Introduction to Polymer Science and Chemistry

A Problem-Solving Approach

SECOND EDITION

Introduction to Polymer Science and Chemistry

A Problem-Solving Approach

Manas Chanda

CRC Press
Taylor & Francis Group
Boca Raton   London   New York

CRC Press is an imprint of the
Taylor & Francis Group, an **informa** business

CRC Press
Taylor & Francis Group
6000 Broken Sound Parkway NW, Suite 300
Boca Raton, FL 33487-2742

© 2013 by Taylor & Francis Group, LLC
CRC Press is an imprint of Taylor & Francis Group, an Informa business

---

Library of Congress Cataloging-in-Publication Data

Chanda, Manas, 1940-
    Introduction to polymer science and chemistry : a problem-solving approach / Manas Chanda. -- Second edition.
        pages cm
    Includes bibliographical references and index.
    ISBN 978-1-4665-5384-2 (hardback)
    1. Polymers. 2. Polymerization. 3. Polymers--Problems, exercises, etc. 4. Polymerization--Problems, exercises, etc. I. Title.

QD381.C473 2013
547'.7--dc23   2012031493

Visit the Taylor & Francis Web site at
http://www.taylorandfrancis.com

and the CRC Press Web site at
http://www.crcpress.com

Dedicated to the memory
of my beloved father and mentor
Narayan Chandra Chanda

Contents

Preface

A question asked during discussion, or even ahead of it, excites a student's mind and rouses his eagerness to probe, thus making the process of learning more thorough. During my teaching of polymer science and chemistry over a period of nearly four decades, I have thus always believed that learning becomes much easier if problem solving with a question-and-answer approach is intimately integrated with the text. It was this belief that motivated me to embark on writing this new text on polymer science and polymer chemistry, even though I was fully aware that the field was already crowded with more than a dozen well-written polymer texts. Adopting a distinctly different and innovative approach, the text in this new polymer book has been laced with questions and answers at every step of the development of a theory or concept in each chapter. The book thus features a significantly large number (286) of solved problems interspersed with the text that is spread over 720 pages. In addition, a large number (277) of problems are included as end-of-chapter exercises and these are fully worked out in a separate *Solutions Manual*. As my experience in teaching has shown me the value of dealing with numbers to deepen one's understanding, most of the problems with which the text is studded are numerical. The same is true for exercise problems appended at the end of each chapter and each such problem is provided with numerical answers that the reader can compare with his own.

To describe the present book briefly, it is a revised and enlarged second edition of *Introduction to Polymer Science and Chemistry: A Problem Solving Approach* and it contains a total of 563 text-embedded solved problems and chapter-end exercise problems. It has evolved from the first edition by retaining all the latter's ten chapters, which, however, have been fully relaid and restructured to afford greater readability and understanding, and adding two new and large chapters to deal with two recent topics that are said to have ushered in a renaissance in polymer chemistry, namely, living/controlled radical polymerization and application of "click" chemistry in polymer synthesis. The book thus has twelve chapters that fall into three distinct groups. The first four chapters introduce the reader to polymers and their basic characteristics, both in solid state and in solution, and the next six chapters are concerned with various polymerization reactions, mechanisms, and kinetics, while the last two chapters are devoted to two recent topics, as cited above, of great interest and importance in polymer chemistry. This division into groups is, however, notional and is not made explicit by numbering these groups of chapters separately. Instead, for convenience, a single sequence of numbers is used throughout the book.

Chapter 1 is devoted to introductory concepts and definitions, while Chapter 2 deals with physical and molecular aspects of polymers, that is, those relating to molecular shape and size, distinctive characteristics, conformational and configurational behavior, structural features, morphology, thermal transitional phenomena, and relaxation properties. Chapter 3 discusses polymer solution behavior, the emphasis being on thermodynamics, phase equilibria, solubility, swelling, frictional properties, and viscosity. Molecular weight determination, which is one of the first steps of polymer characterization and a centrally important topic of polymer science, mostly involves analysis of polymers in solution. The next chapter, Chapter 4, is therefore devoted to polymer molecular weights with focus on the fundamentals of molecular weight statistics and methods of measurement, their origins, and significance.

The chemistry part of the book focusing on polymerization reactions, mechanisms, and kinetics starts with Chapter 5. Five main types of polymerization reactions — condensation (step), free radical (chain), ionic (chain), coordination (chain), and ring-opening — are dealt with separately in five essentially self-contained chapters. Copolymerization that may involve any of these polymerization mechanisms is included in respective chapters, an exception being free-radical chain copolymerization which, in view of its great practical importance and considerable theoretical development that has taken place in this field, has been accorded the space of one full chapter. While polymerization reactions have been characterized on the basis of mechanisms and kinetic features, emphasis has been placed on understanding the reaction parameters which are important in controlling polymerization rates, degree of polymerization, and structural features, such as branching and crosslinking.

The development of living/controlled radical polymerization (CRP) methods, which started only in 1985, has been a long-standing goal in polymer chemistry because a radical process is more tolerant of functional groups and impurities and is the leading industrial method to produce polymers, while the livingness of polymerization allows unprecedented control of polymer types, architecture, end-functionalities, molecular weght, and distributions. CRP is thus among the most rapidly developing areas of polymer chemistry, with the number of publications nearly doubling each year in the initial phase of development. Presently, the most popular CRP methods are nitroxide-mediated polymerization (NMP), atom transfer radical polymerization (ATRP), and reversible addition/fragmentation chain transfer (RAFT) techniques, all of which are described elaborately in the newly added Chapter 11, using once again the unique problem solving approach, which is a hallmark of the book.

Since being introduced only in 2001, click chemistry, which may thus be called a new 21st century technique, has made great advances in the realm of polymer chemistry over the last 10 years, giving access to a wide range of complex polymers (dendrimers, dendronized linear polymers, block copolymers, graft copolymers, star polymers, etc.) and new classes of functionalized monomers in a controlled fashion, which would be inaccessible or difficult to synthesize via conventional chemistry. Chapter 12, which is the last chapter of this new edition, is fully devoted to application of click chemistry in polymer synthesis, using the unique problem solving approach.

In writing the first ten chapters, which deal with conventional polymer science and chemistry, I have received much inspiration and valuable guidance from the many well-known polymer texts that are currently available. However, I should make particular mention of George Odian's *Principles of Polymerization* (McGraw-Hill, New York, 1970), the first edition of which appeared when I was still a student and it made a marked impression on me. Another book which influenced me greatly was Rudin's *The Elements of Polymer Science and Engineering* (Academic Press, Orlando, Florida, 1982), a prescribed text at the University of Waterloo, Canada, where I taught during a sabbatical year (1985–1986).

For writing Chapters 11 and 12 on the two recent topics, living/controlled radical polymerization and polymer synthesis by click chemistry, which have not yet made a significant appearance in polymer chemistry textbooks, I have depended exclusively on original articles that appeared, especially in the last ten years, in many reputed journals. Most of the articles have, however, appeared in journals published by the American Chemical Society and John Wiley & Sons. I am grateful to them for granting permission to reproduce some material in the book from these journals.

While SI units are being used increasingly in all branches of science, non-SI units like the older cgs system are still in common use. This is particularly true of polymer science and chemistry.

In this book, therefore, both SI and non-SI units have been used. However, in most places where non-SI units have been used, equivalent values are given in SI terms. A suitable conversion table is also provided as an appendix.

Synthesized polymers are utilized increasingly in our daily life, and a myriad of industrial applications have contributed to their phenomenal growth and expansion. As this requires polymer chemists and specialists in polymers, many universities throughout the world have set up teaching programs in polymer chemistry, science, and engineering. Their students are drawn from various disciplines in science and engineering. The present book is designed primarily for both undergraduate and graduate students and is intended to serve specially as a classroom text for a one-year course in polymer science and chemistry. Moreover, as two chapters have been added in the new edition focusing on recent advances in polymer chemistry over the last two decades, the book will also be useful to students doing research in the area of polymers.

Polymer industry is the single largest field of employment for students of both science and engineering. However, most workers entering the field have little background in polymer science and chemistry and are forced to educate themselves in its basic principles. This book, with its easy style and a large number of illustrative, worked-out problems, will be useful to them as a self-contained text that guides a beginner in the subject to a fairly advanced level of proficiency.

The manuscript of the book originated from a course in polymers that I offered to graduate students of chemistry and chemical engineering during my sabbatical year (1985-1986) at the University of Waterloo, Ontario, Canada, where I have also been a summer-term visiting faculty spanning over two decades (1980-2000). The manuscript has been tested since then and improved year after year to its present state as the course has been offered every year to a mixed class of students from various disciplines including chemistry, chemical engineering, metallurgy, civil engineering, electrical engineering, electronics, and aerospace engineering at the Indian Institute of Science, Bangalore, where I have served as a permanent faculty. A basic knowledge of mathematics, chemistry, and physics is assumed on the part of the reader, while the book has been written to be self-contained, as far as possible, with most equations fully derived and any assumptions stated.

In the interest of time, I took up the onerous task of preparing the entire book electronically. While I did all the (LaTex) typesetting, formatting, and page designing, I received valuable help from two colleagues, Dr. Ajay Karmarkar and Ms. B. G. Girija, who prepared computer graphics for all diagrams, chemical structures, and chemical formula-based equations. I thank both of them. I am deeply indebted to Dr. P. Sunthar, an acknowledged software expert on the campus, for guiding me patiently in the use of word processing softwares during this difficult venture and to Shashi Kumar of Cenveo Publisher Services, Noida, India, for performing the necessary conversions to font-embedded PDF for printing.

Several academicians have contributed, directly or indirectly, to the preparation of this book. Among them I would like to mention Prof. K. F. O'Driscoll, Prof. G. L. Rempel, and Prof. Alfred Rudin, all of the University of Waterloo, Waterloo, Ontario (Canada), Prof. Kenneth J. Wynne of the Virginia Commonwealth University, Richmond, Virginia (USA), Prof. Harm-Anton Klok of Ecole Polytechnique Fédérale de Lausanne, Lausanne (Switzerland), Prof. Premamoy Ghosh of the Univeristy of Calcutta, Calcutta (India), and Prof. S. Ramakrishnan and Prof. M. Giridhar, both of the Indian Institute of Science, Bangalore (India). I express my gratitude to all of them.

Interaction with students whom I met over the years during my long academic career, both in India and abroad, contributed greatly to the evolution of the book to its present form featuring a unique problem solving approach. It is not possible to thank them individually as the number is too large. However, I should mention, gratefully, two of my erstwhile students, Dr. Amitava Sarkar and Dr. Ajay Karmarkar, who were closely associated with me during the last few years of my service at the Indian Institute of Science, Bangalore, and provided help in many ways in the making of this book. Finally, a word of appreciation and gratitude is due to three persons very close to me, namely, my wife Mridula, daughter Amrita, and little granddaughter Mallika, who showed remarkable understanding and patience, and gladly sacrificed their share of my time to facilitate my work.

Manas Chanda

Author

Manas Chanda has been a professor and is presently an emeritus professor in the Department of Chemical Engineering, Indian Institute of Science, Bangalore, India. He also worked as a summer-term visiting professor at the University of Waterloo, Ontario, Canada with regular summer visits from 1980 to 2000. A five-time recipient of the International Scientific Exchange Award from the Natural Sciences and Engineering Research Council, Canada, Professor Chanda is the author or coauthor of more than 100 scientific papers, articles, and books, including *Plastics Technology Handbook*, 4th Edition (CRC Press, Boca Raton, Florida). His biographical sketch is listed in Marquis' *Who's Who in the World* Millennium Edition (2000) by the American Biographical Society. A Fellow of the Indian National Academy of Engineers and a member of the Indian Plastics Institute, he received B.S. (1959) and M.Sc. (1962) degrees from Calcutta University, and a Ph.D. (1966) from the Indian Institute of Science, Bangalore.

Chapter 1

Introductory Concepts

1.1 Basic Definitions

Many of the terms, definitions, and concepts used in polymer science are not encountered in other branches of science and must be understood in order to fully discuss the synthesis, characterization, structure, and properties of polymers. While most of these are discussed in detail in subsequent chapters, some are of such fundamental importance that they must be introduced at the beginning.

1.1.1 Polymer

The term *polymer* stems from the Greek roots *poly* (many) and *meros* (part). The word thus means "many parts" and designates a molecule made up by the repetition of some simpler unit called a *mer*. Polymers contain thousands to millions of atoms in a molecule that is large; they are also called *macromolecules*. Polymers are prepared by joining a large number of small molecules called *monomers*.

The structure of polystyrene, for example, can be written as

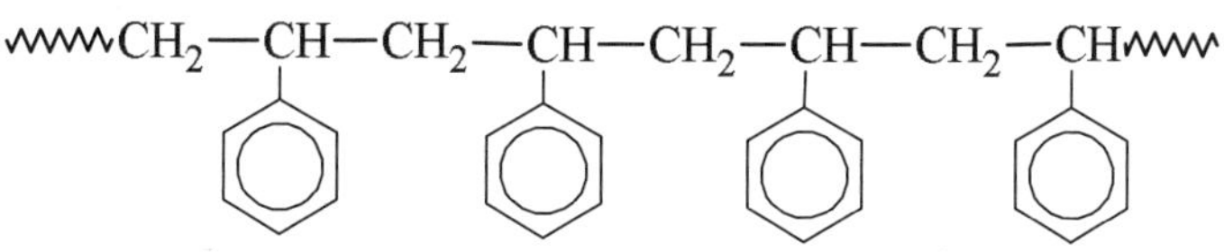

$$\text{wwwCH}_2-\text{CH}-\text{CH}_2-\text{CH}-\text{CH}_2-\text{CH}-\text{CH}_2-\text{CHww}$$

(I)

or, more conveniently, as **(II)**, which depicts the *mer* or *repeating unit* of the molecule within parentheses with a subscript, such as *n*, to represent the number of repeating units in the polymer molecule.

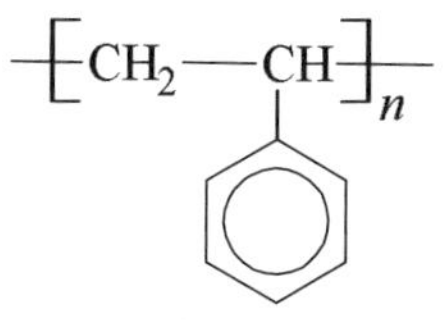

$$\left[\text{CH}_2-\text{CH}\right]_n$$

(II)

The value of *n* usually ranges from a few hundred to several thousand, depending on the molecular weight of the polymer. The polymer molecular weight may extend, on the higher side, to several millions. Often the term *high polymer* is also used to emphasize that the polymer under consideration is of very high molecular weight.

1.1.2   Monomer

Monomers are generally simple organic molecules from which the polymer molecule is made. The structure of the repeating unit of a polymer is essentially that or closely related to that of the monomer molecule(s). The formula of the polystyrene repeating unit (**II**) is thus seen to be essentially the same as that of the monomer styrene $CH_2 =CH\text{-}C_6H_5$.

The repeating unit of a linear polymer is a small portion of the macromolecule such that linking together these units one after another gives rise to the formula of the whole molecule. A repeating unit may be a single component such as (**II**) for the polymer (**I**), or it may consist of the residues of several components, as in poly(ethylene terephthalate), which has the structure:

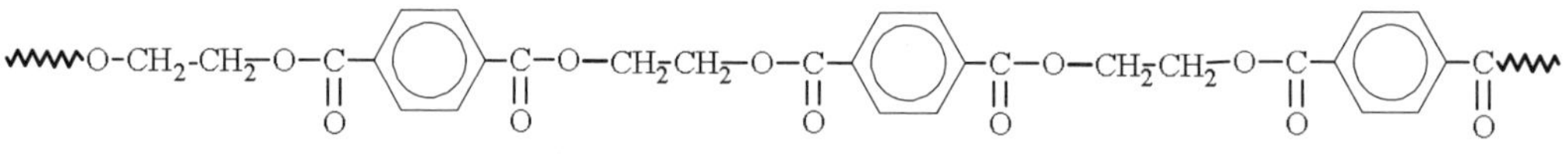

(III)

The repeating unit in (**III**) may be written as

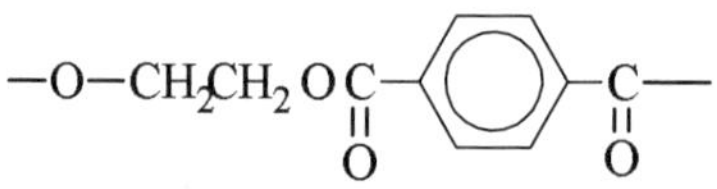

(IV)

Thus, the whole molecule of (**III**) can be built by linking the left-hand atom shown in (**IV**) to the right-hand atom, and so on.

Though it has been stated above that structures of repeating units are essentially those of the monomers from which the polymers are made, this is not always the case. Considering, for example, poly(vinyl alcohol):

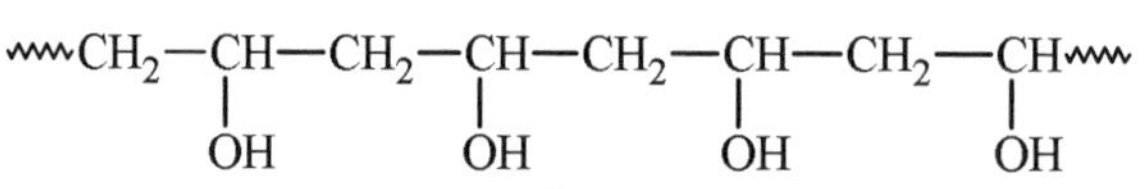

(V)

the obvious precursor monomer for this polymer is vinyl alcohol, $CH_2=CH\text{-}OH$, which is an unstable tautomer of acetaldehyde and does not exist. Poly(vinyl alcohol) is instead made by alcoholysis of poly(vinyl acetate),

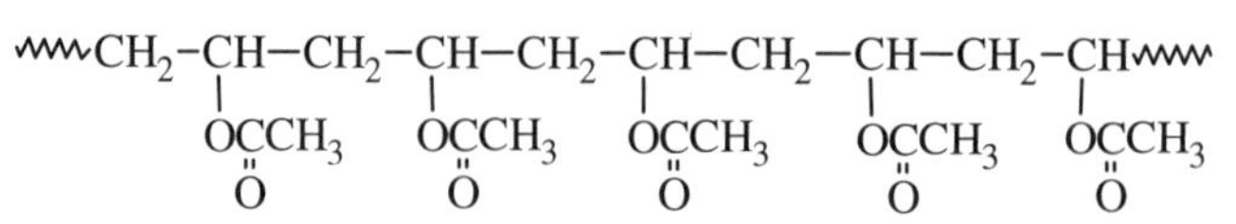

(VI)

which, in turn, is synthesized by polymerization of the monomer vinyl acetate, $CH_2=CHOOCCH_3$.

Another example is cellulose, which is a carbohydrate with molecular formula $(C_6H_{10}O_5)_n$, where n is a few thousand. The structure is

(VII)

Complete hydrolysis of cellulose by boiling with concentrated hydrochloric acid yields D-glucose, $C_6H_{12}O_6$, in 95% yield. Cellulose can thus be considered chemically as a polyanhydroglucose, though it cannot be synthesized from glucose. The concept of the repeating unit is most useful for linear homopolymers (see later).

1.1.3 Molecular Weight and Molar Mass

The term 'molecular weight' is frequently used in practice instead of the term 'molar mass', though the former can be somewhat misleading. Molecular weight is really a dimensionless quantity given by the sum of the atomic weights in the molecular formula. The atomic weights, in turn, are dimensionless ratios of the masses of the particular atoms to 1/12 of the mass of an atom of the most abundant carbon isotope $^{12}C_6$ to which a mass of 12 *atomic mass units* (AMU) is assigned. (AMU, a unit used for expressing the atomic masses of individual isotopes of elements, is approximately 1.6604×10^{-24} g.) The *molar mass* of a substance, on the other hand, is the mass of 1 mol of the substance and usually is quoted in units of g/mol or kg/mol. The numerical value of molecular weight is multiplied by the specific units, such as g/mol, to convert it into an equivalent value of molar mass in dimensions of g/mol. Thus, a molecular weight of 100,000 is equivalent to a molar mass of 100,000 g/mol or 100 kg/mol.

In this book, we shall retain the term 'molecular weight' because of its widespread use in the polymer literature. However, when using in numerical calculations we shall substitute the molecular weight by its numerically equivalent molar mass (g/mol) to facilitate dimensional balancing.

1.1.4 End Groups

In none of the above examples of the structural representation of polymers have the end groups been shown. This is partly because the exact nature of the end groups of polymer molecules is often not known and partly because end groups constitute an insignificant fraction of the mass of high molecular weight polymer and so usually have negligible effect on polymer properties of major interest.

Problem 1.1 Calculate the end group content (weight fraction) of polystyrene of molecular weight 150,000, assuming that phenyl (C_6H_5-) groups constitute both the end groups of an average polymer molecule.

Answer:

Molar mass of phenyl group $= 6 \times 12 + 5 \times 1 = 77$ g mol^{-1}

Wt. fraction of end groups $= \dfrac{2(77 \text{ g mol}^{-1})}{(1.5 \times 10^5 \text{ g mol}^{-1})} = 0.001$

1.1.5 Degree of Polymerization

This term refers to the number of repeating units that constitute a polymer molecule. We shall use the abbreviation *DP* for the degree of polymerization defined in this way. The subscript *n* used on the parentheses in the foregoing structural formulas for polymers represents this *DP*. The relation between degree of polymerization and molecular weight *M* of the same macromolecule is given by

$$M = (DP)M_o \tag{1.1}$$

where M_o is the formula weight of the repeating unit.

1.1.6 Copolymers

If a macromolecule is made from only one species of monomer, the product is a *homopolymer*, referred to simply as a polymer. The word homopolymer often is used more broadly to describe polymers whose structure can be represented by repetition of a single type of repeating unit containing one or more species. Thus, a hypothetical polymer $+AB\}_n$ made from A and B species is also a homopolymer, e.g., poly(ethylene terephthalate) **(III)**.

The formal definition of a *copolymer* is a polymer derived from more than one species of monomer. The copolymer with a relatively random distribution of the different mers or repeating units in its structure is commonly referred to as a *random copolymer*. Representing two different mers by A and B, a random copolymer can be depicted as

$$-ABBABBBAABBAABAAABBA-$$

There are three other copolymer structures: *alternating*, *block*, and *graft* copolymer structures (Fig. 1.1). In the *alternating copolymer*, the two mers alternate in a regular fashion along the polymer chain:

$$-ABABABABABABABABABAB-$$

A *block copolymer* is a linear polymer with one or more long uninterrupted sequences of each mer in the chain:

$$-AAAAAAAAAABBBBBBBBBB-$$

Block copolymers may have a different number of blocks in the molecule. Thus, A_xB_y, $A_xB_yA_x$, $A_xB_yA_xB_y$, $(A_xB_y)_n$ are referred to as AB *diblock*, ABA *triblock*, ABAB *tetrablock*, and AB *multiblock* copolymers, respectively. Since there is a distribution of block lengths and number of blocks along the copolymer chain, *x* and *y* as well as *n* represent average values.

A *graft copolymer*, on the other hand, is a branched copolymer with a backbone of one type of mer and one or more side chains of another mer:

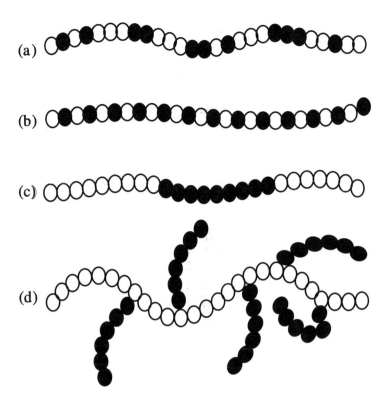

Figure 1.1 Copolymer arrangements : (a) Two different types of mers (denoted by open and filled circles) are randomly placed. (b) The mers are alternately arranged. (c) A block copolymer. (d) A graft copolymer.

Copolymerization, which, in its objective, may be compared to alloying in metallurgy, is very useful for synthesizing polymer with the required combination of properties.

1.2 Polymerization and Functionality

Polymerization may occur only if the monomers involved in the reaction have the proper functionalities. *Functionality* is a very useful concept in polymer science. The *functionality* of a molecule is the number of sites it has for bonding to other molecules *under the given conditions of the polymerization reaction* (Rudin, 1982). Thus, a bifunctional monomer, i.e., a monomer with functionality 2, can link to two other molecules under suitable conditions. Styrene, $C_6H_5CH=CH_2$, for example, has functionality 2 because of the presence of a carbon-carbon double bond. The minimum functionality required for polymerization is 2.

A *polyfunctional* monomer is one that can react with more than two molecules under the conditions of the polymerization reaction. Thus, divinyl benzene **(VIII)** is tetrafunctional in reactions involving additions across carbon-carbon double bonds, while glycerol **(IX)** is trifunctional and pentaerythritol, $C(CH_2OH)_4$, is tetrafunctional in polyesterification reactions.

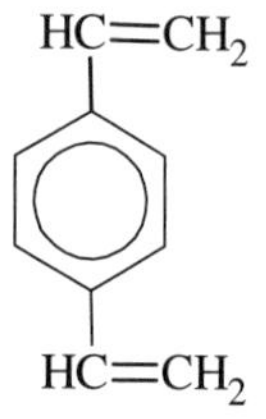

$$HO-CH_2-\underset{\underset{OH}{|}}{CH}-CH_2OH$$

(IX)

(VIII)

Functionality in polymerization is, however, defined only for a given reaction (Rudin, 1982). Thus, a glycol, HOROH, has a functionality of 2 in esterification or ether-forming reactions, but its functionality is zero in amidation reactions.

Problem 1.2   What is the functionality of the following monomers in reactions with (i) styrene, $C_6H_5CH=CH_2$ and (ii) adipic acid, $HOOC(CH_2)_4COOH$?

(a) $H_2C=\underset{\underset{O=\overset{|}{C}-OCH_3}{\overset{|}{C}}}{\overset{CH_3}{C}}$

(b) $HOCH_2CH_2OH$

(c) $H_2C=CH-CH_2OH$

(d) $H_2C=\underset{\underset{CH_3}{|}}{C}-\underset{\overset{||}{O}}{C}-O-CH_2-CH_2O-\underset{\overset{||}{O}}{C}-\underset{\underset{CH_3}{|}}{C}=CH_2$

(e)

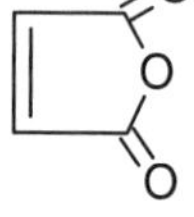

Answer:

(i) In reaction with styrene, the functionalities of the monomers are :

   (a) 2   (one reactive carbon-carbon double bond)

   (b) 0   (-OH groups do not take part in addition reactions)

   (c) 2   (one reactive carbon-carbon double bond; –OH group nonreactive)

   (d) 4   (two reactive carbon-carbon double bonds)

   (e) 2   (one reactive carbon-carbon double bond)

(ii) In reaction with adipic acid, the functionalities of the monomers are :

   (a) 0   (carbon-carbon double bond nonreactive)

   (b) 2   (two reactive -OH groups)

(c) 1 (one reactive -OH group; carbon-carbon double bond nonreactive)

(d) 0 (carbon-carbon double bond nonreactive)

(e) 2 (one reactive anhydride group, which is equivalent to two carboxylic acid groups, each of functionality 1)

1.3 Polymerization Processes

There are two fundamental polymerization mechanisms. Classically, they have been differentiated as *addition polymerization* and *condensation polymerization*. In the addition process, no by-product is evolved, as in the polymerization of vinyl chloride (see below); whereas in the condensation process, just as in various condensation reactions (e.g., esterification, etherification, amidation, etc.) of organic chemistry, a low-molecular-weight by-product (e.g., H_2O, HCl, etc.) is evolved. Polymers formed by addition polymerization do so by the successive addition of unsaturated monomer units in a *chain reaction* promoted by the active center. Therefore, addition polymerization is called *chain polymerization*. Similarly, condensation polymerization is referred to as *step polymerization* since the polymers in this case are formed by *stepwise*, intermolecular condensation of reactive groups. Another polymerization process that has now appeared as a new research area of considerable interest is *supramolecular polymerization* (see Section 1.3.3).

1.3.1 Addition or Chain Polymerization

In chain polymerization, a simple, low-molecular-weight molecule possessing a double bond, referred to in this context as a monomer, is treated so that the double bond opens up and the resulting free valences join with those of other molecules to form a polymer chain. For example, vinyl chloride polymerizes to poly(vinyl chloride):

$$H_2C\!=\!\underset{\underset{Cl}{|}}{CH} \xrightarrow{\text{Polymerization}} \left(\!CH_2\!-\!\underset{\underset{Cl}{|}}{CH}\!\right)_{\!n} \tag{1.2}$$

Vinyl chloride   Poly(vinyl chloride)

It is evident that no side products are formed; consequently the composition of the mer or repeating unit of the polymer ($-CH_2-CHCl-$) is identical to that of the monomer ($CH_2=CHCl$). The identical composition of the repeating unit of a polymer and its monomer(s) is, in most cases, an indication that the polymer is an addition polymer formed by chain polymerization process. The common addition polymers and the monomers from which they are produced are shown in Table 1.1.

Chain polymerization involves three processes: *chain initiation*, *chain propagation*, and *chain termination*. (A fourth process, *chain transfer*, may also be involved, but it may be regarded as a combination of chain termination and chain initiation.) Chain initiation occurs by an attack on the monomer molecule by a free radical, a cation, or an anion; accordingly, the chain polymerization processes are called *free-radical polymerization*, *cationic polymerization*, or *anionic polymerization*. A free radical is a reactive substance having an unpaired electron and is usually formed by the decomposition of a relatively unstable material called an *initiator*. Benzoyl peroxide is a common free-radical initiator and can produce free radicals by thermal decomposition as

$$\underset{\substack{\displaystyle \\ \text{R-C-O-O-C-R}}}{\overset{\substack{O \quad\quad\ O\\ \parallel \quad\quad \parallel}}{}} \longrightarrow \ \underset{\substack{\displaystyle \\ \text{R-C-O}^{\bullet}}}{\overset{\substack{O\\ \parallel}}{}} + \ R^{\bullet} + CO_2 \tag{1.3}$$

(R = Phenyl group for benzoyl peroxide initiator)

Free radicals are, in general, very active because of the presence of unpaired electrons (denoted by dot). A free-radical species can thus react to open the double bond of a vinyl monomer and add to one side of the broken bond, with the reactive center (unpaired electron) being transferred to the other side of the broken bond:

$$\underset{}{\overset{O}{\underset{\parallel}{R-C-O^{\bullet}}}} + \ H_2C\!=\!\overset{H}{\underset{X}{C}} \longrightarrow \ \underset{}{\overset{O}{\underset{\parallel}{R-C-O-CH_2-}}}\overset{H}{\underset{X}{C^{\bullet}}} \tag{1.4}$$

(X = CH$_3$, C$_6$H$_5$, Cl, etc.)

The new species, which is also a free radical, is able to attack a second monomer molecule in a similar way, transferring its reactive center to the attacked molecule. The process is repeated, and the chain continues to grow as a large number of monomer molecules are successively added to propagate the reactive center:

$$\underset{O}{\overset{H}{R-C-O-CH_2-C^{\bullet}}} \ \xrightarrow[\text{of monomer}]{\text{Successive addition}} \ \underset{O}{\overset{H}{R-C-O-(-CH_2-C)^{\bullet}_{m}}} \tag{1.5}$$

This process of *propagation* continues until another process intervenes and destroys the reactive center, resulting in the *termination* of the polymer growth. There may be several termination reactions depending on the type of the reactive center and the reaction conditions. For example, two growing radicals may combine to annihilate each other's growth activity and form an inactive polymer molecule; this is called termination by *combination* or *coupling*:

$$R-\underset{O}{C}-O\!\!-\!\!\left(CH_2-\underset{X}{\overset{H}{C}}\right)^{\bullet}_m + \ {}^{\bullet}\!\!\left(\underset{X}{\overset{H}{C}}-CH_2\right)_n\!\!-\!\!O-\underset{O}{C}-R \longrightarrow$$

$$R-\underset{O}{C}-O\!\!-\!\!\left(CH_2-\underset{X}{\overset{H}{C}}\right)_m\!\!-\!\!\left(\underset{X}{\overset{H}{C}}-CH_2\right)_n\!\!-\!\!O-\underset{O}{C}-R \tag{1.6}$$

Inactive polymer molecule

Table 1.1 Typical Addition Polymers (Homopolymers)

	Monomer	Polymer	Comments
1.	Ethylene $CH_2{=}CH_2$	Polyethylene (PE) $-(CH_2-CH_2)_n-$	High density polyethylene (HDPE) and low density polyethylene (LDPE); molded objects, tubing, film, electrical insulation, used for household products, insulators, pipes, toys, bottles, e.g., Alkathene, Lupolan, Hostalen, Marlex.
2.	Propylene $CH_2{=}CH$ $\quad\;\; CH_3$	Polypropylene (PP) $-(CH_2-CH)_n-$ $\qquad\;\; CH_3$	Lower density, stiffer, and higher temperature resistance than PE; used for water pipes, integral hinges, sterilizable hospital equipment, e.g., Propathene, Novolen, Moplen, Hostalen, Marlex.
3.	Styrene $CH_2{=}CH$	Polystyrene (PS) $-(CH_2-CH)_n-$	Transparent and brittle; used for cheap molded objects, e.g., Styron, Carinex, Hostyren, Lustrex. Modified with rubber to improve toughness, e.g., High impact Polystyrene (HIPS) and acrylonitrile-butadiene-styrene copolymer (ABS). Expanded by volatilization of a blended blowing agent (e.g., pentane) to make polystyrene foam, e.g., Styrocell, Styrofoam.
4.	Acrylonitrile $H_2C{=}CH-CN$	Polyacrylonitrile $-(CH_2-CH)_n-$ $\qquad\quad CN$	Widely used as fibers; best alternative to wool for sweaters, e.g., Orlon, Acrilan.
5.	Vinylacetate $CH_2{=}CH-O-\underset{\underset{O}{\|}}{C}-CH_3$	Poly (vinyl acetate) $-(CH_2-CH)_n-$ $\qquad\quad O$ $\qquad\quad C{=}O$ $\qquad\quad CH_3$	Emulsion paints, adhesives, sizing, chewing gum, e.g., Flovic, Mowilith, Mowicoll.
6.	Vinyl chloride $CH_2{=}CH-Cl$	Poly (vinyl chloride) (PVC) $-(CH_2-CH)_n-$ $\qquad\quad Cl$	Water pipes, bottles, gramophone records, plasticized to make PVC film, leather cloth, raincoats, flexible pipe, tube, hose, toys, electrical cable sheathing, e.g., Benvic, Darvic, Geon, Hostalit, Solvic, Vinoflex, Welvic.
7.	Tetrafluoroethylene $CF_2{=}CF_2$	Polytetrafluoroethylene (PTFE) $-(CF_2-CF_2)-$	High temperature resistance, chemically inert, excellent electrical insulator, very low coefficient of friction, expensive; moldings, films, coatings; used for non-stick surfaces, insulation, gaskets; e.g., Teflon, Fluon.
8.	Methyl methacrylate $\qquad CH_3$ $CH_2{=}C$ $\qquad C{=}O$ $\qquad OCH_3$	Poly(methyl methacrylate) (PMMA) $\qquad\;\; CH_3$ $-(CH_2-C)_n-$ $\qquad\;\; C{=}O$ $\qquad\;\; OCH_3$	Transparent sheets and moldings; more expensive than PS; known as *organic glass*, used for aeroplane windows; e.g., Perspex, Plexiglass, Lucite, Diakon, Vedril.
9.	Isobutylene $\qquad CH_3$ $CH_2{=}C$ $\qquad CH_3$	Polyisobutylene (PIB) $\qquad\;\; CH_3$ $-(CH_2-C)_n-$ $\qquad\;\; CH_3$	Lubricating oils, sealants, copolymerized with 0.5-2.5 mol% isoprene to produce Butyl rubber for tire inner tubes and inner liners of tubeless tires.
10.	Isoprene $CH_2{=}C-CH{=}CH_2$ $\qquad CH_3$	cis-1,4-Polyisoprene $-(CH_2-C{=}CH-CH_2)_n-$ $\qquad\quad CH_3$	Tires, mechanical goods, footwear, sealants, caulking compounds, e.g., Coral, Natsyn, Clariflex I.
11.	Butadiene $CH_2{=}CH-CH{=}CH_2$	cis-1,4-Polybutadiene $-(CH_2-CH{=}CH-CH_2)_n-$	Tires and tire products, e.g., Cis-4, Ameripol-CB, Diene.

A second termination mechanism is *disproportionation,* shown by the following equation:

$$(1.7)$$

In chain polymerization initiated by free radicals, as in the previous example, the reactive center, located at the growing end of the molecule, is a free radical. As mentioned previously, chain polymerizations may also be initiated by ionic systems. In such cases, the reactive center is ionic, i.e., a carbonium ion (in cationic initiation) or a carbanion (in anionic initiation). Regardless of the chain initiation mechanism—free radical, cationic, or anionic—once a reactive center is produced it adds many more molecules in a chain reaction and grows quite large extremely rapidly, usually within a few seconds or less. (However, the relative slowness of the initiation stage causes the overall rate of reaction to be slow and the conversion of all monomers to polymers in most polymerizations requires at least 30 minutes, sometimes hours.) Evidently, at any time during a chain polymerization process the reaction mixture will consist only of unreacted monomers, high polymers and unreacted initiator species, but no intermediate sized molecules. The chain polymerization will thus show the presence of high-molecular-weight polymer molecules at all extents of conversion (see Fig. 1.2).

Problem 1.3   Styrene monomer containing 0.02% (by wt.) benzoyl peroxide initiator was reacted until all the initiator was consumed. If at this stage 22% of the monomer remained unreacted, calculate the average degree of polymerization of the polymer formed. Assume 100% efficiency of the initiator (i.e., all initiator molecules are actually consumed in polymer formation) and termination of chain radicals by coupling alone.

Answer:

Basis: 1000 g styrene

Molar mass of styrene (C_8H_8) = $8{\times}12 + 8{\times}1$ = 104 g/mol

Molar mass of initiator, $(C_6H_5CO)_2O_2$ = 242 g/mol

Amount of initiator   =   0.20 g
  $\equiv$ $(0.20 \text{ g}) / (242 \text{ g mol}^{-1})$ or $8.26{\times}10^{-4}$ mol

Initiator fragment forming
   polymer end groups   = $2 (8.26{\times}10^{-4} \text{ mol})$ $= 1.652{\times}10^{-3}$ mol

Polymer formed by termination
          by coupling   = $(1.652{\times}10^{-3} \text{ mol}) / 2$ $= 8.26{\times}10^{-4}$ mol

Styrene reacted   = $(100 - 22) (1000 \text{ g}) / 100$
          =   780 g
          $\equiv$ $(780 \text{ g}) / (104 \text{ g mol}^{-1})$ or 7.50 mol

$$DP = \frac{(7.50 \text{ mol})}{(8.26 \times 10^{-4} \text{ mol})} = 9080$$

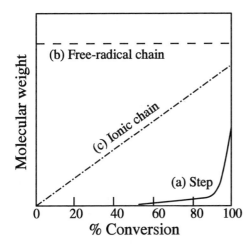

Figure 1.2 Variation of molecular weight with conversion in (a) step polymerization, (b) free-radical chain polymerization, and (c) ionic chain polymerization. (Adapted from Odian, 1991.)

1.3.2 Step Polymerization

Step polymerization occurs by stepwise reaction between functional groups of reactants. The reaction leads successively from monomer to dimer, trimer, tetramer, pentamer, and so on, until finally a polymer molecule with large *DP* is formed. Note, however, that reactions occur at random between the intermediates (e.g., dimers, trimers, etc.) and the monomer as well as among the intermediates themselves. In other words, reactions of both types, namely,

$$n\text{-mer} + \text{monomer} \longrightarrow (n+1)\text{-mer}$$
$$n\text{-mer} + m\text{-mer} \longrightarrow (n+m)\text{-mer}$$

occur equally. Thus, at any stage the product consists of molecules of varying sizes, giving a range of molecular weights. The average molecular weight builds up slowly in the step polymerization process, and a high-molecular-weight product is formed only after a sufficiently long reaction time when the conversion is more than 98% [see Fig. 1.2(a)]. In contrast, polymerization by chain mechanism proceeds very fast, a full-sized polymer molecule being formed almost instantaneously after a chain is initiated; the polymer size is thus independent of reaction time [Fig. 1.2(b)]. In certain ionic chain polymerizations, which feature a fast initiation process coupled with the absence of reactions that terminate the propagating reactive centers, molecular weight increases linearly with conversion [Fig. 1.2(c)].

Since most (though not all) of the step polymerization processes involve polycondensation (repeated condensation) reactions, the terms "step polymerization" and "condensation polymerization" are often used synonymously. Consider, for example, the synthesis of a polyamide, i.e., a polymer with amide (–CONH–) as the characteristic linkage. If we start with, say, hexamethylenediamine and adipic acid as reactants, the first step in the formation of the polymer (nylon) is the following reaction producing a monoamide:

$$H_2N\!\!-\!\!(CH_2)_6\!\!-\!\!NH_2 \ + \ HO\!\!-\!\!\underset{O}{\overset{\|}{C}}\!\!-\!\!(CH_2)_4\!\!-\!\!\underset{O}{\overset{\|}{C}}\!\!-\!\!OH \longrightarrow$$

$$H_2N\!\!-\!\!(CH_2)_6\!\!-\!\!NH\!\!-\!\!\underset{O}{\overset{\|}{C}}\!\!-\!\!(CH_2)_4\!\!-\!\!\underset{O}{\overset{\|}{C}}\!\!-\!\!OH \ + \ H_2O$$

Monoamide

$$(1.8)$$

The reaction continues step-by-step to give the polyamide nylon-6,6. The overall reaction may thus be represented as

$$n \text{ H}_2\text{N}\text{---}(\text{CH}_2)_6\text{---NH}_2 \ + \ n \text{ HO}\text{---}\underset{\text{O}}{\overset{\text{O}}{\text{C}}}\text{---}(\text{CH}_2)_4\text{---}\underset{\text{O}}{\overset{}{\text{C}}}\text{---OH} \longrightarrow$$

$$\text{H}\text{---}\Big[\text{NH}\text{---}(\text{CH}_2)_6\text{---NH}\text{---}\underset{\text{O}}{\overset{\text{O}}{\text{C}}}\text{---}(\text{CH}_2)_4\text{---}\underset{\text{O}}{\overset{}{\text{C}}}\Big]_n\text{---OH} \ + \ (2n\text{-}1)\,\text{H}_2\text{O}$$

$$(1.9)$$

Poly(hexamethylene adipamide)

We see that the composition of the repeating unit (enclosed in square brackets) equals that of two monomer molecules minus two molecules of water. Thus a condensation polymer may be defined as one whose synthesis involves elimination of small molecules or whose repeating unit lacks certain atoms present in the monomer(s).

Problem 1.4 Poly(hexamethylene adipamide) (Nylon-6,6) was synthesized by condensation polymerization of hexamethylenediamine and adipic acid in 1:1 mole ratio. Calculate the acid equivalent of the polymer whose average *DP* is 440.

Answer:

Polymer : $\text{H}\text{---}\Big[\text{NH---(CH}_2)_6\text{---NH---}\underset{\text{O}}{\overset{\text{O}}{\text{C}}}\text{---(CH}_2)_4\text{---}\underset{\text{O}}{\overset{}{\text{C}}}\Big]_n\text{---OH}$

$n \ = \ 440$

Molar mass of repeating unit = 226 g/mol

Average molar mass = 440×226 + 18   or 99,458 g/mol
 Wt. of polymer (in grams) containing
             one acid equivalent   =   99,458

With the development of polymer science and the synthesis of new polymers, the previous definition of condensation polymer is inadequate. For example, in polyurethanes (Table 1.2), which are classified as condensation polymers, the repeating unit has the same net composition as the two monomers (i.e., a diol and a diisocyanate), which react without eliminating any small molecule. To overcome such problems, chemists have introduced a definition which describes condensation polymers as consisting of structural units joined by internal functional groups such as :

ester ($\text{---}\underset{\text{O}}{\overset{}{\text{C}}}\text{---O---}$),   amide ($\text{---}\underset{\text{O}}{\overset{}{\text{C}}}\text{--- NH---}$),   imide ($\text{---N}\overset{\diagup \text{CO---}}{\diagdown \text{CO---}}$),

urethane ($\text{---O---}\underset{\text{O}}{\overset{\text{O}}{\text{C}}}\text{---NH---}$),   sulfide ($\text{---S---}$),   ether ($\text{---O---}$),

carbonate ($\text{---O---}\underset{\text{O}}{\overset{}{\text{C}}}\text{---O---}$),   and   sulfone ($\text{---}\underset{\text{O}}{\overset{\text{O}}{\text{S}}}\text{---}$) linkages.

A polymer satisfying either or both of the above definitions is classified as a condensation polymer. Phenol-formaldehyde, for example, satisfies the first definition but not the second. Some condensation polymers along with their repeating units and condensation reactions by which they can be synthesized are shown in Table 1.2. Some high-performance polymers prepared by poly-condensation are listed in Table 1.3.

The ring-opening polymerizations of cyclic monomers, such as propylene oxide,

$$H_3C-CH \overset{O}{-} CH_2 \longrightarrow -[CH_2-\underset{CH_3}{CH}-O]_n- \tag{1.10}$$

or ϵ-caprolactam

$$(CH_2)_5 \overset{O}{\underset{C}{\parallel}} NH \longrightarrow -[NH-(CH_2)_5-CO]_n- \tag{1.11}$$

proceed either by chain or step mechanisms, depending on the particular monomer, reaction conditions, and initiator employed. However, the polymers produced in Eqs. (1.10) and (1.11) will be structurally classified as condensation polymers, since they contain functional groups (e.g., ether, amide) in the polymer chain. Such polymerizations thus point out very clearly that one must distinguish between the classification based on polymerization mechanism and that based on polymer structure. The two classifications cannot always be used interchangeably. Both structure and mechanism are usually needed in order to clearly classify a polymer.

1.3.3 *Supramolecular Polymerization*

Supramolecular polymers are a relatively new class of polymers in which mono-meric repeating units are held together with directional and reversible (noncovalent) secondary interactions (Lehn, 2000), unlike conventional macromolecular species in which repetition of monomeric units is mainly governed by covalent bonding. A schematic comparison of a covalent polymer and a supramolecular polymer is shown in Fig. 1.3.

The directionality and strength of the supramolecular bonding, such as hydrogen bonding, metal coordination, and $\pi - \pi$ interactions, are important features resulting in polymer properties in dilute and concentrated solutions, as well as in the bulk. It should be noted that supramolecular interactions are not new to polymer science, where hydrogen bonding and other weak reversible interactions are important in determining polymer properties and architectures (Sherrington and Taskinen, 2001). However, for linear supramolecular polymers to form, it is a prerequisite to have strong and highly directional interactions as a reversible alternative for the covalent bond. Hydrogen bonds between neutral organic molecules, though they hold a prominent place in supramolecular chemistry because of their directionality and versatility, are not among the strongest noncovalent interactions. Hence, either multiple hydrogen bonds with cooperativity must be used or hydrogen bonds should be supported by additional forces like excluded volume interactions (Brunsveld et al., 2001). Though the concept has been known for years, it was not known how to incorporate such sufficiently strong but still reversible interactions. However, in the past decade following the development of strong hydrogen-bonding dimers, several research groups have applied these dimers for the formation of hydrogen-bonded supramolecular polymers. Thus the finding by Sijbesma et al. (1997) that derivatives of 2-ureido-4[1H]-pyrimidinone (UPy, **1** in Fig. 1.4) are easy to synthesize and they dimerize strongly (dimerization constant $> 10^6$ M^{-1} in $CHCl_3$) by self-complementary quadrupole (array of four) hydrogen bonding (**2** in Fig. 1.4) prompted them to use this functionality as the associating end group in reversible self-assembling polymer systems.

Table 1.2 Typical Condensation Polymers

Polymer type	Polymerization reaction*	Comments
Polyamide (PA)	$n\ H_2N-R-NH_2 + n\ HOC-R'-COH \rightarrow H-\!\!\left(\!NH-R-NHC-R'-C\!\right)_{\!n}\!-OH + (2n-1)\ H_2O$ $n\ H_2N-R-NH_2 + n\ Cl-C-R'-C-Cl \rightarrow H-\!\!\left(\!NH-R-NHC-R'-C\!\right)_{\!n}\!-Cl + (2n-1)\ HCl$ $n\ H_2N-R-COH \rightarrow H-\!\!\left(\!NH-R-C\!\right)_{\!n}\!-OH + (n-1)\ H_2O$	Moldings, fibers, tirecord; poly(hexamethylene adipamide) (Nylon 6,6)e.g., Ultramid A; polycaprolactam (nylon-6), e.g., Ultramid B, Akulon, Perlenka, poly(hexamethylene sebacamide) (Nylon-6,10), e.g., Ultramid S, Zytel.
Polyester	$n\ HO-R-OH + n\ HOC-R'-C-OH \rightarrow HO-\!\!\left(\!R-OC-R'-CO\!\right)_{\!n}\!-H + (2n-1)\ H_2O$ $n\ HO-R-OH + n\ R''OC-R'-C-OR'' \rightarrow HO-\!\!\left(\!R-OC-R'-CO\!\right)_{\!n}\!-R'' + (2n-1)\ R''OH$ $n\ HO-R-COH \rightarrow HO-\!\!\left(\!R-CO\!\right)_{\!n}\!-H + (n-1)\ H_2O$	Textile fibers, film, bottles; poly(ethylene terephthalate) (PET) e.g., Terylene, Dacron, Melinex, Mylar.
Polyurethane (PU)	$n\ HO-R-OH + n\ OCN-R'-NCO \rightarrow H-\!\!\left(\!O-R-OC-NH-R'-NHC\!\right)_{\!(n-1)}\!-O-R-OC-NH-R'-NCO$	Rubbers, foams, coatings; e.g., Vulkollan, Adiprene C, Chemigum SL, Desmophen A, Moltopren.
Polysulphide	$n\ Cl-R-Cl + n\ Na_2S_x \rightarrow \left(\!R-S_x\!\right)_{\!n} + 2n\ NaCl$	Adhesives, sealants, binders, hose, e.g., Thiokol.
Polysiloxane	$n\ HO-\underset{R}{\overset{R}{Si}}-OH \rightarrow HO-\!\!\left(\!\underset{R}{\overset{R}{Si}}-O\!\right)_{\!n}\!-H + (n-1)\ H_2O$	Elastomers, sealants, fluids, e.g., Silastic, Silastomer, Siloprene.
Phenol-formaldehyde (PF)		Plywood adhesives, glass-fiber insulation, molding compound, e.g., Hitanol, Sirfen, Trolitan.
Urea-formaldehyde (UF)	$n\ H_2N-C-NH_2 + n\ CH_2=O \rightarrow \left(\!NH-C-NH-CH_2\!\right)_{\!n} + n\ H_2O$	Particle-board binder resin, paper and textile treatment, molding compounds, coatings, e.g., Beetle, Resolite, Cibanoid.
Melamine-formaldehyde (MF)		Dinnerware, table tops, coatings, e.g., Formica, Melalam, Cymel.

* R, R', R'' represent aliphatic or aromatic ring. The repeating unit of the polymer chain is enclosed in parentheses.

Table 1.3 Some High-Performance Condensation Polymers

Polymer type and polycondensation reaction	Comments
Polycarbonate (PC)	Moldings and sheets; transparent and tough: used for safety glasses, screens and glazings, electrical and electronics, appliances, compact discs, e.g. Merlon, Baylon, Jupilon.
Polyethersulfone (PES)	Moldings, coatings, membranes; rigid, transparent, self-extinguishing, resistant to heat deformation: used for electrical components, molded circuit boards, appliances operating at high temperatures, e.g., Victrex PES.
Polyetheretherketone (PEEK)	Moldings, composites, bearings, coatings; very high continuous use temperature (260^0C): used in coatings and insulation for high performance wiring, composite prepregs with carbon fibers, e.g., Victrex PEEK.
Poly(phenylene sulphide) (PPS)	Moldings, composites, coatings; outstanding in heat resistance, flame resistance, chemical resistance and electrical insulation resistance: used for electrical components, mechanical parts, e.g., Ryton, Tedur, Fortron.
Poly(p-phenylene terephthalamide)	High modulus fibers; as strong as steel but have one-fifth of weight, ideally suited as tire cord materials and for ballistic vests, e.g., Kevlar, Twaron.
Polyimide	Films, coatings, adhesives, laminates; outstanding in heat resistance, flame resistance, abrasion resistance, electrical insulation resistance, resistance to oxidative degradation, high energy radiation and most chemicals (except strong bases): used in specialist applications, e.g., Kapton, Vespel.

(a)   (b)

Figure 1.3 Schematic representation of (a) a covalent polymer and (b) a supramolecular polymer. (After Brunsveld et al., 2001.)

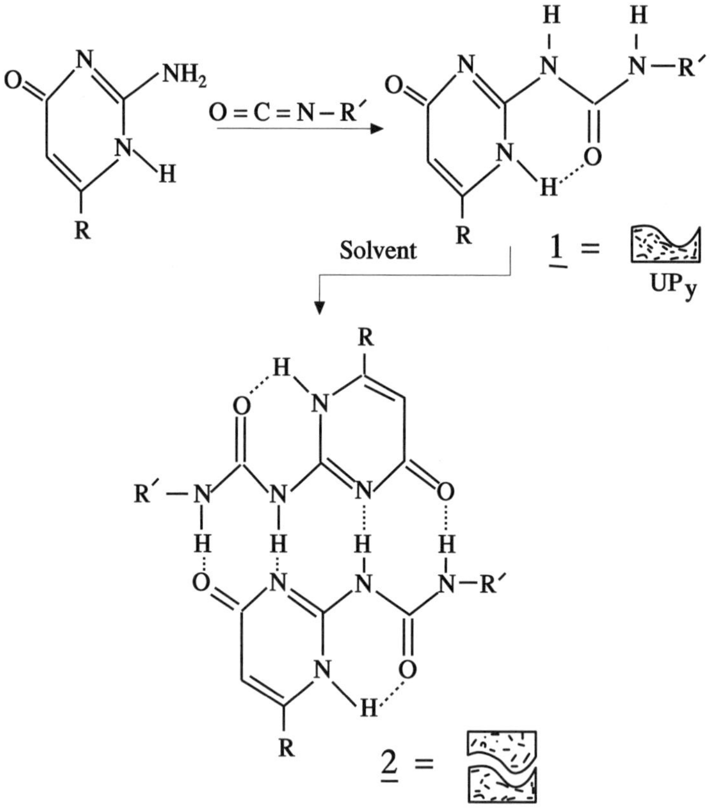

Figure 1.4 Synthesis of a monofunctional 2-ureido-4[1H]-pyrimidinone (UPy) (<u>1</u>) and dimerization of <u>1</u> in solution forming a quadrupole hydrogen-bonded unit. (After Sijbesma et al., 1997.)

A difunctional UPy compound, **4** in Fig. 1.5, possessing two UPy units can be easily made in a one step procedure, from commercially available compounds, methylisocytosine ($R = CH_3$) and hexyldiisocyanate ($R'' = C_6H_{12}$). The compound forms very stable and long polymer chains (**5** in Fig. 1.5) in solution as well as in the bulk (Brunsveld et al., 2001). Dissolving a small amount of the compound in chloroform gives solutions with high viscosities, while calculations show that polymers with molecular weights of the order of 10^6 can be formed. Deliberate addition of small amounts of monofunctional compounds (**1** in Fig. 1.4) results in a sharp drop in viscosity, proving that linkages between the building blocks are reversible and unidirectional and that the monofunctional compounds act as chain stoppers. For the same reason, the supramolecular polymers show polymer-like viscoelastic behavior in bulk and solution, whereas at elevated temperatures they exhibit liquid-like properties (Brunsveld et al., 2001).

The quadrupole hydrogen-bonded unit can be employed in the chain extension of telechelic oligomers such as polysiloxanes, polyethers, polyesters, and polyacrbonates (Folmer et al., 2000). Thus the electrophilic isocyanate group (–NCO) of 'synthon' (**3** in Fig. 1.5) can be reacted with common nucleophilic end groups (–OH or –NH$_2$) of telechelic oligomers, resulting in supramolecular polymers by chain extension (Fig. 1.6). Thus the material properties of telechelic polymers have been shown to improve dramatically upon functionalization with synthon, and materials have been obtained that combine many of the mechanical properties of conventional macromolecules

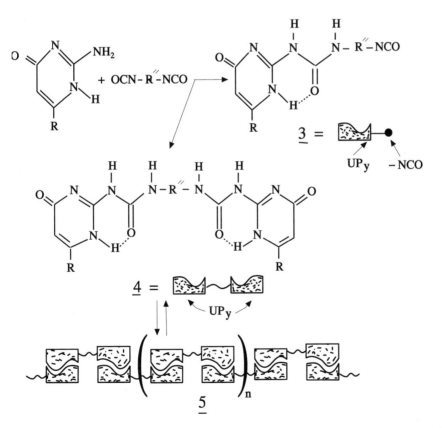

Figure 1.5 Preparation of (a) UPy possessing an isocyanate functional group (**3**) and (b) a difunctional UPy compound (**4**) which forms a supramolecular polymer by hydrogen bonding (cf. Fig. 1.4). (After Folmer et al., 2000; Brunsveld et al., 2001.)

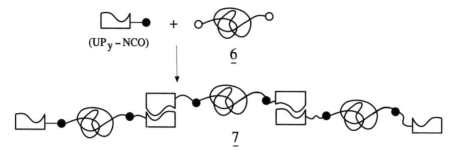

Figure 1.6 Schematic representation of the formation of supramolecular polymer (<u>7</u>) by chain extension of reactive telechelic oligomer with UPy. (From Folmer et al., 2000. With permission from *John Wiley & Sons, Inc.*)

with the low melt viscosity of oligomers (Brunsveld et al., 2001). In contrast to conventional high-molecular-weight polymers, supramolecular (reversible) polymers with a high "virtual" molecular weight show excellent processability due to the strong temperature dependency of the melt viscosity (Folmer et al., 2000). Moreover, hybrids between blocks of covalent macromolecules and supramolecular polymers can be easily made.

1.4 Molecular Architecture

Polymers can be classified, based on structural shape of polymer molecules, as *linear*, *branched*, or *network* (cross-linked). Schematic representations are given in Fig. 1.7.

Linear polymers have repeating units linked together in a continuous length [Fig. 1.7(a)]. In a linear polymer each repeating unit is therefore linked only to two others. When branches protrude from the main polymer chain at *irregular* intervals [Fig. 1.7(b)], the polymer is termed a *branched polymer*. Branches may be short, forming a comblike structure [Fig. 1.7(b)] or may be long and divergent [Fig. 1.7(c)]. *Branched polymers* are thus those in which the repeating units are not linked solely in a linear way.

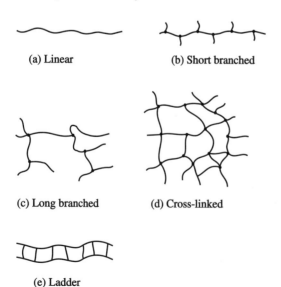

(a) Linear

(b) Short branched

(c) Long branched

(d) Cross-linked

(e) Ladder

Figure 1.7 Schematic representation of various types of polymer molecules. The branch points and junction points are indicated by heavy dots.

Branched polymers may be formed either because at least one of the monomers has functionality greater than 2 or because the polymerization process itself creates branching points on the polymer chain. An example of the first type is the polymer made, for instance, from styrene and a very small amount of divinyl benzene (**VIII**). A segment of such a macromolecule might look like (**X**):

(**X**)

A good example of the second type of branched polymer is the polyethylene that is made by free radical polymerization at high temperatures (100-300°C) and pressures (1,000-3,000 atm). The extent of branching varies considerably depending on reaction conditions and may reach as high as 30 branches per 500 monomer units. Branches in polyethylene are mainly short branches (ethyl and butyl) and are believed to result from intramolecular chain transfer during polymerization (described later in Chapter 6). This branched polyethylene, also called *low-density polyethylene* (LDPE), differs from linear polyethylene (*high-density polyethylene*, HDPE) of a low-pressure process so much so that the two materials are generally not used for the same application.

The term *branched* commonly implies that the polymer molecules are discrete, which means that they can generally be dissolved in a solvent and their sizes can be measured by some of the methods described in Chapter 4.

A *network polymer* [Fig. 1.7(d)], on the other hand, can be described as an interconnected branched polymer. For example, a three-dimensional or space network structure will develop, instead of the branched structure (**X**), if styrene is copolymerized with higher concentrations of divinyl benzene. In a network structure, all polymer chains are linked to form one giant molecule and the molecular weight is "infinite" in the sense that it is too high to be measured by standard techniques (see Problem 1.5). Because of their network structure such polymers cannot be dissolved in solvents and cannot be melted by heat; strong heating only causes decomposition.

Problem 1.5　For a network polymer sample in the form of a sphere of 1 cm diameter with a density of 1.0 g/cm³, estimate the molecular weight assuming that the sample constitutes a single molecule. (Avogadro number = 6.02×10^{23} molecules/mol).

Answer:

$$\text{Molar mass} = \left(\frac{\text{Mass}}{\text{Molecule}}\right)(\text{Avogadro number})$$

$$= (\text{Vol. of sphere})(\text{Density})(\text{Avogadro number})$$

$$= \left[\frac{(3.14)(1.0 \text{ cm})^3}{6}\right](1.0 \text{ g cm}^{-3})(6.02 \times 10^{23} \text{ mol}^{-1})$$

$$= 3.15 \times 10^{23} \text{ g mol}^{-1}$$

$$\text{Molecular weight} = 3.15 \times 10^{23}$$

If the average functionality of a mixture of monomers is greater than 2, copolymerization reaction to a sufficiently high conversion yields network structures (see Chapter 5). Network polymers can also be made by chemically linking already formed linear or branched polymers and the process is called *crosslinking*. *Vulcanization* is an equivalent term that is commonly used for rubbers. Sulfur crosslinks are introduced into a rubber by heating it with sulfur (1-2% by weight) and accelerating agents. Sulfur reacts with the double-bonded carbon atom to produce a network structure, as shown schematically in Fig. 1.8.

Problem 1.6   Assuming that each crosslink produced by vulcanization [Fig. 1.8(b)] contains an average of two sulfur atoms, calculate the sulfur content of vulcanized natural rubber that is 50% crosslinked. (Neglect sulfur other than that is part of the crosslink.)

Answer:

Molar mass of isoprene [Fig. 1.8(a)], C_5H_8 = 5×12 + 8×1 = 68 g/mol. According to representation in Fig. 1.8(b), each sulfide crosslink joins two isoprene units. With $x = 2$, one sulfur atom, on the average, is required for crosslinking per isoprene repeating unit. Therefore,

$$\text{Sulfur content} \quad = \quad \frac{(0.5)(32 \text{ g})}{(0.5)(32 \text{ g}) + (68 \text{ g})} \times 100 \quad = \quad 19\%$$

Problem 1.7   A rubber contains 60% butadiene, 30% isoprene, 5% sulfur, and 5% carbon black. If each sulfide crosslink contains an average of two sulfur atoms, what fraction of possible crosslinks are joined by vulcanization ? (Assume that all the sulfur is present in crosslinks.)

Answer:

Molar mass of butadiene (C_4H_6) = 54 g/mol. Molar mass of isoprene (C_5H_8) = 68 g/mol.

Since one sulfur atom, on the average, is required for crosslinking per repeating unit (cf. Problem 1.6),

$$\text{fraction of cross-links} \quad = \quad \frac{(5 \text{ g})/(32 \text{ g mol}^{-1})}{(60 \text{ g})/(54 \text{ g mol}^{-1}) + (30 \text{ g})/(68 \text{ g mol}^{-1})}$$
$$= \quad 0.101 \quad \text{or} \quad 10.1\%.$$

A *ladder polymer* consists of two parallel strands with regular crosslinks [Fig. 1.7(e)] in between, as in polybenzimidazopyrrolone (**XI**). This polymer is made by polycondensation of pyromellitic dianhydride (**XII**) and 1,2,4,5-tetraminobenzene (**XIII**). The polymer is nearly as resistant as pyrolytic graphite to high temperatures and high energy radiation. It does not burn or melt when heated but forms carbon char without much weight loss.

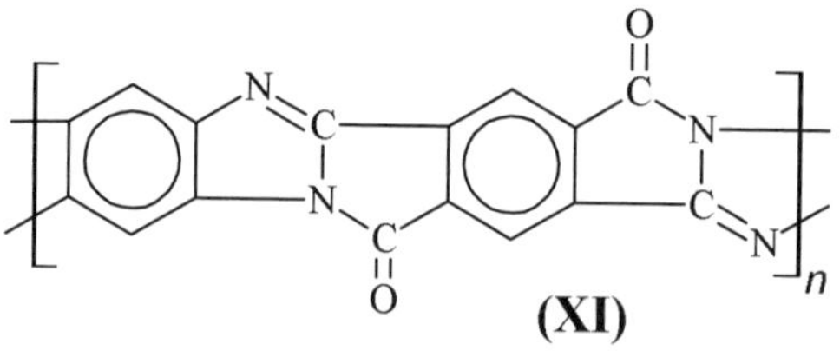

(**XI**)

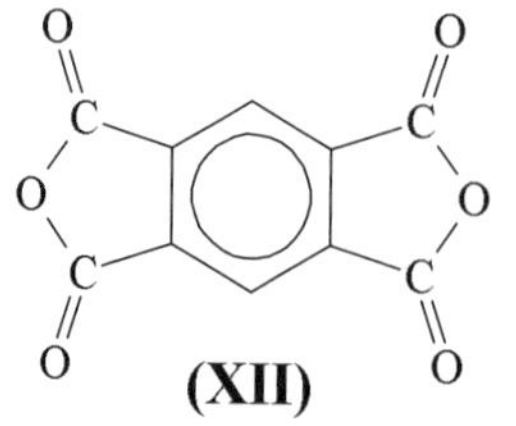

(**XII**)

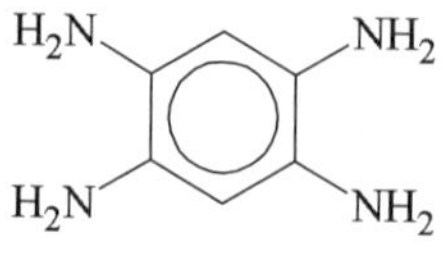

(**XIII**)

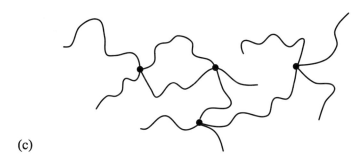

(a)

(b)

(c)

Figure 1.8 Vulcanization of natural rubber with sulfur. (a) Linear polyisoprene (natural rubber). (b) An idealized structure produced by vulcanization with sulfur. The number (x) of sulfur atoms in sulfide crosslinkages is 1 or 2 in efficient vulcanization systems but may be as high as 8 under conditions where cyclic and other structures are also formed in the reaction. (c) The effect of crosslinking is to introduce points of linkage or anchor points between chain molecules, restricting their slippage.

A ladder (or two-strand) structure, as shown above, is one that has an uninterrupted sequence of rings joined one to another at two connecting atoms. A *semiladder* structure, on the other hand, is one in which there are single bonds interconnecting some of the rings. Polyimide (**XIV**) obtained by polycondensation of pyromellitic anhydride (**XII**) and m-phenylenediamine (**XV**) is an example of semiladder polymer.

(XIV)

(XV)

1.5 Classification of Polymers

Polymers can be classified in many ways, such as by source, method of synthesis, structural shape, thermal processing behavior, and end use of polymers. Some of these classifications have already been considered in earlier sections. Thus, polymers have been classified as natural and synthetic according to source, as condensation and addition (or step and chain) according to the method of synthesis or polymerization mechanism, and as linear, branched, and network according to the structural shape of polymer molecules. According to the thermal processing behavior, polymers are classified as thermoplastics and thermosets, while according to the end use it is convenient to classify polymers as plastics, fibers, and elastomers (Rudin, 1982).

1.5.1 Thermoplastics and Thermosets

A *thermoplastic* is a polymer that softens and hardens reversibly on changing the temperature. Both linear and branched polymers are thermoplastic. Thus, they can be softened and made to flow by application of heat. Fabrication processes like injection molding, extrusion molding, and blowing take advantage of this feature to shape thermoplastic resins. The rigidity of thermoplastic resins at low temperatures is attributed to the existence of secondary forces between the polymer chains. These bonds are destroyed at higher temperatures, thereby causing fluidity of the resin.

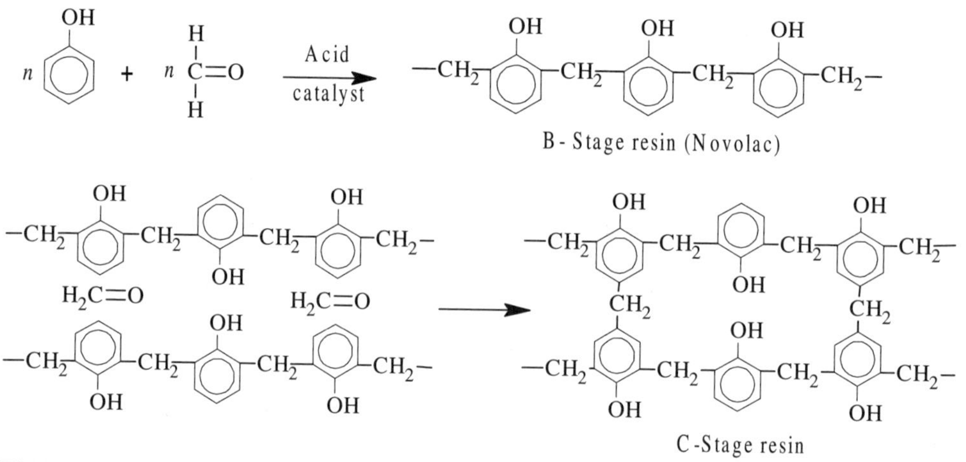

Figure 1.9 Equation (idealized) for the production of phenol-formaldehyde resins.

A thermosetting plastic is a polymer that can be caused to undergo cross-linking to produce a network polymer, called a *thermoset* polymer. Quite commonly, thermosetting resins are prepared, by intent, in only partially polymerized states (*prepolymers*), so that they can be deformed in a heated mold and then hardened by *curing* (crosslinking).

The most important thermosetting resins in current commercial applications are phenolic resins (Fig. 1.9), amino-resins (Fig. 1.10), epoxy resins (Fig. 1.11), unsaturated polyester resins (Fig. 1.12), urethane foams (Fig. 1.13), and the alkyds (Fig. 1.14). The conversion of an un-crosslinked thermosetting resin into a crosslinked network is called *curing*. For curing, the resin is mixed with an appropriate hardener and heated. However, with some thermosetting systems (e.g., epoxies and polyesters), the curing occurs even with little or no application of heat. Epoxies are often preferred to polyesters because they have superior corrosion resistance, mechanical properties, and high-temperature properties, but they are more difficult to handle due to higher viscosities. *Vinyl esters*, which are obtained by reacting epoxies with an unsaturated acid such as acrylic acid

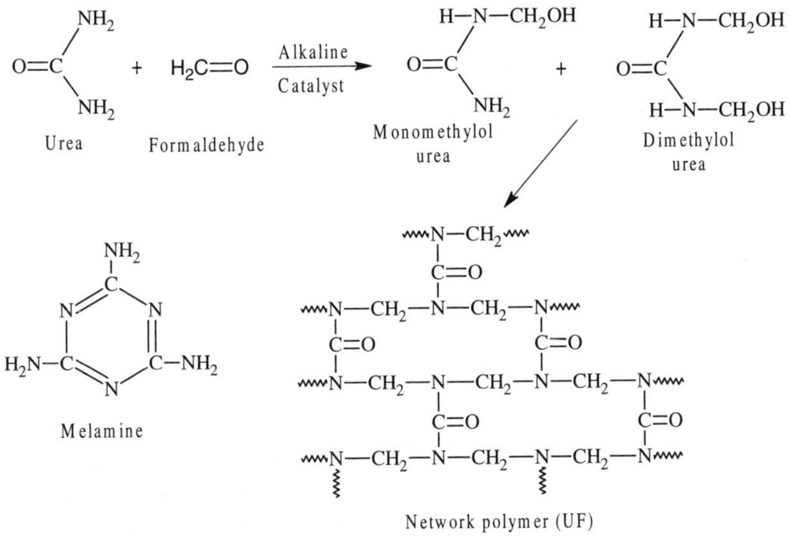

Figure 1.10 The two important classes of amino-resins are the products of condensation reactions of urea and melamine with formaldehyde. Reactions for the formation of urea formaldehyde amino-resins (UF) are shown. Preparation of melamine-formaldehyde resins (not shown) is similar.

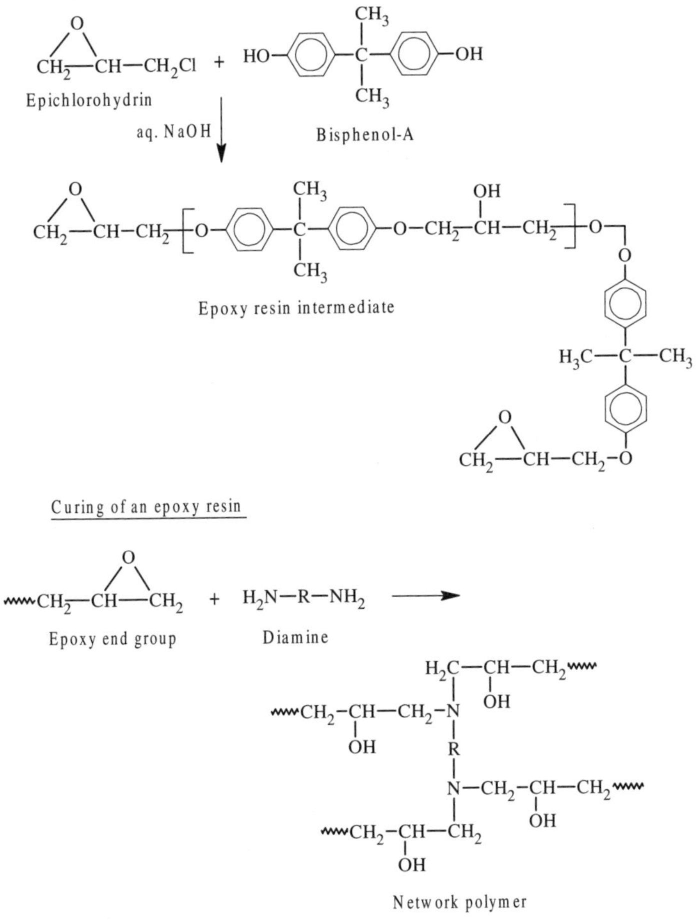

Figure 1.11 Epoxy monomers and polymer and curing of epoxy resins Polyamines such as diethyl enetriamine, ($H_2NC_2H_4NH$- $C_2H_4NH_2$), are widely used for the production of network polymers by room temperature curing.

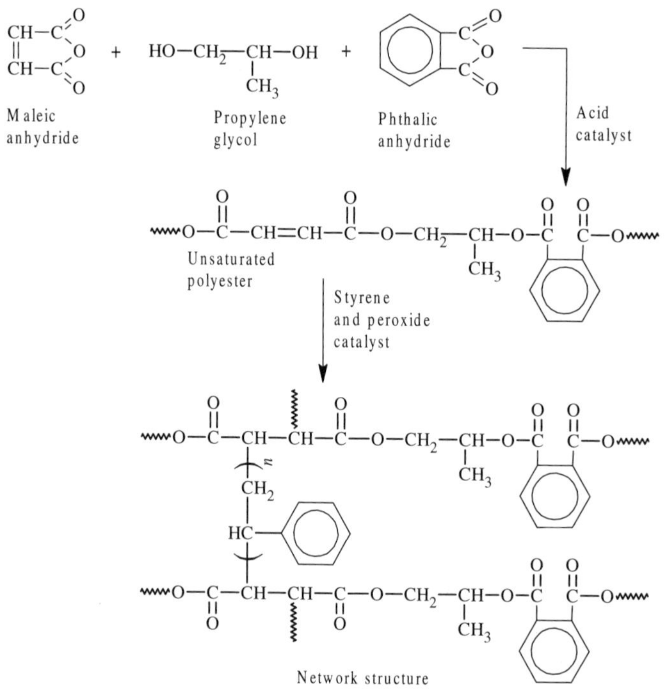

Figure 1.12 Equations for preparation and curing of an unsaturated polyester resin. The presence of ethylenic unsaturation provides sites for crosslinking by a chain reaction mechanism in the presence of styrene. Phthalic anhydride increases flexibility by increasing spacing of crosslinks.

(a) Prepolymer formation:

$$O{=}C{=}N{-}R{-}N{=}C{=}O \ + \ HO{-}P{-}OH \ \longrightarrow \ O{=}C{=}N{-}R{-}NH{-}\underset{O}{\overset{\parallel}{C}}{-}O{-}P{-}O{-}\underset{O}{\overset{\parallel}{C}}{-}NH{-}R{-}N{=}C{=}O$$

Diisocyanate   Glycol

Urethane prepolymer

(b) Chain extension of prepolymer:

(i) With water   ⋯NCO + H_2O + OCN⋯ $\longrightarrow$ ⋯NH$-\overset{O}{\overset{\parallel}{C}}-$NH⋯ + CO_2

Prepolymer   Prepolymer   Urea link

(ii) With glycols   ⋯NCO + HO$-$R$-$OH + OCN⋯ $\longrightarrow$ ⋯NH$-\overset{O}{\overset{\parallel}{C}}-O-R-O-\overset{O}{\overset{\parallel}{C}}-$NH⋯

Urethane link

(iii) With amines   ⋯NCO + H_2N$-$R$-$NH$_2$ + OCN⋯ $\longrightarrow$ ⋯NH$-\overset{O}{\overset{\parallel}{C}}-NH-R-NH-\overset{O}{\overset{\parallel}{C}}-$NH⋯
Double urea link

(c) Crosslinking of chain-extended polyurethane:

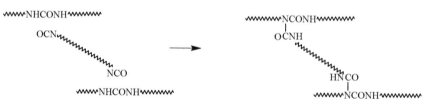

Figure 1.13 Equations for the preparation, chain extension, and curing of polyurethanes.

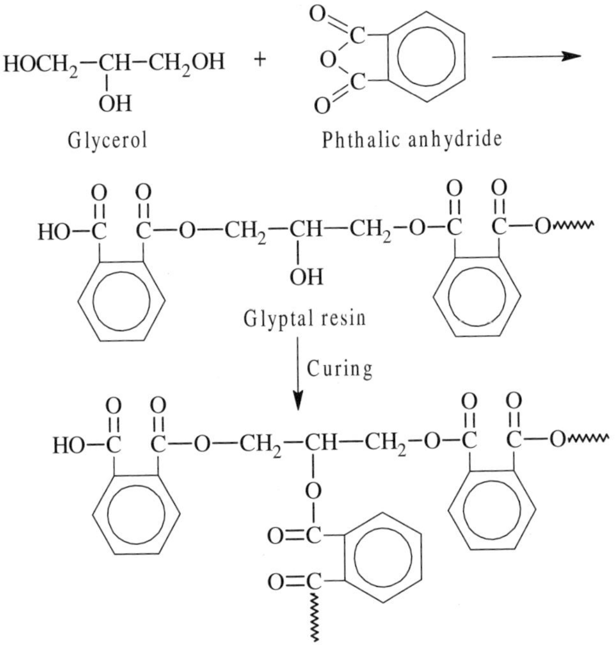

Figure 1.14 Equations for preparation of network glyptal resin.

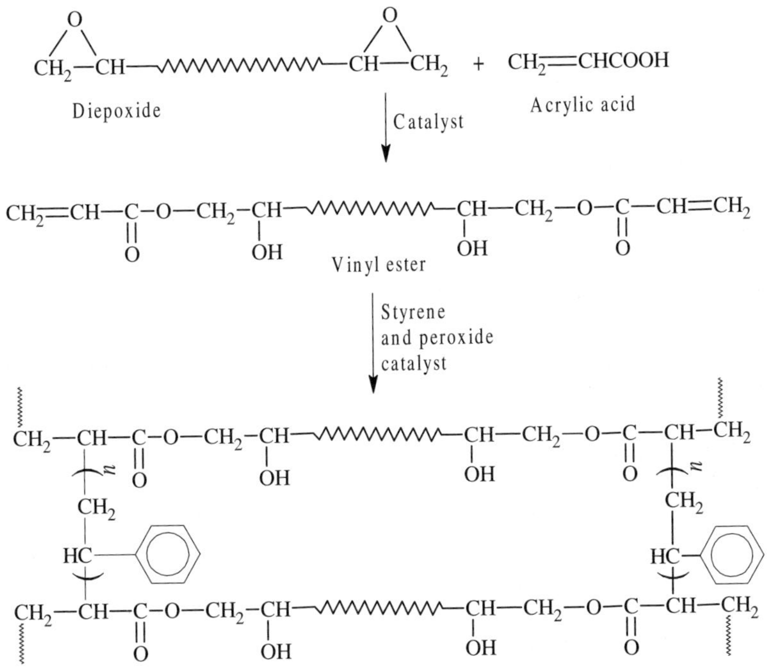

Figure 1.15 Equations for preparation and curing of a vinyl ester.

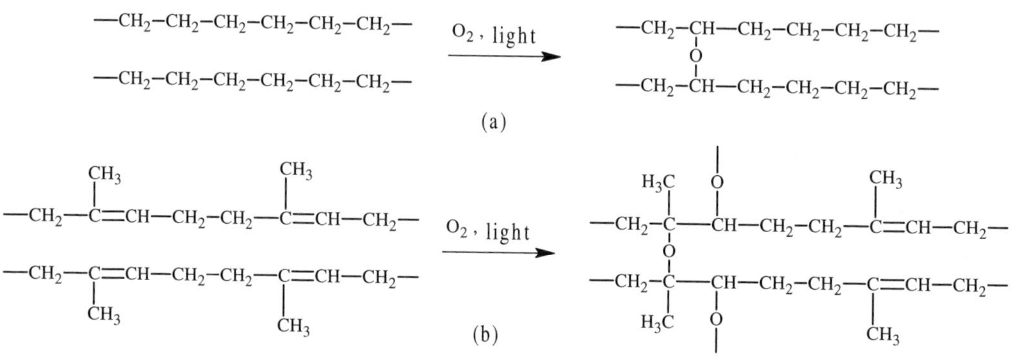

Figure 1.16 Aging of (a) polyethylene and (b) natural rubber by oxidative crosslinking.

or methacrylic acid, combine the benefits of epoxy resins with the lower viscosity and faster curing of unsaturated polyesters. Since vinyl esters contain a double bond (instead of the epoxide group) at each end of the molecule, they can be cured in the same way as polyesters after being dissolved in styrene (Fig. 1.15).

Aging of polymers is often accompanied by crosslinking due to the effect of surroundings. Such crosslinking is undesirable because it greatly reduces the elasticity of the polymer, making it more brittle and hard. The well-known phenomenon of aging of polyethylene with loss of flexibility is due to crosslinking by oxygen under the catalytic action of sunlight [Fig. 1.16(a)]. Natural rubber undergoes a similar loss of flexibility with time due to oxidative crosslinking [Fig. 1.16(b)]. This action may be discouraged by adding to the polymer an antioxidant, such as a phenolic compound, and an opaque filler, such as carbon black, to prevent entry of light (Rudin, 1982).

1.6 Plastics, Fibers, and Elastomers

The distinction between plastics, fibers, and elastomers is most easily made in terms of the tensile stress-strain behavior of representative samples (Rudin, 1982). The curves shown in Fig. 1.17 are typical of those obtained in tension for a constant rate of loading. The parameters of each curve are normal stress (force applied on the specimen divided by the original cross-sectional area), nominal strain (increase in length divided by original length), and the modulus (slope of stress-strain curve). The slope of the curve near zero strain gives the initial modulus.

It may be seen that while fibers have high initial tensile moduli in the range $3 \times 10^3 - 14 \times 10^3$ MN/m^2 ($3 \times 10^4 - 14 \times 10^4$ kgf/cm^2), elastomers have low initial moduli in tension, typically up to 7 MN/m^2 (71 kgf/cm^2), but they generally stiffen (as shown by higher moduli) on stretching. Plastics, in general, have intermediate tensile moduli, typically $3 \times 10^2 - 3 \times 10^3$ MN/m^2 ($3 \times 10^3 - 3 \times 10^4$ kg/cm^2), and their elongation at break varies from a few percent for brittle materials like polystyrene to about 400% for tough, semicrystalline polyethylene.

If the polymer molecule is rigid, it will have less tendency to coil up on itself, and most segments of a given molecule will therefore contact segments of other molecules rather than those of its own. A prime example is the aromatic polyamide fiber, poly(p-phenylene terephthalamide) **(XVI)** marketed as *Kevlar*. It is wet-spun from a solution in concentrated sulfuric acid into fibers which can be stretched to two or three times their original length. The products are as strong as steel but have one-fifth the weight, and can be heated without decomposition to temperatures exceeding 500°C.

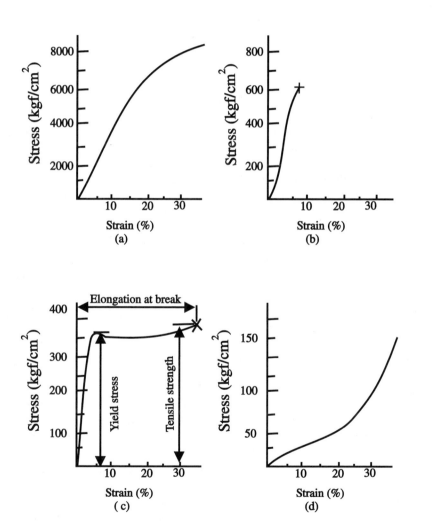

(XVI)

Figure 1.17 Stress-strain curves. (a) Synthetic fiber, like nylon-6,6. (b) Rigid, brittle plastic, like polystyrene (c) Tough plastic, cellulose acetate. (d) Elastomer, like lightly vulcanized natural rubber. (1 kgf/cm^2 = 0.098 MN/m^2) (After Rudin, 1982.)

The requirements that must be satisfied by polymers to be used as textile fibers can thus be summarized as (Rudin, 1982): (1) linear molecules, (2) high molecular weight, and (3) a permanent orientation of the molecules parallel to the fiber axis. The molecules must have a high degree of order and/or secondary forces to permit orientation and crystallization. The chain orientation necessary to develop sufficient strength by crystallization is achieved by a process known as *cold drawing*, in which the initially formed filaments (unoriented or only slightly oriented) are drawn at a temperature above the glass transition temperature (see Chapter 2), which is the temperature at which sufficient energy is available to the molecular segments to cause them to begin to rotate.

Some polymers can be used both as fibers and as plastics, depending on the extent of macromolecular alignment in them. Polyamides, polyesters, and polypropylene are prime examples of such polymers used in both areas. In the fiber making process, known as *cold drawing*, the polymer molecules are aligned in the fiber direction, thereby increasing the intermolecular forces and crystallization. This increases the tensile strength and stiffness and reduces the elongation at break. Thus, typical poly(hexamethylene adipamide) (nylon-6,6) fibers have tensile strengths around 700 MN/m^2 (7.1×10^3 kgf/cm^2) and breaking elongations about 25%. Without cold drawing, the same polymer, used as plastics, has tensile strengths only around 70 MN/m^2 (7.1×10^2 kgf/cm^2) and breaking elongations near 100%. The macromolecules in such articles are randomly oriented and much less extended than in the fiber.

Not all plastics can be converted into practical fibers, however, because the intermolecular forces or crystallization tendency may be too weak to attain high strength by axial orientation. Synthetic fibers are therefore made generally from polymers whose chemical composition and geometry enhance intermolecular attractive forces and crystallization. Such polymers can be converted from plastics to fibers by suitable treatment to cause axial alignment, as explained above.

Elastomeric materials, like thermoplastic resins and fibers, are essentially linear polymers. But certain distinctive features in their molecular structure give rise to rubberlike elasticity. Conventional elastomers consist of highly flexible nonpolar macromolecules that coil up on themselves in the unstressed condition, thereby reducing drastically the extent of intermolecular contacts. A large deformation is possible merely by reorienting the coiled molecules. When elongated, the macromolecular coils partially open up and become aligned more or less parallel to the direction of elongation. This results in an increase in the number of intermolecular contacts and in the sum of intermolecular attractions. In the stressed condition, a rubber is thus harder than in the unstressed condition. The aligned configuration, however, represents a less probable state or a state of lower entropy than a random arrangement. The aligned polymer chains therefore have a tendency to return to their randomly coiled state.

The large deformability of elastomeric materials is due to the presence of a certain internal mobility that allows rearranging the chain orientation, the absence of which in linear chain plastic materials (at normal temperature) constitutes the essential difference between the two groups. Polyethylene, which has weak intermolecular forces but a high degree of crystallinity due to its highly symmetrical and regular molecular structure, is not capable of elastomeric behavior because the crystallites prevent easy coiling or uncoiling of the macromolecules (Rudin, 1982).

Although the aforesaid requirements are necessary conditions for ensuring a large extent of deformability, the remarkable characteristic of the rubbery state—namely, nearly complete recovery—cannot be obtained without a permanent network structure, since permanent deformation rather than elastic recovery will occur. A small amount of crosslinkage is necessary to provide this essential network structure. However, the amount of crosslinkage must be as small as possible to retain the structure; excessive crosslinkages will make the internal structure too stiff to permit even the required rearrangement of chain orientation during deformation and recovery—in other words, it will destroy the rubbery state. An example of this is best furnished by ebonite, which is a rigid plastic made by vulcanizing (crosslinking) natural rubber with large quantities of sulfur.

1.7 Polymer Nomenclature

Polymer nomenclature has been largely a matter of custom without any one system being universally adopted. Though a systematic IUPAC nomenclature now exists for polymers, this nomenclature is rarely used because a common naming system is widely accepted through the force of usage. The common naming system, based either on the source of the polymer, the structure of the polymer, or trade names, usually works without difficulty because the number of polymers that are of interest to the average worker in the field does not exceed a few dozen, and so are not difficult to remember. However, as new polymers enter the area of common usage the number may become too large eventually, necessitating a wider usage of the IUPAC nomenclature. In the following, however, only the common naming system (Rudin, 1982) will be described.

For the common nomenclature the usual practice is to name a polymer according to its source, i.e., the monomer(s) used in its synthesis, and the generic term used is poly"monomer", whether or not the monomer is real. The prefix 'poly' is added on to the name of the monomer to form a single word, e.g., polyethylene, polystyrene, and polyacrylonitrile (see Table 1.1). However, when the monomer has a multiworded name, the name of the monomer after the prefix 'poly' is enclosed in parentheses, e.g., poly(vinyl chloride), poly(vinyl alcohol) and poly(methyl methacrylate) (Table 1.1).

A few polymers are given names based on the repeating unit without reference to the parent monomer. The primary examples are silicones, which possess the repeating unit

$$\left(\!\!\begin{array}{c} R \\ | \\ Si\!-\!O \\ | \\ R \end{array}\!\!\right)$$

(XVII)

Thus, if R = CH$_3$, the polymer is named as poly(dimethyl siloxane).

The nomenclature of random copolymers includes the names of the monomers separated by the interfix *-co-*. Thus **(XVIII)** is named as poly(styrene-*co*-methyl methacrylate) or poly(methyl methacrylate-*co*-styrene), the major component being named first. For alternating copolymers, the interfix *-alt-* is used, e.g., poly (styrene-*alt*-maleic anhydride) **(XIX)**

(XVIII)

(XIX)

Graft copolymers of A and B monomers are named poly(A-*g*-B) or poly A–*graft*-poly B with the backbone polymer –(A–)$_n$– mentioned before the branch polymer. Some examples are poly(ethylene-*g*-styrene) or polyethylene-*graft*-polystyrene and starch-*graft*-poly(methyl methacrylate). In the nomenclature of block copolymers, *b* or *block* is used in place of *g* or *graft*, e.g., poly(A-*b*-B) or poly A-*block*-poly B, poly(A-*b*-B-*b*-A) or poly A-*block*-poly B-*block*-poly A, poly(A-*b*-B-*b*-C) or poly A-*block*-poly B-*block*-poly C, and so on. Thus the triblock polymer (**XX**) is called poly(styrene-*b*-butadiene-*b*-styrene) or polystyrene-*block*-polybutadiene-*block*-polysty-rene. For commercial products, such polymers are usually designated by the monomer initials; thus, structure (**XX**) is named SBS block copolymer.

$$-\left(CH_2-\underset{\underset{\bigcirc}{|}}{CH}\right)_x\left(CH_2-CH=CH-CH_2\right)_y\left(\underset{\underset{\bigcirc}{|}}{CH}-CH_2\right)_z$$

(**XX**)

Condensation polymers are frequently named from the internal linking group between hydro-carbon portions. Thus, (**III**) is a *polyester* which can be written as $-(O\text{-}R\text{-}O\text{-}CO\text{-}R'\text{-}CO)_n-$ and (**XVI**) is a *polyamide* written as $-(HN\text{-}R\text{-}NH\text{-}CO\text{-}R'\text{-}CO)_n-$. Similarly, $-(O\text{-}R\text{-}O\text{-}CO\text{-}NH\text{-}R'\text{-}NH\text{-}CO)_n-$ is a polyurethane, and $-(R\text{-}SO_2)_n-$ is a polysulfone.

In the common naming system, condensation polymers are named by analogy with the lower-molecular-weight products produced by condensation. Thus, since all esters are named by adding the suffix 'ate' to the name of the parent acid (e.g., ethyl acetate), polymer (**III**) is named poly(ethy-lene terephthalate) according to the parent acid, terephthalic acid, which is a para diacid. The word 'ethylene' here implies 'ethylene glycol' because the alcohol used must be a glycol if the polymer is to be linear. Similar reasoning is also followed in naming polyamides. Thus, the word 'hexamethylene' in poly(hexamethylene adipamide) obviously implies hexamethylene diamine because the polymeric structure could be made by the condensation reaction of hexamethylene diamine $H_2N(CH_2)_6NH_2$, and adipic acid, $HOOC(CH_2)_4COOH$.

An alternative naming system is often used for synthetic polyamides derived from unsubsti-tuted nonbranched aliphatic monomers. Thus, a polyamide made from either an amino acid or a lactam is called *nylon-x*, where *x* is the number of carbon atoms in the repeating unit. A nylon made from a diamine and a dibasic acid is designated by two numbers, the first representing the number of carbons in the diamine chain and the second the number of carbons in the dibasic acid.

Problem 1.8   Name the polyamides made from the following monomers and draw their structural formulas (one repeating unit).

(a) Caprolactam; (b) ω-aminoundecanoic acid; (c) dodecyl lactam; (d) hexamethylene diamine and sebacic acid,   and (e) hexamethylene diamine and decanedioic acid.

Answer:

(a) Nylon-6 : $-[NH-(CH_2)_5-CO]_n-$

(b) Nylon-11 : $-[NH-(CH_2)_{10}-CO]_n-$

(c) Nylon-12 : $-[NH-(CH_2)_{11}-CO]_n-$

(d) Nylon-6,10 : $-[NH-(CH_2)_6-NH-CO-(CH_2)_8-CO]_n-$

(e) Nylon-6,12 : $-[NH-(CH_2)_6-NH-CO-(CH_2)_{10}CO]_n-$

There are a few polymers for which the commonly used names convey relatively little informa-tion about the repeating unit structure. The primary examples are polycarbonate, poly(phenylene oxide), polyamide-imides, polysulfones, and polyether ketones (see Table 1.3).

REFERENCES

Allcock, H. R. and Lampe, F. W., "Contemporary Polymer Chemistry", Prentice Hall, Englewood Cliffs, N. J., 1990.

Billmeyer, Jr., F. W., "Textbook of Polymer Science", John Wiley, New York, 1994.

Bosman, A. W., Sijbesma, R. P., and Meijer, E. W., *Materials Today*, p. 34, April (2004).

Brunsveld, L., Folmer, B. J. B., Meijer, E. W., and Sijbesma, R. P., *Chem. Rev.*, **101**, 4071 (2001).

Brydson, J. A., "Plastics Materials", 3rd ed., Chap. 2, Butterworths, London, 1975.

Flory, P. J., "Principles of Polymer Chemistry", Chap. 2, Cornell University Press, Ithaca, N. Y., 1953.

Folmer, B. J. B., Sijbesma, R. P., Versteegen, R. M., van der Rijt, J. A. J., and Meijer, E. W., *Adv. Mater.*, **12**(12), 874 (2000).

Hiemenz, P. C., "Polymer Chemistry. The Basic Concepts", Chap. 1, Marcel Dekker, New York, 1984.

Lehn, J.-M., in "Supramolecular Polymers" (A. Ciferri, ed.), Marcel Dekker, New York, 2000.

Odian, G., "Principles of Polymerization", 3rd ed., Chap. 1, John Wiley, New York, 1991.

Rudin, A., "The Elements of Polymer Science and Engineering", Chap. 1, Academic Press, New York, 1982.

Seymour, R. B. and Carraher, C. E., Jr. "Polymer Chemistry. An Introduction", Marcel Dekker, New York, 1992.

Sherrington, D. C. and Taskinen, K. A., *Chem. Soc. Rev.*, **30**, 89 (2001).

Sijbesma, R. P., Beijer, F. H., Brunsveld, L., Folmer, B. J. B., Hirschberg, J. H. K., Lange, R. F. M., Lowe, J. K. L., and Meijer, E. W., *Science*, **278**, 1601 (1997).

EXERCISES

1.1 Represent, by showing a repeating unit, the structure of the polymer which would be obtained by polymerization of the following monomers:

(a) ω-aminolauric acid; (b) lauryl lactam; (c) ethylene oxide; (d) oxacyclobutan; (e) ethylene glycol and terephthalic acid; (f) hexamethylene diamine and sebacic acid; (g) ethylene glycol and phenylene diisocyanate; (h) *m*-phenylene diamine and isophthaloyl chloride

1.2 Draw the structural formula (one repeating unit) for each of the following polymers: (a) poly(4-methylpent-1-ene); (b) poly(chlorotrifluoroethylene); (c) poly(vinyl ethyl ether); (d) poly(vinylidene chloride); (e) polyethyleneimine; (f) poly(methyl-2-cyano-acrylate); (g) polychloroprene; (h) poly(butylene terephthalate); (i) poly(1,2-propylene oxalate); (j) poly(dihydroxymethylcyclohexyl terephthalate); (k) poly-caprolactam (nylon-6); (l) polyformaldehyde; (m) poly-oxymethylene; (n) poly(propylene oxide); (o) poly (propylene glycol); (p) poly(*p*-phenylene sulfone); (q) poly(dimethyl siloxane); (r) poly (vinyl butyral); (s) poly (*p*-phenylene); (t) poly(*p*-xylylene); (u) polycaprolactone

1.3 What is the degree of polymerization of each of the following polymers with molar mass 100,000 g/mol?

(a) polyacrylonitrile
(b) polycaprolactam
(c) poly(trimethylene ethylene-urethane)
[*Ans.* (a) 1887; (b) 885; (c) 532]

1.4 What is the functionality of the following monomers in reaction with (a) methyl methacrylate and (b) ethylene glycol?

(i) Divinyl benzene
(ii) Maleic anhydride
(iii) Phthalic anhydride

(iv) Acrylic acid

[*Ans.* (i)(a) 4, (b) 0; (ii) (a) 2, (b) 2; (iii) (a) 0, (b) 2; (iv) (a) 2, (b) 1]

1.5   What is the functionality of the monomer shown

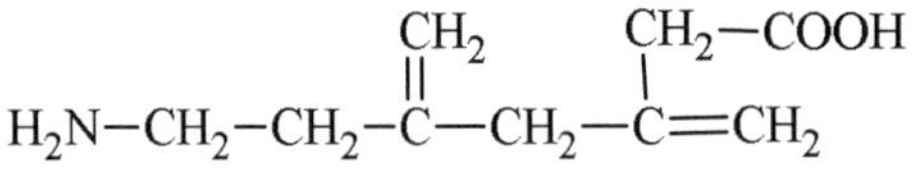

$$H_2N-CH_2-CH_2-\underset{\underset{CH_2}{\|}}{C}-CH_2-\underset{\underset{\|}{CH_2-COOH}}{C}=CH_2$$

(a) in a free radical or ionic addition reaction through $C = C$ double bonds,

(b) in a reaction that produces amide linkages,

(c) in a reaction that produces ester linkages ?

[*Ans.* (a) 4; (b) 2; (c) 1]

1.6   What is the acid value of polycaprolactam (nylon-6) with average *DP* 500 ?   [*Note*: Acid value or acid number is defined as the number of milligrams of KOH required to neutralize 1 g of polymer.]

[*Ans.* 1]

1.7   How would you determine experimentally whether the polymerization of an unknown monomer was proceeding by a step or a chain mechanism ?

1.8   (a) Referring to the epoxy-amine reaction shown in Fig. 1.7, determine the functionality of the diglycidyl ether of bisphenol A (**I**) in a hardening reaction with diethylene triamine (**II**).

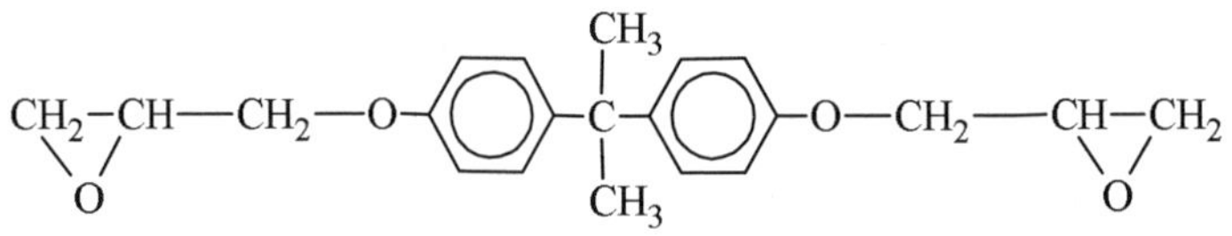

$$CH_2-CH-CH_2-O-\bigcirc-\underset{\underset{CH_3}{|}}{\overset{\overset{CH_3}{|}}{C}}-\bigcirc-O-CH_2-CH-CH_2$$

(I)

$$H_2N-CH_2-CH_2-NH-CH_2-CH_2-NH_2$$

(II)

(b) What is the functionality of (**II**) in this reaction ?

[*Ans.* (a) 2; (b) 5]

1.9   Each of the following polymers can be synthesized from different monomers

(a) $-[-O(CH_2)_5CO-]_n-$

(b) $-(-OCH_2CH_2CH_2-)_n-$

(c) $-[-NH-(CH_2)_7-CO-]_n-$

Show by equations the overall chemical reactions involved in the synthesis of these polymers from different monomers.

1.10 Classify the polymers in Exercise 1.1 as to whether they are condensation or addition polymers. Classify the polymerizations as to whether they are step, chain, or ring opening polymerizations.

1.11 Name the polymers obtained in Exercise 1.1 according to their source, i.e., the monomer(s) used in their synthesis.

1.12 Name the following condensation polymers according to the common nomenclature

(a)

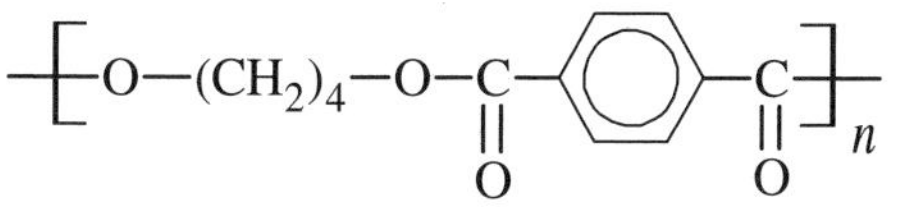

(b)

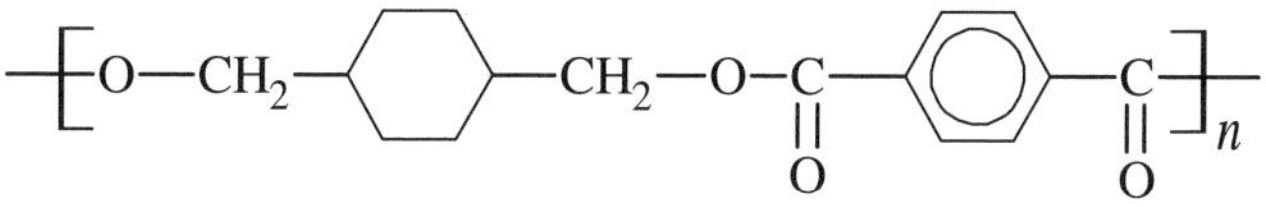

(c)

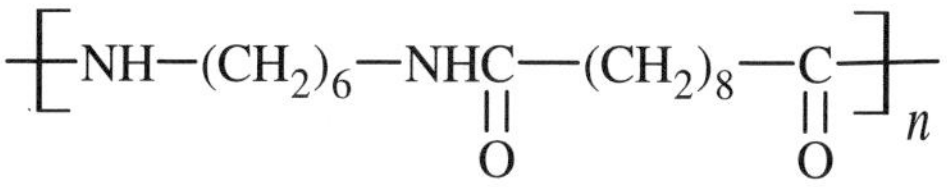

(d)

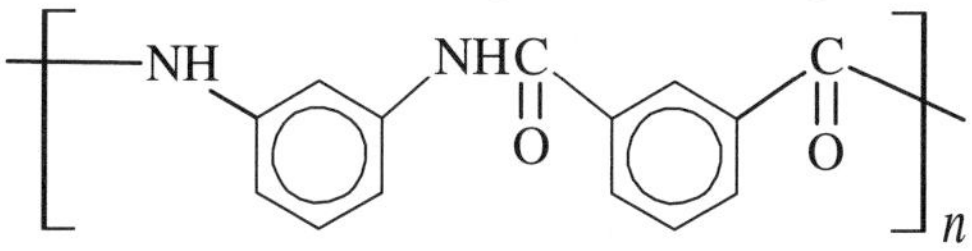

[*Ans.* (a) Poly(tetramethylene terephthalate) or poly(butylene terephthalate); (b) polycyclohexylene terephthalate) or poly(dihydroxymethylcyclohexyl terephthalate); (c) poly(hexamethylene sebacamide); (d) poly(*m*-phenylene isophthalamide)]

1.13 Write repeating formulas and names based on common nomenclature (non-IUPAC) for (a) Nylon-6; (b) Nylon-6,6; (c) Nylon-11; (d) Nylon-6,10; (e) Nylon-5,7.

Chapter 2

Chain Dimensions, Structures, and Transitional Phenomena

2.1 Introduction

The size and shape of a polymer chain are of considerable interest to the polymer scientist, as they influence the physical behavior of a polymer. We shall, however, confine ourselves to the models of the random coil for a polymer chain, as this is usually believed to be most appropriate for synthetic polymers.

The unordered (amorphous) state of aggregation in which the polymer chains also assume random conformations represents one extreme in the physical state of the polymer. This is the state that exists in such amorphous states as solution, melts, or some solids, the randomness being induced by thermal fluctuations. The other extreme is the case where the molecules are able to pack closely in perfect parallel alignment as is found in those polymers that exhibit fibrous behavior—that is, in those possessing a high degree of crystallinity and crystal orientation. In between these two extremes of amorphous and crystalline polymers there is a wide spectrum of polymeric materials with different degrees of crystallinity and amorphous character. These are called *semicrystalline*.

2.2 Polymer Chains: Structures and Dimensions

Before trying to answer questions about the dimension and shape of polymer molecules, we should first consider a simple molecule such as butane and examine the behavior when the molecule is rotated about the bond joining two adjacent carbons. This rotation produces different conformational states of the molecule.

2.2.1 Conformational Changes

The term *conformation* (Koenig, 1980) is used to describe the spatial arrangements of various atoms in a molecule that may occur because of rotations about single bonds. The *planar cis* and *planar trans* conformations of a *n*-butane molecule produced by rotations about the bond joining carbons 2 and 3 are illustrated in Fig. 2.1, while the Newman and "saw horse" projections of the same molecule in these two conformations are shown in Fig. 2.2. Figure 2.3, on the other hand, shows the potential energy of the molecule as a function of the angle ϕ through which the C_3–C_4

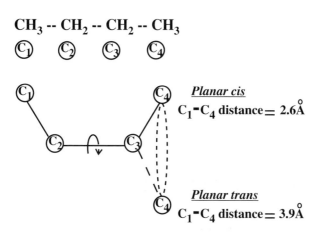

Figure 2.1 Conformations produced by rotation about C_2-C_3 bond of *n*-butane molecule. Each carbon atom in this molecule, being sp^3 hybridized, is tertrahedral with bond angles of 109.50°. The *planar cis* conformation, corresponding to the closest approach of the two bulky methyl groups, is the least stable. Conversely, the *planar trans* conformation, where the bulky groups are farthest apart, is the most stable. (After Young and Lovell, 1990.)

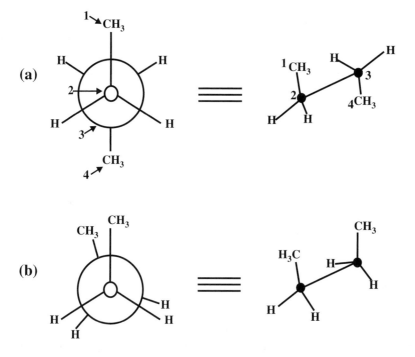

Figure 2.2 Newman and "saw horse" projections for *n*-butane. (a) A staggered state (planar *trans*) with angle of bond rotation $\phi = 0$ and (b) an eclipsed state (planar *cis*) with $\phi = 180°$.

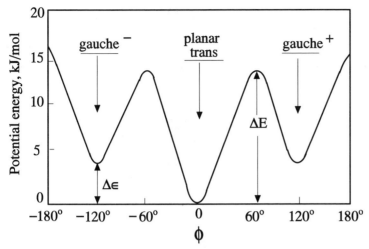

Figure 2.3 Potential energy of *n*-butane as a function of the angle of bond rotation. (After Young and Lovell, 1990.)

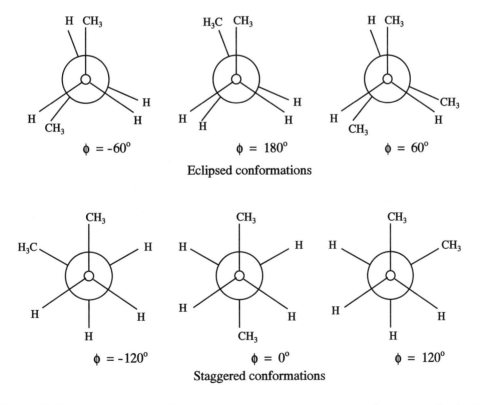

Figure 2.4 Newman projections of the eclipsed and staggered conformations of *n*-butane molecule. (After Young and Lovell, 1990.)

bond is rotated from the *planar trans* conformation ($\phi = 0$) about the plane of C_1–C_2–C_3. The minima correspond to the staggered conformations, namely, one *planar trans* ($\phi = 0$) and two *gauche* ($\phi = \pm 120°$) conformations. All three can be regarded as discrete rotational states. The three staggered conformations are shown as Newman projections in Fig. 2.4 along with the three eclipsed conformations that give rise to the maxima in Fig. 2.3. The higher potential energy of the eclipsed states impose restrictions on the molecular conformation.

For a polymer, the above type of steric interactions occur at a local level in short sequences of chain segments, the conformation about a given chain segment being, however, dependent upon the conformations about the segments on either side to which it is directly connected. Such inter-dependent steric restrictions influence the local chain conformations all along the polymer chain and thus affect its shape and size.

The short-range interactions, as shown above, are important in determining the relative probabilities of existence of different conformations in a polymer chain. Thus the ratio of the number of *trans* (n_t) to *gauche* (n_g) states is given by

$$n_g/n_t = 2 \exp(-\Delta\epsilon/kT) \tag{2.1}$$

where k is the Boltzmann constant, $\Delta\epsilon$ is the energy difference between the energy minima in the *trans* and *gauche* states, and the factor of 2 arises because of the $\pm$ *gauche* states available. When $\Delta\epsilon$ is less than the thermal energy kT provided by collisions of segments, none of the three possible staggered forms will be preferred and all will be equally probable at the given temperature. If this occurs, the overall conformation of an isolated polymer molecule will be a random coil. However, when $\Delta\epsilon > kT$, there will be preference for the *trans* state.

2.2.1.1 Polyethylene

The potential energy diagram for polyethylene $-(CH_2-CH_2-)_n-$ will be expected to be very similar to that for butane shown above, since the conformations can be obtained simply by substituting the $-CH_3$ groups on C_2 and C_3 in Fig. 2.1 by two segments of the polymer chain adjoining the bond of rotation. The distribution of *trans* (t) and *gauche* (g) states along a chain will be a function of the temperature and the relative stability of these states. Consequently, there is an unequal distribution of each (see Problem 2.1).

Problem 2.1   For polyethylene, the energy difference between the *gauche* and *trans* states is about 3.34 kJ/mol. Calculate the ratio of the number of *trans* and *gauche* states along a chain at 100, 200, and 300°K.

Answer:

$\Delta\epsilon = 3.34$ kJ mol^{-1}, $k = 1.38 \times 10^{-23}$ J K^{-1}

$\Delta\epsilon = (3.34 \times 10^3$ J mol$^{-1})/(6.02 \times 10^{23}$ molecules mol$^{-1})$

$\quad = 0.55 \times 10^{-20}$ J (molecule)$^{-1}$

From Eq. (2.1):

$\quad n_g/n_t = 2 \exp[(-0.55 \times 10^{-20})/(1.38 \times 10^{-23} \times T)]$

T :	100°K	200°K	300°K
n_g/n_t :	0.036	0.264	0.524

At low temperatures, *trans* states are thus preponderant.

Since the backbone of polyethylene molecule is composed of a chain of tetrahedral carbon atoms, the molecule in all-*trans* conformation has a linear zigzag structure (like corrugated sheets) shown in (**I**). (The dotted lines denote bonds below and the wedge signifies bonds above the plane of the page.)

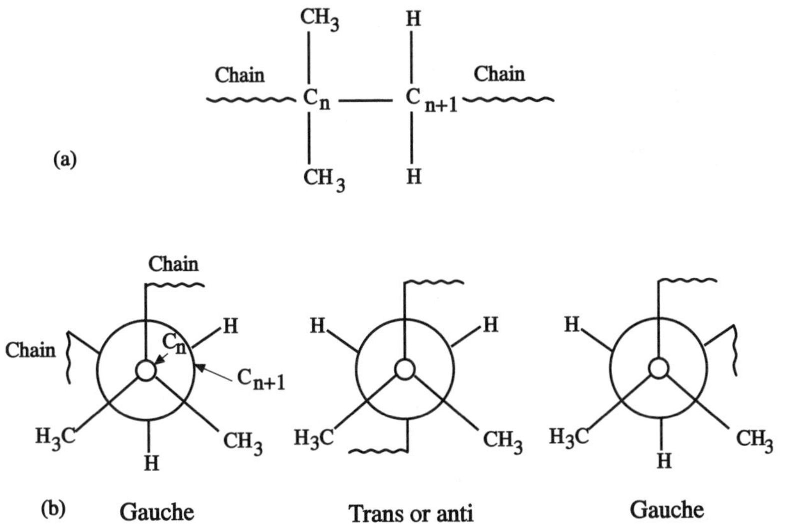

Figure 2.5 (a) Schematic diagram showing two adjacent carbons, C_n and C_{n+1}, in the main chain of polyisobutylene. (b) Newman projections of staggered conformations of adjacent carbons in the main chain.

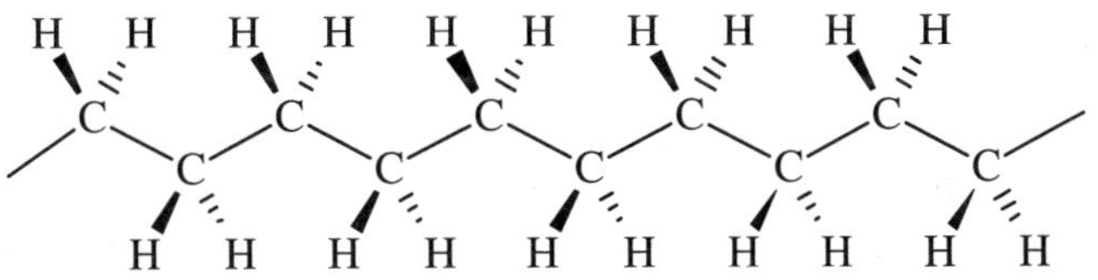

(I)

Linear polyethylene is thus capable of close-packing into tight unit cell and is highly crystalline, despite its low cohesive energy. The all-*trans* zigzag form is the shape of the molecule in crystalline regions of polyethylene. However, at higher temperature the n_g/n_t ratio increases (see Problem 2.1) showing that the chain becomes less extended and more coiled as the temperature increases.

2.2.1.2 Polyisobutylene

To describe the effects of steric restrictions in another polymer, polyisobutylene, consider the Newman projections of the staggered conformations of two adjacent carbons in its repeat unit, as shown in Fig. 2.5. Here the chain substituent on the rear carbon is either between a methyl group and polymer chain or between two methyl groups on the front carbon. There is no significant energy difference between the conformers. Since no conformation is favored, polyisobutylene will tend to form both a helix (*gauche* conformers) and a zigzag (**II**), thereby assuming a highly flexible random coil formation. The polymer thus exhibits rubbery properties at room temperature.

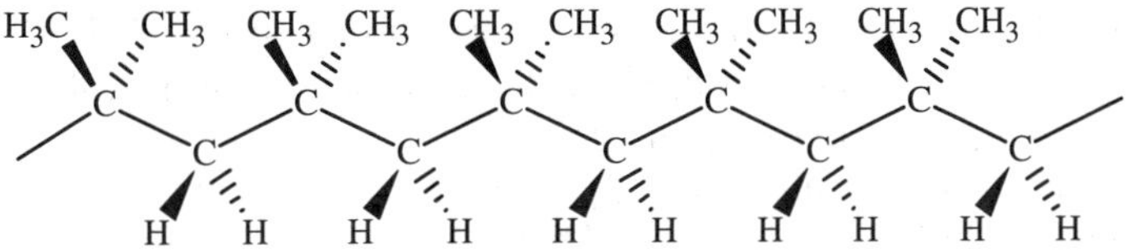

(II)

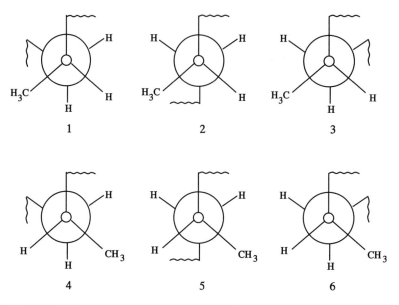

Figure 2.6 Newman projections of the six possible conformations of polypropylene.

2.2.1.3 Polypropylene

Considering again two adjacent carbons in the main chain of the polymer, six conformations are now possible because of the presence of an asymmetrically substituted carbon atom, as shown in Fig. 2.6. Forms **1** and **6** can be neglected for steric reasons; so four different conformations are still possible for the polymer. Atactic polypropylene (see **Stereoisomerism**) has two *trans* forms (**2** and **5**) in the fully extended state (**III**) and so, unlike polyisobutylene, is incapable of crystallizing upon being stretched. Isotactic polypropylene, however, having all the methyl groups on one side, crystallizes easily.

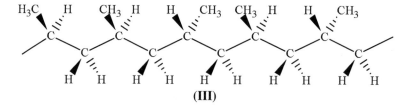

(**III**)

2.2.2 Polymer Conformations in Crystals

The conformation of a polymer in its crystal will generally be that with the lowest energy consistent with regular placement of structural units in the unit cell. Helical conformations occur frequently in polymer crystals. Helices are characterized by a number f_j where f is the number of monomer units per j number of complete turns of the helix. Thus, polyethylene could be characterized as a 1_1 helix in its unit cell with an all-*trans* conformation. The arrangement of the molecules in the polyethylene crystal structure is illustrated in Fig. 2.7.

Isotactic polypropylene crystallizes as a 3_1 helix because the bulky methyl substituents on every second carbon atom in the polymer backbone force the molecule from an all-*trans* con-formation into a *trans/gauche/trans/gauche...* sequence with angles of rotation of $0°$(*trans*) fol-

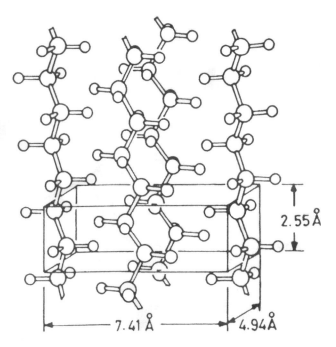

Figure 2.7 Model of the packing of polymer chains in the crystal structure of polyethylene in which $a = 7.41$ Å, $b = 4.94$ Å, and successive pendant atoms are 2.55 Å apart along the chain axis.

lowed by a 120°(*gauche*) twist. In syndiotactic polymers (see **Stereoisomerism**) the substituents are farther apart because the configurations of successive asymmetric carbons alternate and so the *trans/trans/trans*... planar zigzag conformation is generally the lowest energy form. This conformation is observed in syndiotactic poly(vinyl chloride) and 1,2-polybutadiene. Syndiotactic polypropylene can also crystallize in this conformation but a *trans/trans/gauche/gauche*.... sequence forming a 2_1 helix is slightly favored energetically.

In polymers having polar groups, intermolecular electrostatic attractions exert strong influence on chain conformation in their crystals. In polyamides, hydrogen bonds form between the carbonyls and NH groups of neighboring chains (Fig. 2.8) and influence the crystallization of the polymer in the form of sheets, with the macromolecules themselves packed in planar zigzag conformations [see Fig. 2.8(b)].

Problem 2.2 Explain the fact that conversion of the amide groups –CONH– in nylon to methylol groups –CON(CH$_2$OH)– by reaction with formaldehyde, followed by methylation to ether groups – CON(CH$_2$OCH$_3$)– results in the transformation of the fiber to a rubbery product with low modulus and high elasticity.

Answer:

Replacing the –NH– hydrogen in polyamides by an ether group curtails the intermolecular hydrogen bonding. Hence at low degrees of substitution the modulus is reduced and a more elastic fiber is obtained. As the substitution increases, the crystallinity is completely destroyed and rubbery property appears.

Our discussion above on polymer conformations in single chains and in crystals has assumed regularity of macromolecular structure. However, irregularities such as inversions of monomer placements (head-to-head instead of head-to-tail), branches, and changes in configuration may occur. These irregularities, which are considered in a later section, may inhibit crystallization and have a profound effect on polymer properties.

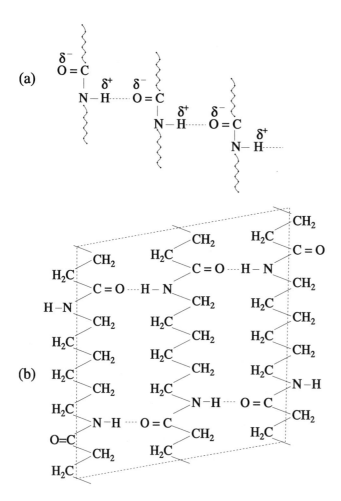

Figure 2.8 (a) Hydrogen bonds between neighboring chains of polyamide. (b) Alignment of chains in hydrogen-bonded sheets in the crystal structure of nylon-6,6. (After Holmes et al., 1955.)

2.2.3 Polymer Size in the Amorphous State

The development of the random coil by H. F. Mark and many further developments by P. J. Flory led to a description of the conformation of chains in the bulk amorphous state. Neutron scattering studies revealed that the conformation in the bulk is close to that found in solution in Θ-solvents (see Chapter 3), thus strengthening the random coil model. On the other hand, some workers suggested that the chains have various degrees of either local or long range order.

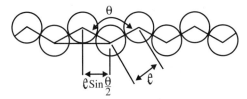

Figure 2.9 Skeletal representation of polyethylene chain in planar zigzag form.

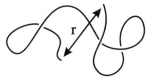

Figure 2.10 Schematic representation of a coiled polymer chain showing the end-to-end distance.

Using the concept of free rotation of the carbon-carbon bond, Guth and Mark (1934) developed the idea of the "random walk" or "random flight" of the polymer chain, which eventually led to the famous relationship between the end-to-end distance of the main chain and the square root of the molecular weight, described below.

2.2.3.1 Freely Jointed Chains

A simple measure of chain dimensions is the length of the chain along its backbone. It is known as the *contour length*. For a chain consisting of n backbone bonds each of length l, this contour length is nl. However, because of the fixed bond angle (109.5°) of carbon, the maximum end-to-end distance of the polymer chain will be somewhat less than nl (see Fig. 2.9 and Problem 2.3). For linear flexible chains that are more like random coils, the distance separating the chain ends, i.e., the *end-to-end distance r* (Fig. 2.10), will be even considerably less than nl.

Problem 2.3   For a linear molecule of polyethylene of molecular weight 1.4×10^5 what would be the end-to-end distance of the polymer molecule in the extended (all-*trans*) state, as compared to the contour length of the molecule ?

Answer:

The polyethylene molecule may be represented skeletally in a planar zigzag form as shown in Fig. 2.9, where $l = 0.154$ nm and $\theta = 109.5°$. In order to perform the calculation, the number n of backbone bonds is required. It can be obtained from the molar mass of the molecule since there is only one methylene group per backbone bond (neglecting chain ends, which make an insignificant contribution to molecular weight for long chains):

$$n \ = \ 1.4 \times 10^5/14 \ = \ 10,000$$

The contour length is the length of the molecule along its backbone and so is given by

Contour length $= \ nl \ = \ 10,000 \ (0.154 \text{ nm}) \ = \ 1540 \text{ nm}$

Since each bond in the fully extended molecule has a projection of $l \sin(\theta/2)$, the end-to-end distance r is given by

$$r \ = \ (10,000) \ (0.154 \text{ nm}) \sin (109.5°/2) \ = \ 1258 \text{ nm}$$

(Magnified a million times, the chain could be represented by a piece of wire 1.26 m long and 0.3 mm in diameter.)

For an isolated polymer molecule it is not possible to assign a unique value of r because the chain conformation (and hence r) changes continuously due to rotation of backbone bonds. Since the single polymer chain can take any of an infinite number of conformations, an average magnitude of r over all possible conformations is therefore computed from the mean of the squares of end-to-end distances and is called the *root mean square* (RMS) end-to-end distance, represented by $\langle r^2 \rangle^{1/2}$, where $\langle \rangle$ means that the quantity is averaged over time. It is given by a rather simple equation (see Appendix 2.1):

$$\langle r^2 \rangle_f^{1/2} = n^{1/2} l \tag{2.2}$$

where the subscript f indicates that the result is for a freely-jointed chain.

Equation (2.2) reveals that $\langle r^2 \rangle_f^{1/2}$ is a factor of $n^{1/2}$ smaller than the contour length nl. Since n is large, this highlights the highly coiled nature of flexible polymer chains.

The dimensions of polymer molecules are also described often in terms of the RMS distance of a chain segment from the center of mass of the molecule. It is defined as the square root of the average squared distance of all the repeating units of the molecule from its center of mass and is known simply as the RMS *radius of gyration*, $\langle S^2 \rangle^{1/2}$. It is thus given by (Orfino, 1961)

$$\langle S^2 \rangle = \langle (1/n) \sum_{i=1}^{n} s_i^2 \rangle \tag{2.3}$$

where s_i is the vector distance from the center of mass to the ith unit of the chain in one particular conformation of the molecule and n is the total number of units. The angular brackets denote a linear average over all possible conformations of the molecular chain.

The radius of gyration has the advantage that it also can be used to characterize the dimensions of branched macromolecules (with more than two chain ends) and cyclic macromolecules (with no chain ends). Moreover, properties of dilute polymer solutions that are dependent on chain dimensions are controlled by $\langle S^2 \rangle^{1/2}$ rather than by $\langle r^2 \rangle^{1/2}$.

The radius of gyration is directly measurable by light scattering (see Chapter 4), neutron scattering, and small angle scattering experiments, whereas the end-to-end distance is not directly observable and has no significance for branched species as they have more than two ends. However, for high-molecular-weight linear macromolecules that have random coil shapes, $\langle S^2 \rangle^{1/2}$ is uniquely related to $\langle r^2 \rangle^{1/2}$ by

$$\langle S^2 \rangle^{1/2} = \frac{\langle r^2 \rangle^{1/2}}{\sqrt{6}} \tag{2.4}$$

and so in the theoretical treatment of linear flexible chains it is usual to consider only $\langle r^2 \rangle^{1/2}$.

2.2.3.2   Real Polymer Chains

The freely jointed chain is a simple model for predicting chain dimensions. It is, however, physically unrealistic. Since each carbon atom in a real polymer chain is tetrahedral with fixed valence bond angles of 109.5°, the links are subject to bond angle restrictions. Moreover, the links do not rotate freely because, as we have seen earlier, there are energy differences between different conformations (cf. Fig. 2.3). Both of these effects cause $\langle r^2 \rangle^{1/2}$ to be larger than that predicted by the freely jointed chain model.

The first modification to the freely jointed chain model is the introduction of bond angle restrictions while retaining the concept of free rotation about bonds. This is called the *valence angle model*. For a polymer chain with all backbone bond angles equal to θ, this leads to Eq. (2.5) for the mean square end-to-end distance

$$\langle r^2 \rangle_{fa} = nl^2 \left(\frac{1 - \cos \theta}{1 + \cos \theta} \right) \tag{2.5}$$

where the subscript fa indicates that the equation is applicable for chains in which the bonds rotate freely about a fixed bond angle.

Since $180° > \theta > 90°$, $\cos \theta$ is negative and $\langle r^2 \rangle$ is greater than nl^2 of the freely jointed chain model [Eq. (2.2)]. For polymers having C–C backbone bonds with $\theta \simeq 109.5°$ for which $\cos \theta \simeq -\frac{1}{3}$, the equation becomes

$$\langle r^2 \rangle_{fa} = 2nl^2 \tag{2.6}$$

Thus for polymers such as linear polyethylene, bond angle restrictions cause the RMS end-to-end distance to increase by a factor of $\sqrt{2}$ from that of the freely-jointed chain.

Problem 2.4   For a linear molecule of polyethylene of molecular weight 1.4×10^5 what would be the RMS end-to-end distance according to the valence angle model as compared to that according to the freely-jointed chain model and the end-to-end distance of a fully extended molecule? Comment on the values obtained, indicating which one is a more realistic estimate of chain dimensions.

Answer:

Length of each bond, $l = 0.154$ nm (see Fig. 2.9); number of bonds, $n = 10,000$ (see Problem 2.3)

End-to-end distance of a fully extended molecule = 1258 nm (see Problem 2.3)

The RMS end-to-end distance, according to freely jointed chain model, is obtained from Eq. (2.2):

$$\langle r^2 \rangle_f^{1/2} = (10,000)^{1/2} (0.154 \text{ nm}) = 15.4 \text{ nm}$$

The RMS end-to-end distance according to valence angle model is obtained from Eq. (2.5):

$$\langle r^2 \rangle_{fa}^{1/2} = (10,000)^{\frac{1}{2}} (0.154 \text{ nm}) \left\{ \frac{1 - \cos(109.5^o)}{1 + \cos(109.5^o)} \right\}^{\frac{1}{2}} = 21.8 \text{ nm}$$

Comment: The most realistic of the above three estimates of chain dimensions is that afforded by the valence angle model. This is because it takes into account coiling of the molecule and also restriction due to fixed valence bond angle.

The valence angle model, though more realistic than the freely jointed model, still underestimates the true dimensions of polymer molecules, because it ignores restrictions upon bond rotation arising from short-range steric interactions. Such restrictions are, however, more difficult to quantify theoretically. A simpler procedure is to assume that the conformations of each sequence of three backbone bonds are restricted to the *rotational isomeric states* that correspond to the potential energy minima such as those shown for *n*-butane in Fig. 2.3. For the simplest case of polyethylene and for vinylidene-type polymers, the application of the rotational isomeric state theory yields the following equation

$$\langle r^2 \rangle_{ha} = nl^2 \left(\frac{1 - \cos\theta}{1 + \cos\theta} \right) \left(\frac{1 + \overline{\cos\phi}}{1 - \overline{\cos\phi}} \right) \tag{2.7}$$

where the subscript *ha* indicates that the result is for a polymer chain with hindered rotation about a fixed bond angle. The quantity $\overline{\cos\phi}$ represents the average value of $\cos\phi$, where ϕ is the angle of bond rotation as defined for *n*-butane (see Figs. 2.3 and 2.4). If the bond rotation is free or unrestricted, all values of ϕ are equally probable causing the positive and negative values of $\cos\phi$ to cancel each other out, so that $\overline{\cos\phi} = 0$ and Eq. (2.7) then reduces to Eq. (2.5). However, due to short-range steric restrictions, values of $|\phi| < 90°$ are more probable which means that $\overline{\cos\phi}$ is positive and $\langle r^2 \rangle_{ha}$ is greater than $\langle r^2 \rangle_{fa}$. Moreover, if bulky side groups occur on the polymer chain, e.g., phenyl group in polystyrene, there will be additional steric interactions due to side groups, producing further hindrance to bond rotation. Since these effects are difficult to evaluate theoretically, Eqn. (2.7) is usually written in a semi-empirical form (Young and Lovell, 1990) as

$$\langle r^2 \rangle_0 = \sigma^2 nl^2 \left(\frac{1 - \cos\theta}{1 + \cos\theta} \right) \tag{2.8}$$

where σ is a *steric parameter* and is the factor by which $\langle r^2 \rangle_0^{1/2}$ exceeds $\langle r^2 \rangle_{fa}^{1/2}$. It is usually evaluated from values of $\langle r^2 \rangle_0^{1/2}$ measured experimentally (the subscript o indicating that the result is observed and not calculated) and typically has values between 1.5 and 2.5.

One can obtain an idea of the stiffness of a polymer chain from the ratio $\langle r^2 \rangle_0^{1/2} / \langle r^2 \rangle_f^{1/2}$, that is, the square root of the ratio, $C_\infty = \langle r^2 \rangle_0 / nl^2$, called the *characteristic ratio* (Young and Lovell, 1990). The value of $C_\infty^{1/2}$ indicates how much greater the RMS end-to-end distance of a real polymer chain is, compared to that of the freely jointed chain.

Problem 2.5   A real polymer chain consisting of n bonds each of length l may be usefully represented by an *equivalent freely jointed chain* of N links each of length b such that it will have the same end-to-end distance and the same contour length. Obtain N and b in terms of the characteristic ratio C_∞ of the polymer chain.

Answer:
If both chains have the same end-to-end distance, then

$$\langle r^2 \rangle_0 = Nb^2 \qquad\qquad (P2.5.1)$$

Also, if both have the same contour length, then

$$nl = Nb \qquad\qquad (P2.5.2)$$

From Eqs. (P2.5.1) and (P2.5.2),

$$b = \frac{\langle r^2 \rangle_o}{nl} = \frac{\langle r^2 \rangle_o l}{nl^2} = C_\infty l \qquad\qquad (P2.5.3)$$

and

$$N = \frac{n^2 l^2}{\langle r^2 \rangle_o} = \frac{n}{C_\infty} \qquad\qquad (P2.5.4)$$

When an average end-to-end distance of a macromolecular coil is $\langle r^2 \rangle_{ha}^{1/2}$ as given by Eq. (2.7) or more generally $\langle r^2 \rangle_0^{1/2}$ as given by Eq. (2.8), the polymer is said to be in its "unperturbed" state (see below). Henceforth the subscript 'o' will be used to indicate the unperturbed state of a polymer molecule.

The model considered above does not impose any restriction on the relative positions of the bonds widely separated in the chain, or, in other words, does not prevent parts of the chain occupying the same space. In a real situation, however, this cannot happen as each part of an isolated polymer molecule excludes other more remotely connected parts from its volume. Because of these long-range steric interactions, the true RMS end-to-end distance, $\langle r^2 \rangle^{1/2}$, is greater than the unperturbed dimension $\langle r^2 \rangle_o^{1/2}$. These effects are usually considered in terms of an *excluded volume* (see Chapter 3) and the extent to which unperturbed dimensions are perturbed in real chains is defined by an expansion factor α such that

$$\langle r^2 \rangle^{1/2} = \alpha \langle r^2 \rangle_0^{1/2} \qquad\qquad (2.9)$$

Perturbations in chain dimensions are also caused by interactions of the chain with its molecular environment, e.g., with solvent molecules or other polymer molecules. The effects of these interactions are also reflected in the expansion parameter α. In a *theta solvent* (see Chapter 3), α is reduced to unity and the end-to-end distance of a polymer chain is then the same as it would be in bulk polymer at the same temperature.

Problem 2.6   Assuming that the RMS end-to-end distance is an approximation to the diameter of the spherical, coiled polymer in dilute solution, calculate the volume occupied by one molecule of polystyrene (molecular weight 10^6) in a *theta* solvent at 25°C. (Carbon-carbon bond length = 1.54×10^{-8} cm; tetrahedral bond angle $\simeq 109.5°$; steric parameter, σ, for polystyrene at 25°C = 2.3)

Answer:

Molar mass of styrene $= 104 \text{ g mol}^{-1}$

Degree of polymerization $= (10^6 \text{ g mol}^{-1})/(104 \text{ g mol}^{-1}) = 9615$

Number of backbone C–C bonds $= 2 \times 9615 = 19230$

From Eq. (2.8): $\langle r^2 \rangle_0 = \sigma^2 n l^2 \left(\dfrac{1 - \cos\theta}{1 + \cos\theta} \right)$

$\sigma = 2.3, \quad n = 19230, \quad l = 1.54 \times 10^{-8} \text{ cm}, \quad \theta = 109.5°, \quad \cos\theta \simeq -1/3.$

Therefore,

$$\langle r^2 \rangle_0 = (2.3)^2 (19230)(1.54 \times 10^{-8} \text{ cm})^2 (2)$$

$$= 2.4 \times 10^{-11} \text{ cm}^2$$

Volume in a *theta* solvent (i.e., $\alpha = 1$) $= (\pi/6)(2.4 \times 10^{-11} \text{ cm}^2)^{3/2}$

$$= 4.4 \times 10^{-16} \text{ cm}^3$$

2.3   Constitutional and Configurational Isomerism

If different isomers of a molecule have different *configurations*, that is, spatial arrangements that cannot be interchanged (without at least momentary breaking of bonds), then it is referred to as *configurational isomerism*. Another type of isomerism, which involves constitutional variations of a molecule, is referred to as *constitutional isomerism*.

2.3.1   Constitutional Isomerism

Except in monomers like ethylene and tetrafluoroethylene where the substituents on the two carbons are identical, the two carbons of the double bond in a vinyl type monomer are distinguishable. One of them can be arbitrarily labeled the *head* and the other the *tail* of the monomer, as shown below for vinyl chloride (**IV**). During polymerization, the monomer in principle can be joined by head-to-tail or head-to-head/tail-to-tail (**V**) additions, as shown by Eqs. (P2.7.1) and (P2.7.2), respectively. However, head-to-tail enchainment is the predominant constitution of most vinyl monomers (see Problem 2.7).

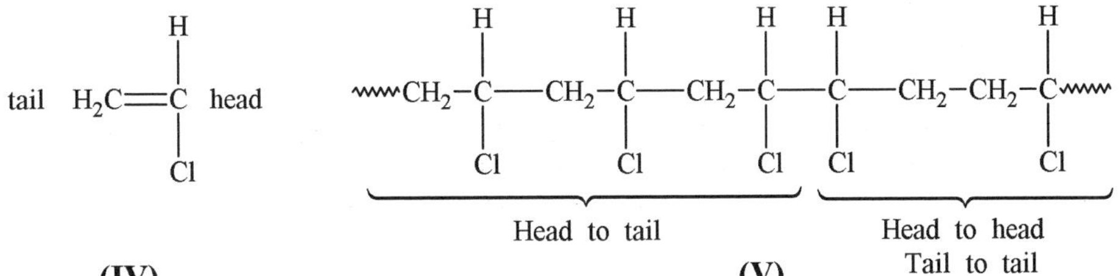

Problem 2.7   Explain the fact that in all polymerizations the head-to-tail addition is usually the predominant mode of propagation.

Answer:

Vinyl monomers polymerize by attack of an active center (**VI**) on the double bond. Equation (P2.7.1) shows the propagation step in head-to-tail enchainment and Eq. (P2.7.2) that in head-to-head or tail-to-tail enchainment:

$$\text{wwCH}_2-\overset{\overset{\displaystyle X}{|}}{\underset{\underset{\displaystyle Y}{|}}{C}}{}^* \;+\; \text{H}_2\text{C}{=}\overset{\overset{\displaystyle X}{|}}{\underset{\underset{\displaystyle Y}{|}}{C}} \longrightarrow \text{wwCH}_2-\overset{\overset{\displaystyle X}{|}}{\underset{\underset{\displaystyle Y}{|}}{C}}-\text{CH}_2-\overset{\overset{\displaystyle X}{|}}{\underset{\underset{\displaystyle Y}{|}}{C}}{}^*$$

(VI)   **(VII)**   (P2.7.1)

$$\text{wwCH}_2-\overset{\overset{\displaystyle X}{|}}{\underset{\underset{\displaystyle Y}{|}}{C}}{}^* \;+\; \text{H}_2\text{C}{=}\overset{\overset{\displaystyle X}{|}}{\underset{\underset{\displaystyle Y}{|}}{C}} \longrightarrow \text{wwCH}_2-\overset{\overset{\displaystyle X}{|}}{\underset{\underset{\displaystyle Y}{|}}{C}}-\overset{\overset{\displaystyle X}{|}}{\underset{\underset{\displaystyle Y}{|}}{C}}-\text{CH}_2^*$$

(VI)   **(VIII)**   (P2.7.2)

The active center involved in the propagation reaction may be a free-radical, ion, or metal-carbon bond (see Chapters 6-10). A propagating species will be more stable if the unpaired electron or ionic charge at the end of the chain can be delocalized across either or both substituents X and Y. Such resonance stabilization is possible in **(VII)** but not in **(VIII)**. Moreover when X and/or Y is bulky there will be more steric hindrance in reaction of Eq. (P2.7.2) than in the reaction of Eq. (P2.7.1). So, in general, head-to-tail addition as in Eq. (P2.7.1) is considered to be the predominant mode of propagation in all polymerizations.

Problem 2.8   A chemical method of determining head-to-head structures in poly(vinyl alcohol) is by means of the following difference in diol reactions (Flory, 1950):

$$\text{wwC}\overset{\overset{\displaystyle H}{|}}{\underset{\underset{\displaystyle OH}{|}}{}}{-}\overset{\overset{\displaystyle H}{|}}{\underset{\underset{\displaystyle OH}{|}}{C}}\text{ww} \;+\; \text{HIO}_4 \longrightarrow \text{wwC}\overset{\overset{\displaystyle H}{|}}{\underset{\underset{\displaystyle O}{\|}}{}} \;+\; \overset{\overset{\displaystyle H}{|}}{\underset{\underset{\displaystyle O}{\|}}{C}}\text{ww} \;+\; \text{HIO}_3 \;+\; \text{H}_2\text{O}$$

(1, 2 Diol)

$$\text{wwC}\overset{\overset{\displaystyle H}{|}}{\underset{\underset{\displaystyle OH}{|}}{}}{-}\text{CH}_2{-}\overset{\overset{\displaystyle H}{|}}{\underset{\underset{\displaystyle OH}{|}}{C}}\text{www} \;\xrightarrow[\text{Soln.}]{\text{HIO}_4}\; \text{No reaction}$$

(1, 3 Diol)

Poly(vinyl acetate) of number-average molecular weight 250,000 is hydrolyzed by base-catalyzed transesterification with methanol to yield poly(vinyl alcohol). Oxidation of the latter with periodic acid yields a poly(vinyl alcohol) with number-average degree of polymerization 485. Calculate the percentages of head-to-tail and head-to-head linkages in poly(vinyl acetate).

Answer:

Repeat unit of poly(vinyl acetate) : $-\!\!\left[\text{CH}_2{-}\text{CH(OCOCH}_3)\right]\!\!-$

Molar mass of repeat unit = 86 g mol^{-1}

$$\overline{X}_n \;=\; \frac{(250{,}000 \text{ g mol}^{-1})}{(86 \text{ g mol}^{-1})} \;=\; 2907$$

Assume that poly(vinyl acetate) is completely hydrolyzed so that $\overline{X}_n$ of the resulting poly (vinyl alcohol) is also 2907. The fact that $\overline{X}_n$ is reduced from 2907 to 485 upon treatment with periodic acid means that each poly(vinyl alcohol) molecule is, on the average, cleaved to yield 2907/485 or 6 smaller molecules each of

$\overline{X}_n = 485$ and so the number of cleavages $= 6 - 1 = 5$. Therefore, the average polymer molecule has 5 head-to-head linkages out of a total of $2907 - 1 = 2906$ linkages in the polymer chain.

$$\% \text{ head-to-head} = \frac{6 \times 100}{2906} = 0.20 \%$$

$$\% \text{ head-to-tail} = 100 - 0.2 = 99.8 \%.$$

Problem 2.9 Polymers of dienes (hydrocarbons containing two carbon-carbon double bonds), such as butadiene and isoprene, have the potential for head-to-tail and head-to-head isomerism and variations in double-bond position as well. How many constitutional isomers can form in the polymerization of (a) polybutadiene and (b) polyisoprene ?

Answer:

The conjugated diene butadiene can polymerize to produce 1,4 and 1,2 products (Rudin, 1982):

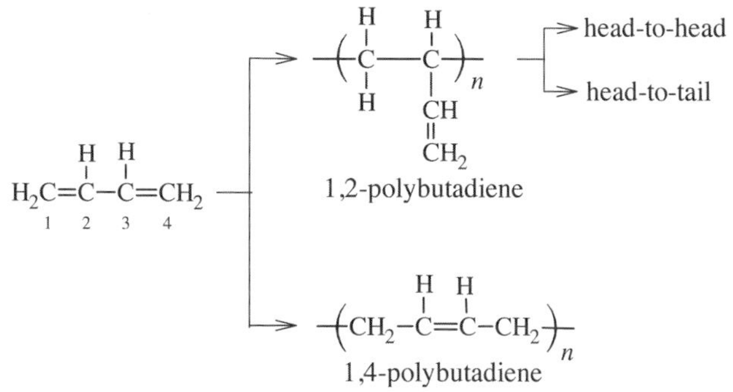

1,2-polybutadiene

1,4-polybutadiene

Thus, there are three possible constitutional isomers and, in addition, there is the possibility of mixed structures. Note that there is no 3,4-polybutadiene because it is identical with 1,2-polybutadiene. This is not the case with 2-substituted conjugated butadienes like isoprene and chloroprene. Thus, as shown below, there are six possible constitutional isomers of isoprene or chloroprene, to say nothing of the potential for mixed structures (Rudin, 1982) :

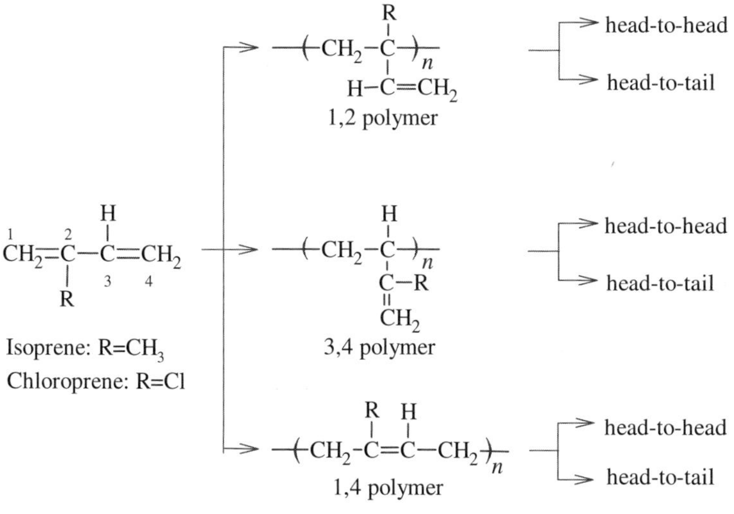

Isoprene: R=CH$_3$
Chloroprene: R=Cl

1,2 polymer

3,4 polymer

1,4 polymer

Elastomeric behavior is shown by 1,4-polymer, particularly if the polymer structure is *cis* about the residual double bond (see Section 2.3.2.1).

2.3.2 *Configurational Isomerism*

The *configuration* of a molecule specifies the relative spatial arrangement of bonds in the molecule (of given constitution) irrespective of the changes in molecular shape which can arise because of rotations about single bonds. A given configuration can thus be changed only by breaking and reforming of chemical bonds. There are two types of configurational isomerism in polymers and these are analogous to *geometrical* and *optical isomerisms* in micromolecular chemistry.

2.3.2.1 *Geometrical Isomerism*

Since rotation cannot take place about a double bond between two carbon atoms, two nonsuperimposable configurations (geometrical isomers) are possible if the two substituents on each carbon differ from each other. For example, the two monomers maleic acid (**IX**) and fumaric acid (**X**) are geometrical isomers, designated *cis* and *trans*, respectively.

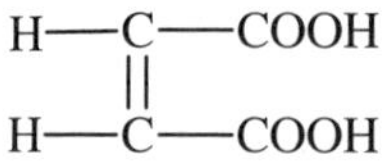

$$
\begin{array}{ll}
\text{H}-\text{C}-\text{COOH} & \text{H}-\text{C}-\text{COOH} \\
\quad\quad \| & \quad\quad \| \\
\text{H}-\text{C}-\text{COOH} & \text{HOOC}-\text{C}-\text{H} \\
\quad\quad \textbf{(IX)} & \quad\quad\quad \textbf{(X)}
\end{array}
$$

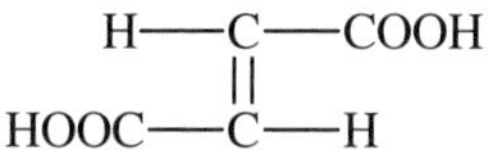

In solid, the molecules of *trans* isomers pack more closely and crystallize more readily than those of *cis* isomers. This is reflected in the fact that the melting point of fumaric acid is about 160°C higher than that of maleic acid. Similarly, the differences in the properties of *cis* and *trans* isomers of polymers are also significant, as shown below with the example of poly(isoprene).

Natural rubber is 1,4-polyisoprene and the polymer configuration is *cis* at each double bond in the chain, as shown in (**XI**). Consequently, the polymer molecule has a bent and less symmetrical structure. Natural rubber does not crystallize at room temperature and is amorphous and elastomeric. Balata (guttapercha) is also 1,4-polyisoprene, but the polymer configuration is *trans* at the double bond (**XII**). The molecule is more extended and has symmetrical structure. The *trans* isomer is thus a nonelastic, hard, and crystalline polymer. It is used as a thermoplastic.

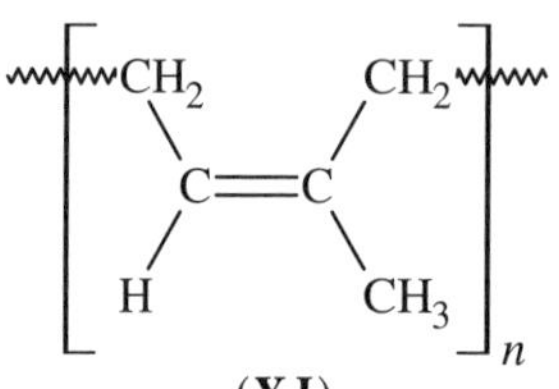

$$
\textbf{(XI)} \quad\quad\quad\quad \textbf{(XII)}
$$

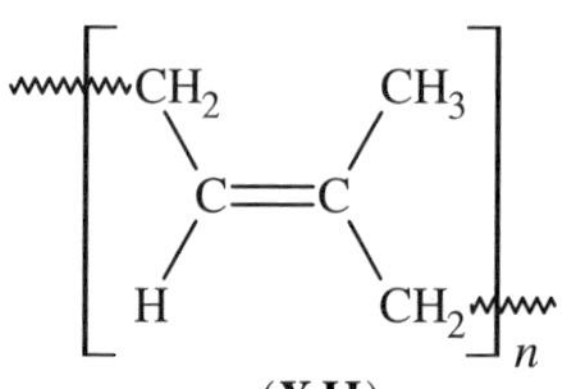

2.3.2.2 *Stereoisomerism*

Optical Activity in Polymers   Stereoisomerism in polymers is formally similar to the optical isomerism of organic chemistry. In a vinyl polymer with the general structure shown in (**XIII**) every other carbon atom in the chain, labeled C^*, is a site of steric isomerism, because it has four different substituents, namely, X, Y, and two sections of the main chain that differ in length (Rudin, 1982).

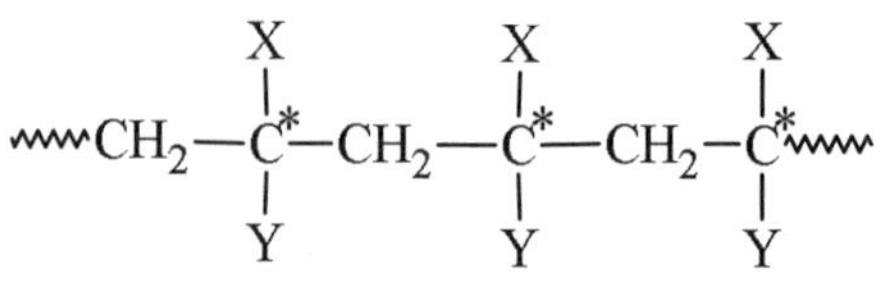

$$
\textbf{(XIII)}
$$

Problem 2.10 Explain why polypropylene of relatively high molecular weight is optically inactive despite having an asymmetric center at every other carbon, while, on the other hand, poly(propylene oxide) is optically active.

Answer: Optical activity is influenced only by the first few atoms around an asymmetric carbon ($C^\star$). For the two sections of the main chain, these will be identical regardless of the length of the whole polymer chain. The carbons marked $C^\star$ in (**XIII**) are thus not truly asymmetric and are termed *pseudoasymmetric* or *pseudochiral* carbons. Only those $C^\star$ centers near the ends of a polymer molecule will be truly asymmetric, but since there are too few chain ends in a high molecular-weight polymer such centers do not confer any significant optical activity on the molecule as a whole. Polypropylene is thus optically inactive.

In poly(propylene oxide), $-\!\!\lfloor CH_2 C^\star(H)(CH_3)O\rfloor\!\!-$, the $C^\star$ is true asymmetric center, as it is surrounded by –H, –CH$_3$, –CH$_2$, and –O–. The polymer is therefore optically active.

It is easy to see that there are two distinct configurational arrangements of the repeat unit of (**XIII**), viz., (**XIV**) and (**XV**), where the wedge and the dotted lines denote bonds which are extending above and below the plane of the paper, respectively. These two *stereoisomers* of the repeat unit cannot be interchanged by bond rotation and they exist because the substituted carbon atom, labeled $C^\star$, is attached to four different groups. Thus, every $C^\star$ may have one or other of the two configurations. One of the configurations is designated as D (or d) and the other as L (or l). The configuration is fixed when the polymer molecule is formed and is independent of any rotations of the main chain carbons about the single bonds connecting them.

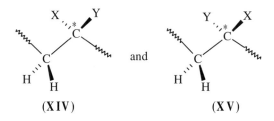

and

(**XIV**)   (**XV**)

Tacticity in Polymers As explained above, the $C^\star$ atoms in (**XIII**) are pseudochiral and hence do not give rise to optical activity; the two mirror image configurations remain distinguishable, however. The different possible configurations or spatial arrangements that occur as a consequence are called the *tacticity* of the polymer (Rudin, 1982; Cowie, 1991).

The usual way to picture the configurational nature of a vinyl polymer is to consider the polymer backbone stretched out so that the bonds between the main chain carbons form a planar zigzag pattern in the plane of the paper and the X and Y substituents on successive pseudoasymmetric or pseudochiral carbons lie either above or below the plane of the backbone, as shown in Fig. 2.11.

If the substituents, say X, are then all above or all below the plane [Fig. 2.11(a)], the configurations of successive pseudoasymmetric carbons are the same; the polymer is termed *isotactic* and designated as *dddd*... or *llll*.... If, however, a given substituent appears alternately above and below the reference plane in the planar zigzag conformation [Fig. 2.11(b)], the polymer is termed *syndiotactic* and designated as *dldl*... or *ldld*.... In both cases, configurations of successive pseudochiral carbons are *regular*, and the polymer is said to be *stereoregular* or *tactic*. When the configurations of pseudoasymmetric carbons are more or less random [Fig. 2.11(c)], the polymer is not stereoregular and is said to be *atactic*. Those polymerizations that yield tactic polymers are called *stereospecific*. It is useful at this point to reiterate that stereoisomerism does not exist if the substituents X and Y are the same.

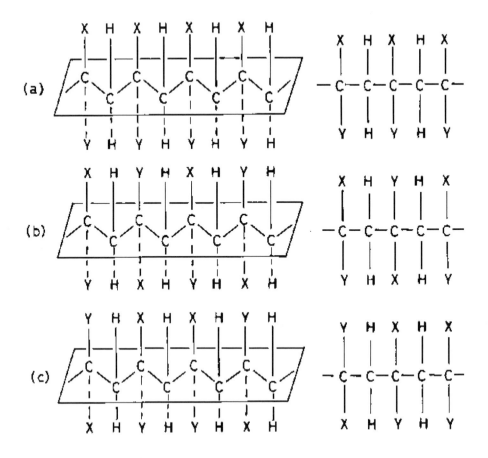

Figure 2.11 Diagrams of (a) isotactic, (b) syndiotactic, and (c) atactic configurations of $-(CH_2\text{-}CXY)_n-$ polymer. The corresponding Fisher projections are shown on the right.

Stereoregularity in vinyl polymers has a profound influence on the crystallizability of the material. Polymer chains with stereoregularity are better able to pack together in a regular manner and crystallize. The increase in crystallinity is reflected in higher melting point, greater rigidity, and less solubility compared to amorphous species with the same constitution. A striking example is provided by polypropylene. Isotactic polypropylene has a crystal melting point of 176°C and is widely used as fiber and filament and in automotive, furniture, and appliance applications, while the atactic polymer is a rubbery amorphous material that finds no important use. Syndiotactic polymers, as we have seen above, are stereoregular and so are crystallizable. They, however, do not have the same mechanical properties as isotactic polymers. Thus most of the highly stereoregular polymers of commercial importance are isotactic, and relatively few syndiotactic polymers are made (Rudin, 1982).

It is important to note at this point that completely tactic and completely atactic polymers represent extremes of stereoisomerism and polymers often possess intermediate degrees of tacticity. The most powerful tool for measuring the extent of stereoregularity in polymers is nuclear magnetic resonance (NMR) spectroscopy (see below).

Meso- and Racemic Placements The Fisher projections in Fig. 2.11 show that placement of the groups in isotactic structures corresponds to a *meso* (same) or *m*-placement of a pair of consecutive pseudochiral centers, whereas the syndiotactic structure corresponds to a *racemic* (opposite) or *r*-placement of the corresponding pair of pseudochiral centers. It is important to note here that the *m* or *r* notation refers to the configuration of one pseudochiral center relative to its neighbor. Thus, the meso dyad is designated *m*, and the racemic dyad *r*. The system of nomenclature can be extended to sequences of any length. Thus, an isotactic triad is *mm*, a heterotactic triad is *mr*, and a syndiotactic triad *rr*. Several such configurational sequences are illustrated in Table 2.1. Each of these, and even more complicated combinations, can be designated through NMR studies.

Let us assume that the probability of generating a meso sequence when a new monomer unit is added at the end of a growing chain can be denoted by a single parameter, P_m (and similarly P_r for a racemic sequence). Implied in it is an assumption that the polymer obeys Bernoullian statistics, that is, the probability of forming an *m* or *r* sequence is independent of the stereochemical configuration of the chain already formed. Since addition can be only *m* or *r* type, it is obvious that $P_m + P_r = 1$.

A triad sequence involves two monomer additions. Thus the Bernoullian probabilities of *mm*, *mr*, and *rr* are P_m^2, $2P_m(1 - P_m)$ and $(1 - P_m)^2$, respectively. A plot of these relations is shown in Fig. 2.12. It will be noted that the proportion of *mr*, that is, heterotactic units, rises to a maximum at $P_m = 0.5$, corresponding to random propagation. For a random polymer ($P_m = 0.5$) the proportion *mm* : *mr* : *rr* will be 1:2:1. The P_m values of 0 and 1 correspond to completely syndiotactic and isotactic polymers, respectively.

Problem 2.11 Derive the Bernoullian probabilities for triad sequences in terms of the parameter P_m.

Answer:

Denoting the rates of meso and racemic dyad placements by R_m and R_r, respectively, the probabilities P_m and P_r can be defined by

$$P_m = R_m/(R_m + R_r), \quad P_r = R_r/(R_m + R_r), \quad P_m + P_r = 1$$

Since P_m and P_r are synonymous with isotactic and syndiotactic dyad fractions *m* and *r*, respectively, probabilities for isotactic, syndiotactic, and heterotactic triad sequences will be given by

$$mm = P_m^2, \quad rr = P_r^2 = (1 - P_m)^2, \quad mr = 2P_m(1 - P_m)$$

The coefficient of 2 for the heterotactic triad arises from the fact that the heterotactic triad can be produced in two ways (as *mr* and *rm*).

For any given polymer following simple Bernoullian statistics, the *mm*, *mr*, *rr* sequence frequencies, as estimated from the relative areas of the appropriate peaks in NMR spectra (see below), should lie on a single vertical line in Fig 2.12, corresponding to a single value of P_m (see Problem below). If this is not the case, then the polymer's configurational sequence deviates from the Bernoullian statistics.

Plots such as Fig. 2.12 can be made (Bovey and Tiers, 1960) for tetrad and other higher order probabilities (or fractions) as a function of P_m using the relationships given in Table 2.1. These are useful for peak assignments in NMR spectra, aided by certain necessary relationships (Bovey, 1969) among the frequencies of occurrences of sequences, which must hold regardless of the configurational statistics (Bernoullian or not).

Table 2.1 Configurational Sequences (Bovey, 1969)

	Designation	Projection	Bernoullian probability
Dyad	m (meso)		P_m
	r (racemic)		$(1-P_m)$
Triad	mm (Isotactic)		P_m^2
	mr (heterotactic)		$2P_m(1-P_m)$
	rr (syndiotactic)		$(1-P_m)^2$
Tetrad	mmm		P_m^3
	mmr		$2P_m^2(1-P_m)$
	rmr		$P_m(1-P_m)^2$
	mrm		$P_m^2(1-P_m)$
	rrm		$2P_m(1-P_m)^2$
	rrr		$(1-P_m)^3$

	Designation	Projection	Bernoullian probability
Pentad	mmmm (Isotactic)		P_m^4
	mmmr		$2P_m^3(1-P_m)$
	rmmr		$P_m^2(1-P_m)^2$
	mmrm		$2P_m^3(1-P_m)$
	mmrr		$2P_m^2(1-P_m)^2$
	rmrm (heterotactic)		$2P_m^2(1-P_m)^2$
	rmrr		$2P_m(1-P_m)^3$
	mrrm		$P_m^2(1-P_m)^2$
	rrrm		$2P_m(1-P_m)^3$
	rrrr (syndiotactic)		$(1-P_m)^4$

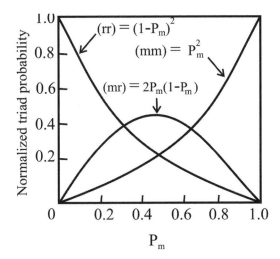

Figure 2.12 The probabilities (or fractions) of isotactic (*mm*), heterotactic (*mr*), and syndiotactic (*rr*) triads as a function of P_m, the probability of isotactic monomer placement during propagation. (Bovey and Tiers, 1960.)

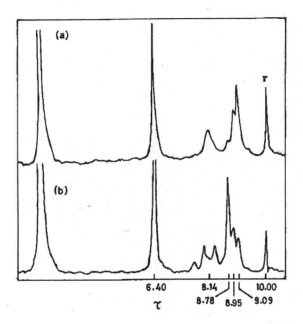

Figure 2.13 NMR spectra of poly(methyl methacrylate) (15% solution in chloroform, *r* = tetramethyl silane reference peak). (a) Mainly syndiotactic; (b) mainly isotactic. The methyl ester group appears at 6.40τ in both spectra and is unchanged by chain configuration. (Bovey, 1969.)

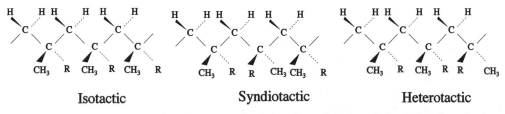

Figure 2.14 Isotactic, syndiotactic, and heterotactic triad configurations for poly(methyl methacrylate).

NMR Measurement of Tacticity   High resolution NMR spectroscopy makes the quantitative measurement of sequence distribution feasible (see Problem below) and allows one to compare actual polymer chains with theoretical predictions.

Problem 2.12   Two typical spectra of poly(methyl methacrylate) samples (Bovey, 1969), known to be predominantly isotactic and syndiotactic, are shown in Fig. 2.13. Justify the assignment of various peaks in these spectra.

Answer:

Since the three protons in an α-methyl group are equivalent, they will absorb radiation at a single frequency, but this frequency will be different for each of the three types of triads, because the environment of α-methyl groups in each is different (see Fig. 2.14). The α-methyl groups give the set of three peaks at 8.78τ, 8.95τ, and 9.09τ. For a sample that is mainly isotactic, the peak at 8.78τ is much larger than the other two. So it is assigned to the α-methyl groups which are flanked on each side by a monomer unit with similar configuration (see Fig. 2.14). Such a group of three units constitutes an isotactic (*mm*) triad. For a mainly syndiotactic sample the peak at 9.09τ is the most prominent of the three peaks. It is assigned to α-methyl protons in the middle of syndiotactic (*rr*) triads. The remaining peak at 8.95τ is therefore attributed to the α-methyl group in the middle of heterotactic (*mr*) triads. The numbers of isotactic, heterotactic, and syndiotactic triads in any particular sample, and hence *mm*, *mr*, and *rr*, are proportional to the areas under the appropriate peaks.

The method of identifying stereoregular structures by NMR can also be applied to polymers that are not predominantly stereoregular, to obtain information about the probability P_m that a monomer adding on to the end of a growing chain will have the same configuration as the unit it is joining. The fraction of each configuration — *mm*, *rr*, and *mr* — is measured from the isotactic, syndiotactic, and heterotactic peak areas, respectively, in the NMR spectra. The results are then fitted to the probability curves by placing one of them, say, the fraction *rr* triad on the curve and letting the other fractions, viz., *mm* and *mr*, fall where they may. If the latter fractions also fall satisfactorily close to the calculated curves, then the configurational statistics is Bernoullian and can be described by a single value of P_m.

Problem 2.13   Determination of the individual peak areas in the α-methyl region of the NMR spectra of poly(methyl methacrylate) shown in Fig. 2.13 yielded the following values (Bovey and Tiers, 1960):

Polymer No.	Polymerization conditions	Iso	Hetero	Syndio
			Area %	
1	Benzoyl peroxide in bulk, 100°C	8.6	37.5	53.9
2	n-BuLi in toluene, −62°C	63	19	18

Determine whether the polymerization in each case can be determined by a single value of P_m in agreement with Bernoullian statistics.

Answer:

We might test for Bernoullian fitting by placing the (*mm*), (*mr*), and (*rr*) triad intensities (that is, the fractional areas representing polymer fractions in iso, hetero, and syndio configurations, respectively) on the probability curves of Fig. 2.12. This is done in Fig. 2.15 for polymer 1 by placing the syndiotactic (*rr*) point (0.54) on the curve to obtain the corresponding P_m value (0.27) and then letting the other two points fall where they may on the vertical line at this P_m. It is observed that the fractions of *mr* (heterotactic) and *mm* (isotactic) units fall satisfactorily close to the calculated curves. This may be interpreted as indicating that the free radical propagation in the given case is Bernoullian and can be described by a single value of P_m (= 0.27).

For polymer 2, made by anionic polymerization, the results are fitted similarly to the probability curves in Fig. 2.15 by placing the (*mm*) point on the curve and letting the other points fall where they may. It is observed that the (*mr*) and (*rr*) points are significantly away from the calculated curves, an indication that the polymer is non-Bernoullian. Thus the anionic polymerization mechanism in the given case cannot be described by a single value of P_m.

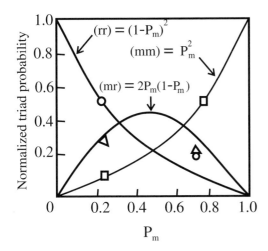

Figure 2.15 Test of Bernoullian model with data for polymer 1 and polymer 2 of Problem 2.13.

2.4 *Crystallinity in Polymers*

Polymers crystallized from melt are seldom fully crystalline. This is due to the presence of extensive chain entanglements in the melt that make it impossible for the polymer chains to be fully aligned to form a 100% crystalline polymer during solidification. The degree of crystallinity is of considerable importance as it influences the properties, and hence applications, of polymers. Several methods have been devised to measure the degree of crystallinity, two widely used among them being the *density method* and the *wide-angle* x-ray scattering (WAXS) *method* (Young and Lovell, 1990). The crystalline regions in the polymer have a higher density than the non-crystalline amorphous regions and this difference in densities, which is relatively large (up to 20%), provides the basis of the density method for the determination of the degree of crystallinity (see Problem 2.15).

Problem 2.14 Derive an equation relating the degree of crystallinity of a semicrystalline polymer to the sample density and densities of the crystalline and amorphous components.

Answer:

Let v_c = total volume of the crystalline components; v_a = total volume of the amorphous components; v = total volume of the specimen; m_c, m_a, and m are the corresponding masses and ρ_c, ρ_a, and ρ are the corresponding densities. Then

$$v = v_c + v_a \tag{P2.14.1}$$

$$m = m_c + m_a \tag{P2.14.2}$$

$$\text{or,} \quad \rho v = \rho_c v_c + \rho_a v_a \tag{P2.14.3}$$

Substituting v_a in Eq. (P2.14.3) from Eq. (P2.14.1) and rearranging leads to

$$\epsilon_c = \frac{v_c}{v} = \left(\frac{\rho - \rho_a}{\rho_c - \rho_a}\right) \tag{P2.14.4}$$

where ϵ_c is the volume fraction of crystalline components. The mass fraction μ_c of the crystalline components is similarly defined as

$$\mu_c = m_c/m = \rho_c v_c/\rho v \tag{P2.14.5}$$

Combination of Eqs. (P2.14.4) and (P2.14.5) then gives

$$\mu_c = \frac{\rho_c}{\rho}\left(\frac{\rho - \rho_a}{\rho_c - \rho_a}\right) \tag{P2.14.6}$$

In terms of specific volumes (reciprocal density), $\bar{v}_c$ and $\bar{v}_a$, of the crystalline and amorphous components, respectively, Eq. (P2.14.6) becomes (Young and Lovell, 1990) :

$$\mu_c = \frac{\bar{v}_a - \bar{v}}{\bar{v}_a - \bar{v}_c} \tag{P2.14.7}$$

Problem 2.15   Estimate the fraction of crystalline material in a sample of polyethylene of density 0.983 g/cm^3.

Data :
Density of amorphous polyethylene = 0.866 g/cm^3
Unit cell (Fig. 2.7) dimensions of polyethylene crystal containing 4 CH_2 groups are :
$a = 7.41$ Å, $b = 4.94$ Å, c (fiber axis) $= 2.55$ Å, $\alpha = \beta = \gamma = 90°$

Answer:
Volume of unit cell $= 7.41 \times 4.94 \times 2.55 \times 10^{-24}$ cm^3
               $= 93.34 \times 10^{-24}$ cm^3

Cell contains 4 CH_2 groups or $\dfrac{4 \times (14 \text{ g mol}^{-1})}{(6.02 \times 10^{23} \text{ mol}^{-1})}$ $= 9.3 \times 10^{-23}$ g

Density of crystalline material $= \dfrac{(9.3 \times 10^{-23} \text{ g})}{(93.34 \times 10^{-24} \text{ cm}^3)}$ $= 0.996$ g cm^{-3}

Using Eq. (P2.14.6),

$$\% \text{ Crystallinity} = 100 \left[\frac{(0.996 \text{ g cm}^{-3})}{(0.983 \text{ g cm}^{-3})}\right]\left[\frac{(0.983 - 0.866) \text{ g cm}^{-3}}{(0.996 - 0.866) \text{ g cm}^{-3}}\right]$$
$$= 91\%.$$

WAXS is a powerful method of determining the degree of crystallinity. Since the scattering from the crystalline regions gives relatively sharp peaks compared to a broad 'hump' produced by scattering from noncrystalline areas (see Fig. 2.16), the degree of crystallinity can, in principle, be determined from the relative areas under the crystalline peaks and the amorphous hump. To a first approximation, the mass fraction of crystalline regions, μ_c, is given by

$$\mu_c = A_c/(A_c + A_a) \tag{2.10}$$

where A_a is the area under the amorphous hump and A_c is the remaining area under the crystalline peaks.

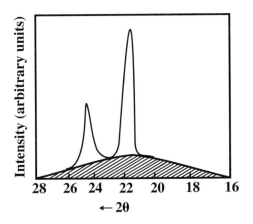

Figure 2.16 A typical WAXS curve for semicrystalline polyethylene where the intensity of scattering is plotted against diffraction angle 2θ. The amorphous hump is shown shaded.

2.4.1 Structure of Bulk Polymers

Most polymers are partially crystalline. The degree of crystallinity of polymers may, however, range very widely from 0 percent for noncrystallizable polymers, through intermediate crystallinities, up to nearly 100 percent for polytetrafluoroethylene and linear polyethylene. A direct evidence of the crystallinity in polymers is obtained from x-ray diffraction studies. The x-ray patterns of many crystalline polymers show both sharp features characteristic of ordered regions (called *crystallites*) and diffuse features characteristic of a molecularly disordered phase like liquids. X-ray scattering and electron microscopy have shown that the crystallites are made up of *lamellae* which, in turn, are built-up of folded polymer chains as explained below.

Lamellae are thin, flat platelets on the order of 100 to 200 Å (0.01-0.02 micron) thick and several microns in lateral dimensions, while polymer molecules are generally on the order of 1000 to 10,000 Å long. Since the polymer chain axis is perpendicular to the plane of the lamellae, as revealed by electron diffraction, the polymer molecules must therefore be folded back and forth within the crystal. This arrangement has been shown to be sterically possible. In polyethylene, for example, the molecules can fold in such a way that only about five chain carbon atoms are required for the fold, that is, for the chain to reverse its direction. Each molecule folds up and down in a regular fashion to establish a *fold plane*. As illustrated in Fig. 2.17(a), a single fold plane may contain many polymer chains. The height of the fold plane is known as the *fold period*. It corresponds to the thickness of the lamellae.

Figure 2.17(b) shows an idealized model of lamellae structure with ideal stacking of lamellar crystals. A more useful model, however, is that of stacks of lamellae interspersed with and connected by amorphous regions that consist of disordered chain segments of polymer molecules. Such a model, referred to as *interlamellar amorphous model* [Fig. 2.17(c)], helps explain the ductility and strength of polymers as a direct consequence of the molecular links between the lamellae forming interlamellar ties. For semicrystalline polymers with amorphous regions to the tune of 20-50%, it is often more advantageous to adopt a *fringed micelle* or *fringed crystalline model* [Fig. 2.17(d)]. It pictures polymers as two-phase systems in which the amorphous regions are interspersed between the randomly distributed crystallites.

2.4.1.1 Spherulites

The most prominent structural entity in a material crystallized from a polymer melt is the *spherulite*. Electron microscopic evidence clearly shows that spherulites are aggregates of lamellar crystallites and the lamellar structure persists throughout the body of spherulites. Spherulites are spherical aggregates, their sizes ranging from microscopic to a few millimeters in diameter. They are recognized by their characteristic appearance in the polarizing microscope (Price, 1958), where they are seen as circularly birefringent areas possessing characteristic Maltese cross optical patterns (Fig. 2.18).

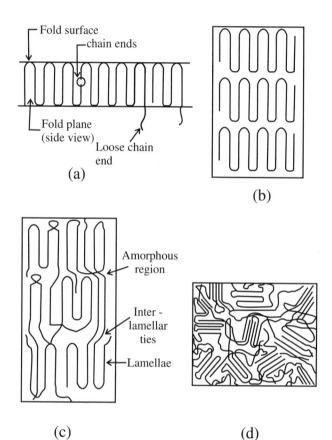

Figure 2.17 Schematic representation of (a) fold plane showing regular chain folding, (b) ideal stacking of lamellar crystals, (c) interlamellar amorphous model, and (d) fringed micelle model of randomly distributed crystallites.

Figure 2.18 Spherulites in a silicone-like polymer, observed in the optical microscope between crossed polarizer. The large and small spherulites were grown by crystallization at different temperatures. (Adapted from Price, 1958.)

The *Avrami equation* [Eq. (2.11)], which was derived in the general context of phase changes in metallurgy, has provided the starting point for many studies of polymer crystallization and spherulitic growth. The equation relates the fraction of a sample still molten, θ, to the time, t, which has elapsed since crystallization began, the temperature being held constant:

$$\theta = \exp(-Zt^n) \tag{2.11}$$

For a given system under specified conditions, Z and n are constants and, in theory, they provide information about the nature of the crystallization process. Taking logarithm twice in succession gives

$$\ln(-\ln\theta) = \ln Z + n\ln t \tag{2.12}$$

A plot of $\ln(-\ln\theta)$ against $\ln t$ should thus be a straight line of slope n, making an intercept of $\ln Z$ with the vertical axis. In practice, this method of evaluating n and Z is very prone to error. So curve fitting methods using Eq. (2.12) and the raw data are usually preferred.

The Avrami exponent, n, has a theoretical value of 3 when crystallization takes the form of spherulitic growth of nuclei that came into being at the same instant in time. Integral values of n ranging from 1 to 4 can be attributed to other forms of nucleation and growth (Hay, 1971). The spherulites formed from a melt have different sizes and degrees of perfection, and they completely fill the volume of a well-crystallized material.

Problem 2.16   Using the Avrami equation calculate approximately the extent of crystallization of polyethylene during cooling from the melt, if the melt is cooled in 1 s by quenching. Use $n = 3$ and $Z = 5$ (for time in seconds) as the average rate constant over the range of temperature in question.

Answer:

From Eq. (2.11),

fractional completion of crystallization $= 1 - \theta$

$$= 1 - \exp(-Zt^n)$$

$$= 1 - \exp(-5) = 0.99$$

This means that crystallization occurs to 99% of its equilibrium extent (which is about 90% in high-density polyethylene).

2.5   Thermal Transitions in Polymers

The term "transition" refers to a change of state induced by changing the temperatures or pressure. Two major thermal transitions are the glass transition and the melting, the respective temperatures being called T_g and T_m.

2.5.1 T_g and T_m

The different types of thermal response in the transition of a thermoplastic polymer from the rigid solid to an eventually liquid state can be illustrated in several ways. One of the simplest and most satisfactory is to trace the change in specific volume, as shown schematically in Fig. 2.19.

The volume change in amorphous polymers follows the curve ABC. In the region C–B, the polymer is a glassy solid and has the characteristics of a glass, including hardness, stiffness, and brittleness. But as the sample is heated, it passes through a temperature T_g, called the *glass transition temperature*, above which it softens and becomes rubberlike. This is an important temperature

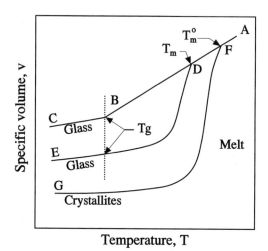

Figure 2.19 Schematic representation of the changes of specific volume of a polymer with temperature for (a) a completely amorphous sample (A–B–C), (b) a semicrystalline sample (A–D–E), and (c) a perfectly crystalline material (A–F–G).

and marks the beginning of movements of large segments of the polymer chain due to available thermal energy (RT energy units/mol). This is reflected in marked changes in properties, such as specific volume, refractive index, stiffness, and hardness. Above T_g, the material may be more easily deformed. A continuing increase in temperature along B–A leads to a change of the rubbery polymers to a viscous liquid without any sharp transition.

In a perfectly crystalline polymer, all the chains would be contained in regions of three dimensional order, called *crystallites*, and no glass transition would be observed. Such a polymer would follow the curve G–F–A, melting at T_m^0 to become a viscous liquid.

Perfectly crystalline polymers are, however, rarely seen in practice and real polymers may instead contain varying proportions of ordered and disordered regions in the sample. These *semicrystalline* polymers usually exhibit both T_g and T_m (not T_m^0) corresponding to the disordered and ordered regions, respectively, and follow curves similar to E–H–D–A. T_m is lower than T_m^0 and more often represents a melting range, because the semicrystalline polymer contains crystallites of various sizes with many defects which act to depress the melting temperature.

Both T_g and T_m are important parameters that serve to characterize a given polymer. While T_g sets an upper temperature limit for the use of amorphous thermoplastics like poly(methyl methacrylate) or polystyrene and a lower temperature limit for rubbery behavior of an elastomer-like SBR rubber or 1,4-*cis*-polybutadiene, T_m or the onset of the melting range determines the upper service temperature for semicrystalline thermoplastics. Between T_m and T_g, these polymers tend to behave as a tough and leathery material. They are generally used at temperatures between T_g and a practical softening temperature that lies above T_g and below T_m.

The onset of softening is usually measured as the temperature at which a given polymer deforms a given amount under a specified load. These values are known as *heat deflection temperatures*. Such data do not have any direct relation with T_m, but they are widely used in designing with plastics.

Problem 2.17 The polyesters, polycaprolactone and poly(ethylene terephthalate), have T_g values of -60°C and +60°C and T_m values of 60°C and 250°C, respectively. Which of the two polymers would be more suitable for a study of the effect of the degree crystallinity on biodegradability at 30°C?

Answer:

A polymer can crystallize whenever it is at a temperature above its T_g. Since at 30°C, polycaprolactone is above T_g, it may crystallize. On the other hand, poly(ethylene terephthalate) can be quenched to 30°C at various rates to produce various degrees of crystallinity which will not change since the temperature is below its T_g. So this polymer will be more suitable for the biodegradability study.

2.5.2 First- and Second-Order Transitions

In first order transitions, such as melting, there is a discontinuity in the volume-temperature plot (see Fig. 2.19) or enthalpy–temperature plot at the transition temperature. In second-order transitions, only a change in slope occurs and there is thus a marked change in the first derivative or temperature coefficients, as illustrated in Fig. 2.20. The glass transition is not a first-order transition, as no discontinuities are observed at T_g when the specific volume or entropy of the polymer is measured as a function of temperature. However, the first derivative of the property-temperature curve, i.e., the temperature coefficient of the property (e.g., heat capacity and volumetric coefficient of expansion), exhibits a marked change in the vicinity of T_g; for this reason it is sometimes called a *second-order transition*.

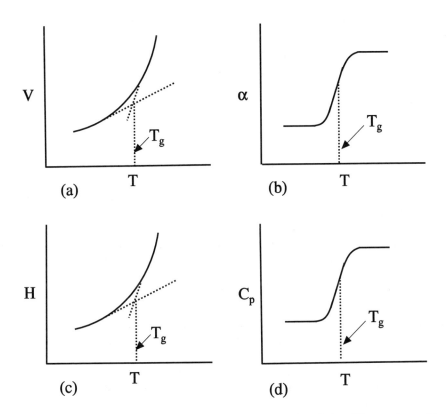

Figure 2.20 Schematic representations of volume (V) and enthalpy (H) variations with temperature. Also shown are variations with temperature of the volume coefficient of expansion (α) and the heat capacity (C_p), which are, respectively, the first derivatives of V and H with respect to temperature (T).

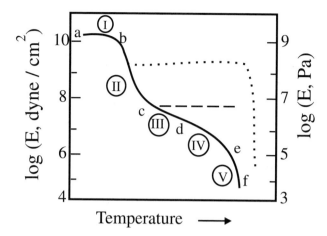

Figure 2.21 Five regions of viscoelastic behavior for a linear, amorphous polymer: I (a to b), II (b to c), III (c to d), IV (d to e), and V (e to f). Dotted line ($\cdots$) shows the effect of crystallinity and dashed line ($----$) that of crosslinking. (After Sperling, 1986.)

2.6 Regions of Viscoelastic Behavior

Broadly, there are five regions of viscoelastic behavior for linear amorphous polymers (Aklonis, 1981; Sperling, 1986) as shown in Fig. 2.21. In region I (a to b), described as the *glassy region*, the polymer is usually brittle. Common examples are polystyrene and poly(methyl methacrylate) at room temperature. Young's modulus for many polymers in this region just below the glass transition temperature is typically of the order of 3×10^{10} dyne/cm^2 (3×10^9 Pa).

Region II (b to c) in Fig. 2.21 is the *glass transition region* where the modulus drops typically by a factor of about one thousand over 20-30°C. In this region, polymers exhibit tough leather-like behavior. The glass transition temperature, T_g, is often taken at the maximum rate of decline of the modulus, i.e., where d^2E/dT^2 is at a maximum.

Region III (c to d) in Fig. 2.21 is described as the *rubbery plateau region*. The modulus after a sharp drop, as described above, again becomes nearly constant in this region with typical values of 2×10^7 dyne/cm^2 (2×10^6 Pa) and polymers exhibit significant rubber-like elasticity.

In region III (c to d) of Fig. 2.21, three cases can be distinguished: (1) If the polymer is linear, the solid line is followed with the modulus decreasing slowly with increasing temperature and the length of the plateau depends significantly on the molecular weight of the polymer, i.e., the higher the molecular weight, the longer the plateau. (2) If the polymer is semicrystalline, the dotted line in Fig. 2.21 is followed and the height of the plateau (i.e., the modulus) depends on the degree of crystallinity. (3) If the polymer is crosslinked, the dashed line in Fig. 2.21 is followed, improved rubber elasticity is observed with the creep portion suppressed (a common example being the ordinary rubber band at room temperature), and the region III remains in effect at higher temperatures up to the decomposition temperature. For cross-linked polymers, therefore, regions IV and V (described below) do not occur.

Region IV is the *rubbery flow region*, which is exhibited by un-crosslinked linear polymers when heated past the rubbery plateau region. In this region, the polymer exhibits both rubber elasticity and flow properties depending on the time scale of the experiment. When subjected to stress for a short time, the entanglements of polymer chains are not able to relax and the material still shows rubbery behavior, but over a longer duration under stress at the increased temperature, the chains can move, resulting in a visible flow.

Region V (e to f) in Fig. 2.21 is the *liquid flow region*, which is reached at still higher temperatures where the increased kinetic energy of the chains permits them to wriggle out through entanglements rapidly and move as individual molecules, often producing highly viscous flow. This is the melting temperature and it is always above the glass transition temperature.

Problem 2.18   A new polymer is reported to soften at 60°C. Describe a very simple experiment to determine whether this softening is a glass transition or a melting point.

Answer:

If 60°C is a glass transition, then heating the polymer slowly past 60°C would take it to the rubbery plateau region (region III in Fig. 2.21), where the modulus E, and hence hardness, would remain fairly constant with increase of temperature. For a melting transition, however, the modulus would drop rapidly and the polymer would become increasingly softer in a similar experiment (Sperling, 1986). The penetration of a weighted needle into the polymer as the temperature is raised can be used as a very simple test.

Problem 2.19   Show schematically the results that would be expected from the following experiments with the new polymer of Problem 2.18 if the reported softening temperature (60°C) is indeed a melting transition (Sperling, 1986):

(a) Specific volume (v) as a function of temperature (T);   (b) differential scanning calorimetry;   (c) Young's modulus (E) as a function of temperature (T);   and (d) x-ray diffraction.

The experiments are carried out in the temperature range 50°C-70°C.

Answer:

(a) The specific volume would show a first order transition. However, for reasons explained earlier, the melting transition for polymers usually occurs over a range of temperature.

(b) DSC would show an endothermic peak.

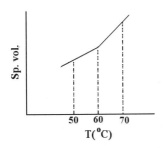

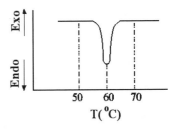

(c) A plot of log$E$ *versus* T would show a sharp downturn at about 60°C.

(d) If the polymer has a high degree of crystallinity, x-ray diffraction at 50°C would show sharp lines but only diffuse rings or halo at 70°C.

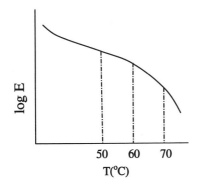

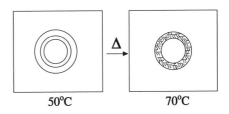

Problem 2.20 What type of results would be obtained if the reported softening temperature (60°C) for the polymer in Problem 2.18 were a glass transition (Sperling, 1986) ?

Answer:

(a) The specific volume would show a second order transition.

(b) Heat capacity of the material changes at the glass transition. T_g is taken as the temperature at which one-half of the change in heat capacity ΔC_p has occurred.

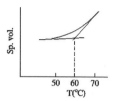

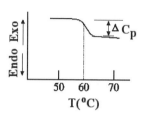

(c) In quasistatic measurements, maximum rate of decrease of the modulus occurs at T_g.

(d) A glass is a material that has lost most of its ductility but has not changed otherwise on cooling; on the basis of x-ray diffraction it will be indistinguishable from the corresponding rubber on the other side of T_g.

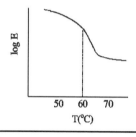

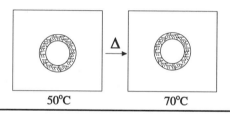

2.7 Factors Affecting T_g

As T_g marks the onset of molecular motion, a number of factors that affect rotation about links (necessary for movement of polymer chains) will also influence the T_g of a polymer. These include (a) chain flexibility, (b) molecular structure (steric effects), (c) molecular weight, and (d) branching and crosslinking.

The flexibility of the chain may be considered to be the most important factor that influences the T_g of a polymer. If the chain is highly flexible, T_g will generally be low and if the chain is rigid, the T_g value will be high. The chain flexibility depends on the rotation or torsion of skeletal bonds. Chains made up of bond sequences which are able to rotate easily are flexible, and hence polymers containing $+CH_2-CH_2+$, $+CH_2-O-CH_2+$, or $+Si-O-Si+$ links will have correspondingly low values of T_g. For example, poly(dimethyl siloxane), which has Si–O bonds in the backbone chain, has one of the lowest T_g values known (−123°C).

The value of T_g increases markedly with the insertion of groups that stiffen the chain by impeding rotation, since more thermal energy is then required to set the chain in motion. Particularly effective in this respect is the p-phenylene ring,

$$-\langle\bigcirc\rangle- \equiv -C_6H_4-$$

Some examples of such polymers are poly(xylylene), $+CH_2-C_6H_4-CH_2+_n$, $T_g \simeq 280°C$ and poly(phenylene oxide), $+C_6H_4-O+_n$, $T_g = 83°C$, as compared to polyethylene $+CH_2CH_2+_n$, $T_g = -93°C$ and poly(ethylene oxide) $+CH_2CH_2-O+_n$, $T_g = -67°C$, respectively.

Problem 2.21   Presence of flexible pendant groups reduces the glass transition of the polymer, whereas bulky or stiff side groups increase it. Why?

Answer:

In general, factors that increase the energy required for the onset of molecular motion increase T_g and those that decrease the energy requirement lower T_g. Flexible pendant groups act as "inherent diluents" and lower the frictional interaction between chains, reducing the T_g thereby. Bulky or stiff side groups, on the other hand, increase interchain friction and energy requirement for molecular motion, thereby increasing T_g.

Problem 2.22   Account for the differences in glass transition temperatures for the following pairs of iso-meric polymers (Young and Lovell, 1990) :

(a) Poly(but-1-ene) $(-24^\circ C)$ and poly(but-2-ene) $(-73^\circ C)$.
(b) Poly(ethylene oxide) $(-67^\circ C)$ and poly(vinyl alcohol) $(85^\circ C)$.
(c) Poly(methyl acrylate) $(6^\circ C)$ and poly(vinyl acetate) $(28^\circ C)$
(d) Poly(ethyl acrylate) $(-24^\circ C)$ and poly(methyl methacrylate) $(105^\circ C)$

Answer:

(a)

$$\left(\!-CH_2-CH-\!\right)_n$$
$$\overset{|}{\underset{\overset{|}{CH_3}}{CH_2}}$$

$T_g = -24^0 C$

The long and bulky side group makes rotation difficult.

$$\left(\!-CH-CH-\!\right)_n$$
$$\overset{|}{CH_3} \; \overset{|}{CH_3}$$

$T_g = -73^0 C$

The side groups are shorter; this makes rotation easier.

(b)

$$\left(\!-CH_2-CH_2-O-\!\right)_n$$

$T_g = -67^0 C$

The backbone chain is flexible; so rotation is easy.

$$\left(\!-CH_2-CH-\!\right)_n$$
$$\overset{|}{OH}$$

$T_g = 85^0 C$

The steric effect of the -OH group restricts rotation. Superimposed on this is the effect of polarity which increases the lateral forces in the bulk state and leads to higher T_g .

(c)

$$\left(\!-CH_2-CH-\!\right)_n$$
$$\overset{|}{\underset{O}{C}}\overset{\diagup\!\diagdown}{} OCH_3$$

$T_g = 6^0 C$

The bulky group hinders rotation causing T_g to increase.

$$\left(\!-CH_2-CH-\!\right)_n$$
$$\overset{|}{\underset{O}{O}}$$
$$\overset{\diagup\!\diagdown}{C} CH_3$$

$T_g = 28^0 C$

The bulky part of the side group is farther from the chain, making rota-tion more difficult.

(d)

$$\left(\!-CH_2-CH-\!\right)_n$$
$$\overset{|}{\underset{O}{C}}\overset{\diagup\!\diagdown}{} OC_2H_5$$

$T_g = -24^0 C$

The side group is long but flexible, so rotation is easier.

$$\left(\!-CH_2-\overset{\overset{\textstyle CH_3}{|}}{\underset{|}{C}}-\!\right)_n$$
$$\overset{C}{\underset{O}{}}\overset{\diagup\!\diagdown}{} OCH_3$$

$T_g = 105^0 C$

The presence of two side groups makes rotation more difficult, leading to a higher T_g

Table 2.2 T_m, T_g, and T_g/T_m for Some Selected Polymers

Polymer	T_m, °C	T_g, °C	T_g/T_m, °K / °K
Silicone rubber	− 58	− 123	0.70
Polyethylene	135	− 68	0.50
Polypropylene	176	− 8	0.59
Polystyrene	240	100	0.73
Poly(methyl methacrylate)	200	105	0.69
Poly(vinyl chloride)	180	82	0.78
Poly(vinylidene fluoride)	210	− 39	0.48
Polyisoprene	28	− 70	0.67
Nylon-6,6	265	50	0.60

Source: Williams, 1971.

2.8 Factors Affecting T_m

Useful generalizations of factors affecting T_m can be derived from the application of macroscopic thermodynamics. At T_m, the free energy change is zero, i.e., $\Delta G_m = \Delta H_m - T_m \Delta S_m = 0$, whence $T_m = \Delta H_m / \Delta S_m$. Thus a high melting point can be the result of a high value of the enthalpy change ΔH_m and/or a small value of the entropy change ΔS_m in melting. The former corresponds to stronger binding of adjacent but unbonded units in the polymer lattice and thus to higher degree of crystallinity. The factors that affect crystallinity and hence T_m can be classified as symmetry, intermolecular bonding, tacticity, branching, and molecular weight.

Chain flexibility has a direct bearing on the melting point. Insertion of groups that stiffen the chain increases T_m, while introducing flexible groups into the chain lowers the value of T_m (cf. **Factors Affecting** T_g). Also, if the chain is substantially branched, reducing the packing efficiency, the crystalline content is lowered and hence the melting point. A good example is low-density polyethylene where extensive branching lowers the density and T_m of the polymer.

2.9 Relation Between T_m and T_g

While T_m is a first order transition, T_g is a second order transition and this precludes the possibility of a simple relation between them. There is, however, a crude relation between T_m and T_g. Boyer (1954) and Beamen (1952) inspected data for a large number of semicrystalline polymers, some of which are shown in Table 2.2. They found that the ratio T_g/T_m ranged from 0.5 to 0.75 when the temperatures are expressed in degrees Kelvin. The ratio is closer to 0.5 for symmetrical polymers such as polyethylene and polybutadiene, but closer to 0.75 for unsymmterical polymers, such as polystyrene and polyisoprene. The difference in these values may be related to the fact that in unsymmterical chains with repeat units of the type $-(CH_2-CHX)-$ an additional restriction to rotation is imposed by steric effects causing T_g to increase, and conversely, an increase in symmtery lowers T_g.

Problem 2.23   A new atactic polymer of the type $-(CH_2-CHX)_n-$ has a T_m of 80°C. What is its T_g likely to be ?

Answer:
The polymer being unsymmetrical, T_g/T_m may be assumed to be about 0.75.

$T_g/T_m = 0.75$, $T_m = 80°C = 353°K$

$T_g = 265°K = − 8°C$

2.10   Theoretical Treatment of Glass Transition

According to the hole theory of liquids (Eyring, 1936), molecular motion in liquids depends on the presence of holes or voids, i.e., places where there are vacancies, as illustrated in Fig. 2.22. For real materials, however, Fig. 2.22 has to be visualized in three dimensions. A similar model can also be constructed for the motion of polymer chains, the main difference being that more than one "hole" will now be required to be in the same locality for the movement of polymer chain segments. On this basis, the observed specific volume of a sample, v, can be described as a sum of the volume actually occupied by the polymer molecules, v_0, and the free volume (empty spaces), v_f, in the system [see Fig. 2.23(a)], i.e.,

$$v = v_0 + v_f \tag{2.13}$$

The glass transition that takes place at a higher temperature can be visualized as the onset of coordinated segmental motion made possible by an increase of the free space in the polymer matrix to a size sufficient to allow this type of motion to occur. Conversely, if the temperature is decreased, the free volume will contract and eventually reach a critical value when the available space becomes insufficient for any large scale segmental motion. The temperature at which this critical value is reached is the *glass transition temperature* (T_g). As the temperature decreases further below T_g, the free volume, i.e., v_f in Eq. (2.13), will remain essentially constant at $v_f^\star$ [see Fig. 2.23(a)], since the chains have now been immobilized ('frozen') in position. However, the occupied volume v_o will change with temperature due to change in vibrational amplitudes. To the first approximation, this change will be a linear function of temperature. The temperature coefficient of the specific volume v will therefore change markedly at T_g.

According to the *iso-free-volume state* theory of Fox and Flory (1950, 1954), same free volume exists at the respective glass temperature (T_g) independent of molecular weight and this same free volume is retained at all temperatures below T_g. This can be explained by considering that at temperatures below T_g the segmental rotations become frozen and the holes that are present at the glass temperature become immobilized. Because the holes are no longer able to diffuse out of the structure, the free volume, which is the sum of the holes, remains nearly unchanged at all temperatures below T_g, though the internuclear separation can still adjust itself below T_g as the thermal vibrations of the atoms become reduced. At temperatures above the glass transition, both the internuclear separation between segments of neighboring chains and the number and size of holes adjust themselves continuously with changing temperature (provided sufficient time is allowed for segmental diffusion). The coefficient of expansion above T_g is therefore higher than that below T_g [see Fig. 2.23(b)].

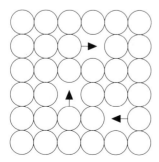

Figure 2.22   A quasicrystalline lattice of molecules (circles) exhibiting vacancies or holes.   The arrow indicates molecular motion.

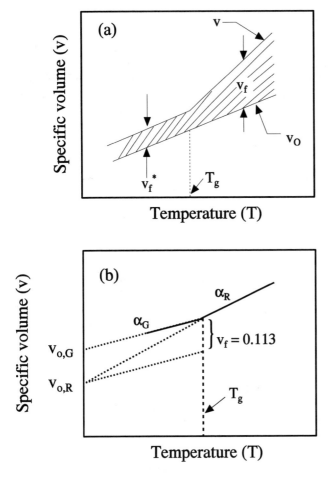

Figure 2.23 (a) Schematic representation of the variation of specific volume with temperature. The free volume (shaded area) is assumed to be constant at $v_f^\star$ below T_g and to increase as the temperature is raised above T_g. (b) Schematic illustration of free volume (Simha and Boyer, 1962).

Simha and Boyer (1962) postulated that the free volume at $T = T_g$ should be defined as

$$v_f = v - v_{0,R}(1 + \alpha_G T) \tag{2.14}$$

where v is the specific volume (i.e., volume per unit mass), v_f is the specific free volume, and $v_{0,R}$ is the extrapolated specific volume. The use of α_G in Eq. (2.14) follows from the conclusion that expansion in the glassy state occurs at nearly constant free volume; $\alpha_G T$ is thus proportional to the occupied volume. Figure 2.23 illustrates these quantities. It is seen that $v_{0,R}$ and $v_{0,G}$ are the hypothetical specific volumes extrapolated from melt or rubbery state to $0°\text{K}$ using α_R and α_G as the coefficients of expansion. Thus at $T = T_g$,

$$v = v_{0,R}(1 + \alpha_R T) \tag{2.15}$$

and

$$v = v_{0,G}(1 + \alpha_G T) \tag{2.16}$$

Substitution of v from Eq. (2.15) into Eq. (2.14) at $T = T_g$ leads to the following expression for free-volume fraction (Simha and Boyer, 1962):

$$(\alpha_R - \alpha_G)T_g = v_f/v_{0,R} \simeq v_f/v \tag{2.17}$$

[since $v_f \ll v_{0,R}$ and $\alpha_G T \ll 1$ in Eq. (2.14)]

Simha and Boyer (1962) postulated that the free-volume fraction defined by Eq. (2.17) is the same for all polymers, that is,

$$(\alpha_R - \alpha_G)T_g = \text{constant} = K_1 \tag{2.18}$$

From the values of $(\alpha_R - \alpha_G)T_g$ based on data of α_R and α_G determined for various polymers, Simha and Boyer (1962) concluded that K_1 in Eq. (2.18) is 0.113, i.e.,

$$(\alpha_R - \alpha_G)T_g = 0.113 \tag{2.19}$$

This implies that the free volume fraction at the glass transition temperature is the same for all polymers and constitutes 11.3% of the total volume in the glassy state. (It is interesting to note that many simple organic compounds show a 10% volume increase on melting.) This is the largest of the theoretical values of free volume derived, while other early estimates yielded a value of about 2%.

A less exact but simpler relationship is

$$\alpha_R T_g = \text{constant} = K_2 \tag{2.20}$$

It thus follows that the fractional free volume defined as [cf. Eq. (2.15)]:

$$(v - v_{0,R})/v = \alpha_R T/(1 + \alpha_R T) = \alpha_R T + O[(\alpha_R T)^2] \tag{2.21}$$

is constant at $T = T_g$.

Equation (2.21) neglects the contribution of $\alpha_G T$ to the total expansion. From the values of α_R determined for various polymers (Simha and Boyer, 1962) Eq. (2.20) is given as

$$\alpha_R T_g = K_2 = 0.164 \tag{2.22}$$

The quantities K_1 and K_2 provide a criterion for the glass transition temperature. They are especially useful for approximate estimation of T_g for new polymers or semicrystalline polymers, where T_g may be obscured.

Problem 2.24 For the polymers for which $\alpha_f = 4.8 \times 10^{-4}$ °K^{-1}, estimate a value of T_g, assuming that these polymers also obey the S-B relationship given by Eq. (2.19).

Answer:

Since the free volume is assumed constant below T_g, the volume change with temperature in the glassy region is only due to the occupied volume, while the volume change in the rubbery region above T_g has contributions from both the occupied volume and the free volume. Therefore, one can write, approximately,

$$\alpha_R - \alpha_G \simeq \alpha_f = 4.8 \times 10^{-4} \text{ °K}^{-1}$$

Substituting this in Eq. (2.19),

$$T_g \simeq 0.113/(4.8 \times 10^{-4} \text{ °K}^{-1}) = 235 \text{ °K} \equiv -38°\text{C}.$$

This value is in the range of the T_g's observed for many polymers.

Problem 2.25   A new linear amorphous polymer is found to have glass transition at 90°C and its cubic coefficient of thermal expansion is 5.6×10^{-4} °K^{-1} at 120°C. Can this polymer be used for an application that requires the cubic coefficient of thermal expansion to be less than 5×10^{-4} °K^{-1} at 60°C (Sperling, 1989) ? [Assume ($\alpha_R - \alpha_G$) to be roughly constant at 3.2×10^{-4} °K^{-1} (Sharma et al., 1972).]

Answer:

Above T_g, the expansion coefficient is α_R. Hence $\alpha_R = 5.6 \times 10^{-4}$ °K^{-1}.

$$T_g = 90°C \equiv 363°K$$

$$\alpha_R T_g = (5.6 \times 10^{-4} \, °K^{-1})(363°K) = 0.203$$

From Eq. (2.19):

$$\alpha_G T_g = 0.203 - 0.113 = 0.09$$

$$\alpha_G = 0.09/(363°K) = 2.5 \times 10^{-4} \, °K^{-1}$$

Also from the approximate relationship $\alpha_R - \alpha_G = 3.2 \times 10^{-4}$ °K^{-1}. Therefore,

$$\alpha_G = (5.6 - 3.2) \times 10^{-4} = 2.4 \times 10^{-4} \, °K^{-1}$$

Hence the polymer is likely to satisfy the thermal expansivity requirement.

As can be seen from Fig. 2.23(a), above T_g there is an important contribution to v_f from the expansion of the polymer in the rubbery or molten state. The free volume above T_g can thus be expressed by

$$v_f = v_f^\star + (T - T_g)(\partial v / \partial T) \tag{2.23}$$

Dividing through by v gives

$$f_T = f_g + (T - T_g)\alpha_f \tag{2.24}$$

where the symbol f_T is now used to represent the free volume fraction v_f/v at T and f_g that at T_g; α_f is the thermal expansion coefficient of free volume. (In a region above T_g but sufficiently close to it, α_f will be given by the difference between the thermal expansion coefficients of the rubbery and the glassy polymer.)

Williams, Landel, and Ferry (1955) proposed that the logarithm of viscosity varies linearly with $1/f$ above T_g, so that

$$\ln\left(\frac{\eta_T}{\eta_g}\right) = B\left(\frac{1}{f_T} - \frac{1}{f_g}\right) \tag{2.25}$$

where η_T and η_g are viscosities at T and T_g, respectively, and B is a constant. Substituting Eq. (2.23) in Eq. (2.25) gives

$$\ln\left(\frac{\eta_T}{\eta_g}\right) = B\left[\frac{1}{f_g + \alpha_f(T - T_g)} - \frac{1}{f_g}\right] \tag{2.26}$$

$$= -\frac{B\alpha_f(T - T_g)}{f_g[f_g + \alpha_f(T - T_g)]} \tag{2.27}$$

Dividing both numerator and denominator by α_f, Eq. (2.27) is rewritten as

$$\ln\left(\frac{\eta_T}{\eta_g}\right) = -\frac{(B/f_g)(T - T_g)}{(f_g/\alpha_f) + (T - T_g)} \tag{2.28}$$

This is one form of the WLF equation. Since the amount of flow is proportional to the flow time t and density ρ and inversely proportional to viscosity, one may write (Cowie, 1991):

$$\frac{\rho_T \cdot t_T}{\eta_T} = \frac{\rho_g \cdot t_g}{\eta_g}$$

and hence

$$\left(\frac{\eta_T}{\eta_g}\right) = \left(\frac{\rho_T \cdot t_T}{\rho_g \cdot t_g}\right) \simeq \left(\frac{t_T}{t_g}\right) \tag{2.29}$$

neglecting small differences in density. Substituting Eq. (2.29) in Eq. (2.28) one then obtains

$$\log_{10}\left(\frac{t_T}{t_g}\right) = \frac{-(B/2.303 f_g)(T - T_g)}{(f_g/\alpha_f) + (T - T_g)} \tag{2.30}$$

This can be compared with another form of the WLF equation written as

$$\log_{10} a_T = \frac{-C_1(T - T_g)}{C_2 + (T - T_g)} \tag{2.31}$$

where a_T is called the *reduced variables shift factor* (defined below), and C_1 and C_2 are constants that can be obtained from experimental data. For many linear polymers, irrespective of chemical structure, $C_1 = 17.44$ and $C_2 = 51.6$ when T_g is the reference temperature. The WLF equation is then written as

$$\log_{10} a_T = \frac{-17.44(T - T_g)}{51.6 + (T - T_g)} \tag{2.32}$$

Problem 2.26   A new linear amorphous polymer has a T_g of $+10°C$. At $27°C$ it has a melt viscosity of 4×10^8 poises. Estimate its viscosity at $50°C$.

Answer:

Eqs. (2.26)-(2.32):

$$T = 27°C, \quad \eta = 4\times10^8 \text{ poise}, \quad T_g = 10°C, \quad T - T_g = 17°C$$

$$\log\left[\frac{4\times10^8}{\eta_g}\right] = \frac{-17.44\times17}{51.6 + 17} = -4.32$$

Solving, $\eta_g = 8.3\times10^{12}$ poise

At $T = 50°C$, $T - T_g = 40°C$. Therefore,

$$\log\left[\frac{\eta}{8.3\times10^{12}}\right] = \frac{-17.44\times40}{51.6 + 40} = -7.61$$

Solving, $\eta = 2.0\times10^5$ poise

A more general form of the WLF equation reads

$$\log_{10} a_T = \frac{-8.86(T - T_s)}{101.6 + (T - T_s)} \tag{2.33}$$

where T_s is the arbitrary reference temperature usually located about $50°C$ above T_g. The constants C_1 and C_2 now have different values, and the shift factor a_T is expressed as a ratio of relaxation times, τ, at T and T_s, that is,

$$a_T = \tau(T)/\tau(T_s) \simeq \eta_T/\eta_s \tag{2.34}$$

It is evident from the derivation given above that while the WLF equation is based on the free-volume theory of glass transition, it also serves to introduce some kinetic aspects into the quantitative theory of glass transition.

Problem 2.27   Assuming for the constant B in Eq. (2.25) a value of unity, calculate the free volume fraction at T_g and the expansion coefficient of free volume, α_f.

Answer:

Comparing Eqs. (2.30) and (2.32),

$$B/2.303 f_g = 17.44 \tag{P2.27.1}$$
$$f_g/\alpha_f = 51.6 \tag{P2.27.2}$$

Considering $B = 1$, the two equations are solved for the two unknowns f_g and α_f, giving $f_g = 0.025$ (i.e., free volume = 2.5%) and $\alpha_f = 4.8 \times 10^{-4} \, °K^{-1}$.

Problem 2.28   The experimental value of T_g depends on the time or frequency frame of the experiment. Calculate from the WLF equation the change that would be expected in the T_g value if the time frame of an experiment is decreased by a factor of 100.

Answer:

From Eq. (2.32):

$$\lim_{T \to T_g} \left(\frac{\log a_T}{T - T_g} \right) = -17.44/51.6 = -0.338$$

Since the time frame of the experiment is decreased by a factor of 100, the shift factor a_T is 1/100. Therefore,

$$T - T_g = \frac{-2.0}{-0.338} \simeq 6°K$$

So the glass transition temperature would be raised by about $6°C$. This is in agreement with the experiment.

2.10.1   *Quantitative Effects of Factors on* T_g

The lower the molecular weight of a polymer sample, i.e., the greater the number of chain ends, the higher becomes the contribution to the free volume when these begin moving; consequently T_g decreases as the molecular weight of the sample is lowered. This behavior can be approximated to an equation of the form (Fox and Flory, 1950, 1954)

$$T_g = T_{g,\infty} - K/M \tag{2.35}$$

relating T_g to molecular weight M; $T_{g,\infty}$ in this equation is the value of T_g for a polymer sample of infinite molecular weight and K is a constant.

In a polymer sample of molecular weight $\overline{M}_n$ and density ρ, the number of chains per unit volume is given by $\rho N_{Av}/\overline{M}_n$, where N_{Av} is Avogadro's number, and so the number of chain ends per unit volume is $2\rho N_{Av}/\overline{M}_n$. If θ is the contribution of one chain end to the free volume then the total *fractional* free volume due to all chain ends, f_c, will be given by

$$f_c = 2\rho N_{Av}\theta/\overline{M}_n \tag{2.36}$$

It can be argued that f_c is the extra amount of free volume in a low molecular weight compound, as compared to a polymer of infinite molecular weight, that causes lowering of the glass transition temperature from $T_{g,\infty}$ to T_g. So f_c will be equivalent to the free volume expansion between T_g and $T_{g,\infty}$, which means that

$$f_c = \alpha_f (T_{g,\infty} - T_g) \tag{2.37}$$

where α_f is the thermal expansion coefficient of the free volume. Combining Eqs. (2.36) and (2.37) and rearranging one then gets

$$T_g = T_{g,\infty} - \frac{2\rho N_{Av}\theta}{\alpha_f \overline{M}_n} \tag{2.38}$$

which is exactly of the same form as Eq. (2.35) when $K = 2\rho N_{Av}\theta/\alpha_f$.

Problem 2.29 A polydisperse polystyrene sample was fractionated into four components of various molecular distributions and T_g of each fraction was measured:

Component	Wt. fraction (w)	$\overline{M}_n$	T_g (°K)
1	0.05	1.5×10^6	378.9
2	0.41	4.8×10^5	378.5
3	0.39	1.2×10^5	377.2
4	0.15	3.7×10^4	373.3

(a) Obtain a relation between T_g and molecular weight.
(b) Calculate the T_g value of the polydisperse polystyrene.

Answer:
(a) T_g is plotted against $1/\overline{M}_n$ in Fig. 2.24 according to Eq.(2.35). From the intercept and slope of the straight line plot,

$$T_{g,\infty} = 379°K, \quad K = 2.1 \times 10^5 \; °K \; g \; mol^{-1}$$

(b) From the definition of number average molecular weight (*see* Chapter 4):

$$\frac{1}{\overline{M}_n} = \frac{w_1}{M_1} + \frac{w_2}{M_2} + \frac{w_3}{M_3} + \frac{w_4}{M_4}$$

Using Eq. (2.35),

$$\frac{T_{g,\infty} - T_g}{K} = \frac{w_1(T_{g,\infty} - T_{g,1})}{K} + \frac{w_2(T_{g,\infty} - T_{g,2})}{K}$$
$$+ \frac{w_3(T_{g,\infty} - T_{g,3})}{K} + \frac{w_4(T_{g,\infty} - T_{g,4})}{K}$$

Simplifying,

$$
\begin{aligned}
T_g &= w_1 T_{g,1} + w_2 T_{g,2} + w_3 T_{g,3} + w_4 T_{g,4} \\
&= (0.05)(378.9°K) + (0.41)(378.5°K) + (0.39)(377.2°K) + (0.15)(373.3°K) \\
&= 377.2°K
\end{aligned}
$$

Therefore the T_g of unfractionated polymer sample is 104°C.

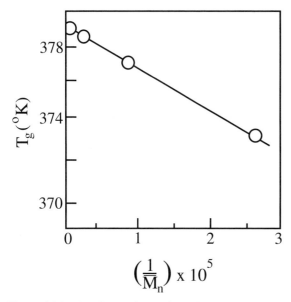

Figure 2.24 Plot of T_g against reciprocal molecular weight (data of Problem 2.29).

Branches present in small numbers on a polymer chain are known to decrease T_g. This effect also can be explained using the free volume concept. Since branches give rise to chain ends, the above analysis of the effect of molecular weight on T_g can be extended to branching. Thus, if the total number ends per chain is y, then by the analysis as given above one may write

$$T_g = T_{g,\infty} - \frac{y\rho N_{Av}\theta}{\alpha_f \overline{M}_n} \qquad (2.39)$$

where $T_{g,\infty}$ is again the glass transition temperature of a linear chain of infinite molecular weight. Since a linear chain itself has two ends, the number of branches per chain is $(y - 2)$. Equation (2.39) is applicable if the number of branches is low. As the density of branching increases, it produces the same effect as side groups in restricting chain mobility and hence T_g rises.

Problem 2.30   The T_g of a linear polymer with $\overline{M}_n = 2500$ is 120°C, while for a sample of the same linear polymer with $\overline{M}_n = 10,000$, T_g is 150°C. On the other hand, a branched product of the same polymer with $\overline{M}_n = 6,000$ has a T_g of 114°C. Determine the average number of branches per chain of the branched polymer.

Answer:

If $\rho N_{Av}\theta/\alpha_f$ is assumed to be constant (K) over the temperature range defined by the T_g's, Eq. (2.39) can be written as

$$T_g = T_{g,\infty} - Ky/\overline{M}_n$$

Linear polymer: $\overline{M}_n = 2500$ g mol^{-1}, $T_g = 393°$K, $y = 2$

$$393°\text{K} = T_{g,\infty} - 2K/(2300 \text{ g mol}^{-1}) \qquad (P2.30.1)$$

Linear polymer: $\overline{M}_n = 10,000$ g mol^{-1}, $T_g = 423°$K, $y = 2$

$$423°\text{K} = T_{g,\infty} - 2K/(10000 \text{ g mol}^{-1}) \qquad (P2.30.2)$$

Solving Eqs. (P2.29.1) and (P2.29.2) simultaneously gives

$$K = 4.48 \times 10^4 \text{ °Kg mol}^{-1}; \quad T_{g,\infty} = 432°\text{K}$$

Branched polymer: $\overline{M}_n = 6{,}000 \text{ g mol}^{-1}$, $T_g = 387°\text{K}$, $y = ?$

$$387°\text{K} = 432°\text{K} - \left[y \frac{(4.48 \times 10^4 \text{ g °K mol}^{-1})}{(6000 \text{ g mol}^{-1})} \right] \implies y = 6.02$$

Therefore, the average number of branches per chain $= 6 - 2 = 4$

The effect of crosslinking on T_g can be treated along similar lines to the effects of molecular weight and branching. When two polymer chains are connected by introducing a crosslink at intermediate points, the chains are pulled closer together at these points and the free volume is decreased. This reduction in the free volume increases T_g since molecular motion becomes more difficult.

According to the free-volume theory, as we have seen earlier, the free-volume fraction f in a polymer at a temperature T above T_g can be expressed in a linear form [cf. Eq. (2.17)]:

$$f = f_g + (\alpha_R - \alpha_G)(T - T_g) \tag{2.40}$$

where f_g is the free volume fraction at the glass transition; α_R and α_G are, respectively, the coefficients of thermal expansion above and below T_g. Considering 1 cm^3 of polymer-diluent mixture comprising ϕ_p cm^3 of polymer and ϕ_d cm^3 of diluent, the free volume f_p associated with ϕ_p cm^3 of polymer at the glass temperature T_g of the mixture is, therefore,

$$f_p = \phi_p[f_g + (\alpha_R - \alpha_G)(T_g - T_{gp})] \tag{2.41}$$

where T_{gp} is the glass transition temperature of the pure polymer. Similarly for ϕ_d cm^3 of the diluent the free volume f_d is

$$f_d = \phi_d[f_g + \alpha_d(T_g - T_{gd})] \tag{2.42}$$

since for a simple liquid the whole of the thermal expansion may, to a first approximation, be regarded as contributing to the free volume; T_{gd} is the glass transition temperature of the diluent. Equations (2.41) and (2.42) are based on the postulate that the free volume fraction has a critical value f_g which is the same for the pure polymer, the diluent and their mixtures at the respective glass temperatures (Kelley and Bueche, 1961). Since for 1 cm^3 of the mixture the free volume fraction at T_g, i.e., f_g, is given by $(f_p + f_d)$, adding Eqs. (2.41) and (2.42) and rearranging gives

$$T_g = \frac{\phi_p T_{gp}(\alpha_R - \alpha_G) + \phi_d T_{gd} \alpha_d}{\phi_p(\alpha_R - \alpha_G) + \phi_d \alpha_d} \tag{2.43}$$

The value of $(\alpha_R - \alpha_G)$ is roughly constant for many polymers, a typical value being 3.2×10^{-4} °K^{-1} (Sharma et al., 1972).

Problem 2.31 In plasticization of poly(methyl methacrylate) ($T_g = 105°\text{C}$ $\alpha_R - \alpha_G = 3.2 \times 10^{-4}$ °K^{-1}) with diethyl phthalate ($T_g = -65°\text{C}$, $\alpha_d = 10 \times 10^{-4}$ °K^{-1}), how much plasticizer would have to be added in order for T_g to be 50°C?

Answer:
Substituting $\phi_p = 1 - \phi_d$; $T_{gp} = 105°\text{C} \equiv 378$ °K; $\alpha_R - \alpha_G = 3.2 \times 10^{-4}$ °K^{-1}; $T_{gd} = -65°\text{C} \equiv 208$ °K; $\alpha_d = 10^{-3}$ °K^{-1}; $T_g = 50°\text{C} \equiv 323$ °K, Eq. (2.43) becomes

$$323 = \frac{(1 - \phi_d) \times 378 \times 3.2 \times 10^{-4} + \phi_d \times 208 \times 10^{-3}}{(1 - \phi_d) \times 3.2 \times 10^{-4} + \phi_d \times 10^{-3}}$$

Solving, $\phi_d = 0.13$. So, 13% (by vol.) of plasticizer is to be added to the polymer.

A relationship between T_g and the composition of a random or statistical copolymer can be derived by assuming that each type of monomer unit has a characteristic free volume and that it is the same in a copolymer or homopolymer. Denoting the two types of monomer units in a copolymer by subscripts 1 and 2, a linear relationship between T_g and the weight fraction composition is thus obtained as

$$T_g = w_1 T_{g,1} + w_2 T_{g,2} \tag{2.44}$$

This equation usually predicts T_g too high.

Using the iso-free-volume criterion for the glass transition and assuming that each type of monomer unit retains its characteristic free volume in the copolymer above T_g, the following relationship has been derived (Mandelkern et al., 1957):

$$\frac{1}{T_g} = \frac{1}{(w_1 + Bw_2)} \left[\frac{w_1}{T_{g,1}} + \frac{Bw_2}{T_{g,2}} \right] \tag{2.45}$$

Here B is a constant for the pair of monomers. With the assumption that B, which is never far from unity, is exactly unity, Eq. (2.45) reduces to the very simple, but nonlinear form

$$\frac{1}{T_g} = \frac{w_1}{T_{g,1}} + \frac{w_2}{T_{g,2}} \tag{2.46}$$

This is the familiar *Fox equation* (Fox, 1956). This equation predicts the typically convex relationship obtained when T_g is plotted against w_1 or w_2.

Equation (2.45) can be rearranged to a more convenient form:

$$w_1(T_g - T_{g,1}) + kw_2(T_g - T_{g,2}) = 0 \tag{2.47}$$

where $k = B(T_{g,1}/T_{g,2})$. This is known as the *Wood equation*, which fits data for many copolymers very well.

Though Eqs. (2.44) to (2.47) are derived for random copolymers, these are also applicable to those cases of polymer blending and plasticization with low molecular weight compounds where phase separation does not take place and one phase is retained.

Problem 2.32   From the following data for T_g of compatible blends of polyphenyleneoxide (PPO) and polystyrene (PS) as a function of weight fraction (w_1) of PS:

w_1	0	0.2	0.4	0.6	0.8	1.0
T_g (°K)	489	458	431	413	394	378

Estimate the value of T_g for a PPO/PS blend with PS weight fraction of 0.3.

Answer:

The T_g values of the PPO/PS blends are plotted against the weight fraction w_1 of PPO in Fig. 2.25. The curves for the linear equation (2.44), the Fox equation (2.46), and the Wood equation (2.47) are drawn on the same graph. Best fit is seen to be obtained for the Wood equation with $k = 0.679$.

From Eq. (2.47):

$$T_g = \frac{w_1 T_{g,1} + kw_2 T_{g,2}}{w_1 + kw_2}$$

With $w_1 = 0.3$, $w_2 = 0.7$, $T_{g,1} = 378°K$, $T_{g,2} = 489°K$,

$$T_g = \frac{0.3 \times (378°K) + 0.679 \times 0.7 \times (489°K)}{0.3 + 0.679 \times 0.7} = 446°K$$

Problem 2.33   How much of a low molecular weight plasticizer with $T_g = -80°C$ should be added to a film of nylon-6,6 in order that T_g is reduced from 50°C to 25°C ?

Answer:

Assuming that single phase is retained on plasticization, equations derived for random polymers can be used

$T_{g,1} = 50°C \equiv 323°K$, $T_{g,2} = -80°C \equiv 193°K$, $T_g = 25°C \equiv 298°K$

From Eq. (2.44) :

$$(298°K) = (1 - w_2)(323°K) + w_2(193°K) \implies w_2 = 0.19$$

From Eq. (2.46) :

$$\frac{1}{(298°K)} = \frac{1 - w_2}{(323°K)} + \frac{w_2}{(193°K)} \implies w_2 = 0.17$$

The difference is due to higher T_g value predicted by the linear equation.

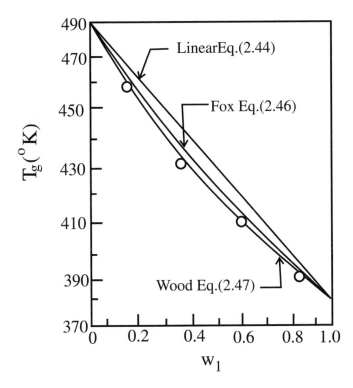

Figure 2.25 Plot of T_g against composition for PPO/PS blends according to (a) the linear equation,   (b) the Fox equation, and (c) the Wood equation. Circles represent experimental points (data of Problem 2.32).

2.11   Chain Movements in Amorphous State

The random chain arrangements in amorphous or molten polymer create extra empty space, called the *free volume*, which essentially consists of holes in the matrix. When sufficient thermal energy is available, the vibrations can cause a segment to jump into a hole by cooperative bond rotation and a series of such jumps can result in the polymer chain eventually changing its position. However, the complete movement of a chain cannot remain unaffected by the surrounding chains since considerable entanglement exists in the melt and any motion will be retarded by other chains.

According to Bueche (1962), the polymer molecule may drag along several others during flow, causing energy dissipation due to friction. The length of the polymer chains in the sample would thus be expected to play a significant role in determining the friction and resistance to flow. The effect of chain length on melt viscosity (η), measured at low shear rates to ensure Newtonian flow, is illustrated in Fig. 2.26. The plot (Fox and Flory, 1951) shows the dependence of melt viscosity on chain length Z defined as the *weight average number* of chain atoms in the polymer molecules. (For vinyl polymers Z is twice and for diene polymers four times the weight-average degree of polymerization.) Below the *critical chain length* Z_c the melt viscosity is given by

$$\eta = K_L Z^{1.0} \tag{2.48}$$

and above Z_c, the melt viscosity is given by

$$\eta = K_H Z^{3.4} \tag{2.49}$$

where K_L and K_H are constants for low and high degrees of polymerization. They are temperature dependent.

While the 1.0 power dependence in Eq. (2.48) represents the simple increase in viscosity as the chain gets longer, the dependence of the viscosity on the 3.4 power of the chain length as shown in Eq. (2.49) arises from entanglement and diffusion considerations. Thus the critical chain length Z_c may be looked upon as the threshold below which the chains are too short to cause significant increase in viscosity by entanglement effects and above which the chains are large enough to cause entanglement, large increase in viscosity and retard melt flow. A few approximate Z_c values (Sperling, 1986) are: polyisobutylene 610, polystyrene 730, poly(dimethyl siloxane) 950, and poly(decamethylene adipate) 290. Relatively flexible polymer chains have a high Z_c, while more rigid-chain polymers have a relatively low Z_c.

Problem 2.34   A plastics extruder is known to work best at a melt viscosity of about 20,000 poises. The vinyl polymer of $\overline{DP}_w = 750$ and $T_g = 80°C$, earlier used with the extruder, had this viscosity at 150°C. However, a batch of the same polymer received subsequently has $\overline{DP}_w = 500$. At what temperature should the extruder now be run so that the optimum melt viscosity for processing is achieved (Sperling, 1986) ?

Answer: For a vinyl polymer, $Z = 2(\overline{DP}_w)$. Since both samples have $Z > 700$, Eq. (2.49) can be used for both.

For polymer with $\overline{DP}_w = 750$ and at 150°C,

$$2 \times 10^4 = K_H (2 \times 750)^{3.4}$$

$$K_H = 3.18 \times 10^{-7} \text{ at } 150°C$$

For polymer with $\overline{DP}_w = 500$,

$$\eta = 3.18 \times 10^{-7} (2 \times 500)^{3.4} = 6023 \text{ Poise at } 150°C$$

The WLF equation (cf. Eq. (2.32)) can be written as

$$\log \frac{\eta_T}{\eta_g} = \frac{-17.44(T - T_g)}{51.6 + (T - T_g)}$$

Since at high molecular weight T_g is essentially constant, both the polymer samples may be assumed to have the same T_g. Therefore,

$$\log \frac{6023}{\eta_g} = \frac{-17.44(150 - 80)}{51.6 + (150 - 80)} \implies \eta_g = 6.6 \times 10^{13} \text{ Poise}$$

Applying the WLF equation again

$$\log \frac{2 \times 10^4}{6.6 \times 10^{13}} = \frac{-17.44(T - 80)}{51.6 + (T - 80)}. \quad \text{Solving,} \quad T = 142°C$$

2.11.1 The Reptation Model

Since independent chain mobility is not possible for polymer chains longer than $\mathcal{Z}_c$ because of the onset of entanglement, a modified model is required to account for the ability of long chains to translate and diffuse through the polymer matrix. The *reptation model* of De Gennes consists of a single polymeric chain, trapped inside a three-dimensional network of entangled chains, which may be considered to constitute a set of fixed obstacles, as shown in Fig. 2.27. The chain is not allowed to cross any of the obstacles, but it may move in a snake-like fashion among them. This snake-like motion is called *reptation*.

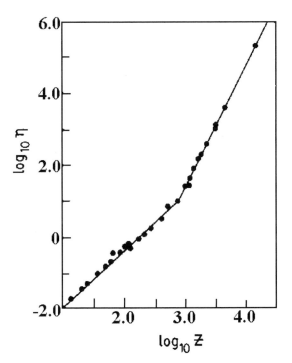

Figure 2.26 Plot of melt viscosity versus chain length $\mathcal{Z}$ for polyisobutylene fractions at 217°C. (After Fox and Flory, 1951.)

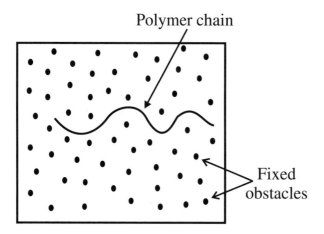

Figure 2.27 A model for reptation. The polymer chain moves among fixed obstacles, but cannot cross any of them. (After De Gennes, 1971; Sperling, 1986.)

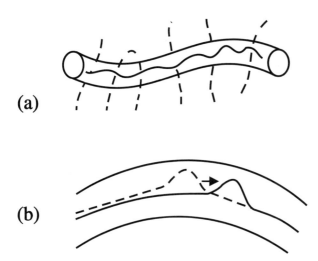

Figure 2.28 (a) Schematic representation of a polymer chain confined in a hypothetical tube contoured by fixed obstacles. (b) Movement of a "kink" along the chain. (After Cowie, 1991.)

In the De Gennes approach, the polymer chain is assumed to be confined in a hypothetical tube [Fig. 2.28(a)], the contours of which are defined by the position of the entanglement points in the network. Mechanistically, reptation can be viewed as the movement of a kink in the chain along its length [see Fig. 2.28(b)] until it reaches the end of the chain and leaves it. Successive motions of this kind translate the chain through the tube and eventually take it completely out of the tube. Using scaling concepts, de Gennes (1971) found that the diffusion coefficient, D, of a chain in the entangled polymer matrix depends on the molecular weight M as

$$D \propto M^{-2} \tag{2.50}$$

Polyethylene of molecular weight 10,000 has a value of D near 10^{-8} cm^2/s at 176°C and polystyrene of molecular weight 100,000 has a D value of about 10^{-12} cm^2/s at 175°C (Tirrell, 1984).

When two blocks of the same polymer are pressed together at a temperature above the T_g for a relatively short time t, interdiffusion of polymer chains takes place (by reptation) across the interface to produce a significant number of entanglements, thereby joining the blocks together (Sperling, 1986). The strength of the junction formed will, however, depend on time t.

Problem 2.35   Mills et al. (1984) related the diffusion coefficient of polystyrene at 170°C to the weight-average molecular weight by the equation:

$$D = 8 \times 10^{-3} (\overline{M}_w)^{-2} \text{ cm}^2/\text{s}$$

In the welding of polystyrene ($\overline{M}_w = 10^5$) at a temperature of 170°C, calculate the flux of polymer molecules across the interface for diffusion over a 100 Å distance. (Density of polystyrene = 1.0 g/cm^3.)

Answer:

Bulk concentration of polymer $= \dfrac{(1.0 \text{ g cm}^{-3})}{(10^5 \text{ g mol}^{-1})} = 10^{-5} \text{ mol cm}^{-3}$

Concentration gradient over 100 Å or 10^{-6} cm distance from the bulk concentration to a zero concentration $= [(10^{-5} - 0) \text{ mol cm}^{-3}]/(10^{-6} \text{ cm}) = 10 \text{ mol cm}^{-4}$

$$D = 8 \times 10^{-3} \times 10^{-10} \text{ cm}^2 \text{ s}^{-1} = 8 \times 10^{-13} \text{ cm}^2 \text{ s}^{-1}$$

$$\text{Flux} = (8 \times 10^{-13} \text{ cm}^2 \text{ s}^{-1})(10 \text{ mol cm}^{-4})(6 \times 10^{23} \text{ molecules mol}^{-1})$$

$$= 48 \times 10^{11} \text{ molecules cm}^{-2} \text{ s}^{-1}$$

In one second, 48×10^{11} molecules will diffuse through a 1-cm^2 area of the interface.

2.12   *Thermodynamics of Rubber Elasticity*

Elastomer deformation is particularly amenable to analysis using thermodynamics since an ideal elastomer behaves essentially as an 'entropy spring', resembling an ideal gas in this respect. In the following we shall derive relationships for internal energy (U) and entropy (S) changes accompanying deformation of an elastomer. More particularly, we shall be concerned with derivation of relationships between U and S and the state variables — force (f), length (l), and temperature (T) — from the first and second laws of thermodynamics and use the analogy of an ideal elastomer and an ideal gas since both can be considered as entropy springs.

The first law of thermodynamics states that the change in internal energy (ΔU) of an isolated system is equal to the total heat (Q) absorbed by the system less the work (W) done <u>by</u> the system:

$$\Delta U = Q - W \tag{2.51}$$

In differential form,

$$dU = dQ - dW \tag{2.52}$$

If the work is done <u>on</u> the system, such as when a gas is compressed or an elastomer is deformed by applying force, the corresponding equation is

$$dU = dQ + dW \tag{2.53}$$

According to the second law of thermodynamics, the heat absorbed by the system dQ at temperature T is related to the increment of entropy dS by

$$dS = dQ/T \tag{2.54}$$

For a reversible process, Eqs. (2.53) and (2.54) thus yield

$$dU = TdS + dW \tag{2.55}$$

If the length of an elastic specimen is increased a small amount dl by a tensile force f then the amount of work fdl will be done <u>on</u> the system (i.e., the specimen). There is also a change in volume dV during elastic deformation. So if a hydrostatic pressure P is acting in addition to the tensile force f, the work PdV is done <u>by</u> the system against the pressure, P. The total work done <u>on</u> the system is therefore

$$dW = fdl - PdV \qquad (2.56)$$

One experimental observation which allows the analysis to be simplified is that the deformation of elastomers takes place approximately at constant volume, i.e., the deformation is nearly *isovolume*. For such a deformation at ambient pressure ($P = 1$ atm), the contribution of PdV to dW will be small and so the work done <u>on</u> the system in creating an elongation dl is

$$dW = fdl \qquad (2.57)$$

Substitution in Eq. (2.55) gives

$$dU = fdl + TdS \qquad (2.58)$$

A useful thermodynamic quantity to characterize changes at constant volume of the working substance is Helmholtz free energy. It is defined as

$$A = U - TS \qquad (2.59)$$

Differentiating at constant temperature,

$$dA = dU - TdS \qquad (2.60)$$

From a comparison of Eqs. (2.58) and (2.60),

$$dA = fdl \qquad (\text{constant } T) \qquad (2.61)$$

Therefore,

$$f = (\partial A/\partial l)_T \qquad (2.62)$$

Combining Eqs. (2.60) and (2.62), the force f can be expressed as

$$f = (\partial U/\partial l)_T - T(\partial S/\partial l)_T \qquad (2.63)$$

The first term in Eq. (2.63) refers to the change in internal energy with extension and the second to the change in entropy with extension. Thus the contribution to the total force consists of an internal energy component and an entropy component. While the internal energy term is dominant in the case of most materials, for elastomers, however, the change in entropy makes the largest contribution to the force. This can be shown after modifying the entropy term in Eq. (2.63). From Eq. (2.59), for any general change,

$$dA = dU - TdS - SdT \qquad (2.64)$$

Combining this equation with Eq. (2.58) gives

$$dA = f\,dl - S\,dT \qquad (2.65)$$

and partial differentiation first at constant temperature and then at constant length gives

$$\left.\begin{array}{rcl} (\partial A/\partial l)_T & = & f \\ (\partial A/\partial T)_l & = & -S \end{array}\right\} \qquad (2.66)$$

Applying the standard relation for partial differentiation, viz.,

$$\frac{\partial}{\partial l}\left(\frac{\partial A}{\partial T}\right)_l = \frac{\partial}{\partial T}\left(\frac{\partial A}{\partial l}\right)_T \qquad (2.67)$$

to Eq. (2.66) gives

$$(\partial S/\partial l)_T = -(\partial f/\partial T)_l \qquad (2.68)$$

Equation (2.63) then becomes

$$f = (\partial U/\partial l)_T + T(\partial f/\partial T)_l \qquad (2.69)$$

This equation is sometimes called the *thermodynamic equation of state for rubber elasticity*. For an ideal elastomer $(\partial U/\partial l)_T = 0$; Equations (2.63) and (2.69) then reduce to

$$f = -T(\partial S/\partial l)_T = T(\partial f/\partial T)_l \qquad (2.70)$$

For real elastomers, however, the internal energy term $(\partial U/\partial l)_T$ cannot be exactly zero, since chain uncoiling would require that the bond rotational energy barriers are overcome.

The values of $(\partial U/\partial l)_T$ and $(\partial S/\partial l)_T$ for real elastomers can be determined by measuring how the tensile force required to hold an elastomer at constant length varies with the temperature, as shown schematically in Fig. 2.29. Above the glass transition temperature, f increases as T increases indicating a contraction of the elastomer. The slope of the curve at any point $(\partial f/\partial T)_l$ gives, according to Eq. (2.68), the variation of entropy with extension at that temperature and enables the value of $(\partial U/\partial l)_T$ to be calculated from Eq. (2.69). It is known from various measurements on polybutadiene and *cis*-polyisoprene elastomers that $(\partial U/\partial l)_T$ is only about 1/10 to 1/5 of f, which means that there is very small change in internal energy during extension and it is the change in entropy that dominates the deformation. Therefore, these elastomers are essentially, but not entirely, entropy springs.

2.12.1 Stress-Strain Behavior of Crosslinked Elastomers

Consider a cube of crosslinked elastomer subjected to a tensile force f. The ratio of the increase in length to the unstretched length is the *nominal strain*, ϵ,

$$\epsilon = (\ell - \ell_0)/\ell_0 \qquad (2.71)$$

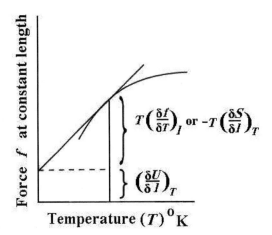

Figure 2.29 An analysis of the thermodynamic equation of state [Eq. (2.69)] for rubber elasticity using a general experimental curve of force versus temperature at constant length. The tangent to the curve at T is extended back to $0°K$. For an ideal elastomer, the quantity $(\partial U/\partial \ell)_T$ is zero, and the tangent goes through the origin. The experimental line is, however, straight in the ideal case. (After Flory, 1953.)

where ℓ and ℓ_0 are the stretched and unstretched specimen lengths, respectively. The deformation, expressed as the extension ratio λ, is thus related to ϵ by

$$\lambda = \ell/\ell_0 = 1 + \epsilon \tag{2.72}$$

For a cube of unit initial dimensions, $d\ell = \ell_0 d\lambda = d\lambda$ and the stress σ is equal to the force f. Equation (2.70) is thus equivalent to

$$\sigma = -T(\partial S/\partial \lambda)_T \tag{2.73}$$

From statistical mechanical calculations (Treloar, 1975) it can be shown that the entropy of deformation is given by

$$\Delta S = -(nR/2N_{Av}) [\lambda^2 + (2/\lambda) - 3] \tag{2.74}$$

where n is the number of chain segments between crosslinks (Fig. 2.30) per unit volume, R is the universal gas constant, and N_{Av} is Avogadro's number. This equation is derived for an ideal network.

Combination of Eqs. (2.73) and (2.74) gives

$$\sigma = (nRT/N_{Av}) (\lambda - 1/\lambda^2) \tag{2.75}$$

For an ideal network the quantity $\rho N_{Av}/n$ (where ρ is the elastomer density, gram per unit volume) is defined as the average molecular weight of chain lengths between crosslinks, denoted by $\overline{M}_c$. So,

$$n = \frac{\rho N_{Av}}{\overline{M}_c} \tag{2.76}$$

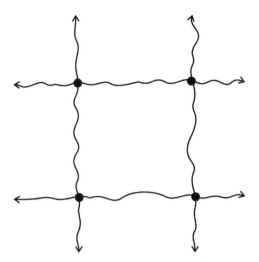

Figure 2.30 Idealized network structure of a crosslinked polymer. • indicates a crosslink (junction) and → signifies continuation of the network structure. Wavy lines between crosslinks are active network chain segments. (Note that for a tetrafunctional cross-link, as shown here, the number of crosslinks is one-half the number of active network chain segments.)

Equation (2.75) can thus be written as

$$\sigma = (\rho RT/\overline{M}_c)(\lambda - 1/\lambda^2) \tag{2.77}$$

$\overline{M}_c$ is a basic variable characterizing the network structure. A high value of $\overline{M}_c$ implies a more tightly crosslinked structure. Hence $1/\overline{M}_c$ may be taken as a measure of the *degree of crosslinking.*

Equation (2.77) thus predicts that the stress required to maintain a given strain will increase with temperature, density, and degree of crosslinking of the elastomer.

Substituting for λ from Eq. (2.72), Eq. (2.77) is approximately given by

$$\sigma = (\rho RT/\overline{M}_c)(3\epsilon + 3\epsilon^2 + \cdots) \tag{2.78}$$

At low strains, Young's modulus E is

$$E \equiv d\sigma/d\epsilon = 3\rho RT/\overline{M}_c \tag{2.79}$$

Thus, the higher the degree of crosslinking (i.e., lower $\overline{M}_c$) of the elastomer, the higher will be its modulus. It will therefore require more force to extend the elastomer a given amount at fixed temperature. Also, for a given elastomer the modulus will increase with temperature, which is characteristic of an entropy spring.

Problem 2.36 A rubber band, 5 cm long, is stretched to 10 cm. The stress at this length is found to increase by 1.2×10^5 Pa when the temperature is raised by 5°C. Assuming ideal behavior, calculate the modulus at 2% elongation of the rubber band at 25°C. Neglect any change in volume with temperature.

Answer:
From Eq. (2.72): $\lambda = \ell/\ell_0 = 10/5 = 2.0$; so, $\lambda - 1/\lambda^2 = 2 - 1/4 = 1.75$.

From Eq. (2.75): $\sigma = (nR/N_{Av})(T)(\lambda - 1/\lambda^2)$. Therefore, at T_1 and T_2:

$$\sigma_1 = (nR/N_{Av})(T_1)(1.75) \quad \text{and} \quad \sigma_2 = (nR/N_{Av})(T_2)(1.75)$$

Hence, $\sigma_2 - \sigma_1 = 1.2 \times 10^5$ Pa $= (nR/N_{Av})(5.0)(1.75)$ which yields $(nR/N_{Av}) = 1.37 \times 10^4$.

At 2% elongation (i.e., $\lambda = 1.02$) and at $T = 298°$K,

$$
\begin{aligned}
E = \sigma/\epsilon = \sigma/(\lambda - 1) &= \frac{(nR/N_{Av})(298)(\lambda - 1/\lambda^2)}{(\lambda - 1)} \\
&= \frac{(1.37 \times 10^4)(298)(0.0588)}{0.02} \\
&= 1.20 \times 10^7 \text{ Pa } (= 12.0 \text{ MPa})
\end{aligned}
$$

Problem 2.37 A styrene-butadiene rubber with 23.5 mol% styrene in the polymer is vulcanized with sulfur. (a) Calculate the stress at 20% elongation of the vulcanizate in which 1.4% of the butadiene units are crosslinked. (b) What would be the corresponding stress if 2% of the butadiene units are crosslinked? Assume random distribution of styrene and butadiene units in the polymer chain. [Density of vulcanizate (without filler) = 0.98 g/cm³ at 25°C.]

Answer:
Molar mass of styrene (C_8H_8) = 104 g/mol. Molar mass of butadiene (C_4H_6) = 54 g/mol. Molar ratio of styrene to butadiene in polymer = 23.5/(100 − 23.5) = 0.307.

(a) Average molar mass of chain lengths between cross-links, $\overline{M}_c$ (cf. Fig. 1.8 in Chapter 1) is

$$\overline{M}_c = \frac{(100 - 1.4)}{1.4}[(54 \text{ g mol}^{-1}) + (0.307)(104 \text{ g mol}^{-1})] = 6052 \text{ g mol}^{-1}$$

For 20% elongation, $\lambda = 1.2$. So from Eq. (2.77):

$$
\begin{aligned}
\sigma &= \frac{(0.98 \text{ g cm}^{-3})(8.37 \times 10^7 \text{ ergs mol}^{-1} \text{ °K})}{(6052 \text{ g mol}^{-1})} \times \left[1.2 - \frac{1}{(1.2)^2}\right] \\
&= 2 \times 10^6 \text{ dyne cm}^{-2} = 0.2 \times 10^6 \text{ N m}^{-2} \quad (= 0.2 \text{ MPa})
\end{aligned}
$$

(b) With 2% of the butadiene units crosslinked,

$$\overline{M}_c = \frac{(100 - 2.0)}{2.0}\left[(54 \text{ g mol}^{-1} + (0.307)(104 \text{ g mol}^{-1}\right] = 4210 \text{ g mol}^{-1}$$

The stress increases in proportion to the degree of crosslinking. Therefore,

$$
\begin{aligned}
\sigma &= \frac{(6052 \text{ g mol}^{-1})}{(4210 \text{ g mol}^{-1})}(2 \times 10^6 \text{ dyne cm}^{-2}) \\
&= 3 \times 10^6 \text{ dyne cm}^{-2} = 0.3 \times 10^6 \text{ N m}^{-2} \quad (= 0.3 \text{ MPa})
\end{aligned}
$$

Problem 2.38 A strip of natural rubber of volume 5 cm³ is slowly and reversibly stretched to twice its length. If the initial slope of the stress-strain curve is 2.5 MPa, how much work (in joules) is done in stretching? Assume that the rubber behaves ideally.

Answer:
Let w = work done, A_0 = initial cross-sectional area, ℓ_0 = initial length, V_0 = initial volume.

$$W = \int f d\ell, \quad \sigma = f/A_0, \quad \lambda = \ell/\ell_0, \quad V_0 = A_0\ell_0 = 5 \times 10^{-6} \text{ m}^3$$

$$(W/V_0) = \int_1^2 \sigma d\lambda = \left(\frac{nRT}{N_{Av}}\right)\int_1^2 \left(\lambda - 1/\lambda^2\right) d\lambda = \left(\frac{nRT}{N_{Av}}\right) \tag{P2.38.1}$$

From Eqs. (2.72) and (2.75):

$$\sigma = \left(\frac{nRT}{N_{Av}}\right)\left[1 + \epsilon - \frac{1}{(1 + \epsilon)^2}\right]$$

$$\left(\frac{d\sigma}{d\epsilon}\right)_{\epsilon=0} = 3\left(\frac{nRT}{N_{Av}}\right) = 2.5 \, \text{MPa} = 2.5 \times 10^6 \, \text{J/m}^3 \tag{P2.38.2}$$

Combining Eqs. (P2.38.1) and (P2.38.2),

$$(W/V_0) = (2.5/3) \times 10^6 \, \text{J/m}^3 \implies W = (2.5/3)(5) \, \text{J} = 4.17 \, \text{J}$$

Problem 2.39   For an ideal rubber strip (20 cm long, 1 cm thick, and 2 cm wide), 20 J of work is expended in stretching it to twice its length at 25°C. Calculate the chain density in mol/cm^3.

Answer:

Energy/volume $= 20 \, \text{J}/40 \, \text{cm}^3 = 0.5 \, \text{J/cm}^3$

Energy/volume $= \int_1^2 \sigma d\lambda = (nRT/N_{Av})\left[\int_1^2 \lambda d\lambda - \int_1^2 \lambda^{-2} d\lambda\right] = (nRT/N_{Av})$

So, $n/N_{Av} = \dfrac{(0.5 \, \text{J cm}^{-3})(10^7 \, \text{erg J}^{-1})}{(8.37 \times 10^7 \, \text{erg mol}^{-1} \, °\text{K}^{-1})(298°\text{K})} = 2 \times 10^{-4} \, \text{mol cm}^{-3}$

Problem 2.40   In a sample of natural rubber (*cis*-polyisoprene) vulcanizate about one of every 200 chain carbon atoms is crosslinked. Estimate the Young's modulus of the sample at low extensions. (Density of vulcanizate $= 0.94 \, \text{g/cm}^3$ at 25°C; gas constant $R = 8.3 \times 10^7$ ergs mol^{-1} °K^{-1} $= 1.987$ cal mol^{-1} °K^{-1}.)

Answer:

There are 200 carbon atoms between crosslinks on the average. This corresponds to 200/4 = 50 monomer units between crosslinks (see Fig. 1.8 in Chapter 1). Since the formula weight of isoprene monomer (C_5H_8) $= 68 \, \text{g mol}^{-1}$, $\overline{M}_c = 50 \times 68 = 3400 \, \text{g mol}^{-1}$

From Eq. (2.79): $E = \dfrac{3(0.93 \, \text{g cm}^{-3})(8.3 \times 10^7 \, \text{dyne.cm mol}^{-1} \, °\text{K}^{-1})(298°\text{K})}{(3400 \, \text{g mol}^{-1})}$

$= 2.0 \times 10^7 \, \text{dyne cm}^{-2} = 2.0 \times 10^6 \, \text{N m}^{-2}$ $(= 2 \, \text{MPa})$

$[\, 1 \, \text{dyne cm}^{-2} = 0.1 \, \text{N m}^{-2} = 0.1 \, \text{Pa}\,]$

Problem 2.41   Stress-strain measurements on a lightly vulcanized natural rubber sample yielded a value of 1.9×10^7 dyne/cm^2 for Young's modulus at low extensions. (a) Calculate the molecular weight of an average chain length between crosslinks. (b) Assuming ideal network structure obtain an estimate of the crosslinked isoprene units in the vulcanized rubber.

Answer:

(a) From Eq. (2.79):

$\overline{M}_c = 3\rho RT/E$

$= \dfrac{3(0.94 \, \text{g cm}^{-3})(8.3 \times 10^7 \, \text{dyne.cm mol}^{-1} \, °\text{K}^{-1})(298 \, °\text{K})}{1.9 \times 10^7 \, \text{dyne cm}^{-2}}$

$= 3671 \, \text{g mol}^{-1}$

(b) Molar mass of isoprene (C_5H_8) $= 68 \, \text{g mol}^{-1}$

According to the idealized structure shown in Fig. 1.4 (Chapter 1), fraction of isoprene units crosslinked $= (68 \, \text{g mol}^{-1})/(3671 \, \text{g mol}^{-1}) = 0.02$

It may be recalled that Eq. (2.77) was derived for an ideal network. The actual behavior of real cross-linked elastomers, however, shows much better accord with the Mooney-Rivlin equation (Mooney, 1948; Rivlin, 1948):

$$C_1 + C_2/\lambda = \frac{1}{2}\sigma(\lambda - 1/\lambda^2)^{-1} \tag{2.80}$$

where C_1 and C_2 are empirical constants.

Equation (2.80) is sometimes written in a form similar to Eq. (2.77) as

$$\sigma = (2C_1 + 2C_2/\lambda)(\lambda - 1/\lambda^2) \tag{2.81}$$

It has been shown that $2C_1$ is related to the degree of crosslinking ($1/\overline{M}_c$), to a fair approximation, by

$$2C_1 = \rho RT/\overline{M}_c \tag{2.82}$$

The parameter C_2 is regarded as a factor correcting for the oversimplification of the elasticity-theory model.

According to Eq. (2.77), the *reduced stress* $\sigma/(\lambda - 1/\lambda^2)$ should be a constant. Equation (2.81), on the other hand, predicts that this quantity depends on λ. Plots of $\sigma/(\lambda - 1/\lambda^2)$ versus $1/\lambda$ are, in fact, found to be linear, especially at low elongation and thus allow for the determination of both C_1 and C_2 (see Problem 2.42)

The parameter C_2 may be considered to be an approximate measure of the departures from the ideal behavior. The magnitude of this parameter is large in dry and lightly swollen rubbers but decreases to zero at high degrees of swelling (see also 'Theory of Swelling' in Chapter 3). A highly swollen rubber thus obeys the ideal rubber Eq. (2.77) more closely.

Problem 2.42   Consider the following stress-strain data (Sperling, 1986) for a rubber band subjected to increasing load at 25°C:

Stress (σ), 10^5 Pa						
	3.6	5.3	7.1	11.1	15.1	19.4
Strain (ϵ)	0.4	0.7	1.2	2.5	3.5	4.2

Make a Mooney-Rivlin plot and determine the constants C_1 and C_2. Obtain an estimate of the degree of crosslinking of the rubber band. (Density of rubber band = 1.09 g/cm^3.)

Answer:

Extension λ is calculated from strain ϵ by the relation (cf. Eq. (2.72): $\lambda = 1 + \epsilon$ and $\sigma/(\lambda - \lambda^{-2})$ is plotted against $1/\lambda$ (Fig. 2.31). The upturn in the curve at high extensions (i.e., low $1/\lambda$) would be attributed to non-Gaussian behavior and limited extensibility of the chain molecules. The values of $2C_1$ and $2C_2$ are obtained from the intercept and slope, respectively, of the linear portion of the plot at low extensions. This yields $C_1 = 1.1 \times 10^5$ Pa $= 1.1 \times 10^6$ dyne/cm^2 and $C_2 = 1.4 \times 10^5$ Pa $= 1.4 \times 10^6$ dyne/cm^2.

From Eq. (2.82):

$$1/\overline{M}_c = 2C_1/\rho RT = \frac{2(1.1 \times 10^6 \text{ dyne cm}^{-2})}{(1.09 \text{ g cm}^{-3})(8.31 \times 10^7 \text{ erg mol}^{-1} \text{ °K}^{-1})(298 \text{ °K})}$$

$$= 8.1 \times 10^{-5} \text{ mol g}^{-1}$$

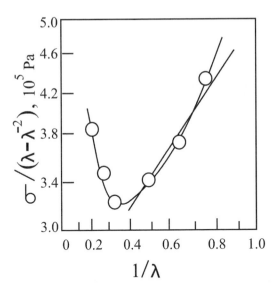

Figure 2.31 A Mooney-Rivlin plot with the data of Problem 2.42. (After Sperling, 1986.)

2.12.2 Nonideal Networks

2.12.2.1 Network Defects

The elastic properties of a vulcanizate is largely influenced by the initial molecular weight of the raw rubber, i.e., before vulcanization. This is because the number of chain ends depends on the molecular weight and each chain end constitutes a flaw in the final network structure. The portion of a molecule from one end to its first crosslinkage along its path, such as the chain AB in Fig. 2.32(a), makes no contribution toward sustaining a stress as it is always free to assume any conformation whatever, as the end A is free. Similarly, the two chains EC and CF of molecule EF are inactive, and so the crosslinkage at C is not a point of constraint on the chain BD. The portion of the network from B to D is thus to be considered a single (active) chain, as if the crosslinkage at C were not present. The crosslink junctions B and D are *elastically active*, while C is not.

It follows from the above discussion that a junction is elastically active if at least three paths leading away from it are independently attached to the network. A polymer chain segment, also called a *strand*, is elastically active if it is bound at each end by elastically active junctions. Loops in the network structure [Fig. 2.32(b)] do not contribute to the elasticity of the network as they have the two ends of a chain segment connected to the network at the same point.

2.12.2.2 Elastically Active Chain Sections

As explained above, a number of chain sections, such as loops and chain ends, do not contribute to the elasticity of the network. It is therefore necessary to develop an expression for the number n_e of active chain sections in a real structure containing a total number n of both active and inactive chain sections.

Suppose that the material consisted originally of N individual molecules of average molecular weight $\overline{M}_n$ per unit volume. The minimum number of links per unit volume which would be required to create a continuous system forming a giant molecule is $N - 1$. It would be incorrect to call it a network, inasmuch as it contains no net-like structure, i.e., it possesses no circuitous

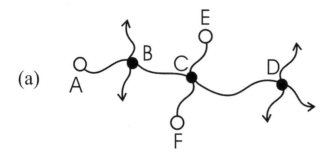

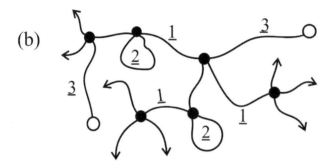

Figure 2.32 (a) Effect of ends of molecules on network structure. (b) Network structure and defects: ● cross-links; ○ terminus of a molecule; (1) elastically active chain; (2) inactive loop; (3) inactive loose end. An arrow (⟶) signifies continuation of the network structure. (After Flory, 1944.)

connections within its structure. As additional crosslinks are added, the structure acquires the character of a network. Only connections in excess of $(N - 1)$ effectively fix the structure so that it responds elastically to deformation. The number of crosslinks per unit volume in a perfect network will be $(n/2)$ (see Fig. 2.30). The effective number of crosslinks per unit volume in a real structure is therefore given by

$$\frac{n_e}{2} = \frac{n}{2} - (N - 1) \simeq \frac{n}{2} - N \tag{2.83}$$

in which 1 has been neglected as being small compared with N. Now N may be expressed by

$$N = \rho N_{Av}/\overline{M}_n \tag{2.84}$$

Combining Eq. (2.83) with Eqs. (2.84) and (2.76) one obtains

$$n_e = \left(\frac{\rho N_{Av}}{\overline{M}_c}\right)\left(1 - \frac{2\overline{M}_c}{\overline{M}_n}\right) \tag{2.85}$$

This expression for n_e should be substituted for n in Eq. (2.75). The combination of Eqs. (2.75) and (2.85) yields

$$\sigma = \left(\frac{\rho RT}{\overline{M}_c}\right)\left(\lambda - \frac{1}{\lambda^2}\right)\left(1 - \frac{2\overline{M}_c}{\overline{M}_n}\right) \tag{2.86}$$

For $\overline{M}_n \to \infty$, Eq. (2.86) reduces to Eq. (2.77).

Equation (2.86) has been shown to be a suitable representation of stress in an elastomer subjected to deformation (Rivlin, 1948). Difficulties arise, however, in accurately measuring or controlling $\overline{M}_c$. In practice, $\overline{M}_c$ may also change during experimentation or application because of oxidative or mechanical degradation.

Problem 2.43   Given a rubber that has $\overline{M}_n$ = 100,000 g/mol before crosslinking. After crosslinking, it is stretched to 2 times its unstretched length at 25°C supporting a stress of 0.84 MN/m². After aging 10 days in air at 125°C, the stress at 2 times its unstretched length at 25°C is only 0.48 MN/m². (a) Calculate the degree of crosslinking of the rubber before aging. (b) What fraction of crosslinks are lost on aging ? The density of the crosslinked rubber (without filler) is 0.98 g/cm³.

Answer:

(a) T = 298°K, λ = 2, σ = 0.84 MN m^{-2} = 8.4x10⁶ dyne cm^{-2}.

From Eq. (2.86) :

$$(8.4 \times 10^6 \text{ dyne cm}^{-2}) = \frac{(0.98 \text{ g cm}^{-3})(8.37 \times 10^7 \text{ erg mol}^{-1} \text{ °K}^{-1})(298°\text{K})}{\overline{M}_c}$$

$$\times \left[2 - \frac{1}{2^2}\right]\left[1 - \frac{2\overline{M}_c}{(10^5 \text{ g mol}^{-1})}\right]$$

Solving, $\overline{M}_c$ = 4621 g mol^{-1}.   (Note that use of Eq. (2.77) gives $\overline{M}_c$ = 5092 g mol^{-1}.)   Degree of crosslinking = $(\overline{M}_c)^{-1}$ = 1/(4621 g mol^{-1}) = 2.2x10^{-4} mol g^{-1}.

(b) After aging, σ = 0.48 MN m^{-2} = 4.8x10⁶ dyne cm^{-2} at T = 25°C (298°K). From Eq. (2.86), $\overline{M}_c$ = 7563 g mol^{-1}. Since $(\overline{M}_c)^{-1}$ is proportional to the number of crosslinks,

$$\text{loss of crosslinks on aging} = \frac{(4621 \text{ g mol}^{-1})^{-1} - (7563 \text{ g mol}^{-1})^{-1}}{(4621 \text{ g mol}^{-1})^{-1}} = 0.39 \text{ or } 39\%$$

REFERENCES

Alkonis, J. J., *J. Chem. Educ.*, **58**(11), 892 (1981).

Beamen, R. G., *J. Polym. Sci.*, **9**, 470 (1952).

Bovey, F. A.,"Polymer Conformation and Configuration", Academic Press, New York, 1969.

Bovey, F. A. and Tiers, G. V. D., *J. Polym. Sci.*, **44**, 173 (1960).

Boyer, R. F., *J. Appl. Phys.*, **25**, 825 (1954).

Bueche, F., "Physical Properties of Polymers", Interscience Publishers, New York, 1962.

Cowie, J. M. G., "Polymers: Chemistry and Physics and Modern Materials", Blackie, Glasgow, 1991.

De Gennes, P. G., *J. Chem. Phys.*, **55**, 572 (1971).

Eyring, H., *J. Chem. Phys.*, **4**, 283 (1936).

Flory, P. J., *Chem. Revs.*, **35**, 51 (1944).

Flory, P. J., "Principles of Polymer Chemistry", Cornell University Press, Ithaca, New York, 1953.

Flory, P. J. and Leutner, F. S., *J. Polym. Sci.*, **3**, 880 (1948); **5**, 267 (1950).

Fox, T. G., *Bull. Am. Phys. Soc.*, **1**, 123 (1956).

Fox, T. G. and Flory, P. J., *J. Appl. Phys.*, **21**, 581 (1950).

Fox, T. G. and Flory, P. J., *J. Phys. Chem.*, **55**, 221 (1951).

Fox, T. G. and Flory, P. J., *J. Polym. Sci.*, **14**, 315 (1954).

Guth, E. and Mark, H., *Monatsch. Chem.*, **65**, 93 (1934).

Hay, J. N., *Brit. Polym. J.*, **3**, 74 (1971).

Holmes, D. R., Bunn, C. W., and Smith, D. J., *J. Polym. Sci.*, **17**, 159 (1955).

Kelley, F. N. and Bueche, F., *J. Polym. Sci.*, **50**, 549 (1961).

Koening, J. L., "Chemical Microstructure of Polymer Chains", chaps. 6-8, Wiley Interscience, New York, 1980.

Mandelkern, L., Martin, G. M., and Quinn, F. A., *J. Res. Natl. Bur. Stand.*, **58**, 137 (1957).

Mills, P. J., Green, P. F., Palmstrom, C. S., Mayer, J. W., and Kramer, E. J., *Appl. Phys. Lett.*, **45**(9), 957 (1984).

Mooney, M., *J. Appl. Phys.*, **19**, 434 (1948).

Odian, G., "Principles of Polymerization", 3rd ed., John Wiley, New York, 1991.

Orfino, T. A., *Polymer*, **2**, 305 (1961).

Price, F. P. in R. H. Doremus, B. W. Roberts, and Turnbull, D., Eds., "Growth and Perfection in Crystals", p. 466, Wiley, New York, 1958.

Rivlin, R. S., *Trans. Roy. Soc. (Lond.)*, **A241**, 379 (1948).

Rudin, A., "The Elements of Polymer Science and Engineering", Academic Press, Orlando, 1982.

Sharma, S. C., Mandelkern, L., and Stehling, F. C., *J. Polym. Sci.*, **B10**, 345 (1972).

Simha, R. and Boyer, R. F., *J. Chem. Phys.*, **37**, 1003 (1962).

Sperling, L. H., "Introduction to Physical Polymer Science", Wiley-Interscience, New York, 1986.

Tirrell, M., *Rubber Chem. Tech.*, **57**, 523 (1984).

Williams, M. L., Landel, R. F., and Ferry, J. D., *J. Am. Chem. Soc.*, **77**, 3701 (1955).

Young, R. J. and Lovell, P. A., "Introduction to Polymers", Chapman and Hall, London, 1990.

EXERCISES

2.1   (a) Calculate the root mean square end-to-end distance and the radius of gyration for a molecule in molten polypropylene of molecular weight 10^5. [Data: carbon-carbon bond length = 1.54×10^{-8} cm; tetrahedral bond angle = $109.5°$; steric parameter, $\sigma = 1.6$ at $140°C$.]   (b) How extensible is the molecule? (*Hint:* Calculate the ratio of the extended chain length to the average chain end separation.) [*Ans.* (a) 2.4×10^{-6} cm; 9.8×10^{-7} cm.   (b) 24 times]

2.2   Assuming that the RMS end-to-end distance is an approximate measure of the diameter of the spherical, coiled polymer molecule in dilute solution, compare the volume occupied by one molecule of polyisobutylene of molecular weight 10^6:

   (a) In a solid at $30°C$ (density = 0.92 g/cm^3).

   (b) In a *theta*-solvent.

Take the value of the steric parameter σ for the polymer as 2.0 and the carbon-carbon bond length as 1.54×10^{-8} cm.

[*Ans.* (a) 1.81×10^6 Å^3; (b) 2.92×10^8 Å^3]

2.3   The RMS end-to-end distance is $2.01 \sqrt{n}$ Å for *cis*-polyisoprene and $2.90 \sqrt{n}$ Å for *trans*-polyisoprene, n being the total number of bonds in each real chain. Compare the chain stiffness of the two polymers. [*Ans. Trans*-polyisoprene is 1.4 times stiffer than *cis*-polyisoprene.]

2.4   From the data given above calculate the values of σ and C_∞ for *cis*- and *trans*-polyisoprene, given that the length of an isoprene repeating unit is 4.60×10^{-10} m. [*Ans. Cis*-polyisoprene: $\sigma = 1.2$; $C_\infty = 3.1$.   *Trans*-polyisoprene: $\sigma = 1.8$; $C_\infty = 6.4$]

2.5   Given that the RMS end-to-end distance $<r^2>_o^{1/2}$ is $(2.90\sqrt{n})10^{-8}$ cm for *trans*-polyisoprene, where n is the total number of bonds in the real chain, and the length of an isoprene repeating unit is 4.60×10^{-8} cm, calculate the number of monomer units a *trans*-polyisoprene molecule has per equivalent random (freely-jointed) chain link. [*Ans.* 1.59]

2.6 The α-methyl resonance in poly(α-methyl styrene) is found to be split into three peaks which are assigned to isotactic, heterotactic, and syndiotactic triads. Fractions of the polymers in the three configurations determined by the area of these peaks are given below for poly(α-methyl styrene) prepared with two different catalysts [Brownstein, S., Bywater, S., and Worsfold, D. J., *Makromol. Chem.*, **48**, 127 (1961)]:

Polymer	Catalyst	Fraction of polymer		
		Iso	Hetero	Syndio
1	BF$_3$	–	0.11	0.89
2	Na-Naphthenide	0.13	0.48	0.39

Determine the probability P_m and test for consistency with Bernoullian statistics. What do the values of P_m obtained in the two cases signify?
[*Ans.* Polymer 1: P_m = 0.06; Polymer 2: P_m = 0.4]

2.7 The density of crystalline polyethylene is 0.996 g/cm^3, while the density of amorphous polyethylene is 0.866 g/cm^3. Calculate the percentage of crystallinity in a sample of linear polyethylene of density 0.970 g/cm^3 and in a sample of branched polyethylene of density 0.917. Why do the two samples have considerably different crystallinities?
[*Ans.* 42.6%; 82.1%]

2.8 Chemically both wax and polyethylene can be described as polymethylene and represented by $-(CH_2)_n-$, but while the former is a brittle solid the latter is a tough plastic. What are the reasons for this difference in mechanical behavior?

2.9 Draw a $\log E$ versus temperature plot for a linear, amorphous polymer and indicate the position and name the five regions of viscoelastic behavior. How is the curve changed if (a) the polymer is semicrystalline, (b) the polymer is crosslinked, and (c) the experiment is run faster?

2.10 To be a useful plastic, an amorphous polymer must be below its T_g at the ambient temperature, while crystalline polymers may be either above or below their T_g's. Why?

2.11 Explain the following
(a) By dipping into liquid nitrogen, an adhesive tape loses its stickiness.
(b) A hollow rubber ball when cooled in liquid nitrogen and thrown hard against the wall breaks into pieces.
(c) A dinner bell coated with latex paint and kept in a freezer makes louder noise than the one coated and kept at room temperature.
(d) Molding or extrusion of plastics too close to T_g can result in a stiffening of the material.
(e) The T_g of a semicrystalline polymer is often higher than the same polymer in a completely amorphous state.
(f) In emulsion polymerization of styrene, often carried out at 80°C, the reaction does not proceed quite to 100% conversion.
(g) When nylon shirts are washed and hung up to drip-dry, creases straighten out by themselves.
(h) The postage stamp, which is coated with linear poly(vinyl alcohol) adhesive, needs to be moistened with water (or saliva) before applying.

2.12 When polyethylene is chlorinated, chlorine replaces hydrogen at random. The softening point of this chlorinated product depends on the chlorine content. It is found that small amounts of chlorine (10 to 50 wt% Cl) lowers the softening point while large amounts (~ 70%) raise the softening point. Rationalize this observation on the basis of intermolecular forces.

2.13 Account for the fact that the coefficient of volume expansion below T_g is smaller than that above T_g. Why is it said that in the glassy state the polymer is not in a true thermodynamic equilibrium?

2.14 A newly prepared thermoplastic polymer has a T_g of 110°C and a melt viscosity of 1.2×10^6 poises at 140°C. The polymer, however, degrades above 160°C. Can this polymer be processed with an extruder which performs best when the melt viscosity is about 2×10^3 poises ? If not, what can be done to make the polymer processable ?

[*Ans.* Cannot be used as $\eta_{160°C} > \eta$ of processability. Plasticizer may be added.]

2.15 The following data were obtained [L. Mandelkern, G. M. Martin, and F. A. Quinn, *J. Res. Natl. Bur. Stand.*, **58**, 137 (1957)] for the glass transition temperature T_g of the copolymers of vinylidene fluoride and chlorotrifluoroethylene, as a function of the weight fraction w_1 of the first comonomer.

w_1	0	0.14	0.35	0.40	0.54	1.0
T_g (°C)	46	19	-3	-8	-15	-38

What would be the T_g of a copolymer with $w_1 = 0.75$?
[*Ans.* −28°C]

2.16 Two polystyrene standards (monodisperse) of molecular weights 1.8×10^6 and 1.6×10^5 have T_g values of 105.9°C and 104.7°C, respectively. Calculate the T_g value of a highly polydisperse polystyrene made up of four low polydisperse components as shown below:

Component	Wt. fraction (w)	$\overline{M}_n$
1	0.05	1.5×10^6
2	0.41	4.8×10^5
3	0.39	1.2×10^5
4	0.15	3.7×10^4

[*Ans.* 104.3°C]

2.17 A new polymer with a weight average degree of polymerization of 1400 and five atoms in the repeating unit has a melt viscosity of 1500 poises at 190°C. What will be the viscosity at the same temperature if its molecular weight is doubled ?
[*Ans.* 15834 poise]

2.18 A vinyl polymer with a Z value of 200 has a melt viscosity of 100 poises. What would the viscosity of this polymer be if $Z = 800$?
[*Ans.* 798 poise]

2.19 A polymer with a T_g of 105°C and a Z value of 400 is found to have a melt viscosity of 500 poises at 170°C. What will be the melt viscosity at 150°C if $Z = 800$? Assume $Z_c = 600$. (*Hint:* Combine the DP dependence with the WLF equation.)
[*Ans.* 7.5×10^5 poise]

2.20 Account for the fact that if a strip of rubber is brought into contact with the temperature-sensitive lips and stretched rapidly, an instantaneous warming of the strip can be easily perceived.

2.21 A polyisoprene rubber has 2.5% of its repeating units crosslinked by sulfur vulcanization. Estimate the modulus of the sample at low extensions. (Density of vulcanizate = 0.94 g/cm^3 at 25°C.)
[*Ans.* 2.7 MN m^{-2} (= 2.7×10^7 dyne cm^{-2})]

2.22 A synthetic rubber containing 60% butadiene and 40% isoprene by weight was vulcanized with 5% (w/w) sulfur. Stress-strain measurements on this vulcanizate at low extensions yielded a value of 2.8×10^7 dyne/cm^2 for Young's modulus. (a) Calculate the molecular weight of the chain lengths between crosslinks. (b) Assuming that all sulfur is used in crosslinking, determine how many sulfur atoms on average are used in each crosslink. (Density of vulcanizate = 0.93 g/cm^3 at 25°C.)
[*Ans.* (a) 2464 g mol^{-1}; (b) 3.8]

2.23 The synthetic rubber in the above problem was vulcanized such that 1.2% of the monomer units in the polymer were crosslinked. (a) Calculate the stress at 25% elongation of the vulcanizate. (b) What would be the corresponding stress if 2.4% of the units are crosslinked ? Assume random distribution of both monomers in the chain. (Density of vulcanizate = 0.93 g/cm^3 at 25°C.)
[*Ans.* (a) 0.29×10^6 N m^{-2} (= 0.29 MPa); (b) 0.59×10^6 N m^{-2} (= 0.59 MPa)]

2.24 An interesting experiment was described [G. V. Henderson, D. O. Cambell, V. Kuzmicz, and L. H. Sperling, *J. Chem. Ed.*, **62**, 269 (1985)] to determine the extent of hydrogen bonding in gelled gelatin by observing the depth of indentation of a sphere into the surface of the gel. In a typical experiment 22 g gelatin was dissolved in 600 ml water and allowed to set in refrigerator at 5°C. A steel ball 4 cm in diameter and weighing 262 g made an indentation 1.20 cm deep when placed on the gelled gelatin. Assuming a molecular weight of 30,000 for the gelatin, calculate the number of hydrogen bonds per molecule.

Young's modulus (E) may be determined from indentation by using the equation

$$E = \frac{3(1 - v^2)F}{4h^{3/2}r^{1/2}}$$

where F represents the force of sphere against the gelatin surface ($= mg$ dynes), h represents the depth of indentation of sphere (cm), r is the radius of sphere (cm), g is the gravity constant, and v is Poisson's ratio (may be assumed to be 0.5).

[*Ans.* 0.46]

2.25 A 6.0 cm long rubber band is stretched to 15.0 cm. It is found that to keep the rubber band at this length when the temperature is raised from 27°C to 32°C, an additional stress of 1.5×10^5 Pa is to be applied. Calculate the modulus of this rubber at 1% elongation at 27°C. (Neglect any changes in volume with temperature.)

[*Ans.* 11.4 MPa]

2.26 Two identical rubber bands, A and B, each 10 cm long, are tied together at their ends and stretched to a total length of 30 cm. The rubber bands A and B are at 25°C and 140°C, respectively. Calculate the distance of the knot from the untied end of A.

[*Ans.* 16.6 cm]

APPENDIX 2.1

End-to-End Distance of a Freely Jointed Chain

The simplest mathematical model of a polymer chain is the freely jointed chain. It has n links, each of length l, joined in a linear sequence with no restrictions on the angles between successive bonds. The length of the chain along its backbone is known as the *contour length* and is given by nl. However, for a linear flexible chain, it is more realistic to consider its dimensions in terms of the distance between the two chain ends, that is the end-to-end distance r [Fig. A2.1(a)].

The analysis of this model is similar to that of the well-known *random-walk model*, which was first developed to describe the random movement of molecules in an ideal gas. The only difference now is that for the freely jointed chain, each step is of equal length l. To analyze the model one end of the chain may be fixed at the origin **O** of a three-dimensional rectangular coordinate system, as shown in Fig. A2.1(b), and the probability, $P(x, y, z)$, of finding the other end within a small volume element $dx.dy.dz$ at a particular point with coordinates (x, y, z) may be calculated. Such calculation leads to an equation of the form (Young and Lovell, 1990):

$$P(x, y, z) = W(x, y, z)\,dx\,dy\,dz \tag{A2.1}$$

where $W(x, y, z)$ is the *probability density function*, i.e., probability per unit volume. When n is large and $r \ll nl$, $W(x, y, z)$ is given by

$$W(x, y, z) = (\beta/\pi^{1/2})^3 \exp[-\beta^2(x^2 + y^2 + z^2)] \tag{A2.2}$$

where $\beta = [3/(2nl^2)]^{1/2}$. Since $r^2 = x^2 + y^2 + z^2$, Eq. (A2.2) simplifies to

$$W(x, y, z) = (\beta/\pi^{1/2})^3 \exp(-\beta^2 r^2) \tag{A2.3}$$

The value of $W(x, y, z)$ corresponds to an end-to-end distance r in a particular direction specified by the set of coordinates (x, y, z). However, there may be many such coordinates each of which gives rise to the same end-to-end distance r, but in a different direction. The probability of finding out chain end at a distance r in *any* direction from the other chain end located at the origin is equal to the probability $W(r)dr$ of finding the chain end in a spherical shell of thickness dr at a radial distance r from the origin. Since the volume of the spherical shell is $4\pi r^2 dr$ and the probability density function is $W(x, y, z)$, the probability $W(r)dr$ is given by

$$W(r)dr = W(x, y, z)4\pi r^2 dr \tag{A2.4}$$

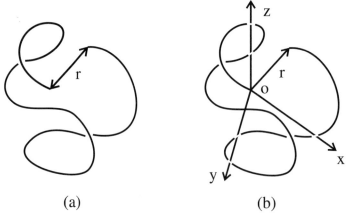

(a)   (b)

Figure A2.1 (a) Schematic representation of a coiled polymer molecule showing the end-to-end distance. (b) Diagram showing a coiled polymer molecule of end-to-end distance r in a rectangular coordinate system with one chain end fixed at the origin. (After Young and Lovell, 1990.)

Substitution of $W(x, y, z)$ from Eq. (A2.3) then gives the following expression for the *radial distribution function* $W(r)$:

$$W(r) = 4\pi(\beta/\pi^{1/2})^3 r^2 \exp(-\beta^2 r^2) \tag{A2.5}$$

This function has a maximum value at a certain value of r. It can be shown by simple differentiation that this maximum occurs at $r = 1/\beta$. It is also possible to show that $W(r)$ normalizes to unity, i.e.,

$$\int_0^\infty W(r)dr = 1 \tag{A2.6}$$

The mean square end-to-end distance, $< r^2 >$, is the second moment of the radial distribution function and so is defined by the integral (Young and Lovell, 1990)

$$< r^2 > = \int_0^\infty r^2 W(r)dr \tag{A2.7}$$

Substituting for $W(r)$ from Eq. (A2.5) and integrating leads to

$$< r^2 > = 3/(2\beta^2) \tag{A2.8}$$

Since $\beta^2 = 3/(2nl^2)$, the RMS end-to-end distance is given by

$$< r^2 >_f^{1/2} = n^{1/2} l \tag{A2.9}$$

where the subscript f is used to indicate a freely jointed chain.

Equation (A2.9) reveals that $< r^2 >_f^{1/2}$ is smaller than the contour length nl by a factor of $n^{1/2}$. The value of n being large, this difference is considerable and it indicates the highly coiled nature of flexible polymer chains. A polymer chain in dilute solution can thus be pictured as a coil, continuously altering its shape and size due to conformational changes under the action of random thermal motions. The angular brackets of $< r^2 >$ denote averaging over all possible conformational sizes available to polymer molecules of the same molecular weight. Size differences become more significant for polymer samples which are polydisperse and contain chains of a wide variety of chain lengths. This calls for another averaging. Chain dimensions of polydisperse linear polymers are thus given by the average RMS end-to-end distance $< \overline{r^2} >^{1/2}$, where the bar indicates averaging due to chain polydispersity, whereas $< r^2 >^{1/2}$ without bar represents the RMS end-to-end distance of a single chain or a monodisperse polymer sample.

Chapter 3

Polymers in Solution

3.1 Introduction

A number of important characteristics of polymers, such as molecular weight, chain length, branching, and chain stiffness, can be explored when the individual molecules are separated from each other. Such studies therefore employ dilute solutions of polymers. However, the dissolution of a polymer also brings with it many new problems. For a correct interpretation of the behavior of polymer solutions it is essential to understand the thermodynamics of polymer-solvent interaction. We will therefore explore some of the basic underlying thermodynamic principles of polymer solutions in this chapter. A major part of the chapter will be concerned with methods of studying polymer solutions that deal with equilibria and can be fully described by thermodynamic relations. These include vapor pressure, osmotic pressure, and phase separation in polymer-solvent systems.

3.2 Thermodynamics of Liquid Mixtures

To characterize the thermodynamic behavior of the components in a solution, it is necessary to use the concept of *partial molar* or *partial specific functions*. The partial molar quantities most commonly encountered in the thermodynamics of polymer solutions are partial molar volume $\overline{V}_i$ and partial molar Gibbs free energy $\overline{G}_i$. The latter quantity is of special significance since it is identical to the quantity called *chemical potential*, μ_i, defined by

$$\mu_i \equiv \left(\frac{\partial G}{\partial n_i}\right)_{P,T,n_{j \neq i}} \equiv \overline{G}_i \tag{3.1}$$

Since any partial molar property of a *pure* substance is simply the corresponding molar property, the chemical potential of a component i in pure form, denoted by μ_i^0, is evidently equal to the molar Gibbs free energy G_i^0 of pure component i at the same temperature and pressure.

The chemical potentials are the key partial molar quantities. The μ_i's determine reaction and phase equilibrium. Moreover, all other partial molar properties and all thermodynamic properties of the solution can be found from the μ_i's if we know the chemical potentials as functions of T, P, and composition.

The free energy of mixing in the formation of a solution, ΔG_{mix}, is given by

$$\Delta G_{\text{mix}} = G - \sum_i G_i^0 \tag{3.2}$$

where G is the free energy characterizing the state of the solution and G_i^0 represents the free energies of the individual (pure) components i forming the solution. Recognizing that chemical potential μ_i is equal to the *partial* molar Gibbs free energy, as shown by Eq. (3.1), and that the chemical potential of pure component i, μ_i^0 is equal to simply the molar Gibbs free energy of pure component i, i.e., G_i^0 at the pressure and temperature of the system, it is easy to show from Eq. (3.2) that

$$\mu_i - \mu_i^0 = \left(\frac{\partial(\Delta G_{\text{mix}})}{\partial n_i}\right)_{P,T,n_{j\neq i}} \tag{3.3}$$

The partial derivative on the right hand side of Eq. (3.3) is the *partial molar Gibbs free energy change*, denoted by $\overline{\Delta G_i}$ (Young and Lovell, 1990).

It may be noted that all known thermodynamic relations are also valid for the partial molar functions. Thus, the relation $G = H - TS$ in terms of partial molar functions is

$$\overline{G}_i = \overline{H}_i - T\overline{S}_i \tag{3.4}$$

Therefore, from Eq. (3.3)

$$\mu_i - \mu_i^0 = \overline{\Delta G_i} = \overline{\Delta H_i} - T\overline{\Delta S_i} \tag{3.5}$$

For ideal solutions that form with a zero heat effect, $\overline{\Delta H_i} = 0$. Hence for ideal solutions

$$\mu_i - \mu_i^0 = \overline{\Delta G_i} = -T\overline{\Delta S_i} \tag{3.6}$$

It will be shown later that ideal entropy of mixing of two components, say, 1 and 2, is given by

$$\Delta S_{\text{mix}} = -R(n_1 \ln x_1 + n_2 \ln x_2) \tag{3.7}$$

where n_i and x_i are, respectively, number of moles and mole fractions of component i. Partial ideal entropies of mixing can be obtained by differentiating Eq. (3.7). Thus for component i, in general,

$$\overline{\Delta S_i} = \frac{\partial(\Delta S_{\text{mix}})}{\partial n_i} = -R \ln x_i \tag{3.8}$$

Substituting in Eq. (3.6),

$$\mu_i - \mu_i^0 = RT \ln x_i \tag{3.9}$$

Thus, the change in chemical potential of the ith component in an ideal solution depends only on the mole fraction of that component in solution.

In the case of a real solution,

$$\mu_i - \mu_i^0 = RT \ln \frac{p_i}{p_i^0} \tag{3.10}$$

where p_i and p_i^0 are the partial vapor pressures of the ith component above the solution and above the pure component, respectively. Since for a thermodynamically stable solution $p_i < p_i^0$, it follows from Eq. (3.10) that $\Delta \mu_i < 0$ or $\mu_i < \mu_i^0$, i.e., the chemical potential of each component in the solution is smaller than that of the pure component.

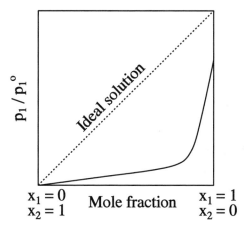

Figure 3.1 Dependence of vapor pressure over a polymer solution on mole fraction of components.

Comparing Eqs. (3.9) and (3.10) we obtain, for an ideal solution,

$$p_i/p_i^0 = x_i \qquad (3.11)$$

Equation (3.11) was obtained empirically by Raoult and is called *Raoult's law*. Hence in ideal solutions Raoult's law holds over the entire range of compositions, this being represented by a straight line in the vapor pressure composition curve (Fig. 3.1). In reality, most solutions do not obey Eqs. (3.9) and (3.11). Such solutions are called *nonideal*, or *real*. Polymer solutions characteristically display sharp negative deviations from ideality, as can be seen in Fig. 3.1.

A method of estimating the deviation of solutions from ideality, suggested by Lewis, consists of substituting a certain function called *activity* for the mole fractions in all thermodynamic equations which are not valid for real solutions. Thus, instead of Eqs. (3.9) and (3.11) we may write

$$\mu_i - \mu_i^0 = \Delta\mu_i = RT \ln a_i \qquad (3.12)$$

and

$$p_i/p_i^0 = a_i \qquad (3.13)$$

Theoretical procedures for deriving expressions for ΔG_{mix} usually start with the construction of a model of the mixture. The most popular *Flory-Huggins theory*, which was developed in the early 1940s, is based on the pseudolattice model and many approximations (Flory, 1953). Though extremely simple, the theory explains correctly (at least qualitatively) a large number of experimental observations, and serves as a basis for other more sophisticated theories.

Flory-Huggins theory is based on splitting ΔG_{mix} into an enthalpy term and an entropy term according to Eq. (3.14) and evaluating these two terms separately

$$\Delta G_{\mathrm{mix}} = \Delta H_{\mathrm{mix}} - T\Delta S_{\mathrm{mix}} \qquad (3.14)$$

The entropy of mixing ΔS_{mix} is computed from the number of possible arrangements of the molecules on a lattice, while the enthalpy of mixing, ΔH_{mix}, is calculated from the change in interaction energies among molecular surfaces during the process of mixing. Before turning to Flory-Huggins theory, it is convenient to develop first the theory for mixtures of low molecular weight solvents to get a better grasp of the principles involved.

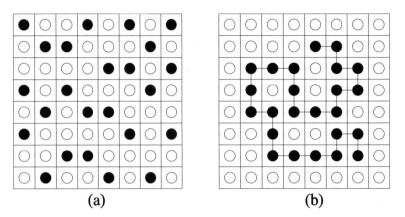

Figure 3.2 Schematic representation of quasicrystalline lattice model for solutions. (a) Simple solution: mixture of molecules of equal size, white circles representing the solvent molecules and filled circles the solute molecules. It is assumed that solvent molecules can exchange sites with solute molecules. This results in an increase in the number of ways they can be arranged, and hence in an increase in entropy. (b) Polymer solution: mixture of solvent molecules (unfilled circles) with a polymer molecule composed of chain segments (each segment represented by a filled circle) tied with chemical bonds. It is assumed that solvent molecules can exchange sites with polymer chain segments. This results in an increase in entropy. (After Flory, 1953.)

3.2.1   Low-Molecular-Weight Mixtures: van Laar Model

Let us consider a mixture of two substances A and B satisfying the conditions that (i) molecules of A and B are identical in size and shape and (ii) the energies of like (i.e., A-A or B-B) and unlike (i.e., A-B) molecular interactions are equal. Such mixtures form *ideal solutions*. Condition (ii) leads to *athermal* mixing (i.e., $\Delta H_{\mathrm{mix}} = 0$), which also means that there are no changes in the rotational, vibrational, and translational entropies and in the entropy of intermolecular inter-actions of the components upon mixing. Thus, ΔS_{mix} depends only upon the *configurational* (or *combinatorial*) entropy change, ΔS_{conf}, which is positive because the number of distinguishable spatial arrangements of the molecules increases when they are mixed. Hence ΔG_{mix} of Eq. (3.14) is negative and formation of an ideal solution is favorable.

The methods of statistical mechanics can be used to derive an equation for ΔS_{conf} by assuming that the molecules are placed randomly into cells that are of molecular size and are arranged in the form of a three-dimensional lattice (represented in two dimensions in Fig. 3.2(a)). The entropy S of an assembly of molecules and the total number Ω of distinguishable degenerate (i.e., of equal energy) arrangements of the molecules are related by the well-known Boltzmann equation:

$$S \; = \; k \ln \Omega \tag{3.15}$$

where k is the Boltzmann constant. Since ΔS_{mix} is equal to ΔS_{conf}, applying Eq. (3.15) to the formation of an ideal solution one obtains

$$\Delta S_{\mathrm{conf}} \; = \; k[\ln \Omega_{\mathrm{mix}} - (\ln \Omega_A + \ln \Omega_B)] \tag{3.16}$$

where Ω_A, Ω_B, and Ω_{mix} are, respectively, the total numbers of distinguishable spatial arrange-ments of the molecules in the pure substance A, pure substance B, and the ideal mixture. All the molecules of a pure substance being identical, there can be only one distinguishable spatial arrangement of them, that is, of placing N indistinguishable molecules in a lattice containing N cells. Thus $\Omega_A = 1$ and $\Omega_B = 1$ and Eq. (3.16) reduces to

$$\Delta S_{\text{conf}} = k \ln \Omega_{\text{mix}} \tag{3.17}$$

The total number of distinguishable spatial arrangements of N_A molecules of A and N_B molecules of B in a lattice of $(N_A + N_B)$ cells is equal to the number of permutations of $(N_A + N_B)$ objects consisting of N_A identical objects of type A and N_B identical objects of type B, that is,

$$\Omega_{\text{mix}} = \frac{(N_A + N_B)!}{N_A! N_B!} \tag{3.18}$$

Substitution of Eq. (3.18) into Eq. (3.17) gives

$$\Delta S_{\text{conf}} = k \ln \left[\frac{(N_A + N_B)!}{N_A! N_B!} \right] \tag{3.19}$$

Applying Stirling's approximation,

$$\ln N! = N \ln N - N \tag{3.20}$$

(for large N), Eq. (3.19) leads to

$$\Delta S_{\text{conf}} = -k \left[N_A \ln \left(\frac{N_A}{N_A + N_B} \right) + N_B \ln \left(\frac{N_B}{N_A + N_B} \right) \right] \tag{3.21}$$

It is more convenient to deal with equations in terms of numbers of moles, n, and mole fractions, x. In the present case, these are given by

$$n_A = N_A / N_{\text{Av}}, \quad n_B = N_B / N_{\text{Av}}, \quad n = n_A + n_B \tag{3.22}$$

$$x_A = n_A / (n_A + n_B), \quad x_B = n_B / (n_A + n_B) \tag{3.23}$$

where N_{Av} is Avogadro's number. Thus, Eq. (3.21) takes a simple form

$$\Delta S_{\text{conf}} = -R [n_A \ln x_A + n_B \ln x_B] \tag{3.24}$$

in which $R (= k N_{\text{Av}})$ is the universal gas constant. (It may be noted that since the configurational entropy of unmixed components is zero, ΔS_{conf} is in fact equal to S_{conf} of the mixture.)

Since for the formation of an ideal solution, $\Delta S_{\text{mix}} = \Delta S_{\text{conf}}$ and $\Delta H_{\text{mix}} = 0$, Eq. (3.14) becomes

$$\Delta G_{\text{mix}} = RT [n_A \ln x_A + n_B \ln x_B] \tag{3.25}$$

Equation (3.25) provides the basis for derivation of standard thermodynamic relationships for ideal solutions (e.g., Raoult's law). To make it more generally applicable, this equation has to be modified for nonideality of solutions when mixing is nonathermal (i.e., $\Delta H_{\text{mix}} \neq 0$) due to nonequivalent intermolecular interactions caused by the so-called dispersion forces. The decrease in energy due to interactions is called the *interaction energy*. This energy is larger for polar molecules than for nonpolar molecules.

In the liquid lattice defined above, each molecule has z contact points, z being the coordination number of the lattice. In a mixture of A and B there are three types of contact, namely, A-A, B-B, and A-B with interaction energies ω_{AA}, ω_{BB}, and ω_{AB}, respectively. (The model implies that all contact points of a molecule are equivalent.) The enthalpy of mixing is given by

$$\Delta H_{\mathrm{mix}} = H - H_A - H_B \tag{3.26}$$

where H, H_A, and H_B represent the contributions of the contact energy to the enthalpies of the mixture, pure component A, and pure component B, respectively. H_A and H_B can be calculated (Munk, 1989) from

$$H_A = zN_A\omega_{AA}/2 \tag{3.27}$$

$$H_B = zN_B\omega_{BB}/2 \tag{3.28}$$

where the factor 2 accounts for the fact that each contact involves two contact points.

Assuming that the number of heterogeneous contacts is N_{AB}, we can thus write (Munk, 1989) for H:

$$H = N_{AB}\omega_{AB} + (zN_A - N_{AB})\omega_{AA}/2 + (zN_B - N_{AB})\omega_{BB}/2 \tag{3.29}$$

Substitution of H_A, H_B, and H from Eqs. (3.27)-(3.29) into Eq. (3.26) then yields

$$\Delta H_{\mathrm{mix}} = N_{AB}[\omega_{AB} - (\omega_{AA} + \omega_{BB})/2] \equiv N_{AB}\Delta\omega_{AB} \tag{3.30}$$

where $\Delta\omega_{AB} = \omega_{AB} - (\omega_{AA} + \omega_{BB})/2$, representing the change in energy for the formation of an AB contact, is the excess contact energy of an AB contact.

The number of heterogeneous contacts N_{AB} can be easily estimated. It is equal to the number of contact points associated with A molecules multiplied by the probability p_B that a neighbor of molecule A is molecule B (Munk, 1989), that is,

$$N_{AB} = zN_A p_B \tag{3.31}$$

For simplicity, p_B is assumed to be the probability that *any* molecule (not necessarily a neighbor of molecule A) in the solution is a molecule B. This latter probability is equal to the mole fraction of component B in the mixture. Its use is justified only when the arrangement of molecules on the lattice is completely random.

Combining Eqs. (3.30) and (3.31) and replacing p_B by the mole fraction x_B of component B, which is equal to $N_B/(N_A + N_B)$, one obtains

$$\Delta H_{\mathrm{mix}} = N_A x_B z\Delta\omega_{AB} \tag{3.32}$$

Substitution of N_A from Eq. (3.22) then yields

$$\Delta H_{\mathrm{mix}} = x_B n_A N_{\mathrm{Av}} z\Delta\omega_{AB} \tag{3.33}$$

The product $N_{\mathrm{Av}}z\Delta\omega_{AB}$ has a physical meaning. It represents an enthalpy change for the transfer of 1 mol of A (or B) from its pure state to an infinitely dilute solution in B (or A).

Combining Eqs. (3.14), (3.24), and (3.33), and noting that $\Delta S_{\mathrm{mix}} = \Delta S_{\mathrm{conf}}$, one obtains

$$\Delta G_{\mathrm{mix}} = RT(n_A \ln x_A + n_B \ln x_B + n_A x_B N_{\mathrm{Av}} z\Delta\omega_{AB}/RT) \tag{3.34}$$

The last term in this expression is commonly simplified by introducing the *interaction parameter* χ_{AB}, defined (Munk, 1989) as

$$\chi_{AB} = N_{Av} z \Delta\omega_{AB}/RT = z\Delta\omega_{AB}/kT \tag{3.35}$$

According to the model, χ_{AB} is thus inversely proportional to temperature, whereas $\Delta\omega_{AB}$ is assumed to be independent of temperature. Note that $\Delta\omega_{AB}$, and hence χ_{AB}, can be either positive or negative.

Combination of Eq. (3.34) with Eq. (3.35) yields

$$\Delta G_{mix} = RT(n_A \ln x_A + n_B \ln x_B + n_A x_B \chi_{AB}) \tag{3.36}$$

Equation (3.36) describes the *van Laar model* of solvent mixtures and is applicable only to mixtures of low-molecular-weight components with approximately the same molar volume.

The van Laar model fails to give realistic predictions of the thermodynamic properties of polymer solutions. This arises from the assumption made in this model that the solvent and solute molecules are identical in size. Flory (1942) and Huggins (1942) proposed, independently, a modified lattice theory that takes into account the large differences in size between solvent and polymer molecules, in addition to intermolecular interactions.

3.2.2 Polymer-Solvent Mixtures: Flory-Huggins Model

In order to place both the solvent molecules and polymer molecules onto the same pseudolattice, it becomes necessary to consider the polymer molecules to be chains of *segments*, each segment being *equal in size to a solvent molecule*. The number, σ, of these segments in the chain defines the size of a polymer molecule and is given by the ratio of molar volumes V_2/V_1 (from now on we will assign the subscript 1 to the solvent and 2 to the polymer solute). Hence σ is not necessarily equal to the degree of polymerization. The model assumes that each lattice cell is occupied by either a solvent molecule or a chain segment, and each polymer molecule is placed in the lattice so that its chain segments occupy a continuous sequence of σ cells (as indicated in Fig. 3.2(b)).

Problem 3.1   Calculate, in terms of the lattice model of the Flory-Huggins theory, the number of segments per polystyrene molecule of molecular weight 290,000 dissolved in (a) toluene and (b) methyl ethyl ketone (MEK) at 25°C. [Data: polymer density = 1.083 g/cm^3 in toluene and 1.091 g/cm^3 in MEK, calculated on the assumption of additivity of volumes; toluene density = 0.861 g/cm^3; MEK density = 0.799 g/cm^3, all at 25°C.]

Answer:

(a) Polystyrene/toluene:

$$V_2 = \frac{M_2}{\rho_2} = \frac{2.90 \times 10^5 \text{ g mol}^{-1}}{1.083 \text{ g cm}^{-3}} = 2.678 \times 10^5 \text{ cm}^3 \text{ mol}^{-1}$$

$$V_1 = (92 \text{ g mol}^{-1})/(0.861 \text{ g cm}^{-3}) = 106.852 \text{ cm}^3 \text{ mol}^{-1}$$

$$\sigma = \frac{V_2}{V_1} = \frac{(2.678 \times 10^5 \text{ cm}^3 \text{ mol}^{-1})}{(106.852 \text{ cm}^3 \text{ mol}^{-1})} = 2506$$

(b) Polystyrene/MEK :

$$V_2 = \frac{(2.90 \times 10^5 \text{ g mol}^{-1})}{(1.091 \text{ g cm}^{-3})} = 2.658 \times 10^5 \text{ cm}^3 \text{ mol}^{-1}$$

$$V_1 = \frac{(72 \text{ g mol}^{-1})}{(0.799 \text{ g cm}^{-3})} = 90.11 \text{ cm}^3 \text{ mol}^{-1}$$

$$\sigma = \frac{(2.658 \times 10^5 \text{ cm}^3 \text{ mol}^{-1})}{(90.11 \text{ cm}^3 \text{ mol}^{-1})} = 2950$$

In comparison, the degree of polymerization of the polymer = $(2.90 \times 10^5 \text{ g mol}^{-1})/(104 \text{ g mol}^{-1}) = 2788$.

In the first stage, the Flory-Huggins theory derives an expression for ΔS_{mix} when $\Delta H_{\text{mix}} = 0$. This involves the application of Eq. (3.16), but this time with $\Omega_2 > 1$ because each molecule can adopt many different *conformations* (i.e., distinguishable spatial arrangements of the sequence of segments). Hence for polymer solutions (considering the configurational entropy change ΔS_{conf} to be the only source of ΔS_{mix}) one can write :

$$\Delta S_{\text{mix}} = k \ln(\Omega_{12}/\Omega_2) \tag{3.37}$$

The evaluation of Ω_{12}, Ω_2, and ΔS_{mix} of Eq. (3.37) is rather complex. The calculations, considered in outline below, are based on the formation of a polymer solution by mixing N_2 polymer molecules, each comprising σ segments in the chain, with N_1 solvent molecules in a lattice containing $N_1 + \sigma N_2 (= N)$ cells. The polymer molecules are added to the lattice one-by-one before adding the solvent molecules.

When adding a polymer molecule it is assumed, using a *mean field approximation*, that the segments of the previously added polymer molecule are distributed uniformly in the lattice. The N_1 solvent molecules are added to the lattice after the addition of N_2 polymer molecules. However, since the solvent molecules are identical, there is only one distinguishable spatial arrangement of them obtained by placing one solvent molecule in each of the remaining empty cells. Considering further the fact that the N_2 polymer molecules are identical, the total number of distinguishable spatial arrangements of the mixture is given by

$$\Omega_{12} = (1/N_2!) \, \Pi_{i=1}^{N_2} \, u_i \tag{3.38}$$

where u_i is the total number of possible conformations of the polymer molecule i and Π is the symbol for a continuous product, which in this case is $u_1 \times u_2 \times u_3 \times u_4 \times \cdots \times u_{N_2}$. The expression for Ω_{12} given in a useful form by Flory is

$$\Omega_{12} = \left[(z-1)^{(\sigma-1)N_2}\right]\left[\left(\frac{\sigma}{e^{\sigma-1}}\right)^{N_2}\right]\left[\left(\frac{N}{N_1}\right)^{N_1}\left(\frac{N}{\sigma N_2}\right)^{N_2}\right] \tag{3.39}$$

The product of the quantities in the first two pair of square brackets of Eq. (3.39) represents Ω_2 and so Eq. (3.37) takes the form

$$\Delta S_{\text{mix}} = k \ln \left[\left(\frac{N}{N_1}\right)^{N_1}\left(\frac{N}{\sigma N_2}\right)^{N_2}\right] \tag{3.40}$$

After simplification, Eq. (3.40) can be written as

$$\Delta S_{mix} = -k[N_1 \ln \phi_1 + N_2 \ln \phi_2] \tag{3.41}$$

where ϕ_1 and ϕ_2 are the *volume fractions* of solvent and polymer given by the fractions of the total number of lattice points occupied by solvent molecules and polymer segments, respectively:

$$\phi_1 = N_1/(N_1 + \sigma N_2) \quad \text{and} \quad \phi_2 = \sigma N_2/(N_1 + \sigma N_2) \tag{3.42}$$

In terms of number of moles, Eq. (3.41) becomes

$$\Delta S_{mix} = -R[n_1 \ln \phi_1 + n_2 \ln \phi_2] \tag{3.43}$$

This expression for polymer solutions is seen to be very similar to the corresponding expression for ideal mixing of small molecules, i.e., Eq. (3.24). The only difference is that in place of the mole fractions in Eq. (3.24) we now use the corresponding volume fractions. It should be noted that Eq. (3.43) reduces to Eq. (3.24) when $\sigma = 1$.

Problem 3.2   Using the entropy change equation of the lattice model, calculate the change in entropy when equal amounts (100 g) of two components are mixed for each of the following systems: (a) toluene and methyl ethyl ketone (MEK); (b) toluene and polystyrene of molecular weight 1.2×10^5; and (c) polystyrene and poly(methyl methacrylate) each of molecular weight 10,000. Take the density of each component as 1.0 g/cm^3. Comment on the comparative entropy changes of mixing for the three systems.

Answer:

(a) To use Eq. (3.24) let A and B represent toluene and MEK, respectively.

Molar masses of A and B: M_A (C_7H_8) = 92 g mol^{-1}; M_B (C_4H_8O) = 72 g mol^{-1}

Moles of A and B: $n_A = \dfrac{(100 \text{ g})}{(92 \text{ g mol}^{-1})} = 1.087 \text{ mol}$, $n_B = \dfrac{(100 \text{ g})}{(72 \text{ g mol}^{-1})} = 1.389 \text{ mol}$

Mole fractions of A and B: $x_A = 1.087/(1.087 + 1.389) = 0.439$, $x_B = 1 - 0.439 = 0.561$

$$\begin{aligned}\Delta S_{mix} &= -(8.314 \text{ J mol}^{-1} \text{ K}^{-1})[(1.087 \text{ mol}) \ln 0.439 + (1.389 \text{ mol}) \ln 0.561] \\ &= 14.11 \text{ J K}^{-1}\end{aligned}$$

(b) Eq. (3.43) is used for calculating ΔS_{mix} of polymer solution.

$$n_1 = n_A = 1.087 \text{ mol}; \quad n_2 = (100 \text{ g}) / (1.2 \times 10^5 \text{ g mol}^{-1}) = 8.33 \times 10^{-4} \text{ mol}$$

The polymer consists of σ segments, each of which can displace a single solvent molecule from a lattice site. Thus, $\sigma = M_2/(\rho_2 V_1^0)$, where M_2 is the molecular weight of the polymer that would have density ρ_2 in the corresponding amorphous state at the solution temperature and V_1^0 is the molar volume of the solvent.

Using appropriate data the calculated values are

$$\sigma = \frac{(1.2 \times 10^5 \text{ g mol}^{-1})}{(1.0 \text{ g cm}^{-3})\left[\dfrac{(92 \text{ g mol}^{-1})}{(1.0 \text{ g cm}^{-3})}\right]} = 1.30 \times 10^3$$

$$\phi_1 = \frac{(1.087 \text{ mol})}{[(1.087 \text{ mol}) + 1.30 \times 10^3 (8.33 \times 10^{-4} \text{ mol})]} = 0.50$$

$$\phi_2 = 1.0 - 0.50 = 0.50$$

$$\begin{aligned}\Delta S_{mix} &= -(8.314 \text{ J mol}^{-1} \, ^\circ\text{K}^{-1})[(1.087 \text{ mol}) \ln 0.50 + (8.33 \times 10^{-4} \text{ mol}) \ln 0.50] \\ &= 6.27 \text{ J K}^{-1}\end{aligned}$$

(c) Eq. (3.43) applies also if two polymers are being mixed. In this case, the number of segments σ_i in the ith component of the mixture is calculated from $\sigma_i = M_i/\rho_i V_r$, where V_r is now a reference volume equal to the molar volume of the smallest polymer repeating unit in the mixture. The corresponding volume fraction is $\phi_i = n_i\sigma_i/\sum n_i\sigma_i$.

For the given problem, $n_1 = n_2 = (100 \text{ g})/(10000 \text{ g mol}^{-1}) = 0.01$ mol. The molar volume of methyl methacrylate repeating unit ($\equiv C_5H_8O_2$) is taken as the reference volume (V_r). Thus,

$$V_r = \frac{M_{C_5H_8O_2}}{\text{Density}} = \frac{100 \text{ g mol}^{-1}}{1.0 \text{ g cm}^{-3}} = 100 \text{ cm}^3 \text{ mol}^{-1}$$

$$\sigma_1 = \sigma_2 = \frac{(10000 \text{ g mol}^{-1})}{(1.0 \text{ g cm}^{-3})(100 \text{ cm}^3 \text{ mol}^{-1})} = 100$$

$$\phi_1 = \frac{0.01 \times 100}{0.01 \times 100 + 0.01 \times 100} = 0.5, \quad \phi_2 = 0.5$$

$$\Delta S_{\text{mix}} = -(8.314 \text{ J mol}^{-1} {}^{\circ}\text{K}^{-1})[(0.01 \text{ mol}) \ln 0.5 + (0.01 \text{ mol}) \ln 0.5]$$
$$= 0.12 \text{ J} {}^{\circ}\text{K}^{-1}$$

Comment: The entropy of mixing of polymers with solvent is small compared to that of micromolecules because there are fewer possible arrangements of solvent molecules and polymer segments than there would be if the segments were not connected to each other. The entropy gain per unit of mixture is further reduced drastically if two polymers are mixed than if one of the components is a low-molecular-weight solvent, because both n_1 and n_2 are very small in the former case.

Having obtained an expression for ΔS_{mix}, the second stage in the development of Flory-Huggins theory is to derive an expression for the enthalpy change, ΔH_{mix}, for the polymer solution. This follows almost exactly the outline depicted for low-molecular-weight mixtures in Eqs. (3.26)-(3.33). The only difference now is in probability p_2 [i.e., p_B in Eq. (3.31)] that a given lattice point is occupied by a polymer segment. In this case, the probability must be approximated by the volume fraction ϕ_2, instead of mole fraction n_2. Thus, for a polymer solution the expression for ΔH_{mix} reads [cf. Eqs. (3.33) and (3.35)]

$$\Delta H_{\text{mix}} = n_1\phi_2 N_{\text{Av}} z \Delta\epsilon_{12} = RT n_1\phi_2\chi_{12} \tag{3.44}$$

where χ_{12} is the Flory-Huggins *interaction parameter* defined by Eq. (3.35). The interaction parameter χ plays a very important role in the theory of polymer solutions. It follows from Eq. (3.44) that χ_{12} is a dimensionless parameter equal to the ratio of the energy of interaction of the polymer with the solvent molecules to the kinetic energy RT, which should not depend on the concentration of a solution. However, a more rigorous derivation reveals that χ_{12} should be concentration dependent. This dependence is a slowly varying function and the absolute value of χ_{12} increases with increasing ϕ_2.

A combination of the entropy term of Eq. (3.43) and the enthalpy term of Eq. (3.44) according to Eq. (3.14) gives the following expression for ΔG_{mix} :

$$\Delta G_{\text{mix}} = RT (n_1 \ln\phi_1 + n_2 \ln\phi_2 + n_1\phi_2\chi_{12}) \tag{3.45}$$

This remarkably simple expression is the famous *Flory-Huggins equation* for the Gibbs free energy of mixing. It has remained as the cornerstone of polymer thermodynamics for more than six decades. Using Flory-Huggins theory it is possible to account for the equilibrium thermodynamic properties of polymer solutions. However, the theory is only able to predict general trends and fails to achieve precise agreement with experimental data.

In dilute polymer solutions, the polymer molecules are isolated from each other by regions of pure solvent, i.e., the polymer segments are not uniformly distributed in the lattice. In view of this, the Flory-Huggins theory is least satisfactory for dilute polymer solutions and only applies to concentrated solutions or mixtures. Furthermore, the interaction parameter introduced to account for the effects of polymer-solvent contact interactions is not a simple parameter and should contain both enthalpy and entropy contributions. Additionally, as noted earlier, it has also been shown to be dependent on the solution concentration.

3.2.2.1 Flory-Huggins Expressions for Thermodynamic Functions

Once the analytical expression for ΔG_{mix} is known, the calculation of chemical potentials and other thermodynamic functions (activities, activity coefficients, virial coefficients, etc.) is straightforward. For polymer solutions, we must apply Eq. (3.3) to Flory-Huggins equation (3.45), keeping in mind that volume fractions ϕ_i are functions of the number of moles, as given by

$$\phi_1 = n_1/(n_1 + \sigma n_2) \tag{3.46}$$

$$\phi_2 = \sigma n_2/(n_1 + \sigma n_2) \tag{3.47}$$

[These relations result from Eqs. (3.42) by dividing numerator and denominator by the Avogadro number N_{Av}.] Thus partial differentiation of Eq. (3.45) with respect to n_1 (note that both ϕ_1 and ϕ_2 are functions of n_1) leads to the following relation for the solvent (see Problem 3.3):

$$\mu_1 - \mu_1^0 = \overline{\Delta G_1} = RT\left[\ln\phi_1 + \left(1 - \frac{1}{\sigma}\right)\phi_2 + \chi\phi_2^2\right] \tag{3.48}$$

and partial differentiation of Eq. (3.45) with respect to n_2 similarly leads to the following relation for the polymer

$$\mu_2 - \mu_2^0 = \overline{\Delta G_2} = RT\left[\ln\phi_2 + (1 - \sigma)\phi_1 + \sigma\chi'\phi_1^2\right] \tag{3.49}$$

where

$$\chi = \chi_{12} - \phi_1(\partial\chi_{12}/\partial\phi_2)_{P,T} \tag{3.50}$$

$$\chi' = \chi_{12} + \phi_2(\partial\chi_{12}/\partial\phi_2)_{P,T} \tag{3.51}$$

Problem 3.3 Starting from Flory-Huggins equation (3.45) show that for the formation of a solution from a monodisperse polymer the partial molar Gibbs free energy of mixing, $\overline{\Delta G_1}$, for the solvent is given by Eq. (3.48).

Answer:

The Flory-Huggins equation (3.45) is

$$\Delta G_{\text{mix}} = RT[n_1\ln\phi_1 + n_2\ln\phi_2 + n_1\phi_2\chi_{12}]$$

and from Eq. (3.1): $\overline{\Delta G_1} = \left[\dfrac{\partial(\Delta G_{\text{mix}})}{\partial n_1}\right]_{P,T,n_2}$

Since $\phi_1 = n_1/(n_1 + \sigma n_2)$ and $\phi_2 = n_2\sigma/(n_1 + \sigma n_2)$, then

$$\overline{\Delta G_1} = RT \left\{ \frac{\partial}{\partial n_1} [n_1 \ln n_1 - n_1 \ln(n_1 + \sigma n_2)] \right.$$

$$+ \frac{\partial}{\partial n_1} [n_2 \ln \sigma n_2 - n_2 \ln(n_1 + \sigma n_2)]$$

$$\left. + \frac{\partial}{\partial n_1} \left[\chi_{12} n_1 \frac{\sigma n_2}{(n_1 + \sigma n_2)} \right] \right\}_{P,T,n_2}$$

$$\frac{\overline{\Delta G_1}}{RT} = \frac{n_1}{n_1} + \ln n_1 - \ln(n_1 + \sigma n_2) - \frac{n_1}{(n_1 + \sigma n_2)}$$

$$+ 0 - \frac{n_2}{n_1 + \sigma n_2} + \sigma n_2 \chi_{12} \left[\frac{-n_1}{(n_1 + \sigma n_2)^2} + \frac{1}{(n_1 + \sigma n_2)} \right]$$

$$+ \frac{\sigma n_1 n_2}{(n_1 + \sigma n_2)} \cdot \frac{d\chi_{12}}{d\phi_2} \cdot \frac{d\phi_2}{dn_1}$$

Simplifying and substituting for $d\phi_2/dn_1$,

$$\frac{\overline{\Delta G_1}}{RT} = 1 + \ln \phi_1 - \frac{(n_1 + n_2)}{(n_1 + \sigma n_2)} + \sigma n_2 \chi_{12} \left[\frac{-n_1 + n_1 + \sigma n_2}{(n_1 + \sigma n_2)^2} \right]$$

$$+ \frac{\sigma n_1 n_2}{(n_1 + \sigma n_2)} \cdot \frac{d\chi_{12}}{d\phi_2} \cdot \frac{-\sigma n_2}{(n_1 + \sigma n_2)^2}$$

$$\frac{\overline{\Delta G_1}}{RT} = \ln \phi_1 + \frac{n_1 + \sigma n_2 - n_1}{(n_1 + \sigma n_2)} + \chi_{12} \phi_2^2 - \phi_1 \phi_2^2 \left(\frac{\partial \chi_{12}}{\partial \phi_2} \right)$$

$$\frac{\overline{\Delta G_1}}{RT} = \ln \phi_1 + \frac{\sigma n_2}{(n_1 + \sigma n_2)} \left[1 - \frac{1}{\sigma} \right]$$

$$+ \left[\chi_{12} - \phi_1 \left(\frac{\partial \chi_{12}}{d\phi_2} \right) \right] \phi_2^2$$

$$\overline{\Delta G_1} = RT \left[\ln \phi_1 + \frac{\sigma n_2}{(n_1 + \sigma n_2)} \left(1 - \frac{1}{\sigma} \right) + \chi \phi_2^2 \right]$$

where $\chi = \left[\chi_{12} - \phi_1 \left(\partial \chi_{12}/\partial \phi_2 \right) \right]$

In the simplest form of the Flory-Huggins theory, the parameter χ_{12} is independent of concentration, and does not depend on ϕ_1 and ϕ_2. Consequently, the functions χ_{12}, χ, and χ' are equal and they are usually represented by the same symbol χ. Further, replacing the factor σ by

$$\sigma = V_2/V_1 \tag{3.52}$$

where V_1 and V_2 are the molar volumes of solvent and solute (polymer), respectively, Eqs. (3.48) and (3.49) are rewritten as

$$\mu_1 - \mu_1^0 = RT \left[\ln \phi_1 + \left(1 - \frac{V_1}{V_2} \right) \phi_2 + \chi \phi_2^2 \right] \tag{3.53}$$

$$\mu_2 - \mu_2^0 = RT \left[\ln \phi_2 + \left(1 - \frac{V_2}{V_1} \right) \phi_1 + \left(\frac{V_2}{V_1} \right) \chi \phi_1^2 \right] \tag{3.54}$$

These equations can be combined with Eq. (3.12) to obtain expressions for the corresponding activities of the components. For example, the activity of the solvent is given by

$$\ln a_1 = \frac{\mu_1 - \mu_1^0}{RT} = \ln(1 - \phi_2) + \left(1 - \frac{V_1}{V_2} \right) \phi_2 + \chi \phi_2^2 \tag{3.55}$$

It can be shown that in the limit of extremely small volume fractions, the right-hand side of Eq. (3.55) is equal to $\ln x_1$ and Eq. (3.55) thus reduces to the equation valid for ideal solutions with activities a_i equal to mole fractions x_i.

3.2.2.2 Colligative Properties and Interaction Parameter χ

Thermodynamic equations relating μ_1 and a_1 to various colligative properties are found in any textbook on thermodynamics. The following relationships are obtained on using the expression for μ_1:

$$\ln \frac{p_1}{p_1^0} = \frac{\mu_1 - \mu_1^0}{RT} = \ln a_1 \tag{3.56}$$

$$\frac{\Delta H_b \Delta T_b}{R(T_b^0)^2} = \frac{-(\mu_1 - \mu_1^0)}{RT_b^0} = -\ln a_1 \tag{3.57}$$

$$\frac{\Delta H_f \Delta T_f}{R(T_f^0)^2} = \frac{\mu_1 - \mu_1^0}{RT_f^0} = \ln a_1 \tag{3.58}$$

$$\frac{\Pi V_1}{RT} = \frac{-(\mu_1 - \mu_1^0)}{RT} = -\ln a_1 \tag{3.59}$$

In these equations, ΔH_f and ΔH_b are the latent heats of freezing and vaporization; T_b and T_f are the boiling and freezing points of the solution; T_b^0 and T_f^0 are the boiling and freezing points of the pure solvent; ΔT_b $(= T_b - T_b^0)$ is the elevation in the boiling point; ΔT_f $(= T_f - T_f^0)$ is the depression in the freezing point; p_1 is the vapor pressure at temperature T of the solution and p_1^0 the vapor pressure over pure solvent at the same temperature; Π is the osmotic pressure of the solution.

Combining Eqs. (3.56)-(3.59) with Eq. (3.55) all the above colligative properties can be related to the interaction parameter χ. For the vapor pressure, for example,

$$\ln \frac{p_1}{p_1^0} = \ln(1 - \phi_2) + \left(1 - \frac{V_1}{V_2}\right)\phi_2 + \chi\phi_2^2 \tag{3.60}$$

The interaction parameter can thus be determined from experimental measurement of colligative properties.

The parameter χ provides a measure of thermodynamic affinity of a solvent for the polymer, or a measure of the quality of the solvent. *The smaller the χ, the better is the solvent thermodynamically.* For very poor solvents, χ may be higher than unity, and for very good ones, χ may be negative. Based on χ values, solvents may be classified as: ideal solvents ($\chi = \frac{1}{2}$); good solvents ($\chi < \frac{1}{2}$); and poor solvent ($\chi > \frac{1}{2}$).

Relative Vapor Pressure and χ Equation (3.60), in which p_1/p_1^0 is the relative vapor pressure of solvent over solution, can be used for the determination of χ. For a polymer of very high molecular weight, V_1/V_2 is small and may be neglected. Then Eq. (3.60) becomes simpler:

$$\ln\left(p_1/p_1^0\right) = \ln(1 - \phi_2) + \phi_2 + \chi\phi_2^2 \tag{3.61}$$

Measuring the value of p_1/p_1^0 over a wide range of concentrations and plotting $[\ln(p_1/p_1^0) - \ln(1 - \phi_2) - \phi_2]$ against ϕ_2^2, we get a straight line the slope of which is equal to χ.

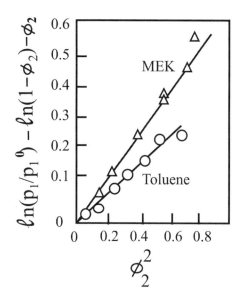

Figure 3.3 Plot of Eq. (3.61) with data of Problem 3.4.

Problem 3.4   Measurements were made of the vapor pressure of polystyrene (mol. wt. 290,000) solutions in toluene and methyl ethyl ketone (MEK) (Bawn et al., 1950). The table below shows the ratios of the vapor pressure of the solvent over the solution, p_1, to the vapor pressure of the pure solvent, p_1^0, against the corresponding weight fraction w_1 of the solvent at 25°C.

Toluene		Methylethyl ketone	
w_1	p_1/p_1^0	w_1	p_1/p_1^0
0.156	0.523	0.108	0.610
0.236	0.704	0.153	0.726
0.304	0.791	0.206	0.824
0.380	0.866	0.208	0.833
0.476	0.920	0.298	0.920
0.599	0.969	0.401	0.968
0.744	0.997	0.529	0.990

Data of Bawn et al., 1950.

For the calculation of volume fractions of the polymer, its density was determined in each solvent at 25°C by measuring the densities of solutions of known concentrations and assuming additivity of volumes. The determinations gave a polymer density of 1.083 g/cm³ in toluene and 1.091 g/cm³ in MEK. Densities of toluene and MEK at 25°C are 0.8610 and 0.7996 g/cm³, respectively.

Determine the parameter χ for polystyrene in the two solvents.

Answer:

Definitions: w_1, w_2 = weights of solvent and polymer, respectively; M_0, M_1, M_2 = molecular weight of the repeating unit (mer) in the polymer, of the solvent, and of the polymer, respectively; ρ_1, ρ_2 = densities of solvent and polymer, respectively; σ = number of segments in the polymer chain in solution; X = degree of polymerization of polymer, i.e., M_2/M_0.

In the lattice theory, the volumes of polymer segments and solvent molecules are assumed to be of equal size, i.e., $(M_2/\rho_2)/\sigma = M_1/\rho_1$. Substituting $M_2 = X M_0$ and rearranging, we get $X/\sigma = (\rho_2 M_1)/(\rho_1 M_0)$.

The ratio X/σ can thus be evaluated from the measured densities: In toluene, $X/\sigma = (1.083 \times 92)/(0.861 \times 104) = 1.11$. In MEK, $X/\sigma = (1.091 \times 72)/(0.7996 \times 104) = 0.94$. It may thus be concluded that the size of the polymer segment is very close to that of the repeat unit of the polymer. The number of segments in the polymer may thus be equated to the number of repeat units in the present case. Then from Eq. (3.47):

$$
\begin{aligned}
\phi_2 &= \frac{(w_2/M_0)}{(w_1/M_1) + (w_2/M_0)} \\
&= \frac{w_2}{(M_0/M_1)w_1 + w_2}
\end{aligned}
$$

Values of $[\ln(p_1/p_1^0) - \ln(1 - \phi_2) - \phi_2]$ are thus calculated and plotted against ϕ_2^2 according to Eq. (3.61) in Fig. 3.3. From the slopes, χ is found to be 0.45 for polystyrene-toluene and 0.70 for polystyrene-MEK. Thus, toluene is a good solvent and MEK is a poor solvent for polystyrene.

Osmotic Pressure and χ A relationship between the osmotic pressure Π of a solution and the activity (a_1) of the solvent in it can be derived (Glasstone, 1947) as

$$
\Pi \overline{V}_1 = -RT \ln a_1 \tag{3.62}
$$

where $\overline{V}_1$ is the molar volume of the solvent in the solution.

Combining Eq. (3.62) and Eq. (3.55) we obtain (note that the partial molar volume $\overline{V}_1$ of a pure solvent is the same as the molar volume V_1):

$$
\Pi = -\frac{RT}{V_1} \ln a_1 = -\frac{RT}{V_1} \ln \phi_1 - \frac{RT}{V_1}(1 - \frac{V_1}{V_2})\phi_2 - \frac{RT}{V_1}\chi\phi_2^2 \tag{3.63}
$$

Assuming, as in the lattice model, that there is no change in volume on mixing, volume fraction ϕ_i and concentration c_i are related as

$$
\phi_i = c_i/d_i \tag{3.64}
$$

where d_i is the density of component i. Similarly, molar volume V_i is related to molecular weight M_i by

$$
V_i = M_i/d_i \tag{3.65}
$$

Now developing $\ln \phi_1$ into a Taylor series

$$
\ln \phi_1 = \ln(1 - \phi_2) = -\phi_2 - \frac{\phi_2^2}{2} - \frac{\phi_2^3}{3} - \cdots
$$

and employing Eqs. (3.64) and (3.65), we can transform Eq. (3.63) into

$$
\frac{\Pi}{c_2} - \frac{RT d_1 c_2^2}{3M_1 d_2^3} = \frac{RT}{M_2} + \frac{RT d_1}{M_1 d_2^2}(\frac{1}{2} - \chi)c_2 = \frac{RT}{M_2} + RT A_2 c_2 \tag{3.66}
$$

where A_2 is the second virial coefficient (see later) represented by

$$
A_2 = \frac{d_1}{M_1 d_2^2}(\frac{1}{2} - \chi) \tag{3.67}
$$

At low concentrations, the second term of LHS of Eq. (3.66) is small and may be neglected. The graphical representation of the resulting equation, plotted as Π/c_2 vs. c_2, is a straight line (see **Membrane Osmometry** in Chapter 4), the slope of which is RTA_2. Hence it is possible to determine A_2 and from it the value of χ using Eq. (3.67).

Problem 3.5   The osmotic pressure data for polystyrene of molecular weight 1.6×10^6 yielded, according to Eq. (3.67), the following values for the second virial coefficient: (a) 2.88×10^{-4} mol cm^3 g^{-2} in dichloroethane and (b) -0.37×10^{-4} mol cm^3 g^{-2} in cyclohexane, both at 22°C. Determine χ for the polymer-solvent systems. Which is a better solvent?

[Data: density of polystyrene = 1.05 g cm^{-3}; density of dichloroethane = 1.24 g cm^{-3}; density of cyclohexane = 0.77 g cm^{-3}.]

Answer:

From Eq. (3.67): $\chi = \frac{1}{2} - \dfrac{A_2 M_1 d_2^2}{d_1}$

Polystyrene-dichloroethane system:

Molar mass of $C_2H_4Cl_2$: $M_1 = 2 \times 12 + 4 \times 1 + 2 \times 35.5 = 99$ g/mol

$$\chi = \frac{1}{2} - \frac{(2.88 \times 10^{-4} \text{ mol cm}^3 \text{ g}^{-2})(99 \text{ g mol}^{-1})(1.05 \text{ g cm}^{-3})^2}{(1.24 \text{ g cm}^{-3})} = 0.475$$

Polystyrene-cyclohexane system:

Molar mass of C_6H_{12}: $M_1 = 6 \times 12 + 12 \times 1 = 84$ g/mol

$$\chi = \frac{1}{2} - \frac{(-0.37 \times 10^{-4} \text{ mol cm}^3 \text{ g}^{-2})(84 \text{ g mol}^{-1})(1.05 \text{ g cm}^{-3})^2}{0.77 \text{ g cm}^{-3}} = 0.504$$

Thermodynamically, dichloroethane is thus a better solvent than cyclohexane for dissolving polystyrene.

Problem 3.6   When analyzing the thermodynamic properties of polymer solutions, it is sufficient to consider either the solvent or the solute, and for reasons of simplicity the former is normally considered. Moreover, for many physicochemical calculations, especially when dealing with dilute solutions, it is convenient to express the solvent activity as a power series in terms of polymer concentrations c_2 in mass/volume units. Therefore, starting with Eq. (3.12), express the difference in chemical potential of the solvent in the solution and in the pure state, i.e., $(\mu_1 - \mu_1^0)$, in terms of the mass concentrations of the solute in dilute solutions.

Answer:

With the solvent labeled as component 1 and solute as component 2, Eq. (3.12) gives for a bicomponent solution:

$$\mu_1 - \mu_1^0 = RT \ln a_1 = RT \ln \gamma_1 x_1 = RT (\ln \gamma_1 + \ln x_1) \tag{P3.6.1}$$

where γ_1 is the activity coefficient and x_1 is the mole fraction of the solvent in the solution. (For the pure solvent, $\gamma_1 = 1$.)

If the solute in solution is neither associated nor dissociated, then $x_1 = 1 - x_2$, where x_2 is the solute mole fraction. For dilute micromolecular solutions it is generally a good approximation to take $\gamma_1 = 1$ and $\ln \gamma_1 = 0$, that is, we assume an ideally dilute solution. Equation (P3.6.1) is accordingly approximated by

$$\mu_1 - \mu_1^0 = RT \ln(1 - x_2) \simeq RT (-x_2 - x_2^2/2 - \cdots) \tag{P3.6.2}$$

In dilute solution, the total number of moles of solute and solvent in unit volume will approach C_1, the molar concentration of solvent. Then the mole fraction x_2 of solute can be expressed as $x_2 = C_2/(C_1 + C_2) \simeq C_2/C_1$, where C_2 is the molar concentration of solute. If the mass/volume concentration of solute is c_2 and M_2 is the molecular weight of solute, then $c_2 = C_2 M_2$ and $x_2 = c_2/C_1 M_2$. If the molar volume of solvent is V_1, with the same volume unit as is used to express the concentration c_1 and C_1 (e.g., liters), then for dilute solution $C_1 = 1/V_1$ and $x_2 = c_2 V_1/M_2$. Substituting all these in Eq. (P3.7.2) gives the following relation for the difference in chemical potential of the solvent in the solution and in the pure state :

$$\mu_1 - \mu_1^0 = -RT V_1 \left[\frac{c_2}{M_2} + \left(\frac{V_1}{2M_2^2} \right) c_2^2 + \left(\frac{V_1^2}{3M_2^3} \right) c_2^3 + \cdots \right] \tag{P3.6.3}$$

3.2.2.3 Virial Coefficients

Equation (P3.6.3) shows the difference in chemical potential of the solvent in the solution and in the pure state to be a power series in solute concentration. Such equations are called *virial equations* (Rudin, 1982).

Equation (P3.6.3) is the key to the application of colligative properties to polymer molecular weights. Thus, insertion of experimental values of c_2, V_1, and $(\mu_1 - \mu_1^0)$ into Eq. (P3.6.3) would provide a measure of the solute molecular weight M_2 and we have seen in the preceding section how the colligative properties are related to $(\mu_1 - \mu_1^0)$, as in Eqs. (3.56) - (3.59).

Equation (P3.6.3) derived above for ideal solutions is invalid for real solutions and polymer solutions are rarely ideal even at the highest dilution that can be used in practice. It is, however, useful to retain the form of the ideal equation and express the deviation of real solutions in terms of empirical parameters. Thus, for a real solution Eq. (P3.6.3) is expressed in a parallel form as

$$\mu_1 - \mu_1^0 = -RT V_1 \left[\frac{c_2}{M_2} + A_2 c_2^2 + A_3 c_2^3 + \cdots \right] \tag{3.68}$$

where A_2 and A_3 are the second and third virial coefficients.

In a bicomponent solution the subscript 2, referring to solute, is often deleted. Equation (3.68) may thus be written as (Rudin, 1982)

$$\mu_1 - \mu_1^0 = -RT V_1 \left[\frac{c}{M} + A_2 c^2 + A_3 c^3 + \cdots \right] \tag{3.69}$$

To obtain virial expressions for colligative properties, Eq. (3.69) may be combined with respective Eqs. (3.56)-(3.59). If, for example, osmotic data are used, Eq. (3.69) is combined with Eq. (3.59) to write the virial equation as

$$\frac{\Pi}{c} = RT \left[\frac{1}{M} + A_2 c + A_3 c^2 + \cdots \right] \tag{3.70}$$

There is, however, no uniformity in the exact form of the virial equations used in polymer science. Alternatives to Eq. (3.70) include (Rudin, 1982):

$$\frac{\Pi}{c} = \left(\frac{RT}{M} \right) \left[1 + \Gamma_2 c + \Gamma_3 c^2 + \cdots \right] \tag{3.71}$$

and

$$\frac{\Pi}{c} = \frac{RT}{M} + Bc + Cc^2 + \cdots \tag{3.72}$$

The three forms are seen to be equivalent if

$$B = RTA_2 = (RT/M)\Gamma_2 \tag{3.73}$$

Virial coefficients are often reported in the literature without specifying the equation to which they apply. However, this can be deduced by inspecting the units of the virial coefficient.

Virial Coefficient and χ In the lattice model for polymer solutions it is assumed that there is no change in volume on mixing. Volume fractions ϕ_i and concentrations c_i are related as

$$\phi_i = c_i v_i \tag{3.74}$$

where the v_i's are specific volumes. Developing $\ln(1 - \phi_2)$ into a Taylor series and employing Eq. (3.74), we can transform (Munk, 1989) Eq. (3.55) into

$$\begin{aligned}
\ln a_1 = \frac{\Delta\mu_1}{RT} &= \left(-\phi_2 - \frac{\phi_2^2}{2} - \frac{\phi_2^3}{3} - \cdots\right) + \left(1 - \frac{V_1}{V_2}\right)\phi_2 + \chi\phi_2^2 \\
&= -\left[c_2 v_2 \frac{V_1}{V_2} + c_2^2 v_2^2(\tfrac{1}{2} - \chi) + \frac{c_2^3 v_2^3}{3} + \cdots\right]
\end{aligned} \tag{3.75}$$

Noting that $V_2 = v_2 M_2$, we can modify (Munk, 1989) this relation further as

$$\begin{aligned}
\ln a_1 = \frac{\Delta\mu_1}{RT} &= -\left[\frac{c_2 V_1}{M_2} + (\tfrac{1}{2} - \chi)c_2^2 v_2^2 + \frac{c_2^3 v_2^3}{3} + \cdots\right] \\
&= -V_1\left[\frac{c_2}{M_2} + A_2 c_2^2 + A_3 c_2^3 + \cdots\right]
\end{aligned} \tag{3.76}$$

or

$$\Delta\mu_1 = -c_2 V_1 RT\left[\frac{1}{M_2} + A_2 c_2 + A_3 c_2^2 + \cdots\right] \tag{3.77}$$

where the *second virial coefficient* A_2 given by

$$A_2 = (\tfrac{1}{2} - \chi)v_2^2/V_1 \tag{3.78}$$

and the *third virial coefficient* A_3 are defined as coefficients of the power series in Eq. (3.76). This equation gives the *virial expansion* of the activity and relative chemical potential of the solvent in a polymer solution.

Equation (3.78) shows that $A_2 = 0$ when the interaction parameter $\chi = \frac{1}{2}$. When A_2 happens to be zero and so also A_3, Eq. (3.77) simplifies greatly and many thermodynamic measurements become much easier to interpret. We should also recall that χ, according to its definition given by Eq. (3.35), is inversely proportional to temperature T. Since χ is positive for most polymer-solvent systems, it should acquire the value $\frac{1}{2}$ at some specific temperature.

3.2.2.4 Modification of Flory-Huggins Theory

It will be recalled that the term $\frac{1}{2}$ appearing in $(\frac{1}{2} - \chi)$, which is the key factor for the calculation of A_2 from Eq. (3.78), originated from the Taylor expansion of $\ln(1 - \phi_2)$ and so could be traced back to the entropy of mixing, while the χ term was of enthalpic origin. Since in real systems, molecular contacts may change the contribution of individual molecules to entropy, χ should also have an entropic component. Similarly, in arriving at the entropy term that led to the value $\frac{1}{2}$, possible contributions from changes in the volume in mixing and from contact interactions were neglected and only the configurational entropy of the lattice model was included. A modification of the theory therefore abandons the particular values $\frac{1}{2}$ and χ for the two terms and adopts more general unspecified values, namely, ψ for the entropic term and κ for the enthalpic term (Flory, 1953). Considering that the enthalpic term must be inversely proportional to temperature, κ is replaced by $\psi\theta/T$, where θ is a new parameter with the dimension of temperature. The factor $(\frac{1}{2} - \chi)$ is thus replaced by $(\psi - \kappa)$, that is,

$$\frac{1}{2} - \chi = \psi - \kappa = \psi(1 - \theta/T) \tag{3.79}$$

In terms of the new symbols, the second virial coefficient is then given by the expression [cf. Eq. (3.78)]:

$$A_2 = \psi(1 - \theta/T)v_2^2/V_1 \tag{3.80}$$

Thus, at a special temperature $T = \theta$, A_2 becomes equal to zero and the solution therefore becomes pseudoideal. Such solutions are also called *theta solutions*. The second virial coefficient is positive at temperatures higher than θ and negative at lower temperatures.

Essentially in the above modification, one interaction parameter χ has been replaced with two new parameters ψ and θ, thus adding flexibility to the treatment. In doing so, a phenomenological approach of modeling has been adopted in place of the pseudolattice model we started with.

Problem 3.7 Osmotic measurements on a high-molecular-weight polystyrene sample in cyclohexane at various temperatures yielded the following values for the second virial coefficient (A_2):

Temperature, °C	$A_2 \times 10^4$, cm^3 mol g^{-2}
25	− 0.334
30	− 0.146
35	0.036
40	0.212

Determine for the polymer-solvent system, (a) the temperature at which theta conditions are attained, (b) the entropy of dilution parameter ψ, and (c) the heat of dilution parameter κ at 27°C. [Specific volume of polymer = 0.96 cm^3/g; molar volume of cyclohexane at 27°C = 108.7 cm^3/mol.]

Answer:

According to Eq. (3.80), the plot of A_2 vs. $1/T$ should be linear over a narrow temperature range and A_2 should be zero at $T = \theta$. The plot shown in Fig. 3.4 thus gives $\theta = 34$°C.

From Eq. (3.80), $\psi = (A_2 V_1)/[v_2^2(1 - \theta/T)]$. At $T = 27$°C $\equiv 300$°K, A_2 (from Fig. 3.4) = -0.26×10^{-4} cm^3 mol g^{-2}. Therefore,

$$\psi = \frac{(-0.26 \times 10^{-4} \text{ cm}^3 \text{ mol g}^{-2})(108.7 \text{ cm}^3 \text{ mol}^{-1})}{(0.96 \text{ cm}^3 \text{ g}^{-1})^2 \left[1 - \frac{(307°\text{K})}{(300°\text{K})}\right]} = 0.13$$

$$\kappa = \frac{\psi\theta}{T} = \frac{(0.13)(307°\text{K})}{(300°\text{K})} = 0.133$$

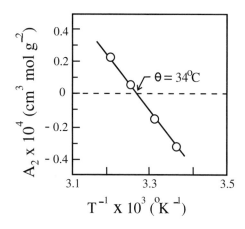

Figure 3.4 Plot of second virial coefficient vs. temperature (data of Problem 3.7).

Significance of ψ and κ The values of ψ, κ, and θ for several polymer-solvent systems are given in Table 3.1. The parameter ψ, which is a measure of the entropy of mixing, appears to be related to the spatial or geometrical character of the solvent. For those solvents which have cyclic structures and are relatively compact and symmetrical (e.g., benzene, toluene, and cyclohexane), ψ has relatively higher values than for the less symmetrical acyclic solvents capable of assuming a number of different configurations. Cyclic solvents are thus more favorable from the standpoint of entropy than acyclic ones. A solvent for which the entropy of mixing of a given polymer (as measured by ψ) is high may be considered as a "geometrically good" solvent for the polymer. In such a solvent, the viscosity of the polymer solution will be relatively high, or will approach a relatively high value at higher temperatures. In a "geometrically poor" solvent (where ψ is small), on the other hand, the viscosity will be relatively low (Flory, 1953).

A solvent for which the heat of mixing (as measured by κ) is low is a "thermally good" solvent, while a solvent with relatively large κ is a "thermally poor" solvent. In an *athermal* solvent ($\kappa = 0$), the volume expansion of the polymer molecule will be independent of temperature, while in a thermally poor solvent it will increase with temperature. The change of viscosity with temperature will thus be low in a solvent with low κ and relatively high in a solvent with relatively large κ.

Table 3.1 Thermodynamic Parameters for Some Polymer-Solvent Systems

Polymer	Solvent	θ in °K	ψ	κ at 25°C
Polyisobutylene	Benzene	297	0.15	0.15
	Toluene	261	0.14	0.12
	Ethylbenzene	251	0.14	0.117
	Cyclohexane	126	0.14	0.059
	n-Heptane	0	0.035	0
Polystyrene	Cyclohexane	307	0.13	0.134
	Benzene	100	0.09	0.03
	Toluene	160	0.11	0.06
	Ethyl acetate	222	0.03	0.02
	Methyl ethyl ketone	0	0.006	0

Data from Fox and Flory, 1951.

Methods to Determine Theta Solvents Solvents in which, *at a given temperature*, a polymer molecule is in the so-called *theta-state*, are called *theta* (θ) *solvents*. The temperature is known as the *theta-temperature* or the *Flory temperature* (as P. J. Flory was the first to show the importance

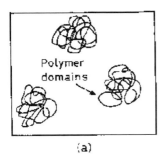 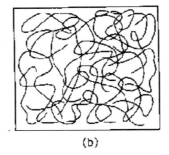

(a)   (b)

Figure 3.5 Schematic illustration of (a) dilute and (b) concentrated polymer solutions.

of the theta-state for a better understanding of molecular and technological properties of polymers). In the theta-state, the polymer solution behaves as thermodynamically ideal at low concentrations.

Several methods can be used to determine theta solvents. These include phase equilibria studies (see **Phase Equilibria in Poor Solvents**), determination of second virial coefficient (see Problem 3.7), viscosity-molecular weight relationship, and cloud point titration.

3.2.2.5 Flory-Krigbaum Theory

A deficiency of the Flory-Huggins theory is its use of a mean-field approximation by which it is assumed that the segments of polymer molecules are distributed uniformly in the lattice. This assumption is valid only at relatively high concentrations of the polymer, since the polymer molecules then interpenetrate each other and the solution thus appears quite homogeneous on a macroscopic scale. At very low concentrations, the molecules lie in regions as far separated from each other as possible, thus making the solution inhomogeneous as illustrated in Fig. 3.5. Then the mean-field approximation is invalid. However, in such solutions, the density of segments within an elemental volume in the polymer domain can still be assumed to be constant, and the results of the Flory-Huggins theory applied to it. Flory and Krigbaum (1953) analyzed dilute solutions in this manner and found that this leads, on proper integration, to the equation:

$$\mu_1 - \mu_1^0 \;=\; RT \left[-\frac{\phi_2}{\sigma} + (\chi - \tfrac{1}{2})F(X)\phi_2^2 + \cdots \right] \tag{3.81}$$

where

$$F(X) \;=\; 1 - \frac{X}{2!\,2^{3/2}} + \frac{X^2}{3!\,3^{3/2}} - \cdots \tag{3.82}$$

$$X \;=\; 4C_M\,\psi\left(1 - \frac{\theta}{T}\right)\frac{M_2^{1/2}}{\alpha_s^3} \tag{3.83}$$

C_M is a parameter depending on the properties of the polymer-solvent system and α_s is an expansion parameter. Both the terms are defined in later sections. At $T = \theta$, X becomes zero, $F(X)$ becomes equal to 1, and the Flory-Krigbaum theory reduces to the Flory-Huggins theory.

Problem 3.8   For polystyrene of molecular weight 100,000, calculate approximately the solution concentration below which the Flory-Krigbaum theory should be used. Use the following relationship between the RMS radius of gyration $\langle S^2 \rangle^{1/2}$ and the polymer molecular weight M: $\langle S^2 \rangle^{1/2} \simeq 2 \times 10^{-9}\sqrt{M}$ cm.

Answer:

The radius of a sphere enclosing the segments of a polymer molecule is taken approximately as twice the value of $\langle S^2 \rangle^{1/2}$, that is, $R \simeq 4 \times 10^{-9} \sqrt{M}$ cm. The volume, V, of the solution containing n polymer molecules is given by : $V \simeq n(4/3)\pi R^3$ cm^3. In terms of the weight, w, of the polymer, n may be written as : $n = (w/M) N_{Av}$ molecules cm^{-3}. So,

$$V \simeq \left(\frac{w}{M}\right) N_{Av} \left(\frac{4}{3}\right) \pi R^3 \simeq 0.16 w \sqrt{M} \text{ cm}^3$$

The concentration at the point of overlap is, therefore,

$$c_{\text{overlap}} = w/V = (0.16\sqrt{M})^{-1} = (0.16\sqrt{10^5})^{-1} = 0.02 \text{ g cm}^{-3}$$

For polystyrene of molecular weight 10^5 the overlap therefore occurs above about 2% concentration. Below this concentration, one should thus use the Flory-Krigbaum theory.

3.2.2.6 Excluded Volume Theory

As noted above, the basis of derivation of the Flory-Huggins equation is satisfactory at higher concentrations of the polymer, but is invalid for dilute solutions, in which the domains inhabited by the individual polymer molecules are far apart. Yet in deriving the chemical potentials and their virial expansion in the preceding sections, this forbidden assumption of low concentration of the polymer was employed. The derivation therefore had to lead to discrepancies. For example, according to Eq. (3.78), the second virial coefficient A_2 should be independent of molecular weight of the polymer, but the results of experiments show that it decreases significantly with increasing molecular weight. To deal with the problem, a different approach is required and the concept of excluded volume assumes importance.

The *excluded volume* of a solute molecule is the volume that is not available (because of exclusion forces or for other reasons) to the centers of mass of other similar solute molecules. This steric interference influences entropy but not enthalpy (Smith, 1963).

The entropy of mixing is based on the concept that the contribution of a solute molecule to entropy depends on the number of ways in which the molecule can be placed as it is added into the solution. Evidently, the latter quantity is proportional to the total volume of the system minus the *excluded* volume of solute molecules already added to the solution. A derivation based on this consideration gives (Smith, 1963):

$$\mu_1 - \mu_1^0 = -RT c_2 V_1 \left(\frac{1}{M_2} + c_2 \frac{u N_{Av}}{2M_2^2} \right) \tag{3.84}$$

where u is the excluded volume of each solute molecule, N_{Av} is Avogadro's number, and other terms are as defined earlier.

Comparison of Eqs. (3.77) and (3.84) yields for the second virial coefficient

$$A_2 = \frac{u N_{Av}}{2M_2^2} \tag{3.85}$$

The molar excluded volume, $u N_{Av}$, for particles of a given shape, will be proportional to the molecular weight of the particles, M_2. Hence, according to Eq. (3.85), derived from the excluded volume model, the second virial coefficient of particles of a given shape decreases with increasing molecular weight, being inversely proportional to the latter.

Problem 3.9   At 27°C, the osmotic pressure of an aqueous solution of globular protein increases from 1.64 mm water to 8.31 mm water as the concentration is increased from 0.1% (w/v) to 0.5% (w/v). The number average molecular weight of the protein is 1,56,500 and its specific volume is 0.75 cm^3/g. Calculate the excluded volume of the protein, assuming it to be spherical in shape, and compare the result with the partial molar volume. Density of water at 27°C is 0.996 g/cm^3.

Answer:

The second virial coefficient (A_2) can be calculated from values of Π/c (omitting the subscript 2 for solute) measured at two concentrations c_1 and c_2 of dilute solution using [cf. Eq. (3.66)]:

$$A_2 = \frac{(\Pi/c)_2 - (\Pi/c)_1}{RT(c_2 - c_1)}$$

Pressure of h cm column of water at 27°C $= (h$ cm$)(0.996$ g cm$^{-3}) \equiv (0.996h$ g cm$^{-2})/(1033$ g cm^{-2} atm$^{-1})$ or $9.64 \times 10^{-4}h$ atm. (Note that 1 atm $= 1033$ g cm^{-2}.)

$c_1 = 1$ mg/cm$^{-3} = 1$ g/L^{-1}; $\Pi_1 = 0.164 \times 9.64 \times 10^{-4}$ or 1.58×10^{-4} atm
$c_2 = 5$ mg/cm$^{-3} = 5$ g/L^{-1}; $\Pi_2 = 0.831 \times 9.64 \times 10^{-4}$ or 8.01×10^{-4} atm

$$A_2 = \frac{(8.01 \times 10^{-4} \text{ atm})/(5 \text{ g L}^{-1}) - (1.58 \times 10^{-4} \text{ atm})/(1 \text{ g L}^{-1})}{(0.082 \text{ L atm mol}^{-1}{}^\circ\text{K}^{-1})(300^\circ\text{K})(5 \text{ g L}^{-1} - 1 \text{ g L}^{-1})}$$

$$= 2.24 \times 10^{-8} \text{ mol L g}^{-2} \quad (= 2.24 \times 10^{-5} \text{ mol m}^3 \text{ kg}^{-2})$$

Excluded volume per mol, uN_{Av}, $= 2A_2M_2^2 = 2(2.24 \times 10^{-8}$ mol L g$^{-2})(1.565 \times 10^5$ g mol$^{-1})^2 = 1.1 \times 10^3$ L mol^{-1} $(= 1.1$ m^3 mol$^{-1})$

Partial molar volume, $\overline{V}_2 = \overline{v}_2 M_2 = (0.75 \times 10^{-3}$ L g$^{-1})(1.565 \times 10^5$ g mol$^{-1}) = 0.12 \times 10^3$ L mol^{-1} $(= 0.12$ m^3 mol$^{-1})$.

The excluded volume (per mol) of the globular protein calculated above is about 9 times the molar volume. This compares well with the case of spherical particles for which the excluded volume equals eight times the particle volume.

Since most polymer molecules have relatively flexible backbone, they tend to be highly coiled and can be represented as random coils. A molecular coil, being very loose, inhabits a region significantly larger than the actual volume of the coil. Each part of the molecule excludes other more remotely connected parts from its volume. These long-range steric interactions cause the RMS end-to-end distance, $\langle r^2 \rangle^{1/2}$, to be greater than $\langle r^2 \rangle_0^{1/2}$. Chain dimensions which correspond to $\langle r^2 \rangle_0^{1/2}$ are unperturbed by the effects of *volume exclusion* and so are called the *unperturbed dimensions*.

Similarly the RMS radius of gyration for perturbed dimensions in real chains, $\langle S^2 \rangle^{1/2}$, is greater than that for unperturbed dimensions, $\langle S^2 \rangle_0^{1/2}$. The perturbed dimensions will differ from unperturbed dimensions by the expansion α of the molecule arising from the long-range effects. Thus, we may write

$$\langle r^2 \rangle^{1/2} = \alpha_r \langle r^2 \rangle_0^{1/2} \tag{3.86}$$

$$\langle S^2 \rangle^{1/2} = \alpha_s \langle S^2 \rangle_0^{1/2} \tag{3.87}$$

No simple relationship exists between the two expansion parameters α_r and α_s. However, for small expansions α_r and α_s are close and one can use an average expansion factor α instead of α_r and α_s for simplicity.

Expansion Factor   The second virial coefficient, A_2, is a measure of solvent-polymer compatibility. Thus, *a large positive value of A_2 indicates a good solvent for the polymer favoring expansion of its size, while a low value (sometimes even negative) shows that the solvent is relatively poor.* The value of A_2 will thus tell us whether or not the size of the polymer coil, which is dissolved in a particular solvent, will be perturbed or expanded over that of the unperturbed state, but the extent of this expansion is best estimated by calculating an expansion factor α. As defined by Eqs. (3.86) and (3.87), α represents the ratio of perturbed dimension of the polymer coil to its unperturbed dimension.

According to Flory (1953), the change in Gibbs free energy with increasing extension, i.e., with increasing α at highly elastic deformation, is given by the equation

$$\frac{\partial \Delta G_{el}}{\partial \alpha} = -3kT\left(\alpha - \frac{1}{\alpha}\right) \tag{3.88}$$

Apart from this excluded volume effect, coil size in solutions is affected by interaction between a coil and a solvent. Solvent molecules penetrate inside the coil, which thus swells and increases in size. The change in Gibbs free energy with coil size increase, caused by swelling, is expressed by the equation (Flory, 1953)

$$\frac{\partial \Delta G_s}{\partial \alpha} = \frac{6C_M kT\psi(1 - \theta/T)M_2^{1/2}}{\alpha^4} \tag{3.89}$$

where k is the Boltzmann constant; C_M is a parameter depending in a complicated way on properties of the polymer-solvent system, and is given by

$$C_M = \left(\frac{27}{2^{5/2}\pi^{3/2}}\right)\left(\frac{v_2^2}{N_{Av}V_1}\right)\left(\frac{\langle r^2 \rangle_0}{M_2}\right)^{-3/2} \tag{3.90}$$

where v_2 is specific volume of polymer and $\langle r^2 \rangle_0$ is mean square end-to-end distance of polymer molecule in ideal solvent. Note that since $\langle r^2 \rangle_0$ is directly proportional to number of backbone bonds in the polymer chain (see Problem 2.5) and hence to molecular weight M_2 of polymer, C_M is independent of polymer molecular weight.

The total change in Gibbs free energy accompanying the change in coil size is, from Eqs. (3.88) and (3.89):

$$\frac{\partial \Delta G}{\partial \alpha} = \frac{6C_M kT\psi(1 - \theta/T)M_2^{1/2}}{\alpha^4} - 3kT\left(\alpha - \frac{1}{\alpha}\right) \tag{3.91}$$

At equilibrium, $\partial \Delta G/\partial \alpha = 0$, and hence

$$\frac{2C_M \psi(1 - \theta/T)M_2^{1/2}}{\alpha^4} = \alpha - \frac{1}{\alpha} \tag{3.92}$$

Rearranging,

$$\alpha^5 - \alpha^3 = 2C_M \psi(1 - \theta/T)M_2^{1/2} \tag{3.93}$$

Equation (3.93) indicates that the size of a polymer coil increases with increasing molecular weight of the polymer. According to this equation, at a temperature equal to the Flory temperature (i.e., $T = \theta$), $\alpha^5 - \alpha^3 = 0$, i.e., $\alpha = 1$. Thus, there is a temperature for each dilute polymer solution, at which it behaves like an ideal solution, the factor α being equal to unity, i.e., long-range interactions have no effect on the size of the macromolecules. The coil is then in an unperturbed state and the chain size depends only on short range interactions that occur between neighboring atoms and groups. All the above arguments apply, however, to very dilute solutions in which chains do not interact with each other.

Problem 3.10 Deduce from Eq. (3.93) that the molecular expansion factor α should increase with increase in temperature in a poor solvent, decrease with increase in temperature in a very good solvent, and be independent of temperature in an athermal solvent.

Answer:

From Eq. (3.79), $\theta = \kappa T / \psi$. In a poor solvent, where both κ and ψ generally are positive, θ also will be positive. Therefore according to Eq. (3.93), $\alpha^5 - \alpha^3$, and hence α, should increase with increase in temperature.

In a very good solvent, where the heat of dilution (κ) is negative and ψ is normal (i.e., positive), θ will be negative, and according to Eq. (3.93), α should decrease with increase in temperature.

In an athermal solvent, $\kappa = 0$, $\theta = 0$, and hence α should be independent of temperature.

Problem 3.11 For a fractionated polyisobutylene sample of molecular weight 1.5×10^6 the intrinsic viscosity was measured in cyclohexane at 30°C (good solvent) and in benzene at 24°C (theta solvent). The ratio of these two intrinsic viscosities yielded the factor α^3, by which the volume of the molecule in cyclohexane at 30°C is enlarged relative to the volume of the unperturbed molecule, as 4.03. Evaluate the total thermodynamic interaction ($\psi - \kappa$) for polyisobutylene-cyclohexane at 30°C. Use the observed relation for polyisobutylene between molecular weight (M_2) and the unperturbed root-mean-square end-to-end distance $\langle r^2 \rangle_0^{1/2}$, given by

$$\langle r^2 \rangle_0^{1/2} = 0.75 \times 10^{-8} M_2^{0.5} \text{ cm}$$

to calculate the parameter C_M as required. [Polymer density = 0.92 g/cm^3; density of cyclohexane at 30°C = 0.772 g/cm^3.]

Answer:

From the given data: $\alpha = (\alpha^3)^{1/3} = (4.03)^{1/3} = 1.591$; $v_2 = 1/(0.92 \text{ g cm}^{-3}) = 1.087 \text{ cm}^3 \text{ g}^{-1}$; M_1 (for C$_6H_{12}$) = 84 g mol^{-1}; $V_1 = (84 \text{ g mol}^{-1})/(0.772 \text{ g cm}^{-3}) = 108.8 \text{ cm}^3 \text{ mol}^{-1}$.

From Eq. (3.90):

$$C_M = \left[\frac{27}{2^{5/2} \pi^{3/2}} \right] \left[\frac{(1.087 \text{ cm}^3 \text{ g}^{-1})^2}{(6.02 \times 10^{23} \text{ mol}^{-1})(108.8 \text{ cm}^3 \text{ mol}^{-1})} \right]$$
$$(0.75 \times 10^{-8} \text{ cm mol}^{1/2} \text{ g}^{-1/2})^{-3}$$

$$= 0.0366 \text{ mol}^{1/2} \text{ g}^{-1/2}$$

From Eqs. (3.79) and (3.93):

$$\psi - \kappa = \psi(1 - \theta/T)$$
$$= \frac{\alpha^5 - \alpha^3}{2C_M M_2^{1/2}} = \frac{(1.591)^5 - 4.03}{2(0.0366 \text{ mol}^{1/2} \text{ g}^{-1/2})(1.5 \times 10^6 \text{ g mol}^{-1})^{0.5}} = 0.068$$

This quantity may be resolved into its entropy and energy components, if the temperature coefficient of the intrinsic viscosity is known (Flory, 1953).

Problem 3.12   Show that the perturbed dimensions of highly expanded polymer coils are proportional to $n^{3/5}$, where n is the number of backbone bonds in the polymer chain.

Answer:

For large expansion, $\alpha^5 \gg \alpha^3$ and so from Eq. (3.93): $\alpha^5 \approx K M_2^{1/2}$, where K is a constant for the polymer-solvent system. Therefore, $\alpha \propto M_2^{1/10}$ and hence, $\alpha \propto n^{1/10}$.

From Eq. (2.2): Unperturbed dimension, $\langle r^2 \rangle_0^{1/2} \propto n^{1/2}$.

From Eq. (3.86): Perturbed dimension, $\langle r^2 \rangle^{1/2} \propto (n^{1/10})(n^{1/2})$, i.e., $\langle r^2 \rangle^{1/2} \propto n^{3/5}$.

3.3   Phase Equilibria in Poor Solvents

The conditions for equilibrium between two phases in a binary system are given by the equality of the chemical potentials in the two phases; that is,

and
$$\mu_1' = \mu_1'' \quad \text{or} \quad \Delta\mu_1' = \Delta\mu_1'' \tag{3.94}$$
$$\mu_2' = \mu_2'' \quad \text{or} \quad \Delta\mu_2' = \Delta\mu_2'' \tag{3.95}$$

where the single and double primes are used to indicate, respectively, the dilute and concentrated phases in thermodynamic equilibrium and subscripts 1 and 2 indicate solvent and polymer solute, respectively. Using these conditions we may now examine the requirements for the occurrence of incomplete miscibility and deduce the corresponding concentrations of the two phases.

Equations (3.48) and (3.49), derived previously, express the chemical potentials μ_1 and μ_2 as functions of the volume fraction ϕ_2 of the polymer (note that $\phi_1 = 1 - \phi_2$) and a single parameter χ (note that χ and χ' are equal) occurs in these fractions. Fulfillment of the conditions represented by Eqs. (3.94) and (3.95) requires that there be two concentrations at which the chemical potential μ_1 or $\Delta\mu_1$ has the same value, and so also the chemical potential μ_2 or $\Delta\mu_2$. Since μ_1 and μ_2 are derived by differentiating the same free energy function, Eq. (3.45), it suffices to consider only one of them.

Problem 3.13   A polymer solution was cooled slowly until phase separation took place. The volume fractions of polymer in the two phases at equilibrium were found to be 0.01 and 0.89, respectively. Using the Flory-Huggins equation for $\Delta\mu_1$ [cf. Eq. (3.48)] and the equilibrium condition $\Delta\mu_1' = \Delta\mu_1''$, obtain a value of the polymer-solvent interaction parameter χ for the conditions of phase separation.

Answer:

In order to use the Flory-Huggins equation for $\Delta\mu_1$, the number-average value of σ (the number of segments per polymer molecule) should be known. However, for most polymers, σ is sufficiently large for $(1 - 1/\sigma) \approx$ 1.0. Making this assumption and applying the condition $\Delta\mu_1' = \Delta\mu_1''$ leads to

$$\ln(1 - 0.01) + (1.0 \times 0.01) + \chi(0.01)^2 = \ln(1 - 0.89) + (1.0 \times 0.89) + \chi(0.89)^2$$

yielding, $\chi = 1.66$.

Figure 3.6 shows plots of the relative chemical potential of the solvent, $-(\mu_1 - \mu_1^0)/RT$, versus the volume fraction of the polymer solute, ϕ_2, calculated according to Eq. (3.48) for several values of χ, considering a polymer of size $\sigma = 1000$. It is seen that for small values of χ, the relative chemical potential decreases [i.e., $-(\mu_1 - \mu_1^0)/RT$ increases] monotonically with ϕ_2 throughout the concentration range, indicating total miscibility. However, as the value of χ is increased, either

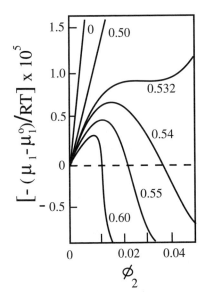

Figure 3.6 Plot of Eq. (3.48) for various values of the interaction parameter, χ, as indicated in the diagram. A value of 1000 has been assumed for σ, the number of segments in the polymer. (Drawn following the method of Flory, 1942.)

by decreasing the temperature or by altering the nature and composition of the solvent, a *critical value* χ_c (in this case 0.532) is reached at which the previously monotonic curve begins to show a point of inflection. Its appearance signifies the onset of immiscibility of polymer and solvent and it is marked by an incipient phase separation (*opalescence*). Since both the first and second derivatives of a function are zero at the inflection point, the conditions for this incipient phase separation can be described (Flory, 1953) by

$$(\partial\mu_1/\partial\phi_2)_{T,P} = 0 \tag{3.96}$$

$$(\partial^2\mu_1/\partial\phi_2^2)_{T,P} = 0 \tag{3.97}$$

It is readily seen from Fig. 3.6 that for each value of χ above χ_c there are two different values of ϕ_2 at which the chemical potential of the solvent in the two phases is the same. This means that solutions with concentrations represented by these two values of ϕ_2 can be in thermodynamic equilibrium for $\chi > \chi_c$. It also implies that a solution which has an intermediate value of ϕ_2 will spontaneously separate into two stable *liquid* phases with these two concentrations and a concomitant decrease in the free energy. Thus, if χ is increased by decreasing the temperature, a fully miscible system at higher temperatures is transformed to one of limited miscibility at the lower temperature. At some *critical temperature* T_c, during this process, incipient phase separation (as indicated by the onset of an opalescence) is encountered, followed by separation into distinct liquid phases at still lower temperatures.

Application of the critical conditions of Eqs. (3.96) and (3.97) to the chemical potential as given by Eq. (3.48) yields (Flory, 1953):

$$1/(1 - \phi_2) - (1 - 1/\sigma) - 2\chi\phi_2 = 0 \tag{3.98}$$

and

$$1/(1 - \phi_2)^2 - 2\chi = 0 \tag{3.99}$$

Eliminating χ, one then obtains for the critical composition,

$$\phi_{2c} = 1/(1 + \sqrt{\sigma}) \tag{3.100}$$

Substituting Eq. (3.100) into either of the Eqs. (3.98) and (3.99) yields

$$\chi_c = (1 + \sqrt{\sigma})^2/2\sigma \tag{3.101}$$

For large σ, Eqs. (3.100) and (3.101) reduce to

$$\phi_{2c} = 1/\sqrt{\sigma} \tag{3.102}$$

and

$$\chi_c = 1/2 + 1/\sqrt{\sigma} \tag{3.103}$$

Equation (3.102) predicts that the critical concentration at which phase separation first occurs on entering the two-phase region is small for a high molecular weight polymer. Thus for a polymer having $\sigma = 1000$, $\phi_{2c} \cong 0.032$. The critical value of χ, according to Eq. (3.103), is 0.50 at infinite molecular weight and will exceed 1/2 by small increments as the chain length decreases.

According to the initial definition given by Eq. (3.35), χ is inversely proportional to the temperature. Also in the later phenomenological approach, $(\chi - \frac{1}{2})$ has been replaced by $\psi(\theta/T - 1)$ [cf. Eq. (3.79)], implying that χ should be a linear function of $1/T$. Accordingly, if χ_c in Eq. (3.101) is replaced by $[\frac{1}{2} + \psi(\theta/T_c - 1)]$, it can be readily shown (Flory, 1953) that

$$\psi(\theta/T_c - 1) = 1/\sqrt{\sigma} + 1/2\sigma \tag{3.104}$$

$$1/T_c = (1/\theta)[1 + (1/\psi)(1/\sqrt{\sigma} + 1/2\sigma)] \tag{3.105}$$

which for large σ can be written simply as

$$1/T_c = (1/\theta)[1 + 1/(\psi\sqrt{\sigma})] \tag{3.106}$$

A plot of $1/T_c$ versus $(1/\sqrt{\sigma} + 1/2\sigma)$ or $1/\sqrt{\sigma}$ (for large σ) should thus be linear. The slope and intercept (at infinite σ) of such plots should yield values of ψ and θ which should depend only on the nature of the polymer and the solvent and not on the degree of polymerization. The critical temperature, T_c, of a given polymer in a given solvent may be determined by observing the precipitation temperature (T_p) as a function of the volume fraction (ϕ_2) of the solute (see Fig. 3.7).

Noting that $\sigma = Mv_2/V_1$ (cf. Problem 3.1), where v_2 is the specific volume of the polymer of molecular weight M and V_1 is the molar volume of the solvent, Eq. (3.106) can be written as

$$1/T_c = (1/\theta)(1 + b/\sqrt{M}) \tag{3.107}$$

where the constant b is given by $b = \sqrt{(V_1/v_2)}/\psi$. Thus according to the above theory, the reciprocal of the critical temperature (in °K) for the onset of opalescence should vary linearly with the reciprocal of the square root of the molecular weight in a given polymer-solvent system. The corresponding θ may now be identified as the *critical miscibility temperature in the limit of infinite molecular weight*.

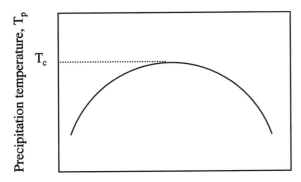

Figure 3.7 The critical temperature, T_c, is determined as the maximum point of the plot (phase diagram) of T_p versus ϕ_2, also called the *cloud point curve* or the *precipitation curve*. T_p is the temperature at which detectable turbidity can first be observed visually upon slow cooling (1-2°C per 10 min).

Problem 3.14 Cyclohexane solutions of polystyrene in varying concentration were slowly cooled to determine the precipitation temperature at which the first detectable turbidity appeared. Four polystyrene samples differing in molecular weight were used. From the maximum point of the precipitation temperature versus volume fraction of polymer plot for each polystyrene sample the critical temperature T_c was determined. The polymer molecular weights and the respective T_c values so obtained are given below.

$M \times 10^{-3}$	T_c, (°C)
920	31
182	28
64	24
31	19

Assuming that the volume of one monomer unit is the same as that of a cyclohexane molecule, estimate the theta temperature θ and the entropic dilution parameter ψ.

Answer:

On the assumption that the volume of mer (repeating unit) is the same as that of a cyclohexane molecule, σ is equal to the degree of polymerization, M/M_o, where M_o is the molar mass of the mer, i.e., 104 g/mol. (Alternatively, σ can be taken as the ratio of the molar volume of the polymer to that of the solvent, the molar volume in each case being obtained from molar mass and density.)

From the plot of $1/T_c$ versus $(1/\sqrt{\sigma} + 1/2\sigma)$ in Fig. 3.8:

$1/\theta$ = Intercept = 3.254×10^{-3} °K^{-1}

θ = 307 °K ≡ 34 °C

$1/\theta\psi$ = Slope = 2.876×10^{-3} °K^{-1}

ψ = $(3.254 \times 10^{-3}$ °K$^{-1})/(2.876 \times 10^{-3}$ °K$^{-1})$ = 1.131

3.3.1 Upper and Lower Critical Solution Temperatures

Cloud-point curves or precipitation curves for different polymer-solvent systems have different shapes (Figs. 3.9 and 3.10). The maxima and minima on these curves indicate the *upper critical*

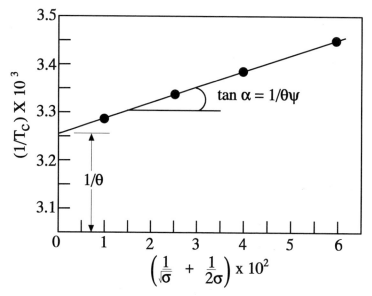

Figure 3.8 A plot of the reciprocal of the critical temperature against the molecular size function occurring in Eq. (3.127) from data in Problem 3.14.

solution temperature (UCST) and the *lower critical solution temperature* (LCST), respectively. As indicated in Figs. 3.9 and 3.10, the phase diagram of a polymer solution has two regions of limited miscibility: (i) below UCST associated with the theta temperature (see Problem 3.14) and (ii) above LCST.

Different polymer-solvent systems may have completely different phase diagrams. For some systems, such as polystyrene-cyclohexanone, UCST < LCST [Fig. 3.10(a)] but for others, e.g., highly polar systems like polyoxyethylene-water, UCST > LCST and closed solubility loop is found [Fig. 3.10(b)].

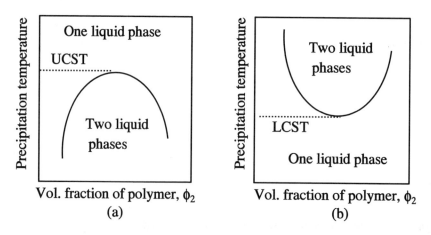

Figure 3.9 Schematic phase diagrams for polymer-solvent systems showing (a) UCST and (b) LCST.

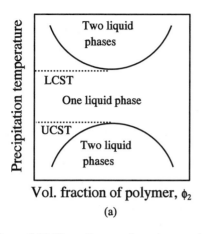
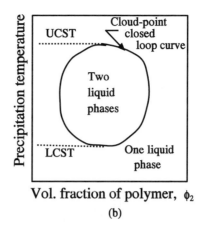

Figure 3.10 Phase diagrams for polymer-solvent systems where (a) UCST < LCST and (b) UCST > LCST.

3.4   Solubility Behavior of Polymers

If we denote the force of attraction between the molecules of one material A by F_{AA}, that between the molecules of another material B by F_{BB}, and represent that between one A and one B molecule as F_{AB}, then the system will be compatible and a solution will result if $F_{AB} > F_{BB}$ and $F_{AB} > F_{AA}$. On the other hand, if F_{AA} or $F_{BB} > F_{AB}$, the system will be incompatible and the molecules will separate, forming two phases. In the absence of any specific interaction (e.g., hydrogen bonding) between solvent and solute, we can reasonably assume the intermolecular attraction forces between the dissimilar molecules to be approximately given by the geometric mean of the attraction forces of the corresponding pairs of similar molecules; that is, $F_{AB} = (F_{AA}.F_{BB})^{1/2}$. Consequently, if F_{AA} and F_{BB} are equal, F_{AB} will also be similar and the materials should be soluble.

A measure of the intermolecular attraction forces in a material is provided by the *cohesive energy*. Approximately, this equals the heat of vaporization (for liquids) or sublimation (for solids) per mol. The cohesive energy density in the liquid state is thus $\Delta E_v/V$, in which ΔE_v is the molar energy of vaporization and V is the molar volume of the liquid. The square root of this cohesive energy density is known as the *solubility parameter* (δ), that is,

$$\delta = (\Delta E_v/V)^{1/2} \tag{3.108}$$

If the vapor behaves approximately like an ideal gas, Eq. (3.108) can be written as

$$\delta = [(\Delta H_v - RT)/V]^{1/2} = [(\Delta H_v - RT)\rho/M]^{1/2} \tag{3.109}$$

where ΔH_v is the molar enthalpy of vaporization and ρ is the density of liquid with molecular weight M. For a volatile liquid, cohesive energy density and, hence, δ can be determined experimentally by measuring ΔH_v and ρ.

Problem 3.15   Calculate an estimate of the solubility parameter for water at 25°C from its heat of vaporization at the same temperature, given by

$$H_2O \, (\ell) = H_2O \, (g), \qquad \Delta H_{25°C} = 10.514 \text{ kcal}$$

Answer:

From Eq. (3.109) :

$$\delta^2 = \left[(10514\,\text{cal}\,\text{mol}^{-1}) - (1.987\,\text{cal}\,\text{mol}^{-1}\,{}^\circ\text{K}^{-1})(298{}^\circ\text{K})\right](1\,\text{g}\,\text{cm}^{-3})/(18\,\text{g}\,\text{mol}^{-1})$$

$$= 551.2\,\text{cal}\,\text{cm}^{-3} \equiv 2.3 \times 10^9\,\text{J}\,\text{m}^{-3}$$

$$\delta = 23.5\,(\text{cal}\,\text{cm}^{-3})^{1/2} = 48.0 \times 10^3\,(\text{J}\,\text{m}^{-3})^{1/2} = 48\,\text{MPa}^{1/2}$$

[*Conversion factors*: 1 cal cm^{-3} = 4.184×10^6 J m^{-3} = 4.184×10^6 Pascal (Pa) = 4.184 MPa. Hence, 1 (cal cm^{-3})$^{1/2}$ = 2.045 MPa$^{1/2}$]

Hildebrand (Hildebrand et al., 1970) first used the solubility parameter approach for calculating estimates of the enthalpy of mixing, ΔH_{mix}, for mixtures of liquids. The equation employed (for derivation see Appendix 3.1) is

$$\Delta H_{\text{mix}} = V_{\text{mix}}\phi_1\phi_2(\delta_1 - \delta_2)^2 \tag{3.110}$$

where V_{mix} is the molar volume of the mixture, and δ_1 and δ_2 are the solubility parameters of components 1 and 2, respectively.

A necessary requirement for solution and blending compatibility is a negative or zero Gibbs free energy change (ΔG_{mix}) when the solution or blend components are mixed, that is,

$$\Delta G_{\text{mix}} = \Delta H_{\text{mix}} - T\Delta S_{\text{mix}} \leq 0 \tag{3.111}$$

Since the ideal entropy of mixing, according to Eq. (3.24), is always positive (the ln of a fraction being negative), the components of a mixture can be assumed to be miscible only if $\Delta H_{\text{mix}} \leq T\Delta S_{\text{mix}}$. Solubility therefore depends on the existence of a zero or small value of ΔH_{mix}, only positive (endothermic) heats of mixing being allowed, as in Eq. (3.110). Miscibility or solubility will then be predicted if the absolute value of the $(\delta_1 - \delta_2)$ difference is zero or small (Burrell, 1955). Specific effects such as hydrogen bonding and charge transfer interactions can lead to negative ΔH_{mix} but these are not taken into account by Eq. (3.110), and separate considerations must be applied in order to predict their effect on miscibility and solubility (Nelson et al., 1970).

Solubility parameters of solvents can be correlated with the density, molecular weight, and structure of the solvent molecule. According to the additive method of Small (1953), the solubility parameter is calculated from a set of additive constants, F, called *molar attraction constants*, by the relationship

$$\delta = \frac{\rho}{M}\sum F \tag{3.112}$$

where $\sum F$ is the molar attraction constants summed over the groups present in the compound; ρ and M are the density and the molar mass of the compound. The same procedure is applied to polymers and Eq. (3.112) is used, wherein ρ is now the density of the amorphous polymer at the solution temperature, $\sum F$ is the sum of all the molar attraction constants for the repeat unit, and M is the molar mass of the repeat unit. Values of molar attraction constants for the most common groups in organic molecules were estimated by Small (1953) from the vapor pressure and heat of vaporization data for a number of simple molecules. A modified version of a compilation of molar attraction constants (Hoy, 1970) is reproduced in Table 3.2. An example of the use of the tabulated molar attraction constants is given in the problem worked out below.

Problem 3.16 Calculate an estimate of the solubility parameter for the epoxy resin DGEBA (diglycidyl ether of bisphenol A) having the repeat unit structure as shown below and density 1.15 g/cm^3.

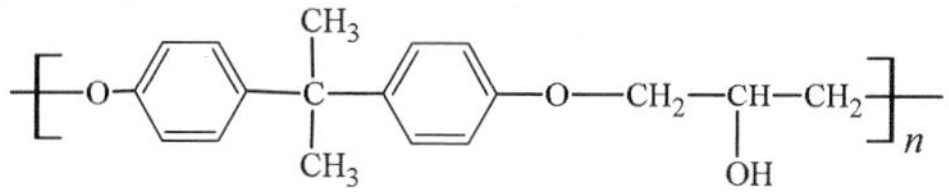

Answer:

M (for repeating unit) = 284 g mol^{-1}

Group	F $\dfrac{(\text{cal cm}^3)^{1/2}}{\text{mol}}$	No. of groups	ΣF $\dfrac{(\text{cal cm}^3)^{1/2}}{\text{mol}}$
–CH$_3$	147.3	2	294.60
–CH$_2$–	131.5	2	263.00
>CH–	85.99	1	85.99
>C<	32.03	1	32.03
–O– (ether)	114.98	2	229.96
–OH	225.84	1	225.84
–CH= (aromatic)	117.12	8	936.96
–C= (aromatic)	98.12	4	392.48
6-membered ring	−23.44	2	−46.88
Para substitution	40.33	2	80.66

			2494.64

$$\delta = \frac{(1.15 \text{ g cm}^{-3})(2494.64 \text{ cal}^{1/2} \text{ cm}^{3/2} \text{ mol}^{-1})}{(284 \text{ g mol}^{-1})}$$

$$= 10.1 \text{ (cal cm}^{-3})^{1/2} = 20.7 \text{ MPa}^{1/2}$$

Table 3.2 Group Molar Attraction Constants

Group	Molar attraction, F (cal cm^3)$^{1/2}$/mol	Group	Molar attraction, F (cal cm^3)$^{1/2}$/mol
–CH$_3$	147.3	–OH→	225.84
–CH$_2$–	131.5	–OH aromatic	170.99
>CH–	85.99	–NH$_2$	226.56
>C<	32.03	–NH–	180.03
CH$_2$= (olefin)	126.54	–N–	61.08
–CH= (olefin)	121.53	–C≡N	354.56
>C= (olefin)	84.51	–N=C=O	358.66
–CH= (aromatic)	117.12	–S–	209.42
–C= (aromatic)	98.12	Cl$_2$	342.67
–O– (ether, acetal)	114.98	–Cl (primary)	205.06
–O– (epoxide)	176.20	–Cl (secondary)	208.27
–COO–	326.58	–Cl (aromatic)	161.0
>C=O	262.96	–Br	257.88
–CHO	292.64	–Br (aromatic)	205.60
(CO)$_2$O	567.29	–F	41.33
Structure feature		Structure feature	
Conjugation	23.26	6-membered ring	−23.44
Cis	−7.13	Ortho substitution	9.69
Trans	−13.50	Meta substitution	6.6
5-membered ring	20.99	Para substitution	40.33

Source: Hoy, 1970; Brandrup and Immergut, 1975; Rudin, 1982.

Table 3.3 Solubility Parameters for Some Common Solvents[a,b]

(i) Poorly H-bonded (hydrocarbons and their halo-, nitro-, and cyano- products)		(ii) Moderately H-bonded (esters, ethers, ketones)	
Solvent	δ, (cal/cm^3)$^{1/2}$	Solvent	δ, (cal/cm^3)$^{1/2}$
n-Hexane	7.3	Isoamyl acetate	7.8
Carbon tetrachloride	8.6	Dioctyl phthalate	7.9
Toluene	8.9	Tetrahydrofuran	9.1
Benzene	9.2	Methyl ethyl ketone	9.3
Chloroform	9.3	Acetone	9.9
Methylene chloride	9.7	1,4-Dioxane	10.0
Nitrobenzene	10.0	Diethylene glycol	
Acetonitrile	11.9	monoethyl ether	10.2
(iii) Strongly hydrogen-bonded (acids, alcohols, aldehydes, amides, amines)			
Piperidine	8.7	Propylene glycol	12.6
Acetic acid	10.1	Methanol	14.5
Meta-cresol	10.2	Ethylene glycol	14.6
t-Butanol	10.6	Glycerol	16.5
1-Butanol	11.4	Water	23.4

[a]*Source:* Burrell, 1975; Rudin, 1982.   [b]SI value of δ in MPa$^{1/2}$ is obtained by multiplying the value in (cal/cm^3)$^{1/2}$ by 2.045.

Table 3.4 Solubility Parameters for Some Common Polymers[a,b]

Polymer	δ, (cal/cm^3)$^{1/2}$	H-bonding group[c]
Polytetrafluoroethylene	6.2	Poor
Polyethylene	8.0	Poor
Polypropylene	9.2	Poor
Polyisobutylene	8.0	Poor
Polybutadiene	8.4	Poor
Polyisoprene	8.1	Poor
Polystyrene	9.1	Poor
Poly(methyl methacrylate)	9.5	Medium
Poly(vinyl acetate)	9.4	Medium
Poly(vinyl chloride)	9.7	Medium
Cellulose diacetate	11.0	Strong
Poly(vinyl alcohol)	12.6	Strong
Polyacrylonitrile	12.7	Poor
Nylon-6,6	13.7	Strong

[a]*Source:* Burrell, 1975; Rudin, 1982.   [b]SI value of δ in MPa$^{1/2}$ is obtained by multiplying the value in (cal/cm^3)$^{1/2}$ by 2.045.   [c]The hydrogen-bonding group of each polymer has been taken as equivalent to that of the parent monomer.

The units of the solubility parameter δ are in (energy/volume)$^{1/2}$. The δ values for some common solvents and polymers, listed in Tables 3.3 and 3.4, have units of cal$^{1/2}$ cm$^{-3/2}$, called *hildebrands*. The SI value in MPa$^{1/2}$ may be obtained by multiplying the δ value in hildebrand by 2.045. Most tabulated solubility parameters refer to 25°C. However, over the temperature range normally encountered in industrial practice, the temperature dependence of δ can be neglected.

While the solubility parameter of a homopolymer can be calculated from the molar attraction constants as illustrated in Problem 3.16, the solubility parameter of random copolymers, δ_c, may be calculated from

$$\delta_c = \sum \delta_i w_i \tag{3.113}$$

where δ_i is the solubility parameter of the homopolymer that corresponds to the monomer i in the copolymer and w_i is the weight fraction of repeating unit i in the copolymer (Krause, 1972).

Problem 3.17 Calculate the solubility parameter for a methyl methacrylate-butadiene copolymer containing 25 mol % methyl methacrylate. The solubility parameter values for poly(methyl methacrylate) (PMMA) and polybutadiene (PB) homopolymers, calculated from molar attraction constants, are, respectively, 9.3 and 8.4 $(cal\ cm^{-3})^{1/2}$.

Answer:

M (butadiene) $=$ 54 g mol^{-1}; M (methyl methacrylate) $=$ 100 g mol^{-1}

Weight fraction of methyl methacrylate in copolymer

$$= \frac{0.25\ (100\ g\ mol^{-1})}{0.25\ (100\ g\ mol^{-1}) + 0.75\ (54\ g\ mol^{-1})} = 0.38$$

From Eq. (3.113): $\delta_c = 0.38(9.3\ cal^{1/2}\ cm^{-3/2}) + (1 - 0.38)(8.4\ cal^{1/2}\ cm^{-3/2})$

$$= 8.7\ (cal\ cm^{-3})^{1/2} = 17.8\ MPa^{1/2}$$

Solubility would be expected if the absolute value of $(\delta_1 - \delta_2)$ is less than about unity and there are no strong polar or hydrogen-bonding interactions in either the polymer or the solvent. To allow for the influence of hydrogen-bonding interactions, solvents have been characterized qualitatively as *poorly*, *moderately*, or *strongly hydrogen bonded*. The solvents listed in Table 3.3 are grouped according to this scheme. It is a useful practice to match both solubility parameter and hydrogen-bonding tendency for predicting mutual solubility.

Hansen (1967) developed a three-dimensional solubility parameter system based on the assumption that the energy of evaporation, i.e., the total cohesive energy ΔE_t which holds a liquid together, can be divided into contribution from dispersion (London) forces ΔE_d, polar forces ΔE_p, and hydrogen-bonding forces ΔE_h. Thus,

$$\Delta E_t = \Delta E_d + \Delta E_p + \Delta E_h \tag{3.114}$$

Dividing this equation by the molar volume of a solvent, V, gives

$$\frac{\Delta E_t}{V} = \frac{\Delta E_d}{V} + \frac{\Delta E_p}{V} + \frac{\Delta E_h}{V} \tag{3.115}$$

or

$$\delta_t^2 = \delta_d^2 + \delta_p^2 + \delta_h^2 \tag{3.116}$$

where δ_d, δ_p, and δ_h are solubility parameters due to dispersion forces, dipole forces, and hydrogen-bonding (or, in general, due to donor-acceptor interactions), respectively. The three parameters, called *Hansen parameters*, were determined (Hansen, 1967; Barton, 1983) empirically on the basis of many experimental observations for a large number of solvents (Table 3.5). Hansen's total cohesion parameter, δ_t, corresponds to the Hildebrand parameter δ, although the two quantities may not be identical because they are determined by different methods. Once the three component parameters for each solvent were evaluated, the set of parameters could then be obtained for each polymer (Table 3.6) from solubility ascertained by visual inspection of polymer-solvent mixtures (at concentrations of 10% w/v).

Problem 3.18 Calculate the composition of a blend of *n*-hexane, 1-butanol, and dioctyl phthalate that would have the same solvent properties as tetrahydrofuran. (Take appropriate data from Table 3.5.)

Answer:

Since δ_d values do not vary greatly, at least among common solvents, only δ_p and δ_h values may be matched. Let ϕ_1, ϕ_2, and ϕ_3 be the volume fractions of *n*-hexane, 1-butanol, and dioctyl phthalate, respectively, in the mixture.

By definition of volume fractions, $\phi_3 = 1 - \phi_1 - \phi_2$. The solubility parameter δ_{mix} of a mixture of two liquids of solubility parameters δ_1 and δ_2, respectively, can be approximated from: $\delta_{mix} = \phi_1\delta_1 + \phi_2\delta_2 + \phi_3\delta_3$. Using the data of Table 3.5,

$$\delta_p(\text{tetrahydrofuran}) = 2.8 = \phi_1(0) + \phi_2(2.8) + (1 - \phi_1 - \phi_2)(3.4)$$
$$\delta_h(\text{tetrahydrofuran}) = 3.9 = \phi_1(0) + \phi_2(7.7) + (1 - \phi_1 - \phi_2)(1.5)$$

Simplifying, $0.6 = 3.4\phi_1 + 0.6\phi_2$ and $2.4 = -1.5\phi_1 + 6.2\phi_2$. Simultaneous solution of these two equations yields $\phi_1 = 0.10$ and $\phi_2 = 0.41$. Hence, $\phi_3 = 0.49$. [Check on δ_d of the mixture: $0.10(7.3) + 0.41(7.8) + 0.49(8.1) = 7.9$. From Table 3.5: δ_d (tetrahydrofuran) = 8.2. On the other hand, simultaneous solution of three equations to match all three parameters δ_d, δ_p, and δ_h yields $\phi_1 = 0.14$, $\phi_2 = 0.41$, and $\phi_3 = 0.48$.]

When plotted in three dimensions, the Hansen parameters provide an approximately spherical volume of solubility for each polymer in δ_d, δ_p, δ_h space. *The scale on the dispersion axis is usually doubled to improve the spherical nature of this volume.* The distance of the coordinates $(\delta_d^i, \delta_p^i, \delta_h^i)$ of any solvent *i* from the center point $(\delta_d^j, \delta_p^j, \delta_h^j)$ of the solubility sphere of polymer *j* is

$$d = [4(\delta_d^i - \delta_d^j)^2 + (\delta_p^i - \delta_p^j)^2 + (\delta_h^i - \delta_h^j)^2]^{1/2} \qquad (3.117)$$

This distance can be compared with the radius R of the solubility sphere of the polymer (Table 3.6), and if $d < R$, the likelihood of the solvent *i* dissolving the polymer *j* is high. This works well, despite the limited theoretical justification of the method.

Table 3.5 Hansen Parameters[a,b] for Solvents at 25°C

Liquid	δ_d $(\text{cal/cm}^3)^{1/2}$	δ_p $(\text{cal/cm}^3)^{1/2}$	δ_h $(\text{cal/cm}^3)^{1/2}$
Acetic acid	6.8	6.0	9.2
Acetone	6.3	4.8	5.4
Benzene	7.9	4.2	2.0
1-Butanol	7.8	2.8	7.7
Chloroform	5.4	6.7	3.1
Cyclohexane	8.0	1.5	0.0
1,4-Dioxane	8.0	4.9	3.9
Dioctyl phthalate	8.1	3.4	1.5
Ethyl acetate	6.5	4.2	4.3
Ethylene glycol	4.9	7.4	14.6
Glycerol	4.5	7.5	15.3
n-Hexane	7.3	0.0	0.0
Methyl ethyl ketone	7.8	4.4	2.5
Methanol	7.4	6.0	10.9
Nitrobenzene	8.6	6.8	0.0
Tetrahydrofuran	8.2	2.8	3.9
Toluene	8.0	3.9	0.8
Water	5.9	11.1	19.7
m-Xylene	8.1	3.5	1.2

[a]*Source:* "Tables of Solubility Parameters", 3rd Ed., Chemicals and Plastics Research and Development Dept., Union Carbide Corporation, Tarrytown, N.Y., 1975. [b]SI value of a parameter in $\text{MPa}^{1/2}$ is obtained by multiplying the value in $(\text{cal/cm}^3)^{1/2}$ by 2.045.

Table 3.6 Hansen Parameters and Interaction Radius of Some Polymers and Resins[a,b]

Polymer	δ_d $(cal/cm^3)^{1/2}$	δ_p $(cal/cm^3)^{1/2}$	δ_h $(cal/cm^3)^{1/2}$	R $(cal/cm^3)^{1/2}$
Acrylonitrile-buta-				
diene elastomer	9.1	4.3	2.0	4.7
Cellulose acetate	9.1	6.2	5.4	3.7
Epoxy resin	10.0	5.9	5.6	6.2
Nitrocellulose	7.5	7.2	4.3	5.6
Polyamide	8.5	−0.9	7.3	4.7
Polyisoprene	8.1	0.7	−0.4	4.7
PMMA	9.1	5.1	3.7	4.2
Polystyrene	10.4	2.8	2.1	6.2
Poly(vinyl acetate)	10.2	5.5	4.7	6.7
Poly(vinyl chloride)	8.9	3.7	4.0	1.7
SBR	8.6	1.7	1.3	3.2

[a]Data from Hansen and Beerbower, 1971. [b]SI value in $MPa^{1/2}$ is obtained by multiplying the value in $(cal/cm^3)^{1/2}$ by 2.045.

Problem 3.19 Using Hansen parameters (Tables 3.5 and 3.6) determine if polystyrene is expected to dissolve in a solvent mixture of 60/40 v/v methyl ethyl ketone/n-hexane.

Answer:

Denote the solubility parameter components of MEK, n-hexane, and polystyrene using superscripts i, j, and k, respectively. From Table 3.5 :

MEK : $\delta_d^i = 7.8$, $\delta_p^i = 4.4$, $\delta_h^i = 2.5$ all in $(cal\ cm^{-3})^{1/2}$

n-hexane : $\delta_d^j = 7.3$, $\delta_p^j = 0$, $\delta_h^j = 0$ all in $(cal\ cm^{-3})^{1/2}$

The Hansen parameters are combined on a 60/40 volume fraction basis :

$$\delta_d^{ij} = 0.6 \times 7.8 + 0.4 \times 7.3 = 7.6 \ (cal\ cm^{-3})^{1/2}$$
$$\delta_p^{ij} = 0.6 \times 4.4 + 0.4 \times 0.0 = 2.6 \ (cal\ cm^{-3})^{1/2}$$
$$\delta_h^{ij} = 0.6 \times 2.5 + 0.4 \times 0.0 = 1.5 \ (cal\ cm^{-3})^{1/2}$$

For polystyrene (Table 3.6): $\delta_d^k = 10.4$, $\delta_p^k = 2.8$, $\delta_h^k = 2.1$, $R = 6.2$ all in $(cal\ cm^{-3})^{1/2}$

From Eq. (3.117): $d = [4(10.4 − 7.6)^2 + (2.8 − 2.6)^2 + (2.1 − 1.5)^2]^{1/2} = 5.6 \ (cal\ cm^{-3})^{1/2}$

As this value is less than the radius of the polymer solubility sphere ($6.2 \ cal^{1/2}\ cm^{-3/2}$), the polymer is expected to be soluble. This is found to be the case for a 10% w/w solution.

Three-dimensional presentations of solubility parameters are not easy to use and it is more convenient to transform the Hansen parameters into fractional parameters as defined by (Teas, 1968) :

$$
\left.
\begin{aligned}
f_d &= \delta_d / (\delta_d + \delta_p + \delta_h) \\
f_p &= \delta_p / (\delta_d + \delta_p + \delta_h) \\
f_h &= \delta_h / (\delta_d + \delta_p + \delta_h)
\end{aligned}
\right\}
\tag{3.118}
$$

The fractional parameters represent, in effect, the quantitative contribution of the three types of forces to the dissolving abilities for each solvent and can be represented more conveniently in a triangular diagram (Teas, 1968) for the prediction of solubility of polymers.

3.5   Swelling of Crosslinked Polymers

A three-dimensional network polymer, such as vulcanized rubber, does not dissolve in any solvent. It may nevertheless absorb a large quantity of a suitable liquid with which it is placed in contact and undergo swelling. The swollen gel is essentially a solution of solvent in polymer, although unlike an ordinary polymer solution it is an *elastic* rather than a viscous one.

Equilibrium swelling for three-dimensional gel structures occurs when the diluting force equals the elastic reaction of the network chains. Flory and Rehner (1943, 1950) analyzed this problem in terms of the thermodynamic behavior of high-polymer systems. Their statistical mechanical treatment leads to an expression (see below) for the elastic free-energy change ΔG_{el} due to the reaction of the network to swelling (expansion). The other free-energy change accompanying swelling is that due to dilution of the polymer by the solvent and referred to as free energy of mixing ΔG_{mix}. The sum of these two quantities is equal to the total free energy change ΔG involved in the mixing of pure solvent with the initially unstrained, network polymer. Thus we may write

$$\Delta G = \Delta G_{mix} + \Delta G_{el} \tag{3.119}$$

A suitable expression for ΔG_{mix} in the present case may be obtained from Eq. (3.45) by equating the number of moles of polymer molecules, n_2, to zero since the molecular weight of a network polymer is infinitely large (cf. Problem 1.5):

$$\Delta G_{mix} = RT(n_1 \ln \phi_1 + n_1 \phi_2 \chi) \tag{3.120}$$

A suitable expression for ΔG_{el} derived (Flory and Rehner, 1943, 1950) by analogy with the deformation of rubber under the condition of isotropy is

$$\Delta G_{el} = (RT\nu_e/2)(3\alpha^2 - 3 - \ln \alpha^3) \tag{3.121}$$

where α is the deformation factor and ν_e is the total number (mol) of *active sections* in the network. If n_e is the number of chain sections per unit volume and V is the total volume, ν_e will be given by

$$\nu_e = n_e V / N_{Av} \tag{3.122}$$

where N_{Av} is Avogadro's number. Let us now recall Eq. (2.85) from Chapter 2, namely,

$$n_e = \left(\frac{\rho N_{Av}}{\overline{M}_c}\right)\left(1 - \frac{2\overline{M}_c}{\overline{M}_n}\right) \tag{3.123}$$

where ρ is the density, $\overline{M}_n$ is the primary molecular weight (number average), and $\overline{M}_c$ is the average molecular weight per cross-linked unit. [$\overline{M}_c$ can be interpreted as the average molecular weight of chains between crosslinks, and hence, $1/\overline{M}_c$ can be taken as a measure of the degree of crosslinking. Further, it may be noted that the factor $(1 - 2\overline{M}_c/\overline{M}_n)$ expresses the correction for network imperfections or flaws resulting from chain ends. For a perfect network, $\overline{M}_n = \infty$ and the factor reduces to unity.] Substituting Eq. (3.123) into Eq. (3.122) now yields the following expression:

$$\nu_e = \left(\frac{\rho V}{\overline{M}_c}\right)\left(1 - \frac{2\overline{M}_c}{\overline{M}_n}\right) = \left(\frac{V}{v_2 \overline{M}_c}\right)\left(1 - \frac{2\overline{M}_c}{\overline{M}_n}\right) \tag{3.124}$$

where v_2 is the specific volume of the polymer.

Problem 3.20 The structure of a three-dimensional random network may be described quantitatively by two quantities: the density of crosslinking designated by the fraction ϵ of the total structural units engaged in crosslinkages and the fraction ϵ_t of the total units which occurs as terminal units or free chain ends (i.e., which are connected to the structure by only one bond). Alternative quantities, such as the number (mole) N of primary molecules and the number (mole) ν of crosslinked units, in addition to $\overline{M}_n$ and $\overline{M}_c$, defined above, are also used to characterize a random network structure. Relate N and ν to these other quantities.

Answer:

The quantities N and ν are related to ϵ_t and ϵ by

$$N = N_0 \epsilon_t / 2 \tag{P3.20.1}$$

and $\quad \nu = N_0 \epsilon$ $\hfill$ (P3.20.2)

where N_0 is the total number (mole) of units. N and ν can be further related to $\overline{M}_n$ and $\overline{M}_c$ by

$$N = N_0 M_0 / \overline{M}_n = V/(v_2 \overline{M}_n) \tag{P3.20.3}$$

and $\quad \nu = N_0 M_0 / \overline{M}_c = V/(v_2 \overline{M}_c)$ $\hfill$ (P3.20.4)

where M_0 is the (mean) molecular weight per structural unit; V is the total volume and v_2 the specific volume of the polymer.

The chemical potential of the solvent in the swollen gel is obtained by adding Eqs. (3.120) and (3.121) and partial differentiation with respect to n_1 in order to apply Eq. (3.3). Thus,

$$\mu_1 - \mu_1^0 = \left(\frac{\partial \Delta G}{\partial n_1}\right)_{T,P} = \left(\frac{\partial \Delta G_{\mathrm{mix}}}{\partial n_1}\right)_{T,P} + \left(\frac{\partial \Delta G_{\mathrm{el}}}{\partial \alpha}\right)_{T,P}\left(\frac{\partial \alpha}{\partial n_1}\right)_{T,P} \tag{3.125}$$

In order to evaluate $(\partial \alpha / \partial n_1)$, we note that

$$\alpha^3 = \frac{V_s}{V} = \frac{1}{\phi_2} \tag{3.126}$$

where V is the volume of the unswollen polymer when the crosslinkages were introduced into it and V_s is the volume of the swollen gel; ϕ_2 is the volume fraction of polymer in the swollen gel. Assuming further that mixing occurs without change in the total volume of the system (polymer + solvent), it follows that

$$\alpha^3 = \frac{1}{\phi_2} = \frac{V + n_1 V_1}{V} \tag{3.127}$$

and hence

$$(\partial \alpha / \partial n_1)_{T,P} = (V_1 / 3\alpha^2 V) \tag{3.128}$$

where V_1 is the molar volume of the solvent.

Evaluating the other two derivatives occurring in Eq. (3.125) after substitution for ΔG_{mix} and ΔG_{el} from Eqs. (3.120) and (3.121), respectively, we obtain (cf. Eq. (3.48)) the Flory-Rehner equation (Flory and Rehner, 1943, 1950):

$$\mu_1 - \mu_1^0 = \overline{\Delta G}_1 = RT\left[\ln(1 - \phi_2) + \phi_2 + \chi\phi_2^2 + \left(\frac{V_1 \nu_e}{V}\right)\left(\phi_2^{1/3} - \phi_2/2\right)\right] \tag{3.129}$$

Substituting for v_e from Eq. (3.124), we may also write the following expression:

$$\mu_1 - \mu_1^0 \;=\; \overline{\Delta G}_1 \;=\; RT \Big[\ln(1 - \phi_2) + \phi_2 + \chi\phi_2^2$$
$$+ \; (V_1/v_2\overline{M}_c)\,(1 - 2\overline{M}_c/\overline{M}_n)\,(\phi_2^{1/3} - \phi_2/2) \Big] \qquad (3.130)$$

The first three terms in the right hand side of Eq. (3.130) correspond to $(\mu_1 - \mu_1^0)$ according to Eq. (3.48) for a polymer of infinite molecular weight (i.e., $\sigma = \infty$), while the last part of the equation is the modification of the chemical potential due to the elastic reaction of the network structure. The activity of the solvent a_1 can be obtained from Eq. (3.130), since a_1 is related to chemical potential by Eq. (3.56). Thus,

$$\ln a_1 \;=\; \frac{\mu_1 - \mu_1^0}{RT} \;=\; \ln(1 - \phi_2) + \phi_2 + \chi\phi_2^2 + \left(\frac{V_1}{v_2\overline{M}_c}\right)\!\left(1 - \frac{2\overline{M}_c}{\overline{M}_n}\right)\!(\phi_2^{1/3} - \phi_2/2) \qquad (3.131)$$

The polymer concentration (ϕ_{2m}) at which $\mu_1 = \mu_1^0$, or the activity of the solvent is unity, represents the composition at *swelling equilibrium*. This is the concentration at which the relative chemical potential $(\mu_1 - \mu_1^0)$, or the relative partial molar free energy $\overline{\Delta G}_1$, plotted against ϕ_2, reaches a maximum. To locate this composition we equate $(\mu_1 - \mu_1^0)$ of Eq. (3.131) to zero, obtaining thereby

$$- \Big[\ln(1 - \phi_{2m}) + \phi_{2m} + \chi\phi_{2m}^2\Big] \;=\; \left(\frac{V_1}{v_2\overline{M}_c}\right)\!\left(1 - \frac{2\overline{M}_c}{\overline{M}_n}\right)\!(\phi_{2m}^{1/3} - \phi_{2m}/2) \qquad (3.132)$$

For a perfect network $(\overline{M}_n = \infty)$ and Eq. (3.132) simplifies to

$$- \Big[\ln(1 - \phi_{2m}) + \phi_{2m} + \chi\phi_{2m}^2\Big] \;=\; \left(\frac{V_1}{v_2\overline{M}_c}\right)\!(\phi_{2m}^{1/3} - \phi_{2m}/2) \qquad (3.133)$$

The left-hand member in Eqs. (3.132) and (3.133) represents the lowering of the chemical potential owing to mixing of polymer and solvent, while the right-hand side represents the increase of the chemical potential due to the elastic reaction of the network. If χ, V_1, and M_c are known, ϕ_{2m} is uniquely determined from Eq. (3.132). However, a more direct method of expressing the degree of swelling is as the maximum imbibition Q_m, defined as the volume of solvent imbibed per unit volume of polymer. Q_m is related to ϕ_{2m} by

$$Q_m \;=\; (1 - \phi_{2m})/\phi_{2m} \qquad (3.134)$$

Problem 3.21   A neoprene rubber vulcanizate containing 90% (by wt.) chloroprene has 1% of the latter units crosslinked. Given that the interaction parameter (χ) value of the polymer in *n*-hexane is 1.13, calculate an estimate of the maximum swelling expressed as volume of solvent imbibed per unit volume of the vulcanizate. [Data: Rubber density = 1.23 g/cm^3;   molar volume of *n*-hexane at 25°C= 131.6 cm^3/mol.]

Answer:

Molecular weight of chloroprene (C$_4$H$_5$Cℓ), M_o = 4x12 + 5x1 + 1x35.5 = 88.5 g mol^{-1}.   Fraction of chloroprene units engaged in cross-linkages, $\epsilon = 0.01$

From Eqs. (P3.20.2) and (P3.20.4), $\overline{M}_c = M_o/\epsilon = (88.5\,\text{g mol}^{-1})/0.01 = 8850\,\text{g mol}^{-1}$

Neglecting the effect of chain ends, Eq. (3.133) is applicable for swelling equilibrium. Referring to this equation,

$$\text{RHS} \;=\; \frac{(131.6\,\text{cm}^3\,\text{mol}^{-1})(1.23\,\text{g cm}^{-3})}{(8850\,\text{g mol}^{-1})}\,(\phi_{2m}^{1/3} - \phi_{2m}/2) \;=\; (0.0183)(\phi_{2m}^{1/3} - \phi_{2m}/2)$$

$$\text{LHS} \;=\; -\big[\ln(1 - \phi_{2m}) + \phi_{2m} + 1.13\phi_{2m}^2\big]$$

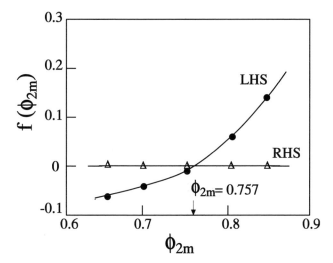

Figure 3.11 Graphical solution of Eq. (3.133) with data of Problem 3.21.

Both RHS and LHS may be evaluated for various values of ϕ_{2m} between 0 and 1 and plotted against ϕ_{2m} (Fig. 3.11). This gives $\phi_{2m} = 0.757$ at the point of intersection of the plots. Hence, maximum swelling (imbibition) = $(1 - 0.757)/0.757 = 0.32$ vol/vol and the polymer swells to $1/0.757$ or 1.32 times its original volume.

Problem 3.22 A crosslinked sample of polyisobutylene swells to 10 times its original volume in cyclohexane. What volume will it swell to in toluene ? [Data: χ (polyisobutylene-cyclohexane) = 0.436; molar volume of cyclohexane at 25°C = 108.7 cm^3/mol; χ (polyisobutylene-toluene) = 0.557; molar volume of toluene at 25°C = 106.9 cm^3/mol.]

Answer:

Assuming perfect network for the crosslinked polymer, Eq. (3.133) can be used and rearranged to give

$$(v_2 \overline{M}_c) = \frac{(\phi_{2m}^{1/3} - \phi_{2m}/2)V_1}{-[\ln(1 - \phi_{2m}) + \phi_{2m} + \chi\phi_{2m}^2]} \tag{P3.22.1}$$

In cyclohexane, $\phi_{2m} = 1/10 = 0.1$. Substituting $\phi_{2m} = 0.1$ and $V_1 = 108.7$ cm^3 mol^{-1} in Eq. (P3.22.1) gives $(v_2 \overline{M}_c) = 44996$ cm^3 mol^{-1}.

For polyisobutylene-toluene system, Eq. (3.133) becomes

$$-[\ln(1 - \phi_{2m}) + \phi_{2m} + 0.557\phi_{2m}^2] = \frac{(106.9 \text{ cm}^3 \text{ mol}^{-1})}{(44996 \text{ cm}^3 \text{ mol}^{-1})}(\phi_{2m}^{1/3} - \phi_{2m}/2)$$

By plotting the LHS of this equation against the RHS with various assumed values of ϕ_{2m} between 0 and 1, the value of ϕ_{2m} at the intersection of the plots is found to be 0.21. Therefore, the polymer swells to $1/0.21$ or 4.75 times its original volume in toluene.

Problem 3.23 Many polymers undergo crosslinking by gamma radiation. One such polymer was subjected to different doses of gamma radiation and the maximum swelling or imbibition (Q_m) of the resulting products in *n*-hexane at 25°C was measured which yielded the following data:

Dose, Mrad	Q_m, vol/vol
1	10.0
2	8.1
4	6.2
8	4.6
16	3.4

Assuming that the crosslink is tetrafunctional and its density is directly proportional to the radiation dose, deduce the crosslink density at the radiation dose of 10 Mrad. Also obtain an estimate of the interaction parameter for the polymer in *n*-hexane. Molar volume of *n*-hexane at 25°C $= 131.6$ cm^3/mol.

Answer:

Let $y =$ dose (Mrad), $Z =$ crosslink density, mol cm^{-3}, and hence, $Z = k.y$. Since each crosslink (tetrafunctional) joins four chains, the number of chains is $4 \times \frac{1}{2}$ or 2 times the number of crosslinks in a given volume. So, $(v_2 \overline{M_c})^{-1} = 2Z = 2ky$, where v_2 is the specific volume of the polymer. Substituting in Eq. (3.133) and rearranging gives: $Y = 2kV_1 X + \chi$, where $Y = -[\ln(1 - \phi_{2m}) + \phi_{2m}] / \phi_{2m}^2$ and $X = y(\phi_{2m}^{1/3} - \phi_{2m}/2) / \phi_{2m}^2$.

The value of ϕ_{2m} at each value of y is obtained from Q_m using Eq. (3.134). Y is plotted against X. The intercept of the straight line plot gives $\chi = 0.501$, while the slope gives $2kV_1 = 5.7 \times 10^{-4}$ (Mrad)$^{-1}$. Hence,

$$\text{Crosslink density, } (Z)_{y=10} = \frac{(5.7 \times 10^{-4} \text{ Mrad}^{-1})(10 \text{ Mrad})}{2(131.6 \text{ cm}^3 \text{ mol}^{-1})} = 2.2 \times 10^{-5} \text{ mol cm}^{-3}$$

3.5.1 Determination of χ from Swelling

A necessary preliminary to obtaining reliable values of the interaction parameter χ from swelling measurements is the determination of the degree of crosslinking of the network, that is, the crosslinked polymer must be calibrated before use. A method used is to determine the volume fraction of polymer present in the swollen gel when a sample of the *crosslinked* polymer has swelled to equilibrium in an excess of a solvent in which the χ value is known for the polymer. The relation between the ratio of initial to final weight and the volume fraction is given by

$$\phi_2 = \frac{\rho_1 W_b}{\rho_1 W_b + \rho_2 W_a - \rho_2 W_b}$$

$$= \frac{1}{1 + \frac{\rho_2}{\rho_1}\left(\frac{W_a}{W_b}\right) - \frac{\rho_2}{\rho_1}} \tag{3.135}$$

where W_b and W_a represent the weights before and after swelling, and ρ_1 and ρ_2 are the densities of the solvent and the polymer.

Problem 3.24 Osmotic pressure measurements at 25°C for poly(vinyl chloride) in dioxane solution yielded a molecular weight value of 4.6×10^5 g/mol and a χ-value of 0.48. Crosslinking was introduced into thick films of this material by heating at temperatures of about 140°C for 40 hours. Extraction with cyclohexanone was used to remove molecules not linked to the network. Samples cut from the adjacent positions in the crosslinked polymer film were used for swelling measurements at 25°C in dioxane ($V_1 = 85.7$ cm^3/mol, $\rho = 1.034$ g/cm^3) and cyclohexanone ($V_1 = 104.2$ cm^3/mol, $\rho = 0.947$ g/cm^3). From the equilibrium swelling data given below calculate the χ value for poly(vinyl chloride) in cyclohexanone at 25°C. (Polymer density, $\rho_2 = 1.41$ g/cm^3 at 25°C.)

| | Weight, g | |
Solvent	Before swelling	After swelling
Dioxane	0.0455	0.286
Cyclohexanone	0.0410	0.542

Answer:

From Eq. (3.135) for equilibrium swelling in dioxane:

$$\phi_{2m} = \frac{1}{1 + \left(\dfrac{1.41 \text{ g cm}^{-3}}{1.034 \text{ g cm}^{-3}}\right)\left(\dfrac{0.286 \text{ g}}{0.0455 \text{ g}}\right) - \left(\dfrac{1.41 \text{ g cm}^{-3}}{1.034 \text{ g cm}^{-3}}\right)} = 0.122$$

Rearranging Eq. (3.132) to solve for $\overline{M}_c^{-1}$ and replacing v_2 by $1/\rho_2$ gives

$$\overline{M}_c^{-1} = \frac{-[\ln(1 - \phi_{2m}) + \phi_{2m} + \chi\phi_{2m}^2]}{\rho_2 V_1 (\phi_{2m}^{1/3} - \phi_{2m}/2)} + \frac{2}{\overline{M}_n}$$

$$= \frac{-[\ln(1 - 0.122) + 0.122 + 0.48\,(0.122)^2]}{(1.41 \text{ g cm}^{-3})(85.7 \text{ cm}^3 \text{ mol}^{-1})[(0.122)^{1/3} - (0.122/2)]} + \frac{2}{(4.6 \times 10^5 \text{ g mol}^{-1})}$$

$$= 2.27 \times 10^{-5} \text{ mol g}^{-1}$$

$$\overline{M}_c = 4.4 \times 10^4 \text{ g mol}^{-1}$$

For equilibrium swelling in cyclohexanone, calculating as above from Eq. (3.135) : $\phi_{2m} = 0.0521$.

Rearranging Eq. (3.132) gives:

$$\chi = -(1/\phi_{2m}^2)\left[(\rho_2 V_1/\overline{M}_c)(1 - 2\overline{M}_c/\overline{M}_n)(\phi_{2m}^{1/3} - \phi_{2m}/2) + \ln(1 - \phi_{2m}) + \phi_{2m}\right].$$

Substituting cyclohexanone data,

$$\chi = -\frac{1}{(0.0521)^2}\left[\frac{(1.41 \text{ g cm}^{-3})(104.2 \text{ cm}^3 \text{ mol}^{-1})}{(4.40 \times 10^4 \text{ g mol}^{-1})}\left(1 - \frac{2(4.40 \times 10^4 \text{ g mol}^{-1})}{(4.6 \times 10^5 \text{ g mol}^{-1})}\right)\times \right.$$
$$\left. ((0.0521)^{1/3} - 0.0521/2) + \ln(1 - 0.0521) + 0.0521\right] = 0.17$$

3.6 *Frictional Properties of Polymer Molecules in Dilute Solution*

A polymer molecule moving in a dilute solution undergoes frictional interactions with solvent molecules due to its motion relative to the surrounding medium. The effects of these frictional interactions are related to the size and shape of the polymer molecule. Thus, the chain dimensions of polymer molecules can be evaluated from measurements of their frictional properties (Flory and Fox, 1950).

At a definite shear rate in a laminar flow, different parts of the polymer molecule move at different rates depending on whether they are in a zone of rapid or relatively slow flow, with the result the polymer molecule experiences a couple of forces which makes it rotate in the flow. Rotational and translational movements of polymer molecules cause friction between their chain segments and the solvent molecules. This is manifested in an increase in viscosity of the solution relative to the viscosity of the pure solvent.

In a dilute solution, the long flexible polymer molecule forms a coil. One can identify two extremes of the frictional behavior of such polymer coils in solution, namely free-draining and nondraining. In the case of a *free-draining polymer*, the solvent molecules can pass through the coil-form of the molecule, flowing past each segment of the chain with equal ease, but a *nondraining polymer* molecule in coil-form retains a definite amount of solvent enclosed by it and moves together with this solvent. This difference in the draining behavior leads to different dependencies of the frictional coefficient, f_0, of a polymer molecule on chain length.

A *free-draining* polymer molecule, referred to as the *free-draining coil*, is considered by dividing it into identical segments each of which has the same frictional coefficient ζ. Assuming that solvent molecules permeate all regions of the polymer coil with equal ease (or difficulty), each segment makes the same contribution to f_0 and one can thus write (Young and Lovell, 1990)

$$f_0 = n\zeta \tag{3.136}$$

where n is the number of segments in the chain.

A *nondraining* polymer molecule, also referred to as the *impermeable coil*, can be represented by an equivalent impermeable hydrodynamic sphere of radius R_h. The frictional coefficient of this sphere which represents the frictional coefficient of the nondraining polymer coil can thus be written, according to Stokes' law, as

$$f_0 = 6\pi \eta_0 R_h \tag{3.137}$$

where η_0 is the viscosity of the pure solvent. With the reasonable assumption that the unperturbed root-mean-square radius of gyration $\langle S^2 \rangle_0^{1/2}$ multiplied by the expansion factor α (which accounts for osmotic swelling of the polymer coil by solvent-polymer interactions) can be taken as a measure of R_h, Eq. (3.137) can be rewritten in the form

$$f_0 = K_0 \alpha \langle S^2 \rangle_0^{1/2} \tag{3.138}$$

where K_0 is a constant for a given system. Since $\langle S^2 \rangle_0^{1/2}$ is proportional to $n^{1/2}$ and for highly expanded coils α is approximately proportional to $n^{1/10}$ (see Problem 3.12), Eq. (3.138) indicates that

$$f_0 = K_0' n^{a_0} \tag{3.139}$$

where K_0' is another constant and $0.5 \leq a_0 \leq 0.6$. Comparison of Eqs. (3.136) and (3.139) shows that while for both the free-draining polymer and nondraining polymer the frictional coefficient increases with polymer chain size, the effect however is less pronounced for the nondraining polymer molecule.

The frictional behavior of real polymer molecules is made of contributions of both free-draining and non-draining polymer molecules represented by Eqs. (3.136) and (3.139), respectively. The free-draining contribution dominates for very short chain or elongated rodlike molecules, while for flexible chain molecules (coils) it decreases rapidly with the increase in chain length (Young and Lovell, 1990). As most polymers consist of long flexible chain molecules, their frictional behavior approximates to that of nondraining or impermeable coils. In the following section, the nondraining behavior is therefore considerd further, specifically in relation to the viscosity of dilute polymer solutions.

3.6.1 Viscosity of Dilute Polymer Solutions

Einstein derived the following equation for the viscosity of suspensions of rigid, uncharged, spherical particles which do not interact with the suspension medium:

$$\eta = \eta_0 \left[1 + \frac{5}{2}\phi_2 \right] \tag{3.140}$$

where η and η_0 are the viscosities of the suspension and the suspension medium, respectively, and ϕ_2 is the volume fraction of the spherical particles.

Equation (3.140) can be rewritten as

$$\eta_{sp} = (5/2)\phi_2 \tag{3.141}$$

where $\eta_{sp} = (\eta - \eta_0)/\eta_0$ is known as the *specific viscosity*.

It may be assumed that under the action of a shear stress the polymer coil with the solvent enclosed by it behaves like an Einsteinian sphere, and hence, the viscosity of polymer solution should obey Eq. (3.141) for noninteracting spherical particles. Noninteraction of the polymer coils requires infinite dilution. Mathematically this is achieved by defining a quantity called the *intrinsic viscosity*, $[\eta]$, according to equation (Young and Lovell, 1990):

$$[\eta] = \lim_{c \to 0} (\eta_{sp}/c) \tag{3.142}$$

If V_h is the hydrodynamic volume of each polymer molecule (it is assumed that all polymer molecules are of the same molecular weight) then

$$\phi_2 = (c/M)N_{Av} V_h \tag{3.143}$$

where c is the polymer concentration (mass per unit volume), M is the molar mass of the polymer (mass per mol), and N_{Av} is Avogadro's number. Substituting Eq. (3.143) into Eq. (3.141) gives

$$\eta_{sp} = (5/2)(c/M)N_{Av} V_h \tag{3.144}$$

A more satisfactory form of this equation to satisfy the condition of noninteraction of polymer coils is given in terms of intrinsic viscosity, $[\eta]$, defined by Eq. (3.142). Thus,

$$[\eta] = (5/2)(N_{Av}/M)(V_h)_{c \to 0} \tag{3.145}$$

This equation can be rearranged to give an equation for the hydrodynamic volume of an impermeable (nondraining) polymer molecule in infinitely dilute solution:

$$(V_h)_{c \to 0} = (2/5)[\eta]M/N_{Av} \tag{3.146}$$

Thus, the quantity $[\eta]M$ is proportional to $(V_h)_{c \to 0}$.

Problem 3.25   The relative viscosity of a benzene solution of natural rubber of concentration 0.05 g/100 cm^3 is measured to be 1.18 at 25°C. The molecular weight of the rubber is 200,000. Calculate an estimate of the hydrodynamic volume of the rubber molecule. State the assumptions made for the calculation.

Answer:

In order to use Eq. (3.144) it is necessary to assume that the polymer coils are impermeable and noninteracting. Since the natural rubber is a high-molecular-weight flexible polymer and the solution is very dilute, these conditions may be assumed to be closely approximated. Further it should be assumed, though incorrectly, that all molecules are of the same molecular weight (i.e., homodisperse).

$$\eta_{sp} = \frac{\eta - \eta_0}{\eta_0} = \frac{\eta}{\eta_0} - 1 = 1.18 - 1 = 0.18$$

$$c = 0.0005 \text{ g cm}^{-3}$$

From Eq. (3.144):

$$V_h = \frac{2\eta_{sp}M}{5cN_{Av}} = \frac{2(0.18)(2.0 \times 10^5 \text{ g mol}^{-1})}{5(0.0005 \text{ g cm}^{-3})(6.02 \times 10^{23} \text{ mol}^{-1})} = 4.8 \times 10^{-17} \text{ cm}^3$$

In dealing with linear polymers, the root-mean-square end-to-end distance $\langle r^2 \rangle^{1/2}$ can be taken as a measure of the size. Therefore, assuming that $(V_h)_{c \to 0}$ is proportional to $\langle r^2 \rangle^{3/2}$, we obtain

$$[\eta] = \Phi\left(\frac{\langle r^2 \rangle^{3/2}}{M}\right) \tag{3.147}$$

where Φ is a constant. If $\langle r^2 \rangle$ is determined from dissymmetry measurements on the light scattered by dilute solutions of a polymer fraction, the results being extrapolated to infinite dilution, and if the intrinsic viscosity is determined in the same solvent and at the same temperature, then it is possible to calculate Φ from Eq. (3.147). It is important to use well-fractionated samples for these measurements, since Eq. (3.147) is actually derived for monodisperse samples. So when measurements are performed with a heterodisperse polymer sample having a considerable range in molecular weight, a number average of $\langle r^2 \rangle^{3/2}$ over the molecular weight distribution should be used in conjunction with number average molecular weight $\overline{M}_n$ in Eq. (3.147).

Analysis of experimental data for $[\eta]$ and $\overline{\langle r^2 \rangle}^{1/2}$ by light scattering of solution led Flory to the conclusion that the parameter Φ should have a universal value for all linear polymers in all solvents. For nondraining polymer molecules, Φ, known as *Flory constant*, is independent of chain structure and chain length, and is dependent only on the spatial distribution of segments about the center of gravity in the molecular coil. For randomly coiled linear polymers this distribution is approximately Gaussian. For nondraining polymer coils with Gaussian distribution of segments the most acceptable value of Φ determined on the basis of theoretical calculations and experimental measurements is 2.5×10^{23} (cm^3 g^{-1})(g mol^{-1} cm^{-3}), or simply mol^{-1}, when $[\eta]$ is expressed in cm^3/g and $\langle r^2 \rangle^{1/2}$ in cm. The value is 2.5×10^{21} (dL g^{-1})(g mol^{-1} cm^{-3}) when $[\eta]$ is expressed in dL/g and $\langle r^2 \rangle^{1/2}$ in cm. This is often quoted as Φ for all polymer-solvent systems. In practice, Φ decreases (Krigbaum and Carpenter, 1955) from its theoretical value of 2.87×10^{23} valid for unperturbed gaussian coils (2.87×10^{21} if $[\eta]$ is expressed in dL/g) to about 2.1×10^{23} for polymers dissolved in good solvents (Munk, 1989).

According to Eq. (3.147), the contribution of a high polymer molecule to the viscosity should be that of an equivalent sphere having a volume proportional to $\langle r^2 \rangle^{3/2}$. Since for a randomly coiled linear chain the mean square radius of gyration $\langle S^2 \rangle$ is related to mean square end-to-end distance $\langle r^2 \rangle$ by $\langle S^2 \rangle = \langle r^2 \rangle/6$ [cf. Eq. (2.4)], Eq. (3.147) can be rewritten as

$$[\eta] \;=\; \Phi' \left(\frac{\langle S^2 \rangle^{3/2}}{M} \right) \tag{3.148}$$

where Φ' is another Flory constant given by $\Phi' = 6^{3/2}\Phi$ and hence has the value of 3.6×10^{24} mol^{-1} when $[\eta]$ is expressed in cm^3/g and $\langle S^2 \rangle^{1/2}$ in cm.

It is convenient to separate $\langle r^2 \rangle^{1/2}$ into its component factors, viz., the unperturbed dimension $\langle r^2 \rangle_0^{1/2}$ and a linear expansion factor α_η for hydrodynamic chain dimensions [cf. Eq. (3.86)]. Equation (3.147) may thus be rewritten as follows:

$$[\eta] \;=\; \Phi \left(\frac{\langle r^2 \rangle_0}{M} \right)^{3/2} \alpha_\eta^3 M^{1/2} \tag{3.149}$$

Since $\langle r^2 \rangle_0$ is directly proportional to the number, n, of chain segments, it is also directly proportional to M. Therefore, $\langle r^2 \rangle_0/M$ is a constant and Eq. (3.149) is more commonly written in the form of the *Flory-Fox equation* (Flory and Fox, 1951):

$$[\eta] \;=\; K_0 \alpha_\eta^3 M^{1/2} \tag{3.150}$$

where
$$K_0 \;=\; \Phi \left(\frac{\langle r^2 \rangle_0}{M} \right)^{3/2} \tag{3.151}$$

In view of the preceding analysis, K_0 should be a constant independent of both polymer molecular weight and solvent. To some extent, it may, however, vary with temperature, considering that the unperturbed dimension $\langle r^2 \rangle_0^{1/2}$ is modified by hindrances to free rotation whose effects are, in general, temperature-dependent.

The Flory-Fox equation predicts that the intrinsic viscosity would depend on (i) the stiffness of the polymer chain (through the term $\langle r^2 \rangle_0/M$), (ii) the molecular weight of the polymer (through $M^{1/2}$), and (iii) the solvent-polymer-temperature combination (through α_η). Under theta conditions, α_η is unity and Eq. (3.149) predicts a dependence of $[\eta]$ on the square root of the molecular weight. This behavior has been confirmed experimentally for several polymer-solvent systems. The exponent 0.5 can be considered as a lower limit since much poorer solvents will not dissolve the polymer. On the other hand, in very good solvents, where there is large expansion of polymer coil, α_η is proportional to $M^{1/10}$ (see Problem 3.12). Hence the intrinsic viscosity will vary as $M^{(1/2+3/10)}$, i.e., as $M^{0.8}$. This is an upper limit. For other solvents and non-theta conditions, the behavior is intermediate between these two limits. The Flory-Fox equation (3.149) thus suggests a general relationship of the form

$$[\eta] \;=\; KM^a \tag{3.152}$$

where K and a are empirical constants for a given polymer-solvent-temperature combination; a increases (in the range $0.5 \le a \le 0.8$) with the degree of expansion of the molecular coils from their unperturbed dimensions under theta conditions, where $a = 0.5$. This important relationship between $[\eta]$ and M is commonly known as the *Mark-Houwink equation* or *Mark-Houwink-Sakurada equation* and was first proposed on the basis of experimental data. The Mark-Houwink constants K and a are evaluated from $[\eta]$ and M data of polymer samples with narrow molecular weight distributions. Generally, a plot of $\log[\eta]$ against $\log M$ is fitted to a straight line from which K and a are obtained (see Chapter 4).

In an "ideal" solvent (i.e., a poor solvent at the theta temperature for which the second virial coefficient vanishes), also known as *theta solvent*, $\alpha_\eta = 1$ and Eq. (3.150) reduces to

$$[\eta]_\theta = K_o M^{1/2} \tag{3.153}$$

It has been established that K_0 normally is independent of the solvent and the molecular weight of the polymer, though it often depends to some extent on temperature. Values for the expansion factor in good solvents can thus be derived from intrinsic viscosities measured in them. From Eqs. (3.150) and (3.153) the linear expansion factor α_η, which is a measure of long range interactions and pertains to hydrodynamic chain dimensions, is thus given by

$$\alpha_\eta = \left([\eta]/[\eta]_\theta\right)^{1/3} \tag{3.154}$$

where $[\eta]$ is the intrinsic viscosity in the given solvent and $[\eta]_\theta$ is that of the same polymer in a theta solvent. From the ratio of the intrinsic viscosity of the polymer in a good solvent to that in an "ideal" solvent at (nearly) the same temperature one may thus obtain the factor by which the molecule is enlarged.

It has been suggested that α thus obtained indirectly from viscosity is related to the more directly measured α of Eqs. (3.86) and (3.87) by

$$\alpha_\eta = \alpha^{0.81} \tag{3.155}$$

There is considerable experimental evidence to support this conclusion.

3.6.1.1 *Determination of Polymer Molecular Dimensions from Viscosity*

The size of a polymer chain in any solvent can be obtained directly from light scattering measurements, if the polymer coil is large enough to scatter in an asymmetric manner (see **Light Scattering Method** in Chapter 4). However, when the chain is too short for the light scattering method an alternative technique has to be used.

If Φ is considered to be a universal constant, the average dimensions of polymer molecules in solution can be drived simply from their intrinsic viscosities and molecular weights. More specifically, the natural, or unperturbed, dimensions of the polymer chain can be estimated from the knowledge of intrinsic viscosity in a theta solvent (Kurata and Stockmayer, 1963; Stockmayer and Fixman, 1963).

Problem 3.26   The intrinsic viscosity of polystyrene of molecular weight 3.2×10^5 in toluene at 30°C was determined to be 0.846 dL/g. In a theta solvent (cyclohexane at 34°C) the same polymer had an intrinsic viscosity of 0.464 dL/g. Calculate (a) unperturbed end-to-end distance of the polymer molecule, (b) end-to-end distance of the polymer in toluene solution at 30°C, and (c) volume expansion factor in toluene solution. ($\Phi = 2.5 \times 10^{23} \text{ mol}^{-1}$)

Answer:

$[\eta]_\theta = 0.464 \text{ dL g}^{-1} = 46.4 \text{ cm}^3 \text{ g}^{-1}$.

From Eq. (3.153):

$$K_0 = [\eta]_\theta / M^{1/2}$$
$$= (46.4 \text{ cm}^3 \text{ g}^{-1})/(3.2 \times 10^5 \text{ g mol}^{-1})^{1/2} = 0.082 \text{ cm}^3 \text{ mol}^{1/2} \text{ g}^{-3/2}$$

From Eq. (3.151):

$$\langle r^2 \rangle_0^{3/2} = K_0 M^{3/2} / \Phi$$

$$= \frac{(0.082 \text{ cm}^3 \text{ mol}^{1/2} \text{ g}^{-3/2})(3.2 \times 10^5 \text{ g mol}^{-1})^{3/2}}{(2.5 \times 10^{23} \text{ mol}^{-1})} = 5.94 \times 10^{-17} \text{ cm}^3$$

$$\langle r^2 \rangle_0^{1/2} = 3.90 \times 10^{-6} \text{ cm} = 390 \text{ Å}$$

(b) From Eq. (3.147):

$$\langle r^2 \rangle^{3/2} = \frac{[\eta]M}{\Phi} = \frac{(84.6 \text{ cm}^3 \text{ g}^{-1})(3.2 \times 10^5 \text{ g mol}^{-1})}{(2.5 \times 10^{23} \text{ mol}^{-1})} = 108.28 \times 10^{-18} \text{ cm}^3$$

$$\langle r^2 \rangle^{1/2} = 4.76 \times 10^{-6} \text{ cm} = 476 \text{ Å}$$

(c) Neglecting the effect on K_0 of small temperature difference between $[\eta]$ and $[\eta]_\theta$ measurements, the hydrodynamic chain dimension expansion factor, α_η, can be obtained from Eq. (3.154). Thus,

$$\alpha_\eta = \left[\frac{(0.846 \text{ dL g}^{-1})}{(0.464 \text{ dL g}^{-1})} \right]^{1/3} = 1.22$$

From Eq. (3.155):

Average linear expansion factor, $\alpha = (\alpha_\eta)^{1/0.81} = 1.28$

Volume expansion factor, $\alpha^3 = (1.28)^3 = 2.1$

Problem 3.27   For a fractionated sample of *cis*-1,4-polybutadiene of molecular weight 1.23×10^5 intrinsic viscosities were measured (Moraglio, 1965) in three different solvents at respective theta temperatures. From the results given below determine the variation of the unperturbed dimensions of the polymer molecule with temperature.

Solvent	θ-Temperature (°C)	$[\eta]_\theta$, dL/g
n-Heptane	− 1.0	0.670
Isobutyl acetate	20.5	0.656
n-Propyl acetate	35.5	0.645

Answer:

The K_0 values for the temperatures −1, 20.5, and 35.5°C are calculated from Eq. (3.153) using $[\eta]_\theta$ values and $M = 1.23 \times 10^5$ g mol^{-1}. These are plotted as $\ln K_0$ vs. T°K. From the slope, $d \ln K_0 / dT = 1.09 \times 10^{-3}$.

Differentiation of Eq. (3.151) with respect to T yields

$$\frac{d \ln \langle r^2 \rangle_0}{dT} = \frac{2}{3} \frac{d \ln K_0}{dT} = 0.73 \times 10^{-3}$$

Note: The parameter $d \ln \langle r^2 \rangle_0 / dT$ appears in the theoretical treatments of thermoelasticity of macromolecular substances.

It is not always possible to find a suitable theta solvent for a polymer and methods have been developed which allow unperturbed dimensions to be estimated in nonideal (good) solvents.

For small expansions, the hydrodynamic chain expansion parameter, α_η, of Gaussian polymer chains is given by a closed expression of the form (Young and Lovell, 1990):

$$\alpha_\eta^3 = 1 + b.z \tag{3.156}$$

where z is an excluded volume parameter (which can be shown to be proportional to the square root of the polymer molecular weight) and b is a constant. (The value of b is uncertain, e.g., values of 1.55 and 1.05 have been obtained from different theories.) Combination of this relation with the Flory-Fox equation (3.150) gives

$$[\eta] = K_0 M^{1/2} + b K_0 z M^{1/2} \tag{3.157}$$

which upon rearrangement, recognizing that $z \propto M^{1/2}$, leads to

$$[\eta] M^{-1/2} = K_0 + B' M^{1/2} \tag{3.158}$$

where B' depends upon the chain structure and polymer-solvent interaction; it is a constant for a given polymer-solvent pair at a given temperature. Thus $[\eta]$ and M data of calibration samples with narrow molecular weight distributions can be plotted as $[\eta] M^{-1/2}$ against $M^{1/2}$ to give K_0 as the intercept. The value of K_0 can then be used to evaluate unperturbed chain dimensions and the expansion factor for the chain dimensions.

Problem 3.28   Viscosity measurements on polystyrene fractions of different molecular weights in benzene at 20°C yielded the following values for intrinsic viscosities.

Molecular weight M	$[\eta]$, dL/g
44,500	0.268
65,500	0.356
262,000	1.07
694,000	2.07
2,550,000	5.54

Evaluate for polystyrene of molecular weight 10^6 (a) the unperturbed end-to-end chain lengths and (b) the volume expansion factor in benzene at 20°C. Assume $\Phi = 2.5 \times 10^{23}$ mol^{-1}.

Answer:

In Fig. 3.12, $[\eta] M^{-1/2}$ is plotted against $M^{1/2}$ and a straight line is fitted. The intercept at $M^{1/2} \to 0$ yields $K_0 = 0.00117$ dL mol$^{1/2}$ g$^{-3/2}$ = 0.117 cm^3 mol$^{1/2}$ g$^{-3/2}$.

(a) From Eq. (3.151)

$$\langle r^2 \rangle_0^{1/2} = \left(\frac{K_0 M^{3/2}}{\Phi} \right)^{1/3}$$

$$\langle r^2 \rangle_0^{1/2} = \left[\frac{(0.117 \text{ cm}^3 \text{ mol}^{1/2} \text{ g}^{-3/2})(10^6 \text{ g mol}^{-1})^{3/2}}{(2.5 \times 10^{23} \text{ mol}^{-1})} \right]^{1/3}$$

$$= 7.76 \times 10^{-6} \text{ cm } (776 \text{ Å})$$

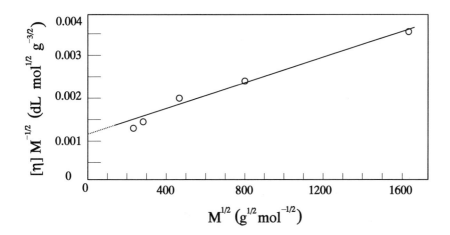

Figure 3.12 Plot of $[\eta]M^{-1/2}$ against $M^{1/2}$ for polystyrene solutions in benzene at 20°C. (Data of Problem 3.28.)

(b) From Eqs. (3.153)–(3.155) :

$$\alpha^3 = \left(\frac{[\eta]M^{-1/2}}{K_0}\right)^{1/0.81}$$

From Fig. (3.12), $[\eta]M^{-1/2} = 0.00275$ dL mol$^{1/2}$ g$^{-3/2}$ for $M = 10^6$ g mol^{-1}. Therefore,

$$\alpha^3 = \left[\frac{(0.275 \text{ cm}^3 \text{ mol}^{1/2} \text{ g}^{-3/2})}{(0.117 \text{ cm}^3 \text{ mol}^{1/2} \text{ g}^{-3/2})}\right]^{1/0.81} = 2.87$$

The interpretation of $[\eta]$ for branched polymers and copolymers is much more complicated. Branching increases the segment density within the molecular coil leading to a smaller size for a molecule of the same molecular weight. Thus, a branched polymer molecule has a smaller hydrodynamic volume, and hence a lower intrinsic viscosity, than a similar linear polymer of the same molecular weight.

REFERENCES

Barton, A. F. M., "Handbook of Solubility Parameters and Other Cohesion Parameters", CRC Press, Boca Raton, Florida, 1983.

Bawn, C. E., Freeman, R. F., and Kamaliddin, A. R., *Trans. Farady. Soc.*, **46**, 677 (1950).

Brandrup, J. and Immergut, E., Eds., "Polymer Handbook", 2nd ed., Wiley Interscience, New York, 1975.

Burrell, H., *Official Digest*, **27**, 726 (1955).

Burrell, H., "Solubility Parameter Values", pp. IV-337-359 in "Polymer Handbook", (J. Brandrup and E. Immergut, Eds.), 2nd ed., Wiley Interscience, New York, 1975.

Flory, P. J., *J. Chem. Phys.*, **10**, 51 (1942).

Flory, P. J., "Principles of Polymer Chemistry", Cornell Univ. Press, New York, 1953.

Flory, P. J. and Fox Jr., T. G., *J. Polym. Sci.*, **5**, 745 (1950).

Flory, P. J. and Fox Jr., T. G., *J. Am. Chem. Soc.*, **73**, 1904 (1951).

Flory, P. J. and Krigbaum, W. P., *J. Am. Chem. Soc.*, **75**, 1775 (1953).

Flory, P. J. and Rehner, J., *J. Chem. Phys.*, **11**, 521 (1943); **18**, 108 (1950).

Fox, T. G. and Flory, P. J., *J. Am. Chem. Soc.*, **73**, 1909, 1915 (1951).

Glasstone, S., "Thermodynamics for Chemists", p. 373, Van Nostrand, Princeton, 1947.

Hansen, C. M., *J. Paint Technol.*, **39**, 104, 511 (1967).

Hansen, C. M. and Beerbower, A., "Kirk Othmer Encycl. Chem. Technol." (A. Standen, Ed.), 2nd ed., Suppl., 889, 910, Wiley Interscience, New York, 1971.

Hildebrand, J. H., Prausnitz, J. M., and Scott, R. L., "Regular and Related Solutions", Van Nostrand Reinhold, New York, 1970.

Hoy, K. L., *J. Paint Technol.*, **42**, 76 (1970).

Huggins, M. L., *J. Phys. Chem.*, **46**, 151 (1942); *J. Am. Chem. Soc.*, **64**, 1712 (1942).

Krigbaum, W. R. and Carpenter, D. K., *J. Phys. Chem.*, **59**, 1166 (1955).

Krause, S., *J. Macromol. Sci.-Macromol. Rev.*, **C7**, 251 (1972).

Kurata, M. and Stockmayer, W. H., *Fortschr. Hochpolym. Forsch.* (Advances in Polymer Science), **3**, 196 (1963).

Moraglio, G., *Europ. Polym. J.*, **1**, 103 (1965).

Munk, P., "Introduction to Macromolecular Science", John Wiley, New York, 1989.

Nelson, R. C., Hemwall, R. W., and Edwards, G. D., *J. Paint Technol.*, **42**, 636 (1970).

Rudin, A., "The Elements of Polymer Science and Engineering", Academic Press, Orlando, 1982.

Small, P. A., *J. Appl. Chem.*, **3**, 71 (1953).

Smith, R. P., *J. Chem. Phys.*, **38**, 1463 (1963).

Stockmayer, W. H. and Fixman, M., *J. Polym. Sci.*, *C1*, 137 (1963).

Teas, J. P. *J. Paint Technol.*, **40**, 519 (1968).

Young, R. J. and Lovell, P. A., "Introduction to Polymers", Chapman and Hall, London, 1990.

EXERCISES

3.1 Calculate (a) the enthalpy of mixing and (b) Gibbs free energy change in dissolving 10^{-5} mol of poly(methyl methacrylate) of $\overline{M}_n = 10^5$ and $\rho = 1.20$ g/cm^3 in a 150 g of chloroform ($\rho = 1.49$ g/cm^3) at 25°C. The value of χ is 0.377. Assume that the volumes are additive.
[*Ans.* (a) 25.5 J; (b) −16.1 J]

3.2 Use the Flory-Huggins equation for $\Delta\mu_1$ to calculate an estimate of χ for solutions of natural rubber ($\overline{M}_n = 2.5 \times 10^5$) in benzene, given that vapor pressure measurements show that the activity of the solvent in a solution with $\phi_2 = 0.250$ is 0.989.
[*Ans.* 0.426]

3.3 Osmotic measurements on polyisobutylene (mol. wt. 1.46×10^6) in toluene at various temperatures yielded the following values for the second virial coefficient (A_2):

Temperature, ^{o}C	$A_2 \times 10^4$, cm^3 mol g^{-2}
−5	0.41
0	0.68
5	0.95

Determine for the polymer-solvent system (a) the temperature at which theta conditions are attained, (b) the entropy of dilution parameter ψ, and (c) the heat of dilution parameter κ at 0°C. [*Data*: Specific volume of polymer at 0°C = 1.0 cm^3/g; molar volume of toluene at 0°C= 106.1 cm^3/mol.]
[*Ans.* (a) −12°C; (b) 0.16; (c) 0.15]

3.4 Values of second virial coefficient (A_2) of a polydimethylsiloxane gum of molar mass (M) 4.43×10^5 g/mol in several solvents are given below along with data of molar volumes (V_1) at room temperature.

Solvent	V_1, cm^3/mol	A_2, mol cm^3 g^{-2}
[(CH$_3$)$_2$SiO]$_4$	307.8	6.35×10^{-4}
n-Heptane	147.5	6.47×10^{-4}
Toluene	106.9	4.06×10^{-4}

It has been found that mean square end-to-end distance of polymer molecule and molar mass are related by $\left(\langle r^2\rangle_o/M\right)^{1/2} = 7.30\times10^{-9}$ cm mol$^{1/2}$ g$^{-1/2}$.

Calculate an estimate of the polymer-solvent interaction parameter (χ) and the relative volumetric expansion (α^3) of the polymer in each solvent. [Polymer density = 0.960 g/cm^3]

Ans.

Solvent	χ	α	α^3
[(CH$_3$)$_2$SiO]$_4$	0.319	1.44	3.02
n-Heptane	0.412	1.43	2.92
Toluene	0.458	1.35	2.46

3.5 The second virial coefficient (Γ_2) of a test sample of polystyrene ($\overline{M}_n = 4.8\times10^6$) in toluene ($\rho = 0.845$ g/cm^3 at 25°C) was found to be 219 cm^3/g at 25°C. The partial specific volume of the polymer in the solution at this temperature was found to be 0.91 cm^3/g. Evaluate the interaction parameter χ. [*Ans.* 0.494]

3.6 A solution of poly(methyl methacrylate) ($\rho = 1.20$ g/cm^3, $\overline{M}_n = 3.5\times10^5$) in chloroform ($\rho = 1.49$ g/cm^3 at 20°C) has been prepared by dissolving 100 mg of the polymer in 200 mL of the solvent. Estimate the osmotic pressure of the resulting solution. [$\chi = 0.377$]
[*Ans.* 4×10^{-5} atm]

3.7 The second virial coefficient Γ_2 of a sample of polyisobutylene ($\overline{M}_n = 428,000$) in chlorobenzene at 25°C is 94.5 cm^3/g. Calculate the osmotic pressure in g/cm^2 of a 7.0×10^{-6} mol/L solution of this polymer in chlorobenzene at 25°C and compare with the value calculated for an ideal solution. [Density of chlorobenzene at 25°C = 1.11 g/cm^3.]
[*Ans.* 0.23 g cm^{-2} (ideal 0.18 g cm^{-2})]

3.8 What is the "excluded volume effect"? Show the dependence of the expansion factor α on the molecular weight of the polymer. How does the mean square end-to-end distance $\langle r^2\rangle$ vary with the molecular weight for theta and better solvents?

3.9 Osmotic pressure determinations were made at 27°C on a series of solutions of a globular protein in water. The following results were obtained:

Protein concentration, mg/cm^3	Osmotic pressure, mm H$_2$O
1	1.64
2	3.29
3	4.95
4	6.62
5	8.31

(a) Calculate the molecular weight of the protein and its second virial coefficient. (b) Calculate the interaction parameter and the excluded volume and compare your result with the partial molar volume assuming that the partial specific volume is 0.75 cm^3/g. Comment on the result of this comparison. [Density of water at 27°C = 0.996 g/cm^3.]
[*Ans.* (a) 1,56,200 and 2.2×10^{-5} mol cm^3 g^{-2}; (b) 0.499, 1.1 m^3 mol^{-1} (cf. 0.12 m^3 mol^{-1})]

3.10 Polyisobutylene of molecular weight 1.46×10^6 was dissolved in toluene at 65°C and the solution was slowly cooled till it became turbid. Calculate an estimate of the volume fraction of polymer in the separated phase. [Polymer density = 0.92 g/cm^3; molar volume of toluene = 106.9 cm^3/mol.]
[*Ans.* 0.0082]

3.11 Make suitable derivations from the Flory-Huggins theory to show that when phase separation takes place in a polymer solution, the proportion of *x*-mer in the polymer-rich phase increases as *x* increases.

3.12 A series of solutions of polyisobutylene fractions of different molecular weights (M) in ethylbenzene were slowly cooled and the temperature (T_c) at which the stirred solution became turbid on slow cooling was recorded for each concentration. The values of T_c together with polymer molecular weights are given below.

T_c, $^\circ$C	M
24.4	8.2×10^6
23.8	1.8×10^6
17.0	1.8×10^5
10.7	5.4×10^4

Calculate from the data an estimate of the theta temperature θ and the entropic dilution parameter ψ. [Polymer density = 0.92 g/cm^3; molar volume of ethylbenzene = 122 cm^3/mol]
[*Ans.* 26°C; 0.849]

3.13 The critical temperature (T_c) at which the solutions of *cis*-1,4-polybutadiene in *n*-heptane underwent separation into two phases were measured for three carefully fractionated polymer samples having different molecular weights (M). From the data, given below, determine the theta temperature (θ) for the polymer in *n*-heptane.

$M \times 10^{-3}$	T_c, $^\circ$C
295	-14.0
940	-8.5
1550	-6.8

[*Ans.* -1°C]

3.14 Calculate an estimate of the solubility parameter for acetone at 20°C from its heat of vaporization at the boiling point (56°C), given as $(\Delta H_v)_{56^\circ C} = 7231$ cal/mol, and heat capacities in liquid and gaseous states, given by

$$C_p \text{(liquid)} = 17.251 + 44.31 \times 10^{-3} T \quad \text{cal mol}^{-1} \, {}^\circ\text{K}^{-1}$$

$$C_p \text{(gas)} = 5.371 + 49.227 \times 10^{-3} T - 15.182 \times 10^{-6} T^2 \quad \text{cal mol}^{-1} \, {}^\circ\text{K}^{-1}$$

Density of acetone at 20°C $= 0.791$ g/cm^3.
[*Ans.* 9.82 (cal cm^{-3})$^{1/2}$ or 20.1 MPa$^{1/2}$]

3.15 Using the values of molar attraction constants from Table 3.2, calculate the solubility parameter values for (a) polystyrene ($\rho = 1.05$ g/cm^3), (b) polyacrylonitrile ($\rho = 1.18$ g/cm^3), and (c) poly(ethylene terephthalate) ($\rho = 1.38$ g/cm^3).
[*Ans.* (a) 8.9; (b) 12.7; (c) 11.2, all in (cal cm^{-3})$^{1/2}$]

3.16 Derive the following expression

$$\chi = V_1^0 (\delta_1 - \delta_2)^2 / RT$$

which relates the Flory-Huggins interaction parameter χ to the solubility parameters δ_1 and δ_2 of polymer and solvent, respectively. What are its limitations ?

3.17 Calculate an estimate of the interaction parameter χ for Buna-N rubber (butadiene-acrylonitrile copolymer) and *n*-hexane at 25°C from the solubility parameter of the solvent (δ_1) and the polymer (δ_2). [*Data :* $\delta_1 = 14.7 \times 10^3$ (J m^{-3})$^{1/2}$; $\delta_2 = 18.7 \times 10^3$ (J m^{-3})$^{1/2}$; molar volume of *n*-hexane at 25°C = 131.6 cm^3/mol.]
[*Ans.* 1.1]

3.18 Dioctyl sebacate ($\delta = 8.7$) which is used as a plasticizer for poly(vinyl chloride) is to be substituted by a mixture of tritolyl phosphate ($\delta = 9.8$) and aromatic oils ($\delta = 8.0$) on the basis of equal δ value. Calculate the composition of the mixture.
[*Ans.* Volume fractions: tritolyl phosphate 0.39, aromatic oils 0.61]

3.19 Using Hansen parameters (Tables 3.5 and 3.6) determine if poly(methyl methacrylate) (PMMA) is expected to dissolve in a solvent mixture of 1:4 (by volume) chloroform/benzene.

3.20 A nitrile rubber (butadiene-acrylonitrile copolymer, Buna-N) vulcanizate containing 34% acrylonitrile, 56% butadiene, and 2% sulfur (by weight) has a χ-value of 2.52 in n-hexane. Estimate (a) fraction of crosslinked units, (b) degree of crosslinking as represented by $\overline{M}_c$, and (c) maximum extent of swelling in hexane, expressed as volume of hexane imbibed per unit volume of vulcanizate. Assume that all sulfur in the vulcanizate is used in crosslinking and that, on average, each crosslink contains 8 sulfur atoms. [*Data*: Rubber density = 1.00 g/cm^3; molar volume of *n*-hexane at 25°C = 131.6 cm^3/mol.] [*Ans.* (a) 0.0046; (b) 1.16x10^4 g mol^{-1}; (c) 0.038]

3.21 A natural rubber vulcanizate contains 96% polyisoprene and 1.5% (by wt.) sulfur. Calculate an estimate of (a) fraction of crosslinked units and (b) degree of crosslinking represented by $\overline{M}_c$. Assume that all sulfur is used in crosslinking and that, on average, one crosslink is formed for every 8 sulfur atoms. [*Ans.* (a) 0.004; (b) 1.64x10^4 g mol^{-1}]

3.22 Equilibrium swelling of a sample of acetone-extracted smoked natural rubber of number average molecular weight 2,23,000 was measured in several nonsolvents giving the following equilibrium values of polymer concentration (expressed as volume fraction of polymer, ϕ_{2m}) in gel:

Nonsolvents	ϕ_{2m}
Ethylacetate	0.443
Dioxane	0.244
Methyl ethyl ketone	0.498
n-Butyl formate	0.140
Isopentane	0.096

Arrange the above nonsolvents in the order of increasing goodness based on the interaction parameter values at the miscibility limit determined from the swelling data.
[*Ans.* Methyl ethyl ketone (0.771), ethyl acetate (0.724), dioxane (0.600), *n*-butyl formate (0.553), isopentane (0.535)]

3.23 The following data were collected on the swelling of polybutadiene vulcanizates of different sulfur contents in toluene at 25°C :

Sulfur content (wt, %)	Maximum imbibition, Q_m (vol./vol.)
2	5.0
4	3.5
6	2.5
12	1.65

Calculate the number of crosslinks per atom of sulfur in the vulcanizates and comment on the results obtained. Assume $\chi = 0.3$ for the polymer-solvent system and $\rho = 1.0$ g/cm^3 for the vulcanizates. Molar volume of toluene at 25°C, $V_1 = 107$ cm^3/mol.
[*Ans.* 0.12, 0.11, 0.13, 0.13. An average value of number of crosslinks per atom of sulfur is 0.123, which indicates that approximately one crosslink is formed for every eight sulfur atoms on the average.]

3.24 Natural rubber of $\overline{M}_n = 2.23\times10^5$ was vulcanized with 2% sulfur. Calculate an estimate of the extent of swelling of the vulcanizate in benzene at equilibrium, given that the interaction parameter of rubber and benzene is 0.41. Assume that all sulfur is used in crosslinking and that there is, on the average, one crosslink for every eight sulfur atoms. [*Data*: Rubber density = 0.91 g/cm^3; molar volume of benzene = 89.4 cm^3/mol.]
[*Ans.* $\phi_{2m} = 0.134$, i.e., swelling to about 7.5 times the dry volume]

3.25 The network polyurethane obtained by reacting a diisocyanate R(N=C=O)$_2$ with pentaerythritol C(CH$_2$ OH)$_4$ contains, according to an elemental analysis, 0.2% (w/w) nitrogen and has a density of 1.05 g/cm^3. The network polymer swells to 6 times its original volume when equilibrated with chlorobenzene ($\rho = 1.10$ g/cm^3, mol. wt. 112.5) at 27°C. Calculate an estimate of the polymer-solvent interaction parameter.
[*Ans.* 0.435]

3.26 A crosslinked sample of butadiene-acrylonitrile copolymer undergoes 4% swelling in *n*-heptane. How much will it swell in cyclohexane if the values of the interaction parameter of the copolymer for *n*-hexane and cyclohexane are, respectively, 2.46 and 0.94. The molar volumes of *n*-hexane and cyclohexane at 25°C are 147.5 and 108.7 cm³/mol, respectively.
[*Ans.* 49%]

3.27 The intrinsic viscosities of polyisobutylene of molecular weight 5.58×10^5 in cyclohexane at 30°C and in benzene at 24°C (theta temperature) are 2.48 dL/g and 0.799 dL/g, respectively. Calculate (a) unperturbed end-to-end chain length of the polymer, (b) end-to-end chain length of the polymer in cyclohexane at 30°C, and (c) volume expansion factor in cyclohexane at 30°C. Take $\Phi = 2.5 \times 10^{23}$ mol^{-1}.
[*Ans.* (a) 562 Å; (b) 821 Å; (c) 4.0]

3.28 Polyisobutylene fraction of molecular weight 540,000 was used for viscosity measurements in cyclohexane and benzene. The intrinsic viscosity values obtained were 2.48 dL/g in cyclohexane at 30°C and 0.80 dL/g in benzene at the theta temperature (24°C). The observed relation between molecular weight and unperturbed end-to-end distance is given by

$$\langle r^2 \rangle_0^{1/2} = 0.76 \times 10^{-8} M^{1/2} \text{ cm}$$

Evaluate the total thermodynamic interaction $(\psi - \kappa)$ for the polyisobutylene-cyclohex-ane system at 30°C. [*Data*: Polymer density = 0.92 g/cm³; molar volume of cyclo-hexane = 109 cm³/mol at 30°C.]
[*Ans.* 0.129]

3.29 Viscosity measurements were made on solutions of fractionated *cis*-1,4-polybutadiene samples in toluene at 30°C and in *n*-heptane at −1°C (theta temperature), yielding the following values of intrinsic viscosities (in dL/g) :

Fraction	$[\eta]$ in toluene at 30°C	$[\eta]_\theta$ in *n*-heptane at −1°C
1	1.056	0.508
2	1.493	0.672
3	2.267	0.890

The viscosity-molecular weight relationship for the polymer in toluene at 30°C is given by

$$[\eta] = 2.27 \times 10^{-4} M^{0.75}$$

Derive a relationship between the unperturbed root-mean-square end-to-end distance of the polymer molecule and the molecular weight. The value of the universal parameter, Φ, in toluene is 2.5×10^{21} dL mol^{-1} cm^{-3}.
[*Ans.* $\langle r^2 \rangle_0^{1/2} = 0.91 \times 10^{-8} M^{1/2}$ cm]

APPENDIX 3.1

Regular Solutions: Solubility Parameter

A so-called "regular solution" is obtained when the enthalpy change (ΔH_{mix}) is nonideal (i.e., non-zero, either positive or negative) but the entropy change (ΔS_{mix}) is still ideal. So on the molecular level, while an ideal solution is one in which the different types of molecules (A and B, for example) behave exactly as if they are surrounded by molecules of their own kind (that is, all intermolecular interactions are equivalent), a *regular* solution can form only if the random distribution of molecules persists even in the presence of A-B interactions that differ from the purely A-A and B-B interactions of the original components A and B. This concept has proved to be very useful in the development of an understanding of miscibility criteria.

The derivation given below follows that given in Rudin (1982). Consider a regular solution containing N_A molecules of component A and N_B molecules of component B, and assume the following conditions: (a) Each molecule of A has a volume v_A and can make z_A contacts with other molecules, the corresponding values for component B being v_B and z_B, respectively. (b) Each A-A contact contributes an interaction energy ω_{AA}, and the corresponding energies for (B-B) and (A-B) contacts are ω_{BB} and ω_{AB}. (c) Only the first-neighbor contacts need to be taken into consideration and the mixing is random. So if a molecule is selected at random, one may assume that the probability that it makes contact with a molecule of a particular component is proportional to the volume fraction of that component. If this randomly selected molecule were of component A, its energy of interaction with its neighbors would thus be ($z_A\omega_{AA}N_Av_A/V_{mix} + z_A\omega_{AB}N_Bv_B/V_{mix}$), where the total volume of the mixture V_{mix} is equal to ($N_Av_A + N_Bv_B$). The energy of the interaction of N_A molecules of component A with the rest of the system is then obtained by multiplying both the terms in the previous sum by N_A and dividing the first term by 2, i.e., ($z_A\omega_{AA}N_A^2v_A/2V_{mix} + z_A\omega_{AB}N_AN_Bv_B/V_{mix}$). (The division of the first term by 2 is necessitated by the fact that it takes two A molecules to make an A-A contact.) Similarly, the interaction energy of N_B molecules of component B with the rest of the system is $z_B\omega_{BB}N_B^2v_B/2V_{mix} + z_B\omega_{AB}N_AN_Bv_A/V_{mix}$. The total contact energy of the system E_t is given by the sum of the interaction energies of A-A, B-B, and A-B contacts in the system.

Interaction energy of A-A contacts $= z_A\omega_{AA}N_A^2v_A/2V_{mix}$ (A3.1)

Interaction energy of B-B contacts $= z_B\omega_{BB}N_B^2v_B/2V_{mix}$ (A3.2)

Interaction energy of A-B contacts $=$

$$\frac{1}{2}\left[\frac{z_A\omega_{AB}N_AN_Bv_B}{V_{mix}} + \frac{z_B\omega_{AB}N_AN_Bv_A}{V_{mix}}\right] \tag{A3.3}$$

Therefore,

$$E_t = [z_A\omega_{AA}N_A^2v_A + \omega_{AB}N_AN_B(z_Av_B + z_Bv_A) + z_B\omega_{BB}N_B^2v_B]/2V_{mix} \tag{A3.4}$$

After substituting V_{mix} by $N_Av_A + N_Bv_B$, Eq. (A3.4) can be written as

$$E_t = N_A\left(\tfrac{1}{2}z_A\omega_{AA}\right) + N_B\left(\tfrac{1}{2}z_B\omega_{BB}\right) + \frac{N_AN_B}{2(N_Av_A + N_Bv_B)}$$
$$\times[\omega_{AB}(z_Av_B + z_Bv_A) - \omega_{AA}z_Av_B - \omega_{BB}z_Bv_A] \tag{A3.5}$$

In order to eliminate ω_{AB} it may be assumed that

$$\tfrac{1}{2}\omega_{AB}\left(\frac{z_A}{v_A} + \frac{z_B}{v_B}\right) = \left[\frac{z_A\omega_{AA}}{v_A}\cdot\frac{z_B\omega_{BB}}{v_B}\right]^{1/2} \tag{A3.6}$$

In effect, this takes ω_{AB} to be equal to the geometric mean of ω_{AA} and ω_{BB}. The geometric mean rule, based partly on theoretical principles and partly on observation, is expected to hold, however, only in situations where dispersion forces provide the only significant interaction energy.

With the help of Eq. (A3.6), Eq. (A3.5) can be written in the form

$$E_t = N_A \frac{z_A \omega_{AA}}{2} + N_B \frac{z_B \omega_{BB}}{2}$$
$$- \frac{N_A N_B v_A v_B}{N_A v_A + N_B v_B} \left[\left(\frac{z_A \omega_{AA}}{2 v_A} \right)^{1/2} - \left(\frac{z_B \omega_{BB}}{2 v_B} \right)^{1/2} \right]^2 \qquad (A3.7)$$

The first two terms on the right hand side of Eq. (A3.7) represent the interaction energies of individual components A and B, while the last term represents the change in internal energy ΔE_{mix} of the system when the species are mixed. Assuming that the contact energies are independent of temperature, the enthalpy change on mixing (ΔH_{mix}) is thus seen to be

$$\Delta H_{mix} = \Delta E_{mix} = \frac{N_A N_B v_A v_B}{N_A v_A + N_B v_B} \left[\left(\frac{z_A \omega_{AA}}{2 v_A} \right)^{1/2} - \left(\frac{z_B \omega_{BB}}{2 v_B} \right)^{1/2} \right]^2 \qquad (A3.8)$$

Since the terms $(z_A \omega_{AA}/2 v_A)^{1/2}$ and $(z_B \omega_{BB}/2 v_B)^{1/2}$ represent solubility parameters δ_A and δ_B, respectively, Eq. (A3.8) can be recast into the form

$$\Delta H_{mix} = \left(\frac{N_A N_B v_A v_B}{N_A v_A + N_B v_B} \right) (\delta_A - \delta_B)^2$$
$$= \left(\frac{N_A v_A}{V_{mix}} \right) \left(\frac{N_B v_B}{V_{mix}} \right) (\delta_A - \delta_B)^2 V_{mix}$$
$$= \phi_A \phi_B V_{mix} (\delta_A - \delta_B)^2 \qquad (A3.9)$$

where ϕ_A and ϕ_B are volume fractions of components A and B in the mixture.

Chapter 4

Polymer Molecular Weights

4.1 Introduction

If the molecular weight and molecular weight distribution for a polymer are known along with a good understanding of its chain conformation, many properties of the polymer can be predicted. While polymer chain conformations were discussed in Chapter 2, polymer molecular weights are dealt with in the present chapter.

Not only are polymer molecular weights very large, typically ranging from a few thousand to a million or more; unlike conventional chemicals, the molecular weight within any polymer sample is not uniform. Owing to this heterogeneity, the numerical value assigned to the molecular weight of a polymer depends on the way in which the heterogeneity is averaged. Thus, if the molecular weight is computed by dividing the total mass by the total number of molecules, which is equivalent to weighting the molecular weight of each species by its mole fraction, then we obtain the *number-average molecular weight* $\overline{M}_n$. Alternatively, the molecular weight of each species may be weighted by its weight fraction to give the *weight-average molecular weight* $\overline{M}_w$. The relation between $\overline{M}_n$ and $\overline{M}_w$ depends on the form of molecular weight distribution. Only for a homogeneous sample of a polymer are the two averages equal, and otherwise $\overline{M}_w > \overline{M}_n$.

The principal methods for the measurement of the number-average molecular weight $\overline{M}_n$ make use of the well-known properties of dilute solution, such as osmotic pressure, elevation of the boiling point, and depression of the freezing point (i.e., the colligative properties of solutions), since these properties are all proportional to the *number* of dissolved solute molecules. Measurement of the turbidity of a dilute polymer solution, i.e., the intensity of light scattered relative to the intensity of the incident beam, is the standard method for obtaining $\overline{M}_w$. Gel permeation chromatography is essentially a process for the separation of polymer molecules according to their size and affords determination of both $\overline{M}_n$ and $\overline{M}_w$. The most widely used characterization procedure is, however, viscosity because it is the easiest of the various methods and requires no complicated instrument. All these methods are described in this chapter.

4.2 Molecular Weight Averages

4.2.1 Arithmetic Mean

The distribution of molecular weights in a polymer sample is commonly expressed as the proportions of the sample with particular molecular weights. The various molecular weight averages

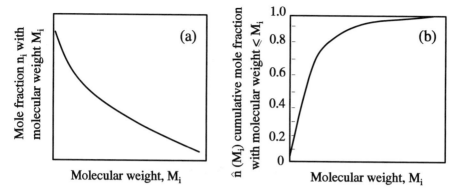

Figure 4.1 (a) A normalized differential number distribution curve. (b) A normalized integral number distribution curve.

used for polymers can be shown to be simply arithmetic means of molecular weight distributions.

Let us assume that unit volume of a polymer sample contains a total of A molecules consisting of a_1 molecules with molecular weight M_1, a_2 molecules with molecular weight M_2, ... a_j molecules with molecular weight M_j. The arithmetic mean molecular weight $\overline{M}$ is then the total measured quantity divided by the total number of molecules:

$$
\begin{aligned}
\overline{M} &= \frac{a_1 M_1 + a_2 M_2 + \cdots + a_j M_j}{a_1 + a_2 + \cdots + a_j} \\
&= \frac{a_1 M_1 + a_2 M_2 + \cdots + a_j M_j}{A} \\
&= \frac{a_1}{A} M_1 + \frac{a_2}{A} M_2 + \cdots + \frac{a_j}{A} M_j
\end{aligned}
\tag{4.1}
$$

The ratio a_i/A represents the proportion of molecules with molecular weight M_i. Denoting this proportion by f_i, the arithmetic mean molecular weight will be given by

$$
\overline{M} = f_1 M_1 + f_2 M_2 + \cdots + f_j M_j = \sum_i f_i M_i
\tag{4.2}
$$

Equation (4.2) gives the arithmetic mean of the distribution of molecular weights. Almost all molecular weight averages can be related to this equation.

4.2.2 Number-Average Molecular Weight

If we substitute the proportion of species f_i, which have molecular weight M_i, by the corresponding mole fraction n_i in Eq. (4.2), we obtain the definition of number-average molecular weight, $\overline{M}_n$, representing the number distribution:

$$
\overline{M} = \sum_i n_i M_i = \overline{M}_n
\tag{4.3}
$$

The mole fraction n_i is also the differential number function, and a plot of n_i versus M_i represents a differential number distribution curve, as shown in Fig. 4.1(a). The distribution being normalized, the scale of the ordinate in this figure goes from 0 to 1, and the area under the curve is unity.

The cumulative number (or mole) fraction is defined as

$$\hat{n}(M_i) = \sum_i n_i \tag{4.4}$$

where n_i is the mole fraction of molecules with molecular weight M_i and $\hat{n}(M_i)$ is the cumulative mole fraction with molecular weight $\leq M_i$. A plot of $\hat{n}(M_i)$ against the corresponding M_i yields an integral number distribution curve, as in Fig. 4.1(b). The units of ordinate are mole fractions and extend from 0 to 1; the distribution is therefore said to be *normalized*.

While Eq. (4.3) gives a simple definition of the number average molecular weight $\overline{M}_n$, we can derive other equivalent definitions following a simple arithmetic. For this let us define the following terms, some of which have already been used above.

n_i : mole fraction of species i (that is, molecules of same size with molecular weight M_i) in a sample
N_i : moles of species i
N : total of all N_i's
w_i : weight fraction of species i
W_i : weight of species i
W : sum of all W_i's

It now follows that

$$n_i = N_i / \sum N_i = N_i/N \tag{4.4a}$$

$$W_i = N_i M_i \tag{4.4b}$$

$$w_i = W_i / \sum W_i = N_i M_i / \sum N_i M_i \tag{4.4c}$$

and

$$\overline{M}_n = \sum n_i M_i = \frac{\sum N_i M_i}{\sum N_i} \tag{4.5}$$

$$= \frac{\sum W_i}{\sum N_i} = \frac{W}{\sum (W_i/M_i)} = \frac{1}{\sum (w_i/M_i)} \tag{4.6}$$

The number-average degree of polymerization, $\overline{DP}_n$, is defined as

$$\overline{DP}_n = \frac{\text{Number-average molecular weight}}{\text{Mer weight}}$$

$$= \frac{\overline{M}_n}{M_0} = \frac{\sum n_i M_i}{M_0} = \sum n_i x_i \tag{4.7}$$

which may also be written as

$$\overline{DP}_n = \sum n_x x \tag{4.7a}$$

where n_x is the number fraction of molecules containing x number of repeating units.

Alternatively, from Eq. (4.6)

$$\overline{DP}_n = \frac{\overline{M}_n}{M_0} = \frac{1}{M_0 \sum (w_i/M_i)} = \frac{1}{\sum (w_i/x_i)} \tag{4.8}$$

which may also be written as

$$\overline{DP}_n = 1 / \sum (w_x/x) \tag{4.8a}$$

where w_x is the weight fraction of molecules containing x number of repeating units.

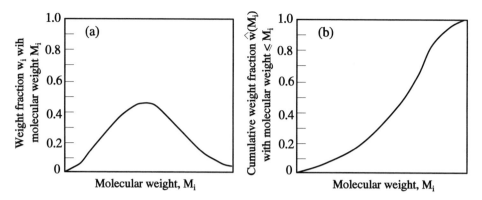

Figure 4.2 (a) A normalized differential weight distribution curve. (b) A normalized integral weight distribution curve.

4.2.3   Weight-Average Molecular Weight

The situation for weight distribution corresponds to that for a number distribution described in the previous section. Thus recording the weight of each species in the sample, instead of the number of molecules of each size, would give a weight distribution. The differential weight fraction is simply the weight fraction w_i, while the integral (cumulative) weight fraction $\hat{w}(M_i)$ is given by

$$\hat{w}(M_i) \;=\; \sum_i w_i \tag{4.9}$$

and is equal to the weight fraction of all species with molecular weight not greater than M_i.

A plot of w_i against M_i yields a normalized differential weight distribution curve, as in Fig. 4.2(a) and that of $\hat{w}(M_i)$ against M_i yields a normalized integral distribution curve, as in Fig. 4.2(b). The scale of the ordinate in both these figures goes from 0 to 1. Substituting w_i for f_i in Eq. (4.2) produces the following expression for the arithmetic mean of the weight distribution :

$$\overline{M} \;=\; \sum_i w_i M_i \;=\; \overline{M}_w \tag{4.10}$$

$\overline{M}_w$ is the weight-average molecular weight. Combining Eqs. (4.10) and (4.4c) it can also be expressed as

$$\overline{M}_w \;=\; \frac{\sum N_i M_i^2}{\sum N_i M_i} \tag{4.11}$$

The weight-average degree of polymerization, $\overline{DP}_w$, is defined as

$$\overline{DP}_w \;=\; \frac{\text{Weight-average molecular weight}}{\text{Mer weight}}$$

$$=\; \frac{\overline{M}_w}{M_0} \;=\; \frac{\sum w_i M_i}{M_0} = \sum w_i x_i \tag{4.12}$$

which may also be written as

$$\overline{DP}_w = \sum w_x x \qquad (4.12a)$$

where w_x is the weight fraction of molecules containing x number of repeating units.

If all species in a polymer sample have the same molecular weight (that is, the polymer is *monodisperse*) then $\overline{M}_n = \overline{M}_w = \overline{M}_z$ ($\overline{M}_z$ denotes the z-average molecular weight). Such monodispersity is, however, not found in synthetic polymers and it is always true that $\overline{M}_z > \overline{M}_w > \overline{M}_n$. The ratio $\overline{M}_w/\overline{M}_n$, called the *polydispersity index* (PDI), is commonly used as a simple measure of the *polydispersity* of the polymer sample, though it is not a sound statistical measure of the distribution breadth. The breadth and shape of the distribution curve are characterized more appropriately with parameters derived from the moments of distribution.

4.3 Molecular Weights in Terms of Moments

In the previous section we have seen that average molecular weights are arithmetic means of distributions of molecular weights. An alternative and generally more useful definition of average molecular weights is, however, obtained in terms of moments of distribution (Rudin, 1982).

A moment in mechanics is generally defined as $U_j^a = Q d^j$, where U_j is the jth moment, about a specified line or plane a of a vector or scalar quantity Q (e.g., force, weight, mass, area), d is the distance from Q to the reference line or plane, and j is a number indicating the power to which d is raised. [For example, the first moment of a force or weight about an axis is defined as the product of the force and the distance of the line of action of the force from the axis. It is commonly known as the *torque*. The second moment of the force about the same axis (i.e., $j = 2$) is the *moment of inertia*.] If Q has elements Q_i, each located a distance d_i from the same reference, the moment is given by the sum of the individual moments of the elements:

$$U_j^a = \sum_i Q_i d_i^j \qquad (4.13)$$

In applying Eq. (4.13), the ordinate at $M = 0$ in the graph of the molecular weight distribution [Figs. 4.1(a) and 4.2(a)] is commonly chosen as the reference, though it is usually not mentioned explicitly. The distance d_i from the reference line is measured along the abscissa in terms of the molecular weight M_i, while the quantity F_i is replaced by the quantity or proportion of polymer with molecular weight M_i, and j assumes a wider range of values (0, 1, 2, 3, ...) than in mechanics. A general definition of a statistical moment of a molecular weight distribution taken about zero is then

$$U_j' = \sum q_i M_i^j \qquad (4.14)$$

where q_i is the quantity of polymer in the sample with molecular weight M_i. The prime superscript (′) used on the symbol indicates that the moment is taken about the $M = 0$ axis. It requires to define q_i appropriately and assign numerical values to j to obtain equations for different moments of distribution. To distinguish between number distribution and weight distribution we shall henceforth use the notation $_nU$ to refer to a moment of the number distribution and $_wU$ to denote a moment of the weight distribution.

Problem 4.1 Write general equations for statistical moments for (a) number distribution and (b) weight distribution of molecular weights.

Answer:

(a) Number distribution

Unnormalized: $_nU_j' = \sum N_i M_i^j$ (P4.1.1)

Normalized: $_nU_j' = \sum n_i M_i^j$ (P4.1.2)

(b) Weight distribution

Unnormalized: $_wU_j' = \sum W_i M_i^j$ (P4.1.3)

Normalized: $_wU_j' = \sum w_i M_i^j$ (P4.1.4)

Weight distributions are commonly encountered during analysis of polymer samples, while number distributions are more useful in consideration of polymerization kinetics.

4.3.1 Ratio of First and Zeroth Moments

In general terms, the ratio of the first moment to the zeroth moment of any distribution gives the arithmetic mean. For an unnormalized number distribution, the zeroth ($j = 0$) and first ($j = 1$) moments of the distribution about zero are given, respectively, by [cf. Eq. (P4.1.1)]:

$$_nU_0' = \sum_i N_i (M_i)^0 = \sum N_i = N$$ (4.15)

$$_nU_1' = \sum_i N_i (M_i)^1 = \sum N_i M_i$$ (4.16)

The ratio of these moments is the arithmetic mean of the number distribution and by comparing with Eq. (4.5) we can write

$$\overline{M} = \frac{_nU_1'}{_nU_0'} = \sum N_i M_i / \sum N_i = \overline{M}_n$$ (4.17)

The arithmetic mean of an unnormalized weight distribution is likewise given by [cf. Eq. (4.10)]:

$$\overline{M} = \frac{_wU_1'}{_wU_0'} = \frac{\sum W_i (M_i)^1}{\sum W_i (M_i)^0} = \frac{\sum W_i M_i}{\sum W_i} = \sum w_i M_i = \overline{M}_w$$ (4.18)

In the above two examples, we have chosen unnormalized distributions. For normalized distributions, the area under the curve for the differential number distribution [Fig. 4.1(a)] or weight distribution [Fig. 4.2(a)] equals unity. That is,

$$_nU_0' = \sum n_i (M_i)^0 = \sum n_i = 1$$ (4.19)

$$_wU_0' = \sum w_i (M_i)^0 = \sum w_i = 1$$ (4.20)

It is now seen from Eqs. (4.17) and (4.18) that the arithmetic mean is numerically equal to the first moment of the normalized distribution.

4.3.2 Ratios of Higher Moments

We may define an average, in general, as the ratio of successive moments of the distribution. We have seen above that $\overline{M}_n$, the number average molecular weight, is equal to the ratio of the first to the zeroth moment of the number distribution. In the same way, the ratios of successively higher moments of the number distribution give other average molecular weights (Rudin, 1982):

$$\frac{_nU_2'}{_nU_1'} = \frac{\sum N_i M_i^2}{\sum N_i M_i} = \overline{M}_w \tag{4.21}$$

$$\frac{_nU_3'}{_nU_2'} = \frac{\sum N_i M_i^3}{\sum N_i M_i^2} = \overline{M}_z \tag{4.22}$$

$$\frac{_nU_4'}{_nU_3'} = \frac{\sum N_i M_i^4}{\sum N_i M_i^3} = \overline{M}_{z+1} \tag{4.23}$$

This process can be continued to obtain other higher averages. The averages commonly used in practice are, however, limited to $\overline{M}_n$, $\overline{M}_w$, $\overline{M}_z$, and the viscosity average molecular weight $\overline{M}_v$. While $\overline{M}_n$, $\overline{M}_w$, and $\overline{M}_v$ may be obtained by direct measurements (see later), it is usually necessary to measure the detailed distribution of molecular weights to estimate $\overline{M}_z$ and other averages. $\overline{M}_v$ is given by

$$\overline{M}_v = \left[\frac{\sum N_i M_i^{a+1}}{\sum N_i M_i} \right]^{1/a} \tag{4.24}$$

where a is a constant. A derivation of Eq. (4.24) is given in a later section. Combination of Eq. (4.24) with Eq. (4.4c) gives

$$\overline{M}_v = \left[\frac{\sum W_i M_i^a}{\sum W_i} \right]^{1/a} = \left[\sum w_i M_i^a \right]^{1/a} \tag{4.25}$$

In terms of moments, $\overline{M}_v$ is given by

$$\overline{M}_v = \left[\frac{_nU_{a+1}'}{_nU_1'} \right]^{1/a} \tag{4.26}$$

and

$$\overline{M}_v = \left[_wU_a' \right]^{1/a} \tag{4.27}$$

corresponding to number distribution and weight distribution, respectively.

Problem 4.2   A sample of poly(vinyl chloride) is composed according to the following fractional distribution:

Weight fraction, w_i	0.04	0.23	0.31	0.25	0.13	0.04
Mean mol. wt., $M_i \times 10^{-3}$	7	11	16	23	31	39

(a) Compute $\overline{M}_n$, $\overline{M}_w$, and $\overline{M}_z$.
(b) How many molecules per gram are there in the polymer?

Answer:

The w_i vs. M_i data are used to make the following table :

w_i	M_i	$w_i M_i$	w_i / M_i	$w_i M_i^2$
0.04	7,000	280	0.57×10^{-5}	1.96×10^6
0.23	11,000	2,530	2.09×10^{-5}	2.78×10^7
0.31	16,000	4,960	1.94×10^{-5}	7.94×10^7
0.25	23,000	5,750	1.09×10^{-5}	1.32×10^8
0.13	31,000	4,030	0.42×10^{-5}	1.25×10^8
0.04	39,000	1,560	0.10×10^{-5}	6.08×10^7
Σ		19,110	6.21×10^{-5}	4.27×10^8

(a) From Eqs. (4.6), (4.10), and (4.22) :

$$\overline{M}_n = \frac{1}{\Sigma(w_i/M_i)} = \frac{1}{6.21 \times 10^{-5}} = 16,100$$

$$\overline{M}_w = \sum w_i M_i = 19,110$$

$$\overline{M}_z = \frac{\sum N_i M_i^3}{\sum N_i M_i^2} = \frac{\sum w_i M_i^2}{\sum w_i M_i} = \frac{4.27 \times 10^8}{19,110} = 22,344$$

(b) Number of molecules per gram $= (\sum w_i/M_i) \times (\text{Avogadro's number}) = (6.21 \times 10^{-5}) \times (6.02 \times 10^{23}) = 3.74 \times 10^{19}$ molecules/g.

4.4 Molecular Weight Determination

The different methods of molecular weight determination can be divided into two categories: absolute methods and secondary methods. Absolute methods give values that provide a direct estimate of the molecular weight. Secondary methods, on the other hand, yield comparisons between the molecular weights of different polymers and must be calibrated with a reference molecular weight that has been studied by one of the absolute methods. Measurements of colligative properties, light scattering, and sedimentation under ultracentrifugation of polymer solutions are several methods for the determination of absolute molecular weights. Ultracentrifugation experiments yield $\overline{M}_z$ values. They are used primarily for biological polymers. Among the absolute methods, only the osmotic method and the light scattering method are given primary consideration in this chapter. Other absolute methods are only briefly discussed.

The secondary methods, such as solution viscosity and gel permeation chromatography, require prior establishment of empirical relationships that relate the molecular weight to the viscosity of the polymer solution or to the retention times in a gel-permeation column. Once such calibration has been done, the secondary methods provide a fast, simple, and accurate way to obtain molecular weights. The solution viscosity and gel permeation chromatography methods are described in a later part of this chapter.

4.4.1 End-Group Analysis

End-group analysis can be used to determine $\overline{M}_n$ of polymer samples if the substance contains detectable end groups, and the number of such end groups per molecule is known beforehand.

End-group analysis has been applied mainly to condensation polymers, since these polymers by their very nature have reactive functional end groups. The end groups are often acidic or basic in nature, as exemplified by the carboxylic groups of polyesters or the amine groups of polyamides; such groups are conveniently estimated by titration. From the experimental data $\overline{M}_n$ is derived according to

$$\overline{M}_n = \frac{f.w.e}{a} \tag{4.28}$$

where f is the functionality or number of reactive groups per molecule in the polymer sample, w is the weight of the polymer, a is the amount of reagent used in the titration, and e is the equivalent weight of the reagent.

Problem 4.3 A sample (3.0 g) of carboxyl terminated polybutadiene (CTPB) required titration with 20 mL 0.1 N KOH to reach a phenolphthalein end point. Calculate $\overline{M}_n$ of the polymer.

Answer:

Here $f = 2$ eq mol^{-1}, $e = 56$ g eq^{-1}, $w = 3.0$ g

$$a = \frac{(20\,\text{mL})(0.1\,\text{eq}\,\text{L}^{-1})(56\,\text{g}\,\text{eq}^{-1})}{(1000\,\text{mL}\,\text{L}^{-1})} = 0.112\,\text{g}$$

From Eq. (4.28), $\overline{M}_n = \dfrac{(2\text{ eq mol}^{-1})(3.0\text{ g})(56\text{ g eq}^{-1})}{(0.112\text{ g})} = 3{,}000 \text{ g mol}^{-1}$

End-group analysis yields the equivalent weight of the polymer, M_e, which is the mass of the polymer per mole of end groups and related to $\overline{M}_n$ by

$$M_e = \overline{M}_n/f \tag{4.29}$$

The functionality f of low-molecular-weight functionalized prepolymers (e.g., polyether polyols used in the preparation of polyurethanes) is often determined by combining the functional group analysis (which yields the equivalent weight of the polymer, M_e) with another suitable method of molecular weight ($\overline{M}_n$) determination, such as vapor phase osmometry. Note that in this respect the functional groups do not have to be end groups.

Problem 4.4 A sample (2.0 g) of polyether polyol prepolymer ($\overline{M}_n = 2048$) dissolved in chlorohydrocarbon solvent was treated with excess succinic anhydride to convert each hydroxyl group in the polyol to a carboxyl group by formation of succinic half-ester. A sample (1.0 g) of this treated polymer recovered from the solution by precipitation (in excess of ethanol) required 12.8 mL of N/10 KOH for carboxyl titration. Determine the hydroxyl functionality of the polyol.

Answer:

One molecule of succinic anhydride is added to each hydroxyl group,

$$\text{—OH} + \begin{array}{c} \text{CH}_2\text{CO} \\ | \quad\quad \searrow \\ \text{O} \\ | \quad\quad \nearrow \\ \text{CH}_2\text{CO} \end{array} \longrightarrow \text{—OCOCH}_2\text{CH}_2\text{COOH}$$

Carboxyl end group in 1 g of succinic anhydride-reacted polymer (SAP) = $(12.8 \text{ mL})(0.1\times10^{-3} \text{ mol mL}^{-1})$ = 1.28×10^{-3} mol

Equivalent weight (mass per mol of carboxyl group) of SAP = $(1 \text{ g})/(1.28\times10^{-3} \text{ mol})$ = 781 g mol^{-1}. Since the molar mass of succinic anhydride = 100 g mol^{-1}, equivalent weight (mass per mol of hydroxyl group) of polyol, M_e = (781 − 100) or 681 g mol^{-1}.

Now from Eq. (4.29),

$$f = \frac{\overline{M_n}}{M_e} = \frac{2048 \text{ g mol}^{-1}}{681 \text{ g mol}^{-1}} \simeq 3.$$

The end-group analysis method for $\overline{M_n}$ determination cannot be used in many cases of practical interest because f in Eq. (4.28) is not known. This is particularly true for branched or crosslinked polymers with variable number of end groups per molecule. There is also the problem of selecting a suitable solvent to dissolve the polymer.

4.4.2 Colligative Property Measurement

4.4.2.1 Ebulliometry (Boiling Point Elevation)

In applying this method, the boiling point of a solution of known concentration is compared to that of the solvent at the same pressure. For ideally dilute solutions, the elevation of the boiling point, $T - T_b$, is related to the normal boiling temperature of the solvent T_b, its molar latent heat of evaporation L_e, and molecular weight M_1, and also to the molecular weight of the solute M_2, and relative weights of solvent and solute W_1 and W_2, respectively, by

$$\Delta T_b = T - T_b = \frac{RT_b^2}{L_e} \cdot \frac{W_2}{W_1} \cdot \frac{M_1}{M_2} \tag{4.30}$$

For convenience, Eq. (4.30) is rewritten as

$$\Delta T_b = \frac{RT_b^2 M_1}{1000 L_e} \cdot \frac{1000 W_2}{W_1 M_2} = k_e m_2 \tag{4.31}$$

where $k_e = (RT_b^2 M_1)/(1000 L_e)$ is the *molal boiling-point elevation constant* of the solvent and m_2 is the solute molality (in units of moles per kilogram), given by $m_2 = (1000 W_2)/(W_1 M_2)$. For benzene, for example,

$$k_e = \frac{(1.987 \text{ cal mol}^{-1}\,^\circ\text{K}^{-1})(353\,^\circ\text{K})^2(78 \text{ g mol}^{-1})}{(1000 \text{ g kg}^{-1})(7,497 \text{ cal mol}^{-1})} = 2.6\,^\circ\text{K kg mol}^{-1}$$

Some other values of k_e calculated in this way are: water 0.51, acetic acid 3.0, benzene 2.5, acetone 1.7, methyl alcohol 0.8, chloroform 3.8, carbon tetrachloride 5.0.

To determine a molecular weight, one measures ΔT_b for a *dilute* solution of solute in solvent and calculates m_2 from Eq. (4.31). The molecular weight M_2 of the solute then equals the value of $(1000 W_2)/(W_1 m_2)$. Ebulliometry, like end-group analysis, is limited to low-molecular-weight polymers. By use of thermistors sensitive to $0.0001°C$, it is possible to measure molecular weight values up to 50,000, although typical limits (using Beckman thermometer with temperature difference measurement accuracy of $0.001°C$) are 5,000.

4.4.2.2 Cryoscopy (Freezing Point Depression)

Calculation of the freezing-point depression of the solvent and hence the molecular weight of the solute by this method proceeds exactly the same way as for the boiling-point elevation. For cryoscopy of ideal solutions, equations corresponding to those for ΔT_b and k_e are $\Delta T_f = -k_f m_2$ and $k_f = (RT_f^2 M_1)/(1000 L_f)$, where $\Delta T_f \equiv T - T_f$ is the *freezing-point depression*, T_f is the freezing point of pure solvent, and L_f is the molar latent heat of fusion. The solvent's *molal freezing-point depression constant* k_f is calculated in the same way as k_e is calculated in ebulliometry. For water, for example,

$$k_f = \frac{(1.987\,\mathrm{cal\,mol^{-1}\,°K^{-1}})(273\,°K)^2(18.01\,\mathrm{g\,mol^{-1}})}{(1000\,\mathrm{g\,kg^{-1}})(1,436\,\mathrm{cal\,mol^{-1}})} = 1.86\,°K\,\mathrm{kg\,mol^{-1}}$$

Some k_f values so obtained are: water 1.8, acetic acid 3.8, benzene 5.1, succinonitrile 20.3, camphor 40. The large k_f of camphor makes it especially useful in molecular weight determinations. Like ebulliometry, the cryoscopic method is also limited to relatively low-molecular-weight polymers with $\overline{M}_n$ up to 50,000.

4.4.2.3 Membrane Osmometry

Osmotic pressure is the most important among all colligative properties for the determination of molecular weights of synthetic polymers (see Table 4.1 for comparison). To explain osmotic pressure, let us imagine a box (Fig. 4.3) divided into two chambers by a semipermeable membrane that allows the solvent to pass through it but not the solute. Suppose that the right chamber is filled with pure solvent A and the left chamber with a solution of B in A such that initially the heights of the liquids in the two capillary tubes are equal. The chambers are thus initially at equal pressures, that is, $P_R = P_L$, where the subscripts stand for right and left. Let us assume that thermal equilibrium is always maintained, that is, $T_R = T_L = T$.

The chemical potential of the solvent A on the right $\mu_{A,R}$ is μ_A^0, representing chemical potential of pure solvent A. If the solution on the left is dilute enough to be considered ideally dilute, then $\mu_{A,L} = \mu_A^0 + RT \ln x_A$, which is less than $\mu_{A,R} = \mu_A^0$, since x_A being mole fraction of A in solution is less than 1. Since $\mu_{A,R} > \mu_{A,L}$, solvent A will flow through the membrane from right to left and the liquid height in the left tube will rise, thereby increasing the pressure in the left chamber. For an ideally dilute solution, the partial molar volume $\overline{V}_A$ for the solvent is the same as for the pure

Table 4.1 A Comparison of Colligative Properties of a 1% (w/v) Solution of Polystyrene of Molecular Weight 20,000 in Benzene

Property	Value
Vapor pressure lowering	0.004 mm Hg
Boiling point elevation	0.0013°C
Freezing point depression	0.0025°C
Osmotic pressure	15 cm solvent

Source: Billmeyer, Jr., 1984

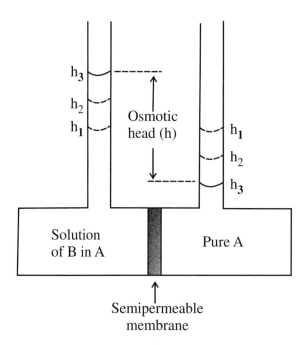

Figure 4.3 Schematic diagram showing the development of osmotic head as a function of time, where h_1 represents the initial liquid levels, h_2 the levels after some time, and h_3 the levels when equilibrium is attained.

solvent, i.e., $\overline{V}_A^0 = V_A^0$, and since, in general, $(\partial \mu_i / \partial P) = \overline{V}_i$, one can write $(\partial \mu_A / \partial P)_T = \overline{V}_A = V_A^0$. Since V_A^0 is positive, the increase in pressure will increase $\mu_{A,L}$ until finally equilibrium is reached, that is, $\mu_{A,L} = \mu_{A,R} = \mu_A^0$. (Note that the membrane being impermeable to solute B, there is no equilibrium relation for μ_B.)

Let the equilibrium pressures in the right and left chambers be P and $P + \Pi$, respectively. The difference in pressures, viz., Π, is the *osmotic pressure*. This extra pressure makes the chemical potential of the solvent (μ_A) in solution equal to that (μ_A^0) in pure solvent. If the solution in the left chamber is dilute enough to be considered as ideally dilute, then at equilibrium $\mu_{A,R} = \mu_{A,L}$, or

$$\mu_A^0(P,T) = \mu_A^0(P+\Pi, T) + RT \ln x_A \tag{4.32}$$

where the right side of the equation follows from the general relation: $\mu_i = \mu_i^0 + RT \ln x_i$. Since $(\partial \mu_i / \partial T)_P = -\overline{S}_i$ and $(\partial \mu_i / \partial P)_T = \overline{V}_i$, it follows that $d\mu_A^0 = -S_A^0 dT + V_A^0 dP$. Thus, at constant T, $d\mu_A^0 = V_A^0 dP$. Integration from P to $P + \Pi$ then gives

$$\mu_A^0(P+\Pi, T) - \mu_A^0(P,T) = \int_P^{P+\Pi} V_A^0 dP \tag{4.33}$$

Substitution of Eq. (4.33) into Eq. (4.32) gives

$$RT \ln x_A = -\int_P^{P+\Pi} V_A^0 dP \tag{4.34}$$

Considering that liquids are rather incompressible, V_A^0 would hardly vary with pressure and we can take V_A^0 as practically constant. The integral in Eq. (4.34) then becomes $V_A^0 \Pi$ and Eq. (4.34) thus gives $\Pi = -(RT/V_A^0)\ln x_A$. With $x_A = 1 - x_B$, we have $\ln x_A = -x_B - x_B^2/2 - \cdots \simeq -x_B$, where, since $x_B \ll 1$, we can neglect x_B^2 and higher powers. Therefore,

$$\Pi = (RT/V_A^0)x_B \qquad (4.35)$$

Since the solution is quite dilute, we have $x_B = (n_B)/(n_A + n_B) \simeq n_B/n_A$ and therefore,

$$\Pi = \frac{RT}{V_A^0}\frac{n_B}{n_A} \qquad (4.36)$$

where n_A and n_B are the number of moles of solvent and solute in the solution that is in equilibrium with pure solvent A across the membrane.

Problem 4.5   A solution containing 1.018 g of a protein per 100 g of water is found to have an osmotic pressure of 10.5 torr at 25°C. Estimate the molecular weight of the protein.

Answer:

The osmotic pressure (in atm) is: $\Pi = \dfrac{10.5 \text{ torr}}{760 \text{ torr/atm}} = 0.0138 \text{ atm.}$

For water: density at 25°C and 1 atm $= 0.997$ g cm^{-3}; molar mass $= 18.0$ g mol^{-1}. So, $V_A^0 = 18.054$ cm^3 mol^{-1} and $n_A = (100 \text{ g})/(18.0 \text{ g mol}^{-1}) = 5.555$ mol.

From Eq. (4.36):

$$
\begin{aligned}
n_B &= \frac{\Pi V_A^0 n_A}{RT} \\
&= \frac{(0.0138 \text{ atm})(18.054 \text{ cm}^3 \text{ mol}^{-1})(5.555 \text{ mol})}{(82.06 \text{ cm}^3 \text{ atm mol}^{-1}\,{}^{\circ}\text{K}^{-1})(298\,{}^{\circ}\text{K})} = 5.659 \times 10^{-5} \text{ mol}
\end{aligned}
$$

Hence, $M = (1.018 \text{ g})/(5.659 \times 10^{-5} \text{ mol}) = 18,000$ g mol^{-1}

Since the solution is quite dilute, the solution volume V is approximately equal to that of the solvent $n_A V_A^0$ and Eq. (4.36) becomes

$$\Pi V = n_B RT \qquad (4.37)$$

$$\text{or} \quad \Pi = C_B RT \qquad (4.38)$$

where the concentration C_B in moles/volume equals n_B/V. Both Eq. (4.37), which has formal resemblance to the equation of state for an ideal gas, $PV = nRT$, where n is the number of moles of gas in volume V, and the equivalent Eq. (4.38) are called the *van't Hoff law*. It is valid in the limit of infinite dilution, where the solution behaves ideally. The osmotic pressure in a nonideally dilute two-component solution is, however, given by

$$\Pi = RT\,(M_B^{-1}c_B + A_2 c_B^2 + A_3 c_B^3 + \cdots) \qquad (4.39)$$

which has formal resemblance to the virial equation for gases. In Eq. (4.39), M_B is the solute molecular weight and c_B is the solute mass concentration defined as $c_B = m_B/V$, where m_B is the mass of solute B in the solution of volume V. The quantities A_2, A_3, ... are related to the solute-solvent interaction and are functions of temperature. In the limit of infinite dilution, c_B approaches zero, and Eq. (4.39) approximates to

$$\Pi = RTc_B/M_B = RTm_B/M_BV = RTn_B/V = C_BRT \tag{4.40}$$

which is the van't Hoff law [Eq. (4.38)].

Problem 4.6 Find the osmotic pressure of a solution of 1.0 g glucose ($C_6H_{12}O_6$) in 1000 cm^3 of water at 1 atm and 25°C.

Answer:

Molar mass of glucose = 180.16 g mol^{-1}; $n_B = (1.0 \text{ g})/(180.16 \text{ g mol}^{-1}) = 0.00555$ mol. Substitution in Eq. (4.37) gives

$$\Pi = \frac{(0.00555 \text{ mol})(82.06 \text{ cm}^3 \text{ atm mol}^{-1} {}^{\circ}K^{-1})(298 {}^{\circ}K)}{(1000 \text{ cm}^3)}$$
$$= 0.135 \text{ atm} = 102.6 \text{ torr}$$

Note the large value of Π even for the very dilute (0.0055 mol/L) glucose solution. Since the density of water is 1/13.6 times that of mercury, an osmotic pressure of 102 torr (102 mm Hg) corresponds to a height of 10.2 cm × 13.6 = 139 cm = 4.5 ft of liquid in the left-hand tube in Fig. 4.3. This large value of Π is required to increase the chemical potential of A in the solution to that of pure A, since the chemical potential of a component of a condensed phase is rather insensitive to pressure.

Though the osmotic pressure, according to Eq. (4.40), is inversely proportional to solute molecular weight, the relatively large, measurable values of Π obtained even for dilute solutions (see Table 4.1) make osmotic pressure measurements valuable in determining molecular weights of substances with high molecular weights like polymers.

Virial Equations The osmotic data of a real solution are expressed, according to Eq. (4.39) and omitting the subscript for solute, as

$$\frac{\Pi}{c} = RT\left[\frac{1}{M} + A_2 c + A_3 c^2 + \cdots \right] \tag{4.41}$$

where A_2 and A_3 are called the second and third virial coefficients. Two alternative forms of Eq. (4.41) are

$$\frac{\Pi}{c} = \frac{RT}{M}\left[1 + \Gamma_2 c + \Gamma_3 c^2 + \cdots \right] \tag{4.42}$$

and

$$\frac{\Pi}{c} = \frac{RT}{M} + Bc + Cc^2 + \cdots \tag{4.43}$$

Obviously, the three forms of the virial equation are equivalent (Rudin, 1982), if

$$B = RTA_2 = (RT/M)\Gamma_2 \tag{4.44}$$

The virial coefficients are often reported in the literature without specifying the equation to which they apply, and this can usually be deduced by inspecting their units.

At low solution concentrations, the c^2 terms in any of the above virial equations, Eqs. (4.41) to (4.43), will be very small, and the data of Π/c versus c will be expected to be linear with intercepts at $c = 0$ yielding values of M^{-1} and slopes giving the second virial coefficient of the polymer solution.

The value of the third virial coefficient Γ_3 in Eq. (4.42) is often assumed to be equal to $(\Gamma_2/2)^2$, so that Eq. (4.42) can be rewritten as

$$\left(\frac{\Pi}{c}\right)^{1/2} = \left(\frac{RT}{M}\right)^{1/2} (1 + \tfrac{1}{2}\Gamma_2 c) \qquad (4.45)$$

This form is sometimes convenient to use for extrapolation of Π/c data to zero concentration because, for solutions in good solvents, plots given by Eqs. (4.41)-(4.43) are not linear.

Problem 4.7   Show that the molecular weight determined from osmotic pressure measurements is the number average molecular weight.

Answer:

Consider a whole polymer to be made up of a series of monodisperse macromolecules i with concentration (weight/volume) c_i and molecular weight M_i. For each monodisperse species i one can write from Eq. (4.41) for very dilute solutions: $\Pi_i = RTc_i/M_i$. For the polydisperse polymer, $\Pi = \sum \Pi_i = RT \sum (c_i/M_i)$ and $c = \sum c_i = \sum (w_i/V)$, where w_i is the weight fraction of species i in volume V. For the osmotic molecular weight one can then write:

$$\begin{aligned}
\overline{M} &= \frac{RTc}{\Pi} \\
&= \frac{RT \sum c_i}{RT \sum (c_i/M_i)} = \frac{1}{\sum(w_i/M_i)} = \overline{M}_n \quad \text{[From Eq. (4.6)]}
\end{aligned}$$

The nonideality of polymer solutions is incorporated in the virial coefficients. Predicting nonideality of polymer solutions means, in reality, predictions of the second virial coefficient. Better solvents generally produce greater swelling of macromolecules and result in higher virial coefficients. Most polymers become more soluble in their solvents as the temperature is increased which is reflected in an increase of the virial coefficient. Conversely, the second virial coefficient reduces as the temperature is reduced and at a sufficiently low temperature it may actually be zero. This is the *Flory theta temperature*, which has been defined in Chapter 3 as that temperature at which a given polymer of infinitely high molecular weight would be insoluble at great dilution in a given solvent.

Practical Aspects of Osmometry   In static osmometers, the heights of liquid in capillary tubes attached to the solvent and solution compartments (Fig. 4.3) are measured. At equilibrium, the hydrostatic pressure corresponding to the difference in liquid heights is the osmotic pressure. The main disadvantage of this static procedure is the length of time required for attainment of equilibrium.

It may be noted that the osmotic pressure is essentially the extra pressure that must be applied to the solution to maintain equilibrium when solution and pure solvent are separated by a semipermeable membrane. This extra pressure can be measured by attaching a counter pressure device to the solution tube (Fig. 4.3).

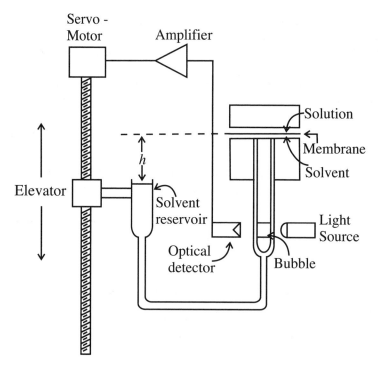

Figure 4.4 Schematic diagram of essential components of a high-speed membrane osmometer. (Hewlett-Packard Corp., Avondale, Pa.)

This method of determining the osmotic pressure is conveniently referred to as the 'dynamic equilibrium' technique. It is especially useful when rapid determinations of osmotic pressure are required. Dynamic osmometers reach equilibrium pressures in 10 to 30 minutes, as compared to hours in the static method, and indicate osmotic pressure automatically. There are several types. Some models employ sensors to measure solvent flow through the membrane and adjust a counteracting pressure to maintain a zero net flow. In a commercially available high-speed membrane osmometer, schematically shown in Fig. 4.4, the movement of an air bubble inside the capillary immediately below the solvent cell is used to indicate this solvent flow. Such movement is immediately detected by a photocell, which in turn is coupled to a servomechanism that controls the flow.

The data obtained by osmotic pressure measurements are pressures (*osmotic heads*) in terms of heights (h) of solvent columns at different concentrations (c) of the polymer solution. In applying the data, h/c is plotted against c and extrapolated to $c = 0$, yielding the value of $(h/c)_0$. The column height h is then converted to osmotic pressure Π by $\Pi = h\rho g$, where ρ is the density of the solvent and g is the gravitational acceleration constant, and $\overline{M}_n$ is calculated from Eqs. (4.41)-(4.43), which in the limit of $c \longrightarrow 0$ reduce to

$$\left(\frac{\Pi}{c}\right)_0 = \frac{RT}{\overline{M}_n} \tag{4.46}$$

The second virial coefficient can be obtained from the slope of the straight line portion of the (Π/c) versus c plot by removing the c^2 terms in Eqs. (4.41)-(4.43). When plotted according to

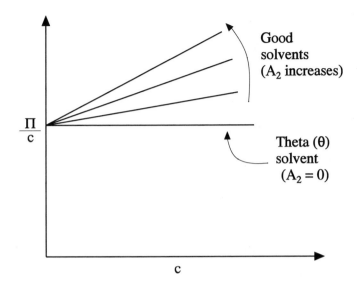

Figure 4.5 Schematic illustration of the effect of solvating power of solvent on reduced osmotic pressure (Π/c) versus concentration (c) plots for the same polymer. (After Rudin, 1982.)

Eq. (4.41), the osmotic pressures of solutions of the same polymer in different solvents should yield plots with the same intercept (at $c = 0$) but with different slopes (see Fig. 4.5), since the second virial coefficient, which reflects polymer-solvent interactions, will be different in solvents of differing solvent power. For example, the second virial coefficient can be related to the Flory-Huggins interaction parameter χ (see p. 110) by

$$A_2 = (\tfrac{1}{2} - \chi)v_2^2/V_1^{\,0} \tag{4.47}$$

where v_2 is the specific volume of the polymer, $V_1^{\,0}$ is the molar volume of the solvent, and χ is an interaction energy per mol of the solvent divided by RT. For a theta solvent, $\chi = 0.5$ and $A_2 = 0$. Better solvents have lower χ values and higher second virial coefficients. Equation (4.47) can be used to determine χ from osmotic measurements.

Problem 4.8 At 20°C, the osmotic pressure of a polycarbonate was measured in chloro-benzene solution with the following results:

Concentration (g/L)	1.95	2.93	3.91	5.86
Osmotic pressure (cm chlorobenzene)	0.20	0.36	0.53	0.98

[Solvent density = 1.10 g/cm³; polymer density = 1.20 g/cm³.]

Estimate: (a) polymer molecular weight, $\overline{M}_n$, (b) second virial coefficients A_2 and Γ_2, and (c) polymer-solvent interaction parameter χ.

Answer:

(a) From the osmotic data we obtain the following table :

$c\left(\dfrac{g}{1000\ cm^3}\right)$	h (cm C_6H_5Cl)	$\left(\dfrac{h}{c}\right)\ \dfrac{(cm\ C_6H_5Cl)(1000\ cm^3)}{g}$
1.95	0.20	0.102
2.93	0.36	0.123
3.91	0.53	0.135
5.86	0.98	0.167

The next step is to plot (h/c) against c. When this is done, we get a straight line (Fig. 4.6) with slope 0.016 and intercept 0.072 on the ordinate, i.e., $(h/c)\mid_{c\to0} = 0.072$ (cm C_6H_5Cl)(1000 cm^3)(g^{-1}) = 72.0 (cm C_6H_5Cl)(cm^3)(g^{-1}).

Pressure exerted by a column of C_6H_5Cl, h cm long : $\Pi = h\rho g = (h\,cm)(1.10\,g\,cm^{-3})(980$ cm s^{-2}) = 1078h g cm^{-1} s^{-2}.

Therefore, $(\Pi/c)_{c\to0} = 72.0 \times 1078\,(g\,cm^{-1}\,s^{-2})(cm^3)(g^{-1}) = 77{,}616\ cm^2\ s^{-2}$. Hence, $\left(RT/\overline{M}_n\right) = 77{,}616\ cm^2\ s^{-2}$ and noting that R should be in ergs mol^{-1} $^\circ$K^{-1} to yield $\overline{M}_n$ in g mol^{-1}, we then obtain

$$\overline{M}_n = \frac{(8.314\times10^7\ ergs\ mol^{-1}\,^\circ K^{-1})(293\,^\circ K)}{(77{,}616\ cm^2\ s^{-2})} = 3.1\times10^5\ g\ mol^{-1}$$

(b) Slope of h/c versus c linear plot (Fig. 4.6) = 0.016 (cm C_6H_5Cl)(10^6 cm^6)(g^{-2})
= 0.016$\times$1078$\times$10^6 (g cm^{-1} s^{-2}) (cm^6)(g^{-2}) = 1.725$\times$10^7 g^{-1} cm^5 s^{-2}. Neglecting c^2 and higher powers of c in Eq. (4.41) for dilute solutions, this slope can be equated to A_2RT. Hence, A_2RT = 1.725$\times$10^7 g^{-1} cm^5 s^{-2} or

$$A_2 = \frac{(1.725\times10^7\ g^{-1}\ cm^5\ s^{-2})}{(8.314\times10^7\ ergs\ mol^{-1}\,^\circ K^{-1})(293\,^\circ K)} = 7.0\times10^{-4}\ mol\ cm^3\ g^{-2}$$

From Eq. (4.44), $\Gamma_2 = A_2\overline{M}_n$. Therefore, $\Gamma = (7.0\times10^{-4}\ mol\ cm^3\ g^{-2})(3.1\times10^5\ g\ mol^{-1}) = 217\ cm^3\ g^{-1}$.

(c) Considering Eq. (4.47),

$v_2 = (1.2\ g\ cm^{-3})^{-1} = 0.833\ cm^3\ g^{-1}$; $V_1^o = (112.5\ g\ mol^{-1}/(1.10\ g\ cm^{-3}) = 102.27\ cm^3\ mol^{-1}$.

Substituting these values in $\chi = \frac{1}{2} - A_2V_1^o/v_2^2$ derived from Eq. (4.47) then gives $\chi = 0.40$.

The practical range of molecular weights that can be measured by membrane osmometry is approximately 30,000 to one million. For measurements of $\overline{M}_n$ less than 30,000 another technique known as vapor-phase osmometry described next is more suitable.

4.4.2.4   Vapor-Phase Osmometry

Vapor-phase osmometry is based on vapor pressure lowering, which is a colligative property. The method therefore gives $\overline{M}_n$. Combining Raoult's law and Dalton's law we have

$$P = P_1 + P_2 = x_1 P_1^o + x_2 P_2^o \tag{4.48}$$

where P is the vapor pressure of the solution ; P_1 and P_2 are the vapor pressures, respectively, of the solvent and solute in solution; P_1^o and P_2^o are the pure component vapor pressures of the   solvent

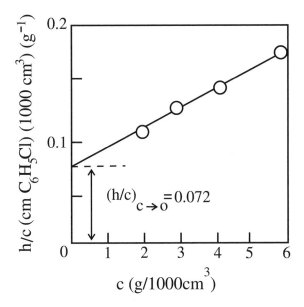

Figure 4.6 Molecular weight determination from osmotic head (h) and concentration (c) data (Problem 4.8).

and solute, respectively; x_1 and x_2 are the mole fractions, respectively, of the solvent and solute in solution.

When the solute is nonvolatile, as is the case with high-molecular-weight polymers, the vapor phase consists only of solvent. Therefore, Eq. (4.48) reduces to

$$P = x_1 P_1^0 = (1 - x_2)P_1^0 \tag{4.49}$$

Rearranging,

$$x_2 = 1 - \frac{P}{P_1^0} = \frac{P_1^0 - P}{P_1^0} = \frac{-\Delta P}{P_1^0} \tag{4.50}$$

Here ΔP is the vapor pressure lowering given by difference between the vapor pressure of the solvent above the solution and the vapor pressure of the pure solvent at the same temperature. Rewriting Eq. (4.50) as

$$\frac{\Delta P}{P_1^0} = -x_2 = \frac{W_2/M_2}{(W_1/M_1) + (W_2/M_2)} \tag{4.51}$$

where W and M are weight and molecular weight, respectively, with the subscript 1 used for solvent and 2 for solute, we obtain for dilute solutions

$$\lim\left(\frac{\Delta P}{P_1^0}\right)_{c_2 \to 0} = -\frac{(W_2/M_2)}{(W_1/M_1)} = -\frac{W_2 M_1}{W_1 M_2} = -c_2 \frac{V_1^o}{M_2} \tag{4.52}$$

in which V_1^0 is the molar volume of the solvent and c_2 is the concentration of the solute in mass per unit volume. The expanded virial form of Eq. (4.52) for vapor pressure lowering is

$$\frac{\Delta P}{P_1^0} = -V_1^0 c_2 \left[\frac{1}{M} + B c_2 + C c_2^2 + \cdots \right] \qquad (4.53)$$

[cf. Eq. (4.43)] where B and C are virial coefficients. As ΔP is inversely proportional to molecular weight of the solute, its magnitude for polymer solutions is quite small and difficult to measure directly. It is more accurate and convenient to convert this vapor pressure difference into a temperature difference. This is done in the method called *vapor-phase osmometry*, also known as *vapor-pressure osmometry*.

The temperature difference ΔT, corresponding to the vapor pressure difference ΔP in Eq. (4.53), can be deduced from the Clausius-Clapeyron equation

$$\Delta T = \frac{\Delta P}{P_1^0} \cdot \frac{RT^2}{\Delta H_v} \qquad (4.54)$$

where ΔH_v is the latent heat of vaporization of the solvent at temperature T. Combining Eq. (4.54) with Eq. (4.53) we obtain

$$\frac{\Delta T}{c_2} = -\frac{RT^2}{\Delta H_v} V_1^0 \left[\frac{1}{M} + B c_2 + C c_2^2 + \cdots \right] \qquad (4.55)$$

The molecular weight of the solute can thus be determined by measuring $\Delta T / c_2$ and extrapolating it to $c_2 = 0$.

Practical Aspects   There is no membrane in a vapor-pressure osmometer. Instead there are two matched thermistors in a thermostated chamber that is saturated with solvent vapor (Fig. 4.7). With a hypodermic syringe a drop of solution is placed on one thermistor and similarly a drop of solvent of equal size on the other thermistor. The solution has a lower vapor pressure than the solvent at the same temperature, and so the solvent vapor condenses on the solution droplet. The solution droplet, therefore, starts getting diluted as well as heated up by the latent heat of condensation of solvent condensing on it. In a steady state, the total rise in temperature ΔT can be related by an analog of Eq. (4.55):

$$\frac{\Delta T}{c_2} = k_s \left[\frac{1}{\overline{M}_n} + B c_2 + C c_2^2 + \cdots \right] \qquad (4.56)$$

where k_s is an instrument constant, which is normally determined for a given solvent, temperature, and thermistor pair, by using solutes of known molecular weight.

The temperature difference between the two thermistors can be measured very accurately as a function of the bridge imbalance output voltage, ΔV. The operating equation is

$$\frac{\Delta V}{c} = \frac{K}{M} + KBc \qquad (4.57)$$

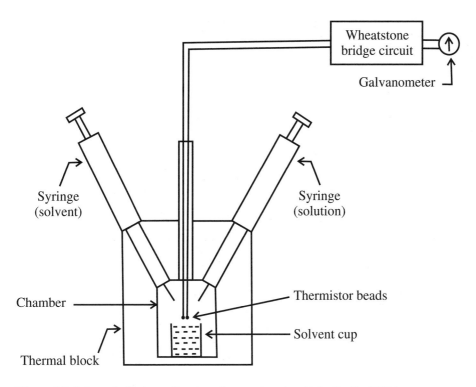

Figure 4.7 Schematic diagram of a vapor-phase osmometer. (After Rudin, 1982.)

[cf. Eq. (4.55)] where K is the calibration constant. A plot of $\Delta V/c$ versus c (where c is the solution concentration) is made and extrapolated to zero concentration to obtain the ordinate intercept $(\Delta V/c)_{c\to 0}$. The calibration constant K can be computed using the equation

$$K = M(\Delta V/c)_{c\to 0} \tag{4.58}$$

where M is the molecular weight of the known standard sample. For determining the molecular weight of an unknown sample, solutions of the sample are made in different concentrations in the same solvent used for the standard sample and the whole procedure is repeated to obtain the ordinate intercept $(\Delta V/c)_{c\to 0}$. The molecular weight of the unknown sample is then given by

$$\overline{M}_n = K/(\Delta V/c)_{c\to 0} \tag{4.59}$$

The upper limit of molecular weights for vapor-phase osmometry is considered to be 20,000. Development of more sensitive machines has extended this limit to 50,000 and higher.

Problem 4.9 Following are the vapor phase osmometry data for a standard polystyrene of known molecular weight and an experimental sample of hydroxyl terminated polybutadiene (HTPB) in toluene solutions at 70°C. Calculate the molecular weight of HTPB.

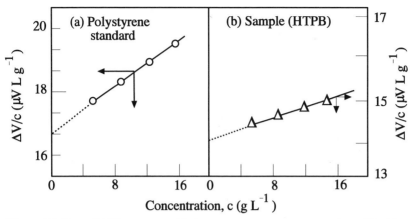

Figure 4.8 Plots of $\Delta V/c$ versus c with the vapor-pressure osmometry data of Problem 4.9.

Polymer	Concentration, c (g/L)	Bridge output, ΔV (μV)
Standard	6	107
polystyrene	9	164
of $\overline{M}_n = 1800$	12	224
	15	287
HTPB of unknown	6	85
molecular weight	9	129
	12	176
	15	225

Answer:

The data in columns 2 and 3 are plotted as $\Delta V/c$ versus c both for the polystyrene standard (Fig. 4.8a) and for HTPB (Fig. 4.8b).

Polystyrene standard (Fig. 4.8a): $M_n = 1800$. $(\Delta V/c)_{c\to 0} = 16.95$.   Therefore, $K = 1800 \times 16.95 = 30,510$.

HTPB sample (Fig. 4.8b): $(\Delta V/c)_{c\to 0} = 13.45$.   Hence, $\overline{M}_n = \dfrac{K}{(\Delta V/c)_{c\to 0}} = \dfrac{30,510}{13.45} = 2268$

4.4.3   *Light-Scattering Method*

The measurement of light scattering by polymer solutions is an important technique for the determination of weight-average molecular weight, $\overline{M}_w$. It is an absolute method of molecular weight measurement. It also can furnish information about the size and shape of polymer molecules in solution and about parameters that characterize the interaction between solvent and polymer molecules. The experimental technique is, however, exacting, mainly because of the large difference in intensity of the incident beam and light scattered by the polymer solution.

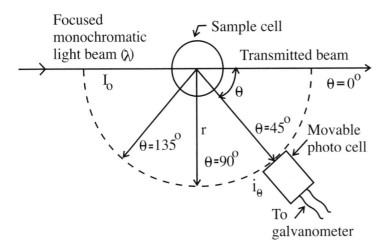

Figure 4.9 Arrangement of the apparatus required to measure light scattered from a solution at different angles with respect to the incident beam. (After Billingham, 1977.)

4.4.3.1 Rayleigh Ratio

A theoretical treatment of the scattering of light by the molecules of a gas was given by Lord Rayleigh in 1871. A schematic diagram that shows the basic features of the apparatus required for a light scattering experiment is given in Fig. 4.9. For a dilute gas, the intensity of scattered light as a function of scattering angle θ is given by the *Rayleigh equation*

$$\frac{i_\theta r^2}{I_0(1 + \cos^2\theta)} \;=\; R_\theta \;=\; \frac{2\pi^2}{\lambda^4 N_{Av}} \frac{(\tilde{n} - 1)^2 M}{c} \tag{4.60}$$

In this equation, I_0 is the intensity of incident light of wavelength λ; i_θ is the intensity of scattered light per unit volume of the system measured at angle θ to the incident beam direction and at a distance r from the center of the system (Fig. 4.9); M is the molecular weight of the gas; $\tilde{n}$ is the refractive index; c is the density or concentration of the gas in mass per unit volume; and N_{Av} is Avogadro's number. The term R_θ is called the *Rayleigh ratio*. It is the reduced relative scattering intensity defined by

$$R_\theta \;=\; \frac{i_\theta r^2}{I_0(1 + \cos^2\theta)} \tag{4.61}$$

and is independent of both r and θ on the basis of the Rayleigh equation (4.60). The Rayleigh ratio is sometimes defined as the quantity $i_{90} r^2/I_0$, which is the reduced scattering intensity at $\theta = 90°$, that is, R_{90}.

The Rayleigh theory of light scattering in dilute gases cannot be applied to liquids, the main reason being that, unlike in gases, strong intermolecular forces are present in liquids. These difficulties are overcome by a different approach to light scattering in liquids, developed by Einstein (1910). In this approach, scattering is considered to be caused by local fluctuations in density due to the thermal motions of the molecules, resulting in local fluctuations in the refractive index and hence in scattering of the incident light. Einstein's theory leads to the following expression for the Rayleigh ratio (Allcock and Lampe, 1990):

$$R_\theta \;=\; \frac{i_\theta r^2}{I_0(1+\cos^2\theta)} \;=\; \frac{2\pi^2}{\lambda^4 N_{Av}} \frac{RT}{\beta}\left(\tilde{n}\frac{d\tilde{n}}{dp}\right)^2 \tag{4.62}$$

where p is the hydrodynamic pressure on the liquid and β is the compressibility, that is, $\beta = -(1/V)(\partial V/\partial p)_T$. The other terms are as given before.

The presence of dissolved solute in a solution causes additional scattering of light, besides that due to the solvent alone. Debye (1944) put forward a treatment in which the additional scattering of light by a solution is viewed as resulting from local fluctuations in the concentration of the solute. Analogous to Einstein's theory of density fluctuations in pure liquids, Debye's model considers that the local fluctuations in solute concentration due to random thermal motion are opposed by the osmotic pressure of the solution and leads to the following expression (Rudin, 1982; Allcock and Lampe, 1990) for the Rayleigh ratio of the scattering due to solute:

$$R'_\theta \;=\; \frac{i'_\theta r^2}{I_0(1+\cos^2\theta)} \;=\; \frac{2\pi^2}{\lambda^4 N_{Av}}\left(\tilde{n}_0\frac{d\tilde{n}}{dc}\right)^2 \frac{RTc}{(\partial\Pi/\partial c)_T} \tag{4.63}$$

where the primes are used to denote the excess scattering from the liquid due to the solute; $\tilde{n}$ and $\tilde{n}_0$ are the refractive indices of the solution and solvent, respectively; $d\tilde{n}/dc$ is the specific refractive index increment with concentration; c is the concentration of solute in mass per unit volume; Π is the osmotic pressure of the solution; and the other terms are as described previously. R'_θ, commonly known as the *excess Rayleigh ratio*, is the difference between the Rayleigh ratios of the solution and the pure solvent:

$$R'_\theta \;=\; R_\theta(\text{solution}) - R_\theta(\text{solvent}) \tag{4.64}$$

The excess Rayleigh ratio R'_θ can be derived from the "raw" galvanometer readings I_g and I_{gs} when the sample cell contains the solution and the solvent, respectively, with the photocell of Fig. 4.9 positioned at an angle θ in both cases. The equation used is

$$R'_\theta \;=\; \frac{k(I_g - I_{gs})\sin\theta}{1+\cos^2\theta} \tag{4.65}$$

where k is the conversion factor. The value of k can be obtained from the following equation for the solvent,

$$R_\theta \;=\; \frac{kI_{gs}\sin\theta}{1+\cos^2\theta} \tag{4.66}$$

using I_{gs} measured at an angle θ and the known value of R_θ for the solvent. [Note that I_g and I_{gs} are multiplied by $\sin\theta$ to correct for the variation that occurs in the effective scattering volume as θ is varied (see Fig. 4.10). Multiplying by the factor $\sin\theta$ one actually computes the value that the galvanometer would have read, if the "viewed" volume had remained constant (Margerison and East, 1967).]

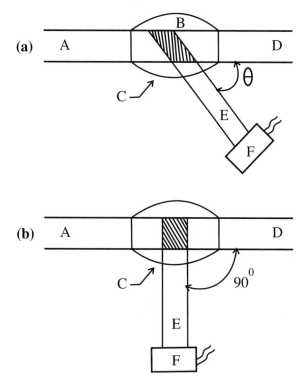

Figure 4.10 The variation of effective scattering volume (shaded) with angle: (a) $\theta < 90°$; (b) $\theta = 90°$. A: incident beam; B: polymer solution in solvent; C: scattering cell; D: light trap; E: scattered light beam; F: movable photocell connected to galvanometer. (After Margerison and East, 1967.)

4.4.3.2 Turbidity and Rayleigh Ratio

If I_o and I are the intensities of the beam before and after passing through a length ℓ of the medium, they can be related in terms of Beer's law for the absorption of light as follows (Rudin, 1982):

$$I/I_0 = e^{-\tau\ell} \tag{4.67}$$

where τ is the measure of the decrease of the incident beam intensity per unit length of the medium and is called the *turbidity*. It has the dimensions of reciprocal length.

Noting that $I = I_0 - I_s$, where I_s is the total intensity of light that is scattered by the solution, Eq. (4.67) may be written in the form (Allcock and Lampe, 1990):

$$e^{-\tau\ell} = \frac{I_0 - I_s}{I_0} = \frac{I_0 - I_s'\ell}{I_0} = 1 - \frac{I_s'\ell}{I_0} \tag{4.68}$$

where I_s' is the total intensity of scattered light per unit path length. Expanding $e^{-\tau\ell}$ in series and neglecting the square and higher powers, since the fraction of light scattered is small, the exponential in Eq. (4.68) can be approximated as

$$e^{-\tau\ell} = 1 - \tau\ell + \tfrac{1}{2}(\tau\ell)^2 - \tfrac{1}{6}(\tau\ell)^3 + \cdots \simeq 1 - \tau\ell \tag{4.69}$$

A comparison of Eqs. (4.68) and (4.69) gives

$$\tau = I'_s/I_o \tag{4.70}$$

as the relationship between the turbidity and the total intensity of scattered light per unit path.

The total of light intensity scattered per unit path length through all angles of polar coordinates is given by (Allcock and Lampe, 1990):

$$I'_s = \int_0^\pi \int_0^{2\pi} r^2 i'_\theta \sin\theta\, d\theta\, d\phi \tag{4.71}$$

From Eq. (6.43), $r^2 i'_\theta = I_o R'_\theta (1 + \cos^2\theta)$. From Eqs. (4.70) and (4.71) one thus obtains

$$\frac{I'_s}{I_o} = \tau = \int_0^\pi \int_0^{2\pi} R'_\theta (1 + \cos^2\theta) \sin\theta\, d\theta\, d\phi \tag{4.72}$$

Substituting for R'_θ from Eq. (4.63), Eq. (4.72) can be written as

$$\tau = \frac{2\pi^2}{\lambda^4 N_{Av}} \left(\tilde{n}_o \frac{d\tilde{n}}{dc} \right)^2 \frac{RT\,c}{(\partial\Pi/\partial c)_T} \int_0^\pi (1 + \cos^2\theta) \sin\theta\, d\theta \int_0^{2\pi} d\phi \tag{4.73}$$

The value of the product of the definite integrals is $16\pi/3$ (Rudin, 1982; Allcock and Lampe, 1990), so that we obtain the following relationship involving turbidity and the total scattered intensity:

$$\tau = \frac{I'_s}{I_o} = \left(\frac{32\pi^3}{3\lambda^4 N_{Av}} \right) \frac{\left(\tilde{n}_o \dfrac{d\tilde{n}}{dc} \right)^2 RT\,c}{(\partial\Pi/\partial c)_T} \tag{4.74}$$

This expression provides us with a way to determine molecular weights of polymers from light scattering of polymer solutions as shown below. Moreover, a comparison of Eqs. (4.74) and (4.63) shows that the relationship between the turbidity and the Rayleigh ratio is

$$\tau = \frac{16\pi}{3} R'_\theta = \left(\frac{16\pi}{3} \right) \frac{i'_\theta r^2}{I_o(1 + \cos^2\theta)} \tag{4.75}$$

This expression enables us to determine the turbidity by measuring the intensity of light scattered at given angles.

4.4.3.3 Turbidity and Molecular Weight of Polymer

The light scattering of solutions may now be related to the solute molecular weight by substituting $(\partial\Pi/\partial c)_T$ into Eq. (4.74). Representing the osmotic pressure for a monodisperse (i.e., single molecular weight) solute by a virial equation [cf. Eq. (4.41)] in the form

$$\frac{\Pi}{c} = RT \left(\frac{1}{M} + A_2 c + A_3 c^2 + \cdots \right) \tag{4.76}$$

where A_2 and A_3 are the second and third virial coefficients, we obtain by differentiation

$$\left(\frac{\partial\Pi}{\partial c} \right)_T = RT \left(\frac{1}{M} + 2A_2 c + 3A_3 c^2 + \cdots \right) \tag{4.77}$$

Substituting in Eq. (4.74) we obtain

$$\tau = \left(\frac{32\pi^3}{3\lambda^4 N_{Av}}\right) c \left(\tilde{n}_0 \frac{d\tilde{n}}{dc}\right)^2 \Big/ \left(\frac{1}{M} + 2A_2 c + 3A_3 c^2 + \cdots\right) \qquad (4.78)$$

Rearranging we have (Rudin, 1982):

$$\frac{Hc}{\tau} = \frac{1}{M} + 2A_2 c + 3A_3 c^2 + \cdots \qquad (4.79)$$

where the function H is a lumped constant given by

$$H = \left(\frac{32\pi^3}{3\lambda^4 N_{Av}}\right)\left(\tilde{n}_0 \frac{d\tilde{n}}{dc}\right)^2 \qquad (4.80)$$

The procedure for determination of molecular weight according to Eqs. (4.79) and (4.80) is therefore as follows. Take a series of polymer solutions (dust free) of different concentrations. Choose a particular angle to the incident beam (usually $90°$) and determine excess Rayleigh ratio R'_θ (or excess scattering intensity i'_θ), and hence turbidity from Eq. (4.75), for each solution. Calculate H from Eq. (4.80) using measured values of $\tilde{n}$ and $\tilde{n}_0$. Evaluate Hc/τ for each concentration and plot this against c. By extrapolation of the linear portion to zero concentration the intercept on the Hc/τ axis gives $1/M$ directly, while the initial slope yields the second virial coefficient (see Problem 4.10).

It is also customary to define another optical constant K such that with substitution from Eq. (4.80) one obtains (Rudin, 1982):

$$K = \frac{3H}{16\pi} = \left(\frac{2\pi^2}{\lambda^4 N_{Av}}\right)\left(\tilde{n}_0 \frac{d\tilde{n}}{dc}\right)^2 \qquad (4.81)$$

It is easy to see by comparing Eqs. (4.75), (4.79), and (4.81) that (Rudin, 1982):

$$\frac{Kc}{R'_\theta} = \frac{3Hc}{16\pi R'_\theta} = \frac{Hc}{\tau} = \frac{1}{M} + 2A_2 c + 3A_3 c^2 + \cdots \qquad (4.82)$$

Hence M and A_2 can be determined from a plot of Kc/R'_θ against c (see Problem 4.12) in the same way as from the plot of Hc/τ versus c using Eq. (4.79).

The molecular weight obtained by application of Eq. (4.82) to a polydisperse polymer will be some *average* over the molecular weight distribution characteristic of the polymer. It can be easily shown that the molecular weight determined from light scattering measurement is $\overline{M}_w$ (see Problem 4.13).

Problem 4.10   Given below are typical light scattering data for solutions of polystyrene in benzene ($\tilde{n}_0 = 1.5010$) with $\lambda = 4358$ Å at $20°C$:

Conc. (g/100 cm^3)	$(\tilde{n} - \tilde{n}_o) \times 10^2$	$\tau \times 10^2$ (cm^{-1})
0.175	0.021	0.093
0.385	0.043	0.144
0.594	0.066	0.182
0.730	0.082	0.198
1.000	0.110	0.226

Determine the molecular weight of the polymer.

Answer:

The value of the refractive index increment $d\tilde{n}/dc$ is needed at infinite dilution, but there is little concentration dependence in the normal concentration range used for light scattering of polymer solutions. The required value can therefore be obtained from

$$\frac{\tilde{n} - \tilde{n}_0}{c} = \frac{\Delta\tilde{n}}{c} \simeq \frac{d\tilde{n}}{dc}$$

Therefore Eq. (4.80) can be written as

$$H = \left(\frac{32\pi^3}{3\lambda^4 N_{Av}}\right)\tilde{n}_0^2\left(\frac{\tilde{n} - \tilde{n}_0}{c}\right)^2 \tag{P4.10.1}$$

[The value of $(d\tilde{n}/dc)$ can also be obtained from the plot of $(\tilde{n} - \tilde{n}_0)$ against c and Eq. (4.80) can then be used directly.]

For $c = 0.1750$ g/100 cm^3, using Eq. (P4.10.1):

$$H = \frac{32(3.143)^3}{3(4358\times10^{-8}\text{ cm})^4(6.02\times10^{23}\text{ mol}^{-1})}(1.501)^2\left[\frac{(0.021\times10^{-2})}{(0.175\times10^{-2}\text{ g cm}^{-3})}\right]^2$$

$$= 4.48\times10^{-6}\text{ mol cm}^2\text{ g}^{-2}$$

So,

$$\frac{Hc}{\tau} = \frac{(4.48\times10^{-6}\text{ mol cm}^2\text{ g}^{-2})(0.175\times10^{-2}\text{ g cm}^{-3})}{(0.093\times10^{-2}\text{ cm}^{-1})}$$

$$= 8.45\times10^{-6}\text{ mol g}^{-1}$$

Similarly, Hc/τ values are calculated for other c values and are plotted against c in Fig. 4.11. From the intercept at $c = 0$,

$$\left(\frac{Hc}{\tau}\right)_{c=0} = 6.8\times10^{-6}\text{ mol g}^{-1} = \frac{1}{M} \Rightarrow M = 1.47\times10^5\text{ g mol}^{-1}$$

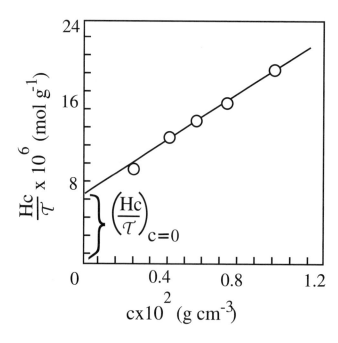

Figure 4.11 Light-scattering determination of the molecular weight of polystyrene. (Data of Problem 4.10.)

Problem 4.11   A solution of polystyrene in benzene at 25°C used for light scattering experiments with light of wavelength 4358 Å has the following values: $\tilde{n} = 1.5130$, $d\tilde{n}/dc = 0.111$ cm^3 g^{-1}.

Determine the optical constant K for the system.

Answer:

From Eq. (4.81):

$$K = \frac{2(3.143)^2 (1.5130)^2 (0.111 \text{ cm}^3 \text{ g}^{-1})^2}{(6.023 \times 10^{23} \text{ mol}^{-1})(4358 \times 10^{-8} \text{ cm})^4} = 2.564 \times 10^{-7} \text{ cm}^2 \text{ g}^{-2} \text{ mol}$$

(It may be noted that K is a constant for a polymer-solvent system and is independent of the concentration and molecular weight.)

Problem 4.12   Measurements of excess scattering for a solution of polystyrene in benzene at 25°C with light of wavelength 4358 Å yielded the following data:

c, g/100 cm^3	0.132	0.197	0.296	0.445	0.667
$R'_{90} \times 10^5$, cm^{-1}	3.88	5.44	7.63	10.1	13.3

Determine (a) the molecular weight of the polymer and (b) the second virial coefficient for this particular polymer-solvent system. (The value of K for the polymer-solvent system is 2.564×10^{-7} cm^2 g^{-2} mol.)

Answer:

For $c = 0.132$ g / 100 cm^3 = 1.32×10^{-3} g cm^{-3},

$$\frac{Kc}{R'_{90}} = \frac{(2.564 \times 10^{-7} \text{ cm}^2 \text{ g}^{-2} \text{ mol})(1.32 \times 10^{-3} \text{ g cm}^{-3})}{(3.88 \times 10^{-5} \text{ cm}^{-1})} = 8.72 \times 10^{-6} \text{ mol g}^{-1}$$

Similarly, Kc/R'_θ values for other concentrations (c) are calculated and these are plotted against c in Fig. 4.12.

(a) Eq. (4.82): From the intercept of the graph on Kc/R'_{90} axis (Fig. 4.12),

$$\frac{1}{M} = \left(\frac{Kc}{R'_\theta}\right)_{c=0} = 7.6 \times 10^{-6}$$

$$M = 1.31 \times 10^5 \text{ g mol}^{-1}$$

(b) Eq. (4.82): From the slope of the graph (Fig. 4.12),

$$2A_2 = 8.0 \times 10^{-4} \text{ mol g}^{-2} \text{ cm}^3 \quad \text{or} \quad A_2 = 4.0 \times 10^{-4} \text{ mol g}^{-2} \text{ cm}^3$$

From Eq. (4.44) another form of the second virial coefficient is

$$\Gamma_2 = A_2 M = (4.0 \times 10^{-4} \text{ mol g}^{-2} \text{ cm}^3)(1.31 \times 10^5 \text{ g mol}^{-1}) = 52 \text{ cm}^3 \text{ g}^{-1}$$

Problem 4.13   Show that the molecular weight of a polydisperse polymer determined by the light scattering method is a weight-average molecular weight.

Answer:

For a solution in the limit of infinite dilution, Eq. (4.79) becomes: $\tau = HcM$.   If the solute molecules are independent agents and contribute additively to the observed turbidity, one can also write: $\tau = \sum \tau_i = H \sum c_i M_i$ and $c = \sum c_i$, where τ_i, c_i, and M_i refer, respectively, to the turbidity, weight concentration, and molecular weight of monodisperse species i, which is one of the components of the mixture that make up the polymer sample.

The average molecular weight $\overline{M}$ is obtained from the overall turbidity τ and concentration c:

$$\overline{M} = \frac{\tau}{Hc} = \frac{H \sum c_i M_i}{H \sum c_i} = \frac{\sum c_i M_i}{\sum c_i} \tag{P4.13.1}$$

Since $c_i = N_i M_i / V$, where N_i is the number of moles of species i having molecular weight M_i in volume V of solution, it is seen on substitution in Eq. (P4.13.1) that

$$\overline{M} = \frac{\sum N_i M_i^2}{\sum N_i M_i} = \overline{M}_w$$

4.4.3.4 Dissymmetry of Scattering

Intraparticle Interference Equation (4.82) is derived on the assumption that each solute molecule is small enough compared to the wavelength of the incident light to act as a point source of secondary radiation, so that the intensity of the scattered light is symmetrically distributed as shown in Fig. 4.13. This condition is satisfied by vinyl polymers having a degree of polymerization less than about 500. If any linear dimension of the solute particle is as great as about $\lambda'/20$ or greater, as in the case of large size polymer molecules, then the secondary radiations from dipoles in various regions of the scattering molecule may vary in phase at a given viewing point. (Note that λ' is the wavelength of the light in solution. For dilute solutions it is equal to $\lambda/\tilde{n}_0$, where $\tilde{n}_0$ is the refractive index of the solvent.) The resulting interference will depend on the size and shape of the molecule and on the observation angle.

Problem 4.14 Using light of wavelength 5461 Å and benzene as solvent (refractive index 1.5014), what is the limiting size (root mean square end-to-end distance) of a polymer coil above which the molecule can no longer be regarded as a point source in refraction ?

Answer:

$$\text{Limiting } <r^2>^{1/2} = \frac{\lambda'}{20} = \frac{(\lambda/\tilde{n}_0)}{20}$$

$$= \frac{5461 \text{ Å}}{20 \times 1.5014} = 182 \text{ Å}$$

This represents the critical value of the diameter or length of the scattering molecule.

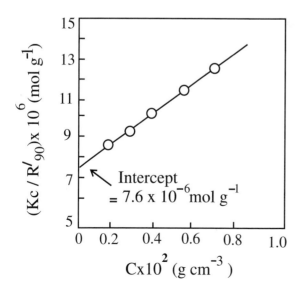

Figure 4.12 Plot of Kc/R'_{90} versus c in the absence of interference effects. (Data of Problem 4.12.)

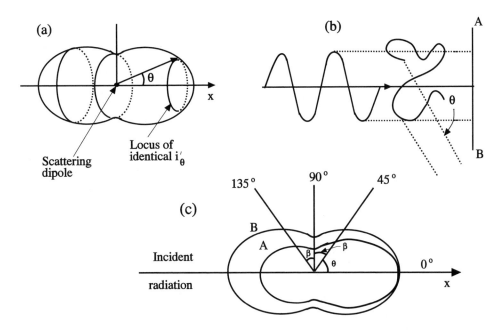

Figure 4.13 (a) Scattering envelope for a point scatterer. (b) Scattering by a random coil which is comparable in size with the wavelength of the incident radiation. (c) Scale diagram of two-dimensional scattering envelope: Curve A shows the effect of interference in relation to curve B for the same polymer solution without interference. Distance from the scattering particle to the boundary of the envelope (which is cylindrically symmetrical about x axis) represents the magnitude of scattered light as a function of angle.

With reference to Fig. 4.13(b), the scattered light received at the plane AB (zero angle) is in phase, no matter what part of the polymer molecule is acting as the source of the secondary radiation. However, as θ increases, there occurs an increasing path difference between the light received from different parts of the molecule. The destructive interference due to phase differences between the scattered rays reduces the intensity of the scattered radiation. This interference effect will be greater the larger the scattering angle θ, and so the radiation envelope will not be symmetrical. Both these effects are illustrated in Fig. 4.13(c). In both cases of interference and noninterference the envelopes are cylindrically symmetrical about the incident ray, but the envelope in the presence of interference effects is no longer symmetrical about a plane through the scatterer and normal to the incident radiation. Thus the scattering is less in the direction for which $\theta = 90° + \beta$ than for $\theta = 90° - \beta$. This effect is called *dissymmetry*. The observed ratio of these intensities is usually referred to as the *dissymmetry coefficient* (Debye, 1947) designated by z_β, i.e.,

$$z_\beta = \frac{i'_{90°-\beta}}{i'_{90°+\beta}} = \frac{i'_\theta}{i'_{\pi-\theta}} = \frac{R'_\theta}{R'_{\pi-\theta}}$$

The root-mean-square distance $\langle r^2 \rangle^{1/2}$ between the ends of the polymer chain is a convenient measure of the diameter of the randomly coiled polymer molecule. The dissymmetry coefficient z_β will be unity for $\langle r^2 \rangle^{1/2} < \lambda'/20$ and will increase as $\langle r^2 \rangle^{1/2}$ increases. In general, macromolecules with a linear structure, in good solvents and with molecular weights between 10^5 and 10^7, have diameters between 200 and 3000 Å. In light scattering measurements using mercury arcs ($\lambda = 4,000\text{-}5,000$ Å) as the source of light, the particle diameter is therefore larger than $\lambda'/20$.

Intraparticle interference effects, described above, diminish as the viewing angle θ to the incident light approaches zero, at which point the scattering is unperturbed and so Eq. (4.82) can be used for determination of molecular weight. This principle is used in the method of *low-angle laser light scattering* (LALLS) following the advent, in recent years, of light scattering photometers based on helium-neon (He-Ne) lasers (λ = 6328 Å) with which scattering can be measured accurately at angles as low as 2° to 10° off the incident beam path. However, the optics of older commercial instruments that are still in wide use are limited to angles greater than about 30° to the incident beam. Appropriate corrections must be applied to scattering intensities measured at such large angles, as discussed below.

If the experimentally observed values of R'_θ and $R'_{\pi-\theta}$ differ significantly from one another, interference effects are present. These effects may be taken into account by defining a parameter $P(\theta)$, called the *particle scattering factor*. It is simply the ratio of the scattering intensity to the intensity in the absence of interference, measured at the same angle θ:

$$P(\theta) \;=\; \frac{(R'_\theta)_{\text{observed}}}{(R'_\theta)_{\text{no interference}}} \;=\; \frac{(i'_\theta)_{\text{observed}}}{(i'_\theta)_{\text{no interference}}} \tag{4.83}$$

The R'_θ in our previous equation (4.82) is $(R'_\theta)_{\text{no interference}}$, since in their derivation the scattering molecules have been considered as point sources. Thus Eq. (4.82) may be rearranged in terms of the observed R'_θ and $P(\theta)$ as

$$\frac{Kc}{(R'_\theta)_{\text{observed}}} \;=\; \frac{Kc}{P(\theta)\,(R'_\theta)_{\text{no interference}}}$$

$$\;=\; \frac{1}{P_\theta}\left[\frac{1}{M} + 2A_2 c + 3A_3 c^2 + \cdots\right] \tag{4.84}$$

According to Zimm, the following approximation may be written

$$\frac{Kc}{R'_\theta} \;=\; \frac{Hc}{\tau} \;=\; \frac{1}{M.P(\theta)} + 2A_2 c \tag{4.85}$$

(omitting the subscript 'no interference' for simplicity)

Since the forward scatter at zero angle is the same whether interference effects are present or not, $P(\theta) = 1$ at $\theta = 0°$. Therefore, if Hc/τ or Kc/R'_θ measured at different values of θ and c is extrapolated to both $\theta = 0$ and $c = 0$, Eq. (4.85) reduces to

$$\left(\frac{Kc}{R'_\theta}\right)_{c\to 0,\ \theta\to 0} \;=\; \left(\frac{Hc}{\tau}\right)_{c\to 0,\ \theta\to 0} \;=\; \frac{1}{M} \tag{4.86}$$

The molecular weight of the solute can thus be obtained without making any assumptions about the shape of the polymer molecule.

Scattering Factor   The scattering function $P(\theta)$ depends on polymer dimension, wavelength of the light, and the refractive index of the solvent. The following function has been derived for *random coil polymer* (Doty and Edsall, 1951):

$$P(\theta) \;=\; \frac{2}{u^2}\left[e^{-u} - (1-u)\right] \tag{4.87}$$

where $u = [(4\pi/\lambda')\sin(\theta/2)]^2 \langle S^2 \rangle$ and $\langle S^2 \rangle$ is the mean square radius of gyration of the polymer molecule (see p. 46). It is related to the mean square end-to-end distance $\langle r^2 \rangle$ by $\langle S^2 \rangle^{1/2} = \langle r^2 \rangle^{1/2}/\sqrt{6}$. Substituting this to replace $\langle S^2 \rangle$ by $\langle r^2 \rangle$ in the above equation for u leads to

$$u = \left(\frac{2}{3}\right)\left(\frac{\langle r^2 \rangle}{\lambda'^2}\right)\left(2\pi \sin \frac{\theta}{2}\right)^2 \tag{4.88}$$

When $u \ll 1$, one can approximate Eq. (4.87) by

$$P(\theta) = 1 - u/3 \tag{4.89}$$

Since $1/(1-x) \approx 1+x$ when $x \ll 1$, one obtains from Eq. (4.89), $1/P(\theta) = 1+u/3$. Substituting for u from Eq. (4.88) gives

$$1/P(\theta) = 1 + \left(\frac{8\pi^2}{9\lambda'^2}\right)\langle r^2 \rangle \sin^2 \frac{\theta}{2} \tag{4.90}$$

Substituting this in Eq. (4.85) one then obtains

$$\frac{Kc}{R'_\theta} = \frac{Hc}{\tau} = \frac{1}{M} + \left(\frac{1}{M}\right)\left(\frac{8\pi^2}{9\lambda'^2}\right)\langle r^2 \rangle \sin^2(\theta/2) + 2A_2 c \tag{4.91}$$

4.4.3.5 Zimm Plots

The double extrapolation to zero θ and zero c required for the use of Eq. (4.91) is effectively done on the same plot by the Zimm method (Zimm, 1948). Zimm plots consist of graphs in which Kc/R'_θ (or Hc/τ) is plotted against $\sin^2(\theta/2) + bc$, where b is a constant arbitrarily chosen to give an open display of the experimental data, and to enable the two extrapolations to $c = 0$ and $\theta = 0$ to be carried out with comparable accuracy. (It is often convenient to take $b = 100$.) In practice, intensities of scattered light are measured at a series of concentrations and at several angles for each concentration. The Kc/R'_θ (or Hc/τ) values are plotted, as shown in Fig. 4.14. The extrapolated points on the $\theta = 0$ line, for example, are the intensities of the lines through the Kc/R'_θ values for a fixed c and various θ values with the ordinates at the corresponding bc values. Similarly, the $c = 0$ line is drawn through the intersections of the lines through the Kc/R'_θ values, for a fixed θ and various c values, with the corresponding $\sin^2(\theta/2)$ ordinates. The $\theta = 0$ and $c = 0$ lines intersect on the ordinate and the intercept equals $1/M$.

Problem 4.15 Scattering from benzene and a series of polystyrene solutions was measured with a light-scattering photometer. The resulting experimental data, given in Table P4.15.1, consist of a series of galvanometer readings I_g and I_{gs}, for solution and solvent, respectively, with a photomultiplier situated at the various angles shown to an incident beam of unpolarized, monochromatic light of wavelength 5461 Å.

Determine the molecular weight of the polymer by constructing a Zimm plot. Use the values: $\tilde{n}_0 = 1.5014$, $d\tilde{n}/dc = 0.106$ cm^3 g^{-1}, $R_{90°}$ of benzene $= 16.3 \times 10^{-6}$ cm^{-1}.

Table P4.15.1 Values of I_g and I_{gs} at various c and θ

c g/cm³ θ	30°	45°	60°	75°	90°	105°	120°	135°	150°
2.0×10^{-3}	1542	917	607	461	408	440	540	755	1235
1.5×10^{-3}	1383	820	550	413	363	384	475	660	1080
1.0×10^{-3}	1158	682	455	343	301	319	392	540	880
0.75×10^{-3}	998	590	396	297	263	275	339	464	755
0.50×10^{-3}	803	477	319	241	214	224	275	376	607
Pure benzene	282	170	128	105	100	105	127	170	285

Source: Margerison and East, 1967.

Answer:

From Eq. (4.66):

$$R_{90°} = k(I_{gs})_{90°}$$
$$k = \frac{R_{90°}}{(I_{gs})_{90°}} = \frac{16.3\times10^{-6}\ \text{cm}^{-1}}{100} = 16.3\times10^{-8}\ \text{cm}^{-1}$$

The values of excess Rayleigh ratio R'_θ calculated from Eq. (4.65) using the above value of k and data of Table P4.15.1 are recorded in Table P4.15.2.

Table P4.15.2 Values of $R'_\theta \times 10^6$ in cm^{-1} calculated from data in Table P4.15.1

c g/cm³ θ	30°	45°	60°	75°	90°	105°	120°	135°	150°
2.0×10^{-3}	58.7	57.4	54.1	52.5	50.2	49.4	46.6	44.9	44.2
1.5×10^{-3}	51.3	49.9	47.6	45.4	42.9	41.2	39.3	37.6	37.0
1.0×10^{-3}	40.8	39.3	36.9	35.1	32.8	31.6	29.9	28.4	27.7
0.75×10^{-3}	33.3	32.3	30.2	28.3	26.6	25.1	23.9	22.6	21.9
0.50×10^{-3}	24.3	23.6	21.6	20.1	18.6	17.5	16.7	15.8	15.0

From Eq. (4.81):

$$K = \frac{2(3.143)^2\,(1.5014)^2\,(0.106\ \text{cm}^3\ \text{g}^{-1})^2}{(6.023\times10^{23}\ \text{mol}^{-1})(5.461\times10^{-5})^4} = 9.34\times10^{-8}\ \text{cm}^2\ \text{g}^{-2}\ \text{mol}$$

The values of Kc/R'_θ calculated using the aforesaid K and data in Table P4.15.2 are recorded in Table P4.15.3. A Zimm plot obtained from these data is shown in Fig. 4.14.

Table P4.15.3 Values of $(Kc/R'_\theta)\times 10^6$ in mol g^{-1} calculated from data in Table P4.15.2

c g/cm³ θ	30°	45°	60°	75°	90°	105°	120°	135°	150°
2.0×10^{-3}	3.18	3.25	3.45	3.56	3.72	3.78	4.01	4.16	4.22
1.5×10^{-3}	2.73	2.81	2.94	3.08	3.26	3.40	3.56	3.72	3.78
1.0×10^{-3}	2.29	2.37	2.53	2.66	2.85	2.95	3.12	3.29	3.37
0.75×10^{-3}	2.10	2.17	2.32	2.47	2.63	2.79	2.93	3.10	3.20
0.50×10^{-3}	1.92	1.98	2.16	2.32	2.51	2.67	2.79	2.95	3.11

Source: Margerison and East, 1967.

From the intercept of the $c = 0$ and $\theta = 0$ lines on the Kc/R'_θ axis (Fig. 4.14): $\left(Kc/R'_\theta\right)_{c\to0,\theta\to0}$ $= 1.37\times10^{-6}$ mol g^{-1}. From Eq. (4.86): $M = 1/(1.37\times10^{-6})$ g mol^{-1} $= 730,000$ g mol^{-1}.

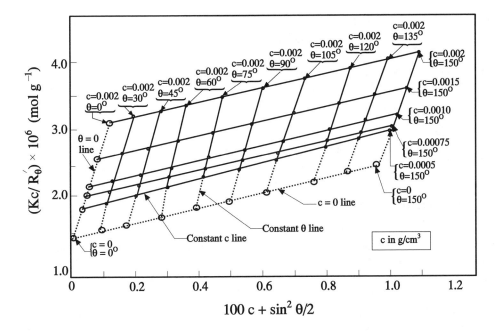

Figure 4.14 Zimm plot for the polystyrene sample of Problem 4.15. The concentration units employed are g/cm³. The symbols ○ represent extrapolated points. (Drawn following the method of Margerison and East, 1967.)

Problem 4.16 For the polystyrene sample in Problem 4.15 calculate (a) the second virial coefficient, (b) the root mean square end-to-end distance, and (c) the root-mean-square radius of gyration.

Answer:

(a) According to Eq. (4.91), the limiting slope of the $\theta = 0$ line of Kc/R'_θ versus c plot is $2A_2$. From the Zimm plot in Fig. 4.14, limiting slope = 8.75×10^{-4} mol g^{-2} cm³. Therefore, $A_2 = 4.375 \times 10^{-4}$ mol g^{-2} cm³.

From Eq. (4.44): Another form of the second virial coefficient is $\Gamma_2 = A_2 M = (4.375 \times 10^{-4}$ mol g^{-2} cm³)(7.3 10^5 g mol^{-1}) = 3.2×10^2 cm³ g^{-1}.

(b) The root-mean-square end-to-end distance of the polymer chains is found from the limiting slope of the $c = 0$ line. From Eq. (4.91): Limiting slope of the $c = 0$ line = $(1/M)(8\pi^2/9\lambda'^2) < r^2 >$, where $\lambda' = \lambda/\tilde{n}_0$.

From the Zimm plot in Fig. 4.14: slope = 1.26×10^{-6} mol g^{-1}. Hence,

$$< r^2 > = \frac{(9)(1.26 \times 10^{-6} \text{ mol g}^{-1})(7.3 \times 10^5 \text{ g mol}^{-1})}{8\pi^2} \left(\frac{5461 \times 10^{-8} \text{ cm}}{1.5014} \right)^2$$

$$= 1.386 \times 10^{-10} \text{ cm}^2$$

$$< r^2 >^{1/2} = 1.18 \times 10^{-5} \text{ cm} = 1180 \text{ Å}$$

(c) From the relation between $< r^2 >^{1/2}$ and $< S^2 >^{1/2}$ given earlier [Eq. (2.4)],

$$< S^2 >^{1/2} = < r^2 >^{1/2} / \sqrt{6} = (1180 \text{Å})/\sqrt{6} = 482 \text{ Å}.$$

Table 4.2 Solution Viscosity Nomenclature

Name	Symbol	Definition[a]
Relative viscosity	η_r	$\eta_r = \eta/\eta_0$
Specific viscosity	η_{sp}	$\eta_{sp} = \eta_r - 1 = \eta/\eta_0 - 1$
Reduced viscosity	η_{sp}/c	$\eta_{sp}/c = (\eta_r - 1)/c$
Inherent viscosity	η_{inh}	$\eta_{inh} = \dfrac{\ln \eta_r}{c}$
Intrinsic viscosity[b]	$[\eta]$	$[\eta] = \lim_{c \to 0} \dfrac{1}{c}\left(\dfrac{\eta}{\eta_0} - 1\right)$

[a] η = solution viscosity; η_0 = solvent viscosity.
[b] Also called the *limiting viscosity number*.

4.4.4 Dilute Solution Viscometry

The dependence of viscosity on size permits estimation of an average molecular weight from solution viscosity. The average molecular weight that is measured is the viscosity average $\overline{M}_v$, which differs from $\overline{M}_n$ and $\overline{M}_w$ described so far in this chapter. The relation between viscosity increase and molecular weight that is used for the calculation of $\overline{M}_v$ is given by the Mark-Houwink-Sakurada (MHS) equation (see below). Before viscosity increase data are used to calculate $\overline{M}_v$ of the polymer, it is necessary to eliminate the effects of polymer concentration. The methods whereby this is achieved are described in a later section. These methods, however, do not remove the effects of polymer-solvent interactions, and so $\overline{M}_v$ of a given polymer sample will depend to some extent on the solvent used in the solution viscosity measurements.

The nomenclature commonly used for solution viscosity is represented in Table 4.2. It may be noted that the viscosity terms listed in it are not viscosities at all. Thus η_r and η_{sp} are actually unitless ratios of viscosities. The intrinsic viscosity $[\eta]$ is a ratio of viscosities divided by concentration and so has the units of reciprocal concentration, commonly quoted in cm^3/g or dL/g.

The Mark-Houwink-Sakurada (MHS) equation for the calculation of $\overline{M}_v$ is written as

$$[\eta] = K\overline{M}_v^a \tag{4.92}$$

where K and a are MHS constants. In terms of molecular weights (M_i) of monodisperse species i, $\overline{M}_v$ is given by [cf. Eqs. (4.24) and (4.25)]

$$\overline{M}_v = \left[\frac{\sum N_i M_i^{a+1}}{\sum N_i M_i}\right]^{1/a} = \left[\sum w_i M_i^a\right]^{1/a} \tag{4.93}$$

Problem 4.17   Assuming that the MHS constants K and a are independent of molecular weight, derive the definitions of $\overline{M}_v$ provided by Eq. (4.93).

Answer:

Consider a polymer sample to be made up of a series (i) of monodisperse macromolecules each with concentration (weight/volume) c_i and molecular weight M_i. Then,

$$c_i = N_i M_i \tag{P4.17.1}$$

where N_i is the concentration in terms of moles/volume. From the definition of intrinsic viscosity $[\eta]$ given in Table 4.2, one can write for the species i of monodisperse macromolecules in very dilute solution

$$(\eta_i/\eta_0) - 1 = c_i[\eta_i] \tag{P4.17.2}$$

Since K and a are independent of molecular weight, one can write for the species i in very dilute solution [cf. Eq. (4.92)]

$$[\eta_i] = KM_i{}^a \qquad (P4.17.3)$$

Substituting Eqs. (P4.17.1) and (P4.17.3) in Eq. (P4.17.2) yields

$$(\eta_i/\eta_0) - 1 = N_i K M_i^{a+1} \qquad (P4.17.4)$$

We may regard the viscosity of the solution of a whole polymer as the sum of the contributions of the monodisperse species that make up the polymer. That is,

$$\left[\frac{\eta}{\eta_0} - 1\right]_{\text{whole}} = \sum_i \left(\frac{\eta_i}{\eta_0} - 1\right) = K \sum_i N_i M_i^{a+1} \qquad (P4.17.5)$$

Substituting Eq. (P4.17.5) in the definition of intrinsic viscosity (see Table 4.2),

$$[\eta] = \lim_{c \to 0} \frac{1}{c}\left(\frac{\eta}{\eta_0} - 1\right)_{\text{whole}} = \lim_{c \to 0} \frac{K}{c} \sum N_i M_i^{a+1} \qquad (P4.17.6)$$

However, $c = \sum c_i = \sum N_i M_i$ and so $[\eta] = \lim_{c \to 0}\left(K \sum_i N_i M_i^{a+1} / \sum N_i M_i\right)$. Substituting this in Eq. (4.92) gives

$$K\overline{M}_v^a = \lim_{c \to 0}\left(K \sum_i N_i M_i^{a+1} / \sum_i N_i M_i\right) \qquad (P4.17.7)$$

So in the limit of infinite dilution,

$$\overline{M}_v = \left[\sum N_i M_i^{a+1} / \sum N_i M_i\right]^{1/a} \qquad (P4.17.8)$$

An alternative definition in terms of weight fraction composition of the polymer can be derived by noting that $\sum N_i M_i = \sum W_i = W$, where W_i and W are the concentrations in weight/vol. Equation (P4.17.8) is then converted to the form

$$\overline{M}_v = \left[\sum w_i M_i^a\right]^{1/a} \qquad (P4.17.9)$$

where w_i is the weight fraction of species i.

Problem 4.18   Show that the intrinsic viscosity of a mixture of polymers is the weight average value of the intrinsic viscosities of the components of the mixture in the given solvent.

Answer:

Let $[\eta_i]$ be the intrinsic viscosity of species i (monodisperse macromolecules) in a particular solvent. The MHS equation [Eq. (4.92)] for this species is

$$[\eta_i] = KM_i^a \qquad (P4.18.1)$$

From Eq. (4.93), for the whole polymer, $(\overline{M}_v)^a = \sum w_i M_i^a$. Substituting this in Eq. (4.92) and then combining with Eq. (P4.18.1) gives $[\eta] = \sum w_i [\eta_i]$.

It is important to note that $\overline{M}_v$, like $\overline{M}_n$ or $\overline{M}_w$, is a function of the molecular weight distribution of the polymer but unlike the latter it is also a function of the solvent (through the exponent a). According to Eq. (4.93), $\overline{M}_v = \overline{M}_n$ for $a = -1$ and $\overline{M}_v = \overline{M}_w$ for $a = 1$. For polymers that assume random coil shapes in solution, it is found that $0.5 \le a \le 0.8$. Since a is thus closer to 1 than to -1, $\overline{M}_v$ will be much closer to $\overline{M}_w$ than to $\overline{M}_n$.

4.4.4.1 Calibration of the Mark-Houwink-Sakurada Equation

The classical method for determining K and a values of the Mark-Houwink-Sakurada (MHS) equation involves fractionation of a whole polymer into subspecies, or fractions, with narrow molecular weight distributions. An average molecular weight can be determined on each such fraction, by osmometry ($\overline{M}_n$) or light scattering ($\overline{M}_w$), and, if the fractions are narrow enough, the measured average can be approximated to $\overline{M}_v$ of monodisperse polymer. The intrinsic viscosities measured at a constant temperature for a number of such fractions of known $\overline{M}_v$ are fitted to the equation

$$\ln[\eta] \;=\; \ln K + a\ln(\overline{M}_v) \tag{4.94}$$

to yield the MHS constants K and a for the particular polymer/solvent system at the temperature of viscosity measurement.   (Since actual fractions are not really monodisperse and $\overline{M}_v$ is closer to $\overline{M}_w$ than to $\overline{M}_n$, it is a better practice to determine the molecular weight by light scattering than by osmometry.)

The two constants K and a are derived from the intercept and slope of a linear least squares fit to $[\eta] - M$ values for a series of fractionated polymers. The method assumes that K and a are fixed for a given polymer type and solvent and do not vary with polymer molecular weight. This is not strictly true, however, and the MHS constants determined for higher-molecular-weight species may depend on the molecular weight range. Tabulations of such constants therefore usually list the molecular weights of fractions for which the particular K and a values were determined. Such a list for some common systems of more general interest is presented in Table 4.3.

Table 4.3 Values of Mark-Houwink-Sakurada Constants (K and a) for Some Polymer-Solvent Systems

Polymer	Solvent	Temp. (°C)	Mol. wt. range ($M \times 10^{-4}$)	$K \times 10^2$ (cm^3 g^{-1})	a
Poly(acryl-amide)	Water	30	2-50	0.63	0.80
Poly(acrylo-nitrile)	Dimethyl formamide	25	3-25	2.43	0.75
Polyisobutylene	Cyclohexane	30	0.05-126	2.65	0.69
	Toluene	25	14-34	8.70	0.56
Poly(methyl methacrylate)	Acetone	30	6-263	0.77	0.70
	Benzene	30	4-73	0.63	0.76
	Chloroform	30	13-263	0.43	0.80
Polystyrene	Benzene	25	3-70	0.92	0.74
	Toluene	25	4-52	0.85	0.75
Poly(vinyl acetate)	Acetone	30	2.7-130	1.02	0.72
	Methanol	30	2.7-130	3.14	0.60
Poly(vinyl alcohol)	Water	30	0.60-16	6.66	0.64
Cellulose triacetate	Chloroform	30	3-18	0.45	0.9

Source: Brandrup and Immergut, 1975.

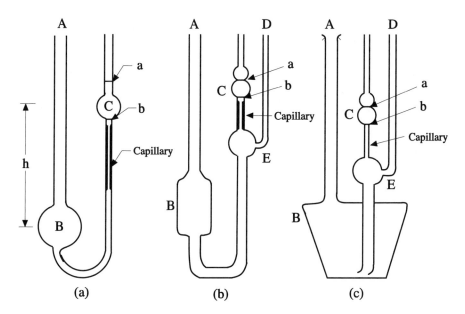

Figure 4.15 Common types of glass viscometers. (a) Ostwald viscometer; (b) Ubbelohde suspended-level viscometer; (c) A modified Ubbelohde suspended-level viscometer (see text for description).

4.4.4.2 Measurement of Intrinsic Viscosity

This determination is performed very easily with simple glass viscometers. Since the viscosity of a liquid depends markedly on temperature, viscosity measurements must be made at a carefully controlled temperature (within $\pm\ 0.1°C$). Before a measurement, the viscometer is therefore equilibrated in a carefully controlled thermostatic bath at the required temperature.

Two popular viscometers are of the Ostwald and Ubbelohde types, shown in Fig. 4.15. To operate the Ostwald viscometer, a given volume of liquid is introduced into bulb B through stem A and is drawn up by suction till it fills the bulb C and moves beyond the fiducial mark *a*. The suction is then released and the time taken by the liquid meniscus to pass between the fiducial marks *a* and *b* across the bulb C is measured. The liquid obviously flows under a varying driving force proportional to the changing difference (*h*) in the levels of the liquids in the two tubes. To ensure that this driving force is the same in all cases, the same amount of liquid must always be taken in bulb B.

The above condition of always using the same volume of liquid does not apply, however, in the case of Ubbelohde suspended level viscometer (Ubbelohde, 1937), shown in Fig. 4.15(b). A modified design of the Ubbelohde viscometer is shown in Fig. 4.15(c). For measurement with the Ubbelohde viscometer a measured volume of polymer solution with known concentration is pipetted into bulb B through stem A. This solution is transferred into bulb C by applying a pressure on A with compressed air while column D is kept closed. When the pressure is released, the solution in bulb E and column D drains back into bulb B and the end of the capillary remains free of liquid. The solution flows from bulb C through the capillary and around the sides of the bulb E into bulb B. The volume of liquid in B has no effect on the rate of flow through the capillary because there is no back pressure on the liquid emerging from the capillary as the bulb E is open to atmosphere. The flow time *t* for the solution meniscus to pass between the fiducial marks *a* and *b* on bulb C above the capillary is noted. Since the volume of solution in B has no effect on the flow time t, the solution in B can be diluted *in situ* by adding a measured amount of solvent through A. The diluted solution, whose concentration is easily calculated from the solvent added, is then raised up into C, as before, and the new flow time is measured. In this way, the concentration of the solution in B can be changed by successive dilution with measured volumes of solvent and the corresponding flow times can be determined.

With the assumption of a constant flow rate, the flow time t can be related to the viscosity η of the liquid by the Hagen-Poiseuille equation (Moore, 1972)

$$\eta = \pi r^4 \Delta P t / 8 V \ell \tag{4.95}$$

where ΔP is the pressure difference between the ends of a capillary tube of length ℓ and radius r, and V is the volume of liquid that flows through the capillary tube in time t. If the capillary tube is in a vertical position, as shown in Fig. 4.15, the driving force is essentially the weight of the liquid itself. The pressure difference, δP, which is the driving force per unit area, will therefore be given by

$$\Delta P = h \rho g \tag{4.96}$$

where h is the average height of the liquid during measurement, ρ the density of the solution, and g the gravitational acceleration constant.

Substitution of Eq. (4.96) in Eq. (4.95) yields

$$\eta = \frac{\pi r^4 h g \rho t}{8 V \ell} \tag{4.97}$$

This equation is applicable if the flow is Newtonian or viscous and all of the potential energy of the driving force is expended in overcoming the frictional resistance. These conditions are usually satisfied (Allcock and Lampe, 1990) for the apparatus used and for measurement of relative viscosities η/η_0 (see Table 4.2) which are of interest in polymer chemistry.

Denoting the terms related to solvent with subscript zero, a ratio of viscosities of solution and solvent in terms of respective flow times for the same volume V through the same capillary is obtained from Eq. (4.97) as

$$\frac{\eta}{\eta_0} = \frac{\rho t}{\rho_0 t_0} \tag{4.98}$$

For dilute solutions ρ is very close to ρ_0 and Eq. (4.98) simplifies to

$$\frac{\eta}{\eta_0} = \frac{t}{t_0} \tag{4.99}$$

Thus, the ratio of viscosities needed for the determination of $[\eta]$ (see Table 4.2) can be obtained from flow times without measuring absolute viscosities.

The intrinsic viscosity $[\eta]$ is defined in Table 4.2 as a limit at zero concentration. Since the η/η_0 ratios are obtained from Eq. (4.99) with measurements made at finite concentrations of the solution, it becomes necessary to extrapolate the data to zero concentration in order to satisfy this definition. There are a variety of ways to carry out this extrapolation. The variation in solution viscosity (η) with increasing concentration (c) can be expressed as a power series in c. The equations usually used are the Huggins equation (Huggins, 1942):

$$\frac{\eta_{sp}}{c} = \frac{1}{c}\left(\frac{\eta}{\eta_0} - 1\right) = [\eta] + k_H [\eta]^2 c + k'_H [\eta]^3 c^2 + \cdots \qquad (4.100)$$

and the Kraemer equation (Kraemer, 1938):

$$\eta_{inh} = \frac{\ln(\eta/\eta_o)}{c} = [\eta] - k_1 [\eta]^2 c - k'_1 [\eta]^3 c^2 - \cdots \qquad (4.101)$$

It is easy to show that both equations should extrapolate to a common intercept equal to $[\eta]$. The usual calculation procedure thus involves a double extrapolation of Eqs. (4.100) and (4.101) on the same plot (see Problem 4.19) to determine $[\eta]$ and hence $\overline{M}_v$ from the MHS equation.

Problem 4.19   The following are data (Schultz and Blaschke, 1941) from viscosity measurements with an Ostwald viscometer ($r = 1.5 \times 10^{-2}$ cm, $\ell = 11$ cm) on a solution of poly(methyl methacrylate) in chloroform at 20°C.

Concentration (g/cm^3) $\times 10^2$	Flow time (s)
0.0000	170.1
0.03535	178.1
0.05152	182.0
0.06484	185.2
0.100	194.3
0.200	219.8
0.400	275.6

(a) Determine $[\eta]$ by plotting η_{sp}/c and η_{inh} against c.

(b) Find $\overline{M}_v$ for this polymer, for which the MHS equation is $[\eta] = 3.4 \times 10^{-3}\, \overline{M}_v^{0.80}$ cm^3/g.

Answer:

Neglecting kinetic energy correction since the error is small, $\eta_{sp}/c = \dfrac{t - t_0}{t_0 c}$ and

$\eta_{inh} = \dfrac{\ln(t/t_0)}{c}$, where $t_0 =$ flow time for solvent $= 170.1$ s and $t =$ flow time for solution.

Both η_{sp}/c and η_{inh} are plotted on the same graph in Fig. 4.16. The common intercept of the plots on the ordinate at $c = 0$ gives $[\eta] = (\eta_{sp}/c)_{c=0} = (\eta_{inh})_{c=o} = 1.325 \times 10^2$ cm^3 g^{-1}. From Eq. (4.92):

$$\overline{M}_v = \left(\frac{1.325 \times 10^2}{3.4 \times 10^{-3}}\right)^{1/0.8} = 5.47 \times 10^5$$

A useful initial concentration for solution viscometry of most synthetic polymers is about 1 g/100 cm^3. High-molecular-weight polymers may require lower concentrations to produce a linear plot (Fig. 4.16) of $(\eta/\eta_0 - 1)/c$ against c, which does not curve away from the c-axis at higher concentrations. At very low concentrations such plots may curve upwards, which is attributed to adsorption of polymer on the capillary walls.

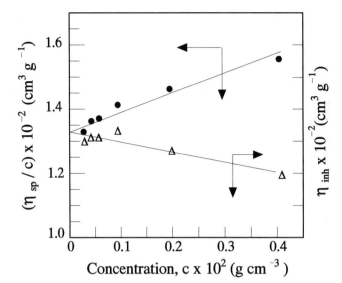

Figure 4.16 Plot of η_{sp}/c and η_{inh} against c (Data of Problem 4.19).

4.4.5 Gel Permeation Chromatography

Gel permeation chromatography (GPC) is an extremely powerful method for determining the complete molecular weight distribution and average molecular weights. It is essentially a process for the separation of polymer molecules according to their size. The separation occurs as a dilute polymer solution is injected into a solvent stream which then passes through a column packed with porous gel particles, the porosity being typically in the range 50-10^6 Å. GPC is also known as *gel filtration, gel exclusion chromatography, size-exclusion chromatography* (SEC), and *molecular sieve chromatography.*

The principle of GPC is simple. It is shown schematically in Fig. 4.17. A schematic layout of a typical GPC system is shown in Fig. 4.18. Consider a stationary column packed with finely divided solid particles, all having the same pore size. The smaller polymer molecules which are able to enter the pores (tunnels) of the gel particles will have longer effective paths than larger molecules and will hence be "delayed" in their passage (elution) through the column. On the other hand, larger polymer molecules with coil size greater than the pore diameter will be unable to enter the pores and will thus be swept along with the solvent front to appear in the exit from the GPC column (Fig. 4.17) and reach the detector (Fig. 4.18) ahead of the smaller molecules. (In reality, the column packing itself has a distribution of pore sizes and this improves the effectiveness of the fractionation process for the whole molecular weight distribution.) The volume of solvent thus required to elute a particular polymer species from the point of injection to the detector is known as its *elution volume*. Molecular weight can be calculated from the GPC data only after calibration of the GPC system in terms of elution volume or retention time with polymer standards of known molecular weights (see below).

The most common type of column packing used for analysis of synthetic polymers consists of polystyrene gels, called *styragel* particles (hence the name of *gel* permeation). These are highly porous polystyrene beads and are highly cross-linked so that they can be packed tightly without clogging the columns when the sample and the elution solvent are pumped through them under pressure. These are used for organic solvents. For GPC in aqueous systems, the common packings are crosslinked dextran (*Sephadex*) and polyacrylamide (*Biogel*). Porous glass beads can be used with both aqueous and organic solvents.

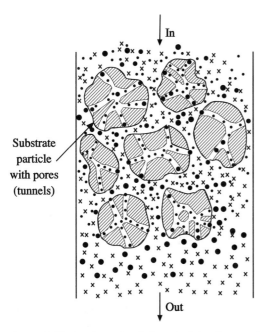

Figure 4.17 A schematic of the principle of separation by gel permeation chromatography. Black circles represent molecules of coil sizes ≤ pore diameter, while crosses represent molecules of coil sizes > pore diameter. If a sample with molecular size distribution enters the column at the same time, the molecules will emerge from the column sequentially, separated according to molecular size, from larger to smaller.

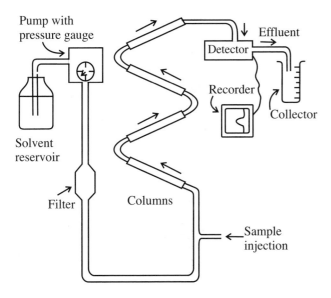

Figure. 4.18 Schematic layout of a typical gel permeation chromatography apparatus. It is, however, a normal practice to use a set of several columns each packed with porous gel particles having a different porosity, depending on the range of molecular sizes to be analyzed. (Adapted from Allcock and Lampe, 1990.)

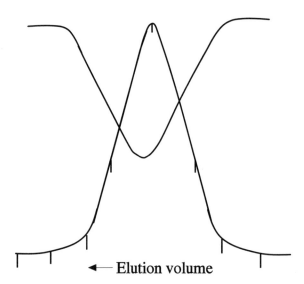

← Elution volume

Figure 4.19 A typical gel permeation chromatogram. The lower trace with short vertical lines is the differential refractive index while the upper curve is an absorption plot at a fixed ultraviolet frequency. The short vertical lines are syphon dumps counted from the time of injection of the sample. The units of the ordinate depend on the detector, while those of the abscissa can be in terms of syphon volumes (counts) or volume of solvent. (After Rudin, 1982.)

After passage through the column(s), the solvent stream (eluant) carrying the size-separated polymer molecules passes through a detector, which responds to the weight concentration of polymer in the eluant. The most commonly used detector is a differential refractometer. It measures the difference in refractive index between the eluted solution and the pure solvent. This difference is proportional to the amount of polymer in solution. Spectrophotometers are also used as alternative or auxiliary detectors.

The *elution volume* (also called the retention volume) is the volume of solvent that has passed through the GPC column from the time of injection of the sample. It is conveniently monitored by means of a small siphon, which actuates a marker every time it fills with eluant and dumps its contents. The raw GPC data are thus available as a trace of detector response proportional to the amount of polymer in solution and the corresponding elution volumes. A typical GPC record (gel permeation chromatogram) is shown in Fig. 4.19.

4.4.5.1 Data Interpretation and Calibration

A GPC chromatogram (Fig. 4.19) can yield a plot of the molecular weight distribution, since the ordinate corresponding to the detector response can be transformed into a weight fraction of total polymer while by suitable calibration the elution volume axis can be transformed into a logarithmic molecular weight scale (explained below). Using a baseline drawn through the recorder trace, the chromatogram heights are measured for equal small increments of elution volume. The weight fraction corresponding to a particular elution voume is taken as the height of the ordinate divided by the sum of the heights of all the ordinates under the trace (see Problem 4.21). This process normalizes the chromatogram.

Let us now consider how the elution volume axis of a chromatogram, such as shown in Fig. 4.19, can be translated into a molecular weight scale. This necessitates calibration of the particular GPC column using monodisperse polymer samples. The main problem encountered in this task is

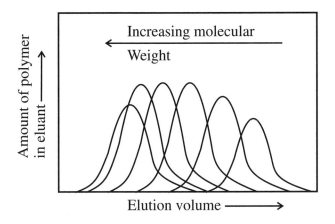

Figure 4.20 Gel permeation chromatography elution curves for polymer standards having very narrow molecular weight distribution. (After Rudin, 1982.)

that monodisperse or very narrow distribution samples of most polymers are not generally available. However, such samples are available for a few specific polymers. For polystyrene, for example, anionically polymerized samples of narrow molecular weight distributions (MWD) with polydispersity index less than 1.15 are commercially available in a wide range of molecular weights (10^3 to 10^6). Using such narrow MWD samples, a polystyrene calibration of molecular weight versus elution volume can be easily obtained for the given GPC column (or columns) and the given GPC solvent. A problem that would then remain is to establish a relationship for the particular GPC column between the elution volume and molecular weight of some chemically different polymer. We now consider these two steps in sequence.

When a narrow MWD polystyrene sample, as described above, is injected into the GPC column, the resulting chromatogram, though narrow, is not a simple spike because of band-broadening effects and because the polymer itself is not truly monodisperse. However, no significant error is committed by assigning the elution volume corresponding to the peak of the chromatogram to the molecular weight of the polystyrene sample, since the distribution is very narrow. Thus, a series of narrow MWD polystyrene samples used with the particular GPC column and the GPC solvent yield a set of GPC chromatograms, as shown in Fig. 4.20. The peak elution volumes and the corresponding molecular weights thus provide a polystyrene calibration curve (Fig. 4.21) for the particular GPC column and solvent used. In the next step, known as *universal calibration*, the polystyrene calibration curve is translated to one that will be effective for another given polymer in the same apparatus and solvent.

To extend the calibration to other polymers, a calibration parameter that is independent of the chemical nature of the polymer, that is, a *universal calibration parameter*, is required. Experimentally, it has been found (Grubisic et al., 1967) that such a parameter could be the product of the intrinsic viscosity and molecular weight (i.e., $[\eta]M$). Thus, as shown in Fig. 4.22, the logarithm of the product $[\eta]M$ plotted against elution volume, with tetrahydrofuran used as the solvent, provides a *single curve* for a wide variety of polymers, which thus suggests that a universal calibration procedure may be possible. (Such a single curve for different polymers is not obtained, however, by simply plotting $\log M$ against elution volume.)

A theoretical validity of the aforesaid experimental observation is obtained from a consideration of the hydrodynamic volume of the polymer, as shown below. As long ago as 1906, it was shown by Einstein (1906) that the viscosity of a dilute suspension relative to that of the suspending

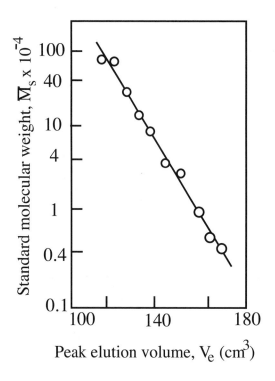

Figure 4.21 A typical polystyrene standard calibration curve (M_s vs. V_e) for GPC.

medium is given by the expression

$$\eta_r = (\eta/\eta_0) = 1 + \omega\phi \tag{4.102}$$

where η_r is the relative viscosity, η the viscosity of the solution, η_0 the viscosity of the pure solvent, ϕ the volume fraction of the suspended material, and ω is a factor that depends on the general shape of the suspended particle. For spherical particles, ω is 2.5 and this equation then becomes

$$\eta/\eta_0 - 1 = 2.5\phi \tag{4.103}$$

If all polymer molecules exist in solution as discrete entities, without overlap and each solvated molecule of a monodisperse polymer has an equivalent volume (or hydrodynamic volume) V and molecular weight M, then the volume fraction ϕ of solvent-swollen polymer coils at a concentration c (mass/volume) is

$$\phi = cVN_{Av}/M \tag{4.104}$$

where N_{Av} is Avogadro's number. Combination of Eqs. (4.103) and (4.104) yields

$$\frac{1}{c}\left(\frac{\eta - \eta_0}{\eta_0}\right) = \frac{2.5VN_{Av}}{M} \tag{4.105}$$

Implicit in this equation is the assumption that the contributions of the individual macromolecules to the viscosity increase are independent and additive, which is true when the polymer solution is

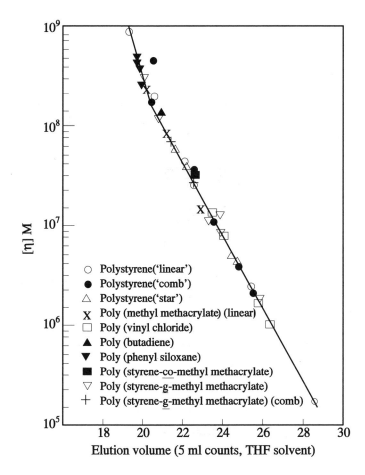

Figure 4.22 A universal calibration curve for several polymers in tetrahydrofuran. (Drawn with data of Grubisic et al., 1967.)

very dilute. Viscosity data are therefore used to extrapolate the left-hand side of Eq. (4.105) to zero concentration conditions, thus giving

$$[\eta] \equiv \lim_{c \to 0} \frac{1}{c} \left(\frac{\eta - \eta_0}{\eta_0} \right) = \left(\frac{2.5 N_{Av}}{M} \right) \lim_{c \to 0} V \qquad (4.106)$$

Multiplying both sides by M and taking logarithm gives

$$\log([\eta]M) = \log(2.5 N_{Av}) + \log\left(\lim_{c \to 0} V \right) \qquad (4.107)$$

The product $[\eta]M$ is thus seen to be a direct function of the hydrodynamic volume of the solute at infinite dilution. Since studies of GPC separations have shown that polymers appear in the eluate in inverse order of their hydrodynamic volumes in the particular solvent, it may thus be stated that *two different polymers that appear at the same elution volume in a given solvent and particular GPC column at a given temperature have the same hydrodynamic volumes* and hence the same

[η]M characteristics; that is,

$$\log \left([\eta]_x M_x\right) \; = \; \log \left([\eta]_s M_s\right) \tag{4.108}$$

where the subscripts x and s indicate the unknown polymer (i.e., polymer with unknown molecular weight) and the standard polymer, respectively. If each intrinsic viscosity term in Eq. (4.108) is replaced by its MHS expression [Eq. (4.92)], one obtains for the two polymers at equal elution volumes:

$$\log \left(K_x M_x^{a_x+1}\right) \; = \; \log \left(K_s M_s^{a_s+1}\right) \tag{4.109}$$

Solving for $\log M_x$ gives

$$\log M_x \; = \; \left(\frac{1}{1+a_x}\right) \log \frac{K_s}{K_x} + \frac{1+a_s}{1+a_x} \log M_s \tag{4.110}$$

This equation describes the elution volume calibration curve for M_x. The elution volume (V_e) that corresponds to a GPC peak in the unknown polymer is used to obtain a value of $\log M_s$ from the polystyrene standard curve (Fig. 4.21) that has been obtained in the same column and solvent, and M_x is then calculated from Eq. (4.110). Alternatively, a number of values of V_e can be chosen and a new calibration curve for the polymer under study can be constructed using a standard curve such as Fig. 4.21 and Eq. (4.110).

The above procedure is applicable only if the MHS constants, K_s, a_s, K_x, and a_x, are known. Values of K_s and a_s are available in the literature for the standard polymer in various solvents (Table 4.3), and in many cases values of K_x and a_x may also be available. However, if the desired MHS constants are not available for the polymer under study in the GPC solvent or for the standard in the same solvent, they can be determined by measuring the intrinsic viscosities, as described earlier.

Problem 4.20   A series of narrow distribution polystyrene standards dissolved in chloroform were injected into a GPC column at 35°C yielding a set of chromatograms. The following data of peak elution volumes and corresponding sample molecular weights were reported (Dawkins and Hemming, 1975):

$\overline{M}_s \times 10^{-3}$ (g/mol)	867	670	411	160	98.2	51	19.8	10.3	3.7
V_e (cm^3)	122.7	126.0	129.0	136.5	141.0	147.0	156.5	162.5	170.0

Using the above data for polystyrene standards, construct a calibration curve for the molecular weight-elution volume of polymer X in chloroform at 35°C. The MHS constants in chloroform at 35°C may be taken as: $K = 4.9 \times 10^{-3}$ cm^3/g, $a = 0.79$ for polystyrene and $K = 5.4 \times 10^{-3}$ cm^3/g, $a = 0.77$ for polymer X.

Answer:

A semilogarithmic plot of M_s versus V_e gives the polystyrene calibration curve (cf. Fig. 4.21) for the given GPC column, solvent, and temperature.

Substitution of the MHS constants in Eq. (4.110) gives the expression: $\log M_x \; = \; -0.0238 + 1.0113 \log M_s$. To construct an elution calibration curve (M_x vs. V_e) for polymer X, various values of V_e are assumed and corresponding to each of them the M_s value is first obtained from the polystyrene calibration curve (Fig. 4.21) and then M_x from the above expression. A semilog plot of M_x versus V_e gives the required calibration curve (Fig. 4.23).

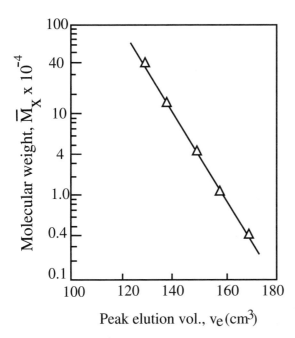

Figure 4.23 Elution calibration curve (M_x vs. V_e) for polymer X (Problem 4.20) derived from polystyrene calibration curve and MHS constants.

Problem 4.21 The GPC column of Problem 4.20 was used for the determination of molecular weight of a sample of the same polymer X. After injecting a chloroform solution of the polymer into the GPC column, the refractive index difference ($\Delta\tilde{n}$) between the eluted solution and pure solvent was measured as a function of elution volume (V_e), which yielded the following data:

$\Delta\tilde{n} \times 10^5$	0.6	3.4	12.4	15.0	9.9	3.0	0.4
V_e (4-mL count)	40	39	38	37	36	35	34

Using the calibration curve obtained in Problem 4.20, calculate $\overline{M}_n$, $\overline{M}_w$ and the polydispersity index of the sample.

Answer:

The molecular weight corresponding to each elution volume is determined from the elution calibration curve for this polymer in Fig. 4.23. The corresponding weight fraction w_i is computed from the refractive index difference by the following relation based on the assumption that $\Delta\tilde{n}$ is proportional to concentration and the proportionality factor is independent of molecular weight: $w_i = \Delta\tilde{n}_i / \sum \Delta\tilde{n}_i$. The results are tabulated below.

Elution volume (4-mL count)	$w_i = \Delta\tilde{n}_i / \sum \Delta\tilde{n}_i$	$M_i \times 10^{-3}$ (from Fig. 4.23)	$w_i M_i$	$(w_i/M_i) \times 10^6$
34	0.009	182	1620	0.049
35	0.067	117	7851	0.573
36	0.221	76	16834	2.914
37	0.336	50	16780	6.712
38	0.277	33	9154	8.406
39	0.076	21	1598	3.624
40	0.013	14	187	0.957
			$\sum = 54 \times 10^3$	$\sum = 23.236$

Therefore, $\overline{M}_w = \sum w_i M_i = 54 \times 10^3$ g mol^{-1}; $\overline{M}_n = 1/\sum (w_i/M_i) = 43 \times 10^3$ g mol^{-1}; polydispersity index = $\overline{M}_w / \overline{M}_n = 1.25$.

REFERENCES

Allcock, H. R. and Lampe, F. W., "Contemporary Polymer Chemistry", 2nd ed., Prentice Hall, Englewood Cliffs, N.J., 1990.

Brandrup, J. and Immergut, E., Eds., "Polymer Handbook", 2nd ed., Wiley Interscience, New York, 1975.

Billingham, N. C., "Molar Mass Measurements in Polymer Science", Kogan Page, London, 1977.

Billmeyer, F. W., Jr., "Textbook of Polymer Science", Wiley, New York, 1984.

Dawkins, J. V. and Hemming, M., *Makromol. Chem.*, **176**, 1777 (1975).

Debye, P., *J. Appl. Phys.*, **15**, 338 (1944).

Doty, P. and Edsall, J. T., "Advances in Protein Chemistry", vol. VI, pp. 35-121, Academic Press, New York, 1951.

Debye, P., *J. Phys. Colloid Chem.*, **51**, 18 (1947).

Einstein, A., *Ann. Physik*, **19**, 289 (1906).

Einstein, A., *Ann. Physik*, **33**, 1275 (1910).

Grubisic, Z., Rempp, P., and Benoit, H., *Polymer Lett.*, **5**, 753 (1967).

Huggins, M. L., *J. Am. Chem. Soc.*, **64**, 2716 (1942).

Kraemer, E. O., *Ind. Eng. Chem.*, **30**, 1200 (1938).

Kronberg, B. and Patterson, D., *Macromolecules*, **12**, 916 (1979).

Margerison, D. and East, G. C., "An Introduction to Polymer Chemistry", Pergamon Press, New York, 1967.

Moore, W. J., "Physical Chemistry", 4th ed., Prentice Hall, Englewood Cliffs, N.J., 1972.

Rudin, A., "The Elements of Polymer Science and Engineering", Academic Press, Orlando, 1982.

Schultz, G. V. and Blaschke, F., *J. Prakt. Chem.*, **158**, 130 (1941).

Ubbelohde, L., *Ind. Eng. Chem., Anal. Ed.*, **9**, 85 (1937).

Zimm. B. H., *J. Chem. Phys.*, **16**, 1099 (1948).

EXERCISES

4.1 If a polymer sample contains an equal number of moles of species with degrees of polymerization $x = 1$, 2, 3, 4, 5, 6, 7, 8, 9, and 10, what are the number-average and weight-average degrees of polymerization ?
[*Ans.* $\overline{DP}_n = 5.5$; $\overline{DP}_w = 7$]

4.2 One gram of polymer A ($\overline{DP}_n = 1000$, $\overline{DP}_w = 2000$) is mixed with 2 g of polymer B ($\overline{DP}_n = 2000$, $\overline{DP}_w = 5000$). Calculate the degree of polymerization of the mixture that would be derived from osmotic pressure measurements at several concentrations.
[*Ans.* $\overline{DP}_n = 1500$; $\overline{DP}_w = 4000$]

4.3 Polyethylenes A, B, and C from three sources are to be blended to achieve a weight-average molecular weight of 210,000 and polydispersity index (PDI) of 3.0. How much of each polymer should be used to obtain 10,000 kg blend ?

Polyethylene	Weight-average mol. wt.	PDI
A	130,000	2.50
B	220,000	2.00
C	400,000	2.50

[*Hint:* $\overline{M}_w = \sum w_i (\overline{M}_w)_i$; $1/\overline{M}_n = \sum w_i/(\overline{M}_n)_i$; PDI $= \overline{M}_w/\overline{M}_n$.]
[*Ans.* A = 1892 kg, B = 5776 kg, C = 2332 kg; (Total 10,000 kg)]

4.4 Polymer samples A and B are monodisperse polyisobutylenes, while sample C is a polydisperse polyisobutylene. Sample A is known to have a molecular weight twice that of B and $\overline{M}_w$ for sample C is

given as 1.8×10^5. Deduce the $\overline{M}_n$ for sample C from the following two measurements on a mixture of all three samples. The mixture contains 30 g of A, 40 g of B, and 30 g of C. Light scattering measurements give a molecular weight of 94,000, while the measurement of osmotic pressure gives a molecular weight of 50,000.
[*Ans.* 48,000 g mol^{-1}]

4.5 By a fractional precipitation a polymer A with most probable distribution ($I = 2.0$) and an osmotic molecular weight of 120,000 is separated into two fractions B and C with molecular weights, 200,000 and 300,000, respectively, determined by light scattering. Calculate the weight of B that would be obtained from 100 g of initial polymer A. If both B and C have the same polydispersity, what is the value of the polydispersity index ?
[*Ans.* $W_B = 60$ g; $I = 1.92$]

4.6 A polymer P having number-average and weight-average molecular weights of 120,000 and 300,000, respectively, was cut into fractions A and B by fractional precipitation. If A and B have number-average molecular weights of 90,000 and 130,000, respectively, what are the weight fractions of A and B obtained from the initial polymer P ? If A and B have the same polydispersity, what is the polydispersity index ?
[*Ans.* $w_A = 0.19$; $w_B = 0.81$; $I = 2.45$]

4.7 By fractional precipitation, 200 g of polymer P with the most probable distribution and a number-average molecular weight $\overline{M}_n = 150,000$ is separated into two fractions, A and B, the former weighing 68.0 g. Light scattering of A gives a molecular weight of 250,000. If the polydispersity index of A is the same as that of B, what is the number-average molecular weight of B ?
[*Ans.* $(\overline{M}_n)_B = 143,300$]

4.8 For end-group analysis, 0.8632 g of a carboxyl terminated polybutadiene (CTPB) sample dissolved in 1:3 mixture of ethanol and toluene consumed 5.2 ml of 0.1240 N alcoholic potassium hydroxide solution in titration using phenolphthalein as the indicator. Calculate the molar mass of the polymer.
[*Ans.* 2672 g mol^{-1}]

4.9 A linear polyester was synthesized from a mixture of diacid and glycol with carboxyl to hydroxyl ratio greater than unity. A sample of the polyester (3.5 g) requires titration with 23 mL of N/50 KOH to reach a phenolphthalein end point. By vapor-pressure osmometry the molecular weight of the polymer was determined to be 12,000. Calculate the average carboxyl functionality of the polymer (that is, average number of carboxyl groups per polymer molecule).
[*Ans.* 1.6]

4.10 About 1 g (dry) of a sample of polyester polyol of $\overline{M}_n = 3,000$ was treated with bromoacetyl bromide ($BrCH_2COBr$) to convert the hydroxyl end groups to bromoacetyl end groups. The treated polymer was found to contain 4.88% Br by elemental analysis. Estimate the average number of hydroxyl groups on each molecule of the polyol.
[*Ans.* 2.0]

4.11 A polyol for polyurethane foam was synthesized by polymerizing propylene oxide by using glycerol as the initiator (see Chapter on "Ring-Opening Polymerization"). For analysis, a sample of this polyol was reacted with phenyl isocyanate to convert the hydroxyl groups to urethanes which were then analyzed for nitrogen. The nitrogen content of the treated polymer was found to be 1.523%. Estimate the hydroxyl equivalent weight and molecular weight of the polyol.
[*Ans.* OH equiv. = 800 g (mol OH)$^{-1}$; $\overline{M}_n = 2,400$]

4.12 The molecular weight of a polymer determined by an osmotic pressure measurement in a theta solvent is 20,000. What osmotic pressure (atm) would be expected at a concentration of 1.20 g/dL and 30°C ? Would there be a difference in molecular weight if the osmotic measurement were made in a good solvent ?
[*Ans.* $\Pi = 0.015$ atm; no difference]

4.13 The osmotic pressure measurement of a 0.22 g/100 cm^3 solution of poly(vinyl chloride) in toluene at 25°C in the apparatus shown in Fig. 4.3 indicated a difference of 7.1 mm in the heights of the solution and solvent levels. (a) What is the osmotic pressure (atm) of the solution ? (b) If the second virial coefficient for poly(vinyl chloride) in toluene is $\Gamma_2 = 200$ cm^3/g, calculate $\overline{M}_n$ of the polymer. [Density of toluene at 25°C $= 0.867$ g/cm^3.]

[*Ans.* (a) 5.96×10^{-4} atm; (b) 1,34,000]

4.14 A polyisobutylene sample has $\overline{M}_n = 400,000$. The second virial coefficient of the polymer in chlorobenzene solution at 25°C is $\Gamma_2 = 94.5$ cm^3/g. Calculate the osmotic pressure in g/cm^2 of 0.30 g/dL solution of this polymer in chlorobenzene at 25°C. Compare this with the value calculated for an ideal solution. [Chlorobenzene density at 25°C is 1.11 g/cm^3.]

[*Ans.* $\Pi = 0.246$ g cm^{-2}; Π (ideal) $= 0.190$ g cm^{-2}]

4.15 A polystyrene sample was dissolved in benzene to obtain solutions of different known concentrations. Measurement of osmotic head for these solutions at 30°C with a static equilibrium osmometer produced the following data :

Concentration, c (g dL^{-1})	Osmotic head, h (cm C$_6$H$_6$)
0.260	0.140
0.540	0.516
0.755	0.966
0.920	1.389
1.025	1.730

Determine for the polymer (a) $\overline{M}_n$, (b) the second virial coefficient, and (c) the χ value in benzene at 30°C. [Density of benzene at 30°C $= 0.868$ g cm^{-3}.]

[*Ans.* (a) $\overline{M}_n = 1,972,000$; (b) $A_2 = 5.0\times10^{-4}$ mol cm^3 g^{-2}; (c) $\chi = 0.45$]

4.16 Calculate the fraction of light of $\lambda = 5460$ Å that would be scattered in a 1-cm path length through a 2.0×10^{-4} g/cm^3 solution of a polymer of $\overline{M}_w = 5.2\times10^4$ in a solvent of refractive index 1.3688. The specific refractive index increment of the polymer solution is 0.120 cm^3/g.

[*Ans.* $I_s/I_o = 1.74\times10^{-5}$]

4.17 For a polymer (0.14 g) dissolved in dimethylsulfoxide (100 mL) the ratio of the intensity of scattering at 90° to that at 0°, i.e., (i_{90}/i_0), is measured to be 0.01 with incident light of wavelength 546 nm. The solvent has a measured refractive index ($\tilde{n}$) of 1.475 at 21°C and a $d\tilde{n}/dc$ of 1.0. The turbidity τ can be obtained from the relation $\tau = k\tilde{n}^2(i_{90}/i_0)$

using the equipment constant value $k = 0.100$ cm^{-1} supplied with the light-scattering photometer. Determine the apparent molecular weight for the polymer sample. [Note : The molecular weight is called "apparent" since it is for a single point and not extrapolated to zero c.]

[*Ans.* 1.15×10^4]

4.18 Light scattering results for a polystyrene sample dissolved in benzene are shown below :

c, g dm^{-3}	1.38	3.29	4.90	7.00	9.42
$R'_{90}\times10^5$, cm^{-1}	4.51	7.78	9.91	11.89	13.30

Given : $K = 2.587\times10^{-7}$ cm^2 mol g^{-2}. Determine the weight-average molar mass of the polymer sample.

[*Ans.* 1,46,000 g mol^{-1}]

4.19 By measurement of light scattering ($\lambda = 436$ nm) from toluene solutions of a polystyrene sample [D. Rahlwes and R. G. Kirste, *Makromol. Chem.*, **178**, 1793 (1977)], the following results were obtained for the Rayleigh ratio, $R'(\theta)$, at various concentrations and scattering angles :

$c\times10^3$	$R'(\theta)\times10^4$ (cm^{-1})		
(g/cm^3)	15°	45°	75°
0.20	2.47	1.53	0.86
0.40	4.40	2.84	1.65
0.60	5.91	3.96	2.37
0.80	7.07	4.91	3.02
1.00	7.98	5.71	3.61

Make a Zimm plot using the data and determine the weight-average molecular weight of the polymer. Determine also the second virial coefficient Γ_2 and the radius of gyration of the polymer in solution. (For toluene, $\tilde{n} = 1.4976$ and for polystyrene-toluene solutions $d\tilde{n}/dc = 0.112$ cm^3/g.)
[*Ans.* $\overline{M}_w = 6.25 \times 10^6$; $\Gamma_2 = 460$ cm^3 g^{-1}; radius of gyration = 1027 Å.]

4.20 The results of light-scattering studies [D. Rahlwes and R. G. Kirste, *Makromol. Chem.*, **178**, 1793 (1977)] for $\lambda = 4360$ Å on a styrene/α-methyl styrene block copolymer in toluene solution were as follows:

$c \times 10^3$	$R'(\theta) \times 10^4$ (cm^{-1})		
(g/cm^3)	15°	45°	75°
0.20	1.91	1.47	1.01
0.40	3.55	2.78	1.95
0.60	4.95	3.94	2.80
0.80	6.16	4.96	3.59
1.00	7.19	5.87	4.31

The refractive index of toluene is 1.4976 and the specific refractive index increment of the polymer in toluene is 0.1263 cm^3/g.

(a) Construct a Zimm plot of the data and evaluate the weight-average molecular weight of the polymer.

(b) Evaluate the second virial coefficient A_2.

(c) Evaluate the root-mean-square end-to-end distance of the polymer in toluene solution.
[*Ans.* (a) $\overline{M}_w = 3.4 \times 10^6$; (b) $A_2 = 0.73 \times 10^{-4}$ cm^3 mol g^{-2}; (c) $\langle r^2 \rangle^{1/2} = 164$ nm]

4.21 From the light scattering data of a polystyrene sample given in Problem 4.20, determine the radius of gyration of the polymer in solution.
[*Ans.* 480 Å]

4.22 The relative flow times (t/t_o) of a poly(methyl methacrylate) polymer in chloroform solution are given below (Rudin, 1982):

Concentration (g/dL)	t/t_o
0.20	1.290
0.40	1.630
0.60	2.026

(a) Determine $[\eta]$ by plotting η_{sp}/c and η_{inh} against c.

(b) Find $\overline{M}_v$ for this polymer, given that $[\eta] = 3.4 \times 10^{-5} \, \overline{M}_v^{0.80}$ (dL/g)
[*Ans.* (a) $[\eta] = 1.32$ dL/g; (b) $\overline{M}_v = 545,000$]

4.23 Show that in the limit of infinite dilution, the reduced viscosity (η_{sp}/c) and inherent viscosity (η_{inh}) are equal.

4.24 The following data on intrinsic viscosities and GPC peak elution volumes at 25°C for standard polystyrene samples in tetrahydrofuran (THF) solution were obtained [M. Kolinsky and J. Janca, *J. Polym. Sci., Chem. Ed.*, **12**, 1181 (1974)]:

$\overline{M}_w \times 10^{-3}$	867	411	173	98.2	51	19.85	10.3	5.0
$[\eta]$ (cm^3/g)	206.7	125.0	67.0	43.6	27.6	14.0	8.8	5.2
V_e (cm^3)	149	157	177	186.5	199.5	219	234	253.5

Construct an elution volume calibration curve for poly(vinyl bromide) in THF at 25°C, given that the MHS constants for this system are $K = 1.59 \times 10^{-2}$ cm^3/g and $a = 0.64$ [A. Ciferri, M. Kryezewski, and G. Weil, *J. Polym. Sci.*, **27**, 167 (1958)]. A sample of the polymer in THF was injected into the same GPC column. The refractive index difference between the eluted solution and pure solvent was measured and plotted against the elution volume. This produced a broad peak, the maximum of which occurred at an elution volume of 180 cm^3. Calculate the $\overline{M}_w$ of the polymer corresponding to the peak maximum.
[*Ans.* 2.20×10^5]

Chapter 5

Condensation (Step-Growth) Polymerization

5.1 Introduction

Condensation polymerizations, also known as *step-growth* or simply *step polymerizations*, are merely classical organic reactions that are used to produce linear macromolecules starting from bifunctional monomers (that is, monomers containing two functional groups per molecule), or to produce polymer networks from mixtures of bifunctional and multifunctional monomers having three or more functional groups per molecule. The polymerization of bifunctional monomers may be described as a stepwise or progressive conversion of monomers with two reactive end groups to higher molecular-weight homologues, which themselves retain two reactive end groups. It may take place either by a *polycondensation reaction*, whereby a low-molecular-weight by-product is formed along with the polymer, as is exemplified by polyesterification:

$$n \, \text{HOOC} \, (CH_2)_x COOH \; + \; n \, \text{HO}(CH_2)_y OH \; \longrightarrow$$
$$\text{HO-}[OC(CH_2)_x COO(CH_2)_y O]_n H \; + \; (2n - 1) \, H_2O$$

or by a *polyaddition reaction* in which the total reactants are incorporated in the polymer chain (and no by-products are formed), as is typified by polyurethane formation:

$$(n + 1) \, \text{OCN} \, (CH_2)_x NCO \; + \; n \, \text{HO}(CH_2)_y OH \; \longrightarrow$$
$$\text{OCN-}[(CH_2)_x NHCOO(CH_2)_y OOCNH]_n (CH_2)_x NCO$$

These equations represent the overall reactions in the respective step-growth polymerizations. The growth of polymer molecules, however, occur by a stepwise intermolecular reaction. Thus two monomer molecules react to form a dimer; a dimer reacts with a monomer to form a trimer or with a dimer to form a tetramer, and so on. In fact, any two species in the reaction mixture can react with each other. Step polymerization can therefore be expressed by the general reaction:

$$n\text{-mer} \; + \; m\text{-mer} \; \longrightarrow \; (n + m)\text{-mer}$$

where n and m can have any value from 1 to very large number. Thus, a polyesterification reaction mixture of diacid and diol at any instance will consist of various-sized diol, diacid, and hydroxy-acid molecules. Any two of these molecules containing OH and COOH groups can react and the chemical reaction in each step is the same, which may be written as

$$\text{\small www-COOH} + \text{HO-\small www} \longrightarrow \text{\small www-COO-\small www} + H_2O \tag{5.1}$$

Similarly, polyamidation and polyurethane-forming reactions can be written as

$$\text{\small www-COOH} + H_2N\text{-\small www} \longrightarrow \text{\small www-CONH-\small www} + H_2O \tag{5.2}$$

$$\text{\small www-NCO} + \text{HO-\small www} \longrightarrow \text{\small www-NH-COO-\small www} \tag{5.3}$$

Implicit in these equations is the assumption that the functional group on the end of a monomer has the same reactivity as a similar group on a *n*-mer of any size and that the reactivities of both functional groups of bifunctional species (e.g., COOH groups of a diacid and OH groups of a diol) in the reaction mixture are the same. These simplifying assumptions are known as the concept of *equal reactivity of functional groups*. Experimental evidence and theoretical justifications have been provided in support of this concept (Flory, 1953; Odian, 1991).

A comparative account of the differences between step polymerization or condensation polymerization on the one hand and chain polymerization or addition polymerization (see Chapter 6) on the other hand is given in Table 5.1.

5.2 Rates of Polycondensation Reactions

Since the reactivity of two functional groups (or rate constant) is independent of the size of the molecule to which they are attached, it is possible to measure the polycondensation reaction rate simply by determining the concentration of functional groups as a function of time.

Let us consider the polyesterification of a diacid and a diol to illustrate the kinetic behavior of a typical step polymerization. Like simple esterification (Sykes, 1986), polyesterification is an acid-catalyzed reaction that can be represented by a sequence of reactions as shown by Eqs. (5.4)-(5.6). In these equations, the wavy lines (www) are used to signify that these equations apply, irrespective of the size of the molecular species. Equation (5.4) represents protonation of oxygen in carbon-oxygen double bond which leads to a more positive carbon atom for subsequent addition of a nucleophile, in this case www-OH [Eq. (5.5)], followed by elimination of H_2O and H^+ and formation of ester group [Eq. (5.6)]. The mechanism is generally referred to as A_{AC2} (Acid-catalyzed, acyl-oxygen cleavage, bimolecular). The rate limiting step is the nucleophilic addition (Neckers and Doyle, 1977).

$$\tag{5.4}$$

$$\tag{5.5}$$

$$\tag{5.6}$$

Table 5.1 Comparison of Step-Growth and Chain-Growth Polymerizations

Step-growth/condensation polymerization	Chain-growth/addition polymerization
Monomers bearing functional groups such as -OH, -COOH, -NH$_2$, -NCO, etc., undergo step polymerization.	Monomers with carbon-carbon unsaturation undergo polymerization when an active center is formed.
The growth of polymer molecules proceeds by a stepwise intermolecular reaction (at a relatively slow rate), normally with the elimination of small molecules as by-products of condensation, such as H$_2$O, HCl, NH$_3$, etc., in each step. The molecule never stops growing during polymerization.	Each polymer molecule/chain increases in size at a rapid rate once its growth has been started by formation of an active center. When the macromolecule stops growing (due to termination reaction) it can generally not react with more monomers (barring side reactions).
Monomer units can react with each other or with polymers of any size. Growth occurs in a series of fits and starts as the reactive species of a monomer or polymer encounters other species with which it can form a link. This can occur even in the absence of an added catalyst.	Growth of a polymer molecule is caused by a kinetic chain of reactions involving rapid addition of monomer to an active center that may be a free radical, ion, or polymer-catalyst bond. The active center is produced by some external source (energy, highly reactive compound, or catalyst).
At any moment during the course of polymerization, the reaction mixture consists of molecules of all sizes ranging from monomer, dimer, trimer, etc., to large polymer depending on the extent of conversion. The average size increases with conversion.	At any moment the reaction mixture essentially consists of full-grown, large polymer molecules, unreacted monomer molecules, and a very low concentration ($10^{-8} - 10^{-3}$ mol L^{-1}) of growing chains (i.e., possessing an active center) of intermediate sizes.
Molecular species in the intermediate stages of growth can be readily isolated. Molecular weight slowly increases throughout the process and it is only at a very high range of conversion that polymer molecules of very high molecular weight are obtained.	Molecular species in the intermediate stages of growth cannot be isolated. Conversion of monomer to polymer increases with time but the molecular weight of the polymer remains more or less unchanged with the progress of reaction.
Control of molecular weight is achieved by using a stoichiometric imbalance of the reactive functional group or a calculated amount of an appropriate monofunctional monomer.	Control of molecular weight is achieved by employing appropriate concentrations of initiator and monomer and temperature of polymerization or adding calculated amount of a chain transfer agent.
Backbone of polymer chains contains heteroatoms such as N, O, S, etc., at regular intervals due to condensed interunit links.	Usually the backbone of polymer chains consists of -C-C- linkages and other kinds of atoms such as O, N, S, etc., may appear in the side groups.

Source: Ghosh, 1990.

The rate of a step polymerization is most conveniently expressed in terms of the rate of disappearance of the reacting functional groups. Thus, the rate of polyesterification, R_p, can be expressed as the rate of disappearance of carboxyl groups:

$$R_p = \frac{-d[\text{COOH}]}{dt} \tag{5.7}$$

where [COOH] represents the concentration of unreacted carboxyl groups. The progress of a polyesterification reaction can thus be followed experimentally by titrating the unreacted carboxyl groups with a base during the reaction.

5.2.1 Irreversible Polycondensation Kinetics

For the usual polyesterification, k_1, k_{-1}, and k_3 are large compared to k_2. So when the reaction is run under nonequilibrium conditions by continuous removal of the by-product water, the polymerization rate can be considered to be synonymous with the rate of the forward reaction in Eq. (5.5) (Moore and Pearson, 1981):

$$R_p = \frac{-d[\text{COOH}]}{dt} = k_2[\text{C}^+(\text{OH})_2][\text{OH}] \tag{5.8}$$

where [OH] and [C$^+$(OH)$_2$] represent the concentrations of hydroxyl and protonated carboxyl groups, respectively. The equilibrium expression for the protonation reaction (Eq. 5.4) is:

$$K_p = \frac{k_1}{k_{-1}} = \frac{[\text{C}^+(\text{OH})_2]}{[\text{COOH}][\text{H}^+]} \tag{5.9}$$

Combination of Eq. (5.8) with Eq. (5.9) gives

$$\frac{-d[\text{COOH}]}{dt} = \frac{k_1 k_2 [\text{COOH}][\text{OH}][\text{H}^+]}{k_{-1}} \tag{5.10}$$

Equation (5.10) leads to two distinct kinetic situations depending on the source of H$^+$, that is, on whether or not a strong acid such as sulfuric acid or p-toluene sulfonic acid is externally added as a catalyst. The former is known as "catalyzed" polyesterification and the latter as "uncatalyzed" or "self-catalyzed" polyesterification (Odian, 1991).

In uncatalyzed or self-catalyzed polyesterification, the diacid monomer acts as its own catalyst for the esterification reaction. Assuming then that [H$^+$] is proportional to [COOH], Eq. (5.10) can be written (Flory, 1953) as

$$-\frac{d[\text{COOH}]}{dt} = k[\text{COOH}]^2[\text{OH}] \tag{5.11}$$

where k is the overall rate constant.

For equimolar initial concentrations of the diacid and diol, [COOH] = [OH] = C and Eq. (5.11) may then be simplified (Ghosh, 1990; Odian, 1991) to

$$-\frac{dC}{dt} = kC^3 \quad \text{or} \quad -\frac{dC}{C^3} = kdt \tag{5.12}$$

Integrating Eq. (5.12),

$$2kt = \frac{1}{C^2} - \frac{1}{C_0^2} \tag{5.13}$$

where C_0 is the initial concentration of functional groups (carboxyl or hydroxyl), that is, $C = C_0$ at $t = 0$.

It is convenient here and for many other purposes as well to introduce a parameter p, called the *extent of reaction* (Ghosh, 1990; Odian, 1991), which represents the fraction of functional groups initially present that have undergone reaction at time t, that is,

$$p = \frac{C_0 - C}{C_0} \quad \text{or} \quad C = C_0(1 - p) \tag{5.14}$$

Equation (5.13) then becomes

$$\frac{1}{(1 - p)^2} = 2C_0^2 kt + 1 \qquad (5.15)$$

A plot of $1/(1 - p)^2$ vs. t should thus be linear. However, when plots are made of experimental values of $1/(1 - p)^2$ versus time, it is found that Eq. (5.15) is not obeyed from $p = 0$ up to $p = 0.80$ (i.e., for the first 80% of the esterification of -COOH and -OH groups) (Solomon, 1967), while above 80% conversion the same equation is obeyed very well. It may be noted that the deviations below 80% conversion are also observed for simple esterifications such as when the dicarboxylic acid is replaced by a monocarboxylic acid. An important reason for deviations from Eq. (5.15) in the initial region below 80% conversion is that there occur large changes in the polarity of the reaction mixture as the polar carboxylic acid groups are converted to the less polar ester linkages. The *true* reaction rate constants are thus to be obtained from the linear plots of $1/(1 - p)^2$ versus time in this high conversion region (Solomon, 1972).

Problem 5.1 Equimolar mixture of 1,10-decanediol and adipic acid was polymerized under mild conditions to 82% conversion of the original carboxyl groups and the resulting product was further polymerized at a higher reaction temperature without any externally added catalyst, yielding the following data (Hamann et al., 1968) for condensation of the residual carboxyl and hydroxyl end groups. The reverse reaction of hydrolysis was avoided by removing the water of condensation by passing a stream of dry nitrogen through the mixture.

Temperature 190°C		Temperature 161°C	
Time (min)	% Reaction	Time (min)	% Reaction
0	0	0	0
30	20.6	20	9.1
60	39.0	40	16.0
90	50.2	100	31.6
150	61.2	150	41.1
225	66.8	210	47.9
300	71.5	270	52.5
370	74.4	330	57.0
465	77.2	390	60.0
510	78.2	450	62.6
550	78.8	510	64.6
600	79.6	550	65.5
660	80.6	700	69.2
730	81.7	840	71.9
800	82.5	880	72.4

Data from Hamann et al., 1968.

Determine the rate constants and the activation energy for the uncatalyzed reaction.

Answer:

Temperature 190°C

At $t = 0$, conversion (p) of the original COOH groups = 0.82 and $1/(1 - p)^2 = 1/(1 - 0.82)^2 = 0.3086 \times 10^2$.

At $t = 30$ min, conversion (p) of the original COOH groups = $0.82 + (1 - 0.82) \times 0.206 = 0.857$ and $1/(1 - p)^2 = 1/(1 - 0.857)^2 = 0.49 \times 10^2$.

Values of $1/(1-p)^2$ are similarly calculated for other t values and plotted against t in Fig. 5.1, which shows that Eq. (5.15) is obeyed very well in the later stages of conversion. Thus, $2C_0^2k = \text{slope} = 1.23 \text{ min}^{-1}$.

Molar mass of adipic acid $(C_6H_{10}O_4) = 146$ g mol^{-1}
Molar mass of 1,10-decanediol $(C_{10}H_{22}O_2) = 174$ g mol^{-1}

Considering a mixture of 1 mol adipic acid and 1 mol 1,10-decanediol, that is, total mass = 0.32 kg, $C_0 = [COOH]_0 = [OH]_0 = (2 \text{ moles carboxyl or hydroxyl group})/(0.32 \text{ kg}) = 6.25$ mol kg^{-1}. Therefore,

$$k_{190^0C} = \frac{(1.23 \text{ min}^{-1})}{2(6.25 \text{ mol kg}^{-1})^2} = 1.57 \times 10^{-2} \text{ kg}^2 \text{ mol}^{-2} \text{ min}^{-1}.$$

Temperature 161°C

The kinetic data for 161°C are similarly plotted in Fig. 5.1, from which $2[COOH]_0^2k = \text{slope} = 0.428 \text{ min}^{-1}$. Hence,

$$k_{161^0C} = \frac{(0.428 \text{ min}^{-1})}{2(6.25 \text{ mol kg}^{-1})^2} = 5.48 \times 10^{-3} \text{ kg}^2 \text{ mol}^{-2} \text{ min}^{-1}$$

From the Arrhenius expression $k = A\exp(-E/RT)$,

$$\ln\left[k_{190^0C}/k_{161^0C}\right] = \left(\frac{1}{434°K} - \frac{1}{463°K}\right) \frac{E}{(8.314 \text{ J mol}^{-1} \text{ °K}^{-1})}$$

$$E = \frac{\ln\left(k_{190^0C}/k_{161^0C}\right)(8.314 \text{ J mol}^{-1} \text{ °K}^{-1})}{\left(\frac{1}{434°K} - \frac{1}{463°K}\right)} = 6.06 \times 10^4 \text{ J mol}^{-1}$$

If a strong acid, such as sulfuric acid or p-toluene sulfonic acid, is added to a polyesterification system, it is a case of *catalyzed polyesterification* and $[H^+]$ in Eq. (5.10) then represents the concentration of this added catalyst. Since the catalyst concentration remains constant during the course of the polymerization [see Eqs. (5.4) and (5.6)], Eq. (5.10) can be written as

$$-\frac{d[COOH]}{dt} = k'[COOH][OH] \qquad (5.16)$$

where the three rate constants k_1, k_{-1}, k_2, and the catalyst concentration have been collected into a single rate constant k'. For a typical polyesterification (Hamann et al., 1968), k' of the catalyzed reaction [Eq. (5.16)] is nearly 2 orders of magnitude larger than k of uncatalyzed reaction [Eq. (5.11)]. Assuming $k' \gg k$, the rate of polymerization in the presence of strong acid catalyst can thus be represented by Eq. (5.16).

For equimolar initial concentrations of diacid and diol, $[COOH] = [OH] = C$ and Eq. (5.16) simplifies to

$$-\frac{dC}{dt} = k'C^2 \qquad (5.17)$$

Integration of Eq. (5.17) between $C = C_0$ at $t = 0$ and $C = C$ at $t = t$ yields

$$k't = \frac{1}{C} - \frac{1}{C_0} \qquad (5.18)$$

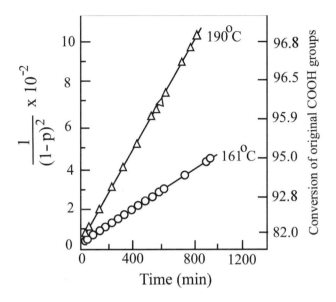

Figure 5.1 Plot of Eq. (5.15) for the later stages of uncatalyzed polyesterification of adipic acid and 1,10-decamethylene glycol (data of Problem 5.1). (Note that the reaction time of zero corresponds to 82% conversion of the original COOH groups present.)

Combination of Eq. (5.18) with Eq. (5.14) leads to

$$\frac{1}{(1-p)} = C_0 k' t + 1 \tag{5.19}$$

This equation provides a quantitative description of the dependence of conversion on reaction time for catalyzed polyesterification.

Problem 5.2   How would you modify Eq. (5.18) for a case of polyesterification where k is not negligible relative to k' ?

Answer:

If k is not negligible relative to k', the polymerization rate will be given by the sum of the rates of both the catalyzed and uncatalyzed polymerizations (Odian, 1991), that is,

$$-d[COOH]/dt = k[COOH]^3 + k'[COOH]^2 \tag{P5.2.1}$$

For equimolar initial concentrations of diacid and diol, $[COOH] = [OH] = C$ and Eq. (P5.2.1) simplifies to

$$-dC/dt = kC^3 + k'C^2 \tag{P5.2.2}$$

Integration of Eq. (P5.2.2) yields (Hamann et al., 1968) :

$$k't = \frac{k}{k'} \ln\left[\frac{C(kC_0 + k')}{C_0(kC + k')}\right] + \frac{1}{C} - \frac{1}{C_0} \tag{P5.2.3}$$

The first term on the right side of Eq. (P5.2.3) represents the contribution of the uncatalyzed polyesterification. For very small value of k/k', Eq. (P5.2.3) reduces to Eq. (5.18) for catalyzed polyesterification.

Problem 5.3   In an externally added acid catalyzed polyesterification, the conversion increased from 98% to 99% in 82 minutes. Calculate an estimate of time that was needed to reach the 98% conversion.

Answer:

From Eq. (5.19), $t = p/[k'C_0(1-p)]$. For $p = 0.98$, $t_{0.98} = 49/(k'C_0)$ and for $p = 0.99$, $t_{0.99} = 99/(k'C_0)$. Therefore, $t_{0.99} - t_{0.98} = 50/(k'C_0) = 82$ min, which yields $k'C_0 = 0.61$ min^{-1}. Therefore, $t_{0.98} = 49/(k'C_0) = 49/0.61 \simeq 80$ min. This is, however, an approximate estimate since Eq. (5.19) is not obeyed in the low conversion region.

Problem 5.4   In another series of experiments similar to those in Problem 5.1 the low-molecular weight polyester corresponding to 82% conversion of the original COOH groups were further polymerized at 161°C in the presence of *p*-toluene sulfonic acid (0.004 mol per mol of polymer) yielding the following conversion data (Hamann, et al., 1968) of the end hydroxyl and carboxyl acid groups:

Time (min)	% Reaction	Time (min)	% Reaction
0	0	40	82.9
5	34.6	50	85.7
10	54.7	60	87.9
15	65.5	75	90.1
20	70.8	90	91.5
30	77.9	105	92.6

Determine the rate constant for the catalyzed polyesterification.

Answer:

Let p = conversion of the original COOH groups.   At $t = 0$, $p = 0.82$ and $1/(1-p) = 1/(1 - 0.82) = 5.550$.   At $t = 5$ min, $p = 0.82 + (1 - 0.82) \times 0.346 = 0.8823$, and $1/(1-p) = 8.49$.

Values of $1/(1-p)$ are similarly calculated for other t values and plotted against t in Fig. 5.2 which shows that Eq. (5.19) is obeyed very well in the later stages of conversion. Also, a comparison with Fig. 5.1 shows that the catalyzed polyesterification is significantly faster than the uncatalyzed reaction. Now from Fig. 5.2, $C_0 k'$ = slope = 0.67 min^{-1}. Since C_0 = [COOH]$_0$ = 6.25 mol kg^{-1} (see Problem 5.1), k' = (0.67 min^{-1})/(6.25 mol kg^{-1}) = 0.107 kg mol^{-1} min^{-1}.

Step polymerizations are usually carried out with nearly stoichiometric amounts (i.e., equal moles) of the two reacting functional groups and Eqs. (5.15) and (5.19) are then applicable. The kinetics of step polymerizations with nonstoichiometric amounts of the groups can also be treated (Lin and Yu, 1978) in a similar way, though the final rate expressions are quite different (see Problem 5.5).

Problem 5.5   A nonstoichiometric mixture of a dicarboxylic acid and a glycol having 20 mol % excess of the latter and a carboxyl group concentration of 5.64 mol/kg was polymerized under nonequilibrium conditions with continuous removal of the by-product water, to about 80% esterification of the carboxylic acid groups. The mixture was then further polymerized in the presence of a strong acid catalyst which yielded the following data of the concentration of carboxylic acid groups in the mixture versus reaction time.

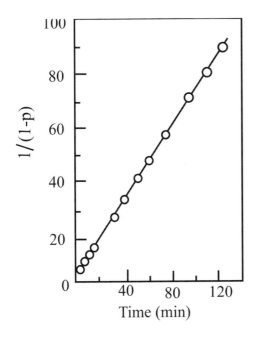

Figure 5.2 Later stages of catalyzed polyesterification of adipic acid and 1,10-decamethylene glycol (Data of Problem 5.2). (Note that the reaction time of zero corresponds to 82% conversion of the original COOH groups present.)

Time (min)	[COOH] (mol/kg)
0	1.24
5	0.83
10	0.48
15	0.40
20	0.31
25	0.21

Determine the reaction rate constant for the catalyzed polyesterification.

Answer:

Stoichiometric imbalance, $r = [COOH]_0/[OH]_0 = 1/1.2 = 0.833$, where $[COOH]_0$ and $[OH]_0$ are the concentrations of the original COOH and OH groups namely, $[COOH]_0 = 5.64$ mol kg^{-1}, and $[OH]_0 = 6.77$ mol kg^{-1}.

It is convenient to define the polymerization rate as the rate of disappearance of the functional groups present in deficient amount. Thus in the present case,

$$-d[COOH]/dt = k'[COOH][OH] \qquad \text{(P5.5.1)}$$

where k' is the rate constant of the catalyzed reaction.

In esterification, COOH and OH groups react in equal molar amounts. Therefore,

$$[COOH]_0 - [COOH] = [OH]_0 - [OH] \qquad \text{(P5.5.2)}$$

Combination of Eqs. (P5.5.1) and (P5.5.2) and integration (Moore and Pearson, 1981) of the differential equation leads to

$$\frac{1}{[OH]_0 - [COOH]_0} \ln\left(\frac{[COOH]_0[OH]}{[OH]_0[COOH]}\right) = k't \qquad \text{(P5.5.3)}$$

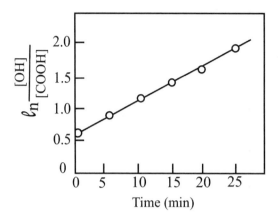

Figure 5.3 Test of Eq. (P5.5.4) for nonstoichiometric polyesterification.

In terms of r, defined above, Eq. (P5.5.3) can be rewritten as

$$\ln \frac{[OH]}{[COOH]} = -\ln r + \left(\frac{1-r}{r}\right)[COOH]_0 k' t \tag{P5.5.4}$$

To test Eq. (P5.5.4), $[OH]$ is calculated from Eq. (P5.5.2) and $\ln([OH]/[COOH])$ is plotted against t (Fig. 5.3). The plot is linear with a positive slope of 0.049 min^{-1}. Therefore,

$$\left(\frac{1-r}{r}\right)[COOH]_0 k' = 0.049 \text{ min}^{-1}$$

$$k' = (0.049 \text{ min}^{-1})\left(\frac{0.833}{1-0.833}\right)\left(\frac{1}{5.64 \text{ mol kg}^{-1}}\right) = 0.043 \text{ kg mol}^{-1} \text{ min}^{-1}$$

5.2.2 Reversible Polycondensation Kinetics

The concentration of a strong acid catalyst added to the reaction mixture for polyesterification remains constant throughout the course of polymerization. For the reversible polyesterification (as in a closed batch reactor) represented by

$$\text{wwwCOOH} + \text{wwwOH} \underset{k'_{-1}}{\overset{k'}{\rightleftarrows}} \text{wwwCOOwww} + H_2O \tag{5.20}$$

in which the initial hydroxyl group and carboxyl group concentrations are both C_0 and the reaction is catalyzed by an externally added strong acid catalyst, we thus have the rate equation

$$\frac{d[COO]}{dt} = -\frac{d[COOH]}{dt} = k'[COOH][OH] - k'_{-1}[COO][H_2O] \tag{5.21}$$

For an equimolar polymerization system, $[COO] = [H_2O] = pC_0$ and $[COOH] = [OH] = (1-p)C_0$, where p is the fractional conversion of carboxyl and hydroxyl groups. The rate equation is then obtained as

$$\frac{1}{C_0} \cdot \frac{dp}{dt} = k'(1-p)^2 - k'_{-1}p^2 \tag{5.22}$$

The integrated form (Levenspiel, 1972) of this equation is

$$\ln\left[\frac{p_E - (2p_E - 1)p}{p_E - p}\right] = 2k'\left(\frac{1}{p_E} - 1\right)C_0 t \tag{5.23}$$

where p_E is the fractional conversion at equilibrium.

At equilibrium, $d[COO]/dt = -d[COOH]/dt = 0$. The equilibrium constant K is then obtained from Eq. (5.21) as (Odian, 1991):

$$K = \frac{k'}{k'_{-1}} = \frac{[COO]_E[H_2O]_E}{[COOH]_E[OH]_E} = \frac{(p_E C_0)^2}{[C_0(1 - p_E)]^2} \tag{5.24}$$

Rearranging Eq. (5.24) gives

$$p_E^2(K - 1) - 2Kp_E + K = 0 \tag{5.25}$$

Solution of this quadratic equation yields 2 roots, namely, $p_E = \sqrt{K}/(\sqrt{K} - 1)$ and $\sqrt{K}/(\sqrt{K} + 1)$. The first root is meaningless since $p_E > 1$ for $K > 1$ and $p_E = $ -ve for $K < 1$. The equilibrium conversion p_E is therefore given by

$$p_E = \sqrt{K}/(\sqrt{K} + 1) \tag{5.26}$$

Combining Eqs. (5.23) and (5.26), p can be obtained (Levenspiel, 1972) as

$$p = \frac{\sqrt{K}[1 - \exp(-2\tau/\sqrt{K})]}{1 + \sqrt{K} - (\sqrt{K} - 1)\exp(-2\tau/\sqrt{K})} \tag{5.27}$$

where $\tau = k'C_0 t$.

Using Eq. (5.27), the extent of conversion in a closed batch reactor below the upper limit of equilibrium conversion p_E can be calculated as a function of time. The equation applies equally to A—B and stoichiometric A—A plus B—B polymerization where none of the reaction products are removed from the system and none is present initially.

Problem 5.6 Calculate the conversion that would be obtained in 1 h for reversible polyesterification in an equimolar system catalyzed by externally added strong acid. It is given that $C_0 k' = 5 \times 10^{-4}$ s^{-1} and $K = 1$. Compare with the conversion that would be obtained if the reaction were carried out in an irreversible manner by removing water from the system.

Answer:

Since $\tau = C_0 k't = (5 \times 10^{-4} \text{ s}^{-1})(3600 \text{ s}) = 1.80$, Eq. (5.27) gives

$$p = \frac{1 - \exp(-2 \times 1.80)}{1 + 1 - 0} = 0.49$$

For irreversible polymerization, Eq. (5.19) can be applied so that $1/(1 - p) = 1.80 + 1 = 2.80$ and hence $p = 0.64$.

Problem 5.7 Polyurethanes having the characteristic linkage –NHCOO– are formed when a diol undergoes condensation polymerization with a diisocyanate (see Table 1.2). The polymerization kinetics is found to be of third order at very low concentrations but of second order at higher concentrations. This led to the proposition (Baker et al., 1949) of the following reaction mechanism:

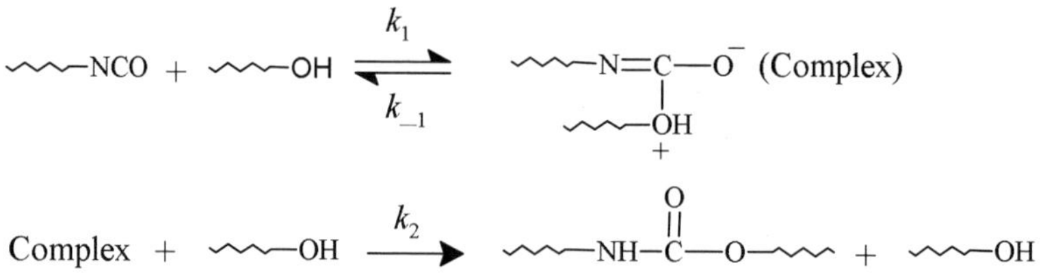

in which the alcohol acts as a weak-base catalyst. Derive a kinetic expression that accounts for the observed reaction orders.

Answer:

Corresponding to the proposed reaction scheme, the following rate expressions can be written:

$$-\frac{d[-NCO]}{dt} = k_1[-NCO][-OH] - k_{-1}[Complex] \tag{P5.7.1}$$

$$\frac{d[Complex]}{dt} = k_1[-NCO][-OH] - k_{-1}[Complex] - k_2[Complex][-OH] \tag{P5.7.2}$$

Application of the steady-state approximation, $d[Complex]/dt = 0$, to Eq. (P5.7.2) yields

$$[Complex] = \frac{k_1[-NCO][-OH]}{k_2[-OH] + k_{-1}} \tag{P5.7.3}$$

Substitution of Eq. (P5.7.3) into Eq. (P5.7.1) then gives

$$-\frac{d[-NCO]}{dt} = \frac{k_1 k_2[-NCO][-OH]^2}{k_{-1} + k_2[-OH]} \tag{P5.7.4}$$

Equation (P5.7.4) conforms to the experimental observation that the kinetics of the reaction changes from second to third order as [–OH] reduces.

5.3 *Number-Average Degree of Polymerization*

The number-average degree of polymerization of the *reaction mixture*, $\overline{X}_n$, is defined (Ghosh, 1990; Odian, 1991) as the total number (N_0) of monomer molecules initially present, divided by the total number (N) of molecules present at time t, that is,

$$\overline{X}_n = \frac{N_0}{N} \tag{5.28}$$

It is easy to see that for hydroxyacids and for stoichiometric mixtures of diol and diacid, there is an average of one carboxyl per molecule at any state of reaction. So Eq. (5.28) can be written as

$$\overline{X}_n = C_0/C \tag{5.29}$$

where C_0 is the initial (at $t = 0$) concentration of hydroxyl or carboxyl groups and C is the concentration at some time t. Now combining Eq. (5.29) with Eq. (5.14) one obtains

$$\overline{X}_n = \frac{1}{1 - p} \tag{5.30}$$

This equation is sometimes referred to as the *Carothers equation* and is applicable to all step polymerizations represented by equations

$$n\text{A}\!-\!\text{B} \;\longrightarrow\; -\!(\text{A}\!-\!\text{B})_{\overline{n}}- \tag{5.31}$$

$$n\text{A}\!-\!\text{A} + n\text{B}\!-\!\text{B} \;\longrightarrow\; -\!(\text{A}\!-\!\text{AB}\!-\!\text{B})_{\overline{n}}- \tag{5.32}$$

in systems containing stoichiometric amounts of A and B groups. In subsequent treatment, a condensation polymer formed from a monomer of the type A—B (i.e., monomer having one A and one B functional groups) will be referred to as *type I polymer* and the polymerization process will be called *type I condensation*, while a polymer formed from a mixture of A—A and B—B type monomers will be termed a *type II polymer* and the process referred to as *type II condensation*.

It is important to note that for a polymerization system, $\overline{X}_n$ given by Eq. (5.29) or (5.30) is the number-average degree of polymerization of the *reaction mixture* (including even the unreacted monomers) and not just of the polymer that has been formed. For a polymer, however, there is a subtle difference between the number average degree of polymerization, $\overline{X}_n$, and the average degree of polymerization, $\overline{DP}_n$. The former quantity is given by the average number of *structural units* per polymer chain, the residue from each monomer in the polymer chain being termed a *structural unit* (or a *monomer unit*). On the other hand, $\overline{DP}_n$ is defined as the average number of repeating units per polymer chain. So, $\overline{X}_n$ is not necessarily equal to $\overline{DP}_n$. For example, for an average polyester chain represented by H$-$(O-R-CO$-$)$_{100}$$-$OH, both $\overline{X}_n$ and $\overline{DP}_n$ are equal to 100, while for an average chain represented by H$-$(O-R-OOC-R$'$-CO$-$)$_{100}$$-$OH, $\overline{X}_n = 200$ and $\overline{DP}_n = 100$.

Problem 5.8 Show that in the high conversion region, $\overline{M}_n$ for uncatalyzed polyesterification is proportional to $t^{1/2}$ and that for catalyzed polyesterification is proportional to t, where t is the reaction time.

Answer:

The number average molecular weight, $\overline{M}_n$, is related to $\overline{X}_n$ by (Odian, 1991) :

$$\overline{M}_n = \overline{X}_n.M_0 + M_{eg} = \frac{M_0}{(1 - p)} + M_{eg} \tag{P5.8.1}$$

where M_0 is the molecular weight of the monomer residue(s) in the repeating unit and M_{eg} is the molecular weight of the end groups of the polymer. For example, in the polyesterification of adipic acid, $HO_2C(CH_2)_4CO_2$ and ethylene glycol, $HOCH_2CH_2OH$, the polymer is $HO-(CH_2-CH_2-COO(CH_2)_4COO)-H$. So, M_0 is 86 and M_{eg} is 18. Since even for a relatively low molecular weight polymer M_{eg} makes negligibly small contribution to $\overline{M}_n$, Eq. (P5.8.1) can be approximated to

$$\overline{M}_n = \overline{X}_n.M_0 = \frac{M_0}{1 - p} \tag{P5.8.2}$$

By combining Eq. (P5.8.2) with Eqs. (5.15) and (5.19) one then obtains

$$\overline{M}_n\,(\text{uncatalyzed}) = M_0(1 + 2C_0^2 kt)^{1/2} \tag{P5.8.3}$$

$$\overline{M}_n\,(\text{catalyzed}) = M_0(1 + C_0 k't) \tag{P5.8.4}$$

The kinetic expressions (5.15) and (5.19) are obeyed for esterifications above $\sim 80\%$. For conversions of this magnitude the values of t are usually sufficiently large that unity in the parentheses of Eqs. (P5.8.3) and (P5.8.4) may be neglected, yielding the approximate equations (Allcock and Lampe, 1990):

$$\overline{M}_n(\text{uncatalyzed}) \simeq M_0\,[\text{COOH}]_0\,(2k)^{1/2}\,t^{1/2} \tag{P5.8.5}$$

$$\overline{M}_n(\text{catalyzed}) \simeq M_0\,[\text{COOH}]_0\,k't \tag{P5.8.6}$$

where $[\text{COOH}]_0$ is the concentration of the carboxyl group at $t = 0$. Evidently, $\overline{M}_n$ increases much faster with t in catalyzed polyesterification than in uncatalyzed polyesterification.

In a closed batch reactor, water produced by the condensation reaction is not removed. As a result, its concentration builds up until the rate of the reverse reaction (*depolymerization*) becomes equal to the forward reaction (polymerization) rate. The maximum polymer molecular weight is determined by the extent of forward reaction when equilibrium is established.

Problem 5.9   Show that in polyesterification of hydroxyacid or equimolar mixture of diacid and diol carried out in a closed vessel, the upper limit of $\overline{M}_n$ that can be obtained is $(\sqrt{K} + 1)M_0$, where K is the equilibrium constant for the esterification reaction and M_0 is the molecular weight of the monomer residue(s) in the repeating unit.

Answer:

The extent of conversion at equilibrium, p_E, is given by [cf. Eq. (5.26)]

$$p_E = \sqrt{K}/(\sqrt{K} + 1) \tag{P5.9.1}$$

Using this expression to substitute for p_E in Eqs. (5.30) and (P5.8.2) at equilibrium yields the equilibrium or limiting values of $\overline{X}_n$ and $\overline{M}_n$ as

$$\overline{X}_n = \sqrt{K} + 1 \tag{P5.9.2}$$

$$\overline{M}_n = (\sqrt{K} + 1)M_0 \tag{P5.9.3}$$

Note: According to Eq. (P5.9.2), even for a high equilibrium constant of 10^3, a degree of polymerization of only about 32 can be obtained in a closed reactor with no removal of by-product water.

Problem 5.10   The equilibrium constant K for the esterification reaction of decamethylene glycol and adipic acid is of the order of unity at $110°C$. If equimolar amounts of the diol and the diacid are used in polycondensation at $110°C$, what weight ratio of dissolved water (of condensation) to polymer would correspond to an equilibrium $\overline{X}_n$ value of 60 at $110°C$?

Answer:

Since the concentration of the by-product water at equilibrium is to be determined, Eq. (5.24) is rewritten in the form

$$K = \frac{[p_E[\text{COOH}]_0]\,[\text{H}_2\text{O}]_E}{[\text{COOH}]_E\,[(1 - p_E)[\text{COOH}]_0]} = \frac{p_E\,[\text{H}_2\text{O}]_E}{(1 - p_E)[\text{COOH}]_E} \tag{P5.10.1}$$

which is then rearranged to

$$\frac{[H_2O]_E}{[COOH]_E} = \frac{K(1 - p_E)}{p_E} \tag{P5.10.2}$$

[It should be noted that Eq. (P5.10.2), derived above for stoichiometric A—A plus B—B polymerization, applies equally to A—B systems.]

Using Eq. (5.30) at equilibrium, we set $\overline{X}_n = 1/(1 - p_E) = 60$ or $p_E = 0.9833$.
From Eq. (P5.10.2), the ratio of dissolved water (of condensation) to polymer at this equilibrium conversion is

$$\frac{[H_2O]_E}{[COOH]_E} = \frac{(1)(1 - 0.9833)}{0.9833} = 0.0170 \text{ mol/mol} \tag{P5.10.3}$$

The repeating unit of the polyester being $\text{-[O-(CH_2)_{10}\text{-}OOC\text{-}(\text{-}CH_2)_4\text{-}CO]-}$, the molar mass of the repeating unit = 284 g mol^{-1}. The average molar mass of two monomer residues in a repeating unit is $M_0 = (284$ g mol$^{-1})/2 = 142$ g mol^{-1}. For $\overline{X}_n = 60$, the average molar mass of polymer, $\overline{M} = 60 \times 142$ or 8520 g mol^{-1}. Due to stoichiometric balance between COOH and OH groups, there is an average of one COOH group per molecule. From Eq. (P5.10.3), the water content of the reaction mixture is then

$$\frac{(0.0170 \text{ mol H}_2\text{O}/\text{mol COOH})(18 \text{ g/mol H}_2\text{O})}{(8520 \text{ g/mol COOH})} = 3.6 \times 10^{-5} \text{ g/g}$$

Note: Step polymerizations are often conducted at temperatures near or above the boiling point of water to facilitate water removal and thus drive the reaction to higher conversions required for a higher molecular weight.

5.4 Control of Molecular Weight

The quantitative dependence of $\overline{X}_n$, and hence molecular weight, on the extent of reaction in condensation polymerization is shown by Eq. (5.30). From the corresponding data presented in Table 5.2 it is evident that to produce a relatively high-molecular-weight polymer ($\overline{X}_n > 100$), it is necessary to allow the reaction to proceed to a very high degree of conversion ($p > 0.98$) or, in other words, to a product that contains a very small number of chain ends. In general, this will be

Table 5.2 Effect of Extent of Reaction on Number-Average Degree of Polymerization

% Reaction ($p \times 100$)	Number average degree of polymerization ($\overline{X}_n$) from Eq. (5.30)
50	2
75	4
90	10
95	20
98	50
99	100
99.9	1,000
99.99	10,000

possible only when equal concentrations of the two reactive functional groups are maintained throughout the course of the reaction.

It is further evident from Table 5.2 that $\overline{X}_n$, and hence molecular weight, increases very rapidly with conversion in the high conversion range ($p > 0.99$). An increase in conversion from 0.990 to 0.999, for example, leads to a tenfold increase in $\overline{X}_n$. It is thus not a realistic proposition to control polymer molecular weight by adjusting the extent of conversion. An alternative procedure to control $\overline{X}_n$ would be to introduce, deliberately, an imbalance in the ratio of the two types of functional groups in the feed.

For example, the use of *excess* diol in the polymerization of a diol with a diacid yields a polyester (**I**) with hydroxyl end groups which are incapable of further reaction, since the diacid has been completely reacted:

$$\text{HOOCRCOOH} + \text{HO R}'\text{OH}\,(\textit{excess}) \longrightarrow$$
$$\text{H}\text{-}(\text{OR}'\text{O OCRCO})_n\text{OR}'\text{OH} \tag{5.33}$$
$$(\textbf{I})$$

The use of excess diacid leads to a similar result, as the polyester (**II**) now has only carboxyl groups at both ends after the diol is completely reacted:

$$\text{HOR}'\text{OH} + \text{HO OCRCOOH}\,(\,\textit{excess})\, \longrightarrow$$
$$\text{HO}\text{-}(\text{OCRC OOR}'\text{O})_n\text{OCRCOOH} \tag{5.34}$$
$$(\textbf{II})$$

Another method of achieving the desired molecular weight is by addition of a small amount of a *monofunctional monomer*. The monofunctional monomer effectively caps the chain end it reacts with and prevents further growth of the chain.

5.4.1   *Quantitative Effect of Stoichiometric Imbalance*

The *stoichiometric imbalance* of the two types of functional groups in a reaction mixture is determined by the ratio of the numbers of the two groups initially present. We consider below two specific cases (Hiemenz, 1984; Manaresi and Munari, 1989; Odian, 1991) of stoichiometric imbalance.

Case 1: *Polymerization of bifunctional monomers A—A and B—B, one being present in excess.*

At any point of time during the polymerization process, let N_{AA} and N_{BB} be the numbers (mol) of monomer species A—A and B—B, and N_A and N_B those of functional groups A and B, respectively, with $N_{BB} > N_{AA}$. Using the same symbols with an additional subscript '$_0$' to denote quantities present initially, we may write $N_{A_0} = 2N_{AA_0}$, $N_{B_0} = 2N_{BB_0}$, and stoichiometric imbalance ratio or parameter, $r = N_{A_0}/N_{B_0} = N_{AA_0}/N_{BB_0}$. (Note that the ratio r is always defined such that it has a value equal to or less than unity, but never greater than unity, i.e., the groups present in excess are denoted as B groups.)

Extent of reaction of A groups, p_A = fraction of A groups which have reacted = $(N_{A_0} - N_A)/N_{A_0}$. Similarly, p_B = fraction of B groups which have reacted = $(N_{B_0} - N_B)/N_{B_0}$. Since at any time, the number of A groups reacted must be equal to the number of reacted B groups, $(N_{A_0} - N_A) = (N_{B_0} - N_B)$. Therefore $p_B = rp_A$.

If we assume for simplicity that any secondary reactions, such as intramolecular reactions leading to cyclic molecules with no end groups, can be neglected (which is justified in many cases), it is easy to see that

Number of A (or B) groups reacted
= Decrease in the number of molecules (5.35)

Equation (5.28) can therefore be written as

$$\overline{X}_n = \frac{N_{AA_0} + N_{BB_0}}{N_{AA_0} + N_{BB_0} - (N_{A_0} - N_A)} \tag{5.36}$$

where the term in parentheses in the denominator represents the number of A functional groups that reacted and, hence, the decrease in the number of molecules.

Introducing the parameter r, Eq. (5.36) may then be written as

$$\overline{X}_n = \frac{1 + r}{1 + r - 2rp_A} \tag{5.37}$$

It must be noted that the extent of reaction (p) is usually defined with respect to functional groups not in excess, i.e., A groups in the present case. Therefore, Eq. (5.37) may be simply written as

$$\overline{X}_n = \frac{1 + r}{1 + r - 2rp} \tag{5.38}$$

omitting the subscript A from p. When polymerization is 100% complete, i.e., $p = 1.00$, Eq. (5.38) becomes

$$\overline{X}_n = \frac{1 + r}{1 - r} \tag{5.39}$$

which gives the limiting value of $\overline{X}_n$ and hence the limiting molecular weight.

When the two bifunctional monomers are present in stoichiometric amounts, i.e., $r = 1.00$, Eq. (5.38) reduces to

$$\overline{X}_n = \frac{1}{(1 - p)}$$

which is the same as Eq. (5.30).

Problem 5.11 What feed ratio of hexamethylene diamine and adipic acid should be employed in order to obtain a polyamide of $\overline{M}_n = 10,000$ at 99% conversion ? Identify the end groups of this product.

Answer:

Formula weight of repeating unit $+HN(CH_2)_6NHCO(CH_2)_4CO+$ $= 226$

$M_0 = \frac{1}{2} \times 226 = 113$; $\overline{X}_n = \overline{M}_n/M_0 = 10,000/113 = 88.5$

From Eq. (5.38), with $p = 0.99$

$$\overline{X}_n = \frac{1 + r}{1 + r - 2r(0.99)} = 88.5 \quad \text{Solving, } r = 0.9974$$

The polymerization is carried out with either COOH/NH$_2$ or NH$_2$/COOH $= 0.9974$. For COOH/NH$_2$ = 0.9974, all end groups will be NH$_2$; for NH$_2$/COOH = 0.9974, all end groups will be COOH.

Case 2: *Addition of small amounts of a monofunctional reactant, for example, B, to an equimolar mixture of A—A and B—B.*

Two approaches may be used to calculate $\overline{X}_n$. (Note that the same arguments and equations as given below will apply to the case where a monofunctional reactant, say B, is added to bifunctional monomer A–B having inherent stoichiometry.)

Approach I uses a stoichiometric imbalance r, which is now redefined (Odian, 1991) as

$$r = \frac{N_{A_0}}{N_{B_0} + 2N'_{B_0}} \tag{5.40}$$

where N'_{B_0} is the number of monofunctional B molecules initially present and other terms are as defined before; thus, $N_{A_0} = 2N_{AA_0}$ and $N_{B_0} = 2N_{BB_0}$. [Note that N'_{B_0} in Eq. (5.40) is multiplied by 2. This is beacuse a monofunctional B molecule here has the same effect on limiting the chain growth as a bifunctional B—B molecule present in excess, since only one of the two B groups can then react.] Thus, for 1 mole percent excess B and $N_{AA_0} = N_{BB_0}$, $r = 100/(100 + 2\times1) = 0.9804$. With r defined as by Eq. (5.40), Eq. (5.38) is applicable for calculation of $\overline{X}_n$ for a given p.

Approach II uses average functionality f_{av} defined (Rudin, 1982) by

$$f_{av} = \frac{\sum N_i f_i}{\sum N_i} \tag{5.41}$$

where N_i is the number of moles of species i with f_i number of functional groups. Note that f_{av} represents the average number of functional groups per molecule in the reaction mixture. Equation (5.41) holds strictly when functional groups of opposite kinds are present in equal concentrations, i.e., for stoichiometric mixtures. In nonstoichiometric mixtures, the excess reactant does not enter the polymerization (in the absence of side reactions) and so it should not be taken into account for calculating f_{av}. Let us consider a polymerization system in which $N_{A_0} < N_{B_0}$, where N_{A_0} and N_{B_0} are number of equivalents of functional groups of types A and B, respectively, present initially. In this case, the number of B equivalents that can react cannot exceed N_{A_0}, and therefore (Rudin, 1982),

$$f_{av} = \frac{2N_{A_0}}{\sum N_{i_0}} \tag{5.41a}$$

(Thus, f_{av} represents the average number of *useful* equivalents of functional groups of all kinds per molecule present initially in the reaction mixture.)

Let

N_0	= total number (mol) of monomers (of all types) present initially.
N	= total number (mol) of molecules (monomers plus polymers of all sizes) when the reaction has proceeded to an extent p.
$N_0 - N$	= number of linkages formed at the extent of reaction p. (This follows from the fact that every time a new linkage is formed the reaction mixture will contain one less molecule.)

Since two functional groups react to form one linkage, moles of functional groups lost in forming $(N_0 - N)$ moles of linkage $= 2(N_0 - N)$. Thus,

$$\text{Extent of reaction, } p = \frac{\text{No. of functional groups used}}{\text{No. of functional groups present initially}}$$

$$= \frac{2(N_0 - N)}{N_0 f_{av}} \tag{5.42}$$

whence $N = \frac{1}{2}(2N_0 - N_0 p f_{av})$.

Therefore, from Eq. (5.28):

$$\overline{X}_n = \frac{N_0}{N} = \frac{N_0}{(2N_0 - N_0 p f_{av})/2} = \frac{2}{2 - p f_{av}} \tag{5.43}$$

Problem 5.12 For some applications of nylon-6,6, it is desirable to have $\overline{M}_n$ below 20,000. For making the polymer by polycondensation of hexamethylene diamine and adipic acid, how much acetic acid per mol of adipic acid should be added so that $\overline{M}_n$ does not exceed this value? Use both Approaches I and II for the calculation.

Answer:

Formula weight of repeating unit $-\!\!\!\!+HN(CH_2)_6NHCO(CH_2)_4CO\!\!\!\!+- = 226$.
$M_0 = 226/2 = 113$; $\overline{X}_n = 20,000/113 = 177$.

Approach I

From Eq. (5.38): $\lim_{p \to 1} \overline{X}_n = \dfrac{1 + r}{1 + r - 2r} = 177$. Solving, $r = 0.9888$.

Let x mol of acetic acid be added for every mol of adipic acid.

From Eq. (5.40): $r = \dfrac{2}{2 + 2 \times x} = \dfrac{1}{1 + x} = 0.9888$. Solving, $x = 0.0113$.

So for every mol of adipic acid 0.0113 mol of acetic acid is to be added.

Approach II

From Eq. (5.43), $\lim_{p \to 1} \overline{X}_n = \dfrac{2}{2 - 1 \times f_{av}} = 177$. Solving, $f_{av} = 1.9887$.

Let x mol of acetic acid be added for every mol of monomer.

Component	mol	f	Equivalents
$H_2N(CH_2)_6NH_2$	1	2	2
$HOOC(CH_2)_4COOH$	1	2	2
CH_3COOH	x	1	x
	Total $2+x$		

Total amine equivalents $= 2$; total acid equivalents $= 2 + x$.

$$f_{av} = \frac{2 \times \text{equivalents of limiting group}}{\text{total moles}} = \frac{2 \times 2}{2 + x} = 1.9887$$

Solving, $x = 0.0114$, in agreement with the result obtained by Approach I.

5.5 *Molecular Weight Distribution (MWD)*

The molecular weight distribution (MWD) has been derived by Flory (1953) by a statistical approach using the principle of equal reactivity of all functional groups of a given chemical type, irrespective of the size of the molecule to which they are attached; that is, the reactivity of each type of group does not change during the course of polymerization. The derivation given below applies equally to A—B (Type I) and stoichiometric A—A plus B—B (Type II) step polymerizations.

Consider, as an example, an A—B type polymerization

$$x\,\text{HO–R–COOH} \longrightarrow \text{H}\text{-(O-R-CO)}_x\text{OH} + (x-1)\,\text{H}_2\text{O}$$

The polymer chain may be written as (Ghosh, 1990):

$$\underset{1}{\text{HO-R-}} \underset{2}{\text{COO-R-}} \underset{3}{\text{COO-R-}} \underset{4}{\text{COO-R-}} \text{COO-} \cdots\cdots \underset{x}{\text{R- COO-H}}$$

(III)

Probability that a given group has reacted = fractional extent of reaction (p).

Probability that a given group has not reacted = $1 - p$

In a polymer consisting of x monomer residues, the number of ester linkages = $(x - 1)$

Probability that the molecule contains $(x - 1)$ ester groups = p^{x-1}

Probability that the xth carboxyl group is unreacted = $(1 - p)$

Therefore, the probability that the molecule in question is composed of exactly x units = $p^{x-1}(1 - p) = n_x$ (5.44)

where n_x is the mole or number fraction of molecules in the polymer mixture which are x-mers, and is also given by

$$n_x = N_x/N$$

Here N_x is the number of molecules which are x-mers and N is the total number of molecules at the extent of reaction p. However, $N = N_0(1 - p)$, which is the same as Eq. (5.43) with $f_{av} = 2$. Equation (5.44) therefore becomes

$$N_x = N_0(1 - p)^2 p^{x-1} \tag{5.45}$$

Let

M_0 = formula weight of a monomer that has reacted at both ends, i.e.,
     –R-COO– (the end groups OH and H having reacted).
Therefore, for an x-mer (structure III) the formula weight = xM_0 (neglecting unreacted ends OH and H).

Total weight of all molecules = N_0M_0 (neglecting unreacted ends OH and H).
Hence the weight fraction w_x of x-mers is

$$w_x = \frac{N_x x M_0}{N_0 M_0} = \frac{N_x x}{N_0} \tag{5.46}$$

Combining Eq. (5.46) with Eq. (5.45) one obtains

$$w_x = x(1 - p)^2 p^{x-1} \tag{5.47}$$

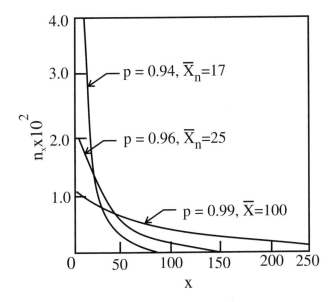

Figure 5.4 Mole fraction or number fraction distribution of reaction mixture in linear step-growth polymerization for conversions of 94%, 96%, and 99% [Eq. (5.44)]. (Drawn following the method of Flory, 1953.)

Equations (5.44) and (5.47) describe the differential number and weight distribution functions, respectively, for linear step polymerizations at the extent of reaction p. These distributions are also known as the *most probable* or *Flory distributions* (Figs. 5.4 and 5.5).

Problem 5.13 Calculate the extent of reaction necessary to obtain the maximum yield by weight of species consisting of 100 monomer units in step-growth polymerization of a stoichiometric mixture. Derive first a general expression to obtain this.

Answer:

To determine the value of p at which w_x reaches a maximum, one has to find the value of p at which the derivation of w_x with respect to p is zero. From Eq. (5.47):

$$dw_x/dp = x(x-1)(1-p)^2 p^{x-2} - 2x(1-p)p^{x-1} = 0$$

Thus the value of p at which w_x is maximum is

$$p = \frac{x-1}{x+1} \qquad\qquad\qquad\qquad (P5.13.1)$$
$$= \frac{100-1}{100+1} = 0.98$$

Problem 5.14 Show mathematically that the maximum of the weight distribution function at high conversions in a most probable distribution is located at the number average degree of polymerization $\overline{X}_n$.

Answer:

To determine the value of x at which w_x reaches maximum, one has to find the value of x at which the derivative of w_x with respect to x is zero. Thus from Eq. (5.47):

$$dw_x/dx = (1-p)^2 \left[p^{x-1} + xp^{x-1} \ln p \right] = 0 \qquad\qquad (P5.14.1)$$

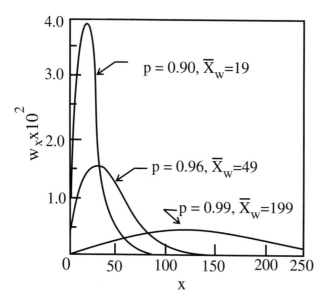

Figure 5.5 Weight fraction distribution of reaction mixture in linear step-growth polymerization for conversions of 90%, 96%, and 99% [Eq. (5.47)]. (Drawn following the method of Flory, 1953.)

At high conversions with p close to unity one can set $q = 1 - p$ and $\ln p = \ln(1 - q) = -q + \cdots \cdots$. For maximum w_x at a given p, one thus derives from Eq. (P5.14.1),

$$x = 1/(1 - p) \tag{P5.14.2}$$
$$= \overline{X}_n \quad [\text{cf. Eq. (5.50) derived below}]$$

which signifies that, at high conversions, maximum yield by weight is obtained for species containing the same number of monomer units as the average degree of polymerization of the reaction mixture.

Note :   An alternative way to derive Eq. (P5.14.2) is to calculate $d\ln w_x/dx$ and set it equal to zero.

5.5.1   Breadth of MWD

The number-average degree of polymerization is given by

$$\overline{X}_n = \frac{\sum xN_x}{\sum N_x} = \frac{\sum xN_x}{N} = \sum x n_x \tag{5.48}$$

where the summations are over all values of x. [Equation (5.48) may be compared with Eq. (4.7a) for $\overline{DP}_n$. Note that n_x in Eq. (4.7a) represents the average number of repeating units in polymer chains, while n_x in Eq. (5.48) represents the average number of monomer residues in polymer chains.]

Combination of Eqs. (5.44) and (5.48) then gives

$$\overline{X}_n = \sum x p^{x-1}(1 - p) \tag{5.49}$$

Evaluation of the series summation in Eq. (5.49) for fixed p yields (see Appendix 5.1):

$$\overline{X}_n = (1 - p) \sum_{x=1}^{\infty} x p^{x-1} = \frac{1 - p}{(1 - p)^2} = \frac{1}{1 - p} \tag{5.50}$$

Note that the same result was also obtained earlier [see Eq. (5.30)].

The weight-average degree of polymerization, $\overline{X}_w$, is given by [cf. Eq. (4.12a)]

$$\overline{X}_w = \sum x w_x \tag{5.51}$$

Substituting for w_x from Eq. (5.47) one gets

$$\overline{X}_w = \sum x^2 p^{x-1} (1 - p)^2 \tag{5.52}$$

Evaluation of the series summation in Eq. (5.52) for fixed p yields (see Appendix 5.1):

$$\overline{X}_w = (1 - p)^2 \sum x^2 p^{x-1} = \frac{(1 - p)^2 (1 + p)}{(1 - p)^3} = \frac{(1 + p)}{(1 - p)} \tag{5.53}$$

The breadth of the molecular weight distribution curve is then obtained as

$$\frac{\overline{M}_w}{\overline{M}_n} = \frac{\overline{X}_w}{\overline{X}_n} = (1 + p) \tag{5.54}$$

The z-average degree of polymerization, $\overline{X}_z$, is given by

$$\overline{X}_z = \frac{\sum x^2 w_x}{\sum x w_x} \tag{5.55}$$

An expression of $\overline{X}_z$ in terms of p is then obtained by combining Eqs. (5.47) and (5.55) and performing series summation (see Problem 5.16):

$$\overline{X}_z = \frac{(1 + 4p + p^2)}{(1 - p^2)} \tag{5.56}$$

Problem 5.15   In a polymerization of $H_2N(CH_2)_{10}COOH$ to form nylon-11, 95% of the functional groups is known to have reacted. Calculate (a) amount of monomer (in terms of weight fraction) remaining in the reaction mixture, (b) weight fraction of the reaction mixture having a number-average degree of polymerization equal to 100, and (c) the extent of reaction at which the 100-mer has the maximum yield by weight.

Answer:

(a) From Eq. (5.47), for $x = 1$ and $p = 0.95$:

$$w_1 = (1)(1 - 0.95)^2 (0.95)^0 = 2.5 \times 10^{-3} \quad \text{(i.e., 0.25\% by weight)}$$

(b) From Eq. (5.47), for $x = 100$, $p = 0.95$:

$$w_{100} = (100)(1 - 0.95)^2 (0.95)^{99} = 1.56 \times 10^{-3} \quad \text{(i.e., 0.156\% by weight)}$$

(c) From Eq. (P5.13.1), for $x = 100$: $p_{100} = (100 - 1)/(100 + 1) = 0.98$

The derivations given in Section 5.5 for A—B type polycondensation (type I) hold also for A—A plus B—B type polycondensation (type II) reactions when there are precisely equivalent proportions of A—A and B—B reactants and the principle of equal reactivity of all functional groups of the same chemical type, irrespective of the size of the molecule, is applicable. Now, however, x will represent the combined number of both types of units in the polymer chain. Therefore, in place of M_0 in Eq. (P5.8.2) one has to use the average weight of units M_A and M_B or one-half of the repeat unit in the polymer chain. If, moreover, there are more than 2 monomer types in the reaction mixture, a weighted average of all units is to be used.

Problem 5.16   In the polymerization of equimolar mixture of $HO_2C(CH_2)_4CO_2H$ and $H_2N(CH_2)_6NH_2$ to form nylon-6,6, 99% of the carboxylic acid groups are known to have reacted. Calculate $\overline{M}_n$, $\overline{M}_w$, and $\overline{M}_z$ of the reaction mixture at this conversion.

Answer:

$M_0 = \frac{1}{2} \times$ Formula weight of repeating unit $\text{--[HN(CH}_2)_6\text{NHCO(CH}_2)_4\text{CO]--} = 113$

From Eq. (5.50):

$$\overline{X}_n = \frac{1}{1 - 0.99} = 100, \quad \overline{M}_n = \overline{X}_n M_0 = 11,300$$

From Eq. (5.53):

$$\overline{X}_w = (1 + 0.99)/(1 - 0.99) = 199; \quad \overline{M}_w = \overline{X}_w M_0 = 22,487$$

From Eq. (5.47):

$$\sum x^2 w_x = \sum x^3 (1 - p)^2 p^{x-1} = (1 - p)^2 \sum x^3 p^{x-1}$$

Using the appropriate series summation from Appendix 5.1,

$$\sum x^2 w_x = \frac{1 + 4p + p^2}{(1 - p)^2} \tag{P5.16.1}$$

From Eq. (5.53):

$$\sum x w_x = \overline{X}_w = (1 + p)/(1 - p) \tag{P5.16.2}$$

Substituting Eqs. (P5.16.1) and (P5.16.2) in Eq. (5.55),

$$\overline{X}_z = \frac{(1 + 4p + p^2)}{(1 + p)(1 - p)}$$

$$= \frac{1 + 4(0.99) + (0.99)^2}{(1 + 0.99)(1 - 0.99)} = 298.50$$

$$\overline{M}_z = \overline{X}_z M_0 = 33,730$$

Problem 5.17 In a synthesis of polyester from 2 moles of terephthalic acid, 1 mol of ethylene glycol, and 1 mol of butylene glycol, the reaction was stopped at 99.5% conversion of the acid. Determine (a) $\overline{M}_n$ and $\overline{M}_w$ of the polyester; and (b) mole fraction and weight fraction of species containing 20 monomer units.

Answer:

$M_A = -OCC_6H_4CO- = 132 \text{ g mol}^{-1}; \quad M_{B_1} = -OCH_2CH_2O- = 60 \text{ g mol}^{-1};$
$M_{B_2} = -O(CH_2)_4O- = 88 \text{ g mol}^{-1}; \quad M_0 = (2M_A + M_{B_1} + M_{B_2})/4 = 103 \text{ g mol}^{-1}.$

(a) From Eq. (5.50),

$$\overline{M}_n = \overline{X}_n M_0 = \frac{M_0}{1-p} = \frac{103 \text{ g mol}^{-1}}{(1-0.995)} = 20,600 \text{ g mol}^{-1}$$

From Eq. (5.53),

$$\overline{M}_w = \overline{X}_w M_0 = \frac{M_0(1+p)}{(1-p)}$$
$$= \frac{(103 \text{ g mol}^{-1})(1+0.995)}{(1-0.995)} = 41,100 \text{ g mol}^{-1}$$

(b) From Eq. (5.44), $n_{20} = (1-0.995)(0.995)^{(20-1)} = 0.0045$

From Eq. (5.47), $w_{20} = (20)(1-0.995)^2(0.995)^{(20-1)} = 0.0005$

Problem 5.18 Consider polymerization of a nonstoichiometric ($r \neq 1$) mixture of A—A and B—B. If the reaction goes to completion (i.e., $p = 1$ for the limiting group), derive expressions for n_x, w_x, $\overline{X}_n$, and $\overline{X}_w$ of the mixture. Assume that only reaction between A and B end groups takes place and all end groups of the same type (A or B) are equally reactive, irrespective of the size of the molecules to which they are attached.

Answer:

Define N_{A_0} = initial number (mol) of A groups; N_{B_0} = initial number (mol) of B groups; N_A, N_B = number (mol) of A and B groups, respectively, at some time after the reaction started; Let $N_{A_0} < N_{B_0}$, i.e., A groups are deficient and can be completely reacted; stoichiometric ratio, $r = N_{A_0}/N_{B_0} < 1$; extent of reaction, $p_A \equiv \left(N_{A_0} - N_A\right)/N_{A_0}$.

Since all A groups have reacted ($p_A = 1$), the reaction mixture can have only odd number species, such as monomer, trimer, pentamer, and so on. Consider one such species, say, pentamer (Rempp and Merrill, 1986):

Probability   B--BA--AB--BA--AB--B
of reaction $r \times 1$ 1×1 $r \times 1$ 1×1 $(1-r)$

The probability that a B group selected at random has reacted (with A) is $p_B = p_A N_{A_0}/N_{B_0} = r$. So the probability that a B group has not reacted $= 1 - r$.

The probability of formation of the pentamer chain shown above can be determined as follows (Rempp and Merrill, 1986): Since all A groups have reacted ($p_A = 1$), the probability of the left end group being B is 1. If we start from this end and move toward the right, the probability that the next B group has reacted is r (as explained above) and that it has reacted with A is $r \times 1$ (since B can react only with A). The reacted A group is followed by another A group which therefore means that the next group is a reacted B and the probability is 1×1. The next group to the right is also B and the probability that it has reacted with an A group is $r \times 1$. Another A group then follows and it is connected to a reacted B with probability 1×1. The probability that the next B group is unreacted is $(1 - r)$. Therefore the probability of selecting a pentamer is

$$P_5 \; = \; (r \times 1)(1 \times 1)(r \times 1)(1 \times 1)(1 - r) \; = \; r^2(1 - r)$$

and the corresponding probabilities for monomer and trimer are

$$P_1 \; = \; (1 - r)$$
$$P_3 \; = \; r(1 - r)$$

In general, $P_x \; = \; r^{(x-1)/2}(1 - r)$. (Note that x can have only odd values as no even number species can exist.)

The probabilities P can be equated to mole fractions (number fractions) n, i.e.,

$$n_x \; = \; r^{(x-1)/2} \cdot (1 - r) \qquad (x \text{ is odd}) \tag{P5.18.1}$$

The number-average degree of polymerization is given by

$$\overline{X}_n \; = \; \frac{\sum_{x=1}^{\infty} x n_x}{\sum_{x=1}^{\infty} n_x} \; = \; \sum_{x=1}^{\infty} x r^{(x-1)/2} \cdot (1 - r) \tag{P5.18.2}$$

If we set $x = 2z - 1$ (Rempp and Merrill, 1986), so that if $x = 1,3,5,\cdots$, $z = 1,2,3,\cdots$, then Eq. (P5.18.2) leads to

$$\overline{X}_n \; = \; \sum_{z=1}^{\infty} (2z - 1) r^{(z-1)} (1 - r)$$

$$= \; (1 - r)\left[2 \sum_{z=1}^{\infty} z r^{(z-1)} - \sum_{z=1}^{\infty} r^{(z-1)} \right]$$

$$= \; (1 - r)\left[\frac{2}{(1 - r)^2} - \frac{1}{1 - r} \right] \qquad \text{(see Appendix 5.1)}$$

$$= \; \frac{1 + r}{1 - r} \tag{P5.18.3}$$

Note that this expression can also be derived from

$$\overline{X}_n \; = \; N_0/N \tag{P5.18.4}$$

where N_0 = number (mol) of mers = $N_{A_0}/2 + N_{B_0}/2 = N_{A_0}(1 + 1/r)/2$ and

$$N \; = \; \text{number (mol) of species on completion of reaction } (p_A = 1)$$
$$= \; \text{number (mol) of pairs of ends}$$
$$= \; (N_{B_0} - N_{A_0})/2 \; = \; N_{A_0}\left(\frac{1}{r} - 1\right)/2$$

Therefore,

$$\overline{X}_n \; = \; \frac{N_0}{N} \; = \; \frac{1 + r}{1 - r} \tag{P5.18.5}$$

which is identical with Eq. (P5.18.3) and with Eq. (5.38) for $p = 1$.

Consider now weight fractions of various species, namely x-mers with $x = 1$, $x = 3$, $x = 5$, and so on. The weight fraction w_x of x-mers is

$$w_x = \frac{N_x x M_0}{N_0 M_0} = \frac{n_x N x}{N_0} \tag{P5.18.6}$$

where N_x = number (mol) of species which are x-mers
            = mole fraction of x-mers (n_x) × total moles of species after reaction (N)
and M_0 = average formula weight of mers.

Substituting for n_x and N/N_0 from Eqs. (P5.18.1) and (P5.18.5) into Eq. (P5.18.6),

$$w_x = x \cdot r^{(x-1)/2} \cdot \frac{(1-r)^2}{(1+r)} \tag{P5.18.7}$$

If we set $x = 2z - 1$ (Rempp and Merrill, 1986), so that for $x = 1, 3, 5, \cdots$ we have $z = 1, 2, 3, \cdots$, and so

$$\sum_{x=1}^{\infty} x w_x = \sum_{z=1}^{\infty} (2z - 1)^2 r^{(z-1)} \cdot \frac{(1-r)^2}{1+r} \tag{P5.18.8}$$

which yields, using appropriate equations from Appendix 5.1,

$$\sum_{x=1}^{\infty} x w_x = \left[\frac{4(1+r)}{(1-r)^3} - \frac{4}{(1-r)^2} + \frac{1}{1-r} \right] \frac{(1-r)^2}{(1+r)}$$

$$= \frac{4}{1-r} - \frac{4}{1+r} + \frac{1-r}{1+r} \tag{P5.18.9}$$

Now since $\overline{X}_w = \sum_{x=1}^{\infty} x w_x$, we have

$$\overline{X}_w = \frac{4}{1-r} - \frac{4}{1+r} + \frac{1-r}{1+r} \tag{P5.18.10}$$

and $\quad \dfrac{\overline{X}_w}{\overline{X}_n} = \dfrac{4}{1+r} - \dfrac{4(1-r)}{(1+r)^2} + \left(\dfrac{1-r}{1+r} \right)^2 \tag{P5.18.11}$

For $r = 1$, this reduces to $\overline{X}_w / \overline{X}_n = 2$ [cf. Eq. (5.54) for $p = 1$].

Problem 5.19   A polyester, poly(ethylene terephthalate), was synthesised by reacting one mole of dimethyl terephthalate and two moles of ethylene glycol, the reaction being carried to 100% exchange of the methyl group forming the ester link between glycol and terephthalate. Calculate (a) number average degree of polymerization, (b) weight-average degree of polymerization, (c) polydispersity index of the product, and (d) mole fractions and weight fractions of monomer, trimer, and pentamer in the reaction product.

Answer:

Consider $A = -COOCH_3$, $B = -OH$. Therefore, for the given composition, $r = \frac{1}{2}$

(a) From Eq. (P5.18.5), $\overline{X}_n = (1 + \frac{1}{2})/(1 - \frac{1}{2}) = 3$

(b) From Eq. (P5.18.10), $\overline{X}_w = \dfrac{4}{1 - \frac{1}{2}} - \dfrac{4}{1 + \frac{1}{2}} + \dfrac{1 - \frac{1}{2}}{1 + \frac{1}{2}} = \dfrac{17}{3}$

(c) $I = \overline{X}_w/\overline{X}_n = 17/9$

(d) From Eq. (P5.18.1): $n_1 = 1/2$, $n_3 = 1/4$, $n_5 = 1/8$
   From Eq. (P5.18.7): $w_1 = 1/6$, $w_3 = 1/4$, $w_5 = 5/24$

Problem 5.20 Assuming that from the polymerization system of Problem 5.19 all excess ethylene glycol is distilled out along with the by-product methanol, what would be (a) $\overline{X}_n$ and (b) $\overline{X}_w$ of the reaction product ?

Answer:

(a)

$$\overline{X}_n = \frac{\sum_{x=3}^{\infty} x n_x}{\sum_{x=3}^{\infty} n_x} = \frac{\sum_{x=1}^{\infty} x n_x - 1(n_1)}{1 - n_1}$$

$$= \frac{(1 + r)/(1 - r) - (1 - r)}{1 - (1 - r)} \quad \text{[cf. Eqs. (P5.18.1) – (P5.18.3)]}$$

$$= 5 \quad (\text{since } r = 1/2)$$

(b)

$$\overline{X}_w = \frac{\sum_{x=3}^{\infty} x w_x}{1 - w_1} = \frac{\sum_{x=1}^{\infty} x w_x - 1(w_1)}{1 - w_1}$$

From Eq. (P5.18.7),

$$w_1 = 1 \times (\tfrac{1}{2})^0 \times \frac{(1/2)^2}{(3/2)} = \frac{1}{6}$$

From Eq. (P5.18.9),

$$\sum_{x=1}^{\infty} x w_x = \frac{4}{1 - r} - \frac{4}{1 + r} + \frac{1 - r}{1 + r} = \frac{17}{3}$$

Therefore, $\overline{X}_w = \dfrac{(17/3) - (1/6)}{1 - (1/6)} = \dfrac{33}{5} = 6.6$

Problem 5.21 Consider the synthesis of nylon-6,6 from hexamethylene diamine and adipic acid, the mole ratio of the diacid to diamine being 0.99. Estimate the degree of polymerization of species which has the maximum yield by weight at the completion of the condensation.

Answer:

Substituting $x = 2z - 1$ (Rempp and Merrill, 1986) in Eq. (P5.18.7),

$$w_x = (2z - 1) r^{(z-1)} (1 - r^2)/(1 + r)$$

To determine the value of z at which the weight fraction w_x reaches a maximum, one has to find the value of z at which the derivation of w_x with respect to z is zero:

$$\frac{dw_x}{dz} = \frac{(1-r)^2}{(1+r)} r^{(z-1)} [2 + 2z\ln r - \ln r] = 0$$

$$z = (\ln r - 2)/(2\ln r)$$

Therefore, $x = 2z - 1 = -2/\ln r$. For $r = 0.99$, $x = 199$.

Proceeding in the same way as in Problem 5.18, derivations can be obtained (Rempp and Merrill, 1986) for the general case of nonstoichiometric mixtures of A—A and B—B with incomplete reactions, i.e., $r \neq 1$ and $p \neq 1$. However, approximate values can also be calculated using the equations for stoichiometric mixtures derived above, if p_A is now replaced by $p_A r^{1/2}$ (see Problem 5.22).

Problem 5.22 Consider synthesis of nylon-6,10 from hexamethylene diamine, $H_2N(CH_2)_6NH_2$, and sebacic acid, $HOOC(CH_2)_8COOH$, where the mole ratio (r) of the diacid to the diamine is 0.99. At 98% conversion of carboxylic acid groups (i.e., $p_A = 0.98$) determine (a) $\overline{M}_n$ and $\overline{M}_w$ of the product, and (b) the sum of the mole fractions and of the weight fractions of species of degrees of polymerization 1 through 6 inclusive. Follow the approximate method of using equations for the stoichiometric case but with p_A replaced by $p_A r^{1/2}$.

Answer:

Let A—A represent the diacid and B—B the diamine.

$M_A = -HN(CH_2)_6NH- \equiv 114$ g mol^{-1}; $M_B = -OC(CH_2)_8CO- \equiv 168$ g mol^{-1}.

$p_A = 0.98$; $r = 0.99$; $p_A r^{1/2} = (0.98)(0.99)^{1/2} = 0.9751$.

$M_0 = (114 + 168)/2$ or 141 g mol^{-1}.

From Eq. (5.50): $\dfrac{\overline{M}_n}{M_0} = \overline{X}_n = \dfrac{1}{1 - 0.9751} = 40.16$. Therefore, $\overline{M}_n = 40.16(141$ g mol$^{-1}) = 5663$ g mol^{-1}.

From Eq. (5.53): $\overline{M}_w = (141$ g mol$^{-1})\left(\dfrac{1 + 0.9751}{1 - 0.9751}\right) = 11{,}184$ g mol^{-1}.

From Eq. (5.44): $\sum_{x=1}^{6} n_x = (1 - 0.9751)\sum_{x=1}^{6}(0.9751)^{x-1} = 0.1404$.

From Eq. (5.47): $\sum_{x=1}^{6} w_x = (1 - 0.9751)^2 \sum_{x=1}^{6} x(0.9751)^{x-1} = 0.012$.

5.6 Nonlinear Step Polymerization

5.6.1 Branching

A special kind of step-growth polymer may be produced by carrying out polymerization of an A—B type bifunctional monomer in the presence of a small amount of a second monomer A_f containing $f(> 2)$ functional groups. The value of f represents the functionality of the monomer;

thus, if $f = 3$, A_f represents a trifunctional monomer, $\bullet(-A)_3$, having three A end groups on the same molecule. If reaction occurs only between A and B groups and neither A groups nor B groups are capable of reacting with each other, the product of polymerization of A—B in the presence of trifunctional A_f will lead to a branched polymer (Odian, 1991) of structure (**IV**).

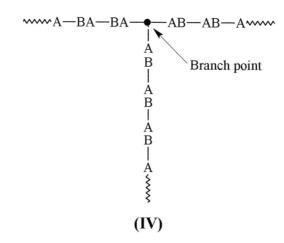

(IV)

Branched polymer molecules of this type cannot react with each other since the reactive group at the end of each branch is A. Thus, the formation of network or crosslinked structures cannot take place. For A—B + $\bullet A_f$ with $f = 2$, the polymer formed will be linear, and for all higher values of f only branched polymers will be formed.

Molecular weight distribution in the above type of nonlinear or multichain step polymerization would be expected to be much narrower than in linear or monochain polymerization. This can be explained as follows. Consider, for example, polymerization of bifunctional monomer A—B with a small proportion of an f-functional substance A_f leading to polymer molecules of the type $\bullet[-A(B-A)_y]_f$, [see (**IV**)] where it is understood that y, the number of units in a chain, may differ for each of the f chains. The sum of the y values of all the chains, say x, gives the size of the multichain polymer molecule. A polymer molecule of size much larger, or much smaller, than the average x is less likely in multichain than in monochain polymerization. Otherwise, the f branches making up a multichain molecule would have to be all very long or all very short, and such cooperation among different components would indeed be very unlikely. Molecular weight distributions for multichain polymers will thus be narrower than for ordinary monochain polymers and this difference will increase as the number of branches, i.e., as the value of f increases.

The number-average degree of polymerization, $\overline{X}_n$, and the weight-average degree of polymerization, $\overline{X}_w$, derived statistically (Shaefgen and Flory, 1948) are expressed as

$$\overline{X}_n = (frp + 1 - rp)/(1 - rp) \tag{5.57}$$

$$\overline{X}_w = \frac{(f-1)^2(rp)^2 + (3f-2)rp + 1}{(frp + 1 - rp)(1 - rp)} \tag{5.58}$$

The ratio of the weight-average degrees of polymerization to number-average degree of polymerization is

$$\frac{\overline{X}_w}{\overline{X}_n} = 1 + \frac{frp}{(frp + 1 - rp)^2} \tag{5.59}$$

In the limit of $r = f = 1$, Eqs. (5.57), (5.58), and (5.59) reduce to Eqs. (5.50), (5.53), and (5.54), respectively, for linear polymerization.

Further, in the limit of $p = 1$, Eq. (5.59) reduces to

$$\frac{\overline{X}_w}{\overline{X}_n} = 1 + \frac{1}{f} \tag{5.60}$$

The size distribution therefore becomes narrower with increasing functionality of A_f. For $f = 1$, Eq. (5.60) gives $\overline{X}_w/\overline{X}_n = 2$, as would be expected for most probable distribution in linear polymerization. Note also that $f = 1$ corresponds to Case 2 of stoichiometric imbalance in linear polymerization (see p. 230).

5.6.2 *Crosslinking and Gelation*

Consider polymerization of a system consisting of

A—B, e.g., hydroxyacid HO–R–COOH

A_f $(f > 2)$, e.g., triol R'(OH)$_3$

and B—B, e.g., dicarboxylic acid HOOC–R''–COOH

In this system, A—B is the chain forming monomer, A_f is the branch inducing monomer, while the monomer B—B is a coupling monomer as it can join two chains having unreacted A groups. A branch joining the two polymer chains is called a *crosslink* (Odian, 1991). Note that branches ending in A groups cannot join by themselves to form a crosslink. The role of B—B, besides that of A_f, is thus critical in crosslink formation, as shown in structure **(V)** (Ghosh, 1990). Crosslinking can also occur in other systems involving multifunctional reactants. A few such systems are (A—A + B—B + A_f), (A—A + B_f), and (A_f + B_f), where $f > 2$. The crosslinking reaction is very important from the point of view of application (see Section 1.5.1).

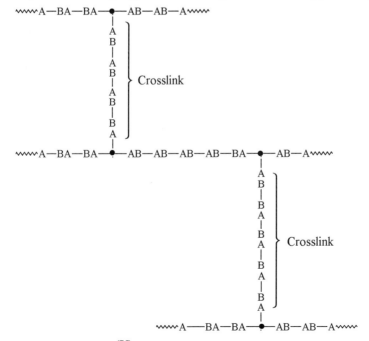

(V)

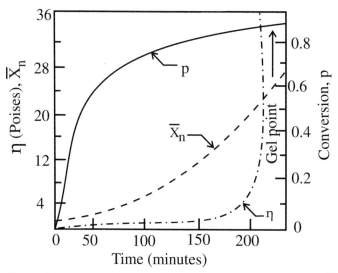

Figure 5.6 Variation of viscosity (η) and degree of polymerization ($\overline{X}_n$) with time and extent of reaction in the system succinic acid + diethylene glycol + 1,2,3-propanetricarboxylic acid. The observed gel point (p_c) is 0.894, while p_c calculated from Eq. (5.72) is 0.843. (From data reported by Flory, 1941, 1946.)

The formation of crosslinks leads to the onset of *gelation* at some point in the polymerization reaction. Phenomenologically, the *gel point* is defined as that point in the reaction where the system loses fluidity and the viscosity becomes so large that an air bubble cannot rise through it. A gel is an insoluble polymer fraction, the insolubility being caused by crosslinks between the polymer chains leading to the formation essentially of a single macroscopic molecule. However, at the gel point not all the material is insoluble. Only the gelled portion (and it represents only a small percentage of the total reaction mixture) is insoluble while the other portion is soluble. The portion that is soluble is referred to as the *sol*, whereas the part that is insoluble is referred to as the *gel*.

The course of polyesterification in an experimental crosslinking system of the type A—A + B—B + A$_f$ is shown in Fig. 5.6. There occurs, at first, a general increase in viscosity and then a sudden big increase as the gel point is approached. Also the reaction, as measured by the conversion of the reactive group, proceeds very slowly near the gel point. These two factors make it difficult to determine the gel point from the measurement of conversion or viscosity.

Since gelation is irreversible and it results in a loss of fluidity of the system, it is obviously important to be able to predict the onset of gelation from the extent of reaction or conversion of functional groups.

Average Functionality Approach   As we derived earlier [see Eq. (5.43)], the number-average degree of polymerization $\overline{X}_n$ of the reaction mixture is given by

$$\overline{X}_n = \frac{2}{2 - p.f_{av}} \tag{5.61}$$

where f_{av} is the average functionality of the reaction mixture. Equation (5.61) can be rearranged to

$$p = \frac{2}{f_{av}} - \frac{2}{\overline{X}_n \cdot f_{av}} \qquad (5.62)$$

Equation (5.61), which reduces to Eq. (5.30) for $f_{av} = 2$ and is also referred to as the *Carothers equation*, relates the extent of reaction and the degree of polymerization to the average functionality of the reaction mixture.

Since the number-average degree of polymerization of gel can be considered to be infinite (see Problem 1.5), i.e., $\overline{X}_n \to \infty$, Eq. (5.61) gives the extent of reaction at the gel point, called the *critical extent of reaction* or *critical conversion* as

$$p_c = \frac{2}{f_{av}} \qquad (5.63)$$

Thus, for a glycerol-phthalic acid (2 : 3 molar ratio) stoichiometric system, $f_{av} = (2 \times 3 + 3 \times 2)/(2 + 3) = 2.4$ and, hence, $p_c = 2/2.4 = 0.833$. For applying Eq. (5.62) to nonstoichiometric mixtures, however, f_{av} must be calculated as the average *useful* functionality of the reaction mixture.

Problem 5.23   Can the following alkyd recipe be reacted to complete conversion of the limiting reactant without gelation ?

Phthalic anhydride $C_6H_4(CO)_2O = 2.0$ mol
Glycerol $CH_2(OH)CH(OH)CH_2(OH) = 0.3$ mol
Pentaerythritol $C(CH_2OH)_4 = 0.6$ mol

Answer:

Component	Moles	Functionality	Equivalents
Phthalic anhydride	2.0	2	4.0
Glycerol	0.3	3	0.9
Pentaerythritol	0.6	4	2.4
Total	2.9		7.3

Total acid equivalents $= 4.0$ and total OH equivalents $= (0.9 + 2.4) = 3.3$

From Eq. (5.41a), since the OH equivalents are in deficient supply, $f_{av} = 2 \times 3.3/2.9 = 2.2759$. (Note that this represents the average **useful** functionality of the reaction mixture, while the overall average functionality is $7.3/2.9 = 2.5172$.)   Therefore,

$$p_c = \frac{2}{f_{av}} = \frac{2}{2.2759} = 0.88$$

Since the OH is the limiting reactant, its conversion at the gel point (Carothers) $= 0.88$. So complete conversion of OH without gelation is not possible.

Problem 5.24   Consider a nonstoichiometric reaction mixture consisting of N_{AA} moles of A—A, N_{BB} moles of B—B, and N_{A_f} moles of f-functional A_f with $f > 2$. Assuming that B groups are in excess, derive an expression for the conversion of the limiting reactant at the gel point.

Answer:

From Eq. (5.41a),

$$f_{av} = \frac{2(2N_{AA} + f.N_{A_f})}{N_{AA} + N_{BB} + N_{A_f}} \tag{P5.24.1}$$

The ratio of A groups to B groups is given by

$$r = \frac{2N_{AA} + f.N_{A_f}}{2N_{BB}} \tag{P5.24.2}$$

The fraction of all A groups which belong to the reactant with f (>2) is given by

$$\rho = \frac{f.N_{A_f}}{2N_{AA} + f.N_{A_f}} \tag{P5.24.3}$$

Combination of Eqs. (P5.24.1)-(P5.24.3) yields

$$f_{av} = \frac{4rf}{f + 2r\rho + rf(1 - \rho)} \tag{P5.24.4}$$

which can be substituted into Eq. (5.63) to yield

$$p_c = \frac{(1 - \rho)}{2} + \frac{1}{2r} + \frac{\rho}{f} \tag{P5.24.5}$$

The extent of reaction p_c at the gel point refers to the extent of reaction of the A functional groups. The extent of reaction of the B groups at the gel point is rp_c. [Note that Eq. (P5.24.5) is applicable only to $A-A + B-B + A_f$ $(f > 2)$ or $B-B + A_f$ $(f > 2)$ systems with B groups being in excess over A groups, while Eq. (5.63) is applicable to all systems with $f_{av} > 2$. Both Eqs. (5.63) and (P5.24.5) are based on Carothers' theory.]

5.6.2.1 Statistical Approach

Systems with one type of branch unit A statistical approach was used by Flory (1941, 1943, 1953) and Stockmayer (1943, 1953) to derive an expression for the prediction of the extent of reaction at the gel point. In this approach, it is assumed that: (1) the reactivity of all functional groups of the same type is the same and independent of molecular size and (2) there are no intramolecular reactions between functional groups of the same molecule.

Consider polymerization of bifunctional molecules $A-A$, $B-B$, and trifunctional molecule $\bullet(-A)_3$ in a mixture, not necessarily in equimolar quantities. This will lead to trifunctionally branched network polymer (**VI**):

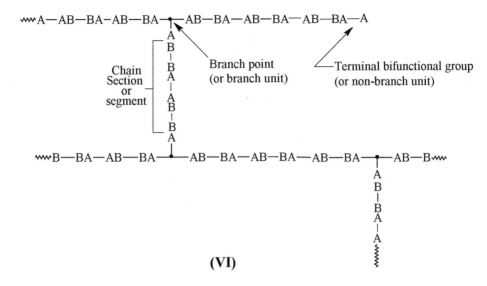

(VI)

A portion of the trifunctionally branched network (**VI**) of gelled polymer is schematically represented in Fig. 5.7, showing a series of contour lines or envelopes on which the branches end, either in a branch point or in a terminal bifunctional group (nonbranch unit). For convenience, the chain sections are shown as being equal in length. However, this will not affect our following analysis.

Consider the chain section in envelope 1 (Fig. 5.7) having a branch point at each end. The four new chains originating from the two branch points on envelope 1 happen to lead further to three new branch points and one nonbranch unit on envelope 2. The resulting six new chain sections that emanate from envelope 2 happen to lead to three branch points and three nonbranch units on

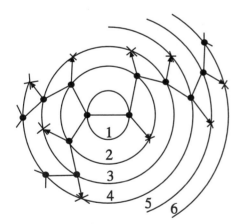

- **● Branch point (or branch unit)**

- **X Terminal nonbranch unit**

- **— Chain section or chain segment**

Figure 5.7 Schematic representation of a portion of trifunctionally branched network polymer. (Drawn following Flory, 1941, 1946.)

envelope 3, and so on. The probability that a branch arising at a branch point leads via bifunctional units to another branch point rather than to a nonbranch unit is referred to as the *branching probability*, α. So $(1 - \alpha)$ is the probability that a branch arising from the branch point leads to a nonbranch unit.

Moving on to an ith envelope from the randomly selected chain section enclosed in envelope 1 in Fig. 5.7, suppose there are Y_i branch points on the ith envelope. If *all* chain sections originating from these branch points ended in branch points on the $(i + 1)$th envelope, then there would be $2Y_i$ branch points on the $(i + 1)$th envelope. However, because of the branching probability α, defined above, the expected number (Y_{i+1}) of branch points on the $(i + 1)$th envelope will be only $2Y_i\alpha$. The criterion for gelation or continuous expansion of the network is that the number of chain sections arising from the $(i + 1)$th envelope, namely, $2Y_{i+1}$, be greater than the number of chain sections arising from the ith envelope, i.e., $2Y_i$. In other words, the criterion is

$$2Y_{i+1} > 2Y_i$$

$$\text{or} \quad 2(2Y_i\alpha) > 2Y_i, \quad \text{that is,} \quad \alpha > \tfrac{1}{2}$$

When $\alpha < \tfrac{1}{2}$, an infinite network cannot form. Clearly, the critical value of α, i.e., α_c, when the branching unit is trifunctional ($f = 3$), as in (**VI**) above, is

$$\alpha_c = \tfrac{1}{2}, \quad \text{or generally,} \quad \alpha_c = \frac{1}{f - 1} \tag{5.64}$$

where f is the functionality of the branch unit, i.e., the number of chain sections meeting at the branch point, which, in turn, is the functionality of the monomer with functionality greater than 2. The quantity α_c is called the *critical branching coefficient* for gel formation. If more than one monomer with $f > 2$ is present, an average value of f of all the monomers with $f > 2$ is used in Eq. (5.64). It should be noted that finite species are present at $\alpha = \alpha_c$, and only when α exceeds α_c by a finite amount does the theory allow the presence of infinite networks.

The actual value of α for a given polyfunctional system must now be computed. Let the fraction of A groups on multifunctional ($f > 2$) units be denoted by ρ [cf. Eq. (P5.24.3)], i.e.,

$$\rho = \frac{\text{Number of all A groups (reacted and unreacted) on trifunctional units}}{\text{Total number of all groups (reacted and unreacted) in mixture}} \tag{5.65}$$

Let us consider the following section (**VII**) of the gel network shown in (**VI**):

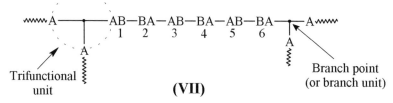

(VII)

Fraction of A groups reacted $= p_A$

Fraction of B groups reacted $= p_B$

Fraction of A groups on trifunctional units $= \rho$

Fraction of A groups on bifunctional units $= 1 - \rho$

The probability that A at position 1 has reacted $= p_A$

Since A can react only with B, the probability that A at position 1 has reacted with bifunctional B $= p_A \times 1 = p_A$

The probability that B at position 2 has reacted $= p_B$

The probability that B at position 2 has reacted with a bifunctional unit and not a trifunctional unit $= (1 - \rho)$

Therefore, the probability that B at position 2 has reacted with A on a bifunctional unit $= p_B(1 - \rho)$

The appropriate probabilities for positions 3, 4, and 5 are similarly obtained.

The probability that B at position 6 has reacted with A on a trifunctional unit $= p_B\rho$

Therefore, the probability α that the chain section shown in (**VII**) has formed $= p_A[p_B(1 - \rho)p_A]^2 p_B\rho$

In general, these sections can be represented (Williams,1971) as

$$A{-}\overset{\displaystyle |}{\underset{\displaystyle A}{\mathsf{T}}}{-}A{\left(\!{-}B{-}BA{-}A\!\right)_x}{-}B{-}BA{-}\overset{\displaystyle |}{\underset{\displaystyle A}{\mathsf{T}}}{-}A \qquad \text{[For the chain section shown in (VII), } x = 2]$$

$$(\textbf{VIII})$$

Therefore, the probability $\alpha(x)$ of obtaining the section shown in (**VIII**), containing x number of $-(B{-}BA{-}A)-$ units, is given by $\alpha(x) = p_A[p_B(1 - \rho)p_A]^x p_B\rho$.

Our objective is to find a total probability, α, that is independent of the length (i.e., the value of x) of the chain section. Accordingly (Williams, 1971),

$$\alpha = \sum_{x=0}^{\infty} \alpha(x) = \sum_{x=0}^{\infty} [p_A p_B(1 - \rho)]^x p_A p_B \rho$$

$$= \frac{p_A p_B \rho}{1 - p_A p_B(1 - \rho)} \tag{5.66}$$

This is the probability that a functional group on a branch unit leads (via bifunctional units) to another branch unit.

Either p_A or p_B can be eliminated by using the ratio r of A groups to B groups. Thus,

$$r = \frac{\text{All A groups}}{\text{All B groups}} = \frac{N_{A_0}}{N_{B_0}} \tag{5.67}$$

$$p_A = \frac{\text{Reacted A groups}}{N_{A_0}}, \quad p_B = \frac{\text{Reacted B groups}}{N_{B_0}} \tag{5.68}$$

Since one A group can react only with one B group, and hence $p_A N_{A_0} = p_B N_{B_0}$, it follows from Eqs. (5.67) and (5.68) that

$$r = p_B/p_A \tag{5.69}$$

Substitution for p_A or p_B from Eq. (5.69) into Eq. (5.66) yields

$$\alpha = \frac{rp_A^2\rho}{1 - rp_A^2(1 - \rho)} = \frac{p_B^2\rho}{r - p_B^2(1 - \rho)} \tag{5.70}$$

Let $p_A = p_c$ at the gel point. So from Eq. (5.70), at the gel point,

$$\alpha_c = \frac{rp_c^2\rho}{1 - rp_c^2(1 - \rho)} \tag{5.71}$$

Combination of Eq. (5.71) with Eq. (5.64) yields a useful expression for the extent of reaction (of the A functional groups) at the gel point:

$$p_c = \frac{1}{[r + r\rho(f - 2)]^{1/2}} \tag{5.72}$$

Equation (5.71) can be used for calculating the conversion at the gel point for a given feed composition, or for the reverse situation, i.e., calculating feed stoichiometry for a desired gel point (p_c).

Problem 5.25   Simplify Eqs. (5.70) and (5.72) for the following special cases of reaction mixtures: (a) two functional groups A and B are present in stoichiometric quantities; (b) feed consists of only A_f ($f > 2$) and B—B with B groups in excess over A groups; (c) feed consists of only A_f ($f > 2$) and B—B with A and B groups present in stoichiometric quantities; and (d) feed consists only of monomers with $f > 2$.

Answer:

(a) Here, $N_{A_0} = N_{B_0}$, $r = 1$, and $p_A = p_B = p$. So Eqs. (5.70) and (5.72) become

$$\alpha = \frac{p^2\rho}{1 - p^2(1 - \rho)} \quad \text{and} \quad p_c = \frac{1}{[1 + \rho(f - 2)]^{1/2}}$$

(b) Here, $\rho = 1$; $r (= N_{A_0}/N_{B_0}) < 1$; Eqs. (5.70) and (5.72) reduce to

$$\alpha = rp_A^2 = p_B^2/r \quad \text{and} \quad p_c = \frac{1}{[r + r(f - 2)]^{1/2}}$$

(c) Conditions of both case (a) and case (b) are present, i.e., $r = \rho = 1$. So Eqs. (5.70) and (5.72) become

$$\alpha = p^2 \quad \text{and} \quad p_c = \frac{1}{[1 + (f - 2)]^{1/2}}$$

(d) Here the probability that a functional group on a branch unit leads to another such unit is simply the probability that it has reacted, i.e., $\alpha = p$. Therefore, from Eq. (5.64),

$$p_c = \frac{1}{f - 1}$$

Use an average f for all monomers with $f > 2$.

Problem 5.26   A crosslinked polyurethane is to be made from diphenylmethane diisocyanate, $OCNC_6H_4$-$CH_2C_6H_4NCO$, and a polymeric tetrol, $R(OH)_4$, without addition of any water (Rempp and Merrill, 1986). Let the stoichiometric ratio of initial concentrations of NCO to OH groups be denoted by r.

(a) Calculate the critical extent of conversion of NCO groups above which gelation would occur for $r = 4$.

(b) What is the minimum value of r below which no gel can ever be obtained?

Answer:

Let A represent the alcohol functions and B the isocyanate functions. All A functions belong to tetrafunctional ($f = 4$) base molecules; therefore $\rho = 1$.

$r = [NCO]_0/[OH]_0 = N_{B_0}/N_{A_0}$; $p_A N_{A_0} = p_B N_{B_0}$; $p_A = p_B(N_{B_0}/N_{A_0}) = r p_B$. Therefore, Eq. (5.66) reduces to $\alpha = r p_B^2$.

The critical value at which gelation occurs is given by $\alpha_c = 1/(f - 1) = 1/3$. To get gelation, α_c must be exceeded.

Denoting the conversion of B functional group (p_B) at the gel point by p_c,

$$r p_c^2 = \alpha_c = 1/3, \quad \text{or} \quad p_c = 1/\sqrt{3r} \quad \text{and} \quad r = 1/(3p_c^2).$$

(a) For $r = 4$, $p_c = 0.29$. Hence gelation occurs above 29% conversion of NCO.

(b) For $p_c = 1$, $r = 0.33$. So gelation can never be obtained for any r value less than 0.33.

Problem 5.27   For a system composed of diethylene glycol (2.0 mol), 1,2,3-propane-tricarboxylic acid (0.6 mol), and adipic acid (1.0 mol), determine the number average degree of polymerization before gelation. Derive first a general expression to obtain this.

Answer:

Consider a condensation polymerization system consisting of A—A, B—B, and A_f ($f > 2$). Let N_{A_0} = total number of A groups present initially; N_{B_0} = total number of B groups present initially; ρ = number fraction of all A groups that belong to the reactant with $f > 2$.

Total number of monomer molecules present initially,

$$N_0 = \frac{N_{A_0}(1 - \rho)}{2} + \frac{N_{A_0}\rho}{f} + \frac{N_{B_0}}{2} \tag{P5.27.1}$$

Since each new linkage formed involves one A group, number of linkages formed at the extent of reaction, p_A, of A groups $= p_A N_{A_0}$.

Since with each linkage formed there will be one molecule less, number of molecules reacted $= p_A N_{A_0}$.

Number of molecules (N) present at the extent of reaction p_A is related by

$$N = N_0 - p_A N_{A_0}$$

Number-average degree of polymerization, $\overline{X}_n = \dfrac{N_0}{N_0 - p_A N_{A_0}}$ \hfill (P5.27.2)

Substituting for N_0 from Eq. (P5.27.1) and using $r = N_{A_0}/N_{B_0}$, Eq. (P5.27.2) yields (Williams, 1971):

$$\overline{X}_n = \frac{f(1 - \rho + 1/r) + 2\rho}{f(1 - \rho - 2p_A + 1/r) + 2\rho} \tag{P5.27.3}$$

This equation is not applicable above the gel point.

For the given system, $f = 3$, $\rho = 3 \times 0.6/(2 \times 1.0 + 3 \times 0.6) = 0.4737$ and $r = (2 \times 1.0 + 3 \times 0.6)/(2 \times 2.0) = 0.950$.

From Eq. (5.72),

$$p_c = \frac{1}{[0.950 + 0.950 \times 0.4737(3 - 2)]^{1/2}} = 0.845$$

From Eq. (P5.27.3), taking $p = 0.845$,

$$\overline{X}_n = \frac{3(1 - 0.4737 + 1/0.95) + 2 \times 0.4737}{3(1 - 0.4737 - 2 \times 0.845 + 1/0.95) + 2 \times 0.4737} = 9.25$$

Comment: The average degree of polymerization of the reaction mixture is thus not large at the gel point. It also does not increase rapidly at the gel point. This merely means that at the gel point many small molecules are still present; it does not preclude the formation of a fractional amount of indefinitely large structures at or beyond the gel point (see **Model for Gelation Process**).

Systems with different types of branch units   Let us consider a more general case where both A and B types of branch units occur and where there are also monofunctional and bifunctional A and B reactants, for example, polymerization of $A + A-A + A_f + B + B-B + B_f$.

To derive a general expression for the conversion of functional groups at the gel point, let symbols f, θ, and ρ denote, respectively, functionality (>2), fraction of functional groups belonging to bifunctional reactant, and fraction of functional groups belonging to multifunctional ($f > 2$) reactant, with subscripts A and B representing the type of functional group. Let r represent the stoichiometric ratio of functional groups such that $r \leq 1$. Thus,

f_A = Functionality of A_f.
θ_A = (A's belonging to A—A) / (Total A's).
ρ_A = (A's belonging to A_f) / (Total A's).
f_B = Functionality of B_f.
θ_B = (B's belonging to B—B) / (Total B's).
ρ_B = (B's belonging to B_f) / (Total B's).

Let N_A and N_B be the total number of A groups and B groups, and N_{A_f} and N_{B_f} be the total number of A_f and B_f molecules, respectively. The probability α that any group A belonging to a branch unit and selected at random is connected via a chain to another branch unit is obtained as follows. This state of affairs may come about in 4 possible ways shown in structures (a)-(d) below (for $f = 3$):

$$A-\!\!\!\overset{\displaystyle|}{\underset{\displaystyle A}{}}\!\!\!-A\left(B-BA-A\right)_n B-BA-\!\!\!\overset{\displaystyle|}{\underset{\displaystyle A}{}}\!\!\!-A \qquad (a)$$

$$A-\!\!\!\overset{\displaystyle|}{\underset{\displaystyle A}{}}\!\!\!-A\left(B-BA-A\right)_n B-\!\!\!\overset{\displaystyle|}{\underset{\displaystyle B}{}}\!\!\!-B \qquad (b)$$

$$B-\!\!\!\overset{\displaystyle|}{\underset{\displaystyle B}{}}\!\!\!-B\left(A-AB-B\right)_n A-\!\!\!\overset{\displaystyle|}{\underset{\displaystyle A}{}}\!\!\!-A \qquad (c)$$

$$B-\!\!\!\overset{\displaystyle|}{\underset{\displaystyle B}{}}\!\!\!-B\left(A-AB-B\right)_n A-AB-\!\!\!\overset{\displaystyle|}{\underset{\displaystyle B}{}}\!\!\!-B \qquad (d)$$

where n is any integer from 0 to ∞.

The probability α is equal to the sum of the 4 individual probabilities of each of these 4 cases occurring in any selection taken at random, each case being equally valid for the purpose of evaluating α. If the extents of reaction, p_A and p_B, are the fractions of A and B groups, respectively, that have reacted, then the 4 probabilities are given by equations written below:

$$\left. \begin{aligned}
\alpha_a &= \sum_0^n p_A [\rho_A N_A/(\rho_A N_A + \rho_B N_B)][\theta_A \theta_B p_A p_B]^n \theta_B p_B \rho_A \\
\alpha_b &= \sum_0^n p_A [\rho_A N_A/(\rho_A N_A + \rho_B N_B)][\theta_A \theta_B p_A p_B]^n \rho_B \\
\alpha_c &= \sum_0^n p_B [\rho_B N_B/(\rho_A N_A + \rho_B N_B)][\theta_A \theta_B p_A p_B]^n \rho_A \\
\alpha_d &= \sum_0^n p_B [\rho_B N_B/(\rho_A N_A + \rho_B N_B)][\theta_A \theta_B p_A p_B]^n \theta_A p_A \rho_B
\end{aligned} \right\} \tag{5.73}$$

[*Note*: The development of these expressions is similar to that of Eq. (5.66) except for the inclusion of the terms $\rho_A N_A/(\rho_A N_A + \rho_B N_B)$ and $\rho_B N_B/(\rho_A N_A + \rho_B N_B)$, which represent the probability of the multifunctional ($f > 2$) unit being of A and B type, respectively.]

The sum of the above four probabilities gives the total probability α:

$$\alpha = \frac{N_A \rho_A^2 \theta_B p_B p_A + N_A \rho_A \rho_B p_A + N_B \rho_A \rho_B p_B + N_B \theta_A \rho_B^2 p_A p_B}{(N_A \rho_A + N_B \rho_B)(1 - \theta_A \theta_B p_A p_B)} \tag{5.74}$$

[*Note*: $\sum_{n=0}^{\infty} y^n = 1/(1 - y)$, y being constant: $0 < y < 1$ (Appendix 5.1).]

At the gel point [cf. (5.64)],

$$\alpha_c = 1/(f - 1) \tag{5.75}$$

where f is the functionality of the branch unit. If there is more than one branch unit, f in Eq. (5.75) must be replaced by the appropriate average. For the present system, the average value of f is $(N_{A_f} f_A + N_{B_f} f_B)/(N_{A_f} + N_{B_f})$ so that instead of Eq. (5.75), we may write

$$\alpha_c = \frac{N_{A_f} + N_{B_f}}{N_{A_f}(f_A - 1) + N_{B_f}(f_B - 1)} \tag{5.76}$$

Combining Eqs. (5.74) and (5.76), the final equation applicable at the gel point is

$$(N_{A_f} + N_{B_f})(N_A \rho_A + N_B \rho_B)(1 - \theta_A \theta_B p_A p_B)$$
$$= [N_{A_f}(f_A - 1) + N_{B_f}(f_B - 1)][p_A p_B (N_A \theta_B \rho_A^2 +$$
$$N_B \theta_A \rho_B^2) + \rho_A \rho_B (N_A p_A + N_B p_B)] \tag{5.77}$$

where p_A and p_B now represent conversions at the gel point. If B functional groups are in excess, $r = N_A/N_B$ and $p_B = rp_A$ (since $N_A p_A = N_B p_B$). It is convenient to express Eq. (5.77) in terms of conversion of the limiting functional group (A in this case). Thus, substituting rp_A for p_B in Eq. (5.77), and replacing p_A by p_c to denote the gel point conversion of the limiting functional group, we have after rearrangement,

$$(XW + \theta_A \theta_B UV)rp_c^2 + WY p_c - UV = 0 \tag{5.78}$$

where

$$U = N_{A_f} + N_{B_f}$$
$$V = N_A\rho_A + N_B\rho_B$$
$$W = N_{A_f}(f_A - 1) + N_{B_f}(f_B - 1)$$
$$X = N_A\theta_B\rho_A^2 + N_B\theta_A\rho_B^2$$
$$Y = 2\rho_A\rho_B N_A \quad \text{if A is the limiting group}$$
$$\text{or } Y = 2\rho_A\rho_B N_B \quad \text{if B is the limiting group}$$

If $(XW + \theta_A\theta_B UV) = 0$,

$$p_c = UV/WY \tag{5.79}$$

When $(XW + \theta_A\theta_B UV) \neq 0$, solving Eq. (5.78) we have

$$p_c = \frac{-WY + \sqrt{W^2Y^2 + 4UVr(XW + \theta_A\theta_B UV)}}{2r(XW + \theta_A\theta_B UV)} \tag{5.80}$$

Note that Eq. (5.80) gives the gel-point conversion p_c of the functional group which is the limiting reactant. The conversion of the other functional group is rp_c, where r is the mole ratio of the two functional groups, such that $r \leq 1$.

Problem 5.28   Predict the conversions of both A and B groups at the point of gelation in the polymerization systems :   (a) A_3 (1 mol) + A_4 (1 mol) + B—B (4 moles);   (b) A (2 moles) + A—A (3 moles) + A_3 (3 moles) + B (2 moles) + B—B (1 mole) + B_4 (3 moles);   (c) A (4 moles) + A—A (30 moles) + A_3 (2 moles) + A_5 (1 mol) + B (5 moles) + B—B (25 moles) + B_4 (5 moles).

Answer:

(a) $\theta_A = 0$, $\rho_A = 1$, $\theta_B = 1$, $\rho_B = 0$

$$f_A = \frac{1 \times 3 + 1 \times 4}{1 + 1} = 3.5$$

$N_{A_f} = 1 + 1 = 2$, $N_{B_f} = 0$, $N_A = 7$, $N_B = 8$, $r = 7/8 = 0.8750$

$$U = 2 + 0 = 2$$
$$V = 7 \times 1 + 8 \times 0 = 7$$
$$W = 2(3.5 - 1) + 0 = 5$$
$$X = 7 \times 1 \times 1^2 + 8 \times 0 \times 0 = 7$$
$$Y = 2 \times 1 \times 0 \times 7 = 0$$

From Eq. (5.80),

$$p_c = \frac{-5 \times 0 + \sqrt{(5 \times 0)^2 + 4 \times 2 \times 7 \times 0.875(7 \times 5 + 0 \times 1 \times 2 \times 7)}}{2 \times 0.875(7 \times 5 + 0)}$$

$$= 0.6761$$

So $p_A = 0.6761$, $p_B = rp_A = 0.5916$

(b) $\theta_A = (3 \times 2)/(2 \times 1 + 3 \times 2 + 3 \times 3) = 6/17 = 0.3529$
    $\theta_B = (1 \times 2)/(2 \times 1 + 1 \times 2 + 3 \times 4) = 2/16 = 0.1250$
    $\rho_A = (3 \times 3)/17 = 0.5294$
    $\rho_B = (3 \times 4)/16 = 0.7500$
    $U = 3 + 3 = 6$

$$V = 17 \times 0.5294 + 16 \times 0.750 = 20.9998$$
$$W = 3(3 - 1) + 3(4 - 1) = 15$$
$$X = 17 \times 0.125(0.5294)^2 + 16 \times 0.3529(0.75)^2 = 3.7717$$
$$r = N_B/N_A = 16/17 = 0.9412$$
$$Y = 2 \times 0.5294 \times 0.750 \times 16 = 12.7056$$

Substituting the above values in Eq. (5.80) yields $p_c = 0.5638$. So $p_B = 0.5638$ and $p_A = rp_B = 0.5306$.

(c) $f_A = (2 \times 3 + 1 \times 5)/(2 + 1) = 3.6667$
$f_B = 4$
$N_{A_f} = 2 + 1 = 3$, $N_{B_f} = 5$
$\theta_A = (30 \times 2)/(4 \times 1 + 30 \times 2 + 2 \times 3 + 1 \times 5) = 60/75 = 0.800$
$\theta_B = (25 \times 2)/(5 \times 1 + 25 \times 2 + 5 \times 4) = 50/75 = 0.6667$
$r = 1$
$\rho_A = (3 \times 3.6667)/75 = 0.1467$
$\rho_B = (5 \times 4)/75 = 0.2667$
$U = 3 + 5 = 8$, $V = 75(0.1467 + 0.2667) = 31.005$
$W = 3(3.6667 - 1) + 5(4 - 1) = 23.00$
$X = 75 \times 0.6667(0.1467)^2 + 75 \times 0.80(0.2667)^2 = 5.3438$
$Y = 2 \times 0.1467 \times 0.2667 \times 75 = 5.8687$

Substituting these values in Eq. (5.80) yields $p_c = 0.7562$. So $p_A = p_B = 0.7562$.

5.6.2.2 *Model for Gelation Process*

In experimental studies, where the loss of fluidity is taken as marking the gel point, the conversion at the observed gel point is almost always found to be higher than that (calculated) at the theoretical gel point. This can be explained by the model proposed by Bobalek et al. (1964) for the gelation process, as shown in Fig. 5.8. According to this model, at the theoretical gel point, a number of macroscopic three-dimensional networks (gel particles) form and undergo phase separation. The gel particles so formed remain suspended in the medium and increase in number as reaction continues. At the experimentally observed gel point, the concentration of gel particles reaches a critical value and causes phase inversion as well as a steep rise in viscosity. The lower value of p_c predicted by the statistical approach is also attributed to the occurrence of some wasteful intramolecular cyclization reactions not taken into account in the derivation and also in some cases to the limited applicability of the assumption of equal reactivity of all functional groups of the same type, irrespective of molecular size.

5.6.2.3 *Molecular Size Distribution*

Expressions for molecular size distributions in multifunctional condensation leading to three-dimensional polymers are derived (Flory, 1941, 1946, 1953; Stockmayer, 1943, 1952, 1953) by following much the same approach as for linear polymers, though with much more difficulty. Only the results of these derivations will, however, be considered here. The derivations are based on three simplifying assumptions of ideal network formation: (1) all functional groups of the same type are equally reactive and independent of size of molecules to which they are attached; (2) all groups react independently of one another; and (3) no intramolecular reactions (cyclization) occur.

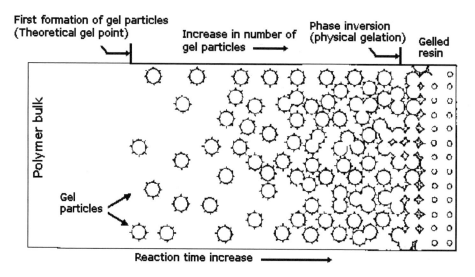

Figure 5.8 A model for the process of gelation. (After Bobalek et al., 1964.)

The simplest possible type of three-dimensional polymer is that formed by stepwise homopolymerization of a multifunctional monomer, such as etherification of pentaerythritol, or by stepwise copolymerization of equimolar amounts of two monomers having the same functionality f, such as the condensation of a trihydric alcohol with an equimolar proportion of a tribasic acid, all three functional groups on each monomer being equally reactive. For such polymerizations, the number N_x, the number or mole fraction n_x, and weight fraction w_x of x-mer molecules in the reaction system are given (Ghosh, 1990; Odian, 1991), respectively, by

$$N_x = N_0 \left[\frac{(fx - x)! f}{x! (fx - 2x + 2)!} \right] \alpha^{x-1} (1 - \alpha)^{fx - 2x + 2} \tag{5.81}$$

$$n_x = \left[\frac{(fx - x)! f}{x! (fx - 2x + 2)! (1 - f\alpha/2)} \right] \alpha^{x-1} (1 - \alpha)^{fx - 2x + 2} \tag{5.82}$$

$$w_x = \left[\frac{(fx - x)! f}{(x - 1)! (fx - 2x + 2)!} \right] \alpha^{x-1} (1 - \alpha)^{fx - 2x + 2} \tag{5.83}$$

where N_0 is the initial number of monomer molecules and α is the branching coefficient or branching probability, defined earlier as the probability that one arm of a branching unit (i.e., unit having a functionality greater than two) leads to another branching unit. For a system containing only monomers with $f > 2$ the branching coefficient α is simply equal to p, the extent of reaction [see Problem 5.25(d)].

Figure 5.9 shows molecular size distribution, calculated from Eq. (5.83), as a function of the extent of reaction in a simple trifunctional condensation ($f = 3$). In contrast to the weight-fraction distribution for linear condensation polymers (see Fig. 5.5), the curve shows a monotonic decrease with increase in size x. Thus, monomers always are present in greater amount, even on a weight basis, than species of any other size. The progressive change in polymer composition as the reaction progresses can be shown in another way by plotting weight fractions of various polymer sizes against the extent of reaction $p = \alpha$. Plots of this type are shown in Fig. 5.10, calculated again from Eq. (5.83).

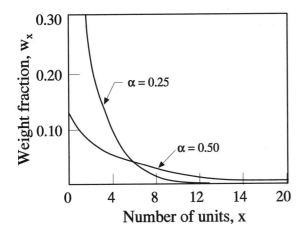

Figure 5.9 Molecular size distribution, w_x vs. x, for step-growth polymerization of trifunctional ($f = 3$) monomers at various stages of reaction, denoted by $\alpha = p$. The curves are calculated from Eq. (5.83) and plotted according to Flory (1941, 1946).

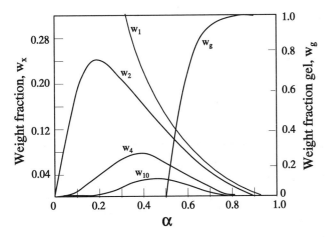

Figure 5.10 Weight fractions of various finite species (w_x) and of gel (w_g) as a function of $\alpha = p$. w_x is calculated from Eq. (5.83) and w_g from Eq. (5.88) and plotted according to Flory (1941, 1946).

The number- and weight-average degrees of polymerization in polymer systems of the above type are given (Ghosh, 1990) by

$$\overline{X}_n = \frac{N_0}{N_0 - N_0 f p/2} = \frac{1}{1 - \alpha f/2} \tag{5.84}$$

$$\overline{X}_w = \sum x w_x = \frac{(1 + \alpha)}{1 - (f - 1)\alpha} \tag{5.85}$$

$$\frac{\overline{X}_w}{\overline{X}_n} = \frac{(1 + \alpha)(1 - \alpha f/2)}{1 - (f - 1)\alpha} \tag{5.86}$$

Note that Eqs. (5.81), (5.82), and (5.83) are similar in form to Eqs. (5.45), (5.44), and (5.47), respectively, derived for linear systems and they simply reduce to the latter on substituting $f = 2$ and $\alpha = p$. Similarly, Eqs. (5.84), (5.85), and (5.86) reduce to Eqs. (5.50), (5.53), and (5.54), respectively, for linear systems on substituting $f = 2$ and $\alpha = p$.

Problem 5.29 The reaction of phenol with formaldehyde to phenolic resins is very complicated. Assuming for simplicity that in phenol-formaldehyde condensations with excess $(3:1)$ formaldehyde, the phenol is a trireactive structural unit and the formaldehyde merely supplies the interunit linkage, calculate the weight-average molecular weight of the phenolic resin at 30% conversion of the functional groups. What is the weight fraction of species containing 5 phenol residues at this conversion?

Answer:

Consider the monomer as
with $M_o(C_9H_{12}O_4) = 184$ g mol^{-1}
and $f = 3$.

From Eq. (5.85), for $\alpha = p = 0.30$, $\overline{X}_w = \dfrac{1 + 0.3}{1 - (3 - 1) \times 0.3} = 3.25$. Therefore, $\overline{M}_w = (3.25)(184$ g mol$^{-1}) - 3.25 \times 3 \times 0.3(18$ g mol$^{-1})/2 = 563$ g mol^{-1} (Note: One water molecule is eliminated by condensation of two methylol groups.)

From Eq. (5.83),

$$w_5 = \left[\frac{(3 \times 5 - 5)! \, 3}{(5 - 1)!(3 \times 5 - 2 \times 5 + 2)!} \right] (0.3)^{(5-1)} (1 - 0.3)^{(3 \times 5 - 2 \times 5 + 2)} = 0.06$$

5.6.2.4 Post-Gel Relations

Gelation in polymerization, which becomes manifested in a sudden loss in fluidity of the reaction mixture, does, in fact, take place long before all the reactant molecules join together forming one giant (hence insoluble) molecule. All but a small portion of the gelled polymer mass, as it exists at the gel point, would thus dissolve in a suitable solvent. The portion that is soluble is referred to as the *sol*, whereas the part that is insoluble is referred to as the *gel*. Only a few large molecules are required to induce gelation.

It may be noted that the curves in Fig. 5.10 representing the weight fractions of various species calculated from Eq. (5.83) continue through the gel point ($\alpha = 0.5$) without discontinuity. This would be expected as the derivations of the distribution equations (5.81)-(5.83) place no restriction on α or p except that it be in the physically real range from zero to unity. These equations are valid only for finite species beyond the gel point. If Eq. (5.83) for w_x is summed over all values of x, the weight fraction of sol (w_s) would result. When the units are trifunctional ($f = 3$), this yields (Flory, 1953):

$$w_s = \sum_{\text{all finite } x} w_x = \frac{(1 - \alpha)^3}{\alpha^3} \tag{5.87}$$

Since the sum of the weight fractions of gel w_g and sol w_s must be unity, we obtain an expression for the gel fraction as

$$w_g = 1 - \frac{(1 - \alpha)^3}{\alpha^3} \tag{5.88}$$

and $\alpha = p$. The weight percentage of gel plotted in Fig. 5.10 is calculated from this equation. Its formation commences abruptly at the critical point ($\alpha = 0.5$), rises rapidly, and proceeds to 100 percent as $\alpha = p$ goes to unity, i.e., as the reaction proceeds to completion.

Problem 5.30 A mixture of equivalent amounts of glycerol and 1,2,3-propanetricarboxy-lic acid is condensed to the extent of 65% conversion. Assuming equal reactivity of functional groups, calculate (a) weight fractions of sol and gel in the mixture, and (b) weight fractions of the monomer and the dimer in the sol fraction.

Answer:

Since both monomers are trifunctional ($f = 3$), $\alpha = p = 0.65$.

(a) From Eq. (5.87),

$$w_s = \frac{(1 - 0.65)^3}{(0.65)^3} = 0.156, \quad w_g = 1 - 0.156 = 0.844$$

(b) From Eq. (5.83),

$$w_1 \text{ (in mixture)} = \left[\frac{2!3}{0!3!}\right](0.65)^0(1 - 0.65)^3 = 0.0429$$

$$w_2 \text{ (in mixture)} = \left[\frac{4!3}{1!4!}\right](0.65)(1 - 0.65)^4 = 0.0293$$

Hence, w_1 (in sol) $= 0.0429/0.156 = 0.275$
and w_2 (in sol) $= 0.0293/0.156 = 0.188$.

Alternatively, take $\alpha = 1 - 0.65 = 0.35$ and calculate w_1 and w_2 directly from Eq. (5.83), to obtain

$$w_1 = \left[\frac{2!3}{0!3!}\right](0.35)^0(1 - 0.35)^3 = 0.275 \text{ and } w_2 = \left[\frac{4!3}{1!4!}\right](0.35)(1 - 0.35)^4 = 0.187.$$

Problem 5.31 Stoichiometric polymerization of 1,3,5-benzenetriacetic acid with 1,10-decanediol was conducted beyond the gel point to measure the mass fraction of sol w_s at different extents of reaction p, yielding the following data (Argyropoulos et al., 1987) :

p	w_s
0.726	0.752
0.809	0.130
0.877	0.036

Calculate the sol fraction by Flory's equation to compare with the experimental w_s values.

Answer:

Let A_3 = 1,3,5-benzenetriacetic acid and B_2 = 1,10-decanediol. Since $\rho = 1$, $r = 1$, and $p_A = p_B = p$, we obtain from Eq. (5.70), $\alpha = p^2$.

For $p = 0.726$, $\alpha = (0.726)^2 = 0.527$. Hence from Eq. (5.87): $w_s = (1 - 0.527)^3/(0.527)^3$
$= 0.723$ (cf. exptl. 0.752).

Similarly, for $p = 0.809$, $w_s = 0.147$ (cf. exptl. 0.130) and for $p = 0.877$, $w_s = 0.027$ (cf. exptl. 0.036).

5.7   Recursive Approach for Average Properties

Molecular size distributions of species as a function of reaction extent in step-growth polymerization, described so far in this chapter, are all based on the probability approach of Flory (1941, 1946, 1953) and Stockmayer (1943, 1952, 1953) and the simplifying assumptions of equal reactivity of all functional groups of the same type and no intramolecular reactions. These distributions are used to calculate the average properties ($\overline{M}_n$, $\overline{M}_w$, and PDI) of the reaction products. However, for cases of practical importance these distribution functions become quite complex.

Derivations based on many alternative approaches have also been made for the molecular weight distributions in both linear and nonlinear step polymerizations (Gordon and Judd, 1971; Burchard, 1979; Durand and Bruneau, 1979; Orlova et al., 1979). However, a knowledge of only the average properties of the distributions ($\overline{M}_n$, $\overline{M}_w$, and PDI) is often sufficient for many practical purposes. Macosko and Miller (Macosko and Miller, 1976; Lopez-Serrano, et al., 1980) developed a most useful approach for obtaining the average properties directly without resorting to the necessity of first calculating the molecular weight distribution. The approach utilizes the recursive nature of a step polymerization and assumes a *Markov* growth process in which the reaction with a polymer chain end depends only on the functional group at the end of that chain and not on any inside group away from the chain end. The method is demonstrated below for a few ideal linear and nonlinear step-growth polymerizations. The method retains the three simplifying assumptions of Flory, namely   (1) all functional groups of the same type are equally reactive;   (2) all groups react independently of one another;   and (3) no intramolecular reactions occur in finite species. A system satisfying the above assumptions is referred to as an "ideal" system. However, some departures from these assumptions, such as unequal reactivity, substitution effects, and some intramolecular reactions, can also be treated by the recursive method (Miller and Macosko, 1976, 1980).

The recursive approach uses an elementary law of conditional expectation. Let A be an event and $\overline{A}$ its complement. Let Y be a random variable, $E(Y)$ its expectation (or average value), and $E(Y \mid A)$ is conditional expectation, given that the event A has occurred. $P(A)$ is the probability that event A occurs. Then the *law of total probability for expectation* is (Case, 1958):

$$E(Y) \;=\; E(Y \mid A)\,P(A) \;+\; E(Y \mid \overline{A})\,P(\overline{A}) \tag{5.89}$$

5.7.1   Linear Step-Growth Polymerization

The simplest step-growth linear polymer is of the AB type. Typical examples of AB polymerization are the step polymerization of HORCOOH and $H_2NRCOOH$. For example, the polymer from HORCOOH is

H$-$(ORCO$-$)$-$(ORCO$-$)$-\cdots\cdots\cdots-$(ORCO$-$)$-$OH

Symbolically, the polymer may be shown as

To derive the molecular weight averages of the polymer, we shall use here the method described by Lopez-Serrano et al. (1980), which is identical in concept to the recursive method of Macosko and Miller (1976) cited above. Selecting an A group (marked by *) at random, the 'in' direction will be defined as the direction from the chosen A toward the B group of the same mer unit. 'Out' is then the opposite direction from the chosen A, i.e., toward the 'A' side of the mer. The 'in' and 'out' directions associated with B groups will also be defined in the same way.

Let us now ask what will be the expected weight of the polymer chain attached to a randomly chosen A group (indicated by *) looking 'out', i.e., $E(W_A^{out})$. Since the A group is chosen at random, W_A^{out} is a random variable. W_A^{out} equals 0 if A has not reacted. If A has reacted (with B of the next mer unit) then W_A^{out} equals W_B^{in}, the weight attached to B looking <u>into</u> B's parent molecule (Macosko and Miller, 1976):

$$W_A^{out} = \begin{cases} 0 & \text{if A does not react} \\ W_B^{in} & \text{if A does react (with B)} \end{cases} \tag{5.90}$$

By Eq. (5.89),

$$\begin{aligned} E(W_A^{out}) &= E(W_A^{out} \mid \text{A does not react})\,P(\text{A does not react}) \\ &\quad + E(W_A^{out} \mid \text{A reacts})\,P(\text{A reacts}) \\ &= 0(1-p) + E(W_B^{in})\,p \end{aligned} \tag{5.91}$$

where p is the extent of reaction of A groups, i.e., the fraction of all A groups that have reacted:

$$p = (N_{A_0} - N_A)/N_{A_0} \tag{5.92}$$

Here N_{A_0} represents the initial moles of A type groups and N_A equals the moles after some reaction time.

The expected weight attached to a B looking 'in' is equal to the weight of an AB mer plus the expected weight attached to an A looking 'out'. Thus,

$$E(W_B^{in}) = M_{AB} + E(W_A^{out}) \tag{5.93}$$

and the repetitive nature of this polymer molecule leads us back to the starting position.

It is obvious that if we had started by asking for the expected weight attached to a B group looking out, we would be led by the same arguments to a parallel set of the following two equations:

$$E(W_B^{out}) = E(W_A^{in})\,p \tag{5.94}$$

$$E(W_A^{in}) = M_{AB} + E(W_B^{out}) \tag{5.95}$$

The recursive nature of these chains is evident from the fact that starting at a point somewhere along the chain and moving along the chain in some direction we eventually reach another position statistically equivalent to the starting position (Lopez-Serrano et al., 1980).

Solving Eqs. (5.92) to (5.95) yields

$$E(W_A^{out}) + E(W_B^{out}) = 2M_{AB}\left(\frac{p}{1-p}\right) \tag{5.96}$$

$$E(W_A^{in}) + E(W_B^{in}) = 2M_{AB}\left(\frac{1}{1-p}\right) \tag{5.97}$$

For calculating average molecular weights we should note that the weight-average molecular weight will be given by the sum of the weight of an AB mer unit plus the expected weights attached to each arm looking 'out.' Thus,

$$\overline{M}_w = M_{AB} + E(W_A^{out}) + E(W_B^{out}) \tag{5.98}$$

Alternatively, $\overline{M}_w$ is obtained by picking an AB mer unit or an A group at random and then finding the expected weight of the molecule of which it is a part; that is,

$$\overline{M}_w = E(W_A^{in}) + E(W_A^{out}) \tag{5.99}$$

and substituting for $E(W_A^{in})$ from Eq. (5.95) leads again to Eq. (5.98).

Combining Eq. (5.98) with Eq. (5.96) gives

$$\overline{M}_w = M_{AB} \left(\frac{1+p}{1-p} \right) \tag{5.100}$$

The weight-average degree of polymerization, $\overline{X}_w$, is given by

$$\overline{X}_w = \frac{\overline{M}_w}{M_{AB}} = \frac{1+p}{1-p} \tag{5.101}$$

which gives the well-known result of Eq. (5.53) derived earlier through size distributions.

In the above calculations, we have chosen molecules at random by picking mer units at random. This is a weight-averaging process for molecules, since the larger the molecules, the proportionately larger chance it has of being chosen. But if we pick chain ends at random to choose molecules, and ask for the expected weight attached to the end group looking 'in,' we obtain a 'number-averaged' quantity, since the smaller the molecule, the proportionately larger chance it has of being chosen (Lopez-Serrano et al., 1980). If we pick end groups, however, we must statistically weight the $E(W^{in})$ by the mole fraction of each type of end group (Lopez-Serrano, 1980), that is, the mole fractions of unreacted A and B (denoted by μ_A and μ_B); thus:

$$\overline{M}_n = \mu_A E(W_A^{in}) + \mu_B E(W_B^{in}) \tag{5.102}$$

Since there are equal numbers of A and B ends in this simple case, $\mu_A = \mu_B = 1/2$, and we get from Eq. (5.97):

$$\overline{M}_n = M_{AB} \left(\frac{1}{1-p} \right) \tag{5.103}$$

$$\text{and} \quad \overline{X}_n = \overline{M}_n/M_{AB} = 1/(1-p) \tag{5.104}$$

This is the same as Eq. (5.50) derived earlier by the method of Flory.

Problem 5.32   Use the recursive approach to derive expressions for $\overline{M}_w$ and $\overline{M}_n$ for a polymerization system composed of equimolar amounts of A—A and B—B. To keep the system as simple as possible take the molecular weights of the two structural units as equal (denoted by M_0).

Answer:

Let the system $A-A + B-B$ react until some fraction p_A of the A's and fraction p_B of the B's have reacted. Since $A-A$ and $B-B$ are in equimolar amounts and $A+B$ is the only type of reaction, $p_A = p_B = p$.

The polymer resulting from the $A-A + B-B$ polymerization is shown below:

By application of Eq. (5.89), the following equations can be written for the system:

$$E(W_A^{out}) = (1 - p_A) \times 0 + p_A E(W_B^{in}) = pE(W_B^{in}) \tag{P5.32.1}$$

$$E(W_B^{in}) = M_{BB} + E(W_B^{out}) = M_0 + E(W_B^{out}) \tag{P5.32.2}$$

$$E(W_B^{out}) = (1 - p_B) \times 0 + p_B E(W_A^{in}) = pE(W_A^{in}) \tag{P5.32.3}$$

$$E(W_A^{in}) = M_{AA} + E(W_A^{out}) = M_0 + E(W_A^{out}) \tag{P5.32.4}$$

Let W_{AA} be the total molecular weight of the molecule to which the randomly chosen $A-A$ belongs. Similarly, let W_{BB} be the weight for a randomly chosen $B-B$. Then,

$$E(W_{AA}) = E(W_A^{in}) + E(W_A^{out}) = M_0 + 2E(W_A^{out}) \tag{P5.32.5}$$

$$E(W_{BB}) = E(W_{AB}^{in}) + E(W_B^{out}) = M_0 + 2E(W_B^{out}) \tag{P5.32.6}$$

Solving Eqs. (P5.32.1) through (P5.32.4) and substituting into Eqs. (P5.32.5) and (P5.32.6) gives

$$E(W_{AA}) = E(W_{BB}) = M_0 + \frac{2pM_0(1 + p)}{1 - p^2} \tag{P5.32.7}$$

To find the weight average molecular weight, we pick a unit of mass at random and compute the expected weight of the molecule of which it is a part [i.e., another application of Eq. (5.89)]. Thus,

$$\overline{M}_w = w_{AA} E(W_{AA}) + w_{BB} E(W_{BB})$$

where w_{AA} and w_{BB} are the weight fractions of $A-A$ and $B-B$ units; $w_{AA} = w_{BB} = 1/2$. Therefore,

$$\overline{M}_w = \tfrac{1}{2} E(W_{AA}) + \tfrac{1}{2} E(W_{BB}) \tag{P5.32.8}$$

which on substitution from Eq. (P5.32.7) yields

$$\overline{M}_w = M_0 + \frac{2pM_0(1 + p)}{1 - p^2} = M_0 \frac{1 + p}{1 - p} \tag{P5.32.9}$$

and $\overline{X}_w = (1 + p)/(1 - p)$ (P5.32.10)

The number average molecular weight is given by

$$\overline{M}_n = \tfrac{1}{2} E(W_A^{in}) + \tfrac{1}{2} E(W_B^{in}) \tag{P5.32.11}$$

Solving Eqs. (P5.32.1)-(P5.32.4) for $E(W_A^{in})$ and $E(W_B^{in})$ and substituting in Eq. (P5.32.11) gives

$$\overline{M}_n = M_0 \left(\frac{1}{1 - p} \right), \quad \text{or} \quad \overline{X}_n = \frac{1}{1 - p} \tag{P5.32.12}$$

Note: The expressions for $\overline{M}_w, \overline{X}_w, \overline{M}_n$, and $\overline{X}_n$ derived above for the simple case of equal molecular weights of structural units are identical with the corresponding expressions for the $A-B$ system. Expressions, though more involved, can be readily derived for a system with unequal structural unit molecular weights.

In addition to being a simpler method for obtaining the average properties such as $\overline{M}_w$ and $\overline{M}_n$, compared to the Flory and similar approaches (Case, 1958), the recursive approach also more easily allows an evaluation of the effect of unequal reactivity and unequal structural unit molecular weights on the average properties (Macosko and Miller, 1976; Lopez-Serrano et al., 1980; Ziegel et al., 1972).

5.7.2   Nonlinear Step-Growth Polymerization

5.7.2.1   Polymerization of A_f ($f > 2$)

The reaction between similar f-functional molecules is the simplest case of nonlinear polymerization. An example is the etherification of pentaerythritol (Macosko and Miller, 1976):

$$C(CH_2OH)_4 \; \longrightarrow \; \text{Branched polyether} + H_2O \tag{5.105}$$

Ignoring for the present the effects of any condensation products, we can schematically represent the polymerization of A_f ($f = 4$) by (Macosko and Miller, 1976):

$$\tag{5.105a}$$

Let the system react until some fraction p of the A groups has reacted. Note that $p = (N_{A_0} - N_A)/N_{A_0}$, where N_A denotes the number of moles of A groups and the zero subscript indicates the value at zero time. Let us pick an A group at random, labeled as A′ in Eq. (5.105) and ask: What is the weight, W_A^{out}, attached to A′ looking *out* from its parent molecule in the direction $\xrightarrow{1}$? Since A′ is chosen at random, W_A^{out} is a random variable. Thus,

$$W_A^{out} = \begin{cases} 0 & \text{if A′ does not react} \\ W_A^{in} & \text{if A′ does react (with A″, say)} \end{cases} \tag{5.106}$$

where W_A^{in} is the weight attached to A″ looking along $\xrightarrow{2}$, into A″'s parent molecule. Now by Eq. (5.89),

$$E(W_A^{out}) = E(W_A^{out} \mid A \text{ does not react}) P(A \text{ does not react})$$
$$+ E(W_A^{out} \mid A \text{ reacts}) P(A \text{ reacts})$$
$$= 0(1 - p) + E(W_A^{in}) p \tag{5.107}$$

$E(W_A^{in})$, the expected weight on any A looking into its parent molecule, will be the molecular weight of A_f plus the sum of the expected weights on each of the remaining $f - 1$ arms which is just $E(W_A^{out})$ for each arm. Thus,

$$E(W_A^{in}) = M_{A_f} + (f - 1) E(W_A^{out}) \tag{5.108}$$

and the recursive nature of the branched molecule leads us back to the starting point.

The weight-average molecular weight is obtained by picking an A_f or an A group at random and then finding the expected weight of the molecule of which it is a part; that is,

$$\overline{M}_w = E(W_A^{in}) + E(W_A^{out}) \tag{5.109}$$

Solving Eqs. (5.107) and (5.108) for $E(W_A^{in})$ and $E(W_A^{out})$ and substituting in Eq. (5.109) yields

$$\overline{M}_w = M_{A_f} \frac{1 + p}{1 - p(f - 1)} \tag{5.110}$$

$$\text{or} \quad \overline{X}_w = \frac{\overline{M}_w(p)}{\overline{M}_w(0)} = \frac{1 + p}{1 - p(f - 1)} \tag{5.111}$$

which is in agreement with Flory's result [cf. Eq. (5.85)] derived by a much longer process involving size distributions (Flory, 1953).

It may be noted that solutions to Eqs. (5.107) and (5.108) exist only when $p(f - 1) < 1$. If $p(f - 1) \geq 1$, then $\overline{M}_w$ diverges and the system forms a gel or infinite network. The critical conversion p_c for gel formation in the polymerization of A_f ($f > 2$) is thus

$$p_c = 1/(f - 1) \tag{5.112}$$

Since only monomer A_f ($f > 2$) is present, the branching coefficient (α), defined earlier, is simply p and the critical condition for gel formation is thus given by

$$\alpha_c = 1/(f - 1) \tag{5.113}$$

This is in agreement with the result of Flory's gel point theory [cf. Eq. (5.64)], described previously. In a situation where more than one multifunctional ($f > 2$) species is present, α_c would be computed using an average f.

If there is a condensation product in polymerization, one needs to subtract the molecular weight (M_c) of the condensate (by-product). In polyetherification, $M_c = 18$. Stockmayer (1943, 1952, 1953) suggested that networks involving formation of by-products could simply be treated by replacing M_{A_f} with $M_{A_f} - f M_c/2$ in the relations derived neglecting formation of by-products. This approach is better than ignoring M_c entirely, but it neglects the unreacted ends of the molecules and can only be strictly valid as p approaches 1.

5.7.2.2   Polymerization of A_f ($f > 2$) + B_2

An example of $A_f + B_2$ system, where $f > 2$, is urethane formation from pentaerythritol and hexamethylenediisocyanate. Consider N_{A_f} moles of monomer A_f reacting with N_{B_2} moles of bifunctional B_2 monomer shown schematically for $f = 4$ (Macosko and Miller, 1976):

$$(5.114)$$

Let the system react until some fraction p_A of the A groups and some fraction p_B of the B groups have reacted. If the reaction takes place only between A and B groups, then the number of A's reacted must equal the number of B's reacted, that is,

$$p_A(fN_{A_f}) = p_B(2N_{B_2})$$

$$\text{or} \quad p_B = \frac{fN_{A_f}}{2N_{B_2}} p_A = rp_A \tag{5.115}$$

Again pick an A group at random, labeled as A′ in Eq. (5.114). What is the weight, W_A^{out}, attached to A′ looking out from its parent molecule, in the direction $\overset{1}{\longrightarrow}$? Since A′ is chosen at random,

$$W_A^{out} = \begin{cases} 0 & \text{if A′does not react} \\ W_B^{in} & \text{if A′does react (with B′)} \end{cases} \tag{5.116}$$

Therefore, according to the law of total probability for expectations [Eq. (5.89)],

$$E(W_A^{out}) = p_A E(W_B^{in}) \tag{5.117}$$

Following the arrows in Eq. (5.114) we can write expected weights, in a similar way as above, until the repetitive nature of the structure leads us back to the starting position, i.e., Eq. (5.117):

$$E(W_B^{in}) = M_{B_2} + E(W_B^{out}) \tag{5.118}$$

$$E(W_B^{out}) = rp_A E(W_A^{in}) \tag{5.119}$$

$$E(W_A^{in}) = M_{A_f} + (f-1)E(W_A^{out}) \tag{5.120}$$

Denoting by W_{A_f} the total molecular weight of the molecule to which a randomly chosen A_f belongs and defining W_{B_2} similarly for a randomly chosen B_2, we may write

$$E(W_{A_f}) = E(W_A^{in}) + E(W_A^{out}) = M_{A_f} + fE(W_A^{out}) \qquad (5.121)$$

$$E(W_{B_2}) = E(W_B^{in}) + E(W_B^{out}) = M_{B_2} + 2E(W_B^{out}) \qquad (5.122)$$

Solving Eqs. (5.117)-(5.120) and substituting the results in Eqs. (5.121) and (5.122) gives

$$E(W_{A_f}) = M_{A_f} + fp_A \left[\frac{M_{B_2} + rp_A M_{A_f}}{1 - rp_A^2(f-1)} \right] \qquad (5.123)$$

$$E(W_{B_2}) = M_{B_2} + 2rp_A \left[\frac{M_{A_f} + p_A(f-1)M_{B_2}}{1 - rp_A^2(f-1)} \right] \qquad (5.124)$$

Picking at random a unit of mass and determining the expected weight of the molecule of which it is a part will give the weight-average molecular weight. This is, in fact, another application of Eq. (5.89). Thus,

$$\overline{M}_w = E(W_{A_f})w_{A_f} + E(W_{B_2})w_{B_2} \qquad (5.125)$$

where w_{A_f} and w_{B_2} are the weight fractions of A_f and B_2, respectively, defined by

$$w_{A_f} = \frac{N_{A_f}M_{A_f}}{N_{A_f}M_{A_f} + N_{B_2}M_{B_2}} \qquad (5.126)$$

$$w_{B_2} = 1 - w_{A_f} \qquad (5.127)$$

Equation (5.125) thus becomes

$$\overline{M}_w = \frac{(2r/f)(1 + rp_A^2)M_{A_f}^2 + 4rp_A M_{A_f}M_{B_2} + [1 + (f-1)rp_A^2]M_{B_2}^2}{(2rM_{A_f}/f + M_{B_2})[1 - r(f-1)p_A^2]} \qquad (5.128)$$

This result is the same as that which Stockmayer (1943, 1952, 1953) obtained by tortuous combinatorial arguments and manipulation of distribution functions. If some of the starting species are oligomers having MWD, average molecular weights (Ziegel et al., 1972) must be used in Eq. (5.128). Note that Eq. (5.128) does not consider the effect of condensation by-products. Their inclusion, though straightforward (Macosko and Miller, 1976), leads to more more unwieldy expressions.

Problem 5.33 Without intramolecular reactions, $\overline{M}_n$ can always be calculated from stoichiometry. Neglecting formation of condensation by-products, derive a general expression to calculate $\overline{M}_n$ for the general system $\sum A_i + \sum B_j$ at the extent of reaction p_A of the A groups.

Answer:

At the extent of reaction p_A, $\overline{M}_n$ is just the total mass, m_t, over the number (mole) of molecules, N, present after reaction, i.e., $\overline{M}_n = m_t/N$. Denoting the initial number (mole) and the molecular weight of the monomers by N and M, respectively, with appropriate subscripts, we have $m_t = \sum_i M_{A_i}N_{A_i} + \sum_j M_{B_j}N_{B_j}$. Note that N, defined above, is just the number (mole) of molecules present initially, N_0, less the number of new bonds formed, N_b (since with the formation of one bond the number of molecules decreases by 1). Further, since the reaction is only between A and B, the number of bonds formed equals the number of either A or B groups reacted. Thus,

$$N = N_0 - N_b = \left(\sum_i N_{A_i} + \sum_j N_{B_j}\right) - p_A \sum_i i N_{A_i}$$

Thus, $\overline{M}_n = \dfrac{\sum_i M_{A_i} N_{A_i} + \sum_j M_{B_j} N_{B_j}}{\sum_i N_{A_i} + \sum_j N_{B_j} - p_A \sum_i i N_{A_i}}$

5.7.2.3 *Polymerization of* $\sum A_i + \sum B_j$

We now consider a more general case of stepwise reaction between A_i and B_j, where $i, j = 2, 3, \cdots$. Addition of bifunctional components increases the chain lengths between the branch points. Consider such a mixture of N_{A_i} moles of A_i and N_{B_j} moles of B_j. Assuming that A groups react only with B groups, this general system can be represented schematically by (Macosko and Miller, 1976):

$$(5.129)$$

For the expected weight along $\xrightarrow{1}$ we must now consider all the possible B_j's with which an A can react. Equation (5.89) is therefore generalized to the form (Macosko and Miller, 1976):

$$
\begin{aligned}
E(W_A^{out}) &= E(W_A^{out} \mid \text{A does not react})\, P(\text{A does not react}) \\
&\quad + \sum_j E(W_A^{out} \mid \text{A reacts with } B_j)\, P(\text{A reacts with } B_j) \\
&= 0(1 - p_A) + \sum_j E(W_{B_j}^{in}) p_A n_{B_j} = p_A \sum_j n_{B_j} E(W_{B_j}^{in})
\end{aligned}
\qquad (5.130)
$$

where n_{B_j} = mole fraction of all B groups that are on B_j molecules and is given by

$$n_{B_j} = \frac{j N_{B_j}}{\sum_j j N_{B_j}} \qquad (5.131)$$

Next considering direction $\xrightarrow{2}$, there will be a relation of the same form as Eq. (5.118) for each B_j molecule

$$E(W_{B_j}^{in}) = M_{B_j} + (j-1)E(W_B^{out}) \tag{5.132}$$

Along directions $\xrightarrow{3}$ and $\xrightarrow{4}$ the expected weights can be derived in the same way as Eqs. (5.130) and (5.132), giving

$$E(W_B^{out}) = p_B \sum_i n_{A_i} E(W_{A_i}^{in}) \tag{5.133}$$

$$E(W_{A_i}^{in}) = M_{A_i} + (i-1)E(W_A^{out}) \tag{5.134}$$

where n_{A_i} = mole fraction of all A groups that are on A_i molecules and is given by

$$n_{A_i} = \frac{iN_{A_i}}{\sum_i iN_{A_i}} \tag{5.135}$$

and p_B is related to p_A as in Eq. (5.115):

$$p_B = \frac{\sum_{i=1} iN_{A_i}}{\sum_j jN_{B_j}} p_A = rp_A \tag{5.136}$$

Solving the system of Eqs. (5.130) and (5.132)-(5.134) yields

$$E(W_A^{out}) = \frac{p_A M_b + p_A p_B (f_B - 1)M_a}{1 - p_A p_B (f_A - 1)(f_B - 1)} \tag{5.137}$$

$$E(W_B^{out}) = \frac{p_B M_a + p_A p_B (f_A - 1)M_b}{1 - p_A p_B (f_A - 1)(f_B - 1)} \tag{5.138}$$

where f_A and f_B are weight average functionalities of A_i and B_j molecules, respectively, defined by

$$f_A = \frac{\sum_i i^2 N_{A_i}}{\sum_i iN_{A_i}} = \sum_i in_{A_i} \tag{5.139}$$

$$f_B = \frac{\sum_j j^2 N_{B_j}}{\sum_j jN_{B_j}} = \sum_j jn_{B_j} \tag{5.140}$$

and M_a and M_b are given by

$$M_a = \sum_i M_{A_i} n_{A_i} \tag{5.141}$$

$$M_b = \sum_j M_{B_j} n_{B_j} \tag{5.142}$$

Let W_{A_i} be the total molecular weight of the molecule to which a randomly chosen A_i belongs, and W_{B_j} the corresponding weight for a randomly chosen B_j. Then, similarly to Eqs. (5.121) and (5.122),

$$E(W_{A_i}) \; = \; M_{A_i} \; + \; iE(W_A^{out}) \tag{5.143}$$

$$E(W_{B_j}) \; = \; M_{B_j} \; + \; jE(W_B^{out}) \tag{5.144}$$

To find $\overline{M}_w$, Eq. (5.125) is generalized as

$$\overline{M}_w \; = \; \sum_i E(W_{A_i}) w_{A_i} \; + \; \sum_j E(W_{B_j}) w_{B_j} \tag{5.145}$$

where w_{A_i} and w_{B_j} are weight fractions of A_i and B_j, respectively, while w_{A_i} is given by

$$w_{A_i} \; = \; \frac{M_{A_i} N_{A_i}}{\sum_i M_{A_i} N_{A_i} \; + \; \sum_j M_{B_j} N_{B_j}} \tag{5.146}$$

and w_{B_j} is defined similarly. Substitution of all terms in Eq. (5.145), followed by rearrangement, gives

$$\overline{M}_w \; = \; \frac{p_B m_a' \; + \; p_A m_b'}{p_B m_a \; + \; p_A m_b}$$
$$+ \; \frac{p_A p_B [p_A(f_A - 1)M_b^2 + p_B(f_B - 1)M_a^2 + 2M_a M_b]}{(p_B m_a + p_A m_b)[1 - p_A p_B(f_A - 1)(f_B - 1)]} \tag{5.147}$$

where

$$m_a \; = \; \sum_i M_{A_i} N_{A_i} / \sum_i i N_{A_i} \; = \; \sum_i M_{A_i} n_{A_i} / i$$

$$m_a' \; = \; \sum_i M_{A_i}^2 N_{A_i} / \sum_i i N_{A_i} \; = \; \sum_i M_{A_i}^2 n_{A_i} / i$$

and m_b and m_b' are analogously defined for the B_j's.

Equation (5.147) is a general relation which holds for nearly all cases of nonlinear stepwise polymerizations. Note that Eqs. (5.110) and (5.128) are only special cases of Eq. (5.147). The extent of reaction at the gel point can be obtained from Eq. (5.147) for $\overline{M}_w \to \infty$; thus,

$$(p_A p_B)_{gel} \; = \; \frac{1}{(f_A - 1)(f_B - 1)} \tag{5.148}$$

Substituting for p_B from Eq. (5.136) and representing $(p_A)_{gel}$ simply by p_c, we obtain

$$p_c \; = \; \frac{1}{\{r(f_A - 1)(f_B - 1)\}^{1/2}} \tag{5.149}$$

Problem 5.34 Calculate the gel-point conversions for the systems cited in Problem 5.28 using the recursive approach for comparison with the corresponding values calculated according to Flory-Stockmayer theory.

Answer:

(a) System: A_3(1 mole) + A_4(1 mole) + B_2(4 moles)

Eq. (5.139): $f_A = \dfrac{3^2 \times 1 + 4^2 \times 1}{3 \times 1 + 4 \times 1} = 3.571$

Eq. (5.140): $f_B = \dfrac{2^2 \times 2}{2 \times 2} = 2$

Eq. (5.136): $r = \dfrac{3 \times 1 + 4 \times 1}{2 \times 4} = 0.875$

Eq. (5.149): $(p_A)_{gel} = p_c = \dfrac{1}{\{0.875(3.571 - 1)(2 - 1)\}^{1/2}} = 0.667$

[cf. Problem 5.28(a): $(p_A)_{gel} = 0.676$]

(b) System: A(2 moles) + A_2(3 moles) + A_3(3 moles) + B_1(2 moles) + B_2(1 mole) + B_4(3 moles)

From Eqs. (5.139), (5.140), and (5.136): $f_A = 2.412$, $f_B = 3.375$, and $r = 1.062$

Then from Eq. (5.149): $(p_A)_{gel} = p_c = 0.530$. [cf. Problem 5.28(b): $(p_A)_{gel} = 0.531$]

(c) System: A_1(4 moles) + A_2(30 moles) + A_3(2 moles) + A_5(1 mole) + B_1(5 moles) + B_2(25 moles) + B_4(5 moles)

Similar calculations as above give: $f_A = 2.227$, $f_B = 2.467$, $r = 1.000$, and $(p_A)_{gel} = p_c = 0.745$ [cf. Problem 5.28(c): $(p_A)_{gel} = 0.756$].

Problem 5.35 Consider the polymerization system (c) in Problem 5.34. Calculate the weight-average molecular weight at 50% conversion of the A groups, given that all the monomers in the system have equal molecular weights of 100.

Answer:

To use Eq. (5.147), one needs to calculte m_a, m'_a, m_b, and m'_b :

$$m_a = \frac{100(4 + 30 + 2 + 1)}{(1 \times 4 + 2 \times 30 + 3 \times 2 + 5 \times 1)} = 49.33; \quad m'_a = \frac{(100)^2(4 + 30 + 2 + 1)}{(1 \times 4 + 2 \times 30 + 3 \times 2 + 5 \times 1)} = 4933.3$$

$$m_b = \frac{100(5 + 25 + 5)}{1 \times 5 + 2 \times 25 + 4 \times 5} = 46.67; \quad m'_b = \frac{(100)^2(5 + 25 + 5)}{1 \times 5 + 2 \times 25 + 4 \times 5} = 4666.7$$

From Problem 5.34(c): $f_A = 2.227$, $f_B = 2.467$, $r = 1.00$. Also, $p_B = rp_A = 1 \times 0.50 = 0.50$, and $M_a = M_b = 100$. Substituting these values in Eq. (5.147) yields: $\overline{M}_w = 417$.

5.7.3 *Post-Gel Properties*

The recursive method of Macosko and Miller (1976) has been described earlier for calculating molecular weight averages up to the gel point in nonlinear polymerization. A similar recursive method (Langley, 1968; Langley and Polmanteer, 1974) can also be used beyond the gel point, particularly for calculating weight-fraction solubles (*sol*) and crosslink density. To illustrate the principles, we consider first the simple homopolymerization, that is, reaction between similar *f*-functional monomers A_f and then a more common stepwise copolymerization, such as reaction of A_f with B_2.

5.7.3.1   Polymerization of A_f

Gel-point Conversion   Let us represent schematically the stepwise homopolymerization of A_f by

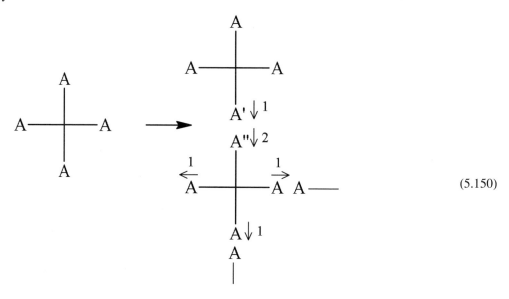

(5.150)

Assume that polymerization has proceeded to an extent that some fraction p of the A's has reacted. Picking an A group at random (say, A' in Eq. (5.150)) we ask: What is the probability that looking *out* from the molecule A_f in the direction $\overset{1}{\longrightarrow}$ we find a finite or dangling chain rather than an infinite network? Let F_A^{out} be the event that $\overset{1}{\longrightarrow}$ is the start of a finite chain; then from Eq. (5.89) it follows that (Macosko and Miller, 1976):

$$
\begin{aligned}
P(F_A^{\text{out}}) &= P(F_A^{\text{out}} \mid A \text{ reacts}) P(A \text{ reacts}) \\
&\quad + P(F_A^{\text{out}} \mid A \text{ does not react}) P(A \text{ does not react}) \\
&= P(F_A^{\text{in}})p + 1(1-p) = pP(F_A^{\text{in}}) + 1 - p
\end{aligned}
$$

(5.151)

where F_A^{in} is the event that $\overset{2}{\longrightarrow}$ in Eq. (5.150), looking *in* from A″, is the start of a finite chain. However, for A″ to lead to a finite chain all the other arms of A_f must be finite, that is,

$$
P(F_A^{\text{in}}) = P(F_A^{\text{out}})^{f-1}
$$

(5.152)

and this leads us back to the starting situation. Solving Eqs. (5.151) and (5.152), one obtains for $P(F_A^{\text{out}})$:

$$
pP(F_A^{\text{out}})^{f-1} - P(F_A^{\text{out}}) - p + 1 = 0
$$

(5.153)

and for $P(F_A^{\text{in}})$:

$$
\left[pP(F_A^{\text{in}}) + 1 - p \right]^{f-1} = P(F_A^{\text{in}})
$$

(5.154)

Physically, $P(F_A^{out}) = 1$ signifies that the system has not yet gelled. In order to derive post-gel relations we therefore desire roots of Eqs. (5.153) and (5.154) between 0 and 1. Since $P(F_A^{out}) = 1$ is a root of Eq. (5.153), and is not of interest, we can factor it out and write (Miller and Macosko, 1976):

$$p \sum_{i=0}^{f-2} P(F_A^{out})^i - 1 = 0 \tag{5.155}$$

which can be readily solved for a given value of f. Thus,

$$\text{for } f = 3, \quad P(F_A^{out}) = (1 - p)/p \tag{5.156}$$

$$\text{for } f = 4, \quad P(F_A^{out}) = (1/p - 3/4)^{1/2} - 1/2 \tag{5.157}$$

Roots between 0 and 1 for higher f are, however, easy to find numerically.

Problem 5.36   Starting with Eq. (5.153) for the simple A_f ($f > 2$) homopolymerization, derive a relation between the critical extent of reaction p_c for gelation and the monomer functionality f.

Answer:

The condition $P(F_A^{out}) = 1$ signifies that an A cannot be found on an infinite chain, or in other words, an infinite network has not started to form. Thus to find the gel point one has to identify the situation in which there is no solution of Eq. (5.153) between 0 and 1. Factoring Eq. (5.153) yields

$$\left[P(F_A^{out}) - 1 \right] \left[p \sum_{i=0}^{f-2} P(F_A^{out})^i - 1 \right] = 0 \tag{P5.36.1}$$

Thus it is enough to determine when

$$g(x) = p \sum_{i=0}^{f-2} x^i - 1 = 0 \tag{P5.36.2}$$

has no roots between 0 and 1. Because $g(x)$ increases monotonically for x between 0 and 1, and $g(0) = -1$ and $g(1) = p(f - 1) - 1$, Eq. (P5.36.2) will have no root between 0 and 1 if and only if $p(f - 1) - 1 < 0$. Thus, if $p \leq 1/(f - 1)$, we have $P(F_A^{out}) = 1$ and if $p > 1/(f - 1)$, we have $P(F_A^{out}) < 1$ and an infinite network has started forming. Thus the critical extent of reaction, p_c, which must be exceeded for the formation of infinite networks to become possible, is given by

$$p_c = 1/(f - 1) \tag{P5.36.3}$$

This is identical with the relation derived from the Flory theory of gelation [see Problem 5.25(d)].

Weight Fraction Solubles and Crosslink Density   Till the gel point is reached, all molecules are finite and the weight fraction of solubles, w_s, is thus unity. Beyond the gel point more and more molecules enter the network with the consequent decrease of w_s. For the homopolymerization of A_f ($f > 2$), a randomly chosen A_f molecule will be part of the sol if all of its f number of arms lead only to finite chains. Thus,

$$w_s = P(F_A^{out})^f \tag{5.158}$$

Using Eq. (5.156) for $f = 3$ gives

$$w_s = (1 - p)^3/p^3 \qquad (5.159)$$

which agrees with Eq. (5.87) derived by Flory for the homopolymerization of trifunctional monomers ($\alpha = p$). Note that these relations neglect the formation of condensation products. If a condensation product forms during polymerization, such as water in etherification of pentaerythritol (p. 286), the above analysis must be modified to accommodate this phenomenon (Macosko and Miller, 1976).

An important property of the polymer network is the crosslink density or concentration of effective junction points. An effective junction point can be identified thus. *An A_f is an effective junction point if three or more of its arms lead out to the infinite network.* (Note that if only one arm is infinite, this A_f unit will be just hanging from the network; if, on the other hand, two arms are infinite, A_f only forms part of a chain connecting two effective junction points and it itself is not an effective junction point.)

Let us consider tetrafunctional A_4 as an example of A_f. The probability that A_4 is an effective crosslink of degree 4 is given by

$$P(X_{4,4}) = \left[1 - P(F_A^{\text{out}})\right]^4 \qquad (5.160)$$

However, A_4 can also be an effective crosslink of degree 3 and the probability that exactly three of the four arms lead out to infinite network is given (Macosko and Miller, 1976) by

$$P(X_{3,4}) = \binom{4}{3} P(F_A^{\text{out}}) \left[1 - P(F_A^{\text{out}})\right]^3 \qquad (5.161)$$

In general, the probability that an A_{f_i} unit will be an effective crosslink of degree m is then

$$P(X_{m,f_i}) = \binom{f_i}{m} P(F_A^{\text{out}})^{f_i - m} \left[1 - P(F_A^{\text{out}})\right]^m \qquad (5.162)$$

The concentration of effective network junctions or crosslink density may be defined (Macosko and Miller, 1976) as "the initial concentration of the appropriate A_{f_i} species, $[A_{f_i}]_0$, times the probability $P(X_{m,f_i})$ summed over $f_i = m$ to the highest functionality f_k", i.e.,

$$[X_m] = \sum_{f_i=m}^{f_k} [A_{f_i}]_0 P(X_{m,f_i}) \qquad (5.163)$$

Summing the individual $[X_m]$'s from $m = 3$ to f_k then gives the total crosslink density $[X]$. At $p = 1$, $P(F_A^{\text{out}})$ becomes zero, so that in the limit of complete reaction, one can write, theoretically, $[X_m] = [A_{f_m}]_0$. The crosslink density is an important parameter as it can be related to the concentration of effective network chains and hence to shear modulus of the crosslinked polymer (Miller and Macosko, 1976; Langley, 1968; Langley and Pollmanteer, 1974).

Problem 5.37 Consider the polyether network formation by the stepwise polymerization of pentaerythritol. Using the recursive method of direct computation, determine the following network properties as a

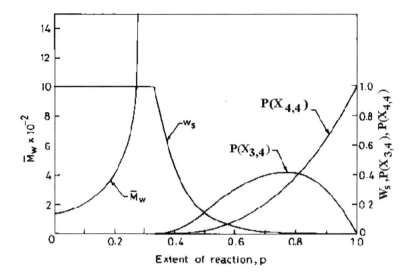

Figure 5.11 Average properties calculated by recursive method for polyether network in stepwise polymerization of pentaerythritol (Data of Problem 5.37).

function of the extent of reaction: (a) weight-average molecular weight, (b) weight fraction of solubles, and (c) crosslink densities. Neglect the effect of condensation products on these properties.

Answer:

(a) M_{A_f} ($C_5H_{12}O_4$) = 136; $f = 4$; from Eq. (P5.36.3), $p_c = 1/(4-1) = 1/3$.

From Eq. (5.110): $\overline{M}_w = (136)\dfrac{1+p}{1-3p}$.

Figure 5.11 shows a plot of $\overline{M}_w$ vs. p.

(b) Combining Eqs. (5.157) and (5.158) for $f = 4$ gives $w_s = \left[(1/p - 3/4)^{1/2} - 1/2\right]^4$ for $p > p_c$.
Figure 5.11 shows a plot of w_s vs. p for $p > 1/3$.

(c) Substituting Eq. (5.157) in Eqs. (5.160) and (5.161) gives

$$P(X_{4,4}) = \left[3/2 - (1/p - 3/4)^{1/2}\right]^4$$
$$P(X_{3,4}) = {}^4C_3\left[(1/p - 3/4)^{1/2} - 1/2\right]\left[3/2 - (1/p - 3/4)^{1/2}\right]^3$$

for $p > p_c$. Figure 5.11 shows a plot of $P(X_{4,4})$ and $P(X_{3,4})$ vs. p for $p > 1/3$. The actual densities, X_4 and X_3, are, however, given by

$$X_4 = [A_f]_0 P(X_{4,4})$$
$$X_3 = [A_f]_0 P(X_{3,4})$$

where $[A_f]_0 = 1/(136 \text{ g mol}^{-1}) = 7.35 \times 10^{-3} \text{ mol g}^{-1}$.

5.7.3.2 Polymerization of $A_f + B_2$

Following the recursive method used above, calculation of $P(F_A^{out})$ can now be extended to a general system of A_i's reacting with B_j's. For illustration, consider, however, a simple case of A_f reacting with B_2, schematically represented (Miller and Macosko, 1976) by

(5.164)

Looking out from A_f in direction $\xrightarrow{1}$ and using Eq. (5.89) gives [cf. Eq. (5.151)]:

$$P(F_A^{out}) = p_A P(F_B^{in}) + 1 - p_A \tag{5.165}$$

where p_A is the fraction of A groups and p_B the fraction of B groups which have reacted. These are related by [cf. Eq. (5.115)]

$$p_B = r p_A \tag{5.166}$$

where r is the initial mole ratio of A groups to B groups. The probability of B′ leading to a finite chain in direction $\xrightarrow{2}$ is just the same as that of B″ in direction $\xrightarrow{3}$, that is,

$$P(F_B^{in}) = P(F_B^{out}) \tag{5.167}$$

Looking out from B″ along $\xrightarrow{3}$ and using Eq. (5.89) gives

$$P(F_B^{out}) = p_B P(F_A^{in}) + 1 - p_B = r p_A P(F_A^{in}) + 1 - r p_A \tag{5.168}$$

For A″ to lead to a finite chain, all of the other arms of A_f must be finite. In the direction $\xrightarrow{4}$ we thus obtain

$$P(F_A^{in}) = P(F_A^{out})^{f-1} \tag{5.169}$$

taking us back to the starting situation of Eq. (5.165).

Combining equations to solve for $P(F_A^{out})$ yields

$$rp_A^2 P(F_A^{out})^{f-1} - P(F_A^{out}) - rp_A^2 + 1 = 0 \tag{5.170}$$

Since $P(F_A^{out}) = 1$ will always be a root of Eq. (5.170), and not of interest as it signifies no gelling, we can factor it out and solve the remaining equation for roots between 0 and 1. This yields

$$\text{for } f = 3, \quad P(F_A^{out}) = (1 - rp_A^2)/rp_A^2 \tag{5.171}$$

$$\text{for } f = 4, \quad P(F_A^{out}) = (1/rp_A^2 - 3/4)^{1/2} - 1/2 \tag{5.172}$$

For higher f the roots are easier to find numerically.

We can readily extend Eq. (5.158) for sol fraction to a mixture of A_f and B_2 by weighting each species by its mass fraction in the mixture. Thus,

$$w_s = w_{A_f} P(F_A^{out})^f + w_{B_2} P(F_B^{out})^2 \tag{5.173}$$

Further generalization to a system of $\sum_i A_i + \sum_j B_j$ ($i, j = 1, 2, 3, ...$) is straightforward (Miller and Macosko, 1976):

$$w_s = \sum_i w_{A_i} P(F_A^{out})^i + \sum_j w_{B_j} P(F_B^{out})^j \tag{5.174}$$

For crosslink density, Eqs. (5.160)-(5.163) are applicable to $A_{f_i} + B_2$ system since only A_{f_i} ($f_i > 2$) can act as an effective junction point in the infinite network.

Problem 5.38 Calculate the sol fraction and the degree of crosslinking in the urethane networks formed by stepwise polymerization of 2-hydroxymethyl-2-ethyl-1,3-propanediol and 1,6-hexamethylene diisocyanate to 90% conversion of the hydroxyl groups (Miller and Macosko, 1976). Compare these properties for two urethane systems with (a) $r = 1$ and (b) $r = 0.75$, where r is the mole ratio of hydroxyl to isocyanate groups.

Answer:

Let $OH \equiv A$, $NCO \equiv B$, and the system be represented by $A_3 + B_2 \longrightarrow$ network.

Substituting Eq. (5.169) in Eq. (5.168) yields

$$P(F_B^{out}) = rp_A P(F_A^{out})^{f-1} + 1 - rp_A \tag{P5.38.1}$$

For $f = 3$, Eq. (5.173) thus becomes

$$w_s = w_{A_3} P(F_A^{out})^3 + w_{B_2} [rp_A P(F_A^{out})^2 + 1 - rp_A]^2 \tag{P5.38.2}$$

M_{A_3} ($C_6H_{14}O_3$) = 134, M_{B_2} ($C_8H_{12}N_2O_2$) = 168.

(a) $r = 1$, $A_3 : B_2$ mole ratio = 2:3, $p_A = 0.90$

$$w_{A_3} = \frac{2 \times 134}{2 \times 134 + 3 \times 168} = 0.347; \quad w_{B_2} = \frac{3 \times 168}{2 \times 134 + 3 \times 168} = 0.653$$

From Eq. (5.72), $p_c = \dfrac{1}{[1 + 1]^{1/2}} = 0.707$. Since $p_A > p_c$, equations for post-gel properties can be used.

From Eq. (5.171), $P(F_A^{out}) = (1 - 0.9)^2/(0.9)^2 = 0.2346$ and then from Eq. (P5.38.2), $w_s = (0.347)(0.2346)^3 + (0.653)[(0.9)(0.2346)^2 + 1 - 0.9]^2 = 0.02$. So at 90% reaction the finite species have essentially disappeared.

Again from Eq. (5.162), $P(X_{3,3}) = [1 - P(F_A^{out})]^3 = (1 - 0.2346)^3 = 0.448$ and from Eq. (5.163), $[X_3] = (0.347)[1/(134 \, \text{g mol}^{-1})](0.448) = 1.16 \times 10^{-3} \, \text{mol g}^{-1}$.

(b) $r = 0.75$, $A_3 : B_2$ mole ratio $= 1:2$, $p_A = 0.90$

$$w_{A_3} = \frac{1 \times 134}{1 \times 134 + 2 \times 168} = 0.285; \quad w_{B_2} = \frac{2 \times 168}{1 \times 134 + 2 \times 168} = 0.715$$

From Eq. (5.72), $p_c = \dfrac{1}{(0.75 + 0.75)^{1/2}} = 0.816$. Since $p_A > p_c$, equations for post-gel properties are applicable.

From Eq. (5.171), $P(F_A^{out}) = (1 - 0.75 \times 0.9^2)/(0.75 \times 0.9^2) = 0.646$ and from Eq. (P5.38.2), $w_s = (0.285)(0.646)^3 + (0.653)[(0.75)(0.9)(0.646)^2 + 1 - 0.9]^2 = 0.17$. So even at 90% reaction there are nearly 17% solubles in the polymer.

Again from Eq. (5.162), $P(X_{3,3}) = (1 - 0.646)^3 = 0.044$ and from Eq. (5.163), $[X_3] = (0.285)[1/(134 \, \text{g mol}^{-1})](0.044) = 0.9 \times 10^{-4} \, \text{mol g}^{-1}$.

<u>Note:</u> Comparison of (a) and (b) show that both the sol fraction and the crosslink density are highly sensitive to the value of r, the former increasing and the latter decreasing markedly with the increase in the proportion of the bifunctional component.

5.8   *Polycondensation of A_xB Monomers*

Flory (1953) described the polycondensation reaction of A_xB monomers from a theoretical point of view. Assuming that (i) the only allowed reaction is between an A group and a B group, (ii) no intramolecular condensation reactions occur, and (iii) the reactivity of a functional group is independent of molecular size, he predicted that condensation of A_xB monomers would give highly branched molecules without network formation and having a multitude of end groups of the same type (Fig. 5.12).

If y molecules of monomer A_xB are reacted, the resulting y-meric species (cf. Fig. 5.12) will contain only a single B group at one end and $(f - 2)y + 1$ A groups at other ends, where $f(= x + 1)$ is the total number of functional groups on the monomer. Consider, for simplicity, an $A_{f-1}B$ monomer with $f = 3$. The probability that an arbitrarily chosen A group has reacted is equal to p_A, the fraction of A groups reacted. The fraction of B groups reacted, p_B, is then given by

$$p_B = p_A(f - 1) \tag{5.175}$$

The probability that a functional group A on a branch point (see Fig. 5.12) leads to another branch point, which defines the *branching coefficient* (α), is evidently equal to the probability that A

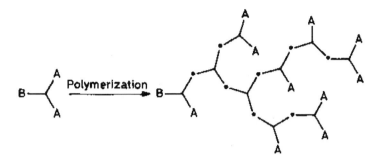

Figure 5.12 Formation of branched polymer by condensation polymerization of A_2B monomer as proposed by Flory (1953).

has reacted, i.e., p_A. Since B is the limiting group, replacing p_B with the conversion p, we thus obtain from Eq. (5.175) an expression for the branching coefficient as

$$\alpha = p/(f - 1) \tag{5.176}$$

The total number of molecules N at the extent of conversion p is $N_0(1 - p)$, where N_0 is the total number of monomer units initially present. Hence the number-average degree of polymerization, $\overline{X}_n$, is

$$\overline{X}_n = \frac{N_0}{N} = \frac{N_0}{N_0(1 - p)} = \frac{1}{1 - p} = \frac{1}{1 - \alpha(f - 1)} \tag{5.177}$$

The critcal value of α being $\alpha_c = 1/(f - 1)$, according to Eq. (5.64), it is obvious from Eq. (5.176) that α may not reach α_c, since p may approach but never reach unity. Flory (1953) therefore concluded that condensation of A_xB monomers would give randomly branched molecules *without* network formation. The inability to form network structure is linked with the limitation of the number of unreacted B groups to one per molecule (Fig. 5.12).

5.8.1 Dendritic and Hyperbranched Polymers

Though Flory (1953) theorized long ago about synthesizing condensation polymers from multi-functional monomers (see above), he did not consider it worthwhile to pursue this line of research, since such polymers being nonentangled and noncrystalline due to their highly branched structure would be expected to have poor mechanical properties. However, a little more than 30 years later, the first papers on synthesis of *dendritic polymers* (*dendron*, Greek for "tree") appeared (Tomalia et al., 1985; Newkome et al., 1985), revealing a number of very unique and different properties of these polymers, compared to their linear analogs. For instance, at high molecular weights the dendritic polymers were found to be globular and, in contrast to linear polymers, they behaved more like molecular micelles (Zeng and Zimmerman, 1997). The descriptors *starburst, dendrimers, arborols, cauliflower, cascade, and hyperbranched*, used for such polymers, all describe specific geometric forms of structure.

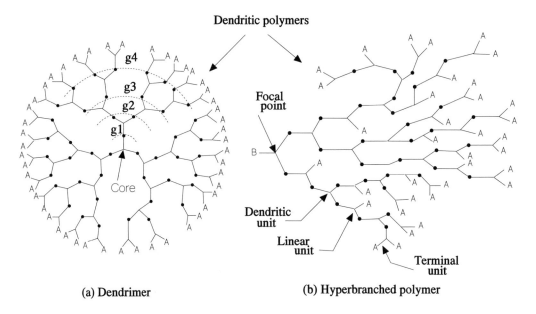

Figure 5.13 Schematic representation of dendritic polymers comprising dendrimers and hyperbranched polymers ('g' level indicates generation number).

Following the first papers of Tomalia et al. (1985) and Newkome et al. (1985) dealing with dendrimers, a large number of dendrimers have been presented in the literature ranging from polyamidoamine (Tomalia et al., 1991), polyethers (Hawker and Fréchet, 1990; Padias et al., 1987), and polyesters (Miller et al., 1992; Ihre et al., 1996) to polysilane (van der Made and van Leeuwen, 1992). Copolymers of linear blocks with dendrimer segments (*dendrons*), block copolymers of different dendrons, and polymers with dendritic side chains (Frauenrath, 2005) have been described.

Dendrimers, as shown in the generalized form in Fig. 5.13(a), are obtained when each ray in a star molecule is terminated by an f-functional branching from which $(f-1)$ rays of the same length again emanate. A next *generation* is created when these $f-1$ rays are again terminated by the branching units from which again rays originate, etc. In recent years, the chemistry of preparing dendrimers has become very successful. However, the synthesis of perfect monodisperse dendrimers is time-consuming and painfully cumbersome (Fréchet, 1994). Another serious drawback is the space filling that occurs with the growth of a dendrimer due to which it has not been possible to prepare more than five generations. Either the reaction to a higher generation stops completely or the outermost shells develop imperfections.

Since the production of perfect monodisperse dendrimers is too complicated and expensive, several researchers at DuPont Experimental Station working on dendritic polymers as rheology control agents and as spherical multifunctional initiators developed a route for a one-step synthesis of dendritic polymers. These polymers were, however, polydisperse and had defects in the form of linear segments between branch points but they were highly branched dendritic molecules. Kim and Webster (1990; 1992) named them *hyperbranched polymers*. (The structures are called hyperbranched, since due to the chemical constraint a very high branching density becomes possible without gelation.) Ever since, a wide variety of hyperbranched polymers have been synthesized. A step change occurred when self-condensing vinyl polymerization was introduced by Fréchet et al. (1995), extending hyperbranched methodologies to vinyl monomers with end group control and living polymerization features (see Section 11.3.8.3). Hyperbranched polymer synthesis based on vinyl monomers (Ambade and Kumar, 2000; Gretton-Watson et al., 2005) was developed further by several research groups and includes free radical polymerization chemistries (Wang et al., 2003), self-condensing vinyl polymerizations (Mori et al., 2004), catalytic chain transfer processes (Heuts et al., 2002), and ionic (Baskaran, 2003) and coordination methodologies (Guan, 2002). Methods of synthesis and various applications of hyperbranched and dendritic polymers have been reviewed recently (Yates and Hayes, 2004; Smith et al., 2005).

Among the various applications suggested for hyperbranched polymers are surface modification, additives, tougheners for epoxy-based composites, coatings, and medicines. It has been demonstrated that hydrophobic, fluorinated, hyperbranched poly(acrylic acid) films can passivate and block electrochemical reactions on metal surfaces thus preventing surface corrosion (Bruening, 1997). The lack of mechanical strength makes hyperbranched polymers more suitable as additives in thermoplast applications. Hyperbranched polyphenylenes, for example, have been shown to act successfully as rheology modifiers when processing linear thermoplastics. A small amount added to polystyrene thus results in reduced melt viscosity (Kim and Webster, 1992).

The use of epoxidized hyperbranched polyesters as toughening additives in carbon-fiber reinforced epoxy composites has been demonstrated (Boogh et al., 1995). Since a hyperbranched resin has a substantially lower viscosity and much shorter drying time than a conventional (less branched) resin of comparable molecular weight, hyperbranched polymers have been used as the base for various coating resins (Pettersen and Sorensen, 1994).

An important application of dendritic polymers being explored in medicine is in advanced drug delivery systems. However, most applications within this field, described in the literature, deal with dendrimers and not with hyperbranched polymers. Hyperbranched polymers, being mostly polydisperse, are unsuitable in vivo applications (Roberts et al., 1996).

A special feature of dendritic polymers is the possibility to combine an interior structure with one polarity, with a shell (end groups) having another polarity, e.g., a hydrophobic inner structure and hydrophilic end groups. Thus hyperbranched polyphenylenes with (anionic) carboxylate end groups have been described (Kim and Webster, 1990), where carboxylate end groups make the polymer water soluble while the hydrophobic interior hosts nonpolar guest molecules. This has led to the development of dendritic polymers with characteristics of *unimolecular micelles* (Hawker and Chu, 1996) and unique guest-host possibility considered suitable for medical applications, such as drug delivery (Uhric, 1997).

REFERENCES

Allcock, H. R. and Lampe, F. W., "Contemporary Polymer Chemistry", Prentice Hall, Englewood Cliffs, N.J., 1990.

Ambade, A. V. and Kumar, A., *Prog. Polym. Sci.*, **25**, 1141 (2000).

Baker, J. W., Davies, N. N., and Gaunt, J., *J. Chem. Soc.*, 24 (1949).

Baskaran, D., *Polymer*, **44**, 2213 (2003).

Bobalek, E. G., Moore, E. R., Levy, S. S., and Lee, C. C., *J. Appl. Polym. Sci.*, **8**, 625 (1964).

Boogh, L., Pettersen, B., Japon, S., and Manson, J-A., *Proceedings of 28th International Conference on Composite Materials*, Whistler, Canada, vol. 4, p. 389 (1995).

Breuning, M. L., Zhou, Y., Aguilar, G., Agee, R., Bergbreiter, and Crooks, R. M., *Langmuir*, **13**, 770 (1997).

Burchard, W., *Polymer*, **20**, 589 (1979).

Case, L. C., *J. Polym. Sci.*, **29**, 455 (1958).

Durand, D. and Bruneau, C. M., *Makromol. Chem.*, **180**, 2947 (1979).

Flory, P. J., *J. Am. Chem. Soc.*, **63**, 3083, 3091, 3096 (1941); *Chem. Rev.*, **39**, 137 (1946).

Flory, P. J., "Principles of Polymer Chemistry", Cornell University Press, Ithaca, N.Y., 1953.

Frauenrath, H., *Prog. Polym. Sci.*, **30**, 325 (2005).

Fréchet, J. M. J., *Science*, **263**, 1710 (1994).

Fréchet, J. M. J., Henmi, M., Gitsov, I., Aoshima, S., Leduc, M. R., and Grubbs, R. B., *Science*, **269**, 1080 (1995).

Ghosh, P., "Polymer Science and Technology of Plastics and Rubbers", Tata McGraw-Hill, New Delhi, 1990.

Gordon, M. and Judd, M., *Nature (London)*, **234**, 96 (1971).

Gretton-Watson, S.P., Alpay, E., Steinke, J. H. G., and Higgins, J. S., *Ind. Eng. Chem. Res.*, **44**, 8682 (2005).

Guan, Z. B., *Chem. Eur. J.*, **8**, 3087 (2002).

Hamann, S. D., Solomon, D. H., and Swift, J. D., *J. Macromol. Sci.-Chem., A2(1)*, 153 (1968).

Hawker, C. J. and Chu, F., *Macromolecules*, **29**, 4370 (1996).

Hawker, C. J. and Fréchet, J. M. J., *J. Am. Chem. Soc.*, **112**, 7638 (1990).

Heuts, J. P. A., Roberts, G. E., and Biasutti, J. D., *Aust. J. Chem.*, **55**, 381 (2002).

Hiemenz, P. C., "Polymer Chemistry", Marcel Dekker, New York, 1984.

Ihre, I., Hult, A., and Soderlind, E., *J. Am. Chem. Soc.*, **27**, 6388 (1996).

Kim, Y. H. and Webster, O. W., *J. Am. Chem. Soc.*, **112**, 4592 (1990).

Kim, Y. H. and Webster, O. W., *Macromolecules*, **25**, 5561 (1992).

Langley, N. R., *Macromolecules*, **1**, 348 (1968).

Langley, N. R. and Polmanteer, K. E., *J. Polym. Sci., Part A-1*, **12**, 1023 (1974).

Levenspiel, O., "Chemical Reaction Engineering", 2nd ed., John Wiley, New York, 1972.

Lin, C. C. and Yu, P. C., *J. Polym. Sci., Polym. Chem. Ed.*, **16**, 1005, 1797 (1978).

Lopez-Serrano, L., Castro, J. M., Macosko, C. W., and Tirrell, M., *Polymer*, **21**, 263 (1980).

Macosko, C. W. and Miller, D. R., *Macromolecules*, **9**, 199 (1976).

Manaresi, P. and Munari, A., "General Aspects", Chap. 2 in "Comprehensive Polymer Science", vol. 5 (G. Allen and J. C. Bevington, eds.), Pergamon Press, Oxford, 1989.

Miller, T. M., Kwoek, E. W., and Neenan, T. X., *Macromolecules*, **25**, 143 (1992).

Miller, D. R. and Macosko, C. W., *Macromolecules*, **9**, 206 (1976); **11**, 656 (1976).

Miller, D. R. and Macosko, C. W., *Macromolecules*, **13**, 1063 (1980).

Moore, J. W. and Pearson, R. G., "Kinetics and Mechanism", 3rd ed., John Wiley, New York, 1981.

Mori, H., Walther, A., Andre, X., Lanzendorfer, M. G., and Müller, A. H. E., *Macromolecules*, **37**, 2054 (2004).

Neckers, D. C. and Doyle, M. P., "Organic Chemistry", John Wiley, New York, 1977.

Newkome, G. R., Yao, Z., Baker, G. R., and Gupta, V. K., *J. Org. Chem.*, **50**, 2004 (1985).

Odian, G., "Principles of Polymerization", 3rd ed., John Wiley, New York, 1991.

Orlova, T. M., Pavlova, S. S. A., and Dubrovina, L. V., *J. Polym. Sci., Polym. Chem. Ed.*, **17**, 2209 (1979).

Padias, B. A., Hall, H. K., Tomalia, D. A., and McConnell, J. R., *J. Org. Chem.*, **52**, 5305 (1987).

Pettersen, B. and Sorensen, K., *Proceedings of the 21st Waterborne, Higher Solids & Powder Coatings Symposium*, New Orleans, Louisiana, p. 753 (1994).

Rempp, P. and Merrill, E. W., "Polymer Synthesis", Huethig & Wepf, Basel, 1986.

Roberts, J., Bhalgat, M., and Zera, R., *J. Biomed. Res.*, **30**, 53 (1996).

Rudin, A., "The Elements of Polymer Science and Engineering", Academic Press, Orlando, 1982.

Shaefgen, J. R. and Flory, P. J., *J. Am. Chem. Soc.*, **70**, 2709 (1948).

Smith, D. K., Hirst, A. R., Love, C. S., Hardy, J. G., Briguell, S. V., and Huang, B., *Prog. Polym. Sci.*, **30**, 220 (2005).

Solomon, D. H., *J. Macromol. Sci. Rev. Macromol. Chem., Cl(1)*, 179 (1967).

Solomon, D. H., "Polyesterification" Chap. 1 in "Step-Growth Polymerizations" (D. H. Solomon, ed.), Marcel Dekker, New York, 1972.

Stockmayer, W. H., *J. Polym. Sci.*, **9**, 69 (1952); **11**, 424 (1953); *J. Chem. Phys.*, **11**, 625 (1964).

Sykes, P., "A Guidebook to Mechanism in Organic Chemistry", 6th ed., Orient Longman, New Delhi, 1986.

Tomalia, D. A., Baker, H., Dewald, J., Hall, M., Kallos, G., Martin, J. R., Ryder, J., and Smith, P., *Polymer J.*, **17**, 117 (1985).

Tomalia, D. A., Hedstrand, D. M., and Ferritto, M. S., *Macromolecules*, **24**, 1435 (1991).

Uhric, K., *Trends in Polym. Sci.*, **5**(12), 388 (1997).

van der Made, A. W. and van Leeuwen, P. W. N. M., *J. Chem. Soc., Chem. Commun.*, 1400 (1992).

Wang, Z. M., He, J. P., Tao, Y. F., Yang, L., Jiang, H. J., and Yang, Y. L., *Macromolecules*, **36**, 7446 (2003).

Williams, D. J., "Polymer Science and Engineering", Prentice Hall, Englewood Cliffs, N.J., 1971.

Yates, C. R. and Hayes, W., *European Polym. J.*, **40**, 1257 (2004).

Zeng, F. and Zimmerman, S. C., *Chem. Rev.*, **97**, 1681 (1997).

Ziegel, K. D., Fogel, A. W., and Pariser, R., *Macromolecules*, **5**, 95 (1972).

EXERCISES

5.1 A mixture of R(COOH)$_2$ and R′(OH)$_2$ in the mole ratio of 1:1.1 was used for studying the kinetics of polyesterification. The mixture having R(COOH)$_2$ concentration of 3.808 mol/kg was initially polymerized under mild conditions to about 80% conversion of the carboxylic acid groups. The mixture was then further polymerized in the presence of an added strong acid catalyst at a sufficiently high temperature to facilitate removal of the water of condensation by passing a stream of nitrogen gas. The polymerization in the second stage yielded the following data:

Time, min	[COOH], mol/kg
0	1.71
5	1.47
10	1.28
20	1.00
30	0.80
40	0.66
60	0.47
90	0.30

Determine the polyesterification rate constant for the monomer system.
[*Ans.* 0.013 kg mol^{-1} min^{-1}]

5.2 Polymeric decamethylene adipate of $\overline{X}_n = 190$ and having only hydroxyl end groups is to be prepared by reacting a small excess of the glycol with the dibasic acid until the condensation was substantially complete. Calculate the mole percent excess glycol that is to be used.
[*Ans.* 1.06 mole%]

5.3 Poly(decamethylene adipate) obtained in Exercise 5.2 was subjected to alcoholysis with 5% of its own weight of (a) ethylene glycol and (b) *n*-octyl alcohol in the presence of *p*-toluene sulfonic acid catalyst till there was no further change in viscosity. Estimate the limiting $\overline{X}_n$ of the mixture. Assume equal reactivity of groups.
[*Ans.* (a) 9; (b) 17.6]

5.4 A sample of phthalic anhydride which contains 5% (w/w) phthalic acid as impurity, is to be polymerized with propylene glycol. What would be the limiting degree of polymerization if 'stoichiometric' polymerization were carried out without taking note of the impurity ?
[*Ans.* $\overline{X}_n = 370$]

5.5 In the polycondensation of H$_2$N(CH$_2$)$_6$NH$_2$ and HOOC(CH$_2$)$_4$COOH, in equimolar quantities, to form nylon-6,6 what is the molecular weight of the species that has the largest yield by weight at 99% conversion ?
[*Ans.* $\overline{M}_n = 11,300$]

5.6 Nylon-11 is poly(ω-aminoundecanoic acid). In the polymerization of ω-aminoundec-anoic acid, $H_2N(CH_2)_{10}COOH$, to produce this polymer, what weight fraction of the reaction mixture will have the structure $-[-NH-(CH_2)_{10}-CO-]_{100}-$ when 99% of the functional groups have reacted ?
[*Ans.* 3.7×10^{-3}]

5.7 Poly(ethylenoxy benzoate) is produced by the following reaction :

$$n\,HOCH_2CH_2O-Ph-COOCH_3 \xrightarrow{\Delta} -(-OCH_2CH_2O-Ph-CO-)_n + nCH_3OH \uparrow$$

(where Ph = p-phenylene). At 75% extent of reaction, what is the weight average degree of polymerization (a) of the reaction mixture and (b) of the polymer ?
[*Ans.* (a) 7.0; (b) 7.4]

5.8 When an equilibrium step-growth polymerization is 99% complete, what fraction of the reaction mixture is still monomer (a) on mole basis and (b) on weight basis ?
[*Ans.* (a) 0.01; (b) 0.0001]

5.9 In the polymerization of $H_2N(CH_2)_{10}COOH$ to form nylon-11, what is the molecular weight of the species that has the largest weight fraction in the reaction mixture at 99% conversion ?
[*Ans.* 18,300]

5.10 Consider synthesis of nylon-6,10 from stoichiometric amounts of hexamethylene diamine, $H_2N(CH_2)_6NH_2$, and sebacic acid, $HOO(CH_2)_8COOH$. At 98% conversion of carboxylic acid groups, determine (a) $\overline{M}_n$ and $\overline{M}_w$ of the product, and (b) the sum of the mole fractions and that of the weight fractions of species containing 1 through 6 (inclusive) monomer units.
[*Ans.* (a) 7,050 and 13,960; (b) 0.1141 and 0.0079]

5.11 In the polymerization of decamethylene glycol and acidic acid in equimolar proportions at 110°C in the presence of an externally added strong acid catalyst, the degree of polymerization showed an increase of 40 in 300 min, the reaction being carried out with continuous removal of water of condensation. (a) Determine the rate constant k' as defined by $-d[COOH]/dt = k'[COOH][OH]$. (b) Suppose in the removal of water from the start of reaction 2% of the initially charged glycol were lost by steam distillation, but no adipic acid. What maximum degree of polymerization could be achieved ?
[*Ans.* (a) 2.2×10^{-3} s^{-1}; (b) 99]

5.12 A nonstoichiometric mixture consists of 1 mol of glycerol and 5 moles of phthalic acid. Can gelation occur in this system ?
[*Ans.* No; $\overline{X}_n$ of mixture at $p = 1$ is 2]

5.13 Compare the gel point calculated from the Carothers equation with that using the statistical approach of Flory for the following mixture: phthalic anhydride, ethylene glycol, and glycerol in the molar ratio 1.50 : 0.70 : 0.50.
[*Ans.* p_c (Carothers) = 0.931, p_c (Flory) = 0.826]

5.14 A polymerization system made of A_f ($f = 3$) and B-B is to be reacted until gelation occurs. What should be the stoichiometric proportions of A_f and B-B if it is desired to effect gelation of the system when 90% of the A groups have reacted ?
[*Ans.* 0.588 : 1 (Carothers); 0.411 : 1 (Flory)]

5.15 Compute the time to reach the gel point for the acid catalyzed system $2A_f$ ($f = 3$) + A—A + 3.5 B—B, where $c_o k' = 4 \times 10^{-4}$ s^{-1}. Also calculate $\overline{X}_n$ at the gel point.
[*Ans.* 1.67 h; 7.7]

5.16 The network polyurethane obtained by reacting a diisocyanate R(N=C=O)$_2$ with pentaerythritol C(CH$_2$OH)$_4$ contains, according to elemental analysis, 0.2% (w/w) nitrogen and has a density of 1.05 g/cm^3. Determine (a) polymer chain segment density in mol/cm^3, and (b) molar mass of chain segments between branch points.
[*Ans.* (a) 7.5×10^{-5} mol cm^{-3}; (b) 14,000 g mol^{-1}]

5.17 Calculate the statistical gel point conversions of A and B groups in the system consisting initially of A (4 moles), A—A (51 moles), A_3 (2 moles), A_4 (3 moles), B (2moles), B—B (50 moles), B_3 (3 moles),

and B_5 (3 moles).
[*Ans.* A groups: 0.7799; B groups: 0.7643]

5.18 Equimolar amounts of trimethylol propane and tricarballylic acid were reacted till 75% esterification took place. Assuming equal reactivity of functional groups, calculate the weight fractions of sol and gel in the polymeric mixture, and the weight fraction of unreacted monomer in the sol.
[*Ans.* $w_s = 0.037$; $w_g = 0.963$; w_1 (sol) = 0.422]

5.19 Use Eq. (5.149) derived by the recursive approach to calculate the gel point conversions of A and B groups for the following step-growth polymerization systems and compare with the corresponding values calculated according to the gel point theory of Flory: (a) A_4 (2 moles) + B—B (4 mols); (b) A—A (3 moles) + A_4 (1 mol) + B—B (4 moles); (c) A_4 (1 mol) + B_4 (1mol); and (d) A (1mol) + A_3 (2 moles) + B_4 (2 moles).
[*Ans.* (a) $p_A = 0.577$ (cf. Flory 0.577), $p_B = 0.577$; (b) $p_A = 0.667$ (cf. Flory 0.667), $p_B = 0.833$; (c) $p_A = 0.333$ (cf. Flory 0.333), $p_B = 0.333$; (d) $p_A = 0.471$ (cf. Flory 0.467), $p_B = 0.412$]

5.20 Recalculate the gel point conversion of the polymerization system of Exercise 5.17 by the simpler recursive approach and compare the results. Also calculate the $\overline{M}_w$ of the polymer formed at 50% conversion of the A groups. Assume that the molecular weights of all the monomers in the polymerization system are equal and have a value of 100.
[*Ans.* $p_A = 0.771$ (cf. Flory 0.779); $\overline{M}_w = 399$]

5.21 Consider the following system:

Phthalic anhydride $C_6H_4(CO)_2O$ 1.5 mol
Tricarballylic acid $CH_2(COOH)CH(COOH)CH_2(COOH)$ 0.4 mol
Pentaerythritol $C(CH_2OH)_4$ 1.0 mol

Can the system be reacted to complete conversion without gelation ? If not, what is the extent of conversion of the acid functionality at the gel point calculated from (a) the Carothers equation, (b) the statistical approach of Flory-Stockmayer, and (c) the recursive method of Macosko-Miller ?
[*Ans.* (a) 0.69; (b) 0.494; (c) 0.497]

5.22 Polyurethane foam was made by stepwise polymerization of methylenediphenyl isocyanate $CH_2(C_6H_4NCO)_2$ and 2-hydroxymethyl-2-ethyl-1,3-propanediol in the mole ratio of 1 : 0.7. Calculate the weight fraction solubles and the total crosslink density in the network if the extent of reaction of the isocyanate is 90%.
[*Ans.* $w_s = 0.03$; $[X_3] = 0.71 \times 10^{-3}$ mol g^{-1}]

5.23 A stoichiometric mixture of A_4 and B_2 type monomers in which only the reaction of A with B is possible was polymerized to 80% conversion of the A groups. Calculate (a) the sol fraction and (b) the ratio of crosslinks of degree 3 and degree 4 in the network polymer. (Molecular weights : $A_4 = 576$, $B_2 = 5000$)
[*Ans.* (a) 0.06; (b) 2.68 : 1]

APPENDIX 5.1

Summations Often Used in Calculation of Molecular Distribution of Polymers

1. $\displaystyle\sum_{x=1}^{\infty} p^{x-1} = \frac{1}{1-p}$, p being a constant $(0 < p < 1)$.

   Derivation

   $$\sum_{x=1}^{n} p^{x-1} = 1 + p + p^2 + \cdots\cdots + p^{n-1}$$

   $$p\sum_{x=1}^{n} p^{x-1} = p + p^2 + p^3 + \cdots\cdots + p^n$$

   Subtracting,

   $$(1-p)\sum_{x=1}^{n} p^{x-1} = 1 - p^n$$

   As $n \to \infty$, $p^n \to 0$; it then follows

   $$\sum_{x=1}^{\infty} p^{x-1} = \frac{1}{1-p} \qquad\qquad (A5.1.1)$$

2. $\displaystyle\sum_{x=1}^{\infty} xp^{x-1} = \frac{1}{(1-p)^2}$, p being a constant $(0 < p < 1)$.

   Derivation

   $$\sum_{x=1}^{n} xp^{x-1} = 1 + 2p + 3p^2 + \cdots\cdots + np^{n-1}$$

   $$p\sum_{x=1}^{n} xp^{x-1} = p + 2p^2 + 3p^3 + \cdots\cdots + np^n$$

   By subtracting

   $$(1-p)\sum_{x=1}^{n} xp^{x-1} = 1 + p + p^2 + p^3 + \cdots\cdots - np^n$$

   Taking into account the result derived in 5.1 and noting that with $p^n \to 0$ as $n \to \infty$

   $$\sum_{x=1}^{\infty} xp^{x-1} = \frac{1}{(1-p)^2} \qquad\qquad (A5.1.2)$$

3. $\displaystyle\sum_{x=1}^{n} x^2 p^{x-1} = \frac{1+p}{(1-p)^3}$

   Derivation

   $$\sum_{x=1}^{\infty} x^2 p^{x-1} = 1 + 4p + 9p^2 + 16p^3 + \cdots\cdots + n^2 p^{n-1}$$

   $$p\sum_{x=1}^{\infty} x^2 p^{x-1} = p + 4p^2 + 9p^3 + 16p^4 + \cdots\cdots + n^2 p^n$$

By subtracting,

$$(1 - p) \sum_{x=1}^{n} x^2 p^{x-1} = 1 + 3p + 5p^2 + 7p^3 + \cdots\cdots - n^2 p^n$$

whence

$$p(1 - p) \sum_{x=1}^{n} x^2 p^{x-1} = p + 3p^2 + 5p^3 + 7p^4 + \cdots\cdots - n^2 p^{n+1}$$

By subtracting these two equations,

$$(1 - p)^2 \sum_{x=1}^{n} x^2 p^{x-1} = 1 + 2p + 2p^2 + 2p^3 + \cdots\cdots n^2 p^{n+1}$$

whence

$$p(1 - p)^2 \sum_{x=1}^{n} x^2 p^{x-1} = p + 2p^2 + 2p^3 + 2p^4 + \cdots\cdots + n^2 p^{n+2}$$

By subtracting again,

$$(1 - p)^3 \sum_{x=1}^{n} x^2 p^{x-1} = 1 + p - n^2 p^{n+2}$$

As $n \to \infty$, $p^{n+2} \to 0$; it follows

$$\sum_{x=1}^{\infty} x^2 p^{x-1} = \frac{1 + p}{(1 - p)^3} \tag{A5.1.3}$$

Similarly, it can be shown that

$$\sum_{x=1}^{\infty} x^3 p^{x-1} = \frac{1 + 4p + p^2}{(1 - p)^4} \tag{A5.1.4}$$

Chapter 6

Free Radical Polymerization

6.1 Introduction

In the previous chapter, step-growth polymerization and its kinetics were considered, while the characteristic features of this polymerization as compared to those of chain-growth polymerization were summarized in Table 5.1. We now turn our attention to this latter type of polymerization processes and especially to those initiated by free-radicals in the presence of monomer. A large number of unsaturated monomers, such as ethylene ($CH_2 = CH_2$, the simplest olefin), α-olefins ($CH_2 = CHR$, where R is an alkyl group), vinyl compounds ($CH_2 = CHX$, where X = Cl, Br, I, alkoxy, CN, COOH, COOR, C_6H_5, etc., atoms or groups), and conjugated diolefins [e.g., butadiene, $CH_2 = CH-CH=CH_2$, and isoprene,$CH_2 = C(CH_3)- -CH=CH_2$] readily undergo chain-growth polymerization, also known as addition or simply chain polymerization. Polymerization of such monomers will be discussed in this and the next three chapters.

Chain-growth polymerization, as well as all other typical chain reactions are usually faster than those in step-growth polymerizations and are typified by three normally distinguishable processes, viz., (i) initiation of the chain, (ii) propagation or growth of the chain, and (iii) termination of the chain. (A fourth process, *chain transfer*, may also be involved.)

The initiation is usually a direct consequence of the generation of a highly active species $R^\star$ by dissociation or degradation of some monomer molecules (M) under the influence of such physical agencies as heat, light, radiation etc., or as a consequence of dissociation or decomposition of some chemical additives commonly known as initiators (I):

$$I \longrightarrow R^\star \tag{6.1}$$

The reactive species $R^\star$ may be a free radical, a cation, or an anion, which adds to the unsaturated monomer molecule by opening the π bond and simultaneously regenerating a reactive center of the same type. The new reactive center then adds to another monomer molecule, M, and the process is repeated in quick succession leading to addition of many more monomer molecules to the same chain, the reactive center being always shifted to the end of the growing chain with each addition. The chain propagation process can thus be represented by

$$R^\star + M \longrightarrow RM^\star \xrightarrow{+M} RMM^\star \xrightarrow{+nM} R(M)_{n+1}M^\star \tag{6.2}$$

The chain growth is terminated when the reactive center is destroyed by some mechanism that depends on the type of reactive center (radical, cation, or anion), nature of the monomer M, and the reaction conditions.

In the present chapter, the basic principles of free-radical chain polymerizations, i.e., chain-growth polymerizations in which the reactive centers are free radicals, will be considered in detail, focusing on the polymerization reactions in which only one monomer is involved, while polymerization reactions involving more than one monomer, referred to as free-radical copolymerization, are considered separately in Chapter 7. Chain-growth polymerizations in which the active centers are ionic are discussed in Chapter 8.

6.2   Scheme of Radical Chain Polymerization

6.2.1   Overall Scheme

Radical chain polymerization, as noted above, is a chain reaction which involves mainly three steps – *initiation*, *propagation*, and *termination*, taking place in sequence. The overall scheme of the polymerization of a vinyl or related monomer M, initiated by the decomposition of a free-radical initiator I, may be schematically represented as follows (Ghosh, 1990):

Initiation

$$I \xrightarrow{k_d} 2R^\bullet \tag{6.3}$$

$$R^\bullet + M \xrightarrow{k_i} RM^\bullet \tag{6.4}$$

Propagation

$$RM^\bullet + M \xrightarrow{k_p} RMM^\bullet \text{ or } RM_2^\bullet$$
$$RM_2^\bullet + M \xrightarrow{k_p} RMMM^\bullet \text{ or } RM_3^\bullet \tag{6.5}$$
$$RM_3^\bullet + M \xrightarrow{k_p} RMMMM^\bullet \text{ or } RM_4^\bullet$$

In general terms,

$$RM_{n-1}^\bullet + M \xrightarrow{k_p} RM_n^\bullet \tag{6.6}$$

Termination

$$RM_n^\bullet + RM_m^\bullet \begin{cases} \xrightarrow{k_{tc}} RM_{(n+m)}R & (6.7a) \\ \xrightarrow{k_{td}} RM_n + RM_m & (6.7b) \end{cases}$$

where $R^\bullet$ is a free-radical generated by the decomposition of the initiator I; $RM_n^\bullet$ and $RM_m^\bullet$ (m, $n = 1,2,3 \cdots$) are the growing polymer chains, each bearing a free-radical center at the chain end and RM_n, RM_m, $RM_{n+m}R$, etc. are dead (inactive) polymer molecules. It is apparent that the group R capping one or both ends of the dead polymer molecule is the initiator fragment in the form of radical $R^\bullet$, trapped as an end group in the polymer structure. A significant feature of the above polymerization scheme is that the active center is retained by the growing chain throughout the propagation process. The aforesaid three steps are discussed below in greater detail.

6.2.2 Chain Initiation

The initiation step consists of two reactions in series. The first is the production of free radicals, which can be accomplished in many ways. The most common method, however, involves the use of a thermolabile compound, called an *initiator* (or *catalyst*), which decomposes to yield free radicals when heated. Thus, the homolytic dissociation of an initiator I yields a pair of radicals R•, as shown by Eq. (6.3), where k_d is the rate constant for initiator dissociation at the particular temperature. Its magnitude is usually of the order of 10^{-4}–10^{-6} s^{-1}. (Being derived from the initiator, R• is referred to as an initiator radical and often as a *primary radical*.) The second step of the initiation process is the addition of the radical R• to a monomer molecule as shown in Eq. (6.4), where RM• is the monomer-ended radical containing one monomer unit and an end group R. The rate constant for the reaction is k_i. For a vinyl monomer, this second step involves opening the π-bond to form a new radical:

$$\text{R}^\bullet \ + \ \text{H}_2\text{C}\!=\!\!\underset{\underset{\text{X}}{|}}{\overset{\overset{\text{H}}{|}}{\text{C}}} \ \longrightarrow \ \text{R}-\text{CH}_2-\underset{\underset{\text{X}}{|}}{\overset{\overset{\text{H}}{|}}{\text{C}}}{}^\bullet \tag{6.8}$$

The rate of radical generation according to Eq. (6.3) is given by

$$d[\text{R}^\bullet]/dt \ = \ 2k_d[\text{I}] \tag{6.9}$$

since each molecular decomposition of the initiator produces two primary radicals, R•. Each R• then attacks a monomer molecule M to produce a chain radical RM• [see Eq. (6.4)] by a fast reaction [Eq. (6.8)]. The decomposition of the initiator [Eq. (6.3)], which is much slower than this initiation reaction, is the rate controlling step (as the primary radicals R• give rise to chain initiation as soon as they are generated in the monomer system). If there is no wastage of the primary radicals by side reactions, the rate of chain initiation, R_i, will thus be the same as the rate of radical generation given by Eq. (6.9), i.e.,

$$R_i \ = \ \left(\frac{d[\text{M}^\bullet]}{dt}\right)_i \ = \ k_i[\text{R}^\bullet][\text{M}] \ = \ 2k_d[\text{I}] \tag{6.10}$$

For practical purposes, however, the expression is modified as

$$R_i \ = \ \left(\frac{d[\text{M}^\bullet]}{dt}\right)_i \ = \ 2fk_d\,[\text{I}] \tag{6.11}$$

where the factor f is the *initiator efficiency* or the efficiency of initiation, representing the fraction of primary radicals R•, which actually contribute to chain initiation as given by Eq. (6.4). This will be discussed later in more detail.

6.2.3 Chain Propagation

Initiation is followed by *chain propagation*, in which the growth of RM•, often represented as $M_1{}^\bullet$, takes place by successive addition of a large number of monomer molecules, according to Eqs. (6.5). The monomer addition takes place in the same way as shown in Eq. (6.8). Following each addition reaction, the chain size increases by one monomer unit, while the radical center is transferred to the end monomer unit.

In writing the sequence of equations (6.5), it was assumed that the radical reactivity is independent of chain length so that all successive propagation steps are characterized by the same rate constant k_p and may thus be represented by the general equation (6.6). The overall rate of propagation is therefore given by

$$R_p = k_p [M][M^\bullet] \qquad (6.12)$$

where [M] is the monomer concentration and [M$^\bullet$] is the total concentration of all chain radicals of size RM$^\bullet$ and larger.

The value of k_p for most monomers is in the range 10^2-10^4 L/mol-s. This is very large compared to rate constant values for step-growth polymerization (typically, 10^{-3} L/mol-s for acid-catalyzed polyesterification). The growth of a chain radical to macromolecular size thus takes place very rapidly. This also explains a distinctive feature of radical chain polymerization that, in sharp contrast to condensation or step-growth polymerization (see Table 5.1), the chain polymerization system even at low conversions would practically consist of full-grown polymer molecules of high molecular weight besides the unreacted monomer and no species in an intermediate stage of growth can be isolated. However, as a consequence of k_d being very small compared to k_p, the time needed for a measurable conversion of the monomer is orders of magnitude greater than the time needed for the full growth of a polymer molecule.

6.2.4 Chain Termination

The sequence of propagation reactions is terminated at some point due to annihilation of the radical center of the propagating chain. Two propagating chains are terminated when two radicals *combine* to form an electron-pair (covalent) bond as in the reaction

(6.13)

This process is called *termination by combination* (or *coupling*). In general terms, this is written as Eq. (6.7a), where k_{tc} is the rate constant for termination by combination (or coupling) or simply the *combination rate constant*.

Alternatively, a pair of radicals can form two new molecules by a *disproportionation* reaction:

(6.14)

In general terms, this is represented as Eq. (6.7b), where k_{td} is the rate constant for *termination by disproportionation* or simply the *disproportionation rate constant*.

Termination of chain radicals can also occur by a combination of coupling and disproportion-ation. However, since both the reactions result in the formation of dead (i.e., without any radical center) polymer molecule(s), the termination step can also be represented by

$$\text{M}_n{}^\bullet \;+\; \text{M}_m{}^\bullet \;\xrightarrow{\;k_t\;}\; \text{dead polymer} \tag{6.15}$$

where k_t is the overall termination rate constant given by

$$k_t \;=\; k_{tc} + k_{td} \tag{6.16}$$

The termination rates, R_t, corresponding to the different modes of termination are

$$\text{From Eq. (6.7a):} \quad R_{tc} = 2k_{tc}\,[\text{M}^\bullet]^2 \tag{6.17}$$

$$\text{From Eq. (6.7b):} \quad R_{td} = 2k_{td}\,[\text{M}^\bullet]^2 \tag{6.18}$$

$$\text{From Eq. (6.15):} \quad R_t = 2k_t\,[\text{M}^\bullet]^2 = 2(k_{tc} + k_{td})\,[\text{M}^\bullet]^2 \tag{6.19}$$

where $[\text{M}^\bullet]$ is the total concentration of chain radicals of all sizes. A factor of 2 is used in the above expressions, in accordance with the generally accepted American convention, since for each incidence of a termination reaction, two chain radicals disappear from the system. Recall that a factor of 2 has also been used previously in Eq. (6.9) for formation of radicals in pairs. It may be noted that in the British convention, this factor of 2 is not used, both in initiation and termination. However, both conventions are nonobjectionable and they lead to the same conclusions about polymerization rates, but the two cannot be mixed.

Equations (6.12) and (6.17)-(6.19) are based on the assumption that both the propagation and termination rate constants are independent of the size of the radical. This assumption facilitates the derivation of a kinetic expression, as shown below, for the overall rate of polymerization, which can be tested experimentally. There is also a considerable experimental evidence to justify the above assumption (Allcock and Lampe, 1990).

Values of k_t (whether k_{tc} or k_{td}) are usually in the range of 10^6–10^8 L/mol-s. Though these values are orders of magnitude greater than k_p, polymerization to high molecular weight still occurs because the concentration of radical species is very small (due to low values of k_d) and because, as shown later, the polymerization rate is proportional to k_p and inversely proportional to $k_t^{1/2}$.

6.2.5 Rate of Polymerization

A radical chain polymerization is started when the initiator begins to decompose according to Eq. (6.3) and the concentration of radicals in the system, $[\text{M}^\bullet]$, increases from zero. The rate of termination or disappearance of radicals, being proportional to $[\text{M}^\bullet]^2$ [cf. Eqs. (6.17)-(6.19)], is thus zero in the beginning and increases with time, till at some stage it equals the rate of radical generation. The concentration of radicals in the system then becomes essentially constant (or "steady"), as radicals are formed and destroyed at equal rates. This condition, described as "steady-state assumption" or "steady-state approximation", can thus be described by the following two equations:

$$R_i = R_t \quad \text{and} \quad d[\text{M}^\bullet]/dt = 0 \tag{6.20}$$

The steady-state approximation is a very useful assumption since, as shown below, it allows one to eliminate the inconvenient radical concentration term [M$^\bullet$] and find an expression for it in terms of known or measurable parameters. The validity of steady-state approximation has been shown experimentally in many polymerizations (see Problem 6.1 below).

Problem 6.1   Experimentally, it is observed that, except in the very earliest stages where the extent of reaction is insignificant, the loss of monomer is accounted for, quantitatively, by the formation of the polymer (Allcock and Lampe, 1990). Justify on this basis the steady-state approximation that all free radicals present in a polymerizing system are at steady-state concentrations.

Answer:

Let [M]$_0$ be the initial concentration of the monomer. Since all the monomer molecules that have reacted must be contained either in the propagating radicals or in the inactive polymer product, the stoichiometry requires that (Allcock and Lampe, 1990):

$$[\text{M}]_0 = [\text{M}] + \sum_n n[\text{P}_n] + \sum_n n[\text{M}_n{}^\bullet] \tag{P6.1.1}$$

Differentiating Eq. (P6.1.1) with respect to time and rearranging one obtains

$$-\frac{d[\text{M}]}{dt} = \sum_n n\left(\frac{d[\text{P}_n]}{dt}\right) + \sum_n n\left(\frac{d[\text{M}_n{}^\bullet]}{dt}\right) \tag{P6.1.2}$$

According to experimental observation, $-\dfrac{d[\text{M}]}{dt} = \sum_n n\left(\dfrac{d[\text{P}_n]}{dt}\right)$.

Hence, $\sum_n n\left(\dfrac{d[\text{M}_n{}^\bullet]}{dt}\right) = 0$. This means that

$$\frac{d[\text{M}_n{}^\bullet]}{dt} = \frac{d[\text{M}^\bullet]}{dt} = 0, \text{ where } [\text{M}^\bullet] = \sum_n [\text{M}_n{}^\bullet] \tag{P6.1.3}$$

Equation (P6.1.3) represents the steady-state approximation.

It can be shown that the steady-state condition is reached soon after polymerization starts (see Problem 6.12). One can, therefore, assume without much error that it applies to the whole course of the polymerization (Rudin, 1982).

Substitution of R_t from Eq. (6.19) into Eq. (6.20) yields

$$R_i = 2k_t[\text{M}^\bullet]^2 \quad \text{or} \quad [\text{M}^\bullet] = \left(\frac{R_i}{2k_t}\right)^{1/2} \tag{6.21}$$

Since monomer disappears both by the initiation reaction [Eq. (6.4)] and by the propagation reactions [Eq. (6.6)], the *rate of monomer disappearance*, which is the same as the *rate of polymerization*, will be given by $-d[\text{M}]/dt = R_i + R_p$. However, the number of monomer molecules consumed in chain initiation is insignificant compared with those consumed in chain propagation in a case where the average chain length of the polymer formed is large. For calculations, the former can thus be neglected and the polymerization rate simply equated to the rate of propagation (*long-chain approximation*), i.e.,

$$-\frac{d[M]}{dt} = R_p = k_p[M][M^\bullet] \tag{6.22}$$

Radical concentrations, being very low ($\sim 10^{-8}$ mol/L), are difficult to measure quantitatively. It is therefore convenient to eliminate $[M^\bullet]$ by substitution from Eq. (6.21). This yields the following expression

$$-\frac{d[M]}{dt} = R_p = k_p[M]\left(\frac{R_i}{2k_t}\right)^{1/2} \tag{6.23}$$

for the rate of polymerization. When initiation takes place by thermal decomposition of initiator, as represented by Eq. (6.3), substitution for R_i from Eq. (6.11) gives

$$-\frac{d[M]}{dt} = R_p = k_p[M]\left(\frac{fk_d[I]}{k_t}\right)^{1/2} \tag{6.24}$$

Equations (6.23) and (6.24) show that the rate of polymerization depends directly on the monomer concentration and on the square root of the rate of initiation. Thus, doubling the monomer concentration doubles the polymerization rate, while doubling the rate of initiation or initiator concentration increases the polymerization rate only by the factor $\sqrt{2}$. It is further evident from Eqs. (6.23) and (6.24) that the polymerizability of a monomer in radical chain polymerization is related to the ratio $k_p/k_t^{1/2}$ and not to k_p alone. As we shall see later, this ratio appears frequently in the equations derived for radical polymerization.

6.2.6 Overall Extent of Polymerization

It should be noted that Eq. (6.24) represents the instantaneous rate of polymerization corresponding to [M] and [I] values at any given instant. Since these values change with conversion, Eq. (6.24) must be integrated over a period of time to determine the overall extent of polymerization.

If the initiator decomposes in a unimolecular reaction [cf. Eq. (6.3)], the rate of initiator disappearance is first order in initiator, represented by

$$-d[I]/dt = k_d[I] \tag{6.25}$$

Integration of Eq. (6.25) between $[I]_0$ at $t = 0$ and $[I]$ at t gives

$$[I] = [I]_0 e^{-k_d t} \tag{6.26}$$

[Integrating between $[I]_0$ at $t = 0$ and $[I] = [I]_0/2$ at $t = t_{1/2}$, gives the half-life ($t_{1/2}$) of the initiator as

$$t_{1/2} = (\ln 2)/k_d \tag{6.27}$$

Being independent of the initial concentration of the initiator, $t_{1/2}$ is a convenient criterion for initiator activities. Several common initiators with their half-lives at various temperatures are listed in Table 6.1.]

Table 6.1 Half-Lives of Some Common Initiators at Different Temperatures

Initiator	Half-life at			
	50°C	70°C	100°C	130°C
Azobisisobutyronitrile	73 h^a	6.1 h^a	7.2 min^b	
Acetyl peroxide	158 h^a	8.1 h^a	15 min^c	
Benzoyl peroxide	364 h^d	16.3 h^a	25 min^d	
	(49.4°C)			
Lauroyl peroxide	87.9 h^a	6.7 h^a	4.8 min^c	
t-Butyl peroxide			220 h^a	6.9 h^c
Cumyl peroxide				1.8 h^a
t-Butyl peracetate			12.5 h^a	20.3 min^a

Solvents used: *a* Benzene;   *b* toluene;   *c* cyclohexene;   *d* styrene.

Source: Calculated from Eq. (6.27) using k_d data from Brandrup et al. (1999).

Inserting Eq. (6.26) into Eq. (6.24) one obtains

$$-\frac{d[M]}{[M]} = \left(\frac{k_p}{k_t^{1/2}}\right)(fk_d\,[I]_0)^{1/2}\left(e^{-k_d}\right)dt \tag{6.28}$$

Integration between $[M]_0$ at $t = 0$ and $[M]$ at t gives

$$-\ln\frac{[M]}{[M]_0} = 2\left(\frac{k_p}{k_t^{1/2}}\right)\left(\frac{f\,[I]_0}{k_d}\right)^{1/2}\left(1 - e^{-k_d t/2}\right) \tag{6.29}$$

Since the extent of monomer conversion, p, is usually defined as

$$p = \frac{[M]_0 - [M]}{[M]_0} \quad\text{or}\quad (1 - p) = [M]/[M]_0 \tag{6.30}$$

Equation (6.29) may also be written as

$$-\ln(1 - p) = 2\left(\frac{k_p}{k_t^{1/2}}\right)\left(\frac{f\,[I]_0}{k_d}\right)^{1/2}\left(1 - e^{-k_d t/2}\right) \tag{6.31}$$

This equation can be used to calculate the amount of polymer produced (that is, moles of monomer converted into polymer) in time t at a given temperature or for determining the time needed to reach different extents of conversion for actual polymerization systems where both [M] and [I] decrease with time.

Problem 6.2   The decomposition of benzoyl peroxide is characterized by a half-life of 7.3 h at 70°C and an activation energy of 29.7 kcal/mol. What concentration (mol/L) of this peroxide is needed to convert 50% of the original charge of a vinyl monomer to polymer in 6 hours at 60°C?   (Data: $f = 0.4$; $k_p^2/k_t = 1.04\times10^{-2}$ L/mol-s at 60°C.)

Answer:

From Eq. (6.27) at 70°C, $k_d = \dfrac{\ln 2}{t_{1/2}} = \dfrac{\ln 2}{(7.3 \times 3600 \text{ s})} = 2.638 \times 10^{-5} \text{ s}^{-1}$

Using Arrhenius-type relationship, $k_d = A_d e^{-E_d/RT}$ at 60°C and 70°C:

$$\ln\left[\frac{(k_d)_{60°}}{(2.638 \times 10^{-5} \text{ s}^{-1})}\right] = \frac{-E_d}{R}\left(\frac{1}{333} - \frac{1}{343}\right) \tag{P6.2.1}$$

From the given value of E_d, $E_d/R = (29.7 \times 10^3 \text{ cal mol}^{-1})/(1.987 \text{ cal mol}^{-1} \text{°K}^{-1}) = 1.495 \times 10^4 \text{ °K}$.

Substituting the value of E_d/R in Eq. (P6.2.1), $(k_d)_{60°} = 7.128 \times 10^{-6} \text{ s}^{-1}$.

Using Eq. (6.31) for 50% conversion at 60°C,

$$-\ln 0.5 = \frac{2(1.04 \times 10^{-2} \text{ L mol}^{-1} \text{ s}^{-1})^{1/2}}{(7.128 \times 10^{-6} \text{ s}^{-1})^{1/2}} (0.4 \, [\text{I}]_0)^{1/2} \times$$
$$\left[1 - e^{-(7.128 \times 10^{-6} \text{ s}^{-1})(6 \times 3600 \text{ s})/2}\right]$$

Solving, $[\text{I}]_0 = 3.75 \times 10^{-2} \text{ mol L}^{-1}$.

Problem 6.3 For a new monomer 50% conversion is obtained in 500 min when polymerized in homogeneous solution with a thermal initiator. How much time would be needed for 50% conversion in another run at the same temperature but with four-fold initial initiator concentration ?

Answer:

Approximating $\left(1 - e^{-k_d t/2}\right)$ by $k_d t/2$, Eq. (6.31) may be written as

$$-\ln(1 - p) = \left(\frac{k_p}{k_t^{1/2}}\right)(f k_d)^{1/2} [\text{I}]_0^{1/2} t$$

Taking ratio of equations for the runs (same conversion, same rate constants),

$$1 = \frac{[\text{I}]_{0,2}^{1/2} t_2}{[\text{I}]_{0,1}^{1/2} t_1}$$

Given $[\text{I}]_{0,2}/[\text{I}]_{0,1} = 4$. Hence, $t_2 = (500 \text{ min})/\sqrt{4} = 250 \text{ min}$.

Problem 6.4 If a 5% solution of a monomer A containing 10^{-4} mol/L of peroxide P is polymerized at 70°C, 40% of the original monomer charge is converted to polymer in 1 h. How long will it take to polymerize 90% of the original monomer charge in a solution containing (initially) 10% A and 10^{-2} mol/L of peroxide P ?

Answer:

Approximating $(1 - e^{-k_d t/2})$ by $k_d t/2$, Eq. (6.31) becomes

$$-\ln(1 - p) = k_p \left(\frac{f k_d}{k_t}\right)^{1/2} [\text{I}]_0^{1/2} t$$

Run 1: $\quad -\ln(0.6) = K\left(10^{-4} \text{ mol L}^{-1}\right)^{1/2} (1 \text{ h})$

Run 2: $\quad -\ln(0.1) = K\left(10^{-2} \text{ mol L}^{-1}\right)^{1/2} (t)$

where K is a lumped constant, same for both runs. Taking ratio to eliminate K, $t = 0.1\,(\ln 0.1/\ln 0.6) = 0.45 \text{ h}$.

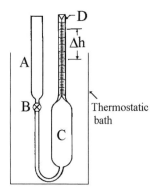

Figure 6.1 Sketch of a simple dilatometer. The change in height of the liquid in the capillary (protruding from the oil in the bath) is measured periodically with a cathetometer (a rigidly mounted, vertically sliding telescope) or from the scale on the dilatometer. The rate of polymerization can then be determined from the rate of change in the height of the liquid. (Adapted from Allcock and Lampe, 1990.)

6.3 Experimental Determination of R_p : Dilatometry

To determine the rate of polymerization, it is merely necessary to measure the monomer concentration [M] as a function of reaction time t and then determine the slope of a plot of [M] versus t. The monomer concentration can be measured by infrared (IR), ultraviolet (UV), and nuclear magnetic resonance (NMR) spectroscopy. It is, however, significantly faster and also more convenient to determine polymerization rates by measuring some physical property of the reaction mixture that changes with the extent of polymerization. Although a number of such methods are available, *dilatometry* is most commonly used in practice. The method is briefly described below.

Because the density of a polymer is usually greater than that of the monomer from which it is formed (for styrene, for example, 1.062 and 0.905 g/cm^3 at 25°C for polymer and monomer, respectively), the rate of addition polymerization can be followed by observing the contraction (Δh) in volume of a fixed weight of monomer as it is polymerized at a constant temperature. The sensitivity of the volume change (shrinkage) with conversion can be increased significantly by measuring the shrinkage in a tube of very narrow diameter. Such a device is called a *dilatometer*. It may be constructed in a variety of ways. A simple apparatus is shown in Fig. 6.1. It has a reservoir (C) to contain a sufficient volume of liquid (for a meaningful volume change with conversion) and is fitted with a capillary tube (D) to facilitate the measurement of volume change due to conversion of monomer to polymer. The apparatus is placed in a thermostatic oil bath maintained at the desired reaction temperature.

Problem 6.5   Show how the rate of a free-radical polymerization can be determined from the measured changes in the capillary liquid height (Δh) in a simple dilatometer (Fig. 6.1).

Answer:

On a weight fraction basis, the fractional yield of polymerization, y, at time t may be written as

$$y = \frac{w_m^0 - w_m}{w_m^0} \tag{P6.5.1}$$

where w_m^0 and w_m are the weights of the monomer initially and at time t, respectively. Assuming that all the monomer is converted to polymer after a sufficiently long time ($t = \infty$) when there is no further decrease in the liquid height (h), w_m and w_m^0 can be related to the change in volume of the reaction system (Allcock and Lampe, 1990) by

$$\frac{w_m}{w_m^0} = \frac{V - V_\infty}{V_0 - V_\infty} \tag{P6.5.2}$$

where V_0 is the initial volume of the reaction system, while V and V_∞ are the volumes at $t = t$ and $t = \infty$, respectively. Then combining Eqs. (P6.5.1) and (P6.5.2), one may write

$$y = \frac{V_0 - V}{V_0 - V_\infty} = \frac{\Delta h(t)}{\Delta h(\infty)} \tag{P6.5.3}$$

where Δh is the fall in liquid height in the capillary tube. An expression for the initial rate of polymerization may then be given (Allcock and Lampe, 1990) as

$$R_p = \frac{y[M]_0}{t} = \frac{\Delta h(t)}{\Delta h(t = \infty)} \frac{[M]_0}{t} \tag{P6.5.4}$$

Since $\Delta h(t = \infty)$ can be calculated (see Problem 6.6), R_p can be determined from the observed decrease in liquid height (Δh) in the capillary of the dilatometer at time t.

Problem 6.6   A dilatometer (total volume 49.0 cm^3, capillary radius 0.1 cm) was filled with a freshly distilled sample of styrene containing 0.1% (by wt) of 2,2′-azo-bis-isobutyro- nitrile. The dilatometer was placed in an oil bath at 70±1°C and was immersed in the oil so that the capillary tube protruded. When the volume of the solution began to decrease after coming to thermal equilibrium, the fall in height of the liquid (Δh) in the capillary tube was determined periodically from the scale on the dilatometer (McCaffery, 1970). This yielded the following data:

Time (s)	Δh (cm)
300	0.83
600	1.68
1800	5.03
3600	10.05
5400	15.10
7200	20.13

Determine the kinetic parameter $k_p/k_t^{1/2}$ at 70°C, given that $f = 0.60$ and $k_d = 4.0 \times 10^{-5}$ s^{-1} at 70°C. [Density (g/cm^3) at 70°C: styrene 0.860; polystyrene 1.046]

Answer:

Initial volume of styrene (V_0) = 49.0 cm^3 at 70°C.   Weight of polystyrene at 100% conversion = initial wt of styrene = (49.0 cm^3)(0.860 g cm^{-3}) = 42.14 g.   Therefore, V_∞ = (42.14 g) / (1.046 g cm^{-3}) = 40.287 cm^3 and $(V_0 - V_\infty)$ = (49.0 − 40.287)   or   8.713 cm^3.

As the capillary tube diameter is known, the fall in liquid height in the capillary can be calculated:

$$\Delta h(t = \infty) = \frac{(8.713 \text{ cm}^3)}{(3.143)(0.1 \text{ cm})^2} = 277.22 \text{ cm}$$

$$\text{Monomer concentration, } [M]_0 = \frac{(1000 \text{ cm}^3 \text{ L}^{-1})(0.860 \text{ g cm}^{-3})}{(104 \text{ g mol}^{-1})} = 8.27 \text{ mol L}^{-1}$$

The yield of polymer (y) can be calculated from Eq. (P6.5.3) and the polymerization rate (R_p) from Eq.

(P6.5.4) using the observed values of Δh. This gives:

Time (s)	% Conversion ($y \times 100$)	$R_p \times 10^5$ (mol L^{-1} s^{-1})
300	0.30	8.40
600	0.62	8.51
1800	1.84	8.49
3600	3.69	8.48
5400	5.55	8.50
7200	7.39	8.49

For low degrees of conversion (generally below 10%) both [I] and [M], and hence R_p, may be assumed to be constant. The average value of R_p obtained from the above data is 8.49×10^{-5} mol L^{-1} s^{-1}.

$$[I]_0 = (1 \ \text{g L}^{-1}) / (164 \ \text{g mol}^{-1}) = 6.1 \times 10^{-3} \ \text{mol L}^{-1}$$

From Eq. (6.24), using initial conditions,

$$
\begin{aligned}
k_p/k_t^{1/2} &= \frac{R_p}{[M]_0 \, (fk_d \, [I]_0)^{1/2}} \\
&= \frac{(8.49 \times 10^{-5} \ \text{mol L}^{-1} \ \text{s}^{-1})}{(8.27 \ \text{mol L}^{-1}) \left[(0.6)(4.0 \times 10^{-5} \ \text{s}^{-1})(6.1 \times 10^{-3} \ \text{mol L}^{-1}) \right]^{1/2}} \\
&= 0.027 \ \text{L}^{1/2} \, \text{mol}^{-1/2} \, \text{s}^{-1/2}
\end{aligned}
$$

6.4   Methods of Initiation

It may be noted that Eq. (6.23) for the rate of polymerization R_p contains a *general term R_i* representing the rate of initiation. The expression for R_i will vary depending on the method used for the generation of primary radicals. For unimolecular thermal decomposition of initiator compounds [Eq. (6.3)], R_i is given by Eq. (6.11). Inserting it in Eq. (6.23) yielded the corresponding expression, Eq. (6.24), for the rate of polymerization.

A variety of other means, besides thermal decomposition of initiator, can be used to produce radicals for chain initiation, such as redox reactions, ultraviolet irradiation, high-energy irradiation, and thermal activation of monomers. The expression for R_i will be different in each case and inserting it into the same equation (6.23) will yield the corresponding expression for the rate of polymerization.

6.4.1   Thermal Decomposition of Initiators

The most common means of generating radicals is the thermal scission of a compound containing a relatively weak bond. It is ideally a unimolecular reaction leading to generation of a pair of radicals, as typified by Eq. (6.3), with a first order rate constant k_d, which is related to the half-life $t_{1/2}$ by Eq. (6.27). It is convenient to use compounds which have bond dissociation energies in the range 100-170 kJ/mol, as compounds with lower dissociation energies will decompose too rapidly and those with higher energies will decompose only too slowly. The major class of compounds with bond dissociation in this range are the organic peroxides. Generally, these may be represented by R–O–O–R′ in which O–O is the peroxide linkage which undergoes scission producing two radicals:

$$\text{RO–OR}' \xrightarrow{k_d} \text{RO}^\bullet + \text{R}'\text{O}^\bullet \tag{6.32}$$

Several types of peroxy compounds that are widely used are listed in Table 6.2. The dissociation of these compounds occurs by reaction of the same type as shown in Eq. (6.32). Besides peroxides, another class of compounds that find extensive use as initiators are the azo compounds. By far the most important member of this class of initiators is 2,2′-azobisisobutyronitrile (AIBN) which generates radicals by the decomposition reaction:

$$\underset{\substack{\text{CH}_3 \\ | \\ \text{H}_3\text{C}-\text{C}-\text{N}=\text{N}-\text{C}-\text{CH}_3 \\ | \quad\quad\quad | \\ \text{CN} \quad\quad\quad \text{CN}}}{} \xrightarrow{\Delta} 2\,\underset{\substack{\text{CH}_3 \\ | \\ \text{H}_3\text{C}-\text{C}^\bullet \\ | \\ \text{CN}}}{} + \text{N}_2 \tag{6.33}$$

In spite of the high dissociation energy ($\sim$ 290 kJ/mol) of the C–N bond, AIBN undergoes facile dissociation because it leads to the formation of very stable nitrogen gas. This initiator has a 10-h $t_{1/2}$ at 64°C.

The temperature of an initiator depends on the rate of decomposition as reflected in its half-life. A good rule-of-thumb in this regard is a $t_{1/2}$ of about 10 h at the particular reaction temperature (Table 6.2). The practical use temperature ranges of some common initiators are: diacetyl peroxide 70-90°C, dibenzoyl peroxide 75-95°C, dicumyl peroxide 120-140°C, and AIBN 50-70°C (Odian, 1991).

6.4.1.1 Initiator Efficiency

A significant proportion of primary radicals that are produced by the initiator decomposition in a reaction system do not actually react with the monomer to form chain radicals, and the initiator efficiency f in Eq. (6.11) usually lies in the range 0.3-0.8. A major cause of this low f is the wastage of primary radicals by the so-called "cage" reactions. To illustrate, the decomposition of diacetyl peroxide could lead to the following reactions (Rudin, 1982; Odian, 1991):

$$\text{CH}_3\text{COO–OOCCH}_3 \rightleftharpoons \boxed{2\text{CH}_3\text{COO}^\bullet} \tag{6.34}$$

$$\boxed{2\text{CH}_3\text{COO}^\bullet} \longrightarrow \boxed{\text{CH}_3\text{COOCH}_3 + \text{CO}_2} \tag{6.35}$$

$$\boxed{2\text{CH}_3\text{COO}^\bullet} \longrightarrow \boxed{\text{CH}_3\text{CH}_3 + 2\text{CO}_2} \tag{6.36}$$

$$\boxed{\text{CH}_3\text{COO}^\bullet} + \text{M} \longrightarrow \text{CH}_3\text{COOM}^\bullet \tag{6.37}$$

$$\boxed{\text{CH}_3\text{COO}^\bullet} \longrightarrow \text{CH}_3\text{COO}^\bullet \tag{6.38}$$

$$\text{CH}_3\text{COO}^\bullet + \text{M} \longrightarrow \text{CH}_3\text{COOM}^\bullet \tag{6.39}$$

$$\text{CH}_3\text{COO}^\bullet \longrightarrow \text{CH}_3{}^\bullet + \text{CO}_2 \tag{6.40}$$

$$\text{CH}_3{}^\bullet + \text{M} \longrightarrow \text{CH}_3\text{M}^\bullet \tag{6.41}$$

Here ☐ represents a "cage" (with solvent and/or monomer molecules comprising the cage wall) enclosing the primary radical(s). According to the concept of "cage effect", the radicals formed on decomposition of an initiator molecule remain surrounded by the "cage" where they may suffer many collisions leading to their recombination [i.e., reverse of Eq. (6.34)] or mutual deactivation

Table 6.2 Some Commonly Used Peroxide Initiators for Radical Polymerization

Type/Example	Formula	Temperature (^{0}C) for $t_{1/2}$ of 10h
Diacyl peroxides		
Diacetyl peroxide		69
Dibenzoyl peroxide		72
Dialkyl peroxides		
Di-t-butyl peroxide		126
Di-cumyl peroxide		117
Peroxyesters		
t-Butyl peroxyacetate		102
t-Butyl peroxybenzoate		105
Hydroperoxides		
Cumyl hydroperoxide		158
Peroxydicarbonates		
Di-isopropyl peroxy-dicarbonates		46
Ketone peroxides		
Methyl ethyl ketone peroxides	and other structures	105

Source: Mageli and Kolczynski, 1968.

[Eqs. (6.35) and (6.36)] before finally reacting with a monomer molecule contained in the wall of the cage [Eq. (6.37)] or diffusing out of the cage [Eq. (6.38)]. Outside the cage, the radical has several possibilities of reaction, as shown by Eqs. (6.39)-(6.41).

The reactions represented by Eqs. (6.37), (6.39), and (6.41) initiate polymerization and so represent an effective use of the initiator. However, the reactions of Eqs. (6.35) and (6.36) leading to stable products, which cannot generate radicals, amount to loss of initiator molecules. Such cage wastage reactions are much more significant than any other in reducing the value of f.

Problem 6.7   Consider the following scheme of reactions for free-radical chain polymerization initiated by thermal homolysis of initiator with cage effect (Jenkins, 1958; Koenig, 1973):

$$\text{I} \xrightarrow{k_d} \boxed{2\text{R}\cdot} \tag{P6.7.1}$$

$$\boxed{2\text{R}\cdot} \xrightarrow{k_r} \text{Q} \tag{P6.7.2}$$

$$\boxed{\text{R}\cdot} + \text{M} \xrightarrow{k_x} \text{M}\cdot \tag{P6.7.3}$$

$$\boxed{\text{R}\cdot} \xrightarrow{k_D} \text{R}\cdot \tag{P6.7.4}$$

$$\text{R}\cdot + \text{R}\cdot \xrightarrow{k_r'} \text{Q}' \tag{P6.7.5}$$

$$\text{R}\cdot + \text{M} \xrightarrow{k_x'} \text{M}\cdot \tag{P6.7.6}$$

$$\text{M}\cdot + \text{M} \xrightarrow{k_p} \text{M}\cdot \tag{P6.7.7}$$

$$\text{M}\cdot + \text{M}\cdot \xrightarrow{k_t} \text{P} \tag{P6.7.8}$$

where $\boxed{\text{R}\cdot}$ represents a primary radical within a cage provided by the medium, Q is the stable product of reaction between a pair of primary radicals within a cage, M$\cdot$ represents chain radicals having an end group R (derived from the initiator), and P is a dead polymer molecule; other species are as defined earlier. Equations (P6.7.2) and (P6.7.3) denote, respectively, the reaction of a cage-confined primary radical with its primary partner within the cage and with a monomer molecule contained in the wall of the cage, while Eq. (P6.7.4) denotes diffusion of the primary radical out of the cage. Reactions (P6.7.5) and (P6.7.6) are analogous to (P6.7.2) and (P6.7.3), but refer to the main body of the solution, that is, they are reactions which may occur only after (P6.7.4).

On the basis of the above scheme, derive an expression for the rate of polymerization including the cage effect but neglecting chain transfer.

Answer:

It should be noted that the rate of reaction (P6.7.2), unlike reaction (P6.7.5), is first order with respect to $\boxed{\boxed{\text{R}\cdot}}$, since the primary radicals are necessarily formed in pairs in independent cages. It is also known that for all practical values of [M], reaction (P6.7.5) is negligible. The net rate of formation of $\boxed{\text{R}\cdot}$ is thus given by (Koenig, 1973):

$$\frac{d\boxed{\boxed{\text{R}\cdot}}}{dt} = 2k_d[\text{I}] - k_r\boxed{\boxed{\text{R}\cdot}} - k_x\boxed{\boxed{\text{R}\cdot}}[\text{M}] - k_D\boxed{\boxed{\text{R}\cdot}}$$

For a steady state concentration of $\boxed{\text{R}\cdot}$, $d\boxed{\boxed{\text{R}\cdot}}/dt = 0$. Therefore,

$$\boxed{\boxed{\text{R}\cdot}} = \frac{2k_d[\text{I}]}{k_r + k_D + k_x[\text{M}]} \tag{P6.7.9}$$

If reaction (P6.7.6) is fast, the rate of initiation will be given by

$$R_i = k_D \boxed{\boxed{R^\bullet}} + k_x \boxed{\boxed{R^\bullet}} [M]$$
$$= \boxed{\boxed{R^\bullet}} (k_D + k_x [M]) \tag{P6.7.10}$$

Combination of Eqs. (P6.7.9), (P6.7.10), and (6.23) yields

$$R_p = k_p \left(\frac{k_d}{k_t}\right)^{1/2} \left(\frac{k_D + k_x[M]}{k_r + k_D + k_x[M]}\right)^{1/2} [I]^{1/2} [M] \tag{P6.7.11}$$

Comparison of Eq. (P6.7.11) with Eq. (6.24) shows that the polymerization rate is reduced by the cage effect.

For $k_D \gg k_x[M]$ and $k_D \gg k_r$, Eq. (P6.7.11) reduces to Eq. (6.24) for ideal kinetics and negligible cage effect.

Determination of Initiator Efficiency   It should be noted that the initiator efficiency is not an exclusive property of the initiator, since it may vary to different extents depending on the condition of polymerization including the identities of monomer and any solvent used. A number of methods can be used for determining the initiator efficiency (Ghosh, 1990; Odian, 1991).

In one method of determining the initiator efficiency, the initiator fragments occurring as end groups in the polymer formed are compared to the amount of initiator consumed. If an initiator labeled with radioactive isotope is used, the polymer formed will possess radioactive end groups. After polymerization the polymer can be separated and subjected to radioactive analysis to provide a sensitive method for determining the number of initiator fragments trapped as end groups in polymer molecules, i.e., number of radicals actually used in chain initiation (Bonta et al., 1976; Berger et al., 1977). Common examples of isotopically labeled initiators are ^{14}C-labeled benzoyl peroxide and other related peroxides, ^{14}C-labeled AIBN and ^{35}S-labeled potassium persulfate.

A variation of the above method involves both radioactive end-group analysis of the polymer and measurement of the polymerization rate (Bevington, 1955). Thus from an assay of the polymer, it is possible to calculate the number of monomer molecules combined in the polymer for each initiator fragment. This number is the *kinetic chain length* (see Section 6.7.1) independent of the mechanism of termination. Dividing the overall rate of polymerization by the kinetic chain length one obtains the rate of initiation, which can then be compared with the rate of production of radicals from the initiator to obtain the efficiency of initiation.

Radical scavengers which are capable of stopping chain growth efficiently and rapidly can be used for counting of radicals to determine the initiator efficiency. The stable free radical diphenylpicrylhydrazyl (DPPH) has been widely used for this purpose (Bonta et al., 1976; Berger et al., 1977). The DPPH radical (purple or deep violet), obtained by oxidation of diphenylpicryl-hydrazine with PbO_2, reacts with other radicals to form a nonradical adduct (light yellow or colorless):

$$\phi_2N-\overset{\bullet}{N}-\underset{O_2N}{\overset{O_2N}{\bigcirc}}-NO_2 \ + \ R^{\bullet} \ \longrightarrow \ R-\bigcirc-N-NH-\underset{O_2N}{\overset{O_2N}{\bigcirc}}-NO_2$$

$$(\phi = C_6H_5) \tag{6.42}$$

The color change associated with the reaction allows one to follow the reaction easily by spectrophotometry. Other radical scavengers have also been used in similar counting of radicals (Cohen, 1945, 1947; Sato et al., 1977). This method of determining f using radical scavengers is, however, not very useful as the reaction between the scavengers and radicals is often not quantitative.

Probably the most useful method for determination of initiator efficiency is the one based on the *dead-end effect* in polymerization technique which is treated in a later section. Using this technique, kinetic data obtained under dead-end conditions can be used to evaluate both k_d and f under the experimental conditions.

6.4.2 Redox Initiation

Free-radicals generated in many oxidation-reduction (or *redox*) reactions can be used to initiate chain poymerization. An advantage of this type of initiation is that, depending on the redox system used, radical production can occur at high rates at moderate (0-50°C) and even lower temperatures. Redox systems are generally used in polymerizations only at relatively low temperatures, a significant commercial example being the production of styrene-butadiene rubber by emulsion copolymerization of butadiene and styrene at 5-10°C ("cold recipe").

A general redox reaction described by

$$A–B \ + \ X \longrightarrow A^{\bullet} + B^{\ominus} + X^{\oplus} \tag{6.43}$$

can occur with any molecule AB if the reducing agent X is sufficiently strong to split the A−B bond. For practical purposes, the A−B bond in redox systems must be relatively weak and this limits the choice of such materials. Some of the common redox systems are described below.

Water soluble peroxides and persulfates, in combination with a reducing agent such as ferrous (Fe^{2+}) and thiosulfate ($S_2O_3^{2-}$) ions, are a common source of radicals in aqueous and emulsion systems. An example of such a redox system has hydrogen peroxide as the oxidant and Fe^{2+} as the reductant:

$$HOOH \ + \ Fe^{2+} \ \overset{k_i}{\longrightarrow} \ HO^{\bullet} \ + \ OH^- \ + \ Fe^{3+} \tag{6.44}$$

Another redox system commonly used consists of persulfate as the oxidant and Fe^{2+} or $S_2O_3^{2-}$ as the reductant (Rudin, 1982; Odian, 1991):

$$^-O_3S-O-O-SO_3^- \ + \ Fe^{2+} \ \longrightarrow \ SO_4^{\bullet-} \ + \ SO_4^{2-} \ + \ Fe^{3+} \tag{6.45}$$

$$^-O_3S-O-O-SO_3^- \ + \ S_2O_3^{2-} \longrightarrow \ SO_4^{\bullet-} + SO_4^{2-} + S_2O_3^{\bullet-} \tag{6.46}$$

An equimolar mixture of $K_2S_2O_8$ and $FeSO_4$ at 10°C produces radicals about 100 times as fast as an equal concentration of the persulfate alone at 50°C. Other redox systems with persulfate include reductants such as HSO_3^- and SO_3^{2-}.

A combination of inorganic oxidant and organic reductant may initiate polymerization by oxidation of the organic component. For example, the oxidation of an alcohol by Ce^{4+} (Mohanty et al., 1980) proceeds according to the reaction

$$Ce^{4+} + RCH_2-OH \xrightarrow{k_d} Ce^{3+} + H^+ + R\overset{\bullet}{C}HOH \qquad (6.47)$$

and the rate of initiation is

$$R_i = k_d[Ce^{4+}][\text{alcohol}] \qquad (6.48)$$

The termination of the propagating radicals in the alcohol-Ce^{4+} system proceeds according to the reaction (Odian, 1991):

$$M_n\bullet + Ce^{4+} \xrightarrow{k_t} Ce^{3+} + H^+ + \text{dead polymer} \qquad (6.49)$$

at high ceric ion concentrations and the rate of termination (neglecting the usual bimolecular termination mechanism) is given by

$$R_t = k_t[Ce^{4+}][M\bullet] \qquad (6.50)$$

The usual steady-state assumption, $R_i = R_t$, then leads to the polymerization rate as

$$R_p = \left(\frac{k_p k_d}{k_t}\right)[M][\text{alcohol}] \qquad (6.51)$$

While some redox systems involve direct electron transfer between oxidant and reductant [see Eq. (6.47)], others involve the intermediate formation of oxidant-reductant complexes (see Problem 6.8).

Problem 6.8   The free-radical polymerization of acrylonitrile initiated by the redox system Mn^{3+}-cyclohexanol (CH) was investigated (Ahmed et al., 1978) in aqueous sulfuric acid in the temperature range 30-45°C. The following reaction mechanism was suggested which involves the formation of a complex between Mn^{3+} and the alcohol that decomposes yielding the initiating free radical, and termination by the mutual combination of polymer chain radicals:

(a) $Mn^{3+} + CH \overset{K}{\rightleftharpoons} \text{complex}$

(b) $\text{complex} \xrightarrow{k_r} R\bullet + Mn^{2+} + H^+$

(c) $R\bullet + Mn^{3+} \xrightarrow{k_0} Mn^{2+} + \text{products}$

(d) $R\bullet + M \xrightarrow{k_i} M_1\bullet$

(e) $M_n\bullet + M \xrightarrow{k_p} M_{n+1}\bullet$

(f) $M_n\bullet + M_m\bullet \xrightarrow{k_t} \text{polymer}$

Derive suitable rate expressions for evaluation of rate parameters from measured initial rates of polymerization.

Answer:

Rate of Mn^{3+} disappearance

Applying the steady-state principle to the primary radical $R\cdot$,

$$[R\cdot] = \frac{k_r K [Mn^{3+}][CH]}{k_0 [Mn^{3+}] + k_i [M]} \tag{P6.8.1}$$

For reaction (a),

$$[complex] = K [Mn^{3+}][CH] \tag{P6.8.2}$$

For total Mn^{3+},

$$[Mn^{3+}]_{tot} = [Mn^{3+}] + [complex]$$

$$= [Mn^{3+}](1 + K[CH]) \tag{P6.8.3}$$

Therefore,

$$-\frac{d[Mn^{3+}]_{tot}}{dt} = -\frac{d[complex]}{dt} - \frac{d[Mn^{3+}]}{dt}$$

$$= k_r K [Mn^{3+}][CH] + k_0 [R\cdot][Mn^{3+}]$$

$$= k_r K [Mn^{3+}[CH] + \frac{k_0 k_r K [Mn^{3+}]^2 [CH]}{k_0 [Mn^{3+}] + k_i [M]} \tag{P6.8.4}$$

If $k_o [Mn^{3+}] \gg k_i[M]$, Eq. (P6.8.4) reduces to the form

$$-\frac{d[Mn^{3+}]_{tot}}{dt} = 2k_r K[Mn^{3+}][CH]$$

$$= \frac{2k_r K [Mn^{3+}]_{tot} [CH]}{1 + K[CH]}$$

$$= 2k_r [Mn^{3+}]_{tot} \left(1 - \frac{1}{[CH]K}\right) \tag{P6.8.5}$$

The values of K and k_r can be computed by plotting $\left(-d[Mn^{3+}]_{tot}/dt\right)$ versus $1/[CH]$ at constant concentration of total Mn^{3+}. From the intercept and the ratio intercept/slope of the plot the rate constant for the unimolecular decomposition of the complex (k_r) and its formation constant (K) can be calculated (Ahmed et al., 1978).

Rate of polymerization

Applying the steady-state principle to the growing chains, i.e., equating the rates of reactions (d) and (f),

$$k_i [R\cdot][M] = k_t [M\cdot]^2$$

$$[M\cdot] = \left(\frac{k_i [R\cdot][M]}{k_t}\right)^{1/2} \tag{P6.8.6}$$

Substituting for [R $\cdot$] in Eq. (P6.8.6) from Eq. (P6.8.1) and then inserting the equation in $R_p = k_p [M\cdot][M]$ yields

$$R_p = \frac{k_p k_i^{1/2} (k_r K [Mn^{3+}][CH])^{1/2} [M]^{3/2}}{k_t^{1/2} (k_0 [Mn^{3+}] + k_i [M])^{1/2}} \tag{P6.8.7}$$

If $k_0 [Mn^{3+}] \gg k_i [M]$, Eq. (P6.8.7) can be approximated to the form

$$R_p = \frac{k_p k_i^{1/2} (k_r K [CH])^{1/2} [M]^{3/2}}{k_o^{1/2} k_t^{1/2}}$$

The composite constant $k_p (k_i / k_0 k_t)^{1/2}$ can be obtained from the slope of the plot of $([M]^3 / R_p^2)$ versus $1/[CH]$, if k_r and K are known from Eq. (P6.8.5).

For fitting experimental rate data, Eq. (P6.8.7) can be conveniently rearranged to the form

$$\left(\frac{[M]}{R_p}\right)^2 = \frac{k_0 k_t}{k_p^2 k_i k_r} \left(\frac{1}{[CH][M]}\right) + \left(\frac{k_t}{k_p^2 k_r K}\right) \frac{1}{[Mn^{3+}][CH]}$$

This equation can be used to evaluate rate parameters using the initial rate of polymerization (R_p) data that may be obtained from the initial slope of the conversion vs. time plots. Since k_r and K are obtained from Eq. (P6.8.5), the value of the composite constant $k_p (k_i / k_0 k_t)^{1/2}$ can be obtained from the intercept of a plot of $([M]/R_p)^2$ against $1/[Mn^{3+}]$ at constant concentration of the monomer [M] and the reducing agent [CH]. The slope gives the parameter $k_p / k_t^{1/2}$ while the quotient of intercept and slope yields k_0/k_i.

6.4.3 *Photochemical Initiation*

Chain-growth polymerization of vinyl monomers can be initiated by exposure to ultraviolet (UV) or visible light generating free radicals by photochemical reaction. An obvious advantage stemming from this method is the avoidance of chemical contamination by initiator residues. Another advantage which has considerable practical significance is that *photochemical initiation* (or simply *photoinitiation*) and polymerization can be spatially controlled to occur only in specific regions and the process can be turned on or off simply by turning the light on or off. Photopolymerization thus finds extensive use in *photolithography* and *photocuring* of paints, adhesives, and inks. Common examples of photoinitiation can be considered under two main categories: (a) direct photoinitiation and (b) sensitized photoinitiation.

6.4.3.1 *Direct Photoinitiation*

There are many examples of monomers which undergo chain polymerization by direct exposure to ultraviolet or visible light. Thus, the absorption of light photons of a specific wavelength by a monomer M may yield an electronically excited molecule $M^\star$, which subsequently decomposes to produce radical fragments

$$M + h\nu \longrightarrow M^\star \longrightarrow R\cdot + R'\cdot \tag{6.52}$$

For example, both alkyl vinyl ketone and vinyl bromide monomers dissociate when irradiated with ultraviolet light (300 nm) by the following reactions:

$$R-\overset{\overset{\text{O}}{\|}}{C}-CH=CH_2 \quad \xrightarrow{h\nu} \quad R-\overset{\overset{\text{O}}{\|}}{C}\cdot \; + \; \cdot CH=CH_2 \tag{6.53}$$

$$CH_2=CH\text{-}Br \xrightarrow{h\nu} Br\cdot \; + \; CH_2=CH\cdot \tag{6.54}$$

The free radicals thus generated from some monomer molecules initiate radical chain polymerization in the remaining monomer.

6.4.3.2 Photosensitization

Even if direct light absorption by monomer as described above does not occur, polymerization can still be initiated by the use of photosensitizers. Being highly photosensitive, a photosensitizer (S) gets readily activated on exposure to light to the excited state $S^\star$:

$$S \xrightarrow{h\nu} S^\star \tag{6.55}$$

which then dissociates to produce chain-initiating radicals [Eq. (6.56)] or transfers energy to monomer or initiator to form the corresponding excited states [Eq. (6.57)]. The excited species thus formed may undergo homolysis to produce chain initiating radicals [Eqs. (6.58)-(6.59)]:

$$S^\star \longrightarrow R\cdot + R'\cdot \tag{6.56}$$

$$S^\star + M \;(\text{or } I) \longrightarrow S + M^\star \;(\text{or } I^\star) \tag{6.57}$$

$$M^\star \;(\text{or } I^\star) \longrightarrow R\cdot + R'\cdot \tag{6.58}$$

The same substances that are used as thermal initiators (I), e.g., azo compounds and peroxides, are also used often as photosensitizers. However, many other substances can be used as photosensitizers even though they do not undergo thermal dissociation to act as thermal initiators. Benzophenone and acetophenone and their derivatives are examples of such substances most commonly used as photosensitizers. When irradiated, these ketones undergo fragmentation generating radicals. For example, benzophenone dissociates as

$$\phi-\overset{\overset{\text{O}}{\|}}{C}-\phi \xrightarrow{h\nu} \left(\phi-\overset{\overset{\text{O}}{\|}}{C}-\phi\right)^* \longrightarrow \phi-\overset{\overset{\text{O}}{\|}}{C}\cdot \; + \; \phi\cdot$$
$$(\phi = C_6H_5-) \tag{6.59}$$

Irradiation in the presence of a hydrogen donor (RH) leads to the production of radicals by hydrogen abstraction:

$$\phi-\overset{\overset{\text{O}}{\|}}{C}-\phi \xrightarrow{h\nu} \left(\phi-\overset{\overset{\text{O}}{\|}}{C}-\phi\right)^* \xrightarrow{RH} \phi-\overset{\overset{\text{OH}}{|}}{\underset{\cdot}{C}}-\phi \; + \; R\cdot$$
$$(\phi = C_6H_5-) \tag{6.60}$$

This reaction is generally more efficient than direct fragmentation [Eq. (6.59)] and occurs at higher wavelength (i.e., lower energy) radiation, which is advantageous in many applications. Amines with α-hydrogens are the most efficient and commonly used hydrogen donors.

6.4.3.3   Rate of Photoinitiated Polymerization

In photochemistry, a mole of light quanta is called an *Einstein*. Thus an Einstein of light quanta of frequency v or wavelength λ has energy $N_{Av}hv$ (or $N_{Av}hc/\lambda$), where N_{Av} is Avogadro's number, h is Planck's constant, and c is the speed of light. The rate of photochemical initiation may then be expressed (Ghosh, 1990) as

$$R_i = 2\Phi I_a \qquad (6.61)$$

simply by replacing k_d [I] of Eq. (6.11) with I_a, the intensity of absorbed light in moles (*Einsteins*) of light quanta per liter per second, and replacing f with Φ for photochemical polymerization. The factor of 2 in Eq. (6.61) is used only for those photoinitiating systems in which two radicals are produced per molecule undergoing photolysis [cf. Eqs. (6.55)-(6.59)]. Referred to as the *quantum yield for photoinitiation*, Φ may be defined as the ratio of the rate of chain initiation to the rate of light absorption; Φ is thus equivalent to f in thermal initiation [Eq. (6.11)], and both describe the fraction of radicals that are actually used in initiating chain polymerization. As in the case of f, the maximum value of Φ is 1. Substituting for R_i in Eq. (6.23) from Eq. (6.61) gives the expression for the rate of photopolymerization as

$$R_p = k_p [M] (\Phi I_a / k_t)^{1/2} \qquad (6.62)$$

Problem 6.9   Methyl methacrylate is polymerized in 10% w/v solution using a photosensitizer and 3130 Å light from a mercury arc lamp. Direct measurement by actinometry shows that light is absorbed by the system at the rate of 1.2×10^5 ergs/L-s. If Φ for the system is 0.60, calculate (a) the rate of initiation and (b) the rate of polymerization. [$k_p/k_t^{1/2}$ at 60°C= 0.102 $L^{1/2}$ $mol^{-1/2}$ $s^{-1/2}$.]

Answer:

$$\text{Energy of 1 photon}\,(hc/\lambda) \;=\; \frac{(6.63 \times 10^{-27}\ \text{erg s})(3 \times 10^{10}\ \text{cm s}^{-1})}{(3.13 \times 10^{-5}\ \text{cm})}$$

$$= 6.35 \times 10^{-12}\ \text{erg}$$

Energy of 1 mol of light quanta

$$= (6.02 \times 10^{23}\ \text{quanta mol}^{-1})(6.35 \times 10^{-12}\ \text{erg quantum}^{-1})$$

$$= 3.82 \times 10^{12}\ \text{ergs mol}^{-1}$$

$I_a = (1.2 \times 10^5\ \text{ergs L}^{-1}\text{s}^{-1})/(3.82 \times 10^{12}\ \text{ergs mol}^{-1}) = 3.14 \times 10^{-8}\ \text{mol L}^{-1}\ \text{s}^{-1}$

$[M] = (100\ \text{g L}^{-1})/(100\ \text{g mol}^{-1}) = 1.0\ \text{mol L}^{-1}$

(a) $R_i = 2\Phi I_a = 2 \times 0.6(3.14 \times 10^{-8}\ \text{mol L}^{-1}\text{s}^{-1}) = 3.77 \times 10^{-8}\ \text{mol L}^{-1}\ \text{s}^{-1}$

(b) $R_p = \left(k_p/k_t^{1/2}\right) [M] (R_i/2)^{1/2}$

$$= \left(0.102\ L^{1/2}\ mol^{-1/2}\ s^{-1/2}\right)\left(1.0\ \text{mol L}^{-1}\right)\left(3.77 \times 10^{-8}\ \text{mol L}^{-1}\ \text{s}^{-1}\right)^{1/2}/\sqrt{2}$$

$$= 1.40 \times 10^{-5}\ \text{mol L}^{-1}\ \text{s}^{-1}$$

Unless very thin reaction vessels are employed for photopolymerization or the absorption of light is very low, the intensity of the absorbed light, I_a, will vary with thickness of the reaction system. From Lambert-Beer's law,

$$I = I_0 e^{-\epsilon[S]\ell} \tag{6.63}$$

where I_0 is the incident light intensity, I is the light intensity at a distance ℓ into the reaction vessel, [S] is the concentration of the species S that absorbs light, and ϵ is the molar absorption coefficient or molar absorptivity (*molar extinction coefficient*) of S at the particular wavelength of light absorption. The light intensity absorbed, I_a, is then given by

$$I_a = I_0 - I = I_0 \left(1 - e^{-\epsilon[S]\ell}\right) \tag{6.64}$$

where ℓ now represents the thickness of the reaction vessel. An expression for the photopolymerization rate can be obtained by combining Eq. (6.64) with Eqs. (6.62) to yield (Odian, 1991):

$$R_p = k_p[M] \left[\frac{\Phi I_0 \left(1 - e^{-\epsilon[S]\ell}\right)}{k_t} \right]^{1/2} \tag{6.65}$$

[Note that to ensure dimensional consistency of Eq. (6.65), it is necessary to convert light intensity I_0 into the units of moles (Einsteins) of light quanta per liter per second.] If the exponent in Eq. (6.65) is small (as for dilute solutions), the exponential can be expanded, $e^{-x} = 1 - x + \cdots$, keeping only the leading terms to give

$$R_p = k_p[M] \left(\frac{\Phi \epsilon I_0 [S] \ell}{k_t} \right)^{1/2} \tag{6.66}$$

Thus for the case where there is negligible attentuation of light intensity in traversing the reaction vessel, the photopolymerization R_p is first order in [M], $\frac{1}{2}$-order in light intensity, and $\frac{1}{2}$-order in [S]. If polymerization is initiated by photolysis of initiator I into a pair of radicals, Eqs. (6.65) and (6.66) can be used with [S] substituted by [I]. However, when the photoexcitation involves monomer, i.e., S is M, Eq. (6.66) becomes (Ghosh, 1990):

$$R_p = k_p[M]^{3/2} \left(\frac{\Phi \epsilon I_0 \ell}{k_t} \right)^{1/2} \tag{6.67}$$

that is, the dependence of R_p on [M] becomes 3/2-order. It should be noted, however, that abnormal orders in [M] have been found under certain circumstances (as when M acts as a quencher).

6.4.4 Initiation by High-Energy Radiations

Alpha particles, beta rays (electrons), gamma rays, or high-velocity particles from a particle accelerator have particle or photon energies in the range 10 keV–100 meV, which are much higher than the range of energies (2–6 eV) of visible-ultraviolet photons used in photo-initiated free-radical polymerizations. So practically any monomer that polymerizes by a free-radical mechanism can also be polymerized by these radiations. However, as a consequence of the higher energies of these radiations, the substrate species S, on excitation, may also suffer ionization by ejection of electrons:

$$S + \text{radiation} \longrightarrow S^+ + e^- \qquad (6.68)$$

A series of reactions, such as splitting of the cations [Eq. (6.69)], absorption of electrons by cationic species [Eq. (6.70)] or neutral species [Eq. (6.71)] to form radicals or anions, respectively, and anion splitting to form radical [Eq. (6.72)] and radical plus electron [Eq. (6.73)], may then take place:

$$S^+ \longrightarrow R^{\cdot} + A^+ \qquad (6.69)$$

$$A^+ + e^- \longrightarrow A^{\cdot} \qquad (6.70)$$

$$S + e^- \longrightarrow S^- \qquad (6.71)$$

$$S^- \longrightarrow R'^{\cdot} + B^- \qquad (6.72)$$

$$B^- \longrightarrow B^{\cdot} + e^- \qquad (6.73)$$

Thus, high-energy irradiation of either a pure monomer or a solution of a monomer generates both free radicals and ions. Polymerization may therefore proceed by radical or ionic mechanism or combinations thereof, depending on prevailing conditions.

Gamma radiation is the most convenient type of high-energy radiation for initiating polymerization. Its high penetrating power affords *uniform* irradiation of the system and solid monomers can be polymerized, thus allowing polymerization of many monomers at low temperatures. However, due to higher costs and safety problems of ionizing radiation sources, compared to photochemical sources, radiation-induced polymerization has achieved far less commercial success than photochemical polymerization.

6.4.5 *Thermal Initiation in Absence of Initiator*

A few monomers, even in a highly purified state, are known to undergo self-initiated polymerization when heated in an inert atmosphere in the absence of an initiator or catalyst. Styrene, for example, undergoes such thermal, self-initiated polymerization at temperatures of 100°C or more. Methyl methacrylate also exhibits this behavior, though at a slower rate. [Note that for most other monomers a spontaneous polymerization observed in the absence of any added initiator is often due to the presence of peroxides or hydroperoxides (formed by the reaction of monomer with O_2) which generate radicals by thermal homolysis.] The rates of thermal, self-initiated poymerizations, though much slower than the corresponding polymerizations initiated by the thermal homolysis of an initiator, are not negligible and must be taken into account in any polymerization conducted at a temperature at which self-initiation is significant. Thus, at very low initiator concentrations, self-initiation makes an appreciable contribution to the polymerization rate for styrene.

The mechanism of self-initiation in thermal polymerization of styrene has been shown (Barr et al., 1978) to involve the formation of a Diels-Alder dimer **(I)** of styrene [Eq. (6.74)] followed by the transfer of a hydrogen atom from the dimer to a styrene molecule [Eq. (6.75)]. The existence of the dimer has been confirmed by UV spectroscopy.

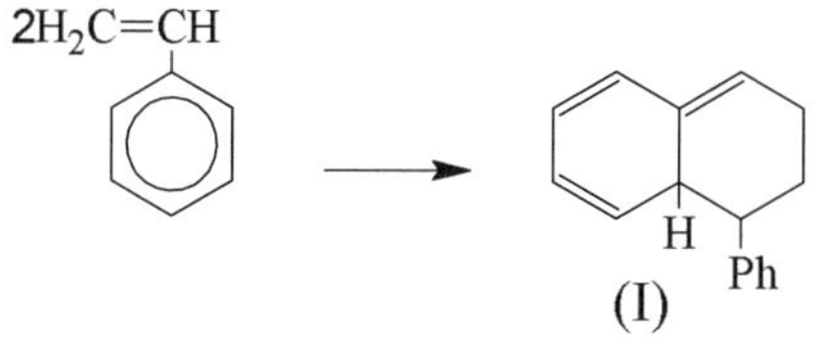

$$(6.74)$$

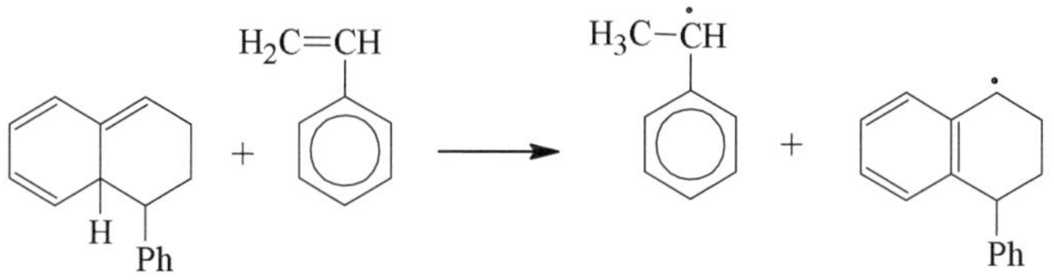

$$(6.75)$$

Problem 6.10   There is evidence (Barr et al., 1978) that thermal self-initiated polymerization of styrene may be of about five-halves order. Show that this is in agreement with the established initiation mechanism involving a Diels-Alder dimer formation [Eqs. (6.74) and (6.75)].

Answer:

The higher than second-order rate observed for thermal conversion of monomer indicates that Eq. (6.74) is the slow step. Representing the concentration of Diels-Alder dimer **(I)** by [D] and that of styrene by [M],

$$R_i = k_i[D][M] = k_i K[M]^3$$

where K is the equilibrium constant for the reaction of Diels-Alder dimer formation [Eq. (6.74)] and k_i is the rate constant for the initiation reaction [Eq. (6.75)].

Using steady-state approximation ($R_i = R_t$),

$$k_t[M\bullet]^2 = k_i K[M]^3 \implies [M\bullet] = \left(\frac{k_i K}{k_t}\right)^{1/2}[M]^{3/2}$$

Substituting this into Eq. (6.22) gives

$$R_p = k_p[M]^{5/2}\left(\frac{k_i K}{k_t}\right)^{1/2}$$

showing five-halves order in monomer concentration.

6.5 Dead-End Polymerization

If the initiator concentration used in a free-radical polymerization system is low and insufficient, leading to a large depletion or complete consumption of the initiator before maximum conversion of monomer to polymer is accomplished, it is quite likely to observe a limiting conversion p_∞ which is less than the maximum possible conversion p_c, as shown in Fig. 6.2. This is known as the *dead-end* effect and it occurs when the initiator concentration decreases to such a low value

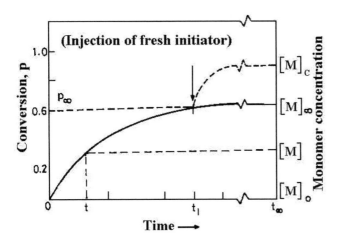

Figure 6.2 Schematic representation of dead-end effect in radical polymerization showing a limiting conversion (p_∞) of monomer to polymer. Injection of the initiator in adequate amounts in the system at time t_1 immediately causes formation of more polymers up to a maximum conversion (p_c) corresponding to the equilibrium monomer concentration $[M]_c$. (After Ghosh, 1990.)

that the half-life of the kinetic chains approximates that of the initiator. However, if there is *autoacceleration effect* or *gel effect* (described later) leading to a sharp rise in rate of polymerization, viscosity of medium, and degree of polymerization, pure dead-end effect cannot be observed.

Recalling Eq. (6.31) for radical chain polymerization initiated by thermal homolysis of an initiator:

$$- \ln(1 - p) \;=\; 2\left(\frac{k_p}{k_t^{1/2}}\right)\left(\frac{f[I]_0}{k_d}\right)^{1/2}\left(1 - e^{-k_d t/2}\right) \tag{6.76}$$

where p is the fractional extent of monomer conversion at time t and $[I]_0$ is the initial concentration of the initiator, let p_∞ be the limiting conversion attained at long reaction times ($t \longrightarrow \infty$). Equation (6.76) then becomes

$$- \ln(1 - p_\infty) \;=\; 2k_p\left(\frac{f[I]_0}{k_d k_t}\right)^{1/2} \tag{6.77}$$

Dividing Eq. (6.76) by Eq. (6.77) and taking logarithm yields (Ghosh, 1990):

$$\ln\left[1 - \frac{\ln(1 - p)}{\ln(1 - p_\infty)}\right] \;=\; -\frac{k_d t}{2} \tag{6.78}$$

A plot of the left side of Eq. (6.78), which is equivalent to the expression $\ln[(\ln[M]_\infty - \ln[M])/(\ln[M]_\infty - \ln[M]_0)]$, versus time ($t$) permits evaluation of k_d (see Problem 6.11). Once k_d is determined, f can be obtained from either Eq. (6.24) or (6.77) using the value of $k_p/k_t^{1/2}$ available from other studies. The thermal dissociation rate constants and activation energy values for several commonly used initiators are listed in Table 6.3.

Table 6.3 Thermal Decomposition Parameters of Some Initiators

Initiator	Solvent	$T°C$	$k_d \times 10^5$	E_d
2,2'-Azobisisobutyronitrile	Styrene	50	0.297	127.6
	Benzene	50	0.216	
Benzoyl peroxide	Styrene	75	1.83	
	Benzene	75	1.48	128.0
Lauroyl peroxide	Styrene	60	1.20	
	Benzene	60	1.51	127.2
2,4-Dichlorobenzoyl peroxide	Styrene	49.4	2.39	
	Benzene	50	1.08	117.6

The units of k_d are s^{-1} and those of E_d are kJ / mol.

Source: Brandrup et al., 1999.

Problem 6.11   Isoprene was polymerized in bulk at several temperatures using AIBN at an initial concentration of 0.0488 mol/L in dead-end polymerization experiments (Gobran et al., 1960). In every case, the conversion increased with time until a limiting value was obtained beyond which no further polymerization was observed. No autoacceleration effect was observed in this system. The data of fractional degree of conversion (p) with time, including the limiting value of conversion (p_∞), determined at each temperature are shown in the table below:

60°C		70°C		80°C	
Time (h)	p	Time (h)	p	Time (h)	p
8	0.054	4	0.055	2	0.100
16	0.086	6	0.102	4	0.172
30	0.167	12	0.180	6	0.202
48	0.220	24	0.235	8	0.240
72	0.280	30	0.273	16	0.290
96	0.315	48	0.310	24	0.305
144	0.360	72	0.310	48	0.305
240	0.390	96	0.325	∞	0.305
300	0.390	150	0.325		
∞	0.390	∞	0.325		

Source: Data from Gobran et al., 1960.

Determine the kinetic parameters k_d and $(k_p / k_t^{1/2}) f^{1/2}$ for isoprene-AIBN system and the activation energy of the initiator dissociation.

Answer:

The left side of Eq. (6.78) is plotted against time t in Fig. 6.3 using the conversion-time data of the isoprene-AIBN system. The slope of the linear plot yields $k_d/2$, and hence k_d, of AIBN in isoprene. $(k_p / k_t^{1/2}) f^{1/2}$ is then calculated from Eq. (6.77). This yields

Temperature (°C)	k_d, s^{-1}	$\left(k_p/k_t^{1/2}\right) f^{1/2} \times 10^3$, $L^{1/2}$ mol$^{-1/2}$ $s^{-1/2}$
60	8.54×10^{-6}	3.27
70	3.08×10^{-5}	4.94
80	9.84×10^{-5}	8.17

It is possible to calculate f from the above values of $(k_p/k_t^{1/2}) f^{1/2}$ if $k_p/k_t^{1/2}$ is known from other studies. An Arrhenius plot of the k_d values as $-\ln k_d$ vs. $1/T$ yields from the slope, $E_d = 122.6$ kJ mol^{-1}.

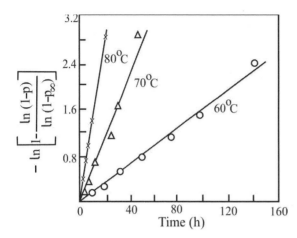

Figure 6.3 Test of Eq. (6.78) for dead-end polymerization of isoprene-AIBN system at different temperatures with data of Problem 6.11. (After Gobran et al., 1960; Odian, 1991.)

6.6 Determination of Absolute Rate Constants

Three different types of rate constants are used in ideal polymerization kinetics described by Eq. (6.24), namely, those for initiation (k_d), propagation (k_p), and termination (k_t). Till now we have discussed only polymerization kinetics under steady-state conditions and how these allow evaluation of k_d (see Dead-end polymerization) and the ratio $k_p/k_t^{1/2}$ or k_p^2/k_t (but not k_p and k_t individually) using measurable quantities like R_p, R_i, and [M]. As we shall see below, employing nonsteady-state polymerization with photoinitiation and intermittent illumination (Flory, 1953; Walling, 1957), it is possible to evaluate k_p/k_t so that by combining it with $k_p/k_t^{1/2}$ obtained from steady-state experiments, one can determine k_p and k_t individually.

6.6.1 Nonsteady-State Kinetics

The radical concentration is zero at the very start of the polymerization and reaches the steady-state level only after a finite time. During this intervening nonsteady period, the rate of change of the concentration of radicals will be given by the difference of the rates of their formation and termination, i.e.,

$$\frac{d[\text{M}^\bullet]}{dt} = R_i - R_t = R_i - 2k_t[\text{M}^\bullet]^2 \qquad (6.79)$$

The advantage of photoinitiation lies in that the generation of radicals can be commenced (or stopped) instantly by turning on (or off) light to the polymerization cell and, moreover, time needed for temperature equilibration (when thermal initiators are used) can be avoided. Using Eq. (6.61) for the rate of photoinitiation, Eq. (6.79) becomes

$$d[\text{M}^\bullet]/dt = 2\Phi I_a - 2k_t[\text{M}^\bullet]^2 \qquad (6.80)$$

At the end of nonsteady period, $d[\text{M}^\bullet]/dt = 0$, and Eq. (6.80) leads to

$$2\Phi I_a = 2k_t[\text{M}^\bullet]_s^2 \qquad (6.81)$$

where the subscript s denotes steady-state condition. Using Eq. (6.81), Eq. (6.80) may be written as (Odian, 1991):

$$\frac{d[\text{M}^\bullet]}{dt} = 2k_t\left([\text{M}^\bullet]_s^2 - [\text{M}^\bullet]^2\right) \tag{6.82}$$

For mathematical convenience, Flory (1953) defined a new parameter τ_s, called the *average lifetime of a growing radical* under steady-state conditions, as

$$\tau_s = \frac{\text{No. of radicals present at steady state}}{\text{No. of radicals disappearing per unit time at steady state}}$$

$$= \frac{[\text{M}^\bullet]_s}{2k_t[\text{M}^\bullet]_s^2} = \frac{1}{2k_t[\text{M}^\bullet]_s} \tag{6.83}$$

Multiplying both the numerator and denominator of Eq. (6.83) by $k_p[\text{M}]$, τ_s is more conveniently expressed as (Ghosh, 1990; Odian, 1991):

$$\tau_s = \frac{k_p[\text{M}]}{2k_tk_p[\text{M}][\text{M}]_s} = \frac{k_p[\text{M}]}{2k_t(R_p)_s} \tag{6.84}$$

As shown below, using intermittent illumination for photoinitiated radical chain polymerization, the parameter τ_s can be evaluated. Since $[\text{M}]$ and $(R_p)_s$ are measurable, determination of τ_s leads to the evaluation of k_p/k_t from Eq. (6.84) and thence to individual k_p and k_t values, by combining with $k_p/k_t^{1/2}$ obtained from steady-state measurements.

Integration of Eq. (6.82) yields (Flory, 1953; Ghosh, 1990; Odian, 1991):

$$\ln\left[\frac{(1 + [\text{M}^\bullet]/[\text{M}^\bullet]_s)}{(1 - [\text{M}^\bullet]/[\text{M}^\bullet]_s)}\right] = 4k_t[\text{M}^\bullet]_s(t - t_0) \tag{6.85}$$

where t_0 enters as a constant of integration such that $[\text{M}^\bullet] = 0$ at $t = t_0$.

Since $\tanh^{-1}u = (1/2)\ln[(1 + u)/(1 - u)]$, combining Eqs. (6.83) and (6.85) one may write

$$\tanh^{-1}\left(\frac{[\text{M}^\bullet]}{[\text{M}^\bullet]_s}\right) = \frac{t - t_0}{\tau_s} \tag{6.86}$$

The t in these equations may be identified with the duration of the illumination.

Problem 6.12 Typical τ_s values determined from photoinitiated radical chain polymerization with intermittent illumination are in the range 0.1-10 s. Calculate from this the duration of the nonsteady-state period and comment on the validity of steady-state assumption made in radical chain polymerization.

Answer:

$[\text{M}^\bullet]_0 = 0$ at $t = 0$. Therefore, $t_0 = 0$.

From Eq. (6.86), $\dfrac{[\text{M}^\bullet]}{[\text{M}^\bullet]_s} = \tanh\left(\dfrac{t}{\tau_s}\right)$

For a steady-state condition, assuming, for example, 99.9999% attainment of the steady-state radical concentration, $t/\tau_s = 7$. Thus the time required for $[\text{M}^\bullet]$ to reach its steady-state value is 70, 7, and 0.7 s, respectively, for τ_s values of 10, 1, and 0.1 s. The time required to reach the steady-state condition is thus negligibly small compared to the time for a typical polymerization study. This shows the validity of the steady-state assumption.

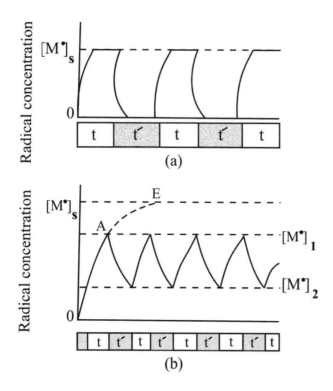

Figure 6.4 Schematic representation of variation of chain radical concentration [M·] over (a) cycles of long illumination period (t) and dark period (t') and (b) cycles of short (intermittent) illumination period (t) and dark period (t'). (After Ghosh, 1990.)

Now suppose the polymerization is being conducted with intermittent illumination, that is, with alternate light and dark periods (Ghosh, 1990). Initially the radical concentration [M·] is zero and it builds up in the period of illumination till it reaches a steady-state value [M·]$_s$, as shown in Fig. 6.4(a). If the light source is switched on and off and held for long but equal time periods of light (t) and darkness (t'), the radical concentration will alternately build up from zero to [M·]$_s$, the steady-state value, and decay from [M·]$_s$ to zero with the repetition of the sequence t and t' of illumination and darkness, respectively [Fig. 6.4(a)]. The intensity of illumination is I_0 during t and zero during t' with short zones of transition in between. The radical concentration during illumination essentially corresponds to I_0, but radicals are present only half of the time in intermittent illumination and hence the average rate of polymerization $(\overline{R}_p)$ observed is actually one-half of the rate $(R_p)_s$, that would be observed for the same intensity on the basis of continuous illumination. Thus for slow blinking, $\overline{R}_p/(R_p)_s = 1/2$.

At high blinking frequency, that is with very short but equal light and dark periods [Fig. 6.4(b)], even if one starts with [M·] = 0 at the beginning of the first light period, the radical build up [curve OAE in Fig. 6.4(b)] is interrupted due to stoppage of light even before the radical concentration reaches [M·]$_s$ (at point E). [M·] reaches a value, say, [M·]$_1$ which is less than [M·]$_s$. The decay of radical concentration from the [M·]$_1$ value begins immediately thereafter as the dark period commences. However, since the dark period (t') is short, radical concentration does not decay to zero, but drops to a minimum value, say, [M·]$_2$ > 0 and starts rising as the illumination is on again. With frequent blinking, the concentration of radicals thus alternates between [M·]$_1$ and [M·]$_2$ as the system passes through the end of a light period to the end of a dark period in successive cycles of illumination and darkness [see Fig. 6.4(b)].

With continuing increase in the frequency of blinking, the difference between $[M^\bullet]_1$ and $[M^\bullet]_2$ will be smaller and smaller till in the limiting case of very fast blinking the radical concentration would reach a constant or plateau value, $[M^\bullet]_c$, below the steady value, $[M^\bullet]_s$. This constant value $[M^\bullet]_c$ would effectively be that corresponding to a continuous illumination of intensity $I_0/2$, because only 50% of the irradiation is received by the system on the average (considering both light and dark periods). It then follows from a combination of Eqs. (6.21), (6.61), and (6.64) that $[M^\bullet]_c$ will be proportional to $(I_0/2)^{1/2}$. Since the rate of polymerization is proportional to the concentration of radicals [cf. Eq. (6.22)], the average rate of polymerization $(\overline{R_p})$ for very fast blinking will thus be proportional to $(I_0/2)^{1/2}$. This may be compared with Eq. (6.66) for steady conditions under *continuous* illumination (i.e., no blinking), indicating that $(R_p)_s \propto (I_0)^{1/2}$.

According to the analysis and consideration detailed above for photoinitiated polymerization, it may now be concluded that R_p or $[M^\bullet]$ can be varied by varying the frequency of blinking. Under otherwise comparable conditions, the average rate $(\overline{R_p})$ at different flashing conditions with the same I_0 is related to the steady-state rate $(R_p)_s$ as

$$\frac{(\overline{R_p})}{(R_p)_s} = \frac{1}{2} \quad \text{(for slow blinking)} \tag{6.87}$$

$$\text{and} \quad \frac{(\overline{R_p})}{(R_p)_s} = \frac{1}{2^{1/2}} \quad \text{(for very fast blinking)} \tag{6.88}$$

For unequal light and dark periods such that $t' = rt$, i.e., the dark period is r times longer than the period of illumination in all successive cycles (note that $r = 1$ for the systems considered above), the expressions for the relative rates may be written, in general, as:

$$\frac{(\overline{R_p})}{(R_p)_s} = \frac{1}{(r+1)} \quad \text{(for slow blinking)} \tag{6.89}$$

$$\frac{(\overline{R_p})}{(R_p)_s} = \frac{1}{(r+1)^{1/2}} \quad \text{(for very fast blinking)} \tag{6.90}$$

Considering, as an example, a case where the ratio (r) of the length (t') of the dark period to the length (t) of a light period is 3, the relative polymerization rate equals 1/4 for slow blinking and 1/2 for very fast blinking. Thus the average rate $(\overline{R_p})$ increases from 1/4 to 1/2 of the steady-state rate $(R_p)_s$ as the cycle time (time for one light period and one dark period) decreases from a much higher value to a much lower value in comparison with the average lifetime of a growing chain (τ_s), or as the frequency of blinking, $1/(t + rt)$, increases from a much lower value to a much higher value in comparison with $1/\tau_s$.

As explained above, in relation to Fig. 6.4(b), the radical concentration, after a number of cycles, will oscillate between two constant values, namely, $[M^\bullet]_1$ at the end of time t of each light period and $[M^\bullet]_2$ at the end of time $t' (= rt)$ of each dark period (Brier et al., 1926). Considering the first light period, if $[M^\bullet] = 0$ at $t = 0$, then from Eq. (6.86), $t_0 = 0$; but for $[M^\bullet] = [M^\bullet]_2 > 0$ at the beginning of light period $(t = 0)$, Eq. (6.86) yields

$$-t_0/\tau_s = \tanh^{-1} ([M^\bullet]_2/[M^\bullet]_s) \tag{6.91}$$

and hence

$$\tanh^{-1}([M^\bullet]/[M^\bullet]_s) - \tanh^{-1}([M^\bullet]_2/[M^\bullet]_s) = t/\tau_s \qquad (6.92)$$

At the end of light period t, $[M^\bullet] = [M^\bullet]_1$ [see Fig. 6.4(b)] and Eq. (6.92) becomes

$$\tanh^{-1}\left(\frac{[M^\bullet]_1}{[M^\bullet]_s}\right) - \tanh^{-1}\left(\frac{[M^\bullet]_2}{[M^\bullet]_s}\right) = \frac{t}{\tau_s} \qquad (6.93)$$

On the other hand, during the dark period radical decay occurs according to

$$d[M^\bullet]/dt' = -2k_t[M^\bullet]^2 \qquad (6.94)$$

which on integration, with $[M^\bullet] = [M^\bullet]_1$ at $t' = 0$, yields

$$\frac{1}{[M^\bullet]} - \frac{1}{[M^\bullet]_1} = 2k_t t' \qquad (6.95)$$

Multiplying Eq. (6.95) through by $[M^\bullet]_s$ and combining with Eq. (6.83),

$$\frac{[M^\bullet]_s}{[M^\bullet]} - \frac{[M^\bullet]_s}{[M^\bullet]_1} = \frac{t'}{\tau_s} \qquad (6.96)$$

Since $t' = rt$ and $[M^\bullet] = [M^\bullet]_2$ at the end of dark period t', one obtains from Eq. (6.96),

$$\frac{[M^\bullet]_s}{[M^\bullet]_2} - \frac{[M^\bullet]_s}{[M^\bullet]_1} = \frac{rt}{\tau_s} \qquad (6.97)$$

Equations (6.93) and (6.97) permit evaluation of the maximum and minimum radical concentration ratios $[M^\bullet]_1/[M^\bullet]_s$ and $[M^\bullet]_2/[M^\bullet]_s$ for given values of $r(= t'/t)$ and t/τ_s. The average radical concentration $\overline{[M^\bullet]}$ over a cycle of light and dark periods is given by (Ghosh, 1990)

$$\overline{[M^\bullet]} = \left(\int_0^t [M^\bullet]dt + \int_0^{rt} [M^\bullet]dt'\right)/(t + rt) \qquad (6.98)$$

The value of $[M^\bullet]$ in the first integral covering the period of illumination is obtained from Eq. (6.92) and that in the second integral covering the dark period is obtained from Eq. (6.96). Evaluation of the integrals yields the following expression (Noyes and Leighton, 1941; Ghosh, 1990):

$$\frac{\overline{[M^\bullet]}}{[M^\bullet]_s} = \frac{1}{(1+r)}\left[1 + \left(\frac{\tau_s}{t}\right)\ln\left(\frac{[M^\bullet]_1/[M^\bullet]_2 + [M^\bullet]_1/[M^\bullet]_s}{1 + [M^\bullet]_1/[M^\bullet]_s}\right)\right] \qquad (6.99)$$

Using this equation, for a given ratio (r) of dark period to light period, values of $\overline{[M^\bullet]}/[M^\bullet]_s$ or $(\overline{R_p})/(R_p)_s$ may be calculated for different assumed values of t/τ_s. A semilog plot of relevant data for $r = 3$ is shown in Fig. 6.5. The plot shows that the average radical concentration falls from one-half of the steady-state value for fast blinking (low t/τ_s) to one-fourth of the same for slow blinking (large t/τ_s), in full conformity with Eqs. (6.89) and (6.90).

A rotating sector method is conveniently used for the determination of τ_s value. The name is derived from the fact that a rotating disc from which a sector-shaped portion is cut out is used.

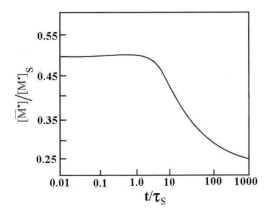

Figure 6.5 Semilog plot of $\overline{[M^\bullet]}/[M^\bullet]_s$ versus t/τ_s for $r = 3$. (Adapted from Matheson et al., 1949.)

It is interposed between the system and the source of light to cause periodic interruption of the light. The cut out portion determines the value r. (For example, a rotating sector with 1:3 opening corresponds to $r = 3$.) The blinking frequency as well as t and t' are determined by the speed of sector rotation. The polymerization rate is conveniently determined by dilatometric technique (p. 321). The rate measured with the rotating sector present is the average polymerization rate $(\overline{R_p})$ while that measured without the sector present is the steady-state rate of polymerization $(R_p)_s$.

A number of rate ratios $(\overline{R_p})/(R_p)_s$ are obtained for a given sector ratio (r) by varying the sector speed and hence t. The rate ratios are then plotted against $\log t$. The theoretical curve for the same r value, such as the one shown in Fig. 6.5 for $r = 3$, is placed above the experimental curve and moved along the abscissa until a best fit is obtained. The required displacement of one curve relative to the other along the abscissa gives the value of $\log \tau_s$. This is obvious from the fact that the theoretical curve represents $(\log t - \log \tau_s)$.

Evaluation of τ_s allows calculation of k_p/k_t from Eq. (6.84) and combining the latter with $k_p/k_t^{1/2}$ obtained from Eq. (6.24) by measuring polymerization rates under steady-state conditions, the absolute rate constants k_p and k_t can be evaluated individually. Table 6.4 lists the k_p and k_t values and the corresponding activation energies for some common monomers.

Similar in principle to the aforesaid rotating sector method is a method known as "spatially intermittent polymerization" (O'Driscoll and Mahabadi, 1976). It uses a tubular glass reactor encased in a metal tube that has regularly spaced slots to provide alternating light and dark segments. A solution of monomer and initiator (or photosensitizer) is passed through the reactor so that it polymerizes effectively under periodic exposure to light.

6.7 Chain Length and Degree of Polymerization

Let us now examine if the chain length or degree of polymerization $(\overline{DP}_n)$ of a polymer product can be calculated from the measured rate of polymerization coupled with a knowledge of the relevant kinetic parameters.

6.7.1 Kinetic Chain Length

The average number of monomer molecules converted into polymer per radical is known as the *kinetic chain length* (ν). This quantity is thus given by the ratio of the rate of propagation (R_p) to

Table 6.4 Some Values of Propagation Rate Constant, k_p, and Termination Rate Constant, k_t, and Activation Energies in Radical Chain Polymerization

Monomer	$k_p \times 10^{-3}$	E_p	$k_t \times 10^{-7}$	E_t
Methyl acrylate (60°C)	1.97	29.7	9.91	22.2
Acrylonitrile (60°C) (Solvent: DMF)	1.96	16.2	78.2	15.5
Methyl methacrylate (60°C)	0.290	26.4	2.1	11.9
Vinyl chloride (50°C) (Solvent: THF)	12.38	16	85.5	17.6
Vinyl acetate (50°C)	1.5-2.6	18	2.5	21.9
Styrene (60°C)	0.187	26	2.9	8.0
Ethylene (83°C) (Solvent: Benzene)	0.47	18.4	105	1.3
Acrylamide (25°C) (Solvent: Water)	18.0	-	1.45	-

k_p and k_t values have the units of L mol^{-1} s^{-1}. The units of E_p are kJ / mol of polymerizing monomer and those of E_t are kJ / mol of propagating radicals.

Source: k_p and k_t values are from Brandrup et al. (1999); E_p and E_t values are from Odian, 1991.

the rate of initiation (R_i), i.e., $v = R_p/R_i$. Using Eq. (6.20) for the steady-state condition, Eq. (6.12) for R_p and Eq. (6.19) for R_t, one can write (Rudin, 1982; Odian, 1991):

$$v = R_p/R_i = R_p/R_t = \frac{k_p[M]}{2k_t[M\bullet]} \tag{6.100}$$

Elimination of [M•] with the help of Eq. (6.12) then gives

$$v = \frac{k_p^2}{2k_t} \frac{[M]^2}{R_p} \tag{6.101}$$

For polymerization initiated through radicals, generated by thermal decomposition of initiator, Eq. (6.24) may be combined with Eq. (6.101) to give an alternative expression for v:

$$v = \left[\frac{k_p}{2(fk_d k_t)^{1/2}}\right] \frac{[M]}{[I]^{1/2}} \tag{6.102}$$

This equation shows that the chain length of polymer is directly proportional to the monomer concentration and inversely proportional to the square root of the initiator concentration. Thus a four-fold increase in initiator concentration would result in halving the size of polymer molecules, though the rate of polymerization [Eq. (6.24)] would be doubled by this change in initiator concentration.

Problem 6.13 In peroxide-initiated polymerization of a monomer, which follows the simple kinetic scheme represented by Eq. (6.24), it is desired to double the initial steady rate of polymerization without changing the molecular weight. How could this be achieved by altering the initial concentrations of the monomer and the initiator (Rudin, 1982)?

Answer:

From Eq. (6.24),

$$[M]_2 [I]_2^{1/2} = 2[M]_1 [I]_1^{1/2}$$

or $\quad [M]_2/[M]_1 = 2[I]_1^{1/2}/[I]_2^{1/2}$ $\hspace{3cm}$ (P6.13.1)

From Eq. (6.102), since there is no change in ν,

$$[M]_2/[I]_2^{1/2} = [M]_1/[I]_1^{1/2}$$

or $\quad [M]_2/[M]_1 = [I]_2^{1/2}/[I]_1^{1/2}$ $\hspace{3cm}$ (P6.13.2)

From Eqs. (P6.13.1) and (P6.13.2), $[I]_2/[I]_1 = 2:1$. Therefore, $[M]_2/[M]_1 = \sqrt{2}:1$

6.7.2 Mode of Chain Termination

The number-average degree of polymerization $\overline{DP}_n$ of the polymer product formed by bimolecular termination is related to the kinetic chain length. Thus if the propagating radicals terminate by coupling or combination [Eq. (6.13)], the resulting dead polymer molecule will be made of two kinetic chain lengths, that is,

$$\overline{DP}_n = 2\nu \hspace{3cm} (6.103)$$

In this case, there will be an initiator fragment at each end of the polymer molecule. On the other hand, if the termination occurs by disproportionation [Eq. (6.14)], the dead polymer molecules formed will be made of one kinetic chain each, that is,

$$\overline{DP}_n = \nu \hspace{3cm} (6.104)$$

In this case, the polymer molecule will have an initiator fragment only at one end. So, if termination takes place by both coupling and disproportionation, the number of initiator fragments per polymer molecule will be between 1 and 2.

Problem 6.14 For a radical chain polymerization with bimolecular termination, the polymer produced has on the average 1.60 initiator fragments per polymer molecule. Calculate the relative extents of termination by disproportionation and by coupling, assuming that no chain transfer reactions occur. Derive first a general relation for this calculation.

Answer:

Let
n = number of propagating chains
ϵ_{tc} = fraction of propagating chains terminating by coupling
$1 - \epsilon_{tc}$ = fraction of propagating chains terminating by disproportionation
b = average number of initiator fragments per polymer molecule

Number of polymer molecules formed $= \dfrac{n\epsilon_{tc}}{2} + n(1 - \epsilon_{tc})$

$$b \;=\; \frac{\text{Total number of initiator fragments}}{\text{Total number of polymer molecules}}$$

$$\;=\; \frac{n}{n\epsilon_{tc}/2 + n(1 - \epsilon_{tc})} \;=\; \frac{2}{2 - \epsilon_{tc}} \tag{P6.14.1}$$

Rearranging, one can write (Ghosh, 1990; Odian, 1991) :

$$\epsilon_{tc} \;=\; (2b - 2)/b \tag{P6.14.2}$$

$$(1 - \epsilon_{tc}) \;=\; (2 - b)/b \tag{P6.14.3}$$

For the given problem,

From Eq. (P6.14.2): fraction of coupling $=$ $(2 \times 1.60 - 2)/1.60 = 0.75$
From Eq. (P6.14.3): fraction of disproportionation $= (2 - 1.6)/1.6 \;=\; 0.25$

If b is the average number of initiator fragments per polymer molecule and ϵ_{tc} is the fraction of propagating chains which undergo termination by coupling, the two being related by Eq. (P6.14.1), the number-average degree of polymerization $\overline{DP}_n$ will then be related to the kinetic chain length v by

$$\overline{DP}_n \;=\; \left(\frac{2}{2 - \epsilon_{tc}}\right) v \tag{6.105}$$

Combining Eqs. (6.101) and (6.105) one obtains

$$\overline{DP}_n \;=\; \frac{1}{(2 - \epsilon_{tc})} \frac{k_p^2}{k_t} \frac{[\text{M}]^2}{R_p} \tag{6.106}$$

Though the termination of propagating chains occurs predominantly by coupling, varying extents of disproportionation are observed depending on the monomer and the reaction conditions employed.

Problem 6.15   Using carbon-14 labeled AIBN as the initiator, a sample of methyl methacrylate was polymerized at $80°C$ to an average degree of polymerization of 1.5×10^3. Measured in a scintillating counter, the AIBN showed an activity (per mol) of 9.1×10^7 counts per minute. If 1.0 g of the poly(methyl methacrylate) had an activity of 337 counts per minute in the same scintillating counter, what was the mode of termination in methyl methacrylate at $80°C$?

Answer:

Mer weight of methyl methacrylate $= 100$; moles of mers in 1.0 g poly(methyl methacrylate) $= 10^{-2}$.

1 mol of AIBN $\equiv 9.1 \times 10^7$ counts min^{-1}.   Therefore, 337 counts/min $\equiv 337/(9.1 \times 10^7)$ or 3.70×10^{-6} mol AIBN $\equiv 7.4 \times 10^{-6}$ mol chain radicals, since 1 mol of AIBN gives rise to 2 moles of chain radicals (assuming 100% initiator efficiency).

Let x fraction of chain radicals terminate by coupling and $(1 - x)$ fraction by disproportionation. Therefore 1 mol of chain radicals give rise to $0.5x + (1 - x)$ or $(1 - 0.5x)$ mol polymer molecules. Hence,

$$\overline{DP}_n \;=\; \frac{10^{-2} \text{ mol}}{(7.4 \times 10^{-6} \text{ mol})(1 - 0.5x)} \;=\; 1.5 \times 10^3$$

Solving, $x = 0.2$. So 20% of the chain radicals terminate by coupling and 80% by disproportionation.

6.7.3 Average Lifetime of Kinetic Chains

The average lifetime (τ) of the kinetic chain is given by the ratio of the steady-state radical concentration to the steady-state rate of radical disappearance:

$$\tau = \frac{[\text{M}^{\bullet}]}{2k_t [\text{M}^{\bullet}]^2} = \frac{1}{2k_t [\text{M}^{\bullet}]} \tag{6.107}$$

Substituting for [M$^{\bullet}$] from Eq. (6.100) yields (Rudin, 1982):

$$\tau = \frac{\nu}{k_p [\text{M}]} \tag{6.108}$$

For termination by disproportionation [cf. Eq. (6.104)],

$$\tau = \overline{DP}_n / k_p [\text{M}] \tag{6.109}$$

and for termination by coupling or combination [Eq. (6.103)],

$$\tau = \overline{DP}_n / 2k_p [\text{M}] \tag{6.110}$$

Problem 6.16   In an experiment, styrene polymerization in bulk at 60°C is initiated by 1×10^{-3} M benzoyl peroxide. The density of liquid styrene is 0.909 g/cm^3 at the reaction temperature. What is the average radical lifetime and what is the steady-state radical concentration? [Data at 60°C: k_t (styrene) = 6.0×10^7 L mol^{-1} s^{-1}; k_d (benzoyl peroxide) = 7.1×10^{-6} s^{-1}; $f = 0.5$.]

Answer:

Combining Eq. (6.108) with Eq. (6.102),

$$
\begin{aligned}
1/\tau &= 2(f k_d k_t [\text{I}])^{1/2} \\
&= 2\left[(0.5)(7.1 \times 10^{-6} \text{ s}^{-1})(6.0 \times 10^7 \text{ L mol}^{-1} \text{ s}^{-1})(1 \times 10^{-3} \text{ mol L}^{-1})\right]^{1/2} \\
&= 0.923 \text{ s}^{-1} \quad \Rightarrow \tau = 1.1 \text{ s}
\end{aligned}
$$

From Eqs. (6.21) and (6.11),

$$
[\text{M}^{\bullet}] = \left(\frac{f k_d [\text{I}]}{k_t}\right)^{1/2} = \left[\frac{(0.5)(7.1 \times 10^{-6} \text{ s}^{-1})(1 \times 10^{-3} \text{ mol L}^{-1})}{(6.0 \times 10^7 \text{ L mol}^{-1} \text{ s}^{-1})}\right]^{1/2}
$$

$$= 7.7 \times 10^{-9} \text{ mol L}^{-1}$$

6.8   Chain Transfer

It is often observed that the measured molecular weight of a polymer product made by free-radical chain polymerization is lower than the molecular weights predicted from Eq. (6.102) for termination by either coupling [Eq. (6.103)] or disproportionation [Eq. (6.104)]. Such an effect, when the mode of termination is known to be disproportionation, can be due to a growing polymer chain terminating prematurely by transfer of its radical center to other species, present in the reaction mixture. These are referred to as *chain transfer reactions* and may be generally written as

$$M_n\cdot \ + \ TA \ \xrightarrow{k_{tr}} \ M_nA \ + \ T\cdot \tag{6.111}$$

where the chain transfer agent TA may be monomer, initiator, solvent, polymer, or any other substance present in the reaction mixture, and A is the atom or species transferred; k_{tr} is the chain transfer rate constant.

The new radical $T\cdot$, which results from chain transfer, can reinitiate polymerization by the reaction

$$T\cdot \ + \ M \ \xrightarrow{k_r} \ TM\cdot \ (\text{or } M_1\cdot) \tag{6.112}$$

where k_r is the rate constant for addition of monomer to $T\cdot$, leading to chain reinitiation by the process,

$$T\cdot \ + \ M \ \xrightarrow{k_r} \ M_1\cdot \ \xrightarrow{k_p} \ M_2\cdot \ \xrightarrow{k_p} \ \cdots \ \xrightarrow{k_p} \ M_n\cdot \tag{6.113}$$

where k_p is the normal propagation rate constant.

While chain transfer to monomer is negligible for most monomers, it may, however, be significant for some monomers, such as vinyl acetate, vinyl chloride, and α-methyl substituted vinyl monomers, e.g., propylene and methyl methacrylate (MMA). For MMA the chain transfer proceeds by the reaction:

$$\text{wwwCH}_2\text{-}\overset{\overset{\displaystyle CH_3}{|}}{\underset{\underset{\displaystyle COOCH_3}{|}}{C}}\cdot \ + \ H_2C=\overset{\overset{\displaystyle CH_3}{|}}{\underset{\underset{\displaystyle COOCH_3}{|}}{C}} \ \xrightarrow{k_{tr,M}} \ \text{wwwCH}_2\text{-}\overset{\overset{\displaystyle CH_3}{|}}{\underset{\underset{\displaystyle COOCH_3}{|}}{CH}} \ + \ H_2C=\overset{\overset{\displaystyle \cdot CH_2}{|}}{\underset{\underset{\displaystyle COOCH_3}{|}}{C}} \tag{6.114}$$

where $k_{tr,M}$ is the rate constant for chain transfer to monomer. In general, the rate of transfer to monomer is given by

$$R_{tr,M} \ = \ k_{tr,M}\,[M]\,[M\cdot] \tag{6.115}$$

where $[M\cdot]$ represents growing chain radicals of all sizes.

Many peroxide initiators have significant chain transfer reactions. Dialkyl and diacyl peroxides undergo chain transfer due to breakage of the O–O bond, e.g.,

$$\text{wwwCH}_2\text{-}\overset{\overset{\displaystyle H}{|}}{\underset{\underset{\displaystyle X}{|}}{C}}\cdot \ + \ R\text{-}\overset{\overset{\displaystyle O}{||}}{C}\text{-}O\text{-}O\text{-}\overset{\overset{\displaystyle O}{||}}{C}\text{-}R \ \xrightarrow{k_{tr,I}} \ \text{wwwCH}_2\text{-}\overset{\overset{\displaystyle H}{|}}{\underset{\underset{\displaystyle X}{|}}{C}}\text{-}O\text{-}\overset{\overset{\displaystyle O}{||}}{C}\text{-}R \ + \ R\text{-}\overset{\overset{\displaystyle O}{||}}{C}\text{-}O\cdot \tag{6.116}$$

where $k_{tr,I}$ is the rate constant for chain transfer to initiator. Usually the strongest transfer agents among the initiators are the hydroperoxides. In general, the rate of chain transfer to initiator is given by

$$R_{tr,I} \ = \ k_{tr,I}\,[I]\,[M\cdot] \tag{6.117}$$

In some polymerizations, the solvent itself may act as the chain transfer agent. For example, during vinyl polymerization in solvent CCl_4 chain transfer takes place by the reaction

$$
\begin{array}{ccc}
\text{www}CH_2-\overset{\overset{\displaystyle H}{|}}{\underset{\underset{\displaystyle X}{|}}{C}}\cdot \; + \; CCl_4 & \xrightarrow{k_{tr,\,S}} & \text{www}CH_2-\overset{\overset{\displaystyle H}{|}}{\underset{\underset{\displaystyle X}{|}}{C}}-Cl \; + \; \cdot CCl_3
\end{array}
\qquad (6.118)
$$

where $k_{tr,S}$ is the rate constant for chain transfer to solvent. However, in industrial free-radical polymerizations the use of solvent is usually avoided for economic reasons, while some ingredients may be deliberately added as chain transfer agents to limit the molecular weight of the polymer. In general, the rate of chain transfer to solvents and/or added chain transfer agents is given by

$$
R_{tr,S} = k_{tr,S}[S][M\cdot] \qquad (6.119)
$$

Since chain transfer terminates a growing chain, its effect is a lower molecular weight than would result in its absence. The effect of any chain transfer on the rate of polymerization will, however, vary depending on the relative rates of the transfer [Eq. (6.111)] and reinitiation [Eq. (6.112)], as compared to the rate of normal propagation reaction [Eq. (6.6)].

Problem 6.17 What would be the effect of chain transfer reactions on the polymerization rate and polymer molecular weight in each of the following cases (Rudin, 1982; Odian, 1991): (a) $k_p \gg k_{tr}$, $k_r \simeq k_p$; (b) $k_p \ll k_{tr}$, $k_r \simeq k_p$; (c) $k_p \simeq k_{tr}$ or $k_p > k_{tr}$ and $k_r < k_p$; (d) $k_p \ll k_{tr}$, $k_r \ll k_p$; (e) $k_p \ll k_{tr}$, $k_r < k_p$?

Answer:

(a) This is the case of normal chain transfer. Since $k_r \simeq k_p$, the new radical ($T\cdot$) formed by the transfer reaction [Eq. (6.111)] will add to a monomer molecule (to initiate a new chain) within about the same time period as that required for addition of a monomer-ended radical to a monomer molecule; the rate of polymerization R_p is therefore not altered. Though the molecular weight is reduced by chain transfer, the stated condition $k_p \gg k_{tr}$ ensures that the product is still macromolecular. This is the case of normal chain transfer.

(b) If $k_p \ll k_{tr}$ and $k_r \simeq k_p$, there will be a large number of transfer reactions compared to propagation reactions, and only a very low-molecular-weight polymer will be formed.

(c) As the reactivity of the radical formed by chain transfer is lower than that of propagating chain radicals ($k_r < k_p$), the rate of polymerization will be reduced (but not to zero); because of chain transfer the polymer molecular weight will also be reduced. Such cases are examples of *retardation*.

(d) In this case, the reduction in polymerization rate is so severe as to make it effectively nil. It is a case of *inhibition*. Retardation and inhibition are described more elaborately in a later section.

(e) The chain transfer is predominant as ($k_{tr} \gg k_p$) and the new radical formed by this reaction reinitiates poorly since ($k_r < k_p$), with the result that there is a large decrease both in polymerization rate and molecular weight. This is the case of *degradative chain transfer*.

Problem 6.18 When a vinyl monomer of molecular weight 132 was polymerized by a free-radical initiator in the presence of dodecyl mercaptan ($C_{12}H_{25}SH$), the rate of polymerization was not depressed by the mercaptan. The purified polymer showed a sulfur content of 0.02% (w/w) and its $\overline{DP}_n$ was determined to be 450. If 80% of the kinetic chains are terminated by coupling and 20% by disproportionation in this polymerization, what should be the extent of terminal unsaturation in the chains?

Answer:

Molar mass of repeat unit = 132 g mol^{-1}; $\overline{M}_n = \overline{DP}_n \times M_o = 450 \times 132 = 5.94 \times 10^4$; polymer chains = $(5.94 \times 10^4)^{-1} = 1.68 \times 10^{-5}$ mol g^{-1}; Sulfur = $(0.02 \times 10^{-2}) / (32$ g g-atom$^{-1}) = 6.25 \times 10^{-6}$ g-atom g^{-1}.

In the presence of mercaptan, the following reactions take place (Flory, 1953):

$$M_n\boldsymbol{\cdot} + RSH \longrightarrow M_n - H + RS\boldsymbol{\cdot}$$

$$RS\boldsymbol{\cdot} \xrightarrow{+M} RSM\boldsymbol{\cdot} \xrightarrow{+M} \text{etc.} \longrightarrow RSM_x\boldsymbol{\cdot}$$

Since the rate of polymerization is not reduced by the mercaptan, the transfer radical RS $\boldsymbol{\cdot}$ evidently reacts readily with the monomer to start a new kinetic chain. Accordingly, it may be assumed that for one mol of mercaptan consumed one mol of dead polymer chains are formed and one mol of new kinetic chains are initiated. Therefore 6.25×10^{-6} moles of polymer chains are produced by transfer to mercaptan. Remaining $(1.68 \times 10^{-5} - 6.25 \times 10^{-6})$ or 1.055×10^{-5} mol of polymer chains result from termination by disproportionation and coupling.

Out of every 100 kinetic chains, 80 terminate by coupling to produce 40 polymer molecules and 20 terminate by disproportionation to produce 20 polymer molecules of which 10 contain terminal unsaturation [see Eq. (6.14)]. Therefore, the amount of terminal unsaturation = $(1.055 \times 10^{-5}$ mol g$^{-1})\,(10/60)$ = 1.76×10^{-6} mol g^{-1}, which means that $(1.76 \times 10^{-6})\,(100) / (1.68 \times 10^{-5})$ or 10.5% of polymer chains have a terminal unsaturation.

6.8.1 Degree of Polymerization

Equations (6.105) and (6.106) for $\overline{DP}_n$ apply to free-radical polymerization following ideal kinetics in which termination of chain growth occurs only by mutual reaction (coupling and disproportionation) of chain radicals. Combining Eqs. (6.100) and (6.105) one may write

$$\overline{DP}_n \;=\; \left(\frac{2}{2 - \epsilon_{tc}}\right)\frac{R_p}{R_t} \;=\; \frac{R_p}{zR_t} \tag{6.120}$$

where $z = (1 - \epsilon_{tc}/2)$. In Eq. (6.120), R_t is the rate of termination of chain radicals and zR_t is the rate of production of dead polymer molecules by bimolecular termination mechanism. (If combination is the sole mechanism, $\epsilon_{tc} = 1$ and $z = 1/2$, while for termination by disproportionation $\epsilon_{tc} = 0$ and $z = 1$.) Since, as we have seen above, chain radicals can also be terminated by chain transfer reactions, Eq. (6.120) will now be modified to include transfer reactions. This can be easily done by redefining $\overline{DP}_n$ as the ratio of rate of polymerization to the rate of formation of dead polymer molecules by all reactions, namely, by the normal bimolecular termination and various transfer reactions. Thus Eq. (6.120) takes the form (using $R_i = R_t$ for steady state) (Ghosh, 1990):

$$\overline{DP}_n \;=\; \frac{R_p}{zR_t + R_{tr,M} + R_{tr,I} + R_{tr,S}}$$

$$=\; \frac{R_p}{zR_i + k_{tr,M}[M\boldsymbol{\cdot}][M] + k_{tr,I}[M\boldsymbol{\cdot}][I] + k_{tr,S}[M\boldsymbol{\cdot}][S]} \tag{6.121}$$

Noting from Eq. (6.12) that $R_p = k_p[M][M^\bullet]$, Eq. (6.121) is conveniently rearranged to the form

$$\frac{1}{\overline{DP}_n} = \frac{zR_i}{R_p} + C_M + C_I \frac{[I]}{[M]} + C_S \frac{[S]}{[M]} \tag{6.122}$$

where C_M, C_I, and C_S are the *chain transfer constants* for monomer, initiator, and solvent/chain transfer agent, respectively, defined as the ratio of the respective chain transfer rate constant to the propagation rate constant, that is,

$$C_M = \frac{k_{tr,M}}{k_p}, \quad C_I = \frac{k_{tr,I}}{k_p}, \quad C_S = \frac{k_{tr,S}}{k_p} \tag{6.123}$$

The term C_S refers to solvent used in polymerization or any component specially added as chain transfer agent. Equation (6.122), often referred to as thc *Mayo equation*, shows the quantitative effect of various transfer reactions on the number average degree of polymerization. Note that the chain transfer constants, being ratios of rate constants with same dimensions, are dimensionless quantities dependent on the types of both the monomer and the substance causing chain transfer as well as on the temperature of reaction. Various methods can be employed, based on the Mayo equation, to determine the values of the chain transfer constants. (See Problems 6.20 and 6.21 for some of these methods.)

Problem 6.19   Vinyl acetate has a relatively high monomer chain transfer constant (2×10^{-4} at $60°$C). Calculate an upper limit for the molecular weight of poly(vinyl acetate) made by radical polymerization at $60°$C.

Answer:

The upper limit of molecular weight corresponds to the lower limit of ($1/\overline{DP}_n$) in Eq. (6.122). The lower limit of the first, third, and fourth terms on the right side of Eq. (6.122) is 0. Hence,

$$\frac{1}{\overline{DP}_n}\Big|_{min} = C_M = 2\times10^{-4} \implies \overline{DP}_n = 5,000$$

$$\overline{M}_n = (5,000)(86\text{ g mol}^{-1}) = 4,30,000\text{ g mol}^{-1}$$

Problem 6.20   Vinyl acetate is to be polymerized in benzene solution at $60°$C using 2,2′-azobisisobutyronitrile (AIBN) as the initiator and carbon tetrachloride as a chain transfer agent. The initial monomer concentration is 200 g/L and solution density is 0.83 g/cm^3. What concentrations of initiator and chain transfer agent should be used to obtain poly(vinyl acetate) with an initial molecular weight (assuming termination by coupling) of 15,000 and 50% conversion of monomer in 2 h? Neglect chain transfer to initiator in calculations. [Data at $60°$C: $k_d = 8.45\times10^{-6}$ s^{-1}; $k_p = 2.34\times10^3$ L/mol-s; $k_t = 2.9\times10^7$ L/mol-s; $f = 1.0$; $C_M = 2.3\times10^{-4}$; C_S(benzene) $= 1.2\times10^{-4}$; C_S(CCl$_4$) $= 1.07$.]

Answer:

Molecular weights: vinyl acetate 86;   AIBN 164;   benzene 78;   CCl$_4$ 154.

$[M]_0 = (200\text{ g L}^{-1})/(86\text{ g mol}^{-1}) = 2.32\text{ mol L}^{-1}$

Neglecting initiator mass, $[S]_{benzene} = (830\text{ g} - 200\text{ g})/(78\text{ g mol}^{-1}) = 8.08\text{ mol L}^{-1}$.

Assuming that chain transfer has no effect on the polymerization rate, Eq. (6.31) can be used. Using relevant data and $p = 0.5$, this equation gives

$$[I]_0 = 6.16\times10^{-3}\text{ mol L}^{-1} = 6.16\times10^{-3}\times164 \text{ or } 1.0\text{ g L}^{-1}$$

From Eqs. (6.102) and (6.103),

$$(\overline{DP}_n)_0 = \frac{k_p [M]_0}{(f k_d k_t [I]_0)^{1/2}} \tag{P6.20.1}$$

Substituting appropriate values in Eq. (P6.20.1), $(\overline{DP}_n)_0 = 4340$.

Desired $\overline{DP}_n = 15,000/86$ or 174. From Eq. (6.122),

$$\frac{1}{\overline{DP}_n} = \frac{1}{(\overline{DP}_n)_0} + C_M + C_S(\text{benzene})\frac{[S]_{\text{benzene}}}{[M]_0} + C_S(\text{CCl}_4)\frac{[S]_{\text{CCl}_4}}{[M]_0}$$

Substituting appropriate values and solving, $[S]_{\text{CCl}_4} = 0.0106$ mol $L^{-1} = 0.0106 \times 154$ or 1.63 g L^{-1}.

Problem 6.21 The following data of rate of polymerization and degree of polymerization at low conversion were obtained in bulk polymerization of monomer M (initial concentration 8.3 mol/L) using different concentrations of thermal initiator I at 60°C:

$[I] \times 10^2$, mol/L	$R_p \times 10^3$, mol/L-s	$\overline{DP}_n$
0.018	0.005	8267
0.072	0.010	5495
0.280	0.020	3296
1.74	0.050	1300
4.48	0.086	714
7.80	0.115	495
13.20	0.15	352

Calculate (a) C_M, (b) $k_p/k_t^{1/2}$, and (c) C_I at 60°C. Assume termination by coupling.

Answer:

Consider polymerization in the absence of a solvent or added chain transfer agent, so that $[S] = 0$. For steady-state polymerization, Eq. (6.23) can be used to express R_i in terms of R_p as

$$R_i = \frac{2 R_p^2 k_t}{k_p^2 [M]^2} \tag{P6.21.1}$$

and Eq. (6.24) can be rearranged to express $[I]$ as

$$[I] = \frac{R_p^2 k_t}{k_p^2 [M]^2 f k_d} \tag{P6.21.2}$$

Substitution of Eqs. (P6.21.1) and (P6.21.2), and $[S] = 0$ into Eq. (6.122) gives (Rudin, 1982; Odian, 1991):

$$\frac{1}{\overline{DP}_n} = \frac{2 z k_t R_p}{k_p^2 [M]^2} + C_M + C_I \frac{k_t R_p^2}{k_p^2 f k_d [M]^3} \tag{P6.21.3}$$

Since Eq. (P6.21.3) is quadratic in R_p, the plot of $1/\overline{DP}_n$ vs. R_p will be curved, the extent of which will depend on C_I, that is, the nature of the initiator. In the present case, the plot of $1/\overline{DP}_n$ vs. R_p (Fig. 6.6) is seen to be linear at low R_p (i.e., the first 3 points). Therefore the values of the intercept and slope are determined from a least-squares calculation using the first 3 data points, yielding (a) $C_M = $ Intercept $= 6.03 \times 10^{-5}$ and (b)

$(2zk_t)/(k_p^2[M]^2)$ = slope = 12.16. Since termination occurs only by coupling, $z = 1/2$. Also $[M] = 8.3$ mol L^{-1}. So, $k_p/k_t^{1/2} = 3.46 \times 10^{-2}$ $L^{1/2}$ mol$^{-1/2}$ s$^{-1/2}$.

(c) Equation (P6.21.3) can be rearranged and divided through by R_p to yield (Rudin, 1982; Odian, 1991):

$$\left(\frac{1}{\overline{DP}_n} - C_M\right)\frac{1}{R_p} = \frac{2zk_t}{k_p^2[M]^2} + C_I\frac{k_tR_p}{k_p^2 f k_d[M]^3} \tag{P6.21.4}$$

It is easy to show from Eq. (6.24) that

$$\frac{k_t}{k_p^2 f k_d[M]^3} = \frac{[I]}{[M]R_p^2} \tag{P6.21.5}$$

Since in the present case, the value of C_M is negligibly small, assuming $C_M \simeq 0$, one can rearrange Eq. (P6.21.4) and combine with Eq. (P6.21.5) to yield (Rudin, 1982; Odian, 1991):

$$\left(\frac{1}{\overline{DP}_n} - \frac{2zk_tR_p}{k_p^2[M]^2}\right) = C_I\frac{[I]}{[M]} \tag{P6.21.6}$$

Turning to the present case, since the termination is only by coupling, $z = 1/2$ and Fig. 6.7 shows a plot of $(1/\overline{DP}_n - k_tR_p/k_p^2[M]^2)$ vs. $[I]/[M]$. From the slope of the linear plot $C_I = 0.066$.

Values of C_M for some common monomers are listed in Table 6.5. These being representative values, C_M is seen to be small (typically, in the range of 10^{-6} to 10^{-4}). Since chain transfer to monomer cannot be avoided, the highest molecular weight that can be obtained for a polymer, in the absence of all other transfer reactions, will be limited by the value of C_M (see Problem 6.19). Table 6.6 lists some values of initiator transfer constants (C_I).

Table 6.5 Some Values of Monomer Chain Transfer Constants

Monomer	$C_M \times 10^4$	Monomer	$C_M \times 10^4$
Acrylamide	0.651	Methyl methacrylate	0.07–0.18
Acrylonitrile	0.26–0.3	Styrene	0.07–1.37
Ethylene	0.4–4.2	Vinyl acetate	1.75–2.8
Methyl acrylate	0.036–0.325	Vinyl chloride	10.8–12.8

All C_M values are for 60°C.
Source: Brandrup et al., 1999; Ghosh, 1990.

Problem 6.22 Weighed amounts of styrene (M) and *n*-butyl mercaptan (S) in sealed glass ampoules were heated at 60°C for different periods of time. The polymers were then precipitated in methanol, dried in oven, and degrees of polymerization evaluated by intrinsic viscosity measurements. From the data given below calculate the chain transfer constant (C_S) for the styrene/dodecyl mercaptan system at 60°C.

$([S]/[M]) \times 10^5$	Time (h)	% Conversion of M	$\overline{DP}_n \times 10^{-3}$	$10^4/\overline{DP}_n$
1.41	12	1.12	2.62	3.82
2.70	13	1.24	1.56	6.4
4.54	14	1.45	0.97	10.3
6.43	16	1.58	0.76	13.1

Answer:

If a sufficiently strong transfer agent is used so that the fourth term of the right side of Eq. (6.122) makes the biggest contribution to $\overline{DP}_n$, the polymerization may be adjusted so as to considerably simplify Eq. (6.122).

Thus, for a series of measurements one may keep $[I]^{1/2}/[M]$ constant, so the first term on the right side of Eq. (6.122) becomes constant and carry out an uncatalyzed polymerization or use an initiator for which C_I is negligibly small (e.g., AIBN) so that the third term becomes nonexistent or negligible. Under these conditions, Eq. (6.122) takes the form (Rudin, 1982; Odian, 1991):

$$\frac{1}{\overline{DP}_n} = \left(\frac{1}{\overline{DP}_n}\right)_0 + C_S \frac{[S]}{[M]} \tag{P6.22.1}$$

where $(1/\overline{DP}_n)_0$ is the value of $(1/\overline{DP}_n)$ in the absence of solvent or the chain transfer agent, and it represents the sum of the first three terms on the right side of Eq. (6.122). The slope of the linear plot of $(1/\overline{DP}_n)$ vs. $[S]/[M]$ then gives the measure of C_S. Figure 6.8 shows such a plot for the data given above, yielding $C_S =$ slope = 21.6.

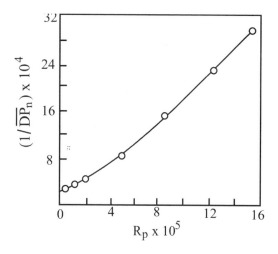

Figure 6.6 Plot of Eq. (P6.21.3) for the determination of C_M with the data of Problem 6.21.

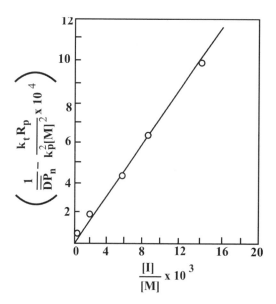

Figure 6.7 Plot of Eq. (P6.21.6) for the determination of C_I with the data of Problem 6.21.

Problem 6.23   In a free-radical polymerization of styrene in solution at 60°C, carbon tetrabromide was used as a chain transfer agent. The initial concentrations of styrene and CBr$_4$ were 1 mol/L and 0.01 mol/L, respectively. In 1 h, these concentrations dropped to 0.85 mol/L and 0.007 mol/L, respectively. What is the chain transfer constant C_S for styrene/CBr$_4$? (Neglect chain transfer to monomer, initiator, and solvent.)

Answer:

Dividing the rate expression for transfer [Eq. (6.119)] by that for propagation [Eq. (6.12)] yields

$$\frac{-d[S]/dt}{-d[M]/dt} = \frac{k_{tr,S}[S]}{k_p[M]} \quad \text{or} \quad \frac{d[S]}{d[M]} = C_S\frac{[S]}{[M]} \tag{P6.23.1}$$

Rearranging to $d[S]/[S] = C_S(d[M]/[M])$ and integrating between t_0 and t gives

$$\ln\left(\frac{[S]}{[S]_0}\right) = C_S\ln\left(\frac{[M]}{[M]_0}\right) \tag{P6.23.2}$$

Substituting appropriate values in Eq. (P6.23.2),

$$C_S = \frac{\ln(0.007/0.01)}{\ln(0.85/1.0)} = 2.19$$

The transfer constants for a number of solvents/additives for polymerization of styrene, methyl methacrylate, and vinyl acetate are listed in Table 6.7. It is obvious from the data that the values of chain transfer constants depend on the chemical structure of both the chain transfer agent and the monomer. Thus the same substance may have different values of C_S for different monomers. Transfer agents with large C_S are especially useful since they can be used in small concentrations (see Problem 6.20).

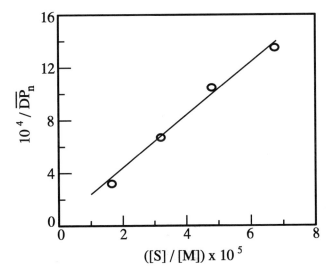

Figure 6.8 Plot of Eq. (P6.22.1) for the determination of C_S with the data of Problem 6.22.

Table 6.6 Some Values of Initiator Chain Transfer Constants

Initiator	$C_I \times 10^4$ for polymerization of		
	Styrene	Methyl metha- crylate	Acrylamide
2,2′-Azobisisobutyronitrile	0	0	–
Dibenzoyl peroxide	480–1010	200	–
t-Butyl hydroperoxide	350	12,700	–
Di-t-Butyl peroxide	2.3–13	–	–
Cumyl hydroperoxide	630	3,300	–
Lauroyl peroxide (70°C)	240	–	–
Persulfate (50°C)	–	–	258

All C_I values are for 60°C except where otherwise noted.
Source: Brandrup et al., 1999; Ghosh, 1990.

Table 6.7 Some Values of Transfer Constants for Solvents and Chain Transfer Agents

Transfer agent (Solvent/Additive)	$C_S \times 10^4$ for polymerization of		
	Styrene	Methyl methacrylate	Vinyl acetate
Benzene	0.018-1.92	0.04-0.83	1.07-2.96
Toluene	0.105-2.05	0.17-0.45	17.8-35
Ethylbenzene	0.67-0.83	0.766	55.2
Isopropyl benzene	0.8-3.88	1.9-2.56	89.9-100
t-Butylbenzene	0.04-0.06	0.26 (80°C)	3.61
n-Butyl chloride	0.04	1.20 (80°C)	5.6-8.4
n-Butyl bromide	0.06	–	8.0-10.0
n-Butyl iodide	1.85	–	800
Chloroform	0.41-3.4	0.454-1.77	125-170
Carbon tetrachloride	69-148	0.925-20	6,700-10,000
Carbon tetrabromide	17,800-22,000	2,700	7390000
n-Butyl mercaptan	220,000	6,700	480,000

All values are for 60°C unless otherwise noted.
Source: Brandrup et al., 1999; Ghosh, 1990.

6.8.2   Chain Transfer to Polymer

In discussions above, we have ignored the possibility that chain transfer may take place to polymer molecules present in the system. At low conversions, the polymer concentration is low and so the extent of transfer to polymer is negligible. However, transfer to polymer cannot be neglected at high conversions. Chain transfer to polymer is also significant with very reactive propagating radicals like those in the polymerizations of vinyl chloride, vinyl acetate, ethylene, and other monomers in which there is no significant resonance stabilization.

Chain transfer to polymer produces a radical on the polymer chain and polymerization of monomer from this site results in the formation of a branch, for example,

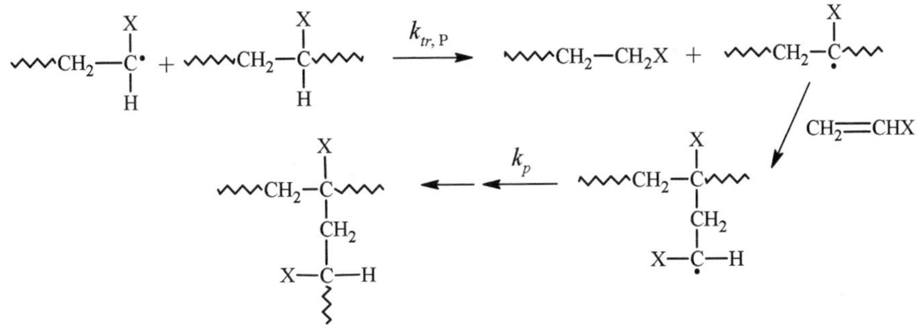

$$(6.124)$$

While long branches are formed by the "normal" chain transfer to polymer, as shown above [Eq. (6.124)], reactive radicals like those of polyethylene can also undergo self-branching by a "backbiting" intramolecular transfer reaction (see Fig. 6.9) in which the chain-end radical abstracts a hydrogen atom from a methylene unit of the same chain resulting in the formation of short branches (as many as 30-50 branches per 1000 carbon atoms in the main chain) that outnumber the long branches by a factor of 20-50.

Chain transfer to polymer does not change either the number of monomer molecules which have been polymerized or the number of polymer molecules over which they are distributed. Chain transfer to polymer thus has no effect on $\overline{DP}_n$ and it is not included in Eq. (6.121). It, however, broadens the molecular weight distribution because the polymers which are already large are more likely to suffer transfer reactions and become yet bigger due to branching.

6.8.3 *Allylic Transfer*

Chain transfer to monomer is particularly favored with allylic monomers (e.g., allylic acetate) which have the structure $CH_2=CH-CH_2X$ with a C-H bond alpha to the double bond described as an allylic C-H. While the propagating radical is very reactive, the radical formed by transfer to allylic C–H in the monomer is particularly stable and unreactive due to resonance stabilization [Eq. (6.125)] and does not initiate new chains. Allylic transfer is thus variously known as *degradative chain transfer* to monomer, *autoinhibition*, or *allylic termination*.

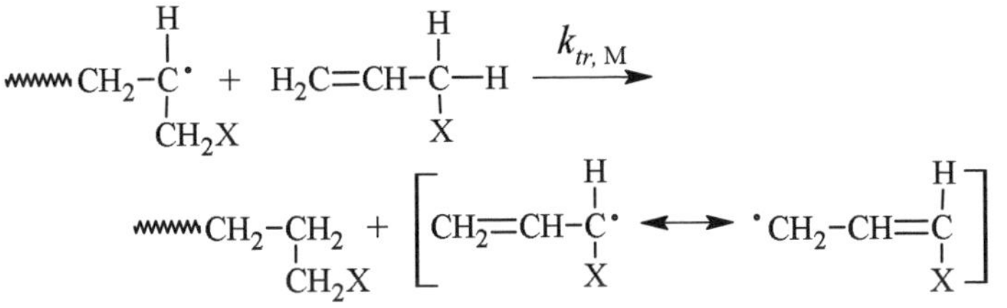

$$(6.125)$$

Problem 6.24 Suggest a kinetic scheme to account for the following characteristics of free-radical polymerization of allylic monomers:

(a) R_p is very low with first order dependence in initiator concentration.

(b) $\overline{DP}_n$ is very low and independent of monomer and initiator concentrations.

(c) Deuterated allylic monomer possessing allylic C-D bond has significantly higher R_p than the normal allylic monomer.

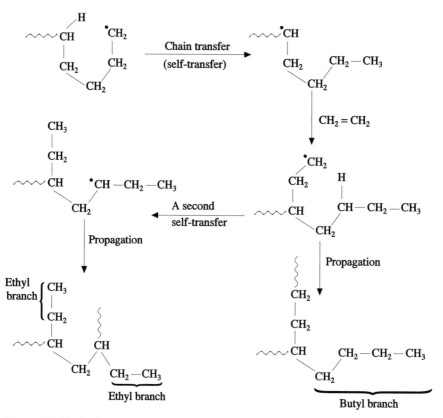

Figure 6.9 Mechanism of ethyl and butyl branching in free-radical polymerization of ethylene. (After Ghosh, 1990.)

Answer:

(a) Since the allylic radical formed [Eq. (6.125)] has high resonance stability, it does not initiate new chains. Chain termination occurs predominantly by chain transfer. At steady state,

$$\frac{d[\mathrm{M}\cdot]}{dt} = R_i - k_{tr,\mathrm{M}}[\mathrm{M}\cdot][\mathrm{M}] = 0 \tag{P6.24.1}$$

Combination with $R_p = k_p[\mathrm{M}\cdot][\mathrm{M}]$ and $R_i = 2fk_d[\mathrm{I}]$ leads to

$$R_p = \frac{k_p R_i}{k_{tr,\mathrm{M}}} = \frac{2fk_p k_d[\mathrm{I}]}{k_{tr,\mathrm{M}}} \tag{P6.24.2}$$

For degradative chain transfer, $k_p \ll k_{tr,\mathrm{M}}$. Therefore R_p is very low. Equation (P6.24.2) shows that R_p is first order in [I].

(b) $\overline{DP}_n = \nu = \dfrac{R_p}{R_i} = \dfrac{R_p}{R_t} = \dfrac{k_p[\mathrm{M}\cdot][\mathrm{M}]}{k_{tr,\mathrm{M}}[\mathrm{M}\cdot][\mathrm{M}]} = \dfrac{k_p}{k_{tr,\mathrm{M}}}.$

Thus $\overline{DP}_n$ is constant and very low.

(c) The C-D bond being stronger than the C-H bond due to its lower zero point energy, the degradative chain transfer is less in the deuterated monomer and R_p is higher.

6.9 Deviations from Ideal Kinetics

A given polymerization conforming strictly to the reaction scheme represented by Eqs. (6.3) to (6.7) is commonly considered as an ideal polymerization. Any system deviating from this pattern of reaction is to be considered as a case of *nonideal polymerization*. The ideal behavior requires constancy of the term $(R_p^2/[I][M]^2)$, as according to Eq. (6.24) it is expressed as

$$\frac{R_p^2}{[I][M]^2} = \left(\frac{k_p^2}{k_t}\right) f k_d \tag{6.126}$$

Practical free-radical polymerizations often deviate from Eq. (6.126) because the assumptions made in the ideal kinetic scheme are not fully satisfied by the actual reaction conditions or because some of these assumptions are not valid. For example, according to the ideal kinetic scheme that leads to Eq. (6.126), the initiation rate (R_i) and initiator efficiency (f) are independent of monomer concentration in the reaction mixture and primary radicals (i.e., radicals derived directly from the initiator) do not terminate kinetic chains, though in reality R_i may depend on [M], as in the case of cage effect (see Problem 6.7) and, at high initiation rates, some of the primary radicals may terminate kinetic chains (see Problem 6.25). Moreover, whereas in the ideal kinetic scheme, both k_p and k_t are assumed to be independent of the size of the growing chain radical, in reality k_t may be size-dependent and diffusion-controlled, as discussed later.

6.9.1 Primary Radical Termination

Primary radicals under ideal conditions would contribute to chain initiation only. But in certain systems and under special conditions in certain others, they may also contribute to chain termination (described as *primary radical termination* or *primary termination*), partly or exclusively, giving rise to significant deviations (Ghosh, 1990) from the ideal kinetics (see Problem 6.25).

Problem 6.25 Show that in the case of radical polymerization where both bimolecular and primary radical terminations occur simultaneously, the parameter $(R_p^2/[I][M]^2)$ is not constant but is dependent on $(R_p/[M]^2)$. Hence suggest a method of analysis of the primary radical termination effect from the experimental R_p data.

Answer:

The reaction scheme (Ghosh, 1990) may be written as:

(a) $I \overset{k_d}{\longrightarrow} 2R\cdot$

(b) $R\cdot + M \overset{k_i}{\longrightarrow} M_1\cdot$

(c) $M_n\cdot + M \overset{k_p}{\longrightarrow} M_{n+1}\cdot$ $(n \geq 1)$

(d) $M_n\cdot + M_m\cdot \overset{k_t}{\longrightarrow} P$

(e) $R\cdot + M_n\cdot \overset{k_{tp}}{\longrightarrow} P$

where P denotes a dead polymer molecule.

If bimolecular and primary radical terminations, (d) and (e), occur simultaneously, then considering steady concentrations of $[M\cdot]$ and $[R\cdot]$ separately, namely,

$$\frac{d[M\cdot]}{dt} = k_i[R\cdot][M] - 2k_t[M\cdot]^2 - k_{tp}[M\cdot][R\cdot] = 0 \qquad \text{(P6.25.1)}$$

$$\frac{d[R\cdot]}{dt} = 2fk_d[I] - k_i[R\cdot][M] - k_{tp}[R\cdot][M\cdot] = 0 \qquad \text{(P6.25.2)}$$

one obtains

$$\frac{k_t[M\cdot]^2}{fk_d[I]} = \frac{k_i[R\cdot][M] - k_{tp}[R\cdot][M\cdot]}{k_i[R\cdot][M] + k_{tp}[R\cdot][M\cdot]} \qquad \text{(P6.25.3)}$$

Substituting $(R_p/k_p[M])$ for $[M\cdot]$ and rearranging gives

$$\frac{R_p^2}{[I][M]^2} = \left(\frac{k_p^2}{k_t}\cdot fk_d\right)\left(\frac{1 - \dfrac{k_{tp}}{k_ik_p}\dfrac{R_p}{[M]^2}}{1 + \dfrac{k_{tp}}{k_ik_p}\dfrac{R_p}{[M]^2}}\right) \qquad \text{(P6.25.4)}$$

Unlike in ideal kinetics [Eq. (6.126)] where $R_p^2/[I][M]^2$ is constant, Eq. (P6.25.4) shows that $R_p^2/[I][M]^2$ is no longer constant but depends on R_p or $R_p/[M]^2$.

Making use of the fact that for vinyl polymerization, $(k_{tp}/k_ik_p)(R_p/[M]^2) \ll 1$, Eq. (P6.25.4) can be simplified to the form :

$$\frac{R_p^2}{[I][M]^2} = \left(\frac{k_p^2}{k_t}fk_d\right)\exp\left(-\frac{2k_{tp}}{k_ik_p}\frac{R_p}{[M]^2}\right) \qquad \left[\text{Note}: \frac{1-a}{1+a} \simeq e^{-2a}\right]$$

Taking logarithm,

$$\ln\frac{R_p^2}{[I][M]^2} = \ln\left(\frac{k_p^2}{k_t}fk_d\right) - \left(\frac{2k_{tp}}{k_ik_p}\right)\frac{R_p}{[M]^2} \qquad \text{(P6.25.5)}$$

A negative slope of the linear plot of the left hand side of Eq. (P6.25.5) against $(R_p/[M]^2)$ is thus indicative of primary radical termination. The magnitude of the slope also gives a measure of the parameter (k_{tp}/k_ik_p).

6.9.2 Initiator-Monomer Complex Formation

There are many reports on polymerization of different monomers showing different degrees of kinetic complexity where the observed effects are well explained and understood on the basis of equilibrium complex formation between the initiator used and the monomer (Walling and Heaton, 1965; Ghosh and Billmeyer, 1969; Ghosh and Banerjee, 1974).

Problem 6.26 Formation of complex (C) between initiator (I) and monomer (M) leads to nonideal kinetic behavior in many cases. A simplified scheme of chain initiation based on the concept of equilibrium complex formation between initiator and monomer may be given as

(a) $I + M \overset{K}{\rightleftharpoons} C$ (initiator-monomer complex)

(b) $C \overset{k_d}{\longrightarrow} 2R\cdot$

(c) $R\cdot + M \overset{k_i}{\longrightarrow} RM\cdot$

where K is the equilibrium constant of the initiator-monomer complexation reaction (a) and the complex C is the true source of initiating radicals. Assuming that the usual bimolecular termination is the only prevailing mode of chain termination, derive an expression for the rate of polymerization R_p and show how K can be evaluated from the experimental R_p vs. [M] data.

Answer:

For reaction (a), since usually $[M] \gg [C]$,

$$K = \frac{[C]}{[I][M]} = \frac{[C]}{([I]_0 - [C])[M]} \tag{P6.26.1}$$

where $[I]_0$ is the initial concentration of the initiator.

$$\text{Rearranging Eq. (P6.26.1),} \quad [C] = \frac{K[M][I]_0}{1 + K[M]} \tag{P6.26.2}$$

The rate of initiation is given by $R_i = 2k_d[C]$. Inserting Eq. (P6.26.2) in this equation and then combining with Eq. (6.23) gives

$$R_p = k_p \left(\frac{k_d}{k_t}\right)^{1/2} \left(\frac{K}{1 + K[M]}\right)^{1/2} [M]^{3/2} [I]_0^{1/2} \tag{P6.26.3}$$

Equation (P6.26.3) describes a change in reaction order with respect to monomer from 1.5 to 1 with increasing [M]. Equation (P6.26.3) may be further transformed to :

$$\frac{[M]^3}{R_p^2} = \frac{k_t}{k_p^2 k_d K[I]_0} + \frac{k_t[M]}{k_p^2 k_d[I]_0} \tag{P6.26.4}$$

According to Eq. (P6.26.4), a plot of $[M]^3/R_p^2$ vs. [M] should give a straight line such that the quotient of the slope and the intercept is equal to K.

6.9.3 *Degradative Initiator Transfer*

Initiators under ideal conditions would contribute only to chain initiation by dissociation [Eq. (6.3)] into primary radicals (R$\cdot$). But in certain systems they also contribute to chain termination

(Ghosh, 1990), partly or exclusively, giving rise to significant deviations from the ideal kinetics (see Problem 6.27). The degradative chain transfer to initiator (I) may be written as

$$M_n^\bullet + I \xrightarrow{k_{tr,I}} M_n + I^\bullet \tag{6.127}$$

where $I^\bullet$ is a radical product of I, which is inactive or less active than the primary radical $R^\bullet$ derived from I.

Problem 6.27   Consider a case of degradative chain transfer to initiator (I) as given by Eq. (6.127). Derive suitable expressions for the rate of polymerization for (a) an extreme case where chain termination occurs exclusively by the degradative initiator transfer and (b) a more general case where chain termination takes place by simultaneous occurrence of the degradative initiator transfer and the usual bimolecular mechanism. In both cases, assume that the radical $I^\bullet$ formed by the chain transfer to initiator I is too inactive to reinitiate polymerization.

Answer:

(a) For termination of the kinetic chains exclusively by the degradative initiator transfer process, the rate of termination is

$$R_t = k_{tr,I}[M^\bullet][I]$$

Using the steady-state concept for the chain radicals,

$$\frac{d[M^\bullet]}{dt} = R_i - R_t = 2fk_d[I] - k_{tr,I}[M^\bullet][I] = 0$$

$$[M^\bullet] = \frac{2fk_d}{k_{tr,I}} \tag{P6.27.1}$$

The rate of polymerization is then given by

$$R_p = k_p[M][M^\bullet] = \left(\frac{k_p}{k_{tr,I}}\right)(2fk_d)[M] \tag{P6.27.2}$$

which shows that R_p is independent of initiator concentration but it still retains first order dependence on monomer concentration.

(b) Under steady state, the following relationship (Ghosh, 1990) holds good when chain termination occurs both by the degradative process and the usual bimolecular mechanism:

$$\frac{d[M^\bullet]}{dt} = 2fk_d[I] - k_{tr,I}[M^\bullet][I] - 2k_t[M^\bullet]^2 = 0$$

On elimination of $[M^\bullet]$ by substituting $[M^\bullet] = R_p/k_p[M]$ and on further rearrangement one obtains

$$\frac{R_p^2}{[I][M]^2} = \left(\frac{k_p^2}{k_t}\right)(fk_d) - \left(\frac{k_p^2}{2k_t}\right)\left(\frac{k_{tr,I}}{k_p}\right)\frac{R_p}{[M]} \tag{P6.27.3}$$

Equation (P6.27.3) allows a plot of $R_p^2/[I][M]^2$ vs. $R_p/[M]$. A negative slope of the linear plot is indicative of termination by the degradative initiator transfer reaction in addition to bimolecular termination. From the slope of the plot, the parameter $(k_{tr,I}/k_p)$ can be obtained and from the intercept fk_d can be calculated, if the kinetic parameter (k_p^2/k_t) for the monomer is known. When R_p becomes independent of [I] for a given [M], Eq. (P6.27.3) is unsuitable for the analysis of the degradative effect and Eq. (P6.27.2) may then be employed.

6.9.4 Autoacceleration

Deviations from ideal kinetics due to size-dependence and diffusion control of termination produce relatively weak effects at low conversions. However, at high conversions these effects are very significant in most radical polymerizations. Thus, instead of the reaction rate falling with time, as would be expected from Eq. (6.24) since the monomer and initiator concentrations decrease with conversion, an exact opposite behavior is observed in many polymerizations where the rate of polymerization increases with time. A typical example of this phenomenon is shown in Fig. 6.10 for the polymerization of methyl methacrylate in benzene solution at 50°C (Schulz and Haborth, 1948). An acceleration is observed at relatively high monomer concentrations and the curve for the pure monomer shows a drastic autoacceleration in the polymerization rate. This type of behavior observed under isothermal conditions is referred to as the *gel effect*. It is also known as the *Tromsdorff effect* or *Norrish-Smith effect* in honor of the early researchers in this field.

The gel effect is generally attributed to a decrease of the termination rate constant k_t due to increased viscosity at higher conversions. From theoretical considerations, the rate constant for reaction between two radicals in low viscosity media (such as bulk monomer) would be very large, about 8×10^9 L/mol-s. Experimentally determined k_t values for radical polymerizations, however, are considerably lower, usually by two orders of magnitude or more (see Table 6.4). Thus diffusion is the rate-determining process for termination. At higher conversions when the polymer concentration becomes high enough, the chain radicals become more crowded and entangled with segments of other polymer chains, attended by increase in viscosity. Consequently, the rate of diffusion of the polymer radicals and the frequency of their mutual encounters decrease. The rate of termination thus becomes increasingly slower.

While termination involves reaction of two large polymer radicals, propagation involves the reaction of a large radical with small monomer molecules. High viscosity thus affects the termination reaction much more than the propagation reaction, that is, k_t decreases much more than k_p, the net result being that there is an increase in the ratio $k_p / k_t^{1/2}$ in Eq. (6.24) and hence an increase in the rate of polymerization. As vinyl polymerizations are exothermic (see later), this increased rate can cause a temperature rise and faster initiator decomposition, leading finally to runaway reaction conditions.

Problem 6.28   The bimolecular chain termination in free-radical polymerization is a diffusion-controlled reaction that can be treated as a three-stage process (North and Reid, 1963; Odian, 1991), described below.

Stage 1. Translational diffusion of the centers of gravity of two macroradicals to such close proximity that certain segments of each chain can be considered to be in contact :

$$M_n{}^\bullet + M_m{}^\bullet \underset{k_2}{\overset{k_1}{\rightleftharpoons}} (M_n{}^\bullet....M_m{}^\bullet) \tag{P6.28.1}$$

Stage 2. Segmental diffusion (movement of parts or segments of a polymer chain relative to its other parts) of the two chains bringing their radical ends sufficiently close for chemical reaction :

$$(M_n{}^\bullet....M_m{}^\bullet) \underset{k_4}{\overset{k_3}{\rightleftharpoons}} (M_n{}^\bullet \mid M_m{}^\bullet) \tag{P6.28.2}$$

Stage 3. Chemical reaction of two radical ends :

$$(M_n{}^\bullet \mid M_m{}^\bullet) \overset{k_c}{\rightleftharpoons} \text{dead polymer(s)} \tag{P6.28.3}$$

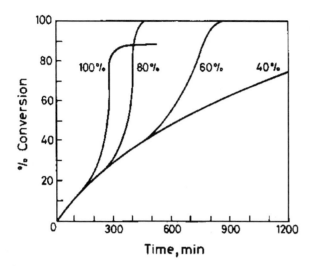

Figure 6.10 Conversion-time plots for the polymerization of methyl methacrylate in benzene at 50°C. The labeled curves are for the indicated monomer concentrations. In the present case, at monomer concentrations less than about 40 wt% the rate (slope of conversion vs. time) is approximately as anticipated from the ideal kinetic scheme, but deviation (rate acceleration) occurs at higher monomer concentrations. (Adapted from Schulz and Haborth, 1948.)

Assuming that diffusion is the rate-determining process for termination, obtain an expression for the rate of termination. Simplify the expression for two limiting situations of slow translational diffusion and slow segmental diffusion.

Answer:

For diffusion control of termination, $k_c \gg k_4$ and the reverse reaction in stage 2 [Eq. (P6.28.2)] is neglected. Assuming a steady state for $(M_n {}^\bullet....M_m {}^\bullet)$,

$$\frac{d[(M_n {}^\bullet....M_m {}^\bullet)]}{dt} = k_1[M_n {}^\bullet][M_m {}^\bullet] - k_2[(M_n {}^\bullet....M_m {}^\bullet)] - k_3[(M_n {}^\bullet....M_m {}^\bullet)] = 0$$

Therefore,

$$[(M_n {}^\bullet....M_m {}^\bullet)] = \left(\frac{k_1}{k_2 + k_3}\right)[M_n {}^\bullet][M_m {}^\bullet]$$

and

$$R_t = k_3[(M_n {}^\bullet....M_m {}^\bullet)] = \frac{k_1 k_3 [M {}^\bullet]^2}{k_2 + k_3} \tag{P6.28.4}$$

where $[M {}^\bullet]$ is the concentration of all chain radicals. For slow translational diffusion, $k_3 \gg k_2$ and Eq. (P6.28.4) then reduces to

$$R_t = k_1 [M {}^\bullet]^2 \tag{P6.28.5}$$

For slow segmental diffusion, $k_2 \gg k_3$ and Eq. (P6.28.4) reduces to

$$R_t = \left(\frac{k_1 k_3}{k_2}\right)[M \cdot]^2 \qquad \text{(P6.28.6)}$$

The experimentally observed termination rate constant k_t thus corresponds to k_1 and $k_1 k_3 / k_2$, respectively, for the two limiting situations (Odian, 1991).

6.10 Inhibition/Retardation of Polymerization

Some substances, when present in the monomer or in the polymerization system, suppress chain formation and/or chain growth by reacting with primary radicals or propagating chain radicals (before they could grow to polymeric size) to yield either nonradical products or radicals that are too low in reactivity to undergo propagation. Such substances are called *inhibitors* or *retarders* depending on their effectiveness. An inhibitor prevents polymerization completely as long as it is present in the system, whereas a retarder reduces the rate of polymerization but does not stop it completely. Figure 6.11 compares these effects, schematically, on the rate of free-radical polymerization. Inhibitors are deliberately added to monomers to prevent polymerization during storage or shipment. So before carrying out any polymerization, the inhibitors should be removed or, alternatively, additional amounts of initiators should be added to compensate for the inhibitors.

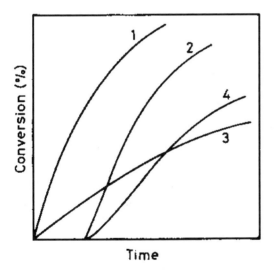

Figure 6.11 Comparison of conversion-time plots for normal, inhibited, and retarded free-radical polymerization. Curve 1: normal polymerization in the absence of inhibitor/retarder. Curve 2: inhibition; polymerization is completely stopped by inhibitor during the initial *induction period* but at the end of this period with the inhibitor having been completely consumed, polymerization proceeds at the same rate as in normal polymerization (curve 1). Curve 3: retardation; a retarder reduces the polymerization rate without showing an induction period. Curve 4: inhibition followed by retardation. (After Ghosh, 1990.)

The difference between inhibitors and retarders is simply one of degree and not kind. Both are either chain transfer agents (TA) or substances (Z) that add to propagating polymer radicals and thus provide an alternative reaction path :

$$M_n^{\bullet} \left\langle \begin{array}{ll} \xrightarrow{\text{(a) Monomer addition}} & M_{n+1}^{\bullet} \\ \xrightarrow{\text{(b) Chain transfer to TA}} & M_nA + T^{\bullet} \\ \xrightarrow{\text{(c) Addition to Z}} & M_nZ^{\bullet} \end{array} \right\} \qquad (6.128)$$

If the rate of reaction in Eq. (6.128b) or (6.128c) is very much greater than that of the reaction in Eq. (6.128a) and the new radicals $T^{\bullet}$ and $M_nZ^{\bullet}$ do not reinitiate by adding to monomer, then high-molecular-weight polymer will not be formed and the rate of polymerization will be practically zero during the period TA or Z is present in the system. This is a case of inhibition (cf. Problem 6.16) and the duration is known as the *induction* or *inhibition period.* On the other hand, if the rate of reaction in Eq. (6.128b) or (6.128c) is comparable to that of the normal propagation reaction in Eq. (6.128a) and the new radicals $T^{\bullet}$ and $M_nZ^{\bullet}$ do not reinitiate (cf. Problem 6.16) or reinitiate only slowly, then the rate of polymerization will be slowed down but not reduced to zero. This is a case of *retardation.*

Quinones are probably the most important class of inhibitors. The following transfer reactions may take place in the presence of quinone :

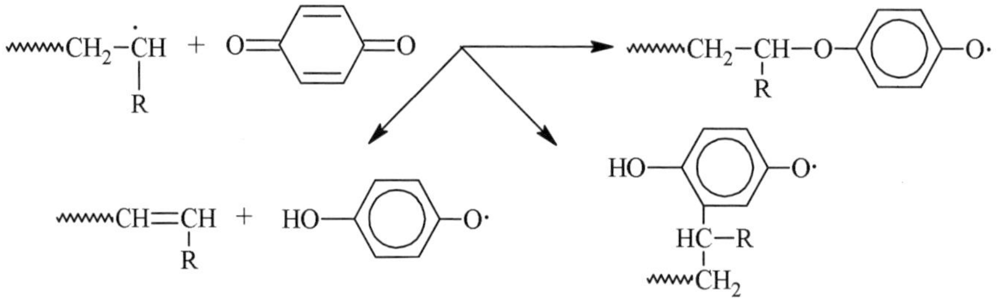

(6.129)

The inhibitor radicals formed in the above reactions are stabilized by resonance to such an extent that they do not add monomer to reinitiate polymerization. They disappear partly through disproportionation (forming quinone and hydroquinone) :

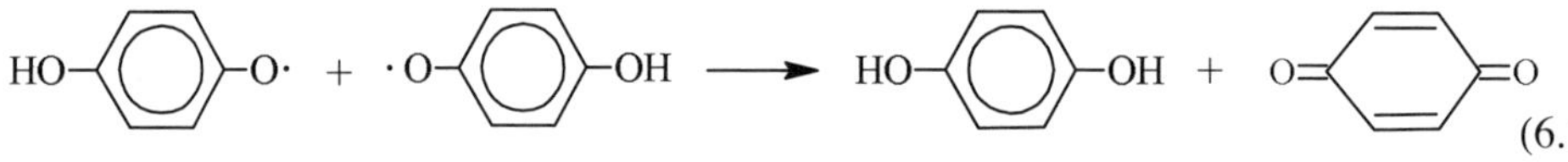

(6.130)

and partly by combining with each other (dimerization) or with new chain radicals.

Reaction (6.130) results in the regeneration of one inhibitor molecule per each pair of inhibitor radicals. Therefore this would lead to a 2:1 stoichiometry between the number of kinetic chains terminated and the number of quinone molecules consumed. Disappearance of inhibitor radicals by dimerization would, however, lead to a 1:1 stoichiometric ratio.

A large number of other substances are also active inhibitors. These include oxygen, NO (one of the most effective inhibitors, so much so that some highly reactive monomers can be distilled only under an atmosphere of NO), aromatic nitro compounds, numerous nitroso compounds, sulfur compounds, amines, phenols, aldehydes, and carbamates. An interesting inhibitor is molecular oxygen. Being a diradical, oxygen reacts with chain radicals to form the relatively unreactive peroxy radical :

$$M_n^{\bullet} + O_2 \longrightarrow M_n\text{—O—O}^{\bullet} \qquad (6.131)$$

6.10.1 Inhibition/Retardation Kinetics

Considering the situation where the inhibitor Z can react either with primary radicals R $^\bullet$, derived from the initiator I, or with propagating radicals M_n $^\bullet$ to give products completely incapable of initiating new polymer chains, the following simple reaction scheme (Eastmond, 1976) may be written:

$$I \xrightarrow{k_d} 2R\bullet \tag{6.132}$$

$$R\bullet + M \xrightarrow{k_i} RM\bullet \ (or \ M_1\bullet) \tag{6.133}$$

$$R\bullet + Z \xrightarrow{k_{zi}} Inactive \ products \tag{6.134}$$

$$M_n\bullet + M \xrightarrow{k_p} M_{n+1}\bullet \tag{6.135}$$

$$M_n\bullet + Z \xrightarrow{k_z} Inactive \ products \tag{6.136}$$

$$M_n\bullet + M_m\bullet \xrightarrow{k_t} Dead \ polymer \tag{6.137}$$

Assuming stationary-state kinetics are applicable, we have for primary radicals and propagating chain radicals the expressions:

$$d[R\bullet]/dt = R_i - k_{zi}[Z][R\bullet] - k_i[M][R\bullet] = 0 \tag{6.138}$$
$$d[M\bullet]/dt = k_i[M][R\bullet] - k_z[Z][M\bullet] - k_t[M\bullet]^2 = 0 \tag{6.139}$$

where $[M\bullet] = \sum_{n=1}^{\infty} M_n\bullet$ and R_i is the rate of initiation in the absence of inhibitor Z, that is, $R_i = 2fk_d[I]$.

From Eq. (6.138),

$$[R\bullet] = \frac{R_i}{k_{zi}[Z] + k_i[M]} \tag{6.140}$$

The rate of formation of chain radicals or rate of initiation in the presence of inhibitor is $(R_i)_z = k_i[M][R\bullet]$. Substituting for $[R\bullet]$ from Eq. (6.140) one may write (Eastmond, 1976)

$$(R_i)_z = \frac{k_i[M]R_i}{k_{zi}[Z] + k_i[M]} = R_i\left\{1 - \frac{k_{zi}[Z]}{k_{zi}[Z] + k_i[M]}\right\} \tag{6.141}$$

If $k_{zi}[Z] \gg k_i[M]$, $(R_i)_z$ will be very small compared to R_i and practically no chain initiation will take place in the presence of inhibitor. Also when $k_{zi}[Z] \gg k_i[M]$, it is likely that $k_z[Z] \gg k_p[M]$, implying that chain growth will be stopped in an early stage and normal bimolecular termination represented by $k_t[M\bullet]^2$ will be absent. Therefore from Eq. (6.139):

$$[M\bullet] = \frac{k_i[M][R\bullet]}{k_z[Z]} \tag{6.142}$$

Substituting for $[M\bullet]$ from Eq. (6.142) into the equation for rate of propagation, $R_p = k_p[M\bullet][M]$ gives

$$R_p = \frac{k_i k_p[R\bullet][M]^2}{k_z[Z]} = \frac{k_p[M](R_i)_z}{k_z[Z]} \tag{6.143}$$

Substituting from Eq. (6.141),

$$
\begin{aligned}
R_p &= \frac{k_p[M]R_i}{k_z[Z]}\left\{1 - \frac{k_{zi}[Z]}{k_{zi}[Z] + k_i[M]}\right\} \\
&= \frac{[M]R_i}{C_Z[Z]}\left\{1 - \frac{k_{zi}[Z]}{k_{zi}[Z] + k_i[M]}\right\}
\end{aligned}
\tag{6.144}
$$

where C_Z is the *inhibition constant* (or *inhibitor constant*) defined as $C_Z = k_z/k_p$.

The average degree of polymerization, $\overline{DP}_n$, which is the average number of monomer molecules consumed per polymer molecule formed, is given (Eastmond, 1976) by

$$
\begin{aligned}
\overline{DP}_n &= \frac{R_p}{k_z[Z][M^\bullet] + k_t[M^\bullet]^2} \\
&\approx \frac{k_p[M][M^\bullet]}{k_z[Z][M^\bullet]} = \frac{k_p[M]}{k_z[Z]} = \frac{[M]}{C_Z[Z]}
\end{aligned}
\tag{6.145}
$$

It is seen from Eqs. (6.144) and (6.145) that in the presence of a powerful inhibitor, that is, under the conditions $k_{zi}[Z] \gg k_i[M]$ and $k_z[Z] \gg k_p[M]$ (hence, $C_Z \gg 1$), any polymerization which occurs even at low [Z] proceeds only at a very low, practically zero, rate and yields only very low-molecular-weight polymer (Eastmond, 1976).

Now considering the situation where the inhibitor reacts with propagating radicals but does not react with primary radicals (i.e., $k_{zi} = 0$), Eq. (6.144) reduces to

$$
-\frac{d[M]}{dt} = R_p = \frac{R_i[M]}{C_Z[Z]}
\tag{6.146}
$$

If x number of chain radicals are terminated per inhibitor molecule, the concentration of inhibitor, [Z], after a time t can be approximately expressed (Odian, 1991) by

$$
[Z] = [Z]_0 - \frac{R_i t}{x}
\tag{6.147}
$$

where $[Z]_0$ is the initial concentration of inhibitor. Substituting Eq. (6.147) into Eq. (6.146),

$$
-\frac{d[M]}{dt} = \frac{R_i[M]}{C_Z\left([Z]_0 - \dfrac{R_i t}{x}\right)}
\tag{6.148}
$$

Rearranging,

$$
-\frac{1}{d\ln[M]/dt} = C_Z\left(\frac{[Z]_0}{R_i}\right) - \left(\frac{C_Z}{x}\right)t
\tag{6.149}
$$

In terms of monomer conversion, $p = ([M]_0 - [M])/[M]_0$, Eq. (6.149) can be written as (Odian, 1991):

$$
-\frac{1}{d\ln(1-p)/dt} = \frac{C_Z[Z]_0}{R_i} - \left(\frac{C_Z}{x}\right)t
\tag{6.150}
$$

A plot of the left side of Eq. (6.150) vs. time should thus be linear and from the intercept and slope of such a plot both C_Z and x can be obtained, if $[Z]_0$ and R_i are known (see Problem 6.29). The method involves measurement of quite small rates of polymerization, especially if C_Z is large [cf. Eq. (6.146)]. However, by use of sufficiently sensitive techniques, such low rates of monomer consumption may be detected (Bevington et al., 1955).

Problem 6.29 Polymerization of vinyl acetate at 45°C in the presence of 0.2 M benzoyl peroxide and 9.3×10^{-4} M 2,3,5,6-tetramethylbenzoquinone (duroquinone), used as inhibitor, yielded the following data (Flory, 1953) of monomer conversion versus time :

Time (min)	% Conversion, $p \times 100$
0	0
100	0.10
200	0.30
300	0.66
400	1.18
500	2.45

Determine the inhibitor constant (C_Z) at 45°C and the number of radicals terminated per inhibitor molecule. [For benzoyl peroxide at 45°C take $k_d = 2.8 \times 10^{-7}$ s^{-1}, $f = 0.25$.]

Answer:
From a polynomial fit of $\ln(1 - p)$ vs. time (t) data, $d\ln(1 - p)/dt$ is obtained by differentiation. The linear plot in Fig. 6.12, according to Eq. (6.150), yields

$$\text{slope} = -\frac{C_Z}{x} = -92.2; \quad \text{intercept} = \frac{C_Z [Z]_0}{R_i} = 5.1 \times 10^4 \text{ min}$$

$$R_i = 2(0.25)(2.8 \times 10^{-7} \times 60 \text{ min}^{-1})(0.2 \text{ mol L}^{-1}) = 1.68 \times 10^{-6} \text{ mol L}^{-1} \text{ min}^{-1}$$

$$C_Z = \frac{(5.1 \times 10^4 \text{ min})(1.68 \times 10^{-6} \text{ mol L}^{-1} \text{ min}^{-1})}{(9.3 \times 10^{-4} \text{ mol L}^{-1})} = 92.1 \quad \text{So, } x = \frac{92.1}{92.2} \simeq 1$$

Table 6.8 shows C_Z values for various systems. The value of C_Z for a compound is seen to depend largely on the monomer being polymerized. A large value signifies that the compound acts as an inhibitor and a small value signifies its role as a retarder. A given compound can thus act as a strong inhibitor for a monomer while acting only as a mild retarder for another monomer.

6.11 Effects of Temperature

6.11.1 Rate of Polymerization

It is clear from Eq. (6.24) that the polymerization rate R_p depends on the combination of three rate constants k_d, k_p, and k_t, which makes the quantitative effect of temperature on R_p rather complex. Expressing the rate constants k_p, k_d, and k_t by an Arrhenius type relationship (Young and Lovell, 1990):

$$k_i = A_i \exp(-E_i/RT) \tag{6.151}$$

where i stands for p, d, and t, A_i is the (nominally temperature-independent) *collision factor*, and E_i

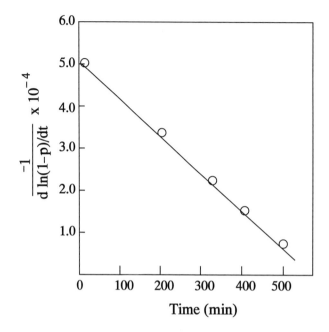

Figure 6.12 Plot of Eq. (6.150) with data of Problem 6.29.

is the activation energy for the individual reaction, the temperature dependence of the group $k_p(k_d/k_t)^{1/2}$ can be expressed by (Ghosh, 1990; Odian, 1991):

$$\ln\left[k_p\left(\frac{k_d}{k_t}\right)^{1/2}\right] = \ln\left[A_p\left(\frac{A_d}{A_t}\right)^{1/2}\right] - \frac{[E_p + (E_d/2) - (E_t/2)]}{RT} \tag{6.152}$$

and hence the temperature dependence of R_p by

$$d(\ln R_p)/dT = E_{RP}/RT^2 \tag{6.153}$$

where E_{RP} is a composite or *overall activation energy* for the rate of polymerization related to E_i ($i = p, d, t$) by

$$E_{RP} = E_p + (E_d/2) - (E_t/2) \tag{6.154}$$

For most monomers E_p is in the range 20-40 kJ/mol and E_t is in the range 8-20 kJ/mol, while for most initiators commonly used in thermal homolysis, E_d is in the range 125-165 kJ/mol. Therefore, E_{RP} is about 80 kJ/mol. A temperature rise of 10°C will thus cause a two- or three-fold increase in the rate of polymerization. It should be noted that E_d dominates in E_{RP}. So as the method of initiation varies, E_{RP} changes significantly. For redox initiation, $E_d \simeq 40 - 60$ kJ/mol and so $E_{RP} \simeq 40$ kJ/mol, which is one-half the value for nonredox initiators. Redox systems are thus generally used for polymerizations at lower temperatures. For photochemical or radiation-induced initiation, $E_d \simeq 0$ and so from Eq. (6.154), $E_{RP} \simeq 20$ kJ/mol. The activation energy being very low, the rate of polymerization in photo- or radiation-initiated systems does not change much with temperature.

Table 6.8 Selected Values of Inhibitor Constants

Inhibitor	Monomer	$C_Z = k_z/k_p$	k_z (L mol^{-1} s^{-1})
Nitrobenzene	Methyl acrylate	0.00464	4.64
	Styrene	0.326	–
	Vinyl acetate	11.2	–
1,3,5-Trinitrobenzene	Methyl acrylate	0.204	204
	Styrene	64.2	–
	Vinyl acetate	404	760,000
p-Benzoquinone	Acrylonitrile	0.91	910
	Methyl methacrylate (44°C)	5.5	2400
	Styrene	518	–
Diphenylpicryl-hydrazyl (DPPH)	Methyl methacrylate (44°C)	2,000	–
FeCl$_3$ in DMF solution	Acrylonitrile (60°C)	3.33	6500
	Styrene (60°C)	536	94000
	Vinyl acetate (60°C)	–	235,000
Oxygen	Methyl methacrylate	33,000	10^7
	Styrene	14,600	10^6-10^7
Phenol	Methyl acrylate	0.0002	< 0.2
	Vinyl acetate	0.012	21

All values are for 50°C unless otherwise noted.
Source: Data from Brandrup and Immergut, 1975.

Problem 6.30 When an organic peroxide is used as the thermal initiator in the polymerization of styrene, the time it takes to convert 20% of the monomer is 70 min at 60°C and 38 min at 70°C in two experiments in which all conditions except the temperature are the same. However, when styrene is photopolymerized, the times for 20% conversion are 18 min and 16 min at 60°C and 70°C, respectively. If the half-life of the peroxide at 60°C is 153 h, what is the half-life at 90°C ?

Answer:

For polymerization initiated by thermal decomposition of initiators, we get from Eq. (6.153) for the two experiments 1 and 2 :

$$\ln \frac{(R_p)_2}{(R_p)_1} = \frac{(E_{RP})_{therm}}{R}\left(\frac{1}{T_1} - \frac{1}{T_2}\right) \tag{P6.30.1}$$

Similarly for photopolymerization [cf. Eq. (6.66)],

$$\ln \frac{(R_p)_2}{(R_p)_1} = \frac{(E_{RP})_{photo}}{R}\left(\frac{1}{T_1} - \frac{1}{T_2}\right) \tag{P6.30.2}$$

Since in photopolymerization, $E_d = 0$, we get from Eq. (6.154),

$$(E_{RP})_{therm} = E_d/2 + (E_{RP})_{photo} \tag{P6.30.3}$$

The rates of polymerization in two experiments are inversely proportional to the times required to attain the same conversion, i.e.,

$$\ln\left[(R_p)_2/(R_p)_1\right] = \ln(t_1/t_2) \tag{P6.30.4}$$

For the peroxide initiated polymerization, we get from Eqs. (P6.30.1) and (P6.30.4),

$$\frac{(E_{RP})_{therm}}{R} = \frac{\ln(t_1/t_2)}{\left(\frac{1}{T_1} - \frac{1}{T_2}\right)} = \frac{\ln(70 \text{ min}/38 \text{ min})}{(1/333°\text{K} - 1/343°\text{K})} = 6,978°\text{K}$$

Similarly for photoinitiated polymerization from Eqs. (P6.30.2) and (P6.30.4),

$$\frac{(E_{RP})_{photo}}{R} = \frac{\ln(18 \text{ min}/16 \text{ min})}{(1/333°\text{K} - 1/343°\text{K})} = 1345°\text{K}$$

From Eq. (P6.30.3),

$$E_d/R = 2(6978°\text{K} - 1345°\text{K}) = 11,266°\text{K}$$

Expressing k_d by $k_d = A_d \exp(-E_d/RT)$ we obtain for experiments at two temperatures T_1 and T_2,

$$\frac{E_d}{R} = \frac{\ln[(k_d)_2/(k_d)_1]}{(1/T_1 - 1/T_2)}$$

Since $t_{0.5} = \ln 2/k_d$ [cf. Eq. (6.27)],

$$\ln[(t_{0.5})_1/(t_{0.5})_2] = \frac{E_d}{R}\left(\frac{1}{T_1} - \frac{1}{T_2}\right)$$

$$\ln\frac{(153 \text{ h})}{(t_{0.5})_2} = (11,266°\text{K})\left(\frac{1}{333°\text{K}} - \frac{1}{363°\text{K}}\right) \implies t_{0.5} \text{ at } 90°\text{C} = 9.3 \text{ h}$$

6.11.2 Degree of Polymerization

It is obvious from Eq. (6.102) for kinetic chain length (ν) that the temperature dependence of ν is determined by the ratio $k_p/(k_d k_t)^{1/2}$. Using the Arrhenius expression (6.151) for the individual rate constants one obtains (Odian, 1991)

$$\ln\left[\frac{k_p}{(k_d k_t)^{1/2}}\right] = \ln[A_p/(A_d A_t)^{1/2}] - E_{DP}/RT \tag{6.155}$$

where E_{DP} is the *activation energy for degree of polymerization*, given by

$$E_{DP} = E_p - (E_d)/2 - (E_t)/2 \tag{6.156}$$

Inserting Eq. (6.103) or (6.104) into Eq. (6.102), applying Arrhenius equations for the rate constants, taking logarithms, and finally differentiating with respect to temperature gives the relation:

$$\frac{d\ln(\overline{DP}_n)}{dT} = \frac{E_{DP}}{RT^2} \tag{6.157}$$

With the usual values of E_p, E_d, and E_t, mentioned above, Eq. (6.156) gives $E_{DP} \simeq -60$ kJ/mol. According to Eq. (6.157), $\overline{DP}_n$ will therefore decrease with increasing temperature. On the other hand, for a purely photochemical polymerization, $E_d \simeq 0$ and hence E_{DP} is positive, being approximately 20 kJ/mol. In photopolymerization, the molecular weight will therefore increase with increasing temperature.

Problem 6.31 When a monomer M was polymerized in solution with peroxide initiator P at two temperatures, using the same concentrations of M and P, the following conversions of M were obtained :

Run	Conversion in 60 min	Initial molecular weight
Run #1 (50°C)	20%	100,000
Run #2 (65°C)	58%	60,000

Calculate the energy of activation (E_d) for the thermal decomposition of P.

Answer:

Approximating $(1 - e^{-k_d t/2})$ in Eq. (6.31) by $k_d t/2$,

$$- \ln(1 - p) = \left(\frac{k_p}{k_t^{1/2}} \right) k_d^{1/2} (f[I]_0)^{1/2} t = K \sqrt{k_d} \sqrt{f[I]_0}\, t \tag{P6.31.1}$$

Only K and $\sqrt{k_d}$ depend on temperature. From Eq. (6.102), assuming termination by coupling and considering initial state,

$$\overline{M}_n = (\overline{DP}_n) M_0 = \frac{K}{\sqrt{k_d}} \frac{[M] M_0}{(f[I]_0)^{1/2}} \tag{P6.31.2}$$

Taking ratio of Eqs. (P6.31.1) and (P6.31.2),

$$- \frac{\ln(1 - p)}{\overline{M}_n} = k_d \frac{f[I]_0}{[M] M_0} t$$

For Run #1 (50°C): $\quad \dfrac{- \ln(0.80)}{100,000} = (k_d)_{50°C} \dfrac{f[I]_0}{[M] M_0} (60 \text{ min}) \tag{P6.31.3}$

For Run #2 (65°C): $\quad \dfrac{- \ln(0.42)}{60,000} = (k_d)_{65°C} \dfrac{f[I]_0}{[M] M_0} (60 \text{ min}) \tag{P6.31.4}$

Taking ratio of Eqs. (P6.31.4) to (P6.31.3) and substituting appropriate values gives

$$(k_d)_{65°C} / (k_d)_{50°C} = \frac{100,000 \ln(0.42)}{60,000 \ln(0.80)} = 6.479$$

$$\ln\left[(k_d)_{65°C} / (k_d)_{50°C} \right] = 1.869 = \frac{E_d}{(8.314 \text{ J mol}^{-1} \,°K^{-1})} \left[\frac{1}{323°K} - \frac{1}{338°K} \right]$$

$$E_d = 1.13 \times 10^5 \text{ J mol}^{-1} \quad (\equiv 113 \text{ kJ mol}^{-1})$$

6.11.3 *Polymerization-Depolymerization Equilibrium*

The free energy of polymerization is given by

$$\Delta G_p = \Delta H_p - T \Delta S_p \tag{6.158}$$

where ΔG_p, ΔH_p, and ΔS_p are the differences in free energy, enthalpy, and entropy, respectively, between 1 mol of monomer and 1 mol of repeating units in the polymeric product. Polymerization occurs if under the prevailing experimental conditions the process leads to decrease in free energy (i.e., ΔG_p is negative). The polymerization of an alkene is exothermic (ΔH_p is negative) because in this process the higher-energy π-bonds of monomer molecules are converted into lower energy σ-bonds in the polymer. The entropy change of polymerization (ΔS_p) is also negative because the conversion of monomer to polymer is accompanied by a decrease in degrees of freedom (randomness). It is thus seen, referring to Eq. (6.158), that ΔH_p is favorable for polymerization, but ΔS_p is not.

According to Eq. (6.158), ΔG_p will become less negative as the contribution of $T\Delta S_p$ increases at higher temperatures and finally a temperature, designated as T_c, may be reached at which $\Delta G_p = 0$, with the result that a high-molecular-weight polymer cannot be produced at temperatures higher than T_c. At T_c, forward (*propagation*) and reverse (*depropagation* or *depolymerization*) reactions proceed at equal rates in a dynamic equilibrium and Eq. (6.6) should thus be written more generally as

$$\text{RM}_{n-1}{}^{\bullet} + \text{M} \underset{k_{dp}}{\overset{k_p}{\rightleftharpoons}} \text{M}_n{}^{\bullet} \tag{6.159}$$

where k_{dp} is the rate constant for depropagation or depolymerization. Accordingly, Eq. (6.12) now becomes

$$R_p = -\frac{d[\text{M}]}{dt} = k_p[\text{M}][\text{M}^{\bullet}] - k_{dp}[\text{M}^{\bullet}] \tag{6.160}$$

At T_c, called the *ceiling temperature*, $-d[\text{M}]/dt = 0$ and, from Eq. (6.160), $k_p[\text{M}]_e = k_{dp}$, or

$$k_p/k_{dp} = K = 1/[\text{M}]_e \tag{6.161}$$

where K is the equilibrium constant and $[\text{M}]_e$ is the equilibrium monomer concentration. The variations of k_p and k_d with temperature are given by the Arrhenius expressions:

$$k_p = A_p \exp(-E_p/RT) \tag{6.162}$$

$$k_{dp} = A_{dp} \exp(-E_{dp}/RT) \tag{6.163}$$

where A_p and A_{dp} are pre-exponential factors and E_p and E_{dp} are the activation energies for propagation and depropagation, respectively; $E_p - E_{dp} = \Delta H_p$ is the enthalpy change for the overall reaction.

The predicted variation of T_c with $[\text{M}]_e$ can be found by inserting the Arrhenius expressions into Eq. (6.161), whence

$$T_c = \frac{\Delta H_p}{R\ln\left(A_p[\text{M}]_e/A_{dp}\right)} \tag{6.164}$$

Equation (6.164) shows that there is a series of ceiling temperatures corresponding to different equilibrium monomer concentrations. For a monomer solution of any concentration taken as $[\text{M}]_e$, there is a temperature at which polymerization will not occur. In fact, for each concentration taken as $[\text{M}]_e$ there will be a plot analogous to Fig. 6.13 showing $k_{dp} = k_p[\text{M}]_e$ at its T_c. Thus there is an

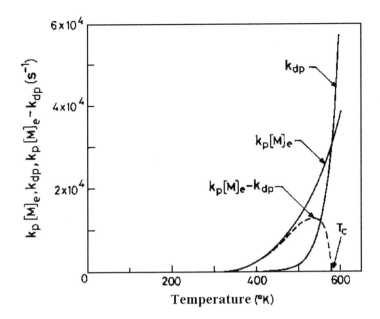

Figure 6.13 Data from Problem 6.31 plotted to show variation of $k_p[M]_e$ and k_{dp} with temperature for pure styrene (Dainton and Irvin, 1958; Ghosh, 1990; Odian, 1991.)

upper temperature limit above which a polymer cannot be produced even from pure monomer at equilibrium. The designation of a singular T_c value, often referred to in the literature as "the ceiling temperature," usually refers to the T_c for the pure monomer or in some cases for the monomer at unit molarity taken as $[M]_e$.

Problem 6.32   For pure styrene (density 0.905 g/cm^3 at 25°C) calculate $k_p[M]_e$ and k_{dp} at different temperatures, and hence determine the ceiling temperature, using the following data (Dainton and Irvin, 1958): $A_p = 10^6$ L mol^{-1} s^{-1}; $A_{dp} = 10^{13}$ s^{-1}; $E_p = 27.2$ kJ mol^{-1}; $E_{dp} = 27.2 + 67.4 = 94.6$ kJ mol^{-1}.

Answer:

For bulk styrene at 25°C, $[M] = \dfrac{(1000 \text{ cm}^3 \text{ L}^{-1})(0.905 \text{ g cm}^{-3})}{(104 \text{ g mol}^{-1})} = 8.7 \text{ mol L}^{-1}$

Assume that $[M]_e = 8.7$ mol L^{-1}. The values of $k_p[M]_e$ and k_{dp} at any given temperature can then be calculated from the following expressions [cf. Eqs. (6.162) and (6.163)] :

$$k_p[M]_e = (10^6 \text{L mol}^{-1}\text{s}^{-1})\exp\left[\frac{-(27.2\times10^3 \text{ J mol}^{-1})}{(8.314\text{J}°\text{K}^{-1}\text{mol}^{-1})(T\,°\text{K})}\right](8.7\,\text{mol L}^{-1}) \qquad \text{(P6.32.1)}$$

$$k_{dp} = (10^{13}\text{ s}^{-1})\exp\left[\frac{-(94.6\times10^3 \text{ J mol}^{-1})}{(8.314 \text{ J}°\text{K}^{-1}\text{mol}^{-1})(T\,°\text{K})}\right] \qquad \text{(P6.32.2)}$$

Values of $k_p[M]_e$, k_{dp}, and $(k_p[M]_e - k_d)$ are calculated for different assumed values of T, starting from room temperature ($T = 300$°K), and these are plotted in Fig. 6.13. The extrapolated curve of $(k_p[M]_e - k_d)$ cuts the temperature axis at $T = 580$°K. Therefore, $T_c = 580 - 273 = 307$°C.

Alternatively, since $\Delta H_p = E_p - E_{dp} = -67.4$ kJ mol^{-1}, one obtains from Eq. (6.164),

$$T_c = \frac{(-67.4 \times 10^3 \text{ J mol}^{-1})}{(8.314 \text{ J }^\circ\text{K}^{-1} \text{ mol}^{-1}) \ln\left[(10^6 \text{ L mol}^{-1} \text{ s}^{-1})(8.7 \text{ mol L}^{-1})/(10^{13} \text{ s}^{-1})\right]}$$

$$= 580^\circ\text{K} \quad (= 307^\circ\text{C})$$

An alternative approach to the problem of determining ceiling temperature is based on the recognition of T_c as the temperature at which $\Delta G_p = 0$, and hence from Eq. (6.158),

$$T_c = \Delta H_p / \Delta S_p \tag{6.165}$$

where ΔH_p and ΔS_p are the heat and entropy changes under the prevailing experimental conditions. Comparing Eqs. (6.164) and (6.165), one may write

$$\Delta S_p = R\ln(A_p/A_{dp}) + R\ln[\text{M}]_e = \Delta S_p^0 + R\ln[\text{M}]_e \tag{6.166}$$

where ΔS_p^0 is the entropy change for $[\text{M}]_e = 1$ mol/L. Therefore,

$$T_c = \Delta H_p / (\Delta S_p^0 + R\ln[\text{M}]_e) \tag{6.167}$$

This equation implies that at the ceiling temperature (T_c) the monomer concentration in equilibrium with long chain polymer is $[\text{M}]_e$. Equation (6.167) can be rewritten as

$$\ln[\text{M}]_e = \frac{\Delta H_p}{R}\left(\frac{1}{T_c}\right) - \frac{\Delta S_p^0}{R} \tag{6.168}$$

showing a linear relationship of $\ln[\text{M}]_e$ with reciprocal T_c. Equation (6.168) can be used to determine ΔH_p and ΔS_p from experimental data of T_c and $[\text{M}]_e$.

The equations derived above can be used to calculate how much monomer will be in equilibrium with high-molecular weight polymer at any temperature (see Problems 6.33 and 6.34).

Problem 6.33   The standard enthalpy of polymerization (ΔH_p^0) of vinyl chloride at 25°C is −72 kJ/mol and the standard entropy of polymerization (ΔS_p^0) can be taken to be approximately −100 J/°K-mol. Can the polymer be safely used at the ambient temperature (25°C), in view of the fact that vinyl chloride is carcinogenic ?

Answer:

For $T_c = 298^\circ$K, Eq. (6.168) gives

$$\ln[\text{M}]_e = \frac{(-72 \times 10^3 \text{ J mol}^{-1})}{(8.314 \text{ J }^\circ\text{K}^{-1} \text{ mol}^{-1})(300^\circ\text{K})} + \frac{(100 \text{ J }^\circ\text{K}^{-1} \text{ mol}^{-1})}{(8.314 \text{ J }^\circ\text{K}^{-1} \text{ mol}^{-1})}$$

$$= -16.84 \implies [\text{M}]_e = 4.9 \times 10^{-8} \text{ mol L}^{-1}$$

So only a negligible concentration (< 0.0005 ppm) of monomer will be in equilibrium with high-molecular-weight polymer at 25°C. If unreacted monomer can be purged from the polymer, no significant concentration will develop thereafter at room temperature because of the polymerization-depolymerization equilibrium.

Problem 6.34 If a free-radical polymerization of 1.0 M solution of methyl methacrylate was being carried out at 100°C, what would be the maximum possible conversion of the monomer to polymer, that is, till the polymerization-depolymerization equilibrium is reached? (Given: $\Delta H_p^0 = -55.2$ kJ/mol and $\Delta S_p^0 = -117$ J/deg-mol at 25°C.)

Answer:

To calculate an approximate value of $[M]_e$, consider the monomer and the polymer in their standard states and assume that there is no significant temperature dependence of enthalpy or entropy of polymerization so that the ΔH_p^0 and ΔS_p^0 data can be used. For $T_c = 373$ °K, Eq. (6.168) then yields,

$$\ln[M]_e = \frac{(-55.2 \times 10^3 \text{ J mol}^{-1})}{(8.314 \text{ J}°\text{K}^{-1}\text{mol}^{-1})(373°\text{K})} - \frac{(-117 \text{ J}°\text{K}^{-1}\text{mol}^{-1})}{(8.314 \text{ J}°\text{K}^{-1}\text{mol}^{-1})}$$

$$= -3.72 \implies [M]_e = 0.024 \text{ mol L}^{-1}$$

Hence, maximum conversion $= (1 - [M]_e) \times 100 = 97.6\%$

6.12 Molecular Weight Distribution

As we have seen in previous sections, the radical chain polymerization involves several possible modes of chain termination — disproportionation, coupling, and various chain transfer reactions. These contribute to the complexity of molecular weight distribution. Further complication arises due to changes in monomer and initiator concentrations with conversion, dependence of termination rate constants on polymer size, and concentration or autoacceleration effects. For polymerizations at low conversions, these additional effects can be neglected and molecular weight distributions can be easily calculated, though at high conversions the distributions may be significantly different.

6.12.1 Low-Conversion Polymerization

A propagating radical may add a monomer molecule to continue propagation or it may undergo chain transfer or bimolecular chain termination reaction. The probability that it will add monomer is then (Rudin, 1982):

$$P = \frac{R_p}{R_p + R_{tr} + R_t} \tag{6.169}$$

where R_p, R_{tr}, and R_t are the rates of propagation, chain transfer, and termination, respectively. Concentrations being approximately constant at low conversions, these rates are given by

$$R_p = k_p[M][M^\bullet] \tag{6.170}$$
$$R_{tr} = k_{tr,M}[M^\bullet][M] + k_{tr,I}[M^\bullet][I] + k_{tr,S}[M^\bullet][S] \tag{6.171}$$
$$R_t = 2(k_{tc} + k_{td})[M^\bullet]^2 \tag{6.172}$$

The last term in Eq. (6.171), using the symbol S for solvent, does in fact include the effects of any chain transfer agent but transfer to polymer is not included in this equation because such transfer can be significant only at high conversions where sufficient polymer is present.

6.12.1.1 Termination by Disproportionation and/or Transfer

Let us first consider the polymerization where each propagating chain results in the formation of one dead polymer molecule, that is, each polymer molecule consists of one kinetic chain. This happens when the chain radical is terminated by disproportionation and/or chain transfer (i.e., k_{tc} = 0). The probability situation in this case is almost identical to that for linear, reversible step-growth polymerization described in Chapter 5. Thus if we select randomly an initiator fragment at the end of a polymer molecule, the probability that the monomer molecule added to this initiator (primary) radical has added another monomer molecule is P. Continuing in this way the probability that x monomer molecules have been added one after another is P^{x-1} (see p. 232). Since the probability that the radical end of a growing chain has terminated is $(1 - P)$, the probability that the polymer molecule under consideration consists of essentially x monomer units is $P^{x-1}(1 - P)$. This probability can be equated to the mole fraction of polymer molecules of this size (Rudin, 1982), that is,

$$n_x = (1 - P)P^{x-1} \tag{6.173}$$

Equation (6.173) represents the number distribution function. The corresponding weight distribution function w_x, by direct analogy to that for step-growth polymerization (p. 232), will be written as

$$w_x = x(1 - P)^2 P^{x-1} \tag{6.174}$$

By analogy with Eqs. (5.50), (5.53), and (5.54) we can then write

$$\overline{DP}_n = 1/(1 - P) \tag{6.175}$$

$$\overline{DP}_w = (1 + P)/(1 - P) \tag{6.176}$$

$$\overline{DP}_w/\overline{DP}_n = 1 + P \tag{6.177}$$

Though Eqs. (6.173) and (6.174) are formally similar to Eqs. (5.44) and (5.47), respectively, there is a very important difference in that the distribution functions in radical chain polymerizations apply only to the polymer which has been formed, while those in step-growth polymerization apply to the whole reaction mixture.

Substituting Eq. (6.169) into Eq. (6.175) and noting that for most addition polymerizations, $R_p \gg (R_{tr} + R_t)$ (as otherwise high-molecular-weight polymer would not be formed),

$$\overline{DP}_n = \frac{R_p + R_{tr} + R_t}{R_{tr} + R_t} \simeq \frac{R_p}{R_{tr} + R_t} \tag{6.178}$$

Insertion of Eqs. (6.170)-(6.172) into Eq. (6.178) and simplification (noting that $k_{tc} = 0$) lead to

$$\frac{1}{\overline{DP}_n} = C_M + C_I \frac{[I]}{[M]} + C_S \frac{[S]}{[M]} + \frac{2k_{td}}{k_p^2[M]^2} R_p \tag{6.179}$$

which can be easily shown to be identical with Eq. (6.122).

A consideration of Eqs. (6.175) and (6.176) indicates that high-molecular-weight polymer (i.e., large $\overline{DP}_n$ and $\overline{DP}_w$) will only be produced if P is close to unity, i.e., if $R_p \gg (R_{tr} + R_t)$. Equation (6.177) indicates that the size distribution $\overline{DP}_w/\overline{DP}_n$ (also referred to as PDI, the *polydispersity index*) has a limiting value of 2 as P approaches unity. The situation is thus analogous to that for linear step-growth polymerization considered in Chapter 5 [cf. Eq. (5.54)].

6.12.1.2 Termination by Coupling

If chain termination occurs only by coupling or combination, each polymer molecule consists of two kinetic chains which grew independently and were joined together following their mutual termination. The polymer formed is analogous to a di-chain ($f = 2$) condensation polymer (Chapter 5). If P represents the probability of continuation of either chain from one of its units to the next, $(1 - P)$ is the probability that a given unit in the chain reacts by termination and the latter can be equated to the ratio of terminated to total units. Since two units are involved in each termination step, one can then write

$$1 - P = R_t/(R_p + R_t) = \frac{2k_{tc}[\text{M}^{\bullet}]}{k_p[\text{M}] + 2k_{tc}[\text{M}^{\bullet}]} \qquad (6.180)$$

where k_{tc} is the rate constant for termination by coupling.

Substitution of $[\text{M}^{\bullet}] = R_p/k_p[\text{M}]$ from Eq. (6.12) into Eq. (6.180) and simplification leads to

$$\frac{1}{P} = 1 + \frac{2k_{tc}R_p}{k_p^2[\text{M}]^2} \qquad (6.181)$$

Then, since each molecule consists of two chains, the number-average degree of polymerization will be given by [cf. Eq. (6.175)],

$$\overline{DP}_n = \frac{2}{(1 - P)} = 2 + \frac{k_p^2[\text{M}]^2}{k_{tc}R_p} \qquad (6.182)$$

For the weight-average degree of polymerization, the following equation can be shown (Tanford, 1961) to be applicable:

$$\overline{DP}_w = \frac{2 + P}{1 - P} \qquad (6.183)$$

Hence,

$$\overline{DP}_w/\overline{DP}_n = (2 + P)/2 \qquad (6.184)$$

The ratio in Eq. (6.184) has a limiting value of 1.5 at high polymer molecular weights when $R_p \gg R_t$ and P approaches 1. This is narrower than the distribution produced in the absence of termination by coupling [cf. Eq. (6.177)], which is, however, quite expected. Since the probability for coupling between same-sized propagating radicals is the same as that between different-sized radicals, polymer molecules of sizes close to the average (and hence of relatively narrow size distribution) are more likely.

The number fraction and weight fraction distributions (Schulz, 1939) are readily shown to be

$$n_x = (x - 1)(1 - P)^2 P^{x-2} \qquad (6.185)$$

$$w_x = \tfrac{1}{2}x(x - 1)(1 - P)^3 P^{x-2} \qquad (6.186)$$

where each initiator radical is counted as a unit. [Note that Eq. (6.186) is obtained simply by multiplying Eq. (6.185) by $x/\overline{DP}_n$.]

6.12.1.3 *Termination by Coupling, Disproportionation, and Chain Transfer*

For polymerization where termination occurs by all three modes, namely, coupling, disproportionation, and chain transfer, one can derive the size distribution as a weighted combination of the above two sets of distribution functions. For example, the weight distribution can be obtained as (Smith et al., 1966):

$$w_x = \omega x (1 - P)^2 P^{x-1} + \tfrac{1}{2}(1 - \omega)x(x - 1)(1 - P)^3 P^{x-2} \tag{6.187}$$

where ω is the fraction of polymer material formed by disproportionation and/or chain transfer reactions. An expression for P can be obtained by inserting Eqs. (6.170)-(6.172) into Eq. (6.169) and using the relation $[\text{M}^\bullet] = R_p/k_p[\text{M}]$. This gives

$$\frac{1}{P} = 1 + C_\text{M} + C_\text{I}\frac{[\text{I}]}{[\text{M}]} + C_\text{S}\frac{[\text{S}]}{[\text{M}]} + \frac{2(k_{tc} + k_{td})}{k_p^2[\text{M}]^2}R_p \tag{6.188}$$

The value of ω is given by

$$\begin{aligned}
\omega &= \frac{k_{tr,\text{M}}[\text{M}^\bullet][\text{M}] + k_{tr,\text{I}}[\text{M}^\bullet][\text{I}] + k_{tr,\text{S}}[\text{M}^\bullet][\text{S}] + 2k_{td}[\text{M}^\bullet]^2}{k_{tr,\text{M}}[\text{M}^\bullet][\text{M}] + k_{tr,\text{I}}[\text{M}^\bullet][\text{I}] + k_{tr,\text{S}}[\text{M}^\bullet][\text{S}] + 2(k_{tc} + k_{td})[\text{M}^\bullet]^2} \\
&= \frac{C_\text{M}[\text{M}] + C_\text{I}[\text{I}] + C_\text{S}[\text{S}] + 2k_{td}R_p/k_p^2[\text{M}]}{C_\text{M}[\text{M}] + C_\text{I}[\text{I}] + C_\text{S}[\text{S}] + 2(k_{tc} + k_{td})R_p/k_p^2[\text{M}]}
\end{aligned} \tag{6.189}$$

To calculate P and ω, and hence w_x, from kinetic and theoretical considerations for any polymer system, it is thus necessary to have the values of transfer constants (C_M, C_I, C_S), k_p^2/k_t, k_t, and $k_{td}/(k_{tc} + k_{td})$.

Problem 6.35   The bulk polymerization of methyl methacrylate (density 0.94 g/cm³) was carried out at 60°C with 0.0398 M benzoyl peroxide initiator (Smith et al., 1966). The reaction showed first order kinetics over the first 10-15% reaction and the initial rate of polymerization was determined to be 3.93×10^{-4} mol/L-s. From the GPC molecular weight distribution curve reported for a 3% conversion sample, the weight fraction of polymer of $\overline{DP}_n = 3000$ is seen to be 1.7×10^{-4}. Calculate the weight fraction from Eq. (6.187) to compare with this value. [Use the following data: $C_\text{I} \simeq 0.02$; $C_\text{M} = 10^{-5}$; $fk_d = 2.7 \times 10^{-6}\,\text{s}^{-1}$; $k_t = 2.55 \times 10^7$ L/mol-s; fraction of termination by disproportionation = 0.85.]

Answer:

From Eq. (6.24),

$$\begin{aligned}
\frac{k_p^2}{k_t} &= \frac{R_p^2}{[\text{M}]^2(fk_d)[\text{I}]} \\
&= \frac{(3.93 \times 10^{-4}\,\text{mol}\,\text{L}^{-1}\,\text{s}^{-1})^2}{(9.4\,\text{mol}\,\text{L}^{-1})^2(2.7 \times 10^{-6}\,\text{s}^{-1})(0.0398\,\text{mol}\,\text{L}^{-1})} = 0.0163\,\text{L}\,\text{mol}^{-1}\,\text{s}^{-1}
\end{aligned}$$

In as much as polymerization rates are measured at less than 10% conversion, it can be assumed that [M] and [I] in Eq. (6.188) are equal to their initial concentrations. Therefore,

$$\begin{aligned}
\frac{1}{P} &= 1 + 10^{-5} + (0.02)(0.0398\,\text{mol}\,\text{L}^{-1})/(9.4\,\text{mol}\,\text{L}^{-1}) \\
&\quad + \frac{(2)(3.93 \times 10^{-4}\,\text{mol}\,\text{L}^{-1}\,\text{s}^{-1})}{(0.0163\,\text{L}\,\text{mol}^{-1}\,\text{s}^{-1})(9.4\,\text{mol}\,\text{L}^{-1})^2} \\
&= 1.0006 \implies P = 0.9994
\end{aligned}$$

Since $k_{td}/(k_{tc} + k_{td}) = 0.85$ and $k_{tc} + k_{td} = k_t = 2.55 \times 10^7$ L mol^{-1} s^{-1}, it follows that $k_{td} = 2.167 \times 10^7$ L mol^{-1} s^{-1}. Therefore,

$$k_p^2/k_{td} = (k_p^2/k_t)(k_t/k_{td})$$

$$= (0.0163 \text{ L mol}^{-1}\text{s}^{-1})(2.55 \times 10^7/2.167 \times 10^7) = 0.0192 \text{ L mol}^{-1}\text{s}^{-1}$$

Substituting the appropriate values in Eq. (6.189) yields $\omega = 0.871$. Therefore, from Eq. (6.187),

$$w_{3000} = (0.871)(3000)(1 - 0.9994)^2 (0.9994)^{3000-1}$$

$$+ \tfrac{1}{2}(1 - 0.871)(3000)(3000 - 1)(1 - 0.9994)^3 (0.9994)^{3000-2}$$

$$= 1.8 \times 10^{-4} \quad (\text{cf. reported value} = 1.7 \times 10^{-4})$$

6.12.2 High-Conversion Polymerization

The size distributions in high-conversion polymerizations are much broader and less predictable than those in low-conversion case. A number of factors combine to make this happen. It is seen from Eq. (6.102) that the kinetic chain length, and hence the polymer molecular weight, depends on the ratio $[M]/[I]^{1/2}$. Since $[I]$ usually decreases faster than $[M]$ as the polymerization proceeds, the molecular weight of the polymer produced at any instant increases with the conversion. This results in broadening of the molecular weight distributions at high conversions with $\overline{DP}_w/\overline{DP}_n$ ratio exceeding the theoretical limiting value of 1.5 or 2. If, moreover, autoacceleration or gel effect also occurs in polymerization, it can lead to even larger broadening. According to Eq. (6.102), the molecular weight also depends on the ratio $k_p/k_t^{1/2}$. Since autoacceleration is caused by reduction in the termination rate resulting in large increases in the $k_p/k_t^{1/2}$ ratio, it is always accompanied by large increases in the average molecular weight and the breadth of the distribution with the $\overline{DP}_w/\overline{DP}_n$ ratio reaching values as high as 5 to 10.

In the case of chain transfer to polymer leading to branching, which assumes greater significance at higher conversions when the polymer concentration is high, a further broadening of the molecular weight distribution can occur and $\overline{DP}_w/\overline{DP}_n$ ratios as high as 20 to 50 can be found in practice. In commercial productions of polymers, efforts are always made to minimize molecular weight broadening.

6.13 Polymerization Processes

Free-radical polymerizations are commonly carried out by four different processes: (a) bulk or mass polymerization, (b) solution polymerization, (c) suspension polymerization, and (d) emulsion polymerization. A new process, termed "grafting-from" polymerization, has recently been added (see later). The processes (c) and (d) are essentially of the heterogeneous type containing a large proportion of nonsolvent (usually water) acting as a dispersion medium for the immiscible liquid monomer. Bulk and solution polymerizations are homogeneous processes, but some of these homogeneous systems may become heterogeneous with progress of polymerization due to the polymer formed being insoluble in its monomer (for bulk polymerization) or in the solvent used to dilute the monomer (for solution polymerization). The kinetic schemes earlier described in this chapter will apply to free radical polymerizations in bulk monomer, solution, or in suspension, but the kinetics of emulsion polymerization, to be discussed later, are different.

Polymerization in bulk, that is, of undiluted monomer, minimizes any contamination of the product. Bulk polymerization is difficult to control, however, due to the high exothermicity and high activation energies of free-radical polymerization and the tendency toward the gel effect in some cases.

By carrying out the polymerization of a monomer in a solvent many of the disadvantages of the bulk process can be avoided. The solvent acting as a diluent reduces the viscosity gain with conversion, allows more efficient agitation or stirring of the medium, thus effecting better heat transfer and heat dissipation. Solution polymerization is advantageous only if the polymer formed is to be applied in solution as, for example, the making of coating (lacquer) grade poly(methyl methacrylate) resins from methyl methacrylate and related monomers.

Suspension polymerization combines the advantages of both the bulk and solution polymerization techniques. It is used extensively in the mass production of vinyl and related polymers. In suspension polymerization (also referred to as *bead* or *pearl* polymerization), the monomer is suspended as droplets by efficient agitation in a large volume (continuous phase) of nonsolvent, commonly referred to as the dispersion or suspension medium. Water is used as the suspension medium for water insoluble monomers because of its obvious advantages. Styrene, methyl methacrylate, vinyl chloride, and vinyl acetate are polymerized by the suspension process.

The size of the monomer droplets usually ranges between 0.1-5 mm in diameter. Suspension is maintained by mechanical agitation and addition of stabilizers. Low concentrations of suitable water-soluble polymers such as carboxymethyl cellulose (CMC) or methyl cellulose, poly(vinyl alcohol), gelatin, etc., are used as suspension stabilizers. They raise the medium viscosity and effect stabilization by forming a thin layer on the monomer-polymer droplets. Water insoluble inorganic compounds such as bentonite, kaolin, magnesium silicate, and aluminum hydroxide, in finely divided state, are sometimes used to prevent agglomeration of the monomer droplets. Initiators soluble in monomer, such as organic peroxides, hydroperoxides, or azocompounds — often referred to as *oil-soluble* initiators — are used. Each monomer droplet in a suspension polymerization thus behaves as a miniature bulk polymerization system and the kinetics of polymerization within each droplet are the same as those for the corresponding bulk polymerization. At the end of the polymerization process, the monomer droplets appear in the form of tiny polymer beads or pearls, and hence, the process is also known as *bead* or *pearl polymerization*.

Problem 6.36   A monomer is polymerized at 80°C (a) in benzene solution and (b) in aqueous suspension in two separate runs, both containing 60 g of the monomer (density 0.833 g/cm^3) and 0.242 g of a peroxide initiator in a total volume of 1 liter. If the initial rate of polymerization for 1 liter of solution is 0.068 mol/h, what is the expected initial rate for 1 liter of suspension ? (Assume that rate constants and the initiator efficiency are same in both cases.)

Answer:

Let M_A = molar mass (g/mol) of monomer A; M_I = molar mass (g/mol) of initiator I.

(a) <u>Solution</u> (1 liter):   [M], mol/L = $60/M_A$;   [I], mol/L = $0.242/M_I$

(b) <u>Suspension</u> (1 liter):

$$[M], \text{mol/L} = \frac{(1000 \text{ cm}^3)(0.833 \text{ g cm}^{-3})}{M_A} = \frac{833}{M_A}$$

$$[I], \text{ mol/L} = \frac{(0.242 \text{ g})(1000 \text{ cm}^3 \text{ L}^{-1})}{(M_1 \text{ g mol}^{-1})(60 \text{ g}/(0.833 \text{ g cm}^{-3})} = \frac{3.36}{M_1}$$

$$(R_p)_a = \left(-\frac{d[M]}{dt}\right)_a = k_p (fk_d/k_t)^{1/2} [M]_a [I]_a^{1/2}$$

$$(R_p)_b = k_p (fk_d/k_t)^{1/2} [M]_b [I]_b^{1/2}$$

$$\frac{(R_p)_b}{(R_p)_a} = \frac{(833/M_A)(3.36/M_1)^{1/2}}{(60/M_A)(0.242/M_1)^{1/2}} = 51.7$$

$$(R_p)_b = 51.7(0.068 \text{ mol h}^{-1} \text{ L}^{-1}) = 3.5 \text{ mol h}^{-1} \text{ L}^{-1}$$

For 1 liter suspension, volume of monomer $= (60 \text{ g})/(0.833 \text{ g cm}^{-3}) = 72 \text{ cm}^3$ monomer; rate $=$ 3.5×0.072 or 0.252 mol h^{-1}.

6.13.1 Emulsion Polymerization

In emulsion polymerization, monomers are polymerized in the form of emulsions and polymerization in most cases involve free-radical reactions. Like suspension polymerization, the emulsion process uses water as the medium. Polymerization is much easier to control in both these processes than in bulk systems because stirring of the reactor charge is easier due to lower viscosity and removal of the exothermic heat of polymerization is greatly facilitated with water acting as the heat sink. Emulsion polymerization, however, differs from suspension polymerization in the nature and size of particles in which polymerization occurs, in the type of substances used as initiators, and also in mechanism and reaction characteristics. Emulsion polymerization normally produces polymer particles with diameters of 0.1-3μ. Polymer *nanoparticles* of sizes 20-30 nm are produced by *microemulsion polymerization* (Antonietti et al., 1999; Yildiz et al., 2003).

6.13.1.1 Qualitative Picture

The original theory of emulsion polymerization is based on the qualitative picture of Harkins (1947) and the quantitative treatment of Smith and Ewart (1948). The essential ingredients in an emulsion polymerization system are water, a monomer (not miscible with water), an emulsifier, and an initiator which produces free radicals in the aqueous phase. Monomers for emulsion polymerization should be *nearly* insoluble in the dispersing medium but not completely insoluble. A slight solubility is necessary as this will allow the transport of monomer from the emulsified monomer reservoirs to the reaction loci (explained later).

Emulsifiers are soaps or detergents and they play an important role in the emulsion polymerization process. A detergent molecule is typically composed of an ionic hydrophilic end and a long hydrophobic chain. Some examples are:

Anionic detergent

Sodium laurate $CH_3(CH_2)_{10}COO^- Na^+$

Sodium alkyl aryl sulfonate C_nH_{2n+1} —⟨◯⟩— $SO_3^- Na^+$

Cationic detergent

Cetyl trimethyl ammonium chloride $C_{16}H_{33} \overset{+}{N}(CH_3)_3 Cl^-$

Anionic and cationic detergent molecules may thus be represented by ——•⁻ and ——•⁺, respectively, indicating hydrocarbon (hydrophobic) chains with ionic (polar) end groups.

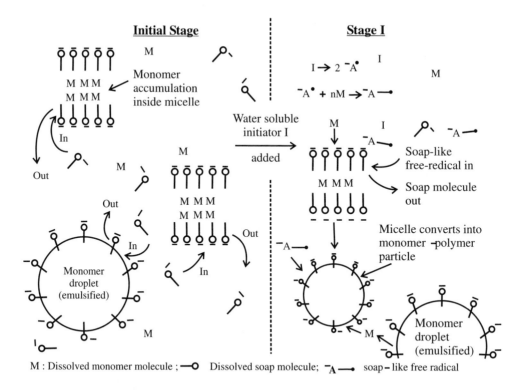

Figure 6.14 Schematic of emulsion polymerization showing three phases present. (After Williams, 1971.)

Let us now consider the locations of the various components in an emulsion polymerization system. A *micelle* of an anionic detergent can be depicted as a cluster of detergent molecules ($—\!\!•^-$) with the hydrocarbon chains directed toward the interior and their polar heads in water (see Fig. 6.14). In the same way, detergent molecules get adsorbed on the surface of an oil droplet suspended in water. Such materials are therefore said to be surface active and are also called *surfactants*.

When a relatively water-insoluble vinyl monomer, such as styrene, is emulsified in water with the aid of an anionic surfactant and adequate agitation, three phases result (see Fig. 6.14): (1) an aqueous phase in which small amounts of both monomer and surfactant are dissolved (i.e., they exist in molecular dispersed state); (2) *emulsified monomer droplets* which are supercolloidal in size ($> 10,000$ Å), stability being imparted by the reduction of surface tension and the presence of repulsive forces between the droplets since a negative charge overcoats each monomer droplet; (3) submicroscopic (colloidal) micelles which are saturated with monomer. This three-phase emulsion represents the initial state for emulsion polymerization (Fig. 6.14).

Stage I (see Fig. 6.14) begins when a free-radical producing water-soluble initiator is added to the three-phase emulsion described above. The commonly used initiator is potassium persulfate, which decomposes thermally to form water-soluble sulfate radical ions:

$$S_2O_8^{2-} \xrightarrow{50-60^{\circ}C} 2SO_4^{\cdot-} \qquad (6.190)$$

The rate of radical generation by an initiator is greatly accelerated in the presence of a reducing agent. Thus an equimolar mixture of $FeSO_4$ and $K_2S_2O_8$ at 10°C produces radicals by the reaction:

$$S_2O_8^{2-} + Fe^{2+} \longrightarrow Fe^{3+} + SO_4^{2-} + SO_4^{-\bullet} \tag{6.191}$$

about 100 times as fast as an equal concentration of the persulfate alone at 50°C (Rudin, 1982). (Redox systems generally find use for polymerizations only at lower temperatures.)

The sulfate radical ions generated from persulfate react with the dissolved monomer molecules in the aqueous phase to form ionic free radicals ($^-\bullet\!-\!-\bullet$):

$$SO_4^{-\bullet} + (n+1)M \longrightarrow {}^-SO_4(M)_nM^\bullet \tag{6.192}$$

These ionic free radicals can be viewed as soaplike anionic free radicals as they are essentially made up of a long hydrocarbon chain carrying an ionic charge at one end and a free-radical center at the other (represented as $^-A\!-\!-\bullet$ in Fig. 6.14). They thus behave like emulsifier molecules and because of the existence of a dynamic equilibrium between micellar emulsifier and dissolved emulsifier they also can at some stage be implanted in some of the micelles.

Once implanted in a micelle, a soap-like anionic free radical initiates polymerization of the solubilized monomer in the micelle. The micelle, thus "stung," grows in size as the solubilized monomer in the micelle is used up and to replenish it more monomer enters the micelle from monomer droplets via the aqueous dispersion phase. The 'stung' micelle is in this way transformed into a monomer-polymer (M/P) particle (see Fig. 6.14). Thus in Stage I, the system will consist of an aqueous phase containing dissolved monomer, dissolved soap-like free radicals, micelles, 'stung' micelles, M/P particles, and monomer droplets. The rate of overall polymerization increases continuously since nucleation of new particles (i.e., conversion of a micelles into M/P particles) and particle growth occur simultaneously. For the same reason a particle size distribution occurs during stage I. However, at 13-20% monomer conversion, nearly all the emulsifier will be adsorbed on the M/P particles and the micelles will disappear. Since new particles mostly originate in micelles, with the disappearance of micelles the nucleation of new M/P particles essentially ceases. This marks the end of Stage I.

Problem 6.37   In the model for emulsion polymerization it is assumed that most of the soap-like free radicals produced in the aqueous phase enters the micelles rather than the emulsified monomer droplets. How would you justify this assumption?

Answer:

The assumption that the soaplike free radicals produced in the aqueous phase enters the micelles rather than the emulsified monomer droplets can be justified because of two reasons:

(a) Micelles have a much higher surface area to volume ratio than the monomer droplets since the former are much smaller in size than the latter. Micelles will therefore have a greater probability of receiving the soap-like anionic free radicals.

(b) The number of micelles per unit volume of aqueous phase is much more than the number of monomer droplets per unit volume, the typical values of these numbers being 10^{18} versus 10^{11}.

The monomer droplets can thus be considered to serve primarily as reservoirs of monomer and the number of free radicals that might be captured by them can be conveniently ignored.

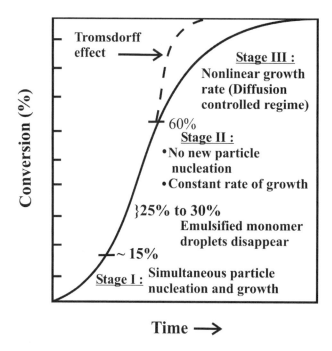

Figure 6.15 Schematic conversion-time curve for a typical emulsion polymerizations showing three main stages of the polymerization process. (After Williams, 1971.)

In Stage II that follows there occurs a continued growth of the existing M/P particles in the absence of any new particle nucleation. Free radicals enter only the M/P particles where polymerization takes place as the particles are supplied with monomer from the emulsified monomer droplets via the aqueous phase. Stage II thus features a constant overall rate of polymerization (Fig. 6.15). It is followed by Stage III which begins when the overall rate of polymerization begins to deviate from linearity and nonlinear growth rate is observed. The nonlinearity may appear (a) as a decrease in rate due to dwindling monomer concentration inside M/P particles or (b) as an increase if the Tromsdorff effect becomes important with the monomer reservoirs having already disappeared in Stage II and the ratio of monomer to polymer within the particle reduced to a point where the reaction becomes diffusion-controlled.

6.13.1.2   Kinetics of Emulsion Polymerization

The kinetic analysis here is based on quantitative considerations of the ideal emulsion polymerization systems which have been described qualitatively in the preceding sections. The treatment centers only around Stage I and Stage II (Fig. 6.15), as no general theory for Stage III is available.

As described above, Stage I is the nucleation period. Ideally, all M/P particles are generated in Stage I and they grow in volume and adsorb surfactant molecules, resulting in diminution of micelles. For simplicity of calculations, it may be assumed that once polymerization is initiated in a particle its volume grows linearly with time at rate u. It may, however, be noted that particles are nucleated at different times during Stage I. Hence at time t in Stage I, the volume of a particle which is nucleated at time τ will be $u(t - \tau)$, if the volume of the initial micelle is negligible. The area of this spherical particle, $a_{t,\tau}$, will thus be (Smith and Ewart, 1948; Rudin, 1982):

$$a_{t,\tau} = [(4\pi)^{1/2} 3u(t - \tau)]^{2/3} \tag{6.193}$$

Assuming that radicals nucleate micelles at a constant rate v (i.e., number of radicals/cm^3 aqueous phase/s), $vd\tau$ particles will be generated in the period $d\tau$ and the area A_t of all particles at time t will be given by the sum of the areas of all particles generated till time t. Thus from Eq. (6.193) one obtains (Smith and Ewart, 1948; Rudin, 1982):

$$A_t = [(4\pi)^{1/2} 3u]^{2/3} \int_o^t (t - \tau)^{2/3} v d\tau = 0.6[(4\pi)^{1/2} 3u]^{2/3} vt^{5/3} \tag{6.194}$$

Clearly, no micelles can remain when $A_t = a_s w_s$, where a_s is the area occupied by unit weight of surfactant and w_s is the weight concentration of surfactant. Suppose this condition is reached at time t_c, which then marks the end of Stage I, that is,

$$a_s w_s = 0.6[(4\pi)^{1/2} 3u]^{2/3} vt_c^{5/3} \tag{6.195}$$

Primary radicals are produced by thermal decomposition of the initiator I in the aqueous phase. In units of number of radicals/cm^3 aqueous phase/s, the rate of radical generation, R_r, is

$$R_r = 2N_{Av} k_d [I] \tag{6.196}$$

where N_{Av} is Avogadro's number. The fraction of radicals that enter particles may be taken as the ratio of the area of all particles to that of the total surfactant/water interface, that is, $A_t/a_s w_s$. Since nucleation is caused by radicals that enter micelles rather than particles, one may write (Rudin, 1982):

$$v = R_r[1 - A_t/a_s w_s] \tag{6.197}$$

Combining Eqs. (6.195) and (6.197) and solving for t_c with the assumption that $A_t/(a_s w_s) \ll 1$, one obtains

$$t_c = 0.53(a_s w_s/R_r)^{0.6}/u^{0.4} \tag{6.198}$$

Thus, the number of particles per unit volume of aqueous phase, N_p, at the end of Stage I is (Smith and Ewart, 1948):

$$N_p = vt_c = 0.53(a_s w_s)^{0.6} (R_r/u)^{0.4} \tag{6.199}$$

This equation indicates that the particle number depends on the 0.6 power of the surfactant concentration and on the 0.4 power of the initiator concentration (through R_r). Typical values (Rudin, 1982) pertaining to emulsion recipes are: $N_p \sim 10^{15} - 10^{16}$ per cm^3 of aqueous phase, $R_r \sim 10^{12} - 10^{14}$ radicals cm^{-3} s^{-1}, $a_s w_s \sim 10^5$ cm^2/cm^3 aqueous phase, $u \sim 10^{-20}$ cm^3 s^{-1}.

Problem 6.38   An emulsion polymerization recipe has the following composition :

Styrene (density 0.9 g/cm^3)	100 g
Water (density 1.0 g/cm^3)	200 g
$K_2S_2O_8$	0.3 g
Sodium lauryl sulfate	3 g

Calculate the number of polymerizing particles per liter of water using the following additional data :   surface area per surfactant molecule $= 4\times10^{-15}$ cm^2; rate of volume increase of latex particle $= 5\times10^{-20}$ cm^3/s; k_d of $K_2S_2O_8$ at 50°C $= 1.7\times10^{-6}$ s^{-1}.

Answer:

Molar mass of sodium lauryl sulfate = 288 g mol^{-1}

$$a_s = \frac{(4\times10^{-15} \text{ cm}^2 \text{ molecule}^{-1})(6.02\times10^{23} \text{ molecules mol}^{-1})}{(288 \text{ g mol}^{-1})}$$

$$= 0.836\times10^7 \text{ cm}^2 \text{ g}^{-1}$$

$$w_s = \frac{(3.0 \text{ g})(1000 \text{ cm}^3 \text{ L}^{-1})}{(200 \text{ cm}^3)} = 15 \text{ g L}^{-1}$$

$$(a_s w_s)^{0.6} = 7.23\times10^4 \text{ cm}^{1.2} \text{ L}^{-0.6}$$

$$[I] = \frac{(0.3 \text{ g})(1000 \text{ cm}^3 \text{ L}^{-1})}{(270 \text{ g mol}^{-1})(200 \text{ cm}^3)} = 0.0056 \text{ mol L}^{-1}$$

$$R_r = 2k_d [I] N_{Av}$$

$$= 2(1.7\times10^{-6} \text{ s}^{-1})(0.0056 \text{ mol L}^{-1})(6.02\times10^{23} \text{ mol}^{-1})$$

$$= 1.146\times10^{16} \text{ L}^{-1} \text{ s}^{-1}$$

$$\left(\frac{R_r}{u}\right)^{0.4} = \left[\frac{(1.146\times10^{16} \text{ L}^{-1} \text{ s}^{-1})}{(5\times10^{-20} \text{ cm}^3 \text{ s}^{-1})}\right]^{0.4} = 1.393\times10^{14} \text{ L}^{-0.4} \text{ cm}^{-1.2}$$

$$N_p = 0.53(a_s w_s)^{0.6}\left(\frac{R_r}{u}\right)^{0.4}$$

$$= 0.53\left(7.23\times10^4 \text{ cm}^{1.2} \text{ L}^{-0.6}\right)\left(1.393\times10^{14} \text{ cm}^{-1.2} \text{ L}^{-0.4}\right)$$

$$= 5.34\times10^{18} \text{ L}^{-1}$$

Here L^{-1} implies per liter of aqueous phase.

In Stage II the conversion is linear with time (see Fig. 6.15) as micelles have already disappeared and no new particle is nucleated. The number of M/P particles N_p in Stage II is fixed at the value formed in Stage I and is given by Eq. (6.199).

Rate of Polymerization   For estimating the rate of polymerization in Stage II, one may generally assume that free radicals enter the particles singly. For example, considering typical values pertinent to emulsion polymerization, if the rate of generation of free radicals (R_r) in the aqueous phase is 10^{14} per second per milliliter and the value of the number of polymer particles is 10^{15} per milliliter, then assuming that all the radicals generated eventually enter M/P particles (since there

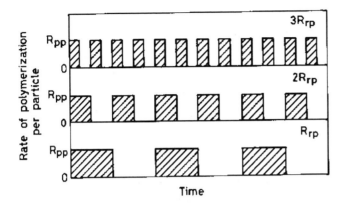

Figure 6.16 A bar diagram showing the effect of rate of radical entry into particle on rate of polymerization per particle (R_{pp}). The total period of activity (sum of shaded regions) is not changed even if the rate of radical entry into particle is increased from R_{rp} to $2R_{rp}$ and further to $3R_{rp}$. (After Williams, 1971.)

are no micelles) the rate of radical entry into a particle will average out to about 0.1 radical per second, i.e., one radical every 10 seconds.

When a (soap-like) free radical enters a M/P particle, it initiates the polymerization of monomer in the particle, but the polymerization is terminated when another free radical enters the same particle (since, as a simple calculation using known k_t values shows, two radicals cannot coexist in the same polymer particle and they would terminate mutually within a few thousands of a second). The particle thereafter remains inactive till another free radical enters and initiates the polymerization afresh. Thus, if a radical enters a M/P particle every 10 seconds as calculated above, the particle will experience alternate periods of activity (growth) and inactivity (no growth), each of 10 seconds duration. In other words, each particle will remain active for half of the total time (and inactive for the other half) and this situation will remain unchanged even if there is a change in the rate of radical entry into the particle. This feature becomes apparent from the bar diagram of Fig. 6.16, where the rate of polymerization within a particle, R_{pp}, for various rates of free radical entry into a particle is shown. One would notice that the total area of active period (shaded in Fig. 6.16) is the same in each case. The rate of polymerization per particle can thus be considered to be independent of the rate at which radicals enter into it. The aforesaid concepts are known collectively as the *Smith-Ewart Theory, Case II*.

The rate of polymerization in a M/P particle, R_{pp}, is given by

$$R_{pp} = k_p[M](\bar{n}/N_{Av}) \qquad (6.200)$$

where [M] denotes the monomer concentration in the M/P particles and $\bar{n}$ the average number of radicals per particle; N_{Av} is Avogadro's number.

Since according to the Smith-Ewart theory, the growing particle contains a free radical only half the time, $\bar{n}$ can be taken as 1/2 and the rate of polymerization per particle is then

$$R_{pp} = \frac{k_p[M]}{2N_{Av}} \qquad (6.201)$$

If R_{pp} is constant and the number of particles per unit volume, N_p, is constant, then the overall rate of emulsion polymerization per unit volume, R_p, is simply given by

$$-\frac{d[M]}{dt} = R_p = \frac{N_p k_p [M]}{2N_{Av}}$$

(6.202)

Substituting for N_p from Eq. (6.199) gives

$$R_p = 0.53 (a_s w_s)^{0.6} \left(\frac{R_r}{u}\right)^{0.4} \frac{k_p [M]}{2N_{Av}}$$

(6.203)

According to the simple Smith-Ewart model described above, the rate of polymerization in Stage II will depend on the 0.6 power of the surfactant concentration (w_s) and the 0.4 power of the rate of radical generation in aqueous phase, R_r, which in turn is related to the aqueous phase initiator concentration, [I], through Eq. (6.196). It is significant that although Eq. (6.203) gives the rate of polymerization in Stage II, the value of [I] that applies to the R_r term in this equation does in fact refer to Stage I where the particles are formed. Therefore, if the above model is valid, adding initiator again in Stage II would have no effect on the rate of polymerization.

Problem 6.39   The experimental value of dynamic concentration of styrene in polymer latex particles under the conditions of constant rate in emulsion polymerization has been found to be 5.2 mol/liter. Assuming this value to be applicable, calculate the rate of polymerization per liter of aqueous phase in Stage II of the reaction of the emulsion polymerization recipe given in Problem 6.38. (Data : k_p for styrene at 50°C= 131 L mol^{-1} s^{-1}.)

Answer:

From Eq. (6.202),

$$-\frac{d[M]}{dt} = N_p k_p [M]/2N_{Av}$$

$$= \frac{(5.34 \times 10^{18} \text{ L}^{-1})(131 \text{ L mol}^{-1} \text{s}^{-1})(5.2 \text{ mol L}^{-1})}{2(6.02 \times 10^{23} \text{ mol}^{-1})}$$

$$= 3.0 \times 10^{-3} \text{ mol L}^{-1} \text{s}^{-1}$$

Here L^{-1} implies per liter of aqueous phase.

Degree of Polymerization   Once inside the M/P particle, a radical propagates at a rate r_p given by

$$r_p = k_p [M]$$

(6.204)

where [M] is the concentration of monomer inside the M/P particle. The rate r_e at which a radical enters a polymer particle in Stage II is given by

$$r_e = R_r / N_p$$

(6.205)

where R_r is the rate of generation of free radicals per unit volume in the aqueous phase [Eq. (6.196)] and N_p is the number of M/P particles per unit volume. Equation (6.205) is based on the assumption that all radicals generated in the aqueous phase eventually enter the M/P particles, that is, radical capture efficiency is 1. Since termination reaction takes place as soon as a radical enters an active or "living" M/P particle (i.e., in which a polymer chain is propagating), r_e is also equal to the rate of termination of a polymer chain. Hence the degree of polymerization will be given by

$$\overline{DP}_n = \frac{r_p}{r_e} = \frac{N_p k_p [M]}{R_r} \tag{6.206}$$

assuming, however, that chain transfer of any kind is negligible. Equation (6.206) shows that the degree of polymerization $\overline{DP}_n$, like the rate of polymerization R_p [Eq. (6.202)], is directly dependent on the number of particles. Thus, *unlike polymerization by the bulk, solution, and suspension techniques, that by the emulsion technique permits simultaneous increase in rate and degree of polymerization* by increasing the number of polymer particles (N_p), that is, by increasing the surfactant concentration, at a fixed rate of initiation. This possibility of combining high molecular weight with high polymerization rate is one reason for the popularity of the emulsion technique.

Substituting Eqs. (6.196) into Eq. (6.206) one obtains

$$\overline{DP}_n = \frac{N_p k_p [M]}{2N_{Av} k_d [I]} \tag{6.207}$$

The initiator concentration here is that which exists at a given instant in Stage II.

Problem 6.40 A particular emulsion polymerization yields polymer with $\overline{M}_n = 200,000$. How would you adjust the operation of an emulsion process to produce polymer with $\overline{M}_n = 100,000$ in Stage II without, however, changing the rate of polymerization, particle concentration, or reaction temperature ?

Answer:

According to Eq. (6.207), $\overline{DP}_n \propto 1/[I]$, since N_p is constant. So $\overline{DP}_n$ can be reduced to half by adding more initiator in Stage II so that [I] is doubled. The rate of polymerization [Eq. (6.203)] and the particle concentration [Eq. (6.199)] will not be altered because the [I] value which is operative in these relations is that for Stage I. An alternative procedure would involve addition of chain transfer agents in appropriate concentrations.

For making large-particle-size latexes by the emulsion technique one can resort to a *seeded polymerization* process. In this a completed "seed" latex is diluted to give the desirable value of N_p particles per liter of emulsion but no additional surfactant is added (so no new polymer particles are formed). When monomer is fed and initiator is added, polymerization occurs in the previously formed particles which grow further as monomer diffuses into them and is converted (see Problem 6.41).

Problem 6.41 A polystyrene latex produced by emulsion polymerization contains 10% by weight of polymer particles (average diameter 0.20 μm). It is decided to grow these particles to a larger size by slowly adding to the latex 2 kg of monomer per kilogram of polymer as polymerization proceeds at 60°C without any new addition of emulsifier. The reaction is to be carried out until all monomer has been added to the reactor and then till the weight ratio of monomer to polymer has come down to 0.2. The unreacted monomer is then to be removed by steam stripping (Rodriguez, 1989).

Determine (a) the time required to reach the desired conversion and (b) the final particle diameter. [Data: k_p at 60°C = 165 L mol^{-1} s^{-1}. Density: monomer = 0.90 g/cm³; polymer = 1.05 g/cm³; dynamic solubility of monomer in polymer = 0.6 g monomer per gram polymer.]

Answer:

To calculate the time of reaction, two consecutive periods have to be considered: (i) period with zero order in monomer (until the monomer reserve is exhausted) and (ii) period with first order (until the required conversion is reached).

Basis: 1000 g latex (900 g water + 100 g polymer)

    Monomer to be added = 2×100 or 200 g

    Monomer to be converted = x g

    Reaction to be stopped when $\left(\dfrac{200-x}{100+x}\right) = 0.2 \implies x = 150$

    Monomer to be removed = (200 − 150) or 50 g

Monomer concentration when reaction is to be stopped is

$$\frac{[(50 \text{ g monomer})/(104 \text{ g mol}^{-1})](1000 \text{ cm}^3 \text{ L}^{-1})}{(250 \text{ g polymer})/(1.05 \text{ g cm}^{-3}) + (50 \text{ g monomer})/(0.90 \text{ g cm}^{-3})} = 1.64 \text{ mol L}^{-1}$$

Dynamic concentration of monomer in particles is

$$\frac{[(0.6 \text{ g monomer})/(104 \text{ g mol}^{-1})](1000 \text{ cm}^3 \text{ L}^{-1})}{(1 \text{ g polymer})/(1.05 \text{ g cm}^{-3}) + (0.6 \text{ g monomer})/(0.90 \text{ g cm}^{-3})} = 3.56 \text{ mol L}^{-1}$$

Let the amount of monomer remaining unconverted at the end of zero-order period be y gram. Therefore,

$$\frac{[(y \text{ g})/(104 \text{ g mol}^{-1})](1000 \text{ cm}^3 \text{ L}^{-1})}{[(300-y)/1.05] \text{ cm}^3 \text{ polymer} + (y/0.9) \text{ cm}^3 \text{ monomer}} = 3.56 \text{ mol L}^{-1}$$

    Solving, $y = 112.4$ g

Number of particles (N_p') in 1 kg of original latex:

    Initial diameter of particle = 0.2 μm = 0.2×10^{-4} cm

    Volume/particle = $(\pi/6)(0.2\times10^{-4} \text{ cm})^3$ = 4.19×10^{-15} cm³

$$N_p' = \frac{(100 \text{ g polymer})}{(1.05 \text{ g cm}^{-3})} \cdot \frac{1}{(4.19\times10^{-15} \text{ cm}^3)}$$

$$= 22.7\times10^{15}$$

Rate of conversion in zero-order period is:

$$N_p' k_p [\text{M}]/2N_{Av} = \frac{(22.7\times10^{15} \text{ particles kg}^{-1})(165 \text{ L mol}^{-1} \text{ s}^{-1})(3.56 \text{ mol L}^{-1})}{2(6.02\times10^{23} \text{ particles mol}^{-1})}$$

$$= 11.1\times10^{-6} \text{ mol s}^{-1} \text{ kg}^{-1} \ (= 4.0\times10^{-2} \text{ mol h}^{-1} \text{ kg}^{-1})$$

(i) Zero-order period:

Monomer converted per 1 kg original latex = 200 − 112.4 or 87.6 g (= 0.84 mol). Therefore,

$$\text{Time for conversion} = (0.84 \text{ mol kg}^{-1})/(4.0\times10^{-2} \text{ mol h}^{-1} \text{ kg}^{-1}) = 21 \text{ h}$$

(ii) First-order period:

From Eq. (6.202),

$$-d\ln[M]/dt = N_p k_p / 2N_{Av}$$

Integrating and solving for t,

$$t = \frac{2\ln([M]_0/[M])N_{Av}}{N_p k_p}$$

Here N_p is the number of particles per liter of latex. It can be taken, without any significant error, to be equal to the number of particles in 1 kg of latex, N_p'. Thus,

$$t = \frac{2\ln[(3.56 \text{ mol L}^{-1})/(1.64 \text{ mol L}^{-1})](6.02 \times 10^{23} \text{ particles mol}^{-1})}{(22.7 \times 10^{15} \text{ particles L}^{-1})(165 \text{ L mol}^{-1}\text{ s}^{-1})(3600 \text{ s h}^{-1})} = 69 \text{ h}$$

(a) Total time = $(21 + 69)$ or 90 h.

(b) Final particle diameter, d:

$$(d/0.2 \,\mu\text{m})^3 = \frac{(100 \text{ g}) + (150 \text{ g})}{(100 \text{ g})} = 2.5 \implies d = 2.7 \,\mu\text{m}$$

6.13.1.3 Other Theories

A number of workers have suggested that emulsion polymerization may not occur homogeneously throughout a polymer particle but either at the particle surface (Sheinker and Medvedev, 1954) or within an outer monomer-rich shell surrounding an inner polymer-rich core (Grancio and Williams, 1970). The latter has been referred to as the *shell* or *core-shell model* and has been proposed to explain the apparent anomaly between the observed constant rate behavior up to about 60 percent conversion, which according to Eq. (6.202) requires [M] in the M/P particle to be constant, and the considerable experimental evidence indicating that emulsified monomer droplets (which serve as monomer reservoirs) disappear at 25 to 30 percent conversion and the monomer concentration drops thereafter.

According to the core-shell model, the growing particle has a heterogeneous rather than a homogeneous composition, and it consists of an expanding polymer-rich (monomer-starved) core surrounded by a monomer-rich (polymer-starved) outer spherical shell (Grancio and Williams, 1970). It is the outer shell that serves as the major locus of polymerization by the Smith-Ewart (on-off) mechanism, while practically no polymerization occurs in the core as it is starved of monomer. Reaction within an outer shell or at the particle surface would be most likely to be operative for those polymerizations in which the polymer is insoluble in its own monomer or under conditions where the polymerization is diffusion-controlled such that a propagating radical cannot diffuse into the center of the particle.

6.13.2 Photoemulsion Polymerization

A combination of photopolymerization and conventional emulsion polymerization has been used recently (Guo et al., 1999) for the synthesis of well defined polyelectrolyte brushes (Fig. 6.17). [The term "brush" denotes a layer of polymeric chains strongly attached with one end to the surface of a substrate and with significant overlap of the chains (Israelachvili, 1992).] The process described for the synthesis of spherical polyelectrolyte brushes consisting of a polystyrene (PS) core and a shell of poly(acrylic acid) chains, attached to the PS-surface by covalent bonds, proceeds in three steps, as shown in Fig. 6.18(a). In the first step, a PS latex is prepared by conventional emulsion polymerization using a surfactant such as sodium dodecyl sulfonate and a water soluble initiator such as potassium persulfate. In the second step, a thin polymeric shell is generated on the core particles by the polymerization of a vinyl group containing photoinitiator, such as 2-[p-(2-hydroxy-2-methyl propiophenone)]-ethylene glycol-methacrylate (HMEM) (Fig. 6.18b), which is, however, added under starved conditions in order to avoid formation of new particles and to ensure a well-defined core-shell morphology of the resulting particles. The polymer of HMEM formed on the surface during this second emulsion polymerization acts as a photoinitiator in the third step, in which the HMEM-covered particles are irradiated with UV light.

The emulsion polymerization in the third step is carried out in the presence of a water soluble monomer, such as acrylic acid. The radicals formed by the photolysis of HMEM (Fig. 6.18b) on the surface start radical chain polymerization by a "grafting-from" technique (see Section 6.13.3) thus generating chains of poly(acrylic acid). The polymer chains remain bound to the surface by an ester bond which can be cleaved by hydrolysis to obtain the polymer for analysis. Thus the molecular weight of the bound polymer chains can be determined which gives their contour length L_c. The thickness L of the brush (Fig. 6.17) attached to the surface of the particles can be deduced from the hydrodynamic radius as measured by dynamic light scattering.

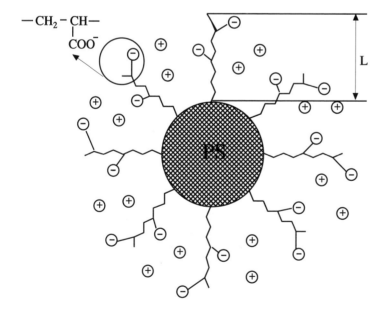

Figure 6.17 Schematic representation of a spherical polyelectrolyte brush in which linear chains of poly(acrylic acid) (PAA) are chemically grafted onto the surface of a colloidal polystyrene (PS) particle. L denotes thickness of the PAA brush. (From Guo and Ballauff, 2000. With permission from *Americam Chemical Society*.)

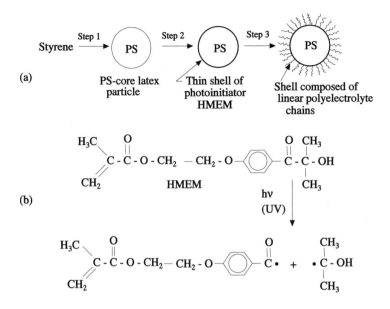

Figure 6.18 Schematic description of the method of preparation of spherical polyelectrolyte brushes by photoemulsion polymerization : (a) general scheme; (b) photodecomposition of initiator HMEM. (After Guo and Ballauff, 2000.)

It should be noted that since two radicals are generated for each decomposed initiator molecule (Fig. 6.18b), only half of the generated polymer will be attached to the surface, while the other half will grow in solution and must subsequently be removed from the latex, such as by extensive serum replacement against pure water (Guo et al., 1999; Guo and Ballauff, 2000).

The method in the third step producing a photochemical reaction in an emulsion medium has been referred to as *photoemulsion polymerization*. The strong turbidity of the latex system presents no obstacle for photoinitiation, however, because the incoming light is scattered elastically. The light that is not absorbed by the photoinitiator on the surface of the particles is scattered, eventually reaching other particles to start a new radical. This method allows one to obtain narrowly distributed polymer brushes by photoemulsion polymerization despite the strong turbidity of these suspensions (Guo and Ballauff, 2000). While the radius of the core latex particles depends on the conventional emulsion polymerization in the first step, the method of photoemulsion polymerization allows one to vary the contour length L_c of the attached chains, and hence the thickness L of the polymer brush, independently.

6.13.3 *"Grafting-From" Polymerization*

In order to modify the surface properties of inorganic materials (e.g., silica gels, silicon wafers), they are often coated with ultrathin films prepared from a large variety of polymers (Halperin et al., 1992). Till recently, most such systems were based on the physisorption of already formed polymers. However, the interaction between the polymer and the surface is usually not strong as in most cases it is caused only by van der Waals forces or hydrogen bonding. A much stronger adhesion between the polymer chains and the substrate results if the macromolecules are covalently bound to the surface. A common way to achieve this is to synthesize end-functionalized polymers and react them with appropriate surface sites ("grafting-to" technique). However, this technique is intrinsically limited to low graft densities and low film thicknesses, as only very small amounts of polymer (typically, less than 5 mg/m^2) can be immobilized to the substrates (Prucker and Rühe, 1998).

A more promising approach is to use surface immobilized initiators for *in situ* generation of grafted polymers ("grafting-from" polymerization). In most such systems described in the literature (Engel, 1980), an anchor molecule is immobilized at the surface of the substrate and the initiating species (usually azo compounds for polymerization of vinyl monomers) is linked to this anchor molecule in one or several additional reaction steps. This procedure of stepwise generation of the monolayers usually leads to low densities of the surface-attached initiator (and polymer, subsequently) besides giving rise to the possibilities of incomplete conversion and formation

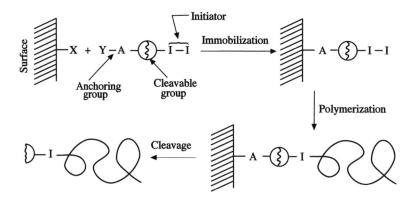

Figure 6.19 Schematic representation of the "grafting-from" technique for the preparation of terminally attached polymer monolayers by radical chain polymerization. (After Prucker and Rühe, 1998.)

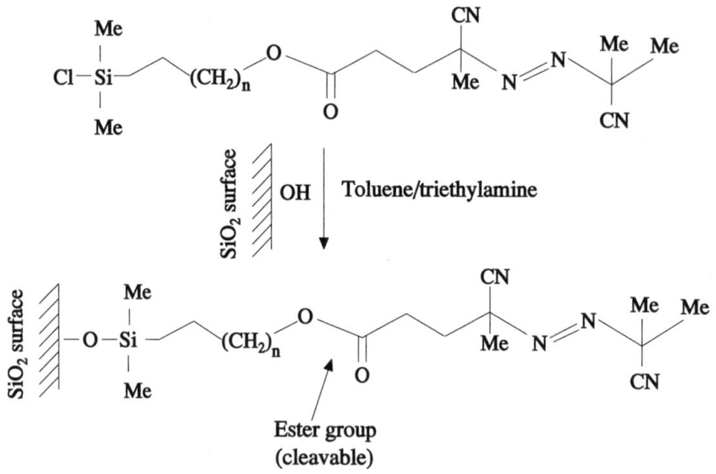

Figure 6.20 Self-assembling of an azomonochlorosilane containing ester group on the surface of silica gel. (After Prucker and Rühe, 1998.)

of structures other than the initiator. In these systems, moreover, there is no simple way to detach the grafted polymers quantitatively to make them available for standard techniques of analysis or other purposes.

The aforesaid problems are avoided in a more recently developed technique (Prucker and Rühe, 1998) in which the initiator containing a cleavable group and an anchoring group is self-assembled on the surface of the substrate (Fig. 6.19) and then activated in the presence of a monomer to initiate *in situ* polymerization at the surface. The presence of the cleavable group then allows for degrafting of the generated macromolecules for analytical purposes. In a typical example, an azomonochlorosilane containing a cleavable ester group is linked to the surface of the silica substrate through a base-catalyzed condensation reaction (Fig. 6.20). Polymer monolayers covalently attached to the surface of the silica gel can then be generated by radical polymerization *in situ*.

The "grafting-from" polymerization technique has attracted a great deal of attention because polymer brushes (Fig. 6.17) can be generated with high grafting densities of surface-attached neutral macromolecules as well as polyelectrolytes (Guo and Ballauff, 2000). While the aforesaid techniques produce grafting-from polymers or polymer brushes on flat substrates, a novel procedure to generate polyelectrolyte brushes by grafting-from polymerization inside a hollow-capsule microreactor has been recently reported (Choi et al., 2005). Figure 6.21 illustrates the fabrication of polyelectrolyte hollow capsules with initiator bound on the inner wall and subsequent grafting-from polymerization inside the capsules. In this procedure, hollow capsule reactors coated inside with initiator are fabricated by the layer-by-layer (LBL) assembly technique (Antipov et al., 2001) on melamine formaldehyde (MF) colloidal particles using poly(allylamine hydrochloride) (PAH) and poly(styrene sulfonate) (PSS) as the positive and negative polyelectrolytes and potassium peroxodisulfate as the water-soluble initiator, followed by removal of the template MF core by treatment with 0.15 M HCl (pH < 1). Poly(styrene sulfonate) bound to the inner wall of the hollow capsules has then been formed by dispersing the water soluble monomer styrene sulfonate (SS) in a suspension of the initiator-bound hollow capsules at 70°C. The hollow capsule can also be used as a novel submicroreactor to control polymerization behavior (Choi et al., 2005).

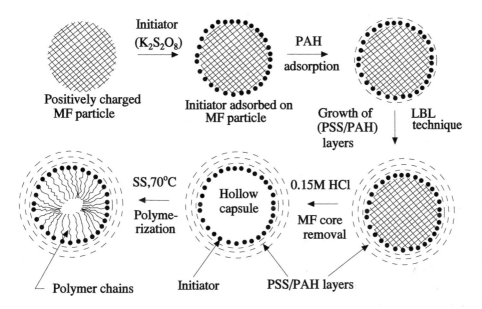

Figure 6.21 Schematic description of the steps used for fabricating hollow capsules (coated inside with potassium persulfate initiator) and grafting-from polymerization of styrene sulfonate (SS) inside the capsules. (From Choi et al., 2005. With permission from *John Wiley & Sons, Inc.*)

6.14 Living Radical Polymerization

Living polymerization was discovered in anionic system by Szwarc (see p. 476) in 1950, which, as we shall see in Chapter 8, offers many benefits including the ability to control molecular weight and polydispersity and to prepare block copolymers and other polymers of complex architecture. Many attempts have then been made to develop a living polymerization process with free-radical mechanism so that it could combine the virtues of living polymerization with versatility and convenience of free-radical polymerization. Considering the enormous importance and application potential of living/controlled radical polymerization techniques, these will be considered in detail in another chapter (Chapter 11) with a state-of-the art discussion on the subject.

REFERENCES

Ahmed, K. R., Natarajan, L. V., and Anivaruddin, Q., *Makromol. Chem.*, **179**, 1193 (1978).

Allcock, H. R. and Lampe, F. W., "Contemporary Polymer Chemistry", Prentice Hall, Englewood Cliffs, N.J., 1990.

Antipov, A. A., Sukhorov, G. B., Donath, E., and Möhwald, H., *J. Phys. Chem. B,* **105**, 2281 (2001).

Antonietti, M., Caruso, R. A., Goltner, C. G., and Weissenberger, M. C., *Macromolecules*, **32**, 1383 (1999).

Bamford, C. H., Barb, W. G., Jenkins, A. D., and Onyon, R. F., "The Kinetics of Vinyl Polymerization by Radical Mechanisms", Butterworths, London, 1958.

Barr, N. J., Bengough, W. I., Beveridge, G., and Park, G. B., *Eur. Polym. J.*, **14**, 245 (1978).

Berger, K. C., Deb, P. C., and Meyerhoff, G., *Macromolecules*, **10**, 1075 (1977).

Bevington, J. C., *Trans. Faraday Soc.*, **51**, 1392 (1955).

Bonta, G., Gallo, B. M., Russo, S., and Uliana, C., *Polymer*, **17**, 217 (1976).

Brandrup, J. and Immergut, E. H., Eds., "Polymer Handbook", 2nd ed., Wiley Interscience, New York, 1975.

Brandrup, J., Immergut, E. H., and Grulke, E. A., Eds., "Polymer Handbook", 4th ed., Wiley Interscience, New York, 1999.

Briers, F., Chapman, D. L., and Walters, E., *J. Chem. Soc.*, 562 (1926).

Choi, W. S., Park, J-H., Koo, H. Y., Kim, J-Y., Cho, B. K., and Kim, D-Y., *Angew. Chem.*, **117**, 1120 (2005).

Cohen, S. G., *J. Am. Chem. Soc.*, **67**, 17 (1945); **69**, 1057 (1947).

Dainton, F. S. and Ivin, K. J., *Qtly. Rev.*, **12**, 61 (1958).

Eastmond, G. C., "Chain Transfer, Inhibition and Retardation", Chap. 2 in "Comprehensive Chemical Kinetics", Vol. 14A (C. H. Bamford, and C. F. H. Tipper, eds.), American Elsevier, New York, 1976.

Engel, P. S., *Chem. Rev.*, **80**, 99 (1980).

Flory, P. J., "Principles of Polymer Chemistry", Cornell Univ. Press, Ithaca, N.Y., 1953.

Ghosh, P., "Polymer Science and Technology of Plastics and Rubbers", Tata McGraw-Hill, New Delhi, 1990.

Ghosh, P. and Billmeyer, F. W., *Advances in Chemistry Series*, **91**, 75 (1969).

Ghosh, P. and Banerjee, A. N., *J. Polym. Sci. Polym. Chem. Ed.*, **12**, 375 (1974).

Gobran, R. H., Berenbaum, M. B., and Tobolsky, A. V., *J. Polym. Sci.*, **46**, 431 (1960).

Grancio, M. R. and Williams, D. J., *J. Polym. Sci. A-1*, **8**, 2617 (1970).

Gregg, R. A. and Mayo, F. R., *Disc. Faraday Soc.*, **2**, 328 (1947).

Guo, X. and Ballauff, M., *Langmuir*, **16**, 8719 (2000).

Guo, X., Weiss, A., and Ballauff, M., *Macromolecules*, **32**, 6043 (1999).

Halperin, A., Tirrell, M., and Lodge, T. P., *Adv. Polym. Sci.*, **100**, 31 (1992).

Harkins, W. D., *J. Am. Chem. Soc.*, **69**, 1428 (1947).

Israelachvili, J. N., "Intermolecular and Surface Forces", 2nd Ed., Academic Press, London, 1992.

Jenkins, A. D., *J. Polym. Sci.*, **29**, 245 (1958).

Koenig, T., "The Decomposition of Peroxides and Azoalkanes", Ch. 3 in "Free Radicals" (J. K. Kochi, ed), Vol. 1, Wiley, New York, 1973.

Le, T. P., Moad, G., Rizzardo, E., and Thang, S. W., PCT Int. Appl. Wo 9801478 A1 980115; Chem. Abstr., **128**, 115390 (1998).

Mageli, O. L. and Kolczynski, J. R., in Vol. 9, "Encyclopedia of Polymer Science and Technology", Mark, H. F., Gaylord, N. G., and Bikales, N. M., Eds., Wiley-Interscience, New York, 1968.

Matheson, M. S., Auer, E. E., Bevilacqua, and Hart, E. H., *J. Am. Chem. Soc.*, **71**, 497 (1949).

McCaffery, E. D., "Laboratory Preparation for Macromolecular Chemistry", McGraw-Hill, New York, 1970.

Mohanty, N., Pradhan, B., and Mahanta, *Eur. Polym. J.*, **16**, 451 (1980).

North, A. M. and Reed, G. A., *Trans. Faraday Soc.*, **57**, 859 (1961); *J. Polym. Sci., A1*, 1311 (1963).

Odian, G., "Principles of Polymerization", John Wiley, New York, 1991.

O'Driscoll, K. F. and Mahabadi, H. K., *J. Polym. Sci. Polym. Chem. Ed.*, **14**, 869 (1976).

Otsu, T., Ogawa, T, and Yamamoto, T., *Macromolecules*, **19**, 2087 (1986).

Prucker, O. and Rühe, J., *Macromolecules*, **31**, 592 (1998).

Rudin, A., "The Elements of Polymer Science and Engineering", Academic Press, Orlando, FL, 1982.

Sato, T., Abe, M., and Otsu, T., *Makromol. Chem.*, **178**, 1951 (1977).

Schulz, G. V. and Haborth, G., *Makromol. Chem.*, **1**, 106 (1948).

Sheinker, A. and Medvedev, S. S., *Dokl. Akad. Nauk SSR*, **97**, 111 (1954).

Smith, W. V. and Ewart, R. W., *J. Chem. Phys.*, **16**, 592 (1948).

Smith, W. B., May, J. A., and Kim, C. W., *J. Polym. Sci., A2*, 365 (1966).

Tanford, C., "Physical Chemistry of Macromolecules", John Wiley, New York, 1961.

Walling, C., "Free Radicals in Solution", Chaps. 3-5, Wiley-Interscience, New York, 1957.

Williams, D. J., "Polymer Science and Engineering", Prentice Hall, Englewood Cliffs, N. J., 1971.

Yildiz, U., Landfester, K., and Antonietti, M., *Macromol. Chem. Phys.*, **204**, 1966 (2003).

EXERCISES

6.1 When a peroxide P is heated to $60°C$ in an inert solvent it decomposes by a first order process and 20% of the peroxide decomposes in 60 min. A bulk monomer is polymerized using this initiator at $60°C$, the initial concentration of the letter being 4.0×10^{-4} mol/L. What fractions of the monomer and the initiator should remain unconverted after 10 min ? At $60°C$, the system parameters are $k_p^2/k_t = 22.34$ L mol^{-1} s^{-1}, $f = 0.8$.
   [*Ans.* Monomer 0.67; Initiator 0.963]

6.2 A solution of 100 g/L acrylamide in methanol is polymerized at $25°C$ with 0.1 mol/L isobutyryl peroxide whose half life is 9.0 h at this temperature and efficiency in methanol is 0.3. For acrylamide, $k_p^2/k_t = 22$ L mol-1 s^{-1} at $25°C$ and termination is by coupling alone. (a) What is the initial steady state rate of polymerization ? (b) How much polymer has been made in the first 10 min of reaction in 1 L of solution ?
   [*Ans.* (a) 0.37 g L^{-1} s^{-1}; (b) 89.4 g L^{-1}]

6.3 A dilatometer which has a 50 cm long capillary (diameter 0.2 cm) has a total volume of 50 cm^3 (including the volume of capillary). The dilatometer was filled with a freshly distilled sample of methyl methacrylate (MMA) containing 0.25 wt% benzoyl peroxide and then immersed in a water bath (at $50°C$) so that the capillary tube protruded from the water. When the volume of the solution began to decrease after coming to thermal equilibrium and overflowing the capillary, the fall in liquid level in the capillary was determined periodically from the scale on the dilatometer. This yielded the following data :

Time (s)	Δh (cm)	Time (s)	Δh (cm)
480	1.48	4080	13.56
1200	4.09	8280	27.24
1920	6.46	9600	31.91

Determine the kinetic parameter $k_p/k_t^{1/2}$ at 50°C, given that $f = 0.80$ and $k_d = 1.11 \times 10^{-6}$ s^{-1} at 50°C. [Density (g/cm^3 at 50°C): MMA 0.893; PMMA 1.160]
[*Ans.* 0.0948 mol$^{1/2}$ L$^{-1/2}$ s$^{-1/2}$]

6.4   Photoinitiated polymerization of methyl methacrylate (1.0 M in benzene) is carried out using a photosensitizer and 3130 Å light from a mercury lamp. If the quantum yield for radical production in this system is 0.50 and light is absorbed by the system at the rate of 10^5 ergs/L-s, calculate the rate of initiation.
[*Ans.* 2.6×10^{-8} mol L^{-1} s^{-1}]

6.5   The peroxide (thermal homolysis) initiated polymerization of a monomer follows the simplest kinetic scheme represented by Eq. (6.24). For a polymerization system with $[M]_0 = 4$ mol/L and $[I]_0 = 0.01$ mol/L, the limiting conversion $p_\infty = 0.10$. To increase p_∞ to 0.20,
(a) would you increase or decrease $[M]_0$ and by what factor?
(b) would you increase or decrease $[I]_0$ and by what factor?
(c) would you increase or decrease the reaction temperature?
[*Ans.* (a) no effect; (b) increase by a factor of 4.5; (c) decrease temperature]

6.6   One hundred liters of methyl methacrylate containing 10.2 moles of an initiator ($t_{1/2} = 50$ h) in solution is polymerized at 60°C. Calculate (a) the kinetic chain length in this polymerization and (b) the amount of polymer formed in the first 1 h of reaction. [Data: monomer density 0.94 g/cm^3; $k_p = 515$ L mol^{-1} s^{-1}; $k_t = 2.55 \times 10^7$ L mol^{-1} s^{-1}; $f = 0.3$.]
[*Ans.* (a) 1397; (b) 70 kg]

6.7   Calculate the time needed to convert half of charge of methyl methacrylate (10 g per 100 mL solution) to polymer using benzoyl peroxide (0.1 g per 100 mL solution) as initiator in benzene at 60°C. What number-average degree of polymerization will be expected initially? What fraction of the initiator will remain unused after 50% conversion of the monomer? [At 60°C, $k_d = 4.47 \times 10^{-6}$ s^{-1}; $k_p^2/k_t = 10^{-2}$ L/mol-s; $f = 0.4$; termination occurs by both disproportionation (58%) and coupling (42%).]
[*Ans.* 24.7 h; 739; 0.67]

6.8   Polymerization of two monomers M_1 and M_2 in homogeneous solutions with the same concentration of peroxide initiator and initial monomer concentrations of 0.100 and 0.200 mol/L, respectively, yielded 5% conversion of the original monomer charge to polymer in 6 min and 18 min, respectively. In another series of experiments at the same temperature, monomers M_1 and M_2 have the same concentration initially. What should be the ratio of concentration of initiator for M_2 compared with that for monomer M_1, if the polymers produced initially are required to have the same degree of polymerization? [Assume $f = 1$ and chain termination by coupling.]
[*Ans.* $[I]_2/[I]_1 = 1/9$]

6.9   Initiator I_1 has half the half-life that initiator I_2 has at 80°C. Monomer M_1 polymerizes 4 times as fast as monomer M_2 at 80°C, when initiator I_1 is used for both and all the concentrations are the same. What is the ratio of degree of polymerization for M_1 and M_2 if they are polymerized with initiator I_1 and I_2, respectively, the ratio of monomer concentrations being 1:2 and that of initiator concentrations being 1:5? It can be assumed that both polymers terminate exclusively by coupling and that the initiator efficiencies are equal.
[*Ans.* 3.16]

6.10 When bulk styrene (density 0.905 g/cm^3) containing a dissolved peroxide initiator is heated at 60°C for 90 min, 3% of the styrene undergoes polymerization and the polymer recovered from the solution is found to have a number average molecular weight of 67,800. From this information determine $k_p/k_t^{1/2}$ at 60°C, assuming that termination occurs by coupling alone.
[*Ans.* 0.021 L$^{1/2}$ mol$^{-1/2}$ s$^{-1/2}$]

6.11 Starting from Eq. (6.86), i.e., the relation $[M^\bullet] = [M^\bullet]_s \tanh(t/\tau_s)$, and assuming low conversion, show that a plot of the fraction of the monomer polymerized versus time t ($\gg \tau_s$) yields a straight line that cuts the time axis at $t = \tau_s \ln 2$.

6.12 A vinyl polymer with a number-average degree of polymerization of 10,000 was produced by polymerization at 70°C using a peroxide initiator concentration of 4×10^{-4} mol/L. If 4.0% of the initial monomer present was converted to polymer in 60 min, what was the initial monomer concentration? The mode of termination is coupling and the initiator is known to have a half-life of 1.0 h at 90°C and an activation energy of 30.0 kcal/mol. Assume $f = 1.0$.
[*Ans.* 6.0 mol L^{-1}]

6.13 Using carbon-14 labeled AIBN as an initiator, a sample of styrene is polymerized to an average degree of polymerization of 1.28×10^4. The AIBN has an activity of 8.97×10^7 counts per minute per mol in a scintillation counter. If 5.0 grams of the polystyrene show an activity of 315 counts per minute, determine the mode of termination of polystyryl radicals.
[*Ans.* Coupling 93%, disproportionation 7%]

6.14 In the bulk polymerization of methyl methacrylate at 60°C with azo-bis-isobutyronitr-ile as the initiator the initial rates of initiation and polymerization are 1.7×10^{-6} mol/L-s and 8.8×10^{-4} mol/L-s, respectively. Predict the initial molecular weight of the polymer formed in this system, if the extent of disproportionation is 70% at 60°C. Neglect chain transfer reactions for the calculation. [*Ans.* 60,800]

6.15 Determine the concentrations (g/L) of initiator (AIBN) and chain transfer agent (*n*-butyl mercaptan) that will give poly(vinyl acetate) with an initial molecular weight (assuming coupling) of 15,000 and 50% conversion of monomer (initial concentration 250 g/L) at 60°C in 30 min. [System parameters (all at 60°C): $k_p^2/k_t = 0.1824$ L/mol-s; $t_{1/2}$ of AIBN = 22 h; $f = 1$; $C_S = 48$.]
[*Ans.* 15.24 g L^{-1}; 0.027 g L^{-1}]

6.16 The molecular weight of polymer when styrene is polymerized in benzene is 400,000. With all other conditions the same, addition of 4.23 mg/L of *n*-butyl mercaptan decreases the molecular weight to 85,000. What concentration (mg/L) of *n*-butyl mercaptan will give a molecular weight of 50,000 (other conditions remaining the same)?
[*Ans.* 8 mg L^{-1}]

6.17 In the polymerization of styrene (1.0 mol/L) in benzene initiated by di-*t*-butylperoxide (0.01 mol/L) at 60°C, the initial rates of initiation and polymerization are 4.0×10^{-11} mol/L-s and 1.5×10^{-7} mol/L-s, respectively. (a) Calculate the initial kinetic chain length. (b) Calculate the initial polymer molecular weight, assuming that termination of chain radicals takes place only by coupling, besides chain transfer to various species. (c) Indicate how often, on the average, chain transfer occurs per each initiating radical from the peroxide. [Data (all at 60°C): $C_M = 5.0 \times 10^{-5}$; $C_I = 7.6 \times 10^{-4}$; $C_S = 2.3 \times 10^{-6}$; density of benzene = 0.85 g/mL; density of styrene = 0.91 g/mL.]
[*Ans.* (a) 3.75×10^3; (b) $\overline{M}_n = 4.9 \times 10^5$; (c) 0.60]

6.18 Consider a free radical polymerization initiated by 10^{-3} M AIBN. At 70°C, k_d is 4.0×10^{-5} s^{-1} and f is close to 0.6. If an inhibitor is to be used to suppress polymerization for an hour, what should be its concentration, if every inhibitor molecule accounts for one primary or monomer-ended radical?
[*Ans.* 1.73×10^{-4} M]

6.19 Methyl acrylate (1 mol) is polymerized using 0.001 mol succinic peroxide in 1 liter solution in benzene at 60°C. If the polymerization is carried out adiabatically, how much would the temperature rise in 30 min? [Data (all at 60°C): $t_{1/2}$ of initiator = 19 h; $f = 1$; $k_p^2/k_t = 0.460$ L/mol-s; $\Delta H_p = -18.6$ kcal/mol.]
[*Ans.* 6.0°C]

6.20 A vinyl monomer is photopolymerized in two experiments in which only the temperature is varied. In these experiments, the time to convert 20% of the original charge of monomer to polymer is found to be 30 min at 60°C and 27 min at 70°C. However, when an organic peroxide is used as the initiator, the corresponding times for 20% conversion are 62 min at 60°C and 29 min at 70°C. What is the activation

energy for the dissociation of the organic peroxide ?
[*Ans.* 124 kJ mol^{-1}]

6.21 The half-lives of azobisisobutyronitrile at 50°C and 70°C are 74 h and 4.8 h, respectively. What will be the half-life at 60°C ?
[*Ans.* 18.1 h]

6.22 In the bulk polymerization of styrene by ultraviolet radiation, the initial polymerization rate and degree of polymerization are 1.3×10^{-3} mol/L-s and 260, respectively, at 30°C. What will be the corresponding values for polymerization at 80°C ? The activation energies for propagation and termination of polystyryl radicals are 26 and 8.0 kJ/mol. What assumption, if any, is made in this calculation ?
[*Ans.* 4.48×10^{-3} mol L^{-1} s^{-1}; 896]

6.23 A radical chain polymerization conforming to ideal behavior shows the indicated conversions for specified initial monomer and initiator concentrations and reaction times :

$[M]_0$ (mol/L)	$[I]_0×10^3$ (mol/L)	Temperature (°C)	Reaction time (min)	Conversion (%)
0.80	1.0	60	60	40
0.50	1.0	75	70	80

Calculate the overall activation energy for the rate of polymerization.
[*Ans.* 63.8 kJ mol^{-1} (= 15 kcal mol^{-1})]

6.24 The enthalpy and entropy of polymerization of α-methylstyrene at 25°C are − 35 kJ/mol and − 110 J/°K-mol, respectively. Calculate approximately the equilibrium constant for polymerization at 25°C and 50°C. Comment on the results.
[*Ans.* 2.45, 1.22]

6.25 Calculate $k_p[M]_e$ and k_{dp} at different temperatures and hence determine the ceiling temperature for pure methyl methacrylate (density 0.940 g/cm^3 at 25°C) using the following data : $A_p = 10^6$ L mol^{-1} s^{-1}, $A_{dp} = 10^{13}$ s^{-1} (assumed), $E_p = 26$ kJ mol^{-1}, $E_{dp} = 82$ kJ mol^{-1}.
[*Ans.* $T_c = 485$°K]

6.26 The enthalpy and entropy of polymerization of methyl methacrylate at 25°C are −56 kJ/mol and −117 J/°K-mol, respectively. For a solution of the monomer (1.0 mol/L), calculate the maximum attainable conversion at (a) 25°C, (b) 120°C, and (c) 200°C.
[*Ans.* (a) 99.98%; (b) 95.0%; (c) 15.4%]

6.27 For polymerization of tetrafluoroethylene, ΔH^0 and ΔS^0 values at 25°C are given as −37 kcal/mol and −26.8 cal/°K-mol. Calculate the ceiling temperature (T_c) from these two values. Account for the fact that in practice poly(tetrafluoroethylene) is found to undergo fragmentation well below the calculated T_c.
[*Ans.* 1380°K]

6.28 If a free-radical polymerization of 1.0 M solution of styrene were being carried out at 100°C, what would be the maximum possible conversion of the monomer to polymer, that is, till the polymerization-depolymerization equilibrium is reached ? (Data : $\Delta H_p^0 = -69.9$ kJ/mol; $\Delta S_p^0 = -104.6$ J/°K-mol.)
[*Ans.* 99.995%]

6.29 Styrene (density 0.90 g/cm^3) was polymerized at 60°C with 0.01 M benzoyl peroxide as the initiator [W. B. Smith, J. A. May, and C. W. Kim, *J. Polym. Sci., Part A2*, **4**, 395 (1966)]. The initial rate of reaction was obtained as 3.95×10^{-5} mol/L-s. From the GPC molecular weight distribution curve reported for a 0.79% conversion sample, the weight fraction of polymer of $\overline{DP}_n = 3000$ is seen to be 2.3×10^{-4}. Calculate the weight fraction from theoretical distribution function to compare with this value. [Data : k_p^2/k_t at 60°C = 0.00119 L mol^{-1} s^{-1}; $k_{td}/k_{tc} = 0$.]
[*Ans.* 2.2×10^{-4}]

6.30 Monomer A is polymerized in solution at 60°C using a peroxide initiator I which has half-life of 5.0 h at the same temperature. A 30% conversion of the monomer is obtained in 25 min when the initial concentration of A is 0.40 M and the initial concentration of I is 0.04 M. Polymerization of A in an emulsion of 8×10^{17} particles per liter at 60°C yields a conversion rate of 18.1 mol/h/L when the concentration of A in the particles is constant at 4.0 M. Determine the termination rate constant of A at 60°C.

[*Ans.* 9.73×10^7 L mol^{-1} s^{-1}]

6.31 What happens to (a) rate of emulsion polymerization, (b) number average degree of polymerization, and (c) polymer particle size, if more monomer is added to the reaction mixture during Stage II polymerization ? Explain.

[*Ans.* (a) No change; (b) no change; (c) increases]

6.32 The rate of emulsion polymerization of styrene at 60°C during the constant rate period (Stage II) is 5.6×10^{-5} mol/cm^3-min and the number of M/P particles is 1.40×10^{15} per cm^3. Taking k_p from Table 6.4, calculate the dynamic concentration of monomer in particles under these conditions.

[*Ans.* 4.8 mol L^{-1}]

6.33 In an emulsion polymerization of isoprene with 0.10 M potassium laurate at 50°C the estimated time required for 100% conversion at steady rate is 30 h. The final latex has 40 g of polymer per 100 mL with particles of 450 Å diameter. During Stage II, the growing swollen polymer particles contain 20 g of monomer per 100 mL of swollen polymer. Assuming that there is no change in total volume on polymerization, estimate the polymerization rate constant from these data. Assume that the polymer has a density of 0.90 g/cm^3.

[*Ans.* 2.4 L mol^{-1} s^{-1}]

6.34 Consider a typical reactor charge for the production of polymer latex : monomer(s) 100, water 180, sodium lauryl sulfate (surfactant) 4, potassium persulfate (initiator) 1 (all quantities are in parts by weight). What effects do the following changes have on the polymerization rate in Stage II ? (a) Using 8 parts surfactant; (b) using 2 parts initiator; (c) using 8 parts surfactant and 2 parts initiator; (d) adding 0.1 part butyl mercaptan (chain transfer agent).

[*Ans.* (a) R_p increases by a factor of 1.52; (b) R_p increases by a factor of 1.32; (c) R_p increases by a factor of 2; (d) R_p unchanged]

6.35 A 10% (by weight) latex of poly(methyl methacrylate) produced by emulsion polymerization contains particles that average 0.2 μm in diameter. In order to grow the particles to a larger size it is decided to feed 4 kg of monomer into the latex per kilogram of polymer as polymerization proceeds at 60°C without further addition of emulsifier. The reaction is to be carried on until all monomer is added to the latex and the weight ratio of monomer to polymer has decreased to 0.2. The unreacted monomer is then to be recovered by steam stripping.

Calculate the total time that will be required for reaction and the final particle diameter. [Data : k_p at 60°C = 515 L mol^{-1} s^{-1}; monomer density = 0.9 g/cm^3; polymer density = 1.2 g/cm^3; dynamic solubility of monomer in polymer = 0.5 g monomer per gram polymer.]

[*Ans.* 42 h; 0.64 μm]

Chapter 7

Chain Copolymerization

7.1 Introduction

We have so far considered free-radical polymerizations where only one monomer is used and the product is a homopolymer. The same type of polymerization can also be carried out with a mixture of two or more monomers to produce a polymer product that contains two or more different mer units in the same polymer chain. The polymerization is then termed a *copolymerization* and the product is termed a *copolymer*. Monomers taking part in copolymerization are referred to as *comonomers*. The simultaneous polymerization of two monomers is known as *binary copolymerization* and that of three monomers as *ternary copolymerization*, and so on. The term *multicomponent copolymerization* embraces all such cases. The relative proportions of the different mer units in the copolymer chain depend on the relative concentrations of the comonomers in the feed mixture and on their relative reactivities. This will be the main subject of our discussion in this chapter.

It should be noted that the chain copolymerization may be initiated by any of the chain initiation mechanisms, namely, free-radical chain initiations considered in the preceding chapter, or ionic chain initiations, which will be described in a later chapter. While polymerization of a single monomer is relatively limited regarding the number of different products that are possible, copolymerization enables the polymer engineer to synthesize an almost unlimited number of products with different properties by variations in the nature and relative amounts of the two monomers in the feed mixture and to tailor-make polymers with specific properties. Copolymerization is thus very important from a technological viewpoint.

The classification of copolymers according to structural types and the nomenclature for copolymers have been described previously in Chapter 1. The present chapter is primarily concerned with the simultaneous polymerization of two monomers by free-radical mechanism to produce random, statistical, and alternating copolymers. Copolymers having completely random distribution of the different monomer units along the copolymer chain are referred to as *random copolymers*. Statistical copolymers are those in which the distribution of the two monomers in the chain is essentially random but influenced by the individual monomer reactivities. The other types of copolymers, namely, graft and block copolymers, are not synthesized by the simultaneous polymerization of two monomers. These are generally obtained by other types of reactions (see Section 7.6).

7.2   Binary Copolymer Composition – Terminal Model

Simultaneous polymerization of two monomers by chain initiation usually results in a copolymer whose composition is different from that of the feed. This shows that different monomers have different tendencies to undergo copolymerization. These tendencies often have little or no resemblance to their behavior in homopolymerization. For example, vinyl acetate polymerizes about twenty times as fast as styrene in a free-radical reaction, but the product obtained by free-radical polymerization of a mixture of vinyl acetate and styrene is found to be almost pure polystyrene with hardly any content of vinyl acetate. By contrast, maleic anhydride, which has very little or no tendency to undergo homopolymerization with radical initiation, readily copolymerizes with styrene forming one-to-one copolymers. The composition of a copolymer thus cannot be predicted simply from a knowledge of the polymerization rates of the different monomers individually. The simple copolymer model described below accounts for the copolymerization behavior of monomer pairs. It enables one to calculate the distribution of sequences of each monomer in the macromolecule and the drift of copolymer composition with the extent of conversion of monomers to polymer.

In order to develop a simple model, one has to necessarily assume that the chemical reactivity of a propagating chain (which may be free-radical in a radical chain copolymerization and carbocation or carboanion in an ionic chain copolymerization) is dependent only on the identity of the chain-end monomer unit and independent of the chain composition preceding the end monomer unit. This is referred to as the *first-order Markov* or *terminal model* of copolymerization. Thus in a binary copolymerization of two monomers, M_1 and M_2, by free-radical mechanism, two different types of chain radicals can be identified, namely those (wwwwM$_1$ •) with radical center on M_1 end unit and those (wwwwM$_2$ •) with the radical center on M_2 end unit. In order to distinguish from free-radical homopolymerization, we shall henceforth indicate the free-radical center in copolymerization by an asterisk ($\star$) and represent the two chain radicals by $M_1^\star$ and $M_2^\star$, respectively, omitting the polymer chain symbol wwww. Since each of these chain radicals is capable of adding both the monomers, though not usually or necessarily with equal ease, the system will be characterized by four types of propagation reactions occurring in parallel, as shown below (Rudin, 1982; Hiemenz, 1984; Hamielec et al., 1989; Allcock and Lampe, 1990; Odian, 1991):

$$M_1^\star + M_1 \xrightarrow{k_{11}} M_1^\star \tag{7.1}$$

$$M_1^\star + M_2 \xrightarrow{k_{12}} M_2^\star \tag{7.2}$$

$$M_2^\star + M_1 \xrightarrow{k_{21}} M_1^\star \tag{7.3}$$

$$M_2^\star + M_2 \xrightarrow{k_{22}} M_2^\star \tag{7.4}$$

where $M_1^\star$ and $M_2^\star$ represent the chain radicals of all sizes with the free-radical bearing terminal unit being M_1 and M_2, respectively; k_{11}, k_{12}, k_{21} and k_{22} are rate constants in which the first subscript refers to the active center of the propagating chain and the second to the monomer. Reactions (7.1) and (7.4) where the reactive chain end adds the same monomer are often referred to as *homopropagation* or *self-propagation*. Propagation reactions involving addition of another monomer (Reactions 7.2 and 7.3) are referred to as *cross-propagation* or *cross-over* reaction. It is assumed in the above scheme that the reaction is carried out below the ceiling temperature of both the monomers and that the various propagation reactions are irreversible.

The rates of consumption of the two types of monomers for copolymer formation are given by [see Eqs. (7.1) to (7.4)]:

$$-d[M_1]/dt = k_{11}[M_1^\star][M_1] + k_{21}[M_2^\star][M_1] \tag{7.5}$$

$$-d[M_2]/dt = k_{12}[M_1^\star][M_2] + k_{22}[M_2^\star][M_2] \tag{7.6}$$

The ratio of these two rates, namely $d[M_1]/d[M_2]$, thus gives the relative rates of incorporation of the two monomeric units in the copolymer (Rudin, 1982):

$$\frac{d[M_1]}{d[M_2]} = \frac{k_{11}[M_1^\star][M_1] + k_{21}[M_2^\star][M_1]}{k_{12}[M_1^\star][M_2] + k_{22}[M_2^\star][M_2]} \tag{7.7}$$

To remove the concentration terms $[M_1^\star]$ and $[M_2^\star]$ from this expression, a steady state is assumed for each chain radical type $M_1^\star$ and $M_2^\star$ in the reaction mixture. This assumption requires that the rate of conversion of $M_1^\star$ to $M_2^\star$ must equal that of $M_2^\star$ to $M_1^\star$, or in mathematical terms,

$$k_{12}[M_1^\star][M_2] = k_{21}[M_2^\star][M_1] \tag{7.8}$$

Solving Eq. (7.8) for $[M_1^\star]$ and substituting in Eq. (7.7) then gives

$$\frac{d[M_1]}{d[M_2]} = \frac{\dfrac{k_{11}k_{21}[M_2^\star][M_1]^2}{k_{12}[M_2]} + k_{21}[M_2^\star][M_1]}{k_{21}[M_2^\star][M_1] + k_{22}[M_2^\star][M_2]} \tag{7.9}$$

By defining two parameters (called *monomer reactivity ratios*), r_1 and r_2, as

$$r_1 = k_{11}/k_{12} \quad \text{and} \quad r_2 = k_{22}/k_{21} \tag{7.10}$$

and substituting them into Eq. (7.9) after dividing the numerator and denominator of the right side of this equation by $k_{21}[M_2^\star][M_2]$, the result is

$$\frac{d[M_1]}{d[M_2]} = \frac{[M_1]}{[M_2]} \cdot \frac{r_1[M_1] + [M_2]}{[M_1] + r_2[M_2]} \tag{7.11}$$

This is the so-called *copolymer equation* or the *copolymer composition equation*. The ratio $d[M_1]/d[M_2]$ representing the ratio of the rates at which the two monomers M_1 and M_2 enter the copolymer gives the molar ratio of the two monomer units in the copolymer (being formed at a given instant), and hence is referred to as the copolymer composition. According to Eq. (7.11), the copolymer composition depends on the concentrations of the two types of monomers in the feed, namely, $[M_1]$ and $[M_2]$, and on the kinetic parameters r_1 and r_2, known as the *monomer reactivity ratios* (ratios of propagation rate constants). Since initiation and termination rate constants are not involved in Eq. (7.11), the copolymer composition should be independent of the initiator used and of the absence or presence of inhibitors/retarders or chain transfer agents.

7.2.1   Significance of Monomer Reactivity Ratios

As defined by Eq. (7.10), the monomer reactivity ratio can be looked upon as the relative tendency for homopropagation and cross-propagation. If for a given monomer pair, $r_1 = 0$ and hence $k_{11} = 0$, it would mean that M_1 does not homopolymerize in the presence of M_2. Similarly, $r_1 > 1$, i.e., $k_{11} > k_{12}$ means that $M_1^\star$ preferentially adds M_1 instead of M_2 and $r_1 < 1$, i.e., $k_{12} > k_{11}$ means that $M_1^\star$ preferentially adds M_2. For example, an r_1 value of 0.5 would mean that $M_1^\star$ adds M_2 twice as fast as M_1.

It is evident from Eq. (7.10) that the values of r_1 and r_2 refer only to a pair of monomers undergoing copolymerization. Thus the same monomer can have different values of r_1 in combination with different monomers, e.g., acrylonitrile has (r_1, r_2) values of (0.35, 1.15), (0.02, 1.8), (1.5, 0.84), and (4.2, 0.05) at 50°C in free-radical copolymerization with acrylic acid, isobutylene, methyl acrylate, and vinyl acetate, respectively, each being designated as M_2 and the other monomer, acrylonitrile, as M_1.

Problem 7.1   The above derivation of the copolymer composition equation [Eq. (7.11)] involves the steady-state assumption for each type of propagating species. Show that the same equation can also be derived from elementary probability theory (Melville et al., 1947; Vollmert, 1973; Odian, 1991) without invoking steady-state conditions.

Answer:

Let

P_{11} =   probability that $M_1^\star$ will add M_1 rather than M_2. (Here the first subscript designates the active center and the second the monomer.)

P_{12} =   probability that $M_1^\star$ will add M_2 rather than M_1.

In the formation of high polymer, the termination occurs rarely. So neglecting it in the present case, one may write

$$P_{11} + P_{12} = 1 \tag{P7.1.1}$$

In the same way,

$$P_{22} + P_{21} = 1 \tag{P7.1.2}$$

The probability that propagating species $M_1^\star$ adds an M_1 unit is equal to the rate of this reaction divided by the sum of the rates of all reactions available to this radical (Rudin, 1982; Hiemenz, 1984; Odian, 1991). This is the probability P_{11} that an M_1 unit follows an M_1 unit in the copolymer. Hence

$$P_{11} = \frac{k_{11}[M_1^\star][M_1]}{k_{11}[M_1^\star][M_1] + k_{12}[M_1^\star][M_2]} = \frac{r_1[M_1]}{r_1[M_1] + [M_2]} \tag{P7.1.3}$$

Similarly, the other probabilities are obtained as

$$P_{12} = \frac{[M_2]}{r_1[M_1] + [M_2]} \tag{P7.1.4}$$

$$P_{21} = \frac{[M_1]}{[M_1] + r_2[M_2]} \tag{P7.1.5}$$

$$P_{22} = \frac{r_2[M_2]}{[M_1] + r_2[M_2]} \tag{P7.1.6}$$

Let $\bar{x}(M_1)$ be the number average sequence length of monomer M_1, that is the average number of M_1 monomer units that follow each other consecutively in a sequence uninterrupted by M_2 units but bounded on each end by M_2 units. Similarly, let $\bar{x}(M_2)$ be the number average sequence length of monomer M_2. In order to evaluate $\bar{x}(M_1)$ and $\bar{x}(M_2)$, it is necessary to determine the distribution of sequence lengths of M_1 and M_2 in the copolymer, namely, $n_x(M_1)$ and $n_x(M_2)$.

The probability that a given sequence contains x number of M_1 units is equal to the fraction of all M_1 sequences which contain x units. That is to say, it is the number distribution function $n_x(M_1)$ for M_1-sequence lengths:

$$n_x(M_1) = P_{11}^{x-1}(1 - P_{11}) = P_{12}P_{11}^{x-1} \tag{P7.1.7}$$

The derivation of Eq. (P7.1.7) is analogous to that of Eq. (5.44). The fraction of all M_2 sequences that contain exactly x number of M_2 units is similarly derived, yielding

$$n_x(M_2) = P_{22}^{x-1}(1 - P_{22}) = P_{21}P_{22}^{x-1} \tag{P7.1.8}$$

The number average sequence length of M_1 is then given by

$$\bar{x}(M_1) = \sum_{x=1}^{\infty} n_x(M_1) \cdot x \tag{P7.1.9}$$

[This is completely analogous to the definition of number average molecular weight $\overline{M}_n$ in Eq. (4.3).] Substituting Eq. (P7.1.7) into Eq. (P7.1.9),

$$\bar{x}(M_1) = \sum_{x=1}^{\infty} xP_{12}P_{11}^{x-1} = P_{12}(1 + 2P_{11} + 3P_{11}^2 + 4P_{11}^3 + \cdots\cdots) \tag{P7.1.10}$$

For copolymerization, $P_{11} < 1$; so the expansion series in Eq. (P7.1.10) is $1/(1 - P_{11})^2$ and Eq. (P7.1.10) reduces to

$$\bar{x}(M_1) = \frac{P_{12}}{(1 - P_{11})^2} = \frac{1}{P_{12}} = 1 + r_1\frac{[M_1]}{[M_2]} \tag{P7.1.11}$$

In the same way one obtains

$$\bar{x}(M_2) = \frac{P_{21}}{(1 - P_{22})^2} = \frac{1}{P_{21}} = 1 + r_2\frac{[M_2]}{[M_1]} \tag{P7.1.12}$$

The mole ratio of monomers M_1 and M_2 in the copolymer chain is the same as the ratio of the two number-average sequence lengths (Odian, 1991); thus

$$\frac{d[M_1]}{d[M_2]} = \frac{\bar{x}(M_1)}{\bar{x}(M_2)} = \frac{[M_1](r_1[M_1] + [M_2])}{[M_2]([M_1] + r_2[M_2])} \tag{P7.1.13}$$

which is identical with Eq. (7.11).

7.2.2 Types of Copolymerization

The observed monomer reactivity ratios of different monomer pairs vary widely but can be divided into a rather small number of classes. A useful classification (Rudin, 1982; Odian, 1991) is based on the product of r_1 and r_2, such as $r_1r_2 \to 0$ (with $r_1 \ll 1$, $r_2 \ll 1$), $r_1r_2 \to 1$, $0 < r_1r_2 < 1$, and $r_1r_2 > 1$ (with $r_1 > 1$, $r_2 > 1$), representing, respectively, *alternating*, *random* (or *ideal*), *random-alternating*, and *block* copolymerizations.

7.2.2.1 Alternating Copolymerization

A zero, or a nearly zero, value for the reactivity ratio means that the monomer is incapable of undergoing homopolymerization and its radical prefers to add exclusively to the other monomer. This leads to alteration of the two monomer units along the copolymer chain. For $r_1 = r_2 = 0$, Eq. (7.11) reduces to $d[M_1]/d[M_2] = 1$. Thus, copolymerization of two monomers for which $r_1 \ll 1$ and $r_2 \ll 1$ will tend to produce an *alternating copolymer* (in which the two monomer units alternate in a regular fashion along the chain), irrespective of the composition of the monomer feed.

7.2.2.2 Ideal (random) Copolymerization

A value of unity (or nearly unity) for the monomer reactivity ratio signifies that the rate of reaction of the growing chain radicals towards each of the monomers is the same, i.e., $k_{11} \simeq k_{12}$ and $k_{22} \simeq k_{21}$ and the copolymerization is entirely random. In other words, both propagating species $M_1^\star$ and $M_2^\star$ have little or no preference for adding either monomer. For $r_1 r_2 = 1$,

$$r_1 = 1/r_2 \quad \text{or} \quad k_{11}/k_{12} = \frac{k_{21}}{k_{22}} \tag{7.12}$$

Equation (7.12) means that k_{11}/k_{12} and k_{21}/k_{22} will be simultaneously either greater or less than unity, or in other words, that both radicals prefer to react with the same monomer. All copolymers whose $r_1 r_2$ product equals, or nearly equals, 1 are therefore called *ideal copolymers* or *random copolymers*. Ionic copolymerizations (Chapter 8) are usually characterized by the ideal type of behavior.

For $r_1 = r_2 = 1$, Eq. (7.11) reduces to

$$d[M_1]/d[M_2] = [M_1]/[M_2] \tag{7.13}$$

which means that the copolymer composition will always be the same as the feed composition. The relative amounts of the two monomer units in the copolymer chain are determined by the relative concentrations of the monomer units in the feed.

For $r_1 r_2 = 1$, Eq. (7.11) reduces to

$$d[M_1]/d[M_2] = r_1 [M_1]/[M_2] \tag{7.14}$$

The relative amounts of the two monomer units along the copolymer chain are thus determined not only by the relative concentrations of the monomer units in the feed but also by the relative reactivities of the two monomers. It is thus obvious that if r_1 and r_2 are widely different, while $r_1 r_2 = 1$, copolymers containing appreciable amounts of both M_1 and M_2 cannot be obtained.

7.2.2.3 Random-Alternating Copolymerization

Copolymerizations in which $0 < r_1 r_2 < 1$ are intermediate between alternating and random types and thus can be described as belonging to *random-alternating* type. Most copolymer systems fall in this category. As $r_1 r_2$ product decreases, cross-propagation reactions are favored and the monomer units in the copolymer chain show an increasing tendency toward alternation. On the other hand, as $r_1 r_2$ approaches 1, the copolymer chain composition becomes increasingly random.

A special situation for $0 < r_1 r_2 < 1$ relates to $r_1 \gg 1$ and $r_2 \ll 1$ or vice versa. In this case, the product composition will tend toward that of the homopolymer of the more reactive monomer and copolymerization cannot occur. For example, when $r_1 \gg 1$ and $r_2 \ll 1$, both $M_1^\star$ and $M_2^\star$ will preferentially add monomer M_1 till this monomer is consumed, which will then be followed by homopolymerization of M_2. This is therefore a case of *consecutive homopolymerizations* (Odian, 1991).

7.2.2.4 Block Copolymerization

If $r_1 > 1$, an $M_1^\star$ propagating species would add many units of M_1 in sucession until the growing chain happens to add an M_2 unit changing itself from $M_1^\star$ type to $M_2^\star$ type. Since r_2 is also more than 1, the $M_2^\star$ propagating species would then preferentially add many M_2 units in succession until an M_1 unit happens to add, converting the chain again to the $M_1^\star$ type. A *block copolymer* consisting of long sequences of each monomer in the copolymer chain would thus be expected. However, if both r_1 and r_2 are quite large, the two types of monomers would only undergo *simultaneous homopolymerization* in each other's presence. Such combinations of reactivity ratios are, however, rare in free-radical copolymerizations, but they can be found in other systems.

7.2.3 Instantaneous Copolymer Composition

The copolymer equation (7.11) can be converted to a more useful form by expressing concentrations in terms of mole fractions. Let f_1 and f_2 be the mole fractions of monomers M_1 and M_2 in the feed, that is,

$$f_1 = 1 - f_2 = [M_1] / ([M_1] + [M_2]) \tag{7.15}$$

and F_1 and F_2 be the mole fractions of monomers M_1 and M_2 in the polymer formed at any instant, i.e.,

$$F_1 = 1 - F_2 = \frac{d[M_1]}{d([M_1] + [M_2])} \tag{7.16}$$

Combining Eqs. (7.15) and (7.16) with the copolymer equation (7.11), one obtains

$$F_1 = \frac{r_1 f_1^2 + f_1 f_2}{r_1 f_1^2 + 2 f_1 f_2 + r_2 f_2^2} \tag{7.17}$$

Equation (7.17), which is also called the *copolymer equation*, gives the mole fraction of monomer M_1 in the copolymer whose feed contained f_1 mole fraction of monomer M_1. It is more convenient to use than its previous form [Eq. (7.11)]. It should be noted, however, that F_1 gives the *instantaneous* copolymer composition and both f_1 and F_1 change as the polymerization proceeds.

Figure 7.1 shows a series of curves for ideal copolymerization (i.e., $r_1 r_2 = 1$) calculated from Eq. (7.17). The term *ideal copolymerization* highlights the similarity between these curves (showing no inflection points) and those found in vapor-liquid equilibria for ideal liquid mixtures.

For cases in which $r_1 = r_2 = 1$, the composition of the copolymer (F_1) is always the same as the feed composition (f_1). Consequently, there occurs no drift in composition with conversion. However, for other cases, there occurs a drift in monomer composition as copolymer is formed. Therefore copolymer composition changes with conversion and to obtain a constant copolymer composition, it becomes necessary to maintain a constant feed composition, such as by adding fresh monomer (one that is consumed faster) to the feed.

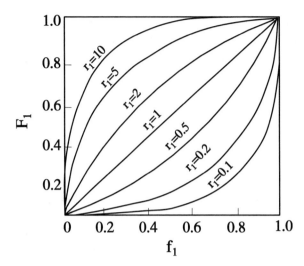

Figure 7.1 Copolymerization diagrams (without inflection points) showing instantaneous composition of copolymer (mole fraction F_1) as a function of monomer composition (mole fraction f_1) for copolymers with the values of $r_1 = 1/r_2$ for ideal copolymerization. (After Ghosh, 1990.)

Problem 7.2   It is desired to form a copolymer of M_1 and M_2 containing twice as many M_1 as M_2. The monomers copolymerize ideally, with monomer reactivity ratios $r_1 = 2.0$ and $r_2 = 0.5$. Describe the feed composition one should use to make this copolymer.

Answer:

For the desired copolymer, $F_1 = 2/3$. To determine the corresponding f_1, rewrite Eq. (7.17) as

$$F_1(r_1 f_1^2 + 2f_1 f_2 + r_2 f_2^2) = r_1 f_1^2 + f_1 f_2$$

and substitute $(1 - f_1)$ for f_2 to obtain a quadratic equation by appropriate rearrangement :

$$f_1^2 r_1 (F_1 - 1) + f_1 (1 - f_1)(2F_1 - 1) + F_1 r_2 (1 - f_1)^2 = 0$$
$$f_1^2 r_1 (F_1 - 1) + f_1^2 (1 - 2F_1) + f_1 (2F_1 - 1) + F_1 r_2 (f_1^2 - 2f_1 + 1) = 0$$
$$f_1^2 [F_1 (r_1 + r_2 - 2) + (1 - r_1)] + f_1 [2F_1 (1 - r_2) - 1] + F_1 r_2 = 0 \qquad \text{(P7.2.1)}$$

Substituting $F_1 = 2/3$, $r_1 = 2$, $r_2 = 0.5$, Eq. (P7.2.1) simplifies to $2f_1^2 + f_1 - 1 = 0$. Solving, $f_1 = 0.5$ as the other root, $f_1 = -1$, is not meaningful. (To check the result, substitution of $f_1 = 0.5$ into Eq. (7.17) gives $F_1 = 0.667$, as required.)

Since the copolymer formed in this case is richer in M_1 as compared to feed, the feed composition will drift toward lower f_1 at higher conversion, leading to progressive change in F_1 and copolymer composition. To obtain copolymer of constant composition $F_1 = 2/3$, calculated amounts of M_1 must therefore be added to the monomer mixture, continuously or periodically, to maintain the feed composition at $f_1 = 0.5$ as the reaction progresses.

Figure 7.2 shows curves for several nonideal cases, that is, where $r_1 r_2 \neq 1$. It is seen that when both r_1 and r_2 are less than 1 there exists some point on the F_1-versus-f_1 curve where the curve crosses the diagonal line representing $F_1 = f_1$, that is, the copolymer composition equals the feed

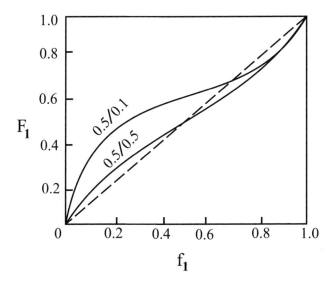

Figure 7.2 Copolymerization diagrams with inflection points showing composition of copolymer F_1 as a function of monomer composition f_1 for the indicated values of the reactivity ratios, r_1/r_2. (After Ghosh, 1990.)

composition. At this point of intersection, polymerization therefore proceeds without change in either feed or copolymer composition. Using distillation terminology, *azeotropic copolymerization* is said to occur at such points and the resulting copolymers are called *azeotropic copolymers*.

Since all azeotropic copolymers must have a point of constant composition, the critical composition $(f_1)_c$ for the azeotrope can be evaluated by solving Eq. (7.11) with $d\,[M_1]/d\,[M_2] = [M_1]/[M_2]$ or Eq. (7.17) with $F_1 = f_1$. By doing so, one obtains

$$\left(\frac{[M_1]}{[M_2]}\right)_c = \frac{1 - r_2}{1 - r_1} \tag{7.18}$$

and

$$(f_1)_c = (1 - r_2)/(2 - r_1 - r_2) \tag{7.19}$$

Note that f_1 in the above equation is physically meaningful ($0 \leq f_1 \leq 1$) only if both r_1 and r_2 are either greater or smaller than unity. (If $r_1 = r_2 = 1$, all values of f_1 are azeotropic compositions.) Since the case of $r_1 > 1$, $r_2 > 1$ is uncommon in free-radical systems, the necessary conditions for azeotropy in such copolymerizations is that $r_1 < 1$, $r_2 < 1$ (see Fig. 7.2). Equation (7.19) then predicts the feed composition that would yield an invariant copolymer composition as the conversion proceeds in a batch reactor.

Problem 7.3   When monomers M_1 and M_2 are copolymerized, an azeotrope is formed at the feed ratio of 1 mol of M_1 to 2 mol of M_2. Monomer M_1 is known *not* to homopolymerize. Will a polymer formed at 50% conversion from an initial mixture of 4 mol of M_1 and 6 mol of M_2 contain more of M_1 or less of M_1 than a polymer formed at 1% conversion ?

Answer:

Since M_1 does not homopolymerize, $r_1 = 0$. Azeotropic feed composition, $f_1 = 1/3 = (1 - r_2)/(2 - 0 - r_2)$. This gives $r_2 = 0.5$. For a feed composition $f_1 = 0.4$ and $f_2 = 0.6$, Eq. (7.17) with $r_1 = 0$ and $r_2 = 0.5$ gives $F_1 = 0.36$. Therefore the feed composition will drift toward higher f_1 at higher conversion and hence F_1 at 50% conversion will be greater than F_1 at 1% conversion.

7.2.4   Integrated Binary Copolymer Equation

It should be noted that the copolymer equations, Eqs. (7.11) and (7.17), give only the *instantaneous* copolymer composition, i.e., the composition of the copolymer that is formed instantly at a given feed composition at very low degrees of conversion (approximately < 5%) such that the composition of the monomer feed may be considered to be essentially unchanged from its initial value. For all copolymerizations except when the feed composition is an azeotropic mixture or where $r_1 = r_2 = 1$, the copolymer product compositions are different from monomer feed compositions. Thus there occurs a drift in the comonomer composition, and correspondingly a drift in the copolymer composition, as the degree of conversion increases. It is important to be able to calculate the course of such changes.

Problem 7.4   A monomer pair with $r_1 = 0.2$ and $r_2 = 5.0$ is copolymerized beginning with a molar monomer ratio $[M_1]/[M_2] = 60/40$. Assuming that the copolymer composition within a 10 mol% conversion interval is constant, calculate instantaneous monomer and copolymer compositions and cumulative average copolymer compositions at 10 mol% conversion intervals up to 100% total conversion. Show the results graphically as change in composition of the copolymer and the monomer mixture during copolymerization.

Answer:

In interval 1: $f_1 = 60/(60 + 40) = 0.60$. From Eq. (7.17), $F_1 = 0.2308$.

At the end of interval 1 (i.e., after 10 mol% conversion), M_1 converted = 2.308 mol; M_1 remaining = 60 − 2.308 = 57.692 mol; $f_1 = 57.692/90 = 0.641$; and $\overline{F_1} = 2.308/10 = 0.2308$.

The residual mixture with $f_1 = 0.641$ will be the starting mixture for the interval 2. From Eq. (7.17) then, $F_1 = 0.2631$. Hence at the end of interval 2 (i.e., 20 mol% conversion), M_1 converted = 2.308 + 2.631 = 4.939 mol; M_1 remaining = 60 − 4.939 = 55.061 mol; $f_1 = 55.061/80 = 0.6883$; and $\overline{F_1} = 4.939/20 = 0.247$.

The results obtained by proceeding in this way are tabulated below:

Interval	mol% Conversion (cumulative)	f_1	F_1	$\overline{F_1}$
1	10	0.60	0.23	0.23
2	20	0.64	0.26	0.25
3	30	0.69	0.31	0.27
4	40	0.74	0.37	0.29
5	50	0.80	0.45	0.32
6	60	0.88	0.58	0.37
7	70	0.95	0.79	0.43
8	80	1.0	1.0	0.50
9	90	1.0	1.0	0.55
10	100	1.0	1.0	0.60

The data in column 3 represent the instantaneous values at the beginning of different intervals. The values are assumed to be constant within the respective intervals. The composition data in column 5 relate to mixtures

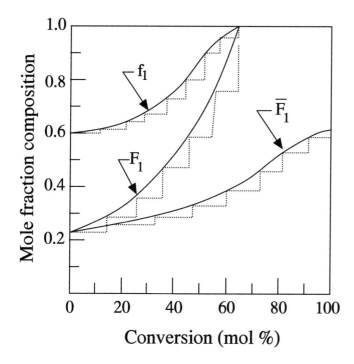

Figure 7.3 Change in the composition of the copolymer and the monomer mixture during copolymerization of a monomer pair with $(f_1)_0 = 0.60$, $r_1 = 0.20$, $r_2 = 5.0$. Data are from Problem 7.4. (After Vollmert, 1973.)

of different copolymers and, at higher conversions, to mixtures of copolymers and some M_1 homopolymer. The results are shown graphically in Fig. 7.3.

Comment: If one makes the conversion interval smaller and smaller, this corresponds to an integration of the copolymer equation (see below).

To follow the drift in composition of both the feed and the copolymer formed one needs to integrate the copolymer equation. The process being rather complex, the numerical or graphical approach of Skeist (1946) based on Eq. (7.17) provides a simple solution to the problem. Consider a system initially containing a total of N moles of the two monomers and choose M_1 as the monomer such that $F_1 > f_1$ (i.e., the polymer being formed contains more M_1 than the feed). Thus at a time when dN moles of the monomer mixture have been converted into polymer, the polymer formed will contain $F_1 dN$ moles of M_1, while the M_1 content in feed will be reduced to $(N - dN)(f_1 - df_1)$ moles. Thus a material balance for monomer M_1 can be written as (Odian, 1991; Billmeyer, Jr., 1994):

$$f_1 N - (N - dN)(f_1 - df_1) = F_1 dN \tag{7.20}$$

Neglecting the small term $df_1 dN$, this equation is rearranged to the form

$$\frac{dN}{N} = \frac{df_1}{F_1 - f_1} \tag{7.21}$$

and then to the integral form

$$\ln \frac{N}{N_0} = \int_{(f_1)_0}^{f_1} \frac{df_1}{(F_1 - f_1)} \tag{7.22}$$

where N_0 and $(f_1)_0$ are the initial values of N and f_1. For given values of r_1 and r_2, the quantities F_1 and $1/(F_1 - f_1)$ are computed from Eq. (7.17) at suitable intervals for $0 < f_1 < 1$. The integration may then be performed graphically or numerically to obtain the degree of conversion $p (= 1 - N/N_0)$ corresponding to a change in feed composition from $(f_1)_0$ to f_1. By a repeated application of this process for suitably chosen values of f_1, it is possible to construct a curve relating f_1 and p. The average overall copolymer composition for any conversion p can be calculated by graphical integration of a plot of F_1 versus f_1 or simply from the amounts of initial and residual monomers (see Problem 7.5).

Problem 7.5   Derive an equation that gives the cumulative or average composition of the copolymer formed at a given overall conversion of the monomers.

Answer:

Consider a batch polymerization mixture containing initially $(N_1)_0$ mol of monomer M_1 and $(N_2)_0$ mol of monomer M_2. Let: after a fraction p of the initial monomers have been polymerized, the unreacted monomers are, respectively, N_1 and N_2 moles; the mole fractions of monomers M_1 and M_2 in the feed after a degree of conversion p are f_1 and f_2, the corresponding initial values being $(f_1)_0$ and $(f_2)_0$. Then,

$$(N_1)_0 + (N_2)_0 = N_0$$

$$(N_1)_0 = (f_1)_0 N_0, \quad (N_2)_0 = (f_2)_0 N_0$$

$$N_1 = f_1(1 - p)N_0, \quad N_2 = f_2(1 - p)N_0$$

Since the average mole fraction of M_1 in copolymer, $\overline{F_1}$, is the ratio of the number of moles of M_1 converted divided by the total number of moles of M_1 and M_2 polymerized in the same interval,

$$\overline{F_1} = \frac{(N_1)_0 - N_1}{[(N_1)_0 - N_1] + [(N_2)_0 - N_2]}$$

Substituting from the above definitions, one obtains

$$\overline{F_1} = \frac{(f_1)_0 - f_1(1 - p)}{p} \tag{P7.5.1}$$

The cumulative average copolymer composition can be calculated in a straightforward manner by entering Eq. (P7.5.1) with the cumulative value of p and the initial value of $(f_1)_0$. [As a check, the value of $\overline{F_1}$ at $p = 1$ must equal $(f_1)_0$.]

Meyer and Lowry (1965) substituted F_1 in Eq. (7.22) using Eq. (7.17) and then integrated Eq. (7.22) analytically to obtain (Odian, 1991; Billmeyer, Jr., 1994):

$$1 - \frac{N}{N_0} = 1 - \left[\frac{f_1}{(f_1)_0} \right]^\alpha \left[\frac{f_2}{(f_2)_0} \right]^\beta \left[\frac{(f_1)_o - \delta}{f_1 - \delta} \right]^\gamma \tag{7.23}$$

which relates the degree of (overall) conversion, $(1 - N/N_0)$, to initial and final feed compositions. The zero subscripts indicate initial state. The quantities α, β, γ, and δ are given by

$$\alpha = \frac{r_2}{(1 - r_2)}, \quad \beta = \frac{r_1}{(1 - r_1)},$$

$$\gamma = \frac{(1 - r_1 r_2)}{(1 - r_1)(1 - r_2)}, \quad \delta = \frac{(1 - r_2)}{(2 - r_1 - r_2)} \quad (7.24)$$

Equation (7.23) can be used to correlate the drift in the feed and copolymer compositions with conversion. This equation is also often used with experimental conversion versus feed composition data to estimate reactivity ratios.

Problem 7.6 A mixture of styrene (M_1) and methyl methacrylate (M_2) was polymerized at 60°C with initial composition $(f_1)_0 = 0.80$, $(f_2)_0 = 0.20$ and the polymer obtained by precipitation at appropriate intervals was analyzed. Some of the conversion-composition data (Meyer, 1966) so obtained are given below :

Conversion (w_c), wt%	Oxygen in polymer (w_{ox}), wt%	f_1
11.74	8.32	0.8091
29.32	7.92	0.8229
46.18	7.61	0.8391
65.88	7.43	0.8769
86.37	7.00	0.9648

With $r_1 = 0.52$ and $r_2 = 0.46$, calculate from Eq. (7.23) the changes in instantaneous monomer and copolymer compositions as a function of conversion and compare the results graphically with the above experimental data. Also calculate the cumulative average copolymer composition at different conversions.

Answer:

The essential procedure for calculating the composition drift with conversion is that f_1 is decreased (or increased) in steps from $(f_1)_0$ to 0 (or to 1.0). For each value of f_1, the corresponding degree of conversion is obtained from Eq. (7.23) and the corresponding instantaneous copolymer composition from Eq. (7.17). With the monomer mixture composition, f_1, and the degree of conversion $p\,(= 1 - N/N_0)$ thus known, it is then easy to also calculate cumulative average copolymer composition $\overline{F_1}$ from Eq. (P7.5.1). For the given monomer system and feed composition, Eq. (7.17) shows that $\overline{F_1} < f_1$, i.e., the polymer is poorer (monomer feed richer) in M_1, as compared to the initial monomer feed; f_1 is therefore to be *increased* in step increments from $(f_1)_0$ in the computation.

The relation between $\overline{F_1}$ and the oxygen content (w_{ox} wt.%) of the copolymer is easily obtained from mass balance, taking note that each MMA unit in the copolymer accounts for 2 oxygen atoms. Taking molar masses of styrene and MMA as 104 and 100 g/mol, the following relation is obtained :

$$\overline{F_1} = (32 - w_{ox})/(32 + 0.04 w_{ox})$$

To convert weight% conversion (w_c) of monomers into mole fraction degree of conversion $(1 - N/N_0)$, the following relation is readily derived :

$$1 - \frac{N}{N_0} = 1 - \left[\frac{4(f_1)_0 + 100}{4 f_1 + 100}\right](1 - w_c/100)$$

[Since for the given system the differences between $(f_1)_0$ and f_1 are small, $1 - N/N_0 \simeq w_c/100$ and mole fraction conversion nearly equals the weight fraction conversion.]

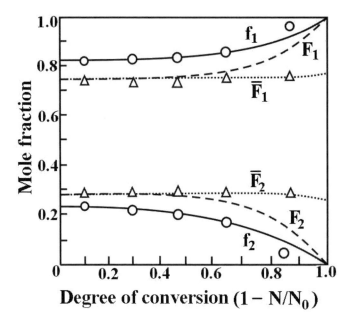

Figure 7.4 Plot of copolymer and monomer feed composition as a function of conversion for styrene (M_1)-methyl methacrylate (M_2) system with $(f_1)_0 = 0.80$, $(f_2)_0 = 0.20$, and $r_1 = 0.53$, $r_2 = 0.46$. Experimental points $\circ$, $\triangle$ are from Problem 7.6. (After Meyer and Chan, 1967.)

The calculated values of f_1, F_1, and $\overline{F_1}$ are plotted against mole fraction degree of conversion along with the given experimental data in Fig. 7.4. The results for f_2 follow from the relation $f_1 + f_2 = 1$, and similarly for F_2 and $\overline{F_2}$.

7.2.5   *Evaluation of Monomer Reactivity Ratios*

The monomer reactivity ratios r_1 and r_2 can be determined from the experimental conversion-composition data of binary copolymerization using both the instantaneous and integrated binary copolymer composition equations, described previously. However, in the former case, it is essential to restrict the conversion to low values (ca. < 5%) in order to ensure that the feed composition remains essentially unchanged. Various methods have been used to obtain monomer reactivity ratios from the instantaneous copolymer composition data. Several procedures for extracting reactivity ratios from the differential copolymer equation [Eq. (7.11) or (7.17)] are mentioned in the following paragraphs. Two of the simpler methods involve plotting of r_1 versus r_2 or F_1 versus f_1.

7.2.5.1   *Plot of r_1 versus r_2*

This method is also known as the *method of intersections*. First described by Mayo and Lewis (1944), the method has been widely used for computing reactivity ratios by fitting experimental data to the differential copolymer equation. In this procedure, Eq. (7.11) is rearranged to the

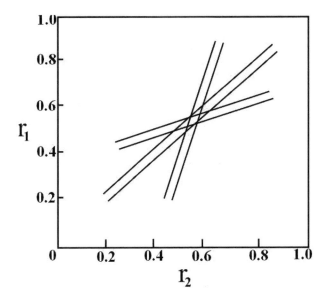

Figure 7.5 Graphical determination of r_1 and r_2 according to Eq. (7.26) for the system styrene/methylmethacrylate (M_1 = styrene; M_2 = methyl methacrylate). (After Mayo and Lewis, 1944.)

form (Ghosh, 1990):

$$r_2 = \frac{[M_1]}{[M_2]}\left[\frac{d[M_2]}{d[M_1]}\left(1 + \frac{[M_1]}{[M_2]}r_1\right) - 1\right] \tag{7.25}$$

or, equivalently, Eq. (7.17) to the form

$$r_2 = \frac{f_1}{f_2}\left[\frac{F_2}{F_1}\left(1 + \frac{f_1}{f_2}r_1\right) - 1\right] \tag{7.26}$$

Knowing experimentally the copolymer composition (F_1) corresponding to a given feed composition (f_1), one can calculate r_2 values corresponding to various assumed values of r_1 and thus obtain a straight line plot of r_1 versus r_2. Each experimental data pair of f_1 and F_1 thus yields one straight line in the $r_1 - r_2$ plane (see Fig. 7.5). Repeating this procedure for different values of F_1 and f_1, a series of straight lines with different slopes are thus obtained. Theoretically, these lines should intersect at a common point representing the actual values of r_1 and r_2. However, because of experimental errors, the lines may not pass through a common point and the small area where most intersections occur is then assumed to represent the most probable r_1, r_2 values.

7.2.5.2 *Plot of F_1 versus f_1*

Fineman and Ross (1950) rearranged Eq. (7.17) to the form

$$\frac{f_1(2F_1 - 1)}{(1 - f_1)F_1} = \left[\frac{f_1^2(1 - F_1)}{(1 - f_1)^2 F_1}\right]r_1 - r_2 \tag{7.27}$$

or (Rudin, 1982; Odian, 1991)

$$G = r_1 H - r_2 \qquad (7.28)$$

where $G = X(Y-1)/Y$, $H = X^2/Y$, $X = [M_1]/[M_2] = f_1/(1 - f_1)$, and $Y = d[M_1]/d[M_2] = F_1/(1 - F_1)$ [see Eqs. (7.15) and (7.16)].

A plot of the term on the left side of Eq. (7.27) or (7.28) against the coefficient of r_1 should thus yield a straight line with slope r_1 and intercept r_2.

Problem 7.7   The initial concentrations of styrene (M_1) and acrylonitrile (M_2) employed in a series of low conversion free-radical copolymerizations are given below together with the nitrogen contents (% N by wt.) of the corresponding copolymer samples produced:

$[M_1]$ mol/L	3.45	2.60	2.10	1.55
$[M_2]$ mol/L	1.55	2.40	2.90	3.45
% N in copolymer	5.69	7.12	7.77	8.45

Determine r_1 and r_2 for the monomer pair by the Fineman-Ross method.

Answer:

Molar mass of repeat unit: styrene (M_1) 104 g mol^{-1}, acrylonitrile (M_2) 53 g mol^{-1}.

$$\%N = \frac{(1 - F_1)(14 \text{ g mol}^{-1})}{F_1(104 \text{ g mol}^{-1}) + (1 - F_1)(53 \text{ g mol}^{-1})} \times 100$$

Solving, $F_1 = [1400 - (\%N \times 53)]/[1400 + (\%N \times 51)]$. Table below gives $f_1 = [M_1]/([M_1] + [M_2])$ and F_1 calculated from %N together with the composite quantities required to make a plot according to Eq. (7.27).

f_1	F_1	$\dfrac{f_1(2F_1 - 1)}{(1 - f_1)F_1}$	$\dfrac{f_1^2(1 - F_1)}{(1 - f_1)^2 F_1}$
0.69	0.6499	1.0268	2.6688
0.52	0.5800	0.2989	0.8499
0.42	0.5501	0.1319	0.4289
0.31	0.5200	0.0346	0.1863

A plot of the data (Fig. 7.6) yields a staight line with slope $(r_1) = 0.40$ and intercept $(-r_2) = -0.04$. Hence, for the given monomer pair: $r_1 = 0.40, r_2 = 0.04$. Analysis of the data may also be done by regression analysis.

The best values of r are obtained from slopes rather than intercepts. While Eq. (7.28) gives r_1 as the slope, it can be rewritten in another form making r_2 the slope :

$$G/H = -r_2/H + r_1 \qquad (7.29)$$

In the aforesaid Mayo-Lewis and Fineman-Ross methods, the experimental composition data are unequally weighted as, for example, at low $[M_2]$ in Eq. (7.28) or low $[M_1]$ in Eq. (7.29) the experimental data have the greatest influence on the slope of a line corresponding to these equations.

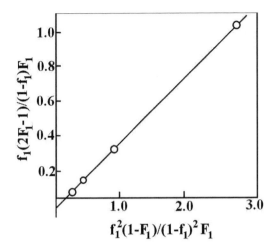

Figure 7.6 Plot according to the Fineman-Ross method (data from Problem 7.7).

The same set of experimental data may thus yield different (r_1, r_2) values depending on which monomer is taken as M_1 and which monomer M_2.

Kelen and Tudos (1975, 1990) modified the Fineman-Ross method by introducing an arbitrary positive constant α into Eq. (7.28) so as to spread the data more evenly thereby giving equal weightage to the data points. Their modification is expressed in the form

$$\eta = \left[r_1 + \frac{r_2}{\alpha}\right]\mu - \frac{r_2}{\alpha} \tag{7.30}$$

where $\eta = G/(\alpha + H)$ and $\mu = H/(\alpha + H)$.

By plotting η against μ one thus obtains a straight line that yields, on extrapolation, $-r_2/\alpha$ and r_1 as η values at $\mu = 0$ and $\mu = 1$, respectively. Using for α the value of $(H_{\min}H_{\max})^{1/2}$, where $H_{\min}$ and $H_{\max}$ are the lowest and highest experimental H values, helps distribute the data symmetrically on the plot.

7.2.5.3 Direct Curve Fitting

An estimate of r_1 and r_2 can be obtained from the slope of the experimental F_1 versus f_1 plot by comparison with curves based on Eq. (7.17) to choose, by trial and error, the values of r_1 and r_2 for which the theoretical curve best fits the data. A limitation of this method is the relative insensitivity of the curves to small changes in r_1 and r_2. Another limitation is the assumption implied in using the differential form of the copolymerization equation [Eq. (7.11) or (7.17)] that the feed composition does not change during the experiment, which is obviously not true. To minimize the error, the polymerization is usually carried out to as low a conversion as possible at which a sufficient amount of the copolymer can still be obtained for direct analysis. The aforesaid limitations can be overcome, however, by the use of an integrated form of the copolymer composition equation, such as Eq. (7.23). In one method, for example, one determines by computational techniques the best values of r_1 and r_2 that fit Eq. (7.23) to the experimental curve of f_1 or f_2 versus $(1 - N/N_o)$.

Some representative values of r_1 and r_2 in radical copolymerization for a number of monomer pairs are shown in Table 7.1. These are seen to differ widely. The reactivity ratios obtained in anionic and cationic copolymerizations are given and discussed in Chapter 8.

Table 7.1 Monomer Reactivity Ratios in Radical Copolymerization

Monomer (M_1)	Monomer (M_2)	T (°C)	r_1	r_2
Acrylic acid	Styrene	60	0.25	0.15
	Vinyl acetate	70	2	0.1
Acrylonitrile	Acrylamide	60	0.8	1.3
	Methyl methacrylate	60	0.13	1.16
	Styrene	60	0.04	0.40
	Vinyl acetate	70	6.0	0.07
Methacrylic acid	Acrylonitrile	70	2.5	0.093
	Styrene	60	0.62	0.2
	Vinyl acetate	70	20	0.01
Methyl methacrylate	Styrene	60	0.46	0.52
	Vinyl acetate	60	20	0.015
	Vinyl chloride	68	10	0.1
Styrene	1,3-Butadiene	50	0.58	1.4
	Maleic anhydride	50	0.04	0.015
	Vinyl acetate	60	56	0.01
	Vinyl chloride	60	15	0.01

Source: Brandrup and Immergut, 1975.

Note: Though only single values are shown for r_1 and r_2, the experimentally reported reactivity ratios often span some range (with a factor of 2 or much larger variation) due to several sources of experimental and statistical uncertainties.

7.2.6   The $Q - e$ Scheme

A useful scheme was proposed by Alfrey and Price (1947) to provide a quantitative description of the behavior of vinyl monomers in radical polymerization, in terms of two parameters for *each* monomer rather than for a monomer *pair*. These parameters are denoted by Q and e and the method is known as the $Q - e$ scheme. An advantage of the method is that it allows calculation of monomer reactivity ratios r_1 and r_2 from the same Q and e values of the monomers irrespective of which monomer pair is used. The scheme assumes that each radical or monomer can be classified according to its reactivity (or resonance effect) and its polarity so that the rate constant for a radical-monomer reaction, e.g., the reaction of M_1 • radical with M_2 monomer, can be written as

$$k_{12} = P_1 Q_2 \exp(-e_1 e_2) \tag{7.31}$$

where P_1 is considered to be a measure for the reactivity of radical M_1 • and Q a measure for the reactivity of monomer M_2; e_1 and e_2, on the other hand, are considered to represent the polar characteristics of the radical and the monomer, respectively.

By assuming that the same e value applies to both a monomer and its radical (that is, e_1 defines the polarities of M_1 and M_1 •, while e_2 defines the polarities of M_2 and M_2 •), one can write expressions for k_{11}, k_{22}, and k_{21} analogous to Eq. (7.31). These can then be appropriately combined to yield the monomer reactivity ratios. Thus for k_{11} one can write by analogy to Eq. (7.31),

$$k_{11} = P_1 Q_1 \exp(-e_1 e_1) \tag{7.32}$$

Table 7.2 $Q-e$ Values

Monomer	e	Q
Butadiene	-1.05	2.39
Styrene (reference standard)	-0.80	1.00
Vinyl acetate	-0.22	0.03
Ethylene	-0.20	0.01
Vinyl chloride	0.20	0.04
Vinylidene chloride	0.36	0.22
Methyl acrylate	0.60	0.42
Methyl methacrylate	0.40	0.74
Acrylic acid	0.77	1.15
Methacrylonitrile	0.81	1.12
Acrylonitrile	1.20	0.60
Methacrylamide	1.24	1.46
Maleic anhydride	2.25	0.23

Source: Brandrup and Immergut, 1975.

Therefore,

$$r_1 = \frac{k_{11}}{k_{12}} = \frac{Q_1}{Q_2} \exp[-e_1(e_1 - e_2)] \tag{7.33}$$

(Note that in any given pair of monomers, the monomer cited first is indexed as M_1 and the other as M_2.) An expression for r_2 is similarly obtained, viz.,

$$r_2 = \frac{k_{22}}{k_{21}} = \frac{Q_2}{Q_1} \exp[-e_2(e_2 - e_1)] \tag{7.34}$$

Thus r_1 and r_2 can be calculated from Q and e values of monomers forming the pair.

Equations (7.33) and (7.34) permit us to calculate the Q and e values for single monomers from the values of r_1 and r_2, provided we have one monomer for which Q and e have been arbitrarily decided. Price chose styrene as the standard monomer with the values $Q = 1$ and $e = -0.8$. Table 7.2 gives a selection of Q and e values for some of most common monomers. As a general rule, monomers with electron-rich double bonds have more negative e values and those that form highly resonance-stabilized radical have higher Q numbers.

Using the tabulated Q and e values for any two monomers, one can calculate the r_1 and r_2 values from Eqs. (7.33) and (7.34) for this monomer pair whether or not they were ever polymerized. The $Q-e$ scheme is of the utmost utility, qualitatively, for predicting copolymerization behavior and for obtaining approximate estimates of r_1 and r_2 values.

Problem 7.8   Calculate the r_1 and r_2 values for the monomer pair styrene(M_1)-acrylonitr-ile(M_2) from the tabulated Q and e values.

Answer:

From Eq. (7.33), $r_1 = \dfrac{1.00}{0.60} \exp[0.80(-0.80 - 1.20)] = 0.336$ (cf. exptl. $r_1 = 0.29$)

From Eq. (7.34), $r_2 = \dfrac{0.60}{1.00} \exp[-1.20(1.20 + 0.8)] = 0.054$ (cf. exptl. $r_2 = 0.03$)

7.2.7   Sequence Length Distribution

It should be noted that the copolymer equation (7.11) describes the instantaneous copolymer composition on a macroscopic scale, that is, composition in terms of the overall mole ratio or mole fraction of monomer units in the copolymer sample produced, but it does not reveal its microstructure, that is, the manner in which the monomer units are distributed along the copolymer chain. Thus for two monomers M_1 and M_2, the ratio $F_1/(1 - F_1)$ gives the overall mole ratio of M_1 and M_2 units in the copolymer but no information about the average lengths (i.e., number of monomer units) of M_1 and M_2 sequences, as illustrated (Allcock and Lampe, 1990) for a typical copolymer by

$$-M_1-M_1-M_1-M_1-M_2-M_1-M_1-M_2-M_2-M_2-M_1-M_1-M_1-$$

where the sequences are underlined. A completely random placement of the two monomer units along the copolymer chain occurs only for the case $r_1 = r_2 = 1$. For all other cases, there will be a definite trend toward a regular placement of monomer units along the chain.

The distributions of the various lengths of the M_1 and M_2 sequences are called the *sequence length distributions*, which define the microstructure of a copolymer. The probabilities of forming M_1 and M_2 sequences of x units, that is, their respective mole fractions $n_x(M_1)$ and $n_x(M_2)$, are given by (see Problem 7.1 for derivation),

$$n_x(M_1) = P_{11}^{x-1} P_{12} \tag{7.35}$$

$$n_x(M_2) = P_{22}^{x-1} P_{21} \tag{7.36}$$

where the probability (P) values are defined (cf. Problem 7.1) by,

$$P_{11} = \frac{r_1[M_1]}{r_1[M_1] + [M_2]} = \frac{r_1 f_1}{r_1 f_1 + f_2} \tag{7.37}$$

$$P_{12} = \frac{[M_2]}{r_1[M_1] + [M_2]} = \frac{f_2}{r_1 f_1 + f_2} \tag{7.38}$$

$$P_{22} = \frac{r_2[M_2]}{[M_1] + r_2[M_2]} = \frac{r_2 f_2}{f_1 + r_2 f_2} \tag{7.39}$$

$$P_{21} = \frac{[M_1]}{[M]_1 + r_2[M_2]} = \frac{f_1}{f_1 + r_2 f_2} \tag{7.40}$$

The mole fractions of different lengths of M_1 and M_2 sequences can be calculated from Eqs. (7.35) through (7.40).

Problem 7.9   The sequence length distribution for a copolymerization system can be described by the mole fractions of sequences with 1, 2, 3, 4, 5, 6, ..., M_1 or M_2 units in a copolymer and plotting against sequence length in a bar graph. Describe such distribution for an ideal copolymerization with equimolar feed composition considering two cases : (a) $r_1 = r_2 = 1$ and (b) $r_1 = 5$, $r_2 = 0.2$.

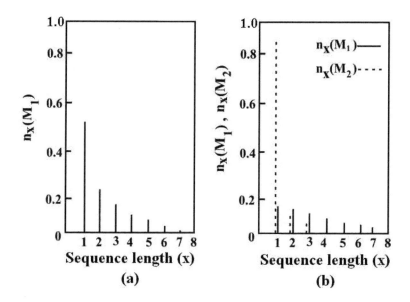

Figure 7.7 Sequence length distribution for an ideal copolymerization with (a) $r_1 = r_2 = 1$, $f_1 = f_2$ and (b) $r_1 = 5$, $r_2 = 0.2$, $f_1 = f_2$. [In (a) the distribution of M_2 sequences is not shown as it is the same as for M_1 sequences. In (b) the plots of $n_x(M_2)$ (- - -) are shown slightly to the left of the actual sequence length for clarity.] (After Vollmert, 1973.)

Answer:

(a) For the system $r_1 = r_2 = 1$, $f_1 = 0.5$, Eqs. (7.37)-(7.40) give $P_{11} = P_{12} = P_{22} = P_{21} = 0.50$. The values of $n_x(M_1)$ calculated from Eq. (7.35) are 0.50, 0.25, 0.125, 0.0625, 0.0313, 0.0156, and 0.0078 for sequences with 1, 2, 3, 4, 5, 6, and 7 M_1 units, respectively. A bar graph is shown in Fig. 7.7(a). Thus, although the most plentiful sequence is single M_1 at 50%, there are considerable amounts of other sequences : 25%, 12.5%, 6.25%, 3.13%, 1.56%, and 0.78%, respectively, of dyad, triad, tetrad, pentad, hexad, and heptad sequences. The distribution of M_2 sequences is exactly the same as for M_1 sequences.

(b) For the system $r_1 = 5$, $r_2 = 0.2$, $f_1 = 0.5$. Equations (7.37)-(7.40) give $P_{11} = P_{21} = 0.8333$ and $P_{12} = P_{22} = 0.1667$. The values of $n_x(M_1)$ calculated from Eq. (7.35) are 0.167, 0.140, 0.116, 0.097, 0.081, 0.067, and 0.056 for sequences with 1, 2, 3, 4, 5, 6, and 7 M_1 units, respectively. A bar graph is shown in Fig. 7.7(b). It is seen that the single M_1 sequences are again the most plentiful but only at 16.7% and the sequence-length distribution is also broader than for $r_1 = r_2 = 1$ of case (a). There are small amounts of relatively long sequences: 3.2% of 10-unit, 1.3% of 15-unit, and 0.4% of 20-unit M_1 sequences. The sequence-length distribution for the less reactive M_2 monomer, calculated from Eq. (7.36) and shown in Fig. 7.7(b), is seen to be much narrower. Single M_2 units are by far the most plentiful (83.3%) with 13.9% dyads, 2.3% triads, and 0.39% tetrads.

Problem 7.10   Consider an alternating copolymerization with $r_1 = r_2 = 0.1$ and $f_1 = 0.5$. What percentage of the alternating copolymer structure is made of $M_1 M_2$ sequence ? Compare the sequence length distribution with that for the ideal copolymer in Problem 7.9(a) which has the identical overall composition.

Answer:

For the system with $r_1 = r_2 = 0.1$ and $f_1 = 0.5$, Eqs. (7.37)-(7.40) give $P_{11} = P_{22} = 0.0910$; $P_{12} = P_{21} = 0.9090$. Therefore, the sequence length distributions for both monomer units are identical. The single M_1

and single M_2 sequences are overwhelmingly the most plentiful at 90.9% each. Thus the M_1M_2 sequence comprises 90.9% of the copolymer structure. From Eqs. (7.35) and (7.36) the dyad and triad sequences are 8.3% and 0.75%, respectively, for both M_1 and M_2. The large difference between this distribution and the distribution in Fig. 7.7(a) for a random copolymer having identical overall composition clearly indicates the difference between alternating and ideal behavior. The ideal copolymer with an overall composition of $F_1 = F_2 = 0.5$ has a microstructure that is very different from that of a predominantly alternating copolymer.

The average sequence lengths $\bar{x}(M_1)$ and $\bar{x}(M_2)$ may also be determined from Eqs. (7.35) and (7.36). We have already derived these expressions in Eqs. (P7.1.11) and (P7.1.12). Thus,

$$\bar{x}(M_1) = 1 + \frac{r_1[M_1]}{[M_2]} = 1 + r_1(f_1/f_2) \tag{7.41}$$

$$\bar{x}(M_2) = 1 + \frac{r_2[M_2]}{[M_1]} = 1 + r_2(f_2/f_1) \tag{7.42}$$

The *run number*, N_R, of the copolymer is defined as the average number of sequences of either type per 100 monomer units (Allcock and Lampe, 1990). Considering, for example, a hypothetical copolymer shown below, in which the sequences are underlined,

$$\underline{M_2}\text{-}\underline{M_1\text{-}M_1}\text{-}\underline{M_2}\text{-}\underline{M_1}\text{-}\underline{M_2}\text{-}\underline{M_2}\text{-}\underline{M_1\text{-}M_1\text{-}M_1}\text{-}\underline{M_2}\text{-}\underline{M_1\text{-}M_1\text{-}M_1}\text{-}\underline{M_2\text{-}M_2\text{-}M_2\text{-}M_2}\text{-}\underline{M_1\text{-}M_1}$$

the number of sequences are 10 and there are 20 monomer units. Hence $N_R = 50$. A larger run number indicates a greater tendency toward alternation. For a perfectly alternating polymer N_R is 100.

Problem 7.11   Assuming (in the case of fairly large molecular weight copolymer) that the rate of sequence formation, regardless of length, is simply the rate at which sequences are ended, derive an expression for *run number*, N_R, in terms of the average sequence lengths $\bar{x}(M_1)$ and $\bar{x}(M_2)$.

Answer:

Since the molecular weight of the polymer is fairly large, the chain termination can be neglected compared to propagation (long-chain approximation). The rate of sequence formation, dS/dt, can then be represented (Allcock and Lampe, 1990) by

$$\frac{dS}{dt} = k_{12}[M_1^\star][M_2] + k_{21}[M_2^\star][M_1] \tag{P7.11.1}$$

and the rate of polymerization by

$$-\frac{d([M_1]+[M_2])}{dt} = k_{11}[M_1^\star][M_1] + k_{12}[M_1^\star][M_2] +$$
$$k_{21}[M_2^\star][M_1] + k_{22}[M_2^\star][M_2] \tag{P7.11.2}$$

Dividing Eq. (P7.11.2) by Eq. (P7.11.1) and making use of the steady-state approximation, as given by Eq. (7.8), one obtains

$$-\frac{d([M_1] + [M_2])}{dS} = \frac{k_{11}[M_1] + k_{12}[M_2]}{2k_{12}[M_2]} + \frac{k_{22}[M_2] + k_{21}[M_1]}{2k_{21}[M_1]}$$

or

$$-\frac{d([M_1] + [M_2])}{dS} = 1 + \frac{r_1}{2}\frac{[M_1]}{[M_2]} + \frac{r_2}{2}\frac{[M_2]}{[M_1]} \tag{P7.11.3}$$

Since the *run number* is the average number of sequences per 100 monomer units, this may now be written (Allcock and Lampe, 1990) as

$$N_R = 100\left(-\frac{dS}{d([M_1] + [M_2])}\right) \qquad (P7.11.4)$$

which after substitution of Eq. (P7.11.3) followed by Eqs. (7.41) and (7.42) yields

$$N_R = \frac{200}{2 + r_1([M_1]/[M_2]) + r_2([M_2]/[M_1])} = \frac{200}{\bar{x}(M_1) + \bar{x}(M_2)} \qquad (P7.11.5)$$

Since for a perfectly *alternating polymer*, $\bar{x}(M_1) = \bar{x}(M_2) = 1$, one obtains from Eq. (P7.11.5), $N_R = 100$.

Problem 7.12 Vinyl acetate (3.0 M) is copolymerized with vinyl chloride (1.5 M) in benzene solution by adding azobisisobutyronitrile to a concentration of 0.1 M and heating to 60°C. Calculate for the copolymer initially formed (a) the probability of forming vinyl acetate and vinyl chloride sequences that are 3 units long, (b) the average sequence lengths of vinyl acetate and vinyl chloride in the copolymer, and (c) the run number of the copolymer. $[r_1 = 0.24, r_2 = 1.80]$

Answer:

(a) $f_1 = (3.0\,M)/(3.0\,M + 1.5\,M) = 0.67$: $f_2 = 1 - 0.67 = 0.33$

From Eqs. (7.37)-(7.40): $P_{11} = 0.3244, P_{12} = 0.6756, P_{22} = 0.4736, P_{21} = 0.5264$

From Eq. (7.35): $n_3(M_1) = (0.3244)^{3-1}(0.6756) = 0.071$

From Eq. (7.36): $n_3(M_2) = (0.4736)^{3-1}(0.5264) = 0.118$

(b) From Eqs. (7.41) and (7.42):

$\bar{x}(M_1) = 1 + (0.24)(0.67/0.33) = 1.48$

$\bar{x}(M_2) = 1 + (1.80)(0.33/0.67) = 1.90$

(c) From Eq. (P7.11.5): $N_R = 200/(1.48 + 1.90) = 59$

7.2.8 Rate of Binary Free-Radical Copolymerization

In deriving an expression for the rate of copolymerization in binary systems the following assumptions (Walling, 1949) will be made: (a) rate constants for the reactions of a growing chain depend only upon the monomer unit at the chain end, and not on other units preceding the end unit; (b) steady-state conditions apply both to the total radical concentration and to the separate concentrations of the two radicals; (c) chain termination is by bimolecular radical reaction.

By assumption (a) the overall rate of monomer disappearance is given by [cf Eqs. (7.5) and (7.6)]:

$$R_p = -\frac{d([M_1] + [M_2])}{dt}$$

$$= k_{11}[M_1{}^\bullet][M_1] + k_{21}[M_2{}^\bullet][M_1] + k_{12}[M_1{}^\bullet][M_2] + k_{22}[M_2{}^\bullet][M_2] \qquad (7.43)$$

where $[M_1]$ and $[M_2]$ are concentrations of the two monomers, $[M_1{}^\bullet]$ and $[M_2{}^\bullet]$ are concentrations of chain radicals ending in M_1 and M_2 units, respectively, and k_{21} is the rate constant for attack of a chain radical ending in M_2 unit upon monomer M_1, etc. By assumptions (b) and (c) and defining an overall termination rate constant k_{to}, two steady-state assumptions may also be written as

$$k_{12}[M_1{}^\bullet][M_2] = k_{21}[M_2{}^\bullet][M_1] \qquad (7.44)$$

$$R_i = R_t = 2k_{to}([M_1{}^\bullet] + [M_2{}^\bullet])^2 \qquad (7.45)$$

where R_i is the overall initiating rate and R_t is the overall termination rate.

Solving Eqs. (7.44) and (7.45) simultaneously for $[M_1{}^\bullet]$ and $[M_2{}^\bullet]$ and substituting into Eq. (7.43) we obtain

$$R_p = -\frac{d([M_1] + [M_2])}{dt} = \frac{R_i^{1/2}(r_1[M_1]^2 + 2[M_1][M_2] + r_2[M_2]^2)}{(2k_{to})^{1/2}(r_1[M_1]/k_{11} + r_2[M_2]/k_{22})} \qquad (7.46)$$

where r_1 and r_2 are the monomer reactivity ratios given by the propagation rate constant ratios k_{11}/k_{12} and k_{22}/k_{21}, respectively.

If the cross-termination rate constant, that is, the rate constant for termination of radical $M_1{}^\bullet$ with radical $M_2{}^\bullet$ is k_{t12}, then the steady-state for the total concentration of radicals can also be written as

$$R_i = R_t = 2k_{t11}[M_1{}^\bullet]^2 + 2k_{t12}[M_1{}^\bullet][M_2{}^\bullet] + 2k_{t22}[M_2{}^\bullet]^2 \qquad (7.47)$$

If we define mole fractions of the respective radicals as $\dot{n}_1$ and $\dot{n}_2$, that is,

$$\dot{n}_1 = \frac{[M_1{}^\bullet]}{[M_1{}^\bullet] + [M_2{}^\bullet]} \text{ and } \dot{n}_2 = 1 - \dot{n}_1 = \frac{[M_2{}^\bullet]}{[M_1{}^\bullet] + [M_2{}^\bullet]} \qquad (7.48)$$

then from Eqs. (7.45), (7.47) and (7.48)

$$\frac{R_t}{([M_1{}^\bullet] + [M_2{}^\bullet])^2} = 2(k_{t11}\dot{n}_1^2 + k_{t12}\dot{n}_1\dot{n}_2 + k_{t22}\dot{n}_2^2) = 2k_{to} \qquad (7.49)$$

where

$$k_{to} = k_{t11}\dot{n}_1^2 + k_{t12}\dot{n}_1\dot{n}_2 + k_{t22}\dot{n}_2^2 \qquad (7.50)$$

A cross-termination factor ϕ defined as

$$\phi = k_{t12}/2(k_{t11}k_{t22})^{1/2} \qquad (7.51)$$

represents the rate constant for cross-termination relative to the geometric mean of the rate constants for chain termination of each monomer alone. Its value > 1 is thus indicative of the preference for cross-termination over homotermination. Equation (7.46) then assumes the following familiar form of the so-called "chemical control" model (Rudin, 1982):

$$-\frac{(d\,[M_1]+d\,[M_2])}{dt} = \frac{\left(r_1[M_1]^2 + 2[M_1][M_2] + r_2[M_2]^2\right)R_i^{1/2}}{\left(r_1\delta_1^2[M_1]^2 + 2\phi r_1 r_2 \delta_1 \delta_2[M_1][M_2] + r_2\delta_2^2[M_2]^2\right)^{1/2}} \qquad (7.52)$$

where δ_1 and δ_2 represent the termination-propagation rate constant ratios given by

$$\delta_1 = \left(2k_{t11}/k_{11}^2\right)^{1/2}, \quad \delta_2 = \left(2k_{t22}/k_{22}^2\right)^{1/2} \qquad (7.53)$$

Equation (7.52) can also be derived from Eq. (7.43) by eliminating radical concentrations with the help of two steady-state assumptions written as Eqs. (7.44) and (7.47) and then using the definitions of r_1 and r_2. Equation (7.52) represents a one-parameter model for the copolymerization rate (Melville et al., 1947) containing the parameter ϕ. Statistically, ϕ is expected to equal unity. However, the values of ϕ obtained in practice by inserting experimental copolymerization rates into Eq. (7.52) are frequently greater than unity. (For styrene-methyl methacrylate, for example, ϕ is 15, while for styrene-butyl acrylate ϕ is 150.) These deviations are ascribed to polar effects that favor cross-termination over homotermination.

Problem 7.13   Show that if reaction probabilities of radicals depend only on encounter rates and are independent of the nature of the radicals, ϕ should equal unity.

Answer:

For simplicity consider a reaction mixture in which $[M_1\cdot] = [M_2\cdot]$. Since $M_1\cdot$ radicals will collide with $M_2\cdot$ and $M_1\cdot$ radicals with equal frequency and since the same applies to $M_2\cdot$ radicals, the frequency of $M_1\cdot - M_2\cdot$ encounters is twice that of $M_1\cdot - M_1\cdot$ or $M_2\cdot - M_2\cdot$ collisions. If the reaction probabilities depend only on encounter rates,

$$\frac{R_{t12}}{(R_{t11}R_{t22})^{1/2}} = 2$$

where $R_{t11} = 2k_{t11}[M_1\cdot]^2$, $R_{t22} = 2k_{t22}[M_2\cdot]^2$, and $R_{t12} = 2k_{t12}[M_1\cdot][M_2\cdot]$.

Thus, $\phi = \dfrac{k_{t12}}{2(k_{t11}k_{t22})^{1/2}} = 1$.

Since segmental diffusivity would be expected to depend on the structure of macroradicals, diffusion controlled termination may depend on copolymer composition. The value of the overall termination rate constant, k_{to}, in copolymerizations may thus be functions of fractions (F_1 and F_2) of the comonomers incorporated in the copolymer. An empirical expression for k_{to} has thus been proposed (Atherton and North, 1962):

$$k_{to} = F_1 k_{t11} + F_2 k_{t22} \qquad (7.54)$$

that represents k_{t0} as the average homotermination rate constants weighted on the basis of mole fractions F_1 and F_2 of the respective monomers in the copolymer. Equation (7.46) in combination with Eq. (7.54) provides a *diffusion-controlled model* with no adjustable parameters (such as ϕ in chemical controlled model). No strong theoretical case can, however, be made for the relationship in Eq. (7.54).

A better fit is often provided by a *combined model* (Chiang and Rudin, 1975) which uses the parameter ϕ of the chemical control model in combination with an empirical formulation for k_{t0} related to copolymer composition. The empirical formulation is thus derived (Rudin, 1982) by substituting mole fractions of each monomer for the radical mole fractions $\dot{n}_1$ and $\dot{n}_2$ in Eq. (7.50):

$$k_{t0} = k_{t11}F_1^2 + k_{t12}F_1F_2 + k_{t22}F_2^2 \tag{7.55}$$

Combination with Eq. (7.51) then yields

$$k_{t0} = k_{t11}F_1^2 + 2\phi(k_{t11}k_{t22})^{1/2}F_1F_2 + k_{t22}F_2^2 \tag{7.56}$$

Equations (7.46) and (7.56) yield the rate of copolymerization, and ϕ may be taken from previous studies of the chemical control model.

Problem 7.14   Bulk polymerization of styrene in the presence of 1 g/L of AIBN initiator at 60°C gave a measured polymerization rate of 5.34×10^{-5} mol L^{-1} s^{-1}. Predict the rate of copolymerization at 60°C of a mixture of styrene (M$_1$) and methyl methacrylate (M$_2$) with 0.579 mole fraction styrene and the same initial concentration of the initiator as in the homopolymerization case. Compare the rates predicted from chemical control, diffusion control, and combined models with the experimental value of 4.8×10^{-5} mol/L-s (Walling, 1949). Use relevant k_p and k_t values for homopolymerization from Table 6.4 and assume $\phi = 15$. [Other data: $r_1 = 0.52, r_2 = 0.46$; monomer density = 0.90 g/cm^3.]

Answer:

Rate constants: $k_{11} = (k_p)_{\text{styrene}} = 165$ L mol^{-1} s^{-1}; $k_{t11} = (k_t)_{\text{styrene}} = 6\times10^7$ L mol^{-1} s^{-1}; $k_{22} = (k_p)_{\text{MMA}}$ = 515 L mol^{-1} s^{-1}; $k_{t22} = (k_t)_{\text{MMA}} = 2.55\times10^7$ L mol^{-1} s^{-1}.

For bulk styrene: [M] = (1000 cm^3 L^{-1})(0.90 g cm^{-3})/(104 g mol^{-1}) = 8.70 mol L^{-1}.   For styrene (M$_1$)–MMA (M$_2$) feed of 0.579 mole fraction styrene: [M$_1$] = 5.09 mol L^{-1} (assuming additivity of monomer volumes).

For styrene homopolymerization [cf. Eq. (6.23)],

$$
\begin{aligned}
R_i &= \frac{2R_p^2 k_{t11}}{k_{11}^2 [\text{M}]^2} \\
&= \frac{2(5.34\times10^{-5}\ \text{mol L}^{-1}\text{s}^{-1})^2(6\times10^7\ \text{L mol}^{-1}\text{s}^{-1})}{(165\ \text{L mol}^{-1}\text{s}^{-1})^2(8.70\ \text{mol L}^{-1})^2} = 1.66\times10^{-7}\ \text{mol L}^{-1}\text{s}^{-1}
\end{aligned}
$$

Chemical Control Model: For copolymerization, Eq. (7.52) for chemical control model is conveniently written (Rudin, 1982) as

$$-\frac{d([\text{M}_1] + [\text{M}_2])}{dt} = \frac{R_i^{1/2}(r_1\alpha^2 + 2\alpha + r_2)[\text{M}_1]/\alpha}{(r_1\delta_1^2\alpha^2 + 2\phi r_1 r_2\delta_1\delta_2\alpha + r_2\delta_2^2)^{1/2}} \tag{P7.14.1}$$

where α = mole ratio of M_1 and M_2 in feed = 0.579/0.421 = 1.375, r_1 = 0.52, r_2 = 0.46, R_i = 1.66×10^{-7} mol L^{-1} s^{-1}, ϕ = 15

$$\delta_1 = \left[\frac{2(6\times 10^7 \text{ L mol}^{-1}\text{ s}^{-1})}{(165 \text{ L mol}^{-1}\text{ s}^{-1})^2}\right]^{1/2} = 66.39 \text{ mol}^{1/2}\text{ s}^{1/2}\text{ L}^{-1/2}$$

$$\delta_2 = \left[\frac{2(2.55\times 10^7 \text{ L mol}^{-1}\text{ s}^{-1})}{(515 \text{ L mol}^{-1}\text{ s}^{-1})^2}\right]^{1/2} = 13.87 \text{ mol}^{1/2}\text{ s}^{1/2}\text{ L}^{-1/2}$$

Substituting the values in Eq. (7.14.1), $-d([M_1] + [M_2])/dt = 5.44\times 10^{-5}$ mol L^{-1} s^{-1}

Diffusion Control Model: Substituting appropriate values in Eq. (7.17) gives $F_1 = 0.5624$, $F_2 = 0.4376$. From Eq. (7.54), one obtains $k_{to} = (0.5624)(6\times 10^7 \text{ L mol}^{-1}\text{ s}^{-1}) + (0.4376)\times (2.55\times 10^7 \text{ L mol}^{-1}\text{ s}^{-1}) = 4.49\times 10^7 \text{ L mol}^{-1}\text{ s}^{-1}$. Writing Eq. (7.46) in the form

$$-\frac{d([M_1] + [M_2])}{dt} = \frac{R_i^{1/2}(r_1\alpha^2 + 2\alpha + r_2)[M_1]}{(2k_{to})^{1/2}(r_1/k_{11} + r_2/\alpha k_{22})\alpha^2} \tag{P7.14.2}$$

where α is the mole ratio of M_1 and M_2 = 1.375 and substituting appropriate values in Eq. (P7.14.2) gives $-d([M_1]+[M_2])/dt = 12.7\times 10^{-5}$ mol L^{-1} s^{-1}.

Combined Model: From Eq. (7.56) with $\phi = 15$, $k_{to} = 31.26\times 10^7$ L mol^{-1} s^{-1} and so from Eq. (P7.14.2), $-d([M_1]+[M_2])/dt = 4.8\times 10^{-5}$ mol L^{-1} s^{-1}.

Comparison

	Chemical control model	$(R_p)_{\text{copolym}} \times 10^5$ Diffusion control model	Combined model
Exptl.	with $\phi = 15$	(no parameter)	with $\phi = 15$
4.8	5.4	12.7	4.8

7.3 Multicomponent Copolymerization: Terpolymerization

Simultaneous copolymerization of three or more types of monomers is generally known as *multi-component copolymerization*, while *terpolymerization* is the term used to designate simultaneous copolymerization of *three* types of monomers. For deriving the copolymer composition equations in these cases, it is assumed (terminal or first-order Markov model) that the reactivity of the propagating species is dependent on the identity of the terminal or end monomer unit and that the various propagation reactions are irreversible. In multicomponent copolymerization of *n* types of monomers, there will thus be *n* different types of growing chains represented by wwwM$_i$ ${}^\bullet$ (i = 1, 2, $\cdots$ *n*). Assuming that each of these can react with any of the *n* types of monomers, there will be n^2 propagation reactions and $n(n-1)$ binary reactivity ratios. The rate of consumption of a monomer *j* in this multicomponent system comprising *n* types of monomers is thus given, in general terms, by

$$-d[M_j]/dt = \left(\sum_{i=1}^{n} k_{ij}[M_i{}^\bullet]\right)[M_j] \tag{7.57}$$

Considering the specific case of terpolymerization of monomers M_1, M_2, and M_3, there are three different types of growing chain ends, viz., wwwM$_1$ ${}^\bullet$, wwwM$_2$ ${}^\bullet$, and wwwM$_3$ ${}^\bullet$ and *nine* different chain propagation reactions (Walling and Briggs, 1945; Hocking and Klimchuk, 1996), as shown below:

Reaction	Rate
$\text{wwww}M_1{}^{\bullet} + M_1 \longrightarrow \text{wwww}M_1{}^{\bullet}$	$-d[M_1]/dt = k_{11}[M_1{}^{\bullet}][M_1]$
$\text{wwww}M_1{}^{\bullet} + M_2 \longrightarrow \text{wwww}M_2{}^{\bullet}$	$-d[M_2]/dt = k_{12}[M_1{}^{\bullet}][M_2]$
$\text{wwww}M_1{}^{\bullet} + M_3 \longrightarrow \text{wwww}M_3{}^{\bullet}$	$-d[M_3]/dt = k_{13}[M_1{}^{\bullet}][M_3]$
$\text{wwww}M_2{}^{\bullet} + M_1 \longrightarrow \text{wwww}M_1{}^{\bullet}$	$-d[M_1]/dt = k_{21}[M_2{}^{\bullet}][M_1]$
$\text{wwww}M_2{}^{\bullet} + M_2 \longrightarrow \text{wwww}M_2{}^{\bullet}$	$-d[M_2]/dt = k_{22}[M_2{}^{\bullet}][M_2]$
$\text{wwww}M_2{}^{\bullet} + M_3 \longrightarrow \text{wwww}M_3{}^{\bullet}$	$-d[M_3]/dt = k_{23}[M_2{}^{\bullet}][M_3]$
$\text{wwww}M_3{}^{\bullet} + M_1 \longrightarrow \text{wwww}M_1{}^{\bullet}$	$-d[M_1]/dt = k_{31}[M_3{}^{\bullet}][M_1]$
$\text{wwww}M_3{}^{\bullet} + M_2 \longrightarrow \text{wwww}M_2{}^{\bullet}$	$-d[M_2]/dt = k_{32}[M_3{}^{\bullet}][M_2]$
$\text{wwww}M_3{}^{\bullet} + M_3 \longrightarrow \text{wwww}M_3{}^{\bullet}$	$-d[M_3]/dt = k_{33}[M_3{}^{\bullet}][M_3]$

$$(7.58)$$

Accordingly, the equations for the rates of monomer consumptions are

$$-d[M_1]/dt = k_{11}[M_1{}^{\bullet}][M_1] + k_{21}[M_2{}^{\bullet}][M_1] + k_{31}[M_3{}^{\bullet}][M_1] \qquad (7.59)$$

$$-d[M_2]/dt = k_{12}[M_1{}^{\bullet}][M_2] + k_{22}[M_2{}^{\bullet}][M_2] + k_{32}[M_3{}^{\bullet}][M_2] \qquad (7.60)$$

$$-d[M_3]/dt = k_{13}[M_1{}^{\bullet}][M_3] + k_{23}[M_2{}^{\bullet}][M_3] + k_{33}[M_3{}^{\bullet}][M_3] \qquad (7.61)$$

We do not know the absolute values of the radical concentrations $[M]_1{}^{\bullet}$, $[M]_2{}^{\bullet}$, $[M]_3{}^{\bullet}$, but we can derive the steady-state relationships for these unknown concentrations. Since in steady state, radicals of the type $\text{wwww}M_1{}^{\bullet}$ are formed just as fast as they are converted to types $\text{wwww}M_2{}^{\bullet}$ and $\text{wwww}M_3{}^{\bullet}$, we may write:

$$k_{12}[M_1{}^{\bullet}][M_2] + k_{13}[M_1{}^{\bullet}][M_3] = k_{21}[M_2{}^{\bullet}][M_1] + k_{31}[M_3{}^{\bullet}][M_1] \qquad (7.62)$$

and similarly for radicals of types $\text{wwww}M_2{}^{\bullet}$ and $\text{wwww}M_3{}^{\bullet}$:

$$k_{21}[M_2{}^{\bullet}][M_1] + k_{23}[M_2{}^{\bullet}][M_3] = k_{12}[M_1{}^{\bullet}][M_2] + k_{32}[M_3{}^{\bullet}][M_2] \qquad (7.63)$$

$$k_{31}[M_3{}^{\bullet}][M_1] + k_{32}[M_3{}^{\bullet}][M_2] = k_{13}[M_1{}^{\bullet}][M_3] + k_{23}[M_2{}^{\bullet}][M_3] \qquad (7.64)$$

Terpolymerization involves $3(3-1)$ or 6 monomer reactivity ratios, viz., $r_{12} = k_{11}/k_{12}$, $r_{13} = k_{11}/k_{13}$, $r_{21} = k_{22}/k_{21}$, $r_{23} = k_{22}/k_{23}$, $r_{31} = k_{33}/k_{31}$, and $r_{32} = k_{33}/k_{32}$. [Note that the symbols of reactivity ratios now require two indices each, since there are three monomers. Thus the reactivity ratios r_1 and r_2 used in Eq. (7.11) would be called r_{12} and r_{21} in this modified notation.] Using these reactivity ratios and combining Eqs. (7.62)-(7.64) with Eqs. (7.59)-(7.61) gives the following composition relations (Walling and Briggs, 1945):

$$\frac{dM_1}{dM_3} = \frac{M_1(M_1 r_{23} r_{32} + M_2 r_{31} r_{23} + M_3 r_{32} r_{21})(M_1 r_{12} r_{13} + M_2 r_{13} + M_3 r_{12})}{M_3(M_1 r_{12} r_{23} + M_2 r_{13} r_{21} + M_3 r_{12} r_{21})(M_3 r_{31} r_{32} + M_1 r_{32} + M_2 r_{31})} \qquad (7.65)$$

$$\frac{dM_2}{dM_3} = \frac{M_2(M_1 r_{32} r_{13} + M_2 r_{13} r_{31} + M_3 r_{12} r_{31})(M_2 r_{21} r_{23} + M_1 r_{23} + M_3 r_{21})}{M_3(M_1 r_{12} r_{23} + M_2 r_{13} r_{21} + M_3 r_{12} r_{21})(M_3 r_{31} r_{32} + M_1 r_{32} + M_2 r_{31})} \qquad (7.66)$$

where M_i reperesents the molar concentration of monomer i ($i = 1, 2, 3$). [Note that in Eqs. (7.65) and (7.66) the usual symbol of $[M_i]$ for a concentration has been abbreviated to M_i in order to facilitate typesetting.]

The above terpolymer composition can be obtained in an equivalent form (Alfrey and Goldfinger, 1946; Hocking and Klimchuk, 1996) as:

$$d[M_1] : d[M_2] : d[M_3] =$$

$$[M_1]\left\{\frac{[M_1]}{r_{31}r_{21}} + \frac{[M_2]}{r_{21}r_{32}} + \frac{[M_3]}{r_{31}r_{23}}\right\}\left\{[M_1] + \frac{[M_2]}{r_{12}} + \frac{[M_3]}{r_{13}}\right\}$$

$$: [M_2]\left\{\frac{[M_1]}{r_{12}r_{31}} + \frac{[M_2]}{r_{12}r_{32}} + \frac{[M_3]}{r_{32}r_{13}}\right\}\left\{[M_2] + \frac{[M_1]}{r_{21}} + \frac{[M_3]}{r_{23}}\right\}$$

$$: [M_3]\left\{\frac{[M_1]}{r_{13}r_{21}} + \frac{[M_2]}{r_{23}r_{12}} + \frac{[M_3]}{r_{13}r_{23}}\right\}\left\{[M_3] + \frac{[M_1]}{r_{31}} + \frac{[M_2]}{r_{32}}\right\} \qquad (7.67)$$

Steady-state relationships have also been expressed (Valvassori and Sartori, 1967) by considering only pairs of radicals. This gives:

$$k_{12}[M_1{}^{\bullet}][M_2] = k_{21}[M_2{}^{\bullet}][M_1]$$

$$k_{23}[M_2{}^{\bullet}][M_3] = k_{32}[M_3{}^{\bullet}][M_2]$$

$$k_{31}[M_3{}^{\bullet}][M_1] = k_{13}[M_1{}^{\bullet}][M_3].$$

Combining these equations with Eqs. (7.59)-(7.61) yields a simpler and shorter version of the terpolymer composition equation (Hocking and Klimchuk, 1996):

$$d[M_1] : d[M_2] : d[M_3] = [M_1]\left\{[M_1] + \frac{[M_2]}{r_{12}} + \frac{[M_3]}{r_{13}}\right\}$$

$$: [M_2]\frac{r_{21}}{r_{12}}\left\{\frac{[M_1]}{r_{21}} + [M_2] + \frac{[M_3]}{r_{23}}\right\}$$

$$: [M_3]\frac{r_{31}}{r_{13}}\left\{\frac{[M_1]}{r_{31}} + \frac{[M_2]}{r_{32}} + [M_3]\right\} \qquad (7.68)$$

Both the conventional Eq. (7.67) and the simplified Eq. (7.68) can be used for predicting terpolymer compositions as they yield similar results (see Problem 7.15).

Problem 7.15   The ternary copolymerization of a monomer mixture containing 35.92, 36.03, and 28.05 mol% of styrene, methyl methacrylate (MMA), and acrylonitrile (AN) at 60°C for 3.5 h yielded at 13.6 wt% conversion a polymer product which analyzed C 78.6% and N 4.68% (by wt.) (Walling and Briggs, 1945). Calculate the initial ternary copolymer composition to compare with the composition obtained by the analysis.

Answer:

Let the mol% composition of the copolymer be x, y, and z for styrene (C_8H_8), MMA ($C_5H_8O_2$) and AN (C_3H_3N), respectively. From C and N mass balance and $x + y + z = 100$, one obtains $x = 45.0$, $y = 25.6$, and $z = 29.4$.

From Table 7.1,

Styrene (M_1)/MMA (M_2): $r_{12} = 0.52$, $r_{21} = 0.46$
Styrene (M_1)/AN (M_3): $r_{13} = 0.40$, $r_{31} = 0.04$
MMA (M_2)/AN (M_3): $r_{23} = 1.16$, $r_{32} = 0.13$

Rewriting Eq. (7.67) as $d[M_1] : d[M_2] : d[M_3] = A : B : C$, where

$$A = f_1 \left[\frac{f_1}{r_{31}r_{21}} + \frac{f_2}{r_{21}r_{32}} + \frac{f_3}{r_{31}r_{23}} \right] \left[f_1 + \frac{f_2}{r_{12}} + \frac{f_3}{r_{13}} \right]$$

$$B = f_2 \left[\frac{f_1}{r_{12}r_{31}} + \frac{f_2}{r_{12}r_{32}} + \frac{f_3}{r_{32}r_{13}} \right] \left[f_2 + \frac{f_1}{r_{21}} + \frac{f_3}{r_{23}} \right]$$

$$C = f_3 \left[\frac{f_1}{r_{13}r_{21}} + \frac{f_2}{r_{23}r_{12}} + \frac{f_3}{r_{13}r_{23}} \right] \left[f_3 + \frac{f_1}{r_{31}} + \frac{f_2}{r_{32}} \right]$$

one can then calculate for a given starting feed composition, viz., f_1, f_2, and f_3, the copolymer compositions as $F_1 = A/(A+B+C)$, $F_2 = B/(A+B+C)$, $F_3 = C/(A+B+C)$.

Equation (7.68) can be similarly written and A, B, C defined accordingly. With $f_1 = 0.3592$, $f_2 = 0.3603$, $f_3 = 0.2805$, and r_{ij} values as given above, the copolymer composition is calculated from both Eqs. (7.67) and (7.68). The results are tabulated below:

Feed composition		Terpolymer composition (mol %)		
Monomer	mol %	Found	Calcd. from Eq. (7.67)	Calcd. from Eq. (7.68)
Styrene	35.92	45.0	45.5	44.7
MMA	36.03	25.6	31.9	31.3
AN	28.05	29.4	22.7	23.9

Knowing the value of the monomer reactivity ratio for each pair of monomers also allows the calculation of sequence distributions. Thus if the monomer mixture is made up of the monomers M_1, M_2, M_3, $\cdots$ M_N, the resulting initial copolymer will consist of sequences of M_1, M_2, $\cdots$, and M_N monomer units. The fraction of all M_1 sequences which possess x number of M_1 units will be given by a distribution function (Alfrey and Goldfinger, 1944):

$$n_x(M_1) = \left(\frac{[M_1]}{[M_1] + \dfrac{[M_2]}{r_{12}} + \dfrac{[M_3]}{r_{13}} + \cdots \dfrac{[M_N]}{r_{1N}}} \right)^{x-1} \times$$

$$\left(1 - \frac{[M_1]}{[M_1] + \dfrac{[M_2]}{r_{12}} + \dfrac{[M_3]}{r_{13}} + \cdots \dfrac{[M_N]}{r_{1N}}} \right) \qquad (7.69)$$

Note that with $[M_i] = 0$ for $i = 3, 4, \cdots N$, Eq. (7.69) reduces to Eq. (7.35) for the binary system.

Problem 7.16 Consider the initial terpolymer in Problem 7.14. Calculate the fractions of styrene, MMA, and AN sequences containing 2 or more monomer units.

Answer:

For a terpolymer Eq. (7.69) becomes

$$
n_x(M_1) = \left(\frac{f_1}{f_1 + \dfrac{f_2}{r_{12}} + \dfrac{f_3}{r_{13}}} \right)^{x-1} \left(1 - \frac{f_1}{f_1 + \dfrac{f_2}{r_{12}} + \dfrac{f_3}{r_{13}}} \right)
$$

For $x = 1$,

$$
\begin{aligned}
n_1(M_1) &= 1 - \frac{f_1}{f_1 + (f_2/r_{12}) + (f_3/r_{13})} \\
&= 1 - \frac{0.3592}{0.3592 + (0.3603/0.52) + (0.2805/0.29)} = 0.822
\end{aligned}
$$

Therefore, fraction of styrene sequences with 2 or more styrene units = $1 - 0.822 = 0.178$.

Similarly, $n_1(M_2) = 1 - \dfrac{f_2}{(f_1/r_{21}) + f_2 + (f_3/r_{23})} = 0.7345$

Therefore, fraction of MMA sequences with 2 or more units $= 1 - 0.7345 = 0.265$.

Again, $n_1(M_3) = 1 - \dfrac{f_3}{(f_1/r_{31}) + (f_2/r_{32}) + f_3} = 0.9865$

Fraction of AN sequences with 2 or more AN units $= 1 - 0.9865 = 0.013$.

7.4 Deviations from Terminal Model

Deviations from the copolymer composition equation [Eq. (7.11)] of the terminal model have been noted for various comonomer pairs and for various polymerization systems. In order to explain some of these deviations, Merz et al. (1946) introduced the concept of the penultimate effect, according to which the reactivity of a propagating species is affected by the penultimate (next-to-last) unit (*penultimate model*). This behavior is referred to as the penultimate or *second-order Markov* behavior. Deviations from the terminal model can also occur because of the formation of monomer complex which undergoes propagation (*complex-participation model*) or because of depropagation reactions becoming significant, which can happen when polymerizations are performed near ceiling temperatures.

7.4.1 Penultimate Model

Since in this model a growing chain is identified by the last two units at the growing chain end, a binary copolymerization system will involve eight propagating reactions which can be represented (Hamielec et al., 1989; Odian, 1991) by

$$
\begin{aligned}
\text{wwwM}_1\text{M}_1\cdot \ + \ \text{M}_1 &\ \xrightarrow{k_{111}} \ \text{wwwM}_1\text{M}_1\text{M}_1\cdot \\
\text{wwwM}_1\text{M}_1\cdot \ + \ \text{M}_2 &\ \xrightarrow{k_{112}} \ \text{wwwM}_1\text{M}_1\text{M}_2\cdot \\
\text{wwwM}_2\text{M}_2\cdot \ + \ \text{M}_1 &\ \xrightarrow{k_{221}} \ \text{wwwM}_2\text{M}_2\text{M}_1\cdot \\
\text{wwwM}_2\text{M}_2\cdot \ + \ \text{M}_2 &\ \xrightarrow{k_{222}} \ \text{wwwM}_2\text{M}_2\text{M}_2\cdot \\
\text{wwwM}_2\text{M}_1\cdot \ + \ \text{M}_1 &\ \xrightarrow{k_{211}} \ \text{wwwM}_2\text{M}_1\text{M}_1\cdot \\
\text{wwwM}_2\text{M}_1\cdot \ + \ \text{M}_2 &\ \xrightarrow{k_{212}} \ \text{wwwM}_2\text{M}_1\text{M}_2\cdot \\
\text{wwwM}_1\text{M}_2\cdot \ + \ \text{M}_1 &\ \xrightarrow{k_{121}} \ \text{wwwM}_1\text{M}_2\text{M}_1\cdot \\
\text{wwwM}_1\text{M}_2\cdot \ + \ \text{M}_2 &\ \xrightarrow{k_{122}} \ \text{wwwM}_1\text{M}_2\text{M}_2\cdot
\end{aligned}
\tag{7.70}
$$

and four reactivity ratios: $r_{11} = k_{111}/k_{112}$, $r_{21} = k_{211}/k_{212}$, $r_{22} = k_{222}/k_{221}$, and $r_{12} = k_{122}/k_{121}$.

Following a procedure similar to that used in deriving Eq. (7.11), the instantaneous copolymer composition equation for the penultimate model is then given (Hamielec et al., 1989) by

$$
\frac{d[\text{M}_1]}{d[\text{M}_2]} = \frac{1 + r_{21}\,([\text{M}_1]/[\text{M}_2])\,(r_{11}[\text{M}_1] + [\text{M}_2])\,/\,(r_{21}[\text{M}_1] + [\text{M}_2])}{1 + r_{12}\,([\text{M}_2]/[\text{M}_1])\,(r_{22}[\text{M}_2] + [\text{M}_1])\,/\,(r_{12}[\text{M}_2] + [\text{M}_1])}
\tag{7.71}
$$

or in terms of mole fraction feed composition (f_1, f_2) and mole fraction polymer composition (F_1, F_2) by

$$
\frac{F_1}{F_2} = \frac{1 + r_{21}\,(f_1/f_2)\,(r_{11}f_1 + f_2)\,/\,(r_{21}f_1 + f_2)}{1 + r_{12}\,(f_2/f_1)\,(r_{22}f_2 + f_1)\,/\,(r_{21}f_1 + f_2)}
\tag{7.72}
$$

7.4.2   Complex-Participation Model

The model for binary copolymerization, in its general form, allows participation of both the comonomers and a donor-acceptor complex of the comonomers in the propagation reactions. The generalized complex-participation model (Cais et al., 1979) can be described by eight propagation reactions and an equilibrium reaction forming the complex from the monomers. The propagation reactions include the same four Eqs. (7.1)-(7.4) which apply in the terminal model, and four additional ones involving the complex (Hamielec et al., 1989; Odian, 1991) which are

$$
\begin{aligned}
\text{wwwM}_1\cdot \ + \ \overline{\text{M}_2\text{M}_1} &\ \xrightarrow{k_{\overline{121}}} \ \text{wwwM}_1\cdot \\
\text{wwwM}_1\cdot \ + \ \overline{\text{M}_1\text{M}_2} &\ \xrightarrow{k_{\overline{112}}} \ \text{wwwM}_2\cdot \\
\text{wwwM}_2\cdot \ + \ \overline{\text{M}_2\text{M}_1} &\ \xrightarrow{k_{\overline{221}}} \ \text{wwwM}_1\cdot \\
\text{wwwM}_2\cdot \ + \ \overline{\text{M}_1\text{M}_2} &\ \xrightarrow{k_{\overline{212}}} \ \text{wwwM}_2\cdot
\end{aligned}
\tag{7.73}
$$

where $\overline{\text{M}_2\text{M}_1}$ and $\overline{\text{M}_1\text{M}_2}$ represent the monomer complex adding to the radical center at the M_2

and M_1 ends, respectively. The equilibrium reaction for the complex formation is

$$M_1 + M_2 \overset{K}{\rightleftharpoons} \overline{M_1 M_2} \qquad (7.74)$$

and the appropriate ratios (Hamielec et al., 1989; Odian, 1991) for this model are: $r_1 = k_{11}/k_{12}, r_2 = k_{22}/k_{21}, r'_1 = k_{\overline{112}}/k_{\overline{121}}, r'_2 = k_{\overline{221}}/k_{\overline{212}}, r''_1 = k_{\overline{121}}/k_{12},$ and $r''_2 = k_{\overline{212}}/k_{21}$. The copolymer composition equation has been derived (Cais et al., 1979) by a statistical approach.

7.5 Copolymerization and Crosslinking

We have considered so far only copolymerization of bifunctional vinyl monomers. However, if any of the monomers in the copolymerization is a divinyl compound or any other olefinic monomer with functionality greater than 2, a branched polymer can result and, furthermore, the growing branches can interconnect to form an infinite crosslinked network known as "gel". It is useful to be able to predict the conditions under which such gel formation will occur.

In copolymerization involving a divinyl compound or diene, crosslinking occurs early or late in the reaction depending on the relative reactivities of the two double bonds in the monomer. If the two double bonds in a monomer are well separated, the reactivity of one is not affected by the polymerization of the other. Ethylene glycol dimethacrylate and allyl acrylate are examples of divinyl monomers of this type. If, on the other hand, the two double bonds are close enough that the reaction of one shields the other sterically, or if they are conjugated as in 1,3-dienes, a difference in reactivity can be expected. In copolymerization with 1,3-dienes, 1,4-polymerization leads to residual 2,3-double bonds of lowered reactivity, which are subsequently used to bring about post-polymerization crosslinking as in vulcanization. Divinyl benzene is in the intermediate category with regard to the effect of the reaction of one double bond on the reactivity of the other. Several different cases are considered in the following.

7.5.1 Vinyl and Divinyl Monomers of Equal Reactivity

Consider the copolymerization of a vinyl monomer X with a divinyl monomer YY where all of the vinyl groups (i.e., the X group and both Y groups) have the same reactivity. [Methyl methacrylate (MMA)–ethyleneglycol dimethacrylate (EGDMA), vinyl acetate–divinyl adipate (DVA), and to an extent styrene–*p*- or *m*-divinylbenzene (DVB) are examples of this type of polymerization system (Odian, 1991). In MMA-EGDMA system,

the unsaturated groups of EGDMA may be confidently assumed to have same reactivity as the identical group of the MMA monomer and the reactivity of one unsaturated group in EGDMA should not depend on whether the other is reacted or not. The vinyl groups in DVB are, however, known to be more reactive than the one in styrene.]

Let the initial molar concentrations of vinyl monomer and monomeric Y groups be $[X]_0$ and $[Y]_0$, respectively, and that of divinyl YY be $[YY]_0$. Thus, $[Y]_0 = 2[YY]_0$. Since the X and Y double bonds are equally reactive, i.e., $r_1 = r_2 = 1$, one obtains from the copolymer equation [Eq.

(7.17)], $F_1 = f_1$. Thus the molar ratio of Y and X groups in the copolymer is simply equal to [Y]/[X].

At the extent of reaction p (defined as the fraction of X and Y groups reacted), the relative number of various monomeric species can be listed as follows:

Unreacted X's : $[X]_0(1-p)$
Reacted X's : $[X]_0 p$
Unreacted YY's : $[YY]_0(1-p)^2$
Singly reacted YY's : $2[YY]_0(1-p)p$
Doubly reacted YY's : $[YY]_0 p^2$

The number of crosslinks can be taken to be the number of YY molecules in which both Y groups are reacted and the number of polymer chains is derived in terms of the degree of polymerization $\overline{DP}$ as:

$$\text{Number of polymer chains} = \frac{\text{Total number of reacted X and Y groups}}{\overline{DP}}$$
$$= ([X]_0 p + [Y]_0 p)/\overline{DP} \tag{7.75}$$

Therefore,

$$\text{Number of crosslinks per chain} = \frac{p^2[YY]_0\overline{DP}}{([X]_0+[Y]_0)p}$$
$$= \frac{p[Y]_0\overline{DP}}{2([X]_0+[Y]_0)} \tag{7.76}$$

It can be generally shown (Rudin, 1982) by probability considerations that *a sufficient condition for gelation occurs when a polymer sample contains one crosslinked unit per weight average molecule or one crosslink per two weight average polymer chains.* Thus, at the critical extent of reaction p_c for the onset of gelation, the number of crosslinks per chain is $\frac{1}{2}$ and the gel point is obtained as

$$p_c = \frac{([X]_0+2[YY]_0)}{2[YY]_0\overline{DP}_w} \tag{7.77}$$

It may be noted that $\overline{DP}$ is now replaced, in view of the aforesaid condition for gelation, by $\overline{DP}_w$ which is the weight average degree of polymerization of "primary molecules". (The term "primary molecule" is used to designate the linear molecule that would exist if all the crosslinks were severed, that is, the polymer chains formed before any crosslinking reactions occurred. In the above problem it may be taken approximately as the weight average degree of polymerization that would be observed in the homopolymerization of monomer A under the particular reaction conditions.)

Equation (7.77) indicates that gelation can be delayed, that is, the extent of reaction at which gelation occurs can be increased by reducing the concentration of divinyl monomer, by reducing the weight average chain length (increase initiator concentration or add chain transfer agents), or by using a divinyl monomer in which one or both the vinyl groups are less reactive than those in the monovinyl monomer (see later).

For $[X]_0 \gg [YY]_0$, Eq. (7.77) reduces to

$$p_c \cong \frac{[X]_0}{2[YY]_0 \overline{DP}_w} \qquad (7.78)$$

Equation (7.78) predicts that extensive crosslinking occurs during copolymerization of X and YY (see Problem 7.17). The equation holds best for systems containing low concentrations of the monomer YY, that is, at higher gel point conversions where the distribution of crosslinks is random.

Problem 7.17 Bulk polymerization of methyl methacrylate (MMA) at 60°C with 0.9 g/L of benzoyl peroxide yielded a polymer with a weight average degree of polymerization of 8600 at low conversions. Predict the conversions of MMA at which gelation would be observed if it is copolymerized with 0.05 mol% of ethylene glycol dimethacrylate (EGDMA) at the same temperature and initiator concentration as in the homopolymerization case.

Answer:

Let MMA be denoted as monomer X and EGDMA as monomer YY. Then $[YY]_0/([X]_0 + [YY]_0) = 0.0005$, or $[X]_0/[YY]_0] \simeq 2000$.

From Eq. (7.78), with $\overline{DP}_w$ (assumed to be the same as that of the homopolymer) = 8600, $p_c \cong 2000/(2 \times 8600) = 0.12$. So gelation would be observed at about 12% conversion of MMA.

Problem 7.18 Calculate the conversion at which gelation should be observed in styrene containing 0.14 mol% p-divinylbenzene (DVB) and 0.04 mol/L benzoyl peroxide initiator at 60°C. Assume for this calculation that the vinyl groups in both styrene and DVB are equally active and that chain termination occurs solely by coupling. [Data at 60°C : $k_d = 2.4 \times 10^{-6}$ s^{-1}; $f = 0.4$; k_p^2/k_t for styrene = 4.54×10^{-4} L mol^{-1} s^{-1}.]

Answer:

[M] for bulk styrene = 8.65 mol L^{-1}

$$\text{Kinetic chain length} = \left(\frac{k_p}{k_t^{1/2}}\right) \frac{1}{2(fk_d)^{1/2}} \frac{[M]}{[I]^{1/2}}$$

$$= (4.54 \times 10^{-4})^{1/2} \frac{1}{2(0.4 \times 2.4 \times 10^{-6})^{1/2}} \cdot \frac{8.65}{(0.04)^{1/2}} = 470$$

$\overline{DP}_n = 2 \times 470 = 940$. For chain termination solely by coupling, $\overline{DP}_w/\overline{DP}_n = (2 + P)/2$ (see p. 384 for derivation). For high polymer, $P \simeq 1$. Therefore, $\overline{DP}_w = 1410$.

Let X $\equiv$ Styrene, YY $\equiv$ DVB. Then $[YY]_0/([X]_0 + [YY]_0) = 0.0014$ or $[X]_0/[YY]_0 = 713$. From Eq. (7.78), $p_c = 713/2(1410) = 0.253 \ (\equiv 25.3\%)$.

Problem 7.19 How would the percentage conversion at the gel point change if the styrene-divinyl benzene mixture of Problem 7.18 contained additionally a mercaptan chain regulator ($C_S = 21$) at a concentration of 2×10^{-4} mol/L ?

Answer:

In the presence of chain regulator (neglecting other transfer reactions) [cf. Eq. (6.122)],

$$\frac{1}{\overline{DP}_n} = \frac{1}{940} + (21)\frac{(2 \times 10^{-4} \text{ mol/L})}{(8.65 \text{ mol/L})} = 1.5 \times 10^{-3} \implies \overline{DP}_n = 645$$

To determine the proportions of unimolecular (chain transfer) and bimolecular (coupling) termination, let y fraction of molecules be terminated by the former mechanism. Therefore,

$$(1 - y)(940) + y(940/2) = 645, \quad \text{or} \quad y = 0.63$$

$\overline{DP}_w/\overline{DP}_n$ is given by a combination of $(1 + P)$ and $(2 + P)/2$ weighted in proportion to the amounts of umimolecular and bimolecular termination, respectively (see pp. 383-384), that is,

$$\overline{DP}_w/\overline{DP}_n = 0.63(1 + P) + 0.37(2 + P)/2 = 1.81 \text{ for } p \rightarrow 1$$

$$DP_w = 1.81 \times 645 = 1170$$

From Eq. (7.78), $p_c = 713/(2 \times 1170) = 0.30 \ (\equiv 30\% \text{ conversion})$

7.5.2   Vinyl and Divinyl Monomers of Different Reactivities

Let us now consider a vinyl (X)-divinyl (YY) system in which the reactivities of the vinyl groups X and Y are not equal, while the two Y groups are equally reactive. If the Y groups are r times as reactive as the X groups, they enter the copolymer r times as rapidly and hence the ratio of Y and X groups in the copolymer, $d[\text{Y}]/d[\text{X}]$, is

$$d[\text{Y}]/d[\text{X}] = r[\text{Y}]/[\text{X}] \tag{7.79}$$

where [X] and [Y] are the concentrations of the X and Y groups in the monomer mixture at any instant. Thus, at the extent of reaction p of X groups, $[\text{X}] = (1 - p)[\text{X}]_0$ and $[\text{Y}] = (1 - rp)[\text{Y}]_0$, where the zero subscript denotes initial concentrations. By a derivation similar to the above, one then obtains an expression for the critical extent of reaction at gelation as

$$p_c = \frac{[\text{X}]_0 + r[\text{Y}]_0}{r^2[\text{Y}]_0\overline{DP}_w} \tag{7.80}$$

For $[\text{Y}]_0 \ll [\text{X}]_0$, Eq. (7.80) reduces to

$$p_c = \frac{[\text{X}]_0}{r^2[\text{Y}]_0\overline{DP}_w} \tag{7.81}$$

An alternative expression to Eq. (7.80) for the gel point conversion can be derived in terms of the reactivity ratios of the two types of vinyl groups in X and YY. Let r_1 represent the relative reactivity of a X monomer and a Y monomeric group, when reacting with a free radical of type X, and r_2, the relative reactivity of a Y group and a X group, when reacting with a Y-type radical, i.e.,

$$r_1 = k_{\text{XX}}/k_{\text{XY}}, \quad r_2 = k_{\text{YY}}/k_{\text{YX}}$$

The composition equation for initial copolymer in terms of monomer group concentration $[\text{Y}]_0$ becomes [cf. Eq. (7.11)]:

$$\frac{d[\text{Y}]}{d[\text{X}]} = \frac{[\text{Y}]_0}{[\text{X}]_0} \frac{(r_2[\text{Y}]_0 + [\text{X}]_0)}{([\text{Y}]_0 + r_1[\text{X}]_0)} \tag{7.82}$$

Ignoring the drift of residual monomer composition with conversion and assuming a random distribution of crosslinks, we may predict gelation to occur at a conversion p_c given (Odian, 1991) by

$$p_c = \frac{(r_1 [X]_0^2 + 2 [X]_0 [Y]_0 + r_2 [Y]_0^2)^2}{\overline{DP}_w [Y]_0 ([X]_0 + [Y]_0)(r_2 [Y]_0 + [X]_0)^2} \tag{7.83}$$

Thus when the double bonds of the divinyl monomer are more reactive than that of the vinyl monomer ($r_2 > r_1$), gelation occurs at lower conversions. On the other hand, gelation is delayed until the later stages, if $r_1 > r_2$.

In the case in which component X is present in considerable excess ($[X]_0 \gg [Y]_0$), Eq. (7.83) reduces to [cf. Eq. (7.81)]:

$$p_c = \frac{r_1^2 [X]_0}{[Y]_0 \overline{DP}_w} \tag{7.84}$$

Problem 7.20 Predict the extent of reaction at which gelation would occur in (a) vinyl acetate-ethylene glycol dimethacrylate and (b) methyl methacrylate-divinyl adipate systems both containing 15 mol% of the divinyl compound. Assume that the reaction conditions for the two systems are such as to yield the same $\overline{DP}_w$ of 1000 for the uncrosslinked polymer. Take r_1 and r_2 values from Table 7.1 for the respective analogous vinyl-vinyl copolymerizations.

Answer:

(a) For vinyl acetate (X) – ethylene glycol dimethacrylate (YY) system, take $r_1 = 0.015$, $r_2 = 20$ (as for vinyl acetate-methyl methacrylate).

Denoting mole fractions of monomer groups X and Y by f_X and f_Y, $f_X = (85)/(85 + 2 \times 15) = 0.74$. So $f_Y = 0.26$.

Equation (7.83) in terms of f_X and f_Y becomes

$$p_c = \frac{[r_1 + 2 f_Y/f_X + r_2 (f_Y/f_X)^2]^2}{(1000)[f_Y/f_X + (f_Y/f_X)^2](r_2 f_Y/f_X + 1)^2} \tag{P7.20.1}$$

Substituting the values, $p_c = 0.3 \times 10^{-3}$ ($\equiv 0.03\%$)

(b) For methyl methacrylate (X) – divinyl adipate (YY) system take $r_1 = 20$, $r_2 = 0.015$ (same as for methyl methacrylate–vinyl acetate). So, $f_X = 0.74$, $f_Y = 0.26$. Substituting in Eq. (P7.20.1) above, $p_c = 0.89$ ($\equiv 89\%$)

7.5.3 One Group of Divinyl Monomer Having Lower Reactivity

In some instances, one group in a divinyl monomer undergoing copolymerization with a vinyl monomer may exhibit much lower reactivity than the other. Consider copolymerization of a vinyl monomer X with a divinyl monomer YZ where groups X and Y have equal reactivities but group Z has lower reactivity. An example (Odian, 1991) of such a system would be methyl methacrylate-allyl methacrylate, where the two methacrylate groups are the X and Y groups and the allyl group is the Z group. The copolymer in this case will consist of copolymerized X and Y groups with pendant unreacted Z groups until later stages of reaction.

Let p be the fraction of X and Y which have polymerized, and r be the reactivity ratio between Z and Y groups ($r = k_{XZ}/k_{XY}$). At conversion p ($< p_c$) the concentrations of unreacted X and Y have been reduced to $[X]_0(1 - p)$ and $[Y]_0(1 - p)$, respectively, while the concentration of unreacted Z groups is still essentially $[Z]_0$. During a small increase in p, given by dp, the following changes in concentrations of X, Y, and Z groups will occur:

$$-d[X] = [X]_0 dp \tag{7.85}$$

$$-d[Y] = [Y]_0 dp \tag{7.86}$$

$$-d[Z] = [X]_0 dp \left(\frac{[Z]_0}{[X]_0 (1 - p)} \right) r = [Z]_0 r \frac{dp}{(1 - p)} \tag{7.87}$$

The decrease in Z concentration up to the conversion p is given by the integral

$$[Z]_0 r \int_0^p \frac{dp}{1 - p} = [Z]_0 r \ln\left(\frac{1}{1 - p} \right) \tag{7.88}$$

The fraction of Z groups which are reacted at the conversion p is therefore

$$\frac{[Z]_0 - [Z]}{[Z]_0} = r \ln\left(\frac{1}{1 - p} \right) \tag{7.89}$$

The average number of reacted Y groups per chain is closely given by $f_{YZ}\overline{DP}$ where f_{YZ} represents the mole fraction of YZ in the original monomer mixture and $\overline{DP}$ is the average degree of polymerization. Since the fraction $r \ln[1/(1 - p)]$ of these are reacted at the Z group, the average number of crosslinks per chain is

$$\text{Number of crosslinks per chain} = f_{YZ}\overline{DP} \, r \ln\left(\frac{1}{1 - p} \right) \tag{7.90}$$

When this reaches the critical value of $\frac{1}{2}$ (with $\overline{DP}$ replaced by $\overline{DP}_w$), gelation occurs. Thus,

$$f_{YZ}\overline{DP}_w \, r \ln\left(\frac{1}{1 - p_c} \right) = \frac{1}{2}$$

$$\ln(1 - p_c) = -\frac{1}{2r f_{YZ}\overline{DP}_w}$$

$$p_c = 1 - \exp[-1/(2r f_{YZ}\overline{DP}_w)] \tag{7.91}$$

Thus gelation is delayed for a lower value of r.

7.6 Block and Graft Copolymerization

Initiating polymerization reactions through active sites bound on an already formed (parent) polymer molecule leads to block or graft copolymerization, the former involving terminal active sites and the latter involving active sites attached either to the backbone or to pendant side groups. Copolymerizations only by free-radical processes are discussed in this section; those involving ionic mechanisms are described in Chapter 8.

Under suitable conditions block and graft copolymers are produced by carrying out free-radical polymerization in a feed mixture containing the parent polymer, the monomer(s) to be grown on this polymer, and initiator. However, the product obtained is usually a mixture of the desired block and graft copolymer, homopolymer of the added monomer, and unreacted parent polymer.

7.6.1 Block Copolymerization

To produce block copolymers by free-radical polymerization, a radical center must be produced at the end of the chain from where fresh chain growth may take place. Two of the ways by which such terminal radicals can be produced are (a) decomposition of peroxide groups introduced as an internal part of a polymer chain backbone or as an end group and (b) breaking of C–C bonds in the polymer chain by mechanical means. More recently, the advent of living or controlled free radical polymerization has opened up a more versatile route to block copolymers by the free radical process.

7.6.1.1 Producing Internal Peroxide Linkages

(a) Peroxide linkages can be introduced in polymer chains by copolymerization of small amounts of oxygen with olefinic monomers. For example, oxygen copolymerizes with styrene, methyl methacrylate, and vinyl acetate to produce peroxide linkages which can then be decomposed to generate free radicals to initiate polymerization of another monomer:

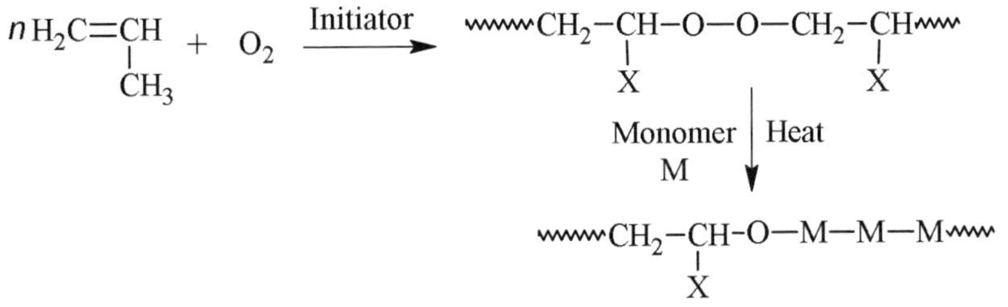

$$n\ H_2C{=}CH \underset{CH_3}{|} +\ O_2 \xrightarrow{\text{Initiator}} \text{wwww}CH_2{-}\underset{X}{\overset{|}{CH}}{-}O{-}O{-}CH_2{-}\underset{X}{\overset{|}{CH}}\text{wwww}$$

$$\downarrow \text{Monomer } | \text{ Heat}$$
$$M$$

$$\text{wwww}CH_2{-}\underset{X}{\overset{|}{CH}}{-}O{-}M{-}M{-}M\text{wwww}$$

(b) Initiation of the parent polymer with polymeric phthaloyl peroxide leads to polymer molecules with peroxide linkages. Phthaloyl polyperoxide has the structure

$$HOOCC_6H_4{-}({-}OOC{-}C_6H_4{-}COO{-})_n{-}C_6H_4COOH$$

The polyperoxide with many internal peroxide linkages decomposes by random cleavage to form shorter diradical species, which then become incorporated into the backbone of the polymer that is produced by the polyperoxide initiation. When fresh monomer is subsequently added and additional heating is supplied, the remaining peroxide linkages in the polymer backbone decompose to form terminal radicals where polymerization is initiated leading to a block copolymer.

7.6.1.2 Introducing Peroxide End Groups

(a) Initiation with *m*-diisopropyl benzene monohydroperoxide produces polymer with isopropylbenzene end groups which can be easily converted into hydroperoxide by reaction with oxygen to initiate polymerization of a second monomer, as shown below:

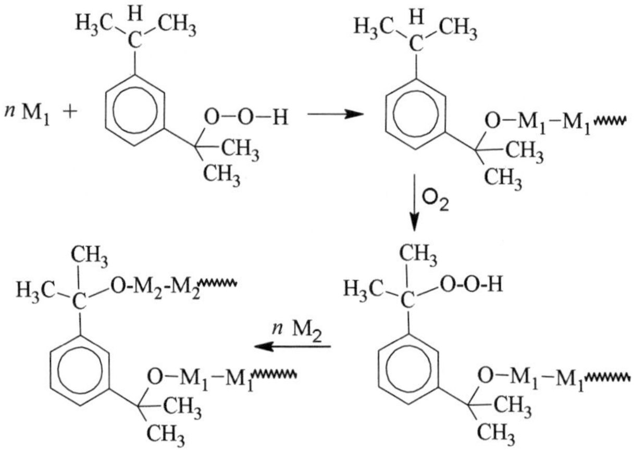

(b) An initiator such as azobiscyanopentanoic acid decomposes to produce free radical species with a carboxyl group at one end:

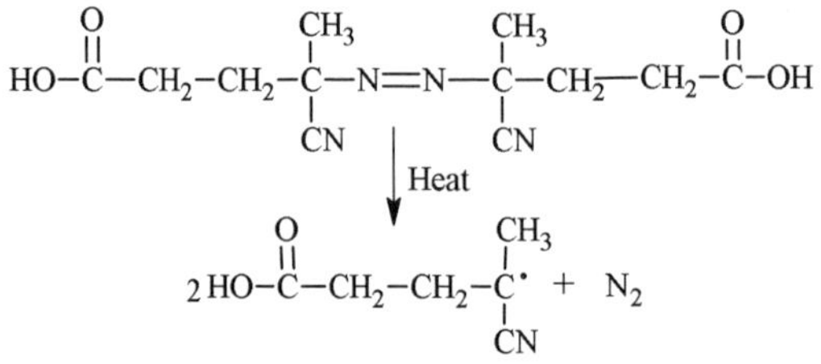

Using such an initiator to produce a polymer, followed by conversion of the carboxylic acid end group to the acid chloride annd then reaction of this with *t*-butyl hydroperoxide results in polymer molecules containing *t*-butyl perester end groups that can be used to initiate polymerization of a second monomer.

7.6.1.3 Mechanical Cleaving of Polymer Chains

Radical centers can be generated at polymer chain ends by subjecting the polymer to high mechanical work (shearing action), a process commonly known as *mastication*. Mastication and mixing are conveniently carried out in two-roll mills or internal mixers. If the shearing forces are sufficiently high, their concentration at individual bonds may cause bond scission, producing radical chain ends. Thus block copolymers are obtained by milling either a mixture of two homopolymers (Odian, 1991):

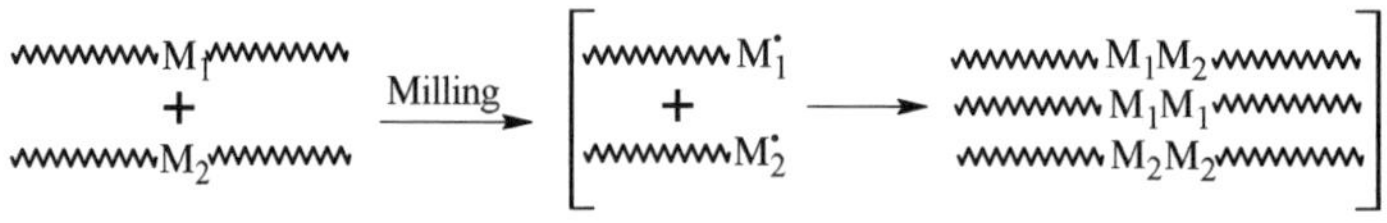

or a mixture of a polymer and a monomer:

In both cases, the process gives rise to mixtures of block copolymer and homopolymer(s).

7.6.1.4 Controlled Radical Polymerization

The methods described above for block copolymer formation by free-radical polymerization are associated with problems such as functionalization of polymer end-groups, relatively low initiator efficiency, and homopolymer formation. However, an improvement to those systems was brought about by the advent of controlled or living radical polymerization (Matyjaszewski, 2000), which can be used for the synthesis of various types of block and graft copolymers with controlled structure (see Section 11.3). Atom transfer radical polymerization (ATRP) is inititiated by alkyl halides (see Section 11.3.1) and, therefore, any polymer which has a sufficiently active alkyl halide end-group could initiate ATRP to afford block copolymers. For example, difunctional polydimethyl-siloxane macroinitiators that contain alkyl bromide end-groups have been used to initiate ATRP of styrene providing control over both chain length and overall composition (Peng et al., 2004). More detailed discussion is provided in Chapter 11.

7.6.2 Graft Copolymerization

We shall consider here graft copolymerization only by free-radical processes. There are three main techniques for preparing graft copolymers via a free-radical mechanism. All of them involve the generation of active sites along the backbone of the polymer chain. These include (i) chain transfer to both saturated and unsaturated backbone or pendant groups; (ii) radiative or photochemical activation; and (iii) activation of pendant peroxide groups.

7.6.2.1 Chain Transfer Methods

The methods depend on the generation of active sites by chain transfer to polymer and reasonable efficiencies of grafting require rather high chain transfer coefficients. Polymers containing carbon-halogen or sulfur-hydrogen bonds are thus susceptible to graft polymerization. The carbon-hydrogen bonds in the α-position to a carbonyl group, as in poly(vinyl acetate), and the carbon-hydrogen bond adjacent to the double bond in unsaturated polymers, such as polybutadiene and polyisoprene, are also suitable for transfer growth.

Unsaturated rubber polymers are especially important for grafting. Consider, for example, the polymerization of styrene in the presence of 1,4-poly(1,3-butadiene) used for producing high-impact polystyrene (HIPS). A method consists of dissolving the polybutadiene rubber (about 5 to 10%) in monomeric styrene containing benzoyl peroxide initiator and applying heat. Polymer radicals, formed by (a) chain transfer between the propagating radical and polymer or (b) addition (copolymerization) of the propagating radical to the double bonds of the polymer, initiate graft polymerization of styrene at the resulting active center on the chain (Odian, 1991):

$$
\text{wwCH}_2\text{-CH=CH-CH}_2\text{ww} + \text{wwCH}_2\overset{\bullet}{\text{CH}}\phi
\begin{cases}
\overset{(a)}{\longrightarrow} & \text{ww}\overset{\bullet}{\text{CH}}\text{-CH=CH-CH}_2\text{ww} \\
& + \text{wCH}_2\text{CH}_2\phi \\[6pt]
\overset{(b)}{\longrightarrow} & \begin{array}{c}\text{wwCH}_2\text{-CH}\phi \\ | \\ \text{wwCH}_2\text{-CH-}\overset{\bullet}{\text{CH}}\text{-CH}_2\text{ww}\end{array}
\end{cases}
$$

The relative amounts of the two processes (a) and (b) depend on the nature of the double bond. For example, the process (b) predominates when the double bond is more reactive. Although the method yields graft copolymer mixed with homopolymer, it has found successful industrial applications, as in the production of HIPS mentioned above and ABS copolymer made by copolymerizing styrene and acrylonitrile in the presence of poly(1,3-butadiene).

7.6.2.2 Irradiation with Ionizing Radiation

Polymer radicals having free radical centers on the backbone chain (to initiate graft copolymerization) can also be produced by irradiation of a polymer-monomer mixture with ionizing radiation. For example, poly(ethylene-*graft*-styrene) can be produced by the irradiation of a monomer-swollen polymer and the initiation reactions can be represented (Odian, 1991) by

$$\text{\scriptsize wwww}CH_2CH_2\text{\scriptsize wwww} \xrightarrow{\text{Radiation}} \text{\scriptsize wwww}CH_2\overset{\bullet}{C}H\text{\scriptsize wwww} + H^{\bullet}$$

$$\text{\scriptsize wwww}CH_2\overset{\bullet}{C}H\text{\scriptsize wwww} + H_2C{=}CH\phi \longrightarrow \begin{array}{c}\text{\scriptsize wwww}CH_2{-}CH\text{\scriptsize wwww}\\ |\\ H_2C{-}\overset{\bullet}{C}H\phi\end{array}$$

A mixture of graft copolymer, parent polymer (polyethylene), and homopolymer (polystyrene) results from the operation.

Table 7.3 Relative Sensitivities of Monomers and Polymers to Ionizing Radiation

Monomers	G value[a]
Styrene	0.66
Acrylonitrile	5.0
Methyl methacrylate	6.1
Vinyl acetate	9.6
Vinyl chloride	~ 10

Polymers	G value[a]
Polystyrene	1.5–3
Polyisoprene	2–4
Polyethylene	6–8
Poly(methyl methacrylate)	6–12
Poly(vinyl acetate)	6–12
Poly(vinyl chloride)	10–15

[a]Number of radicals formed per 100 eV absorbed.
Source: Williams, 1971; Allcock and Lampe, 1990

The grafting efficiency of the radiation method depends on the radiation sensitivity of the monomer to be grafted relative to the parent polymer. Efficient grafting can result only if the monomer is less sensitive than the polymer as in such a case active sites will be generated primarily on the polymer backbone. An indication of the relative sensitivities of monomer and polymer is obtained from a comparison of their *G values*, representing the number of radicals formed per 100 eV absorbed (see Table 7.3). Thus styrene is grafted efficiently onto poly(vinyl chloride) and graftings almost free of polystyrene homopolymer can be obtained. The total radiation dose determines the number of grafted chains and the dose rate determines their length. This is because the dose rate controls the rate of initiation, which, in turn determines the kinetic chain length and hence the molecular weight (Williams, 1971).

Polymer grafting can also be brought about by UV irradiation with or without the presence of a photosensitizer. Most photolytic grafting by UV radiation uses polymers having either pendant carbonyl groups or pendant halogen atoms, since these are easily activated by UV radiation. Some examples are grafting of vinyl acetate, acrylonitrile, or methyl methacrylate onto poly(methyl vinyl ketone) and grafting of styrene or methyl methacrylate on brominated polystyrene. Photolytic grafting resembles radiation grafting, but the depth of penetration by UV is very low as compared to ionizing radiation. The process is thus more suitable for surface modification.

Plasma being a mixture of electrons, ions, excited molecules, radicals and energetic photons (UV light) is able to initiate chemical reactions on polymer surfaces and induce polymerization reactions. The *plasma induced graft polymerization technique* is thus a well known polymer surface modification technique (Tu et al., 2004). The process involves two steps. In the first step, active sites (including carbon radicals or peroxides) are generated on the polymer using plasma, while the second step involves monomer grafting and polymerization at the active sites. Thus the surface properties can be changed without altering the material's bulk properties. Currently, the plasma graft polymerization technique is used to modify the surface of expanded polytetrafluoroethylene membranes to improve its vapor permeation performance.

REFERENCES

Alfrey Jr., T. and Goldfinger, G., *J. Chem. Phys.*, **12**, 115, 205, 332 (1944).

Alfrey Jr., T. and Price, C. C., *J. Polym. Sci.*, **2**, 101 (1947).

Allcock, H. R. and Lampe, F. W., "Contemporary Polymer Chemistry", Prentice Hall, Englewood Cliffs, N.J., 1990.

Atherton, J. N. and North, A. M., *Trans. Faraday Soc.*, **58**, 2049 (1962).

Brandrup, J. and Immergut, E. H., Eds., "Polymer Handbook", 2nd ed., Wiley Interscience, New York, 1975.

Cais, R. E., Farmer, R. G., Hill, D. J. T., and O'Donnell, J. H., *Macromolecules*, **12**, 835 (1979).

Chiang, S. S. M. and Rudin, A., *J. Macromol. Sci. Chem.*, **A9(2)**, 237 (1975).

Fineman, M. and Ross, S. D., *J. Polym. Sci.*, **5**, 259 (1950).

Flory, P. J., "Principles of Polymer Chemistry", Cornell Univ. Press, Ithaca, N.Y., 1953.

Ghosh, P., "Polymer Science and Technology of Plastics and Rubbers", Tata McGraw-Hill, New Delhi, 1990.

Hamielec, A. E., Macgregor, J. F., and Penlidis, A., "Copolymerization", pp 17-31 in "Comprehensive Polymer Science", vol. 3 (G. C. Eastmond, A. Ledwith, S. Russo, and P. Sigwalt, erd.), Pergamon Press, London, 1989.

Hiemenz, P. C., "Polymer Chemistry", Marcel Dekker, New York, 1984.

Hocking, M. B. and Klimchuk, K. A., *J. Polym. Sci.: Part A: Polym. Chem.*, **34**, 2481 (1996).

Kelen, T. and Tudos, F., *J. Macromol. Sci. Chem.*, A9, 1 (1975); *Makromol. Chem.*, **191**, 1863 (1990).

Matyjaszewski, K., *Controlled Radical Polymerization, ACS Symposium Series*, vol. 768, American Chemical Society, Washington, DC (2000).

Mayo, F. R. and Lewis, F. M., *J. Am. Chem. Soc.*, **66**, 1594 (1944).

Melville, H. W., Noble, B., and Watson, W. F., *J. Polym. Sci.*, **2**, 229 (1947).

Merz, E., Alfrey, Jr., T., and Goldfinger, G., *J. Polym. Sci.*, **1**, 75 (1946).

Meyer, V. E., *J. Polym. Sci.,* 4A, 2819 (1966).

Meyer, V. E. and Chan, R. K. S., *Polym. Preprint*, **8**, 209 (1967).

Meyer, V. E. and Lowry, G. G., *J. Polym. Sci.*, **3**, 462 (1948).

Odian, G., "Principles of Polymerization", John Wiley, New York, 1991.

Peng, H., Cheng, S., Feng, L., and Fan, Z., Polm. Int., **53**, 833 (2004).

Rudin, A., "The Elements of Polymer Science and Engineering", Academic Press, Orlando, Florida, 1982.

Skeist, I., *J. Am. Chem. Soc.*, **68**, 1781 (1946).

Tidwell, P. W. and Mortimer, G. A., *J. Polym. Sci.,* A3, 369 (1965).

Tu, C-Y., Chen, C-P., Wang, Y-C., Li, C-L., Tsai, H-A., Lee, K-R., and Lai J-Y., *Eur. Polym. J.*, **40**, 1541 (2004).

Valvassori, A. and Sartori, G., *Adv. Polym. Sci.*, **5**, 28 (1967).

Vollmert, B., "Polymer Chemistry", Springer-Verlag, New York, 1973.

Walling, C., *J. Am. Chem. Soc.*, **71**, 1930 (1949).

Walling, C. and Briggs, E. R., *J. Am. Chem. Soc.*, **67**, 1774 (1945).

Williams, D. J., "Polymer Science and Engineering", Prentice Hall, Englewood Cliffs, N.J., 1971.

EXERCISES

7.1   What values of r_1 and r_2 would yield copolymerization diagrams (F_1 vs. f_1) (a) without inflection points, (b) with inflection points ?
[*Ans.* (a) $r_1 > 1, r_2 < 1$ and $r_1 < 1, r_2 > 1$; (b) $r_1 < 1, r_2 < 1$ and $r_1 > 1, r_2 > 1$ (rare in free-radical copolymerizations, but found in some ionic copolymerizations)]

7.2   A monomer pair with $r_1 = 5.0$ and $r_2 = 0.2$ is copolymerized beginning with a molar monomer ratio $[M_1]/[M_2] = 30/70$. Assuming that the copolymer composition within a 10% conversion interval is constant, calculate instantaneous monomer (f_1) and copolymer (F_1) compositions and cumulative average copolymer compositions at 10 mol% conversion intervals up to 100% conversion. Show the results graphically as change in composition of the copolymer and the monomer mixture during copolymerization.

7.3   On the basis of Q and e values predict the copolymerization behavior of the following pairs of monomers : (a) Vinyl acetate ($Q = 0.03$, $e = -0.22$) and ethyl vinyl ethers ($Q = 0.03$, $e = -1.17$). (b) Styrene ($Q = 1.00$, $e = -0.80$) and vinyl acetate ($Q = 0.03$, $e = -0.22$). (c) Methyl methacrylate ($Q = 0.74$, $e = 0.40$) and acrylic acid ($Q = 1.15$, $e = 0.77$). (d) Styrene ($Q = 1.00$, $e = -0.80$) and acrylonitrile ($Q = 0.60$, $e = 1.20$).

7.4   The following monomer reactivity ratios were determined for the binary copolymerization of ferrocenylmethyl acrylate (FMA) with styrene (STY), methyl acrylate (MA), and vinyl acetate (VA) [C. U. Pittman, Jr., *Macromolecules*, **4**, 298 (1971)] :

M_1	M_2	r_1	r_2
FMA	STY	0.020	2.3
FMA	MA	0.14	4.4
FMA	VA	1.4	0.46

(a) Which of the above comonomer pairs could lead to azeotropic copolymerization ? (b) Is styrene more reactive or less reactive than FMA toward the FMA radical ? By what factor ? (c) List STY, MA, and VA in order of increasing reactivity toward the FMA radical.
[*Ans.* (a) None; (b) STY 50 times more reactive than FMA toward FMA radical; (c) STY > MA > VA]

7.5   Predict the sequence length distributions for an ideal binary copolymerization with $r_1 = r_2 = 1$ for (a) $f_1 = 0.5$, (b) $f_1 = 0.8$, and (c) $f_1 = 0.2$. Compare the distribution patterns and comment on the results.

7.6   Compare the sequence-length distribution (by plotting in a bar graph) in the copolymer from the following monomer pairs with and without azeotrope for $[M_1]/[M_2] = 10/90$: (a) $r_1 = r_2 = 0.1$; (b) $r_1 = 5.0, r_2 = 0.2$.

7.7   When 0.7 mole fraction styrene (M_1) is copolymerized with methacrylonitrile (M_2) in a radical reaction, what is the average length of sequence of each monomer in the copolymer ? ($r_1 = 0.37$, $r_2 = 0.44$).
[*Ans.* $\bar{x}(M_1) = 1.9; \bar{x}(M_2) = 1.2$]

7.8   Acrylonitrile monomer (M_1) is copolymerized with 0.25 mole fraction vinylidene chloride (M_2). What fraction of the acrylonitrile sequences contain 3 or more acrylonitrile units ? ($r_1 = 0.9, r_2 = 0.4$).
[*Ans.* 0.53]

7.9   Styrene (3.0 M) is copolymerized with methacrylonitrile (1.5 M) in benzene solution by adding benzoyl peroxide to a concentration of 0.1 M and heating to 60°C. Calculate for the polymer initially formed (a) composition of the copolymer, (b) probability of forming styrene and methacrylonitrile sequences that are 3 units long, (c) average sequence lengths of styrene and methacrylonitrile in the copolymer, and (d) the run number of the copolymer. [Given : $r_1 = r_2 = 0.25$ at 60°C.]
[*Ans.* (a) M_1 57 mol%, M_2 43 mol%; (b) $n_3(M_1) = 0.0742$, $n_3(M_2) = 0.011$; (c) $\bar{x}(M_1) = 1.50$, $\bar{x}(M_2) = 1.12$; (d) $N_R = 76$]

7.10 The measurement of bulk copolymerization of styrene (M_1) and methyl methacrylate (M_2) at 30°C in a feed of 0.031 mole fraction styrene with initiation by photosensitized decomposition of benzoyl peroxide gave a value of 7.11×10^{-5} mol L^{-1} s^{-1}, while homopolymerization of styrene under the same conditions yielded a polymerization rate of 3.02×10^{-5} mol L^{-1} s^{-1} [H. W. Melville and L. Valentine, *Proc. Roy. Soc., A*, **200**, 337, 358 (1952)]. Calculate the copolymerization rate from (a) chemical control model [Eq. (7.52)] and (b) combined model [Eqs. (7.46) and (7.56)] to compare with the experimental value. Use the homopolymerization rate constants at 30°C for styrene as $k_p = 46$ L mol^{-1} s^{-1} and $k_t = 8.0 \times 10^6$ L mol^{-1} s^{-1} and for MMA as $k_p = 286$ L mol^{-1} s^{-1} and $k_t = 2.44 \times 10^7$ L mol^{-1} s^{-1}. The reactivity ratios at 30°C are $r_1 = 0.485$ and $r_2 = 0.422$. Make the comparison using $\phi = 10$ and $\phi = 13$. Monomer density = 0.90 g/cm^3.

Ans.

	Rate of copolymerization×10^5, mol L^{-1} s^{-1}			
	Chemical control		Combined model	
Experimental	Eq. (7.52)		Eqs. (7.46) and (7.56)	
	$\phi = 10$	$\phi = 13$	$\phi = 10$	$\phi = 13$
7.11	5.57	5.18	8.07	7.58

7.11 Methyl methacrylate (M_1) and vinyl acetate (M_2) constitute a system in which the nature of polyradical ends has no discernible effect on the overall rate of termination (i.e., $\phi = 1$; see Problem 7.13 on p. 451). The copolymerization rate data measured for this system at 60°C are given below [G. M. Burnett and H. R. Gersmann, *J. Polym. Sci.*, **28**, 655 (1958)]:

Mole fraction of M_2 in feed, f_2	Rate × 10^5 (mol L^{-1} s^{-1})
0.756	29.1
0.645	42.4
0.548	60.7
0.453	78.7

Calculate the copolymerization rate using the diffusion control model [Eqs. (7.46) and (7.54)] and combined model [Eqs. (7.46) and (7.56)] to compare with the experimental value. Take the reaction with 0.645 mole fraction vinyl acetate in feed as calibration value to determine R_i. [Kinetic parameters: $r_1 = 28.6$; $r_2 = 0.035$; $k_{11} = 589$, $k_{t11} = 2.9 \times 10^7$, $k_{22} = 3600$, and $k_{t22} = 2.1 \times 10^8$, all in L mol^{-1} s^{-1}; monomer density = 0.90 g/cm^3.]

Ans.

f_2	Rate×10^5, mol L^{-1} s^{-1}		
	Diffusion control	Combined model	Exptl.
0.756	29.7	30.4	29.1
0.645	42.4	42.4	42.4
0.548	53.5	52.6	60.7
0.453	64.2	62.4	78.7

7.12 The ternary copolymerization of a monomer mixture containing 31.24, 31.12, and 37.64 mol% of styrene (St), methyl methacrylate (MMA), and vinylidene chloride (VC) at 60°C for 16 h yielded at 18.2 wt% conversion a polymeric product which analyzed C 68.66% and Cℓ 12.07% (by wt) [C. Walling and E. R. Briggs, *J. Am. Chem. Soc.*, **67**, 1774 (1945)]. Calculate the initial ternary copolymer composition to compare with the composition obtained by the analysis. [Reactivity ratios: St/MMA : $r_1 = 0.52$, $r_2 = 0.46$; St/VC : $r_1 = 1.8$, $r_2 = 0.087$; MMA/VC : $r_1 = 2.4$, $r_2 = 0.36$.]

Ans.

Feed composition		Terpolymer composition (mol%)		
Monomer	Mol%	Found	Calcd. [Eq. (7.67)]	Calcd. [Eq. (7.68)]
St	31.24	43.6	46.8	46.4
MMA	31.12	39.2	38.1	41.9
VC	37.64	17.2	15.1	11.7

7.13 Predict the initial composition of the terpolymer which would be produced from the radical polymerization of a solution containing acrylonitrile (47%), styrene (47%), and 1,3-butadiene 6% (by mol). [*Ans.* M_1 39%, M_2 36%, M_3 25% (by mol)]

7.14 (a) Calculate the mole fraction composition of the initial terpolymer which would be formed from the radical polymerization of a feed containing 0.414 mole fraction methacrylonitrile (M_1), 0.424 mole fraction styrene (M_2), and 0.162 mole fraction α-methyl styrene (M_3). (b) What fraction of styrene sequences in this copolymer contain 2 or more styrene units? [Data M_1/M_2: $r_1 = 0.44$, $r_2 = 0.37$; M_1/M_3: $r_1 = 0.38$, $r_2 = 0.53$; M_2/M_3: $r_1 = 1.124$, $r_2 = 0.627$.] [*Ans.* (a) $F_1 = 0.44$, $F_2 = 0.35$, $F_3 = 0.21$; (b) 0.23]

7.15 What should be the concentration of divinyl benzene in styrene to cause gelation at full conversion of the latter, if styrene were being polymerized under conditions such that the degree of polymerization of the polymer being formed were 1000? Assume that the vinyl groups in divinyl benzene are equally as reactive as those in styrene. [*Ans.* 0.05 mol%]

7.16 Consider the styrene-divinylbenzene system of Problem 7.18 (p. 461). Recalculate the conversion at the gel point taking into consideration the unequal reactivity of styrene and divinylbenzene ($r_1 = 0.3$, $r_2 = 1.0$). [*Ans.* 2.3 mol%]

7.17 (a) How much of the divinyl monomer, ethylene glycol dimethacrylate (EGDMA), should be added to methyl methacrylate (MMA) to cause onset of gelation at 20% conversion when polymerization is carried out at 60°C in the presence of 0.8 g/L benzoyl peroxide. Homopolymerization of MMA under the same conditions is known to yield polymer with $\overline{DP}_w = 1000$. MMA and EGDMA can be reasonably assumed to be of equal reactivity. (b) Recalculate the amount of EGDMA for the case where 1% of a chain regulator is used to bring down $\overline{DP}_w$ to 500. [*Ans.* (a) 0.25 mol%; (b) 0.5 mol%]

7.18 Calculate the conversion for onset of gelation in methyl methacrylate (MMA) containing 0.20 mol% ethylene glycol dimethacrylate (EGDMA) when it is polymerized at 60°C in the presence of 0.04 mol/L AIBN initiator. Take into account the fact that the chain termination in MMA homopolymerization occurs both by disproportionation and coupling, the ratio being 3:1 at 60°C. [Data: k_p^2/k_t (for MMA) $= 1.04 \times 10^{-2}$ L mol^{-1} s^{-1}; $k_d = 8.45 \times 10^{-6}$ s^{-1}; $f = 0.6$; monomer density = 0.90 g/cm^3.] [*Ans.* $p_c = 0.115$]

7.19 How would the percentage conversion at the gel point change if the MMA-EGDMA mixture of Exercise 7.18 contained additionally an effective chain regulator ($C_S = 21$ at 60°C) at a concentration of 10^{-4} mol/L? [*Ans.* $p_c = 0.166$]

7.20 Predict the extent of reaction at which gelation would occur in the following two vinyl-divinyl systems, both containing 1 mol% of the divinyl component: (a) styrene-ethylene glycol dimethacrylate and (b) methyl methacrylate-divinyl benzene. Assume that the reaction conditions for the two systems are such as to yield the same $\overline{DP}_w$ of 1000 for the uncrosslinked polymer. Take the r_1 and r_2 values from Table 7.1 for the analogous vinyl-vinyl copolymerizations. [*Ans.* (a) $p_c = 0.015$; (b) $p_c = 0.012$]

Chapter 8

Ionic Chain Polymerization

8.1 Introduction

Besides free radical mechanisms, discussed in Chapter 6, there are several other mechanisms by which chain or addition polymerization can take place. Prominent among these are ionic mechanisms in which the growing chain end carbon bears a negative charge (*carbanion*) or a positive charge (*carbonium ion*). In the former case, the polymerization is known as *anionic polymerization* and in the latter case as *cationic polymerization*.

Ionic polymerization can, in general, be initiated by acidic or basic compounds. For cationic polymerization, complexes of BF_3, $AlCl_3$, $TiCl_4$, and $SnCl_4$ with water, or alcohols, or tertiary oxonium salts are particularly active initiators, the positive ions in them causing chain initiation. One can also initiate cationic polymerization with HCl, H_2SO_4, and $KHSO_4$. Important initiators for anionic polymerization are alkali metals and their organic compounds, such as phenyllithium, butyllithium, phenyl sodium, sodium naphthalene, and triphenyl methyl potassium.

Several distinctive features of ionic chain polymerization will now be highlighted. Ionic polymerizations are largely selective (as not all olefinic monomers can undergo anionic and/or cationic polymerization) and require stringent reaction conditions including high monomer purity, whereas most olefinic monomers undergo free radical polymerization under much less stringent conditions. Cationic polymerization is essentially limited to those monomers with electron-releasing substituents and anionic polymerization to those possessing electron-withdrawing groups. Also, unlike in free-radical polymerizations where the characteristics of the active centers depend only on the nature of the monomer and are generally independent of the reaction medium, in ionic polymerizations the polarity of the solvent strongly influences the mechanism and rate of ionic polymerization. This can be visualized as follows.

Ionic polymerizations involve successive insertion of monomer molecules between an ionic chain end (positive in cationic and negative in anionic polymerization) and a counterion of opposite charge. The macroion and the counterion form an organic salt which may, however, exist in several forms depending on the nature and degree of interaction between the cation and anion of the salt and the reaction medium (solvent/monomer). Considering, for example, an organic salt A^+B^-, a continuous spectrum of ionicities ("Winstein spectrum") can be depicted as

$$AB \quad \rightleftharpoons \quad A^+B^- \quad \rightleftharpoons \quad A^+/B^- \quad \rightleftharpoons \quad A^+ \| B^- \quad \rightleftharpoons \quad A^+ + B^-$$

Covalent bonding	Contact (tight) ion pair	Solvent separated ion pair	Solvated (loose) ion pair	Free solvated ions
(I)	**(II)**	**(III)**	**(IV)**	**(V)**

$$(8.1)$$

A range of behavior from one extreme of a completely covalent species (**I**) to the other of completely free (and highly solvated) ions (**V**) can be expected, including the intermediate species of *tight* or *contact* ion pair (**II**) and the solvent-separated or loose ion pair (**III**). The contact ion pair always has a *counterion* (or *gegenion*) of opposite charge close to the propagating center and unseparated by solvent, while the solvent-separated ion pair involves ions that are partially separated by solvent molecules.

Ionic polymerizations commonly involve two types of propagating species—an ion pair (**II-IV**) and a free ion (**V**)—coexisting in equilibrium with each other. The relative concentrations of these two types of species, as also the identity of the ion pair (that is, whether of type **II**, **III**, or **IV**), depend on the particular reaction conditions and especially the solvent or reaction medium, which has a large effect in ionic polymerizations. Loose ion pairs are more reactive than tight ion pairs, while free ions are significantly more reactive than ion pairs. In general, more polar media favor solvent-separated ion pairs or free solvated ions. In hydrocarbon media, free solvated ions do not exist, though other equilibria may occur between ion pairs and clusters of ions (Rudin, 1982).

Solvents of high polarity are desirable for solvation of ions. However they cannot be employed for ionic polymerizations. Thus highly polar hydroxylic solvents, such as water and alcohols, react with and destroy most ionic initiators and propagating species. Other polar solvents such as ketones form highly stable complexes with initiators, thus preventing initiation reactions. Most ionic polymerizations are, therefore, carried out in low or moderately polar solvents such as methyl chloride, ethylene dichloride, and pentane.

Though resembling free-radical chain polymerization in terms of initiation, propagation, transfer, and termination steps, ionic polymerizations have significantly different reaction kinetics and rates of ionic polymerization are by and large much faster than in free-radical processes. This is mainly because termination by mutual destruction of active centers, which is prevalent in free-radical systems (see Chapter 6), does not occur in ionic systems as macroions bearing the same charge repel each other and as a result the concentrations of kpropagating species are usually much higher in ionic than in free-radical systems.

As ionic polymerizations with stringent reaction conditions are more difficult to carry out than normal free-radical processes, the latter are invariably preferred where both free-radical and ionic initiations give a similar product. For example, commercial polystyrenes are all free-radical products, though styrene polymerization can be initiated with free radicals as well as with appropriate anions or cations. However, to make research grade polystyrenes with exceptionally narrow molecular-weight distributions and di-block or multi-block copolymers of styrene and other monomers, ionic processes are necessarily employed.

In this chapter, we will review pure ionic polymerizations—first, anionic polymerizations with some of their specific applications and then the polymerization processes which proceed by a cationic mechanism. Coordination polymerizations that are complex polymerizations having partial ionic character and ring opening polymerizations, many of which proceed by anionic and cationic mechanisms, will be reviewed in subsequent chapters.

8.2 *Ionic Polymerizability of Monomers*

The processes whereby a given alkene reacts depend on the inductive and resonance characteristics of the substituent X in the vinyl monomer $CH_2 = CHX$. Electron-releasing substituents,

$$R, \quad RO-, \quad \overset{\overset{\displaystyle H}{|}}{R-C}=\overset{\overset{\displaystyle H}{|}}{C}- \quad \text{and} \quad \langle\!\bigcirc\!\rangle-$$

increase the electron density of the double bond and thus facilitate addition of a cation:

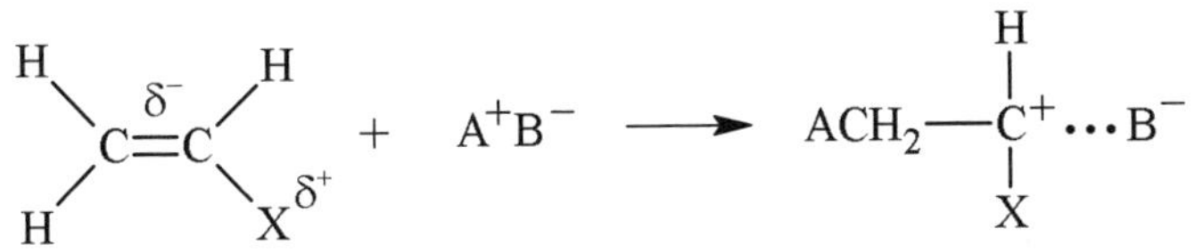

$$
\begin{array}{c}
\underset{H}{\overset{H}{\diagdown}}C\!\!=\!\!C\underset{X^{\delta+}}{\overset{H}{\diagup}} + A^+B^- \longrightarrow ACH_2\!\!-\!\!\underset{X}{\overset{H}{\underset{|}{\overset{|}{C^+}}}}\cdots B^-
\end{array} \tag{8.2}
$$

Thus monomers like isobutylene (**VI**), 3-methylbutene-1 (**VII**), styrene (**VIII**), and vinyl ethers (**IX**) all undergo cationic polymerization.

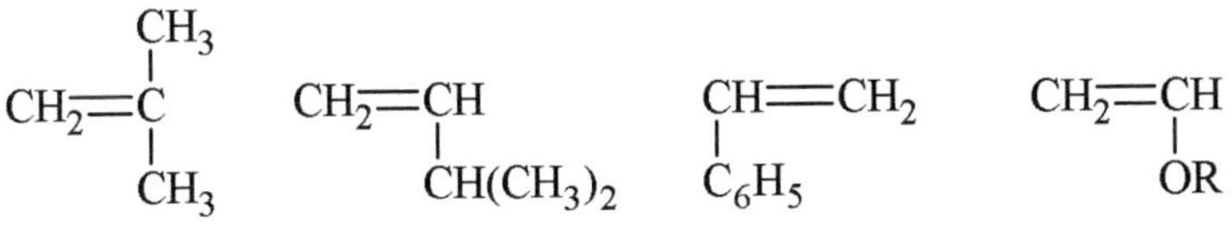

$$
\begin{array}{cccc}
\underset{CH_3}{\overset{CH_3}{\underset{|}{\overset{|}{CH_2\!\!=\!\!C}}}} & \underset{CH(CH_3)_2}{\overset{}{\underset{|}{CH_2\!\!=\!\!CH}}} & \underset{C_6H_5}{\overset{}{\underset{|}{CH\!\!=\!\!CH_2}}} & \underset{OR}{\overset{}{\underset{|}{CH_2\!\!=\!\!CH}}} \\
\textbf{(VI)} & \textbf{(VII)} & \textbf{(VIII)} & \textbf{(IX)}
\end{array}
$$

Electron-withdrawing substituents, such as

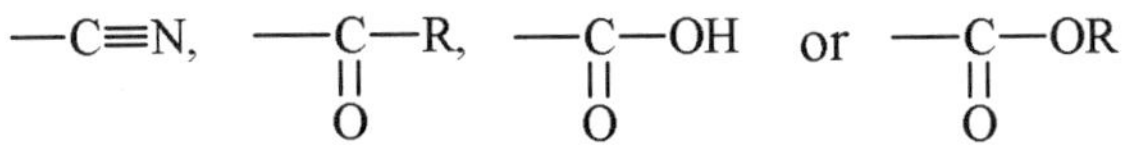

$$
-C\!\!\equiv\!\!N, \quad \underset{O}{\overset{}{\underset{\|}{-C\!\!-\!\!R}}}, \quad \underset{O}{\overset{}{\underset{\|}{-C\!\!-\!\!OH}}} \quad \text{or} \quad \underset{O}{\overset{}{\underset{\|}{-C\!\!-\!\!OR}}}
$$

decrease the electron density of the double bond and thus facilitate attack of an anionic species on the double bond to produce an anion:

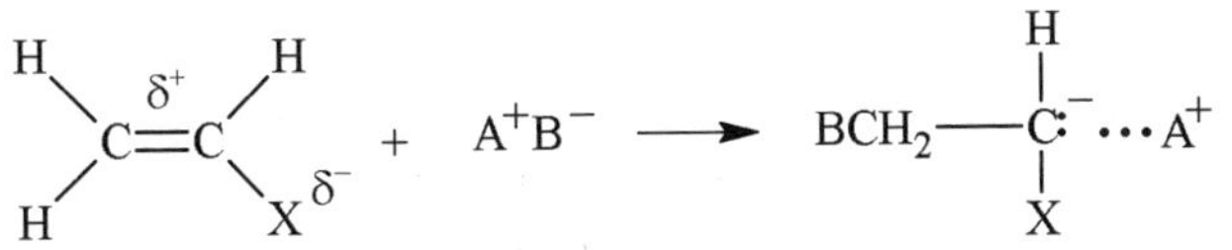

$$
\begin{array}{c}
\underset{H}{\overset{H}{\diagdown}}C\!\!=\!\!C\underset{X^{\delta-}}{\overset{H}{\diagup}} + A^+B^- \longrightarrow BCH_2\!\!-\!\!\underset{X}{\overset{H}{\underset{|}{\overset{|}{C^-}}}}\cdots A^+
\end{array} \tag{8.3}
$$

The electron-withdrawing substituents may also stabilize the anion, e.g., in acrylonitrile polymerization, where the stabilization of the propagating carbanion occurs by delocalization of the negative charge over the α-carbon and the nitrogen of the nitrile group:

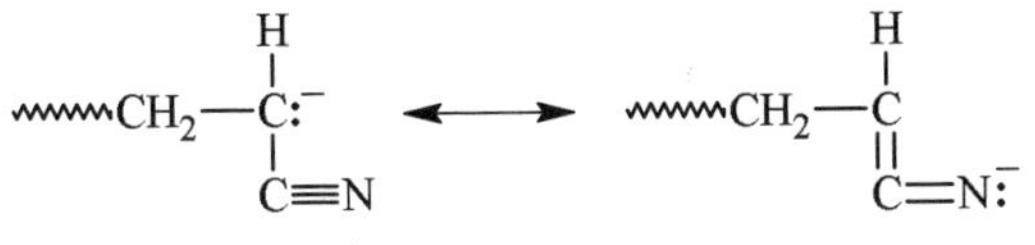

$$
\begin{array}{c}
\text{wwwwCH}_2\!\!-\!\!\underset{C\equiv N}{\overset{H}{\underset{|}{\overset{|}{C:^-}}}} \longleftrightarrow \text{wwwwCH}_2\!\!-\!\!\underset{C=N:^-}{\overset{H}{\underset{\|}{\overset{|}{C}}}}
\end{array} \tag{8.4}
$$

Problem 8.1   Contrary to the high selectivity shown in cationic and anionic polymerization, radical initiators can bring about the polymerization of almost any carbon-carbon double bond. However, aldehydes and ketones are not activated by free radicals. Explain, giving reasons.

Answer:

Radical species are neutral and do not have stringent requirements for attacking the π-bond. Moreover, resonance stabilization of the propagating radical occurs with almost all substituents, for example,

Radical initiation can thus take place with almost any carbon-carbon double bond. Aldehydes and ketones are not activated by free radicals because of the difference in electronegativity of the C and O atoms. Aldehydes and ketones are polymerized only by ionic or heterogeneous catalytic processes.

Phenyl and alkenyl ($-CH=CH_2$) substituents, although electron-pushing inductively, can resonance stabilize the anionic propagating species in the same manner as a cyano group [Eq. (8.4)]. Monomers such as styrene and 1,3-butadiene can therefore undergo both anionic and cationic polymerizations.

The applicability of various types of initiation mechanisms to the polymerization of common olefin monomers (Lenz, 1967) is summarized in Table 8.1. We see that isobutene can be polymerized only by cationic initiation, whereas monomers, such as vinyl chloride, methyl methacrylate, or acrylonitrile with their electronegative substituents, will not yield at all to cationic initiation. Vinyl chloride does not also respond to anionic initiation. Though halogens can withdraw electrons inductively and push electrons by resonance, both effects are relatively weak. Vinyl chloride thus does not undergo either anionic or cationic polymerization.

Table 8.1 Polymerizability of Olefin Monomers by Different Processes[a]

Olefin monomer	Monomer structure	Free radical	Cationic	Anionic	Coordination
Ethylene	$CH_2=CH_2$	+	+	−	+
Propylene	$CH_2=CHCH_3$	−	−	−	+
Butene-1	$CH_2=CHC_2H_5$	−	−	−	+
Isobutene	$CH_2=C(CH_3)_2$	−	+	−	−
Butadiene-1,3	$CH_2=CH-CH=CH_2$	+	+	+	+
Styrene	$CH_2=CHPh$	+	+	+	+
Vinyl chloride	$CH_2=CHCl$	+	−	−	+
Methacrylic esters	$CH_2=C(CH_3)COOCH_3$	+	−	+	+
Vinyl ethers	$CH_2=CHOR$	−	+	−	+
Acrylonitrile	$CH_2=CH-CN$	+	−	+	+

[a]Symbol + signifies that the monomer can be polymerized to high molecular weight polymer by the initiation process indicated. *Source:* Lenz (1967).

8.3   Anionic Polymerization

8.3.1   Anionic Initiation

While initiators for anionic polymerization are all electron donors of varying base strengths, the initiator type required for a particular polymerization depends on the ease with which an anion can be formed from the monomer. In general, the strength of the base required to initiate polymerization decreases with increasing electronegativity of the substituent on the monomer. The electronegativity of some selected substituents is in the following order (Lenz, 1967):

$$-CN \; > \; -COOR \; > \; -C_6H_5 \; \cong \; -CH=CH_2 \; \gg \; -CH_3$$

Since acrylonitrile has a strong electronegative substituent ($-CN$), it can be polymerized by the relatively weak sodium methoxide ($NaOCH_3$) and vinylidene cyanide carrying two $-CN$ groups on the same carbon atom can be polymerized even by weaker bases like water and amines. However, for polymerization of nonpolar monomers such as conjugated olefins very strong bases like metal alkyls should be used as initiators.

The principal anionic initiation processes are (a) nucleophilic attack on the monomer which produces *one-ended* (monofunctional) anions by addition of the initiator across the double bond of the monomer [see Eq. (8.3)] and (b) electron transfer by alkali metals that leads to *two-ended* (bifunctional) anions (see later).

8.3.1.1   Nucleophilic Attack

Nucleophilic attack on an olefinic monomer is essentially addition of a negatively charged entity to the monomer [Eq. (8.3)]. Examples of some reactive bases which can initiate in this manner are n-C_4H_9Li, $C_6H_5CH_2Li$, $NaNH_2$, KNH_2, $C_6H_5CH_2Na$, CH_3ONa, and $EtMgBr$. Alkyllithium compounds are generally low melting and soluble in hydrocarbon solvents. A common example is *n*-butyllithium, which is usually available as a solution in *n*-hexane. Initiation involves addition of the metal alkyl to the olefinic monomer:

$$C_4H_9Li \; + \; CH_2{=}CHY \; \longrightarrow \; C_4H_9{-}CH_2{-}\overset{\displaystyle Y}{\underset{\displaystyle H}{\overset{|}{\underset{|}{C}}}}{:}^- \; Li^+ \qquad (8.5)$$

This type of initiation is known as *monofunctional initiation* as it produces one active (ionic) site. Propagation takes place by the addition of monomer to the ionic site:

$$C_4H_9{-}CH_2{-}\overset{\displaystyle Y}{\underset{\displaystyle H}{\overset{|}{\underset{|}{C}}}}{:}^- \; Li^+ \; + \; n\,CH_2{=}CHY \; \longrightarrow$$
$$C_4H_9{\Large(}CH_2CHY{\Large)}_n CH_2{-}\overset{\displaystyle Y}{\underset{\displaystyle H}{\overset{|}{\underset{|}{C}}}}{:}^- \; Li^+ \qquad (8.6)$$

Alkyllithium initiators find extensive use because of their solubility in hydrocarbon solvents. Alkyls and aryls of the heavier alkali metals, such as Na and K, are poorly soluble in hydrocarbons because of the greater ionic character of the Na$-$C and K$-$C bonds.

Potassium amide initiator is used in liquid ammonia, which has high dielectric constant ($\sim$ 22). This is one of the ionic systems in which the active centers behave kinetically as free ions. Initiation involves the dissociation of potassium amide followed by addition of amide ion to a monomer unit (Ghosh, 1990):

$$KNH_2 \rightleftharpoons K^+ + NH_2^- \tag{8.7}$$

$$NH_2^- + H_2C{=}\overset{\overset{\textstyle H}{|}}{C}\!\!\!\bigcirc \longrightarrow H_2N{-}CH_2{-}\overset{\overset{\textstyle H}{|}}{C}^-\!\!\!\bigcirc \tag{8.8}$$

Styrene and other monomers can be polymerized by KNH_2 in liquid ammonia.

8.3.1.2 Electron Transfer

Alkali metals can donate electrons to the double bonds yielding anion radicals and positively charged, alkali-metal counterions. This may result either from direct attack of the monomer on the alkali metal, or from attack on the metal through an intermediate compound such as naphthalene. Both result in *bifunctional initiation*, that is, formation of species with two carbanionic ends.

Alkali Metals Initiation by direct attack on the alkali metal involves transfer of the loosely held *s* electron from a Group IA metal atom to the monomer. A radical ion (i.e., a species having both ionic and radical centers) is formed:

$$Li^\bullet + CH_2{=}\overset{\overset{\textstyle H}{|}}{\underset{\underset{\textstyle X}{|}}{C}} \longrightarrow {}^\bullet CH_2{-}\overset{\overset{\textstyle H}{|}}{\underset{\underset{\textstyle X}{|}}{C}}{:}^- Li^+ \tag{8.9}$$

The radical ion may dimerize to give a dianion:

$$2\,{}^\bullet CH_2{-}\overset{\overset{\textstyle H}{|}}{\underset{\underset{\textstyle X}{|}}{C}}{:}^- Li^+ \longrightarrow Li^+ {}^-{:}\overset{\overset{\textstyle H}{|}}{\underset{\underset{\textstyle X}{|}}{C}}{-}CH_2{-}CH_2{-}\overset{\overset{\textstyle H}{|}}{\underset{\underset{\textstyle X}{|}}{C}}{:}^- Li^+ \tag{8.10}$$

The initiation process thus results in a bifunctional dicarbanion species capable of propagating at both of its ends.

Free alkali metals may be employed as solutions in certain ether solvents, in liquid ammonia, or as fine suspensions in inert solvents. The latter are prepared by heating the metal above its melting point in the solvent, stirring vigorously to form an emulsion, and then cooling to obtain a fine dispersion of the metal.

Alkali Metal Complexes Polycyclic aromatic compounds can react with alkali metals in ether solution to produce radical ions (Szwarc, 1968). The reaction involves the transfer of an electron from the alkali metal to the aromatic compound. For sodium and naphthalene, for example,

$$
\text{Na}^{\bullet} \ + \ \text{[naphthalene]} \ \longrightarrow \ \text{[naphthalenide]}^{\bullet\,-} \ \text{Na}^{+} \tag{8.11}
$$

(X)

Tetrahydrofuran (THF) is a useful solvent for these reactions. Sodium naphthalenide is formed quantitatively in this solvent, but dilution with hydrocarbons results in precipitation of sodium and regeneration of naphthalene.

The naphthalene anion radical (greenish-blue) rapidly transfers an electron to a monomer such as styrene to form the radical anion (**XI**) ($\phi = -\text{C}_6\text{H}_5$):

$$
\text{[naphthalenide]}^{\bullet\,-} \text{Na}^{+} + \phi\text{HC}{=}\text{CH}_2 \ \rightleftharpoons \ \text{[naphthalene]} + \left[\overset{\text{H}}{\underset{\phi}{\text{C}}}{-}\overset{\text{H}}{\underset{\text{H}}{\text{C}}}{\colon^{-}} \leftrightarrow {}^{-}{\colon}\overset{\text{H}}{\underset{\phi}{\text{C}}}{-}\overset{\text{H}}{\underset{\text{H}}{\text{C}}}{}^{\bullet} \right] \text{Na}^{+} \tag{8.12}
$$

(XI)

The styryl radical ion in (**XI**) is a resonance hybrid of two forms having both anion and radical centers. It dimerizes by reacting at the radical ends to form a red-colored dicarbanion (**XII**):

$$
2 \left[{}^{\bullet}\overset{\text{H}}{\underset{\phi}{\text{C}}}{-}\overset{\text{H}}{\underset{\text{H}}{\text{C}}}{\colon^{-}} \leftrightarrow {}^{-}{\colon}\overset{\text{H}}{\underset{\phi}{\text{C}}}{-}\overset{\text{H}}{\underset{\text{H}}{\text{C}}}{}^{\bullet} \right]\text{Na}^{+} \ \longrightarrow \ \text{Na}^{+}\ {}^{-}{\colon}\overset{\text{H}}{\underset{\phi}{\text{C}}}{-}\text{CH}_2{-}\text{CH}_2{-}\overset{\text{H}}{\underset{\phi}{\text{C}}}{\colon^{-}}\ \text{Na}^{+} \tag{8.13}
$$

(XI)   **(XII)**

The initiation process is thus similar to alkali metal initiation described earlier [cf. Eq. (8.10)]. The dimerization of radical anions is highly probable because of their high concentrations, typically 10^{-3}–10^{-2} M, and the large rate constants (10^6–10^8 L/mol-s) for radical coupling (Odian, 1991). Anionic propagation takes place by monomer addition at both carbanion ends of the styryl dianion:

$$
\text{Na}^{+}\ {}^{-}{\colon}\overset{\text{H}}{\underset{\phi}{\text{C}}}{-}\text{CH}_2{-}\text{CH}_2{-}\overset{\text{H}}{\underset{\phi}{\text{C}}}{\colon^{-}}\ \text{Na}^{+} \ + \ (n{+}m)\,\phi\text{HC}{=}\text{CH}_2 \ \longrightarrow
$$

$$
\text{Na}^{+}\ {}^{-}{\colon}\overset{\text{H}}{\underset{\phi}{\text{C}}}{-}\text{CH}_2{-}\!\left(\!\text{CH}\phi{-}\text{CH}_2\!\right)_{\!n}\!\!\left(\!\text{CH}_2{-}\text{CH}\phi\!\right)_{\!m}\!\!\text{CH}_2{-}\overset{\text{H}}{\underset{\phi}{\text{C}}}{\colon^{-}}\ \text{Na}^{+} \tag{8.14}
$$

Anionic propagation is generally much faster than free-radical reactions.

Problem 8.2   Account for the fact that anionic polymerizations are generally much faster than free-radical reactions although the k_p values are of the same order of magnitude for monomer addition reactions of radicals and solvated ion pairs (free macroanions, however, react much faster).

Answer:

The concentration of radicals in free-radical polymerizations is usually about 10^{-9}–10^{-7} M while that of propagating ion pairs is 10^{-4}–10^{-2} M depending upon initiator concentrations. As a result, anionic polymerizations are 10^3–10^7 times as fast as free radical reactions at the same temperature.

8.3.2   Termination Reactions

To carry out anionic polymerizations, extraneous substances such as water, oxygen, carbon dioxide, or any other impurities that may react with the active ionic centers, must be absent. Since glass surfaces contain adsorbed water, special precautions are taken in laboratory polymerizations, such as flaming under vacuum, to remove this water. In addition, the monomer itself should be very pure and free from inhibitors.

8.3.2.1   Living Polymerization

Following the work of Michael Szwarc in the mid-1950s, it became known that under carefully controlled conditions carbanionic living polymers could be formed using electron transfer initiation. Because the growing chains in anionic polymerization, carrying negative charges, cannot react with each other, there is no compulsory chain termination through recombination. In polymerization systems, especially of nonpolar monomers such as styrene and 1,3-butadiene in perfectly dry inert solvents such as benzene and tetrahydrofuran, initiated by organometallic compounds, termination or transfer is virtually nonexistent and active chain ends can thus have indefinite lifetimes. Such systems are referred to as *living polymers*. Propagation continues till 100% conversion of the monomer is reached, while the propagating anionic centers remain intact and capable of further propagation if more monomer (either same or different) is added.

   One of the first living anionic polymerization systems studied was the polymerization of styrene initiated by sodium naphthalene (p. 481). If the reaction system is highly purified so that all impurities that are liable to react with the carbanions are excluded from the system, propagation continues until all styrene monomer has been consumed, but the color of the polystyryl carbanions remains unchanged indicating that the carbanions are intact. That the resulting polymer chains are still active can be easily demonstrated by adding more styrene to increase the molecular weight or by adding another monomer, such as isoprene, to form a block copolymer. [*Note:* There is no reason in theory why chain growth should not continue indefinitely if more and more monomer is added and chain terminators are absent. In practice, however, small amounts of terminators are invariably produced in the system and chain growth also slows down due to a large increase in viscosity at high molecular weight or due to chains becoming insoluble (Allcock and Lampe, 1990).]

   The living chains can be terminated when desired by adding suitably reactive materials, such as water, alcohol, or ammonia. The unique features of living polymer systems described above provide fascinating possibilities of polymer syntheses, which include making monodisperse polymers (by controlled addition of monomer), structures with specific end groups (by chain termination with appropriate reagents), and block copolymers (by sequential addition of two or more monomers). These are discussed more fully in a later section.

8.3.2.2   Termination by Transfer Agents

Water is an especially effective chain terminating agent. For example, its $C_{tr,S}$ value is approximately 10 in the polymerization of styrene at 25°C with sodium naphthalene. So the presence of even trace quantities of water can significantly reduce the polymer molecular weight and polymerization rate. Water terminates propagating carbanions by proton transfer:

$$\text{wwCH}_2-\overset{\overset{\displaystyle H}{|}}{\underset{\underset{\displaystyle X}{|}}{C}}{:}^- \;+\; H_2O \;\longrightarrow\; \text{wwCH}_2-\overset{\overset{\displaystyle H}{|}}{\underset{\underset{\displaystyle X}{|}}{C}}H \;+\; HO^- \tag{8.15}$$

The hydroxide ion formed is not sufficiently nucleophilic to reinitiate polymerization and the active center is thus effectively destroyed. In contrast, the $C_{tr,S}$ value for ethanol being very small ($\sim 10^{-3}$), its presence in small amounts does not limit the molecular weight.

Oxygen and carbon dioxide from the atmosphere add to propagating carbanions to form peroxy and carboxyl anions:

$$\text{wwCH}_2-\overset{\overset{\displaystyle H}{|}}{\underset{\underset{\displaystyle X}{|}}{C}}{:}^- \;+\; O_2 \;\longrightarrow\; \text{wwCH}_2-\overset{\overset{\displaystyle H}{|}}{\underset{\underset{\displaystyle X}{|}}{C}}-O-O{:}^- \tag{8.16}$$

$$\text{wwCH}_2-\overset{\overset{\displaystyle H}{|}}{\underset{\underset{\displaystyle X}{|}}{C}}{:}^- \;+\; CO_2 \;\longrightarrow\; \text{wwCH}_2-\overset{\overset{\displaystyle H}{|}}{\underset{\underset{\displaystyle X}{|}}{C}}-\overset{\overset{\displaystyle O}{\|}}{C}-O{:}^- \tag{8.17}$$

As these anions are not reactive enough to continue propagation, the chains are effectively terminated. By adding a proton donor subsequently to the polymerization system, the peroxy and carboxyl anions are converted to OH and COOH groups. A notable example of the application of the latter reaction is the preparation of carboxyl ion terminated polybutadiene by anionic polymerization of butadiene with bifunctional initiators, followed by termination with CO_2:

$$\text{Na}^+ \;{:}^-\overset{\overset{\displaystyle H}{|}}{\underset{\underset{\displaystyle H}{|}}{C}}-CH{=}CH-CH_2\text{wwww}CH_2-CH{=}CH-\overset{\overset{\displaystyle H}{|}}{\underset{\underset{\displaystyle H}{|}}{C}}{:}^- \text{Na}^+ \;+\; CO_2 \;\longrightarrow$$

$$\text{Na}^+ \;{:}^-O-\overset{\overset{\displaystyle O}{\|}}{C}-CH_2-CH{=}CH-CH_2\text{wwwww}CH_2-CH{=}CH-CH_2-\overset{\overset{\displaystyle O}{\|}}{C}-O{:}^- \text{Na}^+ \tag{8.18}$$

Reaction with acid then yields carboxyl-terminated polybutadiene (CTPB). Termination with ethylene oxide similarly generates hydroxyl end groups:

$$\text{wwww}\overset{\overset{\displaystyle H}{|}}{\underset{\underset{\displaystyle H}{|}}{C}}{:}^- \text{Li}^+ \;+\; CH_2\overset{O}{\overbrace{\qquad}}CH_2 \;\longrightarrow\; \text{wwww}CH_2-CH_2-CH_2-O{:}^- \text{Li}^+ \xrightarrow{\;CH_3OH\;}$$

$$\text{wwww}CH_2-CH_2-CH_2-OH \;+\; LiOCH_3 \tag{8.19}$$

Hydroxyl-terminated polybutadiene (HTPB) can be produced by such reactions. Such elastomers of low molecular weight (3000–10,000) are used as binders and liquid rubbers. *Telechelic polymers* (that is, polymers with reactive end groups), containing one or more end groups with the capacity to react with other molecules, can be prepared in this way.

8.3.2.3   Spontaneous Termination

As mentioned earlier, living polymers do not, in reality, have infinite life times even in the complete absence of terminating agents as they undergo decay on aging, a process known as *spontaneous termination*. Polystyryl carbanions, known to be the most stable of all anionic chains as they can survive for weeks in hydrocarbon solvents, undergo spontaneous termination by a mechanism known as *hydride elimination*, as shown by the equation (Odian, 1991):

$$\text{wwwCH}_2\text{·CH}\phi\,\text{CH}_2\text{—}\underset{\phi}{\overset{H}{\underset{|}{\overset{|}{C}}}}:^-\text{Na}^+ \longrightarrow \text{wwwCH}_2\text{·CH}\phi\,\text{CH}=\text{CH}\phi + \text{H}:^-\text{Na}^+$$

$$\textbf{(XIII)} \qquad\qquad \textbf{(XIV)}$$

(8.20)

This may be followed by other reactions (Odian, 1991), such as abstraction of an allylic hydrogen from (**XIII**) by a carbanion center (to yield an unreactive anion) or hydrogen abstraction by the sodium hydride (**XIV**) formed in Eq. (8.20).

8.3.3   Polymerization with Complete Dissociation of Initiator

In polymerizations initiated by alkali metals or insoluble organometallic compounds used as fine dispersions in organic media, the initiation step occurs at a phase interface while subsequent propagation reactions may occur in a homogeneous phase. The kinetics of such reactions combining heterogeneous initiation and homogeneous propagation are often very complex and specific to the given systems. However, useful generalizations may be made for systems in which both the initiation and propagation processes are homogeneous. Such systems are discussed next.

There are some cases in homogeneous anionic polymerization in which the initiator dissociates *completely* with quantitative transformation into the active ionic form and the process is also virtually instantaneous (*stoichiometric polymerization*). This is the case, for example, when one uses, as initiators, alkali organic compounds (e.g., phenyllithium, butyllithium, or sodium naphthalene) in solvents which have unshared electron pairs (Lewis bases). The alkali forms stable positively charged complex ions with the Lewis base (Lenz, 1967), while the organic residue becomes negatively charged (carbanion) and can initiate an ionic polymerization [cf. Eqs. (8.11) and (8.12)]:

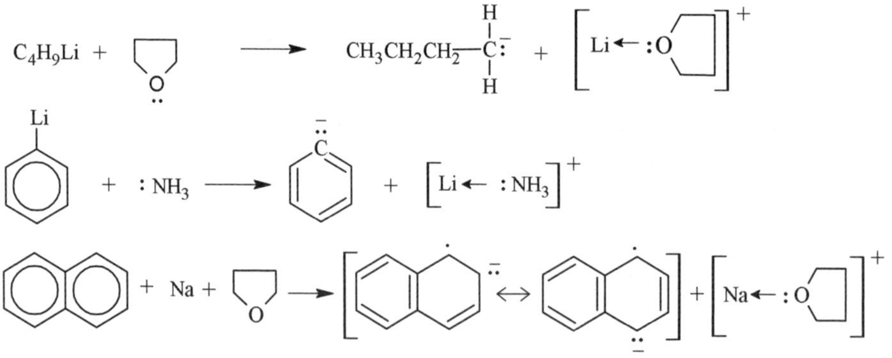

Since the polymerization kinetics in the above cases are extremely simple, ionic polymerization kinetics can be conveniently classified according to whether the initiators are quantitatively and instantaneously dissociated or not.

8.3.3.1 Polymerization Kinetics

For a kinetic analysis, the process of anionic polymerization is divided in the conventional way into three main steps, viz., initiation, propagation, and termination. Representing the initiator by CA (or C^+A^-) and a terminating agent by X, the reactions can be written (Allcock and Lampe, 1990) as :

$$\text{Initiation}: \quad CA + M \xrightarrow{k_i} AM^- C^+ \tag{8.21}$$

$$\text{Propagation}: \quad AM^- C^+ + M \xrightarrow{k_p} AMM^- C^+ \tag{8.22}$$

$$AMM^- C^+ + M \xrightarrow{k_p} AMMM^- C^+ \tag{8.23}$$

$$AM_{x-1}M^- C^+ + M \xrightarrow{k_p} AM_x M^- C^+ \tag{8.24}$$

$$\text{Termination}: \quad AM_x M^- C^+ (+X) \xrightarrow{k_t} AM_x M \tag{8.25}$$

(Depending on the solvent used, the propagating ion may behave as a free ion $AM_x M^-$ or as an ion pair $AM_x M^- C^+$, or as both.) *Living polymerizations* are characterized by the absence of termination (as well as transfer) reactions. The initiators in most cases are reactive enough to give instantaneous initiation, i.e., $k_i > k_p$, which implies that *no initiation takes place during the polymerization and the number of chain (growth) centers to which the monomer molecules may add reaches its maximum value before polymerization begins.* There is, moreover, no change in the number of chain centers during the polymerization as there is no chain termination. Denoting the total concentration of anions (of all degrees of polymerization) by $[M^-]$ (the positive counterion C^+ is not shown for simplicity) and assuming that the initiation reaction (8.21) is both instantaneous and complete, one can write :

$$[M^-] = [CA]_0 \tag{8.26}$$

Hence the rate of polymerization is given by

$$R_p = -d[M]/dt = k_p[M^-][M] = k_p[CA]_0[M] \tag{8.27}$$

The rate constant k_p is, in fact, an apparent rate constant or overall rate constant since there will be both undissociated (ion pair) and dissociated (free ions) species in the polymerization system and their propagation rate constants are different (discussed later). Integrating Eq. (8.27) one obtains the time dependence of the monomer concentration as

$$[M] = [M]_{sp}\exp\left(-k_p[CA]_0 t\right) = ([M]_0 - [CA]_0)\exp\left(-k_p[CA]_0 t\right) \tag{8.28}$$

where the subscript '0' indicates initial value and $[M]_{sp}$ represents the monomer concentration at the start of propagation [Eq. (8.22)], that is, the original monomer concentration less the concentration of the initiator CA which is quantitatively reacted in the initiation step [Eq. (8.21)]. Using $[M]$ measured at various times after the polymerization is started and knowing $[M]_0$ and $[CA]_0$, one may thus determine k_p. The significance of k_p will be discussed in a later section.

8.3.3.2 Experimental Methods

The simple technique of dilatometry, described earlier (p. 324), cannot be used to measure rates of anionic living polymerizations as the reactions proceed too rapidly. For such fast reactions, the so-called *stopped-flow* technique (Sawamoto and Higashimura, 1978, 1979, 1986, 1990) is useful. In stopped-flow, rapid-scan spectroscopy, separate solutions of monomer and initiator are mixed

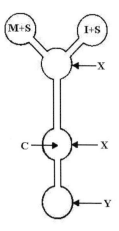

Figure 8.1 Schematic representation of a flow tube for fast polymerization reactions; M monomer; S solvent; I initiator; C chain terminator; X mixing jets; Y receiving container.

instantaneously in a mixing chamber and forced into a capillary tube inside a spectrophotometer where the flow is stopped and the change in absorbance of propagating species or monomer is measured with time.

The rate can also be determined with a modified apparatus (Fig. 8.1) by *short-stopping* the polymerization with a highly active terminating agent, which results in incorporation (into the polymer) of an easily detectable end group derived from the terminating agent. The method involves forcing a rapidly mixed monomer-initiator solution through the flow tube into a solvent containing a terminating agent to quench the reaction mixture. A turbulent flow must prevail in this flow tube (Reynolds number > 10,000) since the chains would have growth times of differing duration if the flow is laminar. As the reaction time is given by the ratio of tube volume to flow rate, very short reaction times (0.005 to 2 s) can be realized in this manner. By analyzing either the polymer or unreacted monomer in the quenched reaction mixture, the extent of monomer conversion, and hence polymerization rate, can then be obtained.

Problem 8.3 Consider the flow tube for rapid polymerization reactions shown schematically in Fig. 8.1. Let V_0 be the volume of the tube (distance between the mixing jets) and V be the volume of the total liquid flowing through in time t. Denoting the total concentration (constant) of polymer chain ends by Eq. (8.26) and the concentration of monomer units in polymer by $[M]_p$ derive an expression for monomer conversion as a function of t.

Answer:

The effective polymerization time is the same as the residence time τ given by

$$\tau = V_o/(V/t) \tag{P8.3.1}$$

Integrating the rate of propagation, given by Eq. (8.27), and assuming $[M]_{sp} \simeq [M]_0$ for high polymers, that is, high monomer to initiator ratio, one obtains

$$-\ln[M] \Big| = k_p[CA]_0 t \Big| \tag{P8.3.2}$$

where $[M]_\tau = [M]_0 - [M]_p$. Equation (P8.3.2) is transposed with the aid of Eq. (P8.3.1) into

$$\ln[1/(1-p)] = \ln\frac{[M]_0}{[M]_\tau} = k_p[CA]_0\tau = k_p[CA]_0\left(\frac{V_0}{V}\right)t \tag{P8.3.3}$$

where p is the monomer conversion given by $p = ([M]_0 - [M]_\tau)/[M]_0$.

8.3.3.3 Average Kinetic Chain Length

As there is no termination step in a true living polymerization, the growth of chains, once initiated, can stop only when the monomer is completely consumed. In a living polymerization with fast and complete dissociation of initiators the average kinetic chain length will thus be given by

$$\nu = \frac{\text{Monomer molecules consumed}}{\text{Number of chain centers}} = \frac{[M]_{sp} - [M]}{[CA]_0} \tag{8.29}$$

where

$$[M]_{sp} = [M]_0 - [CA]_0 \tag{8.30}$$

Substituting [M] from Eq. (8.28), ν is given as a function of time (t) by

$$\nu = \frac{[M]_{sp}}{[CA]_0} \left(1 - e^{-k_p [CA]_0 t}\right) \tag{8.31}$$

The maximum value of ν is obtained at the completion of reaction or in the limit of $t \rightarrow \infty$ as

$$\nu_\infty = \frac{[M]_{sp}}{[CA]_0} \tag{8.32}$$

8.3.3.4 Average Degree of Polymerization

The average degree of polymerization represents the average number of monomer molecules polymerized per polymer molecule formed. In the case of monofunctional initiation [cf. Eq. (8.5)], the number of polymer molecules formed is equal to the number of chain centers (or initiators) and the number average degree of polymerization then is [cf. Eq. (8.29)]

$$\overline{DP} = \frac{[M]_0 - [M]}{[CA]_0} = \frac{\left([M]_{sp} + [CA]_0\right) - [M]}{[CA]_0} = \nu + 1 \tag{8.33}$$

In the case of bifunctional initiation that produces double-ended active species [cf. Eq. (8.10)], the number of polymer molecules formed is 1/2 of the number of chain centers or initiators. The average degree of polymerization is then given by

$$\overline{DP} = \frac{[M]_0 - [M]}{\frac{1}{2}[CA]_0} = \frac{\left([M]_{sp} + [CA]_0\right) - [M]}{\frac{1}{2}[CA]_0} = 2(\nu + 1) \tag{8.34}$$

Problem 8.4 In an experiment (Szwarc et al., 1956), styrene (9.2 g) was added to 60 mL of tetrahydrofuran containing 3.3×10^{-4} mol of sodium naphthalene. Polymerization was carried out at $-80°C$ and after completion (as determined by constant viscosity) an additional 7.7 g of styrene in 50 mL of tetrahydrofuran was added. The final yield was 16.6 g of polystyrene, i.e., about 100% conversion. Calculate the average molecular weight of the final polymer.

Answer:

Total moles of styrene = $(9.2 \text{ g} + 7.7 \text{ g}) / (104 \text{ g mol}^{-1}) = 0.1625$ mol; moles of initiator = 3.3×10^{-4} mol.

Since sodium naphthalene causes bifunctional initiation, Eq. (8.34) is applicable. At complete conversion of monomer,

$$\overline{DP} = \frac{[M]_0}{[CA]_0 / 2} = \frac{\text{Total moles of monomer}}{(\text{Total moles of initiator}) / 2}$$

Since no initiator is added in the second stage of polymerization,

$$\overline{DP} \ = \ (0.1625 \ \mathrm{mol})/\tfrac{1}{2}(3.3 \times 10^{-4} \ \mathrm{mol}) \ = \ 985.$$

Problem 8.5   Sodium (1.15 g) and naphthalene (7.0 g) were stirred together in 50 mL dry tetrahydrofuran to form a dark green solution of sodium naphthalenide. When 1.0 mL of this green solution was introduced into a solution of styrene (208 g) in dry tetrahydrofuran by a rapid injection technique, the latter turned reddish orange. The total final volume of the mixture was 1 L. Assume that the injection of the initiator resulted in instantaneous homogeneous mixing. After 5 seconds of reaction at 25°C, the styrene concentration was determined to be 1.73×10^{-2} mol/L. After 10 seconds the color of the reaction mixture was quenched by adding a few milliliters of methanol. The polymer was then precipitated and washed with methanol. Calculate

(a) Overall propagation rate constant;   (b) initial rate of polymerization;   (c) rate of polymerization after 10 s;   (d) molecular weight of the final polymer.

Answer:

Given quantities: Sodium = 1.15 g ≡ 0.05 mol;   naphthalene = 7.0 g ≡ 0.055 mol;   styrene = 208 g ≡ 2.0 mol

Since naphthalene is in excess, the total number of moles of initiator in 50 mL THF is 0.05 mol. The number of moles transferred with 1 mL of dark green solution = 0.05/50 = 10^{-3} mol. Hence, $[CA]_0 = 10^{-3}$ mol L^{-1}.

$[M]_0 \ = \ 2.0 \ $ mol L^{-1}

$[M]_{sp} \ = \ [M]_0 \ - \ [CA]_0 \ = \ (2.0 \ $ mol L$^{-1}) \ - \ (0.001 \ $ mol L$^{-1}) \ \simeq \ 2.0$ mol L^{-1}.

(a) From Eq. (8.28),

$$
\begin{aligned}
k_p \ &= \ \frac{\ln([M]_{sp}/[M])}{[CA]_0 \, t} \\[2mm]
&= \ \frac{\ln[(2.0 \ \mathrm{mol \ L^{-1}})/(1.73 \times 10^{-2} \ \mathrm{mol \ L^{-1}})}{(10^{-3} \ \mathrm{mol \ L^{-1}})(5 \ \mathrm{s})} \ = \ 950 \ \mathrm{L \ mol^{-1} \ s^{-1}}
\end{aligned}
$$

(b) From Eq. (8.27),

$$(R_p)_0 \ = \ (950 \ \mathrm{L \ mol^{-1} \ s^{-1}})(10^{-3} \ \mathrm{mol \ L^{-1}})(2.0 \ \mathrm{mol \ L^{-1}}) \ = \ 1.90 \ \mathrm{mol \ L^{-1} \ s^{-1}}$$

(c) From Eq. (8.28), at $t = 10$ s

$$
\begin{aligned}
[M] \ &= \ (2.0 \ \mathrm{mol \ L^{-1}}) \exp[-(950 \ \mathrm{L \ mol^{-1} \ s^{-1}})(10^{-3} \ \mathrm{mol \ L^{-1}})(10 \ \mathrm{s})] \\[1mm]
&= \ 1.497 \times 10^{-4} \ \mathrm{mol \ L^{-1}} \\[1mm]
R_p \ &= \ (950 \ \mathrm{L \ mol^{-1} \ s^{-1}})(10^{-3} \ \mathrm{mol \ L^{1}})(1.497 \times 10^{-4} \ \mathrm{mol \ L^{-1}}) \\[1mm]
&= \ 1.42 \times 10^{-4} \ \mathrm{mol \ L^{-1} \ s^{-1}}
\end{aligned}
$$

(d) From Eq. (8.34),

$$\overline{DP}_n \ = \ 2(2.0 \ \mathrm{mol \ L^{-1}} \ - \ 1.497 \times 10^{-4} \ \mathrm{mol \ L^{-1}})/(10^{-3} \ \mathrm{mol \ L^{-1}}) \ \simeq \ 4000$$

8.3.3.5 *Distribution of the Degree of Polymerization*

The distribution of the degree of polymerization in a "living" ionic polymerization is quite different from that in free-radical polymerization, since the former has no termination or transfer mechanism. A derivation of this distribution given below follows that given in Allcock and Lampe (1990) and Cowie (1991).

Consider the reactions (8.21) to (8.24) for an anionic polymerization, with the initiation step [Eq. (8.21)] being instantaneous. Application of mass balance, coupled with the usual assumption that k_p is independent of chain length, then gives the following set of rate equations:

$$\frac{d[\text{AM}^-]}{dt} = -k_p[\text{AM}^-][\text{M}] \tag{8.35}$$

$$\frac{d[\text{AMM}^-]}{dt} = k_p[\text{M}]([\text{AM}^-] - [\text{AMM}^-]) \tag{8.36}$$

$$\frac{d[\text{AMMM}^-]}{dt} = k_p[\text{M}]([\text{AMM}^-] - [\text{AMMM}^-]) \tag{8.37}$$

$$\vdots$$

$$\frac{d[\text{AM}_x\text{M}^-]}{dt} = k_p[\text{M}]([\text{AM}_{x-1}\text{M}^-] - [\text{AM}_x\text{M}^-]) \tag{8.38}$$

Substitution for [M] from Eq. (8.28) into Eq. (8.35) yields

$$\int \frac{d[\text{AM}^-]}{[\text{AM}^-]} = -k_p[\text{M}]_{\text{sp}} \int e^{-k_p[\text{CA}]_0 t} dt \tag{8.39}$$

Since $[\text{AM}^-] = [\text{CA}]_0$ at $t = 0$, integration gives

$$[\text{AM}^-] = [\text{CA}]_0 \exp\left[-\frac{[\text{M}]_{\text{sp}}}{[\text{CA}]_0}\left(1 - e^{-k_p[\text{CA}]_0 t}\right)\right] \tag{8.40}$$

Combining this equation with Eq. (8.31) gives $[\text{AM}^-]$ in terms of kinetic chain length v as

$$[\text{AM}^-] = [\text{CA}]_0 e^{-v} \tag{8.41}$$

In this way the time t is eliminated from Eq. (8.35). Equation (8.31) may also be used to eliminate t from the remainder of the rate equations (8.36) to (8.38). Thus, after differentiation of Eq. (8.31) with respect to t that yields

$$dv = k_p[\text{M}]_{\text{sp}} e^{-k_p[\text{CA}]_0 t} dt \tag{8.42}$$

one may substitute Eqs. (8.28), (8.41), and (8.42) into Eq. (8.36), with elimination of t, to transform the rate equation (8.36) for $[\text{AMM}^-]$ into the differential equation

$$d[\text{AMM}^-]/dv + [\text{AMM}^-] = [\text{CA}]_0 e^{-v} \tag{8.43}$$

The solution of this linear, first-order differential equation of a standard form is given by

$$e^v[\text{AMM}^-] = \int e^v[\text{CA}]_0 e^{-v} dv + z \tag{8.44}$$

where e^v is the integrating factor and z is a constant of integration. Integrating Eq. (8.44) and evaluating z by the condition that at $v = 0$ (i.e., at $t = 0$), $[AMM^-] = 0$, one obtains

$$[AMM^-] = [CA]_0 v e^{-v} \tag{8.45}$$

The above process is then repeated for Eq. (8.37). Thus, elimination of t by substitution of Eqs. (8.28), (8.42), and (8.45) into Eq. (8.37) yields the differential equation

$$d[AMMM^-]/dv + [AMMM^-] = [CA]_0 v e^{-v} \tag{8.46}$$

This equation may be solved in the same manner as Eq. (8.43) using the integrating factor e^v. This yields

$$[AMMM^-] = \tfrac{1}{2} [CA]_0 v^2 e^{-v} \tag{8.47}$$

Proceeding in this way for the other equations in the sequence (8.37) to (8.38), it soon becomes apparent that the concentration of the anion containing x monomer molecules can be expressed by

$$[AM_{x-1}M^-] = [CA]_0 \frac{v^{x-1} e^{-v}}{(x-1)!} \tag{8.48}$$

and at the completion of polymerization by

$$[AM_{x-1}M^-]_\infty = [CA]_0 \frac{v_\infty^{x-1} e^{-v_\infty}}{(x-1)!} \tag{8.49}$$

Since each initial anion leads to one polymer species, the fraction of polymer having degree of polymerization x at the end of reaction is

$$
\begin{aligned}
n_x &= \frac{\text{Number of anions containing } x \text{ monomers}}{\text{Number of anions}} \\
&= \frac{[AM_{x-1}M^-]_\infty}{[CA]_0} \\
&= \frac{v_\infty^{x-1} e^{-v_\infty}}{(x-1)!}
\end{aligned}
\tag{8.50}
$$

Thus the mole fraction of x-mer in the polymer is Poisson's distribution formula (Flory, 1940). Substituting for v_∞ from Eq. (8.32), and using $[M]_{sp} = [M]_0$ for high polymers (i.e., high ratio of initial monomer to initiator), one obtains

$$n_x = \frac{1}{(x-1)!} \left(\frac{[M]_0}{[CA]_0} \right)^{x-1} \exp(-[M]_0/[CA]_0) \tag{8.51}$$

The number distribution of v (or DP) can be predicted from Eq. (8.50) or (8.51) when the average kinetic chain length, v_∞ ($= [M]_0 / [CA]_0$), is known.

Problem 8.6   1-Vinylnaphthalene is polymerized anionically at $25°C$ in a tetrahydrofuran solution containing initially 4×10^{-3} M C_4H_9Li and 0.204 M 1-vinylnaphthalene. Show graphically the number fraction and weight fraction distributions of the degree of polymerization. Show for comparison the corresponding distributions for a polymer of the same average degree of polymerization produced by free-radical reaction (assume termination by disproportionation).

Answer:

From Eq. (8.30), $[M]_{sp} = (0.204 \text{ M}) - (4 \times 10^{-3} \text{ M}) = 0.20 \text{ M}$.

From Eq. (8.32), $\nu_\infty = (0.20 \text{ mol L}^{-1})/(4 \times 10^{-3} \text{ mol L}^{-1}) = 50$.

From Eq. (8.50), $n_x = \dfrac{(50)^{x-1} e^{-50}}{(x-1)!}$, or $\ln n_x = (x-1)\ln 50 - \ln(x-1)! - 50$.

The values of n_x calculated for different assumed values of x ($50 < x < 50$) are plotted against x in Fig. 8.2.

The weight fraction distribution w_x may be obtained by multiplication of the number fraction distribution, Eq. (8.50), by $x/\overline{DP}$ ($\simeq x/\nu_\infty$). When that is done, one has

$$w_x = \frac{x}{(x-1)!} \nu_\infty^{x-2} e^{-\nu_\infty}, \quad \text{or} \quad \ln w_x = \ln x - \ln(x-1)! + (x-2)\ln \nu_\infty - \nu_\infty$$

Figure 8.2 also shows a plot of w_x vs. x for $\nu_\infty = 50$. It should be noted that the observed distributions would usually be somewhat broader than those shown in Fig. 8.2. This anomaly may be attributed to the presence of propagation-depropagation equilibrium:

$$A - M_x - M^- + M \rightleftharpoons A - M_{x+1} - M^-$$

which was not considered in the derivation of Eq. (8.50).

The distribution function for the degree of polymerization of polymer formed by a free-radical chain mechanism in the absence of chain transfer depends only on the kinetic chain length and the ratio of disproportionation to coupling (k_{td}/k_{tc}). The distribution is given (Allcock and Lampe, 1991) by

$$n_x = \frac{1}{\nu}\left(1 + \frac{1}{\nu}\right)^{-x} \frac{(x-1)/\nu + 2(k_{td}/k_{tc})}{1 + 2(k_{td}/k_{tc})} \tag{P8.6.1}$$

For termination solely by disproportionation, i.e., $k_{tc} = 0$, Eq. (P8.6.1) reduces to

$$n_x = \frac{1}{\nu}\left(1 + \frac{1}{\nu}\right)^{-x} \tag{P8.6.2}$$

Equation (P8.6.2) is plotted for $\nu = 50$ in Fig. 8.2 showing number distribution (n_x) of the degree of polymerization. The weight fraction distribution w_x may be obtained by multiplication of Eq. (P8.6.2) by $x/\overline{DP}$ ($= x/\nu$). This yields

$$w_x = \frac{x}{\nu^2}\left(1 + \frac{1}{\nu}\right)^{-x} \tag{}$$

This distribution is also shown graphically in Fig. 8.2.

Polydispersity Index   Using the relation $\overline{DP}_n = \nu_\infty + 1$ [cf. Eq. (8.33)], the number average molecular weight will be given by

$$\overline{M}_n = (\nu_\infty + 1)M_0 \tag{8.52}$$

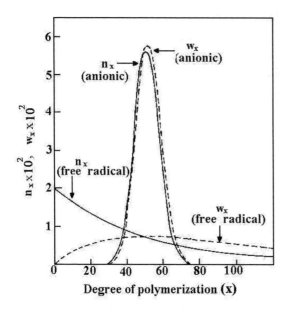

Figure 8.2 Number fraction and weight fraction distribution of the degree of polymerization with $v = 50$ for anionic "living" polymers and for free-radical polymerization (with termination by disproportionation). (Data of Problem 8.6.)

where M_0 is the molecular weight of the monomer unit. On the other hand, the weight fraction distribution can be derived easily by multiplication of the mole fraction distribution [Eq. (8.50)] by $x/\overline{DP}_n$. With $\overline{DP}_n = v_\infty + 1$, this becomes

$$
\begin{aligned}
w_x &= \frac{x}{(v_\infty + 1)} \frac{e^{-v_\infty} v_\infty^{x-1}}{(x-1)!} \\
&= \frac{v_\infty}{(v_\infty + 1)} \frac{x e^{-v_\infty} v_\infty^{x-2}}{(x-1)!}
\end{aligned}
\tag{8.53}
$$

Substituting this for w_x and xM_0 for M_x in $\overline{M}_w = \sum w_x M_x$ gives

$$
\overline{M}_w = M_0 \left(\frac{v_\infty e^{-v_\infty}}{v_\infty + 1} \right) \sum_1^\infty \frac{x^2 v_\infty^{x-2}}{(x-1)!}
\tag{8.54}
$$

Considering first the summation part in this equation, one obtains by rearrangement and simplification (using standard mathematical relations):

$$
\begin{aligned}
\sum_2^\infty \frac{(x^2-1) v_\infty^{x-2}}{(x-1)!} &+ \sum_1^\infty \frac{v_\infty^{x-2}}{(x-1)!} \\
= \sum_3^\infty \frac{v_\infty^{x-2}}{(x-3)!} &+ 3 \sum_2^\infty \frac{v_\infty^{x-2}}{(x-2)!} + \sum_1^\infty \frac{v_\infty^{x-2}}{(x-1)!} \\
= (v_\infty + 3 &+ 1/v_\infty) e^{v_\infty}
\end{aligned}
\tag{8.55}
$$

Table 8.2 Effect of Solvent on Anionic Polymerization of Styrene by 3×10^{-3} M Sodium Naphthalenide

Solvent	Dielectric constant (ϵ)	k_p (L mol^{-1} s^{-1}) at 25°C
Benzene	2.2	2
Dioxane	2.2	5
Tetrahydrofuran (THF)	7.6	550
1,2-Dimethoxyethane (DME)	5.5	3,800

Source: Szwarc and Smid, 1964; Odian, 1991.

Substituting Eq. (8.55) for the summation in (8.54) gives

$$\overline{M}_w = M_0 (v_\infty^2 + 3v_\infty + 1)/(v_\infty + 1) \tag{8.56}$$

Dividing Eq. (8.56) by Eq. (8.52) then yields

$$\overline{M}_w / \overline{M}_n = 1 + \frac{v_\infty}{(v_\infty + 1)^2} \tag{8.57}$$

which can be approximated by

$$\overline{M}_w / \overline{M}_n = 1 + \frac{1}{v_\infty} \tag{8.58}$$

The ratio $\overline{M}_w / \overline{M}_n$ approaches unity asymptotically as v_∞ increases. Narrow molecular weight distributions should thus be obtained in living ionic polymerizations with fast initiation in the absence of depropagation, termination, and chain transfer reactions. Values of polydispersity index (PDI) below 1.1-1.2 are indeed found for many living polymerizations. Molecular weight standards for polystyrene, polyisoprene, poly(α-methylstyrene), and poly(methyl methacrylate) are thus synthesized by living anionic polymerizations. However, the termination reactions in methyl methacrylate polymerizations and depropagation in α-methylstyrene polymerizations tend to broaden the PDI in these systems.

8.3.3.6 *Effects of Reaction Media*

The value of k_p can be determined in a simple manner from Eqs. (8.27) and (8.28) by measuring the extent of reaction or the rate of polymerization. This is enabled by the use of initiators that dissociate quantitatively prior to propagation reactions and by the absence of termination reactions. The rate constants obtained in this way are, however, found to be affected very significantly by the nature of both the solvent and the counterion. The data in Table 8.2 show that while polymerization is much faster in more polar solvents, the dielectric constant is not a quantitative measure of the solvating power (as shown by $\epsilon_{DME} < \epsilon_{THF}$). The higher rate of polymerization in DME may be attributed to a specific solvation effect of two ether groups being in the same molecule (Odian, 1991).

The increase in the overall rate constant k_p with increasing solvent power of the reaction medium can be traced to the corresponding increase in concentration of free ions relative to that of ion pairs. Attention will therefore be directed to the determination of individual propagation rate constants for the free ions and ion pairs and to the relative amounts of these two types of propagating species.

Let us consider for illustration a simple situation that corresponds to an equilibrium between free ions and ion pairs (neglecting other kinds of ion pairs that may exist), propagating at different rates. The reactions involved in the polymerization of monomer M by an initiator CA (C^+A^-) [cf. Eq. (8.22)] can thus be represented (Rudin, 1982) as:

$$AM_x^- C^+ \quad \overset{K}{\rightleftharpoons} \quad AM_x^- \ + \ C^+$$

$$k_p^{\mp} \Big\downarrow M \qquad\qquad\qquad k_p^{-} \Big\downarrow M$$

$$AM_{x+1}^- C^+ \quad \overset{K}{\rightleftharpoons} \quad AM_{x+1}^- \ + \ C^+$$

Ion pairs   Free ions

where $k_p^{\mp}$ and k_p^- are rate constants for ion-pair and free-ion propagation, respectively, and K is the equilibrium constant for dissociation of ion pairs into solvated free ions. K is given by

$$K = \frac{[\sum_x AM_x^-][C^+]}{[\sum_x AM_x^- C^+]} = \frac{[\sum_x AM_x^-]^2}{[\sum_x AM_x^- C^+]} \tag{8.59}$$

since $[C^+]$ must equal $[\sum_x AM_x^-]$ for electrical neutrality, when there is no source of either ion other than C^+A^-.

Equation (8.59) gives

$$\left[\sum_x AM_x^-\right] = K^{1/2}\left[\sum_x AM_x^- C^+\right]^{1/2} \tag{8.60}$$

If there is little dissociation of polymeric ion pairs, $[\sum_x AM_x^- C^+]$ can be set equal to the total concentration of living ends of polymer chain, and Eq. (8.60) then becomes

$$\left[\sum_x AM_x^-\right] = K^{1/2}[M^-]^{1/2} \tag{8.61}$$

If the polymer molecular weight is high, the consumption of monomer in initiation reactions will be negligible and the overall rate of reaction can be expressed as

$$R_p = -\frac{d[M]}{dt} = k_p^{\mp}\left[\sum_x AM_x^- C^+\right][M] + k_p^-\left[\sum_x AM_x^-\right][M] \tag{8.62}$$

Using the approximation of Eq. (8.61) this can be written as

$$R_p = -d[M]/dt = k_p^{\mp}\left[\sum_x AM_x^- C^+\right][M] + k_p^- K^{1/2}[M^-]^{1/2}[M] \tag{8.63}$$

The concentration of polymeric ion pairs is given by

$$\left[\sum_x AM_x^- C^+\right] = [M^-] - \sum_x AM_x^- = [M^-] - K^{1/2}[M^-]^{1/2} \tag{8.64}$$

Combining Eqs. (8.63) and (8.64) and rearranging (note that $[M^-] = [CA]_0$), one obtains

$$\frac{R_p}{[M][CA]_0} = k_p^{\mp} + \frac{(k_p^- - k_p^{\mp})K^{1/2}}{[M^-]^{1/2}} \tag{8.65}$$

In view of Eq. (8.27) for living polymerization, Eq. (8.65) can also be written as

$$k_p = k_p^{\mp} + \frac{(k_p^- - k_p^{\mp})K^{1/2}}{[M^-]^{1/2}} \tag{8.66}$$

where k_p is the overall or apparent rate constant for propagation. Note that k_p is dependent on the living chain ends or initiator concentration.

A plot of the left side of Eq. (8.65) against $1/[CA]_0^{1/2}$ (since $[M^-] = [CA]_0$) gives a straight line having intercept $k_p^{\mp}$ and slope $(k_p^- - k_p^{\mp})K^{1/2}$. Since K can be known by measuring the conductivity of solutions of low molecular weight living polymers, k_p^- can be estimated from the slope. Such measurements show (Szwarc, 1974) that k_p^- values are of the order of $10^4 - 10^5$ L/mol-s compared to $k_p^{\mp}$ values $\sim 10^2$ L/mol-s (see Problem 8.8) which are of the same order as free-radical k_p values. The concentration of free ions that can also be determined by this method (see Problem 8.10) is found to be only about 10^{-3} that of the corresponding ion pairs. However, because of the much larger values of k_p^- compared to $k_p^{\mp}$ as described above, a significant part of polymerization occurs through free ions.

Problem 8.7   A solution of styrene (1.5 M) in tetrahydrofuran is polymerized at 25°C by sodium naphthalene at a concentration of 3.2×10^{-5} M. Calculate the initial polymerization rate and the average degree of polymerization at the completion of the reaction. What fractions of the polymerization rate are due to free ions and ion pairs, respectively ? How will these values change if the sodium naphthalene concentration is increased to 3.2×10^{-2} M ? [Data: $K_d = 1.5 \times 10^{-7}$ mol/L; $k_p^- = 6.5 \times 10^4$ L/mol-s; $k_p^{\mp} = 80$ L/mol-s.]

Answer:

(a) For 3.2×10^{-5} M sodium naphthalene

$$[M^-] = \left(K_d[M^-C^+]\right)^{1/2} \quad [\text{cf. Eq.}(8.60)]$$
$$= [(1.5 \times 10^{-7} \text{ mol L}^{-1})(3.2 \times 10^{-5} \text{ mol L}^{-1})]^{1/2} = 2.19 \times 10^{-6} \text{ mol L}^{-1}$$

From Eq. (8.62),

$$R_p = k_p^-[M^-][M] + k_p^{\mp}[M^-C^+][M]$$
$$= (6.5 \times 10^4 \text{ L mol}^{-1} \text{ s}^{-1})(2.19 \times 10^{-6} \text{ mol L}^{-1})(1.5 \text{ mol L}^{-1})$$
$$+ (80 \text{ L mol}^{-1} \text{ s}^{-1})(3.2 \times 10^{-5} \text{ mol L}^{-1})(1.5 \text{ mol L}^{-1})$$
$$= 0.214 \text{ mol L}^{-1} \text{ s}^{-1} + 0.00384 \text{ mol L}^{-1} \text{ s}^{-1} = 0.218 \text{ mol L}^{-1} \text{ s}^{-1}$$

Fraction of polymerization: due to free ions $= 0.214/0.218 = 0.98$ and due to ion pairs $= 0.00384/0.218 = 0.02$.

At complete conversion of monomer [cf. Eq. (8.34)], $\overline{DP}_n = 2[M]/[M^-C^+] = 2 \times (1.5 \text{ mol L}^{-1})/(3.2 \times 10^{-5} \text{ mol L}^{-1}) = 9.4 \times 10^4$.

(b) For 3.2×10^{-2} M sodium naphthalene,

$$[M^-] = \left[(1.5 \times 10^{-7} \text{ mol L}^{-1})(3.2 \times 10^{-2} \text{ mol L}^{-1})\right]^{1/2} = 6.93 \times 10^{-5} \text{ mol L}^{-1}$$

$$\begin{aligned} R_p &= (6.5 \times 10^4 \text{ L mol}^{-1} \text{ s}^{-1})(6.93 \times 10^{-5} \text{ mol L}^{-1})(1.5 \text{ mol L}^{-1}) \\ &\quad + (80 \text{ L mol}^{-1} \text{ s}^{-1})(3.2 \times 10^{-2} \text{ mol L}^{-1})(1.5 \text{ mol L}^{-1}) \\ &= 6.76 \text{ mol L}^{-1} \text{ s}^{-1} + 3.84 \text{ mol L}^{-1} \text{ s}^{-1} = 10.60 \text{ mol L}^{-1} \text{ s}^{-1} \end{aligned}$$

Fractions of polymerizations due to free ions and ion pairs are thus $6.76/10.60 = 0.64$ and $3.84/10.60 = 0.36$, respectively.

$$\overline{DP}_n = 2(1.5 \text{ mol L}^{-1})/(3.2 \times 10^{-2} \text{ mol L}^{-1}) = 93.8$$

Problem 8.8   The kinetics of sodium naphthalene initiated anionic polymerization of styrene was studied in a less polar solvent dioxane at 35°C.   Using an apparatus which permitted quick mixing of the reaction components in the absence of air and moisture, aliquot samples were withdrawn periodically and deactivated quickly with ethyl bromide. From the residual monomer and the average degree of polymerization of the polymer formed the following data were obtained for four different monomer(M)–initiator(CA) composi-tions. (Reaction times were corrected for the time spent in mixing and deactivation step.)

$[M]_0 = 0.31$ M $[CA]_0 = 4.0 \times 10^{-4}$ M			$[M]_0 = 0.47$ M $[CA]_0 = 7.0 \times 10^{-4}$ M			$[M]_0 = 1.17$ M $[CA]_0 = 11.0 \times 10^{-4}$ M		
Time (s)	Conv. (%)	$\overline{DP}$	Time (s)	Conv. (%)	$\overline{DP}$	Time (s)	Conv. (%)	$\overline{DP}$
101	33	514	107	45	625	80	45	1000
221	58	900	189	65	857	177	74	1564

Determine the propagation rate constant of ion pair, $k_p^{\mp}$.

Answer:

Integrating Eq. (8.27) for constant $[M^-]$ (i.e., total concentration of active end groups is constant throughout the course of polymerization) gives $\ln([M]_0/[M]) = k_p[M^-]t$, where k_p is the apparent or overall rate constant. Defining conversion as $p = ([M]_0 - [M])/[M]_0$, this equation is rearranged to:

$$k_p = \ln[1/(1 - p)]/([M^-]t). \tag{P8.8.1}$$

For fast initiation by quantitative reaction with the initiator CA, $[M^-] = [CA]_0$. To check if this condition is satisfied, $[M^-]$ may be calculated from the equation $[M^-] = 2([M]_0 - [M])/\overline{DP} = 2p[M]_0/\overline{DP}$, the factor of 2 being used because the living polymer chains have two active ends. Calculation of $[M^-]$ from the given data shows that the said condition is satisfied fairly well. Therefore, $[M^-]$ in Eq. (P8.8.1) may be replaced by $[CA]_0$.

Equation (P8.8.1) is used to calculate the overall rate constant k_p corresponding to each value of $[CA]_0$. These values are plotted against $1/[CA]_0^{1/2}$ in Fig. 8.3. The intercept of the linear plot according to Eq. (8.66) gives $k_p^{\mp} = 2.8$ L mol^{-1} s^{-1}. Since the value of $k_p^{\mp}$ is of the same order of magnitude as the overall rate constant k_p, it may be concluded that the polymerization in the given solvent mostly involves ion pairs and the contribution of free ions is small.

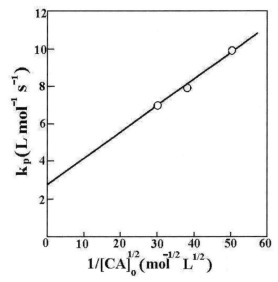

Figure 8.3 The overall rate constant k_p as a function of [CA] plotted according to Eq. (8.66). (Data of Problem 8.8.)

8.3.3.7 Effect of Excess Counterion

Let us now consider how the presence of excess counterion due to the addition of a strongly dissociating salt (e.g., sodium tetraphenyl borate, $NaBPh_4$, to supply excess Na^+) can affect the kinetics of a living anionic polymerization. The retarding effect of a dissociating salt forms the basis of a method by which the dissociation constant of a polymeric ion pair may be determined.

If the polymeric ion pairs are slightly dissociated, $\left[\sum_x AM_x^- C^+\right]$ can be set approximately equal to the total concentration of living ends $[M^-]$ and Eq. (8.59) can be written as

$$K = \frac{\left[\sum_x AM_x^-\right][C^+]}{[M^-]} \tag{8.67}$$

If the added salt is strongly dissociated and a relatively large amount of salt has been added, $[C^+]$ representing the concentration of free C^+ ions in solution can be approximated by the concentration of C^+ produced by the dissociation of the added salt, i.e.,

$$[C^+] \simeq [C^+]_{salt} \tag{8.68}$$

Combining Eqs. (8.67) and (8.68), the concentration of free ions can be given by

$$\left[\sum_x AM_x^-\right] = \frac{K[M^-]}{[C^+]_{salt}} \tag{8.69}$$

and the concentration of ion pairs by

$$\left[\sum_x AM_x^- C^+\right] = [M^-] - \left[\sum_x AM_x^-\right] = [M^-] - \frac{K[M^-]}{[C^+]_{salt}} \tag{8.70}$$

Combination of Eqs. (8.69) and (8.70) with Eq. (8.62) then yields

$$\frac{R_p}{[M][M^-]} = k_p = k_p^{\mp} + \frac{(k_p^- - k_p^{\mp})K}{[C^+]_{\text{salt}}} \tag{8.71}$$

Equations (8.66) and (8.71) allow one to obtain $k_p^{\mp}$, k_p^-, and K from the experimental values of overall or apparent rate constant k_p determined in the absence and presence of added common ion. While a plot of k_p versus $1/[M^-]^{1/2}$ in the absence of added common ion yields a straight line whose slope and intercept are $(k_p^- - k_p^{\mp})K^{1/2}$ and $k_p^{\mp}$, respectively, a plot of k_p versus $1/[C^+]_{\text{salt}}$ in the presence of added common ion yields a straight line [cf. Eq. (8.71)] whose slope and intercept are $(k_p^- - k_p^{\mp})K$ and $k_p^{\mp}$, respectively, thus affording the values of k_p^-, $k_p^{\mp}$, and K, individually (see Problems 8.9 and 8.10). It should be noted that the observed ion pair propagation constant $k^{\mp}$ is, in fact, an apparent rate constant as it is a composite of rate constants for the contact ion pair and the solvent-separated ion pair (see Problem 8.11).

Problem 8.9   Polymerization of styrene with sodium naphthalene initiator was performed at 25°C in tetrahydrofuran (THF) using a static technique (Bhattacharya et al., 1965) that is suitable for monitoring fast reactions. The conversion was determined by monitoring the residual styrene monomer during polymerization and the concentration of living ends $[M^-]$ was determined at the end of the experiment, using spectrophotometry in both cases. In independent experimental series, the overall rate constant k_p was obtained [cf. Eq. (P8.8.1)] both at different concentrations of initiator (and hence $[M^-]$) without addition of electrolyte and at different concentrations of sodium ions from externally added sodium tetraphenyl borate (NaBPh$_4$) salt and constant concentration of initiator. The data are given below:

(a) Without electrolyte		(b) With electrolyte		
$[M^-] \times 10^5$	k_p	$[M^-] \times 10^5$	$[\text{NaBPh}_4] \times 10^5$	k_p
mol/L	L/mol-s	mol/L	mol/L	L/mol-s
1.97	5420	7.1	132.0	114
6.22	3620	16.9	55.5	127
11.0	2450	15.1	58.6	137
16.5	2050	18.5	29.5	163
100	900	11.5	22.0	184
490	538	16.7	13.5	210
1000	396	31.0	7.0	280

Source: Bhattacharya et al., 1965.

Determine the propagation rate constants $k_p^{\mp}$ and k_p^- and the overall dissociation constant K for polystyryl sodium in THF at 25°C. (The dissociation constant of NaBPh$_4$ in THF at 25°C is 8.52×10^{-5} mol/L.)

Answer:

(a) A plot of the observed k_p vs. $1/[M^-]^{1/2}$ is shown in Fig. 8.4(a). The slope of the line, according to Eq. (8.66), gives $(k_p^- - k_p^{\mp})K^{1/2} = 25.2 \text{ L}^{1/2} \text{ mol}^{-1/2} \text{ s}^{-1}$ and its intercept leads to $k_p^{\mp} \sim 150 \text{ L mol}^{-1} \text{ s}^{-1}$.

(b) Let $x = [\text{Na}^+]/[\text{NaBPh}_4]$, where $[\text{NaBPh}_4]$ is the total concentration (c) of the salt initially added. Denoting the dissociation constant of NaBPh$_4$ salt by K_S and noting that $[\text{Na}^+] \simeq [\text{BPh}_4^-]$ if the polymeric ion pairs are slightly dissociated, as compared to salt, one can write $K_S \simeq cx^2/(1-x)$. Solving the quadratic gives: $x.c = [\text{Na}^+] = \frac{1}{2}[(K_S^2 + 4K_S c)^{1/2} - K_S]$. This equation is used to calculate $[\text{Na}^+]$ corresponding to a given $[\text{NaBPh}_4]_0$. A plot of the observed k_p vs. $1/[\text{Na}^+]$ is shown in Fig. 8.4(b). The intercept of the line, according to Eq. (8.71), gives $k_p^{\mp} = 80 \text{ L mol}^{-1} \text{ s}^{-1}$ and its slope $(k_p^- - k_p^{\mp})K = 9.8 \times 10^{-3} \text{ s}^{-1}$. Since

$(k_p^- - k_p^{\mp})K^{1/2} = 25.2$ L$^{1/2}$ mol$^{-1/2}$ s^{-1} from (a), by combining these data one obtains $k_p^- = 6.5 \times 10^4$ L mol^{-1} s^{-1} and $K = 1.52 \times 10^{-7}$ mol L^{-1}.

[The method based on the effect of the salt on the observed k_p gives a much more reliable value of $k_p^{\mp}$ than that obtained from the intercept of the much steeper line in (a).]

Problem 8.10   Calculate the concentrations of polymeric free ions and ion pairs both in the absence of electrolyte and in the presence of electrolyte NaBPh$_4$, considering two polymerization systems of Problem 8.9 that have nearly equal concentrations of living chain ends. What percentage of the free Na$^+$ ions comes from the initiator when the electrolyte is present ?

Answer:

(a) Without electrolyte: The concentrations of polymeric free ions, $\left[\sum AM_x^-\right]$, and that of polymeric ion pairs, $\left[\sum AM_x^- C^+\right]$, can be calculated from Eqs. (8.61) and (8.64). For $[M^-] = 11.0 \times 10^{-5}$ mol L^{-1},

$$\left[\sum_x AM_x^-\right] = K^{1/2}[M^-]^{1/2} = (1.52 \times 10^{-7} \text{ mol L}^{-1})^{1/2}(11.0 \times 10^{-5} \text{ mol L}^{-1})^{1/2} = 0.41 \times 10^{-5} \text{ mol L}^{-1}.$$

$$\left[\sum_x AM_x^- \text{Na}^+\right] = [M^-] - K^{1/2}[M^-]^{1/2} = (11.0 \times 10^{-5} \text{ mol L}^{-1}) - 0.41 \times 10^{-5} \text{ mol L}^{-1} = 10.6 \times 10^{-5} \text{ mol L}^{-1}$$

(b) With electrolyte: Consider the system with $[M^-] = 11.5 \times 10^{-5}$ mol/L and $[\text{NaBPh}_4]_0 = c = 22.0 \times 10^{-5}$ mol/L; then $[\text{Na}^+]_{\text{salt}} = \frac{1}{2}\left[(K_S^2 + 4K_S c)^{1/2} - K_S\right] = 1.01 \times 10^{-4}$ mol L^{-1}.

From Eq. (8.69),

$$\left[\sum_x AM_x^-\right] = \frac{(1.52 \times 10^{-7} \text{ mol L}^{-1})(11.5 \times 10^{-5} \text{ mol L}^{-1})}{(1.01 \times 10^{-4} \text{ mol L}^{-1})} = 1.7 \times 10^{-7} \text{ mol L}^{-1}$$

From Eq. (8.70),

$$\left[\sum_x AM_x^- \text{Na}^+\right] = (11.5 \times 10^{-5} \text{ mol L}^{-1}) - (1.7 \times 10^{-7} \text{ mol L}^{-1}) \simeq 11.5 \times 10^{-5} \text{ mol L}^{-1}$$

(c) Fraction of free Na$^+$ ions coming from the initiator $= (1.7 \times 10^{-7}$ mol L$^{-1})/[(1.01 \times 10^{-4}$ mol L$^{-1}) + (1.7 \times 10^{-7}$ mol L$^{-1})] = 1.68 \times 10^{-3}$ ($\equiv 0.17\%$)

Problem 8.11   Consider propagation by polystyryl sodium ion pairs in a 1 M styrene solution in tetrahydrofuran at 20°C. For an ion pair concentration of 2.0×10^{-3} M, calculate the relative contributions of contact and solvent-separated ion pairs to the propagation process. [Data: At 20°C, rate constant for contact ion pair, $k_c = 24$ L/mol-s; rate constant for solvent-separated ion pair, $k_s = 5.5 \times 10^4$ L/mol-s; equilibrium constant for interconversion between contact ion pair and solvent separated ion pair, $K_{cs} = 2.57 \times 10^{-3}$.]

Answer:

Let K_{cs} be the equilibrium constant for interconversion between solvent-separated and contact ion pairs, i.e.,

$$AM_x^- C^+ \overset{K_{cs}}{\rightleftharpoons} AM_x^-//C^+ \tag{P8.11.1}$$

Let α be the fraction of solvent-separated ion-pairs, i.e.,

$$\alpha = \frac{[AM_x^-//C^+]}{[AM_x^- C^+] + [AM_x^-//C^+]} = \frac{K_{cs}}{1 + K_{cs}} \simeq K_{cs} \quad \text{(since } K_{cs} \ll 1\text{)} \tag{P8.11.2}$$

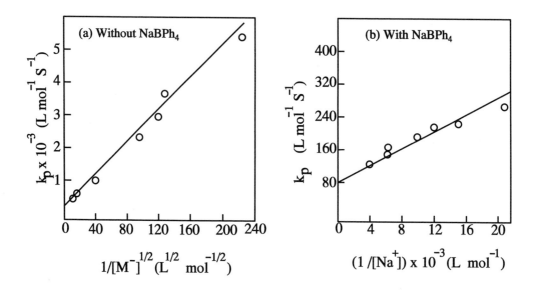

Figure 8.4 The overall rate constant k_p for polystyryl sodium in THF at 25°C as a function of $[M^-]$ and $[Na^+]$ plotted according to Eqs. (8.66) and (8.71). (Data of Problem 8.9.)

Therefore, $[AM_x^-//C^+] = K_{cs}\left([AM_x^- C^+] + [AM_x^-//C^+]\right)$
$$= (2.57 \times 10^{-3})(2.0 \times 10^{-3} \text{ mol L}^-) = 5.14 \times 10^{-6} \text{ mol L}^{-1}.$$

The rate of propagation is given by $R_p = k_c[AM_x^- C^+][M] + k_s[AM_x^-//C^+][M] = (24 \text{ L mol}^{-1} \text{ s}^{-1})(2.0 \times 10^{-3} \text{ mol L}^{-1})(1.0 \text{ mol L}^{-1}) + (5.5 \times 10^4 \text{ L mol}^{-1} \text{ s}^{-1})(5.14 \times 10^{-6} \text{ mol L}^{-1})(1.0 \text{ mol L}^{-1}) = 0.048 + 0.283$ or $0.331 \text{ mol L}^{-1} \text{ s}^{-1}$.

Fractions of propagation carried by solvent-separated and contact ion pairs are thus given by $0.283/0.331 = 0.85$ and $0.048/0.331 = 0.15$, respectively.

The observed ion pair propagation constant $k_p^{\mp}$ is a composite of the rate constants k_c and k_s according to $k_p^{\mp} = \alpha k_s + (1 - \alpha)k_c = (2.57 \times 10^{-3})(5.5 \times 10^4 \text{ L mol}^{-1} \text{ s}^{-1}) + (1 - 2.57 \times 10^{-3})(24 \text{ L mol}^{-1} \text{ s}^{-1}) = 165.3 \text{ L mol}^{-1} \text{ s}^{-1}$.

8.3.4 Polymerization with Incomplete Dissociation of Initiator

If the initiator used in anionic polymerization dissociates completely from the inactive form, CA, to the active form, C^+A^- (or $C^+ + A^-$), before any propagation reactions take place, the kinetics, as we have seen earlier, assumes a very simple form. Some initiators (e.g., lithium alkyls and aryls), however, dissociate only partially and so also the anionic growing chains. In such a case, dissociation equilibria must be considered both for initiation steps, which may now be written as (Allcock and Lampe, 1990):

$$\text{CA} \rightleftharpoons \text{C}^+ \text{A}^- \rightleftharpoons \text{C}^+ + \text{A}^- \tag{8.72}$$

$$\text{C}^+ \text{A}^- + \text{M} \longrightarrow \text{AM}^- \text{C}^+ \rightleftharpoons \text{AMC} \tag{8.73}$$

and for the propagation steps, now to be written as

$$AM^- C^+ + M \longrightarrow AMM^- C^+ \rightleftharpoons AM_2C \tag{8.74}$$

$$\dotsc\dotsc\dotsc\dotsc\dotsc$$

$$AM_{n-1}M^-C^+ + M \longrightarrow AM_nM^- C^+ \rightleftharpoons AM_{n+1}C \tag{8.75}$$

It may be noted that only propagation by ion pairs has been considered in the above scheme. In a more complete scheme, propagation by free ions [cf. Eq. (8.58)] should also be considered. Moreover, the equilibria included in the above scheme may also involve solvation by the solvent, though it has not been shown explicitly. A further complication may also arise, if the initiation depends on the nature of the monomer. In view of these complexities of anionic polymerization, general equations cannot be given for the rate of polymerization, kinetic chain length, or average degree of polymerization (except in simple cases considered above involving complete dissociation of the initiator), and each system should be treated as a "kinetically unique problem" requiring individual solutions (Allcock and Lampe, 1990).

8.3.5 Polymerization with Simultaneous Propagation and Termination

Let us now consider an anionic polymerization where propagation and termination reactions occur simultaneously and polymerization follows in a manner similar to free-radical polymerizations. An example is the potassium amide initiated polymerization in liquid ammonia. This is one of the few anionic systems in which all active centers behave kinetically as free ions. The initiation step in this case consists of dissociation of potassium amide followed by addition of amide ion to monomer (Ghosh, 1990; Odian, 1991):

$$KNH_2 \overset{K}{\rightleftharpoons} K^+ + H_2N:^- \tag{8.76}$$

$$H_2N:^- + CH_2{=}CHR \xrightarrow{k_i} H_2N{-}CH_2{-}\overset{\overset{\displaystyle H}{|}}{\underset{\underset{\displaystyle R}{|}}{C}}:^- \tag{8.77}$$

The dissociation constant K is given by

$$K = \frac{[K^+][NH_2^-]}{[KNH_2]} \tag{8.78}$$

Since the second step [Eq. (8.77)] is slow relative to the first [Eq. (8.76)], the rate of initiation is given by

$$R_i = k_i[NH_2^-][M] \tag{8.79}$$

In general terms, the propagation reactions can be represented by

$$H_2N{-}M_n^- + M \xrightarrow{k_p} H_2N{-}M_nM^- \tag{8.80}$$

with the rate of propagation expressed as

$$R_p = k_p[M^-][M] \tag{8.81}$$

where $[M^-]$ represents the total concentration of the propagating anionic centers.

Extensive chain transfer to solvent occurs by the reaction

$$H_2N-M_n^- \ + \ NH_3 \ \xrightarrow{k_{tr,NH_3}} \ H_2N-M_n-H \ + \ NH_2^- \tag{8.82}$$

with a rate given by

$$R_{tr,NH_3} \ = \ k_{tr,NH_3}\,[M^-]\,[NH_3] \tag{8.83}$$

Assuming a steady state whereby $R_i = R_{tr}$,

$$k_i[NH_2^-][M] \ = \ k_{tr,NH_3}\,[M^-]\,[NH_3] \tag{8.84}$$

$$[M^-] \ = \ \frac{k_i\,[NH_2^-]\,[M]}{k_{tr,NH_3}\,[NH_3]} \tag{8.85}$$

Substituting in Eq. (8.81) we obtain,

$$R_p \ = \ \frac{k_p\,k_i\,[NH_2^-]\,[M]^2}{k_{tr,NH_3}\,[NH_3]} \tag{8.86}$$

When KNH_2 is used as the initiator and no external K^+ is added, the concentrations of K^+ and NH_2^- are equal. Therefore, from Eq. (8.78), $K = [NH_2^-]^2/[KNH_2]$, and Eq. (8.86) can thus be rewritten as

$$R_p \ = \ \frac{k_i\,k_p\,K^{1/2}\,[M]^2\,[KNH_2]^{1/2}}{k_{tr,NH_3}\,[NH_3]} \tag{8.87}$$

Problem 8.12   Justify the steady-state assumption of Eq. (8.84) considering the fact that the chain transfer reaction [Eq. (8.82)] is truly not a termination reaction since the amide ion is regenerated.

Answer:

The second step of initiation [Eq. (8.77)], being slower than the first [Eq. (8.76)], is rate-determining for initiation (unlike in the case of free-radical chain polymerization) and so though the amide ion produced upon chain transfer to ammonia can initiate polymerization, it can be only at a rate controlled by the rate constant, k_i. Therefore, this chain transfer reaction may be considered as a true kinetic-chain termination step and the application of steady-state condition gives Eq. (8.84).

The average kinetic chain length $(\bar{v})$ or degree of polymerization $(\overline{DP})$ is expressed as

$$\overline{DP} \ = \ \bar{v} \ = \ \frac{k_p\,[M]\,[M^-]}{k_{tr,NH_3}\,[M^-]\,[NH_3]} \ = \ \frac{[M]}{C_{NH_3}\,[NH_3]} \tag{8.88}$$

where C_{NH_3} is the chain transfer constant for NH_3. Following the procedure described in Section 6.11, the effects of temperature on R_p and $\overline{DP}$ can be represented by the equations

$$d\ln(R_p)/dT \ = \ \left(E_i + E_p - E_{tr}\right)/RT^2 \tag{8.89}$$

$$d\ln(\overline{DP}_n)/dT \ = \ \left(E_p - E_{tr}\right)/RT^2 \tag{8.90}$$

Since $E_i + E_p - E_{tr} \simeq +38$ kJ mol^{-1} and $E_p - E_{tr} \simeq -17$ kJ mol^{-1}, reducing the reaction temperature decreases R_p but increases $\overline{DP}_n$. However, due to the small magnitude of $(E_p - E_{tr})$ and high concentration of ammonia, chain transfer to ammonia remains highly competitive with propagation and only low molecular weight polymer can be obtained even at low temperatures.

Taking into account also the termination by transfer to adventitious water, besides transfer to NH$_3$, the number-average degree of polymerization is given by

$$\frac{1}{\overline{DP}_n} = \frac{C_{\text{NH}_3}\,[\text{NH}_3]}{[\text{M}]} + \frac{C_{\text{H}_2\text{O}}\,[\text{H}_2\text{O}]}{[\text{M}]} \tag{8.91}$$

where $C_{\text{H}_2\text{O}}$ is the chain transfer constant for water.

8.4 Anionic Copolymerization

For the copolymerization of two monomers by an anionic mechanism, the copolymer composition equation (7.11) or (7.17), derived in Chapter 7, is applicable with the monomer reactivity ratios defined in the same way as before, namely, $r_1 = k_{11}/k_{12}$ and $r_2 = k_{22}/k_{21}$, where k_{11} and k_{22} are the rate constants for the homopropagation reactions:

$$\text{wwwM}_1^- + \text{M}_1 \xrightarrow{k_{11}} \text{wwwM}_1^- \tag{8.92}$$

$$\text{wwwM}_2^- + \text{M}_2 \xrightarrow{k_{22}} \text{wwwM}_2^- \tag{8.93}$$

and k_{12} and k_{21} are rate constants for cross-propagation reactions:

$$\text{wwwM}_1^- + \text{M}_2 \xrightarrow{k_{12}} \text{wwwM}_2^- \tag{8.94}$$

$$\text{wwwM}_2^- + \text{M}_1 \xrightarrow{k_{21}} \text{wwwM}_1^- \tag{8.95}$$

The reactivity ratios r_1 and r_2 can be determined from the composition of the copolymer product. However, a difficulty arises from the fact that the propagation rate constants, k_{ij}, are composite rate constants (Allcock and Lampe, 1990) having contributions from both ion pairs and free ions. Consequently, the reactivity ratios also will be composite quantities composed of free-ion and ion-pair contributions, which are strongly dependent on the reaction conditions. The reactivity ratios can, therefore, be applied only to systems identical to those for which they were determined. This greatly limits the utility of such ratios.

Because of the complicating effects of counterion and solvent associated with anionic polymerization, relatively few reactivity ratios have been determined for anionic systems. Typical reactivity ratios for the anionic copolymerization of styrene and a few other monomers are shown in Table 8.3. Most of the values were determined from the copolymer composition equation [Eq. (7.11) or (7.18)]. A dramatic effect of solvent is seen with styrene-butadiene copolymerization, where a change from the nonpolar hexane to the highly solvating THF reverses the order of reactivity. Again in the case of hydrocarbon solvent, the reaction temperature shows a minimal influence on reactivity ratios, while in the case of polar solvents, such as THF, the reactivity ratios vary considerably, which has been rationalized by considering the solvation of carbon-lithium bond. Thus as the temperature is increased (from $-78°$C to $25°$C), the extent of solvation by THF is expected to decrease, resulting in more covalent carbon-lithium bond.

Table 8.3 Some Typical Values of Monomer Reactivity Ratios for Anionic Copolymerization

Monomer 1	Monomer 2	Initiator	Solvent	Temp. (°C)	r_1	r_2
Styrene	Acrylonitrile[a]	C_6H_5MgBr	C_6H_{12}	−45	0.05	15.0
	Acrylonitrile[b]	RLi	None	–	0.20	12.5
	Methyl metha-crylate[a]	C_6H_5MgBr	Ether	−30	0.05	14.0
		C_6H_5MgBr	Ether	+20	0.30	2.0
	Vinyl acetate[b]	Na	Liq. NH_3	–	0.01	0.01
	Butadiene[b]	s-BuLi	Hexane	0	0.03	13.3
		s-BuLi	Hexane	50	0.04	11.8
		s-BuLi	THF	25	4.0	0.3
		s-BuLi	THF	−78	11.0	0.4
	Isoprene[b]	n-BuLi	C_6H_{12}	40	0.046	16.6
Butadiene[b]	Isoprene	n-BuLi	Hexane	50	3.38	0.47
Methyl metha-crylate[b]	Acrylonitrile	$NaNH_2$	Liq. NH_3	–	0.25	7.9
		RLi	None	–	0.34	6.7
	Vinyl acetate	$NaNH_2$	Liq. NH_3	–	3.2	0.4

Source: [a]Dawans and Smets, 1963; [b]Morton, 1983; Allcock and Lampe, 1990.

8.4.1   Reactivity Groups

An active chain with a carbionic end group will add a second monomer if the reacting carbanion is more basic than the new carbanion that will result from the second monomer. The initiation capacity of a carbanion can thus be correlated approximately with the e-values of the $Q-e$ scheme (Section 7.2.6) as shown in Table 8.4, which lists a number of monomers classified according to their relative reactivities (Fetters, 1969). The carbanion of a monomer in a given reactivity group can generally add any monomer in a group with a higher number, but the reverse is not true. Thus the monomers in group 5 generate the weakest (less basic) anions and cannot initiate monomers in any of the preceding groups. A monomer, however, will generally react with the carbanion generated from any other monomer within the group, though there are exceptions to this rule. For example, the initiation of styrene polymerization by α-methylstyrene carbanion or that of isoprene by styryl carbanion is faster than the reverse initiation.

Problem 8.13   What types of polymers will be obtained from anionic copolymerization of the following monomer pairs: (a) styrene/butadiene, and (b) styrene/methyl methacrylate?

Answer:

(a) Copolymerizations analogous to free radical reactions occur between monomers which are within the same reactivity group in Table 8.4, where each carbanion can initiate either monomer. Since both styrene and butadiene belong to the same reactivity group, they will undergo copolymerization, the copolymer composition depending on the reactivity ratios. From Table 8.3, $r_1 = 4.0$, $r_2 = 0.3$ for s-butyllithium initiated polymerization in THF at 25°C. Therefore ideal copolymerization behavior will be observed and a statistical copolymer containing a larger proportion of styrene in random placement will be obtained. The simple copolymer equation [Eq. (7.11)] can be applied to calculate the copolymer composition.

(b) The styrene-MMA pair contains monomers drawn from different relative reactivity groups in Table 8.4. Polystyryl carbanion will initiate the polymerization of MMA, but the carbanion of the latter monomer is not sufficiently nucleophilic (basic) to cross-initiate the polymerization of styrene.

This is reflected in the r_1, r_2 values (see Table 8.3). For example, the styrene/MMA pair has $r_1 = 0.05$, $r_2 = 14.0$ when initiated by C_6H_5MgBr in ether at −30°C. Thus the polymerization of the mixture in this

Table 8.4 Initiation of Anionic Polymerization by Living Polymer Ends

Reactivity group	Monomer types	e values
1	α-Methylstyrene	−1.2
	Conjugated dienes	−1.0
	Styrene	−0.8
	4-Vinyl pyridine	−0.2
2	Acrylate and methacrylate esters	0.4
3	1,2-Epoxies	−
	Formaldehyde	−
4	Methacrylonitrile	0.8
	Acrylonitrile	1.2
5	Nitroalkenes	−
	Vinylidene cyanide	2.6

Source: Fetters, 1969.

case will cause homopolymerization of MMA followed by that of styrene. However, excess MMA will add to living polystyrene producing a block copolymer of the two monomers.

8.4.2 Block Copolymers

While the simplest vinyl-type block copolymer is a two-segment molecule represented by

(AAAAA·········A)(BBBBBB·········B)   AB block copolymer

other common block copolymer structures are:

(AAAA······A)(BBBB······B)(AAAA······A)   ABA block copolymer

(AAAA······A)(BBBB······B)(CCCC······C)   ABC block copolymer

(AA····A)(BB···B)(AA····A)(BB····B) etc.   Multiblock copolymer

(AAAAAA·········A)(AAAAAAA·········A)   Stereoblock copolymer
Isotactic sequence   Syndiotactic sequence

Block copolymers differ in properties from those of a blend of the corresponding homopolymers or a random copolymer (Chapter 7) with the same overall composition. Let us consider the ABA-type styrene/butadiene/styrene triblock copolymer, which is known to behave as a *thermoplastic elastomer*. While ordinary elastomers, which are crosslinked by covalent bonds, e.g., vulcanization (see Chapter 2) to impart elastic recovery property, are thermosets and so cannot be softened and reshaped by molding, solid thermoplastic styrene/butadiene/styrene tri-block elastomers can be resoftened and remolded. This can be explained as follows. At room temperature, the triblock elastomer has the make-up, as shown in Fig. 8.5, consisting of glassy, rigid, polystyrene domains ($T_g = 100°C$) linked together by rubbery polybutadiene segments. The polystyrene domains thus serve as effective crosslinks and stabilize the structure against moderate stresses. They can be softened sufficiently at $T > 100°C$ to carry out molding but the rigid polystyrene domains will form again on subsequent cooling.

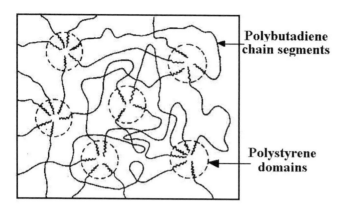

Figure 8.5 Schematic representation of the polystyrene domain structure in styrene-butadiene-styrene tri-block copolymers.

8.4.2.1 Sequential Monomer Addition

The living polymer technique is particularly suitable for preparing block copolymers by sequential addition of monomers to a living anionic polymerization system. Depending on whether mono-functional or difunctional initiators are used, one or both chain ends remain active after monomer A has completely reacted. Monomer B is then added, and its polymerization is initiated by the living polymeric carbanion of polymer A. This method of sequential monomer addition can be used to produce block copolymers of several different types, as classified above.

Monofunctional Initiators   Using monofunctional initiators (e.g., *n*-butyllith-ium), AB, ABA, and multiblock copolymers can be formed. For example, the synthesis of an AB block copolymer can be shown schematically as

$$A \xrightarrow{\text{RLi}} R\text{\tiny wwww}AAA^- \ Li^+ \xrightarrow{B} R\text{\tiny wwww}AAA\text{\tiny wwww}BBB^- \ Li^+ \xrightarrow{H_2O}$$

$$R\text{\tiny wwww}AAA\text{\tiny wwww}BBBH \ + \ LiOH \qquad (8.96)$$

The monomers are to be added in proper order. For example, to prepare an AB type block copolymer of styrene and methyl methacrylate, styrene is polymerized first using a monofunctional initiator and when styrene is fully consumed, the other monomer MMA is added. The copolymer cannot be made by polymerizing MMA first because living poly(methyl methacrylate) is not basic enough to add to styrene. The length of each block in the copolymer is determined by the amount of corresponding monomer added to the reaction mixture. To produce ABA type copolymer by monofunctional initiation, B can be added after A is fully reacted, and A added again when B is fully reacted. Multiblock copolymers can also be made in this way. However, this procedure is possible only if the anion of each monomer sequence can initiate polymerization of the other monomer.

Bifunctional Initiators   Bifunctional initiators like alkali metal complexes of polycyclic aromatic compounds (e.g., naphthalene and biphenyl) can be used to produce ABA triblock copolymers. In these cases, polymerization is started with monomer B to produce a polymeric dianion [see Eq. (8.14)] that initiates the polymerization of monomer A added later. The process is illustrated below for the commercially important styrene-butadiene-styrene (SBS) triblock copolymer:

$$B \xrightarrow{Na} \cdot B{:}^- Na^+ \longrightarrow Na^{+-}{:}BB{:}^- Na^+ \xrightarrow{B} Na^{+-}{:}BBB\cdots BBB{:}^- Na^+$$

$$\xrightarrow{S} Na^{+-}{:}SSSSBBB\cdots BBBSSSS{:}^- Na^+ \qquad (8.97)$$

Thermoplastic elastomers of this type usually have molecular weights in the range 50,000 to 70,000 for the polybutadiene blocks and 10,000 to 15,000 for the polystyrene blocks.

8.4.2.2 Coupling Reactions

A living AB-type block copolymer made by monofunctional initiation can be terminated with a bifunctional coupling agent like a dihaloalkane to make a triblock copolymer. For example, an ABA triblock copolymer can be made as follows:

$$2(AAA\cdots)_x(BBB\cdots)_y{-}B{:}^- Li^+ + Br(CH_2)_z Br \longrightarrow$$
$$(AAA\cdots)_x(BBB\cdots)_{y+1}(CH_2)_z(BBB\cdots)_{y+1}(AAA\cdots)_x + 2LiBr \qquad (8.98)$$

Other coupling agents such as phosgene or dichloromethylsilane are equally effective. If the presence of a linking group, like the $(CH_2)_z$ in the above reaction, is undesirable, a coupling agent like I_2 can be used.

Problem 8.14 In a living polymerization experiment, 8.0 cm^3 of 0.20 mol/L solution of *n*-butyllithium in toluene was added to a solution of 18.0 g styrene in 500 cm^3 toluene. After complete conversion of the styrene, 60 g isoprene was added. When the isoprene had completely reacted, the polymerization was terminated by addition of 8.0 cm^3 of a 0.10 mol/L solution of dichloromethane in toluene.

Write down the reactions taking place in each stage of the above experiment and calculate the molecular weight of the final polymer.

Answer:

The given quantities are:

$$BuLi = (0.20 \text{ mol L}^{-1})(8.0 \text{ cm}^3)(10^{-3} \text{ L cm}^{-3}) = 1.60 \times 10^{-3} \text{ mol}$$

$$Styrene = (18.0 \text{ g})/(104 \text{ g mol}^{-1}) = 0.173 \text{ mol}$$

$$Isoprene = (60 \text{ g})/(68 \text{ g mol}^{-1}) = 0.882 \text{ mol}$$

Dichloromethane $= (0.10 \text{ mol L}^{-1})(8.0 \text{ cm}^3)(10^{-3} \text{ L cm}^{-3}) = 0.80 \times 10^{-3}$ mol

The reactions taking place are as follows (Young and Lovell, 1990):

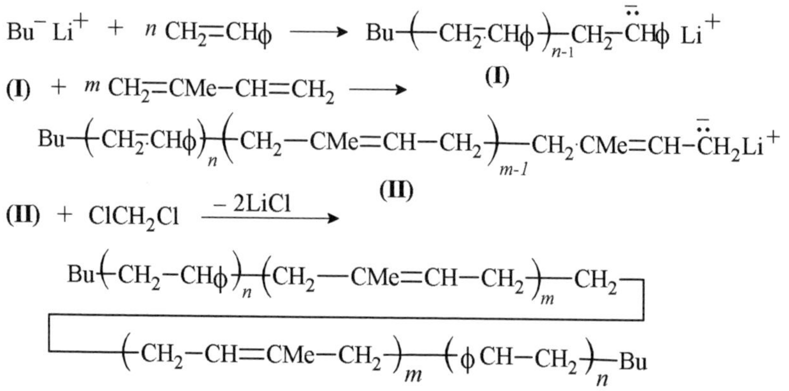

$$\text{Bu}^- \text{Li}^+ + n \text{ CH}_2\!=\!\text{CH}\phi \longrightarrow \text{Bu}\!\!-\!\!\left(\!\text{CH}_2\!\cdot\!\text{CH}\phi\!\right)_{\!n-1}\!\!-\!\!\text{CH}_2\!\cdot\!\overset{..}{\text{C}}\text{H}\phi \text{ Li}^+$$

$$\textbf{(I)}$$

$$\textbf{(I)} + m \text{ CH}_2\!=\!\text{CMe}\!-\!\text{CH}\!=\!\text{CH}_2 \longrightarrow$$

$$\text{Bu}\!\!-\!\!\left(\!\text{CH}_2\!\cdot\!\text{CH}\phi\!\right)_{\!n}\!\!-\!\!\left(\!\text{CH}_2\!-\!\text{CMe}\!=\!\text{CH}\!-\!\text{CH}_2\!\right)_{\!m-1}\!\!-\!\!\text{CH}_2\!\cdot\!\text{CMe}\!=\!\text{CH}\!-\!\overset{..}{\text{C}}\text{H}_2\text{Li}^+$$

$$\textbf{(II)}$$

$$\textbf{(II)} + \text{ClCH}_2\text{Cl} \xrightarrow{\ -2\text{LiCl}\ }$$

$$\text{Bu}\!\!-\!\!\left(\!\text{CH}_2\!-\!\text{CH}\phi\!\right)_{\!n}\!\!-\!\!\left(\!\text{CH}_2\!-\!\text{CMe}\!=\!\text{CH}\!-\!\text{CH}_2\!\right)_{\!m}\!\!-\!\!\text{CH}_2\!\!-\!\!$$
$$\!\!-\!\!\left(\!\text{CH}_2\!-\!\text{CH}\!=\!\text{CMe}\!-\!\text{CH}_2\!\right)_{\!m}\!\!-\!\!\left(\!\phi\text{CH}\!-\!\text{CH}_2\!\right)_{\!n}\!\!-\!\!\text{Bu}$$

The product is an ABA-type triblock thermoplastic elastomer. Styrene is polymerized first since styryl initiation of isoprene is faster than the reverse reaction. The reaction is carried out in a nonpolar solvent with Li^+ as the counterion to enable a block of *cis*-1,4-polyisoprene to be formed in the second growth stage. The living polystyrene-*b*-polyisoprene AB di-block copolymer thus formed is then coupled by a double nucleophilic displacement of Cl^- ions from dichloromethane to give a polystyrene-*b*-polyisoprene-*b*-polystyrene triblock copolymer. (Note that the mole ratio of living diblock chain to dichloromethane is 2:1.)

Polystyrene end-blocks:

$$(\text{M}_n)_{\text{PS}} = \frac{\text{Mass of styrene polymerized}}{\text{Moles of living polymer molecules}}$$

$$= (18.0 \text{ g})/(1.60 \times 10^{-3} \text{ mol}) = 11{,}250 \text{ g mol}^{-1}$$

Polyisoprene center block:

The molecular weight of the polyisoprene block in the final triblock copolymer is twice that of the polyisoprene block in the diblock living copolymer before coupling. Hence,

$$(\text{M}_n)_{\text{PI}} = 2 \times \left(\frac{\text{mass of isoprene polymerized}}{\text{moles of living polymer molecules}}\right)$$

$$= 2(60.0 \text{ g})/(1.60 \times 10^{-3} \text{ mol}) = 75{,}000 \text{ g mol}^{-1}$$

Molecular weight of the final polymer $= 75{,}000 + 2 \times 11{,}250 = 97{,}500$.

Use of polyfunctional linking agents, instead of bifunctional ones, for termination of living polymers can lead to the formation of multibranch polymers. For example, using methyltrichlorosilane and silicon tetrachloride, tribranch and tetrabranch polymers, respectively, can be produced:

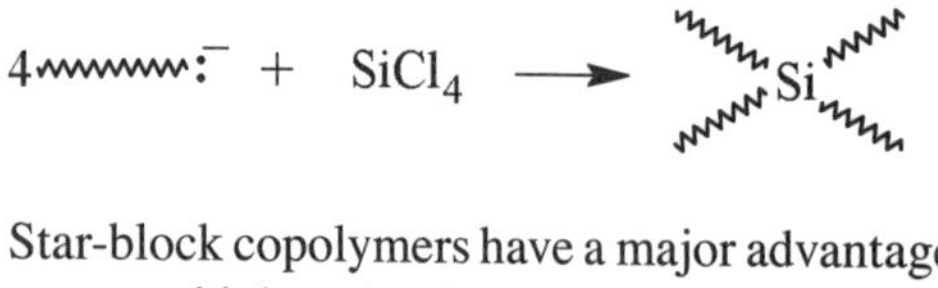

$$4\text{wwwww}{:}^- + \text{ SiCl}_4 \longrightarrow \text{(star-branched Si product)} \tag{8.99}$$

Star-block copolymers have a major advantage in that they exhibit much lower melt viscosities, even at very high molecular weights, than their linear counterparts.

8.5   Cationic Polymerization

As we saw in Section 8.2, the active center in cationic polymerization is a cation and the monomer must therefore behave as a nucleophile (electron donor) in the propagation reaction. Suitability of monomers for cationic polymerization was also discussed in that section and compared in Table 8.1. In short, olefinic monomers with an electron-releasing or electron-donating substituent on the α-carbon can undergo cationic polymerization, while the possibility of resonance stabilization of the carbocationic species increases the reactivity of the monomer (see Problem 8.15).

Problem 8.15   Compare the cationic polymerizability of (a) ethylene, propylene, and isobutylene; (b) styrene, α-methylstyrene, *p*-methoxystyrene, and *p*-chlorostyrene.

Answer:

(a) The electron-releasing alkyl groups cause isobutylene to polymerize readily at low temperatures (yielding high polymer), whereas propylene reacts only with difficulty, yielding low-molecular weight polymers, and ethylene is nearly unreactive. Also from a consideration of carbocation stability (i.e., tertiary most stable, followed by secondary and primary) the polymerizability is in the order

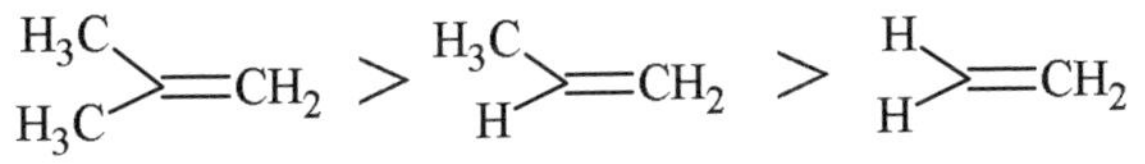

(b) Conjugation when present helps to disperse the positive charge of the carbocation center thus increasing the reactivity of the monomer. The effect becomes stronger when conjugation and electron-donating groups cooperate. On the other hand, substitution of an electron-withdrawing halogen for an *ortho* or a *para* hydrogen decreases the monomer reactivity. Thus the reactivity sequence of the styrene derivatives is as follows:

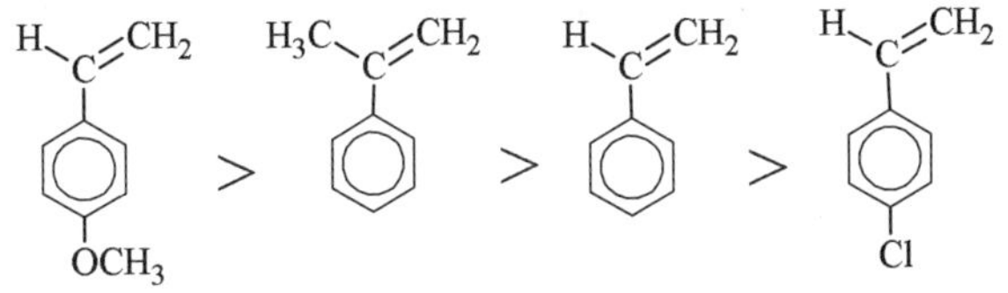

By similar reasoning, electron-releasing substituents, such as RO–, RS–, and aryl at *ortho* or *para* position, increase the monomer reactivity for cationic polymerization.

8.5.1   Cationic Initiation

Two main classes of compounds are known to initiate cationic polymerization of olefins and olefinic derivatives. These are the strong protonic acids and Lewis acids.

8.5.1.1 *Protonic Acids*

The cationic initiation reaction involving a monomer and a protonic acid HA can be written as:

$$\underset{}{H^+ A^-} \;+\; CH_2=\underset{R'}{\overset{R}{\underset{|}{\overset{|}{C}}}} \;\longrightarrow\; CH_3-\underset{R'}{\overset{R}{\underset{|}{\overset{|}{C^+}}}} A^- \tag{8.100}$$

For polymerization to occur, the anion A^-, however, should not be a strong nucleophile, as otherwise it will react with the carbocation to form the nonpropagating covalent compound

$$CH_3-\underset{R'}{\overset{R}{\underset{|}{\overset{|}{C}}}}-A$$

This explains why hydrogen halides with highly nucleophilic halide ions fail to initiate cationic polymerization.

8.5.1.2 *Lewis Acids*

Lewis acids are halides and alkyl halides of Group III metals and of transition metals which have incomplete *d* electron shells. These constitute the most useful group of initiators and include metal halides (e.g., BF_3, $AlCl_3$, $SnCl_4$, PCl_5, $SbCl_5$, $TiCl_4$) and organometallic derivatives (e.g., $RAlCl_2$, R_2AlCl, R_3Al).

Most or perhaps all of the Lewis acids are seldom effective alone as initiators or catalysts; they are used in conjunction with a second compound, called a 'co-catalyst', which very often is water or some other proton donor (*protogen*) such as hydrogen halide, alcohol, and carboxylic acid, or a carbocation donor (*cationogen*) such as *t*-butyl chloride and triphenylmethyl chloride. On reaction with the Lewis acid, they form a catalyst-cocatalyst complex that initiates polymerization. For example, isobutylene is not polymerized by boron trifluoride if both are dry, but immediate polymerization takes place on adding a *small* amount of water. The initiation process is therefore represented by

$$BF_3 + H_2O \;\rightleftharpoons\; H^+(BF_3OH)^- \tag{8.101}$$

$$H^+(BF_3OH)^- + (CH_3)_2C = CH_2 \;\longrightarrow\; (CH_3)_3C^+(BF_3OH)^- \tag{8.102}$$

Initiation by aluminum chloride with *t*-butyl chloride cocatalyst is similarly described by

$$AlCl_3 + (CH_3)_3CCl \;\rightleftharpoons\; (CH_3)_3C^+(AlCl_4)^- \tag{8.103}$$

$$(CH_3)_3C^+(AlCl_4)^- + \phi CH = CH_2 \;\longrightarrow\; (CH_3)_3CCH_2\overset{+}{C}H\phi(AlCl_4)^- \tag{8.104}$$

The initiation steps described above can be generalized as

$$L + IB \;\overset{K}{\rightleftharpoons}\; B^+(LI)^- \tag{8.105}$$

$$B^+(LI)^- + M \;\overset{k_i}{\longrightarrow}\; BM^+(LI)^- \tag{8.106}$$

where L is a Lewis acid catalyst, IB is an ionizable compound acting as the cocatalyst, and M is a monomer. Initiation by the catalyst-cocatalyst system has the advantage over initiation by a Brönsted acid (H^+A^-) that the anion $(LI)^-$ is far less nucleophilic (Odian, 1991) than A^- [cf. Eq. (8.100)].

8.5.2   *Propagation of Cationic Chain*

The propagation of a cationic chain occurs by successive addition of monomer molecules to the initiator ion pair produced in the initiation step [Eq. (8.106)]. In general terms, the propagation can be represented by

$$HM_nM^+ \, (LI)^- \; + \; M \; \xrightarrow{k_p} \; HM_{n+1}M^+ \, (LI)^- \qquad (8.107)$$

which shows that the reaction involves insertion of monomer between the carbocation and its negative counterion. Thus in the polymerization of isobutylene, the cationic chain propagation step is (Odian, 1991):

$$H\!\!+\!\!CH_2C(CH_3)_2\!\!-\!\!\!]_{\overline{n}}\!\!-\!\!CH_2\overset{+}{C}(CH_3)_2 \, (BF_3OH)^- \; + (CH_3)_2C{=}CH_2 \; \longrightarrow$$

$$H\!\!+\!\!CH_2C(CH_3)_2\!\!-\!\!\!]_{\overline{n+1}}\!\!-\!\!CH_2\overset{+}{C}(CH_3)_2 \, (BF_3OH)^- \qquad (8.108)$$

In some cases, the propagation reaction is accompanied by intramolecular rearrangements due to hydride ion ($H:^-$) or methide ion ($CH_3:^-$) shifts. Such polymerizations are referred to as *isomerization polymerizations*. Consider, for example, the polymerization of 3-methyl-1,2-butene in which the carbocation (**XV**), formed initially, isomerizes by a 1,2-hydride shift. The resulting ion (**XVI**), being a tertiary carbocation, is more stable than (**XV**) which is a secondary carbocation.

Propagation thus occurs mostly by (**XVI**), apart from some normal propagation by (**XV**) at higher temperatures. The polymer will therefore contain mostly the rearranged repeating unit (**XVII**) and a smaller number of the first formed repeat unit (**XVIII**). This expectation is supported by the observation that the product contains about 70 and 100% of (**XVII**) at polymerization temperatures of -130 and $-100°C$, respectively (Kennedy et al., 1964; Odian, 1991).

The cationic polymerization of propylene, 1-butene, and higher 1-alkenes yields only very low molecular weight polymers ($DP \leq 10 - 20$) with highly complicated structures that arise due to various combinations of 1,2-hydride and 1,2-methide shifts, proton transfer, and elimination, besides chain transfer during polymerization. In the polymerization of ethylene, initiation involving protonation and ethylation is quickly followed by energetically favorable isomerization:

$$H^+ \; + \; CH_2{=}CH_2 \; \longrightarrow \; CH_3 \overset{+}{C}H_2 \; \overset{CH_2=CH_2}{\longrightarrow}$$

$$CH_3CH_2{-}CH_2{-}\overset{+}{C}H_2 \; \overset{H: \, shift}{\longrightarrow} \; CH_3CH_2{-}\overset{+}{C}H{-}CH_3 \; \longrightarrow \; etc. \qquad (8.109)$$

These transformations are facilitated by the favorable enthalpy differences between primary to secondary (-92 kJ/mol) and secondary to tertiary ($\simeq -138$ kJ/mol) carbocations.

Problem 8.16   Write equations to show the different structural units that may result from intramolecular hydride and methide shifts involving only the end unit in the cationic polymerization of 4-methyl-1-pentene. Which of the resulting repeating units would be the most abundant ?

Answer:

Five different end units (**XIX–XXIII**) may arise from 1,2-hydride and methide shifts. The first-formed carbocation (**XIX**) undergoes hydride shifts to form carbocations (**XX**), (**XXI**), and (**XXII**); (**XXI**) rearranges to (**XXIII**) by a methide shift (Kennedy and Johnston, 1975; Odian, 1991) :

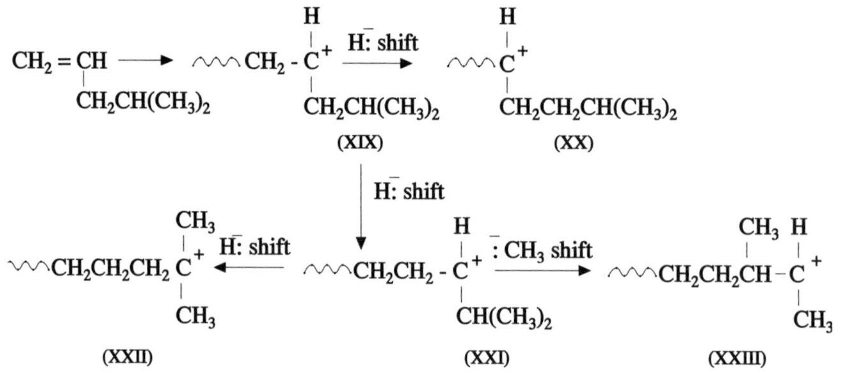

As the tertiary carbocation (**XXII**) is the most stable, the repeating units derived from it will be expected to be most abundant (found 42-51% (Odian, 1991)).

8.5.3   *Chain Transfer and Termination*

While termination of chain growth in cationic polymerization may take place in various ways, many of the termination reactions are, in fact, chain transfer reactions in which the termination of growth of a propagating chain is accompanied by the generation of a new propagating species.

8.5.3.1   *Chain Transfer to Monomer*

In general terms, chain transfer to monomer can be written

$$BM_nM^+ (LI)^- \ + \ M \ \xrightarrow{k_{tr,M}} \ M_{n+1} \ + \ BM^+ (LI)^- \tag{8.110}$$

Chain transfer to vinyl monomer involves transfer of a β-proton from the carbocation to a monomer molecule. Considering, for example, the polymerization of isobutylene, the chain transfer to monomer can be represented by the equation

$$H\text{+}CH_2C(CH_3)_2\text{+}_n\text{--}CH_2 \overset{+}{C}(CH_3)_2 \, (BF_3OH)^- \ + \ CH_2 = C(CH_3)_2 \longrightarrow$$
$$(CH_3)_3C^+ \, (BF_3OH)^- \ + \ H\text{+}CH_2C(CH_3)_2\text{+}_n\text{--}CH_2C(CH_3)=CH_2$$
$$+ \ H\text{+}CH_2C(CH_3)_2\text{+}_n\text{--}CH=C(CH_3)_2 \tag{8.111}$$

which shows the formation of two different unsaturated groups, since there are two different types of β-protons.

As a new propagating species is generated each time a growing chain is terminated by transfer to monomer, many polymer molecules can result for each molecule of catalyst-cocatalyst complex initially formed.

The ratio $k_{tr,M}/k_p$ defines the monomer chain transfer constant C_M and, as in the case of free-radical polymerization (Section 6.8.1), its value determines the polymer molecular weight, in the absence of other chain termination processes.

Another type of chain transfer to monomer involves hydride ion transfer from monomer to the propagating center (Kennedy and Squires, 1965, 1967; Odian, 1991):

$$H \text{+} CH_2C(CH_3)_2 \text{+}_n\text{–}CH_2 \overset{+}{C}(CH_3)_2 \, (BF_3OH)^- \;+\; CH_2{=}C(CH_3)_2 \;\longrightarrow$$

$$CH_2{=}C(CH_3)\overset{+}{C}H_2 \, (BF_3OH)^- \;+\; H\text{+}CH_2C(CH_3)_2\text{+}_n\text{–}CH_2CH(CH_3)_2 \tag{8.112}$$

Note that this transfer reaction results in saturated end groups, unlike proton transfer reaction [Eq. (8.111)] which gives rise to an unsaturated end group. The allyl carbocation formed by the hydride transfer [Eq. (8.112)] is less stable than the tertiary carbocation formed by proton transfer [Eq. (8.111)]. For isobutylene, hydride transfer is thus less likely than proton transfer.

8.5.3.2 Spontaneous Termination

Spontaneous termination, also referred to as *Chain transfer to counterion*, is the unimolecular rearrangement of the ion-pair which results in termination of the growing chain and simultaneous regeneration of the catalyst-cocatalyst complex.

This can be represented in general terms by

$$BM_nM^+ \, (LI)^- \overset{k_{ts}}{\longrightarrow} M_{n+1} \;+\; B^+ \, (LI)^- \tag{8.113}$$

The kinetic chain is thus not terminated and the polymer molecule formed bears a terminal unsaturation. For example, for the system isobutylene/BF_3/H_2O, the reaction is represented (Odian, 1991; Billmeyer, Jr., 1994) by:

$$H\text{+}\text{–}CH_2C(CH_3)_2\text{–}_n\text{–}CH_2 \overset{+}{C}(CH_3)_2 \, (BF_3OH)^- \;\longrightarrow$$

$$H^+ \, (BF_3OH)^- \;+\; H\text{+}\text{–}CH_2C(CH_3)_2\text{–}_n\text{–}CH_2C(CH_3){=}CH_2 \tag{8.114}$$

8.5.3.3 Combination with Counterion

This process, unlike the above mentioned processes, terminates the kinetic chain. In general terms, the reaction can be represented by

$$BM_nM^+ \, (LI)^- \overset{k_t}{\longrightarrow} BM_nMLI \tag{8.115}$$

This type of termination occurs, for example, in the trifluoroacetic acid initiated polymerization of styrene (Throssell et al., 1956; Odian, 1991; Billmeyer, Jr., 1994):

$$(8.116)$$

The propagating ion may also combine with an anionic fragment from the counterion (Kennedy and Feinberg, 1978; Odian, 1991) as in $BX_3.OH_2$ (X = halogen) initiated polymerization or terminate by alkylation or hydridation (Kennedy, 1976; Reibel et al., 1979; Odian, 1991) when aluminium alkyl-alkyl halide initiating sytems are used.

Problem 8.17   Write equations to describe plausible termination reactions in cationic polymerization of isobutylene initiated by (a) $BF_3.OH_2$, (b) $BCl_3.OH_2$, (c) $Al(CH_3)_3$/t-butyl chloride, and (d) $Al(C_2H_5)_3$/t-butyl chloride.

Answer:

(a) For $BF_3.H_2O$ initiated polymerization, chain transfer to monomer is the major mode of chain termination with a minor contribution by combination with OH (Odian, 1991),

$$H\text{--}[CH_2C(CH_3)_2]_n\text{--}CH_2\overset{+}{C}(CH_3)_2\,(BF_3OH)^- \longrightarrow$$
$$H\text{--}[CH_2C(CH_3)_2]_n\text{--}CH_2C(CH_3)_2OH\,+\,BF_3 \qquad\qquad (P8.17.1)$$

(b) Termination in the $BCl_3.H_2O$ initiated polymerization of isobutylene (and styrene) occurs almost exclusively by combination with chloride (Kennedy and Feinberg, 1978; Odian, 1991),

$$H\text{--}[CH_2C(CH_3)_2\text{--}]_n\text{--}CH_2\overset{+}{C}(CH_3)_2\,(BCl_3OH)^- \longrightarrow$$
$$H\text{--}[CH_2C(CH_3)_2]_n\text{--}CH_2C(CH_3)_2Cl\,+\,BCl_2OH \qquad\qquad (P8.17.2)$$

The differences between the reactions (P8.17.1) and (P8.17.2) can be viewed in terms of the order of bond strengths: B–F > B–O > B–Cl. Thus, OH^- is transferred in preference to F^- and Cl^- is transferred in preference to OH^-.

(c) Besides chain transfer to monomer, termination may also occur by *alkylation* (Kennedy, 1976; Odian, 1991), that is, transfer of an alkyl anion to the propagating center:

$$\sim\!\!\sim\!\!CH_2\overset{+}{C}(CH_3)_2\,[(CH_3)_3AlCl]^- \longrightarrow$$
$$\sim\!\!\sim\!\!CH_2\text{--}C(CH_3)_3\,+\,(CH_3)_2AlCl \qquad\qquad (P8.17.3)$$

(d) As the alkyl aluminum in this case contains β-hydrogens, termination by *hydridation*, that is, transfer of a hydride ion from the alkyl anion to the propagating center, occurs in preference to alkylation (Kennedy, 1976; Reibel et al., 1979; Odian, 1991):

$$\sim\!\!\sim\!\!CH_2\overset{+}{C}(CH_3)_2\,[(CH_3CH_2)_3AlCl]^- \longrightarrow$$
$$\sim\!\!\sim\!\!CH_2CH(CH_3)_2\,+\,CH_2\!=\!CH_2\,+\,(CH_3CH_2)_2AlCl \qquad\qquad (P8.17.4)$$

8.5.3.4   *Transfer to Solvents/Reagents*

The termination of growth of individual chains can occur by chain transfer to solvent or impurity present in the system, or to a transfer agent (denoted here by TA or S) deliberately added to the system. In general terms, the reaction can be represented by

$$BM_nM^+(LI)^- + S(TA) \xrightarrow{k_{tr,S}} BM_nMA + T^+(LI)^- \tag{8.117}$$

involving the transfer of a negative fragment A^-. The term $k_{tr,S}$, as in Chapter 6, is used to denote the rate constant for chain transfer to solvent or any other transfer agent. Common examples of such transfer agents are water, alcohols, and acids. Termination by these compounds involves the transfer of HO, RO, or RCOO anion to the chain carbocation (Mathieson, 1963; Odian, 1991), e.g.,

$$\text{wwwCH}_2\overset{+}{\text{C}}(CH_3)_2(BF_3OH)^- + H_2O \longrightarrow$$
$$\text{wwwCH}_2-C(CH_3)_2OH + H^+(BF_3OH)^- \tag{8.118}$$

Water, alcohols, acids, and amines are thus used to quench a cationic polymerization. These are added in excess to terminate the polymerization when desired, usually after complete or maximum conversion of the monomer has been attained.

Problem 8.18   Explain how the following substances act as inhibitors or retarders in cationic polymerization: water, tertiary amines, trialkyl phosphines, and *p*-benzoquinone.

Answer:

In very small concentrations, water acts as a co-catalyst, initiating polymerization in combination with a catalyst (e.g., $SnCl_4$). However, in larger concentrations, it inactivates the catalyst (such as by hydrolysis of $SnCl_4$) or competes successfully with monomer for the catalyst-cocatalyst complex and inactivates the proton by forming hydronium ion [see Eqs. (P8.18.2) and (P8.18.4)] because the basicity of the carbon-carbon double bond is far less than that of water :

$$SnCl_4 + H_2O \rightleftharpoons SnCl_4.OH_2 \overset{H_2O}{\rightleftharpoons} (H_3O)^+(SnCl_4OH)^- \tag{P8.18.1}$$

Termination can also occur by proton transfer to water, e.g.,

$$\text{wwwCH}_2\overset{+}{\text{C}}(CH_3)_2(SnCl_4OH)^- + H_2O \longrightarrow \text{wwwCH}=C(CH_3)_2 + H_3O^+(SnCl_4OH)^- \tag{P8.18.2}$$

or by OH^- transfer to carbocation, e.g.,

$$\text{wwwCH}_2\overset{+}{\text{C}}(CH_3)_2(BF_3OH)^- + H_2O \longrightarrow$$
$$\text{wwwCH}_2C(CH_3)_2OH + H^+(BF_3OH)^- \tag{P8.18.3}$$

$$H^+(BF_3OH)^- + H_2O \longrightarrow (H_3O)^+(BF_3OH)^- \tag{P8.18.4}$$

In the presence of excess water, reaction (P8.18.4) takes place in preference to addition to monomer. So the polymerization rate decreases in the presence of excess water. Alcohols and acids also function similarly as inhibitors or retarders.

Tertiary amines and trialkyl phosphines react with propagating chains to form stable cations that are unable to continue propagation (Biswas and Kamannarayana, 1976; Odian, 1991) :

$$BM_nM^+(LI)^- + :NR_3 \longrightarrow BM_n M\overset{+}{N}R_3(LI)^- \tag{P8.18.5}$$

This causes inhibition or retardation.

p-Benzoquinone acts as an inhibitor by receiving proton from the carbocation and/or catalyst-cocatalyst complex (Odian, 1991):

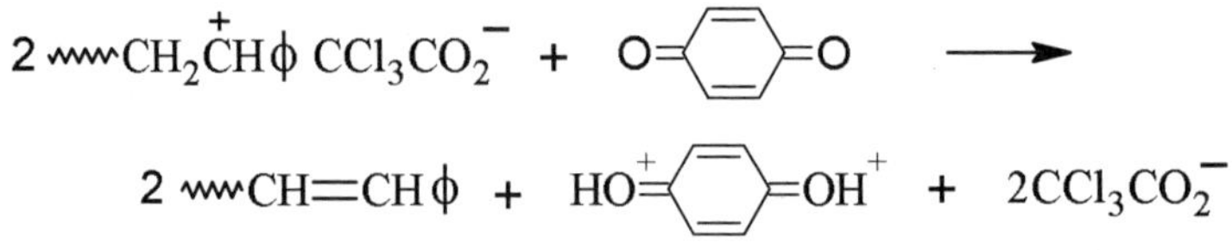

$$2 \text{ } \text{~~~CH}_2\overset{+}{\text{C}}\text{H}\phi \text{ } CCl_3CO_2^- + \text{ } O{=}\langle\overline{}\rangle{=}O \longrightarrow$$

$$2 \text{ } \text{~~~CH}{=}\text{CH}\phi + \text{ } HO\overset{+}{{=}}\langle\overline{}\rangle\overset{+}{{=}}OH + \text{ } 2CCl_3CO_2^-$$

8.5.3.5   Chain Transfer to Polymer

In polymerizations of 1-alkenes such as propylene, intermolecular hydride transfer to polymer can occur giving rise to short chain branches. Such transfer is explained by the fact that propagating carbocations being reactive secondary carbocations can abstract tertiary hydrogens from the polymer (Plesch, 1953; Odian, 1991). For example, reaction (8.119) produces a relatively stable tertiary carbocation from a more reactive propagating secondary carbocation.

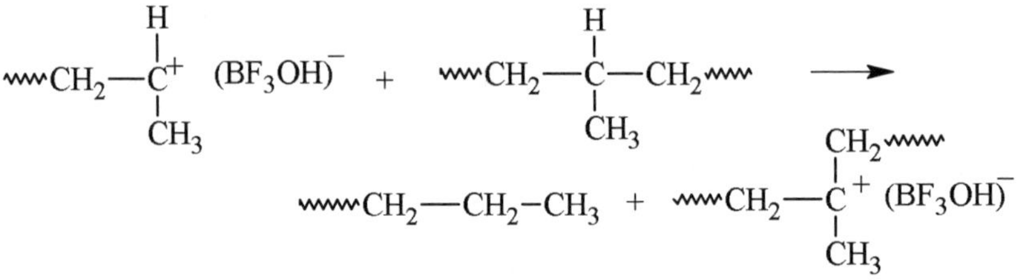

$$\text{~~~CH}_2{-}\underset{CH_3}{\overset{H}{\underset{|}{\overset{|}{C^+}}}} \text{ } (BF_3OH)^- \text{ } + \text{ } \text{~~~CH}_2{-}\underset{CH_3}{\overset{H}{\underset{|}{\overset{|}{C}}}}{-}CH_2\text{~~~} \longrightarrow$$

$$\text{~~~CH}_2{-}CH_2{-}CH_3 \text{ } + \text{ } \text{~~~CH}_2{-}\underset{CH_3}{\overset{CH_2\text{~~~}}{\underset{|}{\overset{|}{C^+}}}} (BF_3OH)^- \qquad (8.119)$$

The reaction contributes to the formation of low-molecular weight products from 1-alkenes.

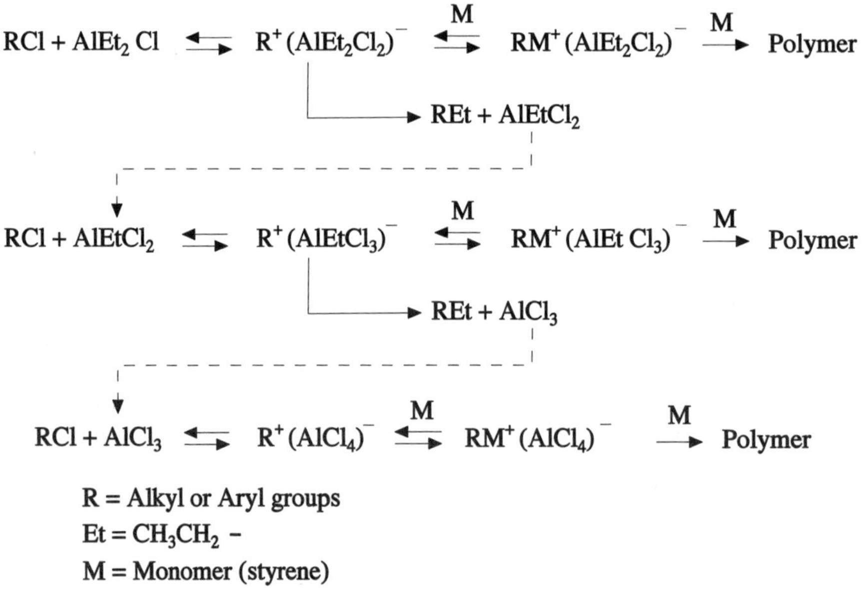

$$RCl + AlEt_2Cl \rightleftharpoons R^+(AlEt_2Cl_2)^- \overset{M}{\rightleftharpoons} RM^+(AlEt_2Cl_2)^- \overset{M}{\longrightarrow} \text{Polymer}$$

$$\longrightarrow REt + AlEtCl_2$$

$$RCl + AlEtCl_2 \rightleftharpoons R^+(AlEtCl_3)^- \overset{M}{\rightleftharpoons} RM^+(AlEtCl_3)^- \overset{M}{\longrightarrow} \text{Polymer}$$

$$\longrightarrow REt + AlCl_3$$

$$RCl + AlCl_3 \rightleftharpoons R^+(AlCl_4)^- \overset{M}{\rightleftharpoons} RM^+(AlCl_4)^- \overset{M}{\longrightarrow} \text{Polymer}$$

R = Alkyl or Aryl groups
Et = CH$_3$CH$_2$ –
M = Monomer (styrene)

Figure 8.6 Equilibria in initiation by aluminum halides and alkyl halides. (Kennedy, 1979; Rudin, 1982.)

8.5.4 Kinetics

Unlike free-radical and homogeneous anionic polymerizations, cationic polymerizations cannot be described by conventional kinetic schemes involving reactions like initiation, propagation, and so on. This is due to the complexity of the cationic initiation and the variable extents of ion-pair formation at the propagating chain end [cf. Eq. (8.1)]. The possibility of the reaction being heterogeneous due to the limited solubility of the initiator in the reaction medium further adds to the complexity of the kinetics.

8.5.4.1 Ions and Ion Pairs

While carbocations must always be accompanied by counterions to maintain electrical neutrality, the distance between the carbocation and counterion largely determines their reactivity. In the majority of cationic polymerization systems reported in the literature, the propagating species are probably associated ion pairs, though an accurate definition of the system ionicity in terms of various species in the solution is almost impossible. Figure 8.6, for example, illustrates the complicated equilibria (Kennedy, 1979) that may exist in initiation by aluminum alkyl halides or aluminum halide Lewis acid which are widely used as initiators (catalysts) in combination with alkyl halides as cocatalysts. Thus each carbocation can add a monomer to initiate polymerization or remove an alkyl (ethyl) group from the counterion producing a saturated hydrocarbon, REt, and a more acidic Lewis acid. The propagating cation can also terminate by the same process to produce ethyl-capped polymers and new Lewis acids. Thus, even if only diethylaluminum chloride is used as the initiator, there may be major contributions to the polymerization from ethylaluminum dichloride or aluminum chloride (Rudin, 1982).

Both the initiation and propagation processes are, moreover, influenced by the degree of association between the active center and its counterion giving rise to a spectrum of ionicities [see Eq. (8.1)]. As a minimum, it is necessary to consider the presence of solvent-separated ion pairs, and free solvated ions. Such a simplified scheme (Ledwith and Sherrington, 1976) is shown in Fig. 8.7. Contact (associated) ion pairs are not included in this scheme, which is justifiable because the concentrations of such species are negligible compared to those of solvated ion pairs in solvents of relatively high dielectric constants (9-15), commonly used in cationic polymerizations.

The observed k_p according to the above simplified scheme is composed of contributions from the ion pairs and free solvated ions (with respective rate constants $k_p^{\pm}$ and k_p^+), while calculations show an order of magnitude ratio of ion-pair and free-ion concentrations (see Problem 8.21). The k_p^+ values are generally at least 100 times as great as the corresponding $k_p^{\pm}$ figures for olefin monomers and, in general, both k_p^+ and $k_p^{\pm}$ *decrease* with increasing solvent polarity since more polar solvents tend to stabilize the initial state of monomer plus ion or ion pair. [In media of low polarity, like bulk monomer, the k_p^+ values for cationic olefin polymerizations are of the order of

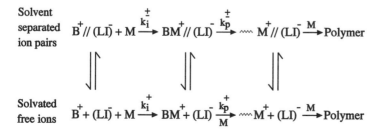

Figure 8.7 A simplified reaction scheme for initiation and propagation in cationic polymerization. (Ledwith and Sherrington, 1976; Rudin, 1982.)

10^6-10^9 L/mol-s, as compared to k_p^- values of the order of 10^3-10^5 L/mol-s for anionic polymerizations under more or less equivalent conditions. Carbocations are thus significantly more reactive than carbanions under similar conditions.]

8.5.4.2 Simplified Kinetic Scheme

It is clear from the above discussions that deriving a general kinetic scheme for cationic polymerizations is rather unrealistic. Nevertheless, we shall postulate here a conventional polymerization reaction scheme based upon the chemistry given in the earlier sections and show where the implied assumptions are not realistic enough. Using an ideal reaction scheme, we may depict a cationic polymerization by the following set of elementary reactions:

Pre-initiation
equilibrium: $L + IB \overset{K}{\rightleftharpoons} B^+ (LI)^-$ (8.120)

Initiation: $B^+ (LI)^- + M \overset{k_i}{\longrightarrow} BM^+ (LI)^-$ (8.121)

Propagation: $BM_nM^+ (LI)^- + M \overset{k_p}{\longrightarrow} BM_{n+1}M^+ (LI)^-$ (8.122)

Termination: $BM_nM^+ (LI)^- \overset{k_t}{\longrightarrow} BM_{n+1}LI$ (8.123)

Transfer: $BM_nM^+ (LI)^- + M \overset{k_{tr,M}}{\longrightarrow} M_{n+1} + BM^+ (LI)^-$ (8.124)

Spontaneous
termination: $BM_nM^+ (LI)^- \overset{k_{ts}}{\longrightarrow} M_{n+1} + B^+ (LI)^-$ (8.125)

Transfer to solvents
or reagents: $BM_nM^+ (LI)^- + S(TA) \overset{k_{tr,S}}{\longrightarrow} BM_nMA + T^+ (LI)^-$ (8.126)

where L, IB, M, S, and TA are as defined previously. The scheme is greatly oversimplified since we have ignored the presence of free solvated ions and ion pairs of various degrees of closeness occurring in equilibria. So all the rate constants appearing in the above scheme are actually composite quantities which will vary with the nature of the medium and the ion pair. In a more nonpolar medium, the values of the rate constants can also vary with the total concentration of reactive species because of the tendency of organic ion pairs to form aggregates having different reactivities than individual ion pairs. Ignoring these complexities in the present discussion, we can write the following expressions (Rudin, 1982) for the rates of initiation (R_i), propagation (R_p), termination by combination with counterion (R_t), transfer to monomer ($R_{tr,M}$), spontaneous termination (R_{ts}), and transfer to reagents or solvents ($R_{tr,S}$):

$$R_i = k_i(K[L][IB])[M] \tag{8.127}$$

$$R_p = k_p[M][M^+] \tag{8.128}$$

$$R_t = k_t[M^+] \tag{8.129}$$

$$R_{tr,M} = k_{tr,M}[M^+][M] \tag{8.130}$$

$$R_{ts} = k_{ts}[M^+] \tag{8.131}$$

$$R_{tr,S} = k_{tr,S}[M^+][S] \tag{8.132}$$

where $[M^+]$ represents $\sum_{n=1}^{\infty}[BM_{n-1}M^+ (LI)^-]$, that is, it is the total concentration of propagating chains of all sizes irrespective of their state of association with counterions.

To proceed further, we may now assume a steady state in the concentration of the active species $(d\,[\text{M}^+]/dt = 0)$. An assumption that this occurs due to $R_i = R_t$ leads to

$$[\text{M}^+] \;=\; \frac{R_i}{k_t} \;=\; \frac{Kk_i\,[\text{L}]\,[\text{IB}]\,[\text{M}]}{k_t} \tag{8.133}$$

$$\text{and} \quad R_p \;=\; \frac{Kk_ik_p\,[\text{L}]\,[\text{IB}]\,[\text{M}]^2}{k_t} \tag{8.134}$$

which show a first order dependence of R_p on R_i or initiator concentration, in contrast to radical polymerizations which show a one-half-order dependence of R_p on R_i. The difference arises from their basically different modes of termination. Thus, whereas termination is second order in the propagating species in radical polymerization, it is only first order in cationic polymerization.

Problem 8.19   Derive expressions for cationic polymerization rate for the following three cases : (a) an added chain transfer agent terminates the kinetic chain [Eq. (8.117)]; (b) the rate-determining step in the initiation process is the forward reaction in Eq. (8.105); and (c) the catalyst L or cocatalyst IB is present in large excess.

How would the order dependence of the polymerization rate change if either monomer or any of the components of the initiating system is involved in solvating the propagating species ?

Answer:

(a) Reaction (8.117) becomes the termination reaction when $\text{T}^+\,(\text{LI})^-$ is unable to initiate new chains (as in the case of a more active agent present in considerable amounts). In that case, at steady state,

$$R_i \;=\; R_t + R_{tr,\text{S}}$$

which yields

$$[\text{M}^+] \;=\; \frac{Kk_i\,[\text{L}]\,[\text{IB}]\,[\text{M}]}{k_t + k_{tr,\text{S}}\,[\text{S}]}$$

From Eq. (8.128),

$$R_p \;=\; \frac{Kk_ik_p\,[\text{L}]\,[\text{IB}]\,[\text{M}]^2}{k_t + k_{tr,\text{S}}\,[\text{S}]}$$

(b) Considering the forward equation in Eq. (8.105), $R_i \;=\; k_1\,[\text{L}]\,[\text{IB}]$. Applying steady-state approximation : $R_i \;=\; R_t$ leads to $[\text{M}^+] \;=\; k_1\,[\text{L}]\,[\text{IB}]/k_t$. Hence $R_p \;=\; k_1\,k_p\,[\text{L}]\,[\text{IB}]\,[\text{M}]/k_t$. Thus, R_p has one order dependence on $[\text{M}]$ instead of 2 order as in Eq. (8.134).

(c) If L is present in large excess, $R_i \;=\; k\,[\text{IB}]\,[\text{M}]$, where k is a composite constant. At steady state, $R_i = R_t$ and $[\text{M}^+] \;=\; k\,[\text{IB}]\,[\text{M}]/k_t$. So, $R_p \;=\; kk_p\,[\text{IB}]\,[\text{M}]^2/k_t$. Thus, R_p is zero order in L. Similarly, if IB is in excess, R_p will be zero order in IB.

If monomer is involved in solvating the propagating species, the R_p will show higher than 2 order in monomer. If L or IB is involved in solvating the propagating species, the propagation rate is then dependent on the concentration of L or IB, while initiation is first order in both L and IB [cf. Eq. (8.127)], thus resulting in higher than first order dependence for R_p on L or IB.

The expressions given above can be employed only if the steady-state conditions exist, at least over some part of the overall reaction. Steady-state is implied if R_p is constant with conversion. However, many, if not most, cationic polymerizations proceed so rapidly that the steady-state is not achieved and even in slower polymerizations, the steady-state may not be achieved if $R_i > R_t$.

The R_p data from nonsteady-state conditions can be used to obtain the value of k_p from Eq. (8.128) when $[M^+]$ is known. In polymerizations initiated by catalyst-cocatalyst systems, $[M^+]$ is taken to be the cocatalyst concentration $[IB]$ when the initiator (catalyst) is in excess or the initiator concentration $[L]$ when the cocatalyst is in excess. Such an assumption holds only if $R_i > R_p$ and the initiator (or the catalyst-cocatalyst system) is highly active (100% efficiency).

The literature contains many instances where the value of the apparent or overall rate constant k_p is assigned to $k_p^\pm$, neglecting the contributions of free ions. This can be erroneous since free ions with their high value of k_p^+ can have a significant effect on k_p even when present in small concentrations. It is also incorrect to write kinetic expressions in terms of only one type of propagating species (usually shown as ion pair) since both ion pairs and free ions are simultaneously present in most systems. The correct expressions for the rate of any step in polymerization should therefore include separate terms for the contributions of both types of propagating species.

Problem 8.20   The term k_p in Eq. (8.128) is only an *apparent* or *overall* propagation rate constant. Show how it could be related to the propagation rate constants k_p^+ and $k_p^\pm$ of the free ions and ion pairs. Suggest methods of obtaining individual k_p^+ and $k_p^\pm$ values.

Answer:

Taking into account the contributions of free ions and ion pairs the propagation rate may be written as

$$R_p = k_p^+ [M^+]_{\text{free}} [M] + k_p^\pm [M^+ (LI)^-][M] \tag{P8.20.1}$$

where $[M^+]_{\text{free}}$ and $[M^+ (LI)^-]$ are the concentrations of free ions and ion pairs, respectively.

From Eq. (8.128),

$$R_p = k_p [M^+][M] \tag{P8.20.2}$$

where $[M^+]$ is the total concentration of cationic ends (comprising free ions and ion pairs).

From Eqs. (P8.20.1) and (P8.20.2),

$$k_p = \frac{k_p^+ [M^+]_{\text{free}} + k_p^\pm [M^+ (LI)^-]}{[M^+]_{\text{free}} + [M^+ (LI)^-]} \tag{P8.20.3}$$

Individual k_p^+ and $k_p^\pm$ values can be obtained by experimental determination of individual concentrations of free ions and ion pairs by a combination of conductivity and short-stop experiments. While conductivity directly yields the concentration of free ions (that is, only free ions conduct), short-stop experiments yield the total concentration of ion-pairs and free ions. For mostly aromatic monomers the total concentration of ion pairs and free ions may also be obtained by UV-visible spectroscopy, assuming that ion pairs show the same UV-visible absorption as free ions since the ion pairs in cationic systems are loose ion pairs (due to large size of the negative counterions).

The overall rate constant k_p is related to k_p^+ and $k_p^\pm$ (see Problem 8.21) by

$$k_p = \alpha k_p^+ + (1 - \alpha)k_p^\pm \qquad (8.135)$$

$$\text{or} \quad k_p = k_p^\pm + k_p^+ K_d^{1/2}/[M^+]^{1/2} \qquad (8.136)$$

where α is the degree of dissociation, K_d the dissociation constant of the propagating ion pair, and $[M^+]$ is the total concentration of propagating cationic ends (both free ions and ion pairs). One can thus safely equate k_p with $k_p^\pm$ only for systems where $[M^+]/K_d = 10^3 - 10^4$ or larger (see Problem 8.21).

Problem 8.21 Consider styrene polymerization by triflic (trifluoroethanesulfonic) acid in 1,2-dichloroethane at 20°C where K_d is 4.2×10^{-7} mol/L (Kunitake and Takanabe, 1979). For experiments performed (using stopped-flow rapid scan spectroscopy) at a styrene concentration of 0.397 M and acid concentration of 4.7×10^{-3} M at 20°C, the maximum concentration of cationic ends (both free ions and ion pairs) was found to be 1.4×10^{-4} M, indicating that the initiator efficiency is 0.030. At 20°C, $k_p^+/k_p^\pm$ is reported to be 12 (Kunitake and Takanabe, 1979).

(a) What is the ratio of free ion and ion pair concentrations ? (b) What would be the relative contributions of free ions and ion pairs to the overall propagation rate ? (c) How much would be the error in assigning the overall rate constant k_p as $k_p^\pm$?

Answer:

(a) For ion-pair dissociation equilibrium,

$$\text{wwww}M^+ (LI)^- \; \overset{K_d}{\rightleftharpoons} \; \text{wwww}M^+ + (LI)^- \qquad (P8.21.1)$$

$$K_d = \frac{[M^+]_{\text{free}} [(LI)^-]}{[M^+]_{\text{pair}}} = \frac{\alpha^2 [M^+]}{(1 - \alpha)} \qquad (P8.21.2)$$

where α is the degree of dissociation of ion-pair and $[M^+]$ is the total concentration of cationic ends, i.e., $\left([M^+]_{\text{free}} + [M^+]_{\text{pair}}\right)$.

Thus,

$$\frac{[M^+]_{\text{free}}}{[M^+]_{\text{pair}}} = \frac{[M^+]_{\text{free}}}{[M^+] - [M^+]_{\text{free}}} = \frac{\alpha}{1 - \alpha} \qquad (P8.21.3)$$

Solving Eq. (P8.21.2) for α,

$$\alpha = (K_d/2[M^+])\left[(1 + 4[M^+]/K_d)^{1/2} - 1\right] \qquad (P8.21.4)$$

Substituting for α in Eq. (P8.21.3) from Eq. (P8.21.4) yields

$$\frac{[M^+]_{\text{free}}}{[M^+]_{\text{pair}}} = \frac{(1 + 4[M^+]/K_d)^{1/2} - 1}{1 + (2[M^+]/K_d) - (1 + 4[M^+]/K_d)^{1/2}} \qquad (P8.21.5)$$

This equation shows that the fraction of free ions decreases rapidly at higher values of $[M^+]/K_d$ (dimensionless). Thus free ions constitute approximately 99, 90, 62, 27, 9, 2, 1, and 0.3% of the propagating species at $[M^+]/K_d$ values of 0.01, 0.1, 1, 10, 10^2, 10^3, 10^4, and 10^5, respectively.

For the given experimental condition,

$$[M^+] = 0.030(4.7 \times 10^{-3} \text{ mol L}^{-1}) = 1.41 \times 10^{-4} \text{ mol L}^{-1}$$

$$[M^+]/K_d = (1.41 \times 10^{-4} \text{ mol L}^{-1})/(4.2 \times 10^{-7} \text{ mol L}^{-1}) = 336$$

From Eq. (P8.21.5), $[M^+]_{\text{free}}/[M^+]_{\text{pair}} = 0.056$; that is, free ions constitute 5.6% of the propagating species.

(b) To obtain a relation between k_p, k_p^+, and $k_p^{\pm}$, Eq. (P8.20.1) can be written as

$$k_p[M^+][M] = k_p^+ \alpha [M^+][M] + k_p^{\pm}(1 - \alpha)[M^+][M]$$

which yields

$$k_p = \alpha k_p^+ + (1 - \alpha)k_p^{\pm} \tag{P8.21.6}$$

If $\alpha \ll 1$, Eq. (P8.21.2) can be approximated to

$$K_d = \alpha^2 [M^+] \tag{P8.21.7}$$

Substitution of Eq. (P8.21.7) into Eq. (P8.21.6) gives, for $K_d/[M^+] \ll 1$,

$$k_p = k_p^{\pm} + k_p^+ K_d^{1/2}/[M^+]^{1/2} \tag{P8.21.8}$$

From Eq. (P8.21.7), for $[M^+]/K_d = 336$, α is approximately 0.054, while Eq. (P8.21.4) yields $\alpha = 0.053$. From Eq. (P8.21.6),

$$
\begin{aligned}
k_p/k_p^{\pm} &= \alpha(k_p^+/k_p^{\pm}) + (1 - \alpha) \\
&= (0.053)(12) + (1 - 0.053) = 0.64 + 0.95 \simeq 1.6
\end{aligned}
$$

Hence, the contribution of free ions to rate = $0.64/(0.64 + 0.95) = 0.40$ (or 40%)

(c) From (b), $k_p/k_p^{\pm} \simeq 1.6$

So there is a 60% error in the value of $k_p^{\pm}$ obtained by equating k_p with $k_p^{\pm}$.

Similarly if $[M^+]/K_d$ is less than 0.1, the system consists of more than 90% free ions and less than 10% less reactive ion pairs (*see* Problem 8.21) and hence one can safely equate k_p with k_p^+. Such conditions may occur in polymerizations initiated by stable carbocation salts such as hexachloroantimonate ($SbCl_6^-$) salts of triphenyl methyl [$(C_6H_5)_3C^+$] and cycloheptatrienyl ($C_7H_7^+$) carbocations. These systems are thus useful for determining free-ion propagation rate constants (Subira et al., 1988). However, since these cations are stable, they can be used only to initiate the polymerization of more reactive monomers like N-vinylcarbazole and alkyl vinyl ethers. The value of the dissociation constant, K_d, for the propagating ion pair can be obtained directly from conductivity measurements or, indirectly, from kinetic data. For the latter a useful method is to study the effect of a common ion salt (e.g., tetra-*n*-butyl ammonium triflate for the triflic acid initiated polymerization) on the kinetics of polymerization in a procedure analogous to that described for anionic polymerization (cf. Problems 8.9 and 8.10). There are relatively few systems where reasonably accurate data for both k_p^+ and $k_p^{\pm}$ are available. However, the difference in reactivity of free ions and ion pairs in cationic polymerization may not generally exceed one order of magnitude (Sauvet and Sigwalt, 1989).

The kinetic influence of water on cationic polymerization is complicated. In some systems, e.g., *i*-butene/TiCl$_4$ in dichloromethane (Biddulph et al., 1965), the initial rate of polymerization increases with concentration of water at low concentrations and becomes independent as this concentration increases. In other systems, the initial rate of polymerization may rise to a maximum and then decline with increasing concentrations of water. Such behavior has been observed in the SnCl$_4$/H$_2$O initiated polymerization of styrene in carbon tetrachloride (Colclough and Dainton, 1958). In view of the complex effects of moisture, as also those of free ions, ion pairs, aggregates, and solvation on the kinetics, the data reported by different workers should be viewed with caution.

Problem 8.22 The problem of reproducibility of rates and the discrepancy between results obtained by various authors is usually attributed to variations in impurity levels, the most important being the traces of water. To determine the minimum concentration levels at which water can still be effective, consider a polymerization where [M] = 0.1 mol/L, k_i = 3 L/mol-s, and the $\overline{DP}_n$ of the polymer formed is 10^3 (these values are reasonably close to a styrene polymerization) and assume that a rate of 1% conversion per 24 h can be determined with sufficient accuracy. Determine the concentration of water in the system containing a Lewis acid (L) catalyst in a relatively large concentration.

Answer:

The rate of polymerization that can be determined in the given case with sufficient accuracy is:

$$R_p = \frac{(0.1 \text{ mol L}^{-1})(0.01)}{(24 \text{ h})(3600 \text{ s h}^{-1})} = 1.16 \times 10^{-8} \text{ mol L}^{-1} \text{ s}^{-1}$$

$$R_p/R_i = \overline{DP}_n = 10^3$$

Therefore, $R_i = 1.16 \times 10^{-11} \text{ mol L}^{-1} \text{ s}^{-1}$

The equation describing ion generation in a L/H$_2$O system, i.e.,

$$\text{H}_2\text{O} + \text{L} \overset{k_i}{\rightleftharpoons} \text{H}^+ + \text{LOH}^-$$

is displaced to the right for a relatively large concentration of L. Therefore,

$$[\text{H}_2\text{O}] = [\text{H}^+]$$

$$R_i = k_i[\text{H}_2\text{O}][\text{M}] = (3 \text{ L mol}^{-1} \text{ s}^{-1})[\text{H}_2\text{O}](0.1 \text{ mol L}^{-1})$$

$$[\text{H}_2\text{O}] = \frac{(1.16 \times 10^{-11} \text{ mol L}^{-1} \text{ s}^{-1})}{(3 \text{ L mol}^{-1} \text{ s}^{-1})(0.1 \text{ mol L}^{-1})} = 3.8 \times 10^{-11} \text{ mol L}^{-1}$$

Note that this water concentration is far below the detection limit of any analytical method and may be present in the equipment, in spite of very careful drying and baking.

8.5.4.3 Degree of Polymerization

Since under practical polymerization conditions the polymer chain growth is more likely to be controlled by chain transfer to monomer [Eq. (8.130)] than by terminaton reactions [Eq. (8.129)], the average chain size is given by

$$\overline{DP}_n \;=\; R_p/R_{tr,M} \;=\; k_p/k_{tr,M} \tag{8.137}$$

When the stoppage of chain growth involves, in addition to chain transfer to monomer, other transfer and termination reactions [Eqs. (8.129), (8.131), and (8.132)] as well, the concentration of the propagating species remains unchanged, and the polymerization rate is still given by Eq. (8.134). However, the degree of polymerization is decreased by all these transfer and termination reactions, and is given by

$$\overline{DP}_n \;=\; \frac{R_p}{R_t + R_{ts} + R_{tr,M} + R_{tr,S}} \tag{8.138}$$

Substituting expressions for the rates given earlier [see Eqs. (8.128)-(8.132)], and rearranging one obtains

$$\frac{1}{\overline{DP}_n} \;=\; \frac{k_t}{k_p\,[M]} + \frac{k_{ts}}{k_p\,[M]} + C_M + C_S\,\frac{[S]}{[M]} \tag{8.139}$$

where C_M and C_S are the chain transfer constants for monomer and chain transfer agent defined by $k_{tr,M}/k_p$ and $k_{tr,S}/k_p$, respectively. Equation (8.139) is the 'cationic polymerization equivalent' (Odian, 1991) of the Mayo equation [Eq. (6.122)] for radical polymerization.

   Since Eq. (8.139) does not depend on either steady-state reaction conditions or a knowledge of $[M^+]$, it is more convenient to calculate the ratios of various rate constants from $\overline{DP}_n$ data than from R_p data. However, the use of $\overline{DP}_n$ data, like the use of R_p data, does require (if the Mayo equation is used) that one employs data at low conversions so that the monomer concentration does not change appreciably.

Problem 8.23   In a low temperature polymerization of isobutylene using $TiCl_4$ as catalyst and H_2O as co-catalyst the following results were obtained (Biddulph et al., 1965) at $-35°C$ showing the effect of monomer concentration on the average degree of polymerization :

$[C_4H_8]$ (mol/L)	0.667	0.333	0.278	0.145	0.059
$\overline{DP}_n$	6940	4130	2860	2350	1030

Using these data, evaluate the rate constant ratios k_{tr}/k_p and k_t/k_p.

Answer:

Since spontaneous termination (chain transfer to counterion) is never a dominant termination reaction compared to chain transfer to monomer, the second term in Eq. (8.139) can be neglected. Further, ignoring the presence of any chain transfer agent (S), Eq. (8.139) can be approximated to

$$\frac{1}{\overline{DP}_n} \;=\; \frac{k_t}{k_p\,[M]} + C_M$$

   Figure 8.8 shows a plot of $1/\overline{DP}_n$ vs. $1/[C_4H_8]$. The slope gives $k_t/k_p = 5.0\times10^{-5}$ mol L^{-1}. From the intercept, $C_M = k_{tr,M}/k_p = 1.1\times10^{-4}$.

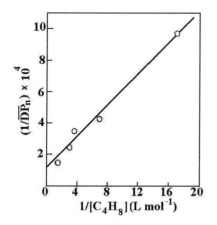

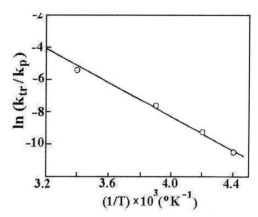

Figure 8.8 Plot of $1/\overline{DP}_n$ vs. $1/[M]$. (Data of Problem 8.23.)

Figure 8.9 Plot of $\ln(k_{tr}/k_p)$ vs. $1/T$. (Data of Problem 8.24.)

Problem 8.24 In studies similar to those in Problem 8.23 but performed over a range of temperatures, the following values were found (Biddulph et al., 1965) for the intercepts of plots of $1/(\overline{DP}_n)$ vs. $1/[C_4H_8]$:

$t(°C)$	+18	−14	−35	−48
$10^3/\overline{DP}_n$	4.37	0.50	0.098	0.027

Evaluate from these data the difference in activation energy between chain propagation and chain transfer to monomer.

Answer:

Expressing the rate constants by the Arrhenius expression

$$\frac{k_{tr}}{k_p} = \left(\frac{A_{tr}}{A_p}\right) e^{(E_p - E_{tr})/RT}$$

where E_p and E_t are the activation energies for propagation and transfer, respectively; A_p and A_{tr} are the respective pre-exponential factors.

Taking logarithms

$$\ln(k_{tr}/k_p) = \ln(A_{tr}/A_p) + \frac{E_p - E_{tr}}{RT}$$

Figure 8.9 shows a plot of $\ln(k_{tr}/k_p)$ *vs.* $1/T$. From the slope of the linear plot, $E_p - E_{tr} = (5.1 \times 10^3 \,°K)(1.987$ cal mol^{-1} $°K^{-1}) = 10.1 \times 10^3$ cal mol^{-1}.

8.5.5 Molecular Weight Distribution

The theoretical molecular weight distributions for cationic chain polymerizations (see Problem 8.25) are the same as those described in Chapter 6 for radical chain polymerizations terminating by disproportionation, i.e., where each propagating chain yields one dead polymer molecule. The polydispersity index (PDI $= \overline{DP}_w/\overline{DP}_n$) has a limit of 2. In many cationic polymerization systems, initiation is rapid, which narrows the molecular weight distribution (MDI) and, in addition, if termination and transfer reactions are very slow or nonexistent, a very narrow MDI with PDI close to unity would result.

Problem 8.25   Consider the cationic polymerization scheme given below:

Initiation: $M + I \xrightarrow{k_i} M_1^+$ (1)

Propagation: $M_x^+ + M \xrightarrow{k_p} M_{x+1}^+$ (2)

Transfer:
  spontaneous $M_x^+ \xrightarrow{k_{ts}} M_x + I$ (3)

  to monomer $M_x^+ + M \xrightarrow{k_{tr,M}} M_x + M_1^+$ (4)

Termination: $M_x^+ \xrightarrow{k_t} M_x$ (5)

Derive expressions for number and weight distributions of chain lengths of cationic polymers for the special case where the monomer concentration is kept constant (which is approximated to in experiments with very low initiator concentration).

Answer:

In the special case where the monomer concentration is kept constant, the rates of all processes are at all times proportional only to the concentration of active species. The *relative* probability, (P), of a growth step as expressed by

$$P = \frac{\text{Rate of growth step}}{\text{Sum of rates of all steps}}$$

$$= \frac{k_p[M][M^+]}{k_p[M][M^+] + k_{ts}[M^+] + k_{tr,M}[M][M^+] + k_t[M^+] + k_{tr,S}[S][M^+]}$$

$$= \frac{k_p[M]}{k_{ts} + k_t + (k_p + k_{tr,M} + k_{tr,S})[M]}$$

is constant for constant [M].

If N chains are started at a time and allowed to grow till all are stopped, then

$$\begin{aligned}
\text{number stopped at length 1 unit} &= N(1-P) \\
2 \text{ units} &= NP(1-P) \\
x \text{ units} &= NP^{x-1}(1-P)
\end{aligned}$$

The number fraction having length x is

$$n_x = N_x/N = P^{x-1}(1-P) \quad \text{[cf. Eq. (6.173)]}$$

giving the number average degree of polymerization:

$$\overline{DP}_n = \frac{\sum xN_x}{\sum N_x} = \frac{\sum xNP^{x-1}(1-P)}{N}$$

$$= (1-P)\sum xP^{x-1} = 1/(1-P)$$

Table 8.5 Representative Monomer Reactivity Ratios in Cationic Polymerization

Monomer 1	Monomer 2	Initiator[a]	Solvent[a]	Temp. (°C)	r_1	r_2
Isobutene[b]	1,3-Butadiene	AlEtCl$_2$	CH$_3$Cl	−100	43	0
		AlCl$_3$	CH$_3$Cl	−103	115	0
	Isoprene	AlCl$_3$	CH$_3$Cl	−103	2.5	0.4
	Cyclopentadiene	BF$_3$.OEt$_2$	ϕCH$_3$	−78	0.60	4.5
	Styrene	SnCl$_4$	EtCl	0	1.60	0.17
		AlCl$_3$	CH$_3$Cl	−92	9.02	1.99
	α-Methylstyrene	TiCl$_4$	ϕCH$_3$	−78	1.2	5.5
Styrene[c]	α-Methylstyrene	SnCl$_4$	EtCl	0	0.05	2.90
		BF$_3$.OEt$_2$	CH$_2$Cl$_2$	−20	0.2-0.5	10-14
	Isobutylene	TiCl$_4$	Toluene	−78	1.20	1.78
		TiCl$_4$	n-Hexane	−20	1.20	0.54
Styrene[d]	Isoprene	SnCl$_4$	EtCl	−30 to 0	0.8	0.1

[a]Et = C$_2$H$_5$, ϕ = phenyl; [b] Kennedy and Marechal, 1982; [c]Tsukamoto and Vogl, 1971; [d]Lipatova et al., 1955.

By direct analogy to low-conversion free-radical polymerization (Section 6.12.1), the weight fraction having length x is

$$w_x = xP^{x-1}(1 - P)^2$$

giving the weight-average degree of polymerization

$$\overline{DP}_w = \sum xw_x = (1 - P)^2 \sum x^2 P^{x-1} = \frac{1 + P}{1 - P}$$

Hence the ratio of weight to number average is

$$\overline{DP}_w/\overline{DP}_n = 1 + P \longrightarrow 2$$

for long chain polymers where $P \longrightarrow 1$.

The cumulative weight fraction of all species up to length x, which is the most convenient function for comparison with an experimental fractionation, will be

$$\sum_1^x w_x = (1 - P)^2 \sum_1^x xP^{x-1} = 1 - P^x(1 + x - xP)$$

Note: Normally, as polymerization progresses and the monomer concentration falls, the value of P will also fall. In such cases, the distribution cannot be formulated by the simple arguments used above. It may, however, be seen qualitatively that the distribution will broaden at higher conversions, its maximum shifting to lower chain lengths.

In polymerizations carried out to high conversions where the concentrations of propagating centers, monomer, and transfer agent as well as rate constants change, the polydispersity index increases considerably. Relatively broad molecular weight distributions are thus generally found in polymerizations by the cationic mechanism.

8.5.6 Cationic Copolymerization

Cationic copolymerization can be treated in identical manner to anionic copolymerization with the mechanistic scheme for propagation obtained simply by replacing the negative signs in Eqs. (8.92) to (8.95) by positive signs. The copolymerization equation derived for free-radical initiation [Eqs. (7.11) and (7.17)] may also be applied to cationic polymerization to determine ratios, several of which are listed in Table 8.5. The situation is complicated, however, due to counterion effects (Ham, 1964; Kennedy and Marechal, 1982). Thus, unlike in free radical copolymerization, different initiators used in cationic polymerization can cause variations in reactivity ratios. Moreover, solvent polarity may have an effect as it governs the degree of chain-end ion-pair dissociation. Another significant difference from free-radical copolymerization is that there is no apparent tendency for alternating copolymers to form in ionic copolymerization. Instead, block copolymers or homopolymer blends are more likely.

The most important of the commercial cationic copolymers is butyl rubber prepared from isobutene and isoprene. Because of its very low air permeability, butyl rubber finds extensive use in tire inner tubes and protective clothing. It is manufactured by low-temperature ($-100°C$) copolymerization of about 97% isobutene and 3% isoprene in chlorocarbon solvents with $AlCl_3$ as initiator (see Table 8.5). More recently, an ozone-resistant copolymer of isobutylene and cyclopentadiene has been marketed.

REFERENCES

Allcock, H. R. and Lampe, F. W., "Contemporary Polymer Chemistry", Prentice Hall, Englewood Cliffs, N.J., 1990.

Bhattacharyya, D. N., Lee, C. L., Smid, J., and Szwarc, M., *J. Phys. Chem.*, **69**, 612 (1965).

Biddulph, R. H., Plesch, P. H., and Rutherford, P. P., *J. Chem. Soc.*, 275 (1965).

Billmeyer, Jr., F. W., "Textbook of Polymer Science", John Wiley, New York, 1994.

Colclough, R. O. and Dainton, F. S., *Trans. Faraday Soc.*, **54**, 886, 894 (1958).

Cowie, J. M. G., "Polymers: Chemistry and Physics of Modern Materials", Blackie, Glasgow, 1991.

Dawan, F. and Smets, G., *Macromol. Chem.*, **59**, 163 (1963).

Fetters, L. J., *J. Polym. Sci. Part C*, **26**, 1 (1969).

Flory, P. J., *J. Am. Chem. Soc.*, **62**, 1561 (1940).

Fontanaille, M., "Carbanionic Polymerization: General Aspects and Initiation," pp 365-386 and "Carbanionic Polymerization: Termination and Functionalization," pp 425-432 in "Comprehensive Polymer Science" (Eastmond, G. C., Ledwith, A., Russo, S., and Sigwalt, P., eds.), vol. 3, Pergamon Press, London, 1989.

Ham, G. E. (ed.), "Copolymerization", Wiley-Interscience, New York, 1964.

Kennedy, J. P., *J. Polym. Sci. Symp.*, **56**, 1 (1976).

Kennedy, J. P., "Cationic Polymerization of Olefins", Wiley-Interscience, New York, 1979.

Kennedy, J. P. and Feinberg, S. C., *J. Polym. Sci., Polym. Chem. Ed.*, **16**, 2191 (1978).

Kennedy, J. P. and Johnston, J. E., *Adv. Polym. Sci.*, **19**, 57 (1975).

Kennedy, J. P. and Marechal, E., "Carbocationic Polymerization", Wiley-Interscience, New York, 1982.

Kennedy, J. P., Minckler, L. S., Wanless, G., and Thomas, R. M., *J. Polym. Sci.*, **A2**, 2093 (1964).

Kennedy, J. P. and Squires, R. G., *Polymer*, **6**, 579 (1965); *J. Macromol. Sci. Chem.*, **A1**, 861 (1967).

Kunitake, T. and Takanabe, K., *Macromolecules*, **12**(6), 1061, 1067 (1979).

Ledwith, A. and Sherrington, D. C., in "Comprehensive Chemical Kinetics" (Bamford, C. H. and Tipper, C. F. H., eds.), vol. 15, chap. 2, Elsevier, Amsterdam, 1976.

Lenz, R. W.,"Organic Chemistry of High Polymers", Wiley-Interscience, New York, 1967.

Lipatova, T. E., Gantmakher, A.R., and Medvedev, S.S., *Dokl. Akad. Nauk. SSSR*, **100**, 925 (1955).

Mathieson, A. R., "Styrene", chap. 6 in "The Chemistry of Cationic Polymerization" (Plesch, P. H., ed.), Macmillan, New York, 1963.

Morton, M.,"Anionic Polymerization: Principles and Practice", Academic Press, New York, 1983.

Odian, G., "Principles of Polymerization", John Wiley, New York, 1991.

Plesch, P. H., *J. Chem. Soc.*, 1653 (1953).

Reibel, L., Kennedy, J. P., and Chung, D. Y. L., *J. Polym. Sci. Polym. Chem. Ed.*, **17**, 2757 (1979).

Rudin, A., "The Elements of Polymer Science and Engineering", Academic Press, Orlando, 1982.

Sauvet, G. and Sigwalt, P., "Carbocationic Polymerization: General Aspects and Initiation," chap. 39 in "Comprehensive Polymer Science", vol. 3, (Eastmond, G. C., Ledwith, A., Russo, S., and Sigwalt, P., eds.), Pergamon Press, Oxford, 1989.

Subira, F., Vairon, J. P., and Sigwalt, P., *Macromolecules*, **21**, 2339 (1988).

Szwarc, M., "Carbanions, Living Polymers, and Electron Transfer Processes", Wiley-Interscience, New York, 1968.

Szwarc, M., "Ions and Ion Pairs in Organic Reactions", Wiley, New York, 1974.

Szwarc, M., Levy, M., and Milkovich, R., *J. Am. Chem. Soc.*, **78**, 2656 (1956).

Szwarc, M. and Smid, J., "The Kinetics of Propagation of Anionic Polymerization and Copolymerization", chap. 5 in "Progress in Reaction Kinetics", (Porter, G., ed.), vol. 2, Pergamon Press, Oxford, 1964.

Throssell, J. J., Sood, S. P., Szwarc, M., and Stannett, V., *J. Am. Chem. Soc.*, **78**, 1122 (1956).

Tsukamoto, A. and Vogl, O., *Progr. Polym. Sci.*, **3**, 199 (1971)

EXERCISES

8.1 Predict the order of reactivity (and justify your prediction) of the given monomers. (a) Styrene, 2-vinylpyridine, and 4 -vinyl pyridine in anionic polymerization. (b) Styrene, *p*-methoxystyrene, *p*-chlorostyrene, and *p*-methylstyrene in cationic polymerization.

8.2 Isobutylene undergoes cationic polymerization in the presence of strong Lewis acids like $AlCl_3$ but it is not polymerized by free radicals or anionic initiators. Acrylonitrile, on the other hand, is polymerized commercially by free radical means and can also be polymerized by anionic initiators like potassium amide but does not respond to cationic initiators. Account for the difference in behavior of isobutylene and acrylonitrile in terms of monomer structure.

8.3 Chlorinated aliphatic solvents are useful for cationic polymerization but cannot be used in anionic polymerization. Conversely, tetrahydrofuran which is a useful solvent in anionic polymerization cannot be used in cationic polymerization. Why ?

8.4 Polymerization of styrene by sodium naphthalene was performed at $35°C$ in dioxane. From the residual monomer and average degree of polymerization of the polymer formed in accurately measured reaction time the following data were obtained for two different monomer (M) - initiator (CA) compositions :

Composition	Time (s)	Conversion (%)	$\overline{DP}$
Run A :	290	69	1063
$[M]_0 = 0.31$ M			
$[CA]_0 = 4.0 \times 10^{-4}$ M			
Run B :	329	90	2079
$[M]_0 = 1.17$ M			
$[CA]_0 = 11.0 \times 10^{-4}$ M			

Calculate the overall rate constant and the total concentration of chain centers corresponding to the two initiator concentrations.

[*Ans.* (a) k_p = 10.1 L mol^{-1} s^{-1}, [M$^-$] = 4.0×10^{-4} M; (b) k_p = 7.0 L mol^{-1} s^{-1}, [M$^-$] = 1.0×10^{-3} M. Note that the overall or apparent rate constant depends on the initiator concentration.]

8.5   The apparent propagation rate constant for polymerization of styrene in THF at 25°C using sodium naphthalene as initiator is 550 L mol^{-1} s^{-1}. If the initial concentration of styrene is 156 g/L and that of sodium naphthalene is 0.03 g/L, calculate the initial rate of polymerization and, for complete conversion of the styrene, the number average molecular weight of the polymer formed. Comment upon the expected value of the polydispersity index ($\overline{M}_w/\overline{M}_n$).
[*Ans.* R_p = 0.165 mol L^{-1} s^{-1}; $\overline{M}_n$ = 1.56×10^6 g mol^{-1}]

8.6   Styrene was added to a solution of sodium naphthalene in tetrahydrofuran at 25°C so that the initial concentrations of styrene and sodium naphthalene in the reaction mixture were 0.2 mol/L and 0.001 mol/L, respectively. Five seconds after the addition of styrene its concentration was determined to be 1.73×10^{-3} mol/L. Calculate   (a) overall propagation rate constant;   (b) initial rate of polymerization;   (c) rate of polymerization after 10 s; and   (d) number average molecular weight of the polymer after 10 s.
[*Ans.* (a) k_p = 950 L mol^{-1} s^{-1};   (b) $(R_p)_0$ = 0.190 mol L^{-1} s^{-1};   (c) $(R_p)_{10s}$ = 1.422×10^{-5} mol L^{-1} s^{-1};   (d) $(\overline{M}_w)_{10s}$ = 20800]

8.7   How much sodium does one need to prepare 1 kg of polystyrene of molecular weight 300,000 by anionic polymerization? How much water is sufficient to completely prevent polymerization in this case?
[*Ans.* Sodium 0.153 g;   water 0.12 g]

8.8   One milliliter initiator solution in scrupulously clean and dry tetrahydrofuran containing 8.6×10^{-4} mol of sodium and 9.4×10^{-4} mol of naphthalene was injected into a solution of styrene (0.048 mol) in 50 mL clean and dry tetrahydrofuran at −70°C. After a few minutes the reaction was complete. A few milliliters methanol was then added to quench the reaction and the reaction mixture was allowed to warm to room temperature. The polymer was recovered by precipitation. What is the $\overline{M}_n$ of the polystyrene formed in the absence of side reactions? What should the $\overline{M}_w$ of the product be if the polymerization were carried out so that the growth of all macromolecules was started and ended simultaneously?
[*Ans.* $\overline{M}_n$ = 11,300; $\overline{M}_w$ = 11,400]

8.9   Polymerization of styrene by sodium naphthalene was performed at 20°C in 3-methyl tetrahydrofuran (3-Me-THF) using a flow apparatus and technique suitable for fast polymerizations. The conversion was determined gravimetrically after precipitation of the polymer with methanol-water or optically by the determination of the residual monomer and the concentration [M$^-$] of the active centers was calculated from the degree of polymerization at a given conversion (see Problem 8.8). In two independent experimental series, the overall rate constant k_p was obtained [cf. Eq. (P8.8.1)] both at different concentrations of initiator (and hence [M$^-$]) without addition of electrolyte and at different concentrations of sodium ions (from added salt, sodium tetraphenyl borate (NaBPh$_4$)) and constant concentration of initiator. From the resulting data given below determine $k_p^{\mp}$, k_p^-, and K for polystyryl sodium in 3-Me-THF at 20°C.

Without electrolyte		With electrolyte	
[M$^-$]×10^3	k_p	[Na$^+$]×10^5	k_p
(mol L^{-1})	(L mol^{-1} s^{-1})	(mol L^{-1})^a	(L mol^{-1} s^{-1})
5.10	1.83	2.78	93
2.20	255	1.69	97
0.73	381	0.75	135
0.40	479	0.45	188
0.26	600	0.37	215

^{a}Calculated from the dissociation constant of NaBPh$_4$ in 3-Me-THF determined by conductivity measurements (1.12×10^{-5} L/mol at 20°C).
[*Ans.* $k_p^{\mp}$ = 66 L mol^{-1} s^{-1}; k_p^- = 1.10×10^5 L mol^{-1} s^{-1}; K = 6×10^{-9} mol L^{-1}]

8.10 (a) Calculate the concentrations of polymeric free ions and ion pairs in the absence of electrolyte taking the polymerization system of Exercise 8.9 in which $[M^-]$ is 0.40×10^{-3} mol/L and no electrolyte has been added. (b) What would be the concentrations of free ions and ion pairs if the polymerization system contained $NaBPh_4$ in concentration equal to the total concentration of living ends? (c) What percentage of free Na^+ ions comes from the initiator in (b)?
[*Ans.* (a) 1.55×10^{-6} mol L^{-1} and 0.40×10^{-3} mol L^{-1}; (b) 1.95×10^{-8} mol L^{-1} and 0.40×10^{-3} mol L^{-1}; (c) 0.016%]

8.11 A polystyrene sample of $\overline{DP}_n = 60$ was made by living polymerization in tetrahydrofuran solvent using butyllithium as the initiator. Show graphically the number fraction distribution of the degree of polymerization for this polymer and, for comparison, similar distribution for a polystyrene of the same average DP produced by free radical polymerization (assuming termination solely by coupling).

8.12 Write reactions to show the type of end group that will result from the use of each of the following reagents to terminate butyllithium initiated living polymers of styrene in THF: (a) Ethyl benzoate; (b) benzyl chloride; (c) bromoaniline; (d) phthalic anhydride; (e) phosgene; (f) carbon disulfide.

8.13 Write reactions illustrating the synthesis of poly(methyl methacrylate)-*block*-polyac-rylonitrile by anionic polymerization. Show how you could synthesize an ABA block copolymer of the two monomers.

8.14 Show by equations the synthesis of the following types of block copolymers (a) ABA and (b) CABAC where A, B, and C represent styrene, butadiene, and isoprene.

8.15 Explain why acrylates (or vinyl acetate) cannot be poymerized by cationic mechanism.

8.16 In the polymerization of oxetane initiated by BF_3 (0.1 M) and H_2O, a simple rate law is found to be obeyed in which the polymerization rate is first order in monomer, initiator, and the total concentration of added and adventitious water. Suggest a kinetic mechanism to explain this behavior.

8.17 If trifluoroacetic acid (mixed with ethylbenzene) is added dropwise to styrene, no polymerization occurs. However, if styrene is added to the acid, high molecular weight polymer forms rapidly [J. J. Throssell, S. P. Sood, M. Szwarc, and V. Stannett, *J. Am. Chem. Soc.*, **78**, 1122 (1956)]. Suggest an explanation.

8.18 Account for the fact that 1-butene can be used to control polymer molecular weight in cationic polymerization of isobutylene.

8.19 Account for the fact that polymerization of propylene by the cationic process yields oligomers that have extremely complicated structure with methyl, ethyl, *n*- and *i*-propyl, and other groups. Why only oligomers are formed by this polymerization?

8.20 Write equations to show the different structural units that may result from intramolecular hydride and methide shifts (involving the end monomeric unit) in cationic polymerization of 4-methyl-1-pentene.

8.21 The cationic polymerization of 3-methyl-1-butene produced a polymer whose NMR spectra consisted of only two singlets. Propose a structure for the polymer consistent with the NMR and suggest a mechanism for this polymerization.

8.22 Under certain conditions, the rate of polymerization of isobutylene using $SnCl_4/H_2O$ initiating system is found to be first-order in $SnCl_4$, first order in water, and second order in isobutylene. The polymer formed initially has the number average molecular weight of 20,000 and contains 3.0×10^{-5} mol OH groups per gram but no chlorine. Suggest plausible reaction schemes to fit these data and derive appropriate expressions for the rate and degree of polymerization. Indicate any assumptions made in the derivations.

8.23 Under what reaction conditions might a cationic polymerization with $SnCl_4/H_2O$ initiating system show a dependence of the polymerization rate which is (a) first order in monomer, (b) zero order in $SnCl_4$ or water, and (c) second order in $SnCl_4$ or water?

8.24 Discuss the general effects of temperature, solvent, and catalyst on the monomer reactivity ratios in ionic copolymerization. How do these compare with the corresponding effects in radical copolymerization?

8.25 What experimental approaches could be used for determining whether the polymerization of a particular monomer by ionizing radiation proceeds by a radical or ionic mechanism ?

8.26 Account for the following facts observed in carbocationic polymerization :

(a) Isobutylene polymerization proceeds rapidly in CH_3Cl but not in CH_3I diluent.

(b) Olefins cannot be polymerized in diethyl ether and the polymerization of vinyl ethers by $BF_3.OEt_2$ proceeds faster in hexane than in Et_2O.

(c) Propylene can be polymerized by $HBr/AlBr_3$ to a very broad molecular-weight-distribution product.

(d) Whereas the initiation rate is enhanced by increasing the solvent polarity, the propagation rate may decrease.

(e) Bulk polymerization of unsaturated hydrocarbons initiated by γ-radiation often shows very high rate constants (10^6 to 10^8 L mol^{-1} s^{-1}).

8.27 Give plausible explanations for the following facts:

(a) Molecular weight does not depend on initiator concentration in ionic polymerization as it does in free radical polymerization.

(b) Polymerization rate is more sensitive to solvent effects in ionic polymerization than in free-radical polymerization.

(c) In cationic polymerization, $\overline{DP} = \bar{\nu}$, but this is not always the case in free radical or anionic polymerization.

(d) Ethyl vinyl ether undergoes cationic polymerization faster than β-chloroethyl vinyl ether under the same conditions.

8.28 The cationic polymerization of styrene initiated by CF_3SO_3H was investigated [T. Kunitake and K. Takanabe, *Macromolecules*, **12**(6), 1067 (1979)] in 1,2-dichloroeth-ane at 20°C by the stopped-flow/rapid scan sectroscopy which was combined with the rapid quenching technique. The reaction was followed by monitoring absorbance at 340 nm due to styryl cation with a rapid-scan spectropho-tometer. The absorbance change was extrapolated to the zero absorbance in order to determine the start time. The residual monomer was determined in the quenched mixture by UV spectroscopy. Representa-tive data of the time course of the formation of styryl cation and conversion measured with styrene and CF_3SO_3H initial concentrations of 0.305 M and 5.4 mM, respectively, are given below:

Reaction time (millisecond)	Conversion (%)	$[M^+] \times 10^4$ mol/L	$[M]$
5	24.9	1.87	0.229
8	31.7	1.97	0.208
10	33.2	1.98	0.204
13	39.3	1.99	0.185
15	40.5	1.98	0.181
20	45.7	1.91	0.166
30	54.1	1.56	0.140
40	60.3	1.27	0.121
50	65.5	1.02	0.105

The initiation rate was found to be proportional to the initiator concentration and roughly first order with respect to the monomer concentration, although the amount of the styryl cation formed was only 1-4% of the total acid concentration.

Analyze the given data to obtain the rate constants of the elementary steps of the polymerization system.

[*Ans.* $k_p^o = 1.0 \times 10^5$ M^{-1} s^{-1}; $k_i = 38$ M^{-1} s^{-1}; $k_t' = k_{ts} + k_t = 0.21 \times 10^3$ s^{-1}]

8.29 In studies of the low temperature polymerization of isobutylene using $TiCl_4$ as coinitiator [R. H. Bid-dulph, P. H. Plesch, and P. P. Rutherford, *J. Chem. Soc.*, 275 (1965)], the following results were obtained at −35°C for the effect of monomer concentration on the average degree of polymerization :

$[C_4H_8]$ (mol/L)	0.667	0.333	0.278	0.145	0.059
$\overline{DP}_n$	6940	4130	2860	2350	1030

From these data, evaluate the rate constant ratios k_{tr}/k_p and k_t/k_p.

[*Ans.* $C_M = k_{tr,M}/k_p = 1.1 \times 10^{-4}$; $k_t/k_p = 5.0 \times 10^{-5}$ mol L^{-1}]

Chapter 9

Coordination Addition Polymerization

9.1 Introduction

The field of coordination polymerization originated in the mid-1950s with the pioneering works of Karl Ziegler in Germany and Giulio Natta in Italy. While Ziegler discovered in the early 1950s that a combination of aluminum alkyls with certain transition metal compounds such as $TiCl_4$ or VCl_4 generated complexes that would polymerize ethylene at low temperatures and pressures producing polyethylene with an essentially linear structure, now referred to as high-density polyethylene (HDPE), Natta's work led to the recognition that the catalytic complexes described by Ziegler were capable of polymerizing 1-alkenes (commonly known as *alpha olefins* in the chemical industry) to yield stereoregular polymers. The range of this type of catalysts, referred to as *Ziegler-Natta catalysts*, was subsequently extended to produce polymers exhibiting a wide range of stereoregular structures including those derived from dienes and cycloalkenes. Many polymers are now manufactured on a commercial scale using Ziegler-Natta catalysts, the most prominent among them being stereoregular (isotactic) polypropylene of high molecular weight. (Recall that propylene cannot be polymerized by either free radical or ionic initiators.) Ziegler and Natta received the 1963 Nobel prize in chemistry jointly in recognition of the scientific and practical significance of their work.

Since the Ziegler-Natta catalyst systems appear to function via formation of a coordination complex between the catalyst, growing chain, and incoming monomer, the process is also referred to as *coordination addition polymerization* and the catalysts as *coordination catalysts*. Other types of complex catalysts that have received attention for stereospecific polymerization are the reduced metal oxides and the *alfin catalysts* (prepared from compounds of sodium). All three types are mainly used in heterogeneous polymerization, although some homogeneous processes are also used. Metallocene-based Ziegler-Natta catalysts appeared on the scene in the early 1980s, following the pioneering studies of Brintzinger (Wild et al., 1982), Ewen (1984), Kaminsky (Kaminsky et al., 1985), and coworkers. Compared to conventional heterogeneous Ziegler-Natta systems in which a variety of active centers with different structures and activities usually coexist, homogeneous metallocene-based catalysts (see Section 9.7) have very uniform catalytically active sites which possess controlled, well-defined ligand environments, thus allowing much greater control on the polymer formed (Tait, 1988).

Table 9.1 Stereoregularity of Polypropylene Produced with Different Catalyst Systems[a]

Catalyst system[b]	Stereoregularity (%)
$(C_2H_5)_3Al + TiCl_4$	48
$(C_2H_5)_3Al + TiCl_3$ (α, γ or δ)	80-92
$(C_2H_5)_3Al + TiCl_3$ (β)	40-50
$(i\text{-}C_4H_9)_3Al + TiCl_4$	30
$(C_6H_5)_3Al + TiCl_3$ (α)	65
$(C_2H_5)_3Al + ZrCl_4$	55
$(C_2H_5)_3Al + VCl_3$	73
$(C_2H_5)_3Al + VCl_4$	48
$(C_2H_5)_3Al + CrCl_3$	36
$R_2AlX + TiCl_3$	90-99
$RAlX_2 + \gamma\text{-}TiCl_3 + $ Amine	>99

[a]Data from Jordan (1967).   [b]R = alkyl, X = halogen

9.2   Ziegler-Natta Catalysts

9.2.1   Catalyst Composition

A Ziegler-Natta catalyst may be simply defined as a combination of two components: (1) a transition metal compound of an element from Groups IVB to VIIIB, and (2) an organometallic compound of a metal from Groups I to III of the periodic table. The transition metal compound is referred to as the *catalyst* and the organometallic compound as the *cocatalyst*. A combination of the catalyst and the cocatalyst is often referred to simply as the catalyst.

The catalyst component usually consists of halides or oxyhalides of titanium, vanadium, chromium, molybdenum, or zirconium, and the cocatalyst component often consists of an alkyl, aryl, or hydride of metals such as aluminum, lithium, zinc, tin, cadmium, beryllium, and magnesium. The catalyst systems may be *heterogeneous* (some titanium-based systems) or *soluble* (most vanadium-containing species). The best known systems are probably those derived from $TiCl_4$ or $TiCl_3$ and an aluminum trialkyl.

The catalysts which are useful for the preparation of isotactic polymers are heterogeneous, i.e., they are insoluble in the solvent, or diluent, in which they are prepared. They are prepared by mixing the components in a dry, inert solvent in the absence of oxygen, usually at a low temperature. The nature of the insoluble Ziegler-Natta catalysts is not well understood and the trial-and-error method is frequently used in developing new catalysts.

Making changes in the catalyst system and mixing with additives affect the rate of polymerization as well as yield, molecular weight, and the degree of stereoregularity of the polymer. Table 9.1 shows how some of the structural variables of Ziegler-Natta catalysts affect the stereoregularity of polypropylene.

9.2.2   Nature of the Catalyst

The nature of the Ziegler-Natta catalyst systems is not precisely known. In the case of insoluble catalyst systems, it is, however, certain that the true catalysts are *not* simple coordination adducts of aluminum alkyl and the metal halide, but complex reactions occur during the initial period of "aging", often needed to achieve high activity. The reactions probably involve exchange of substituent groups (Allcock and Lampe, 1990; Odian, 1991) such as those shown in Eqs. (9.1)-(9.3):

$$AlR_3 + TiCl_4 \rightleftharpoons R_2AlCl + RTiCl_3 \qquad (9.1)$$

$$R_2AlCl + TiCl_4 \rightleftharpoons RAlCl_2 + RTiCl_3 \qquad (9.2)$$

$$AlR_3 + RTiCl_3 \rightleftharpoons R_2AlCl + R_2TiCl_2 \qquad (9.3)$$

leading to the formation of transition metal-carbon (Ti–C) bonds. The organotitanium halides thus formed are unstable and can undergo reductive decomposition processes, such as shown in Eqs. (9.4) and (9.5):

$$RTiCl_3 \longrightarrow R^\bullet + TiCl_3 \qquad (9.4)$$

$$R_2TiCl_2 \longrightarrow R^\bullet + RTiCl_2 \qquad (9.5)$$

(Note that $TiCl_3$ itself can be used as an initial catalyst component in place of $TiCl_4$.) Further reduction may also occur yielding $TiCl_2$:

$$RTiCl_2 \longrightarrow TiCl_2 + R^\bullet \qquad (9.6)$$

$$RTiCl_3 \longrightarrow TiCl_2 + RCl \qquad (9.7)$$

In addition, $TiCl_3$ may be formed by the reaction:

$$TiCl_4 + TiCl_2 \rightleftharpoons 2TiCl_3 \qquad (9.8)$$

Some of the steps in the above sequence of reactions are reduction steps in which the transition metal is reduced to a low valency state possessing unfilled ligand sites. The reduction steps are very important as the low-valency transition metal species are believed to be the real catalysts or precursors of real catalysts. For heterogeneous catalysts, the reactions are, in fact, more complicated than those shown above. Radicals formed in these reactions may be removed by different processes such as combination, disproportionation, or reaction with solvent. Unlike heterogeneous catalysts, the soluble catalysts appear to have well defined structures. For example, the soluble catalyst system that is obtained by the reaction of triethyl aluminum and bis(cyclopentadienyl)titanium dichloride is known by elemental and x-ray analysis to have a halogen-bridged structure (**I**):

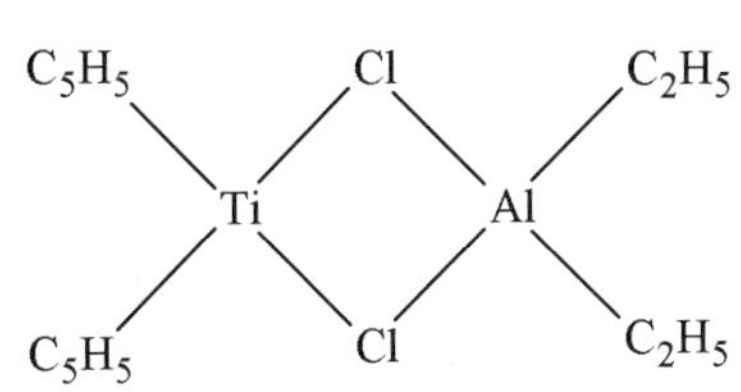

(I)

9.2.3 Evolution of the Titanium-Aluminum System

The efficiency or activity of the early Ziegler-Natta catalyst systems was low. The term *activity* usually refers to the rate of polymerization, expressed in terms of kilograms of polymer formed per gram of catalyst. Thus a low activity meant that large amounts of catalyst were needed to obtain reasonably high yields of polymer, and the spent catalyst had then to be removed from the product to avoid contamination. This problem effectively disappeared with the advent of subsequent generations of catalysts leading to large increases in activity without loss of stereospecificity. This was achieved by increasing the effective surface area of the active component by more than two orders of magnitude through impregnation of the catalyst on a solid support such as $MgCl_2$ or MgO. For example, in contrast to a typical $TiCl_3-AlR_3$ catalyst which yields about 50 to 200 g of polyethylene per gram of catalyst per hour per atmosphere of ethylene, as much as 200,000 g of polyethylene and over 40,000 g of polypropylene per gram titanium per hour may be produced using a $MgCl_2$-supported catalyst, thus obviating the need to remove the spent catalyst (a costly step) from the product. Such catalyst systems are often referred to as *high-mileage catalysts.*

Stereospecificity of the catalyst is kept high (> 90-98% isotactic dyads) by adding electron-donor additives such as ethyl benzoate. Typically, a superactive high-mileage catalyst is made by grinding and mixing magnesium chloride (or the alkoxide) and $TiCl_4$ in a ball mill, followed by the addition of $Al(C_2H_5)_3$ with an organic Lewis base (Hu and Chien, 1988).

9.3   Mechanism of Ziegler-Natta Polymerization

It is generally agreed that the heterogeneous Ziegler-Natta polymerization occurs at localized sites on the catalyst surface, which are activated by the organometallic component by alkylation of a transition metal atom at the surface. Of the various mechanisms that have been proposed, the two that are most generally accepted are the so-called *monometallic* and *bimetallic mechanisms* (Boor, Jr., 1979; Jordan, 1967), the former being favored in heterogeneous processes. In both processes, it is postulated that the monomer is incorporated into a polymer by insertion between a transition metal atom and the terminal carbon of the coordinated polymer chain. These two mechanisms are separately discussed in later sections.

In polymerization with Ziegler-Natta catalysts, a radical mechanism is not acceptable as it cannot explain the stereoregularity and formation of isotactic polymers. An ionic mechanism is, of necessity, widely favored and accepted. Coordination catalysts can thus be visualized as performing two functions (Odian, 1991). First, they provide the species that initiates the polymerization. Second, a fragment of the catalyst, acting as 'gegenion' or counterion of the propagating species, coordinates with both the propagating chain end and the incoming monomer in order to orient the monomer with respect to the growing chain end for a stereospecific addition (see Fig. 9.1). In short, the mechanism can thus be described as a *concerted multi-centered reaction.*

9.3.1   Mechanism of Stereospecific Placement

The prerequisite for a constant specific orientation of the monomer is a fixed location of both the cation and the anion. If either of them is soluble or dispersible, poor stereoregularity or atactic poly(α-olefin) structures would result. Many different mechanisms have been proposed to explain the usual isotactic placement produced by coordination initiators (Tait and Watkins, 1989). Figure 9.1 presents a simple mechanism for an *anionic coordination polymerization* with isotactic placement. According to this mechanism, the polymer chain end bears a partial negative charge, while the catalyst fragment **G** (gegen-ion or counterion) has a partial positive charge. (A *cationic coordination polymerization* mechanism would be similar but the signs of the partial charges would be reversed.) The stereospecific polymerization of α-olefins and other nonpolar alkenes proceeds by π-complexation between monomer and the transition metal in **G**, followed by insertion of a monomer molecule into the metal-carbon bond by a four-center anionic coordination insertion process (Odian, 1991), permitting head-to-tail addition and hair-like growth on the anionic site.

The chain propagation via the anionic coordination mechanism, as shown above, involves insertion of an oriented monomer molecule. In more detail, since the catalyst fragment **G** (containing transition metal) is coordinated with both the propagating chain end and the incoming monomer

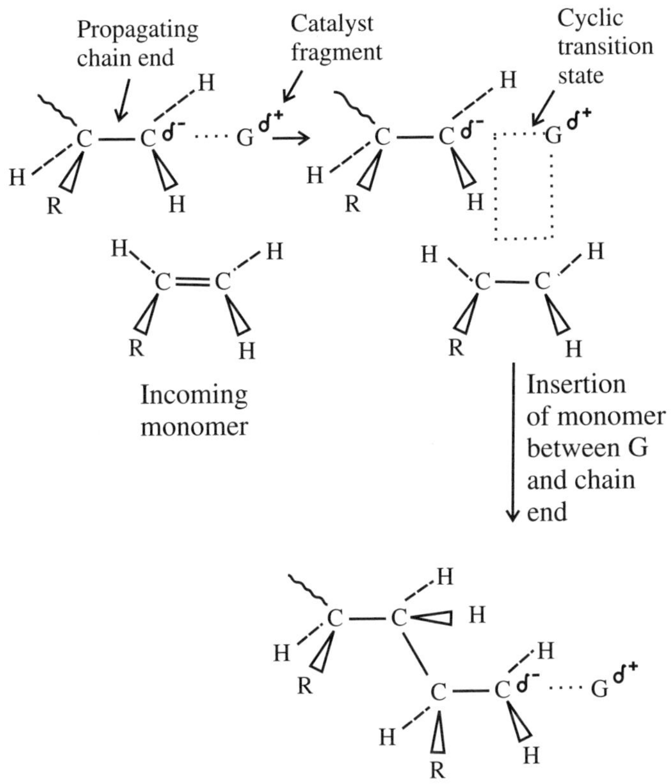

Figure 9.1 Mechanism for anionic coordination polymerization with isotactic placement. (After Odian, 1991.)

molecule, the latter becomes oriented as it is added to the polymer chain. The four-center cyclic transition state results from the formation of bonds between the propagating center and the incoming monomer unit and between the latter and the catalyst fragment **G**. This change is supposedly brought about by the nucleophilic attack of the carbanion polymer chain end on the α-carbon of the alkene monomer together with an electrophiic attack of cationic counterion ($\mathbf{G}^+$) on the alkene π-electrons. The catalyst fragment acts essentially as a 'template' or 'mold' for the orientation and isotactic placement (catalyst site control) of incoming successive monomer units (Odian, 1991).

9.3.2 Bimetallic and Monometallic Mechanisms

The catalyst complex of the $TiCl_3$ / AlR_3 system essentially acts as a template for the successive orientation and isotactic placement of the incoming monomer units. Though a number of structures have been proposed for the active species, they fall into either of two general categories: *monometallic* and *bimetallic*, depending on the number of metal centers (Patat and Sinn, 1958; Natta, 1960; Arlman and Cossee, 1964). The two types can be illustrated by the structures (**II**) and (**III**) for the active species from titanium chloride ($TiCl_4$ or $TiCl_3$) and alkylaluminum (AlR_3 or AlR_2Cl).

(II)   **(III)**

Structure (**II**), representing a bimetallic species, is the coordination complex that arises from the interaction of the original catalyst components (titanium and aluminum compounds) with exchange of R and Cl groups [cf. Eqs. (9.1)-(9.8)]. The placing of R and Cl groups in parentheses signifies that the exact specification of the ligands on Ti and Al cannot be made. Structure (**III**), representing a typical monometallic species, constitutes an active titanium site at the surface of a $TiCl_3$ crystal. Besides the four chloride ligands that the central Ti atom shares with its neighboring Ti atoms, it has an alkyl ligand (received through the aforesaid exchange reactions) and a vacant orbital (□).

9.3.2.1   Bimetallic Mechanism

The truly active bimetallic catalysts are complexes that have an electron-deficient bond, e.g., Ti···C···Al in (**II**). Chain propagation by the bimetallic mechanism (Patat and Sinn, 1958) occurs at two metal centers of the bridge complex as shown in Fig. 9.2(a). The mechanism is similar to that shown in Fig. 9.1 except that the identity of the catalyst fragment **G** is shown explicitly in Fig. 9.2(a). As shown in Fig. 9.2(b), the ion of the transition metal forms a π-complex with the nucleophilic olefin, which is then incorporated, following a partial delocalization of the alkyl bridge, into a six-membered ring transition state [Fig. 9.2 (c)]. The monomer then enters into the growing chain between the Al and the C, thereby regenerating the complex [Fig. 9.2(d)]. While the chain growth takes place always from the metal end, the incoming monomer is oriented, for steric reasons, with the $=CH_2$ group pointing into the lattice and the CH_3 group to one side, with the result that the process leads to the formation of an isotactic polymer.

While a limited amount of experimental evidence does lend support to the bimetallic concept, majority opinion favors the second and simpler alternative, the monometallic mechanism.

9.3.2.2   Monometallic Mechanism

It is generally accepted that the d-orbitals in the transition element are the main source of catalytic activity and that it is the Ti$-$alkyl bond that acts as the polymerization center where chain growth occurs. The function of the aluminum alkyl is thus only to alkylate $TiCl_3$. The monometallic mechanism described below is the one based mainly on the ideas of Cossee and Arlman (1964). The quantum theory developed by Cossee (1967) provided a sound theoretical basis for this mechanism.

According to the mechanism, the active center is formed by the interaction of aluminum alkyl with an octahedral vacancy around Ti. For $\alpha-TiCl_3$ catalyst the formation of active center can be represented as shown in Fig. 9.3. To elaborate, the five-coordinated Ti^{3+} on the surface has a vacant d-orbital, represented by $-\square$, which facilitates chemisorption of the aluminum alkyl followed by alkylation of the Ti^{3+} ion by an exchange mechanism to form the active center $TiRCl_4-\square$. The vacant site at the active center can accommodate the incoming monomer unit, which forms a π-complex with the titanium at the vacant d-orbital and is then inserted into the Ti$-$alkyl bond. The sequence of steps is shown in Fig. 9.4 using propylene as the monomer.

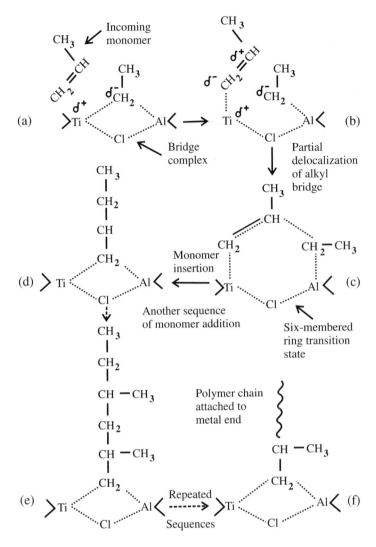

Figure 9.2 Bimetallic mechanism for stereospecific polymerization. (After Patat and Sinn, 1958.)

After the monomer is inserted into the Ti−alkyl bond, the polymer chain migrates back to its initial position, while the vacant site migrates to its original position to accept another monomer molecule. This migration is necessary, as otherwise an alternating position would be offered to the monomer leading to the formation of a syndiotactic polymer instead of an isotactic polymer.

Problem 9.1 From the accurate kinetic data that have been obtained (Natta and Pasquon, 1959) for the polymerization of C_3H_6 with α-$TiCl_3$ and $Al(C_2H_5)_3$ it appears that in the steady state the rate is strictly proportional to the pressure of C_3H_6. The polymerization rate is also proportional to the amount of α-$TiCl_3$ and independent of the concentration of $Al(C_2H_5)_3$. Suggest a kinetic scheme in conformity with these observations. A qualitative use may be made of the fact that an activation energy of 11-14 kcal/mol has been observed for this polymerization and that no stable complex between α-olefins and Ti has been found.

Answer:

The complex formation and the rearrangement can be described schematically as

$$S + M \underset{k_2}{\overset{k_1}{\rightleftharpoons}} SM \quad \text{and} \quad SM \overset{k_3}{\longrightarrow} S', \text{ respectively, where S is the vacant site, SM is the complex between vacant}$$

site and monomer, M is the monomer, and S' is the new vacant site with alkyl group being one unit longer. The following relations hold :

$$-d[M]/dt = k_1[S][M] - k_2[SM] \tag{P9.1.1}$$

$$d[SM]/dt = -k_3[SM] + k_1[S][M] - k_2[SM] \tag{P9.1.2}$$

$$[SM] + [S] = [C] \tag{P9.1.3}$$

where [C] is the total number of active sites (vacant + filled). Applying steady-state conditions, $d[SM]/dt = 0$. One then obtains, after a few eliminations,

$$-d[M]/dt = (k_1 k_3 [C][M])/(k_1[M] + k_2 + k_3) \tag{P9.1.4}$$

According to Eq. (P9.1.4), strict proportionality of the polymerization rate to the monomer pressure requires that either $k_2 \gg k_1[M]$ or $k_3 \gg k_1[M]$. When $k_3 \gg k_1[M]$, one has the situation where every molecule entering the vacant site would react immediately. The rate determining step would then be the complex formation. This, however, seems not very likely, if it is assumed that diffusion of the monomer through the growing polymer is not a limiting factor. It is also difficult to visualize that putting a neutral molecule into a vacant position would require an activation energy of 11-14 kcal/mol that has been observed. This activation energy would preferably be attributed largely to the rearrangement, and the conclusion may thus be drawn that $k_2 \gg k_1[M]$. This also fits the observation that no stable complexes between α-olefins and Ti have been found.

The polymerization rate now becomes

$$R_p = (k_1 k_3 / k_2)[C][M] \tag{P9.1.5}$$

and the measured activation energy is : $\Delta E = \Delta E_R - \Delta H_C$, where ΔE_R is the activation energy for the rearrangement and ΔH_C is the heat of complex formation. Since the latter is probably very small, the activation energy for the rearrangement of the complex will be of the order of 10 kcal/mol.

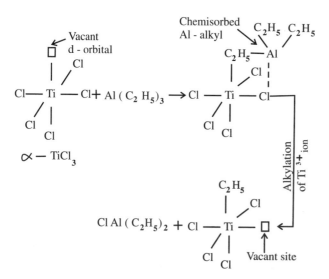

Figure 9.3 Interaction of aluminum alkyl with an octahedral vacancy around Ti in the first stage of monometallic mechanism. (After Cossee, 1967.)

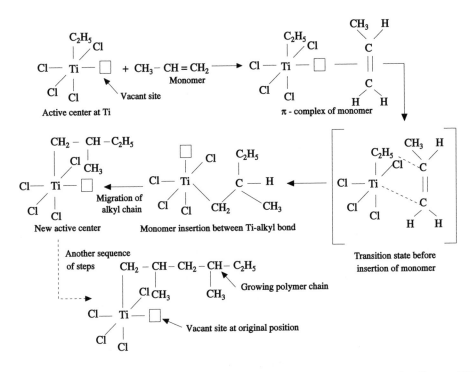

Figure 9.4 Monometallic mechanism for stereospecific polymerization. (After Cossee, 1967.)

Stereoregulation The distinctive characteristic of Ziegler-Natta catalysts is their ability to produce stereoregular polymers. To obtain a stereoregular polymer, according to the monometallic mechanism of Cossee and Arlman, the chemisorption of the monomer should always lead to the same orientation of the molecule on the catalyst surface. It becomes apparent from an examination of models that a molecule such as propylene can fit into the catalyst surface in only one way if a position of closest approach of the double bond to the Ti^{3+} ion is to be achieved. This requires that the $=CH_2$ group of the incoming monomer points into the lattice and, consequently, for steric reasons the orientation of the $-CH_3$ group to one side is preferred. It thus ensures that the configuration of the monomer while complexing at the vacant site, prior to incorporation into the polymer chain, is always the same so that an isotactic polymer is formed. Since migration of the vacant site back to its original position is necessary for isospecific polymerization, this also implies that the tacticity of the polymer formed depends essentially on the rates of both the alkyl shift and the migration. Since both these processes slow down at lower temperatures, a syndiotactic, instead of isotactic, polymer would be formed when the temperature is decreased. In fact, syndiotactic polypropylene can be obtained at $-70°C$.

9.4 Kinetics of Ziegler-Natta Polymerization

9.4.1 Typical Shapes of Kinetic Curves

In Ziegler-Natta polymerizations, the reaction systems are more often heterogeneous than homogeneous. While the relatively few polymerizations that are homogeneous behave in a manner generally similar to ionic polymerizations, described in Chapter 8, the heterogeneous systems

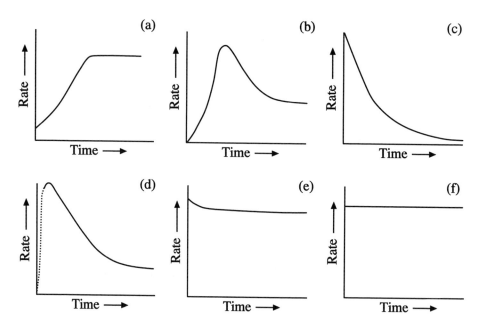

Figure 9.5 Some typical kinetic rate-time profiles. (a) Rate increases in an initial acceleration or settling period to reach a more or less steady value. (b) Rate increases in an initial settling period to reach a maximum and then decreases. (c) No initial settling period and rate decreases rapidly from beginning. (d) Rate rises very rapidly to a maximum value and then decreases rapidly. (e) No settling period and rate decreases slowly. (f) No settling period and rate remains constant. (After Tait and Watkins, 1989.)

usually exhibit complicated behavior, as can be seen from some typical kinetic rate-time profiles of Ziegler-Natta poymerizations, *Types* (a)-(f) in Fig. 9.5. The shapes of these profiles may be characteristic of particular catalysts or catalyst-monomer systems and may be considered to consist of three periods, viz., an *acceleration period*, a *stationary period*, and a *decay period*. Some catalyst systems, however, show all three types.

Type (a) behavior is observed for many first generation catalyst systems, e.g., α-TiCl$_3$, VCl$_3$, etc. with dialkylaluminum halides as cocatalysts in the polymerization of propylene in hydrocarbon media. During an initial *acceleration period*, which is of 20-60 minutes duration for many propylene polymerizations at 1 atm pressure in the temperature range 50-70°C, the rate increases from the beginning to reach a more or less steady value. Natta and Pasquon (1959) attributed this behavior to the breakdown of the α-TiCl$_3$ matrix to smaller crystallites due to the pressure of the growing polymer chains in the initial stages, leading to exposure of fresh Ti atoms and creation of new active centers with consequent increase in the polymerization rate. After all the α-TiCl$_3$ matrix has broken down, the polymerization rate remains steady for a relatively long period of time. The effect of particle size is further discussed in a following section.

Type (b) behavior, in which the rate of polymerization increases in an acceleration period to reach a maximum and then decreases, may be observed for a more active but less stable catalyst, while *Type (c)* or *Type (d)* behavior in which the rate starts at a maximum value or rises very rapidly to a maximum value and then decrease rapidly with time is exhibited by many supported high-activity catalyst systems, e.g., MgCl$_2$/ethylbenzoate/TiCl$_4$–AlEt$_3$ in ethylene or propylene polymerization. *Type (c)* behavior is also shown by many homogeneous catalyst systems, e.g., Cp$_2$TiEtCl–AlEtCl$_2$ in ethylene polymerization (Cp = cyclopentadiene).

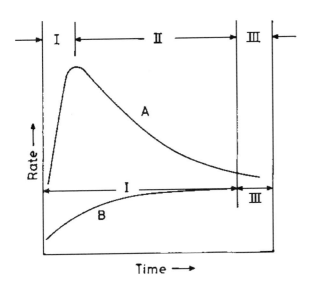

Figure 9.6 Typical kinetic curves observed during propylene polymerization with ground and unground TiCl$_3$. Curve A: decay type; curve B: build-up or acceleration type; zone I: build-up period; zone II: decay period; zone III: stationary period. (After Keii, 1973.)

Type (e) behavior may be observed when there is an almost instantaneous breakdown of porous catalyst particles on treatment with the cocatalyst so that an acceleration or settlement period is nearly nonexistent. This behavior is shown by ether-treated highly porous catalysts in propylene polymerization (Tait and Watkins, 1989). The polymerization rate decreases very slowly with time due to good stability of the catalyst system.

Type (f) behavior, featuring constant polymerization rate, is not often found in practice. One example of this type is the polymerization of 4-methyl-1-pentene on MgCl$_2$-supported catalysts containing phthalate esters. The polymerization rate for this system is found to be almost constant with time (Kissin, 1985).

At higher conversions, the diffusion of monomer to the propagation centers through the formed polymer may be slow enough to become rate-determining and the rate of propagation may then be diffusion controlled. Even at low conversions, diffusion may be rate-controlling in heterogeneous systems at a low degree of agitation. In such systems, the polymerization rate increases with increased rate of stirring (Keii, 1973).

9.4.2 Effect of Catalyst Particle Size

The effect of the catalyst particle size on the rate of polymerization was first studied by Natta and Pasquon (1959), who found that for a constant concentration of the monomer the rate of polymerization changed with time depending on the particle size. While with ground TiCl$_3$ (particle size $\leq 2\mu$) the rate of propylene polymerization quickly reached a maximum and then decreased gradually to an asymptotic stationary value (curve A in Fig. 9.6), with unground particles (size up to 10μ) no maxima were found but the rate accelerated to approach the same asymptotic stationary value (curve B in Fig. 9.6). The former behavior is referred to as the *decay type*, whereas the latter is known as the *build-up* or *acceleration type*. Corresponding to the shapes of the rate curves, the different zones can thus be classified into *build-up*, *decay*, and *stationary periods* (see Fig. 9.6).

Natta and Pasquon (1959) attributed the acceleration type behavior to an increase in surface due to break-up of catalyst particles under mechanical pressure of growing polymer chains (anchored to the catalyst active centers) in the early stages of polymerization. However, as the particle size becomes smaller, greater mechanical energy is required for further size reduction. Consequently, the particle size, and hence the specific surface area, would reach some asymptotic value. The observed stationary polymerization rate would correspond to this particle size of the catalyst.

In some cases, a decay type behavior is observed which may be caused by active site destruction, such as by thermal deactivation or further reduction of the transition metal by the Group I-III metal component. The decay type kinetics is explained later in terms of the deactivation of active sites.

9.4.3   Chain Termination

The termination of a polymer chain growing at an active center may occur by various reactions (Young and Lovell, 1990; Odian, 1991), as shown below with propylene as the example.

1. *Chain Transfer to monomer*:

$$\bullet\!-\!Ti\!-\!CH_2\!-\!CH(CH_3)\!-\!\!\text{\tiny\/}\!\!\!\!\!\! \quad + \quad CH_3CH\!=\!CH_2 \quad \xrightarrow{k_{tr,M}} \quad \bullet\!-\!Ti\!-\!CH_2CH_2CH_3$$
$$+ \quad CH_2\!=\!C(CH_3)\!-\!\!\text{\tiny\/} \tag{9.9}$$

$$\bullet\!-\!Ti\!-\!CH_2\!-\!CH(CH_3)\!-\!\!\text{\tiny\/} \quad + \quad CH_3CH\!=\!CH_2 \quad \xrightarrow{k_{tr,M}} \bullet\!-\!Ti\!-\!CH\!=\!CH\!-\!CH_3$$
$$+ \quad CH_3\!-\!CH(CH_3)\!-\!\!\text{\tiny\/} \tag{9.10}$$

where $\bullet\!-\!Ti$ represents the transition metal active center on the catalyst site at which chain propagation takes place. Note that it is the methylene carbon atom from the monomer that is bonded to the transition metal atom (cf. Fig. 9.2).

2. *Chain transfer to the Group I-III metal alkyl*:

$$\bullet\!-\!Ti\!-\!CH_2\!-\!CH(CH_3)\!-\!\!\text{\tiny\/} + Al(C_2H_5)_3 \quad \xrightarrow{k_{tr,A}} \quad \bullet\!-\!Ti\!-\!CH_2CH_3 \quad +$$
$$(C_2H_5)_2Al\!-\!CH_2\!-\!CH(CH_3)\!-\!\!\text{\tiny\/} \tag{9.11}$$

3. *Spontaneous intramolecular β–hydride transfer*:

$$\bullet\!-\!Ti\!-\!CH_2\!-\!CH(CH_3)\!-\!\!\text{\tiny\/} \quad \xrightarrow{k_s} \quad \bullet\!-\!Ti\!-\!H \quad + \quad CH_2\!=\!C(CH_3)\!-\!\!\text{\tiny\/} \tag{9.12}$$

4. *Chain transfer to an active hydrogen compound such as molecular hydrogen*   (*external agent*):

$$\bullet\!-\!Ti\!-\!CH_2\!-\!CH(CH_3)\!-\!\!\text{\tiny\/} \quad + \quad H_2 \quad \xrightarrow{k_{tr,H_2}} \quad \bullet\!-\!Ti\!-\!H \quad + \quad CH_3\!-\!CH(CH_3)\!-\!\!\text{\tiny\/} \tag{9.13}$$

The above reactions terminating the growth of polymer chains are indeed chain transfer reactions since in each case a new propagating chain is initiated. The relative extents of these reactions depend on various factors such as the monomer, the initiator components, temperature, concentrations, and other reaction conditions. Under normal conditions of polymerization, intramolecular hydride transfer is negligible and termination of propagating chains occurs mostly by chain transfer processes. Being a highly effective chain transfer agent, molecular hydrogen is often used for polymer molecular weight control.

9.4.4 Kinetic Models

In this section several empirical rate expressions for Ziegler-Natta polymerizations will be presented and attempts to model the polymerization will be described. It is found that several models could be proposed to explain the same rate equations. The models are based on the assumption of a fixed geometric center that has a definable identity and activity invariant with time. These assumptions, however, are far too simplistic and only limited general agreement of the models with the observed kinetic behavior or good agreement only in specific cases could be expected.

9.4.4.1 Early Models

A number of simple kinetic models (Natta and Pasquon, 1959; Kissin, 1985) have been developed for polymerization systems that have rate-time profiles of the types shown in Fig. 9.5 (*a* to *f*). These models are based on the assumption that the polymerization consists of three steps, namely, chain initiation, chain propagation, and chain transfer, and that the total concentration of active centers, $C^\star$, remains constant during the polymerization.

Chain Initiation   Chain initiation in Ziegler-Natta polymerization consists of insertion of the first monomer molecule, M, into a transition metal-carbon bond (Cat–R) in an active center, resulting in the formation of a polymerization center (Cat–P):

$$\text{Cat} - \text{R} + \text{M} \xrightarrow{k_i} \text{Cat} - \text{P}_1 \tag{9.14}$$

Chain Propagation   The chain propagation step involves successive insertion of monomer molecul into a transition metal-carbon bond in a polymerization center:

$$\text{Cat} - \text{P}_1 + \text{M} \xrightarrow{k_p} \text{Cat} - \text{P}_2 \tag{9.15}$$

$$\text{Cat} - \text{P}_2 + \text{M} \xrightarrow{k_p} \text{Cat} - \text{P}_3 \tag{9.16}$$

$$\text{Cat} - \text{P}_n + \text{M} \xrightarrow{k_p} \text{Cat} - \text{P}_{n+1} \tag{9.17}$$

For simplicity, all polymerization centers are considered to be equally active with the same propagation rate constant, k_p, irrespective of their geometric location and the degree of polymerization of the growing chains bound to the centers. It is, however, found that all polymerization centers are not equally active and the k_p of the above propagation steps must therefore be regarded as an average.

Chain Transfer   In the absence of any externally added transfer agent, three transfer reactions [Eq. (9.9)-(9.12)] may be considered, viz., chain transfer with monomer, chain transfer with alkylaluminum, and spontaneous chain transfer, represented by

$$\text{Cat} - \text{P}_n + \text{M} \xrightarrow{k_{tr,M}} \text{Cat} - \text{R}' + \text{P}_n \tag{9.18}$$

$$\text{Cat} - \text{P}_n + \text{AlR}_x \xrightarrow{k_{tr,A}} \text{Cat} - \text{R}'' + \text{P}_n - \text{AlR}_{x-1} \tag{9.19}$$

$$\text{Cat} - \text{P}_n \xrightarrow{k_s} \text{Cat} - \text{R}''' + \text{P}_n \tag{9.20}$$

Equations (9.18) and (9.19) are in agreement with the experimental observations that the molecular weight of the polymer formed decreases as the concentration of monomer or alkylaluminum compound is increased. Reactions represented by Eqs. (9.18)-(9.20) lead to the formation of three new initiation centers having different chemical structures. For example, in the polymerization of propylene using either $AlEt_3$ or $AlEt_2Cl$ as the cocatalyst, $R' = n$-propyl, $R'' = $ ethyl, and $R''' = H$. Equation (9.20), reperesenting spontaneous chain transfer can, however, be neglected for most catalyst systems and usual concentrations of monomer and alkylaluminum.

Problem 9.2   Using the kinetic scheme given above, derive suitable expressions for the rate of polymerization (R_p) and average degree of polymerization ($\overline{DP}_n$). Show how chain transfer constants can be evaluated from the measurement of $\overline{DP}_n$.

Answer:

At any instant there will be polymerization centers generally represented by Cat−P_n ($n = 1, 2, \cdots$) and different initiation centers Cat−R′, Cat−R″, and Cat−R‴ [cf. Eqs. (9.18)-(9.20)]. Thus the total concentration of active centers, $C^\star$, is given by the sum of the concentrations of polymerization centers, $C_p^\star$, and the concentrations of different initiation centers, C_i, at which initiation takes place at that particular moment (Kissin, 1985):

$$C^\star = C_p^\star + C_i' + C_i'' + C_i''' \tag{P9.2.1}$$

Under stationary state conditions only the steady state with respect to the concentration of polymerization centers need be considered. A steady state thus implies the following conditions:

$$dC_p^\star/dt = 0 \tag{P9.2.2}$$

$$\sum_j R_{\text{initiation}} = \sum_j R_{\text{transfer}} \tag{P9.2.3}$$

Thus for the above reaction scheme,

$$k_i' C_i' [M] = k_{tr,M} C_p^\star [M] \tag{P9.2.4}$$

$$k_i'' C_i'' [M] = k_{tr,A} C_p^\star [A] \tag{P9.2.5}$$

$$k_i''' C_i''' [M] = k_s C_p^\star \tag{P9.2.6}$$

where [M] and [A] are the concentrations of monomer and alkylaluminum, respectively; k_i', k_i'', and k_i''' are the rate constants for chain initiation involving the Cat−R′, Cat−R″, and Cat−R‴ species, respectively [cf. Eq. (9.14)].

Equation (P9.2.1) can be rewritten as

$$C_p^\star = \frac{C^\star}{1 + (C_i'/C_p^\star) + (C_i''/C_p^\star) + (C_i'''/C_p^\star)} \tag{P9.2.7}$$

Substituting from Eqs. (P9.2.4)−(P9.2.6) then gives

$$C_p^\star = \frac{C^\star}{1 + k_{tr,M}/k_i' + k_{tr,A} [A]/k''[M] + k_s/k_i'''[M]} \tag{P9.2.8}$$

Hence the rate of polymerization is

$$R_p = k_p C_p^\star [M]$$

$$= \frac{k_p C^\star [M]^2}{[M] + k_{tr,M} [M]/k_i' + k_{tr,A} [A]/k_i'' + k_s/k_i'''} \tag{P9.2.9}$$

The number average degree of polymerization, $\overline{DP}_n$, is given by

$$\overline{DP}_n = \frac{k_p[M]}{k_{tr,M}[M] + k_{tr,A}[A] + k_s} \tag{P9.2.10}$$

or,

$$1/\overline{DP}_n = k_{tr,M}/k_p + k_{tr,A}[A]/k_p[M] + k_s/k_p[M] \tag{P9.2.11}$$

Hence a plot of $1/\overline{DP}_n$ vs. $1/[M]$ for polymer samples prepared with constant $[A]$ should be a straight line having $k_{tr,M}/k_p$ as the intercept and $(k_{tr,A}[A]/k_p + k_s/k_p)$ as the slope. Similarly when $1/\overline{DP}_n$ is plotted vs. $[A]$ for polymer samples prepared with constant $[M]$, a straight line should be obtained having $(k_{tr,M}/k_p + k_s/k_p[M])$ as the intercept and $k_{tr,A}/k_p[M]$ as the slope. Consequently, all chain transfer constants can be evaluated (Kissin, 1985) as ratios of chain transfer rate constants $k_{tr,M}$, $k_{tr,A}$, and k_s to the propagation rate constant k_p. Values listed in Table 9.2 are derived in this way.

Some compounds, particularly hydrogen, are effective chain transfer agents [cf. Eq. (9.13)] that cause reduction of molecular weight. For this reason hydrogen is usually added in the commercial production of polyethylene and polypropylene. It is easy to modify Eq. (P9.2.11) to include the effect of hydrogen by adding an extra term $k_{tr,H_2}[H_2]/k_p[M]$ to the right side of Eq. (P9.2.11).

It should be noted that the aluminum alkyls are known to be mainly dimeric in solution, existing in equilibrium with the monomeric species:

$$A_2 \overset{K_d}{\rightleftharpoons} 2A$$

$$[A] = K_d^{1/2}[A_2]^{1/2} \tag{9.21}$$

where the brackets are used to indicate concentrations and K_d represents the dissociation equilibrium constant. Due to the occurrence of dimerization, the actual amount of monomeric AlR_3 is small and $[A]$ in Eqs. (P9.2.9)–(P9.2.11) must be replaced by $K_d^{1/2}[A_2]^{1/2}$. It should be noted that the values of $k_{tr,A}$ for $AlEt_3$ and $AlEt_2Cl$ listed in Table 9.2 are the effective values including the equilibrium constant for dissociation of dimers into monomeric species, i.e.,

$$(k_{tr,A})_{real} = k_{tr,A}/K_d^{1/2} \tag{9.22}$$

Problem 9.3   Derive limiting expressions from Eqs. (P9.2.9) and (P9.2.10) for the case where $[M]$ is high.

Answer:

When $[M]$ is high, Eqs. (P9.2.8)-(P9.2.10) approximate to

$$C_p^\star = \frac{C^\star}{1 + k_{tr,M}/k_i'} \tag{P9.3.1}$$

$$R_p = \frac{k_p C^\star [M]}{1 + k_{tr,M}/k_i'} \tag{P9.3.2}$$

$$\overline{DP}_n = k_p/k_{tr,M} \tag{P9.3.3}$$

Further simplification results when $k_i' \approx k_p \gg k_{tr,M}$. Equations (P9.3.1) and (P9.3.2) then reduce to $C_p^\star \approx C^\star$ and $R_p = k_p C^\star [M]$. Thus at high $[M]$, R_p becomes proportional to $[M]$ and $\overline{DP}_n$ becomes independent of all polymerization variables.

Table 9.2 Chain Transfer Constants (k_{tr}/k_p) for Alkene Polymerization with Heterogeneous Ziegler-Natta Catalysts

Monomer	Catalyst system	Temp. (°C)	$k_{tr,M}/k_p$ ($\times 10^{-4}$)	$k_{tr,A}/k_p$ ($\times 10^{-4}$)	k_s/k_p ($\times 10^{-4}$)
Ethylene	δ-TiCl$_3$-AlEt$_3$	75	0.4	1.7	–
Ethylene	γ-TiCl$_3$-AlEt$_2$Cl	40	1.4	0.04	
Propylene	α-TiCl$_3$-AlEt$_3$	70	6-12	17-34	1.2-2.5
Propylene	VCl$_3$-Al(Bu^i)$_3$	60	10	6	0.25

Source: Erich and Mark, 1956; Zakharov et al., 1977; Grieveson, 1965.

9.4.4.2 Adsorption Models

Ziegler-Natta catalyst systems being mostly heterogeneous in nature, adsorption reactions are most likely to occur in such polymerizations and feature in their kinetic schemes (Erich and Mark, 1956). A number of kinetic schemes have thus been proposed based on the assumption that the polymerization centers are formed by the adsorption of metal alkyl species on to the surface of a crystalline transition metal halide and that chain propagation occurs between the adsorbed metal alkyl and monomer. In this regard the *Rideal rate law* and the *Langmuir-Hinshelwood rate law* for adsorption and reaction on solids assume importance (*see* Problem 9.4).

Problem 9.4 Considering reaction between A and B catalyzed by a solid there are two possible mechanisms by which this reaction could occur. The first is that one of them, say A, gets adsorbed on the solid surface and the adsorbed A then reacts chemically with the other component B, which is in the gas phase or in solution and is not adsorbed on the surface. The second mechanism is that both A and B are adsorbed, and the adsorbed species undergo chemical reaction on the surface. The reaction rate expression derived for the former mechanism is the *Rideal rate law* and that for the second mechanism is the *Langmuir-Hinshelwood rate law*. Obtain simple derivations of these two rate laws.

Answer:

Both the Rideal and Langmuir-Hinshelwood rate laws are based upon the Langmuir adsorption equation which is applicable for gas-solid as well as liquid-solid systems where diffusion of the sorbate to the solid surface is not rate limiting (generally true). The basic assumption of the Langmuir adsorption is that adsorption occurs at adsorption sites and all these sites are equivalent. For gas-solid systems, the rate of adsorption, r_a, of the gas A is proportional to the gas pressure, p_A, and the number of vacant sites, i.e.,

$$r_a = k_a\, p_A\, (n_0 - n_A) \tag{P9.4.1}$$

where n_0 is the total number of adsorption sites and n_A is the number of sites which are occupied by molecules of A. The rate of desorption, on the other hand, is postulated to be

$$r_d = k_d\, n_A \tag{P9.4.2}$$

At equilibrium the rates of adsorption and desorption are equal. Equating Eqs. (P9.4.1) and (P9.4.2) then gives

$$n_A = n_0\, \frac{k_a\, p_A/k_d}{(1 + k_a\, p_A/k_d)} \tag{P9.4.3}$$

Defining the fraction of adsorption sites covered by A as $\theta_a = n_A/n_0$ and the equilibrium constant for the adsorption equilibrium as $K_A = k_a/k_d$, the above equation reduces to

$$\theta_A = \frac{K_A \, p_A}{1 + K_A \, p_A} \tag{P9.4.4}$$

The term θ_A in Eq. (P9.4.4) represents the fraction of total adsorption sites occupied by A. If there are two kinds of molecules, A and B, which are competing for the adsorption sites, one modifies Eq. (P9.4.4) to

$$\theta_A = \frac{K_A \, p_A}{1 + K_A \, p_A + K_B \, p_B} \tag{P9.4.5}$$

$$\theta_B = \frac{K_B \, p_B}{1 + K_A \, p_A + K_B \, p_B} \tag{P9.4.6}$$

where p_A and p_B are the partial pressures of A and B.

If there is a gas molecule A_2 which is adsorbed in the dissociated form, A, Eq. (P9.4.4) is modified to

$$\theta_A = \frac{K_A (K_d \, p_{A_2})^{1/2}}{1 + K_A (K_d \, p_{A_2})^{1/2}} \tag{P9.4.7}$$

where K_d is the dissociation equilibrium constant of A_2.

Equations (P9.4.4)–(P9.4.7) are applicable for gas-solid adsorption. For liquid-solid adsorption, the partial pressures p_A, p_B, and p_{A_2} in these equations are replaced by concentrations [A], [B], and [A_2], respectively.

(a) If the reaction takes place between the adsorbed gas A and the other component B in the gas phase, the rate of reaction is given as

$$R_{AB} = k_s [S] \theta_A \, p_B \tag{P9.4.8}$$

where k_s is the surface reaction rate constant and [S] is the concentration of adsorption sites.

Substituting Eq. (P9.4.4) into Eq. (P9.4.8) gives the *Rideal rate law* as

$$R_{AB} = k_s K_A [S] \frac{p_A \, p_B}{1 + K_A \, p_A} \tag{P9.4.9}$$

For liquid-solid systems, the corresponding rate law is

$$R_{AB} = k_s K_A [S] \frac{[A][B]}{1 + K_A [A]} \tag{P9.4.10}$$

(b) If both A and B are adsorbed before the chemical reaction occurs, the rate of reaction is given as

$$R_{AB} = k_s \theta_A \theta_B [S] \tag{P9.4.11}$$

Substituting for θ_A and θ_B from Eqs. (P9.4.5) and (P9.4.6) one obtains the *Langmuir-Hinshelwood rate law* as

$$R_{AB} = \frac{k_s K_A K_B [S] p_A \, p_B}{(1 + K_A \, p_A + K_B \, p_B)^2} \tag{P9.4.12}$$

The corresponding equation for liquid-solid systems is then

$$R_{AB} = \frac{k_s K_A K_B [S][A][B]}{(1 + K_A [A] + K_B [B])^2} \tag{P9.4.13}$$

For Ziegler-Natta polymerization it may be postulated that the dimeric alkylaluminum molecules are adsorbed on TiCl$_3$ sites to give rise to polymerization centers by the following equilibrium process:

$$(AlR_3)_2 + \text{Active site} \overset{K_A}{\rightleftharpoons} \text{Polymerization center (PC)} \tag{9.23}$$

Equations (P9.4.8)–(P9.4.10) can now be applied and the rate so obtained will be applicable in the stationary zone because steady-state conditions were assumed in deriving these equations. Equation (P9.4.9) is applicable if the PCs react with monomer molecules present in the reaction medium. In this case one has, corresponding to Eq. (P9.4.8),

$$R_{\infty 1} = k_{s1} [S] \theta_{Al_2R_6} [M] \tag{9.24}$$

where [M] is the monomer concentration in the bulk of the reaction mass and $R_{\infty 1}$ is the rate under the assumed conditions. Substituting for $\theta_{Al_2R_6}$ by comparison with Eq. (P9.4.4) one can write

$$R_{\infty 1} = k_{s1} [S] [M] \frac{K_A [A_2]}{1 + K_A [A_2]} \tag{9.25}$$

where K_A is the adsorption equilibrium constant and $[A_2]$ is the concentration of the alkylaluminum dimer.

Applying Henry's law for the dissolution of the gaseous monomer in the solution, [M] is given by

$$[M] = H.p_M \tag{9.26}$$

where H is the Henry constant and p_M is the partial pressure of the monomer. Equation (9.25) then becomes

$$R_{\infty 1} = k_{s1} [S] H p_M \frac{K_A [A_2]}{1 + K_A [A_2]} \tag{9.27}$$

Equations (9.25) and (9.27) are the Rideal rate laws for the Ziegler-Natta polymerization.

If the PCs are assumed to react with the adsorbed monomer molecules, the Langmuir-Hinshelwood rate equation [Eq. (P9.4.13)] is applicable and the rate expression in the stationary zone then becomes

$$R_{\infty 2} = \frac{k_{s2} K_A K_M [S] [A_2] [M]}{(1 + K_A [A_2] + K_M [M])^2} \tag{9.28}$$

where K_M is the adsorption equilibrium constant for the monomer.

Equations (9.27) and (9.28) have been experimentally verified and the values of K_M and K_A determined. Some results (Keii, 1973; Vesley, 1962) are shown in Table 9.3.

If polymerization takes place in the presence of donor-type impurities, such as COS, CS$_2$, H$_2$O, H$_2$S, a general rate equation of the following type can be used (Vesley et al., 1961):

$$R_p = \frac{k_p K_M K_A [S] [A_2] [M]}{(1 + K_A [A_2] + K_M [M] + K_D [D])^2} \tag{9.29}$$

Table 9.3 Experimental Values of Adsorption Equilibrium Constants in Langmuir-Hinshelwood and Rideal Rate Laws for Propylene Polymerization

Catalyst system	Temp. (°C)	K_A (L/mol)	K_M (L/mol)	Rate law used[a]
TiCl$_4$ + AlEt$_3$	32	280	–	R-type
	44	170	–	R-type
	57	60	–	R-type
TiCl$_3$ + AlEt$_3$	50	21.2	0.163	L-H type
VCl$_3$ + AlEt$_3$	40	40-60	6-9	L-H type

[a]R: Rideal; L-H: Langmuir-Hinshelwood.
Source: Vesley, 1962; Keii, 1973; Tait and Watkins, 1989.

where K_D is the equilibrium adsorption constant and [D] is the concentration of donor impurities. The extra term K_D[D] in the denominator of Eq. (9.29) takes into account the donor impurities competing with other species for the sorption sites.

Problem 9.5   Show that under conditions where $K_M[M] \ll K_A[A_2]$ (cf. Table 9.3) the Langmuir-Hinshelwood rate equation becomes indistinguishable from the Rideal expression for the stationary zone rate.

Answer:

For $K_M[M] \ll K_A[A_2]$, Eq. (9.28) can be simplified to

$$R_{\infty 2} \simeq \frac{k_{s2} K_A K_M [S][A_2][M]}{(1 + K_A[A_2])^2} \tag{P9.5.1}$$

or

$$R_{\infty 2} \simeq \frac{k_{s2} K_A K_M [S][A_2][M]}{1 + 2K_A[A_2]} \tag{P9.5.2}$$

where $K_A^2[A_2]^2$ is neglected. If Henry's law is valid for the dissolution of monomer, Eq. (P9.5.2) becomes

$$R_{\infty 2} \simeq \left(\frac{k_{s2}}{2}\right) [S] H K_M p_M \frac{K'[A_2]}{1 + K'[A_2]} \tag{P9.5.3}$$

which is seen to be of the same form as the Rideal expression, Eq. (9.27).

Problem 9.6   Under conditions of high metal alkyl concentrations the propagation rate in Ziegler-Natta polymerization of propylene is found (Saltman, 1960) to be given by

$$R_p = k_p[M][S]$$

where [S] is the concentration of adsorption sites and is proportional to the catalyst weight; [M] is the monomer concentration and k_p is the rate constant for propagation. At low metal alkyl concentrations, the rate is found to be dependent on the metal alkyl content. Suggest a reaction scheme to explain these results with Langmuir-Hinshelwood model, assuming that alkylated transition metal entities form the polymerization centers.

Answer:

Assuming that monomeric metal alkyl A and monomer M are competitors for the adsorption sites S on the catalyst surface, the following adsorption-desorption equilibria exist:

$$S + A \rightleftharpoons S\text{---}A \tag{P9.6.1}$$
$$S + M \rightleftharpoons S\text{---}M \tag{P9.6.2}$$

where the broken line indicates attachment to the surface site. Considering this as a Langmuir-Hinshelwood adsorption, the fractions of surface sites covered with monomeric metal alkyl (θ_A) and monomer (θ_M) are, respectively,

$$\theta_A = \frac{K_A [A]}{1 + K_M [M] + K_A [A]} \quad \text{and} \quad \theta_M = \frac{K_M [M]}{1 + K_M [M] + K_A [A]} \tag{P9.6.3}$$

If use is made of Eq. (9.21), then

$$\theta_A = \frac{K_A K_d^{1/2} [A_2]^{1/2}}{1 + K_M [M] + K_A K_d^{1/2} [A_2]^{1/2}} \tag{P9.6.4}$$

At equilibrium the concentration of the adsorbed monoalkyl is $[S]\theta_A$. Now assuming that the propagation occurs by reaction of monomer (whether adsorbed or not) with the activated adsorbed alkyl complex, i.e.,

$$S\text{---}A + M \longrightarrow S\text{---}AM$$
$$S\text{---}AM + M \longrightarrow S\text{---}AM_2, \text{ etc.} \tag{P9.6.5}$$

where the species AM_n is also an aluminum alkyl, differing from A only in the length of the alkyl group, the propagation rate is

$$R_p = k_p [M][S]\theta_A = \frac{k_p K_A K_d^{1/2} [M][S][A_2]^{1/2}}{1 + K_M [M] + K_A K_d^{1/2} [A_2]^{1/2}} \tag{P9.6.6}$$

If $K_M [M]$ is small and $K_A K_d^{1/2} [A_2]^{1/2}$ large compared to unity (i.e., θ_A approaches 1), the propagation rates become that found experimentally, namely,

$$R_p = k_p [M][S] \tag{P9.6.7}$$

According to Eq. (P9.6.7), at very high monomer concentrations the propagation rate should tend to become independent of monomer concentrations; similarly, at low alkyl concentrations it should become dependent on the metal alkyl content.

Langmuir-type adsorption has been used (Keii et al., 1967) to explain the observed rate behavior both during the initial stage (*build-up period*) and in the stationary state of the polymerization of propylene with $TiCl_3$–$AlEt_3$ which exhibits a *decay-type* behavior (curve A in Fig. 9.6) (see Problem 9.7). The observed rate behavior in the build-up period of the *acceleration-type* curve (curve B in Fig. 9.6) can also be explained in a similar way.

Problem 9.7 The kinetics of propylene polymerization was investigated (Keii et al., 1967) using $TiCl_3$–$AlEt_3$ in the temperature range 30–70°C with propylene pressure (P) in the range 100–700 mm Hg and catalyst concentration 2 g $TiCl_3$/liter ([A]/[Ti] = 0.4–3.0). The polymerization exhibits a decay type rate profile (curve A in Fig. 9.6) with the polymerization rate R_p at first increasing rapidly with the polymerization time t (**Stage I**) and then gradually decreasing to the stationary value R_∞ (**Stage II**). The polymerization rate in **Stage I** in the case where propylene is introduced after the addition of $(AlEt_3)_2$ is found to agree with the rate expression

$$R_p = k_1 P^2 \frac{K[A_2]}{1 + K[A_2]} t \tag{P9.7.1}$$

while the stationary rate R_∞ shows a Langmuir-type dependence expressed by

$$R_\infty = k_p P K[A_2]/(1 + K[A_2]) \tag{P9.7.2}$$

Rationalize the above kinetic behavior assuming Langmuir-type adsorption and formation of polymerization centers from Al_2Et_6 (represented by A_2) and propylene monomer on the surface of $TiCl_3$.

Answer:

Stage I.

Stage I may be considered as the period during which the polymerization center is formed on the surface of titanium trichloride from Al_2Et_6 and the olefin monomer. Al_2Et_6 adsorbs reversibly on the surface S to form a surface complex C. Assuming that the adsorption equilibrium is of the Langmuir type [cf. Eq.(P9.4.4)],

$$[C]_e = [S]\left(\frac{K[A_2]}{1 + K[A_2]}\right) \tag{P9.7.3}$$

where A_2 is Al_2Et_6 (dimer).

The polymerization center $C^\star$ may be formed irreversibly by the attack on the surface complex C by a propylene monomer M from the solution. Then, the formation rate of the center can be expressed by

$$d[C^\star]/dt = k_1'[M]([C]_e - [C^\star]) \tag{P9.7.4}$$

which yields

$$[C^\star] = [C]_e(1 - e^{-k_1'[M]t}) \tag{P9.7.5}$$

Note that the formation of the polymerization center [Eq. (P9.7.4)] is assumed to be irreversible and that of the surface complex [Eq. (P9.7.3)] to be reversible because the polymerization center is more stable than the surface complex.

Initially, Eq. (P9.7.5) can be approximated by

$$[C^\star] = k_1'[M][C]_e t \tag{P9.7.6}$$

If the polymerization occurs by the Rideal mechanism, that is, the rate determining step of the polymerization is the attack of a propylene monomer from solution onto the polymerization center, the rate R_p can be given by

$$R_p = k_p'[M][C^\star] \tag{P9.7.7}$$

Combining Eqs. (P9.7.3), (P9.7.6) and (P9.7.7), the initial polymerization rate is represented by

$$R_p = k_p' k_1'[M]^2[S]\frac{K[A_2]}{1 + K[A_2]} t \tag{P9.7.8}$$

Using Henry's law, $[M] = H.P$, where H is Henry's constant, Eq. (P9.7.8) can also be written as

$$R_p = k_1 P^2 \frac{K[A_2]}{1 + K[A_2]} t \tag{P9.7.9}$$

where $k_1 = k'_p k'_1 H^2 [S]$.

Stage II

The maximum polymerization rate in the stationary state is obtained from Eq. (P9.7.7) by substituting, successively, Eq. (P9.7.5) at $t \to \infty$ to replace $[C^\star]$ and then Eq. (P9.7.3) to replace $[C]_e$. This yields

$$R_{\max} = k'_p [M][C]_e = k'_p [M][S] \frac{K[A_2]}{1 + K[A_2]} = k_p P \frac{K[A_2]}{1 + K[A_2]} \tag{P9.7.10}$$

where $k_p = k'_p H[S]$

The value of R_∞ would, however, be lower as compared to $R_{\max}$ due to the reduction in surface site concentration $[S]$ caused by deactivation.

While most authors consider that the initiation and propagation of chain growth in the Ziegler-Natta polymerization occur by the same type of mechanism, there is a considerable amount of evidence that the rates of these two processes are very different, the rate of chain initiation being much lower than that of chain propagation (Cossee, 1960). Thus two types of active sites may be assumed to exist in these polymerization systems, viz., *polymerization centers* and *potential polymerization centers*. The formation of polymerization centers is, however, more likely to involve alkylated transition metal haides as proposed by Cossee (1960) and not metal alkyl molecules merely adsorbed onto the surface. Thus, models built on the assumption that alkylated transition metal entities form the polymerization centers have been found to give good agreement for several polymerization systems.

Problem 9.8 The kinetic rate behavior of Ziegler-Natta polymerization of ethylene and propylene is given by the empirical equation (Keii, 1973):

$$dR_p/dt = k(R_\infty - R_p)$$

which applies to the build-up period of acceleration-type kinetics (curve B in Fig. 9.6). Show that this equation can be derived on the basis of a reaction scheme which assumes that (a) a potential polymerization center is generated when adsorption of an aluminum ethylate dimer $(AlEt_3)_2$ occurs on an active site; (b) the reaction of the first monomer molecule with this potential polymerization center generates a polymerization center; and (c) a polymerization center undergoes propagation reaction with monomer molecules, the polymerization center being regenerated after monomer addition in each propagation step.

Answer:

The reaction scheme can be represented as (Kumar and Gupta, 1978):

$$S + (AlEt_3)_2 \rightleftharpoons S^\star \tag{P9.8.1}$$

$$S^\star + M \xrightarrow{k_c} C^\star \tag{P9.8.2}$$

$$C^\star + M \xrightarrow{k_p} C^\star \tag{P9.8.3}$$

where S represents a surface site (TiCl$_3$) for adsorption, S* a *potential* polymerization center, and C* a polymerization center. Note that S* is kinetically distinguished from C* in its reaction with monomer M. Reaction in Eq. (P9.8.3) is the propagation reaction.

Let $[C^\star]_\infty$ be the total concentration of polymerization centers present at $t = \infty$, i.e., at the stationary state. In the early stages of the reaction, $[C^\star] < [C^\star]_\infty$ and at any time the following mass balance (Kumar and Gupta, 1978) holds:

$$[S] + [S^\star] + [C^\star] = [S]_0 \qquad (P9.8.4)$$

where $[S]_0$ is the total concentration of the active sites at $t = 0$. At the stationary state, Eq. (P9.8.4) becomes

$$[S]_\infty + [S^\star]_\infty + [C^\star]_\infty = [S]_0 \qquad (P9.8.5)$$

Subtracting Eq. (P9.8.5) from Eq. (P9.8.4) one obtains

$$([S]_\infty - [S]) + \left([S^\star]_\infty - [S^\star]\right) + \left([C^\star]_\infty - [C^\star]\right) = 0 \qquad (P9.8.6)$$

Since S* is an intermediate entity in the formation of the polymerization center, $[S^\star]_\infty$ $\simeq 0$. This follows from the reasoning that after a sufficiently long interval of time, monomer molecules would have reacted completely with all the potential polymerization centers by the irreversible reaction in Eq. (P9.8.2). Moreover, since both [S] and $[S]_\infty$ are large numbers and it may be assumed that only a few of the active sites participate in polymerization, $[S]_\infty - [S] \simeq 0$. Therefore, Eq. (P9.8.6) reduces to

$$[C^\star]_\infty - [C^\star] - [S^\star] \simeq 0$$

or,

$$[S^\star] = [C^\star]_\infty - [C^\star] \qquad (P9.8.7)$$

The rate of formation of polymerization centers [cf. Eq. (P9.8.2)] is given by

$$d[C^\star]/dt = k_c[M][S^\star] \qquad (P9.8.8)$$

Substituting for $[S^\star]$ from Eq. (P9.8.7),

$$d[C^\star]/dt = k_c[M]\left([C^\star]_\infty - [C^\star]\right) \qquad (P9.8.9)$$

The rates of polymerization at time t and in the stationary period are given by

$$R_p = k_p[M][C^\star] \qquad (P9.8.10)$$
$$(R_p)_\infty = k_p[M][C^\star]_\infty \qquad (P9.8.11)$$

If [M] is constant, multiplying Eq. (P9.8.9) by $k_p[M]$ gives

$$d\left(k_p[M][C^\star]\right)/dt = k_c[M]\left(k_p[M][C^\star]_\infty - k_p[M][C^\star]\right)$$

or,

$$dR_p/dt = k\left(R_\infty - R_p\right) \qquad (P9.8.12)$$

where $k = k_c[M]$.

Problem 9.9   For the polymerization system VCl$_3$/AlR$_3$/4-methyl-1-pentene, Burfield et al. (1972) proposed the following reaction scheme :

(a) alkylated transition metal entities form the polymerization centers and chain initiation involves reaction between the polymerization centers and monomer;

(b) there is excess alkylaluminum over and above the amount needed to activate the transition metal sites in (a) and so alkylaluminum molecules are also involved, besides monomer, in competitive adsorption reaction with polymerization centers;

(c) chain propagation involves insertion of the adsorbed monomer into a transition metal-carbon bond (i.e., the process is monometallic); and

(d) chain transfer reactions take place between the growing chain attached to the polymerization center and both adsorbed metal alkyl and adsorbed monomer.

Derive expressions for (i) overall rate of steady state polymerization and (ii) number average degree of polymerization $\overline{DP}_n$. Predict the mode of variation of $\overline{DP}_n$ with reaction conditions.

Answer:

(a) Chain initiation arises from the following reactions :

$$S + Al(Et)_3 \rightleftharpoons S^\star \tag{P9.9.1}$$

$$S^\star + M \xrightarrow{k_i} C^\star \tag{P9.9.2}$$

(b) Reversible competitive adsorption of monomer and alkylaluminum on polymerization centers is represented by

$$C^\star + M \overset{K_M}{\rightleftharpoons} \underline{M} \tag{P9.9.3}$$

$$C^\star + A \overset{K_A}{\rightleftharpoons} \underline{A} \tag{P9.9.4}$$

where $\underline{M}$ amd $\underline{A}$ represent adsorbed monomer and adsorbed alkylaluminum, respectively. The adsorptions are described by Langmuir-Hinshelwood isotherms, viz.,

$$\theta_M = \frac{K_M[M]}{1 + K_M[M] + K_A[A]} \tag{P9.9.5}$$

$$\theta_A = \frac{K_A[A]}{1 + K_M[M] + K_A[A]} \tag{P9.9.6}$$

where θ_M and θ_A are fractions of polymerization centers having adsorbed M and A, respectively. (It should be noted that where the metal alkyl is known to be dimeric, $K^{1/2}[A_2]^{1/2}$ should be substituted for [A]; [A$_2$] is the concentration of metal alkyl dimer and K is the dissociation constant.)

(c) Chain propagation by a monometallic process can be represented by

$$C^\star + \underline{M} \xrightarrow{k_p} C^\star \tag{P9.9.7}$$

The overall rate equation for the period of steady-state polymerization is thus given by (Burfield et al., 1972)

$$R_p = k_p \theta_M [C^\star] \tag{P9.9.8}$$

where k_p is the propagation rate constant with respect to adsorbed monomer, and $[C^\star]$ is the concentration of polymerization centers during this polymerization period. Substituting for θ_M from Eq. (P9.9.5) yields

$$R_p = \frac{k_p K_M [M] [C^\star]}{1 + K_M [M] + K_A [A]} \tag{P9.9.9}$$

(d) The rates of chain transfer with adsorbed metal alkyl, $R_{tr,A}$, and with adsorbed monomer, $R_{tr,M}$, are given by

$$R_{tr,A} = k_{tr,A} \theta_A [C^\star] \tag{P9.9.10}$$

$$R_{tr,M} = k_{tr,M} \theta_M [C^\star] \tag{P9.9.11}$$

(e) The number average degree of polymerization can be given (Burfield, 1972) by

$$\overline{DP}_n = \frac{\int_0^t k_p \theta_M [C^\star] dt}{[C^\star] + \int_0^t k_{tr,M} \theta_M [C^\star] dt + \int_0^t k_{tr,A} \theta_A [C^\star] dt} \tag{P9.9.12}$$

Integrating, inverting, and removing $[C^\star]$ gives

$$\frac{1}{\overline{DP}_n} = \frac{1 + k_{tr,M} \theta_M t + k_{tr,A} \theta_A t}{k_p \theta_M t} \tag{P9.9.13}$$

Substituting for θ_A, θ_M and simplifying yields (Burfield, 1972):

$$\frac{1}{\overline{DP}_n} = \frac{k_{tr,A} K_A [A]}{k_p K_M [M]} + \frac{1}{k_p K_M [M] t} + \frac{K_A [A]}{k_p K_M [M] t} + \frac{(k_{tr,M} + 1/t)}{k_p} \tag{P9.9.14}$$

From Eq. (P9.9.14) one can predict that $\overline{DP}_n$ will: (a) depend initially on the duration of polymerization, becoming invariant at a later stage of the polymerization, i.e., when t is large; (b) decrease with increasing metal alkyl concentration; (c) increase with increasing monomer concentration; and (d) be independent of the metal halide concentration (since $[C^\star] \propto [VCl_3]$).

[**Note:** The model agrees with the observed kinetics and molecular weight dependencies of the $VCl_3/Al(i-Bu)_3$/4-methyl-1-pentene system (Burfield, 1972). It also has general applicability to many other Ziegler-Natta systems.]

To evaluate the kinetic parameters using Eq. (P9.9.9), the concentration of polymerization centers $[C^\star]$ is to be determined. This can be done by experiments in which the $C^\star$ sites are quenched (made inactive) with CH_3O^3H, ^{14}CO, or $^{14}CO_2$ (Tait and Watkins, 1989; Vozka and Mejzlik, 1990). The number average molecular weight (together with polymer yield) and ^{14}C-labeled Group I-III-metal alkyl component can also be used to determine $[C^\star]$. However, each of the techniques has some shortcomings that must be taken into account to ensure reliable results.

Equation (P9.9.9) has been applied successfully to the polymerization systems characterized by kinetic rate-time profiles of the type (a) in Fig. 9.5. Table 9.4 shows comparative values for some of the principal kinetic parameters derived from this equation.

The monometallic mechanism of Cossee and Arlman (1964) for Ziegler-Natta polymerization has found much acceptance in the literature as it has sound quantum mechanical basis. According to this mechanism, as described earlier and shown in Figs. 9.3 and 9.4, the initiation process involves interaction of aluminum alkyl with an octahedral ligand vacancy around Ti, leading to

Table 9.4 Values of Some Kinetic Parameters for Ziegler-Natta
Polymerization of 4-Methyl-1-pentene at 40°C

Catalyst system	K_A (L/mol)	K_M (L/mol)	$k_p \times 10^3$ (min^{-1})
δ-TiCl$_3$·0.33AlCl$_3$-Al(Bu^i)$_3$	16	0.38	1.29
MgCl$_2$/EB/TiCl$_4$-Al(Bu^i)$_3$	30	1.5	4.26
VCl$_3$-Al(Bu^i)$_3$	1.5	0.16	3.10

Source: Tait and Watkins, 1989.

alkylation of the Ti and regeneration of the ligand vacancy. This alkylated Ti with a ligand vacancy constitutes a *polymerization center*. Its formation can thus be described schematically by

$$\bullet\text{Ti}-\square \ + \ \text{AlEt}_3 \ \rightleftharpoons \ \bullet\text{C}^\star-\square \ + \ \text{AlEt}_2\text{Cl} \tag{9.30}$$

where $\bullet\text{C}^\star-\square$ represents a polymerization center.

The propagation reaction proceeds in two steps: (a) adsorption of monomer M at the ligand vacancy of the polymerization center followed by (b) monomer insertion in the polymer chain (via a complex formation) with simultaneous regeneration of the polymerization center with a ligand vacancy (see Fig. 9.4). The propagation reaction can thus be described schematically as

$$\bullet\text{C}^\star-\square \ + \ \text{M(adsorbed)} \ \rightleftharpoons \ \text{Monomer complex} \ \xrightarrow{k'_p} \ \bullet\text{C}^\star-\square \tag{9.31}$$

The polymerization center, according to Cossee, reacts with the monomer molecules which are adsorbed on the catalyst. The process involves migration of the adsorbed molecules on the catalyst surface towards the empty ligand to form the monomer (activated) complex, in which the monomer molecule is bonded between the alkyl group and the transition metal atoms. The alkyl group then breaks off from the titanium ion giving rise to a vacant ligand site in the regenerated polymerization center.

Problem 9.10 Based on the mechanism of Arlman and Cossee, derive a suitable expression for the rate of polymerization and predict the rate behavior that would be expected according to this rate model (Kumar and Gupta, 1978).

Answer:

For the propagation reaction described by Eq. (9.31), the rate in the stationary period may be written as

$$R_\infty \ = \ k'_p \, [\text{Monomer complex}] \tag{P9.10.1}$$

Since the monomer complex is in equilibrium with $\bullet\text{C}^\star-\square$ and the adsorbed monomer molecules,

$$[\text{Monomer complex}] \ = \ K \, [\bullet\text{C}^\star-\square] \, \theta_M \tag{P9.10.2}$$

where θ_M is the fraction of the catalyst surface covered by the monomer molecules. Therefore,

$$R_\infty \ = \ k'_p \, K \, [\bullet\text{C}^\star-\square] \, \theta_M \ = \ k_p \, [\bullet\text{C}^\star-\square] \, \theta_M \tag{P9.10.3}$$

where $k_p = k_p' K$ and θ_M is given by the Langmuir equation [cf. Eq. (P9.4.4)]:

$$\theta_M = \frac{K_M[M]}{1 + K_M[M]} \tag{P9.10.4}$$

Substitution of Eq. (P9.10.4) in Eq. (P9.10.3) then gives

$$R_\infty = k_p[\bullet C^\star - \square] \frac{K_M[M]}{1 + K_M[M]} \tag{P9.10.5}$$

The concentration of $[\bullet C^\star - \square]$ can be evaluated by assuming an equilibrium of the initiation reaction in Eq. (9.30), i.e.,

$$K_A = \frac{[\bullet C^\star - \square][AlEt_2Cl]}{[\bullet Ti - \square][AlEt_3]} \tag{P9.10.6}$$

From stoichiometry, one has

$$[\bullet Ti - \square] = [\bullet Ti - \square]_0 - [\bullet C^\star - \square] \tag{P9.10.7}$$

$$[AlEt_3] = [AlEt_3]_0 - [\bullet C^\star - \square] \tag{P9.10.8}$$

$$[AlEt_2Cl] = [\bullet C^\star - \square] \tag{P9.10.9}$$

where the subscript $_0$ indicates the concentration at $t = 0$. Substituting these in Eq. (P9.10.6) gives

$$[\bullet C^\star - \square]^2 = K_A \left([\bullet Ti - \square]_0 - [\bullet C^\star - \square]\right) \left([AlEt_3]_0 - [\bullet C^\star - \square]\right) \tag{P9.10.10}$$

For low values of $[\bullet C^\star - \square]$, one can use the approximation

$$[AlEt_3] \simeq [AlEt_3]_0 \tag{P9.10.11}$$

and hence

$$\frac{[\bullet C^\star - \square]^2}{K_A[AlEt_3]_0} = [\bullet Ti - \square]_0 - [\bullet C^\star - \square] \tag{P9.10.12}$$

Two possible cases may now be considered (Kumar and Gupta, 1978):

(a) When $K_A[AlEt_3]_0 \gg [\bullet Ti - \square]_0$ and hence $K_A[AlEt_3]_0 \gg [\bullet C^\star - \square]$, the left side of Eq. (P9.10.12) is close to zero and one has

$$[\bullet C^\star - \square] \simeq [\bullet Ti - \square]_0 \tag{P9.10.13}$$

and Eq. (P9.10.5) becomes

$$R_{\infty 1} = k_p[\bullet Ti - \square]_0 \frac{K_M[M]}{1 + K_M[M]} \tag{P9.10.14}$$

(b) When $[\bullet Ti - \square]_0 \gg K_A[AlEt_3]_0$, the term $[\bullet C^\star - \square]$ in the right side of Eq. (P9.10.12) can be neglected, and Eq. (P9.10.12) reduces to

$$[\bullet C^\star - \square]^2 = K_A[\bullet Ti - \square]_0[AlEt_3]_0$$

or,

$$[\bullet C^{\star}-\Box] = [\bullet Ti-\Box]_0 \left[\frac{K_A [AlEt_3]_0}{[\bullet Ti-\Box]_0} \right]^{1/2} \tag{P9.10.15}$$

and hence, from Eq. (P9.10.5),

$$R_{\infty 2} = k_p [\bullet Ti-\Box]_0 \left[\frac{K_A [AlEt_3]_0}{[\bullet Ti-\Box]_0} \right]^{1/2} \left(\frac{K_M [M]}{1 + K_M [M]} \right) \tag{P9.10.16}$$

The adsorption constant K_M in these equations has the same meaning as in the models considered previously and it is small in value (see Tables 9.3 and 9.4). So one should observe that the rates $R_{\infty 1}$ and $R_{\infty 2}$ in Eqs. (P9.10.14) and (P9.10.16) are proportional to the monomer concentrations. These equations also show that the stationary rate would be affected by the concentration of AlEt$_3$ at its lower range ($K_A [AlEt_3]_0/[\bullet Ti-\Box]_0 \ll 1$) and would be independent at its higher range ($K_A [AlEt_3]_0/[\bullet Ti-\Box]_0 \gg 1$), which are in qualitative agreement with experimental observations. The above model based on the Cossee-Arlman mechanism does not, however, conform precisely to all the kinetic features of Ziegler-Natta polymerizations.

9.4.4.3   Average Degree of Polymerization

The number average degree of polymerization at any given time is given by the following general relation

$$\overline{DP}_n = \frac{\text{Number of monomer molecules polymerized in time } t}{\text{Number of polymer molecules produced in time } t} \tag{9.32}$$

The number of monomer molecules polymerized can be obtained by integrating the rate of polymerization. The denominator can, however, be obtained only if the transfer and termination rates are known. If R_t denotes the sum of these rates, then (Kumar and Gupta, 1978):

$$\overline{DP}_n = \frac{\int_0^t R_p\, dt}{[C^{\star}]_t + \int_0^t R_t\, dt} \tag{9.33}$$

where $[C^{\star}]_t$ is the concentration of the polymerization centers at time t. Equation (9.33) can be applied for the stationary state to find $\overline{DP}_n$. Since at the stationary state, R_p and R_t are both constant, Eq. (9.33) can be written as

$$\frac{1}{\overline{DP}_n} = \frac{R_{t\infty}}{R_{p\infty}} + \frac{[C^{\star}]_\infty}{t\, R_{p\infty}} \tag{9.34}$$

In Eq. (9.34), the contribution to the integrals from the transition zone has been ignored for simplicity. For long durations of time, the second term in right side of Eq. (9.33) tends to zero, and the equation reduces to

$$1/\overline{DP}_n = R_{t\infty}/R_{p\infty} \tag{9.35}$$

In order to evaluate $\overline{DP}_n$, the termination and transfer processes must be known. In the case of propylene, for example, these can be described by Eqs. (9.9)–(9.12). The corresponding rates are given by

$$\text{Transfer to monomer}: \quad R_{tr,\text{M}} \;=\; k_{tr,\text{M}}\,[\text{C}^\star][\text{M}] \tag{9.36}$$

$$\text{Transfer to AlEt}_3: \quad R_{tr,\text{A}} \;=\; k_{tr,\text{A}}\,[\text{C}^\star][\text{A}] \tag{9.37}$$

$$\text{Spontaneous transfer}: \quad R_s \;=\; k_s\,[\text{C}^\star] \tag{9.38}$$

This equation has been confirmed experimentally (Natta and Pasquon, 1959). A transfer agent like H_2 is often added to reduce the polymer molecular weight.

The polymers obtained by Ziegler-Natta polymerization generally have very wide molecular weight distributions, e.g., PDI (= $\overline{M}_w/\overline{M}_n$) is 5–20 for polyethylene and 5-15 for polypropylene. While the cause of the wide dispersity is not clearly known, some workers believe that the propagation reaction becomes diffusion controlled after a few percent conversion contributing to the large dispersity and some other workers believe that the rate constants are dependent upon the molecular size.

9.5   Supported Metal Oxide Catalysts

Metal oxides supported on finely divided inert materials have been useful for polymerization of ethylene and other vinyl monomers. The mechanism is presumably similar to that of heterogeneous Ziegler-Natta polymerization with initiation occurring at active sites on the catalyst surface (Boor, 1979). Unlike the traditional Ziegler-Natta catalyst which are two-component systems, the supported metal-oxide catalysts are usually one-component systems. Chromium, vanadium, molybdenum, nickel, cobalt, niobium, tantalum, tungsten, and titanium are among the metals that have been investigated for these catalysts. Materials typically used as supports include alumina, silica, and charcoal.

The most active catalyst is chromium oxide (Tait and Watkins, 1989) with silica (SiO_2) or aluminosilicates (mixed SiO_2/Al_2O_3) used as the support material, often modified with titania (TiO_2). The chromium oxide ($Cr^{VI}O_3$) catalyst is referred to as Phillips catalyst. The best known among other metal oxide catalysts is the molybdenum oxide ($Mo^{VI}O_3$) catalyst, originally developed at Standard Oil of Indiana. The catalysts are prepared by one of two methods – impregnation and coprecipitation.

Much of the high-density polyethylene is manufactured by ethylene polymerization with supported metal oxide catalysts. The supported metal oxides, however, are not as active as Ziegler-Natta catalysts and they do not generate a high degree of stereoregularity. Thus propylene forms partially crystalline polymer and higher 1-alkenes give only amorphous product.

9.5.1   Polymerization Mechanism

The mechanism of polymerization on one-component metal oxide catalysts seems to differ from that on traditional Ziegler-Natta two-component catalysts only in the initiation stage, while the mechanism of continued propagation of polymer chain in the two cases has many common features. The details of initiation mechanism are not precisely understood. It is, however, recognized that the reaction is a surface catalyzed process which requires the monomer to be adsorbed onto the catalyst surface. It is proposed that initiation involves the formation of a metal-carbon σ-bond, which is followed, consecutively, by coordination of an incoming monomer molecule and insertion

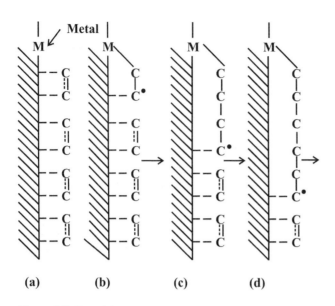

Figure 9.7 Bound-ion-radical mechanism for polymerization on a catalyst surface : (a) adsorbed monomer; (b) initiation; (c), (d) etc. propagation. (After Friedlander and Oita, 1957.)

into the metal-carbon bond. In general, two mechanisms have been proposed to explain polymerization with precipitated catalysts: (a) the bound-ion-radical mechanism and (b) the bound-ion-coordinate mechanism. The *bound-ion-radical mechanism* postulates that chain growth occurs in a chemisorbed layer of monomer molecules after initiation by radicals or ion-radicals bound to the surface of the catalyst, while in the *coordinate mechanism*, chain growth is assumed to take place from a complex ionic center in the catalyst.

9.5.1.1 *Bound-Ion-Radical Mechanism*

According to this mechanism, the catalyst adsorbs monomer on its surface [Fig. 9.7(a)] and initiation occurs when an adsorbed monomer is polarized by some species on the catalyst surface converting it to an ion (or a radical or an ion-radical pair) bound to the surface [Fig. 9.7(b)]. Propagation takes place along the surface [Fig. 9.7(c)] and the polymer chain is eventually terminated and desorbed from the surface, being replaced by fresh monomer. The chain termination may be caused by transfer with monomer or spontaneous transfer or detachment from the surface. The molecular regularity of the polymer formed depends on the surface layer line-up of the adsorbed monomer molecules.

In bound-ion-radical mechanism for ethylene polymerization, initiation can occur either through a chemisorbed ethylene molecule or a chemisorbed hydrogen atom. With chemisorbed ethylene as initiator, polymerization occurs simultaneously at two sites and the growing polymer, which is attached at each end (Fig. 9.7), is converted to a single attachment by transfer or termination at one of the ends. With adsorbed hydrogen as initiator, the initiation is caused by an adsorbed hydrogen atom attaching to a neighboring adsorbed ethylene, following which propagation occurs in the adsorbed layer, with the growing chain adding as an ion-radical to a neighboring adsorbed ethylene molecule (Fig. 9.8). Chain transfer occurs by the shift of a hydrogen atom to a neighbor-

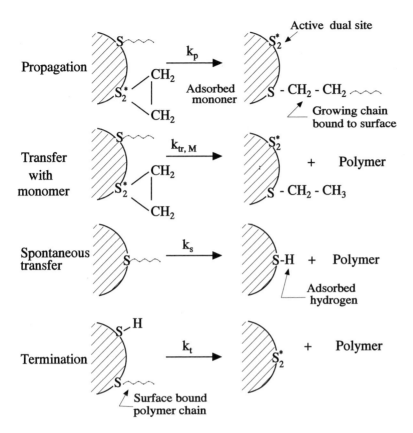

Figure 9.8 Ethylene polymerization in adsorbed layer by adsorbed hydrogen initiation. (After Friedlander, 1959.)

ing adsorbed monomer or spontaneously to a vacant surface site, while chain termination (in the absence of poisons) occurs only by reaction with a chemisorbed hydrogen which frees a dual site for readsorption of ethylene.

Problem 9.11 Metal-oxide catalyzed polymerization of ethylene was carried out in benzene solution in a stirred autoclave with a suspension of hydrogen-reduced molybdena-alumina catalyst (Friedlander and Oita, 1957). The pressure was maintained nearly constant by repressuring the autoclave with ethylene as it was consumed in the polymerization process. Temperatures of 200–275°C were studied. The ethylene concentration in solution was controlled by adjusting the pressure (in the range 625 to 1000 psi) at any particular temperature. The ethylene uptake rate (rate of pressure drop, dP/dt) was measured as a function of the catalyst amount (w_{cat}) and ethylene concentration in solution (calculated from ethylene partial pressure) $[C_2H_4]_s$ at different tmperatures. The experimental data plotted as $(dP/dt)/(w_{cat}[C_2H_4]_s)$ vs. $[C_2H_4]_s$ produce good fit to straight lines whose slopes decrease at higher temperatures. Further, at higher temperatures, the plot of $(dP/dt)/w_{cat}$ vs. $[C_2H_4]_s$ fits to straight lines passing through the origin.

Derive suitable expressions to explain the aforesaid experimental results, considering that the polymerization takes place in an adsorbed layer of ethylene with initiation by adsorbed hydrogen and transfer and termination processes as illustrated in Fig. 9.8.

Answer:

Since chemisorbed ethylene disappears by initiation reaction with adsorbed hydrogen, propagation reaction with surface-bound polymer chain, and polymer chain transfer with monomer (Fig. 9.8), one may write (Friedlander and Oita, 1957):

$$dP/dt = k_i [C_2H_4]_a [H]_a + (k_p + k_{tr,M}) [C_2H_4]_a \sum_n [\text{Polymer}]_a \qquad (P9.11.1)$$

where $[C_2H_4]_a$ is the concentration of ethylene in the adsorbed layer, $[H]_a$ is the concentration of adsorbed hydrogen, and $\sum [\text{Polymer}]_a$ that of all growing chains (bound to the surface). Under steady-state conditions, the growing chains arise in initiation and disappear by spontaneous transfer and termination. Thus at steady state,

$$k_i [H]_a [C_2H_4]_a = (k_s + k_t [H]_a) \sum_n [\text{Polymer}]_a$$

or

$$\sum_n [\text{Polymer}]_a = k_i [H]_a [C_2H_4]_a / (k_s + k_t [H]_a) \qquad (P9.11.2)$$

Substituting in Eq. (P9.11.1),

$$dP/dt = k_i [C_2H_4]_a [H]_a \left\{ 1 + \frac{(k_p + k_{tr,M})}{k_s + k_t [H]_a} [C_2H_4]_a \right\} \qquad (P9.11.3)$$

At low concentrations of ethylene,

$$[C_2H_4]_a = K_E [S_2^\star][C_2H_4]_s \qquad (P9.11.4)$$

where K_E is the Langmuir adsorption equilibrium constant, $[C_2H_4]_s$ is the equilibrium concentration of ethylene in solution, and $[S_2^\star]$ is the concentration of active dual sites (at which ethylene can be adsorbed with two-point adsorption). $[C_2H_4]_s$, in turn, is related to the partial pressure of ethylene, p_E, in the gas phase by

$$[C_2H_4]_s = K_s p_E \qquad (P9.11.5)$$

where K_s is the equilibrium constant for saturation.

Substitution of Eq. (P9.11.4) into Eq. (P9.11.3) yields

$$\frac{(dP/dt)}{[C_2H_4]_s} = K_E k_i [S_2^\star][H]_a \left\{ 1 + \frac{(k_p + k_{tr,M})K_E[S_2^\star]}{k_s + k_t [H]_a} [C_2H_4]_s \right\} \qquad (P9.11.6)$$

which simplifies to

$$\frac{(dP/dt)}{[S_2^\star][C_2H_4]_s} = K_E k_i [H]_a + B [C_2H_4]_s \qquad (P9.11.7)$$

where B is a complex function of the various equilibrium and rate constants. Since $[S_2^\star]$ is proportional to the weight of the solid catalyst, a plot of $(dP/dt)/(w_{cat} [C_2H_4]_s)$ vs. $[C_2H_4]_s$ should yield a straight line at each temperature, as observed experimentally. Since termination and spontaneous transfer become more important at higher temperatures, the value of the slope B would decrease at higher temperatures and if B is small, Eq. (P9.11.7) would simplify to

$$\frac{(dP/dt)}{[S_2^\star]} = K_E k_i [H]_a [C_2H_4]_s \tag{P9.11.8}$$

Thus, a plot of $(dP/dt)/w_{cat}$ vs. $[C_2H_4]_s$ at higher temperatures should fit straight lines passing through the origin, as observed experimentally.

Note: If ethylene initiation is considered, instead of hydrogen initiation, the growing polymer is attached at both the ends and is converted to a single attachment by transfer or termination at one of the ends. The equations describing polymerization take exactly the same form as in the case of initiation by hydrogen but are complicated by extra terms dealing with interconversion of the polymer growing from one or both ends.

Problem 9.12 Derive an expression for the average degree of polymerization corresponding to the reaction scheme assumed in Problem 9.11. Predict from this relation how the molecular weight of the polymer would be affected by (a) increased amount of catalyst, (b) increased amount of hydrogen adsorbed on the catalyst, (c) increased ethylene concentration, and (d) increased temperature.

Answer:

The average degree of polymerization is determined as the sum of all the chain growth reactions divided by all the chain transfer and termination reactions (Friedlander and Oita, 1957):

$$\overline{DP}_n = \frac{k_p [C_2H_4]_a \sum_n [\text{Polymer}]_a}{\left(k_{tr,M} [C_2H_4]_a + k_s + k_t [H]_a\right) \sum_n [\text{Polymer}]_a} \tag{P9.12.1}$$

The sum of the growing polymer chains $\sum_n [\text{Polymer}]_a$ cancels out, and the reciprocal of the simplified $\overline{DP}_n$ becomes

$$\frac{1}{\overline{DP}_n} = \frac{n}{k_{tr,M}} k_p + \frac{k_s + k_t [H]_a}{k_p [C_2H_4]_a} \tag{P9.12.2}$$

Substituting Eq. (P9.11.4) into Eq. (P9.12.2), one obtains

$$\frac{1}{\overline{DP}_n} = \frac{k_{tr,M}}{k_p} + \frac{k_s + k_t [H]_a}{k_p K_E [S_2^\star][C_2H_4]_s} \tag{P9.12.3}$$

Thus the molecular weight should increase slightly with the increased amount of catalyst or increased ethylene concentration, and decrease with an increase in hydrogen adsorbed on the catalyst. (These predictions are in accord with the experimental observation.)

Because k_p and $[C_2H_4]_s$ are much larger than k_s and $k_t [H]_a$, especially at lower temperatures, Eq. (P9.12.3) simplifies (Friedlander and Oita, 1957) to

$$1/\overline{DP}_n \simeq k_{tr,M}/k_p \tag{P9.12.4}$$

Expressing the rate constants in the Arrhenius form,

$$\frac{1}{\overline{DP}_n} = \frac{A_{tr,M}}{A_p} \exp\left[\frac{E_p - E_{tr,M}}{RT}\right] \tag{P9.12.5}$$

The energy of activation for transfer with monomer, $E_{tr,M}$, would ordinarily be greater than the energy of activation for propagation, E_p; so the molecular weight should decrease with increasing temperature.

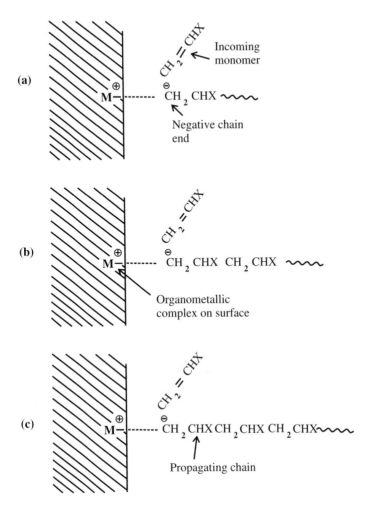

Figure 9.9 The process of chain growth in bound-ion-coordination mechanism for polymerization on a catalyst surface with the growth taking place from a single active site and the monomer being supplied from the liquid phase. Consecutive propagation steps are represented in (a), (b), and (c). (After Natta, 1955.)

9.5.1.2   Bound-Ion-Coordination Mechanism

The chain growth by this mechanism is shown schematically in Fig. 9.9. The reaction is considered anionic because the negative end of the olefin coordinates with an organometallic complex in the surface. Olefin molecules are inserted between the metal ion and the alkyl chain in the complex one at a time, extending the chain by two carbon atoms. The mechanism has the advantage that the ion pair never becomes widely separated. Addition at an electron-deficient bond between the metals in an organo-metallic complex has also been proposed. In either case, the addition of a monomer molecule held in a fixed orientation at the instant of reaction is the most importat step as it decides the stereospecificity of the polymer. Moreover, since the orientation of the monomer molecules can greatly reduce the activation energy necessary for the propagation step, a high rate of reaction would be expected.

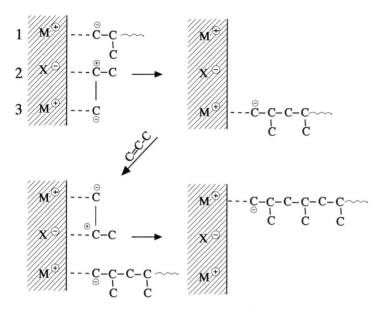

Figure 9.10 The surface-coordinate mechanism of chain growth applied to a surface consisting of three points of contact. (After Friedlander and Resnick, 1958.)

Problem 9.13 The coordinate and bound-ion-radical mechanisms, although apparently quite dissimilar, have many features in common. If, in the bound-radical hypothesis, the surface involved decreases to the limiting case of three points of contact, the two mechanisms would appear to be quite similar (Friedlander, 1959). Explain this similarity by applying the idea of growth on the surface (used in the bound-radical hypothesis) to the coordinate mechanism, considering a surface with only three points of contact.

Answer:

The scheme is illustrated in Fig. 9.10. The growing polymer molecule is in the form of an organometallic compound at position 1, and adsorbed next to it at positions 2 and 3 is an olefin molecule. The growing end of the polymer is transferred to the olefin molecule giving a new organometallic compound in position 3. The adsorption of a new olefin molecule in the free positions 1 and 2, followed by transfer of the organometallic compound back from 3 to 1 in a manner similar to the original olefin addition step, then gives an organometallic compound in the original position but two monomer units longer. This increase can continue back and forth along that portion of the surface that has the proper geometry. Similarity between the surface-coordinate mechanism shown in Fig. 9.10 and the coordinate mechanism shown in Fig. 9.9 is easily seen.

9.6 Ziegler-Natta Copolymerization

Random copolymers of ethylene and α-olefins (1-alkenes) can be obtained with Ziegler-Natta catalysts, the most important being those of ethylene and 1-butene (LLDPE) and of ethylene and propylene (EPM or EPR and EPDM). Some reactivity ratios are listed in Table 9.5. Tha ratios vary with the nature and physical state of the catalyst and in most instances, r_1r_2 is close to unity. However, all these values show that ethylene is much more reactive than higher alkenes. Copolymers produced using Ziegler-Natta catalysts usually have a wide range of compositions. This may

Table 9.5 Some Values of Reactivity Ratios (r) in Ziegler-Natta Copolymerization

Monomer 1	Monomer 2	Catalyst[a]	Reaction type[b]	r_1	r_2
Ethylene	Propylene	$TiCl_3/AlR_3$	H	15.7	0.11
		$TiCl_4/AlR_3$	C	33.4	0.03
		VCl_3/AlR_3	H	5.6	0.14
		VCl_4/AlR_3	C	7.1	0.09
Ethylene	1-Butene	VCl_3/AlR_3	H	27.0	0.04
		VCl_4/AlR_3	C	29.6	0.02
Propylene	1-Butene	VCl_3/AlR_3	H	4.0	0.25
		VCl_4/AlR_3	C	4.3	0.23

[a]R = C_6H_{13}; [b]H = heterogeneous; C = colloidal.
Source: Boor, Jr., 1979.

be due to the presence of different active sites in the catalyst giving rise to different reactivity ratios, or due to encapsulation of active sites contributing to fall in activity.

More homogeneous copolymer compositions are obtained with soluble Zieg-ler-Natta catalysts, especially where monomer compositions are maintained relatively constant during polymerization. Commercially important ethylene-propyl-ene binary copolymers (EPM rubbers) and ethylene-propylene-diene ternary copolymers (EPDM rubbers) are made by this process.

Ziegler-Natta catalysts have been used to prepare a number of block copolymers, though in most cases these are mixed with significant amounts of homopolymers. The Ziegler-Natta method is clearly inferior to anionic polymerization for preparing block copolymers of controlled compositions. Nevertheless, block co-polymers of ethylene and propylene (*Polyallomers*), which are high-impact plastics exhibiting crystallinity characteristics of both isotactic polypropylene and linear polyethylene, have been made by this process as a commercial product.

9.7   *Metallocene-Based Ziegler-Natta Catalysts*

Ziegler-Natta catalysts are mostly heterogeneous and function in a ternary gas-polymer-catalyst or liquid-polymer-catalyst system. The first homogeneous Ziegler-Natta catalyst was discovered in 1957 (Breslow and Newburg, 1957; Natta et al., 1957). The catalyst, *bis*(cyclopentadienyl)titanium dichloride (Cp_2TiCl_2, Cp = η^5-cyclopentadienyl) activated with alkylaluminum chloride (AlR_2Cl) exhibited a low polymerization activity for ethylene ($\approx 10^4$ g polyethylene/mol Ti-h-atm) and none for propylene. It was found later that small amounts of water increased significantly the activity of the catalyst as the reaction between water and aluminum alkyls resulted in the formation of aluminoxanes. In 1980 Kaminsky and coworkers (Sinn and Kaminsky, 1980) used oligomeric methylaluminoxane (MAO) with Group IVB metallocene compounds to obtain ethylene polymerization catalysts having extremely high activities. For instance, a polyethylene *productivity* of 9.3×10^6 g polyethylene/mol Ti-h-atm was obtained with Cp_2TiCl_2/MAO at 20°C and 9×10^7 g polyethylene/mol Zr-h-atm with Cp_2ZrCl_2/MAO at 70°C. However, these catalysts were nonstereospecific and produced only atactic polypropylene.

In the early 1980s Brintzinger and coworkers (Wild et al., 1982; 1985; Huang and Rempel, 1995) synthesized racemic ethylene-bridged *bis* (indenyl) zirconium dichloride, $Et(Ind)_2ZrCl_2$, and racemic ethylene-bridged *bis*(4,5,6,7-tetrahydro-indenyl)zirconium dichloride, $Et(H_4Ind)_2ZrCl_2$, as well as their titanium analogues, $Et(Ind)_2TiCl_2$ and $Et(H_4Ind)_2TiCl_2$, which have both *meso* and racemic configurations. The $Et(Ind)_2ZrCl_2$ and $Et(H_4Ind)_2ZrCl_2$ catalysts, activated with MAO, catalyzed the stereospecific polymerization of propylene showing high productivities. It was the

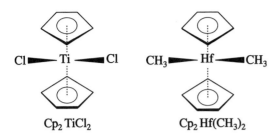

Figure 9.11 Structures of two metallocenes with C_{2v} symmetry.

first time that the isotactic polyolefins were made by homogeneous Ziegler-Natta polymerization (Huang and Rempel, 1995). This discovery was highly significant as it demonstrated the ability of chiral *ansa*-indenyl ligands (Latin *ansa*, a handle) to exert stereochemical control on migratory insertion of a vinyl monomer, in contrast to the *meso*-Et(Ind)$_2$TiCl$_2$/MAO system, which, as expected, produced only atactic polypropylene. A large group of *ansa*-metallocene compounds were developed subsequently, each possessing unique catalytic activity and stereospecificity. These homogeneous metallocene-based catalysts have theoretical significance with regard to studies of Ziegler-Natta polymerization because the polymerization processes in homogeneous systems, often being more simple, lend to greatly simplified kinetic and mechanistic analyses.

The metallocene catalysts have been under development for nearly 25 years and a breakthrough has been achieved for a technical realization of these catalyst systems, culminating in the announcement of "single-site" catalysts (SSC) by different companies. These catalysts permit olefins to react only at single sites on the catalyst molecules with unprecedented control over reactivity and so are being used to produce tailor-made high performance polyolefins. Metallocene catalysts also have opened up possibilities to create novel polymers that have never been produced by conventional Ziegler-Natta catalysts. One such example is the development of hybrid thermoplastic polyolefins having almost any combination of stiffness and impact behavior (Guyot, 1993).

9.7.1 Catalyst Composition

The precursor of homogeneous Ziegler-Natta catalyst systems is the Group IVB transition metallocenes (titanocenes, zirconocenes, and hafnocenes) that are characterized by two bulky cyclopentadienyl (Cp) or substituted cyclopentadienyl (Cp′) ligands. Figure 9.11 shows two such simple metallocenes which have C_{2v} symmetry (the two Cp rings in the molecules are not parallel). The molecular structures of the two famous *Brintzinger catalysts* (chiral *ansa*-metallocenes), Et(Ind)$_2$ZrCl$_2$ and Et(H$_4$Ind)$_2$ZrCl$_2$, having indenyl (Ind) and tetrahydroindenyl (H$_4$Ind) ligands arranged in a chiral way and connected together with chemical bonds by ethylene (Et) bridging groups, are depicted in Fig. 9.12. Since the discovery of the effectiveness of these catalysts, a large number of *ansa*-metallocenes have been synthesized (Huang and Rempel, 1995) by changing the transition metals (Ti, Zr, or Hf) and substituents on the Cp rings, as well as the bridging groups. Among a wide variety of Cp ligands investigated, the most commonly used are methylcyclopentadienyl (MeCp), pentamethylcyclopentadienyl (Me$_5$Cp), indenyl (Ind), tetrahydroindenyl (H$_4$Ind), and fluorenyl (Flu) ligands, while the commonly used bridging groups are ethylene (Et, $-$CH$_2$CH$_2-$), dimethylsilene [Me$_2$Si, (CH$_3$)$_2$Si=], isopropylidene [*i*-Pr, (CH$_3$)$_2$C=], and ethylidene (CH$_3$CH=).

The steric interaction of the Cp type ligands (surrounding the active metal center) with incoming monomer provides a controlling influence on the stereospecificity of polymers made with metallocene catalysts. Consequently, a change in the steric structure of these ligands leads to

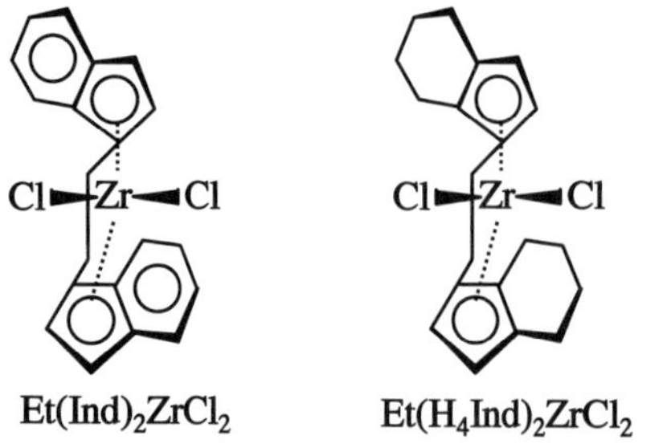

Et(Ind)$_2$ZrCl$_2$ Et(H$_4$Ind)$_2$ZrCl$_2$

Figure 9.12 Structures of two Brintzinger catalysts. (Wild et al., 1985.)

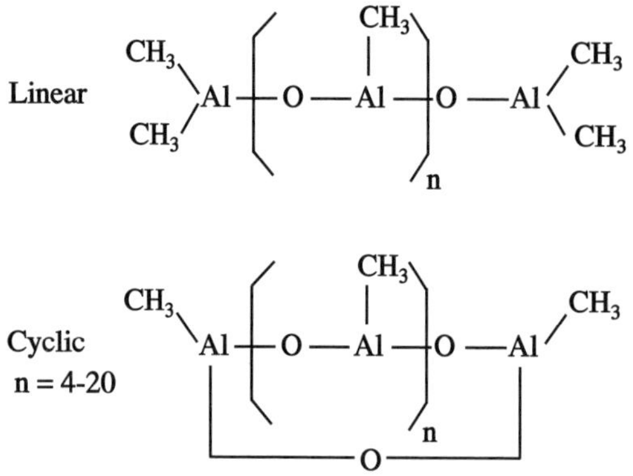

Figure 9.13 Possible structures of MAO. (After Kaminsky et al., 1983.)

changes in steric structures of polyolefin products produced on them. The bridging group provides a stereorigid conformation for the catalyst. It moreover determines the distance between the transition metal atom and the Cp ligands as also the bending angle, thereby influencing the catalyst activity and stereospecificity (i.e., whether isotactic, syndiotactic, or atactic). The stereorigid metallocene (catalyst precursor) can thus be tailored to obtain a desired type of stereospecificity.

While Group IVB transition metallocenes are the main component of homogeneous Ziegler-Natta catalyst systems, the most important cocatalyst which activates them is MAO. Before the discovery of MAO, alkylaluminum chloride was used to activate the homogeneous Ziegler-Natta catalyst Cp$_2$TiCl$_2$, but the activity of the resulting catalyst was poor. The use of MAO cocatalyst, however, increased the catalyst activity by several orders of magnitude. MAO is formed by hydrolysis of trimethylaluminum (TMA). Early workers employed hydrated inorganic salts, such as Al$_2$(SO$_4$)$_3$.18H$_2$O, as the water source. However, commercial producers of MAO add water directly, either by use of moist nitrogen or a marginally wet aromatic solvent (Kaminsky et al., 1983).

MAO is an oligomer with 6-20 [–O–Al(Me)–] repeat units (Huang and Rempel, 1995). A higher degree of oligomerization of MAO is known to have a beneficial effect on the catalyst activity. The structure of MAO is not known precisely. While earlier research suggested that MAO might exist in a linear and/or cyclic form (Fig. 9.13), later investigations indicated that there appears to be no logical structure for MAO with $n > 4$, in which all Al atoms simultaneously possess a coordination of 4. A possible structure with a coordination number of 4 is shown in Fig. 9.14 (Sugano et al., 1993).

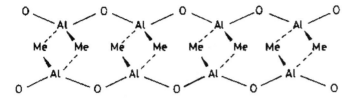

Figure 9.14 A proposed structure of MAO with coordination number of 4. (Sugano et al., 1993.)

9.7.2 The Active Center

The active species in metallocene-MAO systems are believed to be metallocene alkyl cations, i.e., cationic d^0 14-electron complexes of the type $[Cp_2M(R)]^+$ (M = Ti, Zr, Hf). As shown below, the formation of the catalytically active complex involves a series of reactions between metallocenes and MAO (Kaminsky et al., 1992). In the case of halogen-containing metallocenes, a rapid alkylation of metallocene by MAO takes place first, and a methyl transfer reaction between the resulting metallocene alkyls and MAO then generates the active species. Thus, the first step in the reactions after the metallocene Cp_2ZrCl_2 is mixed with MAO is the alkylation of the former via complexation (Huang and Rempel, 1995):

$$Cp_2 \, Zr \, Cl_2 \; + \; [Al \, (CH_3) \text{-} O]_n \; \rightleftharpoons \; Cp_2 \, Zr \, Cl_2 \cdot [Al \, (CH_3) \text{-} O]_n$$

$$Cp_2 \, Zr \, (CH_3) \, Cl + Al_n \, (CH_3)_{n\text{-}1} \, O_n \; Cl \rightleftharpoons Cp_2 \, Zr \, (CH_3) \, Cl \, [\, Al_n \, (CH_3)_{n\text{-}1} \text{-} O_n \; Cl \,]$$

A further alkylation then leads to the formation of $Cp_2Zr(CH_3)_2$:

$$Cp_2Zr(CH_3)Cl \; + \; MAO \; \longrightarrow \; Cp_2Zr(CH_3)_2 \; + \; [Al_n(CH_3)_{n-1}\text{-}O_nCl]$$

The alkylated metallocene, $Cp_2Zr(CH_3)_2$, again reacts (Huang and Rempel, 1995) with MAO, forming a compound (**A**) that features the structural element Zr–O–Al:

Compound **A** is believed to be the active species in the metallocene/MAO systems. Since the Zr–O bond has a polar character and could be ionic in nature, compound **A** possibly exists in two different states (Huang and Rempel, 1995) that are in equilibrium:

The metallocene alkyl cation $[Cp_2Zr(R)]^+$ formed might be the true active center.

Under proper conditions, a homogeneous catalyst can be made as a "single-site" catalyst that has only a single type of active center in a homogeneous system. With such "single-site" stereospecific catalyst, polymers with sharp melting transitions (T_m) and markedly narrow molecular weight distribution ($M_w/M_n \leq 2$) can be produced (Chien and Tsai, 1993).

9.7.3   Polymerization Mechanism

There is now a good understanding of the role played by the active species in the polymerization of α-olefins by homogeneous Ziegler-Natta catalysis. An olefin being a weak base, the cationic metallocene alkyls have a strong tendency to coordinate with olefin molecules and once an olefin molecule coordinates to the metallocene alkyl, it can be readily inserted into the alkyl-metal bond as the energy gained in the transformation of M–R and M–(C=C) bonds into M–C and C–R bonds (M = Ti, Zr, or Hf; R = alkyl, C=C is olefin) provides the necessary driving force. As a result of the insertion, a new d^0 alkyl complex is formed which can coordinate and then insert another olefin molecule. Repetition of this cycle a large number of times leads finally to a polymer. Figure 9.15 shows such an insertion mechanism (Kaminsky and Steiger, 1988) for ethylene polymerization with Cp_2ZrCl_2/MAO.

A monometallic mechanism was proposed by Corradini and Guerra (1991) in which the active center is a metal-carbon bond and the propagation consists of two stages: the coordination of the olefin to the active site, followed by insertion into the metal-carbon bond through a *cis* opening. The active species in this model is the cationic Cp_2MR^+ complex. In addition to aromatic Cp ligands of the precursor metallocenes, an incoming monomer molecule and a growing polymer chain are also coordinated to the metal in the stage preceding monomer insertion.

9.7.4   Kinetic Models

9.7.4.1   Ewen's Model

Ewen's model (Ewen, 1984) was the first kinetic model for propagation in homogeneous systems. This scheme, shown in Fig. 9.16 for the polymerization of propylene with $Cp_2Ti(IV)$, is representative for polymerizations with Group IVB metallocenes and is analogous to the mechanism postulated for ethylene polymerizations with titanocene derivatives and $AlC_2H_5Cl_2$ (Laulier and Hoffman, 1976). Species (1) and (4) in the scheme represent coordinatively unsaturated Ti(IV) complexes that are formally d^0 16-electron pseudotetrahedral species; species (2) represents the interacting catalyst/cocatalyst combination, while intermediate species (3) is shown with the monomer coordinated at an a_1 molecular orbital with the three non-Cp ligands and the transition metal occupying a common equatorial plane. The growing chain is held between two lateral coordination sites accommodating an unidentified non-Cp anion (R') and the monomer.

Under pseudo-first-order conditions, the rate of polymerization of propylene can be expressed (Huang and Rempel, 1995) as

$$R_p = k_{obs}[M][M_T][C] \qquad (9.39)$$

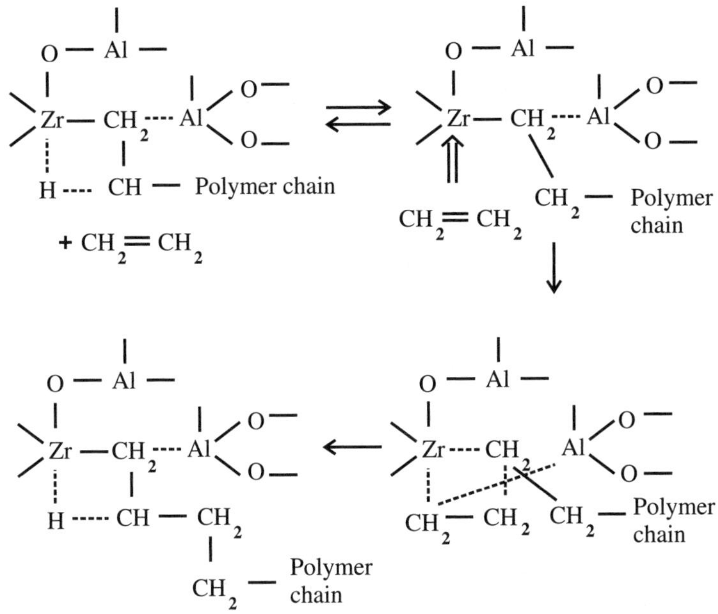

Figure 9.15 Kaminsky's model for ethylene polymerization. (Kaminsky and Steiger, 1988.)

where $\qquad k_{obs} = k_p K_C K_M / (1 + K_M[M] + K_C[C])$

and $\qquad K_M[M] + K_C[C] \ll 1$

The terms k and K in the above equations represent, respectively, the rate constant and the equilibrium constant.

Ewen's model can explain the experimental results that show polymerization rates vary linearly with the product of the monomer [M], metallocene [M_T], and aluminoxane concentrations [C] at low monomer conversions with [Al] over a certain range. At constant [M], [M_T], [C], and temperature, the experimental results (Ewen, 1984) show a linear relationship between polymerization rate and polymerization time (t) according to equation

$$[N] = [C^\star] + k_{tr}[C^\star]t \qquad (9.40)$$

where [N] represents the number of polymer chains produced per Ti center with the assumption of one chain per active center, [$C^\star$] represents the fraction of Ti present as active propagation centers, and k_{tr} is the sum of pseudo-first-order rate constants for chain transfer.

9.7.4.2 Chien's Model

Chien's kinetic model (Chien and Wang, 1990; Chien and Sugimito, 1991), unlike Ewen's model described above, is for the systems in which more than one active species is present. The model assumes the presence of different types of active center, chain transfer to MAO, chain transfer by β-H elimination (see p. 530), and first-order deactivation reactions for active centers. Chien applied the model in the study of ethylene polymerization with Cp_2ZrCl_2/MAO catalyst and propylene polymerization with $Et(Ind)_2ZrCl_2$/MAO and $Et(H_4Ind)_2ZrCl_2$/MAO catalysts.

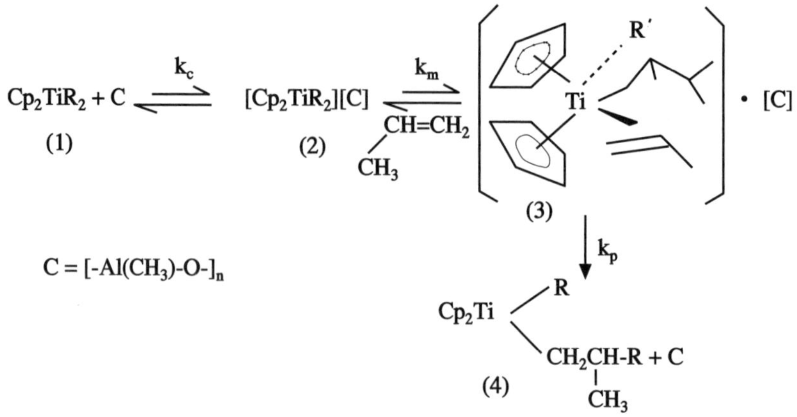

Figure 9.16 Ewen's kinetic model. (Ewen, 1984.)

For a system with i types of active centers, the polymerization rate of the ith species is

$$R_{p,i} = k_{p,i}[C_i^\star][M] \tag{9.41}$$

where $k_{p,i}$ and $[C_i^\star]$ are the propagation rate constant and the concentration of the ith active species, respectively. The overall rate of polymerization is obtained by summing Eq. (9.41):

$$R_p = \sum_i k_{p,i}[C_i^\star][M] \tag{9.42}$$

or, if $[C_i^\star]$ cannot be determined,

$$R_p = k_{p,\text{avg}}[C^\star][M] \tag{9.43}$$

where k_p is the propagation rate constant and $[C^\star]$ is the total concentration of active species.

The total productivity P can be written as

$$P = [M] \sum_i k_{p,i} \int [C_i^\star] dt \tag{9.44}$$

Assuming that the catalytic species deactivate according to first order kinetics, the total productivity at time t is

$$P(t) = [M] \sum_i k_{p,i}[C_i^\star]_0 [1 - \exp(-k_{d,i}t)] \tag{9.45}$$

where $k_{d,i}$ is the deactivation rate constant.

The value of $[C^\star]_0$ can be taken to be equal to the initial metal-polymer bond concentration, $[\text{MPB}]_0$. The metal-polymer bond concentration at time t, $[\text{MPB}]_t$, can be related to the polymerization yield at time t, Y_t, according to (Chien and Sugimoto, 1991):

$$[\text{MPB}]_t = [\text{MPB}]_0 + k_{tr}^{\text{Al}} Y_t / k_p [M] \tag{9.46}$$

Table 9.6 Rate Constants for Et(H$_4$Ind)$_2$ZrCl$_2$/MAO in Propylene Polymerization^a

[Al]/[Zr] ratio	PP fraction	k_p (M-s)$^{-1}$	k_{tr}^{Al} (s^{-1})
3500	T	970	0.015
	C$_7$	1840	0.015
	C$_6$	1370	0.026
	C$_5$	80	0.003
	E	130	0.0078
350	T	1480	0.047
	C$_7$	2550	0.027
	C$_6$	2590	0.041
	C$_5$	97	0.0045
	E	275	0.0027

^{a}[Zr] = 1.0 μM, T_p = 30°C, [C$_3$H$_6$] = 0.47 M (P_p = 1.7 atm).
T: total polymer; C$_5$: *n*-pentane soluble, C$_6$: *n*-hexane soluble, C$_7$: *n*-heptane soluble, E: ether soluble fractions.

Source: Chien and Sugimoto, 1991; Huang and Rempel, 1995.

where k_{tr}^{Al} is the rate constant of chain transfer to MAO. Since [MPB] can be determined by the tritium radiolabelling method (by reacting the polymerization mixture with CH$_3$O^{3}H), [MPB]$_t$ can be plotted against Y_t and extrapolated to $Y_0 = 0$ to obtain [MPB]$_0$ = [$C^\star$]$_0$. However, if the polymers produced with *i*th active species can be separated from others (for example, by solvent extraction (Herwig and Kaminsky, 1983)), [$C_i^\star$]$_0$ can also be determined from the [MPB]–Y plot for the *i*th fraction. The propagation rate constants $k_{p,i}$ and $k_{p,avg}$ can be calculated according to Eqs. (9.41) and (9.43). In addition, k_{tr}^{Al} values can be obtained from the slope of the [MPB]–Y plots. Some rate constant values for propylene polymerization with Et(H$_4$Ind)$_2$ZrCl$_2$/MAO systems, reported by Chien (Chien and Sugimoto, 1991), are listed in Table 9.6.

Problem 9.14 Besides chain transfer to MAO, the chain transfer by β-H elimination (see Section 9.7.4.3) is another main termination process in Chien's model for polymerization with metallocene/MAO catalysts. Show how the rate constant (k_{tr}^{β}) of chain transfer by β-H elimination can be calculated from the polymer molecular weight data.

Answer:
The rates of chain transfer to MAO (R_{tr}^{Al}) and chain transfer by β-H elimination (R_{tr}^{β}) are expressed as:

$$R_{tr}^{Al} = k_{tr}^{Al}[C^\star][MAO] \quad \text{and} \quad R_{tr}^{\beta} = k_{tr}^{\beta}[C^\star]$$

The number-average degree of polymerization is

$$\overline{DP}_n = \frac{R_p}{\sum R_{tr}} = \frac{k_p[C^\star][M]}{k_{tr}^{Al}[C^\star][MAO] + k_{tr}^{\beta}[C^\star]} \quad \text{or} \quad \frac{1}{\overline{DP}_n} = \left(\frac{k_{tr}^{Al}[MAO] + k_{tr}^{\beta}}{k_p}\right)\frac{1}{[M]}$$

The values of k_p and k_{tr}^{Al} having been determined via the [MPB]–Y plot, the value of k_{tr}^{β} can be obtained from the slope of $(1/\overline{DP}_n) - (1/[M])$ plot.

According to the data in Table 9.6, it was suggested (Chien and Sugimoto, 1991) that the system Et(H$_4$Ind)$_2$ZrCl$_2$/MAO has two types of active species (**I** and **II**), species **I** producing

Table 9.7 Fraction of Active Zr in Et(H$_4$Ind)$_2$ZrCl$_2$/MAO^a

	[C*] (mol/mol Zr)				
[Al]/[Zr] ratio	T	E	C$_5$	C$_6$	C$_7$
3,500	0.66	0.15	0.16	0.07	0.28
350	0.13	0.07	0.017	0.01	0.053

^{a}[Zr] = 1.0 μM, T_p = 30°C, [C$_3$H$_6$] = 0.47 M (P_p = 1.7 atm).
T: total polymer; C$_5$: *n*-pentane soluble, C$_6$: *n*-hexane soluble, C$_7$: *n*-heptane soluble, E: ether soluble fractions.
Source: Chien and Sugimoto, 1991.

n-hexane and *n*-heptane soluble fractions and species **II** producing *n*-pentane and ether-soluble fractions. The two types of active species coexist in about equal amounts (see Table 9.7); one has higher selectivity, 10-20 times greater rate constant of propagation, and a factor of 5-15 times faster chain transfer to MAO than the second type of active species (see Table 9.6). Metallocene complexes with different states of coordination with MAO may be responsible for the various active species.

Homogeneous metallocene-based catalyst systems have much simpler kinetics than heterogeneous systems and represent the best systems for kinetic study of Ziegler-Natta polymerization. They also offer a good opportunity to study the durability and deactivation of the catalysts (Herwig and Kaminsky, 1983).

9.7.4.3 Molecular Weight and Chain Transfer

Unlike heterogeneous Ziegler-Natta catalysts, homogeneous matallocene catalysts often produce only low molecular weight polymers, especially in stereospecific polymerizations. The molecular weight is given by

$$\overline{M}_n = \overline{DP}_n \cdot M_0 = \left(R_p / \sum R_{tr,i}\right) M_0 \tag{9.47}$$

where M_0 is monomer molecular weight. Several types of transfer reactions which may occur in homogeneous systems terminating polymer chain growth are described below (Huang and Rempel, 1995).

β-Hydrogen Elimination The metal center abstracts a H atom bonded to the β-C of the growing polymer chain, forming an M–H bond (M = transition metal) and leaving a polymer with an unsaturated end group:

Chain Transfer by Monomer β-Hydrogen elimination takes place simultaneously with monomer insertion at the active center without forming the M–H bond:

Chain Transfer to MAO If MAO is used to activate catalyst, the growing polymer chain attached to an active center can exchange with the methyl group of a MAO molecule, forming the Al-terminated polymer chain and the M–CH_3 bond in the active center:

$$
\underset{}{Zr-CH_2-\overset{\overset{\displaystyle CH_3}{|}}{CH}-\text{\tiny www}} \quad \xrightarrow{\ MAO\ } \quad Al-CH_2-\overset{\overset{\displaystyle CH_3}{|}}{CH}-\text{\tiny www} \ + \ Zr-CH_3
$$

$$
\downarrow H^+
$$

$$
\overset{H_3C}{\underset{H_3C}{}}\!\!\!\diagdown CH-\text{\tiny www}
$$

β-CH_3 Elimination This chain transfer mechanism has been discovered in the polymerization of propylene. The metal center abstracts a CH_3 group, instead of a H atom, at the β-carbon of the growing chain, thus forming a M–CH_3 bond at the active center and leaving a polymer with an allylic end group (Resconi et al., 1992):

$$
Zr-CH_2-\overset{\overset{\displaystyle CH_3}{|}}{CH}-CH_2-\overset{\overset{\displaystyle CH_3}{|}}{CH}-\text{\tiny www} \quad \longrightarrow \quad Zr-CH_3 \ + \ CH_2{=}CH-CH_2-\overset{\overset{\displaystyle CH_3}{|}}{CH}-\text{\tiny www}
$$

Poor molecular weights of polymer products is a major drawback of metallocene-based polymerization catalysts, limiting them from being fully exploited for practical applications. However, technical breakthroughs have been achieved in that specially designed zirconocene catalysts (Spaleck et al., 1992) can produce polymers with molecular weights far above 100,000.

A unique feature of polymer products obtained with homogeneous metallocene-based catalysts is their narrow molecular weight distribution (MWD). Thus the values of $\overline{M}_w/\overline{M}_n$ for polyethylenes and polypropylenes are typically near 2 which would indicate the presence of uniform active species in catalysts. Some researchers have thus suggested that the catalysts could be "single site," while others have explained that two or more kinds of species can have very similar values for the k_p/k_{tr} ratio, giving rise to narrow MWD products.

9.8 Immobilized Metallocene Catalysts

Though metallocene/MAO systems exhibit high activities for olefin poymerization under homogeneous conditions, the polymer formed (fluff) has a low bulk density and consists of very finely divided particles. By making the catalyst heterogeneous via immobilization on suitable carriers or supports, it is possible to increase the polymer bulk density. Immobilization of the homogeneous catalyst on a solid support is also required for adapting the metallocene technology to existing industrial processes. An added advantage of using the supported catalyst is that the extremely high atomic ratios of MAO to zirconium that are usually needed to obtain high activity in homogeneous (solution) polymerization become unnecessary. For example, excellent activities are reported (Turner, 1987) for supported metallocene/MAO catalysts with Al/Zr atomic ratio less than 100. This low MAO requirement has greatly facilitated the commercialization of metallocenes.

The support commonly favored for heterogenizing metallocene/MAO catalysts is amorphous silica containing fairly high pore volume ($> 1\ \text{cm}^3/\text{g}$) and an average particle diameter of about 50 μm. As identified by IR and NMR spectroscopy (Garoff, 1993), there are three kinds of OH

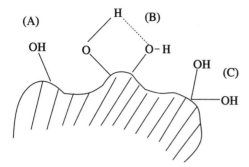

Figure 9.17 Surface hydroxyl groups on silica: (A) isolated groups; (B) hydrogen bonded groups; (C) geminated hydroxyl groups (from the same Si atom). (After Hindryckx et al., 1997.)

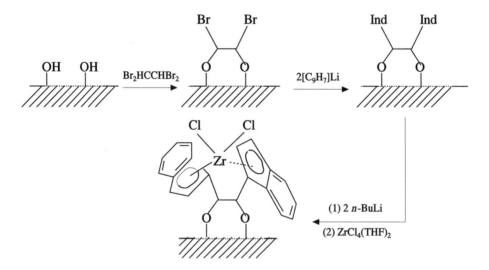

Figure 9.18 Schematic of the method of synthesis of a bridge-anchored *ansa*-metallocene. (The indenyl rings are shown in pseudo-racemic orientation as the high isotacticity of polypropylene produced with the catalyst suggests a structural analogy to *rac*-Et(Ind)$_2$ZrCl$_2$. (From Soga et al., 1994. With permission from *John Wiley & Sons, Inc.*)

groups on the silica surface, viz., isolated, hydrogen bonded, and geminated double hydroxyl groups (Fig. 9.17). In a reported work (Soga et al., 1994), bridged metallocenes are covalently bound to silica through reaction of the surface silanols with reactive carbon halide or silicon halide bonds on the ligand bridge. The synthesis of one such bridge-anchored *ansa*-metallocene is shown schematically in Fig. 9.18. Activities of catalysts based on such surface-bound precursors are reported (Soga et al., 1994) to be greatly superior to the homogeneous analogs.

Besides silica, numerous other materials, including alumina, zeolites, clays, and organic polymers, have been used as supports for metallocene-based catalysts. In some cases, a covalent bond between metallocene and polymer support has been achieved through prior attachment of part of

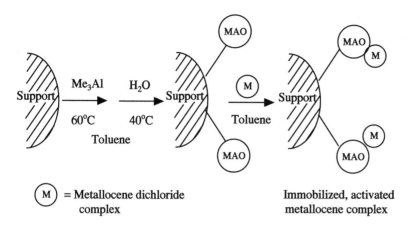

M = Metallocene dichloride complex

Immobilized, activated metallocene complex

Figure 9.19 Synthesis of a heterogeneous metallocene catalyst with immobilized MAO. (Adapted from Köppl et al., 2001.)

the ligand to a polymeric substrate (Stork et al., 1999) and in others through incorporation of a pendent unsaturated group on the ligand into the growing polymer (Jung et al., 1997).

In a different approach to heterogenization, supported MAO cocatalyst has been formed directly by reacting MAO with functional groups on the silica surface (Janiak and Rieger, 1994). It is, however, difficult to attach all the aluminoxane in this way onto the carrier material. In an alternative and significantly more effective procedure (Köppl et al., 2001), trimethylaluminum (TMA) is first attached to silica gel by heating the carrier material with TMA in an inert solvent such as toluene at 60°C until the gas (methane) evolution subsides. In this step, TMA (a strong Lewis acid) presumably blocks all surface Lewis basic centers of the carrier material to prevent them later from poisoning cationic metallocene polymerization centers (strong Lewis acids) and thus lead to higher productivities. The reaction mixture following TMA attachment is cooled down to 40°C and a little amount of water is bubbled through it using a moist argon flow to produce a completely immobilized form of MAO. [Alternatively, using a direct method (Sinn et al., 1988), often preferred, a calculated amount of ice (H_2O:TMA mole ratio = 1:2) of −80°C may be added to the reaction mixture, cooled to −80°C, and stirred vigorously until the ice is crushed.] :

$$Me_3Al + SiO_2/H_2O \xrightarrow[-CH_4]{Toluene} SiO_2 - MAO \qquad (9.48)$$

Finally, a catalyst precursor (metallocene) is added to the mixture at room temperature, the amount being dependent on the desired Al:Zr ratio. The process is shown schematically in Fig. 9.19.

Cellulose, starch, and flour are excellently suited for immobilization of MAO because of the presence of OH groups as anchor points (Fig. 9.20). The arrangement of the OH groups on the surface possibly serves as a template for the formation of aluminoxane structures that exhibit advantageous cocatalyst properties and the activities of the resulting supported catalysts are typically in the range of 270 kg polyethylene/g Zr-h (Köppl et al., 2001).

In the new *polymerization filling technique* (PFT), Ziegler-Natta catalyst is attached onto the surface of an inorganic filler, so that olefin can be polymerized from the filler surface, thus allowing a very high filler loading to be reached in composites with improved mechanical properties

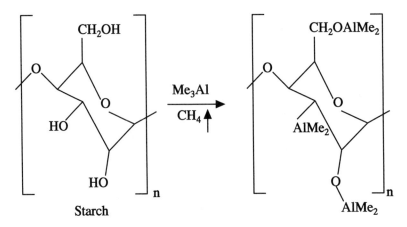

Figure 9.20 Reaction of trimethylaluminum with starch in toluene. (After Köppl et al., 2001.)

(Hindryckx et al., 1997). Various filler/coordination catalyst pairs have been used in PFT. One of the most active catalyst reported among them is $CaSO_4$-MAO-Cp_2ZrMe_2 with a productivity of 400 kg polyethylene/g Zr-h (Kaminsky et al., 1993).

9.9 Oscillating Metallocene Catalysts

The discovery of homogeneous Ziegler-Natta catalysts, in 1980s, based on metallocenes having bridged structures (see Fig. 9.12), opened up new and exciting possibiities to fine-tune the stereochemical structure of polymer products. As described previously (p. 523), chiral *ansa*-metallocenes produce isotactic polyolefins, whereas achiral meso isomers form atactic polyolefins. As discussed in Chapter 2, the stereochemistry of polymers strongly influences their properties. Thus, isotactic polypropylene is a crytalline thermoplastic with a melting point of ~165°C, whereas atactic (stereorandom) polypropylene is an amorphous gum elastomer. On the other hand, polypropylene consisting of blocks of atactic and isotactic stereosequences is a rubbery material with properties of a *thermoplastic elastomer*. To produce such stereoblock polymers Coates and Waymouth (1995) prepared an unbridged metallocene bis(2-phenylindenyl)-zirconium dichloride, (2-PhInd)$_2$ZrCl$_2$, which was designed to isomerize between chiral *rac*-like (isospecific) and achiral *meso*-like (aspecific) geometries by rotation of the indenyl ligands about the metal-ligand bond axis during the course of polymerization in order to generate isotactic and atactic sequences in the same polymer chain (Fig. 9.21). In order to produce isotactic-atactic stereoblock polymers, it is, however, necessary that the rate of ligand rotation is slower than that of monomer insertion and yet faster than the time required to construct one polymer chain. This was achieved by having a phenyl substituent on the indene ligand to inhibit the rate of ligand rotation so that the catalyst can switch between isospecific and aspecific coordination geometries.

The nonbridged metallocene (2-PhInd)$_2$ZrCl$_2$ was isolated in 82% yield from the reaction of 2-phenylindenyllithium and ZrCl$_4$. In the presence of cocatalyst methylaluminoxane (MAO), the catalyst (2-PhInd)$_2$ZrCl$_2$ is active for oscillating stereopolymerization of propylene yielding stereoblock polymers (Coates and Waymouth, 1995) with an isotactic pentad content, [*mmmm*] (defined as fraction of stereosequences containing five adjacent isotactic stereocenters), ranging from 6.3 to 28.1%, molecular weight (M_w) in the range $2.4{\times}10^4$–$8.9{\times}10^5$, and the productivity of $1.9{\times}10^5$–$17.3{\times}10^5$ g polymer per mole of Zr per hour. The MWDs are in the range of $\overline{M}_w/\overline{M}_n = 1.5$ to 2.8, which is characteristic of single-site homogeneous catalysts.

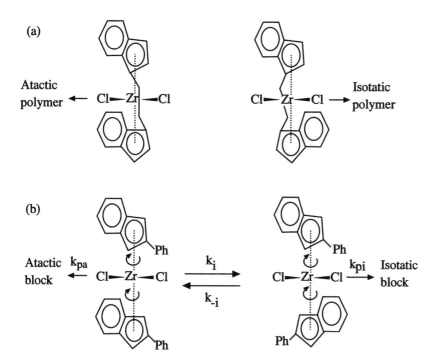

Figure 9.21 (a) Stereorigid *ansa*-metallocenes for the production of isotactic and atactic polymers. (b) Oscillating metallocenes for the production of stereoblock copolymers. (Adapted from Coates and Waymouth, 1995.)

Both the productivity and stereoselectivity of the oscillating metallocene catalyst (2-PhInd)$_2$ ZrCl$_2$ are strongly influenced by the nature of the cocatalyst and reaction conditions, such as temperature, pressure, and monomer concentration. Best performance of the catalyst has been obtained with MAO as the cocatalyst (Wilmes et al., 2002). Both the productivity and [*mmmm*] are found to increase with decreasing temperature or increasing pressure, with the other variable remaining constant. The effect of monomer concentration [M] on the tacticity of the polymer can be qualitatively predicted if it is assumed that the rate of monomer enchainment at both aspecific and isospecific sites is directly proportional to [M], whereas the catalyst isomerization (oscillation) rate is independent of [M] (Coates and Waymouth, 1995):

$$\text{Isotactic block size} \propto [mmmm]$$
$$\propto \frac{\text{Isotactic enchainment rate}}{\text{Catalyst isomerization rate}}$$
$$\propto \frac{k_{pi}[M]}{k_i} \tag{9.49}$$

According to this model, the isotactic block length should increase with [M], and hence propylene pressure, as observed experimentally.

REFERENCES

Arlman, E. J. and Cossee, P., *J. Catal.*, **3**, 99 (1964).

Boor, Jr., J., "Ziegler-Natta Catalysts and Polymerizations", Academic Press, New York (1979).

Breslow, D. S. and Newburg, N. R., *J. Am. Chem. Soc.*, **79**, 5072 (1957).

Burfield, D. R., Mackenzie, I. D., and Tait, P. J. T., *Polymer*, **13**, 302 (1972).

Chien, J. C. W. and Sugimoto, R., *J. Polym. Sci., Polym. Chem. Ed.*, **29**, 459 (1991).

Chien, J. C. W. and Tsai, M. W., *Makromol. Chem., Macromol. Symp.*, **66**, 141 (1993).

Chien, J. C. W. and Wang, B. P., *J. Polym. Sci., Polym. Chem. Ed.*, **28**, 15 (1990).

Coates, G. W. and Waymouth, R. M., *Science*, **267**, 217 (1995).

Corradini, P. and Guerra, G., *Prog. Polym. Sci.*, **16**, 239 (1991).

Cossee, P., *Tetrahedron Lett.*, **17**, 12 1(1960).

Cossee, P., "The Mechanism of Ziegler-Natta Polymerization. II. Quantum-Chemical and Crystal-Chemical Aspects", chap. 3 in "The Stereochemistry of Macromolecules", vol. 6 (A. D. Ketley, ed.), Marcel Dekker, New York, 1967.

Erich, F. and Mark, H. F., *J. Colloid Sci.*, **11**, 748 (1956).

Ewen, J. A., *J. Am. Chem. Soc.*, **106**, 6355 (1984).

Friedlander, H. N., *J. Polym. Sci.*, **38**, 91 (1959).

Friedlander, H. N. and Oita, K., *Ind. Eng. Chem.*, **49**, 1885 (1957).

Friedlander, H. N. and Resnick, W., "Solid Polymers from Surface Catalysts," in "Advances in Petroleum Chemistry and Refining", vol. I (K. A. Kobe and J. J. McKetta, eds.), Interscience, New York, 1958.

Garoff, T., "Techniques and Measurement in Heterogeneous and Homogeneous Catalysis", Univ. of Liverpool, 6-10 September, 1993.

Grieveson, B. M., *Makromol. Chem.*, **84**, 93 (1965).

Guyot, A. *Makromol. Chem., Macromol. Symp.*, **66**, 1 (1993).

Herwig, J. and Kaminsky, W., *Polym. Bull.*, **9**, 464 (1983).

Hindryckx, F., Dubois, P., Jerome, R., Teyssie, P., and Marti, M. G., *J. Appl. Polym. Sci.*, **64**, 423 (1997).

Hu, Y. and Chien, J. C. W., *J. Polym. Sci. Polym. Chem. Ed.*, **26**, 2003 (1988).

Huang, J. and Rempel, G. L., *Prog. Polym. Sci.*, **20**, 459 (1995).

Janiak, C. and Rieger, B., *Angew. Macromol. Chem.*, **215**, 47 (1994).

Jordan, D. O., "The Stereochemistry of Macromolecules", vol. I (A. D. Ketley, ed.), Marcel Dekker, New York, 1967.

Jung, M., Alt, H. G., and Welch, B. M. (to Phillips Petroleum), U. S. Pat. 5,854,363 (priority date Jan. 8, 1997).

Kaminsky, W., Bark, A., and Steiger, R., *J. Mol. Catal.*, **74**, 109 (1992).

Kaminsky, W., Külper, K., Brintzinger, H. H., and Wild, F. R. W. P., *Angew. Chem. Int. Ed. Engl.*, **24**, 507 (1985).

Kaminsky, W., Sinn, H., and Woldt, R., *Makromol. Chem. Rapid Commun.*, **4**, 417 (1983).

Kaminsky, W. and Steiger, R., *Polyhedron*, **7**(22/23), 2375 (1988).

Kaminsky, W. and Zielonka, H., *Polym. Adv. Technol.*, **4**, 415 (1993).

Keii, T., "Kinetics of Ziegler Natta Polymerizations", Halsted Press, New York, 1973.

Keii, T., Soga, K., and Saiki, N., *J. Polym. Sci., Part C*, **16**, 1507 (1967).

Kumar, A. and Gupta, S. K., "Fundamentals of Polymer Science and Engineering", Tata McGraw-Hill, New Delhi, 1978.

Kissin, Y. V., "Isospecific Polymerization of Olefins", Springer-Verlag, Berlin, 1985.

Köppl, A., Alt, H. G., and Phillips, M. D., *J. Appl. Polym. Sci.*, **80**(3), 454 (2001).

Natta, G., *Makromol. Chem.*, **16**, 213 (1955).

Natta, G., *J. Polym. Sci.*, **48**, 219 (1960); *Chim. Ind.* (Milan), **42**, 1207 (1960).

Natta, G. and Pasquon, I., *Advan. Catal.*, **11**, 1 (1959).

Natta, G., Pino, P., Mazzanti, G., and Lanzo, R., *Chim. Ind. (Milan)*, **39**, 1032 (1957).

Odian, G., "Principles of Polymerization", John Wiley, New York, 1991.

Patat, F. and Sinn, H., *Angew. Chem.*, **70**, 496 (1958).

Resconi, L., Piemontesi, F., Franciscono, G., Abis, L., and Fiorani, T., *J. Am. Chem. Soc.*, **114**, 1025 (1992).

Saltman, W. M., *J. Polym. Sci.*, **46**, 375 (1960).

Sinn, H., Bliemeister, J., Clausnitzer, D., Tikwe, L., and Winter, H., "Some New Results on Methyl Aluminoxane", in "Transition Metals and Organometallics as Catalysts for Olefin Polymerization" (W. Kaminsky and H. Sinn, eds.), Springer-Verlag, Berlin, 1988.

Sinn, H. and Kaminsky, W., *Adv. Organomet. Chem.*, **18**, 99 (1980).

Soga, K., Kim, H. J., and Shiono, T., *Macromol. Chem. Phys.*, **195**, 3347 (1994).

Spaleck, W., Antberg, M., Rohrmann, J., Winter, A., Bachmann, B., Kaprof, P., Behm, J., and Herrmann, W. A., *Angew. Chem. Int. Ed. Engl.*, **31**(10), 1347 (1992).

Stork, M., Koch, M., Klapper, M., Müllen, K., Gregorius, H., and Rief, U., *Macromol. Rapid Commun.*, **20**, 210 (1999).

Sugano, T., Matsubara, K., Fujita, T., and Takahasi, T., *J. Mol. Catal.*, **82**, 93 (1993).

Tait, P. J. T., "Transition Metal and Organometallics as Catalysts for Olefin Polymerization" (W. Kaminsky and H. Sinn, eds.), p. 315, Springer, Berlin, 1988.

Tait, P. J. T. and Watkins, N. D., "Monoalkene Polymerization: Mechanisms", chap. 2 in "Comprehensive Polymer Science", Vol. 4 (G. C. Eastmond, A. Ledwith, S. Russo, and P. Sigwalt, eds.), Pergamon Press, Oxford, 1989.

Turner, H. W. (to Exxon), U. S. Pat. 4,752,597 (priority date Feb. 19, 1987).

Vesley, K., *Pure Appl. Chem.*, **4**, 407 (1962).

Vesley, K., Ambroz, J., Vilim, R., and Hamrick, O., *J. Polym. Sci.*, **55**, 25 (1961).

Vozka, P. and Mejzlik, J., *Makromol. Chem.*, **19**, 589 (1990).

Wild, F. R. W. P., Wasincionek, M., Huttner, G., and Brintzinger, H. H., *J. Organomet. Chem.*, **288**, 63 (1985).

Wild, F. R. W. P., Zsolnai, L., Huttner, G., and Brintzinger, H. H., *J. Organomet. Chem.*, **232**, 233 (1982).

Wilmes, G. M., Polse, J. L., and Waymouth, R. M., *Macromolecules*, **35**, 6766 (2002).

Young, R. J. and Lovell, P. A., "Introduction to Polymers", Chapman and Hall, London, 1990.

Zakharov, V. A., Chumaevsky, N. B., Butakov, G. B., and Yermakov, Y. I., *Kinet. Katal.*, **18**, 848 (1977).

EXERCISES

9.1 Account for the fact that atactic structures commonly predominate in free-radical polymerization in contrast to ionic and Ziegler-Natta polymerizations.

9.2 Explain why it is possible to synthesize two diisotactic polymer structures but only one syndiotactic structure.

9.3 What are the basic modes of addition of 1,3-dienes ($CH_2=CR-CH=CH_2$) to a growing polymer chain ? How many stereochemical arrangements are possible for each mode ?

9.4 Give plausible explanations for the fact that nonpolar alkenes generally require the use of heterogeneous catalysts to produce isotactic polymers. What type of catalysts would be needed for isospecific polymerization of methacrylates, vinyl ethers, styrene, and 1,3-dienes ?

9.5 In the build-up period of the decay-type Ziegler-Natta polymerization of propylene, the rates are found to be different when propylene is introduced after $TiCl_3$ and $AlEt_3$ are allowed to equilibrate from the case when $AlEt_3$ is added after the gas is introduced. The rates in the former case are given by the expression

$$R_p = k_1 P^2 \frac{K[AlEt_3]}{1 + K[AlEt_3]} t$$

where P is the propylene gas pressure.

It is assumed that the following equilibrium exists in the former case:

$$S + AlEt_3 \rightleftharpoons S^\star$$

where S represents an active site (i.e., a Ti atom) and $S^\star$ a potential polymerization center. The concentration of $S^\star$ would be given by the Langmuir equation. On introduction of the gas, polymerization centers, $C^\star$, are formed. Show that a proper balance of $C^\star$ yields the aforesaid equation.

9.6   Consider now the case where propylene is introduced before $AlEt_3$ is added. Then, all the reactions in Eqs. (P9.8.1)-(P9.8.3) would occur simultaneously. Derive the following rate expression for small extents of time

$$R_0 = kP[AlEt_3] t$$

where R_0 refers to the rate for small intervals of time, P is the propylene gas pressure, $[AlEt_3]$ is the concentration of $AlEt_3$, and t is the time of the reaction.

9.7   For the measurement of polymerization rate of a gaseous monomer with a heterogeneous Ziegler-Natta catalyst, the catalyst may be mixed with a suitable solvent for the gaseous monomer, and the monomer supplied at constant pressure through a gas meter to the catalyst suspension. Any dependence of the rate of polymerization on the stirring speed would then indicate diffusion control of the propagation reaction.

Using the above type of experimental set-up for propylene polymerization with $TiCl_3$-$AlEt_3$ in n-heptane, the rate of polymerization was measured [T. Keii, "Kinetics of Ziegler-Natta Polymerizations", Halsted Press, New York, 1973] at different speeds of stirring and constant propylene pressure. The results obtained indicated that there were two different steady-state rate curves for the stirring speeds of 400 and 600 rpm. In each case, a steady bulk monomer concentration was reached in about 3-4 hours. Show how the overall process at steady state can be modeled to show dependence of the polymerization rate on stirring speed and to enable determination of both the mass transfer rate constant and polymerization rate constant from rate measurements at different stirring speeds.

Hint: Defining $[M]_0$ = monomer concentration at the gas-liquid interface, $[M]$ = average concentration of monomer in solution, $[cat]$ = concentration of catalyst, k_m = mass transfer constant (dependent on stirring speed), and k_p = polymerization rate constant, the mass balance for the monomer gives $d[M]/dt = k_m([M]_0 - [M]) - k_p[cat][M]$. At the steady state, $d[M]/dt = 0$. One may assume $k_m \gg k_p[cat]$ at very high stirring speeds and hence the rate of polymerization, $R_p = k_p[cat][M]_0$, while at low stirring speeds, $k_m \ll k_p[cat]$ so that $R_p = k_m[M]_0$.

9.8   Considering the reaction scheme of Problem 9.9, derive an expression for the rate of polymerization at steady state and show it can be used to evaluate the initiation rate constant k_i.

9.9   What predictions can be made from Eq. (P9.9.14) about the dependence of the number-average degree of polymerization on the various parameters?

9.10 Using Eq. (P9.9.9) for the rate of polymerization, derive the following Mayo-Walling equation for Ziegler-Natta polymerization

$$\frac{1}{DP_n} = \frac{k_{tr,M}}{k_p} + \frac{k_s}{k_p K_M[M]} + \frac{k_{tr,A} K_A}{k_p K_M[M]}$$

How would this equation be modified if polymerization is carried out in the presence of H_2 used as a terminating agent?

9.11 Hydrogen is used as the chain length regulator in Ziegler-Natta polymerization. A proposed mechanism postulates that there exists a pre-established equilibrium of dissociative adsorption of hydrogen on $TiCl_3$

catalyst surface as

$$H_2 \;\rightleftharpoons\; 2H_{ads}$$

It is this adsorbed hydrogen that participates in the following reaction:

$$Cat{-}P \;+\; 2H_{ads} \;\longrightarrow\; Cat{-}H \;+\; PH$$

The Cat-H reacts with monomer molecules at a different rate in the following fashion:

$$Cat{-}H \;+\; monomer \;\longrightarrow\; Cat{-}P$$

Derive an expression for the rate of polymerization.

9.12 Ethylene was polymerized [Schindler, A., *J. Polym. Sci. Part C*, **4**, 81 (1963)] in *n*-heptane with a catalyst system consisting of diisobutylaluminum hydride (1.5 mmol/L) and titanium tetrachloride (2.5 mmol/L). The catalyst components were reacted at these concentrations at 0°C and the mixture was then aged under a nitrogen blanket for 15 min. Ethylene was fed at 770 mm Hg pressure and the temperature of the catalyst suspension maintained at 40°C. It was found that the experimental data for the rate of polymerization could be represented best by assuming a mixed first and second-order dependence of the form

$$R_p \;=\; Kp_E^2/(1 + K'p_E) \tag{I}$$

where p_E is the ethylene pressure.

Consider a generalized mechanism based on the following assumptions: (a) monomer is added to an organometallic compound forming a polymerization center, (b) chain growth proceeds thereafter through the addition of further units, and that the growth of an individual polymer chain can be terminated either (c) by a transfer reaction with the monomer or (d) by a spontaneous termination reaction involving the transfer of a hydride ion. Show that this reaction scheme leads to a rate expression similar to Eq. (I) if one considers the process (c) as a true transfer reaction and only (d) as a termination step.

How would Eq. (I) be modified if the experiments were conducted with hydrogen addition?

9.13 A major problem in using expressions such as Eq. (P9.8.10) and Eq. (9.33) is the determination of the actual number of polymerization centers $C^\star$. Radioactive quenching techniques have been widely used. The quenching reaction with, for example, radioactive alcohol can be represented by the scheme

$$L_xM_T...CH_2P + ROH^\star \longrightarrow L_xM_TOR + H^\star CH_2P$$

where L_x represents the ligands surrounding the transition metal M_T and $H^\star$ represents the radioactive tritium.

Write an equation for calculating $[C^\star]$ from the measured activity of the polymer yield.

9.14 Name a few solid-support catalyst systems and the monomers for which they can be used. How is the stereoregularity achieved in these polymerizations?

9.15 Compare polymerizations with conventional Ziegler-Natta catalysts and metallocene-based Ziegler-Natta catalysts in respect of (a) reaction mechanism, (b) stereospecificity of polymer product, and (c) polymer molecular weight and distribution.

Chapter 10

Ring-Opening Polymerization

10.1   Introduction

The four types of polymerizations described in the previous chapters can be grouped into two broad categories – condensation and olefin polymerizations. We introduce in this chapter a third category, the ring-opening polymerization (ROP) of cyclic compounds. Examples of commercially important polymers prepared by ROP are given in Table 10.1. Only the polyalkenylenes among them are composed solely of carbon chains, while the most popular among them are polyethers prepared from three-membered ring cyclic ethers (epoxides), polyamides from cyclic amides (lactams), polyesters from cyclic esters (lactones), and polysiloxanes from cyclic siloxanes.

Cyclic compounds are potentially polymerizable as the difunctionality criterion for polymerizability is achieved by a ring-opening process as shown below for ethylene oxide:

$$x \; CH_2\!-\!\!-\!\!CH_2 \longrightarrow -\!\!\left[CH_2\!-\!CH_2\!-\!O \right]\!_x \tag{10.1}$$

Cyclic monomers should thus be able to polymerize provided a suitable mechanism for ring opening is available (*kinetic feasibility*). A more important consideration, however, is the thermodynamic feasibility of polymerization. ROP is thermodynamically favored if the corresponding free-energy change, $\Delta G = \Delta H - T \Delta S$, is negative. Table 10.2, for example, lists some calculated values (Sawada, 1976) of heat (ΔH), entropy (ΔS), and free energy (ΔG) of polymerization of the liquid cycloalkanes to linear polymer at 25°C. The data show that ΔH, which is a fair indicator of the extent of ring strain (caused by either forcing the bonds between the ring atoms into angular distortion or by steric interaction of substituents on the ring atoms), makes the main contribution to ΔG for 3- and 4-membered rings. Thus the most reactive (i.e., thermodynamically least stable) monomers are those containing 3- or 4-membered rings. The data in Table 10.2 further show that cyclohexane is the most resistant to polymerization, since ΔG is positive. For cyclopropane, cyclobutane, cyclopentane, cycloheptane, and cyclooctane, ΔG for polymerization is negative, indicating that the polymerization is feasible. However, thermodynamic feasibility does not always guarantee realization in practice and no high polymers of cyclopropane and cyclobutane are known (Sawada, 1976).

Small changes in the physical conditions and chemical structure can have a large effect on the polymerizability of a cyclic monomer. Thus, whereas 5-membered cyclic ethers such as tetrahydrofuran have negative free-energy change and so are polymerizable, the five-membered cyclic

Table 10.1 Polymers of Commercial Interest Prepared by Ring-Opening Polymerization

Monomer type	Monomer structure	Repeating unit	Polymer type
Cycloalkene	$(CH_2)_x$ ring with CH=CH	$-[CH=CH(CH_2)_x]-$	Poly(alkenylene)
Trioxane	CH_2, CH_2, CH_2 ring with O's	$-[CH_2O]-$	Polyether
Cycle ether[a]	$(CH_2)_x$ O ring	$-[(CH_2)_x - O]-$	Polyether
Lactone (Cyclic ester)	$(CH_2)_x$ ring with $C{=}O$, O	$-[(CH_2)_x \overset{O}{\overset{\|}{C}}O]-$	Polyester
Lactam (Cyclic amide)	$(CH_2)_x$ ring with $C{=}O$, NH	$-[(CH_2)_x \overset{O}{\overset{\|}{C}}NH]-$	Polyamide
Aziridine (Alkyleneimine)	CH_2—CH_2 with NH	$-[CH_2CH_2NH]-$	Polyamine
Cyclic siloxane	$[Si(CH_3)_2]_x$	$-\begin{bmatrix} CH_3 \\ SiO \\ CH_3 \end{bmatrix}-$	Polysiloxane (Silicone)
Hexachloro-cyclotriphos-phazene[b]	ring of P, N with Cl substituents	$-\left[\underset{Cl}{\overset{Cl}{P}}{=}N\right]-$	Polyphosphazene

[a] Epoxide(x=2); oxetane (x=3)

[b] Phosphonitrilic chloride trimer

Note: The ring opening polymerization of cycloalkenes resulting in polymers (polyalkenylenes) with a double bond in the repeating unit is referred to as *ring-opening metathesis polymerization* (ROMP) by analogy to the olefin metathesis reaction. The ROMP takes place in the presence of coordination initiators based on transition metals. The original initiators used for ROMP were two-component systems, composed of a halide or oxide of a transition metal such as W, Mo, Rh, or Ru with an alkylating agent (Lewis acid) such as R_4Sn or $RAlCl_2$, which generate metal-alkylidenes (metal carbenes) in situ that act as the initiating and propagating species.

esters (γ-butyrolactone) have positive free-energy change and are not polymerizable. In marked contrast, 6-membered cyclic ethers do not polymerize while the corresponding cyclic esters do. For cyclic imides and anhydrides, on the other hand, both 5- and 6-membered rings are polymerizable, while for cyclic ethers the reactivity in terms of ring size decreases in the order: 3 > 4 > 8 > 7 > 5 > 6. Substitution in the ring tends to make the free-energy change more positive, thereby decreasing polymerizability. Thus, whereas tetrahydrofuran is polymerizable, 2-methyl tetrahydrofuran is not.

Considering the thermodynamic relation given above, since ΔH and ΔS are negative for polymerization, ΔG becomes less negative as the temperature increases, reaching a value of zero at some temperature (the *ceiling temperature*) above which ΔG is positive and polymerization is no longer favorable. Ceiling temperatures are often quite low in ROP as compared to vinyl polymerizations, particularly where the rings involved are 5- or 6-membered.

Table 10.2 Thermodynamic Parameters for Polymerization of Cycloalkanes at $25°C^a$

$C(CH_2)_x$	ΔH_{lc} (kJ/mol)	ΔS_{lc} (J/mol-deg)	ΔG_{lc} (kJ/mol)	$C(CH_2)_x$	ΔH_{lc} (kJ/mol)	ΔS_{lc} (J/mol-deg)	ΔG_{lc} (kJ/mol)
3	−113.0	−69.0	−92.5	6	+2.9	−10.5	+5.9
4	−105.0	−55.2	−88.7	7	−21.3	−15.9	−16.3
5	−21.7	−42.7	−9.2	8	−34.7	−3.3	−34.3

^{a} Subscript '*lc*' denotes change from liquid to crystalline.
Source: Adapted from Sawada (1976) by conversion of units.

Besides the thermodynamic feasibility, there should also be a kinetic pathway for the ring to open, facilitating polymerization. Cycloalkanes, for example, have no bond in the ring structure that is prone to attack and thus lack a kinetic pathway. This is in marked contrast to the cyclic monomers such as lactones, lactams, cyclic ethers, acetals, and many other cyclic monomers that have a heteroatom in the ring where a nucleophilic or electrophilic attack by an initiator species can take place to open the ring and initiate polymerization. Both thermodynamic and kinetic factors are thus favorable for these monomers to polymerize (Odian, 1991).

10.2 Polymerization Mechanism and Kinetics

The overall process of ring-opening polymerization of cyclic compounds can be schematically represented by the reaction:

$$\left(\begin{array}{c} A\text{---}A \\ | \quad | \\ A\text{---}B \end{array} \right)_x \longrightarrow \big(\!A\text{---}A\text{---}A\text{---}B\big)_x \tag{10.2}$$

One notes that, in contrast to condensation reactions, ring-opening polymerization does not result in the loss of small molecules and, unlike olefin polymerization for which the loss of unsaturation is a powerful driving force, ring-opening polymerization does not involve a change in the type of bonding. Ring-opening polymerizations are primarily initiated by ionic initiators (including coordinate ionic) as well as initiators that are molecular species, e.g., water. This latter class of initiators are generally only effective for the more reactive cyclic monomers. Ionic initiators, however, are usually more reactive. They are typically the same as those described previously for the cationic and anionic polymerizations of monomers containing unsaturation. These ionic ring-opening polymerizations exhibit most of the characteristics of cationic and anionic polymerizations of vinyl monomers (see Chapter 8), such as the effects of solvent and counterion, participation of different species (covalent, ion pairs, and free ions) in propagation, and association phenomena.

Individual ring-opening polymerizations may exhibit the characteristics of step-growth or chain-growth polymerizations or a combination of both these processes depending on the initiator used. Two considerations are generally used to classify ring-opening polymerizations, namely, the kinetic behavior and the molecular weight behavior. In particular, the nature of molecular weight dependence on conversion makes a clear distinction between step growth and chain growth processes. Thus, whereas in the former the molecular weight slowly increases with conversion, in the latter it remains more or less unchanged with conversion. Most, but not all, ring-opening polymerizations, however, show a superficial resemblance to step polymerization in that the molecular weight increases relatively slowly with conversion, and also to chain polymerization as only monomer adds to the growing chains in the propagation step [cf. Eqs. (10.4)-(10.6)] and propagation rates are usually described by equations resembling those of chain growth polymerization. Further, it should be noted that, as explained earlier, ceiling temperatures of ring-opening polymerizations are often quite low and, as a result, the kinetics may be complicated by the occurrence of polymerization-depolymerization equilibria (see later).

10.2.1 Cyclic Ethers/Epoxides

Three-membered cyclic ethers (see Table 10.1) are known as epoxides, the simplest epoxides being ethylene oxide and propylene oxide. Their polymerization is a subject of considerable technological importance. The ring-opening polymerization of epoxides can be initiated by both anionic and cationic methods.

10.2.1.1 Anionic Polymerization

The anionic polymerization of epoxides can be initiated by hydroxides, alkoxides, metal oxides, and some organometallic derivatives. For simplicity the initiator will be generally represented as M^+A^-, e.g., $K^+(Bu^tO)^-$ for potassium butoxide. Thus, the polymerization of ethylene oxide can be represented (Young and Lovell, 1990; Odian, 1991) by the initiation step:

$$\overset{\displaystyle O}{\underset{\displaystyle CH_2\!-\!\!-\!\!-CH_2}{\diagup\!\diagdown}} + M^+A^- \longrightarrow A\!-\!CH_2CH_2O^-\,M^+ \tag{10.3}$$

followed by propagation:

$$A\!-\!CH_2CH_2O^-\,M^+ + \overset{\displaystyle O}{\underset{\displaystyle CH_2\!-\!\!-\!\!-CH_2}{\diagup\!\diagdown}} \longrightarrow$$
$$A\!-\!CH_2CH_2OCH_2CH_2O^-\,M^+ \tag{10.4}$$

or, generally,

$$A\!-\!(CH_2CH_2O)_n^-\!CH_2CH_2O^-\,M^+ + \overset{\displaystyle O}{\underset{\displaystyle CH_2\!-\!\!-\!\!-CH_2}{\diagup\!\diagdown}} \longrightarrow$$
$$A\!-\!(CH_2CH_2O)_{\overline{n+1}}\!CH_2CH_2O^-\,M^+ \tag{10.5}$$

Since the $-O^-\cdots M^+$ bond is more ionic than the $-CH_2-A$ bond, propagation occurs by insertion of monomer molecules into the $-O^-\cdots M^+$ bond, as shown in Eqs. (10.4) and (10.5).

A number of organometallic compounds acting as initiators can cause epoxide polymerization to proceed through anionic coordination mechanism. Propagation in such systems involves a "concerted or coordinated process" in which the epoxide monomer is inserted into an oxygen-metal (O–M) bond (Odian, 1991):

$$\text{wwwCH}_2\text{CH}_2\text{O}\!\!\overset{\displaystyle CH_2}{\underset{\displaystyle\underset{\textstyle -M}{\diagup\diagdown}\ O\!-\!CH_2}{\cdots}} \longrightarrow \text{wwwCH}_2\text{CH}_2\text{OCH}_2\text{CH}_2\text{O}\underset{\displaystyle\underset{\textstyle -M}{\diagup\diagdown}}{|} \tag{10.6}$$

This propagation reaction can be viewed as involving intermediate formation of an alkoxide anion through cleavage of the O–M bond in the growing chain—hence the name *anionic coordination*.

Anionic polymerization of a substituted epoxide such as propylene oxide yields only low molecular weight (< 5000) polymers as chain growth is severely restricted by chain transfer to monomer. The transfer reaction becomes more prevalent as the substituent group possesses hydrogen atoms on the α-carbon atom. The chain transfer to propylene oxide monomer occurs by proton abstraction from the methyl group attached to the epoxide ring (Young and Lovell, 1990):

$$
\begin{array}{c}
\text{CH}_3 \\
| \\
\sim\sim\sim\text{CH}_2\text{-CH-O}^-\,\text{M}^+ \quad + \quad \text{CH}_3\text{-CH-CH}_2 \quad \xrightarrow{\quad k_{tr,M} \quad} \\[1em]
\text{CH}_3 \\
| \\
\sim\sim\sim\text{CH}_2\text{-CH-OH} \quad + \quad \text{CH}_2\text{=CH-CH}_2\text{-O}^-\,\text{M}^+
\end{array}
\tag{10.7}
$$

producing allylic alkoxide ions which reinitiate polymerization to produce polymers with C=C end groups.

Kinetics Most epoxide polymerizations have the characteristoics of living polymerization, that is, initiation is fast relative to propagation and there is an absence of termination processes. The expressions for the rate and degree of polymerization used in living chain polymerizations (see Chapter 8) can thus be applied for epoxide polymerizations. The polymerization rate is given by

$$
R_p \;=\; k_p^0 [\text{M}^-][\text{M}]
\tag{10.8}
$$

where k_p^0 is the overall apparent rate constant and $[\text{M}^-]$ is the total concentration of all living anionic propagating centers (free ions and ion pairs); $[\text{M}]$ is the concentration of monomer.

The number average degree of poymerization at time t during the reaction is given simply by the ratio of the decrease in monomer concentration to the initial concentration of initiator:

$$
\overline{DP}_n \;=\; ([\text{M}]_0 - [\text{M}]_t)/[\text{I}]
\tag{10.9}
$$

where $[\text{M}]_0$ and $[\text{M}]_t$ are the monomer concentrations at times 0 and t.

Problem 10.1 The polymerization of ethylene oxide was studied (Gee et al., 1959) in 1,4-dioxane solution, catalyzed by solutions of sodium in a small excess of a simple alcohol, ROH. (An excess of alcohol is added to increase the solubility of the catalyst.) Their experiments support the suggestion that there is no termination reaction and that both the ion pair, RO^-Na^+, and the ion, RO^-, are catalysts. Derive a simple expression for the rate of polymerization under the experimental conditions.

Answer:

Making a simplifying assumption that all the polymerization steps have the same bimolecular rate constant, that is, k_p^- or $k_p^\mp$, depending on whether free ion or ion pair is involved,

$$
R_p \;=\; \frac{-d[\text{C}_2\text{H}_4\text{O}]}{dt} \;=\; \left[\alpha k_p^- + (1-\alpha)k_p^\mp\right][\text{M}^-][\text{C}_2\text{H}_4\text{O}]
\tag{P10.1.1}
$$

where $[\text{M}^-]$ is the total concentration of all anionic propagating centers and α is the degree of dissociation of ion pairs.

So long as the degree of ionization remains constant, no error will be introduced by using an overall rate constant k_p^0 in place of $[\alpha k_p^- + (1-\alpha)k_p^\mp]$. Moreover, $[\text{M}^-]$ can be replaced by c_0, the total initial alkoxide concentration. Thus,

$$
R_p \;=\; k_p^0 c_0 [\text{C}_2\text{H}_4\text{O}]
\tag{P10.1.2}
$$

Note: A series of proton exchange reactions (see later) arise from the presence of excess alcohol. If these exchange reactions are much faster than the polymerization, the effect on the kinetics of polymerization will be negligible, unless the alcohol added differs markedly in acid strength from the polyether alcohols formed by the exchange reactions.

Problem 10.2   Account for the fact that the lower molecular weight polymers produced by polymerization of epoxides with metal alkoxides in dimethyl sulfoxide (DMSO) solvent shows (Bawn et al., 1969) the presence of sulfur in low concentrations ($< 0.4\%$).

Answer:

A rapid proton transfer takes place between the metal alkoxide and DMSO leading to equilibria of the type, as shown by taking potassium butoxide as an example:

$$Bu^tO^-K^+ + CH_3SOCH_3 \overset{K}{\rightleftharpoons} CH_3SOCH_2^-K^+ + Bu^tOH \tag{P10.2.1}$$

Methyl sulfinyl carbanion, $(CH_3SOCH_2^-)$, has been given the trivial name *dimsyl ion*. It is involved in the majority of the base-catalyzed reactions in DMSO, in spite of the fact that the equilibrium lies far to the left $(K = 1.5\times10^{-7})$. The equilibrium of reaction (P10.2.1) is established quickly and even though the dimsyl ion is present in very low concentrations, it is orders of magnitude more reactive than the *t*-butoxide ion (Ledwith and Mcfarlane, 1964).

Initiation by the dimsyl ion involves the reaction:

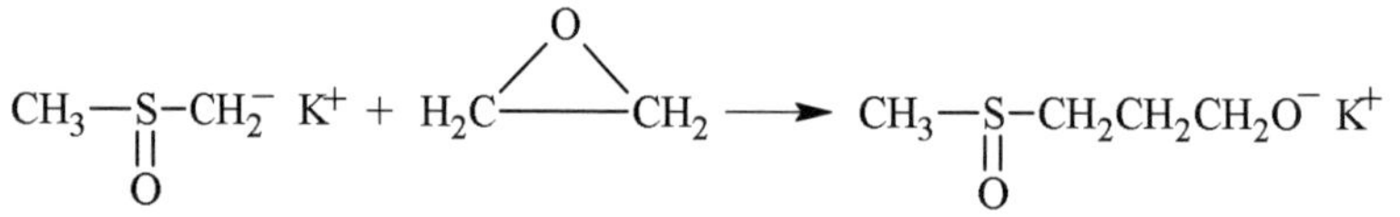

$$CH_3{-}\underset{O}{\overset{O}{S}}{-}CH_2^-\ K^+ + H_2C\overset{O}{\underset{}{\triangle}}CH_2 \longrightarrow CH_3{-}\underset{O}{\overset{O}{S}}{-}CH_2CH_2CH_2O^-\ K^+ \tag{P10.2.2}$$

and propagation proceeds by

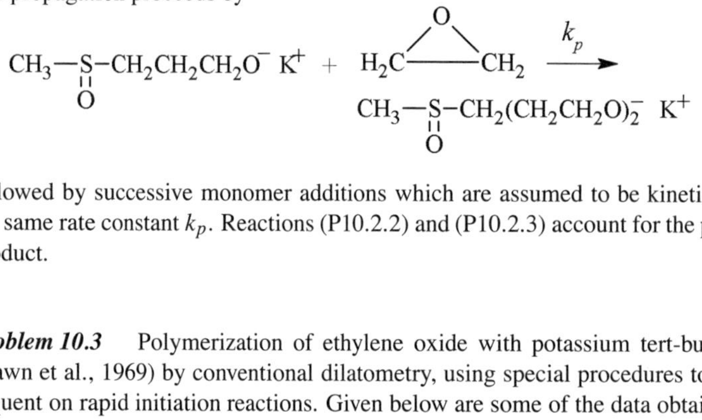

$$CH_3{-}\underset{O}{\overset{O}{S}}{-}CH_2CH_2CH_2O^-\ K^+ + H_2C\overset{O}{\underset{}{\triangle}}CH_2 \xrightarrow{k_p}$$
$$CH_3{-}\underset{O}{\overset{O}{S}}{-}CH_2(CH_2CH_2O)_2^-\ K^+ \tag{P10.2.3}$$

followed by successive monomer additions which are assumed to be kinetically indistinguishable and have the same rate constant k_p. Reactions (P10.2.2) and (P10.2.3) account for the presence of sulfur in the polymer product.

Problem 10.3   Polymerization of ethylene oxide with potassium tert-butoxide in DMSO was followed (Bawn et al., 1969) by conventional dilatometry, using special procedures to eliminate zero time errors consequent on rapid initiation reactions. Given below are some of the data obtained for this system at $50°C$:

| Expt. | Concentration, mol L^{-1} | | Initial rate ($R_p \times 10^3$) |
	Monomer	Initiator$\times10^3$	mol L^{-1} s^{-1}
1	3.14	9.61	6.84
2	3.43	7.81	6.51
3	3.45	1.36	0.99
4	4.81	7.30	7.50
5	2.42	7.60	5.00
6	1.54	7.90	3.65

Calculate an estimate of the apparent rate constant for propagation. Which of the systems given in the table would yield the highest molecular-weight polymer at 90% conversion?

Answer:

Beside the metal oxide, dimsyl ions formed by reaction of metal alkoxide with DMSO [Eq. (P10.2.1)] take part in the initiation of polymerization. The initial rate of polymerization may be given by the expression [cf. Eq. (10.8)]:

$$R_p^0 = -d[M]/dt\,|_0 = k_p^{app}[M]_0 \left[Bu^tO^-K^+\right]_0 \tag{P10.3.1}$$

The values of k_p^{app} calculated for the six cases from Eq. (P10.3.1) are 0.23, 0.24, 0.21, 0.21, 0.27, and 0.30 L mol^{-1} s^{-1}, yielding an average value of 0.24 L mol^{-1} s^{-1}.

According to Eq. (10.9), the highest molecular weight at a given conversion will be obtained for the case which has the highest value of the [Monomer]/[Initiator], that is, for expt. 3 with $[M]_0 = 3.45$ mol L^{-1} and $[I]_0 = 1.36\times10^{-3}$ mol L^{-1}, giving

$$\overline{DP}_n = \frac{(0.90)(3.45\ \text{mol L}^{-1})}{(1.36\times10^{-3}\ \text{mol L}^{-1})} = 2286$$

Exchange Reactions   In epoxide polymerizations initiated by alkali metal alkoxides and hydroxides, protonic substances such as water or alcohol are often added to dissolve the initiator and produce a homogeneous system. However, these substances can contribute to an exchange reaction with the propagating chain. In the presence of alcohol (ROH), for example, the exchange reaction can be represented by

$$R\text{-}OCH_2CH_2\text{-})_\pi O^-Na^+ + ROH \rightleftharpoons$$
$$R\text{-}OCH_2CH_2\text{-})_\pi OH + RO^-Na^+ \tag{10.10}$$

The polymeric alcohol formed in the reaction can also take part in similar exchange reactions with other propagating chains. However, since the total ion concentration remains unchanged by the exchange reactions, they do not affect the observed overall rate (assuming equal reactivity of all ions). The alcohol will, however, affect the molecular weight. Since each alcohol molecule contributes to the generation of a polymeric chain as does the alkoxide initiator, the number-average degree of polymerization will be given by [cf. Eq. (10.9)]:

$$\overline{DP}_n = \frac{[M]_0 - [M]_t}{[I] + [ROH]} \tag{10.11}$$

This equation assumes each initiator and alcohol molecule to be a potential polymer chain. Alcohol or other protonic substances can thus be used to control polymer molecular weight. The molecular weight limitation due to exchange reactions, as represented by Eq. (10.11), does not, however, apply to polymerizations initiated by alkoxides and hydroxides in aprotic polar solvents, nor does it apply to polymerizations initiated by other initiators such as metal alkyls and aryls and the various coordination initiators, since the latter initiators are dissolved in aprotic solvents such as benzene or tetrahydrofuran (Odian, 1991).

Problem 10.4   Discuss the effect of exchange reaction on epoxide polymerizations initiated by metal alkoxides in alcoholic solution, for the cases where the added alcohol is   (a) equally acidic, (b) more acidic, and (c) less acidic than the polymeric alcohol formed by the exchange reaction. What would be the result of the use of HCl or RCOOH in place of ROH (or H_2O)?

Answer:

(a) The exchange reaction will occur throughout the course of the polymerization, if the acidities of the two alcohols are approximately the same. The polymerization rate will be unaffected while the molecular weight will decrease [Eq. (10.11)], but the molecular weight distribution (MWD) will be Poisson.

(b) If the added alcohol ROH is much more acidic than the polymeric alcohol, most of it will undergo reaction with the first-formed propagating species :

$$R'OCH_2CH_2O^-Na^+ + ROH \longrightarrow R'OCH_2CH_2OH + RO^-Na^+ \qquad \text{(P10.4.1)}$$

before polymerization begins. Since ROH is more acidic, reinitiation by RO^-Na^+ would be usually slower, resulting in a decreased polymerization rate and a broadening of the molecular weight.

(c) For the case in which ROH is less acidic than the polymeric alcohol, the rate of polymerization will be relatively unaffected during most of the polymerization and exchange will occur in the later stages of reaction with a broadening of the MWD.

When protonic compounds such as HCl or RCOOH take part in the exchange reaction, the result is not exchange, as occurs with ROH (or H_2O), but inhibition or retardation, since an anion such as Cl^- or $RCOO^-$ possesses little or no nucleophilicity. Reinitiation does not occur or is very slow and the polymeric alcohols are no longer dormant but are dead. Both the polymerization rate and polymer molecular weight thus decrease along with a broadening of the polymer molecular weight.

10.2.1.2 Cationic Polymerization

The initiators used in cationic ring-opening polymerization of cyclic ethers are of the same type as used in cationic chain poymerizations (Chapter 8); for example, strong protonic acids (e.g., H_2SO_4, CF_3CO_2H, CF_3SO_3H) and Lewis acids in conjunction with co-catalysts [e.g., $BF_3.H_2O$, $H^+(SbCl_6)^-$, $Ph_3C^+PF_6^-$]. For simplicity, we shall represent these initiators generally as R^+A^-.

Chain Initiation and Propagation   In general, two mechanisms have been suggested for initiation and chain propagation in ring-opening polymerization by the cationic process. One mechanism involves the formation of an onium ion by interaction of the catalyst system with the monomer, as shown here for the cationic polymerization of tetrahydrofuran :

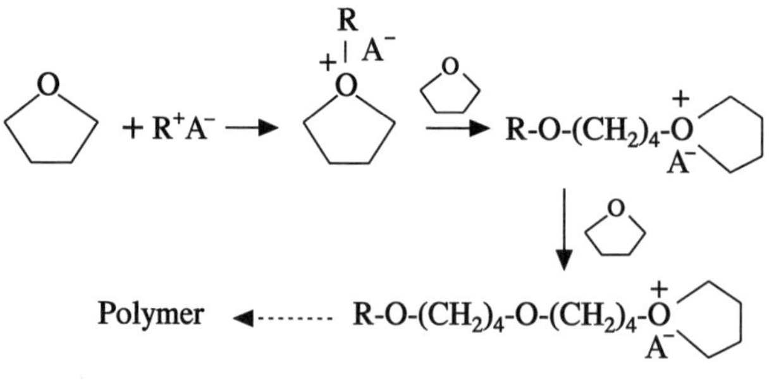

$$\text{(10.12)}$$

The α-carbon of the oxonium ion being electron-deficient because of the adjacent positively charged oxygen, propagation involves a nucleophilic attack of the oxygen of a monomer molecule on the α-carbon of the oxonium ion. In each case, the product of propagation has a terminal cyclic oxonium ion formed from the newly added monomer molecule.

The second mechanism postulates that the catalyst causes monomer ring cleavage forming an ionic species, which is followed by attack of another monomer with ring cleavage and regeneration of ionic active site :

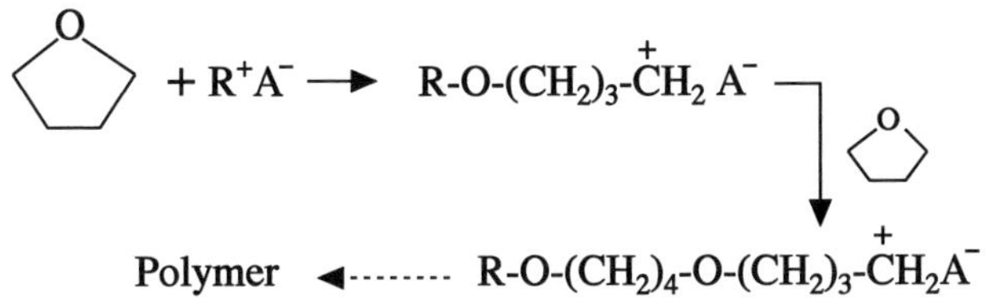

$$(10.13)$$

Though useful polymers can be made by these reactions, their low ceiling temperatures (see p. 599) and consequent tendency to undergo facile depolymerization by an "unzipping" mechanism pose serious limitations. To overcome this problem the technique of *end-capping* or *end-blocking* may be used. Thus polyoxymethylene (polyacetal), an engineering plastic, prepared from the cyclic acetal 1,3,5-trioxane (cyclic trimer of formaldehyde) using boron trifluoride etherate as the initiator (in the presence of small amount of water required for polymerization) can be stabilized by capping the thermolabile hydroxyl end groups of the macromolecule by esterification using acetic anhydride :

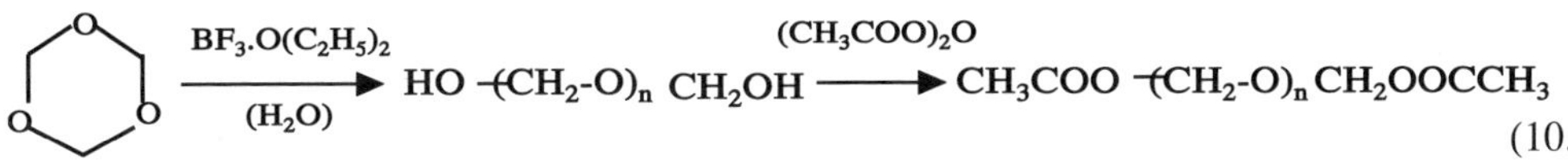

$$(10.14)$$

The end-capping increases the decomposition temperature of polyacetals by a couple of hundred degrees.

Problem 10.5   Like cyclic ethers, cyclic amines can be polymerized by ionic ring-opening method. Thus poly(ethyleneimine) can be prepared by the ring-opening polymerization of aziridine (see Table 10.1) with initiation by protonic acids followed by nucleophilic attack of the monomer. Account for the fact that the process gives rise to a branched polyamine and suggest a method by which the branching could be avoided to obtain a linear polyamine.

Answer:

While protonation of aziridine followed by nucleophilic attack of the monomer on the aziridinium ion gives the dimer, further addition of the monomer and continued chain growth can lead to linear polymer formation :

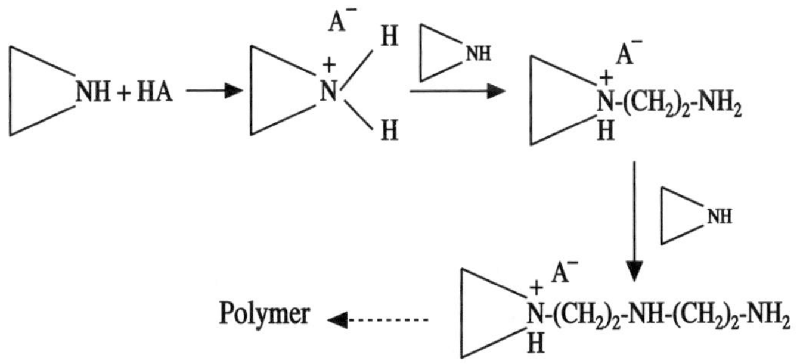

$$(P10.5.1)$$

As the chain growth leads to the formation of secondary amine functions, the latter can react with an aziridinium ion followed by proton transfer to form branched points in the chain:

$$\text{(P10.5.2)}$$

In practice, the final product may contain 20-30% branching.

A linear form of polyethyleneimine can be prepared by using a protecting group such as α-tetrahydropyranyl chloride (Cowie, 1991):

$$\text{(P10.5.3)}$$

According to this scheme, N-(α-tetrahydropyranyl) aziridine is prepared and polymerized cationically to give the poly(iminoether) which is then hydrolyzed with aqueous acid and finally neutralized with alkali.

With some cationic initiators, such as the triphenyl methyl (trityl) salts of the general formula $Ph_3C^+ X^-$, where X^- is a stable anion such as ClO_4^-, $SbCl_6^-$, and PF_6^-, the initiation does not involve direct addition to monomer. The carbocation abstracts a hydride (H^-) ion from the α-carbon of the monomer and the resulting new carbocation initiates polymerization. Thus the reaction of trityl hexafluorophosphate and tetrahydrofuran has been shown to proceed as follows (Cowie, 1991):

$$\text{(10.15)}$$

The reaction takes place below room temperature amd shows the characteristics of living polymerization with no evidence of a termination reaction. The nature of the counterion is an important factor in the elimination of the termination reaction. The rate and degree of polymerizations are then given by expressions similar to Eqs. (10.8) and (10.9).

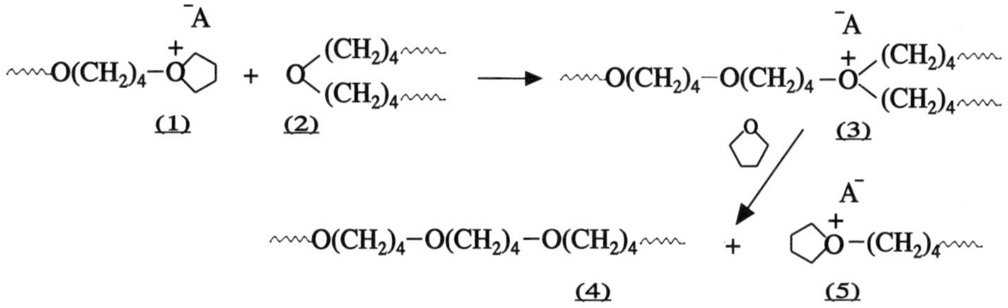

Figure 10.1 Termination by chain transfer to polymer in cationic polymerization of tetrahydrofuran. Nucleophilic attack by an ether oxygen (**2**) in a polymer chain on an oxonium ion propagating center (**1**) forms a tertiary oxonium ion (**3**), which then undergoes nucleophilic attack by monomer leading to a dead polymer (**4**) and a regenerated propagating species (**5**). (After Odian, 1991.)

Chain Transfer and Termination   There are a variety of reactions by which a propagating cationic chain may terminate by transferring its activity. Some of these reactions are analogous to those observed in cationic polymerization of alkenes (Chapter 8). Chain transfer to polymer is a common method of chain termination. Such a reaction in cationic polymerization of tetrahydrofuran is shown as an example in Fig. 10.1. Note that the chain transfer occurs by the same type of reaction that is involved in propagation described above and it leads to regeneration of the propagating species. Therefore, the kinetic chain is not affected and the overall effect is only the broadening of MWD.

Termination of the propagating oxonium ion may also occur by combination with either the counterion or an anion derived from it; for example, in epoxide poymerization initiated by $BF_3.H_2O$:

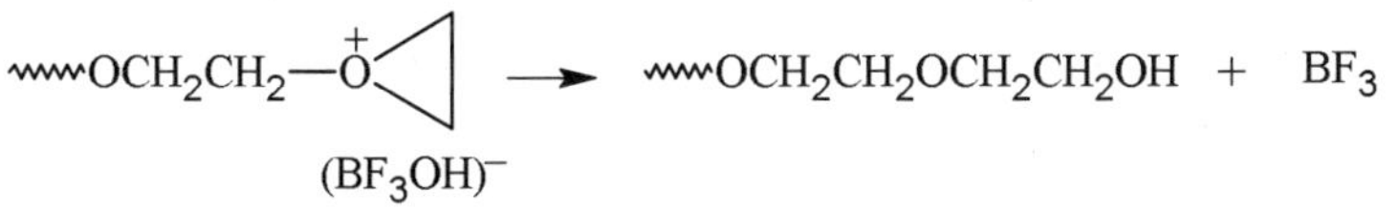

$$\text{(10.16)}$$

Termination may also occur by chain transfer with an externally added chain transfer agent. In this way, desired end groups may be added to the polymer. For example, hydroxyl and amine end groups are obtained by using, respectively, water and ammonia as chain transfer agents [cf. Eq. (8.118)].

Kinetics   The rate of the cationic ring-opening polymerization of cyclic ethers may be expressed in different ways depending on the monomer and the reaction conditions. Polymerizations without termination can be described (Odian, 1991) by kinetic expressions similar to those used in living polymerizations of alkenes (see Chapter 8), e.g.,

$$R_p = k_p[M][M^+] \tag{10.17}$$

where $[M^+]$ is the total concentration of propagating oxonium ions of all sizes.

For reversible ring-opening polymerizations, the propagation-depropagation equilibrium can thus be expressed by

$$M_{n-1}^+ + M \underset{k_{dp}}{\overset{k_p}{\rightleftharpoons}} M_n^+ \tag{10.18}$$

which is analogous to Eq. (6.159) for reversible free-radical polymerization of alkenes and the rate of polymerization is similarly given by :

$$R_p \; = \; -d[M]/dt \; = \; k_p[M][M^+] - k_{dp}[M^+] \tag{10.19}$$

As the rate of polymerization is zero at equilibrium, Eq. (10.19) leads to

$$k_p[M]_e \; = \; k_{dp} \tag{10.20}$$

where $[M]_e$ is the equilibrium monomer concentration, as in Eq. (6.161) for equilibrium alkene polymerization considered in Chapter 6. The derivations given there for $[M]_e$ and ceiling temperature T_c as a function of ΔS^0 and ΔH^0, viz., Eqs. (6.168) and (6.167) are also applicable to ring-opening polymerizations.

Substituting Eq. (10.20) in (10.19) gives the polymerization rate as

$$-d[M]/dt \; = \; k_p[M^+]([M] - [M]_e) \tag{10.21}$$

Integration then gives

$$\ln\!\left(\frac{[M]_0 - [M]_e}{[M] - [M]_e}\right) \; = \; k_p[M^+]t \tag{10.22}$$

where $[M]_0$ is the initial monomer concentration.

Using either Eq. (10.19) or (10.21), k_p can be determined from the monomer conversion versus time data (see Problem 10.6).

Problem 10.6   (a) The kinetics of polymerization of tetrahydrofuran was studied (Vofsi and Tobolski, 1965) with the use of triethyloxonium tetrafluoroborate, $(C_2H_5)_3O^+BF_4^-$, as initiator and dichloromethane as solvent. Conversion versus time was measured at $0°C$ with initial catalyst concentration $[I]_0 = 0.61 \times 10^{-2}$ mol/L   and monomer concentration $[M]_0$ varying from 3 to 9 mol/L. The initial rates, R_p, determined from these data are given in Table A.

(b) In another series of experiments, all with $[M]_0 = 6.1$ mol/L and $[I]_0 = 3.05 \times 10^{-2}$ mol/L, the monomer conversion (p) was measured as a function of time (t) and the number average molecular weight ($\overline{M}_n$) was determined, yielding the data shown in Table B.

Table A   Table B

$[M]_0$, mol L^{-1}	R_p, mol L^{-1} min^{-1}	t, min	p	$\overline{M}_n$
9.15	0.00930	24	0.085	1405
8.00	0.00820	48	0.207	2795
7.00	0.00680	84	0.310	4078
6.10	0.00500	162	0.443	–
5.00	0.00350	210	0.490	–
4.06	0.00170	398	0.565	–
3.05	0.00037	–	–	–

Determine (a) the equilibrium monomer concentration and (b) the propagation rate constant k_p at $0°C$.

Answer:

(a) The polymerization proceeds in the manner shown in Eqs. (P10.6.1) and (P10.6.2).

Initiation:

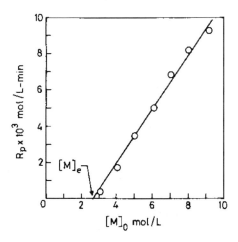

$$\left[\begin{matrix} R \\ | \\ O{-}R \\ | \\ R \end{matrix}\right]^{+} BF_4^- + O \quad \xrightarrow{k_i} \quad \begin{matrix} R \\ \backslash \\ O \\ / \\ R \end{matrix} + R{-}\overset{+}{O} \quad BF_4^- \tag{P10.6.1}$$

Propagation/depropagation:

$$R{-}\overset{+}{O} \underset{BF_4^-}{} + O \quad \underset{k_{dp}}{\overset{k_p}{\rightleftharpoons}} \quad R{-}OCH_2CH_2CH_2CH_2{-}\overset{+}{O} \underset{BF_4^-}{} \tag{P10.6.2}$$

In these equations, k_i, k_p, and k_{dp} are the specific rate constants of initiation, propagation, and depropagation reactions, respectively.

The data from column 2 are plotted against the data from column 1 in Fig. 10.2. From the intercept at R_p = 0, $[M]_e$ = 2.65 mol/L.

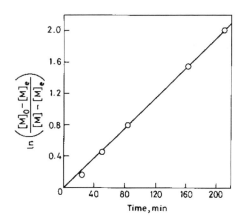

Figure 10.2 Determination of the equilibrium monomer concentration $[M]_e$ from initial rate (R_p) versus initial monomer concentration ($[M]_0$) data. (Data of Problem 10.6.) (After Vofsi and Tobolski, 1965.)

Figure 10.3 Plot of Eq. (10.22) for the determination of $k_p[M^+]$. (Data of Problem 10.6.) (After Vofsi and Tobolsky, 1965.)

(b) Monomer conversion, $p = \dfrac{[M]_0 - [M]}{[M]_0}$ or $[M] = [M]_0(1 - p)$. $\qquad$ (P10.6.3)

With $[M]_e$ = 2.65 mol/L and $[M]$ calculated from Eq. (P10.6.3), the left side of Eq. (10.22) is evaluated and plotted against t in Fig. 10.3. The slope gives $k_p[M^+]$ = 9.5×10^{-3} min^{-1}.

Assuming that the termination of the cationic polymer chain is brought about by the addition of water, the polymer has the formula $C_2H_5(OCH_2CH_2CH_2CH_2)_xOH$ and molecular weight = 46 + 72x. For $\overline{M}_n$ = 1405, $\overline{DP}_n = x$ = (1405 − 46)/(72) = 19. Similarly, for $\overline{M}_n$ = 2795, $\overline{DP}_n$ = 38 and for $\overline{M}_n$ =

4078, $\overline{DP}_n = 56$. Since for the living polymerization, $\overline{DP}_n$ will be given by $\overline{DP}_n = ([M]_0 - [M])/[M^+]$, $[M^+]$ can be calculated from $[M]$ and $\overline{DP}_n$. With $[M]_0 = 6.1$ mol/L, this yields

p	$[M]$, mol/L	$\overline{DP}_n$	$[M^+]$, mol/L
0.085	5.58	19	2.74×10^{-2}
0.207	4.84	38	3.31×10^{-2}
0.310	4.21	56	3.37×10^{-2}

Since the value of $[M^+]$ is nearly constant at monomer conversions $p > 0.20$, it may be assumed with fair approximation that the initiator is completely reacted. Since $k_p[M^+] = 9.5 \times 10^{-3}$, as obtained above,

$$k_p = \frac{(9.5 \times 10^{-3} \text{ min}^{-1})}{(3.37 \times 10^{-2} \text{ mol L}^{-1})} = 0.28 \text{ L mol}^{-1} \text{ min}^{-1} \ (\equiv 4.7 \times 10^{-3} \text{ L mol}^{-1} \text{ s}^{-1}).$$

For ring-opening polymerizations of various cyclic ethers, k_p is in the range 10^{-1}-10^{-3} L/mol-s (Chien et al., 1988; Mijangos and Leon, 1983; Penczek and Kubisa, 1989). These values are seen to be much closer to the rate constants for step-growth polyesterification than to those for various chain polymerizations.

Degree of Polymerization   To derive a relation showing quantitative dependence of the degree of polymerization on reaction parameters for an equilibrium polymerization (Tobolsky, 1957, 1958; Tobolsky and Eisenberg, 1959, 1960), let us assume that an initiator XY brings about the polymerization of a cyclic monomer M in accordance with the following equilibria:

$$XY + M \rightleftharpoons XMY, \qquad K_i = \frac{[XMY]}{[XY][M]} \tag{10.23}$$

$$XMY + M \rightleftharpoons XM_2Y, \qquad K_p = \frac{[XM_2Y]}{[XMY][M]} \tag{10.24}$$

$$XM_2Y + M \rightleftharpoons XM_3Y, \qquad K_p = \frac{[XM_3Y]}{[XM_2Y][M]} \tag{10.25}$$

$$XM_nY + M \rightleftharpoons XM_{n+1}Y, \qquad K_p = \frac{[XM_{n+1}Y]}{[XM_nY][M]} \tag{10.26}$$

where XY is the initiator (which can be both ionic or nonionic), M is the monomer, and K_i and K_p the equilibrium constants for initiation and propagation, respectively. Combining the above equations one obtains the following expression for $[XM_nY]$:

$$[XM_nY] = K_i[XY][M]\big(K_p[M]\big)^{n-1} \tag{10.27}$$

Summation of Eq. (10.27) to include all species from $n = 1$ to $n = \infty$ gives the total concentration, $[N]$, of polymer molecules of all sizes as

$$[N] = \sum_{n=1}^{\infty} XM_nY = \frac{K_i[XY][M]}{1 - K_p[M]} \tag{10.28}$$

One can similarly obtain by summation the total concentration of monomer segments incorporated into the polymer chain. Denoting this quantity by $[W]$, one has

$$[W] = \sum_{n=1}^{\infty} n[XM_nY] = \frac{K_i[XY][M]}{(1 - K_p[M])^2} \tag{10.29}$$

The average degree of polymerization $\overline{DP}_n$ is given by W/N :

$$\overline{DP}_n = \frac{[W]}{[N]} = \frac{1}{1 - K_p[M]} \tag{10.30}$$

The initial concentration of monomer $[M]_0$ is clearly given by

$$
\begin{aligned}
[M]_0 &= [M] + [W] \\
&= [M] \left\{ 1 + \frac{K_i[XY]}{(1 - K_p[M])^2} \right\} \\
&= [M] \left\{ 1 + K_i[XY]\overline{DP}_n^2 \right\} \tag{10.31}
\end{aligned}
$$

The initial concentration of initiator $[XY]_0$ is given by

$$
\begin{aligned}
[XY]_0 &= [XY] + [N] \\
&= [XY] \left\{ 1 + \frac{K_i[M]}{(1 - K_p[M])} \right\} \\
&= [XY] \left\{ 1 + K_i[M]\overline{DP}_n \right\} \tag{10.32}
\end{aligned}
$$

Combination of Eqs. (10.30) through (10.32) gives an expected relationship :

$$\overline{DP}_n = \frac{[M]_0 - [M]}{[XY]_0 - [XY]} \tag{10.33}$$

The quantities $[XY]$ and $[M]$ in the foregoing equations represent the *equilibrium* concentrations of unreacted initiator and unreacted monomer.

The quantities $[XY]_e$, $[M]_e$, $[XY]_0$, $[M]_0$, and $\overline{DP}_n$ are experimentally measurable. To evaluate the constants K_p and K_i one might therefore proceed as follows : obtain K_p from Eq. (10.30) using the experimentally determined values of $[M]$ and $\overline{DP}_n$, and obtain K_i from Eq. (10.31) using the experimentally determined values of $[M]$, $[XY]$, and $\overline{DP}_n$ (see Problem 10.7).

Problem 10.7 A set of experimental data available in the literature on the equilibrium polymerization of caprolactam (CL) are of the type : a known mole ratio m of H_2O to CL is charged in a vessel; the vessel is brought to a definite temperature and polymerization is carried out to equilibrium. When equilibrium is reached, the value of $\overline{DP}_n$ and $[M]$ (or conversion p) are determined.

Two sets of data (Tobolsky and Eisenberg, 1959, 1960) obtained over a sufficiently wide range of variables (temperature and initiator concentration) are given below :

Temp. ($^\circ$C)	Run	m	p	$\overline{DP}_n$
221.5	1	0.060	0.9390	120
	2	0.192	0.9359	60
253.5	1	0.060	0.9210	100
	2	0.132	0.9188	60

Determine the ΔH^0 values for the initiation and propagation steps of the equilibrium polymerization of caprolactam.

Answer:

The polymerization of caprolactam is accompanied by a volume contraction; therefore, the use of concentration units of moles per kilogram is preferred over the application of moles per liter. It is also necessary to be able to compute $[M]_0$ and $[XY]$ from the initial mole ratio m. This is accomplished from the following relationships:

$$[XY]_0 = (1000m)/(18m + 113) \tag{P10.7.1}$$

$$[M]_0 = [M]_{00} - ([M]_{00}/[XY]_{00})\,[XY]_0 = 8.85 - 0.1594[XY]_0 \tag{P10.7.2}$$

where the molecular weight of H_2O is 18 and that of CL is 113; $[M]_{00}$ is equal to the moles per kilogram of pure caprolactam (= 8.85) and $[XY]_{00}$ is equal to the moles per kilogram of pure water (= 55.5).

From a set of values of $\overline{DP}_n$, p, $[XY]_0$, and $[M]_0$ the constants K_i and K_p can be computed by the following steps:

(a) $[M]$ is calculated from $[M] = [M]_0(1 - p)$.

(b) K_p is calculated from Eq. (10.30), rearranged to the form: $K_p = \dfrac{(\overline{DP}_n - 1)}{[M]\overline{DP}_n}$.

(c) $[XY]$ is calculated from Eq. (10.33), rearranged to the form:

$$[XY] = [XY]_0 - \frac{[M]_0 - [M]}{\overline{DP}_n}.$$

(d) K_i is calculated from Eqs. (10.32) and (10.33), combined into the form:

$$K_i = \frac{[XY]_0 - [XY]}{[XY][M]\overline{DP}_n} = \frac{[M]_0 - [M]}{[XY][M]\overline{DP}_n^2}$$

[It should be noted that in the theory outlined above leading to Eqs. (10.30)-(10.33), the presence of cyclic oligomers has not been taken into consideration. These are present in the equilibrium mixture of caprolactam polymerization to the extent of at most 5%.]

The ΔH^0 values can be calculated from the equilibrium constant values at several temperatures. Only one set of values for $[XY]_0$, $[M]_0$, $[M]$, and $\overline{DP}_n$ are necessary in principle to calculate the equilibrium constants; however, both sets at each temperature may be taken to determine the constancy of K_i and K_p. The values are:

$T_1 = 221.5°C\,(\equiv 494.5°K)$			$T_2 = 253.5°C\ (\equiv 526.5°K)$		
Run 1	$K_i = 0.0023$	$K_p = 1.855$	Run 1	$K_i = 0.0026$	$K_p = 1.430$
Run 2	$K_i = 0.0027$	$K_p = 1.787$	Run 2	$K_i = 0.0031$	$K_p = 1.397$
Average	$K_i = 0.0025$	$K_p = 1.821$	Average	$K_i = 0.00285$	$K_p = 1.4135$

From van't Hoff's equation, $\ln[(K_i)_2/(K_i)_1] = (\Delta H^0/R)\,[(T_2 - T_1)/(T_1 T_2)]$, the present data lead to

$$\ln\left(\frac{0.00285}{0.00250}\right) = \frac{\Delta H_i^0}{(1.987\ \text{cal mol}^{-1}\ °K^{-1})}\left[\frac{(526.5 - 494.5)\ °K}{(494.5 \times 526.5)\ °K^2}\right]$$

Solving, $\Delta H_i^0 = 2118$ cal/mol. Similarly, from K_p values, $\Delta H_p^0 = -4095$ cal/mol.

Thus, the results obtained are

$$CL + H_2O \rightleftharpoons H(CL)OH \qquad \Delta H^0 = 2118 \text{ cal/mol}$$
$$H(CL)OH + CL \rightleftharpoons H(CL)_2OH \qquad \Delta H^0 = -4095 \text{ cal/mol}$$
$$H(CL)_nOH + CL \rightleftharpoons H(CL)_{n+1}OH$$

The overall process is thus exothermic.

10.2.2 Lactams

Lactams are cyclic amides formed by the intramolecular amidation of amino acids. The ring-opening polymerization of lactams [Eq. (10.34)], especially caprolactam ($m = 5$), provides a valuable noncondensation route to the synthesis of nylons:

$$(10.34)$$

This reaction can be initiated by strong bases (metal hydrides, alkali metals, metal amides, metal alkoxides, and organometallic compounds), protonic acids, or by water. Water is often used as the initiator (cf. Problem 10.7) for industrial polymerization of lactams and the process is referred to as *hydrolytic polymerization*. Anionic initiation with bases is preferred when polymerization is to be carried out in the mold itself for converting monomer directly into a molded object. Cationic initiation with acids, however, is not so useful because both the extents of conversion and polymer molecular weights are significantly lower (Odian, 1991).

10.2.2.1 Hydrolytic Polymerization

The hydrolytic polymerization (Bertalan et al., 1984; Sekiguchi, 1984) of ϵ-capro-lactam can be carried out, on a laboratory scale, by heating a mixture of the monomer and water (2% by wt.) in a thick-walled polymerization tube with suitable safety precautions (Sorensen and Campbell, 1968). The water initiates the reaction by attacking the carbonyl group in the ring to form the open chain species, ϵ-aminocaproic acid:

$$(10.35)$$

This is followed by step polymerization of the amino acid with itself:

$$\text{wwwCOOH} + \text{H}_2\text{Nwww} \rightleftharpoons \text{wwwCONHwww} + \text{H}_2\text{O} \tag{10.36}$$
$$(\text{S}_n) \qquad (\text{S}_m) \qquad (\text{S}_{n+m}) \qquad (\text{W})$$

and initiation of ring-opening polymerization of lactam by the amino acid, in which the COOH group of the amino acid protonates the lactam followed by nucleophilic attack of amine on the protonated lactam:

$$(10.37)$$

The propagation process follows in the same manner.

The initial ring-opening [Eq. (10.35)] and subsequent propagation steps [Eqs. (10.36) and (10.37)], which include both condensation and stepwise addition reactions, constitute the principal mechanism of the polymerization and may be represented by the following three equilibria (Bertalan, 1984):

$$M + W \rightleftharpoons S_1 \qquad\qquad K_1 = [S_1]/([M][W]) \qquad\qquad (10.38)$$

$$S_n + S_m \rightleftharpoons S_{n+m} + W \qquad\qquad K_2 = ([S_{n+m}][W])/([S_n][S_m]) \qquad\qquad (10.39)$$

$$M + S_n \rightleftharpoons S_{n+1} \qquad\qquad K_3 = [S_{n+1}]/([M][S_n]) \qquad\qquad (10.40)$$

Employing the usual simplifying assumption, that the reactivity of the end groups are equal and independent of the chain length of the respective molecules, K_2 may also be expressed as

$$K_2 = \frac{[S_{n+1}][W]}{[S_n][S_1]} \qquad\qquad (10.41)$$

From Eqs. (10.38), (10.40), and (10.41) it is then obvious that

$$K_3 = K_1 K_2 \qquad\qquad (10.42)$$

Applying the principle of equal reactivities one can also write a more generalized expression for K_2 as:

$$K_2 = \frac{[NHCO][W]}{[NH_2][COOH]} \qquad\qquad (10.43)$$

and since $[NH_2] = [COOH] = [S]$,

$$K_2 = [NHCO][W]/[S]^2 \qquad\qquad (10.44)$$

where

$[NH_2]$	= Concentration of amino end groups
$[COOH]$	= Concentration of acid end groups
$[NHCO]$	= Concentration of amide linkages
$[S]$	= Concentration of polymeric chains

Values for the equilibrium constants, K_1, K_2, and K_3, may be calculated from Eqs. (10.38), (10.44), and (10.42), respectively, by substituting the various terms in these equations with quantities that are experimentally obtainable.

Kinetics   The significant quantities upon which the formulation of kinetic equations may be based are the concentrations of caprolactam (M), polymeric chains (S), and amide groups in linear chains (Z). Neglecting amide groups in cyclic oligomers, the relationship between these quantities is given by (Bertalan, 1984)

$$[Z] = 1 - [M] - [S] \qquad\qquad (10.45)$$

from which it follows that

$$d[Z]/dt = -d[M]/dt - d[S]/dt \qquad\qquad (10.46)$$

We recall that the three principal chemical equations for the considered mechanism [cf. Eqs. (10.38)-(10.40)] were

Ring opening (RO):

$$M + W \; \underset{k_{-1}}{\overset{k_1}{\rightleftharpoons}} \; S_1, \qquad K_1 = k_1/k_{-1} \tag{10.47}$$

Polycondensation (PC):

$$S_n + S_m \; \underset{k_{-2}}{\overset{k_2}{\rightleftharpoons}} \; S_{n+m} + W, \qquad K_2 = k_2/k_{-2} \tag{10.48}$$

or, in general,

$$
\begin{array}{ccccccc}
NH_2 & + & COOH & \rightleftharpoons & NHCO & + & H_2O \\
(S) & & (S) & & (Z) & & (W)
\end{array}
$$

Polyaddition (PA):

$$
\begin{array}{ccccc}
S_n & + & M & \underset{k_3}{\overset{k_3}{\rightleftharpoons}} & S_{n+1}, \qquad K_3 = k_3/k_{-3} \\
(S) & & & & (S-S_1)
\end{array} \tag{10.49}
$$

The kinetic equations representing the contributions of the individual reactions are as follows:

RO: $\quad -d[M]/dt = k_1([M][W] - [S_1]/K_1)$ $\tag{10.50}$

PC: $\quad d[S]/dt = k_1([M][W] - [S_1]/K_1)$
$$- k_2[S]\left\{[S] - \frac{[W]}{K_2}\left(\frac{1 - [M]}{[S]} - 1\right)\right\} \tag{10.51}$$

PA: $\quad -d[M]/dt = k_3[S]\left\{[M] - \frac{(1 - [S_1]/[S])}{K_3}\right\}$ $\tag{10.52}$

Since $[W] = [W]_0 - [S]$, it follows that

$$-d[M]/dt = k_1\left\{[M]([W]_0 - [S]) - \frac{[S_1]}{K_1}\right\}$$
$$+ k_3[S]\left\{[M] - \frac{(1 - [S_1]/[S])}{K_3}\right\} \tag{10.53}$$

$$-d[S]/dt = k_2[S]\left\{[S] - \frac{([W]_0 - [S])}{K_2}\left(\frac{1 - [M]}{[S]} - 1\right)\right\}$$
$$- k_1\left\{[M]([W]_0 - [S]) - \frac{[S_1]}{K_1}\right\} \tag{10.54}$$

Since it has been assumed that the reactivities of all carboxyl and amino groups are equal (that is, independent of the chain length of the molecule), the equations derived for linear macromolecules may also be applied to aminocaproic acid (S_1). For this case, Eqs. (10.38)-(10.40) may be written as

$$M + W \rightleftharpoons S_1, \quad S_n + S_1 \rightleftharpoons S_{n+1} + W, \quad M + S_1 \rightleftharpoons S_2$$

The following rate equation may then be written for S_1:

$$\frac{d[S_1]}{dt} = k_1\left\{[M]([W]_0 - [S]) - \frac{[S_1]}{K_1}\right\}$$
$$- 2k_2\left\{[S][S_1] - \frac{([W]_0 - [S])([S] - [S_1])}{K_2}\right\} - k_3[S_1]\left([M] - \frac{1}{K_3}\right) \tag{10.55}$$

Note that in the last term of this equation, it has been assumed for the sake of simplicity that $[S_2] \simeq [S_1]$.

The set of differential equations comprising Eqs. (10.53)-(10.55) is suitable for the estimation of k_1, k_2, and k_3 from the experimental data of $[M]$, $[S]$, and $[S_1]$, obtained as a function of time.

10.2.2.2 *Anionic Polymerization*

Here we will consider the polymerization of caprolactam initiated by strong bases. They are usually referred to as catalysts and also as initiators by some authors (Sekiguchi, 1984). Strong bases initiate polymerization by the replacement of H in the N–H residue of caprolactam by a cation to give *lactam anion*; e.g., for caprolactam with an alkali metal, M :

$$(CH_2)_5\!-\!NH \;+\; M \;\rightleftharpoons\; (CH_2)_5\!-\!\bar{N}\;M^+ \;+\; \tfrac{1}{2}H_2 \qquad (10.56)$$
$$\textbf{(I)} \qquad\qquad\qquad \textbf{(II)}$$

The alkaline salt of the lactam is preferably prepared *extra situ* and introduced into the mixture to polymerize (Sekiguchi, 1984), or prepared *in situ* by the action of an alkali metal compound (M^+A^-) with the caprolactam monomer :

$$(CH_2)_5\!-\!NH \;+\; M^+A^- \;\longrightarrow\; (CH_2)_5\!-\!\bar{N}\;M^+ \;+\; AH \qquad (10.57)$$
$$\textbf{(I)} \qquad\qquad\qquad\qquad \textbf{(II)}$$

Under conditions of complete dissociation, the concentration of the lactam anion, represented in the following discussions by -CO-N$^-$-, is equal to that of the lactam salt **(II)**, denoted by LS. If LS is weakly dissociated, the concentration $[-CO\text{-}N^-\text{-}]$ varies with $[LS]^{1/2}$, since

$$K \;=\; \frac{[-CO-N^--][M^+]}{[LS]} \;\simeq\; \frac{[-CO-N^--]^2}{[LS]}$$
$$\text{or}\quad [-CO-N^--] \;\simeq\; K^{1/2}[LS]^{1/2} \qquad (10.58)$$

Strongly dissociated catalysts may thus be used in lower concentrations than weakly dissociated catalysts.

The subsequent mechanism of the base-catalyzed polymerization of caprolactam is complicated. Ring opening of the initiated monomer does not occur. Since the lactam anion is stabilized by resonance, viz.,

$$\left[-\overset{\overset{\displaystyle O}{\|}}{C}-\bar{N}- \quad\longleftrightarrow\quad -\overset{\overset{\displaystyle O^-}{|}}{C}=N- \right]$$

its further reaction with caprolactam monomer is energetically unfavorable. It thus attacks the carbonyl atom in a molecule of monomer very slowly, causing scission of the CO–NH bond to produce a highly reactive terminal $-\overline{N}H$ ion (**III**), as shown by Eq. (10.59) (Sekiguchi, 1984; Odian, 1991):

$$
\text{(II)} \quad + \quad \text{(I)} \quad \xrightarrow{\text{Slow}} \quad \text{(III)} \tag{10.59}
$$

(II): $\underset{\text{(II)}}{(CH_2)_5 - N^- \; M^+}$ with ring carbonyl; (I): $\underset{\text{(I)}}{HN - (CH_2)_5}$ with ring carbonyl; (III): $\underset{\text{(III)}}{(CH_2)_5 - N - CO(CH_2)_5 - \overset{H}{N^-} \, M^+}$

The new primary amine anion (**III**), unlike the lactam anion (**II**), is not stabilized by conjugation with a carbonyl group. It is therefore highly reactive and, once it is formed, it undergoes a rapid proton-abstraction reaction with a caprolactam monomer producing the dimer compound *N*-caproylcaprolactam (**IV**), which is a *N*-acyllactam, and regenerating the lactam anion (**II**):

$$
\underset{\text{(III)}}{(CH_2)_5 - N - CO(CH_2)_5 - \overset{H}{N^-} \, M^+} \quad + \quad \underset{\text{(I)}}{(CH_2)_5 - NH} \quad \xrightarrow{\text{Fast}}
$$

$$
\underset{\text{(IV)}}{(CH_2)_5 - N - CO(CH_2)_5 - NH_2} \quad + \quad \underset{\text{(II)}}{(CH_2)_5 - N^- \, M^+} \tag{10.60}
$$

The ring carbonyl carbon atom in the *N*-acyllactam (**IV**) is much more strongly activated towards nucleophilic attack than that in the monomer (**I**) because of the presence of the second carbonyl group bonded to the ring nitrogen atom. Propagation thus occurs through a sequence of reactions (Odian, 1991) similar to those shown in Eqs. (10.61) and (10.62), where monomeric lactam ion attacks the ring carbonyl carbon of *N*-acyllactam (**IV**) causing scission of the CO–N bond to form a $-\overline{N}-$ anion (**V**) which abstracts H^+ from a monomer molecule (**I**) to regenerate the lactam anion (**II**) and the propagating *N*-acyllactam (**VI**). The propagating center in the above mechanism is thus the cyclic amide linkage of *N*-acyllactam as propagation takes place by the unusual process of insertion of lactam anion (and not the monomer lactam) into an N–CO bond of *N*-acyllactam. The lactam anion is thus often referred to as *activated monomer*.

$$
\underset{\text{(II)}}{(CH_2)_5 - N^- \, M^+} \quad + \quad \underset{\text{(IV)}}{(CH_2)_5 - N - CO(CH_2)_5 - NH_2} \quad \longrightarrow
$$

$$
\underset{\text{(V)}}{(CH_2)_5 - N - CO(CH_2)_5 - \overset{M^+}{\overline{N}} - CO(CH_2)_5 - NH_2} \tag{10.61}
$$

$$
\underset{\text{(V)}}{(CH_2)_5 - N - CO(CH_2)_5 - \overset{M^+}{\overline{N}} - CO(CH_2)_5 - NH_2} \quad + \quad \underset{\text{(I)}}{(CH_2)_5 - NH} \quad \longrightarrow
$$

$$
\underset{\text{(VI)}}{(CH_2)_5 - N - CO(CH_2)_5 - NHCO(CH_2)_5 - NH_2} \quad + \quad \underset{\text{(II)}}{(CH_2)_5 - N^- \, M^+} \tag{10.62}
$$

Long induction periods are often observed for this polymerization which is a consequence of the slowness of the reaction in Eq. (10.59). It can be eliminated and the polymerization rate enhanced significantly by inclusion of an acylating agent (e.g., acid chloride or anhydride) that reacts rapidly with the lactam to form an N-acyllactam (**VII**) [see Eq. (10.63)], which then propagates by a sequence of reactions similar to those of Eqs. (10.61) and (10.62), except that the propagating chain now has an acylated end group instead of an amine end group.

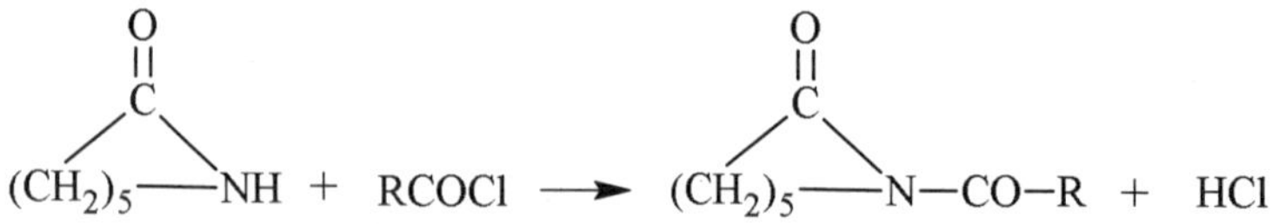

$$(CH_2)_5\text{---}NH + RCOCl \longrightarrow (CH_2)_5\text{---}N\text{---}CO\text{---}R + HCl \qquad (10.63)$$
$$\textbf{(I)} \qquad\qquad\qquad\qquad\qquad \textbf{(VII)}$$

Polymerizations in the presence of an acylating agent are often referred to as *assisted* (or activated) polymerizations. The rate of lactam polymerization depends on the concentrations of the activated monomer and propagating chains, which, for an assisted lactam polymerization, are determined by the concentrations of base and N-acyllactam (or acylating agent), respectively. The degree of polymerization increases with increasing conversion and concentration of monomer and with decreasing concentration of N-acyllactam. Qualitatively, these features are similar to those of living polymerizations.

Problem 10.8   (a) Assuming a constant rate of formation of growing chains and neglecting reverse reactions, derive an expression for the rate of anionic polymerization of lactam with a strongly dissociated catalyst. (b) Given that the propagation rate constant of a lactam polymerization is 8 L mol^{-1} s^{-1} and free lactam ion concentration is 0.01 mol L^{-1}, calculate an approximate value for the induction period. (c) Discuss qualitatively the type of molecular weight distribution one would expect for the polymer at high conversion.

Answer:

(a) As shown above, propagation of anionic polymerization requires the interaction of two active species, lactam anion and N-acyllactam end unit, while monomer (M) is only consumed via its anion. Assuming that the chain length does not influence the reactivity of chain end groups, one can thus write (Sekiguchi, 1984) [cf. Eqs. (10.59) and (10.61)]:

$$-d[M]/dt = k_0[M][-CO-N^--] + k_1[P_n][-CO-N^--] \qquad (P10.8.1)$$

where k_0 is the rate constant of the forward reaction of chain initiation and k_1 is that of propagation; P_n represents a polymer chain of any degree of polymerization n bearing a N-acyllactam end unit.

For a constant rate of formation of growing chains, Eq. (P10.8.1) can be further expressed as

$$\begin{aligned} -d[M]/dt &= k_0[M][-CO-N^--] + k_1(k_0[M][-CO-N^--]t)[-CO-N^--] \\ &= k_0[M][-CO-N^--](1 + k_1[-CO-N^--]t) \end{aligned} \qquad (P10.8.2)$$

(b) The value of time t for which $k_1[-CO-N^--]t$ becomes greater than 1 gives a measure of the induction period. Thus, induction period $\simeq 1/[(8 \text{ L mol}^{-1} \text{ s}^{-1})(0.01 \text{ mol L}^{-1})] = 12.5$ s.

(c) In the aforesaid anionic poymerization, a small number of dimer anions (**III**), once formed, would grow rapidly to give very long chains, while the chain initiation reaction [Eq. (10.59)] continues to add slowly new dimer chains. Consequently, the polymerization in the initial stage would produce a very wide distribution of molecular weights. However, with higher conversions leading to increase in the concentration of linear amide groups and decrease in that of cyclic amide, the self-initiation reaction would be increasingly replaced by the chain multiplication reaction and the lengths of all chains would gradually approach the statistical distribution (Sekiguchi, 1984).

10.2.3 *Lactones*

The ring opening polymerization of lactones, lactides, and glycolides to polyesters can be carried out by both anionic and cationic means using a variety of initiators similar to those used for cyclic ethers. However, cationic polymerization is not as useful as anionic polymerization for synthesizing high-molecular weight polyesters due to detrimental side reactions such as intramolecular transesterification (cyclization) and chain transfer to polymer reactions (including proton and hydride transfer) which lower the molecular weight.

The anionic polymerization of almost all lactones proceeds by nucleophilic attack at the ring carbonyl group followed by scission of the CO–O bond, e.g., for caprolactone the initiation and propagation steps can be depicted as

$$(10.64)$$

Much of the recent activities in anionic ring-opening polymerization involve the use of anionic coordination initiators. For these initiators, the metal *coordinates* with the carbonyl (C=O) oxygen of the monomer, which is followed by cleavage of the acyl-oxygen (CO–O) bond of the monomer and *insertion* into the metal-oxygen (M–O) bond of the initiator. The experimental evidence for this *coordination-insertion* mechanism comes from end group analysis of the polymer formed.

Alkoxides of some metals with *d*-orbitals, including Al and Sn alkoxides, are efficient coordination initiators for ring opening polymerization of lactones. Aluminum alkoxides $[R_{3-n}Al(OR)_n]$ (Mecerreyes and Jérôme, 1999; Penczek et al., 1998) — in particular, Al trialkoxides $[Al(OR)_3]$ and dialkyl Al alkoxides $[R'_2AlOR]$ — are widely used, while tin octoate, $Sn(OOCCH(C_2H_5)C_4H_9)_2$, is often preferred owing to its acceptance as a food additive. The active species in the latter case are, however, believed to be tin(II) alkoxides formed *in situ* by the reaction of tin octoate with either adventitious water and hydroxyacids, or with an alcohol co-initiator. The predominance of Al alkoxides and Sn alkoxides as initiators for ring-opening polymerization is due to their ability to produce stereoregular polymers of narrow molecular weight distribution and predictable molecular weights with well-defined end groups.

The coordination-insertion mechanism is illustrated in Fig. 10.4 taking the example of ε-caprolactone polymerization with aluminum trialkoxide, $Al(OR)_3$. The first step in the mechanism is the coordination of the metal alkoxide to the carboxy of the monomer. This nucleophilic attack causes cleavage of the CO–O bond of the monomer with simultaneous insertion into the metal alkoxide bond, followed by propagation. The hydrolysis of the propagating alkoxide produces poly(ε-caprolactone), a linear polyester with an OR-containing ester as one end group and a hydroxyl group at the other end. There are two possibilities of transesterification side reactions in ring-opening polymerization of lactones, namely, monomolecular (intramolecular) chain transfer leading to formation of macrocyclics and bimolecular (intermolecular) chain transfer, leading to broadening of molecular weight distribution. Though the formation of cyclic compounds is thermodynamically inevitable for polyesters, kinetic conditions can be found for formation of linear macromolecules almost free of ring structures.

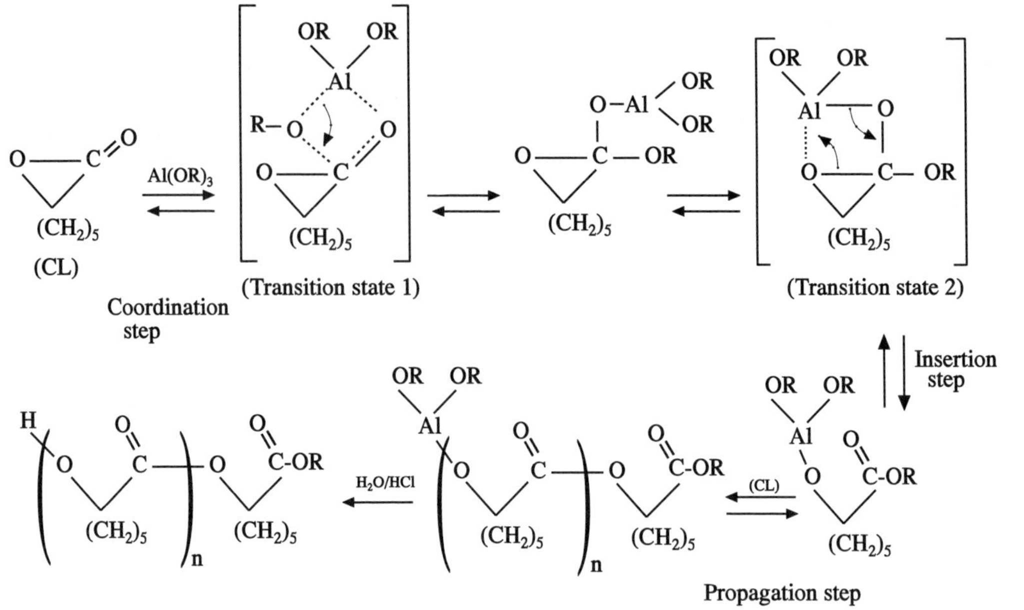

Figure 10.4 The coordination-insertion mechanism for ring-opening polymerization of ϵ-caprolactone with trialkoxy aluminum. (After Deshayes et al., 2003.)

Poly(ϵ-caprolactone), polylactides, and polyglycolides have quite unusual properties of biodegradability and biocompatibility. The majority of polymers used in the biomedical field to develop implants, sutures, and controlled drug-delivery systems are the aforesaid resorbable polyesters produced by ring-opening polymerization of cyclic (di)esters.

REFERENCES

Allcock, H. R. and Lampe, F.W., "Contemporary Polymer Chemistry", Prentice Hall, Englewood Cliffs, New Jersey, 1990.

Bawn, C. E., Ledwith, A., and Mcfarlane, N., *Polymer*, **10**, 653 (1969).

Bertalan, G., Rusznak, I., and Anna, P., *Makromol. Chem.*, **185**, 1285 (1984).

Chien, J. C. W., Cheun, Y. G., and Lillya, C. P., *Macromolecules*, **21**, 870 (1988).

Cowie, J. M. G., "Polymers: Chemistry and Physics of Modern Materials", Blackie, London, 1991.

Deshayes, G., Mercier, F. A. G., Degée, P., Verbrugges, I., Biesemans, M., Willem, R., and Dubois, P., *Chem. Eur. J.*, **9**, 4346 (2003).

Gee, G., Higginson, W. C. E., and Merrall, G. T., *J. Chem. Soc.*, 1345 (1959).

Ledwith, A. and Mcfarlane, N. R., *Proc. Chem. Soc.*, 108 (1964).

Mecerreyes, D. and Jérôme, R., *Macromol. Chem. Phys.*, **200**, 2581 (1999).

Mijangos, F. and Leon, L. M., *J. Polym. Sci. Polym. Lett. Ed.*, **21**, 885 (1983).

Odian, G., "Principles of Polymerization", John Wiley, New York, 1991.

Penczek, S., Duda, A., and Libiszowski, J., *Macromol. Symp.*, **128**, 241 (1998).

Penczek, S. and Kubisa, P. "Cationic Ring-Opening Polymerization: Ethers", chap. 8 in "Comprehensive Polymer Science", vol. 3 (G. C. Eastmond, A. Ledwith, S. Russo, and P. Sigwalt, eds.), Pergamon Press, London, 1989.

Sawada, H., "Thermodynamics of Polymerization", Marcel Dekker, New York, 1976.

Schenck, von H., Ryner, M., Albertsson, Ch. A., and Svensson, M., *Macromolecules*, **35**, 1556 (2002).

Sekiguchi, H., "Lactams and Cyclic Amides", chap 12 in "Ring Opening Polymerization", vol. 2 (K. J. Ivin and T. Saegusa, eds.), Elsevier, London, 1984.

Sorenson, W. R. and Campbell, T. W., "Preparative Methods of Polymer Chemistry", 2nd ed., Wiley Interscience, New York, 1968.

Tobolsky, A. V., *J. Polym. Sci.*, **25**, 220 (1957); **31**, 126 (1958).

Tobolsky, A. V. and Eisenberg, A., *J. Am. Chem. Soc.*, **81**, 2302 (1959); **82**, 289 (1960).

Vofsi, D. and Tobolsky, A. V., *J. Polym. Sci., Part A*, **3**, 3261 (1965).

EXERCISES

10.1 Salts of carbazole are excellent initiators for ethylene oxide polymerization, giving living polymers of predicted molecular weights. From conductivity measurements at 20°C, the dissociation constant of carbazylpotassium (NK) was found [Sigwalt, P. and Boileau, S., *J. Polym. Sci. Polym. Symp.*, **62**, 51 (1978)] to be 7.0×10^{-9} in THF, 1.1×10^{-5} in THF + [2.2.2]cryptand and 7.5×10^{-2} in hexamethylphosphoramide (HMPA). (a) Calculate the fraction of free ions in the three cases with NK concentration 10^{-3} mol/L. (b) Calculate the molecular weights of polyethylene oxide initiated by NK in the three cases at 90% conversion of the monomer of initial concentration 1.4 mol/L. (c) What is the number of initiator residue per molecule of the polymer formed?
[*Ans.* (a) THF 0.003, THF + [2.2.2] 0.11, HMPA 0.99; (b) 55,440; (c) One initiator residue per molecule.]

10.2 The initiation of ethylene oxide polymerization by sodium naphthalene involves direct addition of the monomer to the radical anion and reduction of the adduct by sodium naphthalene producing a dianion. Suggest two tests to support this mechanism.

10.3 Consider the following monomers and initiating systems:

(a) *Monomers*: Propylene oxide, trioxane, oxacyclobutane, γ-butyro-lactam, δ-valerolactam, and ethyleneimine.

(b) *Initiating system*: n-C_4H_9Li, H_2O, $BF_3 + H_2O$, $NaOC_2H_5$, H_2SO_4.

Which initiating system(s) can be used to polymerize each of the various monomers? Write chemical equations for the polymerizations.

10.4 Explain why the polymerization of an epoxide by hydroxide or alkoxide ion is often carried out in the presence of an alcohol. What effects will the presence of alcohol have on the polymerization rate and on the degree of polymerization?

10.5 Explain the following observations:

(a) Anionic polymerization of propylene oxide usually gives rise to a relatively low-molecular-weight polymer.

(b) The rate of polymerization of tetrahydrofuran by BF_3 is greatly enhanced by the addition of a small amount of epichlorohydrin, even though the latter is much less basic than tetrahydrofuran.

(c) The rate of polymerization of oxetane by BF_3 increases on addition of small amounts of water, but the degree of polymerization decreases.

(d) In the presence of an acylating agent, the anionic polymerization of lactams occurs without an induction period.

10.6 Polymerization of ϵ-caprolactam (8.79 mol/L) at 220°C with water (0.352 mol/L) as the initiating agent yielded polymer with $\overline{DP}_n = 152$ at equilibrium conversion of 94.5% (Odian, 1991). Calculate the values of K_p and K_i at equilibrium.
[*Ans.* 2.05 L mol$^-$; 2.5×10^{-3} L mol^{-1}]

10.7   The following data apply to an equilibrium polymerization of tetrahydrofuran (Odian, 1991): $[M]_0 =$ 12.1 mol/L, $[M]^+ = 2.0 \times 10^{-3}$ mol/L, $[M]_e = 1.5$ mol/L, and $k_p = 1.3 \times 10^{-2}$ L/mol-s. Calculate the initial polymerization rate and the polymerization rate at 20% conversion.
[*Ans.* $(R_p)_0 = 2.76 \times 10^{-4}$ mol/L-s, $R_p = 2.13 \times 10^{-4}$ mol/L-s]

Chapter 11

Living/Controlled Radical Polymerization

11.1 Introduction

In polymer chemistry, *living polymerization* or *controlled polymerization* is a form of addition polymerization where the ability of a growing polymer chain to undergo irreversible chain transfer and chain termination has been removed and, in addition, the rate of chain initiation is also much larger than the rate of chain propagation. The result is that the polymer chains grow at a more constant rate than seen in traditional chain polymerization and this is reflected in polymer chain lengths being equal (i.e., the polymer chains have a very low polydipersity index). Unlike in conventional free radical chain polymerization, the molecular weight of polymer chains increases with conversion and the polymer can be synthesized in stages, each stage containing a different monomer that is added to the reaction system, thereby yielding a block copolymer. Living polymerization is thus a popular method for synthesizing block copolymers. Additional advantages are predetermined molar mass and control over end groups.

Though much of the academic and industrial research on living polymerization has focused on anionic, cationic, ring-opening, and coordination polymerization, the development of living/controlled *radical* polymerization (LRP/CRP) methods has been a long-standing goal in polymer chemistry as this could combine the virtues of living polymerization with versatility and convenience of free-radical polymerization. However, because the propagating radical in free-radical polymerization is very short lived in a homogeneous system (since it undergoes bimolecular termination via coupling/disproportionation or chain transfer), LRP or CRP in homogeneous solution had long appeared to be almost impossible. A solution to this problem could be provided by the *reversible* deactivation of growing radicals, $P^{\bullet}$, with scavenger, $S^{\bullet}$, into dormant species, $P{-}S$:

$$P^{\bullet} \; + \; S^{\bullet} \; \underset{k_{act}}{\overset{k_{deact}}{\rightleftarrows}} \; P{-}S \tag{11.1}$$

Reducing the number of growing radicals in this way, their irreversible bimolecular termination could be minimized. Several such techniques are described below. The extensive scientific literature in this field from 1980 has been recently reviewed (Hawker et al., 2001).

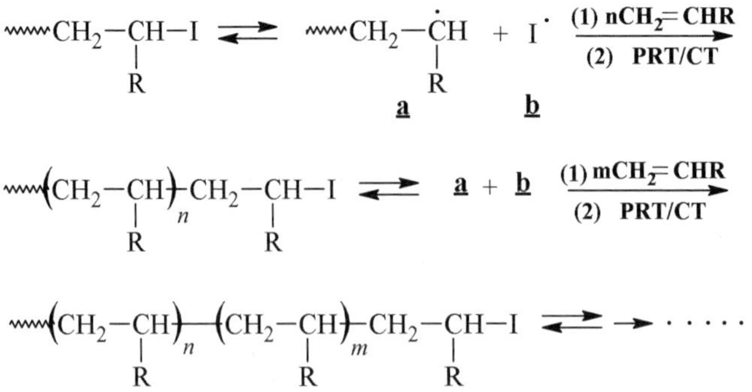

$$\text{wwwCH}_2\text{--CH--I} \rightleftharpoons \text{wwwCH}_2\text{--}\overset{\bullet}{\text{CH}} + \text{I}^\bullet \xrightarrow[\text{(2) PRT/CT}]{\text{(1) nCH}_2\text{= CHR}}$$

$$\underset{\text{R}}{|} \qquad \underset{\text{R}}{|}$$

$$\qquad\qquad\qquad \underline{\textbf{a}} \qquad\qquad \underline{\textbf{b}}$$

$$\text{www}(\text{CH}_2\text{--CH})_n\text{CH}_2\text{--CH--I} \rightleftharpoons \underline{\textbf{a}} + \underline{\textbf{b}} \xrightarrow[\text{(2) PRT/CT}]{\text{(1) mCH}_2\text{= CHR}}$$

$$\underset{\text{R}}{|}\qquad\quad \underset{\text{R}}{|}$$

$$\text{www}(\text{CH}_2\text{--CH})_n(\text{CH}_2\text{--CH})_m\text{CH}_2\text{--CH--I} \rightleftharpoons \rightarrow \cdots\cdots$$

$$\underset{\text{R}}{|}\qquad\quad \underset{\text{R}}{|}\qquad\quad \underset{\text{R}}{|}$$

Figure 11.1 Proposed scheme of radical polymerization through the use of iniferter. (After Otsu et al., 1989.)

The so-called *iniferter* (*ini*tiator-trans*fer* agent-*ter*minator) concept was proposed (Otsu et al., 1982) in 1982 for the design of the polymer chain-end structure that can facilitate radical polymerization showing features of a living mechanism. An iniferter (I–I′) has a very high reactivity for chain transfer and/or primary radical termination. If such a substance is used as an initiator in radical polymerization of a vinyl monomer, ordinary bimolecular termination can be neglected and a polymer bearing two initiator fragments at its chain ends will be obtained as shown below:

$$\text{I--I}' + n\,\text{H}_2\text{C=CH} \longrightarrow \text{I--(CH}_2\text{--CH})_n\text{--I}' \tag{11.2}$$

$$\qquad\qquad\quad \underset{\text{R}}{|} \qquad\qquad\qquad \underset{\text{R}}{|}$$

If the end groups of the polymers formed still have an iniferter function, the radical polymerization will be expected to show features of a living radical mechanism, i.e., both yield and molecular weight of the polymers produced would increase with reaction time (conversion). The proposed iniferter model (Otsu et al., 1989) is shown in Fig. 11.1.

The C–I bond in the propagating chain-end, acting as an iniferter, dissociates into a reactive propagating radical **a** and a nonreactive small radical **b**, which does not initiate but readily undergoes primary radical termination (PRT) with **a** to give the identical C–I bond. Chain transfer (CT) of **a** to the C–I bond may also occur giving a similar **a** and the identical C–I bond. Therefore, if the polymerization proceeds by repetition of dissociation at the C–I bond followed by addition of monomers to **a** and PRT with **b** and/or CT reaction of **a** to C–I bond, such polymerization (Fig. 11.1) may show features of a living radical polymerization.

The polymerization of styrene and methyl methacrylate with organic compounds containing *N,N*-diethyldithiocarbamate groups (Otsu et al., 1986) as photoiniferters and of methyl methacrylate with phenylazotriphenylmethane (Otsu and Tazaki, 1986) as a thermal iniferter are found to proceed via a living radical mechanism similar to the scheme shown in Fig. 11.1, facilitating synthesis of various types of block and graft copolymers with controlled structure. However, despite a number of successes of the application of iniferter technique, one of the goals for complete living polymerization was not reached: polymer with a narrow molecular weight distribution could not be obtained. This provided impetus to search for new and better methods.

Narrow polydispersity, in principle, may be obtained in a free-radical polymerization process, if the process proceeds by a living mechanism, with no premature termination, and if all the propagating chains are initiated at about the same time, similar to what occurs in an anionic polymerization process (Georges et al., 1993). A variety of living radical polymerization systems have been developed in recent years. These are based on either reversible termination or reversible transfer of chain radical for which four principal mechanisms have been put forward: (1) Polymerization with reversible termination (deactivation) of growing chains by coupling, the best example in this class being the alkoxyamine-initiated or *nitroxide-mediated polymerization* (NMP) or *stable free radical polymerization* (SFRP), as first described by Solomon et al. (1985); (2) polymerization with reversible termination by ligand transfer to a metal complex (usually abbreviated as ATRP) (Wang and Matyjaszewski, 1995); (3) polymerization with reversible chain transfer (also termed *degenerative chain transfer*); and (4) polymerization with reversible addition/fragmentation chain transfer (RAFT). All of these methods are, however, based on establishing a rapid dynamic equilibrium between a *minute* amount of growing free radicals and a large majority of dormant species. The methods have attracted much interest as pseudoliving or "living" (often simply called living) radical polymerization for the synthesis of a wide variety of homopolymers and copolymers with controlled molecular weights, low polydispersities, and well defined end groups.

Before various methods of CRP/LRP are described in this chapter, it is important to take a brief look at the criteria that can be used to identify a CRP/LRP process. A CRP/LRP process is characterized by three general features (Qiu et al., 2001).

FEATURE 1 : First Order Kinetics with Respect to Monomer. Figure 11.2 shows typical variations of conversion with time in linear and semilogarithmic coordinates for a living polymerization. A frequently used test for the livingness of polymerization is the linearity of the plot of the rate of polymerization (R_p) versus time (t), the conversion being expressed as $\ln([M]_0/[M])$ where $[M]_0$ and $[M]$ represent the initial concentration of monomer and that at time t, respectively. The linearity, however, shows only that there is no termination reaction, i.e., the number of propagating radicals is constant throughout the polymerization. It does not eliminate the existence of transfer reactions. For constancy of the propagating radical concentration, $[P^\bullet]$, one can write

$$R_p = -d[M]/dt = k_p[M][P^\bullet] = (constant)[M] \tag{11.3}$$

$$\text{or } \ln([M]_0/[M]) = (constant)t \tag{11.4}$$

For a living process (with $R_i \geq R_p$) we have $[P^\bullet] = [I]_0$, i.e., the concentration of the propagating chains is equal to the initial concentration of initiator. The rate of polymerization can then be expressed as

$$-d[M]/dt = k_p[M][I]_0 \text{ or } \ln([M]/[M]_0) = -k_p[I]_0 t \tag{11.5}$$

It should be noted that in conventional radical polymerizations, where termination occurs readily, pseudo first order kinetics could also be observed in the classical steady state, where the concentration of radicals remains constant, because the loss of radicals by termination is compensated by continuous generation of radicals, with a rate equal to that of termination. In contrast, the concentration of radicals in controlled radical polymerizations is established by an equilibrium between the activation and deactivation processes, allowing the rate of initiation (R_i) to be far greater than the rate of propagation (R_P).

FEATURE 2 : Linear Increase of Degree of Polymerization (DP$_n$) with Conversion. Figure 11.3 illustrates schematically the evolution of the molecular weight and polydispersity with conversion for a living polymerization. A linear variation of DP$_n$ with time, and hence conversion, requires that the initiation should be sufficiently fast so that all chains start to grow simultaneously and that no chain transfer occurs to increase the total number of chains initiated by the initiator molecules. Then,

$$DP_n = \frac{\triangle[M]}{[P^\bullet]} = \frac{[M]_0 - [M]}{[I]_0} \tag{11.6}$$

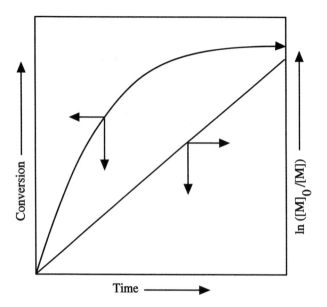

Figure 11.2 Schematic representation of the variation of conversion with time in linear and semilogarithmic coordinates for a living polymerization. (After Matyjaszewski and Xia, 2001.)

Thus,

$$[M] = [M]_0 - DP_n \times [I]_0 \tag{11.7}$$

Features 1 and 2 can be combined into a single equation (Penczek et al., 1991) by substituting for [M] from Eq. (11.7) into Eq. (11.5) to yield

$$\ln\left(1 - \frac{[I]_0}{[M]_0}DP_n\right) = -k_p[I]_0 t \tag{11.8}$$

The linearity of a plot of the left-hand side of Eq. (11.8) with time t is thus a sufficient criterion for $R_t = 0$ (no termination) and $R_{tr} = 0$ (no chain transfer) and from the slope of this plot k_p can be estimated.

FEATURE 3 : Narrow Molecular Weight Distribution. Although narrow molecular weight distribution (MWD) or low polydispersity index (PDI) is a desirable feature, it is not necessarily the result from a living polymerization, which requires only the the absence of chain termination and chain transfer, but ignores the rate of initiation, exchange and depropagation (Qiu et al., 2001). Many studies have indicated that to obtain a narrow MWD each of the following conditions should be fulfilled: (i) initiation is at least as fast as propagation, thus allowing simultaneous growth of all polymer chains; (ii) exchange between species of different reactivities is faster than propagation; (iii) chain transfer or termination is negligible; (iv) polymerization is irreversible; and (v) reaction system is homogeneous and mixing is sufficiently good, thus ensuring availability of all active centers at the onset of polymerization. Under these conditions, a polymer with Poisson distribution may be formed, as quantified in the equation (Qiu et al., 2001):

$$\begin{aligned}
\frac{DP_w}{DP_n} &= \frac{M_w}{M_n} \\
&= 1 + \frac{DP_n}{(DP_n + 1)^2} \cong 1 + \frac{1}{DP_n}
\end{aligned} \tag{11.9}$$

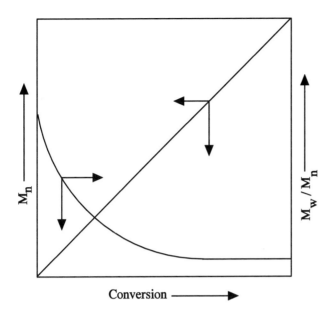

Figure 11.3 Schematic representation of the evolution of molecular weight (M_n) and polydispersity (M_w/M_n) with conversion for a living polymerization. (After Matyjaszewski and Xia, 2001.)

where DP_w (M_w) and DP_n (M_n) represent the weight and number average degree of polymerization (molecular weight), respectively. It follows from Eq. (11.9) that polydispersity index (PDI = $DP_w/DP_n = M_w/M_n$) will decrease with increasing molecular weight. Thus, a polymerization satisfying the aforesaid conditions will be expected to produce a polymer with PDI<1 for $DP_n > 10$. In case of non-compliance with the above conditions, such as polymerizations with slow exchange rate or with termination and transfer reactions, the polydispersity will deviate from Poisson distribution. In such cases, it has been shown that the variation of polydispersity with conversion may be used as a mechanistic criterion to distinguish between various possible mechanisms (Litvinenko and Müller, 1997).

FEATURE 4 : Long Lived Polymer Chains. If chain termination and chain transfer are negligible, all polymer chains retain their ability to grow further after all of the monomer is consumed and resume propagation when fresh monomer is introduced. This unique feature of living polymers is made use of in the preparation of block copolymers by sequential monomer addition.

11.2 Stable Free Radical Polymerization

As living polymerization implies that during the process, side reactions such as irreversible termination and transfer reactions are virtually absent, in free radical polymerization it can be achieved by a *reversible* termination by reaction between the active center and another radical, using photoactivation (Otsu and Kuriyama, 1984) or thermal activation (Bledzki et al., 1983; Crivello et al., 1986; Otsu et al., 1987). An example of the latter is provided by stable nitroxyl radicals like TEMPO (2,2,6,6-tetramethylpiperidinyl-1-oxy) (**I**).

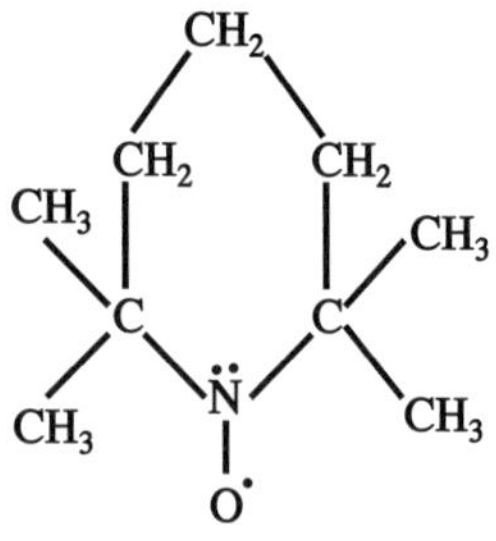

(I)

Thus, in stable free radical polymerization (SFRP), also called nitroxide-mediated polymerization or NMP (which was discovered while using TEMPO as a radical scavenger in investigating the rate of initiation during free radical polymerization), it is believed that reversible combination of a polymer radical, P•, with a stable nitroxyl radical, N•, takes place forming an adduct, P–N, that exists as a dormant species:

$$P\text{–}N \underset{k_c}{\overset{k_d}{\rightleftharpoons}} P\bullet + N\bullet \tag{11.10}$$

$$K = k_d/k_c = \frac{[P\bullet][N\bullet]}{[P\text{–}N]} \tag{11.11}$$

where k_d and k_c are the rate constants of dissociation (activation) and combination (deactivation), respectively. Note that N• is reactive only to P•, while P• is reactive not only to N• and monomer M (propagation) but also to P• (irreversible termination) and neutral molecules S (chain transfer). When the last two reactions of P• (i.e., termination and transfer) are unimportant compared with the first two, the polymerization may proceed via a living radical mechanism (similar to that shown in Fig. 11.1 with I• replaced by N•). Figure 11.4 shows the reversible termination of a polystyryl chain radical by TEMPO radical.

The equilibrium between dormant chains (P–N) and active chains (P•) is designed and the temperature is adjusted so as to heavily favor the dormant state, which effectively reduces the radical (P•) concentration to sufficiently low levels that allow controlled polymerization. For example, the equilibrium constant K in Eq. (11.11) for the polystyrene (PSt)/TEMPO reversible reaction in the bulk polymerization of styrene at 125°C in the presence of a PS-TEMPO adduct

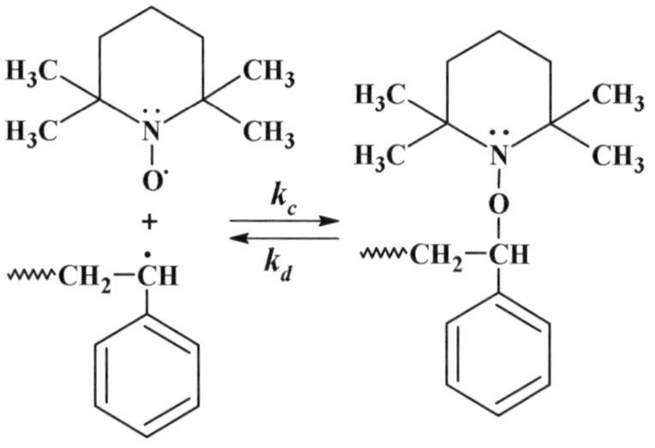

Figure 11.4 Reversible termination of polystyrene chain radical by TEMPO radical. (After Georges et al., 1993.)

was estimated to be 2.1×10^{-11} mol L^{-1} (Fukuda et al., 1996). The value is very low showing that the equilibrium between dormant chains (P–N) and active chains (P$^\bullet$) heavily favors the dormant state, which reduces the radical (P$^\bullet$) concentration sufficiently to limit irreversible bimolecular termination (combination and disproportionation) and transfer reactions.

11.2.1 Monomers

The key to success in synthesizing polymers with narrow polydispersity and well-defined chain end structure by carrying out free-radical polymerization in the presence of nitroxide SFRs such as TEMPO, is the essentially simultaneous initiation and reversible termination of the polymer radical with the SFR (Georges et al., 1994). However, the dissociation such as depicted in Fig. 11.4 for the polystyrene (PSt)-TEMPO adduct is known to occur in a limited number of systems (at high temperatures). A versatile use of the simple TEMPO-based SFRP is therefore not possible. For example, attempts to perform SFRP of monomers such as acrylonitrile (AN), methyl and ethyl acrylates (MA and EA), and 9-vinylcarbazole (VCz) with benzoyl peroxide (BPO) and TEMPO have not been successful. Interestingly, however, styrene has been successfully copolymerized (see Section 11.2.4) with these monomers using BPO initiator and TEMPO under a "living" fashion (Fukuda et al., 1996).

11.2.2 Stable Nitroxide Radicals

Most studies on stable nitroxide radical-mediated polymerization are made with TEMPO or a substituted TEMPO, such as, 4-methoxy-2,2,6,6-tetramethyl-piperidine-1-oxy (MTEMPO) as the stable counter radical. However, the bond formed between the polymer radical and these nitroxides becomes labile around 120°C. This high temperature favors thermal polymerization as also side reactions, such as transfer or termination by dismutation between the growing chain and the stable radical (Jousset et al., 1997). To minimize these reactions, a more hindered radical, namely, di-*tert*-butyl nitroxide (**II**), produced from the compound (**III**) can be used.

(II)   **(III)**

Heated, the compound (**III**) leads to the formation of a di-*tert*-butyl nitroxide radical (**II**) and an alkyl radical, $C_6H_5C^\bullet(H)CH_3$, that exhibits a chemical structure similar to that of the styryl radical. The choice of (**III**) as the capping agent comes from its absence of reactivity toward alkenes and from the low strength of the bond formed with the active site (Catala et al., 1995). It allows free radical polymerization of styrene and substituted styrene monomers at 90°C with complete control of the molecular weight and monomer consumption (Joussel et al., 1997).

11.2.3 Mechanism and Kinetics

The NMP or SFRP is based on capping the active sites located at the chain ends by a stable nitroxide radical (SNR). This termination type reaction is thermoreversible as represented by Eqs.

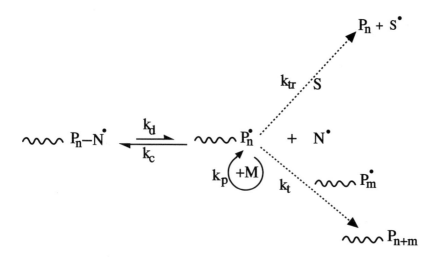

Figure 11.5 General scheme of SNR-mediated controlled/living radical polymerization methods. [cf. Eq. (11.10) and Fig. 11.4.]

(11.10) and (11.11). As shown in the general scheme for SFRP methods in Fig. 11.5, free radicals (P$^\bullet$) generated by the spontaneous thermal process can undergo propagation as well as irreversible bimolecular termination and transfer reactions like in a usual free-radical system. This distinguishes the SFR-mediated polymerization from genuine living polymerization and so the former is often referred to as pseudo-living or "living" radical polymerization (Fukuda et al., 1996).

An SFRP process can be performed by heating a styrenic monomer with polymer- or alkyl-TEMPO adduct, P–N, added as the initiator. However, initiation of a living radical polymerization is generally done by using classical initiators such as AIBN or benzoyl peroxides, which are added to the monomer containing the SNR. Besides thermal radicals, active species P$^\bullet$ are generated by the P–N adduct in the former case and both by the added initiator and P–N adduct in the latter case. SNRs being well-known as free-radical inhibitors and not known to initiate polymerization, there would be little concern with SNRs initiating new chains. Furthermore, SNRs have been shown to promote the dissociation of peroxide initiators and, therefore, could contribute to enabling all the polymeric chains to initiate at the same time (Georges et al., 1993), as required for CRP.

While the SNR-mediated polymerization process comprises heating a mixture of monomer(s) and P–N adduct (that acts both as an initiator and a controlling agent), or a mixture of monomer(s), free-radical initiator, and SNR (or P–N adduct), the best temperature of polymerization, determined by experiments, is the one that leads to (i) a fast initiation rate as compared to the propagation rate; (ii) a fast equilibrium between the active species and the dormant ones; (iii) a low concentration of active species in order to minimize the termination and/or transfer reactions; and (iv) a negligible thermal polymerization (of styrenic monomers). For example, the rate of formation of thermal radicals in styrene is equal to 1.6×10^{-8} mol L^{-1} s^{-1} at 100°C, 0.6×10^{-8} mol L^{-1} s^{-1} at 90°C, and 0.2×10^{-8} mol L^{-1} s^{-1} at 80 °C (Catala et al., 1995). At higher temperature, the formation of thermal radicals which recombine with the growing radicals contribute to a decrease of the number of the active sites with time leading to a higher polydispersity index, which is undesirable. On the other hand, though at a lower temperature the thermal polymerization can be neglected (when compared to the initiator concentration used), the temperature should not be so low that the initiation rate and the equilibrium exchange between the growing radical and the dormant species are not sufficiently fast to provide a good polydispersity index. Thus, for SNRP of styrenic monomers, 90°C appears to be an appropriate temperature.

The three main requirements for an ideal SNR-mediated living/controlled free-radical polymerization are: (i) essentially simultaneous initiation; (ii) reversible reaction of SNR given in Eq. (11.10); and (iii) no important degree of irreversible termination, if any. Under these conditions, one would expect the system to be described by a simple kinetic scheme, as described below.

The free radical polymerization that includes the reversible reaction in Eq. (11.10) may generally be described by the following differential equations (Fukuda et al., 1996):

$$d[P\cdot]/dt \; = \; R_i \; - \; 2k_t[P\cdot]^2 \; + \; k_d[P\text{–}N] \; - \; k_c[P\cdot][N\cdot] \tag{11.12}$$

$$d[N\cdot]/dt \; = \; k_d[P\text{–}N] \; - \; k_c[P\cdot][N\cdot] \tag{11.13}$$

where R_i is the rate of initiation due to an initiator and/or thermal initiation and k_t is the mutual termination rate constant for polymer radicals.

Now let us consider a system composed of PS-TEMPO (denoted as P–N) adduct and monomer styrene with no extra N· added. When the system is heated to a sufficiently high temperature where the adduct dissociation takes place, the concentration of P· and N· will start to increase from 0 up to the values determined by the equilibrium constant K. Since P· /P· biradical termination will continually reduce [P·] relative to [N·], polymerization will eventually stop because of numerous N· radicals accumulated in the system. However, if initiation also takes place in the system, the newly formed radicals will combine with N·preventing its accumulation and a stationary state for both [N·] and [P·] will be reached. Setting $d[P\cdot]/dt = 0$ in Eq. (11.12) and $d[N\cdot]/dt = 0$ in Eq. (11.13), and using Eq. (11.11), one then has

$$[P\cdot] \; = \; (R_i/2k_t)^{1/2} \tag{11.14}$$

$$[N\cdot] \; = \; K[P\text{–}N]/[P\cdot] \tag{11.15}$$

The stationary rate of polymerization, $R_p \; = \; -d[M]/dt$, is given by

$$R_p \; = \; k_p[P\cdot][M] \; = \; \left(\frac{k_p^2}{2k_t}\right)^{1/2} [M]R_i^{1/2} \tag{11.16}$$

where [M] is the monomer concentration and k_p is the propagation rate constant. Equation (11.16) is the same as Eq. (6.23) for the conventional system. It thus implies that the polymerization rate is independent of the adduct concentration and equal to that of the adduct-free system (see Problem 11.1).

Problem 11.1 A TEMPO-mediated polymerization of styrene (St) was carried out (Goto and Fukuda, 1997) as follows: A mixture of benzoyl peroxide (0.072 mol L^{-1}), TEMPO (0.086 mol L^{-1}), and St (0.90 g mol^{-1}), degassed and sealed off under vacuum, was heated at 90°C for 1 h. There was no formation of polymer as shown by GPC. Then the system was heated at 125°C for 5 h to yield a polymer (recovered by precipitation in excess methanol) with $\overline{M}_n = 1700$ and $\overline{M}_w/\overline{M}_n = 1.11$. Analysis showed that this polymer (PSt) had a TEMPO molecule at the chain end and so it was termed a PSt-TEMPO adduct. In chain extension tests, styrene (St) was polymerized at 114°C using in one case (System I) only PSt-TEMPO, 10 wt% (0.049

mol L^{-1}) as the initiator and in another case (System II) *t*-butyl hydroperoxide (BHP) (0.036 wt% or 4.0×10^{-3} mol L^{-1}) in addition to PSt-TEMPO (10 wt%) as initiators in the same reaction mixture.

The concentration of styrene in the two systems and molecular weight of the polymer formed were measured periodically. Selected data derived from these measurements are given below.

Time	System I		System II	
(h)	$\ln([M]_0/[M])$	M_n	$\ln([M]_0/[M])$	M_n
1	0.0615	2727	0.1846	4909
2	0.1230	4000	0.3691	7236
3	0.1999	5091	0.4737	8000
4	0.2415	5455	0.6567	10,000
5	0.3076	6545	0.8215	11,636

Note: The conversion data are read from experimental graph and M_n values are obtained by interpolation of experimental data (Goto and Fukuda, 1997).

(a) Even in the absence of an externally added initiator, styrene undergoes thermal polymerization, the rate of which is given by $(R_p)_0/[M]_0 = 2.4 \times 10^{-5}$ s^{-1} at 114°C (Hui and Hamielec, 1972). Compare this with the initial rate of polymerization in System I. [For thermal polymerization of styrene under initial conditions: $(k_p^2/k_t)_0 = 0.880 \times 10^5 \exp(-6270/T)$ L/mol-s (Hui and Hamielec, 1972).]

(b) Calculate the polymer molecular weight vs. conversion in the two systems (for comparison with the given values) based on the assumption of a constant number of polymer molecules due to PSt-TEMPO adduct throughout the course of polymerization.

(c) Calculate the number of new chains initiated in the two systems, compared to the number of chains introduced by PSt-TEMPO adduct.

(d) Compare the rate of polymerization of styrene in the two systems and discuss its significance with respect to the characteristics of living/controlled free radical polymerization.

Answer:

(a) Approximating $(1 - e^{-k_d t/2})$ by $k_d t/2$, Eq. (6.29) becomes

$$\ln \frac{[M]_0}{[M]} = k_p \left(\frac{f k_d}{k_t} \right)^{1/2} [I]_0^{1/2} t \tag{P11.1.1}$$

Differentiating Eq. (P11.1.1) and comparing with Eq. (6.24) for the initial state,

$$\frac{d \ln([M]_0/[M])}{dt} = \frac{(R_p)_0}{[M]_0} \tag{P11.1.2}$$

Since $\ln([M]_0/[M])$ vs. time is essentially linear, $(R_p)_0/[M]_0 \simeq 0.0615/3600$ or 1.7×10^{-5} s^{-1} for System I (without BHP). Since this value is comparable to (and somewhat less than) the purely thermal polymerization rate, $(R_p)_0/[M]_0 = 2.4 \times 10^{-5}$ s^{-1}, R_p can be considered to be essentially unchanged by the presence of the adduct PSt-TEMPO. [This phenomenon was verified experimentally for styrene polymerization with a PSt-TEMPO (Fukuda et al., 1996) or a low-mass model adduct (Catala et al., 1995).]

(b) If it is assumed that during polymerization, growth takes place only on polymer species of the PSt-TEMPO adduct and no new chains are initiated, polymer molecular weight (M_n) as a function of monomer conversion (p) will be given by

$$M_n = \frac{W_M \times p}{N_A} + (M_n)_0 \tag{P11.1.3}$$

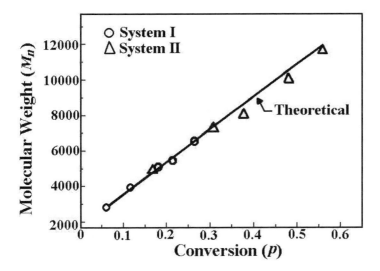

Figure 11.6 Molecular weight, M_n, dependence on monomer conversion for bulk polymerization of styrene at 114°C with only PSt-TEMPO adduct initiator (System I) and with both PSt-TEMPO and *t*-butyl hydroperoxide initiators (System II). The theoretical line represents molecular weight calculated on the assumption of a constant number of polymer molecules (due only to PSt-TEMPO adduct) throughout the course of poymerization (Problem 11.1).

where W_M is the mass (g/L) of monomer, N_A is the number (mol L^{-1}) of PSt-TEMPO adduct molecules, and $(M_n)_0$ is the initial molecular weight of the adduct.

The conversion p is related to $\ln([M]_0/[M])$ by

$$p = \frac{[M]_0 - [M]}{[M]_0} = 1 - \left[\exp\left(\ln\frac{[M]_0}{[M]}\right)\right]^{-1} \tag{P11.1.4}$$

Thus for System I, at $t = 1$ h, $\ln([M]_0/[M]) = 0.0615$, p from Eq. (P11.1.4) is 0.0596. So from Eq. (P11.1.3),

$$M_n = \frac{900 \times 0.0596}{0.049} + 1700 = 2795$$

Similarly, M_n values are calculated at different conversions for both the Systems I and II and are shown in Fig. 11.6 as a 'theoretical line' to compare with the given M_n values. Clearly, the molecular weight increases linearly with conversion, meeting a criterion of living polymerization at least approximately.

(c) From Eq. (11.16) at $t = 0$,

$$(R_i)_0 = 2\left(\frac{(R_p)_0}{[M]_0}\right)^2 \left(\frac{k_p^2}{k_t}\right)_0^{-1} \tag{P11.1.5}$$

At 114°C

$$\left(k_p^2/k_t\right)_0 = 0.880 \times 10^5 \exp(-6270/387)$$
$$= 8.09 \times 10^{-3} \text{ L mol}^{-1} \text{ s}^{-1} \qquad\qquad\qquad\text{(P11.1.6)}$$

System I:

$$(R_i)_0 = 2(1.7 \times 10^{-5} \text{ s}^{-1})^2/(8.09 \times 10^{-3} \text{ L mol}^{-1} \text{ s}^{-1})$$
$$= 0.714 \times 10^{-7} \text{ mol L}^{-1} \text{ s}^{-1}$$

Number of chains initiated in 5 h (maximum estimate)

$$= (0.714 \times 10^{-7} \text{ mol L}^{-1} \text{s}^{-1})(5 \times 3600 \text{ s}) = 1.286 \times 10^{-3} \text{ mol L}^{-1}$$

It is only 2.6% of [PSt-TEMPO] = 0.049 mol L^{-1} added to the system. Its effect on M_n is therefore not significant.

System II:

From Eq. (P11.1.2) and given data,

$$(R_p)_0/[M]_0 = 0.1846/3600 \quad \text{or} \quad 5.128 \times 10^{-5} \text{ s}^{-1}$$

From Eq. (P11.1.5) and (P11.1.6):

$$(R_i)_0 = 2(5.128 \times 10^{-5} \text{ s}^{-1})^2/(8.09 \times 10^{-3} \text{ L mol}^{-1} \text{s}^{-1}) = 6.50 \times 10^{-7} \text{ mol L}^{-1} \text{s}^{-1}$$

Number of chains initiated in 5 h (maximum estimate):

$$= (6.50 \times 10^{-7} \text{ mol L}^{-1} \text{ s}^{-1})(5 \times 3600 \text{ s}) = 1.17 \times 10^{-2} \text{ mol L}^{-1}$$

It is still only about 23% of the initiator adduct [PSt-TEMPO] added to the system.

(d) The conversion versus time data are plotted in Fig. 11.7. These are essentially straight lines, showing that the order in monomer is equal to unity (see Eqs. (6.24) and (P11.1.1)) and indicating that the concentration of the active species remains constant, which is a criterion of a living polymerization. It is suggested (Fukuda et al., 1996) that the constant concentration of active radicals comes from the establishment of a stationary state due to the competition between generation of thermal radicals and termination of growing chains (Jousset et al., 1997).

It may be recalled that the rate of polymerization in System I (without BHP) is comparable to (though somewhat less than) the thermal polymerization rate, as shown in (a). However, the rate of polymerization in System II (with BHP) is much higher than the thermal polymerization rate. This is in agreement with the proposition (Hammouch and Catala, 1996) that the generation of radicals in the medium (through the addition of a radical initiator) controls the rate of polymerization while the amount of PSt-TEMPO or the alkoxyamine controls the molecular weight.

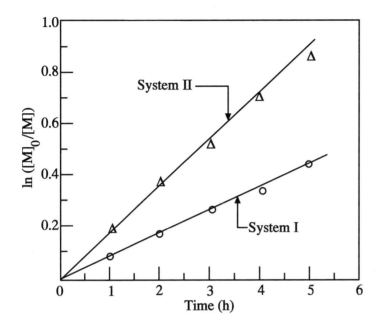

Figure 11.7 Conversion-time plot for polymerization of styrene in bulk at 90°C with only PSt-TEMPO adduct initiator in System I and with both PSt-TEMPO and *t*-butyl hydroperoxide initiators in System II (Problem 11.1).

11.2.4 Copolymerization

Though monomers other than styrene (St) and its derivatives, such as acrylonitrile (AN), methyl and ethyl acrylates (MA and EA), and 9-vinylcarbazole (VCz), give no polymer (or no well-defined polymer) by TEMPO-based SFRP, they are found to take part in copolymerization with styrene. In a typical experiment, styrene mixed with BPO (0.015 mol L^{-1}) and TEMPO (0.018 mol L^{-1}) was preheated at 95°C for 3.5 h to allow BPO to decompose completely and further heated at 125°C for 5.5 h to yield a PSt with $M_n = 2.2 \times 10^4$ and $M_w/M_n = 1.18$ (conversion 31%). This PSt, purified by reprecipitation (with chloroform/methanol) had a TEMPO molecule at the chain end, and was used for block copolymerization in which 22 parts by weight of the same PSt was dissolved in 78 parts of an St/AN azeotropic mixture (mole fraction of St in the feed, $f_1 = 0.63$) and heated at 125°C for 10 h, yielding a diblock copolymer, PSt–PStAN, comprising a PSt and a random PStAN copolymer sequence with composition of $F_1 = 0.63$ (see Section 7.2.3 for definitions of f_1 and F_1). Results of similar experiments with MA, EA, and VCz comonomers are shown in Table 11.1. It is suggested that the penultimate unit effect, which strongly affects the radical reactivity or stability in most systems, may also have something to do with the dissociation of the TEMPO adducts of the otherwise undissociative growing radicals (Fukuda et al., 1996). The success of these "living" radical copolymerizations stimulated further work to synthesize block copolymers comprising random copolymer sequence(s) with narrow polydispersities (Hawker, 1995). Various combinations of the above mentioned monomers are possible, leading to synthesis of a new group of block copolymers that did not exist before.

Table 11.1 TEMPO-mediated Copolymerizations of Styrene with Various Monomers at 125°C

Comono-mer	[BPO] (mol/L)	[TEMPO] (mol/L)	f_1	Time (h)	Conv. (wt%)	F_1	M_n	M_w/M_n
AN	0.07	0.07	0.63	10	69	0.64	16000	1.23
MA	0.03	0.037	0.77	16	61	0.82	13000	1.24
EA	0.03	0.037	0.77	15	49	0.83	23000	1.18
VCz	0.06	0.063	0.79	15	60	0.91	38000	1.27

Source: Fukuda et al. (1996).

As shown above, living free radical system, based on TEMPO as a mediating counter-radical, occurs by a modified free-radical polymerization process which could permit the preparation of well-defined random copolymers. Thus, using the hydroxy derivative of TEMPO, 2-phenyl-2-[(2,2,6,6-tetramethyl-piperidino)oxy]-1-ethanol (**IV**)

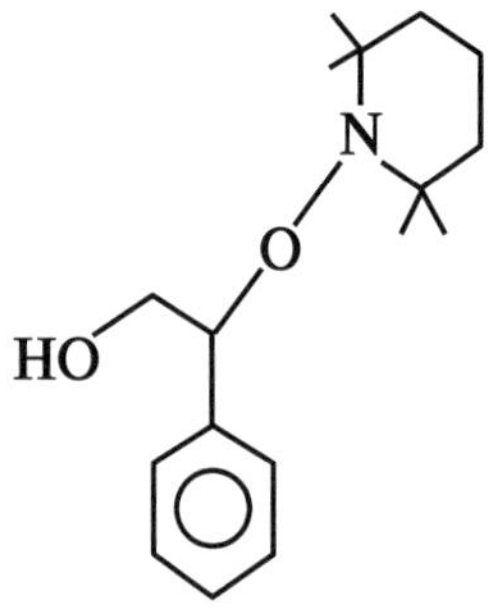

(IV)

as the unimolecular initiator (Hawker, 1995), a variety of monomers were copolymerized (Hawker et al., 1996), e.g., methyl methacrylate, *n*-butyl acrylate, acrylonitrile, hydroxyethyl methacrylate, *p*-(hydroxymethyl)styrene, *p*-(chloromethyl)styrene, and *N*-vinylpyrrolidone, with styrene to give well-defined copolymers. Thus, heating a neat mixture of styrene and butyl acrylate (feed ratio 8:2) at 130°C for 72 h resulted in an 81% yield of copolymer (Fig. 11.8) with M_n of 9000 and a polydispersity of 1.22. Styrene was similarly copolymerized with methyl methacrylate using various feed ratios (Table 11.2). For all the copolymers, proton NMR analysis confirmed that the ratio of repeat units in the copolymer was essentially the same as that expected from the feed ratio. Though the polydispersity is observed to increase as the styrene ratio decreases, for both the methyl methacrylate and *n*-butyl acrylate series, the control over macromolecular structure is significantly greater than with classical free-radical techniques, with the observed polydispersities for all copolymers with greater than 50% styrene being below the theoretical lower limit of 1.50 for a normal free-radical process (Hawker et al., 1996).

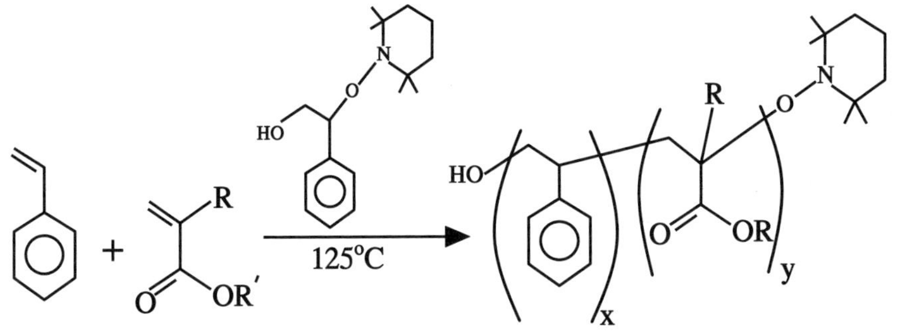

(1) R = Me, R′ = Me

(2) R = H, R′ = n-Bu

Figure 11.8 Random copolymerization of styrene with either (1) methyl methacrylate or (2) butyl acrylate by nitroxide-mediated living free-radical procedures. (Drawn following the synthesis method of Hawker et al., 1996.)

Table 11.2 Random Copolymers of Styrene and either Methyl Methacrylate or *n*-Butyl Acrylate prepared by Living Free-Radical Process

Styrene[a]	Methyl methacrylate		*n*-Butyl acrylate	
(%)	M_n	PD[b]	M_n	PD[b]
90	9,400	1.13	9,500	1.14
80	8,800	1.23	9,000	1.22
70	8,100	1.29	7,800	1.33
60	7,100	1.32	8,000	1.29
50	6,600	1.49	6,900	1.44
40	6,800	1.60	6,400	1.52
30	6,400	1.67	6,500	1.65

Source: Hawker et al. (1996).

[a]Molar percentage of styrene in feed mixture; [b]Polydispersity = M_w/M_n.

The tacticity of the random copolymers prepared by living free radical polymerization is also found to have the same sequence distribution and tacticity as those prepared by normal free-radical methods. Besides low polydispersities, another advantage of living free radical polymerization is that the chain ends can be controlled to a degree previously only obtainable with more demanding techniques. In one example, a pyrene-ended styrene-methyl methacrylate copolymer (**VI**) was prepared by living free radical polymerization using the unimolecular initiator (**V**), which in turn was prepared by esterification of pyrene-1-butyryl chloride with (**IV**) in the presence of 4-(dimethylamino)pyridine (see Fig. 11.9).

While the living free-radical polymerization of styrene from an alkoxyamine can lead to narrow polydispersity polymers, these having the nitroxyl radical at the terminals can be expected to behave as polymeric counter radicals for propagating polymeric radicals. The use of such

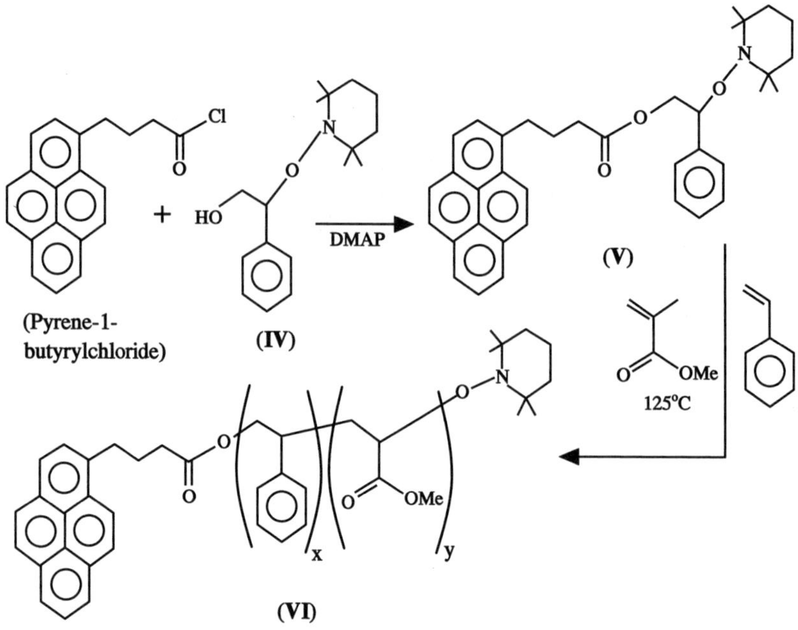

Figure 11.9 Synthesis of a pyrene-ended random copolymer of styrene and methyl methacrylate by nitroxide mediated living free-radical procedure. (Drawn following the synthesis method of Hawker et al., 1996.)

polymeric counter radicals results in the formation of block copolymers by living radical polymerization. While in this case, both the block sequences are made by the same living free-radical procedures, using the hydroxy functionalized unimolecular initiator (**IV**) the living free-radical procedure can be combined with another living polymerization procedure, namely, living ring-opening polymerization. In other words, (**IV**) can be used as a *bifunctional initiator*. This aspect is further elaborated below.

One of the major advantages of living free-radical chemistry, when compared to other living procedures for the polymerization of vinyl monomers, is the stability of the initiating, or propagating, centers (Hawker et al., 1998). This made possible the development of a wide variety of functionalized unimolecular initiators, including (**IV**), for the synthesis of well-defined linear polymers (Hawker, 1994), block copolymers (Fukuda et al., 1996), and other complex polymer structures (Hawker, 1995) by combining living free radical procedures with a wide variety of other living polymerizations. A significant advantage of this synthetic strategy is that novel block copolymers can be prepared using mild conditions and a minimum number of steps without the need for intermediate functionalization reactions to convert the propagation center of one block into an initiating site for polymerization of another monomer to produce a second block. The block copolymers thus obtained have been shown to have low polydispersities and controllable molecular weights for both the blocks. Problem 11.2 presents an example of successful synthesis of diblock copolymers by a sequential two-step method without any transformation or activations of intermediates. Problem 11.3, on the other hand, presents an example, which shows that a living polymer obtained by ROP can be capped by nitroxide and used to initiate living radical polymerization of a second monomer, yielding a di-block copolymer in which the nitroxide moiety serves as a bridge between the blocks.

Problem 11.2 The hydroxy-functionalized alkoxyamine (**IV**) can be used (Hawker et al., 1998) to perform either the living ring opening polymerization (ROP) of ε-caprolactone, or the living free radical polymerization of a vinyl monomer leading to narrow polydispersity polymeric initiators, which can then be directly used to initiate the living polymerization of other monomer systems. With this approach, give outline of a scheme for the preparation of a diblock copolymer, polystyrene-*b*-polycaprolactone, using the aforesaid dual initiator, (**IV**).

Answer:

The dual or double-headed initiator, (**IV**), contains a single primary alcohol, which can be used as the initiating center for the living ROP of cyclic lactones, as well as a secondary benzylic group which is an efficient initiator for the nitroxide-mediated living free radical polymerization of vinyl monomers. The basic outline for the polymerization strategy is shown in Scheme P11.2.1. The ROP of ε-caprolactone with (**P2-I**) (same as (**IV**) as initiator and a catalytic amount of aluminum tris(isopropoxide) as a promoter (Dubois et al., 1996; Duda, 1996), gives polycaprolactone (**P2-II**) having single alkoxyamine chain end per macromolecule. The polycaprolactone, (**P2-II**), can then be used to perform nitroxide-mediated living free radical polymerization of vinyl monomers, such as styrene, to give (requiring no intermediate steps) a low polydispersity block copolymer, (**P2-III**). Alternatively, (**P2-I**) may be used to first carry out the nitroxide-mediated polymerization of styrene, giving hydroxy-terminated polystyrene, (**P2-IV**). The hydroxy chain end can then be used to initiate the living ROP of ε-caprolactone (again without any intermediate transformation steps) to give the block copolymer, (**P2-III**), which is obtained in high yield, has low polydipersity index (~ 1), and its molecular weight can be controlled by monomer ratio.

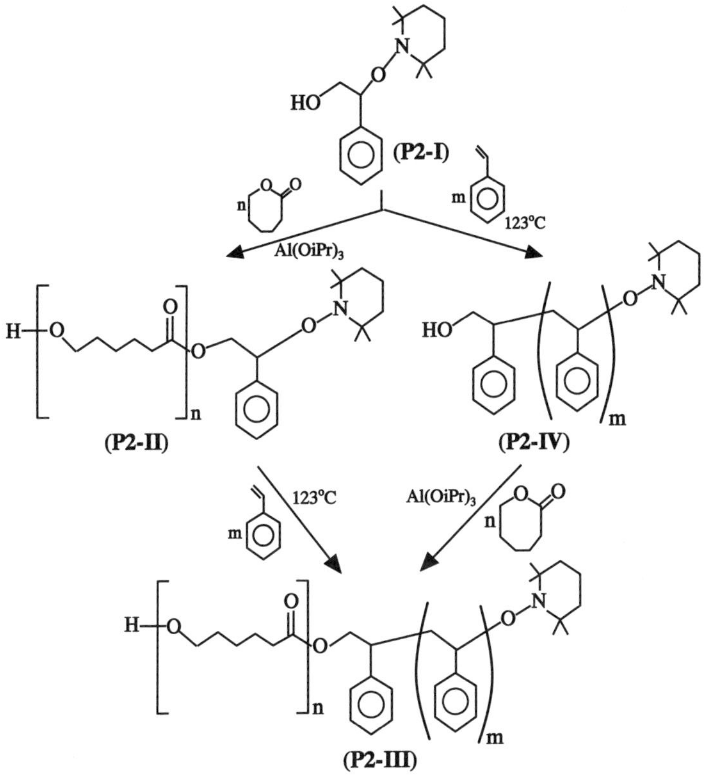

Scheme P11.2.1 Copolymerization of styrene and ε-caprolactone (Problem 11.2). (Drawn following the synthesis method of Hawker et al., 1998.)

Problem 11.3   Suggest a scheme for the preparation of poly(styrene-*b*-tetramethylene oxide) block copolymer using living polymerizations.

Answer:

A strategy for this synthesis could be to first prepare poly(tetrahydrofuran, THF) having a nitroxyl radical
at its chain end (PTN) by the reaction of living poly(THF) with sodium 4-oxy-TEMPO and then apply it
as a counter radical in living radical polymerization (Yoshida and Sugita, 1996). Living poly(THF) can be
prepared (see Section 10.2.1.2) by cationic polymerization of THF using methyl trifluoromethanesulfonate
as an initiator in bulk at room temperature for about 7 min under nitrogen. The living cationic poly(THF)
thus obtained is subjected to the reaction with sodium 4-oxy-TEMPO (which is prepared by the reaction
of 4-hydroxy-TEMPO with sodium hydride in THF). The polymer PTN, isolated and purified by repeated
washing of the hexane solution of the product with water, is highly viscous and red, indicating the presence
of the living nitroxyl radical.

   Radical polymerization of styrene is then performed with BPO as an initiator in the presence of PTN. The
polymerization is carried out in bulk at 125°C after being held at 95°C for 4 h. The outline of the procedure
is shown in Scheme P11.3.1.

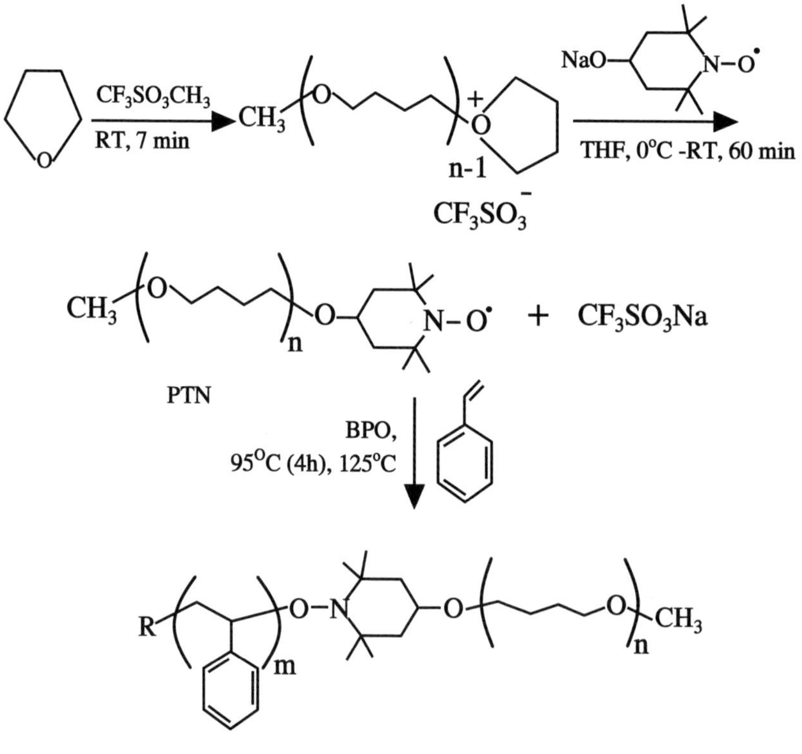

Scheme P11.3.1 Preparation of block copolymer consisting of poly(THF) and polystyrene blocks joined by
a nitroxide bridge (Problem 11.3). (Drawn following the synthesis method of Yoshida and Sugita, 1996.)

Although the traditional method of synthesizing block copolymers by sequential polymerization of corresponding monomers by the same chemistry (e.g., anionic polymerization), may be successful, its extension to comonomers which do not polymerize by the same chemistry is essentially a challenge. Two strategies have been proposed in this case, namely, (a) the coupling of preformed polymers with functional groups at chain ends and (b) using macromolecular initiator for the polymerization of the second monomer. The second strategy usually requires a series of transformation reactions to convert the propagation center of the first block into an initiating site for polymerization of the second monomer. Whichever method is used, two polymerization steps cannot be avoided, however.

A new strategy has been proposed for the *one-step* synthesis of block copolymers, based on living/controlled free-radical process. It involves the use of an asymmetric difunctional initiator that is able to start simultaneous polymerization of two comonomers by different polymerization chemistries in such a way that this initiator remains attached to each type of the growing chain (Mecerreyes et al., 1998). The implementation of one-step synthesis is not simple, however. The two catalysts must be tolerant to each other as also to the two comonomers and the reaction temperature must be closely controlled. Living radical polymerization and ROP by coordination and insertion can meet these requirements.

Problem 11.4 Suggest a scheme for preparing poly(styrene-*b*-ethyleneimine) by a one-step, one-pot simultaneous block copolymerization using a bifunctional initiator for controlled/living polymerization processes.

Answer:

The scheme P11.4.1, outlined below for the simultaneous cationic ROP and controlled free radical polymerizations, is based on the methods that were first used and experimentally verified by Weimer et al. (1998). A multifunctional initiator, such as (**P4-I**), containing orthogonal reactive sites for CRP and ROP is used. This alkoxyamine adduct, namely, benzoic acid 2-(4-chloromethyl)phenyl)-2-(2,2,6,6-tetramethyl piperidin-1-yloxy) ethyl ester, can be prepared (Puts and Sogah, 1997) by heating a solution of benzoyl peroxide and TEMPO in 4-vinylbenzyl chloride at 80°C for 24 h, followed by purification by chromatography.

First, the compound (**P4-I**) is treated with silver triflate (AgOTf, TfOH = CF_3SO_3H) followed by addition of excess 2-phenyl-2-oxazoline (PhOXA) at 25°C. This leads to the formation of the corresponding triflate salt (**P4-II**) in a pool of PhOXA (Scheme P11.4.1). Styrene is then added to the mixture of (**P4-II**) and PhOXA at 25°C, followed by heating the mixture at 125°C for 10 h. Simultaneous polymerization of both monomers occurs yielding (**P4-III**), which is then quenched with methanol to give (**P4-IV**) in good yield. (In the absence of AgOTf, the reaction is found to be too slow.) The polymeric oxazoline segment in the copolymer is then hydrolyzed in aqueous hydrochloric acid to give the corresponding poly(styrene-*b*-ethyleneimine) (**P4-V**).

Though graft and star polymers are traditionally prepared by anionic and cationic polymerization or by group transfer polymerization, these techniques suffer from rigorous synthetic conditions and incompatibility with a wide range of monomer units. In view of this, a free radical approach to the preparation of graft and star polymers with the same degree of macromolecular control as the above techniques but without their synthetic drawbacks has always been a goal of synthetic polymer chemistry. Significant progress toward this goal has been achieved through living free radical methodology which allows the synthesis of star and graft copolymers with

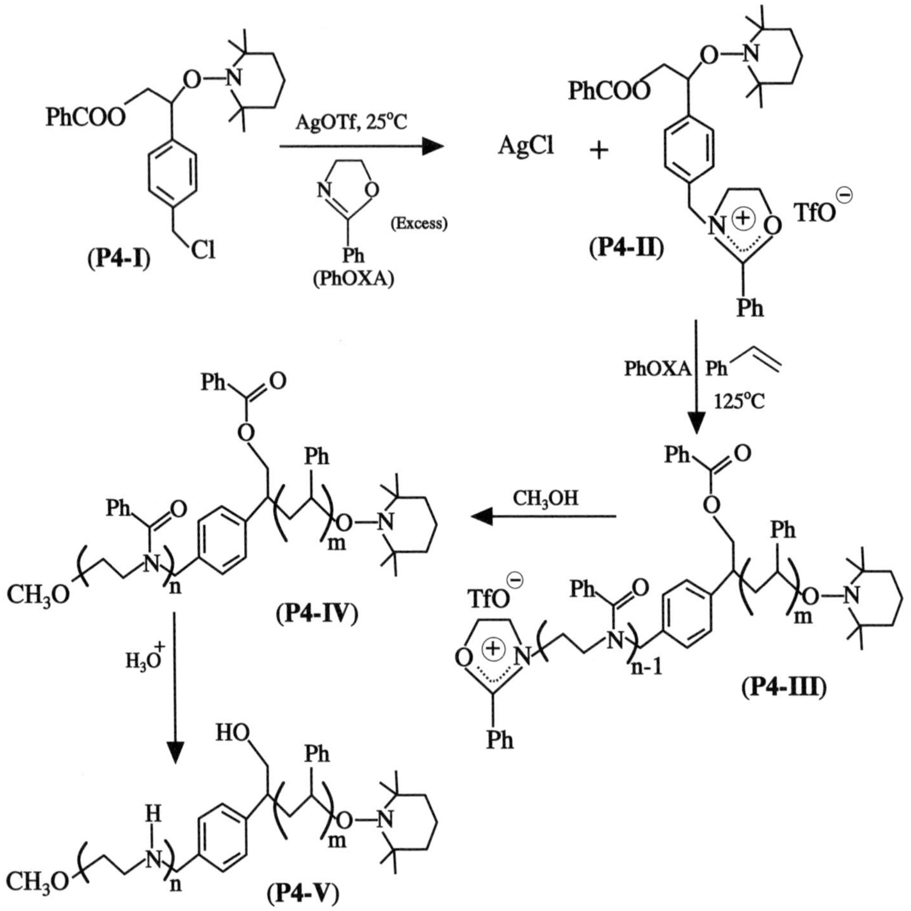

Scheme P11.4.1 One-pot synthesis of poly(styrene-*b*-ethyleneimine) block copolymer (Problem 11.4). (Drawn following the synthesis method of Weimer et al., 1998.)

controlled molecular weights and low polydispersities under mild conditions. Thus, Hawker (1995) synthesized a three-arm star polymer with polystyrene arms (Fig. 11.10), using a tri-functional initiator (**VII**) that contained three initiating styrene-TEMPO groups and was obtained (in 71% yield) by reaction of (**IV**) with 1,3,5-benzenetricarbonyl chloride in the presence of 4-dimethylaminopyridine and pyridine in THF solution. Bulk polymerization of 200 equivalents of styrene with (**VII**) at 130°C for 72 h was found to give polystyrene (**VIII**) in 84% yield with M_n = 16500 and PDI = 1.20, as compared to the theoretical M_n = 21000 with each arm having a M_n of 7000. The discrepancy in molecular weight was attributed to lower hydrodynamic volume of a star polymer as compared to a linear polymer. It may be noted that the polymerization of styrene monomer in the presence of (**VII**), leading to the formation of (**VIII**), takes place by the insertion of styrene units between the TEMPO and styrene units of (**VII**). It was demonstrated (Hawker, 1995) that the TEMPO linkage is stable to hydrolysis conditions, suh as treatment with KOH. Thus, the hydrolysis of (**VIII**) by KOH releases the individual arms as linear polystyrene molecules each with a nitroxide and a hydroxyl end group (**IX**) (see Fig. 11.10).

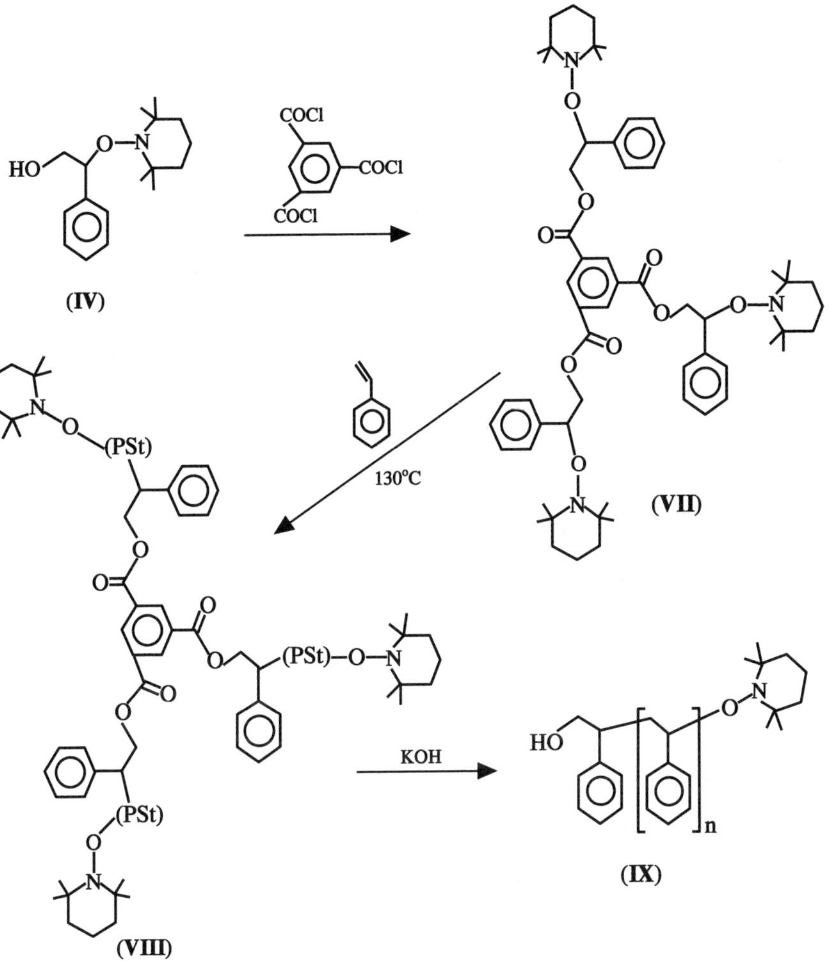

Figure 11.10 Synthesis of three-arm star-shaped polystyrene by TEMPO-mediated living radical polymerization and hydrolysis to TEMPO-capped linear polystyrene. (Drawn following the synthesis method of Hawker, 1995.)

The lack of termination step in SFRP also opens up the possibility of using this polymerization process to prepare graft systems. One approach to synthesis of graft systems using TEMPO-based SFRP involves initial synthesis of a monomer unit incorporating the styrene-TEMPO group, followed by styrene polymerization under normal free radical conditions, and then followed again by a living free radical polymerization at a higher temperature with a second feed of styrene and comonomer (see Problem 11.5) to give grafts of controlled molecular weight and low polydispersity. This ability to conduct one free radical polymerization followed by another by simply increasing the temperature and adding a new monomer, may open new paths to unusual macromolecular architectures. Unusual graft copolymers can be readily prepared by this living free radical technique from monomers which cannot be polymerized by better known anionic or cationic techniques. The molecular weights of the arms can be controlled by varying the equivalents of monomer added while maintaining very low polydispersities. It is believed that the novel polymerization process offers the architectural control obtainable only under synthetically more rigorous anionic or cationic conditions (Hawker, 1995).

Problem 11.5 Devise a synthetic scheme, using TEMPO-based SFRP, to obtain graft copolymers of controlled molecular weight and low polydispersity, in which random copolymers of styrene (St) and ethyl acrylate (EA) containing about 0.2 mole fraction of EA are grafted onto polystyrene (PS) chains.

Answer:

A synthesis procedure based on the strategy that living free radical polymerization occurs at a higher temperature than conventional free radical polymerization, is shown in Scheme P11.5.1. The first step in this Scheme is to synthesize a monomer unit (**P5-II**) containing a styrene-TEMPO group as well as bound styrene. To achieve this, using the same proportions as used by Hawker (1995), the hydroxy derivative of TEMPO, namely, 2-phenyl-2-[(2,2,6,6-tetramethylpiperidine)oxy]-1-ethanol (**P5-I**) (3.6 mmol) in THF (20 mL) is reacted with *p*-chloromethylstyrene (10.0 mmol) in the presence of sodium hydride (5.0 mmol) at room temperature. Copolymerization of (**P5-II**) (1.15 mmol) with styrene (23.0 mmol) is conducted under normal conditions using azobisisobutyronitrile (AIBN) (0.23 mmol) as an initiator in refluxing THF (20 mL) at 65°C. This results in copolymer (**P5-III**), which is obtained as a white solid by evaporating the reaction mixture to dryness, redissolving in dichloromethane, and precipitating into methanol. Containing pendant styrene-TEMPO groups (depending on the relative proportions of (**P5-II**) and styrene in feed, (**P5-III**) acts as the polymeric initiator. A solution of (**P5-III**) (0.085 mmol) in styrene (14.0 mmol) and ethyl acrylate (3.5 mmol) is heated at 125°C with stirring under nitrogen for 72 h. The reaction mixture is dissolved in dichloromethane and precipitated into methanol. (<u>Note</u>: As the results in Table 11.1 show, in TEMPO-mediated copolymerization of St and EA the ratio of repeat units in the copolymer is essentially the same as that expected from the feed ratio.)

A goal of research has been to design a multifunctional reagent that can serve as an initiator for multiple living polymerizations (namely, controlled free radical, anionic, and cationic polymerizations) and also can serve as a source for a variety of macromonomers including those for condensation polymerization and functional polymers. Puts and Sogah (1997) designed and synthesized such a novel multifunctional compound (**X**, Fig. 11.11) containing an alkoxyamine, an oxazoline, and a protected hydroxyl group. The compound (**X**) is 1-{2-((*tert*-butyldimethylsilanyl)oxy)-1-{4-[2-4,5-dihydro-oxazol-2-yl)ethylphenyl}ethoxy}-2,2,6,6-tetramethylpiperidine and has characteristics that enable reactive sites to be accessed independently. Thus alkyl adducts of nitroxyl radicals can initiate controlled free radical polymerization and neutral oxazolines, which are stable toward free radical and anionic reactions, can undergo cationic ring-opening polymerization, while the α-position with respect to the oxazoline, being relatively acidic, can serve as a carbanion source to initiate anionic vinyl polymerization. These possibilities are indicated in Fig. 11.11.

Living free radical polymerization of styrene with (**X**) as initiator can be carried out at 135°C to give oxazoline-ended polystyrene macromonomer/macroinitiator (see Fig. 11.12). Subsequently, cationic ROP of the oxazoline group can be carried out. For example, initiation by methyl triflate, i.e., methyl trifluoromethane sulfonate ($CF_3SO_3CH_3$) in *o*-dichlorobenzene at 140°C produces cationic homopolymerization of oxazoline (see Section 10.2.1.2) yielding a comb polymer with a hydrophilic poly(ethyleneimine) backbone (Fig. 11.12). The comb polymer still possesses protected alcohol groups that can be readily converted to initiator sites for polymerization of such monomers as lactones, lactides, and epoxides. Since the oxazoline and the TBDMS-protected hydroxyl groups are stable under the free radical conditions employed (Puts and Sogah, 1997) and since each initiating site of (**X**) can be addressed selectively and independently, *one-pot* synthesis for a variety of complex functional polymers is possible.

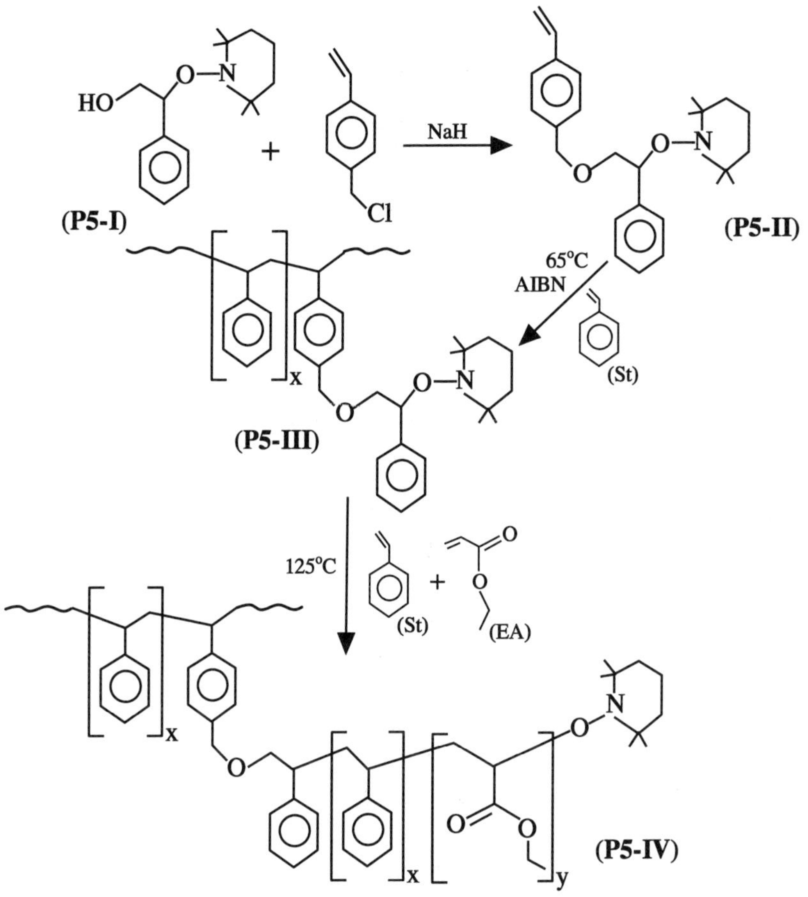

Scheme P11.5.1 Graft copolymer synthesis: random copolymer of styrene and ethyl acrylate grafted onto polystyrene (Problem 11.5). (Drawn following the synthesis method of Hawker, 1995.)

11.2.5 Aqueous Systems

Since functional groups which impart water solubility to monomers and polymers are not compatible with ionic polymerization systems, requiring these groups to be protected prior to polymerization and deprotected thereafter, narrow polydisperse water-soluble polymers cannot be prepared directly by conventional (anionic/cationic) living polymerization systems. In contrast, stable free radical polymerization (SFRP) in water-based systems, which has been studied intensely in recent years (as it combines the environmental and technical advantages of polymerization in aqueous dispersed media), provides a novel, direct route to water-soluble narrow-polydispersity resins of controlled structures. Nearly all aqueous systems have been investigated, ranging from homogeneous solution to heterogeneous dispersed systems, such as aqueous alcoholic dispersion, aqueous suspension, seeded emulsion, ab initio emulsion, and miniemulsion. Various nitroxides used as mediators and various alkoxyamines used as initiators in living radical polymerization in aqueous dispersed systems are summarized in Figs. 11.13 and 11.14, respectively.

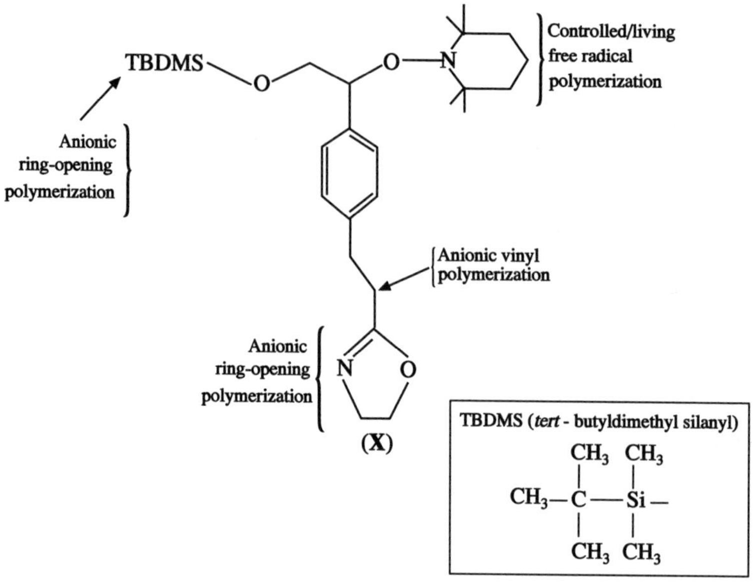

Figure 11.11 A multifunctional reagent containing orthogonal reactive sites for controlled free-radical, anionic, and cationic polymerizations. (From Puts and Sogah, 1997. With permission from *American Chemical Society.*)

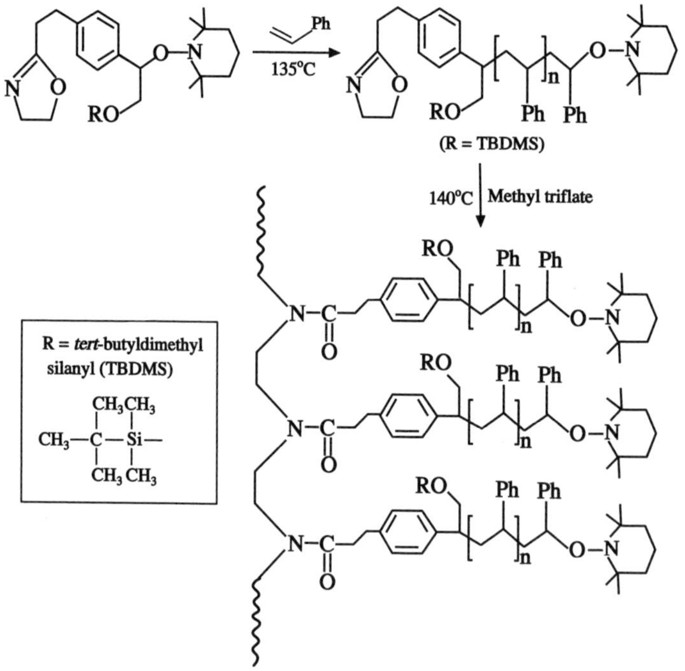

Figure 11.12 Different types of polymerizations with multifunctional reagent (**X**): living radical polymerization of styrene followed by cationic ring-opening homopolymerization of the resulting oxazoline-ended macromonomer. (From Puts and Sogah, 1997. With permission from *American Chemical Society.*)

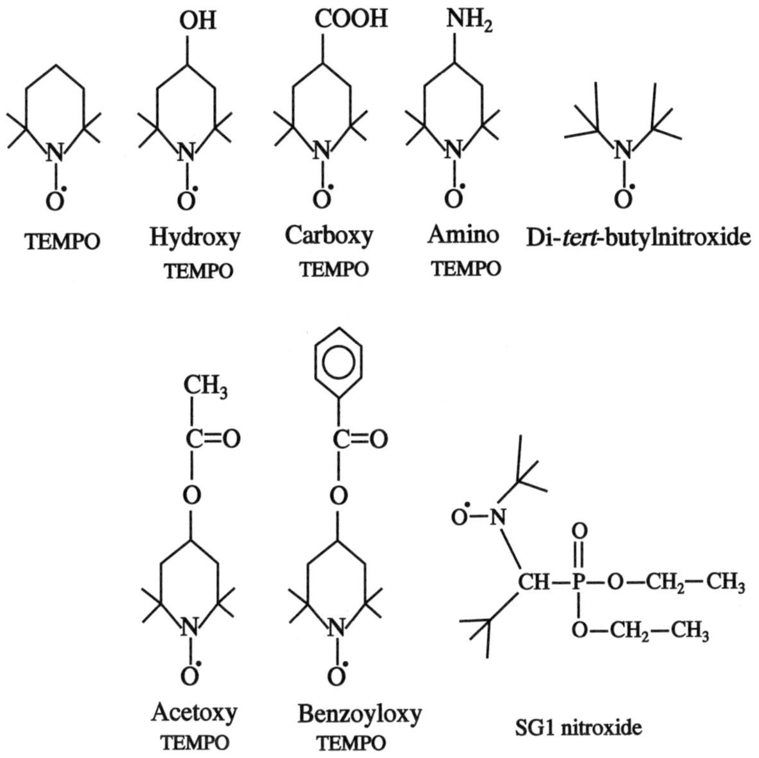

Figure 11.13 Various nitroxides used as mediators in controlled radical polymerization in aqueous dispersed systems.

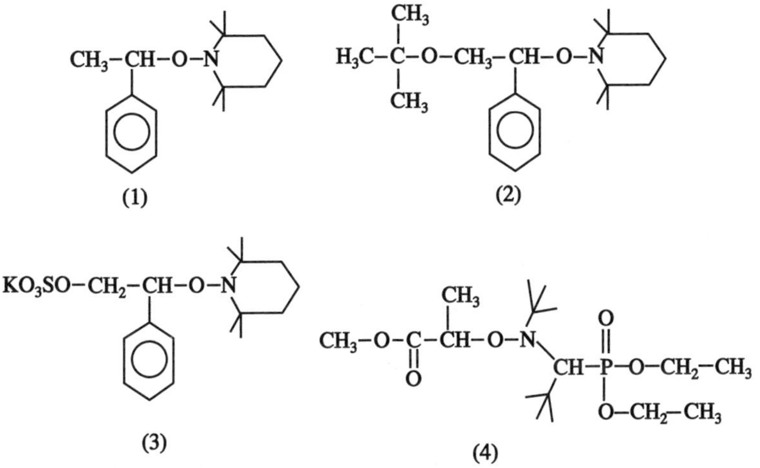

Figure 11.14 Several alkoxyamines used as initiators in controlled radical polymerization in aqueous dispersed systems.

To carry out SFRP in homogeneous aqueous solutions, the radical mediator, as also the monomer and the polymers, should be soluble in the medium. A number of water-soluble monomers have been polymerized by the living radical process (Qiu et al., 2001). They include uncharged monomers, such as 2-hydroxyethyl acrylate, 2-hydroxyethyl methacrylate, oligo(ethylene glycol) methacrylate, 2-(dimethylamino)ethyl methacrylate, acrylic acid, etc., and charged/ionic monomers, such as sodium methacrylate, sodium vinylbenzoate (NaVBA) and sodium styrene sulfonate (NaSS).

Solution polymerizations of NaSS and NaVBA were carried out in aqueous ethylene glycol mixture (75-80% ethylene glycol) at 120-125°C, using TEMPO as the radical mediator and potassium persulfate/sodium bisulfite as the initiator. While homopolymerization of NaSS (Keoshkerian et al., 1995) gave high conversion (>90%) and it was possible to vary the molecular weights from very low (8000) to very high (900,000), maintaining narrow polydispersities (1.1 to 1.3), the homopolymerization of NaVBA only gave relatively low conversion (<30%), though chain extension of TEMPO-terminated poly(NaSS) resins with NaVBA was reasonably efficient, yielding water-soluble block copolymers. However, the acid forms of the monomers, namely styrene sulfonic acid and vinyl benzoic acid, were not polymerizable at all by SFRP due to nitroxide decomposition in the acid medium.

CRP/LRP methods have been applied in miniemulsion polymerization as it offers some unique technical advantages over an emulsion polymerization and can be regarded as a simplified model of emulsion polymerization, in which, ideally, all monomer droplets (of 50-500 nm size) are nucleated by entering aqueous oligoradicals and become polymer particles, while no new particles are nucleated by homogeneous nucleation, thus leading to simpler kinetic behavior.

Macleod et al. (1999) and Prodpran et al. (2000) were the first to report nitroxide-mediated styrene polymerization in miniemulsion. Prodpran et al. used benzoyl peroxide and TEMPO with hexadecane as a costabilizer at 125°C. Stable latexes were produced showing over 90% conversion in 12 h with molecular weight reaching as high as $M_n \sim 40,000$ and polydispersities ranging from ~1.15 to 1.6. The molecular weight varied linearly with conversion, signifying a living polymerization.

A major disadvantage of miniemulsion polymerization is the need to apply very high shear to the initial reaction mixture for creating 50-500 nm droplets of monomer and so also the need to use a hydrophobic costabilizer, usually a volatile organic compound that may eventually show undesirable presence in the final product. Owing to its simplicity, traditional emulsion polymerization (see Section 6.13.1) is usually the technique of choice. This process does not require any special manipulation of the initial monomer-in-water emulsion and particles are generated in the aqueous phase independently from the oil droplets, which only act as a monomer reservoir (see Fig. 6.14). Using differently substituted TEMPO derivatives (Fig. 11.13), all in position 4, Cao et al. (2001) achieved reasonably well-controlled reactions in styrene emulsion polymerization. Using either potassium persulfate (KPS) or azobisisobutyronitrile (AIBN) at 120°C with sodium dodecyl sulfate (SDS) as the surfactant, significant differences were observed for the different nitroxides with different water solubilities. While very low water solubility of the nitroxide resulted in an uncontrolled polymerization, too high water solubility resulted in slow aqueous phase initiation that hindered the polymerization rate. Using 4-acetoxy-TEMPO (Fig. 11.13), 81% conversion was reached in 12 h, giving a polymer with $M_n = 18,000$ and polydispersity ~1.3.

A major problem of CRP in emulsion systems is that the process often leads to unstable latexes, mainly due to the sensitivity of the nucleation step, and is therefore still a challenge (Qiu et al., 2001; Cunningham, 2002; Charleux, 2003). The formation of unstable latexes has been explained by some very specific features of the chain growth in CRP with respect to a classical free-radical polymerization, namely, slow and simultaneous growth of a large number of oligomers in the early stage of polymerization, which completely modifies the mechanism of particle formation.

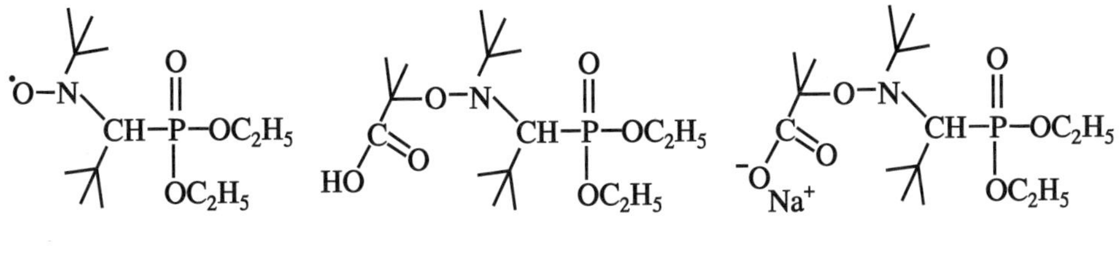

SG1 - nitroxide (A)   (A-H)   (A-Na)

Figure 11.15 Structures of SG1 (*N-tert*-butyl-*N*-(1-diethylphosphono-2,2-dimethyl propyl)) and SG1-based water soluble alkoxyamine initiator. **A**-H represents the acidic form of SG1 and **A**-Na is the sodium salt of **A**-H.

More recently, the CRP method has been improved significantly with the discovery of very efficient acyclic nitroxides such as the stable acyclic phosphonylated nitroxide radical SG1 (Fig. 11.15), which can control the polymerization of much broader range of monomers than cyclic nitroxides like TEMPO. The radical SG1 has a larger equilibrium constant than most other nitroxides, including TEMPO, and is therefore suitable for polymerization temperatures as low as ~90°C (SG1: $K = 1.9 \times 10^{-8}$ mol dm^{-3} at 125°C; TEMPO: $K = 2.1 \times 10^{-11}$ mol dm^{-3} at 125°C).

The very few reports on nitroxide-mediated ab initio emulsion polymerization have mainly concerned the application of a water-soluble radical initiator, such as $K_2S_2O_8$, in conjunction with free nitroxide, either TEMPO or SG1. However, for better control of the chain growth and polymerization kinetics, a monocomponent alkoxyamine initiator, instead of a usual bicomponent initiator system (i.e., conventional radical initiator along with free nitroxide), should be chosen because the bicomponent system suffers from poorly controlled initiator efficiency (Nicolas et al., 2004; Hawker et al., 2001). The main challenge in nitroxide-mediated emulsion polymerization is thus to select an appropriate water-soluble alkoxyamine initiator, and then to find suitable conditions for formation of stable latex with sufficiently high solid content. In this respect, a simple way was reported (Nicolas et al., 2004) for performing nitroxide-mediated emulsion polymerization of *n*-butyl acrylate (BA) and styrene (St), as well as for synthesis of block copolymer of of poly(*n*-butyl acrylate) (PBA) and polystyrene (PSt) with the SG1-based water-soluble alkoxyamine initiator, **A**-Na (Fig. 11.15), the water solubility in this case being imparted by the sodium carboxylate group (for pH > 6), at least during the nucleation step. Due to the hydrophobicity of the SG1 capping agent, **A**-Na exhibits surface activity in aqueous solution and undergoes homolytic dissociation to the highly water-soluble 2-(hydroxycarbonyl)prop-2-yl sodium salt radical, which initiates polymerization in the water phase.

11.3   Atom Transfer Radical Polymerization (ATRP)

Among all LRP methods, ATRP is the most studied and since 1995, when it was first reported, a very large number of articles have appeared on this topic. An excellent review written by the pioneer in the field, Matyjaszewski (Matyjaszewski and Xia, 2001) covers the development in ATRP from 1995 till the end of 2000. ATRP can provide extraordinary control over topologies, compositions, microstructures, and functionalities. This has led to the production of a vast array of polymeric materials with application in nanocomposites, thermoplastic elastomers, bioconjugates, drug delivery systems, etc.

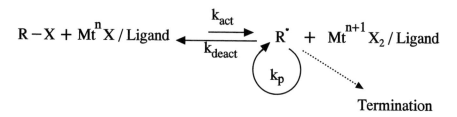

Figure 11.16 General mechanism for ATRP.

ATRP proceeds via standard free-radical polymerization (FRP) mechanism. However, in ATRP the chain initiation is caused by a halogenated organic species in the presence of a transition metal compound and the growing radicals can be reversibly activated or deactivated via a dynamic equilibrium with the transition metal compound by an exchange of halide species between the chain end and metal compound. Using a simple alkyl halide, R–X (X = Cl or Br), as an initiator and a transition metal complex, Mt^n–Y/Ligand (where Y may be a counterion or a ligand that significantly improves the solubility of the transition metal compound), as a catalyst, ATRP of vinyl monomers proceeds in a living fashion, yielding polymers with low polydispersities, $1.1 < M_w/M_n < 1.5$ (Wang and Matyjaszewski, 1995). In these reactions, the catalytic amount of the transition metal compound acts as a carrier of the halogen atom, reversibly releasing growing radicals in a Mt^n/Mt^{n+1} redox process, as shown in Fig. 11.16. The redox process gives rise to the aforesaid equilibrium between dormant (polymer-halide) and active (polymer-radical) chains. The equilibrium is designed to heavily favor the dormant state, which effectively reduces the radical concentration to a sufficiently low level to limit bimolecular coupling. Polymerizations require elevated temperatures (60-120°C).

ATRP is an extension of *atom transfer radical addition* (ATRA), which is a well-known method of carbon-carbon bond formation (catalyzed by transition metal complexes) in organic synthesis. ATRP also has roots in the transition metal catalyzed telomerization reactions (Boutevin, 2000) and connections to the transition metal initiated redox processes as well as inhibition with transition metal compounds (Qin and Matyjaszewski, 1997). ATRP was developed "by designing an appropriate catalyst (transition metal compound and ligands), using an initiator with the suitable structure, and adjusting the polymerization conditions such that the molecular weights increased linearly with conversion and the polydispersities were typical of a living process" (Matyjaszewski and Xia, 2001).

In ATRP, the atom transfer step is the key elementary reaction responsible for the uniform growth of the polymeric chains, which is accomplished through fast initiation (activation rate constant, k_{act}) and rapid reversible deactivation (deactivation rate constant, k_{deact}). Polymer chains grow by the addition of the intermediate radicals (R•) to monomers in a manner similar to a conventional radical polymerization with a rate constant of propagation, k_p. Termination reactions (rate constant, k_t) also occur in ATRP, mainly through radical coupling and disproportionation. With each coupling or disproportionation reaction two equivalents of deactivator, $Mt^{n+1}X_2/L$ (Fig. 11.16), accumulate as persistent radicals. However, in a well-controlled ATRP, the majority of chains exist in a dormant (halogen-capped) state due to the *persistent radical effect* (Fischer, 2001) and irreversible termination of propagating free radicals is statistically suppressed to give polymer synthesized via ATRP its living nature (Chan et al., 2010). (Typically, no more than 5% of the total growing polymer chains undergo irreversible termination during the initial, short, non-stationary period of polymerization.) Though k_{deact} is of a similar order of magnitude as k_t, deactivation becomes the dominant chain ending reaction due to much higher catalyst concentration, as compared to radical concentration. An important aspect of ATRP is that it can tolerate various functional groups. Consequently, well-defined end-functional polymers can be conveniently prepared without the need for additional protecting reactions.

The transition metal catalyst which governs the ATRP equilibrium is traditionally used in stoichiometric or slightly sub-stoichiometric ratio to the initiator. The initiator concentration, in turn, is set in a ratio to the monomer concentration in order to achieve a target molecular weight of the polymer product, the degree of polymerization (DP) being given by the ratio of the concentration of the consumed monomer to that of organic halide (initiator) initially used:

$$\overline{DP}_n = \triangle[\text{M}]/[\text{R–X}]_0 \qquad (11.17)$$

Problem 11.6 Styrene (St) was polymerized by ATRP using a copper(I) bromide (CuBr) catalyst, complexed with N,N,N',N',N''-pentamethyldiethylenetriamine (PMDETA) ligand, and methyl 2-bromopropionate (MBrP) as initiator. Experiments were performed in 1 L mixed vessel at 110°C with excellent temperature control using a monomer to solvent (toluene) ratio of 70:30 wt% and molar ratios of 50:1:1:1 for St/MBrP/CuBr/PMDETA. Under these reaction conditions, only a portion of the catalyst species was soluble and 90% monomer conversion was obtained in 6 h. Calculate a theoretical molecular weight (MW) of the polymer obtained. How would you explain if the experimental MW is found to be higher than the theoretical one ?

Answer:

The catalyst being not fully soluble, only the solubilized catalyst participates in the polymerization. Undissolved catalyst in the system, while not participating in the reaction, acts as a reservoir to maintain the Cu(II) to Cu(I) ratio relatively constant during the polymerization. Assuming 100% initiator efficiency, Eq. (11.17) gives, for 90% monomer conversion,

$$\overline{DP}_n = \frac{[\text{St}]_0 \times 0.90}{[\text{MBrP}]_0} = 45$$

$$M(\text{St}) = 104 \text{ g mol}^{-1}$$

$$\overline{M}_n = (45)(104 \text{ g mol}^{-1}) = 4680 \text{ g mol}^{-1}$$

End groups derived from $CH_3CH(Br)COOCH_3$ are $C_4H_7O_2$ (87) and Br (80). Hence total molar mass of polymer = $4680 + 87 + 80 = 4847$ g mol^{-1}. A higher value of the experimental molar mass would indicate that the initiator efficiency is less than 100%.

Since for coatings applications, the target molecular weight is often quite low, a high concentration of initiator is needed, leading to a high catalyst level in the final polymer (see Problem 11.7). The catalyst can be expensive, especially where it features a specially synthesized ligand. Moreover, the catalyst can be harmful and can add undesirable color to the polymer so that additional steps for catalyst removal and/or recovery may be needed, adding significantly to the production costs.

Problem 11.7 A controlled radical polymerization offers potential advantages if it can be applied to the coatings industry, since the polymer will have lower polydispersity as well as lower viscosity for the same $\overline{M}_n$. ATRP has thus been applied for the production of low molecular-weight solvent-borne acrylic polymers for use in automotive coatings. Poly(ethyl acrylate) with $\overline{M}_n$ of 5000 is to be prepared by ATRP using a low-molecular weight organic halide (R–X) as initiator and a stoichiometric amount of CuX/ligand as catalyst. Estimate the level of initiator required relative to monomer and the level (ppm) of residual copper in the final polymer, assuming 100% initiator efficiency.

Answer:

Formula weight of ethyl acrylate = 100 g/mol. Neglecting contribution of chain ends to polymer molecular weight, $\overline{DP}_n$ of poly(ethyl acrylate) of $\overline{M}_n$ = 5000 is 5000/100 = 50. From Eq. (11.10),

$$[\text{R–X}]_0 \, / \, \triangle [\text{M}] \;=\; 1/\overline{DP}_n \;=\; 0.02$$

Level of initiator required = 0.02×100 = 2 mol% of monomer.

Assuming that the catalyst is used in stoichiometric ratio to the initiator and that the initiator is consumed fully (with one polymer molecule being generated for each initiator molecule), amount of copper (atomic wt = 63) in final polymer

$$= \frac{63 \times 10^6}{5000} \;\simeq\; 10^4 \text{ ppm}$$

11.3.1 ATRP Monomers

Monomers which have been successfully polymerized using ATRP include styrenes, acrylates, methacrylates, and several other relatively reactive monomers such as acrylamides, vinylpyridine, and acrylonitrile, which contain groups (e.g., phenyl, carbonyl, nitrile) adjacent to the carbon radicals that stabilize the propagating chains and produce a sufficiently large atom transfer equilibrium constant. The range of monomers polymerizable by ATRP is thus greater than that accessible by nitroxide-mediated polymerization, since it includes the entire family of methacrylates. However, acidic monomers (e.g., methacrylic acid) have not been successfully polymerized by ATRP and so also halogenated alkenes, alkyl-substituted olefins, and vinyl esters because of their very low intrinsic reactivity in radical polymerization and radical addition reactions (and hence, presumably, a very low ATRP equilibrium constant).

Even under the same conditions of polymerization using the same catalyst, each monomer has its unique atom transfer equilibrium constant for its active and dormant species and the magnitude of this equilibrium constant ($K_{eq} = k_{act}/k_{deact}$) determines the polymerization rate (Matyjaszewski and Xia, 2001). Thus, too large an equilibrium constant will lead to a high radical concentration, resulting in a large amount of termination and hence greater polydispersity of polymer. On the other hand, if the equilibrium constant is too small, ATRP will not occur or occur very slowly.

Too low equilibrium constant may be the main reason that ATRP of olefins, halogenated alkenes, and vinyl acetate has not been successful. Since each monomer has a specific equilibrium constant, optimal conditions for ATRP, which include the type and amount of catalyst, temperature, solvent, and other additives, may be quite different for different monomers (Matyjaszeswki and Xia, 2001).

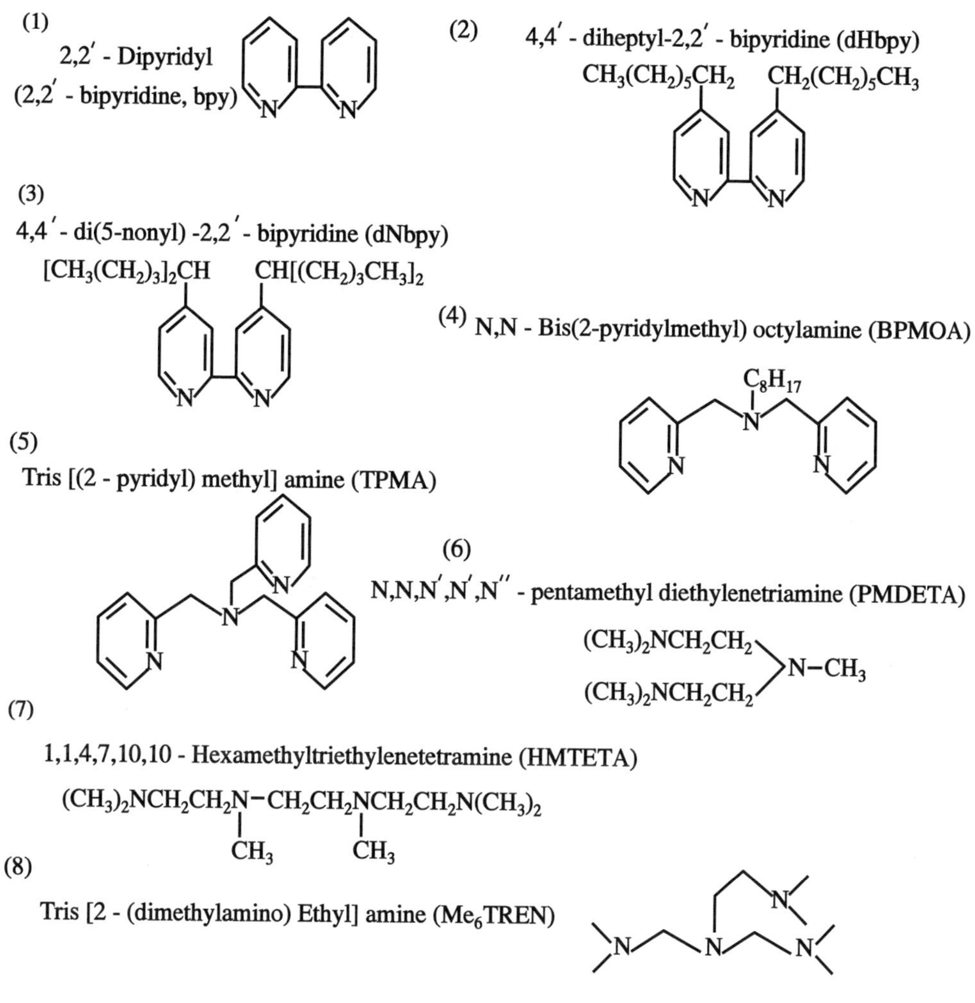

Figure 11.17 Structures of some amine ligands used in copper-mediated ATRP.

While ATRP of styrene and its derivatives has been conducted using copper, iron, ruthenium, and rhenium catalytic systems, in most of the reported cases, copper-based systems have been used, one of the most extensively studied systems being the ATRP of styrene at 110°C with CuBr(dNbpy)$_2$ (see Fig. 11.17) as the catalyst and alkyl bromides as initiators. With the use of a more efficient catalyst such as CuBr/PMDETA (Fig. 11.17) or CuOAc/CuBr/dNbpy, the reaction temperature can be lowered to 80-90°C as better molecular weight control with lower polydispersities is obtained at lower temperatures, presumably due to a lower contribution of the thermal self-initiation. However, temperatures greater than 100°C are usually preferred for styrene ATRP in order to avoid vitrification at high conversion ($T_g \approx 100°C$ for polystyrene), and also to increase the solubility of the catalysts in some cases. Among styrenic monomers, those with electron-withdrawing substituents polymerize faster because the atom transfer equilibrium is more shifted toward the active species side for styrenes with electron-withdrawing groups.

A wide range of acrylates with various side chains have been polymerized using ATRP to obtain well-defined functional polymers, e.g., ATRP of 2-hydroxyethyl acrylate, glycidyl acrylate, and *tert*-butyl acrylate (yielding well-defined poly(acrylic acid) on hydrolysis). Among several transition metal catalysts, viz., copper, ruthenium, and iron-based systems, which have been successfully used for the controlled ATRP of acrylates, copper appears to be superior in producing well-defined polyacrylates with low polydispersities.

A larger range of catalysts are available for ATRP of methacrylates, as compared to acrylates, due to the relative ease of activation of the derived species and the high values of the ATRP equilibrium constants. ATRP of methyl methacrylate (MMA) is usually carried out at 70-90°C using solution polymerization since PMMA has a glass transition temperature $T_g > 100$°C. This also helps keep the concentrations of growing radicals and the catalyst low by dilution, which is needed since the copper-mediated ATRP of MMA has a significantly higher equilibrium constant, as compared to styrene and MA. Besides MMA, other methacrylic esters have also been successfully polymerized by ATRP. These include *n*-butyl methacrylate, 2-(dimethylamino)ethyl methacrylate (DMAEMA), 2-hydroxyethyl methacrylate (HEMA), and silyl-protected HEMA, methacrylic acid in its alkyl-protected form or as its sodium salt, and fluorinated methacrylic esters (Matyjaszewski and Xia, 2001).

For ATRP of acrylonitrile (AN), it is necessary to use a solvent because polyacrylonitrile (PAN) is not soluble in its monomer. Though DMF is a good solvent for PAN, it is not favored because it may also complex with copper. ATRP of AN has been successfully carried out in homogeneous system using ethylene carbonate as the solvent, $CuBr(bpy)_2$ complex as the catalyst and α-bromopropionitrile as the initiator at temperatures ranging from 44°C to 64°C.

According to reports available in the literature on the attempted ATRP of acrylamide, the polymerization has not been successful in all cases. Kinetic studies made using model compounds have revealed that under typical ATRP conditions acrylamide polymerization has a much lower ATRP equilibrium constant than styrene or acrylates. This may be attributed to catalyst inactivation due to complexation of copper by the growing polymer and to displacement of the terminal halogen atom by the amide group. The best results for the ATRP of (meth)acrylamide have been obtained with one of the most powerful catalytic systems, namely, $CuCl/Me_6TREN$ (Fig. 11.17), due to its high equilibrium constant. To minimize side reactions, polymerizations have been carried out at low temperature (20°C) using alkyl chlorides as initiators. Metals other than copper have also been used in the ATRP of acrylamide. For example, dimethylacrylamide has been polymerized with $RuCl_2(PPh_3)_3$ and $Al(OiPr)_3$ in toluene at 60°C using a bromide initiator such as CCl_3Br.

Problem 11.8   Suggest suitable methods for obtaining poly(meth)acrylic acid via ATRP route.

Answer:

Controlled polymerization of acrylic or methacrylic acid by ATRP presents a challenging problem because the acid monomers can poison the catalysts by coordinating to the transition metal (Matyjaszewski and Xia, 2001). Moreover, nitrogen containing ligands can be protonated, which interferes with the metal complexation ability. Poly(meth)acrylic acids can thus be conveniently prepared by polymerization of protected monomers such as trimethylsilyl methacrylate, *tert*-butyl methacrylate, tetrahydropyranyl methacrylate, and benzyl methacrylate, followed by hydrolysis to remove the protective groups (Davis et al., 2000). However, successful ATRP of sodium methacrylate in water, with moderate to good yields, has been reported using $CuBr(bpy)_3$ as the catalyst and a poly(ethylene oxide)-based macroinitiator (Ashford et al., 1999).

11.3.2 ATRP Initiators

In ATRP, alkyl halides (RX) are commonly used as the initiator. Fast initiation is important to obtain well-defined polymers with low polydispersities (PDs). To obtain polymers with low PDs, i.e., narrow molecular weight distributions (MWDs), the halide group, X, must migrate rapidly and selectively between the growing polymer chain and the transition metal complex. Usually, the molecular weight control is found to be the best if X is either bromine or chlorine, though iodine is also found to work well for ATRP of acrylates using copper complex and for styrene polymerization in ruthenium-based ATRP. The C–F bond being too strong to undergo homolytic cleavage, R–F is not used as initiator in ATRP. However, some pseudohalogens, specifically thiocyanates and thiocarbamates, have been successfully used for ATRP of acrylates and styrenes (Nishimura et al., 1999).

Two parameters are important for a successful ATRP initiating system. First, initiation should be fast in comparison with propagation and, second, the probability of side reactions should be minimized. Several general considerations may be used for the initiator choice:

1. Tertiary alkyl halides are better initiators than secondary halides, which, in turn, are better than primary alkyl halides. Sulfonyl chlorides also provide faster initiation than propagaion.

2. The bond strength of the alkyl halides being in the order R–Cl > R–Br > R–I, alkyl chlorides should be the least efficient initiator and alkyl iodides the most efficient. However, alkyl iodides suffer from being light sensitive and have a tendency to form metal iodide complexes. Therefore bromo- and chloro-compounds are most frequently used as initiators. Same halogen is commonly used for the initiator and the metal salt, that is, R–X/Mt–X (X = Cl or Br). In this case, the polymer chains are also terminated by the same X. However, if a mixed halide initiating system is used, i.e., R–X/Mt–Y (X, Y = Cl or Br), majority of polymer chains are terminated by Cl due to the stronger alkyl-chloride bond.

3. Using the same initiator, the success of initiation can depend strongly on the choice of catalyst. For example, CCl_4 is a good initiator for styrene and MMA with $CuBr(bpy)_3$ as the catalyst, but the same is not true using the $CuBr(dNbpy)_2$ catalytic system (Matyjaszewski and Xia, 2001).

4. The method or order of reagent addition (e.g., adding catalyst to the initiator or initiator to the catalyst) can be crucial.

Besides alkyl halides (R–X) and halogenated alkanes (such as $CHCl_3$, CCl_4), many other different types of halogenated compounds are potential initiators, such as benzylic halides (e.g., $PhCHCl_2$, Ph_2CHCl, Ph_2CCl_2), α-haloesters (e.g., α-halopropionates and α-haloisobutyrates), α-haloketones (e.g., CCl_3COCH_3 and $CHCl_2COPh$), α-halonitriles (e.g., 2-bromopropionitrile), and sulfonyl halides. Benzyl substituted halides are useful initiators for the ATRP of styrene and its derivatives because of their structural similarity. However, they are not useful for the ATRP of more reactive monomers, such as MMA, though polyhalogenated benzylic halides, e.g., $PhCHCl_2$ and Ph_2CCl_2, have been used for the ATRP of MMA with $RuCl_2(PPh_3)_3/Al(OiPr)_3$ as the catalyst (Ando et al., 1997). Apparently, two-directional growth is provided by the di-halogenated initiators due to the presence of two Cl groups.

Among α-haloesters, α-haloisobutyrates generally produce initiating radicals faster than the corresponding α-halopropionates due to better stabilization of the radicals. Thus, in the ATRP of methacrylates, α-halopropionates produce slow initiation, though, in contrast, α-bromopropionates are good initiators for the ATRP of acrylates due to their structural similarity (Matyjaszewski and Xia, 2001). Polyhalogenated α-haloesters (e.g., $CCl_3CO_2CH_3$ and $CHCl_2CO_2CH_3$) have also been successfully used as initiators for ATRP of MMA catalyzed by ruthenium complexes, for which, however, polyhalogenated α-haloketones (e.g., CCl_3COCH_3 and $CHCl_2COPh$) are among the best initiators. This faster initiation by ketone, compared with the ester counterparts, can be explained by the stronger electron-withdrawing power of the ketone carbonyl which induces further polarization of the carbon-chlorine bond. The strong electron-withdrawing power of the cyano group is also responsible for fast radical generation by α-halonitriles in ATRP, while the radical formed is also sufficiently reactive, leading to fast initiation by rapid radical addition to monomer.

Well-controlled polymerizations of a large number of monomers have been obtained in copper-catalyzed ATRP with sulfonyl halides as initiators, which yield a much faster rate of initiation than monomer propagation. For example, the apparent rate constants of initiation with sulfonyl chlorides are about four (for styrene and methacrylates) and three (for acrylates) orders of magnitude higher than those for propagation (Matyjaszewski and Xia, 2001).

While the range of available initiators for ATRP is much larger than for other CRP methods, the use of various halo-compounds as initiators described above leads to halogen end groups of the polymers formed. They can be replaced, however, by many synthetic methods to provide more useful functionalities and halogen-free products. Furthermore, as shown later, the use of multi-functional activated halides enables simultaneous growth of chains in several directions, leading to macromolecular stars, combs, and brushes.

11.3.3 ATRP Catalysts

Transition metal complexes functioning as redox catalysts are perhaps the most important components of an ATRP system. (It is, however, possible that some catalytic systems reported for ATRP may lead not only to formation of free radical polymer chains but also to ionic and/or co-ordination polymerization.) As mentioned previously, the transition metal center of the catalyst should undergo an electron transfer reaction coupled with halogen abstraction and accompanied by expansion of the coordination sphere. In addition, to induce a controlled polymerization process, the oxidized transition metal should rapidly deactivate the propagating polymer chains to form dormant species (Fig. 11.16). The ideal catalyst for ATRP should be highly selective for atom transfer, should not participate in other reactions, and should deactivate extremely fast with diffusion-controlled rate constants. Further, it should have easily tunable activation rate constants to meet specific requirements for ATRP monomers. For example, very active catalysts with equilibrium constants $K > 10^{-8}$ for styrenes and acrylates are not suitable for methacrylates.

An efficient ATRP catalyst is described to achieve mainly the following functional attributes (Matyjaszewski and Xia, 2001): (1) initiation is fast and quantitative which ensures that all polymeric chains start to grow simultaneously; (2) the equilibrium between the alkyl halide and the transition metal is heavily shifted toward the dormant species side, thereby ensuring that most of the growing chains remain dormant and the radical concentration is low; (3) deactivation of the active radicals by halogen transfer is fast which ensures that all polymer chains grow at approximately the same rate, leading to low polydispersity; (4) activation of the dormant polymer chains is relatively fast, which provides a reasonable polymerization rate; and (5) there are no side reactions, such as $\beta-H$ abstraction or reduction/oxidation of the radicals. Accordingly, ligands are chosen to adjust the redox potential of the metal center for appropriate reactivity and dynamics for the atom transfer. Another important role of the ligand in ATRP is to solubilize the transition metal salt in the organic media.

A number of transition metal complexes have been applied in ATRP process. It has been successful for molybdenum, chromium, rhenium, ruthenium, and iron, rhodium, nickel, palladium, and copper complexes. Among these, copper catalysts are superior in terms of versatility and cost. Styrenes, (meth)acrylates, (meth)acrylamides, and acrylonitrile have been successfully polymerized using copper-mediated ATRP. The polymerization has been found to be tolerant to a variety of functional groups, such as —OH and —NH$_2$, and insensitive to additives, such as H$_2$O, CH$_3$OH, and CH$_3$CN (Matyjaszewski et al., 1997).

The tacticity of the polymer prepared by ATRP of MMA with copper complexes as catalysts has been found to be similar to that of the PMMA prepared by a conventional free-radical polymerization process.

11.3.4 ATRP Ligands

One main role of the metal-complexing ligands in ATRP is to solubilize the transition metal salt in the organic media. However, when bipyridine is used as a ligand for copper chloride, the solubility of the catalyst in the bulk monomer is limited and the medium remains heterogeneous. Substituted pyridines (Matyjaszewski et al., 1997; Percec et al., 1996), phenanthrolines (Destarac et al., 1997), pyrimines (Haddleton et al., 1997), or multidentate amine ligands (Xia and Matyjaszewski, 1997) must be used in order to get a homogeneous reaction medium. [It may be noted that, when using unsubstituted bipyridine, a homogeneous reaction medium can still be obtained by adding 10 vol% DMF (Cassenbras et al., 1999).] The other main role of ligands is to adjust the properties of the metal center toward the required reactivity and atom transfer dynamics. As the rate constants of activation (k_{act}) and deactivation (k_{deact}) are highly dependent on ligand choice, substantial effort has been made to understand the relationship between ligand structure and catalyst activity in order to design high activity catalysts (Tang and Matyjaszewski, 2006; Tang et al., 2008). It should be noted that ligands may be even more important than metal centers, since they can fine-tune atom transfer equilibium constants and dynamics as well as selectivities.

Nitrogen ligands work particularly well for copper-mediated ATRP, while, in contrast, sulfur, oxygen, or phosphorus ligands are less effective due to inappropriate electronic effects or unfavorable binding constants. Activity of N-based ligands in ATRP decreases with the number of coordinating sites: N4 > N3 > N2 ≫ N1 and with the number of linking C-atoms: C2 > C3 ≫ C4. Also, activity is usually higher for bridged and cyclic systems than for linear analogs (Matyjaszewski and Xia, 2001). Tris[2-(dimethylamino)ethyl]amine (Me$_6$TREN) and tris(2-pyridylmethyl)amine (TPMA) are examples of ligands that form high-activity ATRP complexes with ATRP equilibrium constants several orders of magnitude higher than found for other ligands. Examples of N-based ligands used successfully in Cu-based ATRP are shown in Fig. 11.17.

Phosphorus-based ligands are used to complex most of the transition metals employed in ATRP, including rhenium, ruthenium, iron, rhodium, nickel, and palladium, but not copper. The most frequently used phosphorus ligand is, however, PPh$_3$, which has been used successfully with all the aforementioned transition metals.

11.3.5 ATRP Solvents

ATRP can be performed in bulk, in solution, or in a heterogeneous system (e.g., emulsion, suspension). Various solvents have been used for ATRP in solution for various monomers. These include benzene, toluene, acetone, ethyl acetate, anisole, diphenyl ether, dimethyl formamide, ethylene carbonate, alcohol, water, carbon dioxide, and many others (Matyjaszewski and Xia, 2001). The use of a solvent may become necessary when the polymer formed is insoluble in its monomer. The choice of a solvent is influenced by several factors, such as chain transfer to solvent, interactions between solvent and the catalyst, catalyst poisoning by solvent (e.g., carboxylic acids and phosphine for Cu-based catalyst), and solvent aided side reactions, all of which should be minimum or absent.

The possibility of solvents changing the structure of the catalyst and its ATRP properties should also be taken into consideration (Matyjaszewski and Xia, 2001). A notable example is the ATRP of n-butyl acrylate with $CuBr(bpy)_3$ as the catalyst, which exhibits a much faster rate when the reaction is carried out in ethylene carbonate than in bulk (Matyjaszewski et al., 1998). The higher rate was attributed to a structural change from a dimeric halogen-bridged Cu(I) species in the bulk system to a monomeric Cu(I) species in ethylene carbonate solvent. A similar rate increase in polar media has also been observed by other workers (Wang, et al., 1999; Wang and Armes, 2000; Perrier et al., 2001). Polar media can facilitate dissolution of catalyst to enhance the rate of poly-merization. For example, ATRP can be performed with $CuBr(bpy)_3$ catalyst under homogeneous conditions using DMF solvent (10% v/v).

11.3.6   ATRP Mechanism and Kinetics

The general mechanism of ATRP has been shown in Fig. 11.16, which indicates a radical pathway. The radical nature of the reactive or propagating species in copper-mediated ATRP is proposed on the basis of several experimental observations, which have been summarized by Matyjaszewski and Xia (2001) as :   (1) The ATRP equilibrium can be approached from both sides, i.e., either from $(R–X + Mt^{n+})$ side or (radicals $+ X–Mr^{n+1})$ side (see Fig. 11.16), the latter being termed "reverse ATRP" (see Problem 11.9).   (2) Chemoselectivity for ATRP is similar to that for conven-tional radical polymerization. Thus, ATRP is retarded by the presence of a small amount of oxygen and inhibited by typical radical inhibitors (e.g., TEMPO).   Conventional radical polymerization characteristics are also displayed upon addition of a chain transfer agent, such as octanethiol.   (3) Regioselectivity, stereoselectivity, and tacticity of ATRP polymers are similar to those of polymers made by conventional free-radical process,

In ATRP, while polymer chains propagate by adding new monomer units to the growing chain ends and irreversible termination occurs through combination or disproportionation pathways as in a conventional free-radical process, the termination leads to generation of higher oxidation state stable metal complex, $X–Mt^{n+1}$ (that can be regarded as a persistent metalloradical) and after a sufficient amount of this complex has been built up, the *Persistent radical effect* predominates to reduce the stationary concentration of growing radicals and thereby minimize irreversible radical termination. Besides this small contribution of chain termination, a successful ATRP to obtain well-defined polymers also requires a uniform growth of all chains, which is accomplished through fast initiation and rapid reversible deactivation. However, at high conversions when the monomer concentration becomes very low, propagation slows down but termination may still occur with the usual rate. There is thus a certain 'window of concentrations and conversions' (Matyjaszewski and Xia, 2001) where the polymerization is well controlled to yield low polydispersity polymers.

In ATRP, perhaps the most important kinetic parameters are the rate constants for the activa-tion (k_{act}) and deactivation (k_{deact}) steps (see Fig. 11.16), which determine the magnitude of the equilibrium constant ($K_{eq} = k_{act}/k_{deact}$). In the absence of any side reactions other than radical ter-mination by coupling or disproportionation, K_{eq} determines the polymerization rate. While ATRP does not occur or occurs very slowly if K_{eq} is too small, too large an equilibrium constant, as it has been shown earlier, may actually lead to an apparently slower polymerization. The magnitudes of k_{act} and k_{deact} depend on the structure of the monomer, on the halogens, and on the transition metal complexes. The measured values of these rate constants for some model systems resembling the structure of the dormant/active species, are shown in Tables 11.3 and 11.4.

Table 11.3 Activation Rate Constants(k_{act}) Measured for some Model Systems

RX	Complex[a]	Solvent	k_{act} [M^{-1} s^{-1}]
Benzyl bromide	Cu(I)Br/(dNbpy)$_2$	Acetonitrile	0.043
Phenyl ethyl bromide	Cu(I)Br/(dNbpy)$_2$	Acetonitrile	0.085
Phenyl ethyl bromide	Cu(I)Br/PMDETA	Acetonitrile	0.12
Phenyl ethyl bromide	Cu(I)Br/(dNbpy)$_2$	Ethyl acetate	0.016
Phenyl ethyl chloride	Cu(I)Cl/Me$_6$TREN	Acetonitrile	1.5
Phenyl ethyl chloride	Cu(I)Cl/(dNbpy)$_2$	Acetonitrile	0.000056
Methyl bromopropionate	Cu(I)Br/(dNbpy)$_2$	Acetonitrile	0.052
Methyl bromopropionate	Cu(I)Br/PMDETA	Acetonitrile	0.11

[a] For ligand names and structures see Fig. 11.17.
Source: Matyjaszewski and Xia, 2001; Matyjaszewski et al., 2001.

Table 11.4 Deactivation Rate Constants (k_{deact}) Measured for some Model Systems at 75°C

Radical	Complex[a]	Solvent	k_{deact} [M^{-1} s^{-1}]
Phenyl ethyl	Cu(II)Br$_2$/(dNbpy)$_2$	Acetonitrile	2.5×10^7
Phenyl ethyl	Cu(II)Br$_2$/(dNbpy)$_2$	Ethyl acetate	2.4×10^8
Phenyl ethyl	Cu(II)Br$_2$/PMDETA	Acetonitrile	6.1×10^6
Phenyl ethyl	Cu(II)Br$_2$/Me$_6$TREN	Acetonitrile	1.4×10^7
Phenyl ethyl	Cu(II)Cl$_2$/(dNbpy)$_2$	Acetonitrile	4.3×10^6

[a] For ligand names and structures see Fig. 11.17.
Source: Matyjaszewski and Xia, 2001; Matyjaszewski et al., 2001.

Problem 11.9   While in a normal ATRP, the initiating radicals are generated from an alkyl halide in the presence of a transition metal in its lower oxidation state (e.g., CuBr(dNbpy)$_2$), ATRP can also be initiated by using a thermal free-radical initiator (e.g., AIBN) along with the transition metal compound in its higher oxidation state (e.g., CuBr$_2$(dNbpy)$_2$). Write a general scheme for this latter approach which is named 'reverse ATRP'.

Answer:

The reverse ATRP using AIBN as the initiator has been performed successfully for copper-based heterogeneous (Xia and Matyjaszewski, 1999) and homogeneous (Xia and Matyjaszewski, 1997) systems in solution and in emulsion as well as for iron complexes (Matyjaszewski and Xia, 2001). A general outline of reverse ATRP is shown in Scheme P11.9.1. As shown in this Scheme, the starting materials in reverse ATRP are a thermal free radical initiator (I–I), transition metal halide in the oxidized state (XMt^{n+1}), and monomer (M), while the propagation step resembles a normal ATRP. It may be noted, however, that the reverse ATRP initiated by peroxides sometimes behaves quite differently than that initiated by azo compounds like AIBN. The differences between the benzoyl peroxide (BPO) and AIBN systems possibly arise due to an electron transfer and the formation of a copper benzoate species in the BPO system (Xia and Matyjaszewski, 1999).

For homogeneous ATRP, the reaction mechanism shown in Fig. 11.18 consists of the atom transfer equilibrium followed by the addition of radicals to olefinic monomers at both initiation and propagation steps and finally by the termination step. Assuming a fast initiation, negligible

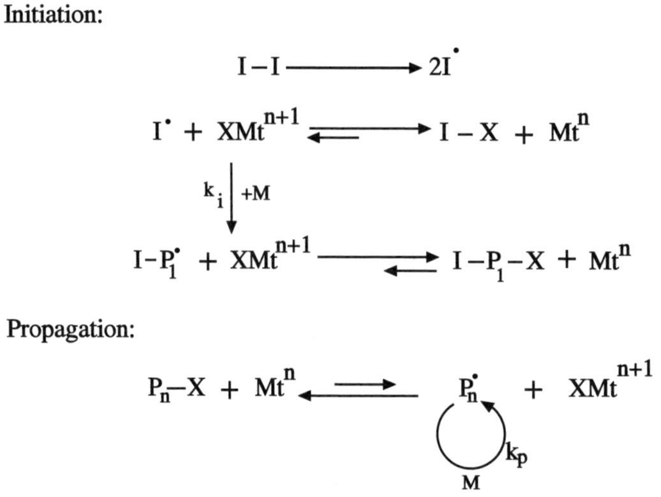

Initiation:

$$I-I \longrightarrow 2I^{\bullet}$$

$$I^{\bullet} + XMt^{n+1} \rightleftharpoons I-X + Mt^n$$

$$k_i \downarrow +M$$

$$I-P_1^{\bullet} + XMt^{n+1} \rightleftharpoons I-P_1-X + Mt^n$$

Propagation:

$$P_n-X + Mt^n \rightleftharpoons P_n^{\bullet} + XMt^{n+1}$$

$$\circlearrowleft k_p$$

$$M$$

Scheme P11.9.1 General scheme of reverse ATRP using AIBN (represented by I–I) thermal initiator (Problem 11.9).

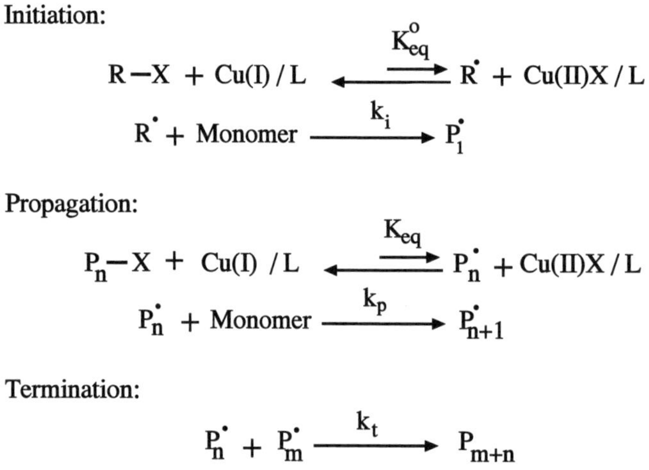

Initiation:

$$R-X + Cu(I)/L \underset{}{\overset{K_{eq}^o}{\rightleftharpoons}} R^{\bullet} + Cu(II)X/L$$

$$R^{\bullet} + Monomer \xrightarrow{k_i} P_1^{\bullet}$$

Propagation:

$$P_n-X + Cu(I)/L \underset{k_p}{\overset{K_{eq}}{\rightleftharpoons}} P_n^{\bullet} + Cu(II)X/L$$

$$P_n^{\bullet} + Monomer \xrightarrow{k_p} P_{n+1}^{\bullet}$$

Termination:

$$P_n^{\bullet} + P_m^{\bullet} \xrightarrow{k_t} P_{m+n}$$

Figure 11.18 Reaction mechanism of homogeneous ATRP.

termination reactions, and a steady concentration of propagating radicals, the rate of polymerization may be written as

$$R_p = k_p[P^{\bullet}][M] \tag{11.18}$$

where k_p is the propagation rate constant, $[P^{\bullet}]$ is the propagating chain concentration, and $[M]$ is the monomer concentration. A fast equilibrium is a necessary condition to obtain low polydispersity in controlled/"living" free radical polymerization. The activation and deactivation rate constants, k_{act} and k_{deact}, combine to form the ATRP equilibrium constant, K_{eq}. For the equilibrium, one can thus write (for fast initiation),

$$K_{eq} = \frac{k_{act}}{k_{deact}} = \frac{[P^{\bullet}][Cu(II)X]}{[Cu(I)][P-X]}$$

$$= \frac{[P^{\bullet}][Cu(II)X]}{[Cu(I)][R-X]} \tag{11.19}$$

where [R–X] is the alkyl halide initiator concentration; [Cu(I)] and [Cu(II)X] are the concentrations of the activator and deactivator species.

Substituting for [P$^{\bullet}$] from Eq. (11.19) into Eq. (11.18) gives

$$R_p = k_p K_{eq}[M][R-X]\frac{[Cu(I)]}{[Cu(II)X]} \tag{11.20}$$

Results from kinetic studies of ATRP under homogeneous conditions for styrene (Matyjaszewski et al., 1997), methyl acrylate (Davis et al., 1999), and methyl methacrylate (Wang et al., 1997) indicate that the polymerization rate is first order with respect to monomer, initiator, and Cu(I) complex concentrations. These observations are in agreement with Eq. (11.20).

Problem 11.10   What would be the effects of temperature increase on the ATRP process ?

Answer:

The rate of polymerization in ATRP will increase with increasing temperature due to the increase of both k_p and K_{eq} (see Eq. (11.20)). Moreover, as a result of the higher activation energy for radical propagation than for radical termination (see Section 6.11.1), higher k_p/k_t ratios and better control ("livingness") may be observed at higher temperatures. However, chain transfer and other side reactions can also become more pronounced at elevated temperatures, thereby reducing the control of polymerization. On the other hand, while a temperature rise will increase the solubility of the catalyst, thereby enhancing the polymerization rate, catalyst decomposition may also occur with the temperature increase.

Equation (11.20) indicates that the rate of polymerization in ATRP is governed by the ratio of activator to deactivator, and not their absolute concentrations. As such, it is theoretically possible to mediate ATRP using a small amount of catalyst with fast activation and deactivation kinetics, However, even if a highly active catalyst is used with ATRP equilibrium constant several orders of magnitude higher than found for other catalysts, it is not possible to reduce the catalyst concentration by the same scale, as irreversible radical termination then leads to accumulation of Cu(II) deactivator as a persistent radical due to which the reaction slows down and eventually stops. To address the aforesaid problem, Matyjaszewski et al. (2006) developed two concepts called *activators regenerated by electron transfer* (ARGET) and *initiators for continuous activator regeneration* (ICAR), through which the catalyst concentration can be reduced to ppm levels with respect to monomer. In both techniques, more oxidatively stable and easier to handle Cu(II) species are used as a starting material, along with an excess of an additional reagent that reduces the accumulated Cu(II) deactivator to generate the Cu(I) activating species, as shown in Fig. 11.19. A wide range of reducing agents, such as stannous 2-ethylhexanoate, ascorbic acid, glucose, hydrazine, phenols, and tertiary amines have been used successfully for an ARGET-ATRP process,

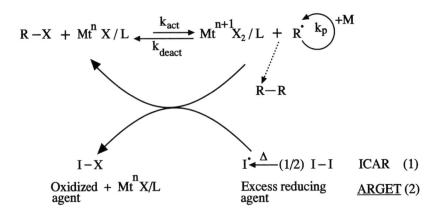

Figure 11.19 Mechanism proposed by Matyjaszewski et al. (2006) for low-catalyst concentration ATRP processes using (1) ICAR and (2) ARGET.

and a variety of polymers with different structures and properties have been prepared (Pietrasik et al., 2006; Dong et al., 2007; Dong and Matyjaszewski, 2008; Pintaner and Matyjaszewski, 2008).

The concentration of catalyst relative to initiator can be significantly decreased when the reducing agent is present in excess relative to the catalyst.   However, the catalyst concentration cannot be decreased indefinitely in any ATRP system, because control over molecular weight distribution in Cu-based ATRP depends on the absolute concentration of the deactivator according to the following relationship (Matyjaszewski, 1995; Matyjaszewski et al., 2006):

$$\text{PDI} = \frac{M_w}{M_n} = 1 + \frac{1}{DP_n} + \left(\frac{[R{-}X]_0 \, k_p}{k_{\text{deact}} \, [\text{Cu(II)X/L}]}\right)\left(\frac{2}{\text{Conv.}} - 1\right) \tag{11.21}$$

where PDI is the polydispersity index; DP_n is the degree of polymerization; subscript 'o' indicates initial state; and 'Conv.' is conversion expressed as a dimensionless fraction. Experimentally, controlled polymer synthesis of acrylates (PDI < 1.2) was realized with catalyst concentrations as low as 50 ppm (Matyjaszewski et al., 2006). The choice of the reducing agent affects the Cu(I)/Cu(II) ratio and hence the polymerization rate (Eq. (11.20)).

A principal disadvantage of the ARGET procedure is the requirement of high ligand concentrations (typically, three to ten times molar excess to metal) to maintain the catalyst complex and protect it from destabilizing side reactions, such as complexation to monomer or to Lewis acid generated from the reduction mechanism. The polymerization is also heavily dependent on ligand choice, since ligand will affect catalyst stability and deactivation kinetics. Both these factors present considerable barriers to large-scale application of ARGET-ATRP, especially since ligands commonly used for ARGET, such as Me$_6$TREN and TPMA (see Fig. 11.17), are expensive and not easily available.

In ICAR-ATRP, a small amount of thermal free radical initiator is used instead of a reducing agent to regenerate activators from accumulated deactivating species, Cu(II). The catalyst is more stable during ICAR and since, unlike in ARGET, side reactions caused by reducing agents do not occur, ligands can be added in a stoichiometric ratio to metal, thus avoiding the aforesaid

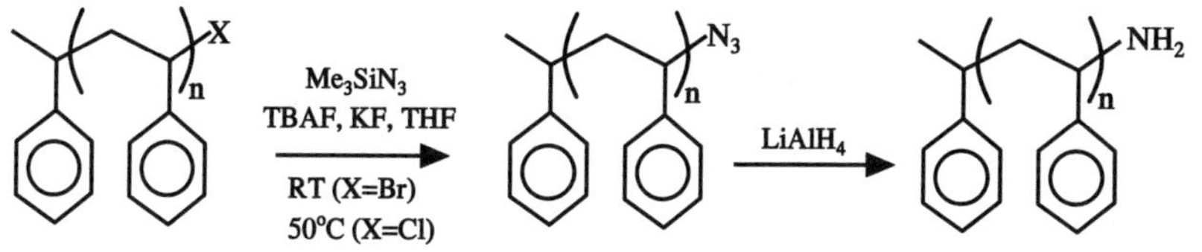

Figure 11.20 Synthesis of amine end-functional polystyrene by displacement of terminal halogen (X) in ATRP polystyrene. (After Matyjaszewski, Nakagawa, and Gaynor, 1997.)

difficulties of excess ligand. A disadvantage of ICAR, however, is that the use of a free radical initiator generates additional chains during the course of the reaction, resulting in broader MWD and formation of unwanted homopolymer when synthesizing block copolymers.

11.3.7 Chain-End Functionality

Besides the obvious aspects of ATRP, namely, control of molecular weight and narrow polydispersity, an important and exciting feature of this types of polymerization is that nearly every chain contains a terminal halogen atom, if termination and transfer are essentially absent. This halogen atom can be successfully replaced through a variety of reactions leading to end-functional polymers. The presence of halogens in the environment being often a cause of concern, a common method of dehalogenation of organic compounds, namely, the reaction with trialkyltin hydrides, $(C_4H_9)_3SnH$, could be applied to ATRP polymers for the removal of terminal halogen species (Coessens and Matyjaszewski, 1999). Thus by the addition of $(C_4H_9)_3SnH$ to the polymeric alkyl halide in the presence of a radical source (such as AIBN, polymer radical, or Cu(I) complex), a saturated hydrogen-terminated polymer is obtained. Similarly, using allyl tri-*n*-butylstannate in place of $(C_4H_9)_3SnH$, allyl-ended polymers are produced (Coessens et al., 2000). Again, starting from halogen-terminated ATRP polystyrenes, nucleophilic substitution of the halogen with azides by reaction with trimethylsilyl azide yields the azide-terminated polymer, which can be followed by reduction with lithium aluminum hydride to produce polystyrene with primary amine-functionalized chain end (Matyjaszewski, 1997) (Fig. 11.20). Halo-terminated polyacrylates have also been transformed similarly to azide- and amine-terminated polymers (Coessens et al., 1998).

The concept of end-functionalization through the addition of a nonpolymerizable monomer resembling telomerization experiments has been applied to ATRP to obtain end-functionalization with different groups. For example, the addition of an excess of allyl alcohol near the end of an acrylate polymerization results in the monoaddition of less reactive monomer. The new alkyl halide chain end does not undergo the ATRP process since the carbon-halogen bond has very low reactivity (Matyjaszewski and Xia, 2001; Coessens and Matyjaszewski, 1999). Similar approach has also been applied to incorporate 1,2-epoxy-5-hexene (Coessens et al., 2000) and maleic anhydride (Koulouri et al., 1999). Taking into account that maleic anhydride can hardly be homopolymerized under normal conditions, ATRP in the presence of an excess of maleic anhydride is performed so as to yield one molecule residing at the chain end. These examples are depicted in Fig. 11.21.

One of the major advantages of living free radical chemistry, when compared to other living procedures for the polymerization of vinyl monomers, is the stability of the initiating, or propagating, centers. This has enabled the development of a wide variety of functionalized unimolecular initiators for the synthesis of well-defined linear polymers, block copolymers, and other complex macromolecular structures. ATRP of monomers with functionalized alkyl halide initiators leads

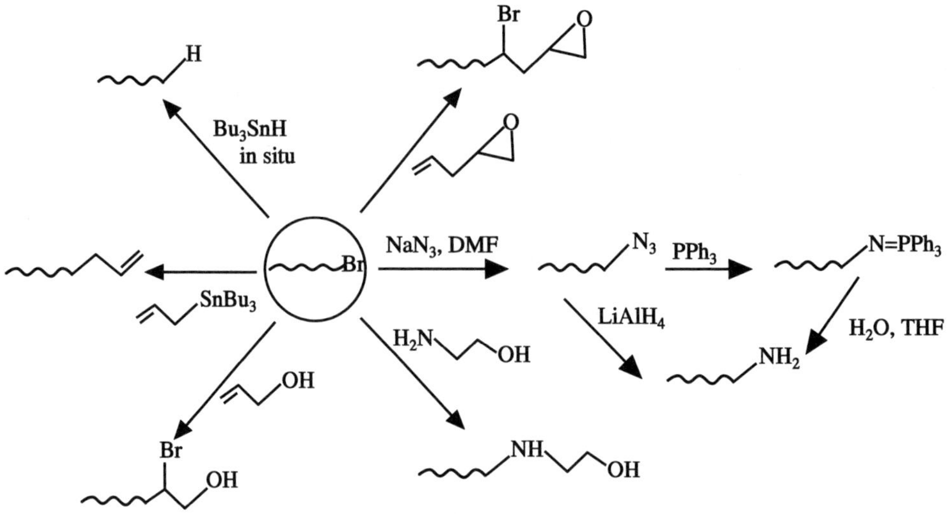

Figure 11.21 Some examples of displacemment of the terminal halogen in ATRP polymers using electrophilic, nucleophilic, and radical (addition) reactions. (Adapted from Matyjaszewski and Xia, 2001.)

to the formation of polymers or macromonomers in which the fragment that forms the α-end of the polymer chain contain the functional groups of the initiators (see Problem 11.11). Polymerizations with initiators containing carboxylic acids are, however, difficult because the acid functionality may poison the catalyst. [**Note**: *Macromonomers* are polymer chains containing a double bond or other polymerizable group at a chain end which can be (co)polymerized in a separate reaction to yield copolymers.]

Problem 11.11 Write the structures of chain end functionalized polymers obtained by ATRP of styrene and methyl acrylate from the following functionalized initiators (I):

For styrene (St): (1) 4-bromobenzyl bromide; (2) 2-bromopropionitrile; (3) vinyl chloroacetate; and (4) glycidol 2-bromopropionate

For methyl acrylate (MA): (1) allyl bromide; (2) hydroxyethyl 2-bromopropionate; (3) α-bromo-γ-butyro-lactone; and (4) 2-bromopropionitrile.

Answer:

In activated alkyl halides, used as initiators for ATRP, the radical stabilizing group (aryl, carbonyl, nitrile, multiple halogens) resides on the α-carbon atom. Initiation is accomplished through homolytic cleavage of the halide and addition of the generated radical at the α-C to alkenes. The functional group of the initiator remains at one end of the polymer chain and the halogen atom at the other end. It may be noted that direct bonding of the halogen to an aryl (e.g., chlorobenzene) or carbonyl group does not facilitate radical generation. ATRP can thus be carried out in chlorobenzene solvent.

The structure of ATRP polystyrene resulting from the cited functionalized initiators are shown in Fig. 11.22 and those of ATRP poly(methyl acrylate) that can be obtained from the cited functionalized initiators are shown in Fig. 11.23. These cases have been reported in the literature (Matyjaszewski et al., 1998).

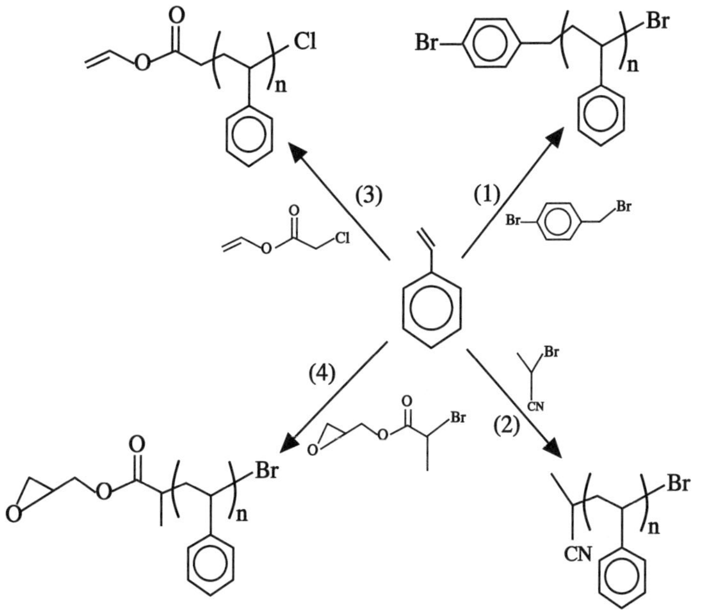

Figure 11.22 ATRP of styrene (St) starting with various functionalized initiators(I). Conditions of polymerization (reported in literature): bulk, $110°C$, $[St]_0 / [I]_0 = 96$, $[I]_0 / [CuBr]_0 / [dNbpy]_0 = 1 / 1 / 2$, time = 3.0 h. (Adapted from Matyjaszewski, Coessens, Nakagawa, Xia, Qui, Gaynor, Coca, and Jasieczek, 1998.)

11.3.8 Copolymerization

11.3.8.1 Block Copolymers

Since the polymer chains obtained by alkyl halide initiated ATRP contain an active halogen end group, they can be used as macroinitiators. Thus a macroinitiator can undergo further extension, either with the monomer used in its synthesis or with a second monomer forming a block copolymer. The growth of subsequent blocks can be achieved from an isolated macroinitiator or by addition of a second monomer *in situ* to an ATRP near completion.

The preparation of hybrid block copolymers, that is, copolymers from monomers that polymerize by different mechanisms has been a challenge facing polymer chemistry for many years. Traditionally, block copolymers are prepared by coupling preformed polymers, by sequential addition of monomers to a living initiator, or by using a mixture of monomers that have substantially different rates of polymerization (Weimer et al., 1998). Where two different polymerization mechanisms are involved, the living end of the first block has to be transformed into an appropriate reactive group capable of initiating polymerization of the second monomer. Much research has been published on the preparation of block copolymers with well-defined molecular structures by mechanism transformation process such as transformation of the ring-opening polymerization (ROP) of one monomer to stable free-radical polymerization (SFRP) (Yoshida and Sugita, 1996, 1998; Yagci et al., 1997), ATRP (Kajiwara and Matyjaszewski, 1998; Xu and Pan, 2000; Guo and Pan, 2001; Guo et al., 2001), or reverse ATRP (Cianga et al., 2002) of another monomer (typically, styrene). It may be recalled that reverse ATRP (see Problem 11.9) is initiated by traditional

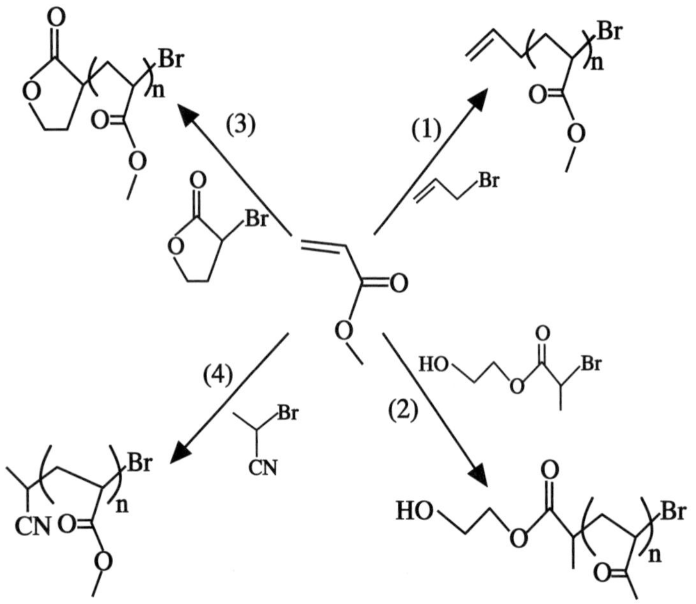

Figure 11.23 ATRP of methyl acrylate (MA) starting with various functionalized initiators (I). Conditions of polymerization (reported in literature): bulk, $110°C$, $[MA]_0 / [I]_0 = 58$, $[I]_0 / [CuBr]_0 / [dNbpy]_0 = 1 / 0.3 / 0.6$, time $= 1.7$ h. (Adapted from Matyjaszewski, Coessens, Nakagawa, Xia, Qui, Gaynor, Coca, and Jasieczek, 1998.)

initiators such as AIBN, with the transition metal compound in its higher oxidation state (e.g. $CuBr_2/L_2$), resulting in polymers with narrow polydispersities.

Another strategy for the preparation of block copolymers by the combination of controlled polymerization techniques consists of the introduction of a *dual initiator* (Bernaerts et al., 2003). Such a dual initiator contains two initiation sites capable of initiating two different polymerization reactions, independently and selectively, for combinations such as ATRP-SFRP (Tunca et al., 2001, 2002), ROP-ATRP (Mecerreyes et al., 1998, 1999; Meyer et al., 2002), and SFRP-ROP (Weimer et al., 1998; Puts and Sogah, 1997; Hawker et al., 1998). Another advantage of dual initiators, in comparison with other methods, is that one or both end groups of the resulting block copolymers can be used for a transformation process to produce ABC triblock (Tao et al., 2003), ABCD tetrablock, and higher block copolymers. Inorganic/organic hybrid ABC triblock copolymers have been synthesized by combining living anionic ROP with ATRP (Miller and Matyjaszewski, 1999), in addition to ABC triblock copolymers synthesized wholly by ATRP (Davis and Matyjaszewski, 2000).

The key to successful synthesis of block copolymers by ATRP is to maintain high chain end functionality, i.e., to limit termination and side reactions, and to balance the reactivity of the end group with that of the monomer, i.e., avoid slow initiation (Davis and Matyjaszewski, 2001). While the latter requirement can be overcome through a careful choice of the block order, radical termination cannot be completely avoided since free radicals are involved in propagation, but can be limited through careful choice of the polymerization conditions and through adjustment of the equilibrium between the active and dormant species, often by adding a "persistent radical" in the form of a higher oxidation state metal (e.g., Cu(II) to Cu(I)) (Davis and Matyjaszewski, 2000).

To develop a general one-pot, one-step method for the synthesis of block copolymers by simultaneous initiation of two or more polymerization mechanisms, without the need for intermediate steps, it is required, among other things, that under similar reaction conditions different monomers polymerize by different mechanisms without crossing over or interfering with each other. In order to eliminate the need of intermediate steps or chemical transformations in between polymerizations by different mechanisms, the concept of dual or double-headed initiator (Hawker et al., 1998) to perform dual living polymerizations from a single initiating molecule with no intermediate activation or transformation was proposed. The compatibility of "living" or controlled free radical procedures (either nitroxide mediated or ATRP) with the living ROP of cyclic esters and vice versa, was demonstrated by the synthesis of a variety of well-defined block copolymers. For example, from a hydroxy-functionalized alkoxyamine, either the living ROP of ε-caprolactone, or the "living" free radical polymerization of styrene can be performed, giving low polydispersity polymeric initiators. These polymeric initiators can then be used to initiate the living polymerization of other monomer systems without the need for modification via intermediate steps. In a similar manner, hydroxy-functionalized ATRP initiators have been used as bifunctional initiators for the polymerization of both ε-caprolactone and a variety of other vinyl monomers (Hawker et al., 1998), leading to novel block copolymers having low polydispersities and controllable molecular weights for all the blocks.

A number of worked examples of block copolymers involving ATRP method of synthesis are given below to provide an overall flavor for the variety of structures that can be produced by combination of ATRP and other methods of polymerization.

Problem 11.12   It was found that while poly(methyl methacrylate) macroinitiator made by ATRP of methyl methacrylate is able to initiate the ATRP of acrylic monomers, for ATRP polyacrylate macroinitiators to effectively initiate the ATRP of methyl methacrylate, the end group should be a bromine atom and the catalyst should be CuCl; that is, halogen exchange should take place (Shipp et al., 1998). Explain this behavior and, taking it into account, propose a strategy to synthesize AB diblock and ABA triblock copolymers, where A = poly(methyl methacrylate) and B = poly(methyl acrylate).

Answer:

Bromo-terminated poly(methyl acrylate) [P(MA)–Br] macroinitiator may first be prepared by ATRP of methyl acrylate (MA) using methyl-2-bromopropionate (MeBrP) as initiator and CuBr/bipyridine (bpy) as the catalyst. The ATRP of methyl methacrylate (MMA) may then be carried out with "halide exchange" (see later), using the obtained P(MA)–Br macroinitiator and a CuCl-based catalyst system, such as CuCl/4,4′-di(5-nonyl)-2,2′-bipyridyl (dNbpy). This was reported to give close to 100% monomer conversion in 16 h at 70°C (Shipp et al., 1998). The Br group initially provides fast initiation, but thereafter the end group of the growing polymer chain is predominantly the less labile Cl with the result that the rate of initiation is higher relative to the propagation rate, thereby improving initiation efficiency.

On the basis of the above approach, the following strategy (Shipp et al., 1998) may be used to synthesize ABA triblock copolymer. Thus P(MMA)–Br macroinitiator may first be synthesized by ATRP of MMA and used to initiate ATRP of MA. The resulting AB diblock copolymer, P(MMA)-*b*-P(MA)–Br, may then be used to initiate ATRP of MMA, with "halide exchange" as described above, forming the desired ABA triblock copolymer. However, obtaining P(MMA) with close to 100% functionality was found (Shipp et al., 1998) to be difficult, possibly because of side reactions that are more prevalent when Br is present as the end group. One way to avoid this difficulty is to first synthesize the central block, P(MA), using a bromide-type difunctional initiator, such as 1,2-bis(bromopropionyloxy) ethane, [CH$_2$OC(=O)CH(CH$_3$)Br]$_2$, and then add MMA by "halide exchange" ATRP to both ends of the resulting bromo-ended difunctional macroinitiator.

Problem 11.13   Suggest ATRP procedures for synthesis of ABC-type triblok copolymers where A is polystyrene, B is poly(*tert*-butyl acrylate), and C is poly(methyl acrylate) in one case (Case 1) and poly(methyl methacrylate) in another case (Case 2).

Answer:

The procedures for the two cases are similar with a slight difference due to high end group reactivity in the second case involving methyl methacrylate (MMA). The synthesis in both cases is carried out in three steps.

Case 1:

ABC triblock copolymer is formed typically through sequential monomer addition in three ATRP steps. **Step One**: polymerization of styrene (St) by mixing the components together in the following order – CuBr, St, PMDETA, and 1-phenyl ethyl bromide, typically at 100°C. The product is a monofunctional macroinitiator, namely, bromo-terminated polystyrene, P(St)–Br, which is isolated from the reaction products mixture and used in **Step Two** for the ATRP of *tert*-butyl acrylate (*t*BA) at 80°C, the order of addition of the components to the reaction mixture being P(St)–Br, CuBr, *t*BA, and PMDETA. The product of step 2 is a monofunctional, bromo-terminated macroinitiator, P(St)-*b*-P(*t*BA)–Br, which is isolated from the reaction products mixture and used in **Step Three** for the synthesis of ABC triblock copolymer by ATRP of methyl acrylate (MA) at 70°C, the order of addition of the components to the reaction mixture being CuBr, P(St)-*b*-P(*t*BA)–Br, MA, and PMDETA. The product of step 3 is the triblock copolymer P(St)-*b*-P(*t*BA)-*b*-P(MA)–Br.

Case 2:

The first two steps leading to the formation of a bromo-terminated monofunctional macroinitiator P(St)-*b*-P(*t*BA)–Br are the same as above. However, some change is needed in step 3 involving chain extension with MMA to obtain the triblock copolymer P(St)-*b*-P(*t*BA)-*b*-P(MMA)–Br. To achieve control of molecular weights and distributions, one needs to use "halide exchange" — a technique that involves the addition of CuCl during the chain extension of bromo-terminated macroinitiator wih MMA. This markedly improves the rate of initiation relative to propagation and gives controlled polymerization ("livingness"). Catalyst solution made by dissolving CuCl in acetone and HMTETA (Fig. 11.17) is added to a solution of P(St)-*b*-P(*t*BA)–Br in MMA and reaction is performed at 70°C. [*Note:* The HMTETA ligand is used as the complexing agent for Cu(I) species as it results in a homogeneous solution in MMA and avoids heterogeneity associated with the corresponding PMDETA complex in MMA. Another important note is that a 1:1 molar equivalent of the complex is used relative to the concentration of *end groups* and not relative to the concentration of initiator.]

Problem 11.14   Explain the mechanism of "halide exchange" in the context of ATRP of methyl methacrylate.

Answer:

The mechanism of "halide exchange" can be described by the illustration (M = methyl methacrylate):

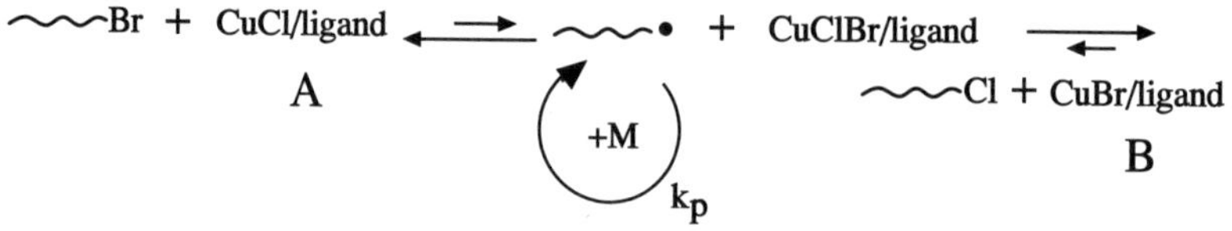

The scheme means that if a bromo-ended (macro)initiator is used in conjunction with a CuCl/ligand catalyst, the formed radical will preferentially abstract the Cl from the deactivator, forming Cl-terminated polymer chains and a CuBr/ligand catalyst. Thus, once cross-propagation occurs and methacrylate chain ends are formed (which is rapid because the monomer is very reactive), the rate of activation of the methacrylate-terminated polymer chain having less labile Cl end group will be lower and, therefore, the equilibrium constant (k_{act}/k_{deact}) will be smaller. The "halide exchange" method would thus enable the use of alkyl halides

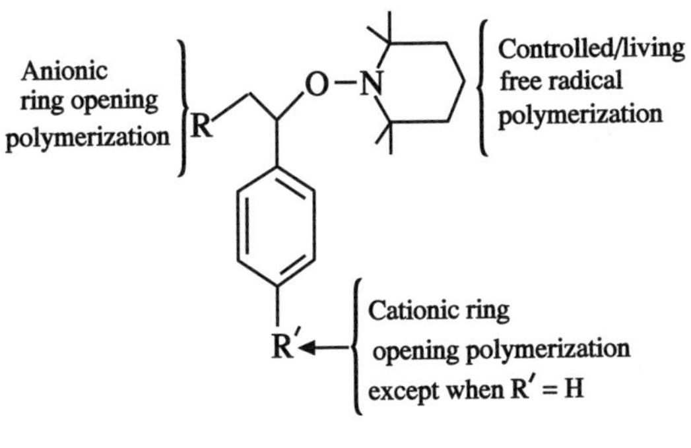

I : R = tBuMe₂SiO, R′ = CH₂CH₂OXA (OXA= oxazoline)
      (TDMS)

II : R = TDMS, R′ = CH₂Cl

III : R = OH, R′ = CH₂Cl

IV : R = C₆H₅COO, R′ = CH₂Cl

V : R = OH, R′ = H

Figure 11.24 Multifunctional initiators possessing initiating sites for more than one type of polymerization. Use of some of these initiators is demonstrated in several worked examples given subsequently. (Adapted from Weimer, Scherman, and Sogah, 1998.)

of apparently lower reactivities in the polymerization of monomers with apparently higher equilibrium constants. (It may be noted that if linear amine ligands are used to complex Cu(I) catalysts, reaction **A** does not participate significantly (~1%). In contrast, if Cu(I)/bpy systems are used, the radical may react with either Br or Cl, resulting in a mixture of halogen chain ends as well as Cu(I) species, so that both **A** (~10%) and **B** (~90%) participate in the activation/deactivation cycle.)

Weimer et al. (1998) reported the synthesis of multifunctional initiators **I–V** (Fig. 11.24) all of which possess initiating sites for more than one type of polymerization. For example, **I** possesses initiating sites that can withstand different polymerization conditions and so can be independently and selectively utilized for living free radical, anionic, anionic ring-opening, and cationic ring-opening polymerizations, thus enabling successful preparation of block and graft copolymers by sequential polymerization of different types of monomers without the need for intermediate steps and chain end transformations, or by one-step, one-pot simultaneous copolymerization involving different polymerization mechanisms.

Problem 11.15 (a) Show how 2,2,2-tribromoethanol, which is a hydroxy-functionalized ATRP initiator, can be used as a dual functional initiator to prepare well-defined block copolymers by dual living free-radical and ring-opening polymerizations. (b) Suggest a method for preparing a triblock copolymer poly(caprolactone)·*b*-poly(methyl methacrylate)-*b*-poly(*n*butyl acrylate) using this dual functional initiator.

Answer:

(a) The compound 2,2,2-tribromoethanol (TBE), Br_3CCH_2OH, has two different types of functional groups (Br and OH) on a simple molecule. This *double-headed initiator* can be used to initiate ATRP of vinyl monomers and also as an initiator for the living ring-opening polymerization (ROP) of cyclic ethers, esters, and amides. Thus, under standard conditions, using CuBr/bpy as the metal complex, the homopolymerization of styrene with TBE (**P15-I**) is an ATRP process with the molecular weight of the product being controlled by the initiator to monomer ratio and polydispersity being low (Hawker et al., 1998). Similarly, homopolymerization of methyl methacrylate (MMA) using TBE and $NiBr_2(PPh_3)_2$ or $RhBr(PPh_3)_3$ as metal catalyst, is a living process leading to low polydispersity, controlled molecular weight materials with one polymer chain being initiated per TBE molecule (see Scheme P11.15.1), in agreement with the finding that polyhalogenated initiators, such as CCl_4, can be used in ATRP.

To demonstrate the use of TBE as an initiator for the living ROP of cyclic esters (e.g., ϵ-caprolactone, CL), aluminum tris(isopropoxide), $Al(OiPr)_3$, can be reacted with 4 equiv of TBE and then added to a solution of CL in toluene and polymerization conducted at room temperature for 1 h (Hawker et al., 1998) to obtain poly(caprolactone) (PCL) derivatives (**P15-III**) with molecular weights similar to theoretical molecular weights and low polydispersities. [Instead of $Al(OiPr)_3$, triethylaluminum (Et_3Al) can also be added at room temperature so as to obtain the activated alkoxide derivative to catalyze the ROP of CL.]

The above two processes can also be combined into a novel synthetic strategy for the synthesis of a wide variety of block copolymers from a readily available double-headed initiator using ATRP. Thus, as shown in Scheme P11.15.1, a short P(MMA) block can be grown from TBE using ATRP to obtain a monohydroxy-terminated P(MMA) derivative, **P15-II**. This, after reacting with Et_3Al at room temperature to generate the activated alkoxide derivative, can be used to initiate the living ROP of CL, leading to the diblock copolymer P(CL)-*b*-P(MMA)–Br (**P15-IV**). Alternatively, **P15-IV** may also be obtained from **P15-III** by ATRP using $NiBr_2(PPh_3)_2$ as metal catalyst.

(b) An added benefit of ATRP is the presence of a dormant initiating group (Br in the present case) at the chain end of a synthesized polymer block. Consequently, additional blocks can be grown by reactivation of the ATRP process. This allows the preparation of novel multiblock copolymers. Thus, using the diblock copolymer **P15-IV** as the macroinitiator, a third block of P(*n*butyl acrylate) can be added by polymerization of *n*-butyl acrylate (*n*BA) under ATRP conditions to obtain the desired triblock copolymer P(CL)-*b*-P(MMA)-*b*-P(*n*BA),

Problem 11.16   A dual initiator, 4-hydroxy-butyl-2-bromoisobutyrate (HBBIB) was synthesized (Bernaerts et al., 2003) by the reaction of 1,4-butanediol with 2-bromoisobutyric acid using *p*-toluene sulfonic acid as catalyst in toluene. Suggest a scheme for the synthesis of well-defined poly(tetrahydrofuran)-*b*-polystyrene block copolymer with controlled molecular weights and low polydispersities using HBBIB.

Answer:

The dual initiator HBBIB is a molecule containing two functional groups (–OH and –Br) capable of initiating two polymerizations occurring by different mechanisms. It can be used (Bernaerts et al., 2003) for the successful two-step synthesis of well-defined block copolymers of styrene (St) and tetrahydrofuran (THF) by ATRP and cationic ring-opening polymerization (CROP), as shown in Scheme P11.16.1. This dual initiator contains a bromoisobutyrate group, which is an efficient initiator for ATRP of styrene in combination with the Cu(0)/Cu(II)/PMDETA catalyst system. This produces polystyrene (PSt) with hydroxyl groups (PSt–OH). On the other hand, the hydroxyl group originating from the dual initiator can be reacted *in situ* with trifluoromethane sulfonic anhydride to produce a triflate ester, which can then be used as the initiating group for the CROP of THF, leading to poly(THF) with a tertiary bromide end group (PTHF–Br). The homopolymers, PSt–OH and PTHF—Br can be applied as macroinitiators for the CROP of THF and ATRP of St, respectively, leading to well-defined PTHF-*b*-PSt block copolymers having low polydispersities (Bernaerts et al., 2003). It may be noted that for PSt–OH to act as a macroinitiator for CROP, the hydroxyl function should stay intact

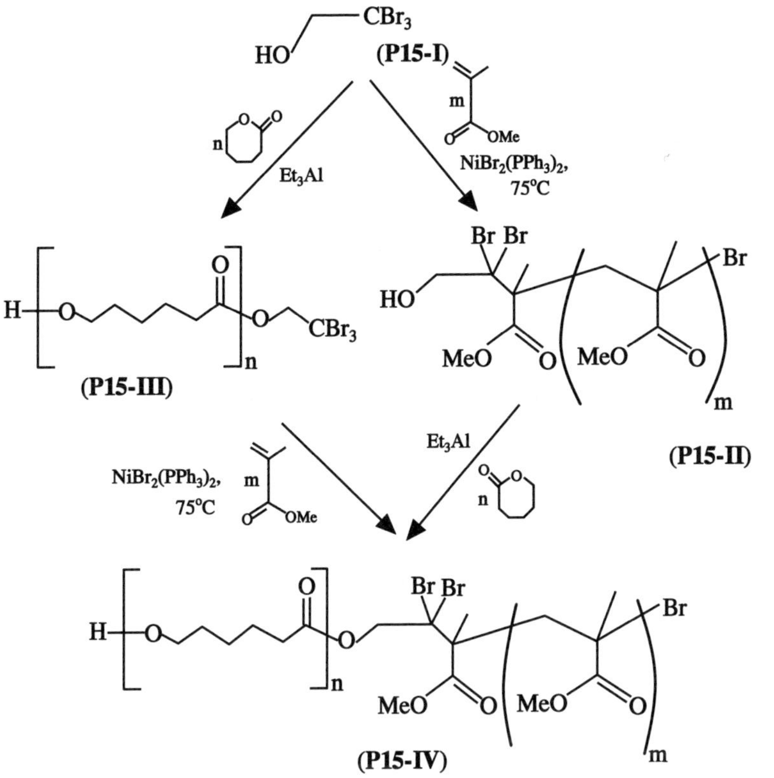

Scheme P11.15.1 Polymerizations by different mechanisms from a multifunctional initiator to obtain block copolymer (Problem 11.15). (From Hawker et al., 1998. With permission from *American Chemical Society.*)

during the ATRP. In fact, it has been demonstrated earlier (Mecerreyes et al., 1998) that the hydroxyl function is compatible with the ATRP of vinyl monomers.

Problem 11.17   Discuss how amphiphilic block copolymers (a) AB diblock poly(acrylic acid)-*b*-polystyrene and (b) ABA triblock poly(acrylic acid)-*b*-polystyrene-*b*-poly(acrylic acid), both asymmetric and symmetric, can be prepared by ATRP.

Answer:

Acrylic acid cannot be polymerized by ATRP which is sensitive to the presence of acids. Moreover, since many of the ligand systems utilized are nitrogen-based, protonation of the nitrogen may occur, disrupting its coordination to the metal center. The solution to this problem is to polymerize protected monomers, followed by a deprotection step to generate the polyacid. Thus Davis and Matyjaszewski (2000) synthesized poly(acrylic acid) (PAA) via hydrolysis of poly(*tert*-butyl acrylate) (P*t*BA), which, in turn, was obtained by ATRP of *t*BA using a CuBr/PMDETA catalyst system in conjunction with an alkyl bromide, such as methyl-2-bromopropionate (MBrP), as the initiator. The monomer conversion was 93% after 320 min at 60°C.

Monofunctional *t*BA Polymerization:

For a typical monofunctional *t*BA polymerization (Davis and Matyjaszewski, 2000) to form P(*t*BA) macroinitiator (with initial concentration ratios *t*BA : MBrP : CuBr : PMDETA : CuBr$_2$ = 50 : 1 : 0.5 : 0.525 : 0.025,

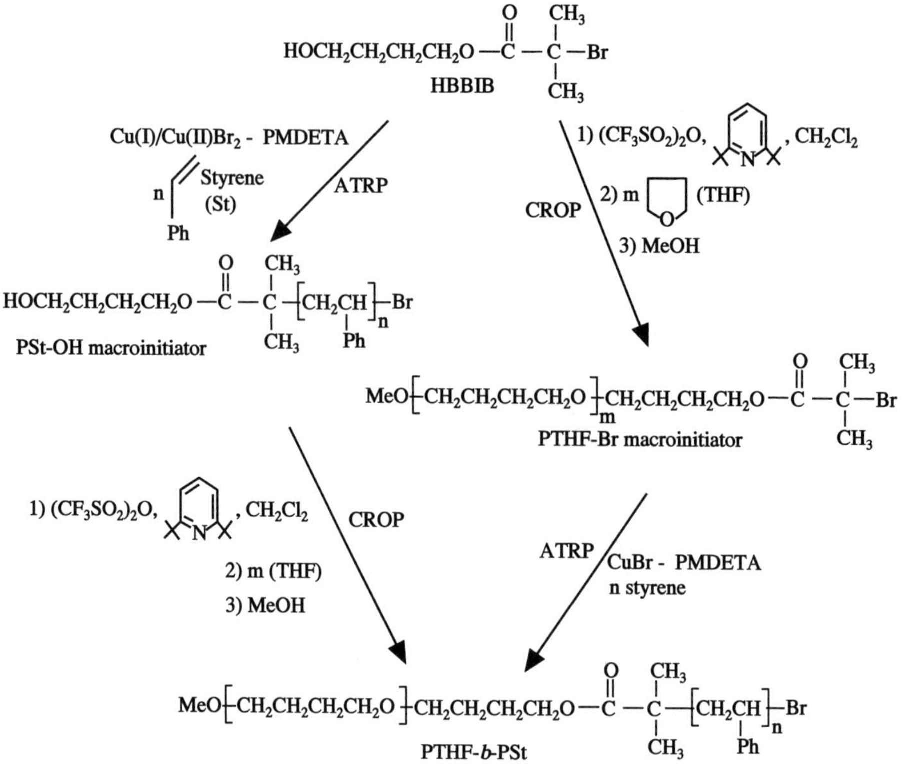
HOCH2CH2CH2CH2O—C(=O)—C(CH3)2—Br
HBBIB
Cu(I)/Cu(II)Br2 - PMDETA
n Styrene (St)
Ph
ATRP
1) (CF3SO2)2O, , CH2Cl2
2) m (THF)
3) MeOH
CROP
PSt-OH macroinitiator
PTHF-Br macroinitiator
CROP
ATRP
CuBr - PMDETA
n styrene
PTHF-b-PSt

Scheme P11.16.1 Synthesis of poly(THF)-*b*-poly(St) block copolymers with the dual initiator HBBIB (Problem 11.16). (From Bernaerts et al., 2003. With permission from *John Wiley & Sons, Inc.*)

25% acetone), an amount of deoxygenated acetone (according to chosen scale of reaction) was added to CuBr and $CuBr_2$ in a nitrogen-filled dry flask, followed by *t*BA and then PMDETA. [*Note:* The CuBr/PMDETA catalyst system being heterogeneous, the addition of acetone (solvent) was necessary to create a homogeneous catalyst system, while the addition of some Cu(II) species was necessary to establish control via the persistent radical effect (Fischer, 1997). The contents of the flask were stirred until the Cu complex had formed (visualized through a change of the solution from cloudy and colorless to clear and light green). After complex formation, the initiator MBrP was added and the flask was heated in a thermostated oil bath at 60°C. The monomer conversion was 96% after 6.5 h. The obtained polymer (i.e., macroinitiator), MP(tBA)$_x$Br (MP = methyl propionate), was recovered from the reaction mixture and purified in several successive steps (Davis and Matyjaszewski, 2000). It had M_n = 7300 and M_w/M_n = 1.11. In comparison, the theoretical M_n = MW of MBrP + (MW of *t*BA)(50/1)(0.96) = 6311.

Block Copolymerization

(a) *AB Diblock Copolymer*

Block copolymerization was performed (Davis et al., 2000) in the same manner as a typical polymerization (see above); however, the order of addition was modified, the macroinitiator and CuBr being added to the flask initially. In a typical procedure, the macroinitiator, MP(tBA)$_x$Br (M_n = 7300, 3.0 g, 4.4×10^{-4} mol) and CuBr (63 mg, 4.4×10^{-4} mol) were added to a dry, nitrogen-filled flask, followed by deoxygenated styrene (5.0 mL, 4.4×10^{-2} mol), which dissolved the macroinitiator. The ligand PMDETA (2 μL, 4.4×10^{-4}

mol) was then introduced and the solution was stirred until the Cu complex had formed (visually confirmed through color change). The flask was then placed in a thermostated oil bath at 100°C for 140 min. GC analysis showed that the monomer conversion was 94%. The diblock copolymer formed, namely, MP(tBA)$_x$-b-(St)$_y$Br, had $M_n = 18150$ and $M_w/M_n = 1.11$; in comparison, theoretical M_n = (MW of macroinitiator) + (MW of St)$(4.4 \times 10^{-2} / 4.4 \times 10^{-4})(0.94) = 7300 + 9776 = 17,076$.

To obtain the amphiphilic copolymer PAA-b-PSt, the diblock copolymer PtBA-b-PSt, synthesized as above, was dissolved in an equal weight of dioxane, and a three-fold excess of concentrated HCl (based on the moles of ester groups present) was added. The mixture was heated under reflux for approximately 4 h. The polymers were then precipitated into water and dried under vacuum (Davis et al., 2000).

(b) *ABA Triblock Copolymer*

An ABA-type triblock copolymer can be obtained by ATRP in two alternative ways, namely, by sequential addition or simultaneous addition of blocks, the former leading to *asymmetric* and the latter to *symmetric* triblock copolymers.

Asymmetric ABA triblock.   The bromo-terminated macroinitiator MP(tBA)$_x$-b-(St)$_y$Br obtained in (a) can be used for ATRP of tBA, which was performed (Davis and Matyjaszewski, 2000) using CuBr/PMDETA as the catalyst system (with the addition of a small amount of CuBr$_2$ for the persistent radical effect) and the initial concentration (mole) ratios of tBA : MP(tBA)$_x$-b-(St)$_y$Br : CuBr : PMDETA : CuBr$_2$ = 100 : 1 : 1 : 1 : 0.05. After 14.5 h at 60°C, the monomer (tBA) conversion was 57%. The triblock copolymer obtained can be represented as MP(tBA)$_x$-b-(St)$_y$-b-(tBA)$_z$Br ($x \neq z$), with theoretical M_n of (tBA)$_z$ = $(100/1)(0.57)$(MW of tBA) = 7296, as compared to theoretical M_n of (tBA)$_x$ = 6311 (calculated earlier).

Symmetric ABA triblock.   In this case, the central P(St) block is first made using a difunctional initiator such as dimethyl 2,6-dibromoheptanedioate (DMDBHD), CH$_2$[CH$_2$CH(Br)CO$_2$CH$_3$]$_2$, and the resulting difunctional Br–P(St)–Br is used as the macroinitiator for simultaneous ATRP of tBA from its two ends, using the CuBr/PMDETA catalyst system. For the synthesis of Br–P(St)–Br, Davis and Matyjaszewski (2000) used the following initial concentration (mole) ratios, St : DMDBHD : CuBr : PMDETA = 20 : 1 : 0.2 : 0.2. In the procedure used, CuBr and DMDBHD were added to a dry nitrogen-filled flask, followed by deoxygenated St and deoxygenated PMDETA. The contents of the flask were stirred to form the copper complex, while the solution changed from cloudy and colorless to dark green but was heterogeneous. The flask was then placed in a thermostated bath at 100°C. The monomer (St) conversion was 59% after 100 min. (The conversion was kept low to insure good chain-end functionality.) The difunctional polymer obtained had $M_n = 1100$ and $M_w/M_n = 1.17$. In comparison, the total theoretical M_n of the two polystyrene segments (represented in the following simply as PSt) = $104(20/1) \times 0.59 = 1227$ (excluding the molar mass of DMDBHD initiator residue between the segments).

To form the symmetric triblock copolymer P(tBA)-b-P(St)-b-P(tBA) from the difunctional macroinitiator Br–P(St)–Br obtained as above, Davis and Matyjaszewski (2000) used the concentration (mole) ratios of tBA : Br–P(St)–Br : CuBr : PMDETA = 70 : 1 : 0.5 : 0.5 and the order of addition was as follows: Br–P(St)–Br and CuBr were added to a dry, nitrogen-filled flask, followed by deoxygenated tBA along with one-third its volume of acetone (which dissolved the polymer). PMDETA was then added and the copper complex formed. The solution was initially heterogeneous, but after heating at 60°C for about 10 min, the solution became homogeneous and after 22.5 h the monomer conversion was 96%. Analysis by ^{1}H NMR showed that the copolymer had the composition 89% tBA, 11% P(St) — based on the ratio of the area for the aromatic protons of P(St) to the total area for the block copolymer — while, in comparison, the theoretical composition (for 100% monomer conversion) is $(70 \times 128) / [(70 \times 128) + 1227] = 88\%$ P(tBA) and 12% P(St).

The *tert*-butyl ester groups, –CO$_2$(C(CH$_3$)$_3$), in both asymmetric and symmetric triblock copolymers obtained above are easily hydrolyzed to –COOH by refluxing with HCl in dioxane solution to yield the corresponding amphiphilic triblock copolymers, containing an interior block of hydrophobic polystyrene and outer blocks of hydrophilic poly(acrylic acid).

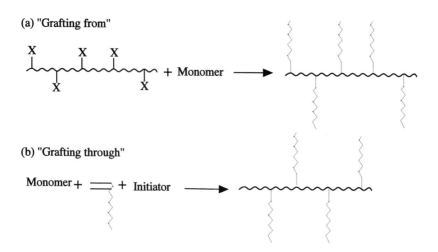

Figure 11.25 Schematic illustration of "grafting from" and "grafting through" processes. Each 'X' represents a chain initiating site. (Adapted from Matyjaszewski and Xia, 2001.)

11.3.8.2   Graft Copolymers

Graft copolymers can be synthesized through one of the three routes (Matyjaszewski and Xia, 2001): (i) "grafting from" processes in which grafts are generated from the pendant functionality of a polymer (macromonomer); (ii) "grafting through" processes which operate by homo- or co-polymerization of a macromonomer (see Fig. 11.25); and (iii) "grafting onto" processes which occur by attachment of growing chains to a polymer backbone. Among the three methods, the first two have been used in combination with ATRP in the design of a variety of graft copolymers.

By extension of the ATRP mechanism of the R–Cl/CuCl/ligand and R–Br/CuBr /ligand systems, various graft copolymers have been accomplished via the "grafting from" process using suitable brominated polymer initiators. An early example of the "grafting from" process using ATRP was the synthesis of graft copolymers of poly(vinyl chloride) (PVC) with styrene and (meth)acrylates (Paik et al., 1998). PVC containing 1% (mol) of pendant chloroacetate groups was used as a macroinitiator in ATRP of styrene, methyl acrylate (MA), n-butyl acrylate (nBA), and methyl methacrylate (MMA) (Fig. 11.26). The molecular weight of the copolymer in each case increased above that of the macroinitiator due to grafting of new chains, yet the polydispersity remained essentially the same (except in the case of MMA). There was a large increase in poly-dispersity for the MMA polymerization, which was ascribed to the slow ATRP initiation of MMA (Paik et al., 1998) from the primary alkyl halide sites. It may be noted that the grafting by ATRP occurs at the chloroacetate sites, while the PVC backbone remains intact because the secondary chlorine on the PVC backbone is too strongly bonded to initiate the polymerization by reaction with Cu(I) complex.

Commercial polymers with suitable pendant functional groups are of great interest as macroinitiators for the synthesis of graft copolymers with controlled architectures and properties. One such product is poly[isobutylene-co-(p-methylstyrene)-co-(p-bromomethylstyrene)] (PIB) (EXXPRO), which is prepared by carbocationic polymerization and can be used for a variety of processes and applications. Using ATRP and a grafting from process, grafts of styrene (Fónagy et al., 1998) and MMA (Hong et al., 2001) on EXXPRO were produced (Fig. 11.27). Thermoplastic elastomeric

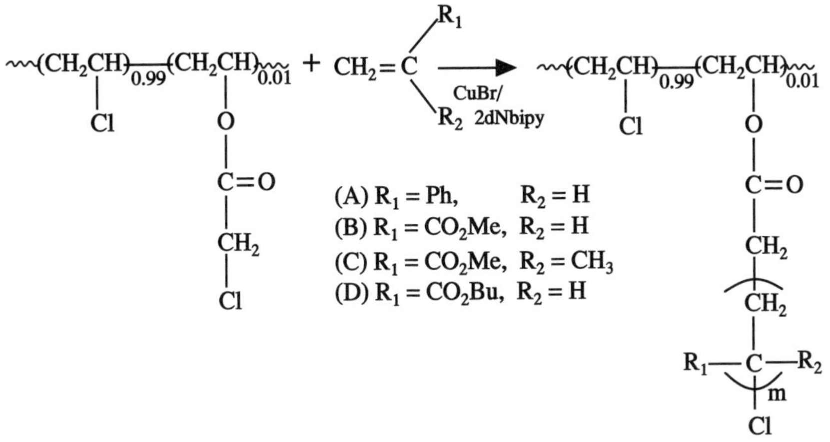

Figure 11.26 Preparation of polystyrene- and poly(metha)acrylates-grafted PVC via ATRP at 110°C and 90°C, respectively, using poly(vinyl chloride)-*co*-(vinyl chloroacetate) as macroinitiator. (From Paik et al., 1998. With permission from *John Wiley & Sons, Inc.*)

behavior with reversible elongations of up to 500% was observed when the copolymer contained 6% (by wt.) polystyrene grafts (Fónagy et al., 1998).

Because of their low chemical reactivity and polarity, high ductility, and good processability, polyethylene (PE) and polypropylene (PP) are often blended with other plastics to improve their performance. However, polyolefins being incompatible with almost all other polymers due to their low surface energy, compatibilizing agents are necessarily added to avoid microphase separation. Such compatibilizers are usually block or graft copolymers containing polyolefin segments. Though free radical graft polymerization onto PE and PP using "grafting from" method with initiation via ionizing radiation (gamma or electron beam), UV with accelerators, or peroxide reaction, is commonly used, this approach leads to ill-defined products due to crosslinking and chain cleavage reactions, besides formation of large quantities of homopolymer. In this respect, a better approach would be the controlled/living free radical techniques, which, as we have noted earlier, allow control of both molecular weights and polydispersities.

The grafting of polymers containing functional groups, such as PMMA, onto ethylene-propylene-diene terpolymer (EPDM) could dramatically increase the interaction of EPDM to a broad range of materials. Such graft copolymers of EPDM could be used as compatibilizers in polymer blends and composites. Earlier attempts to prepare EPDM graft copolymers by the radical or radiation "grafting from" as well as "grafting onto" methods have always resulted in ill-defined products as a result of gel formation (crosslinking), chain degradation, or simultaneous formation of homopolymer. In contrast, EPDM graft copolymers with well-controlled structure can be easily synthesized through controlled/living free radical polymerization (see Problem 11.18).

Problem 11.18 Dienes commonly used in EPDM rubbers include dicyclopentadiene, 4-ethylidene norborn-2-ene, and hexa-1,4-diene. Devise a strategy to prepare (via ATRP and "grafting from" approach) graft copolymers containing EPDM backbone chains and PMMA branches, the diene used in the EPDM being ethylidene norbornene.

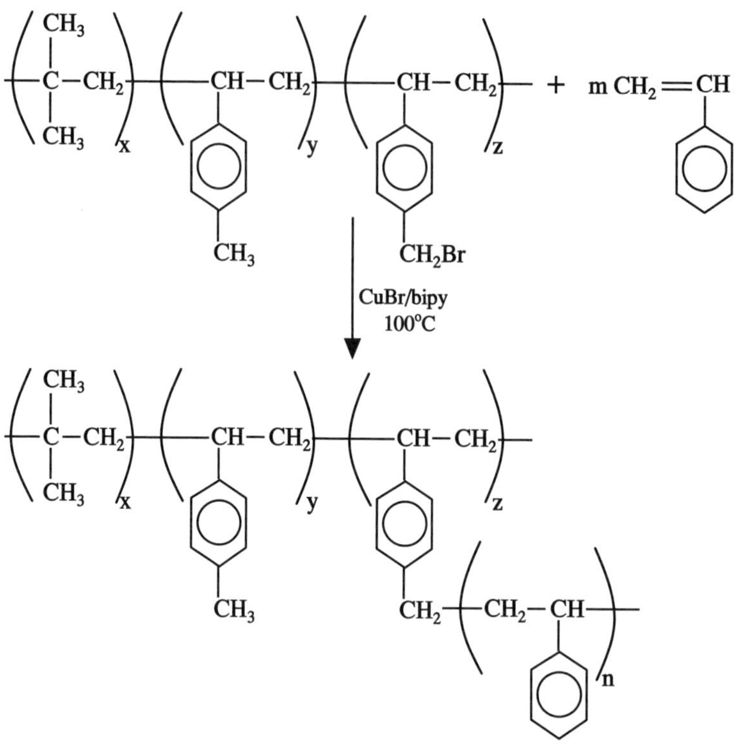

Figure 11.27 ATRP grafting of polystyrene on poly[isobutylene-*co*-(*p*-methyl styrene)-*co*-(*p*-bromomethyl styrene)] ($x = 0.967$, $y = 0.02$, $z = 0.013$) macroinitiator at $100°C$. (Drawn following the synthesis procedure of Fónagy et al., 1998. With permission from *John Wiley & Sons, Inc.*.)

Answer:

The EPDM backbone can be brominated using N-bromosuccinimide (NBS) and 2,2′-azobisisobutyronitrile (AIBN) to introduce allyl bromine on the backbone and then the brominated polymer (EPDM–Br) can be used in conjunction with CuBr/bpy to initiate the polymerization of MMA. Figure 11.28 summarizes the principal stages of the preparation method.

In the procedure employed by Wang et al. (1999), 10 g of a commercial EPDM rubber (Vistollon 2727 from Exxon) was dissolved in 500 mL CCl_4. The allylic bromination was carried out by adding 0.45 g (2.5 mmol) NBS (an excellent allylic bromination agent) and 0.06 g (0.36 mmol) AIBN to the solution and refluxing at $90°C$ for 1.5 h. The bromination efficiency was about 90%. The reaction mixture was cooled, filtered, precipitated in methanol, and the resulting product (EPDM–Br), to be used as the macroinitiator for subsequent ATRP, was dried under vacuum at $60°C$. To perform "grafting from" reaction on this EPDM–Br (0.5 g), the ATRP of MMA (8 mL in 25 mL xylene) was initiated with the concentration (mole) ratio of allylic Br : CuBr : bpy = 1 : *x* : *y*. The maximum grafting efficiency (93%) with 90% monomer conversion was obtained when this mole ratio was 1 : 0.8 : 2.4 and the reaction was conducted at $90°C$ for 20 h. The polymeric product recovered by precipitation into methanol was found to consist of only EPDM-*g*-PMMA and PMMA homopolymer, which was removed by extraction with hexane. The graft copolymers were characterized by solvent extraction, IR, and H^1 NMR spectra.

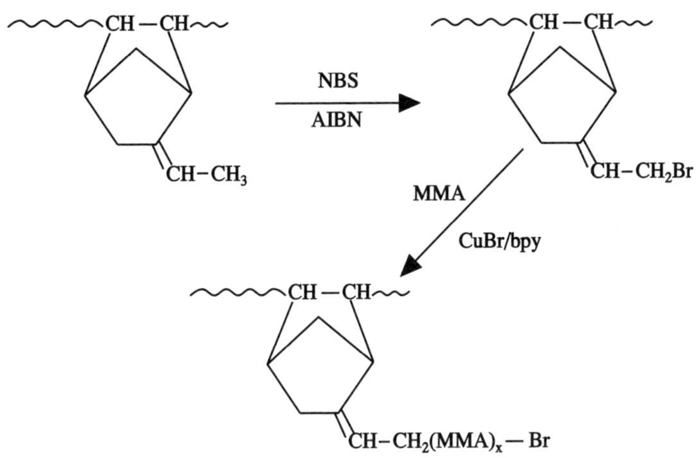

~~~CH—CH~~~

NBS
AIBN →

CH—CH₃

~~~CH—CH~~~

CH—CH₂Br

MMA
CuBr/bpy

~~~CH—CH~~~

CH—CH₂(MMA)ₓ—Br

Figure 11.28 Synthesis of graft copolymers made of EPDM rubber backbone and PMMA branches, using ATRP procedure and "grafting from" approach, the diene of the EPDM rubber being 4-ethylidene norborn-2-ene. (Adapted from Wang et al., 1999.)

Densely grafted copolymers (also called *bottle-brush* copolymers), which have evoked considerable interest in recent years, contain a grafted chain at each repeat unit of the polymer backbone. Several examples of such brush copolymers synthesized by "grafting from" process involving ATRP have been provided in the literature (Beers et al., 1998; Yamada et al., 1999; Boerner et al., 2001). In one study, the copolymers were assembled by grafting from the linear backbone of a macroinitiator, which was synthesized by conventional radical polymerization of 2-(2-bromopropionyloxy) ethyl acrylate in the presence of CBr_4, yielding a polymer (macroinitiator) with $M_n = 27,300$ and high polydispersity $M_w/M_n = 2.3$. ATRP of styrene and *n*-butyl acrylate was conducted with this macroinitiator, leading to the desired densely grafted structures (Fig. 11.29).

The other example of brush copolymers is based on the "grafting through" approach where macromonomers with vinyl terminal are synthesized first and polymerization is carried out through the terminal double bonds to produce densely grafted macromolecules. The "grafting through" approach in combination with ATRP has also been applied in the copolymerization of PMMA macromonomers with *n*-butyl acrylate (Roos et al., 1999, 2000).

11.3.8.3 Star and Hyperbranched Polymers

Among branched polymers, star polymers represent the most elementary way of arranging the subchains since each star contains only one branching point, and as such, they serve as useful models for experimental evaluation of theories about solution properties and rheological behavior of branched polymers (Angot et al., 1998). Star polymers find applications as additives in various areas such as rheology modifiers, pressure sensitive additives, etc. Besides serving as additives, star polymers can also be used as such to achieve specific properties. For instance, star block copolymers with polystyrene-*b*-polybutadiene (PSt-*b*-PB) arms have better processability and me-

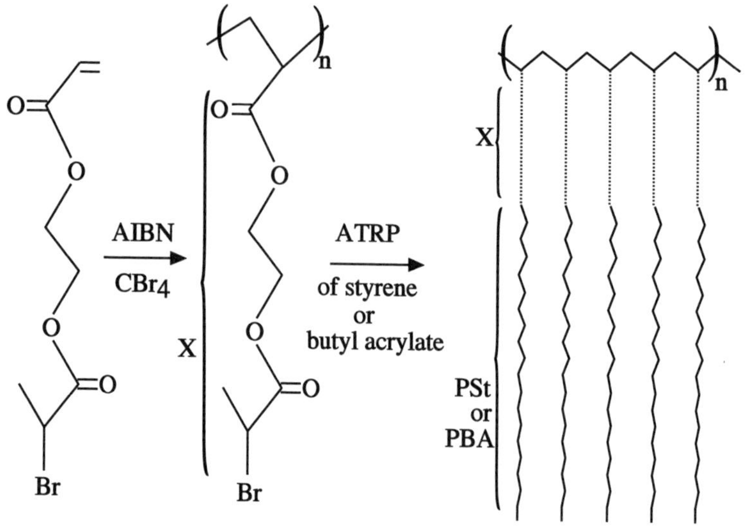

Figure 11.29 Synthesis of densely grafted ("bottle-brush") copolymers using ATRP and "grafting from" process. *Note:* PSt = polystyrene; PBA = poly(n-butyl acrylate). (Adapted from Matyjaszewski and Xia, 2001.)

chanical properties compared to linear PSt-b-PB-b-PSt multiblock copolymers. The polystyrene stars marketed by Philips and star-branched butyls offered by Exxon are examples of commercial star polymers.

There are essentially two methods to obtain star polymers: (a) "arm first" method (linking a given number of linear chains to a central core) (David et al., 1986; Hadjichristidis et al., 1978; Morton et al., 1962) and (b) "Core first" method (growing branches from an active core) (Jacob et al., 1996; Shohi et al., 1992; Schappacher and Deffieux, 1992). To produce star polymers with a precise number of arms, the core-first methodology requires that structurally well-defined plurifunctional initiators be used. Multifunctional initiators of high and yet precise functionality that could serve to polymerize vinylic monomers are scarce. The only examples of such initiators that yield well-defined star polymers are those developed for use in carbocationic polymerization (Jacob et al., 1996). Reports on the use of plurianionic initiators (Eschwey et al., 1973) are much fewer, which might be due to poor solubility of these initiators.

Though ATRP has been applied successfully in macromolecular engineering to synthesize various block and graft copolymers as well as hyperbranched polymers, only a few reports are, however, available on the synthesis of star-shaped polymers using ATRP. Star polymers can be synthesized by ATRP using multifunctional small molecule initiators in core-first methodology. Some examples are the preparation of hexa-arm stars using hexakis(bromomethyl)benzene as the initiators for styrene ATRP (Wang et al., 1995), the synthesis of a tri-arm PMMA (Ueda et al., 1998), and the synthesis of tri-arm liquid-crystalline polycrystals by ATRP using tris(bromomethyl)mesitylene as a trifunctional initiator (Kasko et al., 1998).

The use of *calixarenes* as novel plurifunctional initiators for ATRP has been reported. Well-controlled ATRP of MMA and n-butyl acrylate from dichloroacetate-substituted calixarenes having functionalities of four, six, and eight (Ueda et al., 1998) and that of styrene from octafunctional

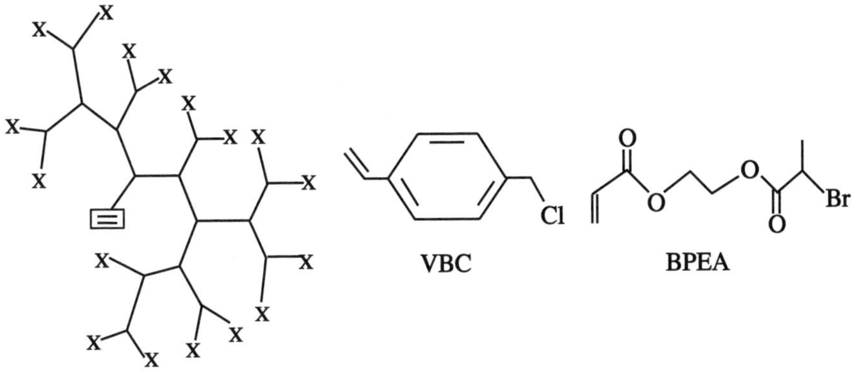

X = Cl, Br

Figure 11.30 Schematic representation of hyperbranched polymer and AB⋆ monomers. (After Matyjaszewski and Xia, 2001.)

2-bromopropionate-modified calixarenes (Angot et al., 1998) have been reported. In a similar way, multifunctional initiators with three, four, six, and eight sulfonyl halide groups have been used to prepare star polymers by ATRP of methacrylates and styrene (Percec et al., 2000). Hexa- and dodeca-functionalized initiators composed of 2-bromoisobutyrates have been synthesized from dendrimer-forming moieties (Heise et al., 1999).

Preparation of hyperbranched polymers using ATRP involves self-condensing vinyl polymerization (SCVP) (Frechét et al., 1995) of AB⋆ monomers, which contain two active species, viz., the double bond A group (polymerizable) and the initiating site B⋆. Two main examples explored in detail within the context of ATRP are *p*-chloromethyl styrene or vinyl benzyl chloride (VBC) and 2-(2-bromopro-pionyloxy) ethyl acrylate (BPEA) (Fig. 11.30). Several other (meth)acrylates with either 2-bromopropionate or 2-bromoisobutyrate groups have also been used.

In the ATRP synthesis with VBC monomer, the structure of the hyperbranched polymer formed was found to depend on the concentration of the catalyst relative to monomer (i.e., initiator). Apparently, in the presence of more catalyst, more deactivator is formed, leading to faster deactivation and a higher degree of branching. However, in the presence of more catalyst, more radicals are also formed, resulting in more termination and additional branching via radical coupling (Weimer et al., 1998). Using multifunctional initiators in an SCVP reaction leads to the possibility of intramolecular cyclization resulting in lower polydispersity.

The terminal halogens in hyperbranched polymers (see Fig. 11.30) can be replaced by more useful functionalities, such as azido, amino, hydroxy, and epoxy. The multifunctional polyols and polyepoxides resulting in the latter two cases can be potentially used to perform thermosetting operations.

Multifunctional initiators, produced by dendrimer techniques, have been used to synthesize "dendrimer-like" star-block copolymers by ATRP method (Hedrick et al., 1998). Starting from a hexafunctional initiator, ε-caprolactone was polymerized and each hydroxyl group was then chemically transformed into two 2-bromo-isobutyrate moieties, which were used to initiate the ATRP of either MMA or a mixture of MMA and hydroxy ethyl methacrylate (HEMA) to produce, in the latter case, dendrimer-like star-block copolymer composed of a poly(ε-caprolactone) core

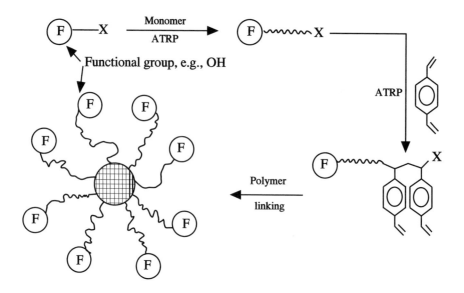

Figure 11.31 Preparation of functional star polymers by ATRP using the "arm first" approach. (Adapted from Matyjaszeswki and Xia, 2001.)

and a poly(MMA-*co*-HEMA) shell (Hedrick et al., 1998).

All of the examples of star polymer formation cited above from the literature represented "core first" approach where the formation of polymeric arms started from a preformed core. The so-called "arm first" approach has also been applied to synthesize star polymers. Thus, linear polymers of polystyrene or poly(*tert*-butyl acrylate) were first prepared by ATRP and then reacted with a cross-linking reagent, such as divinyl benzene (Fig. 11.31), 1,4-butanediol diacrylate, or ethylene glycol dimethacrylate, to generate cross-linked cores (Xia et al., 1999; Zhang et al., 2000). The highest efficiency (~95%) of star formation was observed with 10- to 15-fold excess of the difunctional crosslinking agent over chain ends.

ATRP being highly tolerant to functional groups, functional initiators can be used to directly prepare arms with α-functionalities (Fig. 11.31). End-functional star polymers with diverse groups such as hydroxy, epoxy, amino, cyano, and bromine on the periphery were thus synthesized (Zhang et al., 2000). Radical addition reactions can also be used to incorporate useful functionalities, such as epoxy or hydroxy groups by a chain end transformation process (Coessens et al., 2000). If a difunctional initiator is used for ATRP of a monomer in the "arm first" approach and is followed by reaction with difunctional monomer, it can lead to the formation of fairly homogeneous crosslinked polymer gels (Asgarzadeh et al., 1999).

11.3.9 Aqueous Systems

Application of ATRP to dispersed aqueous systems (emulsion and miniemulsion polymerization) has attracted attention in recent years as it may provide process and economic advantages over the traditional homogeneous bulk and solution polymerizations. However, adaptation of the ATRP process to aqueous dispersions poses several challenges that originate mainly from having more than one phase in the reaction mixture which lead to issues related to phase partitioning and transport of the controlling agent between phases, the role of aqueous phase kinetics, and the phenomena of particle nucleation and colloidal stability. While many of the characteristics of SFRP or NMP also pertain to ATRP, there are critical differences in the phase partitioning behavior of the ATRP activating and deactivating species. With SFRP/NMP, activation of dormant chain (releasing nitroxide) is a unimolecular pocess and deactivation of an active chain (by a free nitroxide) is a bimolecular process. In ATRP, however, a bimolecular pocess is involved in both the activation and deactivation steps. Furthermore, different species (differing in oxidation states) are involved in activation and deactivation steps. They are, therefore, likely to exhibit widely differing phase partitioning. It is possible, for example, that chain deactivation in ATRP occurs only in the organic phase. Consequently, the effects of operating in a heterogeneous environment compared to a homogeneous system can be more complex with ATRP as compared to SFRP/NMP (Cunningham, 2002).

11.4 Degenerative Chain Transfer

In this approach (Matyjaszewski et al., 1995), while the overall polymerization scheme consists of all typical elementary reactions, such as initiation, propagation, and termination, it is moreover supplemented by a reversible chain transfer, also termed *degenerative transfer*, that involves an atom or group transfer from a covalent, dormant species present in large excess over the initiator. The degenerative transfer step in the radical polymerization of alkenes, $CH_2=CHR$, in the presence of a transfer agent, $R'-X$, is shown in Fig. 11.32. Growing radicals, $\sim\!\!\sim\!\!\sim CH_2 \dot{C}HR$, react bimolecularly with $R'-X$ to become dormant species, $\sim\!\!\sim\!\!\sim CH_2 CH(R)X$, through transfer of a group or an atom, X. The new radical $R'\cdot$ reacts with monomer to become a propagating polymer chain which then becomes dormant by transfer of X from another dormant chain. If exchange is fast in comparison with propagation, all R' will become initial end groups in the polymer chains. The total number of chains in the system is then equal to the sum of the chains generated by the initiator and those formed from the transfer agent. When the transfer agent is present in large excess over the initiator, the proportion of chains irreversibly terminated (i.e., those from initiator) should be very low and nearly all chains will terminate with X, which can be activated, thus conferring a living character on the process. The theoretical dependency of the degree of polymerization is given by $DP_n = \Delta[M]/([R'-X]_0 + [I]_0)$, where I is the initiator.

11.5 Reversible Addition-Fragmentation Chain Transfer

Reversible addition-fragmentation chain transfer (RAFT) polymerization is a relatively new method for carrying out controlled/living free-radical polymerization. Since its introduction in 1998, RAFT has rapidly developed into one of the leading techniques of living polymerization. It is arguably the most versatile controlled polymerization process in terms of the variety of monomers for which polymerization can be controlled, reaction conditions (polymerization can be carried out under relatively mild conditions, and in bulk, solution, emulsion or suspension), tolerance to functionalities, and the range of polymeric architectures (e.g., linear block copolymers, comb-like, star, brush polymers, and dendrimers) that can be produced. The main advantage of RAFT over other types of living radical techniques (NMP or ATRP) is that it can be used for a much wider range of functional and nonfunctional monomers — including (meth)acrylates, (meth)acrylamides, acrylonitrile, styrene, and derivatives, butadiene, vinyl acetate, and *N*-vinylpyrrolidone — and with a wider variation of reaction conditions, including aqueous environments at low temperatures.

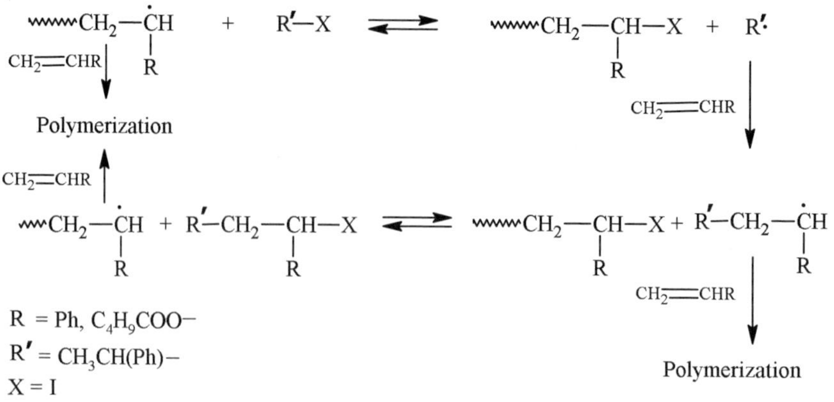

Figure 11.32 Reaction scheme of controlled free-radical polymerization, based on degenerative chain transfer, of butyl acrylate (R = C_4H_9COO-) in the presence of secondary alkyl iodide (R′ = $CH_3CH(Ph)$-, X = I) as the degenerative transfer agent. The latter alone does not initiate polymerization. (Drawn following the method of Matyjaszewski et al., 1995.)

The RAFT technique depends on the use of addition-fragmentation chain transfer agents (called RAFT agents) that possess high transfer coefficients in free-radical polymerization and confer living character on the polymerization via a reversible chain-transfer process. To carry out RAFT polymerization, a chosen quantity of an appropriate RAFT agent is added to a conventional free-radical polymerization, while employing usually the same monomers, initiation, solvents, and temperatures. As in conventional free-radical polymerization, initiation can be classified into three classes: (1) decomposition of organic initiators, (2) initiation via an external source (UV-vis or γ-ray), and (3) thermal initiation. Radical initiators, such as azobisisobutyronitrile (AIBN) and 4,4′-azobis(4-cyanovaleric acid) (ACVA), which is also called 4,4′-azobis(4-cyanopentanoic acid), are widely used as initiators in RAFT polymerization. The overall process of RAFT polymerization is represented by the equation:

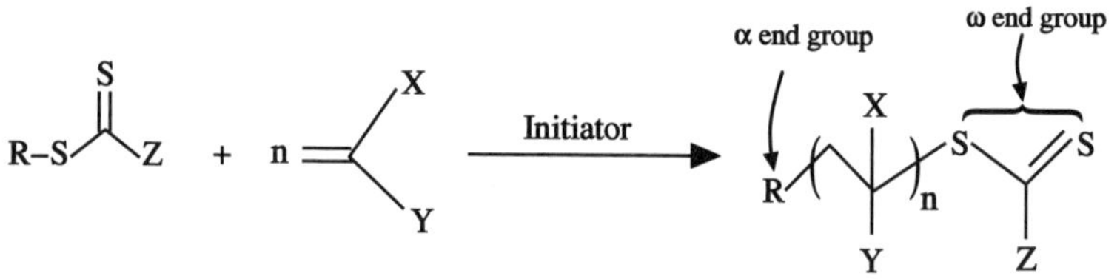

$$(11.22)$$

where R is a group that can initiate polymerization and Z is an activating/stabilizing group. The RAFT process, however, involves a sequence (as shown in Fig. 11.33) in which the chain transfer of the active polymeric chain species ($\text{www}^{\bullet}$) to the RAFT agent takes place to form the intermediate radical ($\text{www}S\text{--}\overset{\bullet}{C}(Z)S\text{--}R$), which then undergoes fragmentation to R$^{\bullet}$ that further reinitiates polymerization. The active moiety [-S-C(Z)=S] from the RAFT agent is now attached to the polymeric chain end, and this makes it a dormant chain. Once the RAFT agent is consumed, equilibrium is established between active and dormant chains based on transfer of S=C(Z)S- moiety between them. This serves to maintain the living character of RAFT polymerization by trapping

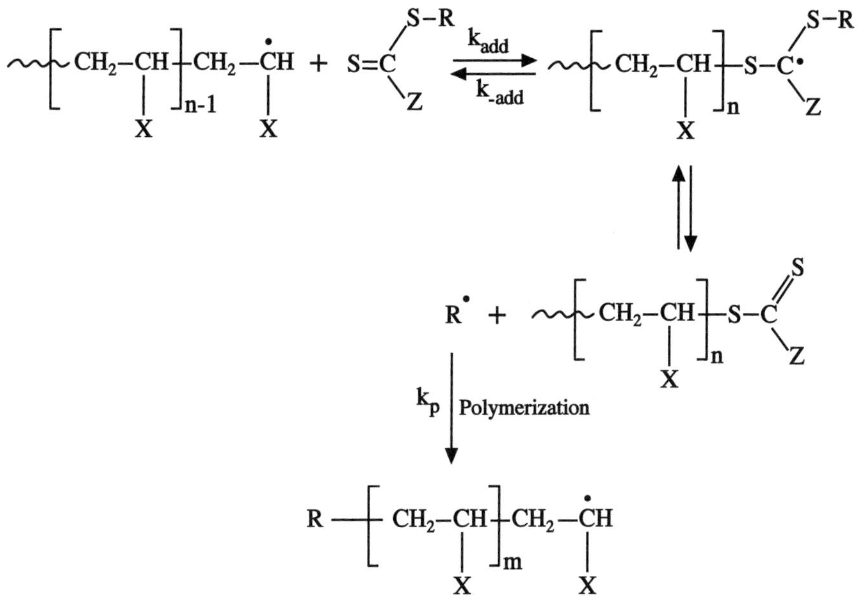

Figure 11.33 Mechanism of RAFT process. Polymerizations can be carried out in bulk, solution, emulsion or suspension, using azo or peroxy initiators as in conventional free-radial polymerization. The moiety S=C(Z)S- remains as the end group.

the majority of the active propagating chains into the dormant thiocarbonyl compound, and thereby limiting the possibility of chain termination through normal processes of radical combination and coupling. The concentration of the active chains is also deliberately kept low relative to the dormant chains by controlling the amounts of the initiator and the RAFT agent in order to limit the termination steps and increase the polymer chain length. The living character of RAFT polymerization has been demonstrated by (a) narrow polydispersity of polymer (usually < 1.2), (b) linear increase of molecular weight with conversion, and (c) ability to produce block copolymers.

It may be mentioned that the RAFT technique was discovered and patented around the same time by Rizzardo and coworkers (Le et al., 1998) at Commonwealth Scientific and Industrial Research Organization (CSIRO, Australia) and by Charmot and coworkers (Charmot et al., 1998, 1999) at Rhodia Chimie (France). The latter group described their invention as *Macromolecular Design via the Interchange of Xanthates* (MADIX) and their patent was restricted to agents which are xanthate-type analogs, whereas the patent of Rizzardo and coworkers covered a much wider range of agents, described as thiocarbonylthio compounds having the generic structure R–S–C(=S)–Z, where Z = aryl, alkyl, NR_2', OR', SR' and R = homolytic leaving group (see later for more details). Xanthates generally have a low reactivity toward acrylate and styrenic monomers, which results in a relatively broad MWD, whereas the the use of highly reactive dithioester type RAFT agents for these systems can lead to formation of polymers with narrow MWD. However, even though the two MWDs are very different, the process for both systems is still considered to be living.

Equation (11.22) shows that the RAFT polymerization leads to the presence of the "R" group (α-functionalization) and residue –SC(=S)Z of the RAFT agent (ω-functionalization) as the two end groups of the resulting polymer. RAFT polymerization can thus be used to synthesize end-functional polymers. However, introducing functionality through the "Z" group is usually not appropriate as the functionality would be lost with cleavage of the thiocarbonylthio group. The alternative of introducing the functionality as part of the "R" group is therefore employed. Though the RAFT process is compatible with a wide range of functional groups including acid, amide, and tertiary amine groups, the process is generally not compatible with unprotected primary or secondary amine groups, since the thiocarbonylthio group reacts rapidly by aminolysis to form, in the first instance, a thiol and a dithiocarbamate (Chiefari et al., 1998). It is therefore necessary to protect amine end groups during RAFT polymerization and to remove the thiocarbonylthio groups before the deprotection step (Postma et al., 2006).

It is evident from the above that while the RAFT agents have the generic structure R–S–C(=S)–Z, the R and Z groups (called the *reinitiating* or *leaving* group and the *stabilizing* or *activating* group, respectively), perform different functions. Thus, the Z group primarily controls the ease with which free-radical species add to the C=S bond and the stability of the resulting carbonyl-thio radical intermediate. The R group, on the other hand, must be a good homolytic leaving group, capable of initiating new polymer chains. Since the Z group strongly influences the stability of the thiocarbonylthio radical intermediate, strong stabilizing groups will favor the formation of the intermediate. However, the stability of the intermediate needs to be fine-tuned so as to favor its fragmentation releasing the reinitiating group R as a free radical. Numerous workers have studied the effect of the Z group on the RAFT polymerization of a variety of monomers (Destarac et al., 2002; Mayadunne et al., 1999; Chiefari et al., 2003; Davis et al., 2003). From these studies, it is the phenyl group that appears to be the ideal candidate for most monomers as it balances the stability of the radical intermediate and its reactivity toward fragmentation. In the case of the benzyl group, the intermediate is less stable, and the fragmentation step occurs more easily, leading to almost no retardation in the polymerization of styrene and faster polymerization for more reactive monomers, such as *N*-isopropyl acrylamide, acrylamide, and methyl acrylate, with good control over the molecular weight (i.e., lower PDI). On the other hand, the less stable intermediate leads to poor control over the polymerization of bulkier propagating radicals such as methyl methacrylate (MMA). Alkyl Z groups also give reasonable control over the polymerization of styrene, butyl acrylate, and MMA. However, for Z with O or N linkage (such as, xanthates and dithiocarbamates), the nonbonded electron pair on the heteroatom is delocalized with the C=S bond. This lowers the reactivity of the double bond toward radical addition, leading to poor control over the molecular weight of the growing polymer chains. However, in the case of fast propagating monomers (e.g., vinyl acetate), chain transfer agents (CTAs) with C=S bond of a lower reactivity are desirable. Indeed, vinyl acetate polymerization can be successfully controlled in the presence of xanthates, whereas it is strongly inhibited by dithiobenzoates. A general classification of Z groups that allow good control over the majority of monomers is as follows: dithiobenzoates > trithiocarbonates > xanthates > dithiocarbamates (Perrier and Takolpuckdee, 2005).

As mentioned earlier, the R group is required to be a good leaving group in comparison with the polymer chain and a good reinitiating species toward the monomer used. It also contributes toward the stabilization of the radical intermediate, although to a lesser extent than the Z group. Experimental data and *ab initio* calculations provide the following general guidelines or the selection of R group for a CTA (Perrier and Takolpuckdee, 2005) : $C[(CH_3)_2]CN \sim C[(CH_3)_2]Ph >$ $CH(CO_2Me)Ph \sim CH(CN)Ph > C[(CH_3)_2]CO_2Et \sim C[(CH_3)_2]CONEt_2 > CH(CH_3)Ph >$ $CH(CH_3)CO_2Et > C(CH_3)_3 \sim CH_2Ph > CH_2CO_2Et$. The first two groups, namely, cyanoisopropyl and cumyl groups thus appear to be the most efficient for the reinitiation step.

CTAs can be categorized into four classes, depending on their activating Z group. These are, as shown in Fig. 11.34: (i) dithioesters (Chong et al., 2003; Le et al., 1998; Tang et al., 2003; An et al., 2005), (ii) xanthates (Corpart et al., 1998; Charmot et al., 2000; Destarac et al., 2002),

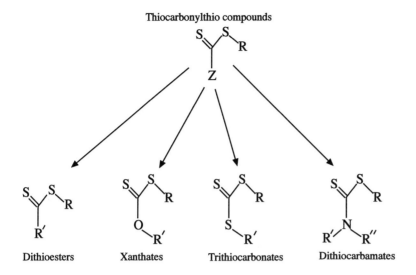

Figure 11.34 General structures of RAFT agents belonging to four classes based on differences in functional groups at the Z position of thiocarbonylthio compounds.

(iii) trithiocarbonates (Mayadunne et al., 2000; Gaillard et al., 2003), and (iv) dithiocarbamates (Schilli et al., 2002; Mayadunne et al., 199; Destarac et al., 2000). A few examples of the above four classes of CTAs used in RAFT synthesis of polymers are shown in Fig. 11.35.

While the Z group mainly influences the rate of addition of radicals to the C=S double bond, it also has effect on stability/reactivity of the macroradical intermediate. In this respect, dithioesters (e.g., with Z = CH₃ or Ph and R = cumyl) are particularly efficient in RAFT polymerization of styrene and alkyl acrylate to produce polymers with narrow MWD, while, in contrast, *O*-ethyl xanthates are relatively inefficient as CTAs, yielding polymers of relatively high polydispersities $(2 < M_w/M_n < 2.2)$ in MADIX polymerization of styrene. This is explained by the existence of slow exchange of the xanthate group among dormant and propagating chains. Trithiocarbonates have an outstanding feature that they can be prepared with either one (e.g., IIIa and IIIb in Fig. 11.35) or two (e.g., IIIc and IIId in Fig. 11.35) good homolytic leaving groups. Since symmetrical trithiocarbonates, such as IIIc and IIId (Fig. 11.35) can grow in two directions, they should yield polymers with active functionality located in the center, e.g.,

$$\text{IIIc (Fig. 11.35)} \longrightarrow \text{PhCH}_2\text{-}\!\!\!\sim\!\!\!\text{S}\!\!-\!\!\overset{\overset{\displaystyle S}{\|}}{\text{C}}\!\!-\!\!\text{S}\!\!\sim\!\!\text{-CH}_2\text{Ph} \xrightarrow{\text{Nucleophile}} 2(\text{PhCH}_2\!\!\sim\!\!\text{SH})$$

One way to confirm this is to cleave the polymer chains at the trithiocarbonate function. This can be achieved readily with mild nucleophiles, such as primary or secondary amines (Mayadunne et al., 2000).

11.5.1 Mechanism and Kinetics

There are five steps in RAFT polymerization: initiation, addition-fragmentation chain transfer, reinitiation, chain equilibrium, and termination (Fig. 11.36).

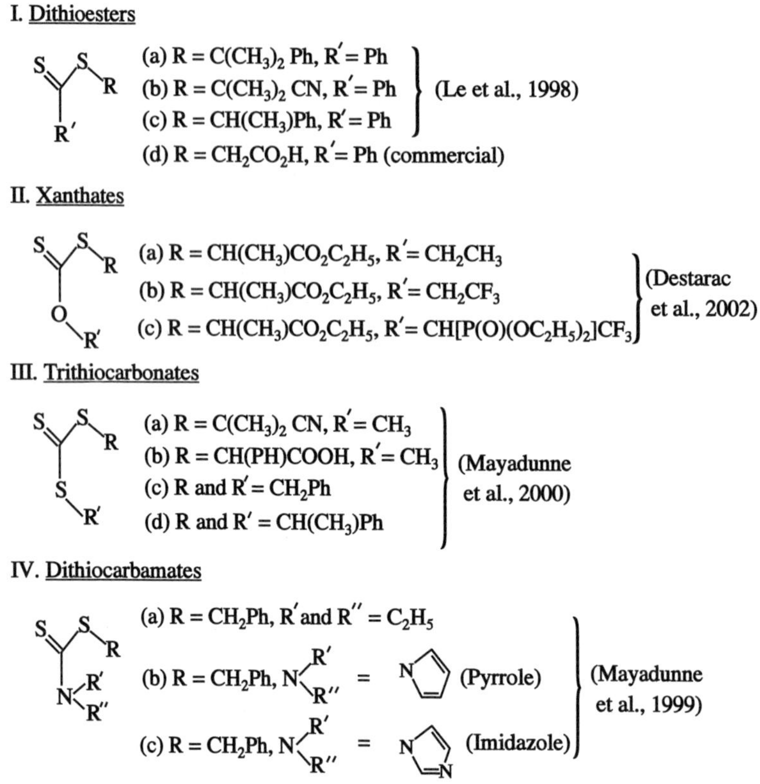

I. Dithioesters

(a) R = C(CH₃)₂ Ph, R′ = Ph
(b) R = C(CH₃)₂ CN, R′ = Ph   (Le et al., 1998)
(c) R = CH(CH₃)Ph, R′ = Ph
(d) R = CH₂CO₂H, R′ = Ph (commercial)

II. Xanthates

(a) R = CH(CH₃)CO₂C₂H₅, R′ = CH₂CH₃
(b) R = CH(CH₃)CO₂C₂H₅, R′ = CH₂CF₃   (Destarac et al., 2002)
(c) R = CH(CH₃)CO₂C₂H₅, R′ = CH[P(O)(OC₂H₅)₂]CF₃

III. Trithiocarbonates

(a) R = C(CH₃)₂ CN, R′ = CH₃
(b) R = CH(PH)COOH, R′ = CH₃   (Mayadunne et al., 2000)
(c) R and R′ = CH₂Ph
(d) R and R′ = CH(CH₃)Ph

IV. Dithiocarbamates

(a) R = CH₂Ph, R′ and R″ = C₂H₅
(b) R = CH₂Ph, N⟨R′/R″ = (Pyrrole)   (Mayadunne et al., 1999)
(c) R = CH₂Ph, N⟨R′/R″ = (Imidazole)

Figure 11.35 Examples of some RAFT agents. *Dithioesters*: Le et al. (1998). *Xanthates*: Destarac et al. (2002). *Trithiocarbonates*: Mayadunne et al. (2000). *Dithiocarbamates*: Mayadunne et al. (1999).

Initiation. The RAFT process is started by radical initiators, such as AIBN. The initiator(I) generates a radical species which starts an active polymerizing chain (P_n •) by reacting with monomer.
Chain transfer (Pre-equilibrium). The active polymer chain P_n • rapidly adds to the reactive C=S bond of the CTA (rate constant, k_{add}) forming an intermediate adduct radical, which undergoes reversible fragmentation either toward the initial growing chain (rate constant, k_{-add}) or to free the group R by β-scission (rate constant, k_β) and simultaneously generate a macro chain transfer agent (macro-CTA), P_n-X.
Re-initiation. The leaving group radical (R •), which is capable of initiation, reacts with monomer to start another active polymer chain (P_m •).
Chain equilibrium (Main equilibrium). The active chain P_m • is able to go through the addition-fragmentation process resulting in equilibrium between the active (radical) and dormant (bound to the CTA) states. This is the main equilibrium that constitutes the fundamental step in the Raft process. It traps majority of the actrive propagating species into the dormant CTA compound, thereby limiting the possibility of conventional radical-radical termination reactions. While one polymer chain (P_m •) is in the dormant stage, the other (P_n •) is active in polymerization.

It is evident from the above mechanism that the thiocarbonylthio group of the original RAFT agent is retained in the polymeric product (via P_n-X and P_m-X). This retention of the thiocarbonylthio moiety is a key feature of the RAFT process and is responsible for the living character of the RAFT polymer.

In chain transfer by addition-fragmentation (Fig. 11.36), the effective rate constant for chain transfer (k_{tr}) is given by the following expression (Moad et al., 1996):

$$k_{tr} = k_{add} \times \frac{k_{fr}}{k_{-add} + k_\beta} \qquad (11.23)$$

where k_{add}, k_{-add}, k_{fr}, and k_β are defined in Fig. 11.36. Equation (11.23) follows from the fact that the transfer constant depends on (a) the reactivity of the C=S double bond for addition of the attacking radical and (b) the partitioning of the fragmentation of the intermediate radical toward either an attacking radical ($P_n \cdot$) or a reinitiating radical ($R \cdot$).

The activity of the transfer agent is usually defined in terms of their chain transfer constants ($C_{tr} = k_{tr}/k_p$). Traditionally, chain transfer constants have been evaluated by applying the Mayo equation [cf. Eq. (6.122)]:

$$\frac{1}{\overline{DP}_n} = C_{tr} \frac{[T]}{[M]} + \frac{1}{\overline{DP}_0} \qquad (11.24)$$

where $\overline{DP}_n$ is the number average degree of polymerization, [T] and [M] are the concentrations of transfer agent and monomer, respectively, and $\overline{DP}_0$ is the number average degree of polymerization obtained in the absence of added transfer agent.

Similar to Eq. (11.23), the effective rate constant for reverse reaction (k_{-tr}) is given by

$$k_{-tr} = k_{-\beta} \times \frac{k_{add}}{k_{-add} + k_\beta} \qquad (11.25)$$

Problem 11.19   Discuss the effect of temperature on chain transfer constant and hence on PDI in RAFT polymerization.

Answer:

Most RAFT agents have C_{tr} values greater than 1. This suggests that as the temperature is reduced, the C_{tr} values will increase because of the higher activation energy of k_p (see Section 6.11.1). Thus one way to lower the PDI by increasing C_{tr} is to simply decrease the temperature. This strategy will work only if C_{tr} is greater than 1. Conversely, when C_{tr} is less than 1, an increase in the temperature will increase the value of C_{tr}, However, high temperatures exacerbate side reactions, such as transfer to monomer and polymer, resulting in deviations from the predicted MWD evolution with conversion. A method for lowering the PDI without changing the temperature is to maintain a constant monomer concentration throughout the polymerization (Monteiro, 2005).

Termination.   Possible termination reactions among propagating and intermediate radicals are shown in Fig. 11.36. If fragmentation occurs fast enough accompanying no side reaction and propagating polymer chains undergo self-termination in conventional ways by coupling (combination) or disproportionation, the RAFT process should have no significant influence on polymerization rate R_p, and R_p may be equated to $R_{p,0}$, the rate for the conventional (RAFT agent-free) polymerization given by (Kwak et al., 2004):

$$R_{p,0} = k_p (R_i/k_t)^{1/2} [M] \qquad (11.26)$$

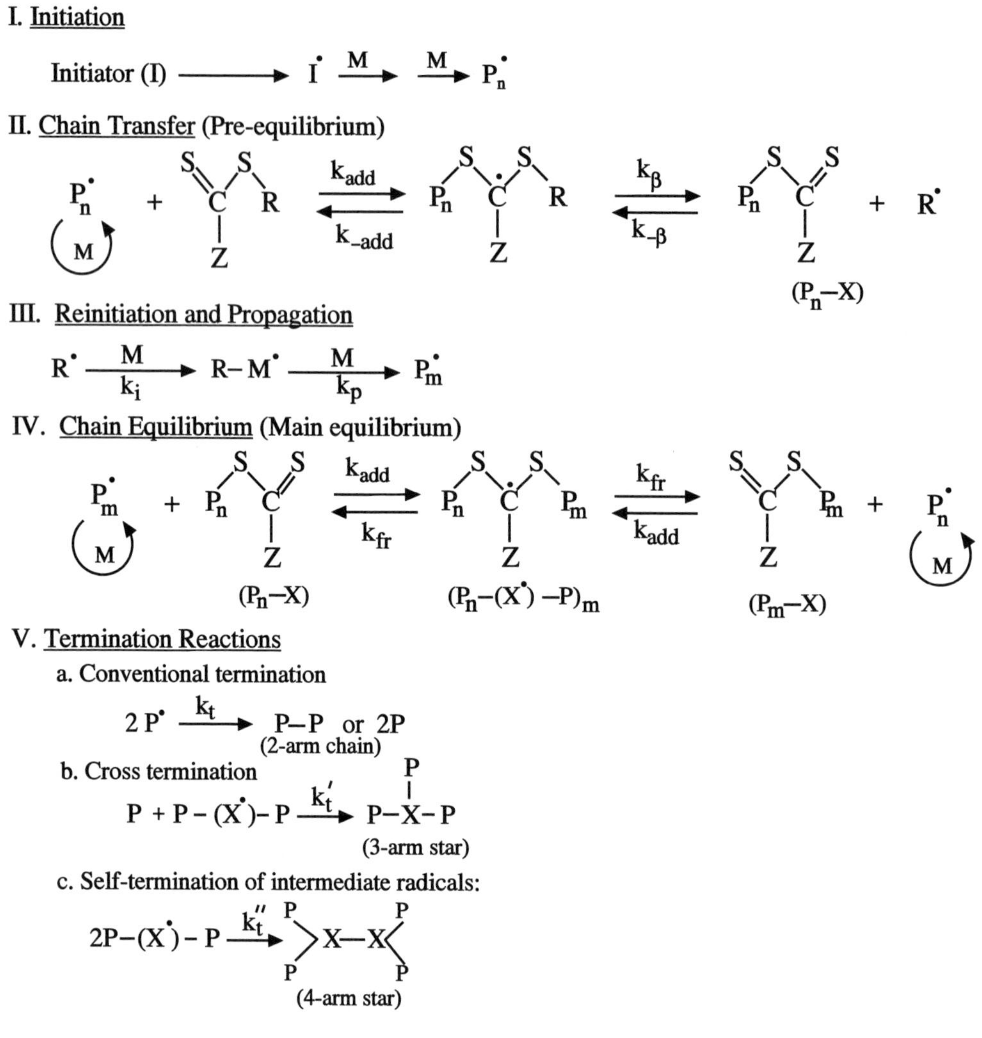

I. Initiation

$$\text{Initiator (I)} \xrightarrow{} \overset{\bullet}{I} \xrightarrow{M} \xrightarrow{M} \overset{\bullet}{P_n}$$

II. Chain Transfer (Pre-equilibrium)

(P$_n$–X)

III. Reinitiation and Propagation

$$R^\bullet \xrightarrow[k_i]{M} R\!-\!M^\bullet \xrightarrow[k_p]{M} \overset{\bullet}{P_m}$$

IV. Chain Equilibrium (Main equilibrium)

(P$_n$–X)   (P$_n$–(X$^\bullet$)–P)$_m$ (P$_m$–X)

V. Termination Reactions

a. Conventional termination

$$2\,P^\bullet \xrightarrow{k_t} \underset{\text{(2-arm chain)}}{P\!-\!P \text{ or } 2P}$$

b. Cross termination

$$P + P\!-\!(\overset{\bullet}{X})\!-\!P \xrightarrow{k_t'} \underset{\text{(3-arm star)}}{\overset{\displaystyle P}{\overset{|}{P\!-\!X\!-\!P}}}$$

c. Self-termination of intermediate radicals:

$$2P\!-\!(\overset{\bullet}{X})\!-\!P \xrightarrow{k_t''} \underset{\text{(4-arm star)}}{\overset{P\quad\quad P}{\underset{P\quad\quad P}{>\!X\!-\!X\!<}}}$$

Figure 11.36 A general scheme for RAFT polymerization with thiocarbonylthio compounds as chain transfer agents. When P$_n$ and P$_m$ are kinetically identical polymers, these need not be distinguished and both may be written as P. (Adapted from Kwak et al., 2004.)

where R_i is the rate of (conventional) initiation, [M] is the monomer concentration, and k_p and k_t are the rate constants of propagation and termination, respectively. Thus, higher apparent rates of polymerization in a RAFT process can be achieved by utilizing CTAs with higher intermediate fragmentation rates, larger monomer and initiator concentrations, and higher temperatures.

In the RAFT polymerization with agents in which Z is a phenyl group (see Fig. 11.35), significant retardation in polymerization rate has, however, been observed. (*Note*: The 'rate retardation' means that R_p is lower in comparison with the corresponding conventional radical polymerization in the absence of RAFT agent and decreases with increasing initial RAFT agent concentrations.) Thus, Monteiro and de Brouwer (2001) found that for a styrene polymerization mediated with cumyl dithioester, as the concentration of the RAFT agent is increased, the rate is increasingly retarded, and the MWD follows ideal living behavior. This is also true for a range of other monomers. Two opposing explanations, among others, have been proposed for the rate retardation. One is slow fragmentation. Thus, Barner-Kowollik et al. (2001) assumed that the intermediate radical, P_n-$(X^\bullet)$-P_m (Fig. 11.36), is stable enough to cause no termination with $P^\bullet$ (i.e., no cross-termination). They concluded from their study that retardation is due to slow fragmentation of the intermediate radicals. However, the biggest drawback to the slow fragmentation model is that the intermediate radical concentration predicted by the model is 2-3 orders of magnitude greater than that found by ESR measurements. Monteiro and de Brouwer (2001), on the other hand, proposed that the only way to account for the rate retardation is through intermediate radical termination (IRT), in which all radicals in the system could terminate with all intermediate radicals to form three- and four-arm stars (see Fig. 11.36). This is in agreement with the experimental and theoretical findings of Kwak et al. (2004).

Problem 11.20   Kwak et al. (2004) presented experimental evidence to show that the rate retardation in the polymerization of styrene (St) with polystyryl dithiobenzoate [PSt–SC(=S)Ph] at $60°C$ is caused by the irreversible cross-termination between the polystyryl radical (PSt$^\bullet$) and the intermediate radical produced by the addition of PSt$^\bullet$ to PSt–SC(=S)Ph. The polymerization rate R_p was found to decrease with the increase of [PSt–SC(=S)Ph] such that a plot of $1/R_p^2$ vs [PSt–SC(=S)Ph] was linear. (a) Show that this is in conformity with the scheme for RAFT polymerization shown in Fig. 11.36, taking into account termination of intermediate radicals and considering that cross-termination to form 3-arm star predominates.   (b) What would be the corresponding rate behavior if termination between two intermediate radicals to form a 4-arm star predominates ?

Answer:

Being kinetically identical, the chains P_n nad P_m in the polymerization system defined in Fig. 11.36 may be represented by the same symbol (P) for the termination reactions. Viewed in a relatively short time scale and considering a stationary state in which all reversible reactions are in equilibrium (or in quasi-equilibrium) and the concentrations of all radical species are (approximately) invariant with time, the following two equations should hold in respect of the propagating and intermediate radical concentrations (Kwak et al., 2004):

$$0 = d[P^\bullet]/dt$$

$$= R_i - k_{add}[P^\bullet][P\text{–}X] + k_{fr}[P\text{–}(X^\bullet)\text{–}P]$$

$$- k_t[P^\bullet]^2 - k_t'[P^\bullet][P\text{–}(X^\bullet)\text{–}P] \tag{P11.20.1}$$

$$0 = d[P\text{–}(X^\bullet)\text{–}P]/dt$$

$$= k_{add}[P^\bullet][P\text{–}X] - k_{fr}[P\text{–}(X^\bullet)\text{–}P]$$

$$- k_t'[P^\bullet][P\text{–}(X^\bullet)\text{–}P] - k_t''[P\text{–}(X^\bullet)\text{–}P]^2 \tag{P11.20.2}$$

Adding Eqs. (P11.20.1) and (P11.20.2),

$$R_i - k_t[P^\bullet]^2 - 2k_t'[P^\bullet][P\text{–}(X^\bullet)\text{–}P] - k_t''[P\text{–}(X^\bullet)\text{–}P]^2 = 0 \tag{P11.20.3}$$

Assuming that the addition-fragmentation equilibrium holds,

$$k_{add}[\text{P}^\bullet][\text{P--X}] = k_{fr}[\text{P--(X}^\bullet\text{)--P}] \tag{P11.20.4}$$

This is a quasi-equilibrium, which is valid when the rates of termination reactions (e.g., k_t' and k_t'' terms in Eq. (P11.20.2)) are negligibly small compared with those of addition and fragmentation (k_{add} and k_{fr} terms in Eq. (P11.20.2)).

In terms of the equilibrium constant K defined by

$$K = k_{add}/k_{fr} \tag{P11.20.5}$$

the concentration of the intermediate radical is obtained from Eq. (P11.20.4) as

$$[\text{P--(X}^\bullet\text{)--P}] = K[\text{P}^\bullet][\text{P--X}] \tag{P11.20.6}$$

Equation (P11.20.3) with Eq. (P11.20.6) is solved for $[\text{P}^\bullet]$ and substituted in the relation $R_p = k_p[\text{P}^\bullet][\text{M}]$ to obtain

$$R_p = \frac{k_p R_i^{1/2} k_t^{-1/2}[\text{M}]}{\left\{1 + 2(k_t'/k_t)K[\text{P--X}] + (k_t''/k_t)K^2[\text{P--X}]^2\right\}^{1/2}} \tag{P11.20.7}$$

Substitution of Eq. (11.26) then yields

$$R_p = \frac{R_{p,0}}{\left\{1 + 2(k_t'/k_t)K[\text{P--X}] + (k_t''/k_t)K^2[\text{P--X}]^2\right\}^{1/2}} \tag{P11.20.8}$$

We may approximate $[\text{P--X}]$ by the initial RAFT agent concentration $[\text{P--X}]_0$, since $[\text{P--(X}^\bullet\text{)--P}]$ and the number of terminated chains are negligibly small compared to $[\text{P--X}]_0$,

(a) When the cross-termination of intermediate radical to 3-arm star formation (Fig. 11.36) is the main cause for the retardation in R_p, then setting $[\text{P--X}] = [\text{P--X}]_0$ and $k_t''/k_t = 0$, we have (Kwak et al., 2002) :

$$R_p^{-2} = R_{p,0}^{-2}\left\{1 + 2(k_t'/k_t)K[\text{P--X}]_0\right\} \tag{P11.20.9}$$

A plot of R_p^{-2} against $[\text{PSt-SC(=S)Ph}]_0$ should thus be a straight line providing a kinetic confirmation of the IRT being the main cause or rate retardation.

(b) If the termination between two intermediate radicals to form a 4-arm star polymer predominates, we may set $k_t'/k_t = 0$ and $[\text{P--X}] = [\text{P--X}]_0$ in Eq. (P11.20.8) to have

$$R_p^{-2} = R_{p,0}^{-2}\left\{1 + (k_t''/k_t)K^2[\text{P--X}]_0^2\right\} \tag{P11.20.10}$$

According to this equation, a plot of R_p^{-2} vs $[\text{P--X}]_0^2$ should be linear.

11.5.2 Theoretical Molecular Weight

It is apparent from the general scheme for RAFT polymerization, shown in Fig. 11.36, that if intermediate radical termination is neglected, most of the RAFT polymers are contributed by P_m–X living chains which have the R group at one end and the thiocarbonylthio moiety at the other. In addition, some dead polymer chains form due to conventional termination of growing chains derived from the free-radical initiator. The theoretical number-average molecular weight (M_n) of a RAFT polymer may therefore be predicted with the following equation:

$$M_n = \frac{[M]_0\, p\, M_{MW}}{([CTA]_0 - [CTA]) + a.f[I]_0 \left(1 - e^{-k_d t}\right)} + CTA_{MW} \qquad (11.27)$$

where $[M]_0$ is the initial monomer concentration, p is the fractional monomer conversion to the polymer, $[CTA]_0$ is the initial RAFT agent concentration, $[CTA]$ is the RAFT agent concentration at p, a is the mode of termination (a equals 1 for termination by combination and 2 for disproportionation), f is the initiator efficiency, and k_d is the initiator dissociation constant. The right hand side of the denominator accounts for radicals derived from initiator with an initial concentration $[I]_0$ in time t. In an ideal RAFT process, polymer directly derived from the initiator is thought to be minimal, and thus the second term in the denominator becomes negligible. Further, if the CTA is fully utilized in the RAFT process, Eq. (11.27) simplifies to

$$M_n = \frac{[M]_0\, p\, M_{MW}}{[CTA]_0} + CTA_{MW} \qquad (11.28)$$

According to this equation, the polymer molecular weight in an ideal RAFT process will increase linearly with increasing monomer conversion, satisfying an important criterion of living polymerization.

Problem 11.21 Bulk polymerization of methyl acrylate (MA) is carried out by the RAFT process at 60°C. The process is mediated by cyanoisopropyl dithiobenzoate (Fig. 11.35, Ib) used as the chain transfer agent (CTA) and initiated by AIBN with the ratio (molar) MA/CTA/AIBN = 1230 : 1 : 0.1, attaining 75% conversion of MA in 7 h.

(a) Obtain an approximate estimate of the average molecular weight (M_n) of the polymer. (b) Calculate theoretically the fraction of polymer molecules that is accounted for by dead polymer chains. (c) Will the proportion of dead chains be higher or lower in the case of a more slowly propagating monomer such as methyl methacrylate (MMA)? [*Data*: monomer density = 0.956 g/cm³; $k_d = 8.54 \times 10^{-6}$ s⁻¹; $a = 2$; $f = 1$.]

Answer:

(a) Assuming that polymer directly derived from the initiator is negligibly small compared to the polymer obtained via CTA and that the CTA is fully utilized in the RAFT process, Eq. (11.28) can be used to derive an approximate value of the molecular weight.

Molar masses: MA 86 g mol⁻¹; CTA 221 g mol⁻¹.

For bulk MA, $[MA]_0 = \dfrac{(1000\ \text{cm}^3\,\text{L}^{-1})(0.956\ \text{g}\,\text{cm}^{-3})}{(86\ \text{g}\,\text{mol}^{-1})} = 11.12\ \text{mol}\,\text{L}^{-1}$

$[CTA]_0 = \dfrac{11.12\ \text{mol}\,\text{L}^{-1}}{1230} = 9.0 \times 10^{-3}\ \text{mol}\,\text{L}^{-1}$

$[AIBN]_0 = \dfrac{11.12 \times 0.1}{1230} = 9.0 \times 10^{-4}\ \text{mol}\,\text{L}^{-1}$

From Eq. (11.28):

$$M_n = \frac{(11.12 \text{ mol L}^{-1})(0.75)(86 \text{ g mol}^{-1})}{(9.0 \times 10^{-3} \text{ mol L}^{-1})} + (221 \text{ g mol}^{-1})$$

$$= 80,000 \text{ g mol}^{-1}$$

(b) The total number of chains produced in a RAFT process will be equal to the number of radicals derived from the initiator and CTA leaving group (R), but the maximum number of living chains will be equal to the number of CTAs. Therefore the proportion of dead chains (x_{dc}) is given by the ratio of the number of initiator-derived radicals to the number of CTAs plus initiator-derived radicals:

$$x_{dc} = \frac{a f ([I]_0 - [I]_t)}{[CTA]_0 + a f ([I]_0 - [I]_t)} \qquad \text{(P11.21.1)}$$

Substituting from Eq. (6.26),

$$x_{dc} = \frac{a f [I]_0 (1 - e^{-k_d t})}{[CTA_0 + a f [I]_0 (1 - e^{-k_d t})} \qquad \text{(P11.21.2)}$$

Using $a = 2$, $f = 1$, $k_d = 8.54 \times 10^{-6}$ s^{-1}, $[I]_0 = 9.0 \times 10^{-4}$ mol L^{-1}, $[CTA]_0 = 9.0 \times 10^{-3}$ mol L^{-1}, and $t = 7 \times 3600$ or 2.52×10^4 s,

$$x_{dc} = 0.011 \text{ (i.e., 1.1\%)}$$

So, only 1.1% of the chains are dead at 75% conversion.

(c) In the case of a more slowly propagating monomer, for similar molecular weights and conversions, t will be greater and hence x_{dc} from Eq. (P11.21.2), other terms being similar, will be larger, signifying that the proportion of dead chains will be higher.

11.5.3   Block Copolymers

One advantage of the RAFT process is its compatibility with a wide range of monomers, including functional monomers. Thus narrow polydispersity block copolymers have been prepared with monomers containing acid (e.g., acrylic acid), hydroxy (e.g., 2-hydroxyethyl methacrylate), and tertiary amino [e.g., 2-(dimethylamino) ethyl methacrylate] functionality (Chiefari et al., 1998). Linear block copolymers are the simplest polymeric architectures achievable via RAFT process. There are two main routes for the synthesis of block copolymers by the RAFT process, viz., (i) sequential monomer addition (chain extension) and (ii) synthesis via macro-CTAs (by R- or Z-group approaches). These are schematically shown in Fig. 11.37. Linear block copolymers are the simplest polymeric architectures achievable via RAFT process.

11.5.3.1   Sequential Monomer Addition

One of the major applications of living polymerization is the synthesis of block copolymers via sequential addition of monomers. In this approach, monomer A is first polymerized quantitatively by the RAFT process, followed by direct addition of a second monomer for chain extension, or alternatively, the RAFT polymer of A from the first step is purified and used as a macro-CTA to

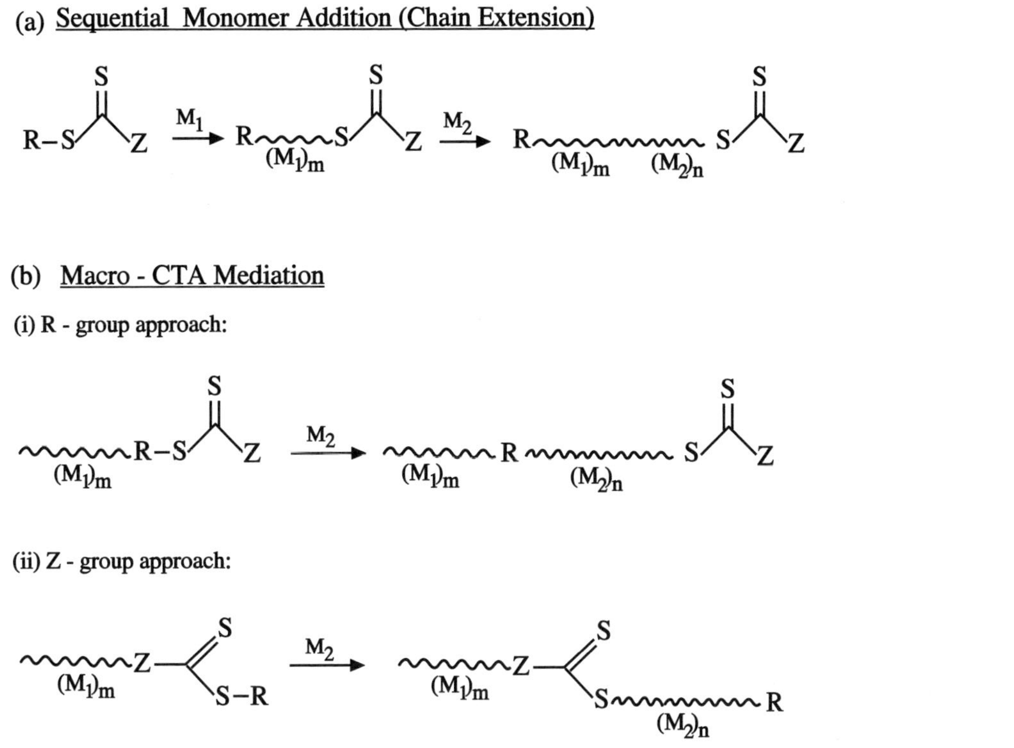

(a) <u>Sequential Monomer Addition (Chain Extension)</u>

(b) <u>Macro - CTA Mediation</u>

(i) R - group approach:

(ii) Z - group approach:

Figure 11.37 General synthetic approaches for the generation of AB diblock copolymers by the RAFT process. (Adapted from Barner, Davis, Stenzel, and Barner-Kowollik, 2007.)

mediate the polymerization of monomer B. If monomer A is polymerized to a high conversion and then B is added directly without purifying the A polymer, the final polymer chain will feature a middle section composed of an AB *gradient polymer* separating the A block from the B block. For the production of block copolymers by sequential monomer addition, it is, however, essential that the first block retains its chain end functionality. This is generally achieved by stopping the polymerization of the first monomer at a conversion below 90%. Moreover, the radical initiator that is added to the RAFT system in order to trigger the degenerative chain transfer also contributes to the formation of homopolymer side products with uncontrolled chain length. A low concentration of the initiator should therefore be used to achieve a high ratio of living chains to (uncontrolled) dead chains. The sequence of monomer addition also needs careful consideration. Thus, to produce a low-dispersity AB diblock copolymer, the first formed polymer with thiocarbonylthio end group [S=C(Z)S–A] should have a high transfer constant to the monomer in the subsequent polymerization step to give the B block. This requires that the leaving ability of propagating radical A• is comparable to, or greater than, that of the propagating radical B• under the reaction conditions (Chong et al., 1999).

Problem 11.22 In preparing the diblock copolymers, polystyrene-*b*-PMMA and polyacrylate-*b*-PMMA, by sequential monomer addition in the RAFT process, what should be the sequence of monomer addition? Answer giving reasons.

Answer:

For forming AB diblock copolymer, when A is a polystyrene or poly(acrylate ester) chain, the transfer constants of S=C(Z)S–A in MMA polymerization are extremely low, which is attributed to the styryl- or acrylyl-propagating radicals being poor leaving groups with respect to a methacrylyl-propagating radical, that causes the adduct radical, B–S–C$^{\bullet}$ (Z)S–A, to partition strongly in favor of starting materials, B$^{\bullet}$ and Z–C(=S)S–A (Chong et al., 1999). So, in preparing a block copolymer, for which one block is based on a methacrylate monomer and the other on a styrene or an acrylate monomer, the methacrylate block should be prepared first, i.e., the PMMA should be taken as the A block.

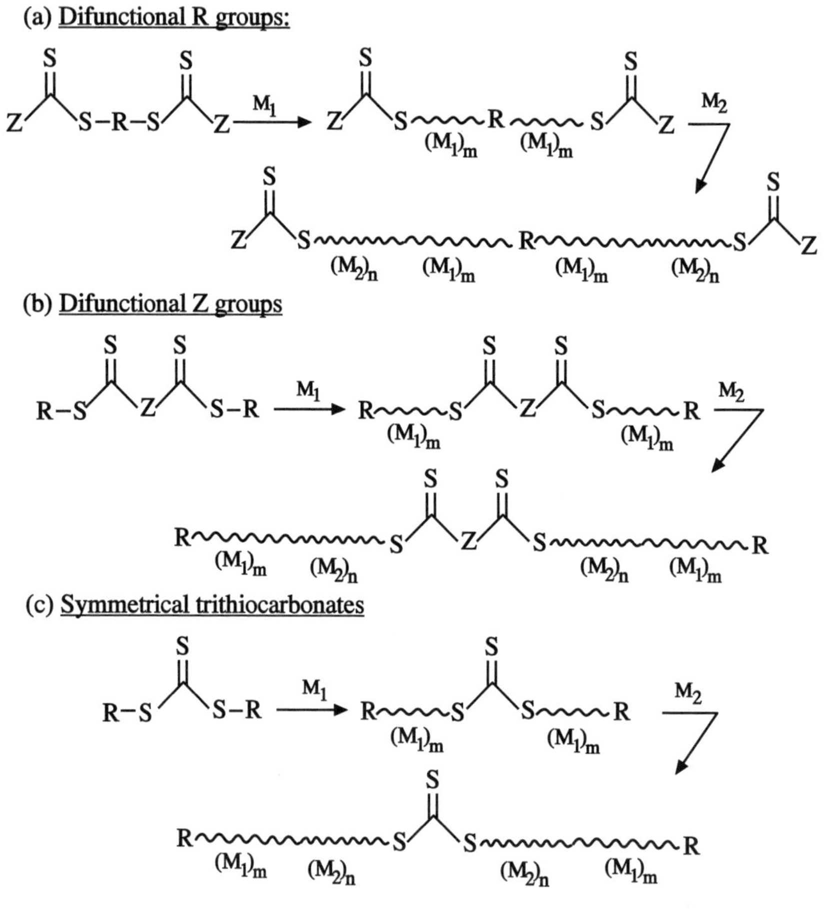

Figure 11.38 RAFT synthesis of ABA type triblock copolymers by sequential monomer addition via mediation by CTAs containing (a) difunctional R groups, (b) difunctional Z groups, and (c) symmetrical trithiocarbonates. (Adapted from Barner, Davis, Stenzel, and Barner-Kowollik, 2007.)

Three techniques may be employed for the RAFT synthesis of ABA type triblock copolymers by sequential monomer addition. As shown schematically in Fig. 11.38, these involve the use of (i) difunctional R groups, (ii) difunctional Z group, and (iii) symmetrical trithiocarbonates. By the use of a CTA with a difunctional R group, telechelic homopolymers and triblock copolymers of type ABA with dithiocarbonylthio functionality as the end groups can be obtained (see Fig. 11.38(a)). Similarly, the use of a CTA with a difunctional Z group can yield telechelic homopolymers and ABA type triblock copolymers with the leaving/reinitiating R group at the chain end (Fig. 11.38(B)). An alternative technique is the use of symmetrial trithiocarbonates, which contain two reinitiating R groups and lead to the formation of ABA type triblok copolymers with the reinitiating group attached to the chain end. A drawback of the last two techniques is the presence of the thiocarbonylthio moiety in the middle of the chain, which might be a weak link depending on the applications. A few examples of the above three types of CTAs that have been synthesized and used are shown in Fig. 11.39.

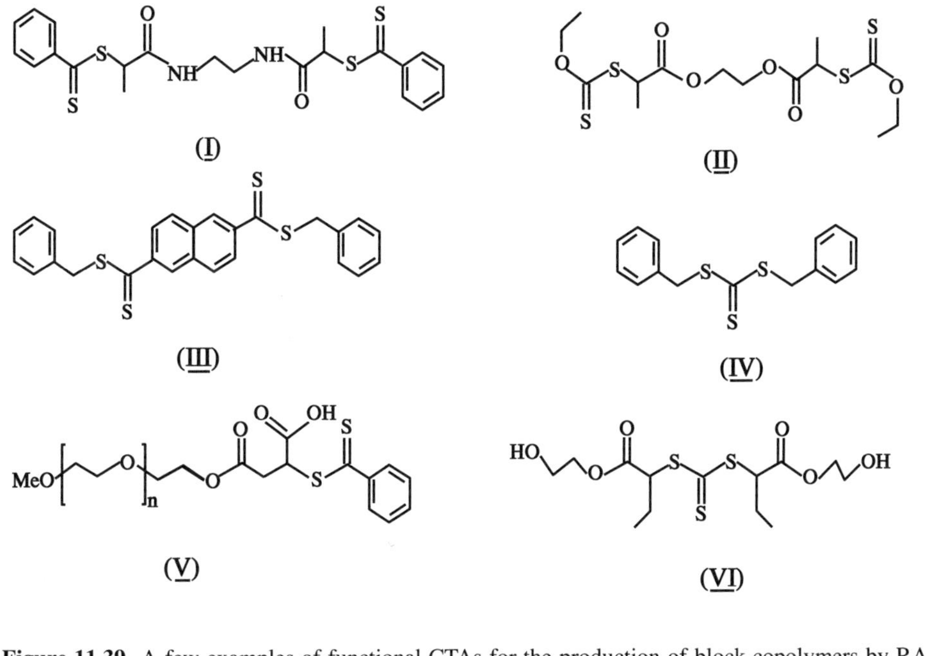

Figure 11.39 A few examples of functional CTAs for the production of block copolymers by RAFT technique: (**I**) a dithioester with difunctional R group, *N,N′*-ethylenebis[2-(thiobenzoylthio) propionamide (Donovan et al., 2003); (**II**) *O*-ethyl xanthate with difunctional R group (Taton et al., 2001); (**III**) a dithioester with difunctional Z group (Dureault et al., 2004); (**IV**) a trithiocarbonate with two identical R (benzyl) groups (Mayadunne et al., 2000); (**V**) a macro-CTA: poly(ethylene oxide) capped with dithiobenzoate group (Hong et al., 2004); (**VI**) S,S′-bis(2-hydroxyethyl-2′-butyrate) trithiocarbonate (You et al., 2004).

Problem 11.23 Mayadunne et al, (2000) prepared an ABA triblock copolymer, comprising polystyrene (PSt) end segments and a central block of poly(*n*-butyl acrylate) (P*n*BA) by first carrying out RAFT polymerization of St (densiy 0.91 g/cm^3) in bulk at 110°C for 20 h using thermal initiation and S,S-di(1-phenylethyl) trithiocarbonate (DPET) (0.017 M) as the RAFT agent to 40% monomer conversion, followed by chain extension of the resulting PSt (0.0016 M) with *n*-BA (2.79 M) using AIBN (0.073×10^{-2} M) initiator at 60°C

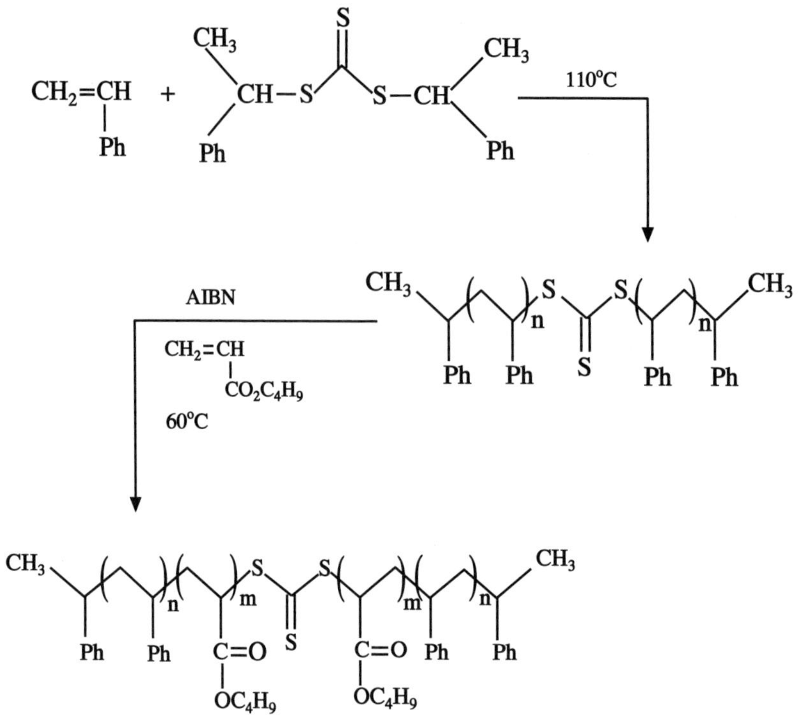

Scheme P11.23.1 Synthesis of PSt-*b*-P*n*BA-*b*-PSt triblock copolymer by RAFT technique. (Problem 11.23.)

for 8 h and 65% monomer conversion to afford the ABA triblock copolymer with M_n = 161,500 and M_w/M_n = 1.16. Calculate the molecular weights of the PSt end segments and P*n*BA central block, and hence the molecular weight of the triblock copolymer to compare with the experimental value.

Answer:

The two sequential steps for the preparation of PSt-*b*-P*n*BA-*n*-PSt triblock copolymer by RAFT polymerization using DPET as the chain transfer agent are shown in Scheme P11.23.1.

Molar masses (M): M_{St} = 104 g mol^{-1}, M_{nBA} = 128 g mol^{-1}

$$[St] = \frac{(1000\,cm^3\,L^{-1})(0.91\,g\,cm^{-3})}{(104\,g\,mol^{-1})} = 8.75\ mol\,L^{-1}$$

[CTA] = [DPET] = 0.0173 mol L^{-1}

Neglecting the relatively small number of chains formed by thermal initiation, M_n of each PSt segment

$$= \frac{[St] \times Conversion}{2 \times [CTA]} \times M_{St}$$

$$= \frac{(8.75\ mol\,L^{-1})(0.40)(104\,g\,mol^{-1})}{2(0.0173\ mol\,L^{-1})} = 10520\ g\,mol^{-1}$$

M_n of PSt = 2×10520 g mol^{-1} = 21040 g mol^{-1}

M_n of P*n*BA (neglecting central trithiocarbonate group)

$$= \frac{[n\text{BA}] \times \text{Conversion}}{[\text{PSt}]} (M_{n\text{BA}})$$

$$= \frac{(2.79 \text{ mol L}^{-1})(0.65)}{(0.0016 \text{ mol L}^{-1})} (128 \text{ g mol}^{-1}) = 1,45,080 \text{ g mol}^{-1}$$

M_n of triblock copolymer $= 1,45,080 + 21,040$ or $1,66,120$ g mol^{-1}

11.5.3.2 Macro-CTA Method

As we have seen earlier, a polymeric CTA (macro-CTA), that is a polymeric chain capped with a dithioester group, is formed as an intermediate in the RAFT synthesis of block copolymers by sequential monomer addition. Macro-CTAs can also be made from already existing macro-molecules via organic synthetic transformation and used as the RAFT agent to synthesize several kinds of block copolymers. The method has the advantage that block copolymers containing poly-mers produced by other (nonradical) mechanisms, e.g., poly(ethylene oxide) (PEO), can easily be synthesized. Moreover, if the initial polymeric chains are modified at both ends, triblock (ABA) or multiblock (CBABC, etc.) copolymers can be obtained. Thus, Hong et al. (2004) produced block copolymers based on PEO by the modification of monohydroxy and dihydroxy PEOs into monofunctional and difunctional CTAs, respectively. The modification of monohydroxy PEO with dithiobenzoic acid led to the monofunctional macro-CTA, which was used to mediate the RAFT polymerization of *N*-isopropylacrylamide to yield the diblock copolymer poly(ethylene oxide-*b*-*N*-isopropylacrylamide) with a controlled molecular weight and PDI < 1.2. A similar proce-dure starting with dihydroxy PEO yielded the triblock copolymer poly(*N*-isopropylacrylamide-*b*-polyethylene oxide-*b*-*N*-isopropylacrylamide).

11.5.4 Star (Co)polymers

A variety of multifunctional CTAs have been used for the preparation of star (co)polymers via RAFT polymerization, the core (or hub) of the star being introduced via functionalization of either the R substituent (R-group approach or R approach) or the Z substituent (Z-group approach or Z approach). Typical examples of functional CTAs used for the production of star (co)polymers in R- and Z-group approaches are shown in Fig. 11.40. The difference between the two approaches is shown schematically in Fig. 11.41, while a mechanism for RAFT star polymerization proposed by Barner et al. (2007) is presented in Fig. 11.42.

In the R-group approach (Fig. 11.42), the RAFT polymerization takes place from a multifunc-tional CTA with the R group serving as the core and the results of the overall process are similar to those obtained from ATRP or NMP (i.e., the polymeric arms grow away from the core). In the Z-group approach, on the other hand, the Z group serves as the core for star formation. The mechanisms of RAFT star polymerization for R- vs. Z-group approach are shown in Fig. 11.42. In the R-group approach, the dithioester moiety leaves the core and the detached moiety mediates the polymerization, while the core itself becomes a radical where monomer addition takes place (the so-called *attached-to process*). As a consequence, the core can undergo radical-radical coupling reactions, thereby preventing the formation of well-defined polymeric material with low polydis-persity (Barner et al., 2007). Therefore, as a general rule, for the polymerization of monomers for which the main termination route is by combination/coupling, conversion should be kept low to limit star-star coupling. Furthermore, the initiator used to trigger the polymerization should be kept at a very low concentration in order to minimize the number of dead polymeric chains.

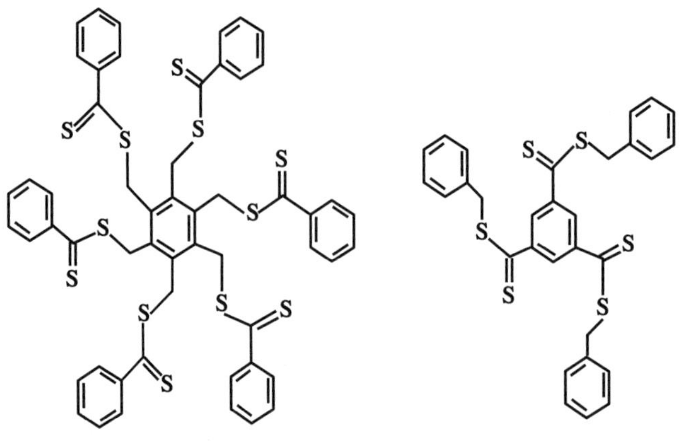

Figure 11.40 Functional CTAs for the production of star (co)polymers: (I) Hexakis (thiobenzoylthiomethyl) benzene (Stenzel-Rosenbaum et al., 2001), used in R-approach; (II) tri(thiobenzoylthiomethyl) benzene (Dureault et al., 2004), used in Z-approach.

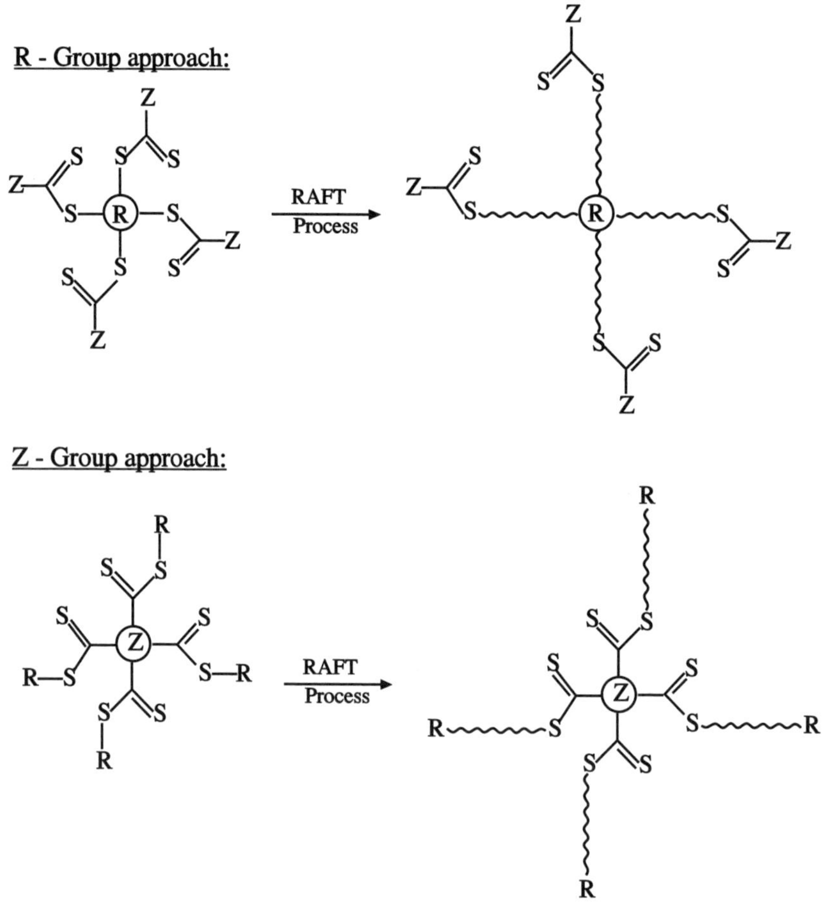

Figure 11.41 Overall processes of star polymer formation by RAFT polymerization with multifunctional CTAs: R- vs. Z-group approach. (Adapted from Barner, et al., 2007.)

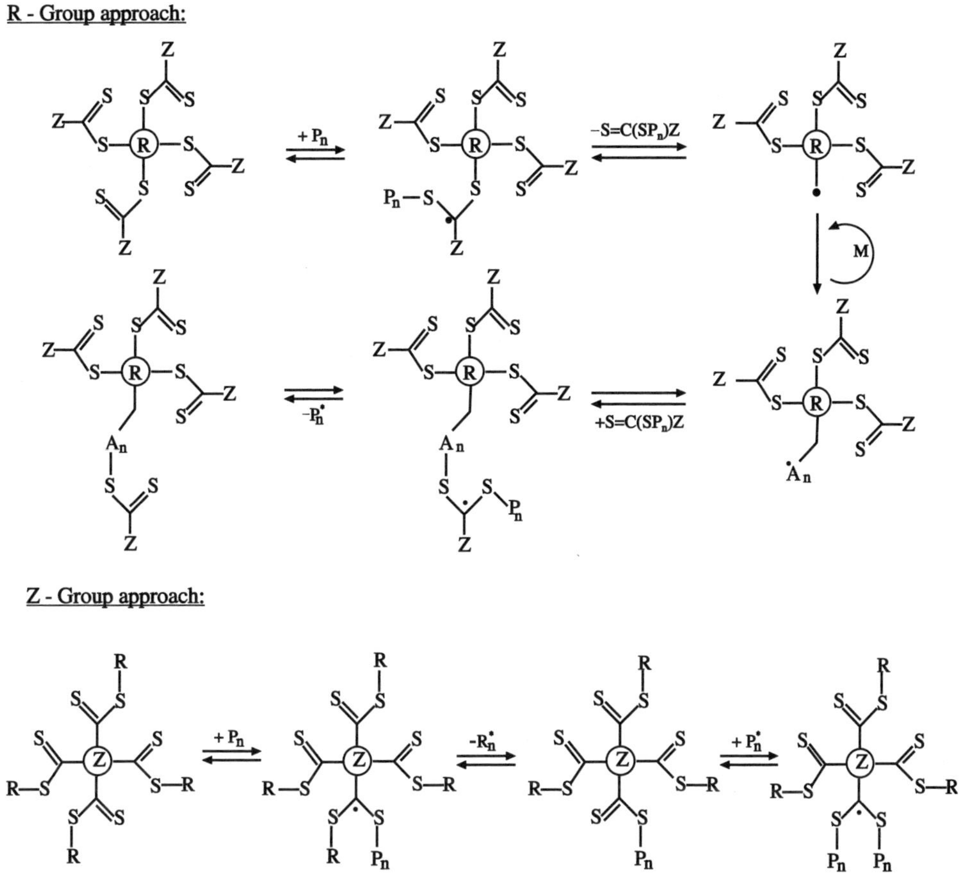

R - Group approach:

Z - Group approach:

Figure 11.42 RAFT star polymerization mechanisms (pre-equilibrium and main equilibrium stages): R- vs. Z-group approach (shown for one arm only). Formation mechanisms of chain radicals $P_n{}^\bullet$ and $P_m{}^\bullet$ are same for R- and Z-group approaches and same as shown in Fig.11.36. (Adapted from Barner et al., 2007.)

On the other hand, when the Z group is used as a core (Fig. 11.42) for star polymers (Z-group approach), the polymeric arms are detached from the core while they grow, and they react back onto the core for the chain transfer reaction (the so-called *away-from process*). The Z-group approach thus implies that the dithioester moiety (RAFT group) remains permanently tethered to the core and consequently the core will not carry any propagating radical functions, thereby avoiding core-core coupling reactions and the formation of higher-order coupling products. The possibility of generating extremely pure star polymer products up to high conversions without the added complexity of cross-coupling reactions is inherent in the Z-group approach of the RAFT process. It thus appears that a Z-group approach is to be preferred over an R-group approach. (It may be noted that in ATRP and NMP, in comparison, the core is always a radical carrying species and so core-core coupling reactions may potentially always occur.) Since in the case of Z-group approach, the dithioester moiety is permanently bonded to the core and the growing macroradicals are detached, with increasing conversion an effective chain equilibration is increasingly hindered due to the shielding effects of the growing polymer chains. However, the access difficulties notwithstanding, the Z-group approach RAFT polymerization is considered to present the better alternative to generate well-defined star polymer material of various monomers (Barner et al., 2007).

Problem 11.24   Besides choosing the Z-approach, what are the key factors that determine the success of the RAFT star polymerization process to achieve large proportion of well-defined star polymers of controlled structures?.

Answer:

The key factors for successful RAFT synthesis of well controlled star (co)polymers are (Barner et al., 2007) : (i) *minimization of linear chain contaminants*, which can be achieved by employing a monomer with a high propagation rate coefficient, while at the same time having a low rate of primary radical delivery into the system, and (ii) *minimization of star-star* (bimolecular radical) *coupling reactions*, which can be achieved by (a) minimizing the number of radicals introduced into the polymerization system to achieve a given conversion after a given time and (b) reducing the propagating radical concentration in the RAFT polymerization.

11.5.5   Branched (Co)polymers

The synthesis of branched (co)polymers can be achieved by (a) grafting from polymeric backbone and (b) using macromonomers. The grafting from approach uses a polymeric chain as a support from which polymeric branches are grown. This can be done in a facile manner by RAFT polymerization using either the R-group (or attached-to) approach or the Z-group (or away-from) approach via synthesis of appropriate polymer-supported RAFT agents. Quinn et al. (2002) thus synthesized comb, star, and graft copolymers of styrene (St) by the RAFT tehnique. The precursors required for these reactions were synthesized readily from RAFT-prepared poly(vinyl benzyl chloride) (PVBC) and poly(St-*co*-VBC) by facile substitution of the chlorine atom with a dithiobenzoate group [–S–C(=S)Ph] to yield intrinsically well-defined star and comb precursors. [Typically, a solution of previously synthesized PVBC and poly(St-*c*-VBC) in THF was refluxed with a 100% excess of sodium dithiobenzoate for 1 h and the sodium cloride formed in the reaction was filtered off.] Comb and star polymers were then synthesized from the precursors, poly(St-*co*-vinyl benzyl dithiobenzate) and poly(vinyl benzyl dithiobenzoate), respectively, by heating the respective solutions in styrene at 60°C in the presence of AIBN.

In the macromonomer approach, polymeric chains (of monomer M_1) end-functionalized with a vinyl group are homopolymerized (with M_1) or copolymerized (with M_2) to produce branched (co)polymers. Thus, poly(ethylene glycol) methyl ether methacrylate (Chen et al., 2003) and poly(dimethyl siloxane) methacrylate (Shinoda and Matyjaszewski, 2001) have been homo- or copolymerized to yield graft (co) polymers with the corresponding pendant chain.

11.5.6   Surface Modification

Surface modification of nanoparticles with synthetic polymers is of great interest due to their potential application in optics, electronics, and engineering. The RAFT polymerization has proved to be a versatile tool to modify nanoparticle surfaces with a variety of functional polymers because the RAFT technique works with the greatest range of vinyl monomers. There are two approaches for covalently attaching polymers to nanoparticles, namely, "grafting to" and "grafting from" approaches. The "grafting to" approach involves reacting an end-functionalized polymer to an activated surface. This method often produces lower grafting density due to steric hindrance caused by grafted chains, though a combination of click chemistry (see Chapter 12) and RAFT could be used to modify surfaces with higher than normal grafting densities of the "grafting to" method, as has been demonstrated for silica particles (Ranjan and Brittain, 2007). Since in the "grafting from" approach, polymer chains grow from the surface and chain extension involves monomer diffusion to the surface, significantly higher graft densities can be achieved.

Among the various methods based on the RAFT technique, surface-initiated RAFT polymerization based on the "grafting from" approach is arguably the most promising one due to its ability to precisely control the structure of the grafted polymer chains, with a low-to-high range of graft densities (Li and Benicewicz, 2008). The "grafting from" approach can be achieved in two different ways for RAFT polymerization : (a) anchoring free radical initiator to the surface, followed by surface-initiated RAFT polymerization in the presence of free RAFT CTA in solution (Baum and Brittain, 2002); (b) surface immobilization of RAFT CTA, followed by surface mediated RAFT polymerization in the presence of free initiator. Baum and Brittain (2002) utilized RAFT polymerization to graft polystyrene and poly(methyl methacrylate) from silica particles using a surface-anchored azo initiator, while Tsujii et al. (2001) reported the first application of surface-initiated RAFT polymerization in the modification of silica particles using a surface-anchored RAFT agent. The latter approach, however, permits more flexible reaction conditions.

The surface immobilization of RAFT CTA can be performed using either (a) the R-group approach where the CTA is attached to the surface via the leaving and reinitiating group or (b) the Z-group approach where the CTA is attached to the surface via the stabilizing Z-group (Perrier et al., 2005). The R-group approach affords higher molecular weight and grafting density of the attached polymer, while the grafting density can be further increased by using trithiocarbonate-type RAFT CTA (Ranjan and Brittain, 2008). Generally, a silyl condensation reaction is used to immobilize a RAFT CTA onto a silica nanoparticles (Li and Benicewicz, 2008). More recently, the click reaction, which is more efficient, has been employed to attach RAFT CTA onto silica nanoparticles, using alkyne-functionalized RAFT CTA and azide functionalized silica nanoparticles to facilitate this coupling (Ranjan and Brittain, 2008). The process will be further elaborated in Chapter 12 (see Section 12.2.2).

11.5.7 Combination of RAFT and Other Polymerization Techniques

Combination of RAFT polymerization with other polymerization techniques allows synthesis of block copolymers containing polymer block segments that are derived by non-free radical processes. For example, You et al. (2004) and Hales et al. (2004) performed ring-opening polymerization initiated by a RAFT CTA, followed by RAFT polymerization to synthesize block copolymers from both vinyl and lactide monomers. Similarly, You et al. (2004) used the hydroxyl functionality of the R group from S,S'-bis(2-hydroxyethyl-2'-butyrate) trithiocarbonate (VI, Fig. 11.39) to initiate the ring-opening polymerization of lactide, followed by the RAFT polymerization of N-isopropylacrylamide to prepare poly(lactide-*b*-isopropylacrylamide-*b*-lactide) with good control of molecular weight and low polydispersity (1.2).

Because of the versatility of the RAFT polymerization involving the use of RAFT agents carrying many useful reactive end groups and the capability to polymerize many monomers that are inherently troublesome for other polymerization techniques, RAFT is also potentially an ideal polymerization technique to combine with cationic polymerization by site transformation (Magenau et al., 2009). This unique combination would allow the synthesis of a variety of block copolymers containing polymer segments derived by cationic polymerization with potentially new or greatly improved properties. As an example, Magenau et al. (2009) thus synthesized block copolymers consisting of polyisobutylene (PIB) and either poly(methyl methacrylate) or polystyrene block segments by combining living cationic and RAFT polymerizations. The initial PIB block was synthesized via quasi-living cationic polymerization of isobutylene and subsequently converted into a hydroxy-terminated PIB and further into a macro-chain-transfer agent (PIB-CTA) which was then employed in a RAFT polymerization of either methyl methacrylate or styrene as model monomers (see Exercise 11.30 and the corresponding answer in *Solutions Manual*). It was envisaged that the technique could be used for the synthesis of PIB-based copolymers with monomers traditionally inaccessible to PIB and cationic polymerization (Magenau et al., 2009).

The click reaction has recently been applied by polymer chemists, who recognized the opportunity of creating new polymer architectures by combining the click approach with other polymerization techniques. A range of complex structures including star polymers and block copolymers have thus been prepared by combining techniques such as ATRP, ROP, RAFT, polycondensation, and polymerization methods with the click chemistry techniques, such as, copper(I) catalyzed azide-alkyne coupling, Diels-Alder reactions, and thiol-ene reactions. These are discussed in the following chapter dealing with applications of click chemistry in polymerization.

11.5.8   *Transformation of RAFT Polymer End Groups*

A common feature of all polymers prepared by RAFT polymerization is that they bear a thiocarbonylthio group at one chain end or both. In addition to the relatively high cost of CTAs, the presence of the thiocarbonylthio end group, which is colored and rather unstable, especially under basic conditions or in the presence of primary or secondary amines (Chong et al., 2003), limits the use of the RAFT process for industrial applications. Much attention has therefore been paid to the removal/transformation of the thiocarbonylthio end groups into colorless groups that are either non-reactive or amenable to controlled modification. The reactions of the thiocarbonylthio group is well-known from small molecule chemistry and much of this knowledge has been shown to be applicable to the thiocarbonylthio group of RAFT-synthesized polymers. Thiocarbonylthio groups react with nucleophiles and ionic reducing agents (e.g., amines, hydroxide, borohydride) producing thiols, and also undergo reaction with various oxidizing agents including NaOCl, H_2O_2, *t*BuOOH, and peracids. However, these reactions may leave reactive groups and hence are not suitable for thiocarbonylthio group removal from RAFT polymers. Other approaches that offer more promise in this respect include (i) thermolysis, (ii) radical-induced reduction, (iii) radical exchange, and (iv) aminolysis/Michael addition sequence.

Thermolysis (Postma et al., 2005) of RAFT polymers results in the cleavage of the thiocabonylthio moieties leading to polymers with unsaturated end groups. Thermolysis has a clear advantage over other methods in that no chemical treatment is required. The method does, however, require that the polymer and any desired functionality are stable to the thermolysis conditions, such as heating to a temperature of 200°C or higher. This limits the range of polymers that can be treated by thermolysis.

Radical-induced reactions (reduction, termination) offer more promise since these processes provide desulfurization by complete end group removal/transfer. Radical-induced reduction of the residing thiocarbonylthio group to hydrogen is an attractive approach since it produces stable saturated end groups (Moad et al., 2005). However, the very effective and most commonly used reducing agent, tri-*n*-butylstannane (Bu_3SnH) is toxic and quantitative removal of excess tributyl stannane and derived byproducts is problematic in that traces of toxic byproducts can remain even after several purification steps. Figure 11.43 shows an idealized mechanism for trirhiocarbonate group removal by radical-induced reduction with Bu_3SnH.

Problem 11.25   Discuss a reaction protocol for the synthesis of a well-defined polystyrene with one end having a primary amine end group, $H–[-CH(C_6H_5)CH_2-]_n–NH_2$, through the use of RAFT techniques.

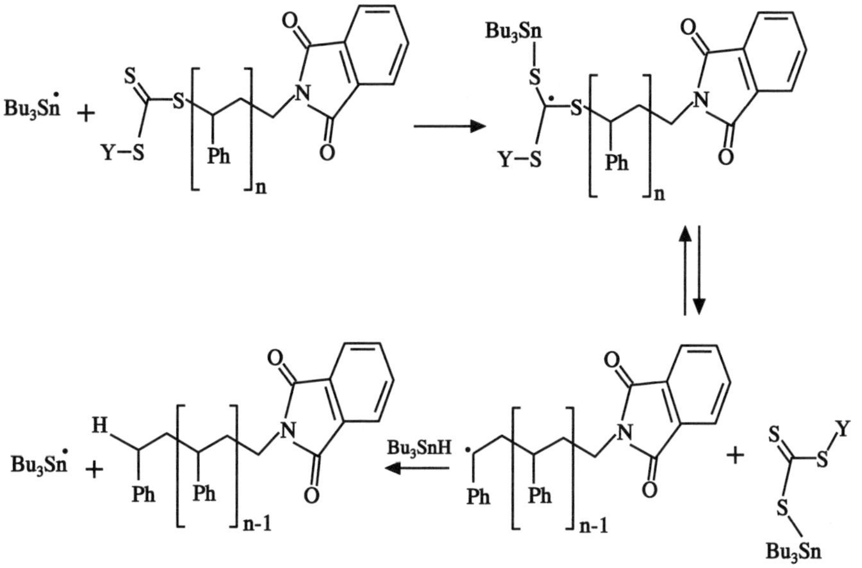

Figure 11.43 An idealized mechanism for trithiocarbonate group removal by radical-induced reduction with Bu₃SnH. (From Postma et al., 2006. With permission from *American Chemical Society*.)

Answer:

Postma et al. (2006) reported the synthesis of well-defined polystyrene (PSt) with primary amine end groups through the use of phthalimide-functional RAFT agents. Styrene (St) polymerization with the RAFT agent, butyl phthalimidomethyl trithiocarbonate (**P25-I**), as shown in Scheme P11.25.1, was conducted at 110°C with thermal initiation in bulk and the following reaction conditions : [RAFT]$_0$ = 0.0288 M and [St]$_0$/[RAFT]$_0$ = 303, achieving 70% monomer conversion in 24 h to obtain RAFT PSt (**P25-II**) with $\overline{M}_n$ = 22,400 g mol^{-1} and $\overline{M}_w/\overline{M}_n$ = 1.12. The polymerization was also successfully conducted at 60°C with AIBN initiation in bulk and reaction conditions : [RAFT]$_0$ = 0.24 M, [St]$_0$/[RAFT]$_0$ = 8.72, [AIBN]$_0$ = 7.69×10^{-2}, achieving 48% monomer conversion in 24 h to obtain $\overline{M}_n$ = 4610 g mol^{-1} and $\overline{M}_w/\overline{M}_n$ = 1.27 of the RAFT polystyrene (**P25-II**).

The clean removal of the thiocarbonylthio end group of (**P25-II**) with retention of the phthalimido end group, as revealed by NMR spectrum, was achieved by radical-induced reduction with tributyl stannane in the presence of AIBN. The replacement of the phthalimido end group from the resulting polymer, (**P25-III**), was then achieved by hydrazinolysis (Scheme P11.25.1). In a typical procedure (Postma et al., 2006), hydrazine hydrate was added to a solution of P25-III in DMF and the resultant mixture was stirred at room temperature overnight. Chloroform was added and the solution was washed with water and brine and dried over anhydrous MgSO₄. The PSt, (**P25-IV**), was isolated by precipitation into methanol and dried in a vacuum.

The radical exchange method which uses a large amount of free radical initiator for end group modification of RAFT polymers is simple and straightforward. The thiocarbonylthio end group is cleaved simply by mixing the RAFT polymer with an excess source of radicals. The process involves *in situ* addition of a radical to the reactive C=S bond of the thiocarbonylthio end group, leading to the formation of an intermediate radical, which can then fragment either back to the original attacking radical or toward a polymeric chain radical (Perrier et al., 2005; Perrier and Takolpuckdee, 2005). In the presence of an excess of free radicals, the equilibrium is displaced toward the chain radical, which can then combine irreversibly with one of the excess free radicals

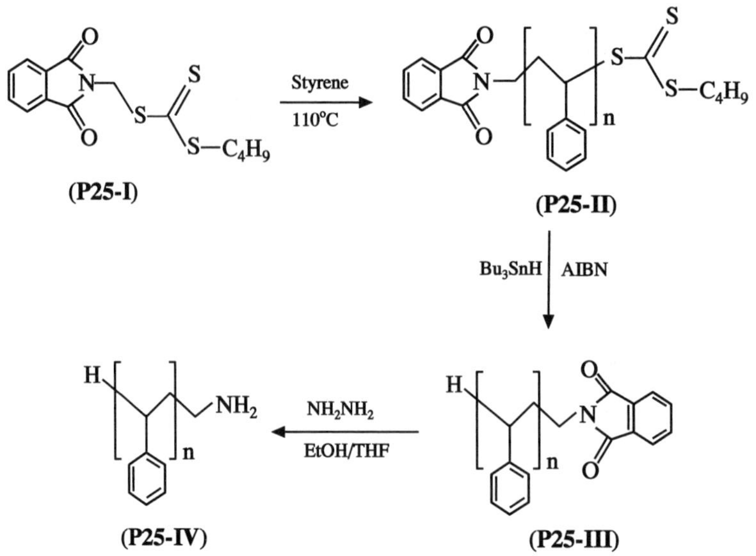

Scheme P11.25.1 Synthesis of polystyrene with primary amine end group through the use of a phthalimide-functional RAFT agent for initiation, removal of trithiocarbonate end group by radical-induced reduction with Bu$_3$SnH and hydrazinolysis of phthalimido end group. (Adapted from Postma et al., 2006.)

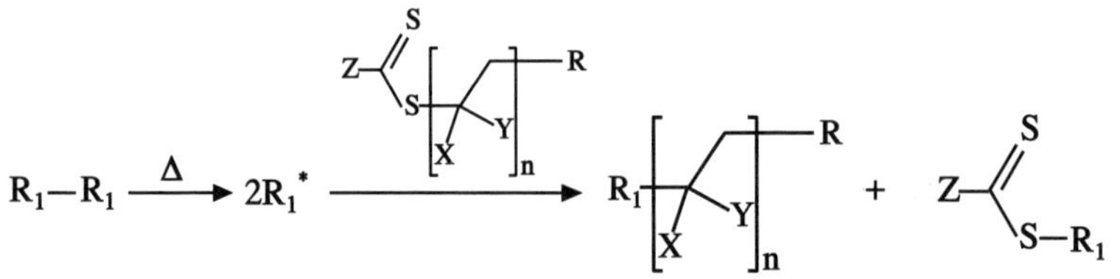

Figure 11.44 End-group modification of RAFT polymers and recovery of CTA by mixing the polymeric chains with an excess source of radicals. (Adapted from Perrier and Takolpuckdee, 2005.)

present in solution, leading to the formation of a dead polymer chain free of sulfur compounds, accompanied by recovery of the CTA, as shown in Fig. 11.44. The process removes the color from the polymer and introduces a new functionality at the end of the polymer chain. Thus, a RAFT-synthesized PMMA which is colored becomes fully decolorized after reaction with AIBN at 80°C for 2.5 h (Perrier and Takolpukdee, 2005).

Both primary and secondary amines, acting as nucleophiles, can convert thiocarbonylthio moieties of RAFT polymers to thiols (Mayadunne et al., 2000) and the latter can be further modified via Michael addition to α,β-unsaturated carbonyl derivative to obtain polymers with hydroxyl termini (see Section 12.2.2), both steps employing mild reaction conditions with good yields (Lima et al., 2005). Considering practical advantages, a one-pot process has also been developed that combines aminolysis of a thiocarbonylthio group and Michael addition of the resulting thiol to an α,β-unsaturated ester under mild conditions, thus transforming the labile thiocarbonylthio moiety into a stable, colorless thioether (Qiu and Winnik, 2006).

REFERENCES

An, Q. F., Qian, J. W., Yu, L. Y., Luo, Y. W, and Lin, X. Z., *J. Polym. Sci. Part A: Polym. Chem.*, **43**, 1973 (2005).

Ando, T., Kamigaito, M., and Sawamoto, M., *Tetrahedron*, **53**, 15445 (1997).

Ando, T., Kato, M., Kamigaito, M., and Sawamoto, M., *Macromolecules*, **29**, 1070 (1996).

Angot, S., Murthy, K. S., Taton, D., and Gnanou, Y., *Macromolecules*, **31**, 7218 (1998).

Asgarzadeh, F., Ourdouillie, P., Beyou, E., and Chaumont, P., *Macromolecules*, **32**, 6996 (1999).

Ashford, E. J., Naldi, V., O'Dell, R., Billingham, N. C., and Armes, S. P., *Chem. Commun.*, 1285 (1999).

Barner, L., Davis, T. P., Stenzel, M. H., and Barner-Kowollik, C., *Macromol. Rapid Commun.*, **28**, 539 (2007).

Barner-Kowollik, C., Quinn, J. F., Morsley, D. R., and Davis, T. P., *J. Polym. Sci. Part A: Polym. Chem.*, **39**, 1353 (2001).

Baum, M. and Brittain, W. J., *Macromolecules*, **35**, 610 (2002).

Beers, K. L., Gaynor, S. G., Matyjaszewski, K., Sheiko, S. S., and Moeller, M., *Macromolecules*, **31**, 9413 (1998).

Bernaerts, K. V., Schacht, E. H., Goethals, E. J., and Prez, F. E. D., *J. Polym. Sci. Part A: Polym. Chem.*, **41**, 3206 (2003).

Bledzki, A., Braun, D., and Titzschkau, K., *Makromol. Chem.*, **184**, 745 (1983).

Boerner, H. G., Beers, K., Matyjaszewski, K., Sheiko, S. S., and Moeller, M., *Macromolecules*, **34**, 4375 (2001).

Boutevin, B., *J. Polym. Sci. Part A: Polym. Chem.*, **38**, 3235 (2000).

Cao, J., He, J., Li, C., and Yang, Y., *Polym. J.*, **33**, 75 (2001).

Cassenbras, M., Pascual, S., Polton, A., Tardi, M., and Vairon, J.-P., *Macromol. Rapid Commun.* **20**, 261 (1999).

Catala, J. M., Bubel, F., and Hammouch, S. O., *Macromolecules*, **28**, 8441 (1995).

Chan, N., Cunningham, M. F., and Hutchinson, R. A., *Macromol. React. Eng.*, **4**, 369 (2010).

Charleux, B., *ACS Symp. Ser.*, **854**, 438 (2003).

Charmot, D., Corpart, P., Michelet, D., Zard, S., Biadatti, T. (Rhodia Chemie). PCT Patent WO 9858974, 1998; Chem. Abstr. **130**, 82018 (1999).

Charmot, D., Corpart, P., Adam, H., Zard, S. Z., Biadatti, T., and Bouhadir, G., *Macromol. Symp*, **150**, 23 (2000).

Chen, Y., Ying, L., Yu, W., Kang, E. T., and Nash, K. G., *Macromolecules*, **36**, 9451 (2003).

Chiefari, J., Chong, Y. K., Ercole, F., Krstina, J., Jefery, J., Le, L. P. T., Mayadunne, R. T. A., Meijs, G. F., Moad, C. L., Moad, G., Rizzardo, E., and Thang, S. H., *Macromolecules*, **31**, 5559 (1998).

Chiefari, J., Mayadunne, R. T. A., Moad, C. L., Moad, G., Rizzardo, E., Postma, A., Skidmore, M. A., and Thang, S. H., *Macromolecules*, **36**, 2273 (2003).

Chong, Y. K., Krstina, J., Le, T. P. T., Moad, G., Postma, A., Rizzardo, E., and Thang, S. H., *Macromolecules*, **36**, 2256 (2003).

Chong, Y. K., Le, T. P. T., Moad, G., Rizzardo, E., and Thang, S. H., *Macromolecules*, **32**, 2071, (1999).

Cianga, I., Hepuzev, Y., Serhatli, E., and Yagci, Y., *J. Polym. Sci. Part A: Polym. Chem.*, **40**, 2199 (2002).

Coessens, V. and Matyjaszewski, K., *Macromol. Rapid Commun.*, **20**, 66, 127 (1999).

Coessens, V., Nakagawa, Y., and Matyjaszewski, K., *Polym. Bull.*, **40**, 135 (1998).

Coessens, V., Pyun, J., Miller, P. J., Gaynor, S. G., and Matyjaszewski, K., *Macromol. Rapid Commun.*, **21**, 103 (2000).

Corpart, P., Charmot, D., Biadatti, T., Zard, S. Z., Michelet, D., PCT Intl. Pat. Appl. WO 9858974, 1998.

Crivello, J. V., Lee, J. L., and Coulon, D. A., *J. Polym. Sci., Polym. Chem. Ed.*, **24**, 1251 (1986).

Cunningham, M. F., *Prog. Polym. Sci.*, **27**, 1039 (2002).

David, B. A., Kinning, D. J., Thomas, E. L., and Fetters, L. J., *Macromolecules*, **19**, 215 (1986).

Davis, T. P., Barner-Kowollik, C., Nguyen, T. L. U., Stenzel, M. H., Quinn, J. F., and Vana, P., *ACS Symp. Ser.*, **854**, 551 (2003).

Davis, K. A., Charleux, B., and Matyjaszewski, K., *J. Polym. Sci. Part A: Polym. Chem.*, **38**, 2274 (2000).

Davis, K. A. and Matyjaszewski, K., *Macromolecules*, **33**, 4039 (2000).

Davis, K. A. and Matyjaszewski, K., *Macromolecules*, **34**, 2101 (2001).

Davis, K. A., Paik, H.-J., and Matyjaszewski, K., *Macromolecules*, **32**, 1767 (1999).

Destarac, M., Bessiare, J.-M., and Boutevin, B., *Makromol. Chem. Rapid Commun.*, **18**, 967 (1997).

Destarac, M., Bzducha, W., Taton, D., Gauthier-Gillaizeau, I., Zard, S. Z., *Macromol. Rapid Commun.*, **23**, 1049 (2002).

Destarac, M., Charmot, D., Franck, X., and Zard, S. Z., *Macromol. Rapid Commun.*, **21**, 1035 (2000).

Doi, Y., Suzuki, S., and Soga, K., *Macromolecules*, **19**(12), 2896 (1986).

Doi, Y., Ueki, S., and Keii, T., *Macromolecules*, **12**(5), 814 (1979).

Dong, H. and Matyjaszewski, K. *Macromolecules*, **41**, 6868 (2008).

Dong, H., Tang, W., and Matyjaszewski, K., *Macromolecules*, **40**, 2974 (2007).

Donovan, M. S., Lowe, A. B., Sanford, T. A., and McCormick, C. L., *J. Polym. Sci. Part A: Polym. Chem.*, **41**, 1262 (2003).

Dubois, P., Ropsen, N., Jérome, R., and Teyssié, P., *Macromolecules*, **29**, 1965 (1996).

Duda, A., *Macromolecules*, **29**, 1399 (1996).

Dureault, A., Taton, D., Destarac, M., Leising, F., and Gnanou, Y., *Macromolecules*, **37**, 5513 (2004).

Eschwey, H., Hallensleben, M. L., and Burchard, W., *Makromol. Chem.*, **173**, 235 (1973).

Feng, X.-S. and Pan, C.-Y., *Macromolecules*, **35**, 4888 (2002).

Fischer, H., *Macromolecules*, **30**, 5666 (1997).

Fischer, H., *Chem. Rev.*, **101**, 3581 (2001).

Fónagy, T., Iván, B., and Szesztay, M., *Macromol. Rapid Commun.*, **19**, 479 (1998).

Frechét, J. M. J., Henmi, M., Gitsov, I., Aoshima, S., Leduc, M., and Grubbs, R. B., *Science*, **269**, 1080 (1995).

Fukuda, T., Terauchi, T., Goto, A., Ohno, K., Tsujii, Y., Miyamoto, T., Kobatako, S., and Yamada, B., *Macromolecules*, **29**, 6393 (1996).

Fukuda, T., Terauchi, T., Goto, A., Tsujii, Y., Miyamoto, T., and Shimizu, Y., *Macromolecules*, **29**, 3050 (1996).

Gaillard, N., Guyot, A., Claverie, J., *J. Polym. Sci. Part A: Polym. Chem.*, **41**, 684 (2003).

Georges, M. F., Veregin, R. P. N., Kazmaier, P. M., and Hamer, G. K., *Macromolecules*, **26**, 2987 (1993).

Georges, M. F., Veregin, R. P. N., Kazmaier, P. M., and Hamer, G. K., *Trends Polym. Sci.*, **2**(2), 66 (1994).

Guo, Y. M., Pan, C., and Wang, J., *J. Polym. Sci. Part A: Polym. Chem.*, **39**, 2134 (2001).

Haddleton, D. M., Jasieczek, C. B., Hannon, M. J., and Shooter, A. S., *Macromolecules*, **30**, 2190 (1997).

Hadjichristidis, N., Guyot, A. N., and Fetters, L. J., *Macromolecules*, **11**, 668 (1978).

Hales, M. Barner-Kowollik, C., Davis, T. P., and Stenzel, M. H., *Langmuir*, **20**, 10809 (2004).

Hammouch, O.S. and Catal, J. M., *Macromol. Rapid Commun.*, **17**, 149 (1996).

Hawker, C. J., *J. Am. Chem. Soc.*, **116**, 11314 (1994).

Hawker, C. J., *Angew. Chem., Int. Ed. Engl.*, **34**, 1456 (1995).

Hawker, C. J., Bosman, A. W., and Harth, E., *Chem. Rev.*, **101**, 3661 (2001).

Hawker, C. J., Elce, E., Dao, J., Volksen, W., Russel, T. P., and Barclay, G. G., *Macromolecules*, **29**, 2686 (1996).

Hawker, C. J., Hedrick, J. L., Malmström, E. E., Trollsas, M., Mecerreyes, D., Moineau, G., Dubois, Ph., and Jerome, R., *Macromolecules*, **31**, 213 (1998).

Hedrick, J. L., Trollsas, M., Hawker, C. J., Atthuff, B., Claesson, H., Heise, A., Miller, R. D., Mecerreyes, D., Jérôme, R., and Dubois, P., *Macromolecules*, **31**, 8691 (1998).

Heise, A., Hedrick, J. L., Trollsas, M., Miller, R. D., and Frank, C. W., *Macromolecules*, **32**, 231 (1999).

Hong, S. C., Pakula, T., and Matyjaszewski, K., *Polym. Met. Sci. Eng.*, **84**, 767 (2001)

Hong, C.-Y., You, Y.-Z., and Pan, C.-Y., *J. Polym. Sci. Part A: Polym. Chem.*, **42**, 4873 (2004).

Jacob, S., Majoros, I., and Kennedy, J. P., *Macromolecules*, **29**, 8631 (1996).

Johnson, C. H. J., Moad, G., Solomon, D. H., Spurling, T. H., and Vearing, D. J., *Aust. J. Chem.*, **43**, 1215 (1990).

Jousset, S., Hammouch, S. O., and Catala, J. M., *Macromolecules*, **30**, 6685 (1997).

Kajiwara, A. and Matyjaszewski, K., *Macromolecules*, **31**, 3489 (1998).

Kasko, A. M., Heintz, A. M., and Pugh, C., *Macromolecules*, **31**, 256 (1998).

Keoshkerian, B., Georges, M. K., and Boils-Boissier, D., *Macromolecules*, **28**, 6381 (1995).

Koulouri, E. G., Kallitsis, J. K., and Hadziioannou, G., *Macromolecules*, **32**, 6242 (1999).

Kwak, Y., Goto, A., Fukuda, T., *Macromolecules*, **37**, 1219 (2004).

Kwak, Y., Goto, A., Tsujii, Y., Murata, Y., Komatsu, K., and Fukuda, T., *Macromolecules*, **35**, 3026 (2002).

Le, T. P., Moad, G., Rizzardo, E., Thange, S. H., PCT Int. Pat. Appl. WO 9801478, 1998; Chem. Abstr., **128**, 115390 (1998).

Li, Y. and Benicewicz, B. C., *Macromolecules*, **41**, 7986 (2008).

Lima, V., Jiang, X., Brokkenzijp, J., Shoenmakers, P. J., Klumperman, B., and Linde, R. V. D., *J. Polym. Sci. Part A: Polym. Chem.*, **43**, 959 (2005).

Litvinenko, G. and Müller, A. H. E., *Macromolecules*, **30**, 1253 (1997).

Liu, S. and Sen, A., *Macromolecules*, **33**, 5106 (2000).

Liu, S. and Sen, A., *Polym. Prepr.* (Am. Chem. Soc., Div. Polym. Chem.), **41**(2), 1573 (2000).

MacLeod, P. J., Keoshkerian, B., Odel, P., and Georges, M. K., *Proc. Am. Chem. Soc.*, Div. Polym. Mater. Sci. Engg., **80**, 539 (1999).

Magenau, A. J. D., Martinez-Castro, N., and Storey, R. F., *Macromolecules*, **42**, 2353 (2009).

Matyjaszewski, K., *J. Phys. Org. Chem.*, **8**, 197 (1995).

Matyjaszewski, K., Coessens, V., Nakagawa, Y., Xia, J., Qiu, J., Gaynor, S., Coca, S., and Jasieczek, C., *ACS Symp. Ser.*, **704**, 16 (1998).

Matyjaszewski, K., Gaynor, S., and Wang, J-S., *Macromolecules*, **28**, 2093 (1995).

Matyjaszewski, K., Goebelt, B., Paik, H.-J., and Horwitz, C. P., *Macromolecules*, **34**, 430 (2001).

Matyjaszewski, K., Jakubowski, W., Min, K., Tang, W., Huang, J., Braunecker, W., and Tsarevsky, N. V., *Proc. Nat. Acad. Sci. (USA)*, **103**(42), 15309 (2006).

Matyjaszewski, K., Nakagawa, Y., and Gaynor, S. G., *Macromol. Rapid Commun.*, **18**, 1057 (1997).

Matyjaszewski, K., Nakagawa, K., and Jasieczek, C. G., *Macromolecules*, **31**, 1535 (1998).

Matyjaszewski, K., Patten, T. E., and Xia, J., *J. Am. Chem. Soc.*, **119**, 674 (1997).

Matyjaszewski, K., Shipp, D. A., Wang, J.-L., Grimaud, T., and Patten, T. E., *Macromolecules*, **31**, 6836 (1998).

Matyjaszewski, K., Teodorescu, M., Miller, P. J., and Peterson, M. L., *J. Polym. Sci. Part A: Polym. Chem.*, **38**, 2440 (2000).

Matyjaszewski, K. and Xia, J., *Chem. Rev.*, **101**, 2921 (2001).

Mayadunne, R. A., Jeffery, J., Moad, G., and Rizzardo, E., *Macromolecules*, **36**, 1505 (2003).

Mayadunne, R. A., Moad, G., and Rizzardo, E., *Tetrahedron Lett.*, **43**, 6811 (2002).

Mayadunne, R. T. A,, Rizzardo, E., Chiefari, J., Chong, Y. K., Moad, G., Thang, S. H., *Macromolecules*, **32**, 6977 (1999).

Mayadunne, R. T. A., Rizzardo, E., Chiefari, J., Kristna, J., Moad, G., Postma, A., and Thang, S. H., *Macromolecules*, **33**, 243 (2000).

Mecerreyes, D., Atthoff, B., Boduch, K. A., Trollsas, M., and Hedrick, J. L., *Macromolecules*, **32**, 5175 (1999).

Mecerreyes, D., Moineau, G., Dubois, P., Jérome, R., Hedrick, J. L., Hawker, C. J., Malmström, E. E., and Trollsas, M., *Angew. Chem. Int. Ed.*, **37**(9), 1274 (1998).

Meyer, U., Palmans, A. R. A., Loontjens, T., and Heise, A., *Macromolecules*, **35**, 2873 (2002).

Miller, P. J. and Matyjaszewski, K., *Macromolecules*, **32**, 8760 (1999).

Moad, G., Chong, Y. K., Rizzardo, E., Postma, A., and Thang, S. H., *Polymer*, **46**, 8458 (2005).

Moad, G., Moad, C. L., Rizzardo, E., Thang, S. H., *Macromolecules*, **29**, 7717 (1996).

Monteiro, M. J., *J. Polym. Sci. Part A: Polym. Chem.*, **43**, 3189 (2005).

Monteiro, M. J. and de Brouwer, H., *Macromolecules*, **34**, 349 (2001).

Morton, M., Helminiak, T. E., Gadkary, S. D., and Bueche, F., *J. Polym. Sci. Part A: Polym. Chem.*, **57**, 471 (1962).

Nicolas, J., Charleux, B., Guerret, O., and Magnet, S., *Angew. Chem. Int. Ed.*, **43**, 6186 (2004).

Nicolas, J., Charleux, B., Guerret, O., and Magnet, S., *Macromoleules*, **37**, 4453 (2004).

Nishimura, M., Kamigaito, M., and Sawamoto, M., *Polym. Prepr.* (Am. Chem. Soc., Div. Polym. Chem.) **40**(2), 470 (1999).

Otsu, T. and Kuriyama, A., *Polym. Bull.* (Berlin), **11**, 135 (1984).

Otsu, T., Matsumoto, A., and Tazaki, T, *Polym. Bull.* (Berlin), **17**, 323 (1987).

Paik, H.-j., Gaynor, S. G., and Matyjaszewski, K., *Macromol. Rapid Commun.*, **19**, 47 (1998).

Penczek, S., Kubisa, P., and Szymanski, R., *Macromol. Rapid Commun.* **12**, 77 (1991).

Percec, V., Barboiu, B., Bera, T. K., van der Sluis, M., Grubbs, R. B., and Frechét, J. M. J., *J. Polym. Sci. Part A: Polym. Chem.*, **38**, 4776 (2000).

Percec, V., Barboiu, B., Neimann, A., Rond, J. C., and Zhao, M., *Macromolecules*, **29**, 3665 (1996).

Perrier, S., Armes, S. P., Wang, X. S., Malet, F., and Haddleton, D. M., *J. Polym. Sci. Part A: Polym. Chem.*, **39**, 1696 (2001).

Perrier, S. and Takolpuckdee, *J. Polym. Sci. Part A: Polym. Chem.*, **43**, 5347 (2005).

Perrier, S., Takolpuckdee, P., and Mars, C. A., *Macromolecules*, **38**, 6770 (2005).

Pietrasik, J., Dong, H., and Matyjaszewski, K., *Macromolecules*, **39**, 6384 (2006).

Pintauer, T. and Matyjaszewski, K., *Chem. Soc. Rev.*, **37**, 1087 (2008).

Postma, A., Davis, T. P., Evans, R. A., Li, G., Moad, G., and O'Shea, M. S., *Macromolecules*, **39**, 5293 (2006).

Postma, A., Davis, T. P., Moad, G., and O'Shea, M. S., *Macromolecules*, **38**, 5371 (2005).

Prodpran, T., Dimonic, V. L., Sudol, E. D., and El-Aasser, M. S., *Macromol. Symp.*, **155**, 1 (2000).

Puts, R. D. and Sogah, D. Y., *Macromolecules*, **30**, 7050 (1997).

Qiu, J., Charleux, B., and Matyjaszewski, K., *Prog. Polym. Sci.*, **26**, 2083 (2001).

Qiu, J. and Matyjaszewski, K., *Acta Polym.*, **48**, 169 (1997).

Qiu, X.-P. and Winnik, F. M., *Macromol. Rapid Commun*, **27**, 1648 (2006).

Quinn, J. F., Chaplin, R. P., and Davis, T. P., *J. Polym. Sci. Part A: Polym. Chem.*, **40**, 2956 (2002).

Quirk, R. P. and Lee, B., *Polym. Intl.*, **27**, 35((1992).

Ranjan, R. and Brittain, W. J., *Macromolecules*, **40**, 6217 (2007).

Ranjan, R. and Brittain, W. J., *Macromol. Rapid Commun.*, **29**, 1104 (2008).

Roos, S. G., Mueller, A. H. E., and Matyjaszewski, K., *Macromolecules*, **32**, 8331 (1999).

Roos, S. G., Mueller, A. H. E., and Matyjaszewski, K., *ACS Symp. Ser.*, **768**, 361 (2000).

Schappacher, M. and Deffieux, A. *Macromolecules*, **25**, 6744 (1992).

Schilli, C., Lanzendoerfer, M. G., Mueller, A. H. E., *Macromolecules*, **35**, 6819 (2002).

Schrock, R. R., Feldman, J., Cannizzo, L. F., and Grubbs, R. H., *Macromolecules*, **20**(5), 1169 (1987).

Scolcard, J. D. and McConville, D. H., *J. Am. Chem. Soc.*, **118** (41), 10008 (1996).

Shinoda, H. and Matyjaszewski, K., *Macromol. Rapid Commun.*, **22**, 1176 (2001).

Shipp, D. A., Wang, J.-L., and Matyjaszewski, K., *Macromolecules*, **31**, 8005 (1998).

Shohi, H., Sawamoto, M., and Higashimura, T., *Makromol. Chem.*, **193**, 2027 (1992).

Stenzel-Rosenbaum, M., Davis, T. P., Chen, V., and Fano, A. G., *J. Polym. Sci. Part A: Polym. Chem.*, **39**, 2777 (2001).

Szwarc, M., *Nature*, **178**, 1168 (1956).

Szwarc, M., Levy, M., and Milkovich, R. J., *J. Am. Chem. Soc.*, **78**, 2656 (1956).

Tang, C. B., Kowaleski, T., and Matyjaszewski, K., *Macromolecules*, **36**, 8587 (2003).

Tang, W., Kwak, Y., Braunecker, W., Tsarevsky, N. V., Coote, M. L., and Matyjaszewski, K., *J. Am. Chem. Soc.*, **130**, 10702 (2008).

Tang, W. and Matyjaszewski, K., *J. Am. Chem. Soc.*, **128**, 1598 (2006).

Tang, W. and Matyjaszewski, K., *Macromolecules*, **39**, 4953 (2006).

Tao, L., Luan, B., and Pan, C., *Polymer*, **44**, 1013 (2003).

Taton, D., Wilczewska, A-Z., Destarac, M., *Macromol. Rapid Commun.*, **22**, 1497 (2001).

Tunca, U., Erdogan, T., and Hizal, G., *J. Polym. Sci.: Part A: Polym. Chem.*, **40**, 2025 (2002).

Tunca, V., Karliga, D., Ertekin, S., Ugur, A. L., Sirkecioglu, O., and Hizal, G., *Polymer*, **42**, 8489 (2001).

Ueda, J., Kamigaito, M., and Sawamoto, M., *Macromolecules*, **31**, 6762 (1998).

Ueda, J., Matsuyama, M., Kamigaito, M., and Sawamoto, M., *Macromolecules*, **31**, 557 (1998).

Wang, X. S. and Armes, S. P., *Macromolecules*, **33**, 6640 (2000).

Wang, J. L., Greszta, D., and Matyjaszewski, K., *Polym. Mater. Sci. Eng.*, **73**, 416 (1995).

Wang, J.-L., Grimaud, T., and Matyjaszewski, K., *Macromolecules*, **30**, 6507 (1997).

Wang, X. S., Luo, N., and Ying, S., *Polymer*, **40**, 4515 (1999).

Wang, J.-L. and Matyjaszewski, K., *Macromolecules*, **28**, 7901 (1995).

Webster, O. W., Hertler, W. R., Sogah, D. Y., Furnham, W. B., and Rajanbabu, T. V., *J. Am. Chem. Soc.*, **105**(17), 5706 (1983).

Weimer, M. W., Frechét, J. M. J., and Gitsov, I., *J. Polym. Sci. Part A: Polym. Chem.*, **36**, 955 (1998).

Weimer, M. W., Scherman, O. A., and Sogah, D. Y., *Macromolecules*, **31**, 8425 (1998).

Xia, J. and Matyjaszewski, K., *Macromolecules*, **30**, 7692 (1997).

Xia, J. and Matyjaszewski, K., *Macromolecules*, **32**, 5799 (1999).

Xia, J., Zhang, X., and Matyjaszewski, K., *Macromolecules*, **32**, 4482 (1999).

Xu, Y. and Pan, C., *J. Polym. Sci. Part A: Polym. Chem.*, **38**, 337 (2000).

Yagci, Y., Baskan, D., and Önen, A., *Polymer*, **38**, 2861 (1997).

Yamada, K., Miyazaki, M., Ohno, K., Fukuda, T., and Minoda, M., *Macromolecules*, **32**, 290 (1999).

Yoshida, E. and Sugita A., *Macromolecules*, **29**, 6422 (1996).

Yoshida, E. and Sugita, A., *J. Polym. Sci. Part A: Polym. Chem.*, **36**, 2059 (1998).

You, Y., Hong, C., Wang, W., Lu, W., and Pan, C., *Macromolecules*, **37**, 9761 (2004).

Zhang, X., Xia, J., and Matyjaszewski, K., *Macromolecules*, **33**, 2340 (2000).

EXERCISES

11.1 Describe the principal criteria that can be used to identify a living polymerization/controlled polymerization (LP/CP). What are the various ways in which living polymerization can be accomplished? What are the advantages of living/controlled radical polymerization (LRP/CRP) compared to other methods of LP/CP? Describe how the main requirements of LP/CP are fulfilled in the following LRP/CRP methods: (a) nitroxide mediated polymerization (NMP) or stable free radical polymerization (SFRP), (b) atom transfer radical polymerization (ATRP), and (c) polymerization by reversible addition fragmentation chain transfer (RAFT)?

11.2 To determine the optimum temperature for the SNR-mediated polymerization of styrene, the process was investigated at three temperatures (80°, 90°, and 100°C) using an alkoxyamine (**III**) as the initiator. This initiator, denoted as S—T, ensured stoichiometry between the growing styryl-type radical S$^\bullet$ and di-*tert*-butyl nitroxide radical (T$^\bullet$) acting as the capping agent. Experimentally, several sealed tubes containing styrene and the initiator (S—T) were prepared (no other radical initiator being added) and submitted to heat during given times. The corresponding polymers were isolated by freeze-drying and the conversion yield calculated from the weight of th samples obtained, while the molecular weight and polydispersity were determined by gel permeation chromatography. Selected data derived from these experiments are given below (Catala et al., 1995).

Table P11.2.1 Valus of polydispersity index (M_w/M_n) at various temperatures and conversions

Temp. (°C)	Monomer conversion (%)	M_w/M_n
80	2.4	1.43
	4.3	1.26
	8.4	1.20
	12.7	1.30
90	7.1	1.15
	11.3	1.16
	13.9	1.16
	17.6	1.17
	22.1	1.17
	26.9	1.17
100	11.1	1.13
	15.7	1.12
	21.0	1.18
	43.2	1.49

Note: The data are read from experimental graph of Catala et al. (1995).

Table P11.2.2 Values of monomer conversion (p) and molecular weight (M_n) at given times (t) in the presence of different concentrations of initiator (S—T)

$[S—T]_0$, (mol/L)	t (h)	p	M_n
3.7×10^{-3}	5.0	0.071	16,000
	8.2	0.113	23,700
	11.2	0.139	31,850
	14.3	0.176	40,000
	17.5	0.221	45,000
	24.7	0.269	63,700
7.4×10^{-3}	3.0	0.049	6,400
	6.0	0.083	11,200
	9.0	0.127	16,000
	12.0	0.159	20,000
	25.0	0.289	35,000
2.2×10^{-2}	15.0	0.221	9,200
	24.5	0.316	13,200
4.4×10^{-2}	20.0	0.265	5,200
	35.0	0.365	7,600

Note: Data are read from experimental graphs of Catala et al. (1995).

Analyze the data to determine (a) the best temperature for living radical polymerization in the presence of S—T alone, (b) the effect of S—T concentration on the polymerization rate and polymer molecular weight.

11.3   In the polymer synthesis considered in **Problem 11.3**, sodium 4-oxy-TEMPO (see Scheme P11.3.1) acts as a transforming agent from living cationic to living radical polymerization. The nitroxyl radical-ended living poly(tetrahydrofuran, THF), denoted as TEMPO-PT, acts as polymeric initiator as well as polymeric counter radical in the polymerization of styrene (St) by benzoyl peroxide (BPO) initiator. In an experimental study [Yoshida, E. and Sugita, A., *Macromolecules*, **29**, 6422 (1996)], radical polymerization of St (8.70 mol/L) was performed with BPO (0.0866 mol/L) in the presence of TEMPO-PT (91.1×10^{-3} mol/L). The polymerization was carried out in bulk at $125°C$ after being held at $95°C$ for 3.5 h. The relation between the molecular weight of the polymer produced and the conversion of St was investigated to determine whether this radical polymerization proceeded in accordance with a living mechanism. The first order time-conversion data and the data on conversion-molecular weight so obtained are shown below, where M_n denotes the molecular weight of the copolymer:

Time (h)	ln($[M]_0/[M]$)	M_n
0	0	3250
2.86	0.4259	5900
6.55	0.7231	6800
17.86	1.907	11000
23.67	2.115	12600

Further, the effect of the concentration of the initiating polymer, TEMPO-PT, on the molecular weight, $(M_n)_P$, of the polystyrene (PSt) block of the diblock copolymer, poly(THF)-*b*-PSt, was investigated, yielding the following results:

[TEMPO-PT]$_0$ (10^{-3} mol/L)	Time (h)	Conversion (%)	$(M_n)_P$
182	163	100	5750
91.1	163	94	11100
47.1	48	97	18700

(a) Based on a previous finding [Yoshida, E. and Sugita, A., *Macromolecules*, **29**, 6422 (1996)] that in the living radical polymerization of St by BPO and 4-methoxy-TEMPO, the concentration of the growing polymer chain is determined not by the initial concentration of the radical initiator BPO but by that of the nitroxyl radical, calculate the copolymer molecular weight vs. conversion for comparison with the experimental values to verify the living character of the system.

(b) Determine the relation between the molecular weight and [TEMPO-PT]$_0$.

11.4   Results from kinetic studies of ATRP for styrene, methyl acrylate, and methyl methacrylate under homogeneous conditions indicate that the rate of polymerization is first order with respect to the concentrations of the initial components, namely, monomer, initiator, and Cu(I)/ligand complex. Derive a rate law that is consistent with this observation. State the assumptions made in the derivation. Predict the variation of monomer conversion with time based on the rate law. Analyze the effect of Cu(II) on the polymerization and the conditions under which living/controlled polymerization will proceed.

11.5   Unlike in a normal ATRP, where the initiating radicals are generated from an alkyl halide in the presence of a transition metal in its lower oxidation state, ATRP can also be initiated using a conventional radical initiator, such as, AIBN. Suggest a general scheme for polymerization with this latter approach.

11.6   Both homogeneous and heterogeneous (or supported) catalytic systems have been used in ATRP. However, polymerization with homogeneous catalysts yields a polymer with much lower polydispersity than that with heterogeneous and supported catalysts. Explain, giving reason.

11.7   Account for the fact that ATRP in aqueous systems is usually more complex than in non-aqueous systems.

11.8   While various monomers, such as styrenes, (meth)acrylates, and acrylonitrile, have been successfully polymerized using ATRP, polymerization of less reactive monomers such as olefins, halogenated alkenes, and vinyl acetate has not been successful. Explain, giving reason. Explain also why optimal conditions of ATRP may be quite different for different monomers.

11.9   Taking into account the fact that acid monomers can poison ATRP catalysts by coordinating to the transition metal, suggest a suitable method for preparing poly(meth)acrylic acids by ATRP.

11.10 Discuss how macromonomers of styrene and methyl acrylate can be prepared and copolymerized by ATRP. Give an outline for the synthesis of allyl end-functionalized macromonomers of styrene and methyl acrylate by ATRP.

11.11 Write the structures of macromonomers that can be obtained by ATRP of styrene using the following initiators: (a) vinyl chloroacetate, (b) allyl chloride, (c) allyl chloroacetate, and (d) vinyl 2-bromoisobutyrate.

11.12 In ATRP, nearly every polymer chain formed should contain a halogen atom at its head-group, which may be a drawback due to increasing concern over the presence of halogens in the environment. Discuss the methods that can be used for the (a) removal and (b) transformation of the terminal halogen in ATRP polymers.

11.13 Explain why it is much easier to form block copolymers by ATRP than by conventional free radical polymerization.

11.14 Suggest a strategy to synthesize AB diblock and ABA triblock copolymers, where B = poly(butyl acrylate) (PBA) and A = poly(methyl methacrylate) (PMMA), taking note of the fact in the polymerization of MMA using ATRP the Cl end group of the growing polymer chain is predominantly less labile that the Br group.

11.15 A bromine-terminated monofunctional poly(*tert*-butyl acrylate) resulting from ATRP of *t*BA catalyzed by the CuBr/N,N,N',N',N''-pentamethyldiethylenetriamine (PMDETA) system (initial mole concentration ratios *t*BA : methyl bromopropionate (MBrP) : CuBr : PMDETA : CuBr$_2$ = 50 : 1 : 0.5 : 0.525 : 0.025, 25% acetone, 60°C; conversion = 96% after 6.5 h) was used as macroinitiator for block copolymerization with styrene (St) with the initial mole concentration ratios of St : P(*t*BA) : CuBr : PMDETA = 100 : 1 : 1 : 1 at 100°C (conversion 94%). The monofunctional bromo-terminated copolymer P(*t*BA)-*b*-P(St) formed was subsequently used as a macroinitiator for a further copolymerization with methyl acrylate (MA). The polymerization was also catalyzed by CuBr/PMDETA (initial concentration ratios MA : P(*t*BA-*b*-P(St) : CuBr : PMDETA = 392 : 1 : 1 : 1), under high dilution in toluene and reached 23% monomer conversion after 3.5 h at 70°C. The experimental molecular weight (M_n) of the resulting triblock copolymer P(*t*BA)-*b*-P(St)-*b*-P(MA) was 24,800 with a PDI = 1.10. Calculate the theoretical M_n to compare with the experimental value.
       [*Ans.* P(*t*BA) macroinitiator: M_n = 6311; P(*t*BA)-*b*-P(St): M_n = 16,087; P(*t*BA)-*b*-P(St)-*b*-PMA: = 23,840 (cf. exptl. M_n = 24,800).]

11.16 Outline a simple ATRP strategy with minimum number of steps for the synthesis of ABCBA pentablock copolymers, where A = poly(methyl acrylate), B = poly(*tert*-butyl acrylate), and C = polystyrene. How would you modify the procedure if A is poly(methylmethacrylate)?

11.17 Within the context of ATRP, discuss the strategies of preparing (a) graft copolymers, (b) star polymers, and (c) hyperbranched polymers.

11.18 Using ATRP techniques in a three-step procedure, a three-arm star copolymer with ABC triblock arms, Y(ABC)$_3$, was prepared, where Y is related to the trifunctional initiator [Davis, K. A. and Natyjaszewski, K., *Macromolecules*, **34**, 2101 (2001)], 1,1,1-tris(4-(2-bromoisobutyry- loxy)phenyl)ethane (TBiBPE), denoted below as I:

and A, B, and C represented homopolymer blocks of stytene (St), *tert*-butyl acrylate (*t*BA), and methyl methacrylate (MMA), respectively. The experimental conditions (initial mole ratios of monomer : initiator : catalyst : ligand, solvent, temperature, time) and results (monomer conversion, $\overline{M_n}$ of polymer product) were as follows (Davis and Matyjaszewski, 2001): Step One St : I : CuBr : PMDETA (Fig. 11.20) = 20 : 1 : 0.2 : 0.2, 25% anisole, 100°C, 360 min, $\overline{57\%}$, 1600. Step Two *t*BA : I : CuBr : PMDETA = 150 : 1 : 0.25 : 0.25, bulk, 60°C, 240 min, 77%, 18,000. Step Three MMA : I : CuCl : HMTETA (Fig. 11.20) = 400 : 1 : 1 : 1, 50%, acetone, 60°C, 48%, 31,900.

Calculate theoretically M_n of the polymer (macroinitiator) obtained by ATRP in each step and compare with the experimental M_n.

[*Ans. Step one*: M_n = 1935 (cf. exptl. M_n = 1600); *Step two*: M_n = 16,719 (cf. exptl. M_n = 18,000); *Step three*: 35,919 (cf. exptl. M_n = 31,900).]

11.19 Functional calixarenes can be used as plurifunctional ATRP initiators for the synthesis of star polymers. Show the structures of 2-bromopropionate and 2-bromoisobutyrate esters of 4-*tert*-butyl calix[*n*] arene used as tetrafunctional (*n* = 4), hexafunctional (*n* = 6), and octafunctional (*n* = 8) initiators for ATRP.

11.20 Suggest a methodology for modular synthesis of an ABC-type miktoarm star polymer consisting of the following three arms: poly(ε-caprolactone) (PCL), polystyrene (PSt), and poly(*tert*-butyl acrylate) (P*t*BA).

11.21 It is desired to synthesize (a) AA$'_2$-type triarm asymmetric polystyrene (PSt) stars with asymmetry in the molar mass of their branches, (b) AB$_2$-type miktoarm star polymer *core*-(PSt)(P*t*BA)$_2$ (where P*t*BA = poly(*tert*-butyl acrylate)), and (c) amphiphilic *core*-(PSt)(PAA)$_2$ (where PAA = poly(acrylic acid)). Suggest a methodology to synthesize these polymers entirely by ATRP processes.

11.22 The research group of U. Tunca [Celik, C., Hizal, G., and Tunca, U., *J. Polym. Sci., Part A: Polym. Chem.*, **41**, 2543 (2003)] synthesized a novel trifunctional initiator, 2-phenyl-2-[(2,2,6,6-tetramethyl)-1-piperidinyloxy] ethyl 2,2-bis[methyl (2-bromopropionato)] propionate from the reaction of 2-phenyl-2-[(2,2,6,6-tetramethylpiperidine)-oxy]-1-ethanol [Hawker, C. J., Barclay, G. G., Orellana, A., Dao, J., and Devonport, W., *Macromolecules*, **29**, 5245 (1996)] and 2.2-bis [methyl (2-bromopropionato)] propionyl chloride [Angot, S., Taton, D., and Gnanou, Y. *Macromolecules*, **33**, 5418 (2000)] in 62% yield and used it to prepare well-defined miktoarm star, *core*-(A)(B)$_2$, and miktoarm star block, *core*-(A)(BC)$_2$, via a combination of SFRP-ATRP route. Using this versatile synthetic approach, give an outline for the synthesis of a miktoarm star polymer, *core*-(PSt)(P*t*BA)$_2$, and miktoarm star block copolymer, *core*-(PSt)(P*t*BA-*b*-PMMA)$_2$, with a controlled molecular weight and a moderate polydispersity. [PSt = polystyrene; P*t*BA = poly(*tert*-butyl acrylate); PMMA = poly(methyl methacrylate).]

11.23 Define RAFT chain transfer agents (or simply RAFT agents) on the basis of their generic structure and functions of R and Z groups. What are the general guidelines for the selection of R- and Z-groups for RAFT agents? How would you classify the RAFT agents on the basis of differences in the Z functional group? Identify a RAFT agent that can yield two identical polymer molecules from one molecule of the agent.

11.24 While trithiocarbonates function effectively as RAFT agents when the substituent on sulfur is a good homolytic leaving group (as compared to the polymer chain), an outstanding feature of trithiocarbonates is that they can be prepared with two good homolytic groups, making these RAFT agents capable of growing in two directions. An important consequence of this is that ABA triblock copolymers can be formed by RAFT polymerization in only two sequential monomer addition steps. To provide proof of this hypothesis, S,S-dibenzyl trithiocarbonate (0.0017 M) was used as the chain transfer agent (CTA) and AIBN (0.073 M) as the initiator to conduct RAFT polymerization of methyl acrylate (MA) (2.22 M) to 54% monomer conversion in 4 h and the resulting polymer (0.0016 M) was chain extended with n-butyl acrylate (nBA) (2.79 M) using AIBN (0.073×10^{-2} M) as the initiator at 60°C for 8 h. If the conversion of monomer in the second step is 70%, calculate the molar mass of the resulting triblock copolymer, poly(MA-b-nBA-b-MA), and that of poly(MA) end segments.
[*Ans.* Triblock copolymer: 2,29,100 g mol^{-1}; poly(MA) end segment (each): 30,322 g mol^{-1}.]

11.25 Define a macro-CTA used in the RAFT process of polymerization. Illustrate R- and Z-group approaches in the context of synthesis of AB diblock and ABA triblock copolymers via macro-CTAs.

11.26 While poly(ethylene oxide) (PEO) has high hydrophilicity and biocompatibility, poly(N-isopropyl acrylamide) (PNIPAM) is a well-known thermosensitive polymer having a low critical solution temperature in water at 32°C. Block copolymers containing PEO and PNIPAM are therefore expected to find critical applications in many areas. Discuss how macro-CTA based RAFT polymerization can be used to produce diblock and triblock copolymers of PEO and PNIPAM.

11.27 What are multifunctional CTAs? How can these be used for the synthesis of star polymers via R- and Z-group approaches?

11.28 In a novel experiment, Feng and Pan [Feng, X.-S. and Pan, C.-Y., *Macromolecules*, **35**, 4888 (2002)] prepared polystyrene (M_n = 5000, M_w/M_n = 1.04) by RAFT polymerization at 80°C using styrene (St)/1-phenylethyl dithiobenzoate/benzoyl peroxide (2000 : 10 : 1 molar ratio). The synthesized poly(St)-S-C(=S)Ph was reacted with maleic anhydride (MAH) in tetrahydrofuran (THF) solution at 80°C, using excess MAH to ensure complete reaction. The resulting product was then used as the RAFT agent to carry out a second RAFT polymerization of methyl acrylate with benzoyl peroxide initiator. Finally, the anhydride group of the reacted MAH was esterified with the terminal hydroxyl group of poly(ethylene glycol) methyl ether. Show schematically the sequence of steps of the copolymer synthesis and describe the nature of the copolymer formed.

11.29 Show schematically how nonsymmetrical trithiocarbonate RAFT agents can be synthesized by a one-pot procedure.

11.30 While the RAFT process is able to polymerize many monomers that are troublesome for other polymerization techniques, it is also possible to synthesize RAFT agents carrying many useful reactive end groups that can later be used for coupling reactions to various macromolecules by site transformation to afford synthesis of various block and graft copolymers, containing polymer blocks created by different polymerization techniques. RAFT is thus potentially an ideal polymerization technique to combine with cationic polymerization by site transformation. An initial example of this prospect was provided by Magenau et al. [Magenau, A. J. D., Martinez-Castro, N., and Storey, R. F., *Macromolecules*, **42**, 2353 (2009)], who synthesized block copolymers consisting of polyisobutylene (PIB) and either poly(methyl methacrylate) (PMMA) or polystyrene (PSt) block segments by a site transformation approach combining living cationic and RAFT polymerizations.

The initial PIB block was synthesized via quasi-living cationic polymerization of isobutylene using TMPCl/TiCl$_4$ initiation system and subsequently converted into a hydroxyl-terminated PIB (HTPIB). Site transformation of HTPIB into a macro-chain-transfer agent (PIB-CTA) was accomplished by DCC/DMAP-catalyzed esterification with a trithiocarbonate-type RAFT CTA carrying a carboxylic acid functionalized R-group. The PIB-CTA was then employed in a RAFT polymerization of either MMA or St, resulting in PIB-b-PMMA or PIB-b-PSt.

Give a schematic outline of the aforesaid synthesis of block copolymers and elaborate the four main steps: (1) PIB synthesis; (2) hydroxyl end-functionalization of PIB; (3) formation of PIB-CTA; and

(4) RAFT polymerization of MMA or St with PIB-CTA as the RAFT agent. [*Note:* TMPCl = 2-chloro-2,4,4-trimethylpentane; DCC = N,N'-dicyclohexylcarbodiimide; DMAP = 4-(dimethylamino) pyridine; CTA = 4-cyano-4-(dodecyl sulfanyl thiocarbonyl sulfanyl) pentanoic acid.]

11.31 A common feature of all polymers prepared by RAFT polymerization is that they bear a thiocarbonylthio group at one chain end or both. This limits the industrial application of the RAFT process because the thiocarbonylthio end group gives color and instability (especially under basic conditions) to the polymer. Discuss and compare the various methods that can be used for removal or transformation of the thiocarbonylthio end group.

Chapter 12

Polymer Synthesis by Click Chemistry

12.1 Introduction

It is the ultimate desire of all chemists to effect reactions of choice with quantitative yields and high functional group tolerance (to avoid protecting group strategies), performing reaction in all solvents (irrespective of protic/aprotic or polar/nonpolar character, with water as the ultimate solvent), under benign conditions (such as ambient temperature), and with no or little amount of side products. The copper(I)-catalyzed azide/alkyne reaction (also called Sharpless "click" reaction), which is a variation of the Huisgen 1,3-dipolar cycloaddition reaction between terminal azide and alkyne groups, fulfills all of the above mentioned requirements. Besides the azide/alkyne reaction that results in the formation of triazoles, two other examples of Huisgen 1,3-dipolar cycloaddition reaction are shown in Fig. 12.1, which involve C–N triple bonds and result in the formation of tetrazoles and oxazoles. These reactions, although known for a long time, are located within a series of reactions (see later) named "click reactions". Critical for the broad application of the 1,3-dipolar cycloaddition pocess as the "click"-type reaction is the discovery that the purely thermal Huisgen reaction can be accelerated tremendously by the addition of Cu(I) species within the reaction system. This transforms the Huisgen reaction from what may be termed a "good" ligation procedure to an extremely "good", fast and general reacton (Binder and Kluger, 2006).

To give a historical perspective, the purely thermal 1,3-dipolar cycloaddition reaction was studied and proposed by Sharpless et al. in 2001 (Kolb et al., 2001) as a candidate of a click reaction on the premise that between alkyl/aryl azides and strongly <u>activated</u> alkynes (viz., acyl and sulfonyl cyanides as well as acyl-alkynes), this purely thermal reaction can proceed at modest temperatures and give high yields (90-100%). In other words, Sharpless recognized the potential of Huisgen's 1,3-dipolar cycloaddition of azides to alkynes yielding triazoles and retooled this reaction as "click" chemistry for the construction of complex molecules. Subsequently, Meldal (Tornoe et al., 2002) discovered that the process can be greatly accelerated by Cu(I) salts, leading to a reaction at 25°C with quantitative yields and high tolerance to various functional groups and solvents. The Cu(I)-catalyzed reaction was also first mentioned by them to have higher regioselectivity (1,4-triazole formation versus 1,5-triazole) compared to the purely thermal process (see Fig. 12.2).

The essence of click chemistry is the idea of facilitation or making synthesis easy by selecting a group of reactions that, when strategically chosen, can facilitate synthesis of a stunning diver-

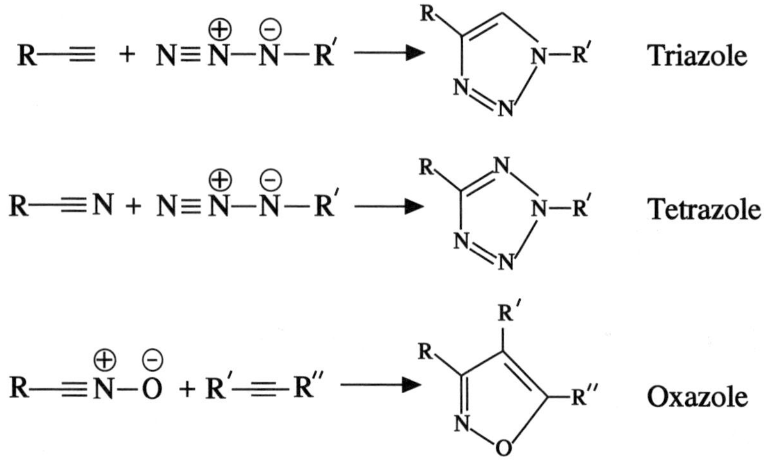

R—≡ + N≡N⁺—N⁻—R' ⟶ [Triazole structure] Triazole

R—≡N + N≡N⁺—N⁻—R' ⟶ [Tetrazole structure] Tetrazole

R—≡N⁺—O⁻ + R'—≡—R'' ⟶ [Oxazole structure] Oxazole

Figure 12.1 Examples of 1,3-dipolar cycloaddition reactions used in click chemistry.

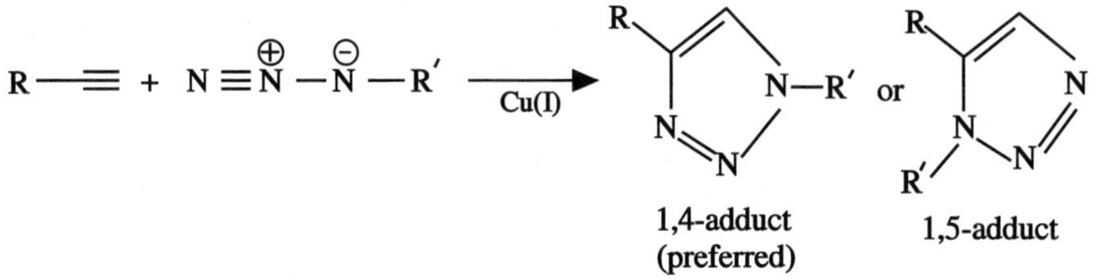

R—≡ + N≡N⁺—N⁻—R' --Cu(I)--> [1,4-adduct structure] or [1,5-adduct structure]

1,4-adduct (preferred)   1,5-adduct

Figure 12.2 The basic process of the Huisgen 1,3-dipolar cycloaddition, [3 + 2] system, between azides and alkynes, generating 1,4- and 1,5-triazoles. Regioselectivity (1,4-triazole versus 1,5-triazole) is higher (Tornoe et al., 2002) in the Cu(I)-catalyzed process, as compared to the purely thermal processes, the former giving 1,4-disubstituted triazoles almost exclusively (Rostovtsev et al., 2002).

sity of polymeric structures and functional materials by the simplest, cleanest, and most efficient means possible (Sumerlin and Vogt, 2010). A central philosophy of click chemistry is avoidance of sophisticated synthetic techniques in favor of simple methods that are modular, reliable, highly specific, efficient, and easy to implement under relatively mild conditions, without being influenced by environments (such as solvents, different functional groups, etc.). A schematic representation of this click concept is shown in Fig. 12.3.

Though the Cu(I)-catalyzed azide-alkyne cycloaddition (CuAAC) has served as the quintessential example of a click reaction since its initial development and has made an enormous impact in the field of macromolecular engineering, click chemistry is not limited to this reaction, but rather "defines a synthetic concept or framework that comprises a range of reactions with different reaction mechanisms, but common reaction trajectories" (Nandivada et al., 2007). The reactions named click reactions are driven by a high thermodynamic force with an enthalpy gain usually greater than 20 kcal mol⁻¹ (Kolb et al., 2001), thus leading to reactions characterized by high yields, high selectivity, fast reaction times, and simple reaction conditions. Besides CuAAC, several other highly efficient reactions also meet the click criteria (Becer et al., 2009). Some examples of such click reactions that are commonly employed in polymer synthesis and functionalization are shown in Fig. 12.4. These include other cycloaddition reactions, such as strain-promoted azide-alkyne coupling (SPAAC), which do not depend on the use of a Cu(I) catalyst (Fig. 12.4b) and Diels-Alder reaction (Fig. 12.4c), many of which possess characteristics of click techniques.

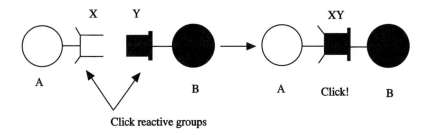

Figure 12.3 Schematic representation of the click concept. The ends **X** and **Y** of molecules **A** and **B** represent functional groups which react quantitatively with high selectivity and extremely high tolerance toward a wide range of functional groups and reaction conditions. (Adapted from Nandivada et al., 2007.)

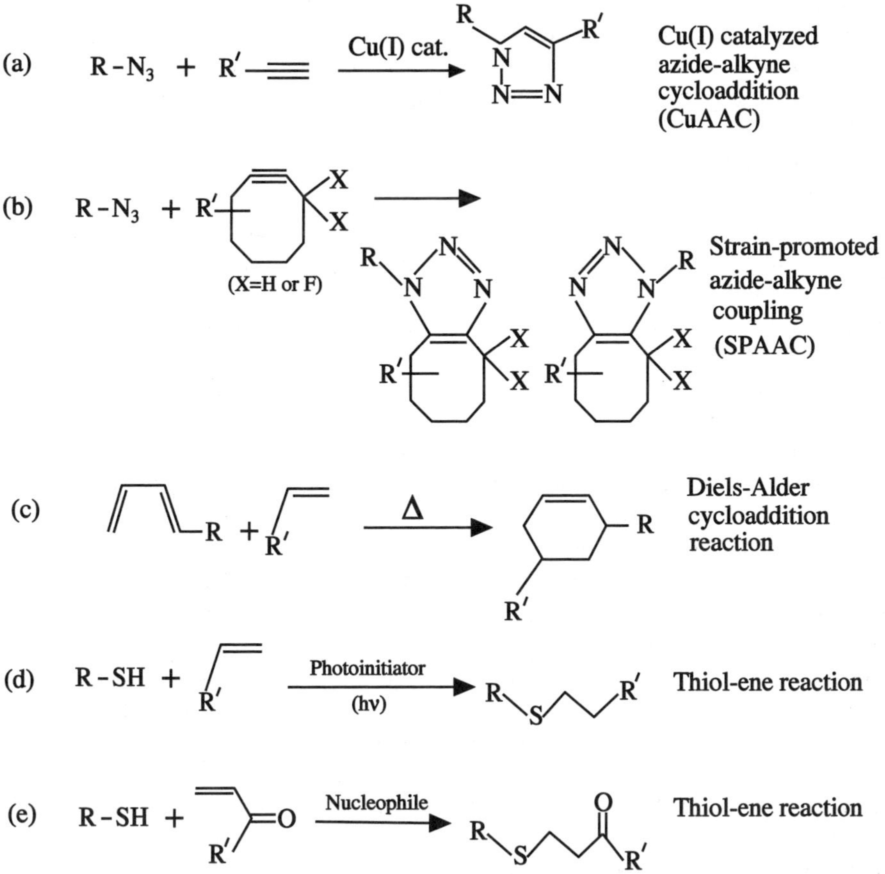

(a) $R\text{-}N_3$ + R'≡ $\xrightarrow{\text{Cu(I) cat.}}$ [triazole structure] Cu(I) catalyzed azide-alkyne cycloaddition (CuAAC)

(b) $R\text{-}N_3$ + R' [cyclooctyne with X] $\longrightarrow$ (X=H or F) [triazole fused cyclooctane structures] Strain-promoted azide-alkyne coupling (SPAAC)

(c) [diene]$\text{-}R$ + [dienophile]$_{R'}$ $\xrightarrow{\Delta}$ [cyclohexene ring with R and R'] Diels-Alder cycloaddition reaction

(d) $R\text{-}SH$ + [alkene]$_{R'}$ $\xrightarrow[\text{(hv)}]{\text{Photoinitiator}}$ $R\diagdown_S\diagup R'$ Thiol-ene reaction

(e) $R\text{-}SH$ + [vinyl ketone]$_{R'}$=O $\xrightarrow{\text{Nucleophile}}$ $R\diagdown_S\diagup \overset{O}{R'}$ Thiol-ene reaction

Figure 12.4 Some common click reactions used in polymer synthesis and functionalization. (Adapted from Sumerlin and Vogt, 2010.)

The Diels-Alder [4 + 2] reaction (Fig. 12.4c), which is a cycloaddition between a conjugated diene (a four π-electron system) and a dienophile (a two π-electron system), is a click reaction that has attracted much attention in macromolecular chemistry, particularly in providing new materials (Jones et al., 1999; Imai et al., 2000; Gheneim et al., 2002; Vargas et al., 2002; Durmaz et al., 2006; and Gacal et al., 2006). Thiol-ene reactions (Fig. 12.4d, e), which also fall within the realm of click chemistry, refer to hydrothiolation of virtually any alkene C=C bond occurring by either radical or nucleophilic mechanisms. These have proven to be practically useful for polymer synthesis under extremely mild conditions, often with no solvent and little or no by-product formation. Some of these reactions will be discussed more elaborately in later sections.

Since their advent in 2001-2002, click chemistry strategies have been integrated rapidly into the field of macromolecules, while the advantage of simplicity, enhanced efficiency, and specificity have allowed the click methods to "flourish vibrantly". The application of click reactions not only facilitates access to well-known polymers, but also allows preparation of complex, previously inaccessible macromolecular structures. Greatly enhanced efficiency and specificity, the two main attributes of click methods, are arguably more important in reactions of polymers than those of small molecules. This is due to the distinctive features of polymers such as high molecular weight and multiplicity of functional groups, which complicate the direct application of standard organic chemistry techniques to polymer synthesis and functionalization, while inefficiency of applied methods leads to products containing reacted units covalently linked to unreacted units due to which no simple method of separation can lead to a pure product. The only method that remains to ensure product purity in such cases is to apply efficient and quantitative transformations as available in click chemistry. High efficiency is especially important in functionalization of polymers involving modification of end groups, since kinetic limitations can lead to low yields because of low concentration and reduced reactivity of end groups present in high-molecular weight linear polymers (Sumerlin and Vogt, 2010).

As the functional groups in a polymer largely affect its properties and potential applications, it is important to introduce quantitatively or preserve carefully all such groups that are needed. Similarly, copolymers containing multiple types of monomer units with widely different functionalities must be reacted with high fidelity in order to avoid any unwanted functionalization or transformation of functional groups. In all such considerations, reaction specificity becomes a critical issue and click methods assume much significance. For example, if multiple transformations are desired on a single polymer, an ideal click reaction procedure with a high level of orthogonality can be used to perform simultaneous parallel or cascade reactions in *one-pot* with little or no possibility of cross-functionalization and interference (see Fig. 12.5). Complex macromolecules typically requiring several reaction steps can thus be prepared conveniently using one-pot procedures. Specificity again is especially important in reactions that involve at least one biological component, since a wide variety of functional groups are present in biological molecules (proteins, peptides, etc.) and modification procedures should be highly specific to limit side reactions. Click techniques have been successfully employed in many such cases.

The advantages of high efficiency and specificity of click techniques have influenced the field of macromolecular synthesis in many ways. Because of their quantitative yields and low susceptibility to side reactions, click procedures are particularly useful in step-growth polymerization processes in which high conversion of functional groups is necessary to achieve high molecular weight. In fact, many of the first examples of click reactions employed in polymer synthesis involved polymers being prepared by step-growth click-type polymerizations (Johnson et al., 2008). While block copolymer synthesis commonly relies on living/controlled polymerization methods to grow one block from another, click reactions can be used to prepare block copolymers in a modular and highly efficient manner by coupling preexisting homopolymers. The modular approach can also be used to synthesize star polymers and graft copolymers by click chemistry (Fig. 12.6).

(a) Simultaneous Reactions

(b) Cascade Reactions

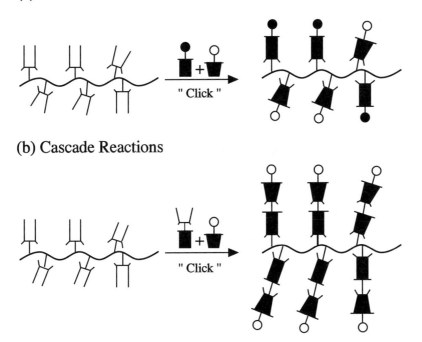

Figure 12.5 Schematic illustration of (a) simultaneous parallel and (b) cascade reactions performed on multi-functional macromolecules using click techniques in one-pot procedure. Such reactions are possible because of orthogonality of many click techniques. (Adapted from Sumerlin and Vogt, 2010.)

Since the concentration of complementary reaction moieties in the preexisting homopolymers is necessarily low and considerable steric hindrance may also be present, the reaction used in the modular approach must be highly efficient. Suitably chosen click reactions thus allow successful ligation of high molecular weight of polymer chains even with low concentrations of reactive groups and significant steric resistance. Multi-block copolymers can also be synthesized in one-pot via a suitable combination of click reactions. Making use of the high efficiency and fidelity in click chemistry approach, diverse dendritic structures have been generated with high purity and excellent yield. The most common click reactions are discussed in this chapter and examples are presented to show how the click approach facilitates access to well-known polymers or allows preparation of previously inaccessible materials.

12.2   *Copper-Catalyzed Azide-Alkyne Cycloaddition*

Sharpless and coworkers published a paper in 2002 (Rostovtsev et al., 2002) where the formation of 1,2,3-triazoles by the Cu(I)-catalyzed Huisgen 1,3-dipolar cycloaddition reaction between alkyl/aryl azides and *non-activated* alkynes was described as a click reaction in contrast to the purely thermal (uncatalyzed) azide-alkyne click reaction which refers to the 1,3-dipolar cycloaddition reaction between alkyl/aryl azides and strongly *activated* alkynes. The Cu(I)-catalyzed

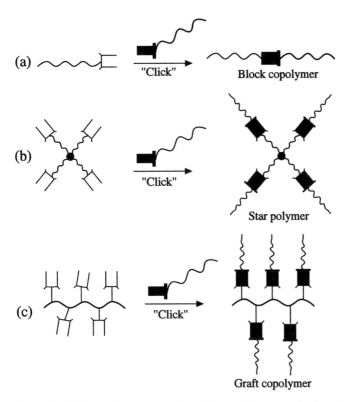

Figure 12.6 Schematic representation of the modular approach of synthesizing (a) block copolymers, (b) star polymers, and (c) graft copolymers by click chemistry. It should be noted that in spite of greatly enhanced efficiency of click reactions, as compared to more conventional methods, quantitative conversion is difficult to achieve in the case of dense grafting. (Adapted from Sumerlin and Vogt, 2010.)

azide-alkyne cycloaddition (CuAAC), which is the most efficient among the click reactions, runs at ambient temperature, is nearly solvent insensitive, and has high tolerance to most functional groups. The main functional groups that interfere with CuAAC reactions are strongly activated azides (i.e., acyl- and sulfonyl azides), cyanides attached to an electron withdrawing group, free thiol-moieties (R–SH), and cyclic alkenes. Double bonds are tolerated to a certain extent, if they are neither electronically activated (i.e., by electron withdrawing substituents) nor embedded into strained ring substrates (Binder and Sachsenhofer, 2007). For example, the thermal 1,3-dipolar cycloaddition reaction rate constants, k (10^7 dm^3 M^{-1} s^{-1}), of phenyl azide towards various substrates having double bonds:

are as follows (Binder and Kluger, 2006): isoprene (0.15, i.e., 0.15×10^7 dm^3 M^{-1} s^{-1}), 1-heptene (0.24), styrene (0.4), cyclopentene (2.4), maleic anhydride (7.20), *N*-phenyl maleimide (27.60), norbornene (188), and 1-pyrrolidinocyclopentene (115,000). The strained and electronically activated alkenes therefore become important competitors in the thermal Huisgen azide-alkyne 1,3-dipolar cycloaddition process.

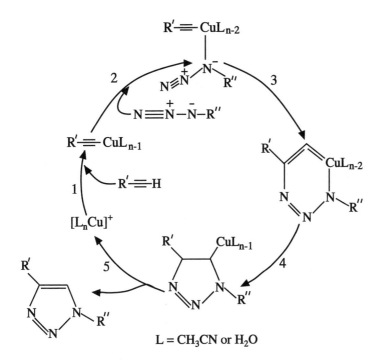

Figure 12.7 Mechanism of the Cu(I)-catalyzed azide-alkyne 1,3-dipolar [3 + 2] cycloaddition reaction. (Adapted from Binder and Kluger, 2006.)

The mechanism of the CuAAC reaction was first proposed by Meldal (Tornoe et al., 2002) and Sharpless (Rostovtsev et al., 2002) and later verified by computational methods by Sharpless (Himo et al., 2005) in a series of papers. The proposed catalytic cycle based on a concerted mechanism via a Cu-acetylide intermediate is shown in Fig. 12.7. The most effective variant of the catalyzed 1,3-dipolar azide-alkyne cycloaddition system uses terminal alkynes in combination with copper sulfate and sodium ascorbate. The sodium ascorbate reduces copper sulfate to Cu(I), which forms a Cu-acetylide by reaction with the terminal alkyne *via* an initial π-complex formation. The copper acetylide formed is considerably more reactive toward the azide so that a rate enhancement of the 1,3-dipolar cycloaddition results (Englert et al., 2005).

Acting as ligand for the azido-moiety, the Cu-acetylide leads to the formation of a 6-membered transition state, in which the Cu(I) plays a central role in the cycloaddition process (Shintai and Fu, 2003). The calculation of relevant energies and activation energies (Himo et al., 2005) revealed that this Cu(I)-mediated cycloaddition results in a lowering of the activation barrier by about 11 kcal/mol with respect to the thermal Huisgen 1,3-dipolar cycloaddition process. This accounts for both the high rate acceleration and greatly enhanced regioselectivity of the catalyzed cycloaddition reaction.

Though in case of the azide-alkyne 1,3-dipolar cycloaddition process, exclusively Cu(I) catalysts have been used (in 0.25-2 molcatalysts (Ru, Ni, Pd, and Pt salts) have also been employed. For Cu(I) catalysts, most methods directly use Cu(I) salts, while other methods generate Cu(I) by reduction of Cu(II) salts with sodium ascorbate or metallic copper. The catalyzed cycloaddition reaction is experimentally simple, perfectly reliable, quantitative, proceeds well in aqueous solutions under ambient conditions without protection from oxygen, requires only stoichiometric amounts of azide and alkyne, and generates virtually no by-products (Wu et al., 2004):

$$
N\equiv\overset{+}{N}-\overset{-}{N}-R \quad + \quad R'-\!\!\!\equiv \quad \xrightarrow[\text{H}_2\text{O/BuOH (1:1), RT, 8h}]{\begin{array}{c}\text{5 mol\% CuSO}_4.5\text{H}_2\text{O,}\\ \text{10 mol \% Sodium ascorbate}\end{array}}
$$

The components and catalysts are simply mixed and stirred. This produces pure products which are simply isolated by filtration or extraction. Typical Cu-systems used in Cu(I)-catalyzed synthesis of 1,2,3-triazoles from azides and alkynes are Cu(II)SO$_4$.5H$_2$O/sodium ascorbate, Cu(II)SO$_4$.5H$_2$O/ copper metal, Cu(I)(MeCN)$_4$PF$_6$, Cu(I)X(X = Br, I), Cu(I)(PPh$_3$)$_3$Br, Cu metal/NEt$_3$,   and   Cu metal/Cu(II)SO$_4$/microwave (Binder and Kluger, 2006). Copper clusters of Cu/Cu oxide nanoparticles with Cu(I)/Cu(II) mole ratio 1:3 and 7-10 mm size, as well as copper clusters of diameter ~2 nm with a specific surface area of 168 m^2/g have also been reported [Molteni et al., 2006; Pachon et al.,2005) to have catalytic activity in the 1,3-dipolar cycloaddition process. Often 1-5 equivalents of base (Binder and Kluger, 2006):

Triethylamine   2,6-lutidine   N,N-diisopropyl
                                              ethylamine

are added to promote Cu(I)-acetylide formation. Besides the Cu(I) catalyst and the base, ligands which lead to a stabilization of the Cu(I) oxidation state are used to prevent Cu(II) catalyzed oxidative coupling reactions. Strong effects of microwave irradiation on the click reaction have been reported. Thus Fokin and coworkers (Appakkuttan et al., 2004) observed triazole formation in 86-93% yield in a single step from benzylic halides, sodium azide, and phenylacetylene under microwave irradiation. The CuAAC reaction, as mentioned earlier, proceeds under aqueous conditions with high yields and complete regioselectivity. Most of the other known solvents are also applicable, such as hexane, toluene, alcohols, halogenated solvents (e.g., dichloromethane, chloroform), dimethyl formamide, dimethyl sulfoxide, tetrahydrofuran, diethyl ether, etc. Biphasic reaction systems, such as water/alcohol and water/toluene, are possible with excellent results.

Tired of the additional work of protective group strategies and faced with insufficient reaction progress, many chemists have long been searching for click-type systems to use as simple and efficient hands-on chemical tools in synthetic strategies. Now, at least a visible step in this direction may be said to have been taken with the advent of the click concept that explains the enormous impact it has made already in material science and chemistry, especially in the field of polymer science, showing a glimmer of an exciting prospect to build complex polymeric structures in a brick-type fashion by simple chemical means. The use of the click concept in diverse applications, such as step-growth polymerization (Fig. 12.8a), synthesis of block copolymers, graft copolymers (Fig. 12.6), dendrimers or hyperbranches, and functionalization/modification of polymers (Fig. 12.8b) with, or by immobilization to, a variety of substrates, is mostly based on linking by CuAAC reactions between azido- and alkyne-termninated entities (see Fig. 12.8).

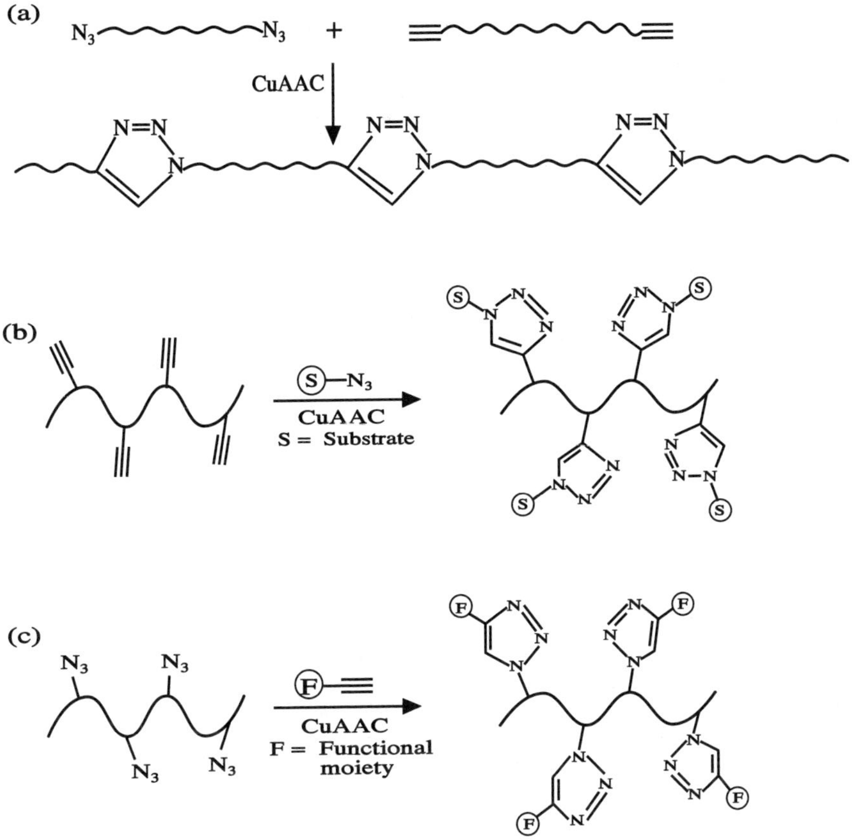

Figure 12.8 Schematic representation of (a) step-growth coupling of bivalent azide and bivalent acety-lene telechelic polymers; (b) polymer modification by CuAAC of pendant alkyne groups of polymers, e.g., poly(vinyl acetylene), with an azide-bearing substrate; and (c) functionalization of polymer by CuAAC of pendant azide with alkyne-bearing functional moiety. Azide terminated dendrimers are similarly subjected to CuAAC with alkyne-derivatized functional moieties to achieve desired functionalization of dendritic macro-molecules.

Most dendrimer syntheses, particularly at later generations, require high monomer loading and need tedious and lengthy chromatographic separations, while simultaneously generating consider-able waste. The high fidelity and efficiency of the CuAAC approach, which provides the exciting possibility of very simple purification due to the use of only stoichiometric amounts of reagents and high conversions, makes the method particularly attractive for the synthesis of dendrimers. In fact, the method first employed by Sharpless and coworkers (Wu et al., 2004) with Frechét's convergent approach for dendrimer preparation clearly marked a breakthrough in the synthesis of dendrimers and dendritic polymeric materials. It was also possibly this application that originally brought click chemistry to the attention of polymer community. Employing monomer structures (**I**) and (**II**) containing azide and alkyne groups (see Fig. 12.9) and creating 1,4-triazole linkages between each generation by CuAAC, the authors succeeded in preparing well-defined dendrimers in a straightforward manner with high yields.

Hawker and Wooley (Joralemon et al., 2005) later extended the CuAAC reaction to prepare dendrimers in a divergent fashion. The divergent method of synthesis generally involves serial

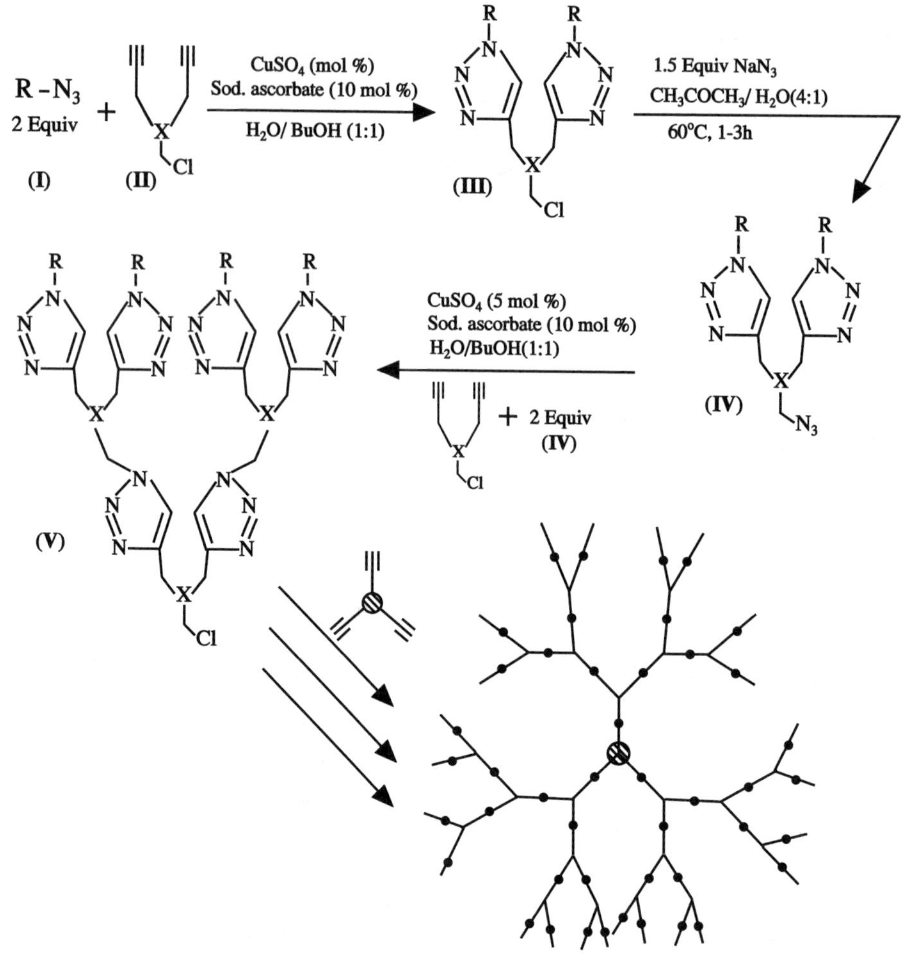

Figure 12.9 Synthesis of dendrimer by repeated CuAAC using Frechét's convergent methodology. Repeated condensation of bivalent acetylene (**II**) with various azides (**I**) yields the core-building block (**III**). Simple nucleophilic substitution of Cl using NaN₃ leads to the 'double edged' building block (**IV**), which in turn is condensed with (**II**) to yield triazole dendrimer (**V**). Individual branches or dendrons can thus be built sequentially and, finally, coupled to a multivalent center piece ("core"), thus leading to a variety of dendrimers with different chain-end groups (**R**) and internal repeat units (**X**). [Adapted from Wu et al. (2004) and Sumerlin and Vogt (2010).]

repetition of two chemical reactions with the growth taking place outward from the core to the dendrimer surface, as the *generation*, and thus the diameter of the dendrimer, increases more or less linearly with the number of cycles. ("Generation" refers to the number of repeated branching cycles that are performed during its synthesis. For example, in the divergent method, if the branching reactions are performed onto the core molecule three times, the dendrimer formed is considered to be a third generation dendrimer.) As 2 or 3 monomers are usually added to each branch point in the reaction cycle, the number of end groups on the surface increases exponentially with each cycle, causing steric crowding and limiting the maximum size of the dendrimer. The convergent method of dendrimer synthesis also involves repetition of several chemical reactions, but here the reaction cycles are used to synthesize *dendrons* or dendrimer branches (instead of successive generation of a dendrimer) which are linked in the last synthetic step to two or more attachment points on a core molecule. While dendrimers synthesized by either method may contain defects, the problem is less pronounced in the convergent method. However, high generation dendrimers (>G8) cannot be prepared efficiently by the convergent approach because of usually low reaction yield between the high generation dendrons and the core. By the divergent approach, in comparison, dendrimers up to G10 can be prepared.

The strategy has been shown to provide facile synthetic routes for obtaining dendritic macromolecules with little or no purification. The divergent method starts with reaction between a core molecule and the monomer. The dendrimer core molecule is a bis-azide, such as 1,2-bis(2-azidoethoxy)ethane (**VI**) (Fig. 12.10) and the monomer is an alkyne-functionalized dihydroxy compound, such as propargyl ether of 1-hydroxybenzene-3,5-dimethanol (**VII**) derived from dimethyl 5-hydroxy isophthalate, as shown in Fig. 12.10. CuAAC reactions between the azide termini of the core and the alkyne terminus of the monomer, as shown in Fig. 12.11, yield the first generation dendrimer in OH-form, i.e., [**G-1**](OH)$_4$, which can then be transformed (Fig. 12.12) into either azide form ([**G-1**](N$_3$)), or alkyne form ([**G-1**](alkyne)$_4$)), demonstrating the versatility of the synthetic approach. Choosing, for example, the azide form for further growth, CuAAC reaction between the first generation azide-terminated dendrimer ([**G-1**](N$_3$)$_4$) and the monomer yields the second generation hydroxyl-terminated dendrimer, namely, [**G-2**](OH)$_8$ (Fig. 12.13). Its reactivation by transformation to [**G-2**](N$_3$)$_8$, followed by CuAAC reaction with the monomer then gives the third generation hydroxyl-terminated dendrimer, namely, [**G-3**](OH)$_{16}$. In this way, employing an iterative sequence that involves CuAAC reaction between azide chain ends of the growing dendrimers and the alkynyl group of the monomer, followed by reactivation via transformation of the hydroxyl groups to azides, one can proceed until the azide-terminated dendrimer of the desired generation number is arrived at (Fig. 12.14). Since alcohol functionalities can be converted to either azide or alkyne groups (see Fig. 12.12), the end groups of the dendrimer of any generation number made by the above divergent method can be either azide, hydroxyl, or alkynyl. Since such azide- and alkyne-functionalized dendrimers are capable of undergoing further click reactions, dendrimers made by the aforesaid divergent click strategy have the potential to find applications as reactive and multifunctional globular macromolecules.

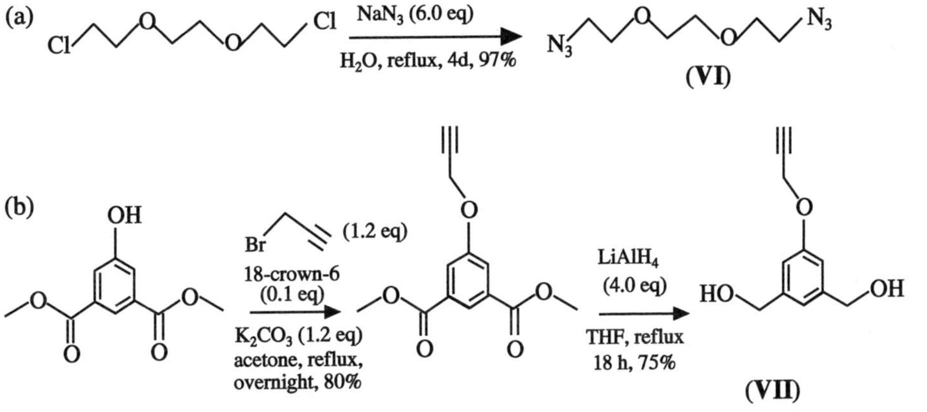

Figure 12.10 Synthesis of 1,2-bis(2-azidoethoxy)ethane (**VI**), used as the bis(azide) core, from 1,2-bis(2-chloroethoxy)ethane. (b) Synthesis of propargyl ether of 1-hydroxybenzene-3,5-dimethanol (**VII**), used as a monomer, from dimethyl 5-hydroxyisophthalate. (Drawn following the synthesis method of Joralemon et al., 2005.)

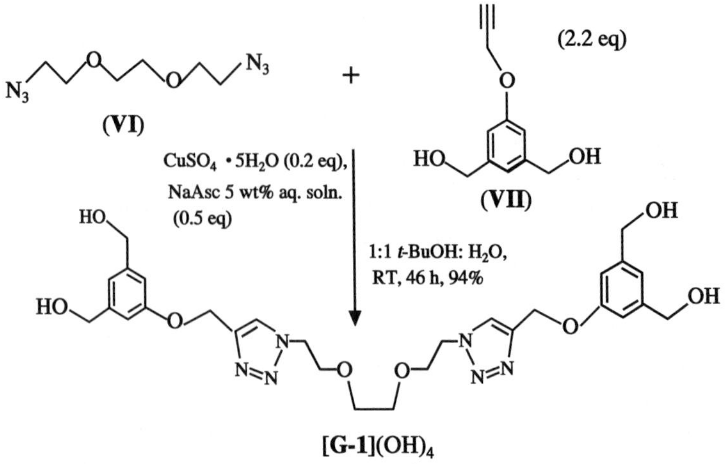

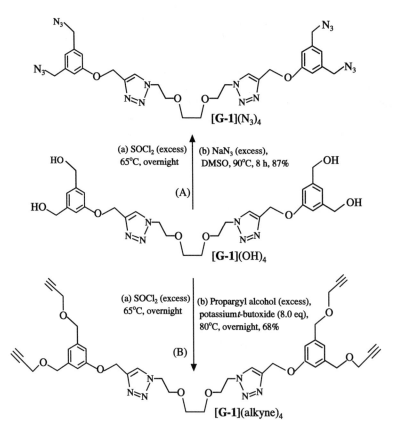

Figure 12.11 Synthesis of first generation dendrimer, **[G-1]**(OH)$_4$, from bis(azide) core and monomer by CuAAC reactions. (NaAsc = Sodium ascorbate.) (From Joralemon et al., 2005. With permission from *American Chemical Society*.)

Figure 12.12 Reactivation of hydroxyl-terminated dendrimer via transformation of hydroxyl groups to azides (Path A) and alkyne groups (Path B). (After Joralemon et al., 2005. With permission from *American Chemical Society*.)

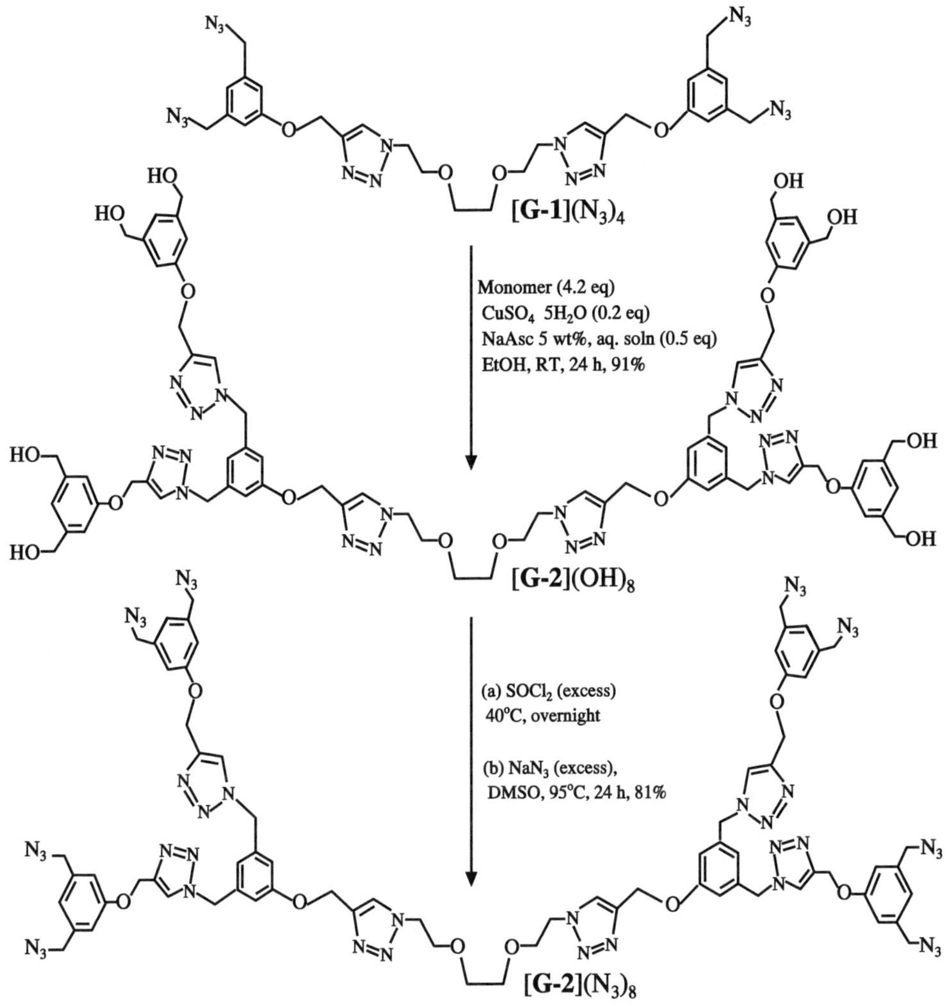

Figure 12.13 Synthesis of generation 2 dendrimer (azide-terminated) by CuAAC reaction using a divergent approach. (After Joralemon et al., 2005. With permission from *American Chemical Society.*)

As we have seen in the preceding pages, click reactions permit C–C bond (or C–N) formation in a quantitative yield without side reactions and without need for an additional purification step. Perhaps these advantages of click reactions have been most thoroughly exploited by combination of the azide/alkyne click chemistry with controlled/living polymerization techniques. For example, the combination of atom transfer radical polymerization (ATRP) and CuAAC has been extensively used to prepare a range of polymers with interesting properties, not easily accessible by other methods. The use of CuAAC has also been very fruitful when combined with other controlled/living polymerization methods, such as ring-opening polymerization (ROP) (Riva et al., 2005; Parrish et al., 2005; Lecomte et al., 2006), ring-opening metathesis polymerization (ROMP) (Yang and Weck, 2008; Murphy et al., 2006; Xia et al., 2008), cationic polymerization (Magenau et al., 2009; Gress et al., 2007; Tasdelen et al., 2008; Luxenhofer and Jordan, 2006), anionic polymerization (Thomsen et al., 2006; Wang et al., 2009; Parrish et al., 2005; Li et al., 2007),

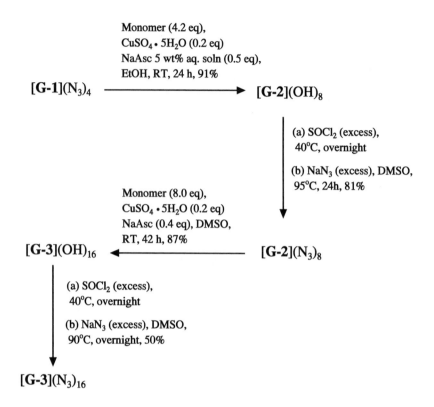

Figure 12.14 Synthesis of generation 3 dendrimer (azide-terminated) by CuAAC reactions using a divergent approach, according to the method of Joralemon et al. (2005). For structures of **G-1** and **G-2** dendrimers see Fig. 12.13.

nitroxide-mediated polymerization (NMP) (Joralemon et al., 2005; Gungor et al., 2007; Binder et al., 2007; Fleischmann et al., 2008; Altintas et al., 2007), and reversible addition-fragmentation chain transfer (RAFT) polymerization (O'Reilly et al., 2006; Vogt and Sumerlin, 2008; Rajan and Brittain, 2007, 2008; Gondi et al., 2007; Sinnwell et al., 2008; Magenau et al., 2009).

Since an azide-alkyne reaction results in the formation of 1,2,3-triazole ring, which acts as the bridge between the parts attached to the reacting azide and alkyne groups, the influence of 1,2,3-triazole moieties on the properties of material prepared by CuAAC cannot be overlooked. The π-electronic system of the triazole ring imparts an aromatic character (and hence stability) that is beneficial (Sumerlin and Vogt, 2010), thus prompting synthesis of new monomers (Thibault et al., 2006; Takizawa et al., 2008), RAFT agents (Akeroyd et al., 2009), and conjugated polymers (Bakbak et al., 2006; van Steenis et al., 2005). Triazoles can serve as ligands for Cu(I) ions producing autocatalytic behavior for many reactions (Chan et al., 2004). This effect is augmented for macromolecules containing multiple closely spaced triazole units. It was hypothesized (Sumerlin et al., 2005) that triazoles formed along the polymer backbone can complex Cu(I), leading to an effective higher local catalyst concentration in the immediate vicinity of neighboring unreacted azide groups to cause acceleration of reaction rate.

12.2.1 Combination of ATRP and CuAAC Reactions

Three different strategies have been employed by various workers to combine ATRP and CuAAC reactions, namely, using (i) azide-telechelic macromonomers, (ii) alkyne-telechelic macromonomers, and (iii) azide or acetylenic moieties within side chains. ATRP has the advantage that the polymers produced by the method contain ω(terminal)-halogen end groups, which can be substituted to contain azide groups. ATRP is thus an attractive technique for the synthesis of well-defined end-functionalized polymers.

The functionalization of ATRP-made polymers by substitution with NaN_3 and subsequent CuAAC with low molecular weight alkynes was first reported by Lutz et al. (2005) and many systematic studies have been conducted since then which have revealed the strong dependence of CuAAC reactions on the end-group structure and catalyst (Golas et al., 2008; Altintas et al., 2006). There has been a rapid proliferation of other methods for combining ATRP and CuAAC to facilitate synthesis of macromonomers (Vogt and Sumerlin, 2006; Topham et al., 2006), functionalized telechelic and high-molecular weight polymers (Gao et al., 2005), multiblock copolymers (Tsarevsky et al., 2005; Kyeremateng et al., 2008), star polymers (Urbani et al., 2007; Altintas et al., 2006; Gao and Matyjaszewski, 2006; Urbani et al., 2008), macrocyclics (Tsarevsky et al., 2008; Eugene and Grayson, 2008; Xu et al., 2007; Ge et al., 2009), networks (Diaz et al., 2004; Johnson et al., 2006; Mespouille et al., 2008), bioconjugates (Lutz and Boerner, 2008; Ladmiral et al., 2006), bionanoparticles (Zeng et al., 2007; Joralemon et al., 2005), and functionalized surfaces (Ranjan and Brittain, 2007).

12.2.1.1 Macromonomer Synthesis

Macromonomers are macromolecules of relatively low molecular weight that contain at least one vinyl group (or other polymerizable group) at a chain end which can be polymerized in a separate reaction. During polymerization, macromonomers serve as building blocks to facilitate synthesis of a variety of branched polymer topologies. While homopolymerization is used to prepare star or densely grafted brush-shaped polymers, copolymerization with low-molecular weight comonomers leads to loosely grafted and statistically distributed graft copolymers (Vogt and Sumerline, 2006).

There are two general synthetic routes to vinyl-capped macromonomers, namely, either an appropriate vinyl-functionalized initiator is used for ATRP of a monomer or the chain ends are modified after polymerization. In principle, post-polymerization modification is much more versatile, allowing access to a large range of different macromonomers. Using this latter approach, ATRP has been combined with CuAAC reaction to prepare well-defined macromonomers from commercially available reagents. Thus poly(*n*-butyl acrylate) (PBA), polystyrene (PSt), and PSt-*b*-PBA were prepared by ATRP and the resulting bromine-terminated polymers were transformed to azide end groups. Reaction of these azido-terminated polymers wih alkyne-containing acrylate and methacrylate monomers resulted in nearly quantitative chain end functionalization, yielding macromonomers with various molecular weights and architectures (Vogt and Sumerline, 2006). A wide range of hydrophilic methacrylic macromonomers with either methacrylic or acrylic terminal groups were also prepared in this way (Topham et al., 2008). Thus, the ATRP of several acrylate monomers, such as 2-amino ethyl methacrylate hydrochloride, 2-(diethylamino) ethyl methacrylate, 2-(dimethylamino) ethyl methacrylate, 2-hydroxyethyl methacrylatre (HEMA), and glycerol monomethacrylate (GMA) was conducted in turn in protic media at room temperature using an azido-functionalized initiator, and the resulting homopolymer precursors were then reacted with propargyl (meth)acrylate to obtain well-defined methacrylic macromonomers.

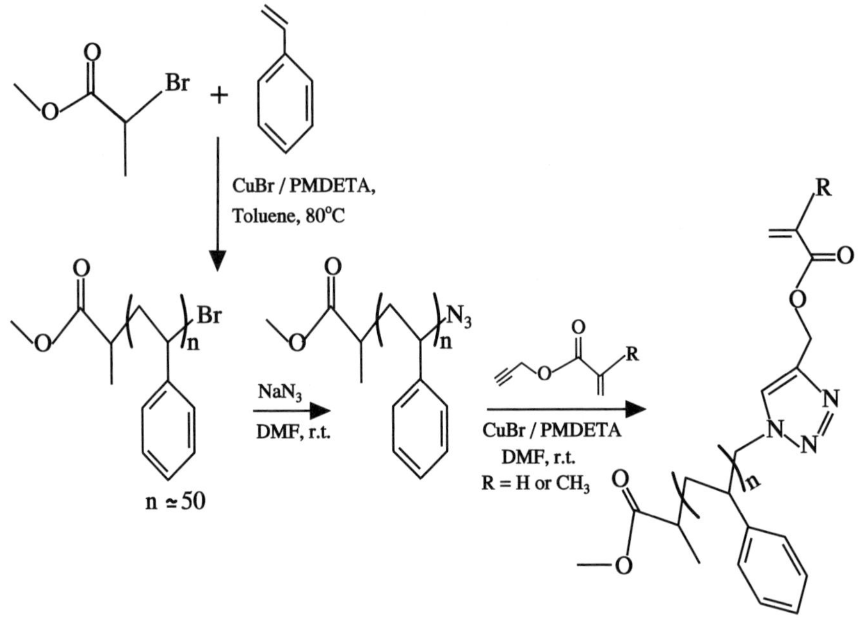

Scheme P12.1.1 Synthesis of PSt$_{50}$-acrylate macromonomer according to the method of Vogt and Sumerlin (2006). For details, see Problem 12.1.

Though typically a stepwise procedure was used in which the azide-terminatred polymer was isolated and purified before reaction with the alkyne species, an in-situ end group transformation process was also conducted in both the cases in which the azide-terminated polymers were reacted in situ with alkyne-functionalized methacrylates. This "one-pot" synthesis is possible due to the orthogonal reactivity inherent in click chemistry.

Problem 12.1   Suggest an efficient route, via ATRP and CuAAC reaction, for the synthesis of a (meth)acrylate terminated polystyrene macromonomer, such as ω-acryloyloxy-polystyrene (DP $\simeq$ 50).

Answer:

The combination of ATRP and postpolymerization modification by CuAAC click chemistry can be employed to prepare well-defined ω-(meth)acryloyl macromonomers in an efficient manner. Thus, polystyrene (PSt) can be prepared by ATRP and subsequently derivatized to contain azide end groups. The azide-terminated polymers can then be reacted with alkyne-containing (meth)acrylate monomers to achieve near-quantitative chain-end functionalization by CuAAC reaction. An efficient synthesis route is shown in Scheme P12.1.1.

For ATRP of St, Vogt and Sumerlin (2006) used methyl 2-bromopropionate (MBP) as the initiator, CuBr as the catalyst, N,N,N',N',N''-pentamethyldiethylene triamine (PMDETA) as the ligand, and propargyl acrylate (PgA) as the alkyne reagent. Briefly, PSt was prepared using the following ATRP conditions: [St]/[MBP]/[CuBr]/[PMDETA] = 100:1:1:1 (mol ratio), 40 vol% toluene, 80°C, and nitrogen blanket. The polymerization was stopped by cooling and exposing the reaction mixture to air. The monomer conversion was limited to less than 60% to ensure end group retention. The reaction mixture was diluted with THF and passed through a neutral alumina column to remove the catalysts, and the polymer (PSt-Br) was isolated by precipitation into methanol. The bromo-terminated polymer was reacted with NaN$_3$ (1.1 equiv) at room temperature in DMF (0.05 M polymer solution) to yield the azido-terminated polymer, PSt-N$_3$, which was isolated by precipitation into methanol. Under a nitrogen atmosphere, CuBr (0.5 equiv) catalyst was added to a solution of polymer (0.1 M in DMF), PMDETA (0.5 equiv), and PgA (1.3 equiv). The reaction was allowed to proceed at room temperature for 24 h. After removal of the catalyst by passing through a column of neutral alumina, the resulting acryloyloxy-end-capped polymer (ω-acryloyloxy-PSt) was isolated by reprecipitation into methanol. [*Note*: While in the above procedure, the end-group transformations required to incorporate the polymerizable functionality were accomplished as a stepwise series of discrete reactions, an in situ process was also successfully used, wherein azidation was immediately followed by azide-alkyne coupling in situ.]

Problem 12.2 An ATRP initiator, propargyl 2-bromoisobutyrate (PgBiB), containing an alkyne functionality was prepared (Tsarevsky et al., 2005) by reacting propargyl alcohol with 2-bromoisobutyric acid in the presence of dicyclohexyl carbodiimide. A heterotelechelic polystyrene with alkyne and bromo end groups was prepared by ATRP of styrene (St), in the presence of PgBiB as the initiator and CuBr/PMDETA as the catalyst in a solvent. The resulting α-alkyne-ω-bromo-terminated polystyrene (ABPSt) was reacted with NaN$_3$ (nucleophilic substitution) in dimethyl formamide (DMF) to yield the corresponding α-alyne-ω-azido-terminated polystyrene (AAPSt).

This macromonomer was then subjected to step-growth click coupling by CuAAC in DMF at room temperature in the presence of CuBr (no additional ligand was necessary as CuBr has sufficient solubility in DMF). Because both ATRP and CuAAC are catalyzed by Cu(I) compounds, a one-pot ATRP-nucleophilic substitution-click coupling process was attempted (Tsarevsky et al., 2005) by (i) initiating ATRP of St with PgBiB ([St] : [PgBiB] : [CuBr] : [PMDETA] = 76 : 1 : 0.25 : 0.25 (mol ratio), 44% toluene, 80°C, 200 min, 13% monomer conversion), (ii) quenching the polymerization by freezing with liquid nitrogen, (iii) adding NaN$_3$, ascorbic acid, and DMF, and (iv) equilibrating to room temperature (116 h) to allow homocoupling of the macromonomer.

(a) Show equations for the sequence of reactions in this process. (b) Calculate the molar mass of the heterotelechelic polymer, ABPSt. (c) Given that the macromonomer conversion in the last step is 90%, determine theoretically the average molecular weight ($\overline{M}_n$) of the polymeric product.

Answer:

(a) The sequence of reactions performed in one-pot leading to the synthesis of α-alkyne-ω-azide-terminated higher molecular weight polystyrene are shown in Scheme P12.2.1.

(b) Molar mass of PgBiB (**P2-I**) = 125 g/mol

Molar mass of St (**P2-II**) = 104 g/mol

$$\text{Molar mass of ABPSt (\textbf{P2-III})} = \frac{[\text{St}] \times 0.13}{[\text{PgBiB}]} (104 \ g \ mol^{-1}) + (125 \ g \ mol^{-1})$$
$$= 1152 \ g \ mol^{-1}$$

(c) Molar mass of macromonomer (**P2-IV**) = M_0 = 1152 − 80 + 42 = 1114 g mol^{-1}. Therefore, molar mass of the structural unit of (**P2-V**) = M_0 = 1114 g mol^{-1}.

Using Eq. (5.30) for the number-average degree of polymerization (n) of (**P2-V**),

$$n = 1/(1 - p)$$

where p = extent of reaction.

For $p = 0.90$, $n = 10$. Therefore,

$$\overline{M}_n = n \times M_0 + M_0 = (n + 1) M_0 = 12254 \ g \ mol^{-1}$$

Problem 12.3 Suggest a method to synthesize, via ATRP and CuAAC reactions, α-(carboxylic acid)-ω-hydroxy-polystyrene (PSt), in which the PSt block has a degree of polymerization (DP) of about 30.

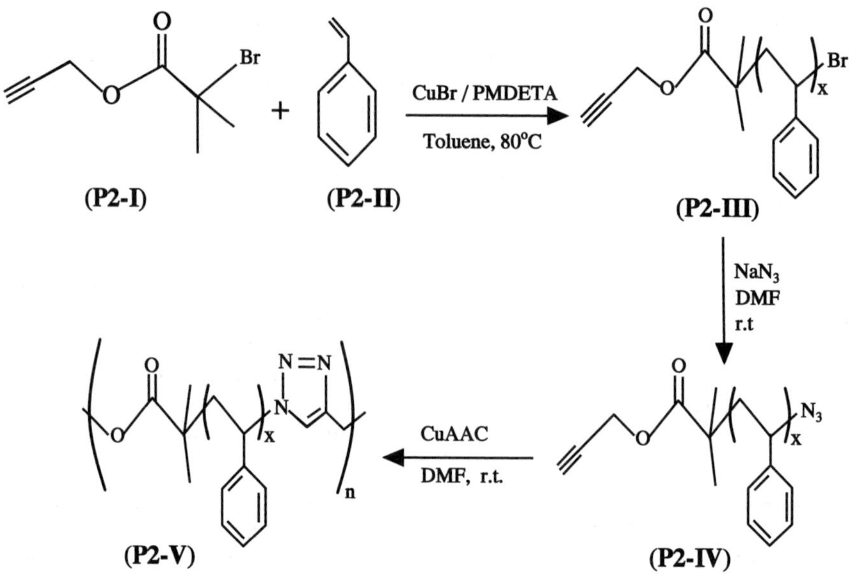

Scheme P12.2.1 Synthesis of α-alkyne-ω-azide-terminated polystyrene (PSt) of high molecular weight by the method of Tsarevsky et al., 2005.

Answer:

The click chemistry approach can be used for the asymmetric functionalization of polymer chains. A convenient strategy in the present case is to use a hetero-difunctional initiator (DFI) containing a bromo group and a protected alkyne group (Opsteen and Van Hest, 2007) so that the bromo group can be used to initiate ATRP of styrene in the presence of Cu/ligand catalyst and the bromo end group of the resulting PSt can be used to introduce the hydroxy group at the ω-end via azidation and click coupling. Thereafter the alkyne group can be deprotected and used to introduce carboxylic acid group at the α-end by CuAAC reaction with an azide functionalized carboxylic acid moiety.

Synthesis of silyl-protected acetylene-functionalized ATRP initiator (Scheme P12.3.1)

Propargyl 2-bromoisobutyrate (see Problem 12.2) with its acetylene functionality protected by a silyl group can be used as the DFI in the present case. Trimethyl silyl (TMS) or triisopropyl silyl (TIPS) protection is commonly used, the latter being preferred when greater stability during ATRP reaction is necessary. (Because of its bulky character, TIPS has significantly higher stability compared with TMS and hence suffers much less loss during ATRP.) The synthesis of TIPS-protected propargyl 2-bromoisobutyrate [3-(1,1,1-triisopropylsilyl)-2-propanyl-2-bromo-2-methyl propanoate] (**P3-I**), involves two steps, namely, silylation of propargyl alcohol (after protecting the acetylene moiety) and esterification of the alcoholic group (Scheme P12.3.1). Adopting the procedure described by Opsteen and Van Hest (2007), a solution of propargyl alcohol (1.13 g, 20.2 mmol) in THF (20 mL) is added dropwise at room temperature to 3.0 M solution (20.0 mL) of ethylmagnesium bromide, which is diluted with 50 mL THF, and then refluxed for 18 h. After cooling the reaction mixture to room temperature, a solution of chlorotriisopropyl silane (TIPS-Cl) (5.57 g, 28.9 mmol) in THF (20 mL) is added dropwise and refluxed for 5 h. The product, 3-(1,1,1-triisopropylsilyl)-2-propyn-1-ol, obtained as a yellow oil by extraction with Et$_2$O and removal of solvent is used without further purification to synthesize the TIPS-protected initiator (**P3-I**) as described below.

A solution of 2-bromoisobutyryl bromide (3.75 mL, 30.4 mmol) in THF (40 mL) is added dropwise to a solution of 3-(1,1,1-triisopropylsilyl)-2-propyn-1-ol (20.2 mmol) and (C$_2$H$_5$)$_3$N (4.23 mL, 30.4 mmol) in THF (60 mL) at 0°C. The reaction mixture is stirred for 2 h at room temperature to complete the reaction forming (**P3-I**).

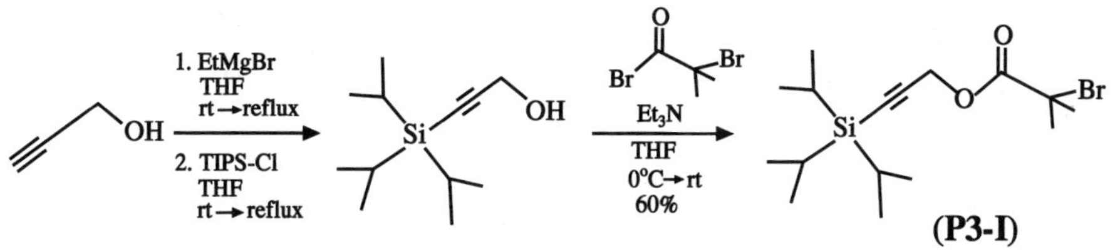

Scheme P12.3.1 Synthesis of triisopropyl silyl (TIPS) protected acetylene functionalized ATRP initiator. (Drawn following the synthesis method of Opsteen and Van Hest, 2007.)

Synthesis of α-(TIPS protected acetylene)-ω-azide functionalized polystyrene (Scheme P12.3.2)

ATRP of styrene is carried out in a Schlenk flask loaded with CuBr to which styrene (much in excess of [St] : [CuBr] = 30 : 1 ratio), phenyl ether (solvent), and N,N,N',N',N''-pentamethyl diethylene triamine (PMDETA) are added and stirred for 15 min to allow catalyst formation. To the reaction mixture, cooled in an ice bath, the difunctional initiator (**P3-I**) is added and heated at 90°C in an oil bath till the monomer conversion (measured by ^{1}H-NMR analysis) corresponding to the target DP of 30 is reached. [The ATRP reactions are stopped at a relatively low (less than 50%) monomer conversion to circumvent loss of bromide end functionality because of termination processes (cf. Problem 12.2).] To quench the polymerization, the reaction mixture is cooled and diluted with CHCl$_3$.

The reaction mixture is then passed through a basic alumina column to remove the catalyst, allowed to concentrate in a vacuum, and the polymer (**P3-II**) is recovered by precipitation in methanol and drying in vacuum at 60°C. For conversion to azide end-functionality, NaN$_3$ (10 equiv) is added to DMF solution (0.01 M) of the above ATRP polymer containing bromide end group and the reaction mixture is stirred for 20 h at room temperature. The heterotelechelic polystyrene (**P3-III**) thus formed is recovered by precipitation in methanol and is dried in vacuum at 60°C.

Coupling of propargyl alcohol (Scheme P12.3.3)

Propargyl alcohol (0.50 mmol) and THF (5 mL) are added to the above heterotelechelic polystyrene (0.05 mmol) in a Schlenk tube, followed by 0.2 mL of a stock solution containing CuBr (0.45 M) and PMDETA (0.45 M) in THF. The reaction mixture is stirred for 18 h at room temperature for completion of the CuAAC reaction (as indicated by the disappearance of FTIR azide signal). The formed hydroxyl-functionalized polystyrene (PSt-OH), (**P3-IV**), is precipitated in methyl alcohol and isolated as a white solid.

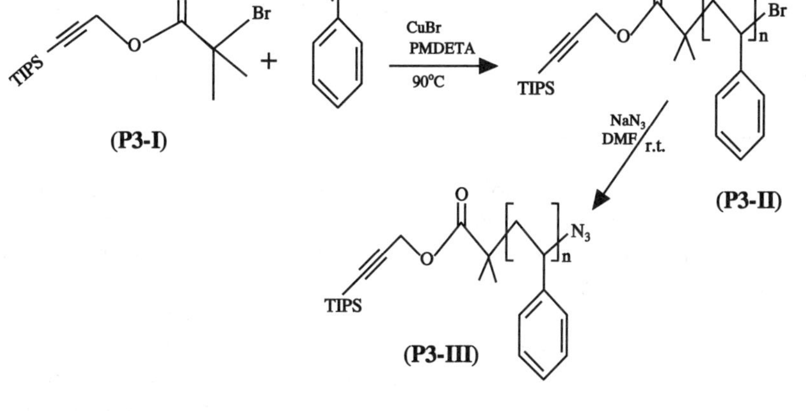

Scheme P12.3.2 Outline for the synthesis of α-(TIPS-acetylene)-ω-azide terminated polystyrene. (Drawn following the synthesis method of Opsteen and Van Hest, 2007.)

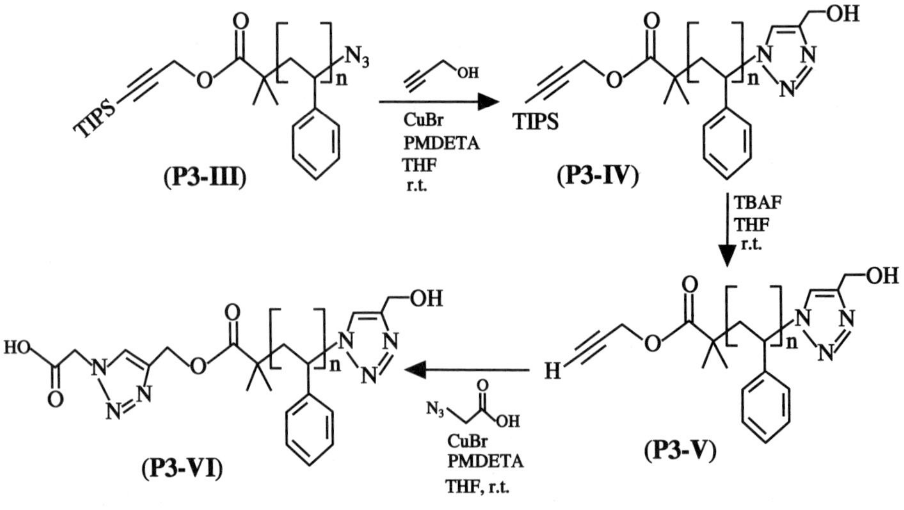

Scheme P12.3.3 Outline for the synthesis of α-(carboxylic acid)-ω-hydroxy-PSt by two consecutive CuAAC reactions on heterotelechelic polystyrene, bearing both azide and TIPS-protected acetytene end groups. (Drawn following the synthesis method of Opsteen and Van Hest, 2007.)

Removal of TIPS group (Scheme P12.3.3)

Tetrabutylammonium fluoride (TBAF), available as 1M solution in THF, is used to remove the TIPS group. To PSt-OH (0.013 mmol), dissolved in THF (1 mL), TBAF (0.13 mL, 0.13 mmol) is added and stirred for 17 h at room temperature. TIPS fluoride, formed as a by-product, is removed by passing the reaction mixture through a basic alumina column and the deprotected hydroxyl functionalized PSt, (**P3-V**) is isolated as a white solid (yield 95%) by precipitation in methanol.

CuAAC reaction with azidoacetic acid (Scheme P12.3.3)

A stock solution (0.1 mL) containing CuBr (0.11 M) and PMDETA (0.11 M) in THF is added to (**P3-V**) (0.011 mmol) and azidoacetic acid (0.11 mmol) in a Schlenk tube and the reaction mixture is stirred for 18 h at room temperature. The reaction mixture is diluted with CH_2Cl_2 (5 mL) and washed with EDTA solution (0.055 M aqueous) and water. The polymeric product, α-(carboxylic acid)-ω-hydroxy-PSt (**P3-VI**), is recovered as a white solid from the organic layer by precipitation in methanol. The yield of (**P3-VI**) from the process is reported to be 82% (Opsteen and Van Hest, 2007).

12.2.1.2   End-Functionalization of (Co)polymer Chains

Since the polymers produced by ATRP have halogen terminal groups and these can be easily and efficiently converted into azide groups by nucleophilic substitution, ATRP is particularly well-suited to subsequent end-group modification via CuAAC reactions to produce a variety of end functional moieties, such as, amino, hydroxy, or carboxy groups (see Problem 12.3). Consequently, CuAAC reactions in combination with ATRP have emerged as a powerful method to modify chain-end and/or chain-side groups for functionalization of various (co)polymers (Gao et al., 2005). This is reflected in the synthesis of a variety of functional (co)polymers via reaction of various alkynes with azide-derivatized telechelic polymers prepared by ATRP (Lutz et al., 2005; Opsteen and Van Hest, 2005; Tsarevsky et al., 2005).

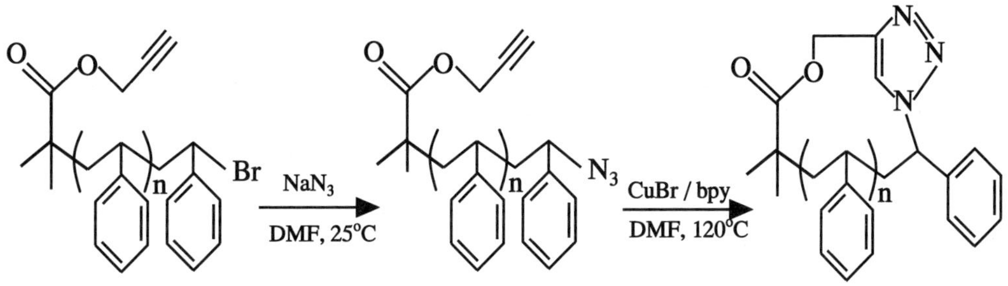

Scheme P12.4.1 Synthesis of macrocyclic polystyrene by terminal azidation and CuAAC reaction on linear polystyrene prepared by ATRP (Tsarevsky, et al., 2005; Laurent and Grayson, 2006.)

12.2.1.3 Cyclization of Linear Polymers

Despite their unique topology and physical properties, cyclic polymers have been a less explored topic, probably due to the challenges encountered in their controlled synthesis. While synthetic routes for the preparation of macrocyclic polymers have long been a goal for polymer chemists, most of the studies made for the preparation of these compounds have involved cyclization of linear polymers that typically suffer from poor yields and competing reactions, requiring tedious purification to isolate pure macrocycles (Laurent and Grayson, 2006). However, in the past years, with the emergence of CuAAC reactions as an alternative approach for the formation of stable covalent linkages due to their high efficiency, quantitative yield and technical simplicity, the possibility of applying click reactions in combination with living radical polymerization, especially ATRP, for the synthesis of macrocyclic polymers by intramolecular ring closure of heterodifunctional linear polymers (bearing alkyne and azide end groups) has attracted much attention. While the bromide end group of an ATRP polymer presents an ideal substrate for a nucleophilic displacement with an azide, incorporation of an alkyne within the ATRP initiator provides the requisite functional groups at opposite (i.e., α, ω) ends of the polymer chain for subsequent cyclization via CuAAC reactions.

An α, ω-heterofunctionalized polymer bearing alkyne and azide end groups can undergo either cyclization or condensation, depending on the concentration of the polymer. Thus, using Cu(I) catalyst (3.6 mM) in THF/water, Laurent and Grayson (2006) found that concentrations of α-alkyne-ω-azido-terminated polystyrene above 0.1 mM favored condensation while dilution below that concentration favored intramolecular cyclization. Since the dilution required to obtain high purity macrocycle would be prohibitively excessive, these workers utilized a continuous addition technique to ensure that the concentration of unreacted linear polymer during the course remained infinitesimally small.

Problem 12.4 Suggest an efficient route, via ATRP and CuAAC reaction, for the preparation of narrow disperse cyclic-polystyrene with an average of 24 styrene units in the ring.

Answer:

α-alkyne-ω-azido heterodifunctional linear polystyrene (PSt) precursor of the required degree of polymerization (DP) is first prepared using a standard ATRP technique. According to the experimental procedure reported by Tsarevsky et al. (2005), propargyl 2-bromo-isobutyrate is used as initiator with Cu(I)Br and N,N,N',N',N''-pentamethyl diethylenetriamine (PMDETA) as catalyst. The reaction is carried out with a [Sty]/[initiator]/[CuBr]/[PMDETA] mol ratio of 40:1:1:1, in 11 vol% phenyl ether at 90°C for 75 min. The

ATRP reactions are stopped at about 60% monomer conversion to correspond to the target DP, yielding narrow disperse heterodifunctional PSt with $\overline{M}_w/\overline{M}_n < 1.2$ (Tsarevsky et al., 2005). The polymer is purified by extraction into CH_2Cl_2 and precipitation into methanol. Azidation of the bromide end group of the polymer is carried out in DMF with sodium azide at 25°C (Scheme P12.4.1).

The cyclization of the α-alkyne-ω-azido PSt linear precursor is carried out under high dilution condition. According to the procedure of Laurent and Grayson (2006), 5 mL of a 2 mM solution of the heterodifunctional PSt in DMF is added over 24 h (i.e., at a rate of 34 μL/10 min) into a 115 mL volume of DMF at 120°C containing Cu(I)Br and bipyridine. After 1 additional hour, DMF is removed under reduced pressure and the product, cyclic-PSt, is recovered by extraction into CH_2Cl_2 and precipitation into methanol. The molecular weight of polymer samples is found to remain unchanged after successful cyclization.

12.2.1.4 Moldular Synthesis of Block Copolymers

As we have noted earlier, ATRP allows the introduction of azide and acetylene end groups, via the use of functional initiators or post-polymerization end group modification methods. Employing both functionalization routes an azide as well as an acetylene end group can be introduced into the same polymer molecule. These heterotelechelic polymers have been used for step-growth intermolecular click coupling (see Problem 12.2) to produce higher molecular weight polymers containing triazole linkages in the repeat units and for the synthesis of macrocyclic polymers by intramolecular click coupling (Problem 12.4). While CuAAC coupling of monofunctional polymer blocks leads to diblock copolymers (Fig. 12.6a), by coupling heterotelechelic polymeric building blocks via CuAAC reactions one can carry out modular synthesis of block copolymers in a controlled fashion, which, otherwise, would be a difficult synthetic task by means of consecutive polymerization of different monomers. Using heterotelechelic polymers bearing protected acetylene moieties and azide functional groups, first CuAAC reactions can be performed using the azide groups to join one polymer block, and, after deprotection of the acetylene groups, a second polymer block can be coupled to the other terminus, yielding ABC type block copolymers. This concept of conducting two consecutive click coupling reactions onto one single polymer chain is versatile and can be used for the asymmetric functionalization of polymer chains with a myriad of distinct moieties (e.g., biomolecules) (Opsteen and van Hest, 2007).

Problem 12.5 Show how an ABC triblock terpolymer, poly(methyl acrylate)-b-polystyrene-b-poly(tert-butyl acrylate) (PMMA-b-PSt-b-PtBA), can be synthesized modularly, in a controlled manner, using ATRP and consecutive CuAAC coupling reactions.

Answer:

For modular synthesis of ABC-type triblock copolymer, two successive CuAAC reactions have to be performed on the central polymer chain (B block). To accomplish this, the B block polymer having both azide and acetylene end groups (heterotelechelic B) has to be used and, moreover, one of the termini has to be protected in order to prevent linear chain extension (cf. Scheme P12.2.1) or formation of cyclic products (Scheme P12.4.1). In a straightforward methodology, the terminal acetylene moiety on B is protected and the azide terminus is used to carry out the first coupling reaction to join the preformed A or C block. Next, the acetylene moiety is to be deprotected to make it available for the second coupling reaction to join the remaining C or A block.

An outline of the synthetic route to prepare PMA-b-PSt-b-PtBA, using the above methodology is given below. Triisopropyl silyl (TIPS) group is used for protecting acetylene functionality. The following end-functionalized polymer blocks made by ATRP, followed by necessary end-group modifications, are employed: α-(TIPS protected acetylene)-ω-azide functionalized polystyrene (TIPS-≡-PSt-N$_3$), terminal acety-

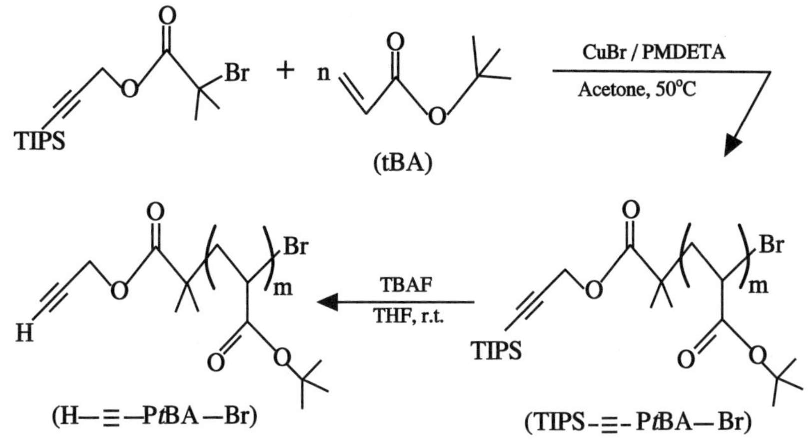

Scheme P12.5.1 Outline for the synthesis of acetylene functionalized poly(*tert*-butyl acrylate). (Drawn following the synthesis method of Opsteen and Van Hest, 2007.)

lene functionalized poly(*tert*-butyl acrylate) (H-≡-PtBA-Br), and α-(TIPS protected acetylene)-ω-azide functionalized poly(methyl acrylate) (TIPS-≡-PMA-N$_3$).

To prepare H-≡-PtBA-Br, tBA polymerization is first carried out under ATRP conditions utilizing a TIPS-protected acetylene-functionalized ATRP initiator (see Scheme P12.3.1) to produce TIPS-≡-PtBA-Br. Its acetylene terminus is then deprotected quantitatively by stirring with tetrabutylammonium fluoride (TBAF) for more than 12 h at room temperature and passing the reaction mixture through a basic alumina column to remove TIPS fluoride. This yields H-≡-PtBA-Br (Scheme 12.3.1) bearing deprotected acetylene functionality as a handle to conjugate to an azide functionalized polymer via CuAAC reaction. TIPS-≡-PSt-N$_3$ and TIPS-≡-PMA-N$_3$ are prepared using protocol similar to that used for TIPS-≡-PtBA-Br, followed by replacement of Br by azide group.

As a first step in the synthesis of triblock terpolymer, PMA-*b*-PSt-*b*-PtBA (Scheme P12.5.2), heterotelechelic PSt (TIPS-≡-PSt-N$_3$) and unprotected acetylene functionalized PtBA (H-≡-PtBA-Br) are clicked together by applying CuBr/Me$_6$TREN as a catalyst with DMF as a solvent. To drive the reaction to completion, a slight excess of PSt (1.13 equiv) is used. After about 18 h of reaction, residual azide groups are reduced via a Staudinger reduction by addition of PPh$_3$ (Brase et al., 2005) so that the resulting amine terminated PSt can be removed by employing 95:5 CH$_2$Cl$_2$/MeOH (Opsteen and van Hest, 2007). The acetylene end group of the diblock copolymer TIPS-≡-PSt-*b*-PtBA-Br thus obtained is then made available for the next coupling reaction by removal of the TIPS group with TBAF in the same manner as for PtBA in Scheme P12.5.1.

As the final step in preparing the triblock terpolymer, TIPS-≡-PMA-N$_3$ (1.5 equiv) is coupled to the free end functionality of the H-≡-PSt-*b*-PtBA-Br diblock copolymer via CuAAC reaction, which is conducted at 50°C with CuBr/Me$_6$TREN as the Cu(I) source and DMF as the solvent (Opsteen and van Hest, 2007). The excess of azide functionalized PMA is again removed by reduction of the azide moiety with PPh$_3$ and extraction of the amine terminated polymer with the mixed solvent CH$_2$Cl$_2$/MeOH (95:5).

12.2.1.5 Nonlinear Polymer Synthesis

Nonlinear polymers primarily include star, miktoarm star, and H-type (co)polymers, hyperbranched dendrimers, and dendrimer-like star polymers. *Homo-arm* (or regular) *star* polymers contain multiple arms of identical chemical composition and similar molecular weight connecting to a central

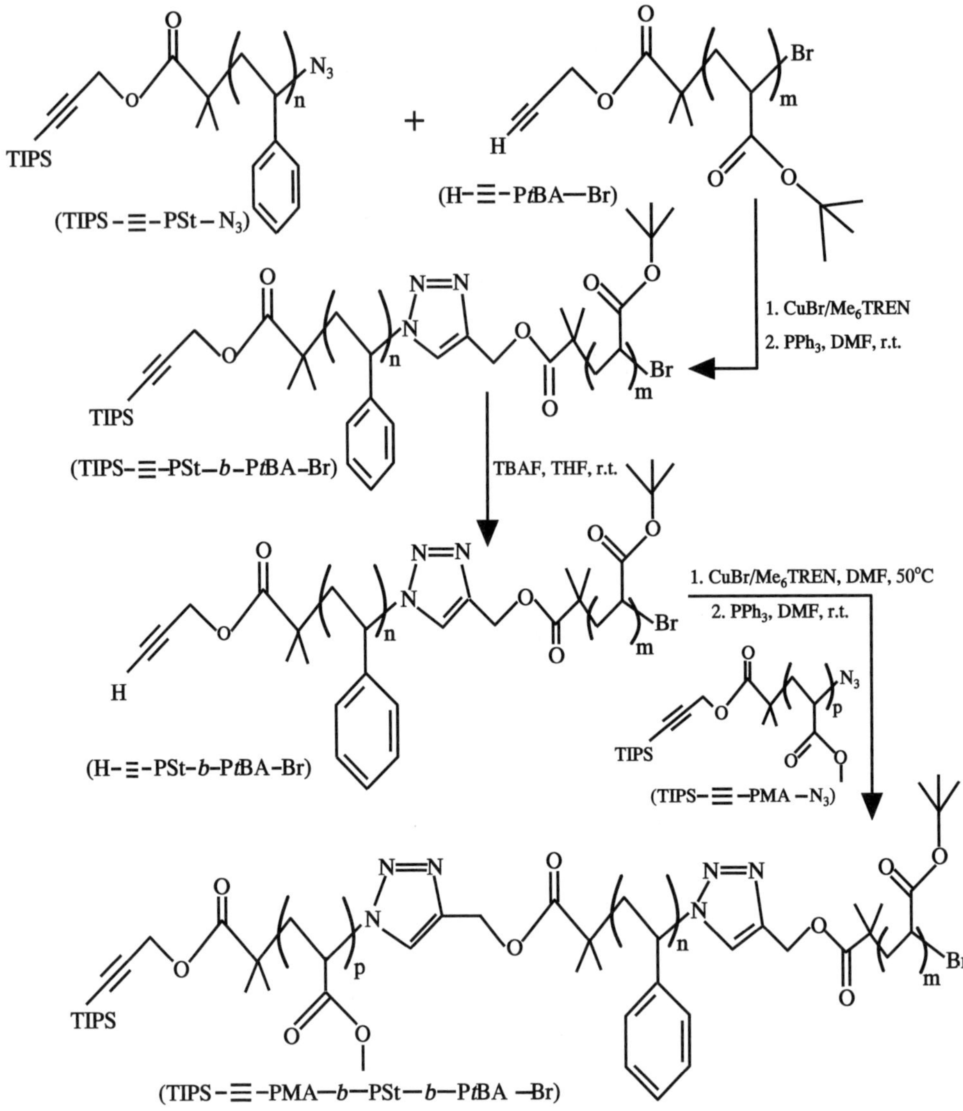

Scheme P12.5.2 Modular formation of an ABC-type triblock terpolymer from heterotelechelic precursors using CuAAC reactions. (Drawn following the synthesis method of Opsteen and Van Hest, 2007.)

point, e.g., three-arm star of monomer A, which we may denote as *core*-(poly A)$_3$. They represent the simplest structure of numerous possible branched topologies (Hadjichristidis et al., 2001). Star polymers in which *each* arm is a multiblock copolymer chain, e.g., *core*-(poly A-*b*-poly B)$_3$ for a diblock 3-arm star, or *core*-(poly A-*b*-poly B-*b*-poly C)$_3$ for a triblock 3-arm star, are called *star-block* copolymers, whereas those in which two or more are species are of different chemical compositions, e.g., (poly A)$_2$-*core*-(poly B), and/or different molecular weights, are known as *miktoarm star* (or heteroarm) copolymers (Fig. 12.15). A special type of nonlinear polymers are *H-shaped* polymers, in which two side chains are attached to each end of a polymer backbone (main chain), e.g., (poly A)$_2$-*core*-poly B-*core*-(poly A)$_2$.

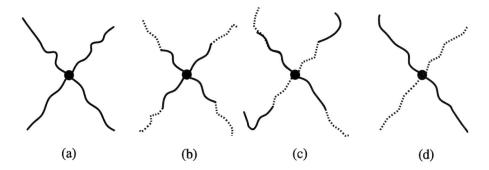

(a)   (b)   (c)   (d)

Figure 12.15 Schematic representations showing the differences between various types of star (co)polymers: (a) star homopolymers; (b) star block copolymers; (c) and (d) miktoarm star copolymers. Solid lines and dotted lines represent polymer chains differing in composition and/or molecular weight.

While *branched polymers* consist of a main (first) chain on which there are a relatively small number of branching points or at least one point where a second chain of monomers branches off, *hyperbranched polymers* have a very large number of branches (Fig. 12.16b). *Dendrimers* are a subset of hyperbranched polymers in that they have no imperfections and they branch at each monomer unit. A dendrimer (Fig. 12.16c) is typically symmetric around the core, and often adopts a spherical three-dimensional morphology. The word 'dendron' is also encountered frequently. A dendron (Fig. 12.16d) usually contains a single chemically addressable group called the *focal* point. The differences between dendrons, dendrimers, amd hyperbranched polymers are illustrated in Fig. 12.16, but the terms are typically encountered interchangeably. The structure of dendrimer-like star polymers is similar to that of dendrimers; however, there are polymeric segments rather than short linkages between the branching points as in dendrimers.

Star polymers with different arm numbers can be prepared in a facile and efficient manner under mild conditions using click chemistry stategy based on 1,3-dipolar cycloaddition between azide and alkyne functional groups (CuAAC reaction) of the polymeric precursors, such as azido-terminated polymers, and alkyne-containing multifunctional compounds (Gao an Matyjaszewski, 2006). Thus, a four-arm star polymer, as shown in Fig. 12.6(b), can be prepared by CuAAC reactions between azido-terminated polymeric precursors forming the arms and a tetraalkyne-containing compound forming the hub or core (see Problem 12.6).

Problem 12.6 Combination of ATRP and the click coupling method provides a simple and efficient route to the synthesis of various types of star polymers. Discuss, in this context, a feasible procedure for the efficient synthesis of polystyrene (PSt) three-arm and four-arm star polymers under mild conditions.

Answer:

Gao and Matyjaszewski carried out the first synthesis of PSt star polymers using a combination of ATRP and click chemistry (Gao and Matyjaszewskki, 2006). PSt linear chains with high azide chain-end functionality were prepared and coupled with tri- and tetra-alkyne-containing coupling agents under mild conditions to produce PSt 3-arm and 4-arm polymers, respectively (Scheme P12.6.1).

A trialkyne-containing coupling agent (**P6-I**) and a tetraalkyne-containing coupling agent (**P6-II**) were prepared, in 85% and 73% yields, by reacting, respectively, 1,1,1-tris(4-hydroxyphenyl) ethane and pentaerythritol with pentynoic acid. The reactions were conducted in methylene chloride in the presence of N-(3-(dimethylamino)propyl)-N'-ethylcarbodiimide hydrochloride and 4-($N.N$-dimethylamino) pyridine for

(a)

(b)

(c)

Generation
number

Core

Branching points

Terminal groups

(d)

Focal point
(Chemically addressable group)

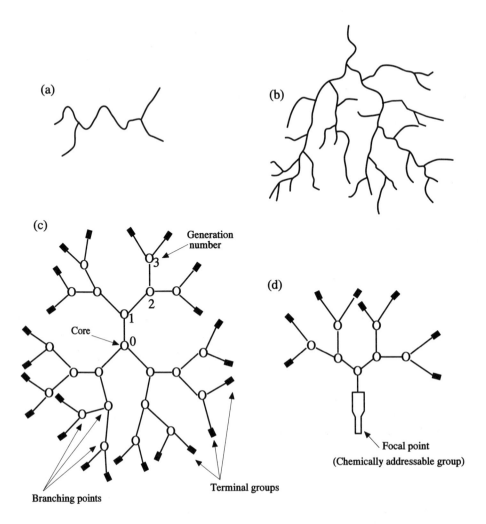

Figure 12.16 Schematic representation of (a) branched polymers, (b) hyperbranched polymers, (c) dendrimers, and (d) dendrons.

40 h at room temperature. To prepare ω-azido-terminated PSt (i.e., PSt-N$_3$), used as PSt precursor, ω-bromo-terminated PSt chain was first prepared by ATRP using CuBr, N, N, N', N', N''-pentamethyldiethylenetriamine (PMDETA) as catalyst and ethyl-2-bromoisobutyrate (EBiB) as initiator, and stopping the reaction at low (< 30%) styrene conversion (via exposure to air and dilution with THF) to circumvent the loss of bromide functionality due to termination processes. The PSt–Br obtained was treated with NaN$_3$ (2 times excess to the mole of bromo group) in DMF at room temperature to introduce azide moieties by nucleophilic substitution, followed by precipitation of the resulting PSt–N$_3$ into MeOH/water mixture (1:1 by volume).

The CuAAC reactions between PSt–N$_3$ and the aforementioned tri- and tetra-alkyne coupling agents were conducted with DMF as solvent and CuBr/PMDETA as catalyst for 10 h at room temperature, the molar ratios used being [PSt–N$_3$]/[**P6-I**]/[CuBr]/[PMDETA] = 1/0.33/0.5/0.5 and [PSt–N$_3$]/[**P6-II**]/[CuBr]/[PMDETA] = 1/0.25/0.5/0.5. The yields of 3-arm and 4-arm star PSt polymers were 90% and 83%, respectively. The addition of a reducing agent Cu(0) [Cu(0)/CuBr = 1/1 by weight] during the coupling reaction to reduce an oxidized Cu(II) species back to Cu(I) was found to have a positive influence on click coupling efficiency, resulting in faster reaction and higher yield of coupling product.

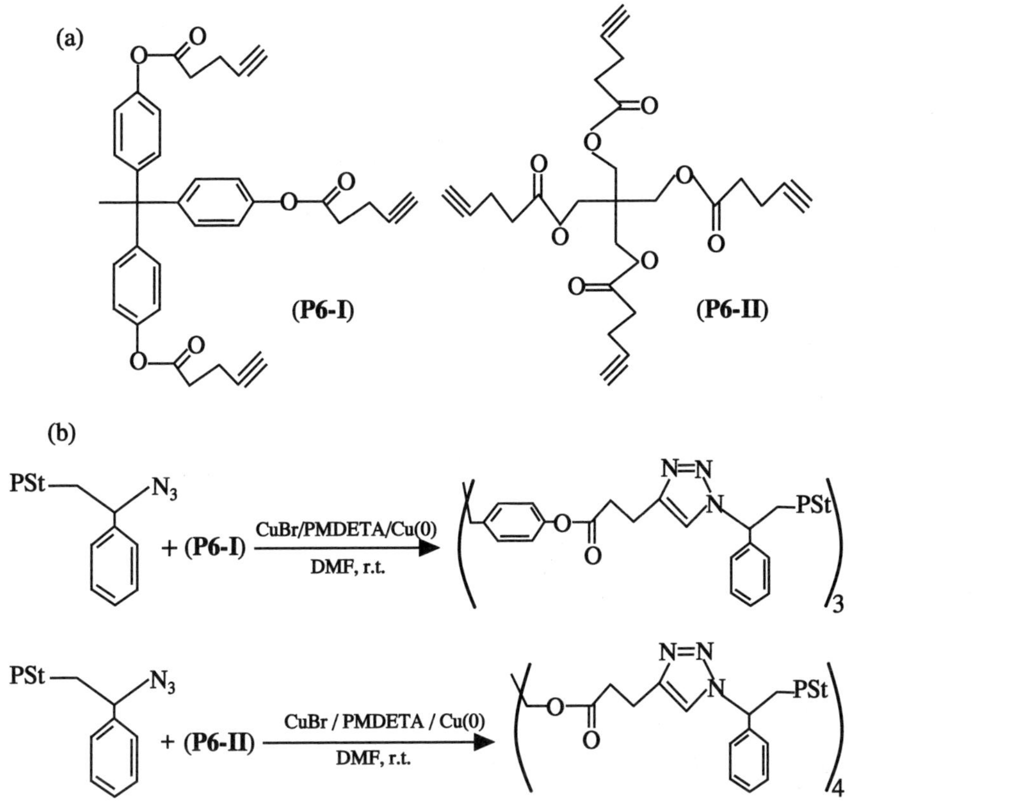

Scheme P12.6.1 (a) Structures of trialkyne containing (**P6-I**) and tetraalkyne containing (**P6-II**) coupling agents. (b) Schematic illustration of synthesis of 3-arm and 4-arm star PSt polymers using a combination of 'core-first' and 'coupling onto' (CuAAC reactions) methodologies. (Adapted from Gao and Matyjaszewski, 2006.)

Problem 12.7 A combination of "core first" and "coupling onto" approaches often offers a simple strategy to synthesize star block copolymers. Suggest, accordingly, a synthetic strategy for the preparation of 3-arm star block copolymers consisting of diblock polymeric arms, polystyrene–*b*–polyethylene oxide (PSt-*b*-PEO).

Answer:

The Scheme P12.7.1 shows an outline for the synthesis of 3-arm star-block copolymer (PSt-*b*-PEO), following the methodologies of Gao, Min, and Matyjaszewski (2007). Using the core-first method, PSt 3-arm star polymers, *core*-(PSt-Br)$_3$, with high bromine chain-end functionality are first synthesized by ATRP of St on a multifunctional core such as 1,1,1-tris[4-(2-bromoisobutyryloxy) phenyl]ethane (TBiBPE) used as initiator with CuBr/PMDETA as catalyst (initial ratio of reagents: [St]/[TBiBPE]/[CuBr]/[CuBr$_2$]/[PMDETA] = 300/1/1.425/0.075/1.5). The reaction is stopped at low St conversion ($\simeq$ 12%) via exposure to air and dilution with THF. The copper complex is removed by passing the solution through a column of neutral alumina before the polymer is precipitated in cold hexane. The chain-end bromo groups of the obtained polymer, *core*-(PSt-Br)$_3$, are transformed to the azide groups by nucleophilic substitution with sodium azide. The resulting product, *core*-(PSt-N$_3$)$_3$, is used for the CuAAC coupling reaction with alkyne-terminated poly(ethylene oxide) chain (PEO-alkyne), the latter being prepared by esterification reaction of the OH end group of poly(ethylene glycol) monomethyl ether (PEO–OH) in CH$_2$Cl$_2$ solution with pentynoic acid. The

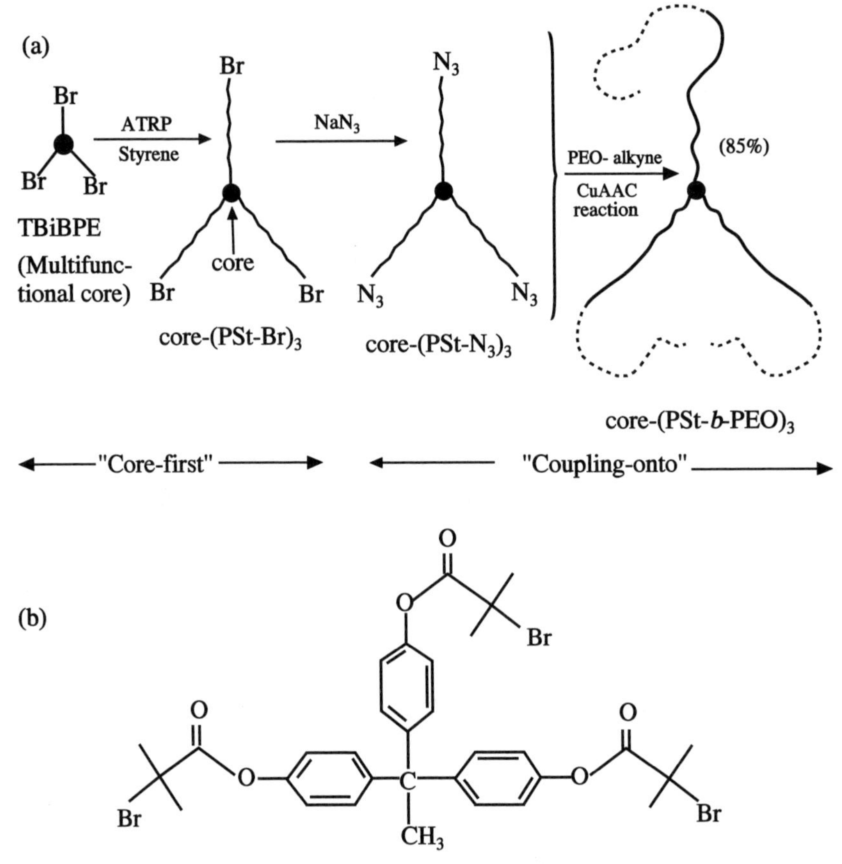

Scheme P12.7.1 (a) Synthesis of *core*-(PSt-*b*-PEO)₃ 3-arm star block copolymers by combination of core-first and coupling-onto methods. (b) Structure of ATRP initiator TBiBPE used as multifunctional core. (Adapted from Gao, Min, and Matyjaszewski, 2007.)

reaction is conducted at room temperature for 40 h and is catalyzed by *N*-(3-dimethylaminopropyl)-*N'*-ethylcarbodiimide hydrochloride in the presence of 4-(*N,N*-dimethylamino) pyridine. The efficiency of the coupling reaction is influenced by the preservation of chain-end functionalities in both (PSt–N₃)₃ and PEO-alkyne precursors. With the azido functionality of *core*-(PSt-N₃)₃ as high as 97% and the alkyne functionality of PEO-alkyne ~100%, the molar fraction of *core*-(PSt-*b*-PEO)₃ was determined (Gao, Min, and Matyjaszewski, 2007) to be 85% with the remaining 15% being accounted for by *core*-(PSt)(PSt-*b*-PEO)₂ (11%) and *core*-(PSt-*b*-PEO)(PSt)₂ (4%).

TBiBPE (see Scheme P12.7.1) is a multifunctional ATRP initiator that can be used for the synthesis of star polymers by ATRP. Several other multifunctional initiators used in the preparation of star (co)polymers are shown in Fig. 12.17. These contain both bromide functionality to initiate ATRP and hydroxyl functionality to initiate ROP.

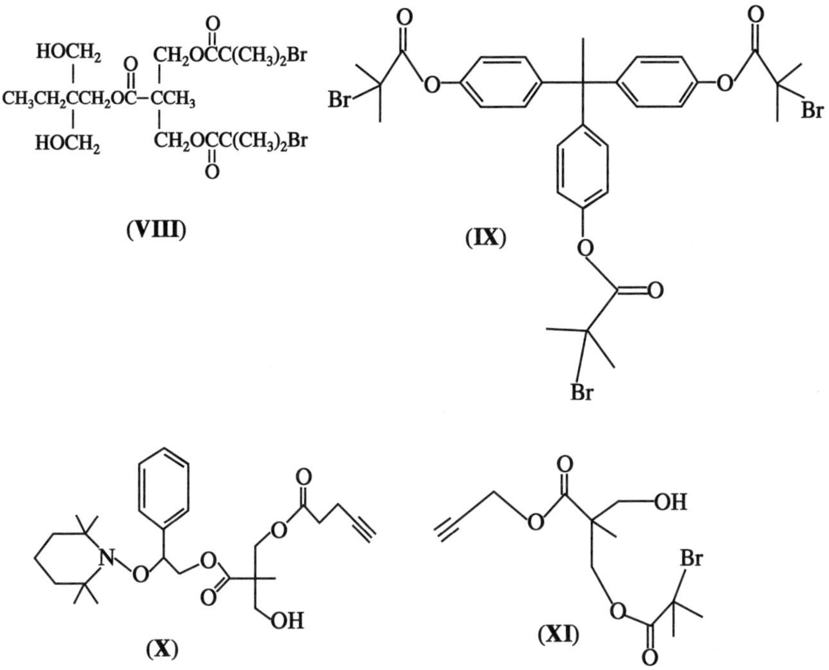

(VIII)

(IX)

(X)

(XI)

Figure 12.17 Several multifunctional initiators used for making star (co)polymers. (**VIII**): 2-((2,2-bis(hydroxymethyl) butoxy) carbonyl)-2-methylpropane-1,3-diyl bis(2-bromo-2-methyl propanoate) (Yang et al., 2008). (**IX**): 1,1,1-tris(4-(2-bromoisobutyryloxy) phenyl) ethane (Matyjaszewski et al., 1999). (**X**): 2-(hydroxymethyl)-2-methyl-3-oxo-3-(2-phenyl-2-(2,2,6,6-tetramethyl piperidin-1-yloxy)ethoxy) propyl pent-4-ynoate (Altintas et al., 2008). (**XI**): propargyl 2-hydroxylmethyl-2-(α-bromoisobutyryloxymethyl)propion-ate (Altintas et al., 2008).

Problem 12.8 Discuss strategies for synthesizing the following copolymers by combination of ATRP and click chemistry : (a) 3-miktoarm star copolymer *core*-(PSt)(P*t*BA)$_2$ and (b) 1st generation miktoarm poly-meric dendrimers (i) *core*-[PSt-*bp*-(P*t*BA)$_2$]$_2$ and (ii) *core*-[PSt-*bp*-(PAA)$_2$]$_2$. [*bp* = branch point, PSt = polystyrene, P*t*BA = poly(*tert*-butyl acrylate), PAA = poly(acrylic acid).]

Answer:

The above syntheses were first reported by Whittaker, Urbani, and Moteiro (2006). The methods used by them are presented below.

(a) PSt-N$_3$ is first synthesized by ATRP of St with an activated bromide initiator, followed by nucleophilic replacement of the bromine end group of the resulting PSt-Br by the azide group (see Scheme P12.8.1). P*t*BA-N$_3$ is also made similarly. Tripropargylamine (in large excess) is coupled to PSt-N$_3$ to form PSt-(-≡)$_2$. A solution of this polymer in DMF is then added to a solution of P*t*BA-N$_3$ in DMF at 80°C containing CuBr/PMDETA catalyst and the CuAAC reaction is allowed to proceed for more than 6 h.

(b) Dendrimers with miktoarm compositions consisting of PSt, P*t*BA, and PAA can be easily synthesized using a combination of ATRP and click reactions, as shown below.

(i) To make *core*-[PSt-*bp*-(P*t*BA)$_2$]$_2$, a difunctional PSt (i.e., Br-PSt-Br) is first synthesized by ATRP of St using a suitable dibromo initiator and CuBr/PMDETA catalyst. Br-PSt-Br is converted to N$_3$-PSt-N$_3$ by reacting with NaN$_3$ and further converted to a tetrafunctional precursor (see Scheme P12.8.2), (≡)$_2$-PSt-(≡)$_2$, by carrying out CuAAC reactions with tripropargylamine (used in large excess) in DMF at 80°C. PtBA-N$_3$ is also made separately by ATRP (using a monobromo initiator) and subsequent azide substitution of Br. Finally, PtBA-N$_3$ is coupled to ($\equiv$)$_2$-PSt-($\equiv$)$_2$ by CuAAC in DMF at 80°C to yield the 1st generation mikto dendritic copolymer *core*-[PSt-*bp*-(PtBA)$_2$]$_2$ (see Scheme P12.8.2).

(ii) A facile conversion of the PtBA arms of *core*-[PSt-*bp*-(PtBA)$_2$] to PAA is achieved by treatment with trifluoroacetic acid at room temperature to yield 3-miktoarm polymeric dendrimers consisting of hydrophobic and hydrophilic segments (see Scheme P12.8.2).

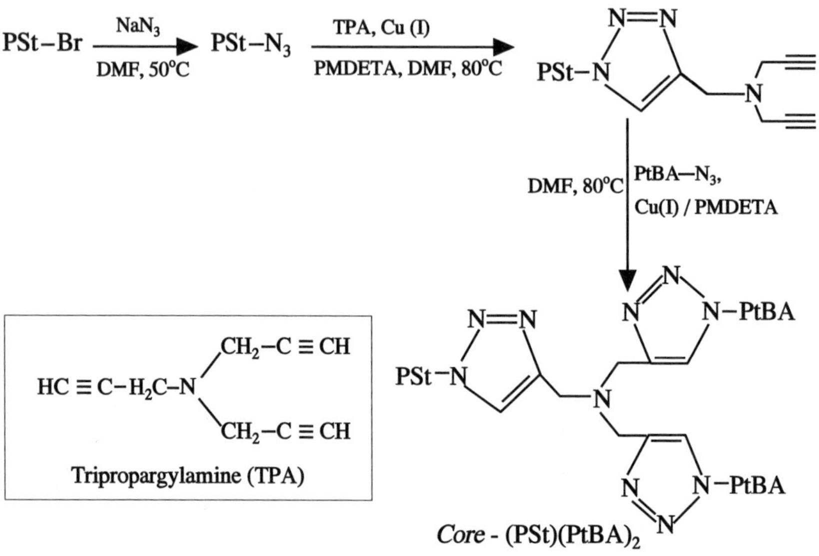

Scheme P12.8.1 Outline for synthesis of 3-miktoarm star copolymer, *core*-(PSt)(PtBA)$_2$. (Adapted from Whittaker, Urbani, and Monteiro, 2006.)

12.2.2 Combination of RAFT Polymerization and CuAAC

While RAFT polymerization, as we have seen in Chapter 11, has in recent years emerged as one of the most versatile controlled radical polymerization (CRP) techniques, its combination with CuAAC has led to creation of versatile postfunctionalization strategies to prepare highly function-alized polymers. These postfunctionalization strategies are commonly utilized in two approaches to prepare end-functionalized and side-functionalized polymers (Li and Benicewicz, 2008). In the first approach, a RAFT agent containing azide or alkyne moiety is prepared and used to mediate the polymerization of various monomers. The polymers obtained containing terminal alkynyl or azido functionalities are used in click reactions with functional azides or alkynes, respectively. In the second approach, a polymer with pendant alkynyl or azide groups is first synthesized by RAFT polymerization and the pendant groups are used subsequently to introduce side-functionalization via click reactions.

Using the first approach, Sumerlin and coworkers (Gondi et al., 2007) synthesized functional telechelic polymers for which two novel azido-functionalized chain trasfer agents (CTAs), namely, (**XII**) and (**XIII**) (see Fig. 12.18), were prepared and employed to mediate the RAFT polymer-ization of styrene (St) and *N,N*-dimethylacrylamide (DMA) under a variety of conditions. Polymerizations of St and DMA were successfully mediated by the aforesaid azido-functionalized CTAs, both the RAFT polymerizations exhibiting pseudo-first-order kinetics, a linear dependence of M_n on conversion, and low polydispersity ($M_w/M_n \leq 1.33$), which are characteristics of controlled polymerization. The obtained homopolymers were found to have retained ω end group functionality, which allowed successful formation of block copolymers using the homopolymers as macro-CTAs (see Problem 12.9). The α-azide terminal polymers and the azide CTAs could be coupled with high efficiency by click chemistry to various alkynes (propargyl acrylate, propargyl methacrylate, and propargyl alcohol) in the presence of Cu(I) catalyst, opening up routes to prepare a range of functional telechelic products and CTAs (see Problem 12.9).

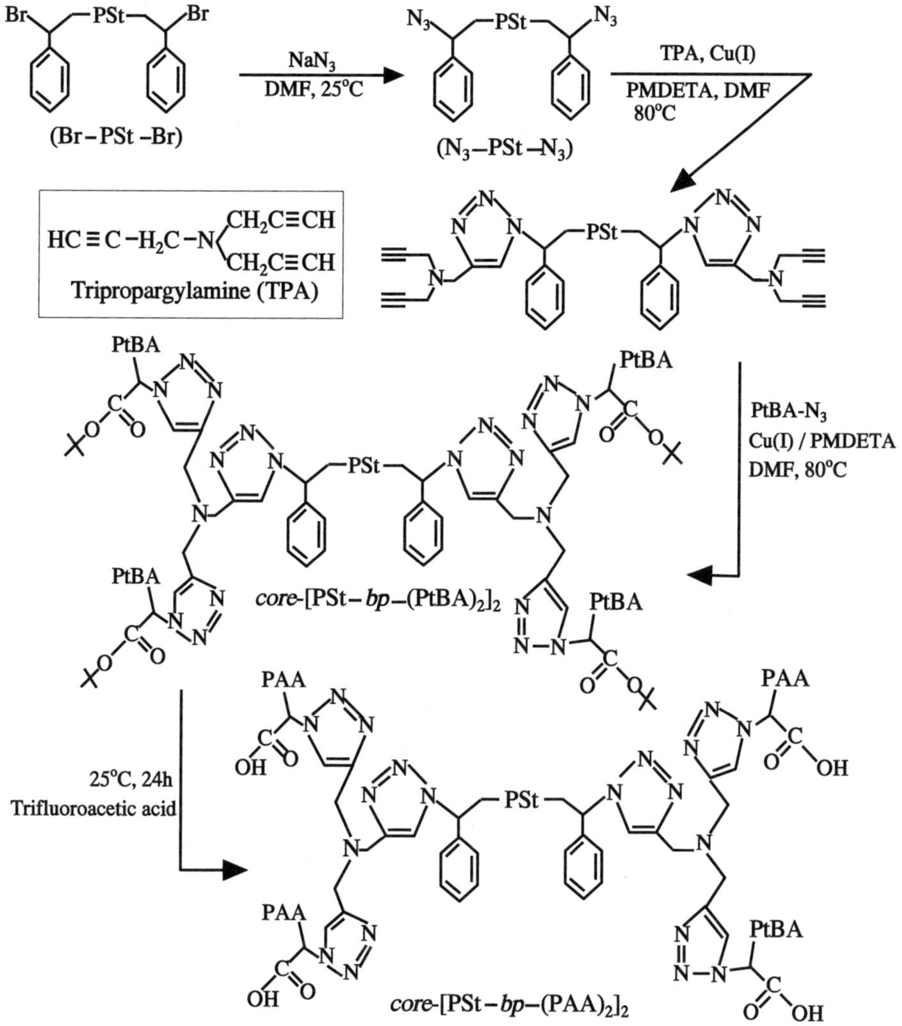

Scheme P12.8.2 Outline for synthesis of 1st generation dendrimers with miktoarm compositions consisting of PSt, P*t*BA, and PAA. (Adapted from Whittaker, Urbani, and Monteiro, 2006.)

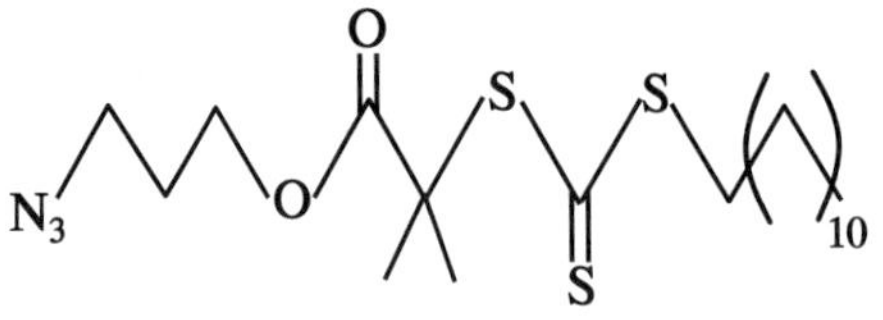

(XII) 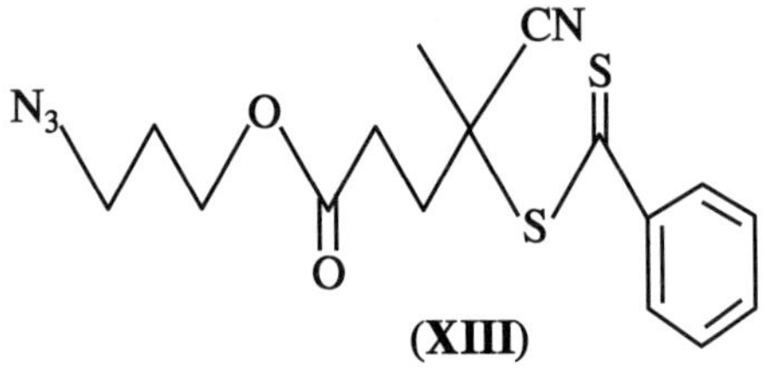   **(XIII)**

Figure 12.18 Examples of azidofuntionalized trithiocarbonate and dithioester type RAFT chain transfer agents: 2-dodecyl-sulfanylthiocarbonylsulfanyl-2-methyl-propionic acid 3-azidopropyl ester (**XII**) and 4-cyano-4-methyl-4-thiobenzoylsulfanyl-butyric acid 3-azidopropyl ester (**XIII**). (Gondi et al., 2007).

Problem 12.9 (a) Describe the methods of synthesis of the new azido functionalized CTA, (**XII**) and (**XIII**). (b) Show how these CTAs can be used to synthesize functional telechelic product by combined RAFT polymerization and click chemistry. Consider for the discussion a specific case of the synthesis of an α, ω-telechelic copolymer having OH group at each end and consisting of polystyrene (PSt) and poly(N,N-dimethhylacrylamide) (PDMA) blocks of approximately 50 and 90 monomer units, respectively.

Answer:

(a) Trithiocarbonate RAFT agents are more active than dithioester type RAFT agents. The azido RAFT agent (**XII**), which is a trithiocarbonate, namely, 2-dodecylsulfanylthiocarbonylsulfanyl-2-methyl-propionic acid 3-azido-propyl ester, can be prepared (Gondi et al., 2007) by mixing together (at room temperature, for 3 h) CH_2Cl_2 solutions of 3-azido-propanol, triethylamine, and 2-dodecylsulfanyl-thiocarbonylsulfanyl-2-methylpropionic acid chloride (DMP-Cl), while the latter can be prepared by reacting 2-dodecylsulfanylthio-carbonylsulfanyl-2-methyl-propionic acid (DMP) in CH_2Cl_2 solution with oxalyl chloride at room temperature for 3 h. The other azido RAFT agent (**XIII**), which is a dithioester, namely 4-cyano-4-methyl-4-thiobenzoylsulfanyl-butyric acid 3-azidopropyl ester, can be prepared (Gondi et al., 2007) by mixing 1-hydroxybenzotriazole and 1-ethyl-3-(3′-dimethylaminopropyl) carbodiimide with 4-cyano-4-((thiobenzoyl) sulfanyl) pentanoic acid (Thang et al., 1999) in CH_2Cl_2 solution at 0°C for 30 min and then reacting with 3-azidopropanol at room temperature for 48 h.

(b) Either of the azido functionalized CTAs, (**XII**) or (**XIII**), along with AIBN can be used. Employing a large molar ratio of CTA to AIBN, e.g., 50:1, so that contribution of the initiator (AIBN) to polymeric chain formation can be neglected, [St]/[CTA] $\simeq$ 50 can be used to closely satisfy the target DP $\simeq$ 50 for the PSt block at nearly complete conversion of monomer by RAFT polymerization at 80°C. Scheme P12.9.1 shows the reaction with CTA (XII). The polymer obtained at this stage, viz., N_3-PSt, can be employed as macro CTA for polymerization of DMA with a molar ratio such as [DMA] : [PSt-macro CTA] : [AIBN] = 200 : 1 : 0.05 at 80°C in DMF and the DMA conversion can be limited to 45% in order to reach the target DP of 90 for the PDMA block. The resulting precursor polymer (PSt-*b*-PDMA) carries 2-methylpropanoic acid 3-azido-propyl ester group on one end and a dodecylsulfanylthiocarbonylsulfanyl group on the other (Scheme P12.9.1). The latter may be converted to a hydroxyl group in a one-pot aminolysis Michael addition sequence (Scheme P12.9.2) carried out in THF, using butylamine as aminolysis agent and hydroxyethyl acrylate (HEA) as Michael addition agent, yielding α-azido ω-hydroxyl telechelic linear precursor. [*Note*: A small amount of a reducing agent, such as tris(2-carboxyethyl) phosphine hydrochloride (TCEP) may be added to the reaction mixture in order to suppress the undesired formation of dimer as a side reaction from the interpolymeric oxidative coupling of the thiol end groups into disulfides (Qiu and Winnik, 2006).] This precursor polymer may then be dissolved in DMF or DMSO and reacted with propargyl alcohol in the presence of CuBr/PMDETA at room temperature to introduce OH group at the α-end via CuAAC reaction (see Scheme P12.9.1).

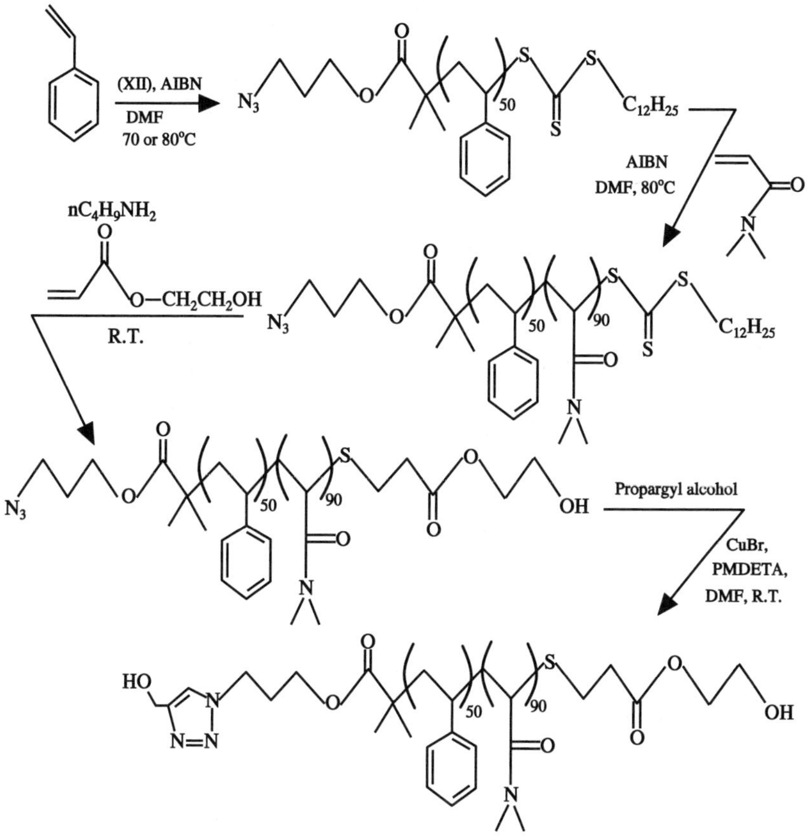

Scheme P12.9.1 Synthesis of hydroxy-ended functional telechelic copolymer PSt-*b*-PDMA.

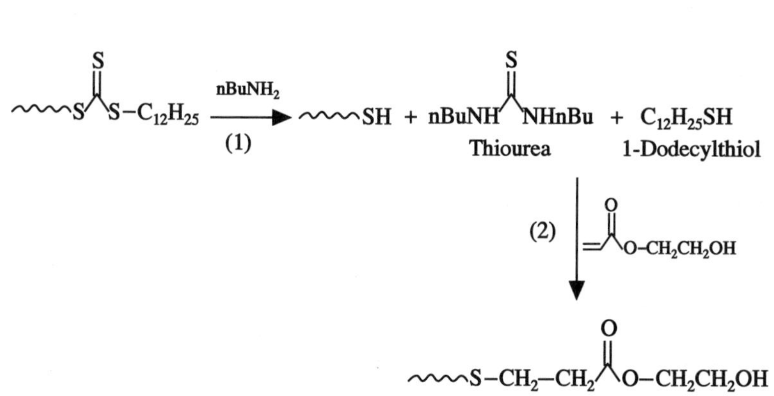

Scheme P12.9.2 Conversion of the trithiocarbonate end group of a RAFT polymer into colorless and stable thioether in a one-pot process combining (1) aminolysis of the trithiocarbonate function and (2) Michael addition of the formed thiol to an α,β-unsaturated carbonyl derivative.

Traditionally, there are two main approaches for surface modification — one is the "grafting to" approach where a preformed, end-functionalized polymer is bound to an activated surface and the other is the "grafting from" approach where polymerization occurs on the surface via propagation from a surface-immobilized initiator. Because the second approach affords a higher grafting density, this is generally preferred. Using a "grafting from" approach, Ranjan and Brittain (2008) performed surface modification of silica nanoparticles by generating high-density, well-defined polymer brushes on the silica surface via combined RAFT polymerization and click chemistry.

RAFT polymerization with a "grafting from" approach can be performed in two different ways: (1) free radical initiator is anchored to the surface and is followed by surface initiated RAFT polymerizatioon in the presence of RAFT CTA; (2) RAFT CTA is anchored to the surface and is followed by surface mediated RAFT polymerization in the presence of a free initiator. Baum and Brittain (2002) used the first option to modify flat silicate surfaces. However, mild reaction conditions that are dictated by the sensitive nature of the initiator group may limit the grafting density. Ranjan and Brittain (2008) therefore used the second option to modify the surface of silica nanoparticles. In this case, however, the selection of RAFT CTA for surface immobilization is crucial to the synthesis of well-defined polymer brushes. The immobilization can be performed by attaching the CTA to the backbone via either the leaving and reinitiating R group (*R-group approach*) or the stabilizing Z group (*Z-group approach*). The R-group approach, however, is known to afford higher molecular weight and grafting density of the attached polymer, thus promoting more efficient ability to synthesize polymer brushes (Ranjan and Brittain, 2008). The grafting density can also be increased further by using the more reactive trithiocarbonate type RAFT CTA instead of the conventional dithioester type RAFT CTA. Though a silyl condensation reaction is commonly used to immobilize a RAFT CTA onto silica nanoparticles, immobilization can be performed with greater efficiency via click reaction between azide and alkyne (see Exercise Problem 12.14 and answer in **Solutions Manual**). This, however, requires click functionalization of both silica nanoparticles and RAFT CTA — for example, azide-functionalization of silica nanopartiles and alkyne functionalization of RAFT CTA — in order to facilitate click coupling (Ranjan and Brittain, 2008).

Though the high efficiency, orthogonality, and simplicity of CuAAC reactions have prompted rapid and extensive adoption of the method in the field of macromolecular engineering, the use of copper is undesirable and a cause of concern in the context of biomaterials synthesis. This has generated interest in new metal-free azide-alkyne cycloaddition reactions (discussed below) which do not suffer from toxicity issues.

12.3   Strain-Promoted Azide-Alkyne Coupling

While CuAAC is called a click reaction, its uncatalyzed counterpart, i.e., the traditional uncatalyzed Huisgen 1,3-dipolar azide-alkyne cycloaddition, is not included into the suite of click chemistry methods because of its reduced yields and demanding reaction conditions. As an alternative approach to lower the activation barrier for [3 + 2] cycloaddition, Codelli et al. (2008) employed the intrinsic ring strain of cyclooctynes. These highly strained alkynes (18 kcal/mol of ring strain) react selectively with azides to form regioisomeric mixtures of tetrazoles (see Fig. 12.4) at ambient temperatures and pressure. More importantly, unlike CuAAC reactions which suffer from toxicity issues that have largely precluded their use for applications *in vivo*, the strain-promoted azide-alkyne cycloaddition (SPAAC) reactions, being metal-free, have no apparent cytotoxicity.

Though nonactivated cyclooctynes react somewhat slowly with azides, incorporating electron-withdrawing groups on the ring leads to dramatically enhanced rates that are characteristic of click reactions. For instance, the presence of a *gem*-difluorogroup adjacent to the strained alkyne (see Fig. 12.4) leads to 30-60 times faster reactions with azides, as compared to non-fluorinated cyclooctynes, and at a rate comparable to CuAAC (Codelli et al., 2008). Such activated SPAAC reactions, showing high efficiency and fidelity, have been employed to label azides on biomolecules in complex lysates and on surfaces of living systems (where R = biomolecule in Fig. 12.4). It may be mentioned that because of its small size, diverse modes of reactivity, and bioorthogonality, azide is an ideal chemical reporter group for biomolecules. Azides can be incorporated into glycans, lipids, and proteins by various modes, such as via metabolic machineries, covalent inhibitors, and enzymatic transfers (Sletten and Bertozzi, 2008).

Looked upon as a Cu-free click reaction with no apparent toxicity issues, the SPAAC reaction would appear to have a rich future. Still, the CuAAC reaction has the advantage of synthetic convenience due to the fact that terminal alkynes can be installed in biomolecules using simple building blocks, suh as commercially available alkynoic acids, while, in contrast, the synthesis of difluorinated cyclooctyne (DIFO) requires 12 steps and the overall yield is only about 1%. Synthetically tractable cyclooctynes having the reactivities of DIFO are therefore much needed in order to expand the use of this Cu free click reaction as an efficient and biocompatible tool (Codelli et al., 2008).

Figure 12.19 shows a panel of DIFO reagents (**XIV**-**XVI**) for Cu-free click chemistry. In contrast to a first generation DIFO reagent (**XIV**) which possesses a C-O bond to a linker substituent at C6, the second generation DIFO reagents, (**XV**) and (**XVI**), while retaining the DIFO core, possess a C-C bond to a linker substituent at C4, which dramatically simplifies the synthesis without affecting the kinetics or bioorthogonality of [3 + 2] cycloaddition with azides. These synthetically tractable second-generation DIFO reagents should facilitate greater application of copper-free SPAAC reactions, especially in biological settings.

While biomolecules labeled with azide can be detected through SPAAC using cyclooctyne as a probe, the intrinsic hydrophobicity of simple cyclooctynes can reduce bioavailability. Sletten and Bertozzi (2008) thus reported the synthesis and evaluation of a novel azacyclooctyne, 6,7-dimethoxyazacyclo-oct-4-yne (DIMAC), reagent. Synthesized in nine steps beginning with methyl 6-bromoglucopyranoside, the DIMAC reagent (**XVII**) has been demonstrated to react with azide-labeled proteins and cells similarly to cyclooctynes. However, its superior polarity and water solubility lead to reduction of non-specific binding and improvement of azide detection sensitivity (Sletten and Bertozzi, 2008).

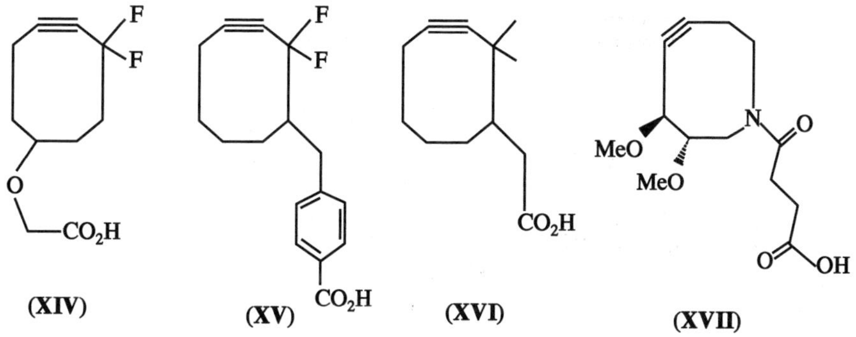

Figure 12.19 Structures of some cyclooctyne derivatives used in Cu-free click chemistry. (**XIV**) first generation cyclooctyne; (**XV**) and (**XVI**) second generation cyclooctynes; (**XVII**) 6,7-dimethoxyazacyclooct-4-yne. (Codelli et al., 2008; Sletten and Bertozzi, 2008.)

12.4   Diels-Alder Click Reactions

The Diels-Alder (DA) reaction, a [4 + 2] system, generally consists of coupling of a diene and a dienophile by intra- or intermolecular reaction. Its reliability, particularly of certain variants such as the furan-maleimide and anthracene-maleimide DA reaction, has included it in the pantheon of click chemistry reactions. As opposed to the majority of click reactions that create new carbon-heteroatom bonds, traditional DA reactions rely on carbon-carbon bond formation between dienes and electron-deficient dienophiles (see Fig. 12.4c). Recently, DA reaction based on the macromolecular chemistry has attracted much attention, particularly for synthesis of new materials. It has been demonstrated that a variety of structures can be synthesized using diene and dienophile-functionalized polymers. Thus, furan-maleimide- and anthracene-maleimide-based DA "click reactions" (Fig. 12.20) have been used as novel route to prepare well-controlled and structured polymers in the synthesis of alternating (Kamahori et al., 1999), block (Durmaz, Karatas, Tunca, and Hizal, 2006; Durmaz, Colakoclu, Tunca, and Hizal, 2006), and graft (Gacal et al., 2006) copolymers, stars (Dag et al., 2007; Dag et al., 2009; Durmaz, Karatas, Tunca, and Hizal, 2006), dendrimers (McElhanon and Wheeler, 2001; Kose et al., 2008; Morgenroth et al., 1997; Szlai et al., 2007), and other highly congested macromolecular architectures. Other applications of DA click reactions in materials synthesis include immobilization of unsaturated DNA strands onto maleimide-coated Au nanoparticles (Proupin-Perez et al., 2005), surface binding of proteins and cell-adhesion ligands (Yeo et al., 2003), and synthesis of polymeric thioxanthone photoinitiators (Gacal et al., 2008).

While the DA cycloaddition "click reaction", like the classical CuAAC "click reaction", has been successfully used, as mentioned above, in the generation of a variety of well-defined polymeric architectures, including block copolymers, graft polymers, and star polymers, both of the strategies have characteristics that may prove to be problematic in certain applications. For

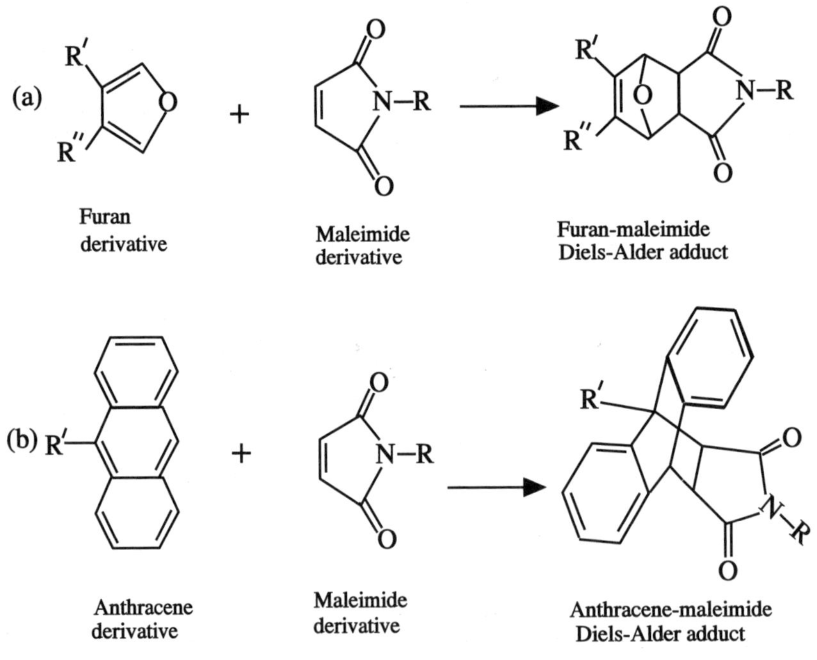

Figure 12.20 Diels-Alder click reactions of maleimide with (a) furan and (b) anthracene.

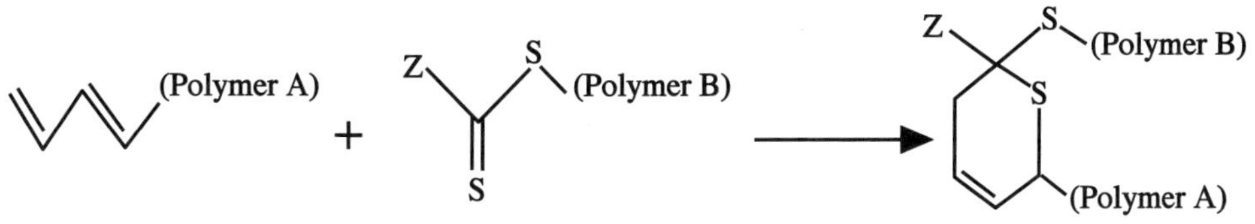

Figure 12.21 Hetero-Diels-Alder (HDA) reaction between diene and electron-deficient dithioesters in the RAFT-HDA approach to make block copolymers.

example, just as CuAAC would have the requirement of a purification stage (to remove Cu) in the development of materials destined for biomedical applications, the high temperatures required for some DA click reactions make it unsuitable for use in forming polymeric conjugates from thermally unstable compounds. Both techniques also require that the compounds to be "click" joined are prefunctionalized with the appropriate complementary moieties in addition to those required for performing the controlled/living (CRP) polymerization. An alternative strategy was, therefore, put forth by Sinwell et al. (2008) for synthesizing polymer conjugates, which involved hetero-Diels-Alder (HDA) cycloaddition in combination with RAFT chemistry. In this process, RAFT controlling agents having thiocarbonyl functionality and bearing electron-withdrawing Z-groups were sequentially used, first for the CRP and then as a reactive heterodienophile in a HDA cycloaddition with an appropriate diene (Fig. 12.21). It was shown that the tendency of electron-deficient dithioesters to undergo HDA cycloadditions could be successfully used to yield modular block copolymers from dissimilar monomer families (Sinwell et al., 2008; Inglis et al., 2009) and star-block copolymers (Inglis et al., 2008; Sinwell et al., 2009), which are elaborated below. While reactions with linear dienes proceed to high conversion in a few hours at 50°C, cyclopentadienes react almost instantaneously at room temperature (Inglis et al., 2009). Depending on sufficient availability of chain transfer agents with electron-withdrawing Z groups, this RAFT-HDA approach can have potential similar to that of the ATRP-CuAAC combination for the synthesis of well-defined macromolecular architectures.

For using DA [4 + 2] cycloaddition reaction in the click chemistry modular approach for the preparation of polymers with different composition and topology – ranging from linear block copolymers to nonlinear macromolecular structures (graft, star, miktoarm star, H-type, dendrimer, dendronized polymer, and network system) – polymers with anthracene or maleimide functional end-groups are prepared separately using living polymerization techniques and then linked covalently together or with maleimide or anthracene functional organic compounds (functional agents) via these functional groups. These end-group functionalities can be introduced into polymer chain through the use of an anthracene or maleimide functionalized initiator for the polymer synthesis or by post-polymerization modification reactions. Several such functionalized initiators and functional agents reported in the literature are shown in Fig. 12.22.

12.4.1 Copolymer Synthesis

A wide range of functionality may be introduced into a polymer chain using a heterofunctional initiator if one or more of the functional groups remain intact during the polymerization. This has made it possible, as we have seen in the preceding chapter, to synthesize well-defined block copolymers by a sequential two-step or one-pot polymerization method, without any chemical transformation or protection of initiating sites. Using this strategy, a number of block copolymers have been prepared by combination of different polymerization mechanisms, such as ATRP-NMP, ATRP-ROP, NMP-ROP, and ATRP-living cationic polymerization. On the other hand, the advent

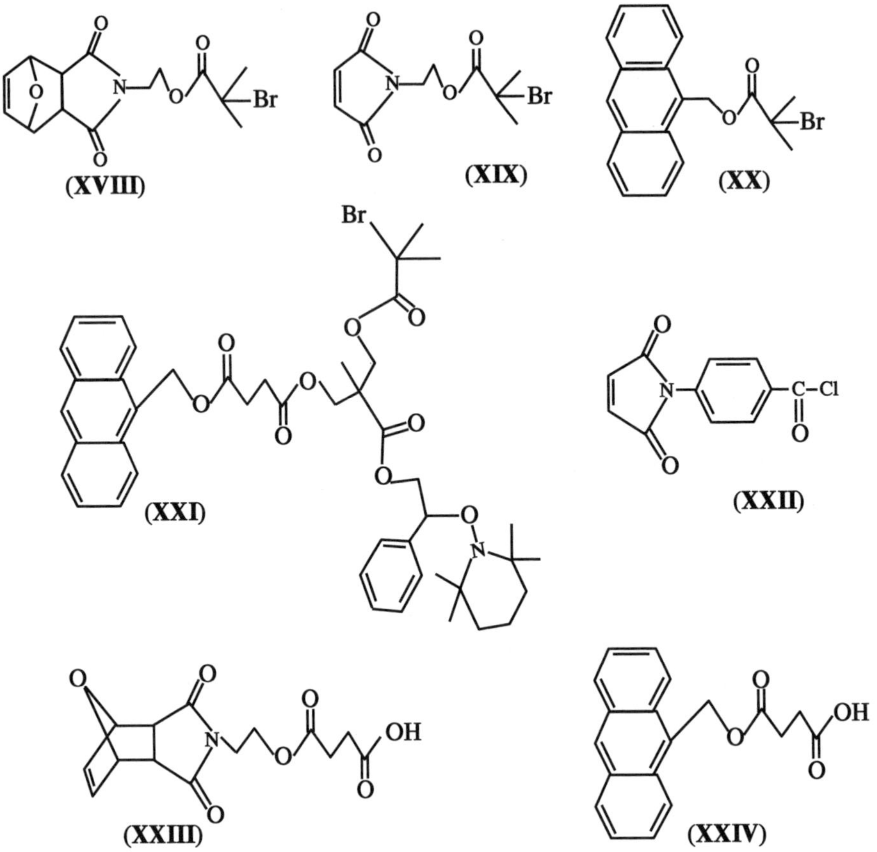

Figure 12.22 Structures of some functional initiators (**XVIII**)-(**XXI**) and post-polymerization functional-izing agents (**XXII**)-(**XXIV**). *Source:* (**XVIII**): Mantovani, et al. (2005); (**XIX**): Mantovani, et al. (2005); (**XX**): Erdogan et al. (2002); (**XXI**): Tunca et al. (2004); (**XXII**): Oishi and Fujimoto (1992); (**XXIII**): Oishi and Fujimoto (1992); (**XXIV**): Lei and Porco (2004). (*Note*: For methods of synthesis of these special compounds refer to answers in **Solutions Manual** to Exercise 12.16.)

of click chemistry has opened up alternative routes for the preparation of block copolymers. One of the routes, as we have noted previously, is based on click reactions between azides and alkynes or nitriles, whih have been applied successfully for the preparation of PSt-*b*-PMMA, PEG-*b*-PMMA, and PEG-*b*-PSt block copolymers, using azide- and alkyne-end functionalized homopolymers pre-pared by ATRP. A second strategy, we consider below, is based on DA reaction, [4 + 2] system, which has been used for the preparation of linear block copolymers containing PEG, PSt, PMMA, or P*t*BA blocks, graft copolymers (PSt-*g*-PEG) and (PSt-*g*-PMMA), star polymers with PEG, PMMA and P*t*BA homopolymer arms, star-block containing PEG-*b*-PSt-*b*-P*t*BA arms, and het-eroarm H-shaped terpolymers (PSt)(P*t*BA)-*bp*-PEO-*bp*(P*t*BA)(PSt). These syntheses have been achieved via DA reaction of maleimide- and anthracene-end functionalized polymers. While such end-functionalized PSt, PMMA, or P*t*BA polymers are directly prepared by polymerization via ATRP or NMP routes utilizing functionalized initiators, maleimide and anthracene functionalities are introduced into PEG by esterification of monohydroxy PEG with carboxylic acid derivatives containing maleimide and anthracene moieties. For example, maleimide-end functionalized PSt,

PMMA, and P*t*BA can be obtained directly by ATRP of respective monomers using (**XVIII**) or (**XIX**) (see Fig. 12.22) as the initiator. Similarly, anthracene-end functionalization can be achieved by using (**XX**) as the ATRP initiator (Durmaz, Colakoglu, Tunca, and Hizal, 2006), while the use of (**XXI**) as initiator facilitates both NMP and ATRP, yielding a di-block copolymer with anthracene functionality in between the blocks. On the other hand, maleimide-end functionalized PEG can be obtained by esterification of monohydroxy PEG with (**XXII**) or (**XXIII**), while esterification with (**XXIV**) leads to anthracene-end functionalized PEG. These methodologies are further elaborated below with the help of several worked out examples.

Problem 12.10 Discuss how the following block copolymers can be prepared via Diels-Alder reaction of maleimide- and anthracene-end functionalized polymers : (a) PMMA-*b*-PSt, (b) PEG-*b*-PSt, (c) P*t*BA-*b*-PSt, and (d) PMMA-*b*-PEG block copolymers.

Answer:

(a) Maleimide-end functionalized PMMA (PMMA-MI) with controlled molecular weight and low polydispersity is prepared by ATRP of MMA in toluene with (**XVIII**) as initiator in the presence of CuCl/PMDETA catalyst at 40°C for 3-4 h. The polymerization mixture, diluted with THF, is passed through a basic alumina column to remove the catalyst and then precipitated in methanol.

Anthracene end-functionalized PSt (PSt-AN) is prepared by ATRP of St with (**XX**) as initiator in the presence of CuBr/PMDETA catalyst at 110°C for 45 min, giving 14% conversion in 45 min (Durmaz et al., 2006). (Low conversion ensures high chain end functionality.) The polymerization reaction mixture is treated as above.

For the preparation of PMMA-*b*-PSt copolymer via DA reaction [Scheme P12.10.1(a)] a solution of PMMA-MI and PSt-AN in toluene is refluxed for 120 h. The efficiency of block copolymerization by this DA reaction is reported to be 98.5% (Durmaz, Colakoglu, Yunca, and Hizal, 2006).

(b) Maleimide end-functionalized PEG (PEG-MI) can be prepared by esterification of monohydroxy PEG (e.g., ethylene glycol monoethyl ether) with 4-maleimido-benzoylchloride (**XXII**), which is added dropwise to a solution of PEG in THF in the presence of Et$_3$N at 0°C and then stirred for 12 h at room temperature. PEG-MI is obtained as a viscous yellow oil. Alternatively, furan-protected maleimide end-functionalized PEG (PEG-FMI) is prepared by reacting PEG with (**XXIII**) for 12 h at room temperature in the presence of *N,N'*-dicyclohexylcarbodiimide (DCC) and 4-dimethylaminopyridine (DMAP). The product PEG-FMI is obtained as a white solid in 80% yield (Dag et al., 2008). (Retro-DA reaction of PEG-FMI is carried out *in situ* in toluene at 110°C to give the deprotected maleimide functionality during the subsequent block copolymerization stage.)

Preparation of PEG-*b*-PSt copolymer is carried out by DA reaction of PEG-MI (or PEG-FMI) and PSt-AN [Scheme P12.10.1(b)]. The two end-functional polymers are dissolved in toluene and the solution is refluxed for 48 h under nitrogen. The efficiency of this block copolymerization (DA reaction) is reported to be 97.5% (Durmaz et al., 2006).

(c) Anthracene end-functionalized PEG (PEG-AN) is made by reacting monohydroxy PEG with (**XXIV**) in CH$_2$Cl$_2$ solution at room temperature for 12 h in the presence of dimethylaminopyridine (DMAP), dimethyl-laminopyridinium 4-toluenesulfonate (DPTS) and DCC. To prepare PMMA-*b*-PEG block copolymer via DA reaction [Scheme P12.10.1(c)], a solution of this PEG-AN and PMMA-MI in toluene is refluxed for 120 h at 110°C. The efficiency of this block copolymerization (DA reaction) is reported to be 93% (Durmaz et al., 2006).

Problem 12.11 Anthracene-maleimide-based DA click reaction provides a novel route to prepare well-defined graft copolymers. Suggest synthetic strategies involving this reaction to prepare well-defined (a) PSt-*g*-PEG and (b) PSt-*g*-PMMA copolymers.

Answer:

As mentioned earlier, graft copolymers can be obtained with three general methods, namely, (i) grafting-onto, in which side chains are preformed and then attached to the backbone; (ii) grafting-from, in which the monomer is grafted from the backbone; and (iii) grafting-through, in which the macromonomers are copolymerized. For availing the anthracene-maleimide-based DA reaction to prepare PSt-*g*-PEG and PSt-*g*-PMMA graft copolymers, Gacal et al. (2006) used the second method (Scheme P12.11.1) by preforming maleimide end-functionalized branches (PEG or PMMA) and creating anthracene functionality on the main chain (PSt).

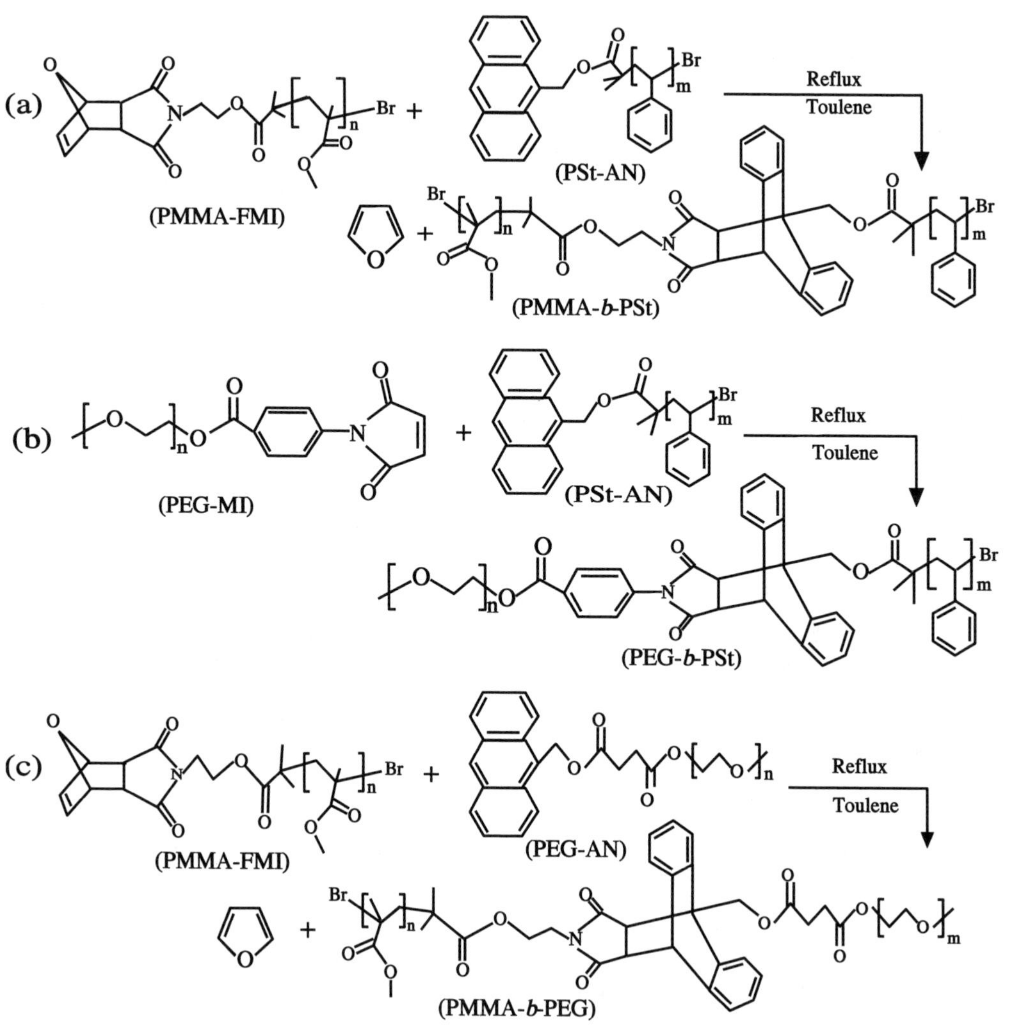

Scheme P12.10.1 Diels-Alder 'click' reactions in the synthesis of (a) PMMA-*b*-PSt, (b) PEG-*b*-PSt, and (c) PMMA-*b*-PEG copolymers. (Adapted from Durmaz, Colakoglu, Tunca, and Hizal, 2006.)

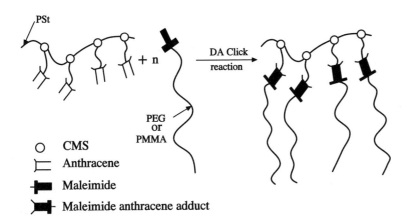

Scheme P12.11.1 Schematic presentation of grafting process by Diels-Alder click reaction. (Adapted from Gacal, Durmaz, Tasdelen, Hizal, Tunca, Yagci, and Damirel, 2006.)

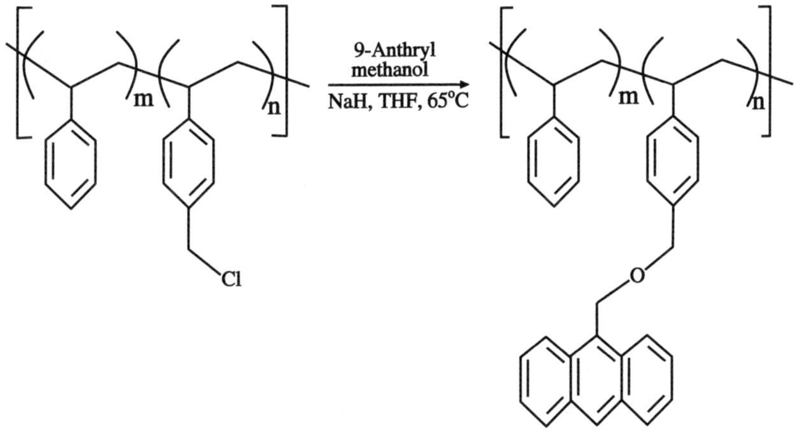

Scheme P12.11.2 Incorporation of anthryl moieties in PSt-*co*-CMS copolymer by etherification of chloromethyl group. (Drawn following the synthesis method of Gacal, Durmaz, Tasdelen, Hizal, Tunca, Yagci, and Damirel, 2006.)

To prepare PEG-maleimide (PEG-MI), PEG methyl ether (Me-PEG) is reacted with (**XXIII**) in the presence of DCC and DPTS (see Problem 12.10) in CH_2Cl_2 solution at room temperature for 12 h. On the other hand, PMMA-maleimide (PMMA-MI) is prepared by ATRP of MMA in toluene using (**XVIII**) as the initiator and Cu(I)Cl/PMDETA catalyst system. After the polymerization at $40°C$ for 3 h, the reaction mixture is passed through a basic alumina column to remove the catalyst and precipitated into hexane to recover the polymer (PMMA-MI).

According to the method used by Gacal et al. (2006) to obtain PSt with anthryl pendant groups (PSt-AN), St, and 4-chloro-methyl styrene (CMS) (which is a functionalizable monomer) are copolymerized and anthracene functionality is attached to the formed (random) copolymer by an etherification procedure, as shown in Scheme P12.11.2. Thus, poly(styrene-*co*-chloromethylstyrene), P(St-*co*-CMS), copolymers containing desired amounts of CMS moieties, are prepared via NMP of St and CMS at $125°C$, while for etherification of the chloromethyl group, the St-CMS random copolymer (1.0 equiv CMS) is reacted (in the dark) with 9-anthrylmethanol (1.1 equiv)/sodium hydride (1.1 equiv) in THF under reflux for 12 h. The product, PSt-*co*-CMS with anthryl pendant groups (denoted as PSt-AN) is recovered by precipitation into methanol. For the preparation of graft copolymer via DA reaction of PEG-MI and PSt-AN, a solution of PEG-MI (1.1 equiv) and PSt-AN (1.0 anthracene equiv) in toluene is refluxed for 48 h at 110°C in the dark. The product, PSt-*g*-PEG, is purified by precipitation into methanol from solution in THF. The copolymer PSt-*g*-PMMA is obtained via DA reaction of PMMA-MI and PSt-AN in a similar way.

Problem 12.12 Durmaz, Karatas, Tunca, and Hizal (2006) synthesized a compound, (**XXI**) (Fig. 12.22), having an anthracene functionality as also a NMP site and an ATRP site. Show how through combination of the DA click reaction, ATRP, and NMP, an ABC type miktoarm star terpolymer with PEG, PSt, and P*t*BA arms [where PSt = polystyrene, P*t*BA = poly(*tert*-butyl acrylate), and PEO = poly(ethylene oxide)] can be synthesized.

Answer:

In the method reported by Durmaz et al. (2006), DA reaction is first carried out between PEG-MI precursor (prepared as in Problem 12.11) and the compound (**XXI**) by refluxing in THF for 48 h (95% conversion). This results in the formation of a maleimide-anthracene adduct having appropriate functionl groups for SFRP/NMP and ATRP. In the second step, the previously obtained adduct is used as a macroinitiator for SFRP of St at 125°C for 15h (85% conversion). In the third step, this PEG-PSt precursor with a bromine functionality in the core is employed as a macroinitiator for ATRP of *t*BA in the presence of Cu(I)Br and PMDETA (Fig. 11.17) at 80°C to give ABC type miktoarm star terpolymer *core*-(PEG)(PSt)(P*t*BA) (see Scheme P12.12.1).

12.4.2   *Thermoresponsive Systems, Dendrons, and Dendrimers*

One of the interesting characteristics of the DA reaction is the thermal reversibility of some of the DA adducts, repetitively regenerating the original reactants. The reversible DA reaction between substituted furans and maleimides (Fig. 12.23) is one of the most often used in the preparation of thermally responsive polymers. This is primarily due to ease of the furan-maleimide adduct formation and dissociation. The adduct formation typically occurs at room temperature, exhibiting fast kinetics and high yields, while adduct dissociation (*retro*-DA or rDA) occurs in solution at temperatures above 60°C (Kwart and King, 1968). This attribute has led to the development of a methodology for protecting the reactive maleimide during polymerization and other transformations that may be affected by its presence. This maleimide can be protected (or masked) by forming a cyclic adduct with furan, while a furan-protected maleimide group is easily deprotected (unmasked) through cycloconversion simply by increasing the temperature (usually above 90°C). Under proper conditions, this retro-DA reaction occurs with virtually 100% efficiency. Thermoreversible crosslinked materials have thus been designed via furan-maleimide chemistry. In fact, most of the approaches to date towards fabrication of self-healing materials have relied on crosslinking of polymers containing furan side chain with bis-maleimide containing small molecules serving as cross-linkers (Fig. 12.24). Alternatively, polymers appended with maleimides could be reacted with polymers or smaller molecules containing multiple furan units to obtain thermoreversible crosslinked materials. The thermoreversibility can be tuned by changing the diene or the dienophile. For example, the use of anthracene as a diene with maleimide dienophile results in adducts that undergo cycloreversion over 200°C. Hence, high yielding stable conjugation can be achieved using the combination of anthracene and maleimide.

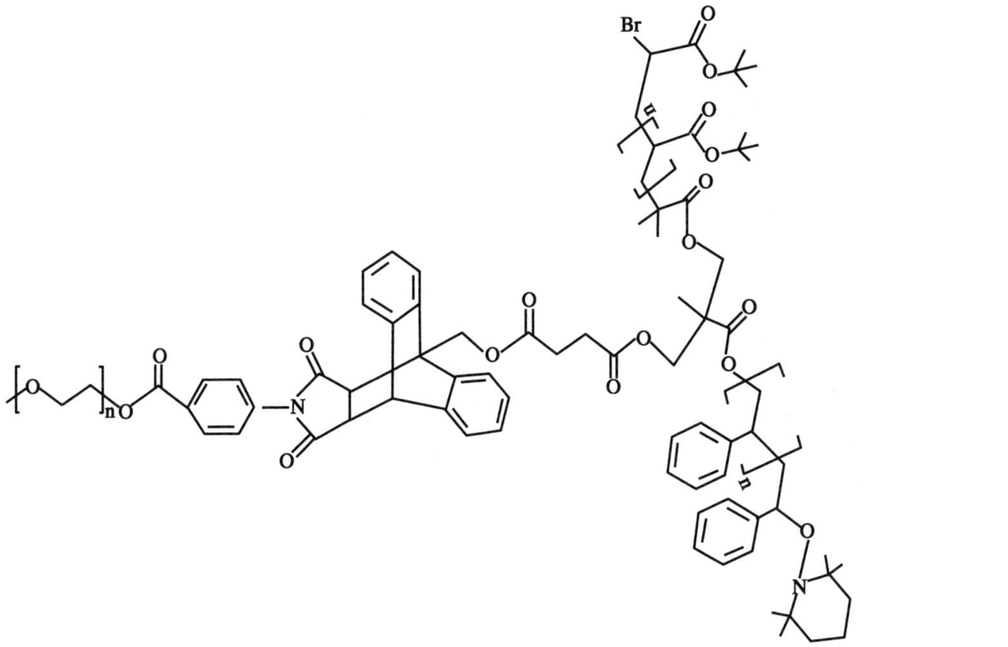

Scheme P12.12.1 Miktoarm star terpolymer, *core*-(PEG)(PSt)(P*t*BA), formed by maleimide-anthracene DA reaction of PEG-MI, NMP of St and ATRP of *t*BA using (**XXI**) as the *core*. (Adapted from Durmaz, Karatas, Tunca, and Hizal, 2006.)

It should be noted that although no distinction is made between the *endo* and *exo* stereoisomers (Fig. 12.25) of furan-maleimide adduct (assuming that their thermal behavior is at least similar if not equal), differences in thermal reversibility (rDA, DA) can have an important influence on the material properties, especially in the case of thermoremendable polymers and networks. For instance, by breaking only the covalent bonds of the *endo* DA-isomers inside the network under mild conditions and reforming them in a less stressed state, stresses can be relaxed. Important material properties such as durability, adhesion to other materials, appearance of micro-voids and micro-cracks, etc., can be improved (Canadell et al., 2010).

The reversibility of the furan-maleimide DA reaction is responsible for its extensive use in a wide variety of thermoremendable polymers and networks, ranging from self-healing (Chen et al., 2003), nonlinear optical (Haller et al., 2004; Kim et al., 2006), hydrogel (Chujo et al., 1990), and

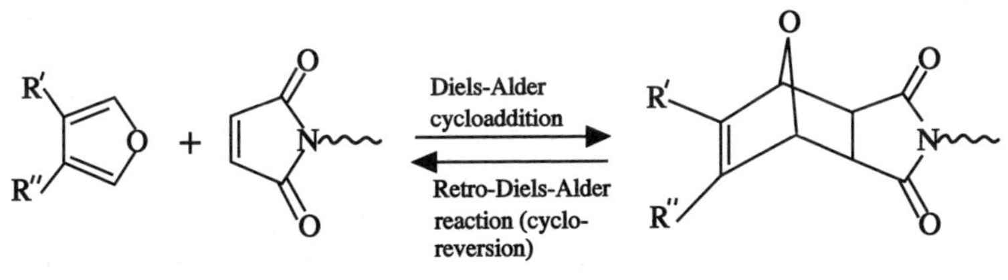

Figure 12.23 Diels-Alder and retro-Diels-Alder reaction sequence.

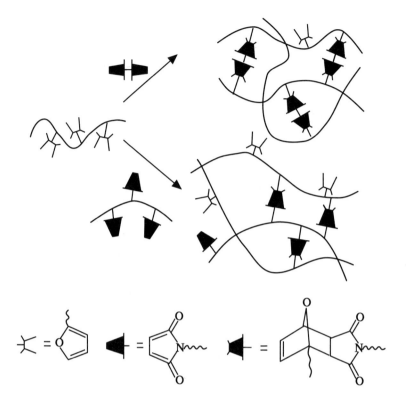

Figure 12.24 Synthetic approaches to self-healing crosslinked materials. (From Sanyal, 2010. With permission from *John Wiley & Sons, Inc.*)

interpenetrating network materials (Imai et al., 2000) to nanoscale lithography (Gotsmann et al., 2006). Furan-maleimide adducts have also been integrated into thermally cleavable encapsulants, foams, and surfactants (McElhanon et al., 2002; 2005) and have been used for the synthesis of thermally labile dendrons and dendrimers (Szalai et al., 2007). A few such examples reported in the literature are cited here.

McElhanon, McGrath, and coworkers (McElhanon et al., 2001; Szalai et al., 2007) reported the preparation and initial investigations of thermally responsive symmetrical dendrimers based on furan-maleimide DA reactions (Fig. 12.26). This represented the first covalent thermally reversible design strategy in a dendritic system. Benzyl aryl ether dendrons and dendrimers containing thermally reversible furan-maleimide DA adducts were prepared up to the third generation (McElhanon and Wheeler, 2001). A convergent approach was employed which involved preparing dendrons with substituted furan-maleimide DA adducts located at the dendron focal point. The process required a disubstituted furan that could serve as an AB_2 monomer and an N-substituted maleimide with a reactive hydroxyl focal point. The convergent approach used by Szalai et al. (2007) for synthesizing thermally labile DA dendrimers was based on reacting appropriately functionalized furan dendrons with a multifunctional maleimide. In the process, first through fourth generation benzyl aryl ether based dendrons that contained furan moieties at their focal point were allowed to react with bismaleimide central linker to provide the corresponding dendrimers (Fig. 12.26). The thermally activated disassembly of these dendrimers was demonstrated by heating

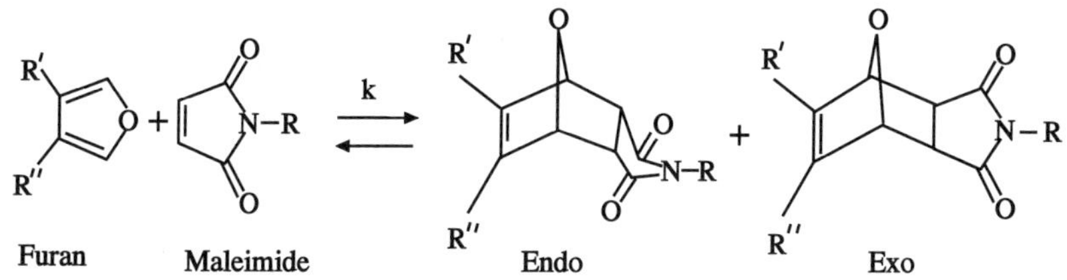

Figure 12.25 Furan-maleimide DA adducts (*endo* and *exo*) formation and dissociation. The adduct formation typically occurs at room temperature and adduct dissociation occurs at elevated temperatures (> 90°C).

them to 110°C, while slow recombination of the disassembled dendrons was accomplished upon thermal treatment.

To obtain segment block dendrimers, Kose et al. (2008) utilized Fréchet type dendrons appended with electron rich furan at the focal point and a complementary set of polyester dendrons containing the electron deficient maleimide group at the focal point. Combination of these two complimentary dendrons was accomplished by heating in benzene to obtain thermoresponsive segment block dendrimers in good yields (Fig. 12.27). The dendrimers fell apart upon heating them to 110°C.

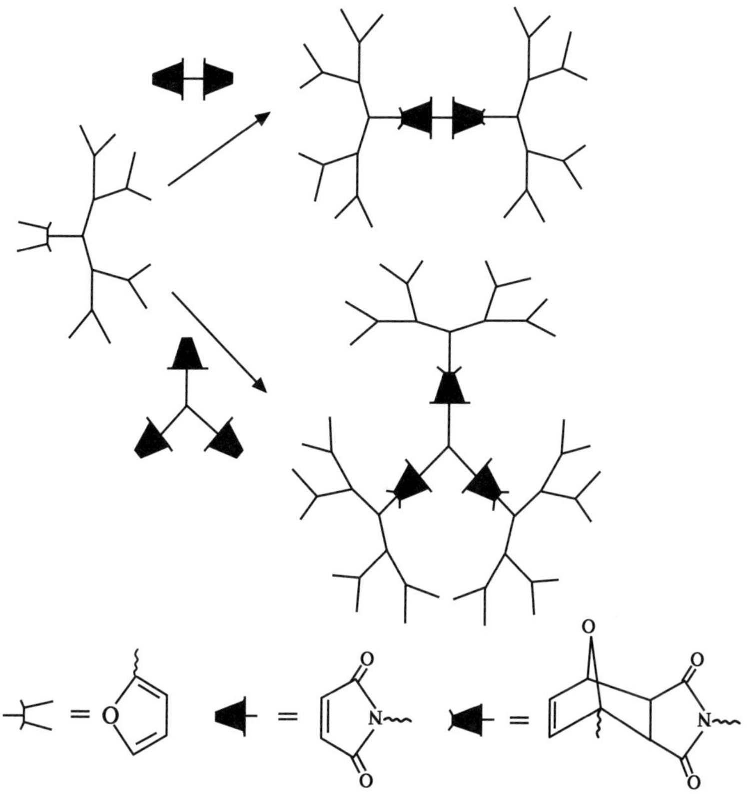

Figure 12.26 Synthesis of thermoreversible symmetrical dendrimers via furan-maleimide DA cycloaddition reactions. (From Sanyal, 2010. With permission from *John Wiley & Sons, Inc.*)

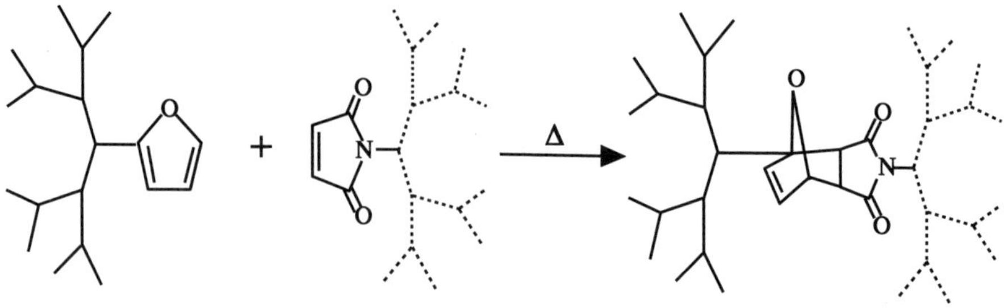

Figure 12.27 Synthesis of thermoresponsive segment-block dendrimers via furan-maleimide DA cycloaddition reaction.

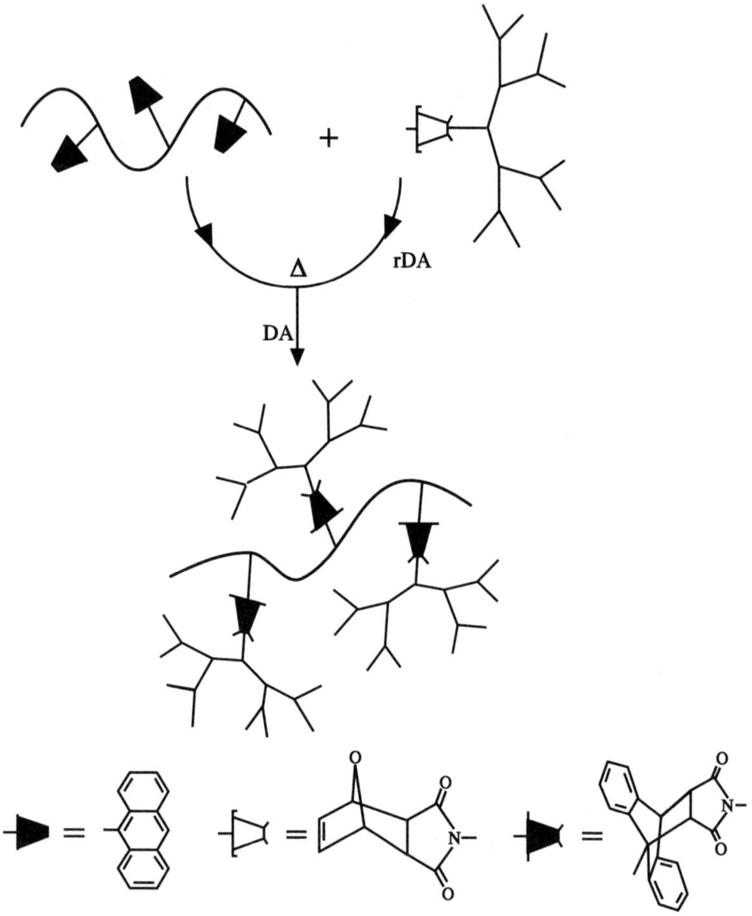

Figure 12.28 Synthesis of dendronized polymers via rDA/DA sequence. (Adapted from Sanyal, 2010.)

Since anthracene and maleimide form high yielding stable adducts (with cycloreversion temperature over 200°C), this combination of diene and dienophile was utilized for the synthesis of dendronized polymers via DA reaction (Tonga et al., 2010). Thus, as shown schematically in Fig. 12.28, polymer appended with anthracene groups was reacted with dendrons containing furan protected (masked) maleimide groups at their focal points by refluxing in toluene. At such high temperatures, the retro-DA takes place, thereby deprotecting (unmasking) the reactive maleimide group, which reacts *in situ* with the anthracene moiety of the copolymer via [4 + 2] cycloaddition reaction resulting in dendron-grafted polymer.

12.4.3 Hetero-Diels-Alder (HDA) Cycloaddition

As we have noted previously, HDA reaction of dienes with terminal thiocarbonylthio groups on RAFT-generated polymers provide the basis of an alternative strategy (Sinnwell et al., 2008) for synthesizing polymer conjugates using RAFT chemistry. Here, chain transfer agents (CTAs) bearing electron-withdrawing Z-groups [benzyl (diethoxyphosphoryl) dithioformate and benzylpyridin-2-yldithioformate] were used in such a way that the thiocarbonyl functionality of the CTAs was sequentially used for the CRP/LRP and as a reactive heterodienophile in a HDA cycloaddition with an appropriate diene, thereby opening a pathway for the synthesis of modular block copolymers and star block copolymers. The HDA cycloadditions are facilitated by the use of a catalyst which enhances the electron-withdrawing nature of the RAFT Z-group (Inglis et al., 2008). The catalysts are selected on the basis of the nature of their interaction with the RAFT Z-group. Thus, in the case of the phosphoryl Z-group, $ZnCl_2$ is used as the catalyst as it chelates with the oxygen on the phosphinyl group, while in the case of the pyridinyl Z-group trifluoroacetic acid (TFA) is used as the catalyst, since the H^+ from it protonates the nitrogen on the pyridinyl group.

Problem 12.13 The combination of RAFT chemistry and the HDA cycloaddition provides a simple and synthetically nondemanding pathway to well-defined macromolecular star-shaped architectures. In support of this contention describe the synthesis of polystyrene (PSt) star polymers with up to 4 arms using the aforesaid two different RAFT end groups, namely, diethoxy-phosphoryldithioformate and pyridin-2-yldithioformate, and HDA coupling reactions.

Answer:

Inglis et al. (2008) used a variant of the "coupling onto" method of star polymer synthesis, as shown in Schemes P12.13.1 and P12.13.2. Thus, RAFT polymerization of St using controlling agents bearing the electron-withdrawing Z-group, benzyl (diethoxyphosphoryl) dithioformate and benzylpyridin-2-yldithioformate, results, respectivly, in diethoxyphosphoryldithioformate and pyridin-2-yldithioformate-terminated PSts (Scheme P12.13.1). When these polymers are submitted to HDA reaction with multidiene coupling agents bearing 2, 3, or 4 diene functional groups, 2-arm star, 3-arm star, and 4-arm star polymers are obtained (Scheme P12.13.2).

According to the methods reported by Inglis et al. (2008), the RAFT polymerization is performed by heating a solution of styrene, RAFT agent (**P13-I**) or (**P13-II**), and AIBN under nitrogen at 60°C for 9 h. The reaction is stopped at low monomer conversion to ensure a high RAFT end-group concentration. Linear PSts, namely, (**P13-III**) and (**P13-IV**) are obtained with number average molecular weights of 3600 and 3500 g mol^{-1}, respectively, and low polydispersity indices (PDI) of 1.10 and 1.15, respectively.

For synthesis of star polymers by HDA cycloaddition of (**P13-III**), for which Z is diethoxyphosphoryl, with multifunctional coupling agent (**P13-V**), a solution of (**P13-III**), (**P13-V**) corresponding to 1.0 equiv of diene, and 1 equiv of $ZnCl_2$ in chloroform is kept at 50°C for 24 h. A similar process is used for HDA cycloaddition of (**P13-IV**) with (**P13-V**), except that 1.2 equiv of TFA is used as the catalyst. The RAFT agents, (**P13-I**) and (**P13-II**), adequately control the polymerization of St in addition to being effective heterodienophiles, thus fulfilling both purposes.

The multifunctional coupling agents (**P13-V**), $m = 2, 3,$ and 4, are obtained by etherification of the corresponding bromomethylbenzenes with *trans*, *trans*-2,4-hexadien-1-ol (Scheme P12.13.3).

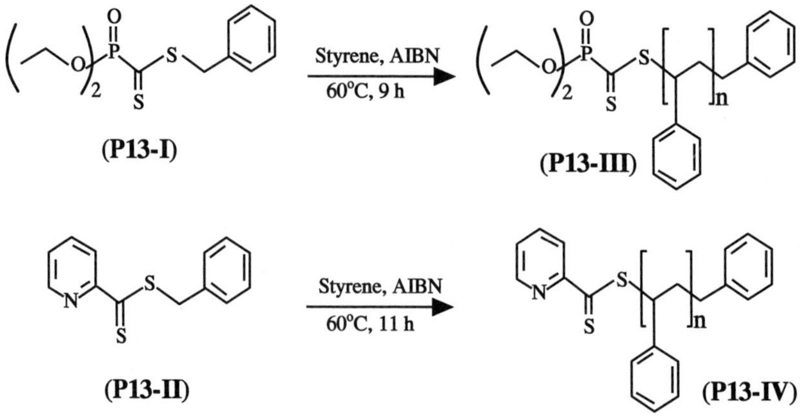

Scheme P12.13.1 Polymerization of styrene with benzyl (diethoxyphosphoryl) dithioformate (**P13-I**) and benzyl pyridin-2-yldithioformate (**P13-II**) used as RAFT chain transfer agents. (After Inglis et al., 2008. With permission from *American Chemical Society*.)

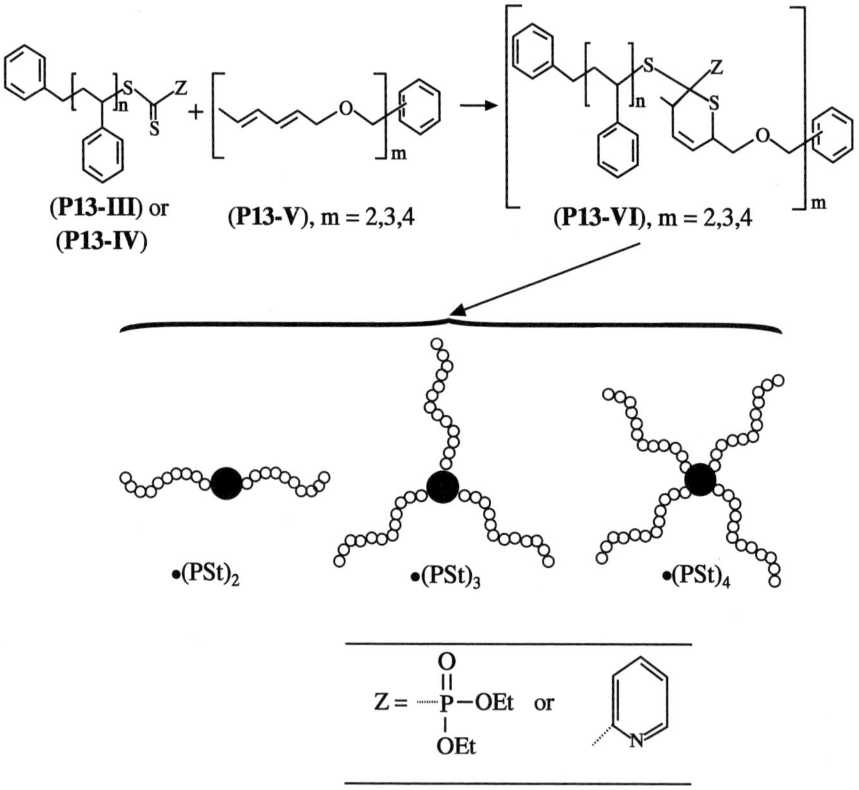

Scheme P12.13.2 Formation of 2-arm, 3-arm, and 4-arm star polymers with PSt arms via HDA cycloadditions of RAFT polymers with dienes. (After Inglis et al., 2008. With permission from *American Chemical Society*.)

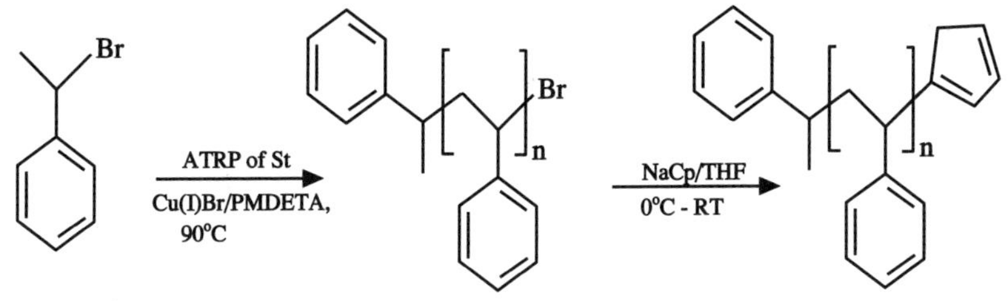

Figure 12.29 Synthesis of cyclopentadienyl-terminated polystyrene. Step 1 : ATRP of St with $C_6H_5CH(Br)CH_3$ as initiator and Cu(I)Br/PMDETA as catalyst. Step 2 : treatment with NaCp (2.0 M in THF) at $0°C$ to room temperature. (Drawn following the synthesis method of Inglis et al., 2009.)

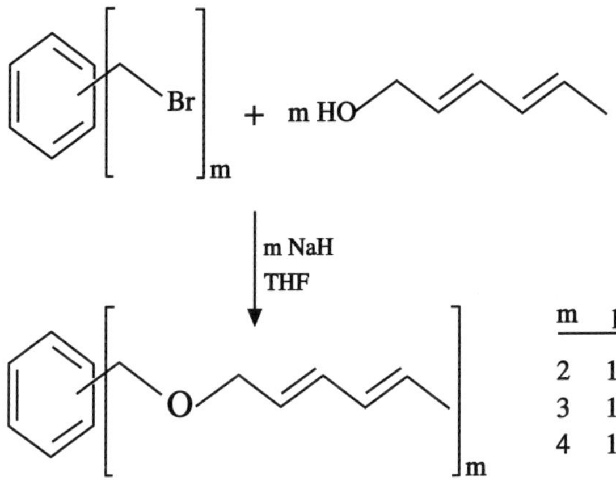

m	position
2	1,4
3	1,3,5
4	1,2,4,5

(P13-V), m = 2,3,4

Scheme P12.13.3 Synthesis of multi-diene coupling agents carrying 2, 3, and 4 diene groups. (Drawn following the synthesis method of Inglis et al., 2008.)

While the CuAAC reaction generally requires reaction times of several hours at temperatures ranging from ambient to 50°C, the HDA cycloaddition between electron-deficient dithioesters and *trans,trans*-2,4-hexadien-1-ol performed at 50°C in the presence of catalysts usually takes between 2 and 24 h to achieve completion. Inglis et al. (2009), however, demonstrated a dramatic reaction-rate improvement of the RAFT-HDA click reaction through the use of novel cyclopentadienyl-functionalized polymers. Thus polystyrene prepared by ATRP and bearing a terminal bromine substituent was treated with sodium cyclopentadienide (NaCp) in THF to achieve a complete substitution of the bromine atom with a cyclopentadienyl moiety (Fig. 12.29), while commercially available poly(ethylene glycol) monomethylether was equipped with a cyclopentadienyl end group through nucleophilic substitution of a tosylated intermediate (Fig. 12.30). These were reacted with **(P13-III)** and **(P13-IV)** in chloroform at ambient temperature and pressure resulting in block copolymers. The reaction was found to be complete within 10 min without the addition of catalyst. This difference was attributed to the high Diels-Alder activity of the cyclopentadienyl end group.

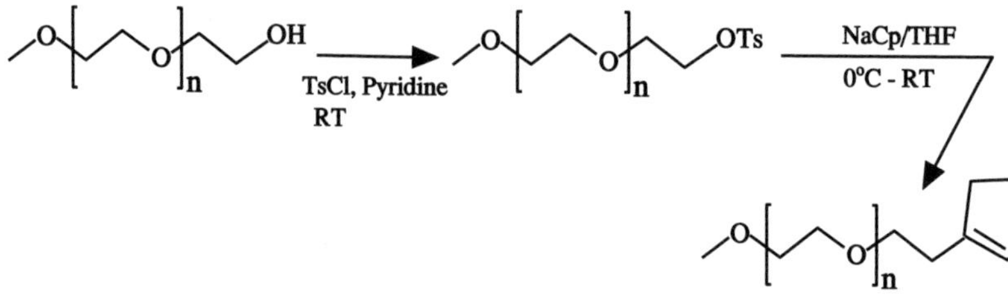

Figure 12.30 Synthesis of cyclopentadienyl-terminated poly(ethylene glycol) (PEG). Step 1:   treatment with p-toluenesulfonyl chloride (TsCl) in pyridine at room temperature.   Step 2:   treatment with NaCp (2.0 M in THF) at $0°C$ to room temperature. (Drawn following the synthesis method of Inglis et al., 2009.)

12.5   Thiol-Ene Reactions

Though tremendous success has been achieved with the development of Cu(I)-mediated Huisgen 1,3-dipolar cycloaddition reaction of azides and acetylenes as a robust and efficient synthetic tool, it has several limitations which include the need for a metal catalyst, an inability to photochemically control the reaction or to conduct the reaction in the absence of solvent. In comparison, the century-old addition of thiols to alkene (the *hydrothiolation* of a C=C bond), which is currently called *thiol-ene coupling* (TEC), has many of the attributes of click chemistry without, however, some of the aforesaid disadvantages of the CuAAC reaction.

It should be noted that the TEC reaction with alkenes can give either of two possible products corresponding to *Markovnikov* (α-product) or *anti-Markovnikov* (β-product) addition:

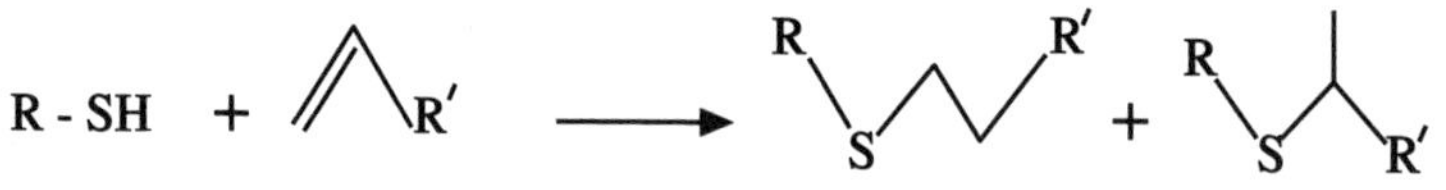

However, thiol-ene addition forming β-products usually predominates. Hence only this addition will be considered in further discussions on TEC reactions. The TEC reaction ocurs by either radical or nucleophilic mechanisms (see Fig. 12.4), depending on the reaction conditions and the unsaturated substrates.

The user-friendliness of the TEC reaction is readily apparent as moisture and air do not need to be excluded, expensive transition metal catalysts are unnecessary, a wide variety of functional groups and solvents (including water) are well tolerated, and high yields and clean products requiring only simple purification are common. Another attribute of the TEC reaction is that the reaction enables the establishment of a robust ligation motif between substrates by virtue of the stability of the thioether linkage in a wide range of chemical environments, such as strong acid and basic media as well as oxidizing and reducing conditions. Moreover, like the classic reaction between azides and alkynes where the starting materials can be easily prepared (e.g., facile conversion of alkyl halides to azides by displacement using NaN$_3$), the wide availability and associated stability of the thiol and alkene starting materials offers a number of advantages from a synthetic viewpoint. Thus, in addition to the range of commercially available thiols, including cysteine-containing peptides, many compounds such as alkyl halides, alcohols, RAFT CTAs, and even alkenes can be readily converted to thiols (David and Kornfield, 2008). It may be recalled that the α-terminus of a (co)polymer chain made by RAFT polymerization is chemically identical to the R-group (assuming no undesirable side reactions) of the RAFT CTA, while the ω terminus bears a thiocarbonylthio functional group, which undergoes facile cleavage to a thiol, typically with a primary or secondary amine (Xu et al., 2006). Sulfhydryl and alkenyl groups are also easily incorporated into polymer or monomer structures to enable postpolymerization modification. Over the years, the TEC reaction has therefore been extensively exploited in polymer chemistry.

To carry out thiol-ene photoreactions between ene-functional polymer and thiol, a general procedure (Campos et al., 2008) consists of dissolving the polymer, 5-10 equiv of thiol (with respect to alkene), and 0.2 equiv of 2,2-dimethoxy-2-phenylacetophenone (DMPA) in a minimal amount of solvent and irradiation with a 365 nm UV lamp for 5-10 min under an inert atmosphere. The corresponding procedure for thiol-ene thermal reactions consists of dissolving the polymer, 5-10 equiv of thiol (with respect to the alkene), and 0.5 equiv of AIBN in a minimal amount of solvent and heating at 80°C in the absence of air (Campos et al., 2008). (*Note*: Though TEC reactions do not require deoxygenation when performed under solvent-free conditions, the requirement for solubilization of polymer in an appropriate solvent necessitates removing oxygen to prevent oxygen inhibition of the radical initiator.)

Problem 12.14 Discuss a possible method of synthesizing an asymmetric telechelic polymer based on polystyrene (DP ~50) with a carboxylic group at one end and a hydroxyl group at the other, using combined thiol-ene and CuAAC click reactions.

Answer:

An alkene-functional bromide-type initiator can be used to prepare by the ATRP method a telechelic polystyrene of desired DP having a single alkene at one end an a single bromine at the other chain end. Campos et al. (2008) used 9-decenyl-2-bromo-2-methylpropanoate (DBMP) as the ATRP initiator for this synthesis (Scheme P12.14.1). In a typical procedure (Campos et al., 2008), to a mixture of DBMP (1.7 mmol), PMDETA (1.70 mmol), styrene (0.170 mol), and chlorobenzene (65 mmol), is added 1.70 mmol of copper(I) bromide and the mixture is heated to 90°C for 15 h giving about 50% conversion of monomer in order to satisfy the DP requirement. The viscous liquid mixture is dissolved in CH_2Cl_2 and washed with water in order to remove all copper. The polymer (e-PSt-Br) is recovered by precipitation in methanol and is the reacted with 10-fold excess sodium azide in DMF solution at 70°C for 24 h to effect replacement (> 90%) of the bromine end unit with azide functionality. The resulting polymer, e-PSt-N_3, is converted to HO_2C-PSt-OH in two steps for which two synthetic pathways are possible, i.e., initial TEC followed by CuAAC of the azide group (Path A) or the reverse strategy where the CuAAC is followed by the TEC reaction (Path B). In a typical procedure following Path A (Campos et al. 2008), e-PSt-N_3 (0.08 mmol), thioglycolic acid (0.80 mmol, 10 equiv), and AIBN (0.04 mmol, 0.5 equiv) are dissolved in benzene (2 mL) and heated to 80°C for 4 h, giving 83% yield of the polymer, HO_2C-PSt-N_3. Then, propargyl alcohol (0.20 mmol, 10 equiv), PMDETA (0.06 mmol, 3 equiv), and dry THF (1 mL) are added to HO_2C-PSt-N_3 (0.02 mmol) and copper(I) bromide (0.06 mmol, 3 equiv). The mixture is stirred at room temperature for 24 h to carry out the CuAAC reaction. The polymer, HO_2C-PSt-OH, is freed of all copper by washing a CH_2Cl_2 solution of the crude product with water and recovered by precipitation into methanol yielding a white solid (44% yield). Campos et al. (2008) found the thermally initiated TEC reaction to be completely orthogonal with the traditional CuAAC reaction allowing the individual chain ends to be quantitatively functionalized without the need for applying protection/deprotection strategies.

Although known since the first half of the last century, the radical-initiated thiol-yne reaction, which like the radical-mediated thiol-ene reaction can proceed extremely rapidly yielding products quantitatively under facile conditions, has been essentially overlooked in the polymer/materials field. However, researchers have recently began to evaluate this potentially highly versatile reaction. Thus, radical mediated thiol-yne step-growth polymerization as a route to highly crosslinked networks has been described (Fairbanks et al., 2009). Polysulfide networks formed via the photopolymerization of a range of alkyldithiols with a series of alkyldiynes have been studied (Chan et al., 2009) and, more recently, a sequential, quantitative nucleophilic thiol-ene/radical thiol-yne process was used for the synthesis of a range of branched structures with potential biomedical significance (Chan et al., 2009).

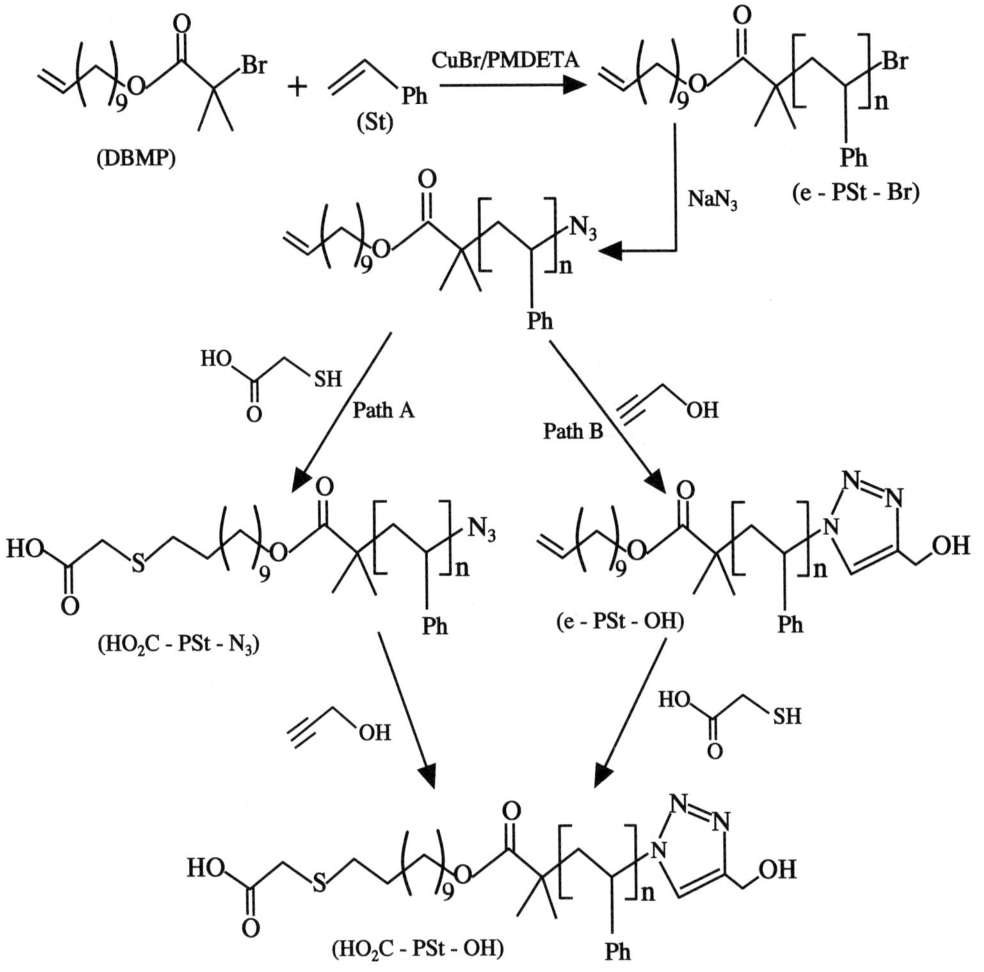

Scheme P12.14.1 Synthesis of HO$_2$C-PSt-OH by ATRP, followed by azidation and combined TEC and CuAAC reactions. (Adapted from Campos et al., 2008.)

Problem 12.15 Suggest a method for the synthesis of poly-N-isopropylacrylamide (PNIPAm) of low polydispersity and DP~50, having either (a) an ene or (b) a yne terminal. Describe the products that would result on performing radical thiol (RSH)-ene and radical thiol (RSH)-yne reactions on these end-functional polymers.

Answer:

Since the thiocarbonylthio end groups of the polymer produced by RAFT polymrization can be easily modified to a variety of species including cleavage to a thiol (typically with a primary or secondary amine) that affords subsequent TEC reaction, it will be convenient to use the RAFT method to make PNIPAm of the desired DP. Yu et al. (2009) used 1-cyano-1-methylethyl dithiobenzoate (CPDB) as the RAFT CTA to make PNIPAm precursors to well-defined ene and yne terminal polymers, as shown in Scheme P12.15.1. The latter are used to carry out thiol-ene and thiol-yne reactions.

In a typical procedure (Yu et al., 2009), a mixture of NIPAm (133 mmol), CPDB (1.75 mmol), and AIBN (0.353 mmol), dissolved in DMF (10 g), is heated at 70°C for 48 h, after which the reaction is quenched by rapid cooling in liquid N_2. (The polymerization is terminated early at about 67% monomer conversion in order to achieve the desired DP and to preserve the thiocarbonylthio end group as also to minimize the presence of terminated products derived from coupling reactions.)

The dithiobenzoate end-group of the homopolymer, PNIPAm, obtained is modified in a one-pot process via primary amine cleavage followed by phosphine-mediated nucleophilic thiol-ene click reactions with either allyl methacrylate (ALMA) or propargyl acrylate (PROPA), yielding ene and yne terminal PNIPAm homopolymers quantitatively. In a typical procedure (Yu et al., 2009), to make allyl end-functionalized PNIPAm (i.e., PNIPAm-S-ALMA), a mixture of PNIPAm (0.424 mmol), ALMA (5.0 mmol), dimethylphenylphosphine (0.072 mmol), CH_2Cl_2, and octylamine (7.5 mmol) is stirred overnight at room temperature. The product, PNIPAm-S-ALMA, is recovered by precipitation into large excess of hexane, followed by dialysis against methanol. The synthesis of propargyl end-functionalized PNIPAm (i.e., PNIPAm-S-PROPA) is achieved following exactly the same protocol as for PNIPAm-S-ALMA.

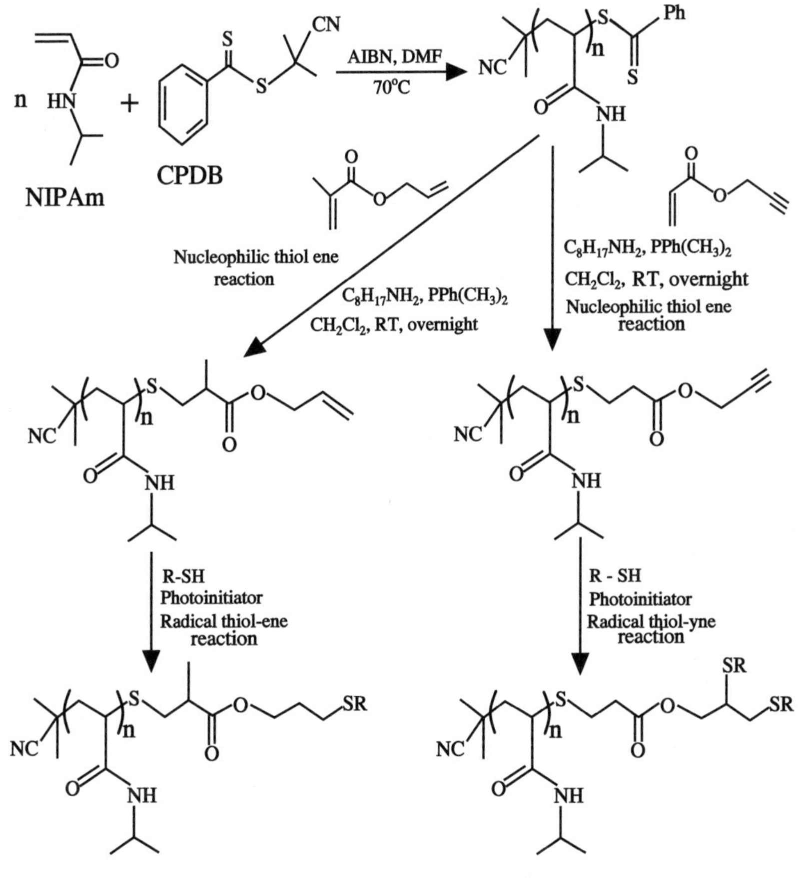

Scheme P12.15.1 RAFT polymerization of NIPAm and subsequent end group modifications via tandem nucleophilic thiol-ene/radical thiol-ene and nucleophilic thiol-ene/radical thiol-yne reactions. (From Yu et at., 2009. With permission from *John Wiley & Sons, Inc.,*)

In a typical procedure for photochemical, radical thiol-ene and thiol-yne reactions (Yu et al., 2009), a mixture of PNIPAm-S-ALMA or PNIPAm-S-PROPA (0.2 g), 100% excess of target thiol (based on the molarity of "ene" or "yne"), benzil dimethyl ketal (Irganure 651) photoinitiator (15.0 mg) in an appropriate volume of THF is exposed to UV light (λ = 350 nm) for 2 h. The sample is dialyzed against methanol. It should be noted that the thiol-ene and thiol-yne reactions yield mono and bis end-functionalized PNIPAm homopolymers, respectively (see Scheme P12.15.1).

Over the years, the TEC reaction has been extensively exploited in polymer chemistry. The UV-induced crosslinking of unsaturated polymers (photocuring) by reaction with multifunctional thiols is employed in surface coating since it has a number of advantages over other curing methods, especially those employing heavy-metal catalysts. Biomaterials for application in medicine, especially dentistry, have been prepared using this process (Carioscia et al., 2005; Dondoni, 2008). Only recently, however, has the click aspect of the TEC reaction been fully appreciated in the field of polymer science. Several promising results have been reported which suggest a significant role for thiol-ene chemistry as a new click reaction, which is compatible with water and oxygen and can be performed in the absence of solvent and under photochemical intiation.

A real synthetic strength of the TEC reaction is its versatility with respect to ene substrates that can be used and include both activated and nonactivaed species such as norbornenes, (meth)acrylates, maleimides, and allyl ethers to name but a few (Yu et al., 2009). Recent examples describing the application of the thiol-ene reaction in the polymer/materials field include the use in thioether dendrimer synthesis (Killops et al., 2008), convergent star synthesis (Chan et al., 2008), synthesis of multifunctional branched organosilanes (Rissing and Son, 2008), and modification/functionalization of wide range of polymers (Justynska and Schlaad, 2004; Killops et al., 2008).

12.5.1 Mechanisms of Thiol-Ene Reactions

The TEC reaction can be accomplished under a variety of experimental conditions, including acid/base catalysis (Li et al., 2008), nucleophilic catalysis (Chan et al., 2009), radical mediation (most commonly induced photochemically) (Dondoni, 2008), and via a solvent-promoted process (Tolstyka et al., 2008). However, the reaction is most commonly performed under radical mediation where it is applicable to many substrates, or under nucleophilic mediation for activated enes where the process proceeds via an anionic chain process.

The radical mediation of the thiol-ene reaction is brought about photochemically or via thermal decomposition of a radical initiator and the reaction is known to proceed by a radical mechanism to give a thioether, as shown in Fig. 12.31(A). Under nucleophile-initiated conditions, such as employing dimethylphenylphosphine (Me$_2$PPh, DMPP) as an initiator, the TEC reaction takes place extremely rapidly as well as quantitatively. The anionic chain mechanism proposed for the amine/phosphine mediated nucleophilic thiol-ene reaction with an acrylate is shown in Fig. 12.31(B).

Under radical-mediated conditions, a thiol also adds to a yne. The radical thiol-yne reaction can be considered as a sister reaction to the radical thiol-ene reaction because it possesses characteristics virtually identical to those of the radical thiol-ene reactions. As with the thiol-ene reaction, the thiol-yne reaction, in general, proceeds rapidly under a variety of experimental conditions in an air atmosphere at ambient temperature and humidity, selectively yielding the mono or bisaddition products with little to no clean up required. In the case of the double addition products formed under radical conditions, the reaction of two equivalents of thiol with a terminal alkyne is itself a two-step process (Fig. 12.32). The first, slower, step 1 involves the addition of thiol to the C≡C bond to yield an intermediate vinylic radical that subsequently undergoes chain transfer to additional thiol, yielding the vinyl thioether and concomitantly generating a new thiyl radical. The vinylthioether product is able to undergo a second, faster, formally thiol-ene, reaction with the newly generated thiyl radical in step 2, yielding the intermediate bisthioether radical that undergoes chain transfer with another thiol molecule quantitatively yielding the target bisthioether (Yu et al., 2009).

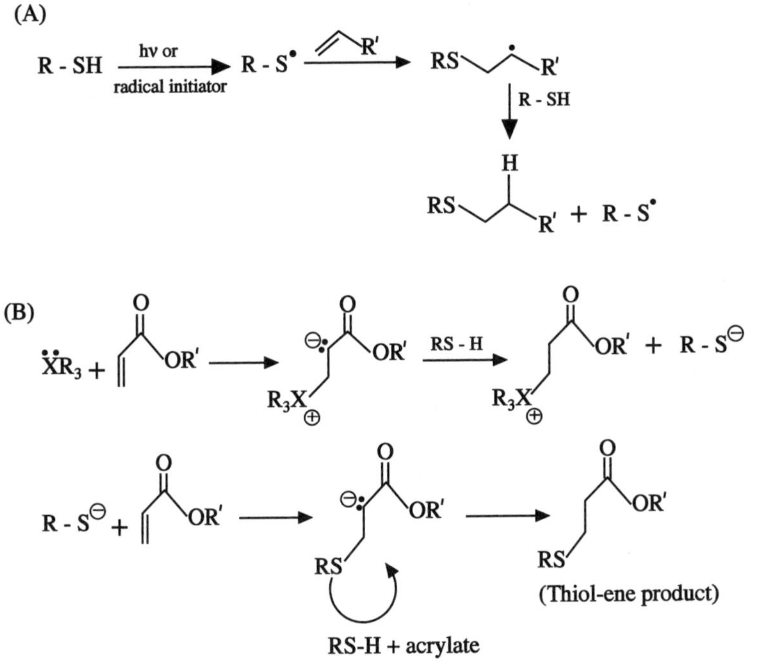

(A)

(B)

Figure 12.31 (A) Mechanism for the radical-mediated thiol-ene reaction with a terminal ene. (B) Anionic chain mechanism for the amine/phosphine mediated nucleophilic thiol-ene reaction with an acrylate. (From Yu et al., 2009. With permission from *John Wiley & Sons, Inc.*)

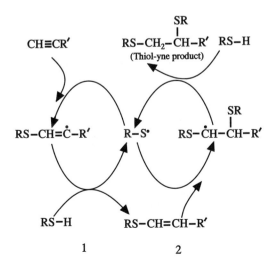

Figure 12.32 Mechanism for the radical mediated thiol-yne reaction with a terminal alkyne. (From Yu et al., 2009. With permission from *John Wiley & Sons, Inc.*)

12.5.2   Synthesis of Star Polymers and Dendrimers

As we have noted earlier, TEC reactions compare favorably with CuAAC reactions, including the use of readily available starting materials and catalysts. However, unlike CuAAC, the TEC reaction is free of metals and can be performed under photochemical initiation. The TEC is also very fast (often quantitative reaction is obseved within a period of seconds at ambient temperature) compared to extended reaction times and elevated temperatures occasionally required for the CuAAC reaction (Binder and Sachsenhofer, 2007). The use of TEC reaction has therefore attracted attention as a means of preparing star polymers and more complex polymers, such as dendrimers and other sterically hindered structures.

Hawker and coworkers (Killops et al., 2008) reported that photochemical TEC reactions allowed robust, efficient, and orthogonal synthesis of dendrimers in a divergent manner (Fig. 12.33). Starting from the tris-alkene core, 2,4,6-triallyloxy-1,3,5-triazine (**A**) and choosing 1-thioglycerol (**B**) as the AB$_2$ monomer for its thiol functionality, two readily functionalizable hydroxy groups, and its miscibility with **A**, they used TEC chemistry to construct both the backbone as well as functionalize the chain ends. The solvent-free TEC reaction between **A** and **B**, in the presence of trace amounts of photoinitiator, 2,2-dimethoxy-2-phenyl-acetophenone (DMPA) and with 1.5 equiv of **B** per alkene, was carried out at room temperature, without deoxygenation, by irradiation for 30 min, giving the first-generation, hexa-hydroxy dendrimer **[G1]**(OH)$_6$ in quantitative yield. Subsequent esterification of **[G1]**(OH)$_6$ with 4-pentenoic anhydride, (H$_2$C=CHCH$_2$CH$_2$CO)$_2$O, in the presence of 4-dimethylaminopyridine (DMAP) furnished the ene-functional dendrimer **[G1]**(ene)$_6$. By repetition of this stepwise procedure, the fourth dendrimer, **[G4]**(OH)$_{48}$, was obtained with unprecedented efficiency.

Various commercially available thiols react photochemically with tetravinylsilane to give the corresponding tetrasubstituted thioether compounds:

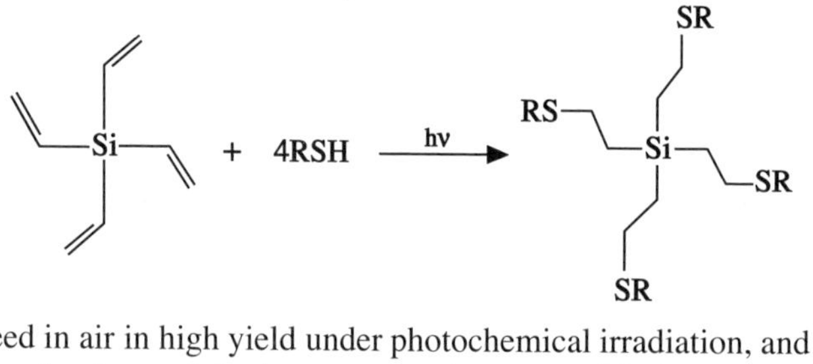

The reactions proceed in air in high yield under photochemical irradiation, and purification steps, if necessary, involve simple precipitation or extraction steps. Rissing and Son (2008) prepared carbosilane-thioether dendrimers in a divergent fashion starting with tetravinylsilane as a core, followed by a succession of alternating thiol-ene and Grignard reactions. While 3-mercaptopropyltrimethoxysilane, HS(CH$_2$)$_3$Si(OMe)$_3$, that is, R = –(CH$_2$)$_3$Si(OMe)$_3$ in the above equation, was used for branching via TEC reaction, the vinylation of the dendrimer ends was performed with vinylmagnesium bromide in THF:

$$Si(OMe)_3 \; + \; CH_2=CHMgBr \; \longrightarrow \; -Si(CH=CH_2)_3$$

Vinyl-terminated dendrimers up to the fifth generation were thus prepared in excellent yields (78-94%).

Since polymers synthesized by RAFT polymerization technique bear thiocarbonylthio end groups that can be reduced to thiol functional species under facile conditions, RAFT-synthesized polymers can serve as masked macromolecular terminal thiol-containing materials capable of undergoing TEC reactions. Primary and secondary amines are convenient and efficient reducing agents for the thiocarbonylthio end groups. Importantly, the thiol-ene reaction also proceeds

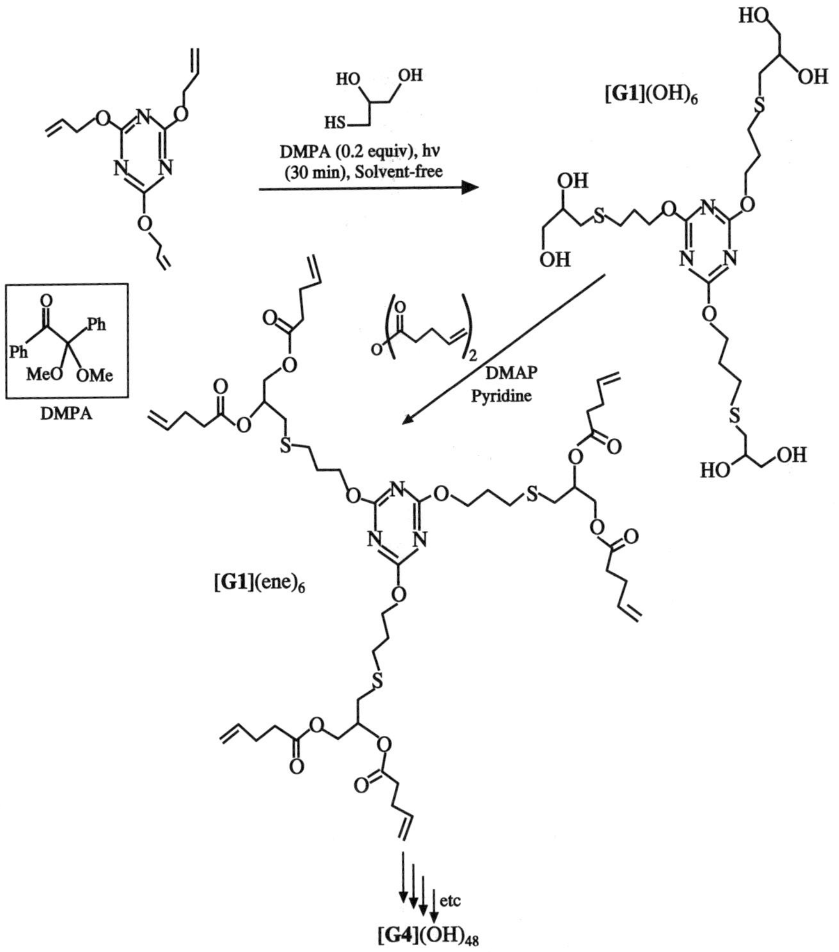

Figure 12.33 Divergent synthesis of a fourth generation hydroxyl-terminated dendrimer with triazine core via thiol-ene click chemistry. (From Killops et al., 2008. With permission from *American Chemical Society.*)

rapidly under nucleophilic catalysis with primary and secondary amines. Consequently, the amine used for the reductive aminolysis can also facilitate addition of the thiol to the activated ene. Moreover, the orthogonal nature of reductive aminolysis and TEC reactions allows the two processes to be conducted in one pot.

Chan et al. (2008) demonstrated the applicability of the aforesaid one-pot concept by performing a novel convergent synthesis of 3-arm star polymers in which a RAFT-prepared homopolymer was subjected to sequential reduction of the thiocarbonylthio end group to a thiol functional group followed by TEC reaction to yield the target star polymers. Thus, a homopolymer of *N,N*-diethylacrylamide, poly(DEA), was prepared under standard RAFT conditions employing 1-cyano-1-methylethyldithiobenzoate, as the RAFT agent in conjunction with AIBN in DMF at 70°C with a target molar mass of 4500 g/mol at quantitative conversion. For one-pot synthesis of 3-arm star polymer (Fig. 12.34), the poly(DEA) homopolymer was mixed with trimethylolpropane triacrylate (TMPTA), hexylamine ($C_6H_{13}NH_2$), and dimethylphenylphosphine (Me_2PPh)

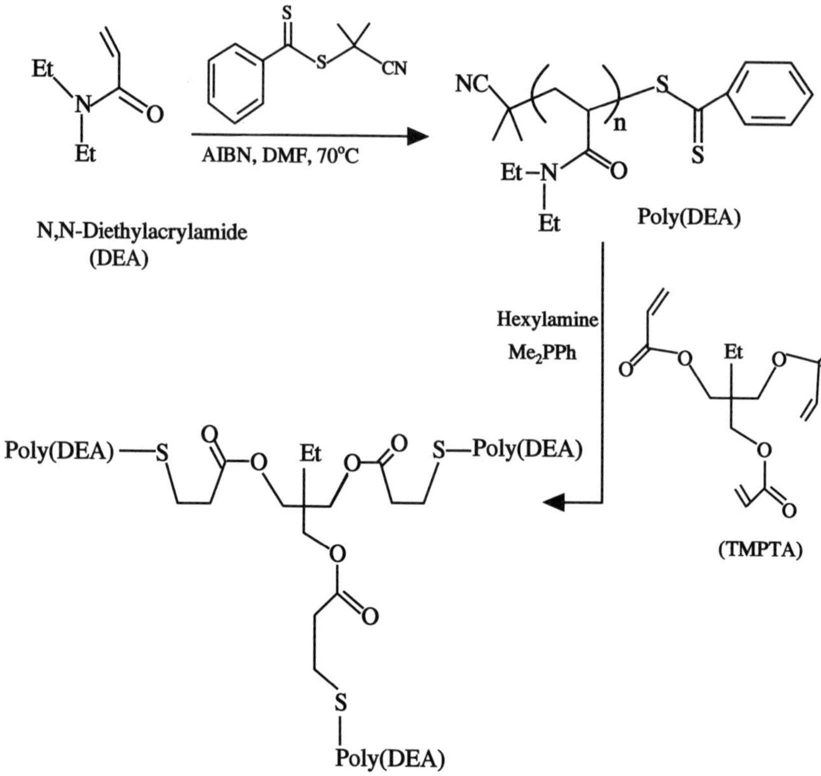

Figure 12.34 Formation of 3-arm star polymers by RAFT polymerization and TEC reactions. [Adapted from Chan et al. (2008) and Sumerlin and Vogt (2010).]

combination at a molar ratio of –SH : ene of 1.5 : 1 (to favor star formation). Since the reaction involves a macromolecular secondary thiol, the primary amine-phosphine combination, as opposed to only amine, was employed (Me$_2$PPh being an extremely potent catalyst for such thiol-ene reactions). The amine-phosphine combination is also beneficial since the phosphine serves a second, important role of eliminating the formation of the polymeric disulfide species that can form after end-group reduction to –SH and often readily occurs in the presence of only amines under a normal air atmosphere (Chan et al., 2008). From FT-IR spectroscopy and ^{1}H NMR analysis of the products it was established that the macromolecular thiol-ene reaction is both fast and quantitative enough to be described as a click reaction.

Employing a similar approach as above, thiol-terminated RAFT polymers have also been reacted with a variety of low and high molecular weight acrylate species for efficient end functionalization (Spruell et al., 2009) and immobilized onto ene-decorated microspheres of poly(divinyl benzene) (Goldmann et al., 2009). Considering the abundance of commercially available activated alkene substrates amenable to nucleophilic addition (e.g., acrylates and maleimides) and the simplicity-cum-robustness of the aforesaid approach, the RAFT/thiol-ene combination can be expected to remain as a valuable tool for macromolecular synthesis (Sumerlin and Vogt, 2010).

REFERENCES

Akeroyd, N., Pfukwa, R., and Klumperman, B., *Macromolecules*, **42**, 3014 (2009).

Altintas, O., Hizal, G., and Tunca, U., *J. Polym. Sci., Part A: Polym. Chem.*, **44**, 5699 (2006).

Altintas, O., Yankul, H., Hizal, G., and Tunca, U., *J. Polym. Sci., Part A: Polym. Chem.*, **45**, 3588 (2007).

Altintas, O., Yankul, B., Hizal, G., and Tunca, U., *J. Polym. Sci., Part A: Polym. Chem.*, **46**, 1218 (2008).

Appakkuttan, P., Dehaen, W., Fokin, V. V., and van der Eyken, E., *Org. Lett.*, **6**, 4223 (2004).

Bakbak, S., Leech, P. J., Carson, B. E., Saxena, S., King, W. P., and Bunz, U. H., *Macromolecules*, **39**, 6793 (2006).

Baum, M. and Brittain, W. J., *Macromolecules*, **35**, 610 (2002).

Becer, C. R., Hoogenboom, R., and Schubert, U. S., *Angew. Chem. Int. Ed.*, **48**, 4900 (2009).

Binder, W. H., Gloger, D., Weinstabl, H., Allmaier, G., and Pittenauer, E., *Macromolecules*, **40**, 3097 (2007).

Binder, W. H. and Kluger, C., *Curr. Org. Chem.*, **10**, 1791 (2006).

Binder, W. H. and Sachsenhofer, R., *Macromol. Rapid Commun.*, **28**, 15 (2007).

Campos, L. M., Killops, K. L., Sakai, R., Paulusse, J. M. J., Damiron, D., Drockenmuller, E., Messmore, B. W., and Hawker, C. J., *Macromolecules*, **41**, 7063 (2008).

Canadell, J., Fischer, H., De With, G., and van Benthem, R. A. T. M., *J. Polym. Sci., Part A: Polym. Chem.*, **48**, 3456 (2010).

Carioscia, J. A., Lu, H., Stanbury, J. W., and Bowman, C. N., *Dent. Mater.*, **21**, 1137 (2005).

Chan, T. R., Hilgraf, R., Sharpless, K. B., and Fokin, V. V., *Org. Lett.*, **6**, 2853 (2004).

Chan, J. W., Yu, B., Hoyle, C. E., Lowe, A. B., *Chem. Commun.*, 4959 (2008).

Chan, J. W., Hoyle, C. E., and Lowe, A. B., *J. Am. Chem. Soc.*, **131**, 5751 (2009).

Chan, J. W., Zhou, H., Hoyle, C. E., and Lowe, A. B., *Chem. Mater.*, **21**, 1579 (2009).

Chen, X., Wudl, F., Mal, A. K., Shen, H., and Nutt, S. R., *Macromolecules*, **36**, 1802 (2003).

Chujo, Y., Sada, K., and Saegusa, T., *Macromolecules*, **23**, 2636 (1990).

Codelli, J. A., Baskin, J. M., Agard, N. J., and Bertozzi, C. R., *J. Am. Chem. Soc.*, **130**, 11486 (2008).

Dag, A., Durmaz, H., Hizal, G., and Tunca, U., *J. Polym. Sci., Part A: Polym. Chem.*, **46**, 302 (2008).

Dag, A., Durmaz, H., Tunca, U., and Hizal, G., *J. Polym. Sci., Part A: Polym. Chem.*, **47**, 178 (2009).

David, R. L. A. and Kornfield, J. A., *Macromolecules*, **41**, 1151 (2008).

Diaz, D. D., Punna, S., Holzer, P., McPherson, A. K., Sharpless, K. B., Fokin, V. V., and Finn, M. G., *J. Polym. Sci., Part A: Polym. Chem.*, **42**, 4392 (2004).

Dandoni, A., *Angew. Chem. Int. Ed.*, **47**, 2, 8995 (2008).

Durmaz, H., Colakoglu, B., Tunca, U., and Hizal, G., *J. Polym. Sci., Part A: Polym. Chem.*, **44**, 1667 (2006).

Durmaz, H., Karatas, F., Tunca, U., and Hizal, G., *J. Polym. Sci., Part A: Polym. Chem.*, **44**, 499, 3947 (2006).

Englert, B. C., Bakbak, S., and Bunz, H. F., *Macromolecules*, **38**, 5868 (2005).

Erdogan, M., Hizal, G., Tunca, U., Hyrabetyan, D., and Pekcan, O., *Tetrahedron*, **43**, 1925 (2002).

Eugene, D. M. and Grayson, S. M., *Macromolecules*, **41**, 5082 (2008).

Fairbanks, B. D., Scott, T. F., Kloxin, C. J., Anseth, K. S., and Bowman, C. N., *Macromolecules*, **42**, 211 (2009).

Fleischmann, S., Komber, H., and Voit, B., *Macromolecules*, **41**, 5255 (2008).

Gacal, B., Akat, H., Balta, D. K., Arsu, N., and Yagci, Y., *Macromolecules*, **41**, 2401 (2008).

Gacal, B., Durmaz, H., Tasdelen, M. A., Hizal, G., Tunca, U., Yagci, Y., and Demirel, A. L., *Macromolecules*, **39**, 5330 (2006).

Gao, H., Louche, G., Sumerlin, B. S., Jahed, N., Golas, P., and Matyjaszewski, K., *Macromolecules*, **38**, 8979 (2005).

Gao, H. and Matyjaszewski, K., *Macromolecules*, **39**, 4960 (2006).

Gao, H., Min, K., and Matyjaszewski, K., *Macromol. Chem. Phys.*, **208**, 1370 (2007).

Ge, Z., Zhou, Y., Xu, J., Liu, H., Chen, D., and Liu, S., *J. Am. Chem. Soc.*, **131**, 1628 (2009).

Gheneim, R., Berumen, C. P., and Gandini, A., *Macromolecules*, **35**, 7246 (2002).

Golas, P. L., Tsarevsky, N. V., and Matyjaszewski, K., *Macromol. Rapid Commun.*, **29**, 1167 (2008).

Goldman, A. S., Walther, A., Nebhani, L., Joso, R., Ernst, D., Loos, K., Barner-Kowollik, C., Barner, L., and Muller, A. H. E., *Macromolecules*, **42**, 3707 (2009).

Gondi, S. R., Vogt, A. P., and Sumerlin, B. S., *Macromolecules*, **40**, 474 (2007).

Gotsmann, B., Duerig, U., Frommer, J., and Hawker, C. J., *Adv. Funct. Mater.*, **16**, 1499 (2006).

Gress, A., Volkel, A., and Schlaad, H., *Macromolecules*, **40**, 7928 (2007).

Gungor, E., Cote, G., Erdogan, T., Durmaz, H., Demirel, A. L., Hizal, G., and Tunca, V., *J. Polym. Sci., Part A: Polym. Chem.*, **45**, 1055 (2007).

Hadjichristidis, N., Pitsikalis, M., Pispas, S., Iatrou, H., *Chem. Rev.*, **101**, 3747 (2001).

Haller, M., Luo, J., Li, H., Kim, T.-D., Liao, Y., Robinson, B. H., Dalton, L. R., and Jen, A. K.-Y., *Macromolecules*, **37**, 688 (2004).

Himo, F., Demko, Z. P., Noodleman, L., and Sharpless, K. B., *J. Am. Chem. Soc.*, **124**, 12210 (2002).

Himo, F., Lovell, T., Hilgraf, R., Rostovtsev, V. V., Noodleman, L., Sharpless, K. B., and Fokin, V. V., *J. Am. Chem. Soc.*, **127**, 210 (2005).

Imai, Y., Itoh, H., Naka, K., and Chujo, Y. *Macromolecules*, **33**, 4343 (2000).

Inglis, A. J., Sinnwell, S., Davis, T. P., Barner-Kowollik, C., and Stenzel, M, *Macromolecules*, **41**, 4120 (2008).

Inglis, A. J., Sinwell, S., Stenzel, M. H., and Barner-Kowollik, C., *Angew. Chem. Int. Ed.*, **48**, 2411 (2009).

Johnson, J. A., Finn, M. G., Koberstein, J. T., and Turro, N. J., *Macromol. Rapid Commun.*, **29**, 1052 (2008).

Johnson, J. A., Lewis, D. R., Diaz, D. D., Finn, M. G., Koberstein, J. T., and Turro, N. J., *J. Am. Chem. Soc.*, **128**, 6564 (2006)

Jones, J. R., Liotta, C. L., Collard, D. M., and Schiraldi, D. A., *Macromolecules*, **32**, 5786 (1999).

Joralemon, M. J., O'Reilly, R. K., Hawker, C. J., and Wooley, K. L., *J. Am. Chem. Soc.*, **127**, 16892 (2005).

Joralemon, M. J., O'Reilly, R. K., Matson, J. B., Nugent A. K., Hawker, C. J., and Wooley, K. L., *Macromolecules*, **38**, 5436 (2005).

Justyuska, J. and Schlaad, H., *Macromol. Rapid Commun.*, **25**, 1478 (2004).

Kamahori, K., Tada, S., Ito, K., and Itsuno, S., *Macromolecules*, **32**, 541 (1999).

Killops, K. L., Campos, L. M., Hawker, C. J., *J. Am. Chem. Soc.*, **130**, 5062 (2008).

Kim, T.-D., Luo, J. D., Tian, Y. Q., Ka, J. W., Tucker, N. M., Haller, M., Kang, J. W., and Jen, A. K. Y. *Macromolecules*, **39**, 1676 (2006).

Kolb, H. C., Finn, M. G., and Sharpless, K. B., *Angew. Chem. Int. Ed.*, **40**(11), 2004 (2001).

Kose, M. M., Yesilbag, G., and Sanyal, A., *Org. Lett.*, **10**, 2353 (2008).

Kwart, H. and King, K., *Chem. Rev.*, **68**, 415 (1968).

Kyeremateng, S. O., Amado, E., Blume, A, and Kressler, J., *Macromol. Rapid Commun.*, **29**, 1140 (2008).

Ladmiral, V., Mantovani, G., Clarkson, G. J., Cauet, S., Irwin, J. L., and Haddleton, D. M., *J. Am. Chem. Soc.*, **128**, 4823 (2006).

Laurent, B. A. and Grayson, S. M., *J. Am. Chem. Soc.*, **128**, 4238 (2006).

Lecomte, P., Riva, R., Schmeits, S., Rieger, J., Van Butsele, K., Jerome, C., and Jerome, R., *Macromol. Symp.*, **240**, 157 (2006).

Lei, X. and Porco, J. A., *Org. Lett.*, **6**, 795 (2004).

Li, Y. and Benicewicz, B. C., *Macromolecules*, **41**, 7986 (2008).

Li, M., De, P., Gondi, S. R., and Sumerlin, B. S., *J. Polym. Sci., Part A: Polym. Chem.*, **46**, 5093 (2008).

Li, M., De, P., Gondi, S. R., and Sumerlin, B. S., *Macromol. Rapid Commun.*, **29**, 1172 (2008).

Li, H., Riva, R., Jerome, R., and Lecomte, P., *Macromolecules*, **40**, 824 (2007).

Lutz, J.-F. and Boerner, H. G., *Prog. Polym. Sci.*, **33**, 1 (2008).

Lutz, J.-F., Boerner, H. G. and Weichenhan, K., *Macromol. Rapid Commun.*, **26**, 514 (2005).

Luxenhofer, R. and Jordan, R., *Macromolecules*, **39**, 3509 (2006).

Magenau, A. J. D., Martinez-Castro, N., and Storey, R. F., *Macromolecules*, **42**, 2353 (2009).

Mantovani, G., Lecolley, F., Tao, L., Haddleton, D. M.,Clerx, J., Cornelissen, J. J. L. M., and Velonia, K., *J. Am. Chem. Soc.*, **127**, 2966 (2005).

Matyjaszewski, K., Miller, P. J., Pyun, J., Kickelbick, G., and Diamanti, S., *Macromolecules*, **32**, 6526 (1999).

McElhanon, J. R., Russick, E. M., Wheeler, D. R., Loy, D. A., and Aubert, J. H., *J. Appl. Polym. Sci.*, **85**, 1496 (2002).

McElhanon, J. R. and Wheeler, D. R., *Org. Lett.*, **3**, 2681 (2001).

McElhanon, J. R., Zifer, T., Kline, S. R., Wheeler, D. R., Loy, D. A., Jamson, G. M., Long, T. M., Rahimian, K., and Simmons, B. A., *Langmuir*, **21**, 3259 (2005).

Mespouille, L., Coulembier, O., Paneva, D., Degee, P., Rashkov, I., and Dubois, P., *J. Polym. Sci., Part A: Polym. Chem.*, **46**, 4997 (2008).

Molteni, G., Bianchi, C. L., Marinoni, G., Santo, N., and Ponti, A., *New J. Chem.*, **30**, 1137 (2006).

Morgenroth, F., Reuther, E., and Müllen, K., *Angew. Chem. Int. Ed.*, **36**, 631 (1997).

Murphy, J. J., Nomura, K., and Paton, R. M., *Macromolecules*, **39**, 3147 (2006).

Nandivada, H., Jiang, X., and Lahann, J., *Adv. Mater.*, **19**, 2197 (2007).

Oishi, T. and Fujimoto, M., *J. Polym. Sci., Part A: Polym. Chem.*, **30**, 1821 (1992).

Opsteen, J. A. and Van Hest, J. M., *J. Polym. Sci., Part A: Polym. Chem.*, **45**, 2913 (2007).

O'Reilly, R. K., Joralemon, M. J., Hawker, C. J., and Wooley, K. L., *J. Polym. Sci., Part A: Polym. Chem.*, **44**, 5203 (2006).

Pachon, L. D., van Maarseveen, J. H., and Rothenberg, G., *Adv. Synth. Catal.*, **347**, 811 (2005).

Parrish, B., Breitenkamp, R. B., and Emrick, T., *J. Am. Chem. Soc.*, **127**, 7404 (2005).

Proupin-Perez, M., Cosstick, R., Liz-Marzan, L. M., Salgucirino-Maccira, V., and Brust, M., *Nuleosides, Nucleotides, Nucleic Acids*, **24**, 1075 (2005).

Qiu, X.-P., Tanaka, F., and Winnik, F. M., *Macromolecules*, **40**, 7069 (2007).

Qiu, X.-P. and Winnik, F. M., *Macromol. Rapid Commun.*, **27**, 1648 (2006).

Ranjan, R. and Brittain, W. J., *Macromolecules*, **40**, 6217 (2007).

Ranjan, R. and Brittain, W. J., *Macromol. Rapid Commun.*, **29**, 1104 (2008).

Rissing, C. and Son, D. Y., *Organometallics*, **27**, 5394 (2008).

Rissing, C. and Son, D. Y., *Organometallics*, **28**, 3167 (2009).

Riva, R., Schmeits, S., Stoffelbach, F., Jerome, C., Jerome, R., and Lecomte, P., *Chem. Commun.*, 5334 (2005).

Rostovtsev, V. V., Green, L. G., Fokin, V. V., and Sharpless, K. B., *Angew. Chem. Int. Ed.*, **41**, 2596 (2002).

Sanyal, A., *Macromol. Chem. Phys.*, **211**, 1417 (2010).

Shintai, R. and Fu, G. C., *J. Am. Chem. Soc.*, **125**, 10778 (2003).

Sinwell, S., Inglis, A. J., Davis, T. P., Stenzel, M. H., and Barner-Kowollik, C., *Chem. Commun.*, 2052 (2008).

Sinnwell, S., Inglis, A. J., Stenzel, M. H., and Barner-Kowollik, C., *Macromol. Rapid Commun.*, **29**, 1090 (2008).

Sinnwell, S., Lammens, M., Stenzel, M. H., Du Prez, F. E., and Barner-Kowollik, C., *J. Polym. Sci., Part A: Polym. Chem.*, **47**, 2213 (2009).

Sletten, E. M. and Bertozzi, C. R., *Org. Lett.*, **10**, 3097 (2008).

Sletten, E. M. and Bertozzi, C. R., *Angew. Chem. Int. Ed.*, **48**, 6974 (2009).

Sumerlin, B. S., Tsarevsky, N. V., Loucho, G., Lee, R. Y., and Matyjaszewski, K., *Macromolecules*, **38**, 7540 (2005).

Sumerlin, B. S. and Vogt, A. P., *Macromolecules*, **43**, 1 (2010).

Szlai, M., McGrath, D. V., Wheeler, D. R., Zifer, T., and McElhanon, J. R., *Macromolecules*, **40**, 818 (2007).

Takizawa, K., Nulwala, H., Thibault, R. J., Lowenhielm, P., Yoshimaga, K., Wooley, K. L., and Hawker, C. J., *J. Polym. Sci., Part A: Polym. Chem.*, **46**, 2897 (2008).

Tasdelen, M. A., Van Camp, W., Goethals, E., Dubois, P., Du Prez, F., and Yagen, Y., *Macromolecules*, **41**, 6035 (2008).

Thang, S. H., Chong, B., Mayadunne, R. T. A., Moad, G., and Rizzardo, E., *Tetrahedron Lett.*, **40**, 2435 (1999).

Thibault, R. J., Takizawa, K., Lowenhielm, P., Helms, B., Mynar, J. L., Frechét, J. M. J., and Hawker, C. J., *J. Am. Chem. Soc.*, **128**, 12084 (2006).

Thomsen, A. D., Malmstroem, E. E., and Hvilsted, S., *J. Polym. Sci., Part A: Polym. Chem.*, **44**, 6360 (2006).

Tolstyka, Z. P., Kopping, J. T., Maynard, H. D., *Macromolecules*, **41**, 599 (2008).

Tonga, M., Cengiz, N., Kose, M. M., Dede, T., and Sanyal, A., *J. Polym. Sci., Part A: Polym. Chem.*, **48**, 410 (2010).

Topham, P. D., Sandon, N., Read, E. S., Modsen, J., Rayan, A. J., and Armes, S. P., *Macromolecules*, **41**, 9542 (2008).

Tornoe, C. W., Christensen, C., and Meldal, M., *J. Org. Chem.*, **67**, 3057 (2002).

Tsarevsky, N. V., Sumerlin, B. S., and Matyjaszewski, K., *Macromolecules*, **38**, 3558 (2005).

Tunca, U., Ozyurek, Z., Erdogan, T., and Hizal, G., *J. Polym. Sci., Part A: Polym. Chem.*, **42**, 4228 (2004).

Urbani, C. N., Bell, C. A., Lonsdale, D. E., Whittaker, M. R., and Monteiro, M. J., *Macromolecules*, **40**, 7056 (2007).

Urbani, C. N., Bell, C. A., Lonsdale, D., Whittaker, M. R., and Monteiro, M. J., *Macromolecules*, **41**, 76 (2008).

Urbani, C. N., Bell, C. A., Whittaker, M. R., and Monteiro, M. J., *Macromolecules*, **43**, 1057 (2008).

Van Steenis, D. J. V. C., David, O. R. P., van Strijdonck, G. P. F., van Maarseveen, J. H., and Reek, J. N. H., *Chem. Commun.*, 4333 (2005).

Vargas, M., Kriegel, R. M., Collard, D. M., and Schiraldi, D. A., *J. Polym. Sci., Part A: Polym. Chem.*, **40**, 3256 (2002).

Vogt, A. P. and Sumerlin, B. S., *Macromolecules*, **39**, 5286 (2006).

Vogt, A. P. and Sumerlin, B. S., *Macromolecules*, **41**, 7368 (2008).

Wang, G., Liu, C., Pan, M., and Huang, J., *J. Polym. Sci., Part A: Polym. Chem.*, **47**, 1308 (2009).

Whittaker, M. R., Urbani, C. N., and Monteiro, M. J., *J. Am. Chem. Soc.*, **128**, 11360 (2006).

Wu, P., Feldman, A. K., Nugent, A. K., Hawker, C. J., Sheel, A., Voit, B., Pyun, J., Frechét, J. M. J., Sharpless, K. B., and Fokin, V. V., *Angew. Chem. Int. Ed.*, **43**, 3928 (2004).

Xia, Y., Verduzco, R., Grubbs, R. H., and Kornfield, J. A., *J. Am. Chem. Soc.*, **130**, 1735 (2008).

Xu, J., He, J., Fan, D., Wang, X., and Yang, Y., *Macromolecules*, **39**, 8616 (2006).

Xu, J., He, J., and Liu, S., *Macromolecules*, **40**, 9103 (2007).

Yang, L.-P., Dong, X.-H., and Pan, C.-Y., *J. Polym. Sci., Part A: Polym. Chem.*, **46**, 7757 (2008).

Yang, S. K. and Weck, M., *Macromolecules*, **41**, 346 (2008).

Yeo, M. N., Yousaf, M. N., Mrksich, M. J., *J. Am. Chem. Soc.*, **125**, 14994 (2003).

Yu, B., Chan, J. W., Hoyle, C. E., and Lowe, A. B., *J. Polym. Sci., Part A: Polym. Chem.*, **47**, 3544 (2009).

Zeng, Q., Li, T., Cash, B., Li, S., Xie, F., and Wang, Q., *Chem. Commun.*, 1453 (2007).

EXERCISES

12.1 Describe the main criteria that should be satisfied for a reaction to be called a "click reaction"? How would you justify the inclusion of the following reactions into the pantheon of click reactions: (a) Copper(I)-catalyzed azide-alkyne cycloaddition (CuAAC) reactions; (b) strain-promoted azide-alkyne coupling (SPAAC) reactions; (c) Diels-Alder (DA) cycloaddition reactions; (d) thiol-ene (TE) reactions; and (e) thiol-yne (TY) reactions?

12.2 The majority of polymer structures today are based on polymerization or functionalization of vinyl monomers derived from a limited range of families, such as styrenic, acrylate, or α-olefin-based systems. The development of a new vinyl monomer family that combines the attractive features of thermal and chemical stability besides having functional handles similar to traditional vinyl systems would represent a significant advance in the area of functionalized materials [Thibault, R. J., Takizawa, K., Lowenheilm, P., Helms, B., Mynar, J. L., Fréchet, J. M. J., and Hawker, C. J., *J. Am. Chem. Soc.*, **128**, 12084 (2006)]. In this respect, 4-vinyl-1,2,3-triazole monomers, which can be easily synthesized by click chemistry techniques, are expected to possess many of the outstanding features of traditional vinyl monomers. Considering this perspective, (a) identify structural similarities between 4-vinyl-1,2,3-triazole (VTZ) monomers and the most commonly used vinyl monomers, such as styrenics, vinyl pyridines, and acrylates; (b) suggest preparation methodologies for two such VTZ derivatives, which can be considered as functional equivalents of styrene and methacrylate, starting from 1-trimethylsilyl-2-vinyl acetylene and 2-methylbut-3-yn-2-ol using click chemistry.

12.3   While in a fairly large number of cases, RAFT-mediated polymerization is used to conduct the controlled polymerization of vinyl monomers, to link such a vinyl polymer to a substrate, esterification via the leaving group of the RAFT agent is used in most cases, though this induces a hydrolyzable link, which is not always desirable. To mitigate this problem, a new RAFT agent with leaving group based on a triazole moiety :

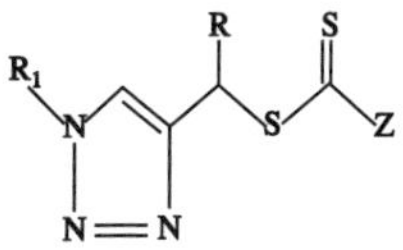

was introduced by Akeroyd et al. (2009) [Akeroyd, N., Pfukwa, R., and Klumperman, B., *Macromolecules*, **42**, 3014 (2009)]. Besides playing an active role in the stabilization of the intermediate radical in RAFT polymerization, comparable to the phenyl group in a benzyl leaving group, triazole provides a link which is aromatic in nature and is therefore extemely stable. Suggest a click procedure to synthesize the above RAFT agent with R = H, R_1 = phenyl, and Z = O-ethyl, i.e.,

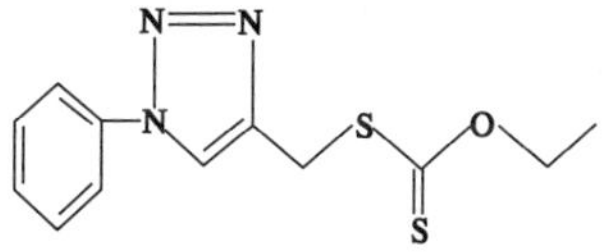

How can additional stabilization of the leaving group radical be provided ?

12.4   Suggest a procedure combining ATRP and CuAAC for "one-pot" synthesis of $(HEMA)_{10}$ methacrylate macromonomer. (HEMA = hydroxyethyl methacrylate.)

12.5   Suggest a route, via combination of ATRP and CuAAC reaction, for efficient synthesis of polystyrene-*b*-poly(*n*-butyl acrylate)-methacrylate macromonomer, where each block in the copolymer chain has a degree of polymerization of about 30.

12.6 α,ω-Dibromo-terminated homo-telechelic polystyrene (DBPSt) was prepared [Tsarevsky, N. V., Sumerlin, B. S., and Matyjaszewski, K., *Macromolecules*, **38**, 3558 (2005)] by ATRP of styrene(St) with a difunctional initiator, dimethyl 2,6-dibromoheptadioate (DM-2,6-DBHD) with the following conditions: [St]/[DM-2,6-DBHD]/[CuBr]/[PMDETA] = 74 : 1 : 0.5 : 0.5, 40 vol% toluene, 80°C, 140 min, 20% monomer onversion. The resulting homotelechelic polymer (DBPSt) was isolated, purified, and reacted with NaN$_3$ in DMF solvent at room temperature to α,ω-diazide-terminated polystyrene (DAPSt). After purification and isolation, this product was subjected to step-growth click coupling with equimolar amount of propargyl ether in DMF at room temperature using a CuBr catalyst.

(a) Show equations for the sequence of reactions in the above process.   (b) Calculate the molar mass of DBPSt.   (c) Given that the macromonomer conversion in the step-growth polymerization was 90%, calculate the number average molecular weight (M_n) of the polymeric product obtained. [*Ans.* (b) 1903 g mol^{-1}; (c) 20,933.]

12.7   Various end-functionalized polymers can be synthesized by reacting alkynes with azide-derivatized polymers prepared by ATRP. Accordingly, suggest a synthetic strategy to prepare α,ω-dihydroxy-terminated polystyrene by a combination of ATRP and subsequent modification via CuAAC reactions.

12.8   Using a combination of ATRP and CuAAC reactions, suggest an efficient route for the synthesis of narrow-disperse cyclic poly(*N*-isopropylacrylamide) with an average of 80 monomer residues in the polymer ring.

12.9   Suggest a modular strategy for the synthesis of the inverse star block copolymer, [poly(ε-caprolactone)-*b*-polystyrene]$_2$-*core*-[poly(ε-caprolactone)-*b*-polystyrene]$_2$.

12.10   Discuss feasible routes for the synthesis of narrow-disperse (a) four-armed star polymer, *core*-(PCL$_{50}$)$_4$ (where CL = caprolactone) and (b) four-armed star diblock copolymer, *core*-(PCL$_{50}$-*b*-PSt$_{50}$)$_4$ (where PSt = polystyrene). Calculate the theoretical molecular weights of the star (co)polymers. [*Ans.* (a) 22,936;   (b) 44,332.]

12.11   Describe, in broad outline, synthetic strategies to make the following dendrimer-like miktoarm star ter-plymers, using a combination of CuAAC click chemistry and controlled/living polymerization methods : (a) *core*-[P*t*BA-*bp*-(PSt)(PCL)]$_3$ and (b) *core*-[PSt-*bp*-(PEG)(P*t*BA)]$_3$. (*Note*:   St = styrene; *t*BA = *tert*-butyl acrylate;   CL = caprolactone;   EG = ethylene glycol.)

12.12   Describe a synthetic strategy for preparation of cyclic poly(*N*-isopropylacrylamide) of polymerization degree approximately 100, based on RAFT polymerization and click chemistry.

12.13   Devise a methodology to synthesize α,ω-dihydroxy telechelic poly(*N*-isopropylacrylamide) by combining click chemistry with (a) ATRP and (b) RAFT polymerization.

12.14   It is proposed to prepare, by combined use of RAFT polymerization and click chemistry, high-density polystyrene-*b*-poly(methyl acrylate) brushes on silica nanopartiles. To this end, discuss an optimum synthetic strategy and give an outline for synthesis.

12.15   Polymers of acrylamide (PAAm) represent an important class of materials because of applications such as in coatings, flocculants, paper making, mining, electrophoresis, and biology. While RAFT polymerization and click chemistry can be used for surface modification by both "grafting to" and "grafting from" approaches, surface modification using the "grafting to" approach is more challenging in achieving higher grafting densities. Considering these points, devise a "grafting to" approach based on RAFT polymerization and click chemistry for surface modification of silica nanoparticles with polymer brushes of tethered PAAm chains.

12.16   Briefly outline the methods of synthesis of (a) the functional initiators (**XVIII**) - (**XXI**) and (b) the post-polymerization functionalizing agents (**XXII**) and (**XXIII**), referring to Fig. 12.22.

12.17   Hizal and coworkers [Durmaz, H., Karatas, F., Tunca, U., and Hizal, G., *J. Polym. Sci., Part A: Polym. Chem.*, **44**, 499 (2006)] synthesized a compound, (**XXI**) (Fig. 12.22), having an anthracene functionality as also a NMP site and an ATRP site. Show how through combination of the DA click reaction, ATRP, and NMP, a heteroarm H-shaped terpolymer (PSt)(P*t*BA)-*bp*-(PEO)-*bp*-(P*t*BA)(PSt), containing PEO as a backbone and PSt and P*t*BA as side arms via a branch point (*bp*) at either end of PEO [where PSt is polystyrene, P*t*BA is poly(*tert*-butyl acrylate), and PEO is poly(ethylene oxide)] can be synthesized.

12.18   Make a suitable choice of click reactions to enable efficient one-pot synthesis of the following ABC tri-block copolymers : (a) PEG-*b*-PSt-*b*-PMMA and (b) PMMA-*b*-PSt-*b*-PCL [where PEG is poly(ethylene glycol), PSt is polystyrene, PMMA is poly(methyl methacrylate), and PCL is poly(ε-caprolactone)].

12.19   Diels-Alder click reactions have been successfully used for the preparation of well-defined star polymers. Devise methodologies using this route to prepare 3-arm star polymers with uniform arms, *core*-A$_3$, where A is (a) poly(ethylene glycol), (b) poly(methyl methacrylate), and (c) poly(*tert*-butyl acrylate).

12.20   How can click chemistry techniques be combined with LRP/CRP methods to carry out surface modification of silica nanoparticles ?

12.21   (a) How will you make polystyrene, poly(methyl methacrylate), and polycaprolactone with multiple alkene functional groups along the backbone chain ? (b) How will you utilize these functional sites to introduce –CO$_2$H groups on the backbone ? (*Note*: By introducing –CO$_2$H groups in this way a significant change in solubility and associated increase in ability to modify the surface of nanoparticles, etc., can be effected.)

12.22 The protected amino acid, *N*-(9-fluorenylmethoxycarbonyl) cysteine (Fmoc-C) is a building block/model for the attachment of peptide fragments to synthetic materials. Devise a stratgy to synthesize low polydispersity polystyrene with multiple Fmoc units along the backbone.

12.23 How can Diels-Alder and retro-Diels-Alder reactions be used for the preparation of thermoresponsive dendrimers and dendronized hyperbranched polymers ?

12.24 Discuss a possible method of synthesizing an asymmetric telechelic polymer based on poly(methyl methacrylate) (DP ~ 50) with a carboxylic group at one end and a hydroxyl group at the other, using combined thiol-ene and CuAAC click reactions.

12.25 Describe the synthesis of a third generation vinyl-terminated dendrimer in a divergent fashion by alternating thiol-ene and Grignard reaction sequence, using tetravinyl silane for forming the core, 3-mercaptopropyl trimethoxysilane for branching, and vinyl magnesium bromide (Grignard reagent) for vinylation of dendrimer ends.

Appendix A

Conversion of Units

SI UNITS AND CONVERSION FACTORS

Physical quantity	Name of SI unit	Symbol for SI unit	Definition of SI unit
Length	Meter	m	Basic unit
Mass	Kilogram	kg	Basic unit
Time	Second	s	Basic unit
Force	Newton	N	$kg\, m\, s^{-2}\ (= J\, m^{-1})$
Pressure	Pascal	Pa	$kg\, m^{-1}\, s^{-2}\ (= N\, m^{-2})$
Energy	Joule	J	$kg\, m^2\, s^{-2}$
Power	Watt	W	$kg\, m^2\, s^{-3}\ (= J\, s^{-1})$

Physical quantity	Customary unit	SI unit	To convert from customary unit to SI units multiply by
Length	in.	m	2.54×10^{-2}
Mass	lb	kg	$4.535\,923\,7 \times 10^{-1}$
Force	dyne	N	1×10^{-5}
	kgf	N	9.806 65
	lbf	N	4.448 22
Pressure	dyne/cm^2	Pa or N m^{-2}	1×10^{-1}
	atm	Pa or N m^{-2}	$1.013\,25 \times 10^5$
	mm Hg	Pa or N m^{-2}	$1.333\,22 \times 10^2$
	lbf/in.^{2} or psi	Pa or N m^{-2}	$6.894\,76 \times 10^3$

SI UNITS AND CONVERSION FACTORS

Physical quantity	Customary unit	SI unit	To convert from customary unit to SI units multiply by
Energy	erg	J	1×10^{-7}
	Btu	J	$1.055\,056 \times 10^3$
	ft-lbf	J	$1.355\,82$
	cal	J	4.187
	eV	J	1.602×10^{-19}
Area	in.^{2}	m^2	$6.451\,6 \times 10^{-4}$
	ft^2	m^2	$9.290\,304 \times 10^{-2}$
Density	lb/ft^3	kg m^{-3}	$1.601\,846\,3 \times 10$
Viscosity	poise (dyne s/cm^2)	kg m^{-1} s^{-1} N s m^{-2}	1×10^{-1}
Viscosity, kinematic	Stoke (cm^2/s)	m^2 s^{-1}	10^{-4}
Surface tension	lbf/ft	N m^{-1}	14.59
	dyne/cm	N m^{-1}	10^{-3}

Additional Conversion Units

$1\ \text{Å} = 10^{-8}\ \text{cm} = 10^{-10}\ \text{m}$	$1\ \text{kgf/cm}^2 = 9.807 \times 10^4\ \text{N/m}^2$
$1\ \text{atm} = 76\ \text{cm Hg (at } 0°\text{C)}$ $= 14.696\ \text{psi}$	$1\ \text{HP} = 550\ \text{ft lbf/s} = 2545\ \text{Btu/h}$ $= 746\ \text{W}$
$1\ \text{eV} = 1.602 \times 10^{-12}\ \text{erg}$	$t°\text{C} = (1.8t + 32)^o\ \text{F}$
$1\ \mu = 10^{-4}\ \text{cm} = 10^{-6}\ \text{m}$	$t^o\ \text{F} = (5/9)(t - 32)°\text{C}$

Appendix B

Fundamental Constants

Constant	CGS system	SI system
Acceleration of gravity (g) (standard value)	980.665 cm s^{-2}	9.8066 m s^{-2}
Normal atmospheric pressure	$1,013,250$ dyne cm^{-2}	1.01325×10^5 N m^2
Volume of 1 mole of ideal gas at at 1 atm and 0°C	22.4136 litre	22.41136×10^{-3} m^3 mole^{-1}
Avogadro's number	6.0220×10^{23} mole^{-1}	6.0220×10^{23} mole^{-1}
Atomic mass unit	1.6604×10^{-24} g	1.6604×10^{-27} kg
Universal gas constant (R)	1.9872 cal deg^{-1} mole^{-1}	8.3143 J deg^{-1} mole^{-1}
Boltzmann constant (k)	1.3807×10^{-16} erg deg^{-1} molecule^{-1}	1.3807×10^{-23} J deg^{-1} molecule^{-1}
Faraday constant	–	$96,487.0$ Coulomb mole^{-1}
Planck constant (h)	6.6262×10^{-27} erg s	6.6262×10^{-34} J s
Velocity of light in vacuum (c)	2.99792×10^{10} cm s^{-1}	2.99792×10^8 m s^{-1}
Electronic charge (e)	4.80325×10^{-10} esu	1.60219×10^{-19} Coulomb
Electron rest mass	9.1095×10^{-28} g	9.1095×10^{-31} kg

FUNDAMENTAL CONSTANTS (continued)

Constant	CGS system	SI system
Proton rest mass	1.67265×10^{-24} g	1.67265×10^{-27} kg
Neutron rest mass	1.67482×10^{-24} g	1.67482×10^{-27} kg
Debye (D)	–	3.336×10^{-30} Coulomb m
Bohr magneton (BM)	9.2731×10^{-21} erg gauss^{-1}	–
Permeability of vacuum ($\mu_s z p$)	–	12.566×10^{-7} N A^{-2}
Permittivity of vacuum ($1/\mu_0 c^2$)	–	8.85419×10^{-12} F m^{-1}

Source: E. R. Cohen and B. N. Taylor, *J. Phys. Chem. Ref. Data*, **2**, 663 (1973).

Index